Neuroscience Intelligence Unit

From Messengers to Molecules: Memories Are Made of These

Gernot Riedel, Ph.D.
Bettina Platt, Ph.D.
School of Medical Sciences
College of Life Sciences and Medicine
University of Aberdeen
Foresterhill, Aberdeen, U.K.

Landes Bioscience / Eurekah.com
Georgetown, Texas
U.S.A.

Kluwer Academic / Plenum Publishers
New York, New York
U.S.A.

From Messengers to Molecules: Memories Are Made of These

Neuroscience Intelligence Unit

Landes Bioscience / Eurekah.com
Kluwer Academic / Plenum Publishers

Printed in the U.S.A.

Kluwer Academic / Plenum Publishers, 233 Spring Street, New York, New York, U.S.A. 10013
http://www.wkap.nl/

Please address all inquiries to the Publishers:
Landes Bioscience / Eurekah.com, 810 South Church Street
Georgetown, Texas, U.S.A. 78626
Phone: 512/ 863 7762; FAX: 512/ 863 0081
www.Eurekah.com
www.landesbioscience.com

From Messengers to Molecules: Memories Are Made of These, edited by Gernot Riedel and Bettina Platt, Landes / Kluwer dual imprint / Landes series: Neuroscience Intelligence Unit

ISBN: 0-306-47862-5

While the authors, editors and publisher believe that drug selection and dosage and the specifications and usage of equipment and devices, as set forth in this book, are in accord with current recommendations and practice at the time of publication, they make no warranty, expressed or implied, with respect to material described in this book. In view of the ongoing research, equipment development, changes in governmental regulations and the rapid accumulation of information relating to the biomedical sciences, the reader is urged to carefully review and evaluate the information provided herein.

Library of Congress Cataloging-in-Publication Data

From messengers to molecules : memories are made of these /
[edited by] Gernot Riedel, Bettina Platt.
p. ; cm. -- (Neuroscience intelligence unit)
Includes bibliographical references and index.
ISBN 0-306-47862-5
1. Neurochemistry. 2. Neurotransmitters. 3. Neurotransmitter receptors. I. Riedel, Gernot. II. Platt, Bettina. III. Series: Neuroscience intelligence unit (Unnumbered)
[DNLM: 1. Memory--physiology. 2. Ion Channels. 3. Learning --physiology. 4. Memory Disorders. 5. Neurotransmitters. 6. Transcription Factors. WL 102 F9308 2004]
QP356.3.F76 2004
612.8'042--dc22

2004001884

Dedication

To our children Daniel and Lisa Sophie,
for wonderful memories.

CONTENTS

Section 3. Neuromodulators

Section 5. Transcription Factors, Genes and Proteins

Section 6. Morphological Changes in Synapses and Neurones

EDITORS

Gernot Riedel, Ph.D.
Bettina Platt, Ph.D.
School of Medical Sciences
College of Life Sciences and Medicine
University of Aberdeen
Foresterhill, Aberdeen, U.K.
Chapters 2.1, 3.1

CONTRIBUTORS

Wickliffe Abraham
Department of Pharmacology
University of Auckland
Auckland, and
Department of Psychology
University of Otago
Dunedin, New Zealand
Chapter 5.2

Daniel L. Alkon
Blânchette Rockefeller Neurosciences Institute
Rockville, Maryland, U.S.A
Chapter 1.1

Fiorenzo Battaini
Department of Neurosciences
University of Roma
Roma, Italy
Chapter 4.4

Pauleen Bennett
Department of Psychology
Monash University
Clayton, Victoria, Australia
Chapter 4.7

Robert W. Berry
Department of Cell and Molecular Biology
Northwestern University Medical School
Chicago, Illinois, U.S.A.
Chapter 6.1

Joyce Besheer
Department of Psychology
University of Nebraska - Lincoln
Lincoln, Nebraska, U.S.A.
Chapter 2.4

Rick A. Bevins
Department of Psychology
University of Nebraska - Lincoln
Lincoln, Nebraska, U.S.A.
Chapter 2.4

Catherine Brandner
Institut de Physiologie
Université de Lausanne
Lausanne, Switzerland
Chapter 3.4

Marie-Christine Buhot
Laboratory of Cognitive Neurosciences
Centre National de la Recherche Scientifique UMR 5106
University of Bordeaux I
Talence Cedex, France
Chapter 2.5

Martín Cammarota
Centro de Memória
Departamento de Bioquímica
Instituto de Ciências Básicas da Saúde
Universidade Federal do Rio Grande do Sul
Porto Alegre, Brasil
Chapter 4.5

Claudio Castellano
Section of Psychopharmacology
Institute of Neuroscience
Rome, Italy
Chapter 2.2

Vincenzo Cestari
Section of Psychopharmacology
Institute of Neuroscience
Rome, Italy
Chapter 2.2

Alessandro Ciamei
Section of Psychopharmacology
Institute of Neuroscience
Rome, Italy
Chapter 2.2

Pauline Curtis
Department of Pharmacology
University of Auckland
Auckland, and
Department of Psychology
University of Otago
Dunedin, New Zealand
Chapter 5.2

Jan P.C. de Bruin
Graduate School Neurosciences
Amsterdam
Netherlands Institute for Brain Research
Amsterdam, The Netherlands
Chapter 2.6

David De Wied
Department of Medical Pharmacology
and Anatomy
Rudolf Magnus Institute
for Neurosciences
University Medical Center
Utrecht, The Netherlands
Chapter 3.3

John F. Disterhoft
Department of Cell and Molecular
Biology
Northwestern University Medical School
Chicago, Illinois, U.S.A.
Chapter 7.2

Mike Dragunow
Department of Pharmacology
University of Auckland
Auckland, and
Department of Psychology
University of Otago
Dunedin, New Zealand
Chapter 5.2

Paul W. Frankland
Programmes in Integrative Biology
and Brain and Behaviour
Hospital for Sick Children
Toronto, Ontario, Canada
Chapter 5.1

Olga T. Ganeshina
Department of Cell and Molecular
Biology
Northwestern University Medical School
Chicago, Illinois, U.S.A.
Chapter 6.1

Yuri Geinisman
Department of Cell and Molecular
Biology
Northwestern University Medical School
Chicago, Illinois, U.S.A.
Chapter 6.1

Robert Gerlai
Department of Psychology
University of Hawaii at Manoa
Honolulu, Hawaii, U.S.A.
Chapter 3.5

Marie Gibbs
Department of Pharmacology
Monash University
Clayton, Victoria, Australia
Chapter 2.7

Karl Peter Giese
Wolfson Institute for Biomedical
Research
University College London
London, U.K.
Chapter 1.2

Maria Grazia Giovannini
Department of Pharmacology
University of Florence
Florence, Italy
Chapter 2.3

Jeffrey Greenwood
Department of Pharmacology
University of Auckland
Auckland, and
Department of Psychology
University of Otago
Dunedin, New Zealand
Chapter 5.2

Jean-Louis Guillou
Laboratory of Cognitive Neurosciences
Centre National de la Recherche Scientifique UMR 5106
University of Bordeaux I
Talence Cedex, France
Chapter 4.1

Rüdiger U. Hasenöhrl
Department of Psychology
University of Hertfordshire
Hatfield Herts, U.K.
Chapter 2.8

Christian Hölscher
Department of Cognitive Neuroscience
University of Tuebingen
Tuebingen, Germany
Chapter 4.2

Joseph P. Huston
Institute of Physiological Psychology and Center for Biological and Medical Research
University of Düsseldorf
Düsseldorf, Germany
Chapter 2.8

Ivan Izquierdo
Departamento de Bioquimica
Instituto de Ciências Básicas da Saúde
Universidade Federal do Rio Grande do Sul
Porto Allegre, Brazil
Chapter 4.3

Robert Jaffard
Laboratory of Cognitive Neurosciences
Centre National de la Recherche Scientifique UMR 5106
University of Bordeaux I
Talence Cedex, France
Chapter 7.1

Sheena A. Josselyn
Programmes in Integrative Biology and Brain and Behaviour
Hospital for Sick Children
Toronto, Ontario, Canada
Chapter 5.1

Tsutomu Kameyama
Department of Chemical Pharmacology
Meijo University, and
Japan Institute of Psychopharmacology
Nagoya, Japan
Chapter 3.2

Ken Kanematsu
Research Institute of Meijo University
Nagoya, Japan
Chapter 3.2

Gábor L. Kovács
Institute of Diagnostics and Management
University of Pécs
Szombathely, Hungary
Chapter 3.3

Barbara Logan
Department of Pharmacology
University of Auckland
Auckland, and
Department of Psychology
University of Otago
Dunedin, New Zealand
Chapter 5.2

Takayoshi Mamiya
Department of Chemical Pharmacology
Meijo University
Nagoya, Japan
Chapter 3.2

Aline Marighetto
Laboratory of Cognitive Neurosciences
Centre National de la Recherche Scientifique UMR 5106
University of Bordeaux I
Talence Cedex, France
Chapter 7.1

Jorge H. Medina
Instituto de Biología Celular y Neurociencias
Universidad de Buenos Aires
Buenos Aires, Argentina
Chapter 4.5

Jacques Micheau
Laboratory of Cognitive Neurosciences
Centre National de la Recherche Scientifique UMR 5106
University of Bordeaux I
Talence Cedex, France
Chapters 2.1, 4.4

Radmila Mileusnic
Department of Biological Sciences
The Open University
Milton Keynes, U.K.
Chapter 5.4

Nicole Mons
Laboratory of Cognitive Neurosciences
Centre National de la Recherche Scientifique UMR 5106
University of Bordeaux I
Talence Cedex, France
Chapter 4.1

Toshitaka Nabeshima
Department of Neuropsychopharmacology and Hospital Pharmacy
Nagoya University Graduate School of Medicine
Nagoya, Japan
Chapter 4.8

Kim T. Ng
Department of Psychology
Monash University
Clayton, Victoria, Australia
Chapter 4.7

M-R. Nikbakht
Institute of Biomedical and Life Sciences
University of Glasgow
Glasgow, U.K.
Chapter 2.9

Xavier Noguès
Laboratory of Cognitive Neurosciences
Centre National de la Recherche Scientifique UMR 5106
University of Bordeaux I
Talence Cedex, France
Chapter 4.4

E. Martin O'Kane
Institute of Biomedical and Life Sciences
University of Glasgow
Glasgow, U.K.
Chapter 2.9

Alessia Pascale
Department of Experimental and Applied Pharmacology
University of Pavia
Pavia, Italy
Chapter 4.4

Giancarlo Pepeu
Department of Pharmacology
University of Florence
Florence, Italy
Chapter 2.3

Ciaran M. Regan
Department of Pharmacology
University College Dublin
Dublin, Ireland
Chapter 6.2

Lianne Robinson
School of Medical Sciences
College of Life Sciences and Medicine
University of Aberdeen
Foresterhill, Aberdeen, U.K.
Chapter 3.1

Carmen Sandi
Brain and Mind Institute
Ecole Polytechnique Federale
de Lausanne
Lausanne, Switzerland
Chapter 3.6

Louis Segu
Laboratory of Cognitive Neurosciences
Centre National de la Recherche
Scientifique UMR 5106
University of Bordeaux I
Talence Cedex, France
Chapter 2.5

Joel C. Selcher
Division of Neuroscience
Baylor College of Medicine
Houston, Texas, U.S.A.
Chapter 4.6

Trevor W. Stone
Institute of Biomedical and Life Sciences
University of Glasgow
Glasgow, U.K.
Chapter 2.9

Oliver Stork
Institute of Physiology
University of Magdeburg
Magdeburg, Germany
Chapter 5.3

Roger Summers
Department of Pharmacology
Monash University
Clayton, Victoria, Australia
Chapter 2.7

Miao-Kun Sun
Blânchette Rockefeller Neurosciences
Institute
Rockville, Maryland, U.S.A
Chapter 1.1

J. David Sweatt
Division of Neuroscience
Baylor College of Medicine
Houston, Texas, U.S.A.
Chapter 4.6

Makoto Ukai
Department of Chemical Pharmacology
Meijo University
Nagoya, Japan
Chapter 3.2

Jeffrey Vernon
Wolfson Institute for Biomedical
Research
University College London
London, U.K.
Chapter 1.2

Monica R.M. Vianna
Departamento de Bioquimica
Instituto de Ciências Básicas da Saúde
Universidade Federal do Rio Grande
do Sul
Porto Allegre, Brazil
Chapter 4.3

Edwin J. Weeber
Division of Neuroscience
Baylor College of Medicine
Houston, Texas, U.S.A.
Chapter 4.6

Hans Welzl
Neuroanatomy and Behavior Group
Institute of Anatomy
University of Zurich
Zurich, Switzerland
Chapter 5.3

Mathieu Wolff
Laboratory of Cognitive Neurosciences
Centre National de la Recherche
Scientifique UMR 5106
University of Bordeaux I
Talence Cedex, France
Chapter 2.5

Wendy W. Wu
Department of Cell and Molecular
Biology
Northwestern University Medical School
Chicago, Illinois, U.S.A.
Chapter 7.2

Kiyofumi Yamada
Laboratory of Neuropsychopharmacology
Department of Clinical Pharmacy
Kanazawa University
Kanazawa, Japan
Chapter 4.8

PREFACE

Memory formation is one of the most important achievements of life, and a main determinant of evolutionary success. For us humans, our present experiences are determined by their relation to our personal past. Recall of memories, new evaluation of their meaning in light of recent achievements, events or problems is therefore a fundamental element of our conscious activities. On a more trivial level, remembering an important phone number or the way to the next shop is essential for our ability to manage day to day life. Failure of the neural mechanisms supporting these functions, as observed in varying levels of severity in different types of dementia, has devastating consequences and leads, in many cases, to a loss of a patient's personality.

This illustrates the importance of memory for the human species, and it also justifies why understanding the mechanisms of memory formation and memory malfunction is in great demand.

While the present book concentrates mostly on pharmacological aspects of memory, we had to neglect the taxonomy of memory, i.e., what forms of memories can be distinguished in humans and what are their counterparts in animals? In the traditional laboratory experiment, the behavioural task is shaped to address specifically one form of memory, for example spatial memory or fear memories in order to avoid confounding influences of other forms of memory, say procedural memory. In a more natural setting, however, forms of memory are mixed and interact, and the recent emergence of neuroecology to understand the brain in relation to native behavior is a clear reflection of this awareness and will be of great benefit in future work. Meanwhile however, we follow the traditional categorization that specific brain regions or neuronal circuits subserve specific forms of memory.

So what are the cellular events underlying memory formation? To put it in simple terms, a learning event will lead to neuronal excitation, activation of ion channels and transmitter receptors in a specific subset of neurones. This will trigger intracellular cascades leading eventually to the activation of transcription factors and genes. The product is the formation of new proteins, which can be used to remodel synapses in their morphology and thus making them more efficient.

While this clearly is an over-simplification, this book follows the general route outlined above and looks at the many different components that are known to contribute to the chain of events, and reveals a number of interactions at different levels.

The book has seven themes. Section one deals with ions and ion channels and concentrates on both calcium and potassium. Section two is dedicated to the principle neurotransmitters and their receptors including excitatory and inhibitory systems. Neuromodulators and their receptor function are summarised in section three. They do not directly activate ion channels and thus impinge on intracellular protein cascades and enzymes via second messengers. These are then covered in section four which looks at various kinases and phosphatases that are crucial for long-term memory formation and can be linked to the activation of transcription factors and genes, as described in section five. Such gene activation should generate novel proteins and these may be incorporated during the formation of new connections between nerve cells, i.e., the process of synaptogenesis, outlined in section six. The final section gives two examples of how pharmacological knowledge can be used to understand malfunction of memory systems, and we return to the outset of this book, namely the roles of ions and ion channels in learning and memory formation.

We are grateful to all our colleagues and friends for contributing to this book despite their tight schedules and multitudes of commitments. With as little interference from us as possible, each chapter is written in such a way that it can be read independently and provides a thorough review of the respective field. We trust that this compendium will appeal to memory researchers, both students and scientists alike. It may hopefully provide a useful overview of the diverse components relevant to memory and other aspects of neuronal plasticity, and serve as a comprehensive introduction for those new in the field and as a source of reference. Finally, we would hope that this summary of cellular mechanisms underlying memory formation may give an impetus for new research in order to strengthen this exciting scientific field.

Gernot Riedel
Bettina Platt

ABBREVIATIONS

2-DG	2-deoxygalactose
5-HETE	5-hydroxyeicosatetraenoic acid
5-HPETE	5-hydroperoxyeicosatetraenoic acid
5-HT	serotonin
6-OHDA	6-hydroxydopamine
7-NI	7-nitroindazole
8-OH-DPAT	8-hydroxy-2(di-n-propyloamino)tetralin
11-HSD	11beta-Hydroxysteroid dehydrogenase
12-HETE	12-hydroxyeicosatetraenoic acid
12-HKETE	12-keto-5,8,10,14-eicosatetraenoic acid ibuprofen
12-HPETE	12-hydroperoxyeicosatetraenoic acid
ACE	angiotensin converting enzyme
ACh	acetylcholine
ACPD	1S,3R-1-amino-cyclopentyl-1,3-dicarboxylic acid
ACTH	adrenocorticotropic hormone
AMPA	α-amino-3-hydroxy-5-methyl-4-isoxazoleproprionic acid
AMPA-R	α-amino-3-hydroxy-5-methyl-4-isoxazolepropionic acid receptor
Ang I	angiotensin I
Ang II	angiotensin II
Ang IV	angiotensin IV
ANP	atrial natriuretic peptide
Anti-Svg-30	antisauvagine-30
AP-1	activating protein
AP5	2-amino-5-phosphonovaleric acid
APP	amyloid precursor protein
ArA	arachidonic acid
Arc	activity-regulated cytoskeleton associated protein
AST	aristolochic acid
ATP	adenosine triphosphate
AVP	$[Arg^8]$-vasopressin
AVP	arginine vasopressin
BC264	$Tyr(SO_3H)$-gNle-mGly-Trp-(NMe)Nle-Asp-Phe-NH_2
BDNF	brain-derived neurotrophic factor
BLA	basolateral nucleus of the amygdala
BNP	brain natriuretic peptide
BOC	tert-butoxycarbonyloxiimino protective group
Ca^{2+}	calcium ion
CalA	calyculin A
CaM	calmodulin
CAM	cell adhesion molecule
CaMk	calmodulin-calcium dependent kinase
CaMKII	Ca^{2+}/calmodulin-dependent protein kinase
CaMK-II	calcium-calmodulin dependent kinase-type II
cAMP	cyclic adenosin monophosphate
CCK	cholecystokinin
CCK-4	C-terminal tetrapeptide of cholecystokinin
CCK-8	C-terminal octapeptide of cholecystokinin
CCK-8s	sulphated C-terminal octapeptide of cholecystokinin

CCK-8US	unsulphated C-terminal octapeptide of cholecystokinin
CEA	central nucleus of the amygdala
CGP42112A	nicotinic acid-Tyr-(N-benzoylcarbonyl-Arg)-Lys-His-Pro-Ile-OH
ChAT	choline acetyl-transferase
CLIP	corticotropin-like intermediate lobe peptide
CNP	C-type natriuretic peptide
CNQX	6-cyano-7-nitroquinoxaline-2,3-dione
CNS	central nervous system
Cort	corticosterone
CREB	cAMP-responsive element-binding protein
CRF	corticotropin releasing factor
CRH	corticotropin releasing hormone
CyA	cyclosporin A
DA	dopamine
DAG	diacylglycerate
DAG	1,2-diacylglycerol
DARPP-32	dopamine- and cAMP-regulated phosphoprotein of 32 kD weight
D-MPRG	D-Met-Pro-Arg-Gly-NH2
DNMTP	delayed non-matching to place
DNMTS	delayed non-matching to sample
DOPAC	3,4-dihydroxyphenylacteic acid
EDRF	endothelium-derived relaxing factor
eNOS	endothelial NOS
EPSPs	excitatory postsynaptic potentials
FF	Fimbria /Fornix
GABA	γ-aminobutyric acid
GAP	GTPase activating protein
GAP-43	growth associated protein of ~50 kD weight
GluR	glutamate receptor
GluR-A	glutamate receptor subtype A
GPII	glycosylphosphatidylinositol
GR 73632	D-ALA-[l-Pro^9,Me-Leu^8]substance P-(7-11)
GR	glucocorticoid receptor
HFS	high frequency stimulation
HMA	hydroxymyristic acid
HODI	homozygous diabetes insipidus
HPA	hypothalamo-pituitary-adrenal
HR	hightened locomotor response
HVA	homovanillic acid
ic	intracerebral
icv	intracerebroventricular
IEG	immediate early gene
IgG	immunoglobulin G
IMHV	intermediate medial hyperstriatum ventrale
INDO	indomethacin
INH-1	inhibitor-1
INH-2	inhibitor-2
iNOS	inducible NOS
IP3	inositol 1,4,5-triphosphate
K^+	potassium ion

L1	a cell adhesion molecule
L-365,260	3R(+)-N-(2,3-dihydroxy-1-methyl-2-oxo-5-phenyl-1-H-1, 4-benzodiazepine-3-yl)
LHRH	luteinizing hormone releasing hormone
L-NA	N^G-nitro-L-arginine
L-NAME	N^G-nitro-L-arginine methyl ester
L-NAME	nomega-nitro-L-arginine methylester-hydrochloride
L-NMMA	N^G-monomethyl-L-arginine acetate
LPH	lipotropin
LPO	lobus parolfactorius
LR	locomotor response
LS	lateral septum
LTD	long-term depression
LTP	long-term potentiation
LTS	long-term sensitisation
M35	galanin-(1-13)-bradykinin-(2-9)amide
MAGUK	membrane-associated guanylate kinase
MAP2	microtubule-associated protein 2
MAPK	mitogen-activated protein kinase
MARCKS	myristoylated alanine-rich C kinase substrate
MeA	methylanthranilate
mGluR	metabotropic glutamate receptor
MK-801	dizocilpine
MR	mineralocorticoid receptor
mRNA	messenger ribonucleic acid
MS	medial septal nucleus
MSB	multiple synapse bouton
MSH	melanocyte stimulating hormone
MWM	Morris water maze
[Nle^1]-Ang IV	norleucine-1-angiotensin IV
NA	noradrenalin
NaCl	sodium chloride, saline
NAME	N^G-nitro-L-arginine methyl ester
NARG	N-nitro-L-arginine
NBM	nucleus basalis magnocellularis
NC-1900	pGlu-Asn-Ser-Pro-Arg-Gly-NH_2 acetate
NCAM	neural cell adhesion molecule
NCDC	2-nitro-4-carboxylphenyl-N,N-diphenylcarbamate
NDGA	nordihydroguaiaretic acid
NGF	nerve growth factor
NK	neurokinin
NKKB	nuclear factor kB
NMDA	N-methyl-D-aspartate
NMDA-R	N-methyl-D-aspartate receptor
nNOS	neuronal NOS
NO	nitric oxide
NOS	nitric oxide synthase
NPY	neuropeptide Y
OA	okadaic acid
OLETF	Otsuka Long-Evans Tokushima fatty rat

ORG 2766	[Met(O_2),D-Lys,Phe[9]]-α-MSH-(4-9)
PAF	platelet activating factor, 1-O-alkyl-2-acyl-sn-3-phosphocholine
PAG	periaqueductal grey matter
PAL	passive-avoidance learning
PAT	passive avoidance task
PG	prostaglandins
PI3	1-Phosphatidylinositol 3-Kinase
PIP_2	phosphoinosytolbisphosphate
PKA	protein kinase A
PKC	protein kinase C
PKG	cGMP-dependent protein kinase, protein kinase G
PLA_2	phospholipase A_2
PLC	phospholipase C
POMC	proopiomelanocortin
PP1	protein phosphatase 1
PP2A	protein phosphatase 2A
PP2B	protein phosphatase 2B (also called calcineurin)
PP2C	protein phosphatase 2C
PPIase	peptidyl prolyl *cis/trans* isomerase (also called immunophilins)
PSA	polysialic acid
PSD	post-synaptic density
PST	polysialyltransferase ST8SiaIV
PTP	protein tyrosine phosphatase
PVN	paraventricular nucleus
RA	retinoic acid
RAM	radial arm maze
RM	reference memory
RT-PCR	reverse transcriptase polimerase chain reaction
SDHACU	sodium-dependent high affinity choline uptake
Ser/Thr	serine/threonine
sGC	soluble guanylyl cyclase
SHS	septo-hippocampal system
SNAP	S-nitroso-N-acetylpenicillamine
SNAP	soluble N-ethylmaleimide-sensitive factor attachment protein
SPRC	synapse-associated polyribosomal complexes
STS	short-term sensitisation
STX	polysialyltransferase ST8SiaII
SVZ	subventricular zone
TF	transcription factor
TRIM	1-(2-trifluoromethylphenyl)imidazole
TX	thromboxanes
US	unconditioned stimulus
VDB	vertical diagonal band
VDCC	voltage-dependent calcium channels
WIN 62577	17-Hydroxy-17-ethynyl-D-4-androstano[3.2-b]pyrimido[1,2]-benzimidazole (non-peptide NK_1 tachykinin receptor antagonist)
WM	working memory

CHAPTER 1.1

Calcium

Miao-Kun Sun and Daniel L. Alkon

Abstract

Ca^{2+} plays an essential role in a variety of intracellular signaling cascades, which underlie mechanisms essential for the dynamic control of cell functions. In cognition, Ca^{2+} participates in control of not only the formation and development of neural structures that cognition depends on, but also signal processing and synaptic plasticity that define learning and memory. The dramatic influence of Ca^{2+} on neural functions relies on the fact that its concentrations and changes are rapidly sensed and recognized by many intracellular molecules, including proteins that trigger neurotransmitter exocytosis and Ca^{2+}-binding enzymes and kinases. Ca^{2+} homeostasis is thus tightly controlled and involves a balance of mechanisms controlling Ca^{2+} entry through the plasma membrane, intracellular storage and release, and sequestration. Each of these mechanisms can be impaired in diseases, by drugs, and in aging, leading to derangement of Ca^{2+} homeostasis. Thus, abnormal Ca^{2+} signaling contributes in important ways to neurological and cognitive disorders. Effective cognitive therapies cannot be achieved without a comprehensive understanding of the roles and mechanisms of Ca^{2+} ions in cognition and without valid strategies for correcting the Ca^{2+} abnormalities. These and other issues are briefly discussed in the chapter.

Introduction

Ca^{2+}, a ubiquitous intracellular messenger, controls almost everything we do (from fertilization to death), including how our minds organize thoughts sufficiently well to investigate our own existence, and for an exceptional few, a clear view of the beginning of our universe. In neurons, for instance, Ca^{2+} regulates development, excitability, secretion, learning, memory, aging, and death.[6,15] Information about the mechanisms regulating Ca^{2+} concentrations and mechanisms regulated by Ca^{2+} is therefore critical for our understanding neural functions and memory.

Intracellular Ca^{2+} signaling is characterized by two phenomena: a broad spectrum of functional roles and precise control of intracellular concentrations. A long-standing question in cell Ca^{2+} signaling is how Ca^{2+}, with its abundant and varied intracellular targets, is able to achieve specificity and activate only a subset of those targets. Temporal and spatial control of Ca^{2+} signaling through the neural networks involved in learning and memory are fundamental for cognitive capacities. The Ca^{2+} signals can not only spread through neurons as global Ca^{2+} waves, but can also be highly localized within micro-domains of sub-cellular compartments such as at close appositions of mitochondria and the endoplasmic reticulum (ER), dendritic spines, or presynaptic terminals.[68,100]

Losing effective control of cytosolic free Ca^{2+} concentration ($[Ca^{2+}]_c$) according to functional demands undoubtedly contributes to neurological and memory disorders, and aging. Abnormally high or low levels of $[Ca^{2+}]_c$ can be cytotoxic. Although high $[Ca^{2+}]_c$ attracts most of attention, there is evidence that neuronal cell injury/death can also be associated with a decrease of $[Ca^{2+}]_c$ (for review see ref. 95). For instance, growth factor deprivation induces cell

From Messengers to Molecules: Memories Are Made of These, edited by Gernot Riedel and Bettina Platt.

death of the sympathetic neurons and the death can be prevented by increasing $[Ca^{2+}]_c$, an effect blocked by Ca^{2+} antagonists or intracellular Ca^{2+} chelators. Blocking L- and N-type voltage-operated Ca^{2+} channel (VOCC) or N-methyl-D-aspartate (NMDA) receptors has also been reported to cause degeneration of neurons.

Aged neurons exhibit a decrease in the maximal rate and amplitude of $[Ca^{2+}]_c$ increase upon depolarization, and a significant decrease in the rate of $[Ca^{2+}]_c$ recovery after neurochemical stimulation. Furthermore, abnormal Ca^{2+} homeostasis contributes to many forms of clinical disorders and offers targets for therapeutic interventions. Moreover, the neuroprotective effects of drugs designed to suppress neuronal cell injury by blocking VOCC may be counterbalanced by the inherent toxicity of these compounds, because a decreased $[Ca^{2+}]_c$ may be sufficient to induce cell injury/death.

Ca^{2+} Influx

The ultimate Ca^{2+} source for neurons exits outside the neurons. Entry of Ca^{2+} across the plasma membrane is known to be important in generating neuronal Ca^{2+} signals, resulting in membrane depolarization and an increased $[Ca^{2+}]_c$. Ca^{2+} channel expression at the cell surface is regulated by intracellular signaling molecules.[13] The latter leads to activation of Ca^{2+}-dependent intracellular signal cascades. There is a large gradient of Ca^{2+} concentration across the plasma membrane: extracellular Ca^{2+} ($[Ca^{2+}]_o$) is slightly above 2 mM, while $[Ca^{2+}]_c$ is approximately 100 nM. Thus, there is a large driving force for Ca^{2+} entry into neurons. Ca^{2+} may enter via either VOCCs (Fig. 1) or receptor-operated Ca^{2+} channels (ROCCs). Ca^{2+} efflux from the ER may also trigger a small, but prolonged Ca^{2+} entry across the plasma membrane through the so-called store-operated Ca^{2+} channels (SOCCs).

Action potentials reliably evoke Ca^{2+} transients in axons and boutons through VOCCs.[35] The VOCCs are involved in providing the Ca^{2+} for neural signals underlying learning and memory in neural networks.[1] Blocking the L-type VOCCs with nimodipine, a 1,4-dihydropyridine, has been reported to dramatically impair learning and memory,[79] limiting their usefulness as therapeutic agents in various brain and cardiovascular disorders, including brain trauma, hypoxia, ischemia, degenerative disorders, memory decline in normal aging, heart failure, and cardiac arrhythmia. Others, however, reported that these substances prevented the performance deficits in spatial memory in rats with a medial septal lesion.[12]

Multiple classes of VOCCs have been distinguished on the basis of their pharmacological and electrophysiological properties and are often termed L, N, P/Q, and T-types. VOCCs are multiple subunit membrane complexes. In the central nervous system, the complexes are comprised of at least α_1, α_2, and β subunits. Transcripts encoding a γ subunit have not been identified in RNA from the brain. The α_1 and β subunits are each encoded by a gene family, including at least six distinct genes for α_1 subunits and four genes for β subunits. Primary transcripts of each of the α_1 genes, the α_2 gene and two of the β genes have been shown to yield multiple, structurally distinct subunits via differential mRNA processing. The α_1 subunits of Ca^{2+} channels contain the Ca^{2+}-selective pore, the essential gating machinery, and the receptor sites for the most prominent pharmacological agents. Some of the cloned α_1 subunits in fact correspond rather well to native L-type or N-type channels. In contrast to the α_1 subunits, Ca^{2+} channel α_2 subunits generally serve as modulatory subunits for the Ca^{2+} channel complex. Although in some cases α_2 subunit coexpression is found also to modulate the rates of activation and inactivation, and the voltage-dependence of inactivation. Functions of the β subunits, on the other hand, more likely depend on their interaction with the α subunits as modulatory subunits, by altering the channel complex properties,[98] such as voltage dependence, rate of activation and inactivation, and current magnitude. Interestingly, calmodulin may mediate two opposing effects on individual channels, initially promoting and then inhibiting channel opening. Both require Ca^{2+}-calmodulin binding to a single 'IQ-like' domain on the carboxyl tail of α_{1A}, but are mediated by different domains of calmodulin. Ca^{2+} binding to the amino-terminal domain selectively initiates channel inactivation, whereas Ca^{2+} sensing by the carboxyl-terminal lobe induces facilitation.[30]

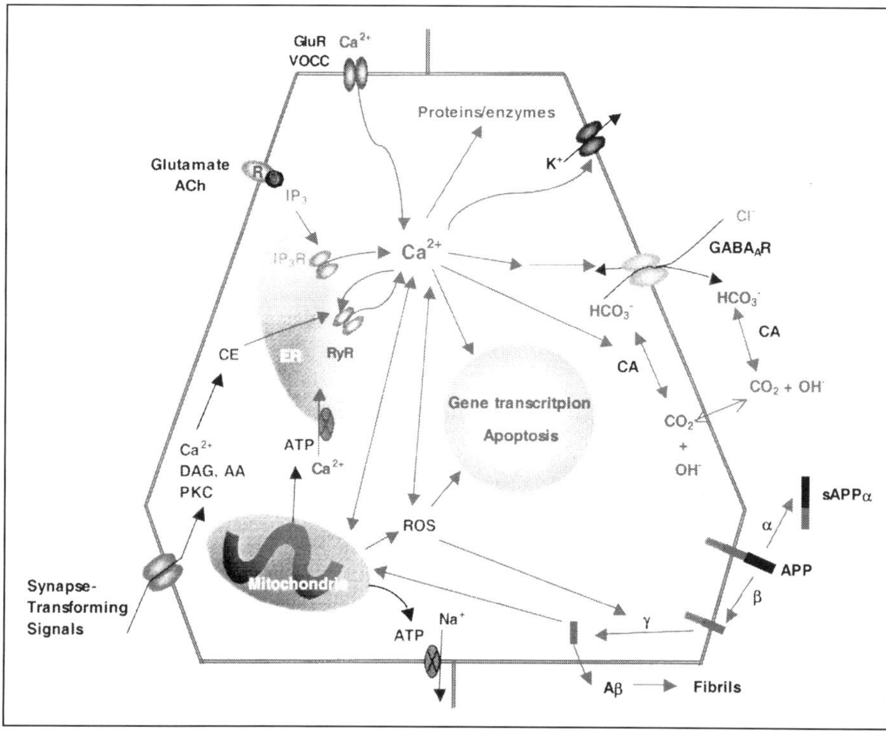

Figure 1. A cartoon to illustrate the features of Ca^{2+} cascades. $[Ca^{2+}]_c$ may increase due to Ca^{2+} influx through plasma membrane channels or intracellular release from ER RyR or IP_3R channels. Ca^{2+} triggers many intracellular responses, such as changes in enzyme activity and receptor/synaptic functions, Ca^{2+} release, mitochondrial functions, gene transcription, and ROS/Aβ formation/apoptosis. Aβdamages neurons and promotes apoptosis by a mechanism involving generation of reactive oxygen species (ROS). ROS promote neuronal apoptosis by damaging various cellular proteins. α, α-secretase; β, β-secretase; γ, γ-secretase; AA, arachidonic acid; ACh, acetylcholine; APP, amyloid precursor protein; ATP, adenosine triphosphate; CA, carbonic anhydrase; CE, calexcitin; DAG, diacylglycerol; ER, endoplasmic reticulum; IP_3R, inositol 1,4,5-triphosphate receptor; PKC, protein kinase C; RyR, ryanodine receptor; sAPPα, α-secretase-derived secreted APP;

L-type Ca^{2+} channels represent a subset of high voltage-threshold Ca^{2+} channel that can generally be distinguished by their persistent activation during a maintained depolarization and by sensitivity to dihydropyridine antagonists and agonists. L-type channels are widely distributed in excitable and nonexcitable cells and are inactivated by Ca^{2+}.[56,59,86] It has been reported that the synaptic transmission between hippocampal CA3 and CA1 neurons does not involve Ca^{2+} from activation of L-type Ca^{2+} channels.

N-type Ca^{2+} channels are found in many central and peripheral neurons and have been proposed to play a role in the release of neurotransmitter at certain synapses. N-type channels can generally be distinguished by the combination of a number of criteria, including activation at potentials more positive than -30 mV (high voltage-threshold), inactivation during a prolonged depolarization, insensitivity to dihydropyridines, and a strong and irreversible block by the neuropeptide toxin ω-conotoxin (ω-CTx)-GVIA. However, this toxin does not block N-type channels exclusively. At micromolar concentrations, ω-CTx-GVIA also reduces currents carried by doe-1, class D L-type channels, and an adrenal chromafin channel that is not the classical N-type.

P-type channels are potently blocked by ω-Aga-IVA, with an IC_{50} of 1-2 nM. In contrast, α_{1A} channels in oocytes are much less sensitive to ω-Aga-IVA, showing an IC_{50} of about 200 nM. However, at submicromolar concentrations, the toxin also strongly inhibits α_{1A} currents. Agonists of metabotropic glutamate receptors (mGluRs) are also found to suppress a large voltage-activated P/Q-type Ca^{2+} conductance in the presynaptic terminal, therefore inhibiting synaptic transmitter release at glutamatergic synapses.

R-type channels in cerebellar granule neurons are resistant to blockade by ω-CTx-GVIA, nimodipine (up to 5 μM), and ω-Aga-IVA (30 nM) at concentrations sufficient to eliminate N-, L-, and P-type channels, respectively.

After blocking N-type channels with ω-Conotoxin GVIA (1-3 μM), much of the synaptic transmission between hippocampal CA3 and CA1 neurons remains. The pharmacological profile of Ca^{2+} channels mediating the remaining transmission resembles that of α_{1A} Ca^{2+} channel subunits expressed in Xenopus oocytes and the Q-type Ca^{2+} channel current in cerebellar granule neurons. Like the R-type channels, Q-type channels are resistant to ω-CTx-GIVA, nimodipine, or ω-Aga-IVA. The Q-type channels appear to be generated by α_{1A} and α_{1E} subunits and are completely blocked by 1.5 μM ω-CTx-MVIIC, and are largely suppressed by ω-Aga-IVA at 1 μM, a concentration 100 to 1000 times that needed to block P-type channels.

N- and P/Q-type Ca^{2+} channels are inhibited by G proteins.[54,57] Ca^{2+} can regulate P/Q-type channels through feedback mechanisms,[41] probably through an association of Ca^{2+}/calmodulin with P/Q type Ca^{2+} channels.[69] Thus, Ca^{2+} entry through P/Q-type channels promotes Ca^{2+}/calmodulin binding to the α_{1A} subunit. The association of Ca^{2+}/calmodulin with the channel accelerates inactivation, enhances recovery from inactivation and augments Ca^{2+} influx by facilitating the Ca^{2+} current so that it is larger after recovery from inactivation.[69]

Low-voltage-activated VOCC channels are called 'T' type because their currents are both transient (owing to fast inactivation) and tiny (owing to small conductance). T-type channels are thought to be involved in pacemaker activity, low-threshold Ca^{2+} spikes, neuronal oscillations and resonance, and rebound burst firing.

ROCCs mediate major classes of signal processing throughout the brain network. L-Glutamate is the major neurotransmitter in the principal pathways that connect the major cell groups in the hippocampus and cortex. Activation of glutamate receptors (GluR) increases Ca^{2+} entry into the neurons. It acts through either mGluRs (coupled to G proteins) or ionotropic receptors (iGluRs; ligand-gated ion channels). iGluR subunits are further subdivided into NMDA, AMPA, and kainate subtypes. When sufficient membrane potential changes are elicited by activation of ROCCs, VOCCs might also be activated, providing additional Ca^{2+} influx. The Ca^{2+} influx initiates intracellular events including intracellular Ca^{2+} release, alterations in gene transcription, and modifications of synaptic strengths. Through the Ca^{2+} signal cascades, glutamatergic activity dramatically alters neuronal activity, which in the hippocampal place cells encode spatial information. Individual hippocampal pyramidal cells demonstrate reliable place field correlates, increasing their discharge rates in selected places within an environment and becoming virtually silent in other places. Excessive activation of the glutamate receptors, however, results in increased Ca^{2+} influx and may cause oxidative stress.

Forming assembling complexes provides a mechanism that ensures specific and rapid signaling through ROCCs. For instance, the β_2-adrenergic receptor is directly associated with one of its ultimate effectors, the class C L-type Ca^{2+} channel $Ca_v1.2$,[26] generating highly localized signal transduction from the receptor to the channel.

Intracellular Release and Storage

Other than Ca^{2+} entry through the plasma membranes, rapid changes in $[Ca^{2+}]_c$ can be induced through Ca^{2+} release from intracellular stores (Fig. 1). Intracellular Ca^{2+} release is generally viewed as a mechanism to amplify and prolong Ca^{2+} influx signals.[48] The release mechanisms are widely used by neurons in signaling. The intracellular Ca^{2+} stores include the ER, mitochondria, and less well defined nuclear store. The involvement of mitochondria in the Ca^{2+} release for Ca^{2+} signaling, however, remains controversial.

The ER is a continuous network that extends throughout the axon, soma, dendrites, and spines and is therefore uniquely placed to generate Ca^{2+} signals in every compartment of a neuron. Ca^{2+} is released from the ER via inositol 1,4,5-triphosphate receptors (IP_3Rs) or ryanodine receptors (RyRs). IP_3Rs are synergistically triggered by IP_3 and Ca^{2+}, while RyRs respond to $[Ca^{2+}]_c$ and the intracellular messenger cyclic ADP ribose. Since the ER has a large capacity, it can function as a Ca^{2+} sink to generate a large number of spikes, but as its load increases the intracellular channels will become increasingly excitable, and Ca^{2+} may be released back into the cytoplasm through the process of Ca^{2+}-induced Ca^{2+} release. Ca^{2+} waves can be generated by first enhancement then inhibition. In Purkinje cells of the cerebellum, Ca^{2+} elevation is required for the IP_3R/channel to open.[16] At Ca^{2+} basal concentrations well below 0.25 μM, increasing $[Ca^{2+}]_c$ increases the open probability of the IP_3R/channel. For $[Ca^{2+}]_c$ higher than 0.25 μM, however, the open probability decreases. The hippocampal pyramidal cells, on the other hand, have complex dendritic arbors, receiving on the order of 10,000 synapses largely on dendritic spines. These dendrites contain a complex ER that reaches into a majority of large spines. In contrast to Purkinje spines, the ER of the hippocampal pyramidal cells is studded with RyRs in dendrites and spines, while IP_3Rs appear to exist largely in dendritic shafts.[107]

The ER can function as an integrator or "memory" depot of neuronal activity. By absorbing and storing the brief pulses of Ca^{2+} associated with each action potential, the ER may keep track of neuronal activity and be able to signal this information to the nucleus through periodic bursts of Ca^{2+}. For example, brief bursts of neuronal activity generate small localized pulses of Ca^{2+} that are rapidly buffered, but prolonged firing may charge up the ER sufficiently for it to transmit regenerative global signals to the nucleus to initiate gene transcription.

IP_3 Receptors

The IP_3Rs consist of three isoforms. Each has a special role in the cell. The IP_3R1 showed a bell-shaped activity in response to $[Ca^{2+}]_c$. This property, however, is not intrinsic to the receptor (its pure form is not inhibited by up to 200 μM Ca^{2+}), rather it is mediated by calmodulin[132] through a negative regulation by binding to calmodulin or a cGMP kinase substrate.[101] The IP_3R3 forms Ca^{2+} channels with single-channel currents that are similar to those of IP_3R1 at low $[Ca^{2+}]_c$; however, the open probability of the IP_3R3 isoform increases monotonically with increased $[Ca^{2+}]_c$ (ref. 50) and channels are more active even at 100 μM $[Ca^{2+}]_c$, whereas the IP_3R1 isoform has a bell-shaped dependence on $[Ca^{2+}]_c$ with maximum channel activity at 250 nM $[Ca^{2+}]_c$ and complete inhibition at 5 μM $[Ca^{2+}]_c$. The properties of IP_3R3 provide positive feedback as Ca^{2+} is released; the lack of negative feedback allows complete Ca^{2+} release from intracellular stores. Thus activation of IP_3R3 in cells that express only this isoform results in a single transient, but globally increased $[Ca^{2+}]_c$, that is better suited to signal initiation. The bell-shaped Ca^{2+}-dependence curve of IP_3R1 is, however, ideal for supporting Ca^{2+} oscillations and the frequency of Ca^{2+} transients can be modulated when IP_3 concentrations are increased.

Ryanodine Receptors

The RyRs correspond to the sarcoplasmic reticulum calcium channels and bind specifically the plant alkaloid ryanodine. All known members of RyR family, namely, skeletal muscle type RyR1, cardiac muscle type RyR2, and brain type RyR3, are abundantly expressed in the central nervous system. They include about 5000 (4872-5037) amino acid residues and are coded by three different genes, which are located on chromosomes 1, 15, and 19, respectively, in humans. The functional receptor is thought to be a homotetramer, which has a quarterfoil shape and a size of 22 to 27 nm on each side. The center of the quarterfoil includes a pore, with a diameter of 1 to 2 nm, which likely represents the Ca^{2+} channel. Near its cytoplasmic end, the channel appears to be blocked by a mass, sometimes referred to as the "plug", which might be involved in the modulation of channel conductance. Hippocampal CA1 pyramidal cells express all three types of RyRs and, compared with other central neurons, have the highest level of the RyR3, in greater abundance than the IP_3Rs. Moreover, in these neurons, RyRs are expressed in the axon, soma, and dendrites, including spines[107] and thus occupy strategically important position for synaptic signaling and integration.

Activation of RyR requires Ca^{2+}, which is therefore thought to be the "physiological" channel activator, since other ligands cannot activate the channel in the absence of Ca^{2+}, or they require Ca^{2+} for maximum effect. Activation of RyR may involve a global conformational change including rotation of channel domain relative to the cytoplasmic domain and appearance of a porelike structure within the channel domain preceding Ca^{2+} release (for review see ref. 58). In the heart cells, a cleft of roughly 12 nm is formed between the cell surface and sarcoplasmic membrane and local Ca^{2+} signal produced by a single opening of an L-type Ca^{2+} channel can trigger about 4-6 RyR receptors to generate a Ca^{2+} spark.[123] The existence of other endogenous RyR activators, such as calexcitin[2,10,17,87,89] or calexcitin-like mammalian proteins, has been proposed. The RyR is activated by caffeine[43] and many other substances.[112,113,116,131] Activation of RyR typically requires large $[Ca^{2+}]_c$ (~1 μM), incompatible with the small bulk NMDAR-mediated Ca^{2+} signals. However, local $[Ca^{2+}]_c$ is more likely to provide sufficient Ca^{2+} for the receptor activation (see below). The RyR is a substrate of several protein kinases, namely cAMP-dependent protein kinase (PKA), cGMP-dependent protein kinase (PKG), protein kinase C (PKC), and calmodulin-dependent protein kinase II (CaMKII). These pathways may be activated in combination to evoke specific functions. The involvement of RyRs in spatial memory is suggested by an increased expression of RyR2 in the rat hippocampus after training.[21,129]

Refilling of the ER is mediated by ER Ca^{2+}-ATPases since it is blocked by cyclopiazonic acid. Even without prior store depletion, the caffeine-induced Ca^{2+} transients disappear after 6-minute exposure to cyclopiazonic acid,[43] suggesting that ryanodine-sensitive Ca^{2+} stores are maintained at rest by continuous Ca^{2+} sequestration. In addition, the store does not refill in Ca^{2+}-free saline, suggesting that the refilling of the stores depends upon Ca^{2+} influx, probably through a 'capacitative-like' transmembrane influx pathway, or store-operated Ca^{2+} channels,[76] at resting membrane potential, a process that depends on a spatial cytoskeleton rearrangement between cell membrane and the ER structures.[47] One possible mechanism underlying neuronal injury by low $[Ca^{2+}]_c$ is a disturbance of ER Ca^{2+} homeostasis. As mentioned previously, low ER Ca^{2+} loading is also neurotoxic. This toxicity may result from other biological activity in the ER that depends on high Ca^{2+} levels. Besides functioning as a major intracellular Ca^{2+} store, the ER plays a pivotal role in the folding, processing, and excretion of membrane and secretory proteins, processes that depend on Ca^{2+} concentration. Depletion of ER Ca^{2+} stores thus is a severe form of stress that blocks the folding and processing of membrane proteins.[73]

The involvement of mitochondria in intracellular Ca^{2+} signaling remains controversial, particularly signaling that requires physiological Ca^{2+} release from mitochondria. It is well established, however, that physiological Ca^{2+} levels are associated with significant movement of Ca^{2+} and Ca^{2+} uptake into mitochondria (Fig. 1). With a bacterial evolutionary origin, mitochondria maintain a modicum of independence from the host cell in some respects (maintaining their own DNA while also deriving many important proteins from the nuclear DNA of the host cell). Nevertheless, they are critical for the life of almost all eukaryotic cells. The primary functions of the mitochondria involve oxidative phosphorylation and ATP supply (Fig. 1). The major targets of mitochondrial Ca^{2+} uptake are the dehydrogenases of the Krebs cycle. Increases in mitochondrial $[Ca^{2+}]$ ($[Ca^{2+}]_m$) participate in activation of the respiratory chain through stimulation of Ca^{2+}-sensitive mitochondrial dehydrogenases (isocitrate, oxoglutarate, and pyruvate dehydrogenases), thereby ensuring adequate ATP synthesis to match the increased energetic demand of stimulated cells.[63] The activation of dehydrogenases stimulates mitochondrial respiration leading to an increase in $\Delta\Psi_m$, driving an increase in ATP production (for review see ref. 33). Thus, $[Ca^{2+}]_c$ oscillations, through their effect on mitochondrial Ca^{2+} uptake, are represented by long-term activation of mitochondrial metabolism. Interestingly, a significant portion of the Ca^{2+} entering mitochondria may not appear as free ionized Ca^{2+} in the matrix, but might rather be present either bound to phosphate or to phospholipids.[27]

Mitochondrial Ca^{2+} uptake may also exert subtle effects on the spatiotemporal characteristics of the $[Ca^{2+}]_c$ in micro-domains through the cell (see below).

Buffering and Sequestration

Buffering and sequestration of Ca^{2+} play an important role in Ca^{2+} homeostasis, involving plasma membrane Na^+-Ca^{2+} exchange, extrusion by plasma membrane Ca^{2+}-ATPase, and uptake into mitochondria and/or the ER. Extrusion through the ATP-dependent Ca^{2+} pump, energized by the mitochondria, across the plasma membrane is the dominant form of Ca^{2+} removal from the bipolar cell synaptic terminals.[127] These mechanisms are, however, vulnerable to energy shortage as occurs in various disease states.

Sequestration of cytosolic Ca^{2+} by intracellular Ca^{2+} stores (ER and mitochondria) contributes substantially to Ca^{2+} clearance in neurons. In permeabilized cells, mitochondria can buffer moderate levels of $[Ca^{2+}]$—the so-called mitochondrial 'set point'—at around 1 μM (for review see ref. 96). The peak $[Ca^{2+}]_m$ of highly responsive mitochondria can be as high as a few hundred μM. Mitochondrial Ca^{2+} accumulation results from the close apposition of the organelles to either ER Ca^{2+} release channels or to plasma membrane Ca^{2+} channels (for review see ref. 100). Mitochondria take up Ca^{2+} primarily through a uniporter,[33] an electrogenic process. The ability to remove Ca^{2+} from local cytosol enables mitochondria to regulate the $[Ca^{2+}]$ in micro-domains close to ER Ca^{2+}-release channels. The sensitivity of the IP_3R/RyR-channels to Ca^{2+} means that, by regulating local $[Ca^{2+}]_c$, mitochondrial Ca^{2+} uptake modulates the rate and extent of propagation of $[Ca^{2+}]_c$ waves in a variety of cell types.

Two observations suggest that intracellular ER Ca^{2+} stores may also act as a buffering system for intracellular Ca^{2+}. First, KCl-induced increase in $[Ca^{2+}]_c$ in bullfrog sympathetic neurons is reported to be substantially attenuated after depletion of ryanodine-sensitive Ca^{2+} stores by prolonged caffeine application. Second, blockers of ER Ca^{2+}-ATPases have been found to prolong the depolarization-induced increases in dendritic $[Ca^{2+}]_c$ in rat neo-cortical layer V pyramidal neurons in slices.[133]

Neurotransmitter Release

VOCC Ca^{2+} entry, a fundamental signaling step in the central nervous system, provides an essential link between membrane depolarization and exocytosis at nerve terminals. $[Ca^{2+}]_c$ thereby profoundly influence neurotransmission that is proportional to the fourth power of $[Ca^{2+}]_c$.[31,85] The central role of Ca^{2+} in transmitter release is that Ca^{2+} triggers the formation of protein complex and drives membrane fusion in neurotransmitter exocytosis[22,118] in less than 1 ms.

Neurotransmitter release at many central synapses is initiated by an influx of Ca^{2+} ion through P/Q-type Ca^{2+} channels,[34,119] which are densely localized in nerve terminals. Intracellular Ca^{2+} does not appear to be involved since depletion of intracellular stores with 1 μM thapsigargin and 1 μM cyclopiazonic acid, two inhibitors of endosomal Ca^{2+}-ATPase activity that deplete all intracellular Ca^{2+} stores, does not affect basal synaptic transmission in the hippocampal CA1 Schaffer collateral pathway inputs.[99] On the other hand, intracellular Ca^{2+}-induced Ca^{2+} release has been shown to contribute to the Ca^{2+} transients in the boutons and to the paired pulse facilitation of excitatory postsynaptic potentials in the hippocampus.[35] Spontaneous transmitter release can occur in the absence of extracellular Ca^{2+} and is largely Ca^{2+} mediated, driven by Ca^{2+} release from internal stores. Boutons display spontaneous Ca^{2+} transients; blocking intracellular Ca^{2+} release reduces the frequency of these transients and of spontaneous miniature synaptic events.[35]

One critical question is: how high must $[Ca^{2+}]_c$ rise during an action potential in order to release a vesicle. In nerve terminals of bipolar cells from goldfish retina, exocytosis requires $[Ca^{2+}]_c$ larger than 100 μM.[83] Such concentrations are unlikely to be reached in the bulk of the cytosol. Thus, vesicles undergoing exocytosis are located within Ca^{2+} micro-domains. The micro-domain Ca^{2+} elevation serves a dual purpose: it permits limited Ca^{2+} elevation to achieve a high, localized maximum regulatory impact for maintaining input specificity of synaptic plasticity and for reducing the risk of excitotoxicity. At fast synapses, step-like elevations to 10 μM $[Ca^{2+}]_c$ have been shown to induce fast transmitter release, deleting around 89% of a pool of available vesicles in less than 3 ms,[102] less than the general assumed 100 μM.[53,72,126] Thus, transient (around 0.5 ms) local elevations of $[Ca^{2+}]_c$ to peak values as low as 25 μM can account for transmitter release during single presynaptic action potentials.

Modulation of Channel Activity

Increases in $[Ca^{2+}]_c$ activate the Ca^{2+}-dependent K^+ channel, either large (BK) or small (SK) conductance, ($K_{Ca}2+$),[7,105,124] limiting the firing frequency of repetitive action potentials. In hippocampal neurons, activation of BK channels underlies the falling phase of the action potential and generation of the fast afterhyperpolarization. In contrast, SK channel activation underlies generation of the slow afterhyperpolarization after a burst of action potentials. The source of Ca^{2+} for BK channel activation is probably N-type channels, which activate the BK channel only, with opening of the two channel types being nearly coincident,[77] suggesting that the N-type Ca^{2+} and BK channels are functionally very close. Direct coupling of NMDA receptors to BK-type Ca^{2+}-activated K^+ channels has also been reported in the inhibitory granule cells of rat olfactory bulb.[60] The slow afterhyperpolarization is blocked by dihydropyridine antagonists, indicating that L-type Ca^{2+} channels provide the Ca^{2+} for activation of SK channels. L-type channels activate SK only and the delay between the opening of L-type channel and SK channels indicates that these two types of channels are 50-150 nm apart.[77] Thus, there exists an absolute segregation of coupling between channels, indicating the functional importance of submembrane Ca^{2+} micro-domains. Some of these effects on K^+ channels may be mediated by Ca^{2+}-binding signal proteins.[88]

Long-Term Changes of Ca^{2+}-Influx via Memory-Specific K^+ Channel Regulation

Memory-related Ca^{2+} signals are decoded through altered operation of membrane channels, including K^+ channels. K^+ channels play an important role in memory formation (for review, see Vernon and Giese in this book). The phosphorylation and dephosphorylation of the Shaker-related fast-inactivating Kv1.4 is regulated by $[Ca^{2+}]_c$.[134] CaMKII phosphorylation of an amino-terminal residue of Kv1.4 leads to N-type inactivated states. Dephosphorylation of this residue induces a fast inactivating mode. Associate learning paradigms in a variety of species have now been closely correlated with long-term changes of voltage-dependent K^+ channels, particularly those in the Shaker family and those that are Ca^{2+}-dependent. Voltage-dependent I_A channels were shown to occur in the single identified type B cells of the Mollusk Hermissenda only when the animal acquired a Pavlovian- conditioned response.[4] The same type of K^+ channel change was demonstrated to last even one month in duration in the post-synaptic dendrites of the cerebellar HVI Purkinje cells only when a rabbit had acquired and retained a Pavlovian-conditioned eye-blink response.[103,104] Similar changes of a post-synaptic K^+ channels were found in the rabbit hippocampus and were correlated with enhanced EPSP summation.[25,74]

These correlated learning-specific changes were found to bear a causal relationship to the acquisition of associative learning using an antisense strategy. Antisense "knockdown" of Shaker postsynaptic Kv1.1 K^+ channels in the hippocampus eliminated retention of a spatial maze learning task[81] while "knockdown" of the presynaptic Kv1.4 K^+ channel did not alter learning or memory of the task.[82]

Such memory-specific reductions of voltage-dependent as well as GABA-mediated K^+ conductance will enhance synaptic depolarization of post-synaptic membranes and thereby enhance opening of VOCC. In this way, learning-specific reduction of K^+ conductance will increase Ca^{2+} influx across the plasma membrane. During learning and even retention, enhanced voltage-dependent Ca^{2+} influx can combine with learning-specific enhancement of intracellular Ca^{2+} release via the RyR and IP_3R to cause further activation of downstream Ca^{2+}-dependent molecular cascades.

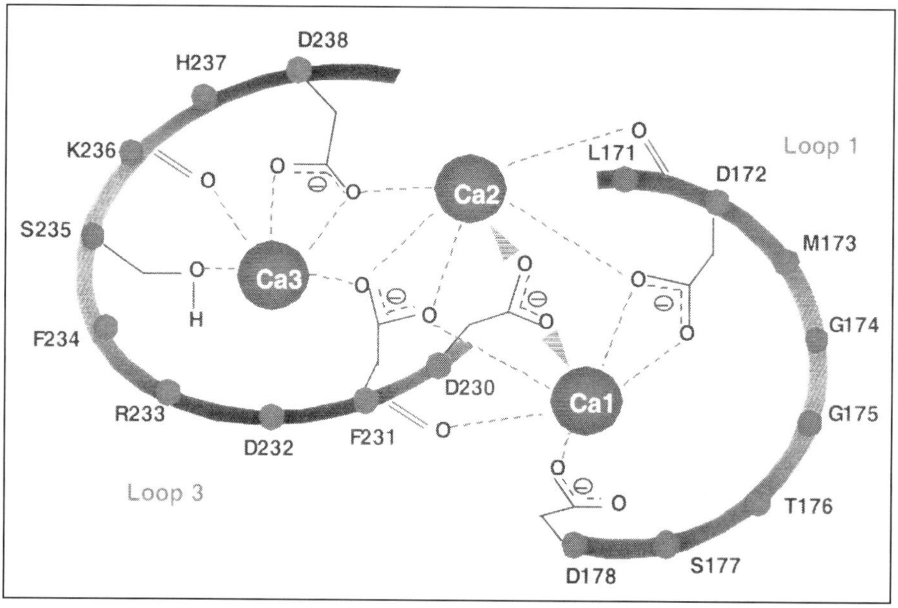

Figure 2. Model of Ca^{2+} binding by C_2 motifs of synaptotagmin I. The Ca^{2+} binding residues are in loop 1 and loop 3. Solid circles represent residues shown in single-letter amino acid code and identified by number (adapted from refs. 39 and 121).

Signal Transduction Cascades

One critical role of Ca^{2+} in neuronal signaling is to couple electrical excitation to the activation of intracellular enzymes, such as various tyrosine protein kinases,[128,130] and signal transduction cascades (Fig. 1). Ca^{2+} regulates a wide variety of biological functions through binding to proteins, so to confine it neatly to one predominant role in mediating effects of signal transduction on synaptic plasticity may be unrealistic.

Most Ca^{2+}-binding proteins can be grouped into families with common structural motifs such as the EF-hand motif[29] or the C_2 modif.[106] The EF-hand motif in L-type Ca^{2+} channels, for instance, is required for initiating Ca^{2+}-sensitive inactivation of the channel.[29] Examples of proteins that contain the C_2 motifs include synaptotagmin and PKC. Synaptotagmin I is a synaptic vesicle protein that involves the coordination of two or three Ca^{2+} ions by five aspartate residues (Fig. 2), one serine residue, and two backbone carbonyl groups located on two separate loops.[39,106,121] Ca^{2+} binding of synaptotagmin initiates vesicle fusion and transmitter release, a basic communication means neurons rely on in information processing for a variety of functions including learning and memory. Ca^{2+}-mediated activation of PKC, on the other hand, plays important roles in associative learning.[3,6] Ca^{2+} also affects a variety of protein kinases and other signal molecules. Many of them play important roles in synaptic plasticity and gene transcription.

Information Coding

Many cellular stimuli result in oscillations in $[Ca^{2+}]_c$. The frequency of such oscillations may encode information and can be important for the induction of selective cellular functions. The frequency, duration, and amplitude of Ca^{2+} oscillations modulate activity of the Ca^{2+}- and calmodulin-dependent protein kinase II (CaM kinase II).[28] A role for repetitive Ca^{2+} spikes has also been suggested for the activation of mitochondrial ATP production,[51] activation of PKC[94] and CaMKII,[28,84] and gene expression.[32,70]

Receptor stimuli that triggered repetitive Ca^{2+} spikes induce a parallel repetitive translocation of PKCγ to the plasma membrane.[94] While Ca^{2+} acts rapidly, diacylglycerol binding to PKCγ is initially prevented by a pseudosubstrate clamp, which keeps the diacylglycerol-binding site inaccessible and delays Ca^{2+}- and diacylglycerol-mediated kinase activation. After termination of Ca^{2+} signals, bound diacylglycerol prolongs kinase activity. The properties of this molecular decoding machine make PKCγ responsive to persistent diacylglycerol increases combined with high- but not low-frequency Ca^{2+} spikes.

Axon Growth

Ca^{2+} transients are environmentally regulated to control axon growth. The motile growth cone at the tip of the axon is sensitive to its $[Ca^{2+}]_c$. Large increases evoked by neurotransmitters or depolarization cause growth cone to collapse, stopping neuritic elongation. NI-35, a growth-inhibitory protein expressed on oligodendrocytes in the CNS, induces growth cones in culture to collapse, associated with a large rise in $[Ca^{2+}]_c$, at least partially due to release from the smooth ER.[11] Growth cones generate transient elevations of $[Ca^{2+}]_c$ and the rate of axon outgrowth is inversely proportional to the frequency of transients.[45] Blockade of Ca^{2+} release prevents collapse. Decreases caused by removing Ca^{2+} from the bathing medium can have similar effects. In some cases, growth cone activity and neuritic elongation can be promoted by elevation of $[Ca^{2+}]_c$ over resting levels; focal changes within the growth cone can produce focal protrusive activity appropriate for changing the direction of growth.[44]

Synaptic Plasticity

Memories are believed to result from changes in synaptic strengths. Synapses are the specialized connections that allow signals to propagate from one nerve cell to the next. Their privileged position and dynamic nature give them a unique role in neural computation. There are about 7-8 x 10^8 synapses in the dentate gyrus of the rat alone. The number of synapses in the human cerebral cortex is undoubtedly many orders of magnitude higher. Activity-dependent changes in the efficacy of synaptic transmission are a basic feature of many synapses in the central nervous system and are believed to underlie memory formation in the brain. Despite the central role for synaptic plasticity in learning and memory, mechanisms underlying synaptic plasticity remain incompletely understood. One of the central challenges of neuroscience is therefore to understand the mechanism of synaptic plasticity.

$[Ca^{2+}]_c$ signals are essential for the induction of synaptic plasticity.[3,6] Ca^{2+} together with diacylglycerol and arachidonic acid then cause PKC activation, which, in turn, is responsible for enhanced synaptic signals.[74] This Ca^{2+} and PKC pathway activated during associative learning in turn activates a series of molecular events such as the release of Ca^{2+} via the RyR, Src-combination with synapsin and synaptophysin, and long-term synthesis of specific proteins such as the RyR itself. Thus, learning-specific initial changes of Ca^{2+} homeostasis are responsible for much longer-lasting molecular changes that themselves are responsible for long-lasting changes of Ca^{2+} homeostasis.[21,129] Many synaptic studies have been performed on neural network in the hippocampus, a major component of the medial temporal lobe, a brain system that plays an important role in declarative or relational memory, those related to personal experience ('episodic memory') and ability to consciously recollect events from everyday experience set within spatiotemporal contexts.

Long-Term Modifications of Synapses

Ca^{2+} plays a crucial role in the induction of all the known forms of synaptic plasticity, long-term potentiation (LTP), depression (LTD see ref. 23), synaptic transformation (LTT see ref. 5,24,62), and enhanced EPSP summation,[74] the putative cellular mechanisms of learning and memory. Ca^{2+} is required to regulate postsynaptic enzymes that trigger rapid modifications of synaptic strengths and also to activate transcription factors that facilitate long-lasting maintenance of these modifications. For instance, in the hippocampal CA1 region, LTP, LTD, and LTT are all blocked by postsynaptic chelators of Ca^{2+} and are thus Ca^{2+}-dependent.

LTP of glutamatergic EPSPs received by the hippocampal pyramidal cells is induced by high frequency (≥100 Hz) stimulation of the presynaptic Schaeffer collateral inputs. High frequency stimulation of this Schaeffer collateral pathway activates NMDA receptors, resulting in an initial Ca^{2+} influx, an event that is believed by many to be essential for LTP induction.[120] The associated Ca^{2+} release from intracellular stores may determine whether LTP or LTD is expressed by activation in the hippocampal CA1 region.[91] Thus, blocking RyR eliminates homosynaptic LTD while blockade or deletion of IP_3R1 leads to a conversion of LTD to LTP and elimination of heterosynaptic LTD.[91] Reduction of Ca^{2+} influx through a partial blockade of NMDA receptors also results in a conversion of LTP to LTD.[91]

LTD can be induced either by low frequency stimulation (1 Hz/15 min) of presynaptic fibers, for instance, the Schaeffer collateral pathway, or in a related manner by asynchronous pairing of presynaptic and postsynaptic activity (for instance asynchronous pairing of postsynaptic action potentials with EPSPs evoked with a delay of 20 ms; 0.3 or 1 Hz for 360s) in slices from young rat brains. According to Reyes and Stanton,[99] induction of LTD by low frequency stimulation alone requires release of Ca^{2+} both from a presynaptic ryanodine pool and from postsynaptic (presumably IP_3-gated) stores. Bath application of ryanodine (10 μM) blocks LTD induction, but impalement of CA1 pyramidal cells with microelectrodes containing ryanodine (2 μM to 5 mM) does not, whereas impalement with microelectrodes containing thapsigarin (500 nM to 200μM) does.[99] Unlike the LTD induced by low frequency stimulation alone, associative LTD induction is independent of NMDA receptors but dependent of mGluR activation and L- and N- VOCC activation.[92]

Central to our understanding learning mechanisms at a synaptic level is the idea that lasting functional change can be driven by the coincidence of multiple signals at a single synaptic site. One candidate for such a change is LTT, a long-term synaptic transformation of GABAergic postsynaptic response from inhibitory to excitatory.[5,24] The induction of LTT requires either cholinergic and GABAergic inputs and/or an associative post-synaptic $[Ca^{2+}]_c$ increase. Its induction by associative activation with calexcitin has been found to be sensitive to RyR blockade,[112] suggesting an essential role of intracellular Ca^{2+} release. Learning-specific up-regulation of the RyR synthesis in this way can facilitate long-term changes of specific GABAergic synapses.[49]

Postsynaptic Switch

Activity-dependent change in the efficacy of transmission through the AMPAR involves alteration in the number and phosphorylation site of postsynaptic AMPARs. Repetitive synaptic activation of Ca^{2+}-permeable AMPARs lacking the GluR2 subunit causes a rapid reduction in Ca^{2+} permeability owing to the incorporation of GluR2-containing AMPARs on cerebellar stellate cells,[135] suggesting a self-regulating mechanism.

Ca^{2+} may mediate a dual function of glutamate and GABA receptors. mGluR activation is generally found to be excitatory. However, depending on the frequency and pattern of afferent input, glutamate can induce an excitation or inhibition by activation of the same mGluR1 receptor.[40] In ventral midbrain dopamine neurons, rapid activation of metabotropic glutamate receptors (mGluR1) induces a pure IPSP, mediated by Ca^{2+} release from ryanodine-sensitive stores,[40] whereas slow and prolonged synaptic activation of the mGluRs may result in a slow EPSP, with suppression of the IPSP. Heterosynaptic interaction of cholinergic and GABAergic synapses may result in a transformation of GABAergic postsynaptic response of the CA1 pyramidal cells from inhibitory to excitatory. The transformation dramatically alters the signal-to-noise ratio and a switch from an excitatory filter to an excitatory amplifier, and thus, the direction of signal transfer through the network.[113-115]

Synaptic Interaction and Associative Learning

Ca^{2+} homeostasis is directly related to learning and memory. First, learning and memory depend on the Ca^{2+}-mediated transmitter release for the associative integration of relevant inputs. Changes in the intensities of neurotransmitter, such as glutamatergic, cholinergic,

GABAergic activities, dramatically alter the signal transfer through the neural network and synaptic plasticity. Excitatory inputs into the hippocampal pyramidal cells, for instance, rapidly change the firing rate of the cells. A large fraction of the pyramidal cells have place fields in any environment. When a rat arrives at a particular location, the 'place field', the firing rate of a particular 'place cell' can exceed 100 Hz from a baseline of < 1 Hz, although during some passes through the place field the cell may not fire at all. Once established, place cells can have the same firing pattern for months.[117] Second, synaptic plasticity that underlies memory formation depends on intracellular Ca^{2+} release (see below). Deficits of cholinergic release/inputs into the hippocampal pyramidal cells are believed to be responsible for the memory decline seen in the AD and elderly.

Oxygen-Sensing and Hypoxic Injury

The brain can be characterized as a metabolically very active organ but has few energy reserves. It must receive adequate and continuous supplies of oxygenated blood and glucose. Normally, as much as 50-60% of the brain cell's energy expenditure may be spent on transporting ions across the cell membranes in order to maintain cellular ion homeostasis,[136,137] including Ca^{2+} homeostasis. Brain ischemia, often resulting in stroke, is a common disorder with a high rate of morbidity and mortality and may be caused by cerebrovascular disruption or hemorrhage, brain tumor, intracranial and/or extracranial inner carotid artery occlusion (e.g., cardiac source embolism or arteriosclerosis), or cardiac arrest. When oxygen supply is halted, there is an initial increase in glycolysis, which is insufficient, however, to make up the energy deficit. The cardiovascular system responds by reorganizing oxygenated blood distribution to the brain.[109] If the insult lasts, after a few minutes there are major perturbations in the energy status of the brain. The efficiency of ion pumps is compromised and there are net movements of ions across the cell membrane down their concentration gradients. Consequently, there is an increase in extracellular K^+, which results in depolarization and an increase in $[Ca^{2+}]_c$.

It is widely believed that disturbances of Ca^{2+} homeostasis play a major role in the pathological process in cell injury of neurons induced by hypoxia/ischemia. An elevation of $[Ca^{2+}]_c$ may result from several factors. First, within minutes following hypoxia-ischemia, neurons are confronted with reduced energy availability, resulting in suppression of the operation of membrane Ca^{2+} pumps. Second, injured cells release K^+, which may depolarize the membrane, resulting in Ca^{2+} influx through the VOCC. Third, Ca^{2+} may be released from intracellular stores. Fourth, there is experimental evidence that the β amyloid protein that accumulates in Alzheimer's disease can potentiate excitotoxic degeneration. Hypoxia/ischemia induces the production of the amyloid β protein, which can form Ca^{2+} channels in bilayer membranes and may contribute to its neurotoxic effects.

Mitochondrial Ca^{2+} may be involved in hypoxic injury. In the progressive transfer of electrons ultimately to molecular oxygen, the respiratory chain also translocates protons across the mitochondrial inner membrane. This process creates and sustains the mitochondrial inner membrane potential ($\Delta\Psi_m$) of some 150 mV negative to the cytosol (together with a low resting concentration of $[Ca^{2+}]_m$, maintained primarily by the mitochondrial Na^+-Ca^{2+} exchanger. Na^+ is then exchanged for protons through a rapid Na^+-H^+ exchange) that provides the energy required to drive the phosphorylation of ADP to ATP. Isolated mitochondria will accumulate Ca^{2+} with impunity in the presence of ATP. A massive influx of Ca^{2+} into the mitochondria leads to production of reactive oxygen species (ROS; Fig. 1), opening of the mitochondrial permeability transition pore and disturbance of energy metabolism. This occurs especially during Ca^{2+} uptake in the absence of ATP or in the presence of pro-oxidants, leading to the release of apoptotic factors from mitochondria. It has been suggested that programmed cell death involves the generation of ROS. Elevations of $[Ca^{2+}]_c$ induce oxidative stress by several mechanisms: activation of nitric oxide synthase (whose product nitric oxide interacts with superoxide anion radical, resulting in production of peroxynitrite), impairment of mitochondrial function (resulting in increased superoxide production by the organelle), and activation of enzymatic cascades that include various oxygenases.[78] Thus, preventing mitochondrial

Ca^{2+} uptake by depolarizing mitochondria with a mitochondrial uncoupler can be neuroprotective.[108]

ER Ca^{2+} stores are also involved in hypoxic injury. This is based on the observations that the stores are depleted after severe hypoxia. As mentioned above, ER Ca^{2+} is required for protein synthesis. Persistent suppression of protein synthesis due to Ca^{2+} depletion induced by hypoxia/ischemia apparently contributes to the pathological process. Following cerebral hypoxia/ischemia, recovery of protein synthesis is closely related to the recovery of cells from metabolic disturbance: protein synthesis recovers in resistant brain regions, but not in areas vulnerable to transient hypoxia/ischemia.

Ca^{2+} is a signal for both life and death. Ca^{2+} can trigger apoptosis.[71] In rat hippocampal CA1 pyramidal cells, hypoxia induces L-glutamate release, Ca^{2+} influx, and Ca^{2+} release probably from IP_3-sensitive stores.[14] Increases in $[Ca^{2+}]_c$ induce mitochondiral Ca^{2+} overload and trigger the production of ROS,[122] which play a central role in hypoxic damage. High $[Ca^{2+}]_m$ plus NO is particularly damaging.[33] NO has been found to activate cardiac RyR by poly-S-nitrosylation in canines.[125] Hypoxia increases Ca^{2+} influx in many types of neurons.[111] In skeletal muscle, the Ca^{2+}-release RyR1 channel has been found to couple the O_2 sensor and NO signaling functions, with most efficient activation at low NO and O_2 concentrations.[37] Ca^{2+} also activates Ca^{2+}-dependent proteases in vulnerable neuronal populations. Ca^{2+}-induced death can be of either the necrotic or apoptotic type. Uncontrolled elevation of $[Ca^{2+}]_c$ has been implicated in neurotoxicological responses and ischemia, by activating phospholipases. Activated phospholipases break down membranes and produce toxic metabolites such as arachidonic acid, proteases. Active proteases break down the cytoskeleton, enzymes, receptors and channels, and endonucleases. The latter induce DNA fragmentation.

Hypoxia/ischemic stroke dramatically impairs learning and memory.[110] Long-term memory decline is evident even after brief episode of hypoxia/ischemia.[90] Therapeutic interventions designed to suppress disturbances of Ca^{2+} homeostasis induced by hypoxia/ischemia must protect mitochondria from Ca^{2+} overload. At the same time, such intervention must prevent the ER from undergoing Ca^{2+} depletion.

Gene Expression

One of the physiological functions of activation of VOCCs[18] and intracellular Ca^{2+} release is to regulate pathways controlling transcription, either through $[Ca^{2+}]_c$ waves and oscillations, or nuclear Ca^{2+} sensor[20,52] that underlies long-lasting cellular events. In hippocampal neurons, electrical activity or K^+ depolarization has been shown to result in rapid translocation of the NF-Atc family of transcription factors from the cytoplasm to the nucleus, activating NF-AT-dependent transcription. These responses require Ca^{2+} influx through L-type VOCCs.[46] GSK-3, a Ser/Thr kinase, can phosphorylate NF-Atc4, promoting its export from the nucleus and antagonizing NF-Atc4-dependent transcription.[46] Induction of the IP_3R1 is also controlled by the Ca^{2+}/NF-Atc pathway.[46] Ultraviolet illumination has been used to release a caged $InsP_3$ analogue after it diffuses into intact cells.[70] The released analogue elicits $[Ca^{2+}]_c$ spikes. Although this study was performed on nonneuronal cells, the findings that the same IP_3 analogue elicits even more gene expression when released by repetitive flashes at 1-minute intervals than at 0.5- or ≥ 2-minute intervals, as a single pulse, or as a slow sustained plateau,[70] may imply general rules for engaging the Ca^{2+}-gene expression cascade. Thus, oscillations in $[Ca^{2+}]_c$ levels at approximately physiological rates may maximize gene expression for a given amount of $InsP_3$, and a well-defined signal-transduction cascade into the nucleus may be tuned to the frequency of $[Ca^{2+}]_c$ spikes.[69] A single burst of IP_3 or excessively frequent oscillations of IP_3 may fail to maintain elevated $[Ca^{2+}]_c$ levels for sufficient periods to trigger gene expression. The lower-frequency oscillations, on the other hand, may allow too much time for rephosphorylation and nuclear exit of NF-Atc between pulses. Slow, steady production of IP_3 is remarkably ineffective at increasing $[Ca^{2+}]_c$ levels for prolonged periods, perhaps because of IP_3R desensitization.[93] The Ca^{2+} waves, however, are more efficient because IP_3 biosynthesis uses many ATP molecules and depletes stores of the scarce lipid phosphatidylinositol-4,5-bisphosphate.

Learning and memory requires gene transcription triggered by synaptically evoked Ca^{2+} signals. Hippocampal neurons are able to convert a burst frequency coded signal in the dendrites into a prolonged nuclear Ca^{2+} amplitude coded signal,[52,80] involving no nuclear import. The frequency-to-amplitude conversion provides a mechanism through which neuronal impulse patterns shape genomic responses.

Alzheimer's Disease

Abnormal Ca^{2+} homeostasis characterizes pathophysiology of Alzheimer's disease (AD). For instance, in AD fibroblasts, bombesin- and bradykinin-induced Ca^{2+} release (through IP_3Rs) is greatly enhanced, compared with those from control groups.[55,61] The Ca^{2+}-mediated acetylcholine release from rat hippocampal slice is potently and acutely inhibited by low concentration (10^{-8} M) of β-amyloid.[64]

β-Amyloid

β-Amyloid (Aβ) causes the death of cortical neurons at micromolar concentrations[38] and can directly form Ca^{2+} channels. It has been suggested that unregulated Ca^{2+} influx via the Aβ-channels may underlie the molecular mechanism of Aβ neurotoxicity and of the AD pathogenesis. Aβ is a hydrophobic peptide and has the intrinsic property of forming aggregates with β-pleated sheet structures. Low concentrations of Aβ increase tyrosine phosphorylation and $[Ca^{2+}]_c$.[75] Incorporation of $A\beta_{1-40}$ into artificial lipid bilayer membranes forms cation-selective (including Ca^{2+}) ion channels[8,9] or $A\beta_{25-35}$ in membranes of acutely dissociated rat cerebral cortical neurons.[42] Formation of Ca^{2+}-conducting channels has been reported in the inside-out membrane patches from immortalized murine hypothalamic neurons within 3-30 min of the addition of $A\beta_{1-40}$ at 4.7 μM,[65] with spontaneous conductance changes over a wide range of 50-500 pS. The channel activity can be inhibited by 250 μM zinc in the bath solution.[65] The secreted form of β-amyloid-precusor protein (APP) is found to attenuate the increase in $[Ca^{2+}]_c$ evoked by L-glutamate in rat cultured hippocampal neurons.[67] APP itself evokes an increase in $[Ca^{2+}]_c$ in 1 or 2 day-cultured hippocampal cells, but not in 7 to 13 day-cultured cells.[67] The APP-induced $[Ca^{2+}]_c$ increase involves an increase in IP_3 and brief intracellular Ca^{2+} release, which triggers a large Ca^{2+} influx[67] and is thus development stage-dependent. On the other hand, intracellular RyR Ca^{2+} release increases the release of Aβ (by 4-fold with 5-10 mM caffeine[97]), whereas thapsigargin, an irreversible inhibitor of Ca^{2+} reuptake from the ER, has been shown to reduce the formation of Aβ.[19]

β-Amyloid also inactivates voltage-dependent K^+ channels in nM concentrations.[36] These same K^+ channels are down regulated in cells of Alzheimer's patients and appear to be diagnostic of the disease. β-Amyloid-reduced K^+ channel activity will also enhance Ca^{2+} influx through VOCC. Furthermore, Alzheimer's-specific enhancement of IP_3-mediated release of intracellular Ca^{2+} has also been observed. Finally, an Alzheimer's gene, presenilin 1, is known to bind a RyR ligand known as calsenilin. This ligand may be responsible for Alzheimer's-specific enhancement of RyR-mediated Ca^{2+} release.

Conclusion

Ca^{2+} is a key regulator of various biological processes, including molecular events related to synaptic plasticity, memory storage and recall. It remains a daunting challenge for memory scientists to elucidate the mechanisms by which memory-related cellular/network events are controlled and the specific roles played by $[Ca^{2+}]_c$ in the critical processes. All the currently known forms of synaptic plasticity that might be involved in memory formation in the brain depend on temporal and spatial increases in $[Ca^{2+}]_c$. Abnormalities of endogenous mechanisms involved in the effective control of $[Ca^{2+}]_c$ are also known to contribute to various neurodegenerative disorders.[66] Elevated $[Ca^{2+}]_c$, however, can induce the synaptic plasticity that underlies memory traces, but also trigger neurodegenerative cascades that lead to the death of the same neurons. The distinction of the two may involve temporal, spatial, and compartmentally specific concentrations of the $[Ca^{2+}]_c$ changes, i.e., temporal changes in Ca^{2+} signal micro-domains, particularly in relation to other signaling events. A comprehensive understanding of the events, then, will be essential for the treatment of memory impairments without evoking neural injury and neurodegeneration. Because of its complexity in space and time, perhaps it is inevitable that many questions remain about the comparative physiology and pathophysiology of Ca^{2+} homeostasis. However, it is our hope that a sufficient understanding of the critical Ca^{2+} homeostatic mechanisms will soon yield therapeutic benefits for the amelioration of neurodegenerative disorders as well as cognitive impairments involving attention, learning, and/or memory.

References

1. Alkon DL. Voltage-dependent calcium and potassium ion conductances: A contingency mechanism for an associative learning model. Science 1979; 205:810-816.
2. Alkon DL. Molecular specificity of synaptic changes responsible for associative memory. Adv Exp Med Biol 1998; 446:1-15.
3. Alkon DL. Ionic conductance determinants of synaptic memory nets and their implications for Alzheimer's disease. J Neurosci Res 1999; 58:24-32.
4. Alkon DL, Lederhendler I, Shoukimas JL. Primary changes of membrane currents during retention of associative learning. Science 1982; 215:693-695.
5. Alkon DL, Sanchez-Andres JV, Ito E et al. Long-term transormation of an inhibitory into an excitatory GABAergic synaptic response. Proc Natl Acad Sci USA 1992; 89:11862-11866.
6. Alkon DL, Nelson TJ, Zhao WQ et al. Time domains of neuronal Ca^{2+} signaling and associative memory: Steps through a calexcitin, ryanodine receptor, K^+ channel cascade. Trends Neurosci 1998; 21:529-537.
7. An WF, Boelby MR, Betty M et al. Modulation of A-type potassium channels by a family of calcium sensors. Nature 2000; 403:553-556.
8. Arispe N, Rojas E, Pollard HB. Alzheimer disease amyloid β protein forms calcium channels in bilayer membranes: Blockade by tromethamine and aluminium. Proc Natl Acad Sci USA 1993; 90:567-571.
9. Arispe N, Rojas E, Pollard HB. Giant multilevel cation channels formed by Alzheimer disease amyloid β protein [AβP-(1-40)] in bilayer membranes. Proc Natl Acad Sci USA 1993; 90:10573-10577.
10. Ascoli GA, Luu KX, Olds JL et al. Secondary structure and Ca^{2+}-induced conformational change of calexcitin, a learning-associated protein. J Biol Chem 1997; 272:24771-24779.
11. Bandtlow CE, Schmidt MF, Hassinger TD et al. Role of intracellular calcium in NI-35-evoked collapse of neuronal growth cones. Science 1993; 259:80-83.
12. Bannon AW, McMonagle-Strucko K, Fanelli RJ. Nimodipine prevents medial septal lesion-induced performance deficits in the Moris water maze. Psychobiol 1993; 21:209-214.
13. Béguin P, Nagashima K, Gonol T et al. Regulation of Ca^{2+} channel expression at the cell surface by the small G-protein kir/Gem. Nature 2001; 411:710-706.
14. Belousov AB, Godfraind J-M, Krnjevi K. Internal Ca^{2+} stores involved in anoxic responses of rat hippocampal neurons. J Physiol Lond 1995; 486:547-556.
15. Berridge MJ, Bootman MD, Lipp P. Calcium – a life and death signal. Nature 1998; 395:645-648.
16. Bezprozvanny I, Watras J, Ehrlich BE. Bell-shaped calcium-response curves of Ins(1,4,5)P3- and calcium-gated channels from endoplasmic reticulum of cerebellum. Nature 1991; 351:751-754.
17. Blackwell KT, Alkon DL. Ryanodine receptor modulation of in vitro associative learning in Hermissenda crasscornis. Brain Res 1999; 822:114-125.
18. Brosenitsch TA, Katz DM. Physiological patterns of electrical stimulation can induce neuronal gene expression by activating N-type calcium channels. J Neurosci 2001; 21:2571-2579.
19. Buxbaum JD, Ruefly AA, Parker CA et al. Calcium rewgulates processing of the Alzheimer amyloid protein precusor in a protein kinase C-independent manner. Proc Natl Acad Sci USA 1994; 91:4489-4493.
20. Carrión AM, Link WA, Ledo F et al. DREAM is a Ca^{2+}-regulated transcription repressor. Nature 1999; 398:80-84.
21. Cavallaro S, Meiri N, Yi CL et al. Late memory-related genes in the hippocampus revealed by RNA fingerprinting. Proc Natl Acad Sci USA 1997; 94:9669-9673.
22. Chen YA, Scales S, Patel SM et al. SNARE complex formation is triggered by Ca^{2+} and drives membrane fusion. Cell 1999; 97:165-174.
23. Christie BR, Schexnayder LK, Johnston D. Contribution of voltage-gated Ca^{2+} channels to homosynaptic long-term depression in the CA1 region in vitro. J Neurophysiol 1997; 77:1651-1655.

24. Collin C, Devane WA, Dahl D et al. Long-term synaptic transformation of hippocampal CA1 gamma-aminobutyric acid synapses and the effect of anandamide. Proc Natl Acad Sci USA 1995; 92:10167-10171.
25. Coulter DA, Lo Turco JJ, Kubota M et al. Classical conditioning reduces amplitude and duration of calcium-dependent afterhyperpolarization in rabbit hippocampal pyramidal cells. J Neurophysiol 1989; 61:971-981.
26. Davare MA, Avdonin V, Hall DD et al. A β_2 adrenergic receptor signaling complex assembled with the Ca^{2+} channel $Ca_v1.2$. Science 2001; 293:98-101.
27. David G. Mitochondrial clearance of cytosolic Ca^{2+} in stimulated lizard motor nerve terminals proceeds without progressive elevation of mitochondrial matrix $[Ca^{2+}]$. J Neurosci 1999; 19:7495-7506.
28. De Koninch P, Schulman H. Sensitivity of CaM kinase II to the frequency of Ca^{2+} oscillations. Science 1998; 279:227-230.
29. De Leon M, Wang Y, Jones L et al. Essential Ca^{2+}-binding motif for Ca^{2+}-sensitive inactivation of L-type Ca^{2+} channels. Science 1995; 270:1502-1506.
30. DeMaria CD, Soong TW, Alseikhan BA et al. Calmodulin bifurcates the local Ca^{2+} signal that modulates P/Q-type Ca^{2+} channels. Nature 2001; 411:484-489.
31. Dodge Jr FA, Rahamimoff R. Cooperative action of calcium ions in transmitter release at the neuromuscular junction. J Phsyiol Lond 1967; 193:419-432.
32. Dolmetsch RE, Xu K, Lewis RS. Calcium oscillations increase the efficiency and specificity of gene expression. Nature 1998; 392:933-936.
33. Duchen MR. Mitochondria and calcium: From cell signaling to cell death. J Physiol Lond 2000; 529:57-68.
34. Dunlap K, Luebke JL, Turner TJ. Exocytotic Ca^{2+} channels in mammalian central neurons. Trends Neurosci 1995; 18:89-98.
35. Emptage NJ, Reid CA, Fine A. Calcium stores in hippocampal synaptic boutons mediate short-term plasticity, storeoeprated Ca^{2+} entry, and spontaneous transmitter release. Neuron 2001; 29:197-208.
36. Etcheberrigarray R, Ito E, Kim CS et al. Soluble beta-amyloid induction of Alzheimer's phenotype for human fibroblast K^+ channels. Science 1994; 264:276-279.
37. Eu JP, Sun JH, Xu L et al. The Skeletal muscle calcium release channel: Coupled O_2 sensor and NO signaling functions. Cell 2000; 102:499-509.
38. Favit A, Grimaldi M, Alkon DL. Prevention of beta-amyloid neurotoxicity by blockade of the ubiquitin-proteasome proteopytic pathway. J Neurochem 2000; 75:1258-1263.
39. Fernández-Chacón R, Könlgstorfer A, Gerber SH et al. Synaptotagmin I functions as a calcium regulator of release probability. Nature 2001; 410:41-49.
40. Fiorillo CD, Williams JT. Glutamate mediates an inhibitory postsynaptic potential in dopamine neurons. Nature 1998; 394:78-82.
41. Forsythe ID, Tsujimoto T, Barnes-Davies M et al. Inactivation of presynaptic calcium current contributes to synaptic depression at a fast central synapse. Neuron 1998; 20:797-807.
42. Furukawa K, Abe Y, Akaike N. Amyloid β protein-induced irreversible current in rat cortical neurons. NeuroReport 1994; 27:2016-2018.
43. Garaschuk O, Yaari Y, Konnerth A. Release and sequestration of calcium by ryanodine-sensitive stores in rat hippocampal neurones. J Physiol Lond 1997; 502:13-30.
44. Goldberg DJ, Grabham PW. Braking news: Calcium in the growth cone. Neuron 1999; 22:423-425.
45. Gomez TM, Spitzer NC. In vivo regulation of axon extension and pathfinding by growth-cone calcium transients. Nature 1999; 397:350-355.
46. Graef IA, Mermelstein PG, Stankunas K et al. L-type calcium channels and GSK-3 regulate the activity of NF-Atc4 in hippocampal neurons. Nature 1999; 401:703-708.
47. Grimaldi M, Favit A, Alkon DL. cAMP-induced cytoskeleton rearrangement increases calcium transients through the enhancement of capacitance calcium entry. J Biol Chem 1999; 274:33557-33564.
48. Grimaldi M, Atzori M, Ray P et al. Mobilization of calcium from intracellular stores, potentiation of neurotransmitter-induced calcium transients, and capacitance calcium entry by 4-aminopyridine. J Neurosci 2001; 21:3135-3143.
49. Gusev PA, Alkon DL. Intracellular correlates of spatial memory acquisition in hippocampal slices: Long-term disinhibition of CA1 pyramidal cells. J Neurophysiol 2001; 86:881-899.
50. Haga RE, Burgstahler AD, Nathanson MH et al. Type III $InsP_3$ receptor channel stays open in the presence of increased calcium. Nature 1998; 396 81-84.
51. Hajnoczky G, Robb-Gaspers LD, Seitz MB et al. Decoding of cytosolic calcium oscillations in the mitochondria. Cell 1995; 82:415-424.
52. Hardingham GE, Arnold FJL, Bading H. Nuclear calcium signaling controls CREB-mediated gene expression triggered by synaptic activity. Nature Neurosci 2001; 4:261-267.

53. Heidelberger R, Heinemann C, Neher E et al. Calcium dependence of the rate of exocytosis in a synaptic terminal. Nature 1994; 371:513-515.
54. Herlitze S, Garcia DE, Mackie K et al. Modulation of Ca^{2+} channels by G protein βγ subunits. Nature 1996; 380:258-262.
55. Hirashima N, Etcheberrigaray R, Bergamaschi S et al. Calcium responses in human fibroblasts: A diagonostic molecular profile for Alzheimer's disease. Neurobiol Aging 1997; 17:549-555.
56. Höfer GF, Hohenthanner K, Baumgartner W et al. Intracellular Ca^{2+} inactivates L-type Ca^{2+} channels with a Hill coefficient of 1 and an inhibition constant of 4 μM by reducing channel's open probability. Biophys J 1997; 73:1875-1865.
57. Ikeda SR. Voltage-dependent modulation of N-type calcium channels by G-protein βγ subunits. Nature 1996; 380:25-258.
58. Ikemoto N, Yamamoto T. Postulated role of inter-domain interaction within the ryanodine receptor in Ca^{2+} channel regulation. Trends Cardiovasc Med 2000; 10:310-316.
59. Imredy JP, Yue DT. Mechanism of Ca^{2+}-sensitive inactivation of L-type Ca^{2+} channels. Neuron 1994; 12:1301-1318.
60. Isaacson JS, Murphy GJ. Glutamate-mediated extrasynaptic inhibition: Direct coupling of NMDA receptors to Ca^{2+}-activated K^+ channels. Neuron 2001; 31:1027-1034.
61. Ito E, Oka K, Etcheberrigaray R et al. Internal Ca^{2+} mobilization is altered in fibroblasts from patients with Alzheimer disease. Proc Natl Acad Sci USA 1994; 91:534-538.
62. Ito E, Oka K, Collin C et al. Intracellular calcium signals are enhanced for days after Pavlovian conditioning. J Neurochem 1994; 62:1337-1344.
63. Jouaville LS, Pinton P, Bastianutto C et al. Regulation of mitochondrial ATP synthase by calcium: Evidence for a long-term metabolic priming. Proc Natl Acad Sci USA 1999; 96:13807-13812.
64. Kar S, Issa AM, Seto D et al. Amyloid β-peptide inhibits high-affinity choline uptake and acetylcholine release in rat hippocampal slices. J Neurochem 1998; 70:2179-2187.
65. Kawahara M, Kuroda Y. Molecular mechanism of neurrodegeneration induced by Alzheimer's β-amyloid protein: Channel formation and disruption of calcium homeostasis. Brain Res Bull 2000; 53:389-3897.
66. Kim CS, Han YF, Etcheberringaray R et al. Alzheimer and beta-amyloid-treated fibroblasts demonstrate s decrease in a memory-associated GTP-binding protein, CP20. Proc Natl Acad Sci USA 1995; 92:3060-3064.
67. Koizumi S, Ishiguro M, Ohsawa I et al. The effect of a secreted form of β-amyloid-precusor protein on intracellular Ca^{2+} increase in rat cultured hippocampal neurons. Br J Pharmacol 1998; 123:1483-1489.
68. Korkotian E, Segal M. Fast confocal imaging of calcium released from stores in dentritic spines. Eur J Neuronsci 1998; 10:2076-2084.
69. Lee A, Wong ST, Gallagher D et al. Ca^{2+}/calmodulin binds to and modulates P/Q-type calcium channels. Nature 1999; 399:155-159.
70. Li W-H, Lliopis J, Whitney M et al. Cell-permeant caged $InsP_3$ester shows that Ca^{2+} spike frequency can optimize gene expression. Nature 1998; 392:936-941.
71. Lipton P. Ischemic cell death in brain neurons. Physiol Rev 1999; 79:1431-1568.
72. Llinás R, Sugimori M, Silver RB. Microdomains of high calcium concentration in a presynaptic terminals. Science 1992; 256:677-679.
73. Lodish HF, Kong N, Wikström L. Calcium is required for folding of newly made subunits of the asialoglycoprotein receptor within the endoplasmic reticulum. J Biol Chem 1992; 267:12753-12760.
74. Lo Turco JL, Coulter DA, Alkon DL. Enhancement of synaptic potentials in rabbit CA1 pyramidal neurons following classical conditioning. Proc Natl Acad Sci USA 1988; 85:1672-1676.
75. Luo YQ, Hirashima N, Li YH et al. Physiological levels of beta-amyloid increase tyrosine phosphorylation and cytosolic calcium. Brain Res 1995; 681:65-74.
76. Ma H-T, Patterson RL, van Rossum DB et al. Requirement of the inositol trisphosphate receptor for activation of storeoperated Ca^{2+} channels. Science 2000; 287:1647-1651.
77. Marrion NV, Tavalin SJ. Selective activation of Ca^{2+}-activated K^+ channels by colocalized Ca^{2+} channels in hippocampal neurons. Nature 1998; 395:900-905.
78. Mattson MP, Guo Q, Furukawa K et al. Presenilins, the endoplasmic reticulum, and the neuronal apoptosis in Alzheimer's disease. J Neurochem 1998; 70:1-14.
79. Maurice T, Bayle J, Privat A. Learning impairment following acute administration of the calcium channel antagonist nimodipine in mice. Beh Pharmacol 1995; 6:167-175.
80. Matzel LD, Lederhendler II, Alkon DL. Regulation of short-term associative memory by calcium-dependent protein kinase. J Neurosci 1990; 10:2300-2307.

81. Meiri N, Ghelardini C, Tesco G et al. Reversible antisense inhibition of Shaker-like Kv1.1 potassium channel expression impairs associative memory in mouse and rat. Proc Natl Acad Sci USA 1997; 94:4430-4434.
82. Meiri N, Sun M-K, Segal Z et al. Memory and long-term potentiation (LTP) dissociated: Normal spatial memory despite CA1 LTP elimination with Kv1.4 antisense. Proc Natl Acad Sci USA 1998; 95:15037-15042.
83. Mennerick S, Matthews G. Ultrafast exocytosis elicited by calcium current in synaptic terminals of retinal bipolar neurons. Neuron 1996; 17:1241-1249.
84. Meyer T, Hanson PI, Stryer L et al. Calmodulin trapping by calcium-calmodulin-dependent protein kinase. Science 1992; 256:1199-1202.
85. Mintz IM, Sabatini BI, Regehr WG. Calcium control of transmitter release at a cerebellar synapse. Neuron 1995; 15:675-688.
86. Neely A, Olcese R, Wei X et al. Ca^{2+}-dependent inactivation of a cloned cardiac Ca^{2+} channel α_1 subunit (α_{1C}) expressed in Xenopus oocytes. Biophys J 1994; 66:1895-1903.
87. Nelson TJ, Collin C, Alkon DL. Isolation of a G protein that is modified by learning and reduces potassium currents in Hermissenda. Science 1990; 247:1479-1483.
88. Nelson TJ, Cavallaro S, Yi C-L et al. Calexcitin: A signal protein that binds calcium and GTP, inhibits potassium channels, and enhances membrane excitability. Proc Natl Acad Sci USA 1996; 93:13808-13813.
89. Nelson TJ, Zhao WQ, Yuan S et al. Calexcitin interaction with neuronal ryanodine receptors. Biochem J 1999; 341:423-433.
90. Newman MF, Kirchner JL, Phillips-Bute B et al. Longitudinal assessment of neurorecognitive function after coronary-artery bypass surgery. New Engl J Med 2001; 344:395-402.
91. Nishiyama M, Hong K, Mikoshiba K et al. Calcium stores regulate the polarity and input specificity of synaptic modification. Nature 2000; 408:584-588.
92. Normann C, Peckys D, Schulze CH et al. Associative long-term depression in the hippocampus is dependent on postsynaptic N-type Ca^{2+} channels. J Neurosci 2000; 20:8290-8297.
93. Oancea E, Meyer T. Reversible desensitization of inositol triphosphate-induced calcium release provides a mechanism for repetitive calcium spikes. J Biol Chem 1996; 271:17253-17260.
94. Oancea E, Meyer T. Protein kinase C as a molecular machine for decoding calcium and diacylglycerol signals. Cell 1998; 95:307-318.
95. Paschen W. Role of calcium in neuronal cell injury: Which subcellular compartment is involved? Brain Res Bull 2000; 53:409-413.
96. Pozzan T, Magalhães P, Rizzuto R. The comeback of mitochondria to calcium signaling. Cell Calcium 2000; 28:279-283.
97. Querfurth HW, Jiang JW, Geiger JD et al. Caffeine stimulates amyloid β-peptide release from β-amyloid precusor protein-transfected HEK293 cells. J Neurochem 1997; 69:1580-1591.
98. Restituito S, Cens T, Barrere C et al. The β_{2a} subunit is a molecular groom for the Ca^{2+} channel inactivation gate. J Neurosci 2000; 20:9046-9052.
99. Reyes M, Stanton PK. Induction of hippocampal long-term depression requires release of Ca^{2+} from separate presynaptic and postsynaptic intracellular stores. J Neurosci 1996; 16:5951-5960.
100. Rizzuto R, Bernardi P, Pozzan T. Mitochondria as all-round players of the calcium game. J Physiol Lond 2000; 529:37-47.
101. Schlossmann J, Ammendola A, Ashman K et al. Regulation of intracellular calcium by a signaling complex of IRAG, IP_3 receptor and cGMP kinase 1β. Nature 2000; 404:197-201.
102. Schneggenburger R, Neher E. Intracellular calcium dependence of transmitter release rates at a fast central synapse. Nature 2000; 406:889-893.
103. Schreurs BG, Tomsic D, Gusev PA et al. Dendritic excitability microzones and occluded long-term depression after classical conditioning of the rabbit's nictitating membrane response. J Neurophysiol 1997; 77:86-92.
104. Schreurs BG, Gusev PA, Tomsic D et al. Intracellular correlates of acquisition and long-term memory of classical conditioning in Purkinje cell dendrites in slices of rabbit cerebellar lobule HVI. J Neurosci 1998; 18:5498-5507.
105. Schumacher MA, Rivard AF, Bächinger HP et al. Structure of the gating domain of a Ca^{2+}-activated K^+ channel complexed with Ca^{2+}/calmodulin. Nature 2001; 410:1120-1124.
106. Shao X, Daletov BA, Sutton RB et al. Bipartite Ca^{2+}-binding motif in C_2 domains of synaptotagmin and protein kinase C. Science 1996; 273:248-251.
107. Sharp AH, McPherson PS, Dawson TM et al. Differential immonohistochemical location of inositol 1,4,5-triphosphate- and ryanodine-sensitive Ca^{2+} release channels in rat brain. J Neurosci 1993; 13:3051-3063.

108. Stout AK, Raphael HM, Kanterwicz BI et al. Glutamate-induced neuron death requires mitochondrial calcium uptake. Nature Neurosci 1998; 1:366-373.
109. Sun M-K. Pharmacology of reticulospinal vasomotor neurons in cardiovascular regulation. Pharmacol Rev 1996; 48:465-494.
110. Sun M-K. Hypoxia, ishcemic stroke, and memory deficits: Prospects for therapy. IUBMB Life 1999; 48:373-378.
111. Sun M-K, Reis DJ. Hypoxia-activated Ca^{2+} currents in pacemaker neurons of rat rostral ventrolateral medulla in vitro. J Physiol Lond 1994; 476:101-116.
112. Sun M-K, Nelson TJ, Xu H et al. Calexcitin transformation of GABAergic synapses: From excitation filter to amplifier. Proc Natl Acad Sci USA 1999; 96:7023-7028.
113. Sun M-K, Nelson TJ, Alkon DL. Functional switching of GABAergic synapses by ryanodine receptor activation. Proc Natl Acad Sci USA 2000; 97:12300-12305.
114. Sun M-K, Dahl D, Alkon DL. Heterosynaptic transformation of GABAergic gating in the hippocampus and effects of carbonic anhydrase inhibition. J Pharmacol Exp Ther 2001; 296:811-817.
115. Sun M-K, Zhao W-Q, Nelson TJ et al. Theta rhythm of hippocampal CA1 neuron activity: Gating by GABAergic synaptic depolarization. J Neurophysiol 2001; 85:269-279.
116. Sutko JL, Airey JA, Welch W et al. The pharmacology of ryanodine and related compounds. Pharmacol Rev 1997; 49:53-98.
117. Thompson LT, Best PJ. Long-term stability of place-field activity of single units recorded from the dorsal hippocampus of freely behaving rats. Brain Res 1990; 509:299-308.
118. Tokumaru H, Umayahara K, Pellegrini LL et al. SNARE complex oligomerization by synaphin/complexin is essential for synaptic vesicle exocytosis. Cell 2001; 104:421-432.
119. Tsien RW, Elinor PT, Horne WA. Molecular diversity of voltage-dependent calcium channels. Trensds Neurosci 1991; 12:349-354.
120. Tsien JZ, Huerta RT, Tonegawa S. The essential role of hippocampal CA1 NMDA receptor-dependent synaptic plasticity in spatial memory. Cell 1996; 87:1327-1338.
121. Ubach J, Zhabg X, Shao X et al. Ca^{2+} binding to synaptotagmin: How many Ca^{2+} ions bind to the tip of a C_2-domain? EMBO J 1998; 17:3921-3930.
122. Vergun O, Sobolevsky AI, Yelshansky MV et al. Exploration of the role of reactive oxygen species in glutamate neurotoxicity in rat hippocampal neurons in culture. J Physiol Lond 2001; 531:147-163.
123. Wang S-Q, Song L-S, Lakatta EG et al. Ca^{2+} signaling between single L-type Ca^{2+} channels and ryanodine receptors in heart cells. Nature 2001; 410:592-596.
124. Xia X-M, Fakler B, Rivard A et al. Mechanism of calcium gating in small-conductance calcium-activated potassium channels. Nature 1998; 395:503-507.
125. Xu L, Eu JP, Meissner G et al. Activation of the cardiac calcium release channel (ryanodine receptor) by poly-S-nitrosylation. Science 1998; 279:234-237.
126. Yamada WM, Zucker RS. Time course of transmitter release calculated from simulations of a calcium diffusion model. Biophys J 1992; 61:671-682.
127. Zenisek D, Matthews G. The role of mitochondria in presynaptic calcium handling at a ribbon synapse. Neuron 2000; 25:229-237.
128. Zhao W, Chen H, Xu H et al. Brain insulin receptors and spatial memory. Correlated changes in gene expression, tyrosine phosphorylation, and signaling molecules in the hippocampus of water maze trained rats. J Biol Chem 1999; 274:34893-34902.
129. Zhao W, Meiri N, Xu H et al. Spatial learning induced changes in expression of the ryanodine type II receptor in the rat hippocampus. FASEB J 2000; 14:290-300.
130. Zhao W, Cavallaro S, Gusev P et al. Nonreceptor tyrosine kinase pp60c-src in spatial learning: Synaptic-specific changes in its gene expression, tyrosine phorphoryation, and protein-protein interactions. Proc Natl Acad Sci USA 2000; 97:8098-8103.
131. Zucchi R, Ronca-Testoni S. The sarcoplasmic reticulum Ca^{2+} channel/ryanodine receptor: Modulation by endogenous effectors, drugs and disease states. Pharmacol Rev 1997; 49:1-51.
132. Michikawa T, Hirota J, Kawano S et al. Calmodulin mediates calcium-dependent inactivation of the cerebellar type 1 inositol 1,4,5-trisphosphate receptor. Neuron 1999; 23:799-808.
133. Markram H, Helm PJ, Sakmann B. Dendritic calcium transients evoked by single back-propagating action potneitals in rat nepcortical pyramidal neurons. J Neurophysiol 1995; 485:1-20.
134. Roeper J, Lorra C, Pongs O. Frequency-dependent inactivation of mammalian A-type K+ channel KV1.4 regulated by Ca2+/calmodulin-dependent protein kinase. J Neurosci 1997; 17:3379-3391.
135. Liu SQ, Cull-Candy SG. Synaptic activity at calcium-permeable AMPA receptors induces a switch in receptor subtype. Nature 2000; 405:454-458.
136. Erecinska M. Silver IA. ATP and brain function. J Cereb Blood Flow Metab 1989; 9:2-19.
137. Silver IA. Erecinska M. Intracellular and extracellular changes of [Ca2+] in hypoxia and ischemia in rat brain in vivo. J Gen Physiol 1990: 95:837-866.

Chapter 1.2

Potassium

Jeffrey Vernon and Karl Peter Giese

Abstract

Potassium-selective channels are ubiquitous constituents of plasma membranes, and are structurally and physiologically diverse. It is considered that their multiplicity of form allows for the fine-tuning of membrane responses to changes in potential or intracellular environment. In neurons, both synaptic transmission and the onset and duration of excitations are governed to a large extent by K^+ channel kinetics. These properties of K^+ channels facilitate higher-order membrane phenomena such as signal integration, spike patterning and synaptic plasticity. It would be surprising if K^+ channels played no part in complex animal behaviours having a neural basis; indeed, across animal models K^+ channel modulation by conserved mechanisms appears to underlie learning and memory. The study of invertebrate and vertebrate models has demonstrated a role for certain K^+ conductances in associative learning and memory. In many reports, learning correlates at the cellular level with a reduction in voltage-sensitive K^+ current amplitude, with a concomitant increase in membrane excitability. Because of the bias in learning and memory studies towards monitoring depolarisations, it is possible that further classes of K^+ channels have been neglected. Protein kinases and phosphatases are implicated in the modulation of channels attending learning. While molecular information is emerging on the channels affected in normal and mutant behaviours, the molecular basis of some learning-associated currents is unknown. We review the fly, mollusc and mammal literature on K^+ currents in learning and memory, and consider some of the problems of the data, as well as possible approaches to outstanding questions. One of these is gene targeting in mice, which has already yielded insights into the relationship between K^+ channel modulations, learning and memory, and membrane phenomena such as long-term potentiation and the slow afterhyperpolarisation.

Introduction

K^+-selective transmembrane channels are the molecular substrate of K^+ currents detected by electrophysiology and modifiable by cytoplasmic components or experimental drugs. The classic work of Hodgkin and Huxley established that one type of outward K^+ flux repolarises the action potential of the squid giant axon[56] and from that time the control of membrane excitability in neural, cardiac and other tissues has remained a major theme of K^+ channel research. Elucidation of the control mechanisms in both native tissue and expression systems has been assisted by the cloning of K^+ channel subunits, the first of which was Shaker.[66,91,93,126] Approximately seventy different alpha subunits have now been cloned, and assigned on sequence criteria to some eleven subfamilies (ten in ref. 20; ref. 109 report a further class). Structure prediction[75] indicates that functional channels of the Shaker type consist of four subunits that form six transmembrane domains plus an ion-permeable pore gated by conformation changes (Fig. 1). These changes depend on the movement of dipoles in voltage-sensing amino acid residues within the electrical field of the membrane. The ensemble activity of many single

From Messengers to Molecules: Memories Are Made of These, edited by Gernot Riedel and Bettina Platt.

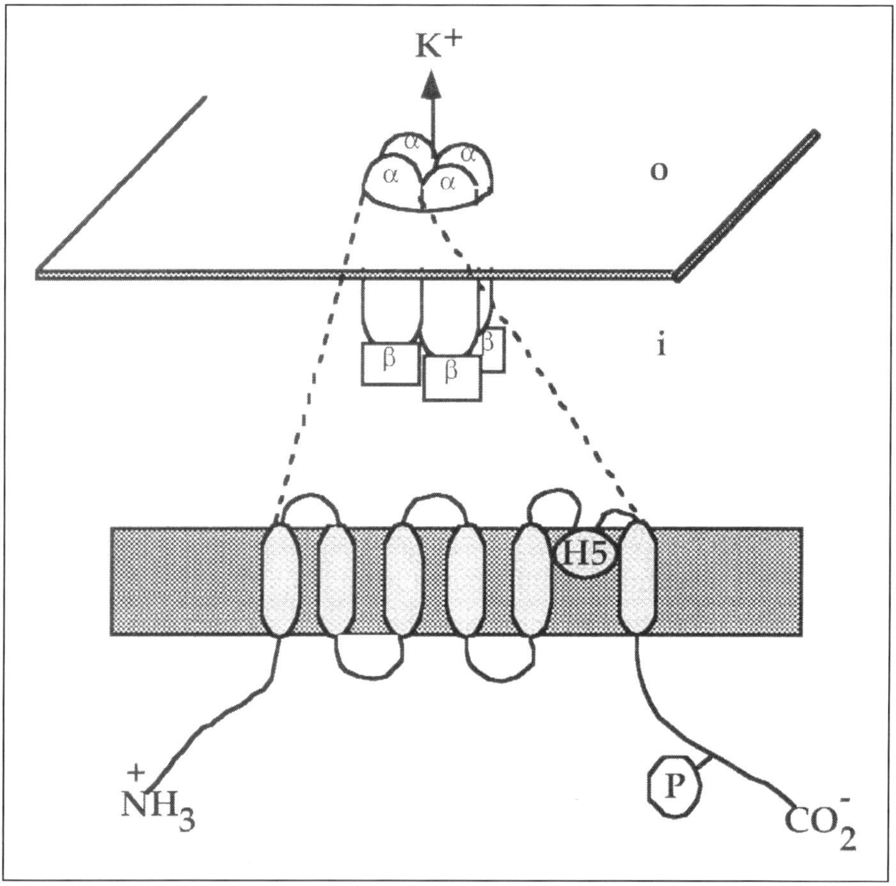

Figure 1. Structure of a 6TM K$^+$ channel. The functional protein is a homo- or hetero-tetramer of alpha-subunits forming six transmembrane domains and a K$^+$ permeable pore (H5). A beta subunit may associate with each alpha peptide. A phosphorylation site has been indicated at a residue near the C-terminus to indicate that phosphorylation events modulate channel activity. o=outside, i=inside the cell membrane.

channels opening at random times under permissive conditions and then inactivating (by intrinsic mechanisms) or deactivating (on membrane repolarisation), gives rise to the macroscopic K$^+$ current.

K$^+$ channels with six, four or two transmembrane domains are known. The structure of Shaker K$^+$ channels is typical of the six-transmembrane (6TM) family. Some but not all 6TM channels are voltage-gated. The 4TM 'leak' K$^+$ channels and the 2TM 'inward rectifiers' are not voltage gated, but their flux is nonetheless sensitive to membrane potential. The heteromerization of channel subunits, alternative splicing and post-transcriptional modifications all generate a further diversity of K$^+$ channels differing in their kinetic and biochemical properties.[20] Under some conditions, K$^+$ channel subunits are known to associate with cytoplasmic β-subunits, of which several have been cloned. For example, $K_v\beta$ subunits,[99] the mammalian homologues of *Hyperkinetic* of *Drosophila melanogaster*, associate with Shaker-related alpha subunits. Beta subunits alter channel opening and closing kinetics, and may act as chaperones during biosynthesis.

The variety of K^+ channels exceeds that of any other ion channel type, implying a role for K^+ channels in the fine-tuning of membrane responses to electrical and molecular stimuli. K^+ channels govern important aspects of membrane physiology, with consequences for resting potential, the spacing between action potentials, the width of an action potential, and subsequent repolarisation. The repertoire of channels in a membrane therefore determines properties such as excitability, and functions such as encoding. Some of the in vitro properties of cloned K^+ channels can be correlated with conductances recorded in vivo; however, in many cases no such correlation can yet be made, in the absence of further discoveries about the effects on currents of subunit heteromerization, channel phosphorylation, and other environmental factors. For example, no cloned subunit appears to correspond to the Na^+-activated K^+ channels detected in mammalian heart and brain.[62] There is evidence for downstream effects of K^+ channel activity beyond the control of membrane excitability, but most of these pathways also remain to be elucidated (see ref. 70 on cell proliferation).

Beyond the detailed kinetic description of K^+ channel opening, closing and inactivation, and the biophysical description of channel function in membranes, there have been attempts to correlate altered K^+ currents with complex phenomena such as disease and behaviour. It is possible that K^+ current modulations are both the cause and the consequence of network changes involved in learning and memory (L&M). Here we will review evidence for the contribution of K^+ channel modification to L&M. So far five identifiable types of 6TM K^+ channel have been implicated in L&M. These channels give rise to 1) the 'transient' A-type currents (I_A) that activate below spiking threshold and inactivate rapidly by intrinsic mechanisms; 2) the 'delayed rectifier' currents that are inactivating or only slowly inactivating, and de-activated by membrane repolarisation; 3) high-conductance 'slowpoke (slo)' or I_{KCa} currents sensitive to Ca^{2+}; and 4) the 'ether-a-go-go (eag)' currents potentially modifiable by cyclic nucleotides. These four currents are all voltage-activated. The fifth is not. This is the apamin-insensitive small-conductance Ca^{2+} sensitive 'SK' current of hippocampal neurons. For some synaptic K^+ channels associated with learning in model organisms, for example the S-K channel in *Aplysia californica*, the molecular identity has not yet been established. Conversely, some K^+ channel subunits underlying biophysical phenomena that are correlated temporally with memory formation (e.g., the hippocampal theta rhythm) or attentiveness (e.g., the M current) are not yet known to be essential in L&M. Finally, there are entire classes of K^+ channel subunits believed to control neuronal excitability that have so far been assigned no role in L&M, for example the 2TM inward rectifiers.

How Can K^+ Channels Contribute to Learning and Memory?

It is generally accepted that synaptic inputs trigger L&M mechanisms. Synaptic receptors activate signalling pathways resulting in the phosphorylation of proteins such as ion channels, modulating their properties. Phosphorylation is a recognised mechanism of channel modulation[74] and the data discussed in this paper show that kinases are implicated in L&M in both invertebrates and vertebrates. K^+ channels may contribute to L&M in three ways, as shown by Figure 2. (1) Modulation of K^+ channels at the synapse can mediate synaptic plasticity. For example, at the presynaptic terminals a reduction of K^+ currents leads to broadening of action potentials enhancing neurotransmitter release. Such a K^+ channel modulation has been shown to underlie the short-term facilitation in *Aplysia californica* that is thought to contribute to L&M. (2) Modulation of K^+ channels can regulate the induction of synaptic plasticity. For example regulation of action potential backpropagation by A-type K^+ channel modification might control the onset of long-term potentiation, a process thought by some workers to underlie mammalian L&M.[14] (3) Modulation of K^+ channels can switch the firing behaviour of neurons without affecting synaptic plasticity. For example, modulation of the slow afterhyperpolarisation (sAHP) is thought to contribute to memory consolidation. It should be noted that such an intrinsic plasticity lacks the input-specificity of a synaptic mechanism.

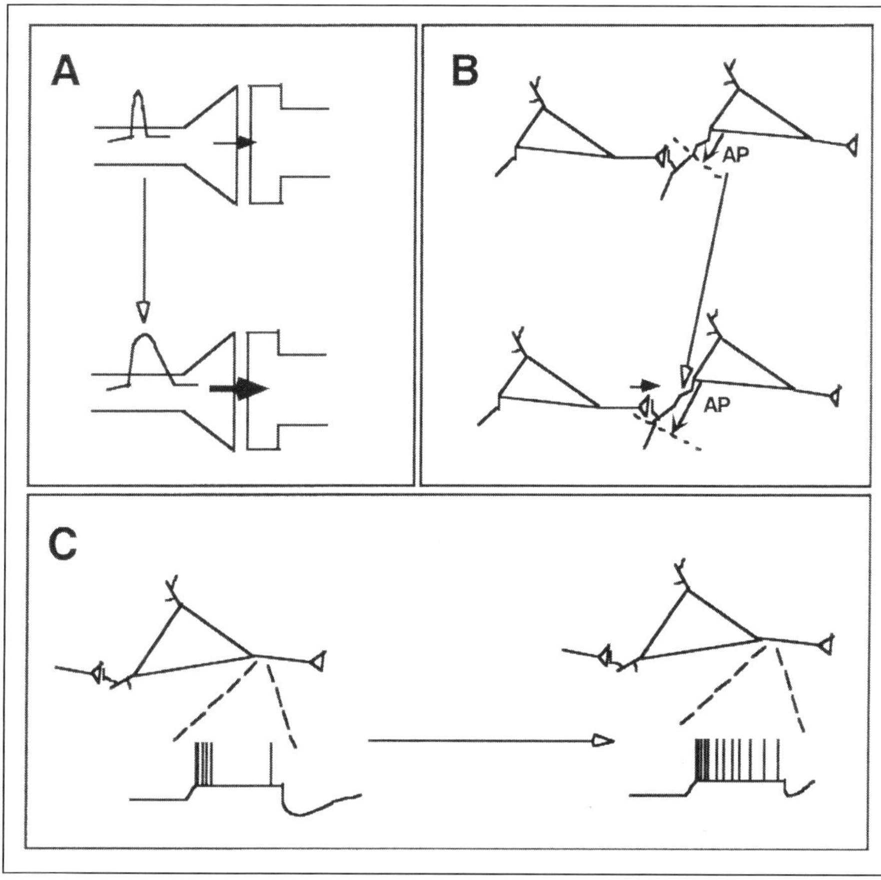

Figure 2. K⁺ channel modulation and the neurophysiology of learning. A) K⁺ channel suppression delays repolarisation. The broadening of the action potential (AP) promotes more neurotransmitter release as observed in short-term facilitation in Aplysia. B) Dendritic K⁺ channels regulate the pairing of inputs to induce synaptic plasticity. Channel inactivation by synaptic inputs (EPSPs or neuromodulatory) allows back-propagation of a temporally paired APs into distal parts of the dendritic tree. This allows coincidence of the APs with incoming EPSPs, which induces long-lasting synaptic plasticity. C) Reduced K⁺ current amplitude alters the firing pattern of neurons. A neuromodulator such as acetylcholine eventually switches off the channels underlying the slow after-hyperpolarisation, altering the response to a given stimulus.

K^+ Channels and Invertebrate Learning and Memory

In the 1960s Eric Kandel and Ladislav Tauc had begun to study habituation of a nonassociative behaviour in single cells of a mollusc, *Aplysia.*[68] Alkon's work with another mollusc, *Hermissenda crassicornis,* showed the neural site of a form of Pavlovian conditioning in invertebrates, and motivated theories of conserved pathways of learning in higher metazoans. The study of memory in *Drosophila melanogaster* turned from ethology to genetics when members of Seymour Benzer's group began to screen for learning mutants having single gene defects, leading to the identification of dunce and other relevant genes.[33,96] Continuing studies of *Hermissenda crassicornis, Aplysia californica* and *Drosophila melanogaster* have revealed the involvement of K^+ channels in L&M (Table 1).

Studies in *Hermissenda crassicornis*

The two eyes of *Hermissenda* consist of five photoreceptors; two 'A' cells, and three 'B' cells. B photoreceptors can support a rhodopsin-mediated light response that depolarises the cell and activates protein kinase C (PKC). Each eye is closely associated with a vestibular (gravity-sensing) organ lined with excitable hair cells. Alkon determined that the vestibular system of *Hermissenda crassicornis* receives signals from the visual system; that an electrical stimulus presented to the animal's photoreceptors elicits an electrical response in the hair cell. *Hermissenda crassicornis* normally approaches a source of light, and clings to the substratum if it detects movement in the aquarium. If the molluscs were trained to associate light with a perturbation, phototaxis would cease and a correlate ought to occur in the visual-vestibular network. Depolarisations recorded in the type B photoreceptor cells confirmed this hypothesis.[2] Conditioning has lasting effects in B cells on evoked spike frequency, input resistance,[89] the shape of action potentials and the amplitude of the after-hyperpolarisation.[43] In B cells isolated from any synaptic input the learning trace can be recorded for weeks after training, suggesting that the photore-

Table 1. Summary of K^+ channel subunits and corresponding currents implicated in learning and memory

	Cloned Subunits	
6TM Group	**Subunit (Number of Known Family Members)**	**Candidate Current**
Shaker	K_v1 (10)	Fly, mammalian I_A Fly PKA-sensitive current $K_v$beta1.1/Hyperkinetic Fly , mammalian I_A
Shab	K_v2 (2)	Fly $I_{K\ major\ component}$; Fly PACAP-activated current Fly CaMKII-sensitive current
Shaw	K_v3 (4)	Fly I_K; Fly c.AMP-sensitive current in mushroom body
Shal	K_v4 (3)	Fly $I_{A\ major\ component}$; Mouse I_A Fly CaMKII-sensitive K^+ current
Eag	Eag (1)	Fly I_A component
K_{Ca}	Slo (1)	Fly, mollusc I_{KCa}
4TM Group		
TRAAK	TRAAK-1 (>1)	Aplysia fast-onset cAMP-sensitive S-K current
	Uncloned Subunits	
SK1 or other subunit	Apamine- insensitive small conductance K^+ channel	Mammalian slow afterhyperpolarisation
Shaker, Shal or other subunits		Mollusc I_A
unknown		Aplysia cAMP-sensitive slow S-K current

Fly data from larval neurons may not apply in adults; mollusc means either Hermissenda or Aplysia.

ceptor cell itself is the encoding site of the altered behaviour. The excitability of the B cell increases with the consolidation of the memory, and decreases after the presentation of light alone causes memory extinction in previously conditioned animals. This excitability increase is thought to depend on decreased amplitude and increased inactivation of two K^+ currents in the soma that can also be reduced in untrained photoreceptors by activation of Ca^{2+}/phospholipid dependent protein kinase C. These two currents are considered to be the transient I_A and the noninactivating composite current $I_{delayed}$, of which the major component is I_{KCa}.[5,6,107] Consistent with the PKC results, inhibitors of PKC antagonise the changes in B cells expected to be induced by learning, by preventing the reduction of I_A and I_{KCa}.[39] Single K^+ channel recordings in B cells[37] resolved two K^+ conductances, of which one, having a 42 pS conductivity, was unchanged by the animal's experience, and the other (64 pS) was modified by learning or by PKC. The 64 pS channel may contribute to a previously detected delayed rectifier current found to participate in repolarisation, and is diminished by application of protein kinase A (PKA);[4] if true, this conductance may be regulated by two parallel kinase pathways. It has not been assigned to a cloned K^+ channel subclass.

How are the data to be interpreted in terms of biochemical pathways within the B-cell? The two modulated K^+ channels may stand at the confluence of two learning-associated pathways that in combination potentiate PKC.[124] For a compatible account but with different emphasis, see refs. 3,13. The effects of PKC may belong to a serotonergic pathway, since light paired with serotonin application to the relevant ganglion in *Hermissenda* produces behavioural suppression.[23] This putative serotonin pathway has not been characterised, although it deserves attention because serotonin is central to synaptic plasticity in *Aplysia*. Finally, we mention the 'crosstalk' possibility that after paired stimuli the response to rotation acquires features of the response to light.[18,90,102]

What is the significance of these K^+ current modulations for L&M of *Hermissenda crassicornis*? The transient I_A current is mediated by A-type K^+ channels that are thought to antagonise depolarisations, and consequently a reduction of I_A current results in increased excitability (see also the role of A-type K^+ channels in mammalian L&M). The I_{KCa} current contributes to repolarization of action potentials, and a reduction of this current broadens action potential duration, which may result in enhanced neurotransmitter release onto the A cell.[38] Conditioning leads to facilitation of the B cell inhibitory synapse onto the A cell, as measured by the increased inhibitory postsynaptic potential (IPSP) amplitude in the A cell.[42] Gandhi and Matzel[43] blocked I_A in B cells to show that spike broadening accounts for the synaptic facilitation.

The importance of B-cell depolarisation per se to *Hermissenda* learning is unclear; some authors report that successive pairings while preventing depolarisation suffice to increase B-cell excitability, meaning that the activation of PKC is the critical event for the memory trace.[81]

Studies in *Aplysia californica*

Synaptic plasticity connected with the sensitisation of the gill-withdrawal reflex in the mollusc *Aplysia californica* is a model for a simple form of learning. The abdominal ganglion, containing sensory cells that synapse onto interneurons and motor neurons controlling the gill muscles, regulates the central component of the gill withdrawal reflex. The abdominal ganglion is connected to a pleural ganglion in the head. When the external siphon of *Aplysia* is touched, the gill within the respiratory mantle retracts. If the head of the organism is given a shock, the subsequent reflex to siphon touching is enhanced.

Electrical stimulation of the sensory neuron elicits an excitatory postsynaptic potential (EPSP) in the motor neuron, giving a cellular representation of the reflex. The motor neuron EPSP can be facilitated by stimulating the connective to the abdominal ganglion from the pleural ganglion (equivalent to shocking the head). Since this facilitation does not fire the presynaptic neuron, and post-synaptic mechanisms were ruled out, stimulating the connective must activate a different pathway that alters the level of neurotransmitter release. This pathway is known to involve a facilitatory interneuron that releases serotonin onto a synapse with the sensory cell, and it is this that brings about facilitation of neurotransmitter release. Depending on the number and timing of serotonin pulses, the facilitation is either short- or long-lasting. Both forms of facilitation appear to be important for behavioural sensitisation.[40]

Short-term facilitation (STF) is associated with two effects on neurotransmitter release, differing in their dependence on action potential spike broadening. It is thought that serotonin release from the interneuron stimulates adenylate cyclase in the sensory neuron, activating a PKA pathway that phosphorylates two types of K^+ channels and reduces their activities.[10,114] The diminished K^+ currents can prolong action potentials via spike broadening, increasing Ca^{2+} entry through N-type Ca^{2+}channels and potentiating glutamate release, leading to enhanced excitability. The two affected channels contribute to a transient voltage dependent K^+ current I_{Kv}, and a cAMP-dependent K^+ current, known as S-K.[10] The first current mediates most of the spike broadening observed in STF, while the second is responsible for most of the spike-duration-independent excitability increase.[16] It has recently been proposed that a fast-onset component of the S-K-current has the properties of TREK-1, a stretch-activated and fatty acid sensitive K^+ 'leak' channel of the 4TM family (ref. 92 and citations therein). This channel would pass net outward potassium flux at the resting potential. A slowly-activating component of I_{SK} has an unknown molecular basis.[47]

PKA is not the only kinase operating on K^+ currents to affect action potential duration in *Aplysia*. Following earlier evidence that PKC also contributed to the STF mediated by serotonin, it was deduced that in synapses depressed by activity, this pathway may lead to a form of spike broadening by slowing the activation of the transient I_{Kv} current and the inactivation of a Ca^{2+}-dependent K^+ channel.[121,122] There are, then, two G-protein pathways coupled to the serotonin receptors on the sensory neuron membrane. The activation of either the PKA or PKC pathway may depend on the excitation history of the synapse, implying cellular discrimination between two kinds of experience (reviewed in ref.16). In addition to sensitisation, forms of associative learning have also been studied in *Aplysia*.[17] Animals receiving a tail shock (the unconditioned stimulus, US) retract the gill for longer periods than in the usual withdrawal reflex to siphon touching (the conditioned stimulus, CS). Following paired presentations of CS and US, *Aplysia* responds to the CS with an enhanced reflex like that produced by the US alone (reviewed in ref. 1). In order to monitor the response of the motor neuron to the stimulations of the sensory neuron and tail, the mechanism of conditioning was studied initially in reduced preparations linking the abdominal and pedal ganglia to a portion of the tail. Later work used synaptic cultures.[34,53] As in STF, conditioning-induced sensorineuron current inhibition increased spike duration, implicating serotonin release from an interneuron, and consequently motor neuron EPSP was enhanced. However, in the case of conditioning the spike broadening and EPSP enhancement were larger than in STF. This suggests that a conditioning paradigm engages the same elements involved in STF, but using a pathway that generates enhanced output.

Studies in *Drosophila melanogaster*

The first K^+ channel subunit, mediating fast transient I_A currents in expression systems, was cloned from the leg-shaking *Drosophila* mutant Shaker (Sh) in 1987.[66,91,93,126] The K^+ channel gene underlying a second leg-shaking mutant, ether-a-go-go (eag), was cloned four years later,[32] and the Slowpoke (slo) calcium-activated K^+ channel gene was cloned from a locus associated with a poor flight and 'sticky feet' phenotype in the same year.[9] Following the sequencing of the Shaker locus, cloning in Drosophila and in vertebrates identified founder members of three other voltage-gated K^+ channel subfamilies, named *Shab*/ K_v2, *Shaw*/ K_v 3, and Shal/ K_v 4.[15]

Physiological studies of the adult giant cervical fibre (GCF) neuron of six Shaker allelic mutants detected protracted action potentials, and additionally in all these genotypes except Sh^5, delayed repolarisations.[125] There are developmental differences in Shaker mutant physiology, so that I_A in Sh^5 larvae has reduced amplitude, while in Sh^5 adults the current inactivates more rapidly.[136] The physiology of larval Shaker may not be relevant to all aspects of K^+ channel modulation in adult flies. In the *Xenopus* oocyte expression system *Drosophila eag* RNA gives rise to a current that reactivates more slowly after increasing hyperpolarisations.[101] This property delays the activation of *eag* during successive depolarizations, narrowing the channel's range of activation and thereby reducing the current amplitude.

Before K^+ channel gene cloning, K^+ current mutants were shown to be defective in acquisition (Sh^5) or retention (*eag*) of conditioned courtship behaviour.[21] Courtship in male *Drosophila* is a ritual involving reorientation, attempted copulation, licking, pursuit and vibration, and is susceptible to mutation at many loci (reviewed in ref. 51). This behaviour depends on the reception of a pheromone secreted from the female cuticle, and is most pronounced in the presence of a virgin female. Its moderation in the presence of a mated female may depend on the secretion of an anti-aphrodisiac by females that have copulated. The identity of this substance has proved elusive, but its existence would account for the behaviour known as courtship conditioning, in which male flies that have courted a mated female subsequently show suppressed courtship of virgins for 3 hours, and of mated females for 24 hours. As if the mated female had given out a 'mixed signal' of mating and anti-mating substances, males later encountering a virgin apparently associate her aphrodisiac pheromone with the anti-aphrodisiac it was previously paired with (summarized in ref. 67). The brain regions engaged by courtship conditioning have been defined in a study of localised CaMKII inhibition.[65] In addition to impaired courtship conditioning Sh^5 flies are also deficient in olfactory conditioning in the paradigm developed by Tully and Quinn.[22,131] In this paradigm wild-type flies are successively exposed to two odours, one of which is paired with presentation of an electric shock. The flies are then transferred to a T-maze in which either arm is scented with the odours used during training. Under these conditions, *Drosophila* avoids the odour previously paired with shock.

K^+ channel mutations can also affect nonassociative learning. Certain *eag* and Shaker alleles enhance habituation to stimulation of a circuit that employs the GCF neuron , controlling a flight-jump response.[35] Conversely, the mutants *slowpoke* (*slo*) and *Hyperkinetic (Hk)* both have reduced habituation. The mutant Sh^{KS133}, having no I_A current in muscle, had no effect on habituation. As the authors point out, this suggests that the habituation observed in the voltage-range and kinetic mutants Sh^{rK0120} and Sh^5 occurs as an indirect consequence of channel modulation. For *slo*, *Hk* and *eag* no significant allelic influences on habituation were reported.

The neural organization of *Drosophila* presents difficulties of access and circuit discrimination, so that K^+ current electrophysiology is generally carried out at the fly neuromuscular junction. These studies may not therefore be relevant to an understanding of mechanisms underlying associative learning. Transgenic flies generated by Ron Davis and expressing a reporter gene confined to mushroom bodies, brain structures essential for chemosensory learning,[55] made possible the in vitro analysis of currents in larval mushroom body neurons.[135] Two composite K^+ currents were recorded; both had a sustained component, but in addition the type 1 currents comprised a 4-aminopyridine (4-AP) insensitive transient, consistent with the presence of Shal/K_v4,[139] while the type 2 currents included a 4-AP sensitive transient. Indicating a role for a PKA pathway in suppressing K^+ channels, cyclic AMP analogues reduced the amplitude of type 1 currents and raised their activation thresholds by +40 mV. Consistent with this result, two well-characterised learning mutants are deficient in the degradation (dunce) and synthesis (rutabaga) of cAMP. Rutabaga encodes a Ca^{2+}/calmodulin-sensitive adenylyl cyclase, which is preferentially expressed in mushroom bodies,[52] while dunce specifies a cAMP phosphodiesterase. In dunce flies, the amplitudes of both I_A and the delayed rectifier I_K are enhanced, whereas in rutabaga a Ca^{2+}-activated K^+ channel shows increased amplitude.[145] In addition, in embryonic neuroblasts of both mutants the firing pattern and spike shape are altered.[142] Furthermore, in dunce mutants, a cAMP-sensitive K^+ current, possibly mediated by a homologue of the Shaw/K_v3 channel, is under-represented in larval mushroom bodies.[25] The morphology and physiology of dunce and rutabaga neuromuscular synaptic boutons are abnormal.[98] Thus, changes in neuronal excitability and synaptic transmission due to abnormal K^+ channel activity might cause the learning deficits observed in dunce and rutabaga.[26]

Two eag mutants, eag^1 and eag^{4pm} have been shown to phenocopy the behaviour of flies transgenic for an inhibitor of CaMKII, suggesting that the predicted phosphorylation sites on the Eag channel may be critical CaMKII substrates. In these experiments, eag mutants courted mated females at up to 1.5 times the frequency of wild type flies.[49] Since the Eag subunit may coassemble with other K^+ channel subunits,[19] the eag mutation might influence neuronal pathways mediated by other K^+ currents, such as delayed rectifier, Shaker and slo currents.[145]

Besides Shaker and Eag, further K^+ channel subunits in Drosophila are altered by perturbations in CaMKII or PKA signalling pathways. Yao and Wu[137] studied giant neuron cultures from *ala1* mutants, which express a specific CaMKII inhibitory peptide, and a mutant DCO^{X4} that has reduced PKA catalytic activity. Both flies showed suppression of a composite K^+ current having a predominant I_A component. DCO^{X4} in addition were inhibited for a second K^+ flux containing a higher proportion of delayed rectifier current. Circumstantial evidence indicates that the K^+ channels differentially modulated by the two kinase pathways are Shal plus Shab (modulated by CaMKII), and Shaker (modulated by PKA). This conclusion relies on the earlier finding that fly neuron I_K is principally a Shab conductance with an admixture of Shaw; while I_A is mostly Shal, together with a Shaker component.[130] Both populations of mutant neurons showed altered firing responses to current injection, with longer spike duration in ala1 cells and erratic firing rates in a subset of DCO^{X4} cells. Moreover, the firing patterns observed in the mutant cells were disorganised in comparison with wild-type responses, suggesting that the K^+ channel activity deregulated by the mutations is responsible.

Two further Drosophila learning mutants carry molecular lesions with consequences for K^+ channels. The *amnesiac* mutant is another fly with perturbed cAMP metabolism. The product of the *amnesiac* gene has limited homology to mammalian PACAP, an activator of adenylyl cyclase in the pituitary. PACAP-38 applied to the neuromuscular junction of wild type flies elicits an outward K^+ current, with a delay of up to 15 minutes; but not in the *rutabaga* mutant (which cannot activate adenylate cyclase). These results imply that the rutabaga protein lies upstream of PACAP effect.[143,144] Currents elicited by PACAP-38 in mammalian systems[63] have quinine and TEA sensitivity reminiscent of the slow activating Shab/K_v2 series of voltage gated channels. In the *Leonardo* (*Leo*) mutant, flies are deficient in one isoform of 14-3-3, a family of proteins with pleiotropic effects that is conserved from yeasts to mammals. *Leonardo* flies have deficits in olfactory learning and short-term memory.[115] At the neuromuscular junction, the leo protein may interact with the K^+ channel Slo subunit via a calcium-binding protein called Slob.[146] There may also be in vivo interaction with the eag subunit. *Leo* is expressed in the mushroom bodies, but its interactions with K^+ channels at this site are unknown.

Studies in Other Invertebrates

In addition to the described studies it may be speculated that K^+ channel modulation also contributes to L&M of other invertebrates.

Behavioural habituation is studied in the worm *C. elegans* and scrutiny of the genome predicts 80 K^+ channel genes, predominantly forming 4TM channels, many of which appear to have highly restricted expression patterns.[108] However, K^+ channel involvement in L&M has not yet been demonstrated. One form of associative learning in *C.elegans*, whereby worms migrate in a thermal gradient after conditioning with food, has been shown to depend on Ca^{2+} signalling mediated by NCS-1/frequenin.[48] In *Drosophila* that over-express frequenin, increased Ca^{2+} fails to enhance larval muscle I_A amplitude in the expected way.[94]

Associative olfactory learning can be studied in the honeybee *Apis mellifera*[84] and network oscillations have been implicated as an underlying mechanism.[118] The $GABA_A$ receptor antagonist picrotoxin disrupts synchrony of oscillations and impairs odour discrimination in that structurally similar odorants cannot be distinguished. This suggests that oscillations enable fine-tuning of sensory inputs and that K^+ channel modulation may be involved. However, as in *C. elegans*, a direct demonstration of K^+ channel involvement in L&M is lacking.

K^+ Channels and Mammalian Learning and Memory

Most mechanistic studies of mammalian L&M concentrate on the hippocampus, because this brain structure is involved in declarative memory and abnormalities in this region contribute to L&M disorders, such as Alzheimer's disease.[85] Modulation of the sAHP and A-type K^+ channel modification in hippocampal pyramidal neurons are thought to contribute to L&M.

Modulation of the Slow Afterhyperpolarization (sAHP) As a L&M Mechanism

Studies on the sAHP have provided the strongest evidence for a contribution of K^+ channel modification to mammalian L&M. The sAHP follows bursts of action potentials, increases with the number of action potentials, and lasts for up to several seconds, preventing further firing of action potentials.[119] In this way the sAHP is a negative feedback for neuronal firing. Depending on the neuron the sAHP is sensitive to the bee venom apamin. For example, apamin blocks the sAHP of hippocampal interneurons,[141] but not the sAHP in hippocampal pyramidal neurons.[119] SK channels, voltage-independent small-conductance Ca^{2+}-activated K^+ channels mediate the sAHP,[105] and three 6TM subunits underlying the apamin-sensitive afterhyperpolarization (AHP) in brain have been cloned.[111,117,133] However, the molecular identity of SK channels mediating the apamin-insensitive sAHP is unknown.

The sAHP amplitude in hippocampal pyramidal neurons can be reduced by signalling pathways triggered by a variety of neurotransmitters, such as acetylcholine (Fig. 2C), which have been implicated in L&M.[105,119] Trace eyeblink conditioning, a hippocampus-dependent L&M task in which the conditioned stimuli are separated from the unconditioned stimuli by a "trace" of hundreds of milliseconds,[116,138] is associated with a reduction of the sAHP in CA1 and CA3 pyramidal cells.[30,88,128] Importantly, the sAHP in hippocampal pyramidal cells is not reduced in nonlearning animals, demonstrating that the sAHP reduction is not a general attention mechanism.[87] In learning animals approximately 45% of the pyramidal cells show a reduced sAHP, indicating that a large number of hippocampal neurons is involved in storing learned information. This conclusion is consistent with the finding that learning of other hippocampus-dependent L&M tasks is associated with transcriptional changes in a large proportion of hippocampal neurons.[50,61] The reduction of the sAHP lasts for about five days, suggesting that this process contributes to memory consolidation. A long-lasting and widespread reduction of the sAHP has also been observed in pyramidal neurons of piriform cortex after olfactory learning[103,104] and in Purkinje cells of lobule HVI of cerebellum after delay eyeblink conditioning.[110]

The functional role of a learning-induced sAHP reduction has been addressed by computer modelling.[12] These simulations suggest that sAHP reductions can move networks from chaotic or zero output into an equilibrium state in which the reproduction of learned patterns is more probable. Interestingly, pharmacological studies showed that a reduction in the sAHP in hippocampal CA1 pyramidal cells is associated with a decreased threshold for induction of long-term potentiation (LTP), a kind of synaptic strengthening and a cellular correlate of L&M.[106] The sAHP in hippocampal CA1 pyramidal neurons has been reported in basal and apical dendrites and the soma so that it could suppress synaptic input in the entire neuron. A reduction of the sAHP leads ultimately to more depolarization (less repression of synaptic input), which is required for the activation of NMDA receptors to induce LTP. Studies with $K_V\beta1.1$ null mutant mice (see below) confirm that some, but not all, sAHP alterations affect the threshold of LTP induction at the CA1 synapse.[45]

The sAHP in hippocampal CA1 pyramidal neurons increases during normal ageing[73,86] due to an enhanced L-type Ca^{2+} channel density allowing more Ca^{2+}-influx during bursts of action potentials, which leads to increased activation of SK channels.[79,127] This sAHP increase could impair the learning-induced reduction of the sAHP, thereby leading to L&M deficits. Thus, the sAHP increase could cause age-related cognitive decline.[31] Consistently, in trace eyeblink conditioning old animals perform poorly. After extensive training old animals can learn this task and these animals have a reduced sAHP which is of a comparable amplitude to the reduced sAHP of young adult animals after less training.[87] A further correlation of increased sAHP leading to decreased L&M abilities derives from studies with animal models of diabetes. These animals have an increased sAHP in hippocampal CA1 pyramidal neurons, impaired LTP at the hippocampal CA1 synapse and spatial learning deficits (see ref. 46 and Gispen, W.H., personal communication).

Apamin has been used to study the requirement for sAHPs in L&M. As well as the previously described effects on sAHPs, apamin also blocks medium AHPs as has been observed in hippocampal pyramidal neurons.[117] In most studies apamin was applied by intraperitoneal injections and the effects on L&M were task-dependent. Apamin does not influence spatial[60] nor passive avoidance L&M,[22,60] it improves memory consolidation in an olfactory discrimination task[41] and operant conditioning,[11] and it enhances learning in an object recognition task.[28]

Studies on the role of the apamin-insensitive sAHP in L&M have been restricted by the lack of specific high-affinity blockers and by the missing molecular identity of the underlying channels. However, indirect blockade experiments confirmed the hypothesis that modification of the apamin-insensitive sAHP is an L&M mechanism. For example, nimodipine, an L-type Ca^{2+} channel blocker, or metrifonate, a cholinesterase inhibitor, reduce the sAHP in aged hippocampal CA1 pyramidal neurons[86,95] and reverse age-related impairments in trace eyeblink conditioning.[29,72]

In addition to the pharmacological studies, experiments with $K_v\beta1.1$ null mutant mice indicate an important role in modulation of the sAHP in L&M.[45] The expression of the *Hyperkinetic*-related β-subunit $K_v\beta1.1$ (for nomenclature, see ref. 36) is restricted to the central nervous system and it is most abundant in the hippocampal CA1 area and in caudate putamen.[15,99] Thus, the $K_v\beta1.1$ null mutation affects a subset and not all of the circuits in brain. $K_v\beta1.1$ confers fast inactivation on otherwise noninactivating $K_v1$subunits.[54,99] Consequently, the loss of $K_v\beta1.1$ transforms fast into noninactivating K^+ channels, as has been analysed in hippocampal CA1 pyramidal neurons.[45] Considering the fact that besides the $K_v\beta1.1$-dependent A-type K^+ channels there are also A-type K^+ channels that are not influenced by $K_v\beta1.1$ (e.g., $K_v4.2$; see below), only a subset of fast inactivating K^+ channels are transformed by the $K_v\beta1.1$ null mutation. In hippocampal CA1 pyramidal neurons the null mutation does not affect single action potentials but it reduces frequency-dependent spike broadening (FSB).[45] FSB is the phenomenon whereby later action potentials in a spike train are longer in their duration. FSB is thought to result mainly from cumulative inactivation of fast-inactivating K^+ channels.[76,112] The reduced FSB leads to decreased Ca^{2+}-influx during the action potentials, causing a smaller sAHP.[45] This indirect reduction of the sAHP does not influence plasticity at the hippocampal CA1 synapse.[44,45] However, in old age the $K_v\beta1.1$-caused reduction of the sAHP compensates for the age-related increase of the sAHP in hippocampal CA1 pyramidal neurons and it restores the threshold of LTP induction at the CA1 synapse.[44] Behavioural studies with young adult $K_v\beta1.1$ null mutants revealed impaired flexibility of L&M.[45] The mutants were impaired in reversal learning in the water maze and in the social transmission of food preferences task. The impaired behavioural flexibility can be explained by an "over-stabilisation" of information resulting from the mutation-caused sAHP reduction. In old age the $K_v\beta1.1$ null mutants do not appear to suffer from age-related impairments in spatial and contextual L&M and the social transmission of food preferences task.[44] Thus, the age-related sAHP increase appears to cause age-related L&M deficits and reductions of the sAHP can overcome these impairments.

Modulation of A-Type K^+ Channels As a L&M Mechanism

Apart from modulation of AHPs, there is evidence that modification of A-type K^+ channels contributes to mammalian L&M. The major characteristic of A-type K^+ channels is their fast inactivation, occurring within milliseconds. Because of this fast inactivation A-type K^+ channels can regulate excitability. For example, action potential propagation can be gated by A-type

K^+ channels, which is only possible when the channels are inactivated.[24,58] Even in a single neuron diverse A-type K^+ channels can be expressed. For example, in the somatodendritic region of hippocampal CA1 pyramidal neurons there are at least two different A-type K^+ channels[45,58] and a third type is located in the presynaptic terminals.[113] In the hippocampus $K_v1.4$,[120] $K_v3.4$,[100] $K_v4.1$, $K_v4.2$[113] and $K_v4.3$[129] have been identified as α-subunits and $K_v\beta1.1$[45] and KChIP2.1[7] as β-subunits of A-type K^+ channels.

It has been proposed that learning is associated with a reduction of A-type K^+ current amplitudes. Pairing of a click (conditioned stimulus) with a local iontophoretic injection of glutamate (unconditioned stimulus) in cat motor cortex results in a reduction of A-type K^+ current.[134] Furthermore, a reduction of A-type K^+ currents has been observed in lobule HVI of cerebellum after delay eyeblink conditioning.[110] Finally, experience-dependent reductions in extracellular spike amplitude in hippocampus may result from enhanced action potential backpropagation in CA1 pyramidal neurons, which is controlled by inactivation of A-type K^+ channels (see ref. 97 and below).

A functional role of A-type K^+ current reduction is the regulation of LTP. The coincidence of incoming EPSPs with action potentials backpropagating from the soma into the dendritic tree, can induce LTP.[77,78] In hippocampal CA1 pyramidal neurons action potential backpropagation is controlled by dendritic A-type K^+ channels.[58] The density of these channels increases with the distance from the soma, making backpropagation more difficult in distal parts of the dendritic tree. PKA, PKC and mitogen-activated protein kinase (MAPK), kinases which have been implicated in L&M (e.g., see ref. 85, 123 and also Vianna and Izquierdo, Nogues et al and Selcher et al in this book), reduce the channel activity by shifting the activation window to the right.[57,64] Thus, kinase signalling could by inactivating dendritic A-type K^+ channels and thereby allowing action potential backpropagation to coincide with incoming EPSPs result in LTP induction. Most of the dendritic A-type K^+ channels are thought to contain the subunits $K_v4.2$ and KChIP2.1, as concluded from the subunit properties and expression patterns.[7,58,113] Consistent with this idea the $K_v4.2$ subunit can be phosphorylated by PKA and MAPK and these modulations vary with the anatomical input in the hippocampus.[8,132] Studies with transgenic mice overexpressing a dominant-negative form of $K_v4.2$ confirm that dendritic A-type K^+ channels are involved in the induction of hippocampal LTP and in L&M.

In addition to dendritic A-type K^+ channels, presynaptic A-type K^+ channels appear also to be required for the induction of LTP at the hippocampal CA1 synapse.[83] Intraventricular application of $K_v1.4$ antisense oligonucleotides blocks this form of synaptic plasticity, probably because it reduces neurotransmitter release, as indicated by a decrease in paired-pulse facilitation. Interestingly, this blockade of LTP does not seem to affect spatial L&M in the water maze (ref. 83; but see ref. 80).

Finally, antisense and pharmacological experiments suggest that $K_v1.1$ containing K^+ channels, which may be A-type K^+ channels, are also involved in L&M.[71,82]

Conclusion

We have reviewed strong evidence that K^+ channel modulation influences neuronal spiking and/or synaptic transmission and appears to contribute to memory formation (and maybe even retrieval). Considering the results from different model organisms, continuities and changes among them are apparent.

Conditioning in two species of mollusc leads to prolonged action potentials. The affected K^+ currents are not identical in each case, perhaps a matter of different circuits. In both species, perturbations of serotonin and PKC signalling can affect K^+ channel activity. However, serotonergic mechanisms of conditioning are not as well characterised for *Hermissenda* photoreceptors as they are in *Aplysia* sensorimotor synapses. In the case of *Aplysia*, PKC and PKA pathways can both suppress the same I_{Kv} transient current, and the adoption of one or the other kinase pathway might depend on the excitation history of the neuron.

In *Drosophila* PKA inhibition in mushroom bodies enhances the amplitude of I_{KCa} currents, while PKA overexpression increases the amplitude of I_A and a delayed rectifier. These PKA results are consistent with the mollusc data, whereas there are no comparable fly data for PKC. CaMKII also reduces the amplitude of *Shal* and *Shab* currents in flies, and rodent CaMKII reduces the I_A and IK_{Ca} currents in *Hermissenda*.[107] Most of our examples concern excitability changes; there is mounting evidence for synaptic mechanisms in *Aplysia* facilitation and this appears to be effected by a reduction of I_A.

The results of the invertebrate learning studies are consistent with the prominent role assigned by other investigators to K^+ channel phosphorylations in brain function.[74] Genetic intervention is not possible for *Aplysia* and *Hermissenda*, a fact that makes *Drosophila* an attractive model for molecular studies of behaviour. However, fly studies do not allow observation of altered neuronal physiology after conditioning. Data from *Drosophila* are obtained from flies with channel deficiencies. These experiments demonstrate that some mutations in channel subunits (eag, Shaker) disrupt courtship conditioning and olfactory conditioning. The molecular lesions can be correlated with channel physiology in culture and in expression systems, providing some insight into the altered neuronal function. A number of caveats may be entered here. Fly channel mutants with gross behavioural phenotypes such as 'sticky feet' may have motor difficulties in task performance. While this objection can be answered by the use of nonchannel mutants such as *rutabaga* and *dunce*, the direct relationship of the K^+ channel to the behaviour becomes harder to address. Indeed data showing normal habitutation in Sh^{KS133} flies lacking I_A in muscle suggest that effects on habituation attributed to other Shaker alleles must be indirect. It may be the case that performance is separable from the acquisition of the task; flies with inhibited PKC form a memory of courtship without showing suppression of courtship during training.[69] As these flies are not inhibited in performance by a motor deficit, they demonstrate that immediate performance of the task may be dissociated from later recall of it.

Even though K^+ channel modulation is likely to contribute to the molecular mechanisms of invertebrate L&M, there are some concerns, which need to be addressed. One is the use of reduced preparations in place of intact animals. It is known that the effect of in vitro conditioning of *Hermissenda* on current amplitudes and voltage dependence is greater than that observed after behavioural conditioning (summarised in ref. 59). This may be an effect only of the relative intensity of the training, or else the mechanisms engaged may be different in vivo and in vitro.

K^+ current effects in the invertebrate studies are detected at sites that are not obviously sufficient for cognitive changes, though they mediate simpler behavioural responses; single photoreceptor cells of *Hermissenda crassicornis*, synapses of cultured neurons of *Aplysia californica*, neuromuscular junctions of *Drosophila melanogaster*. It has been postulated that L&M are a property of local circuits, and of higher order networks, and even of emergent states that are difficult to predict from the cellular organization. Others maintain that present understanding of the signalling of single neurons is too rudimentary to rule out a more reductionist paradigm of theory and research.

Mammalian L&M has not recapitulated all the data from invertebrate studies, but then the same experimental methods are not feasible. Consistent with the invertebrate results, mammalian L&M studies implicate increased excitability due to reductions of I_A. At least four different channel subunits can assemble into functional channels mediating I_A in mammals. Some of these channels are modifiable by protein kinases. Robust data on the role of I_A in mammalian memory are provided by studies on eyeblink conditioning. It has been proposed that transient A type channels can be inactivated by an incoming EPSP, to permit the backpropagation of APs into the dendritic tree and thereby regulate LTP. However, while LTP remains the most plausible cellular analogue of memory, its occurrence after hippocampal learning is questionable in whole animals and LTP impairments may dissociate from spatial L&M abilities (e.g., ref. 140; but see ref. 80).

The sAHP is the best candidate for a mammalian K^+ channel mechanism in learning and memory. Reductions of this Ca^{2+} activated current are associated with memory formation, and increases in the current may mediate cognitive decline in aging animals. Mice lacking $K_v\beta1.1$ in CA1 hippocampal neurons show reduced sAHP with old age compared with controls. The apparent reason for this is that, in the absence of the β subunit, noninactivating K^+ channels are converted into fast inactivators, thereby reducing the time available for Ca^{2+} influx during action potentials. A better understanding of the molecular controls of this process awaits the cloning of the underlying apamin-insensitive SK channel subunit. Perhaps the most important lesson of the $K_v\beta1.1$ null mutant study is that the effect of K^+ channel subunit deletions on downstream physiology can be indirect. The mediations between the sAHP reduction and the observed behavioural changes will also require careful dissection.

Most of the channels identified as L&M correlates are voltage-gated. The two known exceptions are the serotonin-sensitive S-K channel (*Aplysia*) and the SK channel underlying the sAHP. The preponderance of voltage-gated channels might arise because most changes reported are related to AP duration and indices of excitability that are relatively easy to measure, and less likely to be affected by nonvoltage gated channels. Voltage-gated channels are also responsible for altered frequency encoding in another preparation; cultured wild-type neurons of *Drosophila* show four distinct patterns of activity, which depend on the ratios of two distinct currents, and which are disrupted in dunce and rutabaga.[142] As another example of pattern change, evoked spike frequency increases in conditioned *Hermissenda* photoreceptors. The evidence for learning-associated altered spiking is thinner for whole animals, but one example is the conditioned click response of cat cortex.[134] It is not known yet whether training in a behavioural task leads to necessary transcriptional changes in channel profiles that affect membrane excitability. Altered expression might reduce the synthesis of a prelearning channel; or increase the synthesis of channels having a higher activation threshold, or tending to close during the falling phase of an action potential. This question and its relation to memory is potentially solvable by mouse molecular genetics, which permits the manipulation of K^+ channel modulatory molecules in restricted sites within the CNS, or at developmental stages that suit the experimenter.[44]

Acknowledgements

We thank Keiko Mizuno, Anna Need, Boris Rogelj, Steven Rose, and Andy Tinker for commenting on an earlier draft of the manuscript. We are grateful to the British Royal Society and the British Medical Research Council for generous support.

References

1. Abrams TW. Activity-dependent presynaptic facilitation: An associative mechanism in Aplysia. Cell Mol Neurobiol 1985; 5:123-145.
2. Alkon DL. Membrane depolarization accumulates during acquisition of an associative behavioural change. Science 1980; 210:1375-1376.
3. Alkon DL. Ionic conductance determinants of synaptic memory nets and their implications for Alzheimer's disease. J Neurosci Res 1999; 58:24-32.
4. Alkon DL, Acosta-Urquidi J, Olds J et al. Protein kinase injection reduces voltage-dependent potassium currents. Science 1983; 219:303-306.
5. Alkon DL, Farley J, Sakakibara M et al. Voltage-dependent calcium and calcium-activated potassium currents of a molluscan photoreceptor. Biophys J 1984; 46:605-614.
6. Alkon DL, Lederhendler I, Shoukimas JJ. Primary changes of membrane currents during retention of associative learning. Science 1982; 215:693-695.
7. An WF, Bowlby MR, Betty M et al. Modulation of A-type potassium channels by a family of calcium sensors. Nature 2000; 403:553-556.
8. Anderson AE, Adams JP, Qian Y et al. $K_v4.2$ phosphorylation by cyclic AMP-dependent protein kinase. J Biol Chem 2000; 275:5337-5346.
9. Atkinson NS, Robertson GA, Ganetzky B. A component of calcium-activated potassium channels encoded by the Drosophila slo locus. Science 1991; 253:551-555.
10. Baxter DA, Byrne JH. Serotonergic modulation of two potassium currents in the pleural sensory neurons of Aplysia. J Neurophysiol 1989; 62:665-679.

11. Belcadi-Abbassi W, Destrade C. Post-test apamin injection suppresses a Kamin-like effect following a learning session in mice. NeuroReport 1995; 6:1293-1296.
12. Berner J. Functional roles for the slow afterhyperpolarization: Inferences from incorporation of features of the slow gK^+ (Ca) into densely connected artificial neural networks. Network 1991; 2:1-6.
13. Blackwell KT, Alkon DL. Ryanodine receptor modulation of in vitro associative learning in Hermissenda crassicornis. Brain Res 1999; 822:114-125.
14. Bliss TVP, Collingridge GL. A synaptic model of memory: Long-term potentiation in the hippocampus. Nature 1993; 361:31-39.
15. Butler A, Wei AG, Baker K et al. A family of putative potassium channel genes in Drosophila. Science 1989; 243:943-947.
16. Byrne JH, Kandel ER. Presynaptic facilitation revisited: State and time dependence. J Neurosci 1996; 16:425-435.
17. Carew TJ, Walters ET, Kandel ER. Associative learning in Aplysia: Cellular correlates supporting a conditioned fear hypothesis. Science 1981; 211:501-504.
18. Carlson RO, Levitan IB. Role of arachidonic acid in depolarization-induced modulation of ion currents in Aplysia giant neurons. J Neurophysiol 1990; 64:341-350.
19. Chen ML, Hoshi T, Wu CF. Sh and eag K^+ channel subunit interaction in frog oocytes depends on level and time of expression. Biophys J 2000; 79:1358-1368.
20. Coetzee WA, Amarillo Y, Chiu J et al. Molecular diversity of K^+ channels. Ann N Y Acad Sci 1999; 868:233-85.
21. Cowan TM, Siegel RW. Mutational and pharmacological alterations of neuronal membrane function disrupt conditioning in Drosophila. J Neurogenet 1984; 1:333-344.
22. Cowan TM, Siegel RW. Drosophila mutations that alter ionic conduction disrupt acquisition and retention of a conditioned odor avoidance response. J Neurogenet 1986; 3:187-201.
23. Crow T, Forrester J. Light paired with serotonin mimics the effect of conditioning on phototactic behavior of Hermissenda. Proc Natl Acad Sci USA 1986; 83:7975-7978.
24. Debanne D, Guerineau NC, Gähwiler BH et al. Action-potential propagation gated by an axonal I_A-like K^+ conductance in hippocampus. Nature 1997; 389:286-289.
25. Delgado R, Davis R, Bono MR et al. Outward currents in Drosophila larval neurons: Dunce lacks a maintained outward current component downregulated by cAMP. J Neurosci 1998; 18:1399-1407.
26. Delgado R, Latorre R, Labarca P. K^+-channel blockers restore synaptic plasticity in the neuromuscular junction of dunce, a Drosophila learning and memory mutant. Proc R Soc Lond B Biol Sci 1992; 250:181-185.
27. Deschaux O Bizot JC. Effect of apamin, a selective blocker of Ca^{2+}-activated K^+-channel, on habituation and passive avoidance responses in rats. Neurosci Lett 1997; 227:57-60.
28. Deschaux O, Bizot JC, Goyffon M. Apamin improves learning in an object recognition task in rats. Neurosci Lett 1997; 222:159-162.
29. Deyo RA, Straube KT, Disterhoft JF. Nimodipine facilitates associative learning in aging rabbits. Science 1989; 243:809-811.
30. Disterhoft JF, Coulter DA, Alkon DL. Conditioning-specific membrane changes of rabbit hippocampal neurons measured in vitro. Proc Natl Acad Sci USA 1986; 83:2733-7.
31. Disterhoft JF, Thompson LT, Moyer JR et al. Calcium-dependent afterhyperpolarization and learning in young and aging hippocampus. Life Sci 1996; 59:413-420.
32. Drysdale R, Warmke J, Kreber R et al. Molecular characterization of eag: A gene affecting potassium channels in Drosophila melanogaster. Genetics 1991; 127(3):497-505.
33. Dudai Y, Jan YN, Byers D et al. dunce, a mutant of Drosophila deficient in learning. Proc Natl Acad Sci USA 1976; 73:1684-1688.
34. Eliot LS, Hawkins RD, Kandel ER et al. Pairing-specific, activity-dependent presynaptic facilitation at Aplysia sensory-motor neuron synapses in isolated cell culture. J Neurosci 1994; 14:368-383.
35. Engel JE, Wu CF. Genetic dissection of functional contributions of specific potassium channel subunits in habituation of an escape circuit in Drosophila. J Neurosci 1998; 18:2254-2267.
36. England SK, Uebele VN, Kodali J et al. A novel K^+ channel β-subunit ($hK_v\beta1.3$) is produced via alternative splicing. J Biol Chem 1995; 270:28531-28534.
37. Etcheberrigaray R, Matzel LD, Lederhendler II et al. Classical conditioning and protein kinase C activation regulate the same single potassium channel in Hermissenda crassicornis photoreceptors. Proc Natl Acad Sci USA 1992; 89:7184-7188.
38. Farley J, Richards WG, Grover LM. Associative learning changes intrinsic to Hermissenda type A photoreceptors. Behav Neurosci 1990; 104:135-152.
39. Farley J, Schuman E. Protein kinase C inhibitors prevent induction and continued expression of cell memory in Hermissenda type B photoreceptors. Proc Natl Acad Sci USA 1991; 88:2016-2020.

40. Fisher SA, Fischer TM, Carew TJ. Multiple overlapping processes underlying short-term synaptic enhancement. Trends Neurosci 1997; 20:170-177.
41. Fournier C, Kourrich S, Soumireu-Mourat B et al. Apamin improves reference memory but not procedural memory in rats by blocking small conductance Ca^{2+}-activated K^+ channels in an olfactory discrimination task. Behav Brain Res 2001; 121:81-93.
42. Frysztak RJ, Crow T. Enhancement of type B and A photoreceptor inhibitory synaptic connections in conditioned Hermissenda. J Neurosci 1994; 14:1245-1250.
43. Gandhi CC, Matzel LD. Modulation of presynaptic action potential kinetics underlies synaptic facilitation of type B photoreceptors after associative conditioning in Hermissenda. J Neurosci 2000; 20:2022-2035.
44. Giese KP, Peters M, Vernon J. Modulation of excitability as a learning and memory mechanism: A molecular genetic perspective. Physiol Behav 2001; 73:803-810.
45. Giese KP, Storm JF, Reuter D et al. 1998 Reduced K^+ channel inactivation, spike broadening and afterhyperpolarization in $K_v\beta1.1$-deficient mice with impaired learning. Learning Memory 1998; 5:257-273.
46. Gispen WH, Biessels G. Cognition and synaptic plasticity in diabetes mellitus. Trends Neurosci 2000; 23:542-549.
47. Goldsmith BA, Abrams TW. cAMP modulates multiple K^+ currents, increasing spike duration and excitability in Aplysia sensory neurons. Proc Natl Acad Sci USA 1992; 89:11481-11485.
48. Gomez M, De Castro E, Guarin E et al. Ca^{2+} signalling via the neuronal calcium sensor-1 regulates associative learning and memory in C. elegans. Neuron 2001; 30:241-248.
49. Griffith LC, Wang J, Zhong Y et al. Calcium/calmodulin-dependent protein kinase II and potassium channel subunit eag similarly affect plasticity in Drosophila. Proc Natl Acad Sci USA 1994; 91:10044-10048.
50. Guzowski JF, Setlow B, Wagner EK et al. Experience-dependent gene expression in the rat hippocampus after spatial learning: A comparison of the immediate-early genes Arc, c-fos, and zif268. J Neurosci 2001; 21:5089-5098.
51. Hall JC. The mating of a fly. Science 1994; 264:1702-1714.
52. Han PL, Levin LR, Reed RR et al. Preferential expression of the Drosophila rutabaga gene in mushroom bodies, neural centers for learning in insects. Neuron 1992; 9:619-627.
53. Hawkins RD, Abrams TW, Carew TJ et al. A cellular mechanism of classical conditioning in Aplysia: Activity-dependent amplification of presynaptic facilitation. Science 1983; 219:400-405.
54. Heinemann SH, Rettig J, Graak H-R et al. Functional characterization of K_v channel β-subunits from rat brain. J Physiol 1996; 493:625-633.
55. Heisenberg M, Borst A, Wagner S et al. Drosophila mushroom body mutants are deficient in olfactory learning. J Neurogenet 1985; 2(1):1-30.
56. Hodgkin AL, Huxley AF, Katz B. 1952 Measurement of current-voltage relations in the membrane of the giant axon of Loligo. J Physiol 1952; 116:424-448.
57. Hoffman DA, Johnston D. Downregulation of transient K^+ channels in dendrites of hippocampal CA1 pyramidal neurons by activation of PKA and PKC. J Neurosci 1998; 18:3521-3528.
58. Hoffman DA, Magee JC, Colbert CM et al. K^+ channel regulation of signal propagation in dendrites of hippocampal pyramidal neurons. Nature 1997; 387:869-875.
59. Huang H, Farley J. Pp1 inhibitors depolarize Hermissenda photoreceptors and reduce K^+ currents. J Neurophysiol 2001; 86:1297-1311.
60. Ikonen S, Schmidt B, Riekkinen P. Apamin improves spatial navigation in medial septal-lesioned mice. Eur J Pharmacol 1998; 347:13-21.
61. Impey S, Smith DM, Obrietan K et al. Stimulation of cAMP response element (CRE)-mediated transcription during contextual learning. Nat Neurosci 1998; 1:595-601.
62. Isomoto S, Kondo C, Kurachi Y. Inwardly rectifying potassium channels: Their molecular heterogeneity and function. Jpn J Physiol 1997; 47:11-39.
63. Ichinose M, Asai M, Sawada M. Activation of outward current by pituitary adenylate cyclase activating polypeptide in mouse microglial cells. J Neurosci Res 1998; 51:382-90.
64. Johnston D, Hoffman DA, Magee JC et al. Dendritic potassium channels in hippocampal pyramidal neurons. J Physiol (Lond) 2000; 525:75-81.
65. Joiner MA, Griffith LC. Mapping of the anatomical circuit of CaM kinase-dependent courtship conditioning in Drosophila. Learning Memory 1999; 6:177-192.
66. Kamb A, Iverson LE, Tanouye MA. Molecular characterization of Shaker, a Drosophila gene that encodes a potassium channel. Cell 1987; 31:405-413.
67. Kamyshev NG, Iliadi KG, Bragina JV. Drosophila conditioned courtship: Two ways of testing memory. Learning Memory 1999; 6:1-20.

68. Kandel ER, Tauc L. Mechanism of heterosynaptic facilitation in the giant cell of the abdominal ganglion of Aplysia depilans. J Physiol 1965; 181:28-47.
69. Kane NS, Robichon A, Dickinson JA et al. Learning without performance in PKC-deficient Drosophila. Neuron 1997; 18:307-314.
70. Kotecha SA, Schlichter LC. A $K_v1.5$ to $K_v1.3$ switch in endogenous hippocampal microglia and a role in proliferation. J Neurosci 1999; 19:10680-10693.
71. Kourrich S, Mourre C, Soumireu-Mourat B. Kaliotoxin, a $K_v1.1$ and $K_v1.3$ channel blocker, improves associative learning in rats. Behav Brain Res 2001; 120:35-46.
72. Kronforst-Collins MA, Moriearty PL, Schmidt B et al. Metrifonate improves associative learning and retention in aging rabbits. Behav Neurosci 1997; 111:1031-1040.
73. Landfield PW, Pitler TA. Prolonged Ca^{2+}-dependent afterhyperpolarizations in hippocampal neurons of aged rats. Science 1984; 226:1089-1092.
74. Levitan IB. Modulation of ion channels by protein phosphorylation. How the brain works. Adv Second Messenger Phosphoprotein Res 1999; 33:3-22.
75. Li M, Unwin N, Stauffer KA et al. Images of purified Shaker potassium channels. Curr Biol 1994; 4:110-115.
76. Ma M, Koester J. The role of K^+ currents in frequency-dependent spike broadening in Aplysia R20 neurons: A dynamic-clamp analysis. J Neurosci 1996; 16:4089-4101.
77. Magee JC, Johnston D. A synaptically controlled, associative signal for Hebbian plasticity in hippocampal neurons. Science 1997; 275:209-213.
78. Markram H, Lübke J, Frotscher M et al. Regulation of synaptic efficacy by coincidence of postsynaptic APs and EPSPs. Science 1997; 275:213-215.
79. Marrion NV, Tavalin SJ. Selective activation of Ca^{2+}-activated K^+ channels by colocalized Ca^{2+} channels in hippocampal neurons. Nature 1998; 395:900-905.
80. Martin SJ, Grimwood PD, Morris RGM. Synaptic plasticity and memory: An evaluation of the hypothesis. Annu Rev Neurosci 2000; 23:649-711.
81. Matzel LD, Rogers RF. Postsynaptic calcium, but not cumulative depolarization, is necessary for the induction of associative plasticity in Hermissenda. J Neurosci 1993; 13:5029-5040.
82. Meiri N, Ghelardini C, Tesco G et al. Reversible antisense inhibition of Shaker-like $K_v1.1$ potassium channel expression impairs associative memory in mouse and rat. Proc Natl Acad Sci USA 1997; 94:4430-4434.
83. Meiri N, Sun M-K, Segal Z et al. Memory and long-term potentiation (LTP) dissociated: Normal spatial memory despite CA1 LTP elimination with $K_v1.4$ antisense. Proc Natl Acad Sci USA 1998; 95:15037-15042.
84. Menzel R. Searching for the memory trace in a mini-brain, the honeybee. Learning Memory 2001; 8:53-62.
85. Milner B, Squire LR, Kandel ER. Cognitive neuroscience and the study of memory. Neuron 1998; 20:445-468.
86. Moyer Jr JR, Thompson LT, Black JP et al. Nimodipine increases excitability of rabbit CA1 pyramidal neurons in an age- and concentration-dependent manner. J Neurophysiol 1992; 68:2100-2109.
87. Moyer JR, Power JM, Thompson LT et al. Increased excitability of aged rabbit CA1 neurons after trace eyeblink conditioning. J Neurosci 2000; 20:5476-5482.
88. Moyer JR, Thompson LT, Disterhoft JF. Trace eyeblink conditioning increases CA1 excitability in a transient and learning-specific manner. J Neurosci 1996; 16:5536-5546.
89. Muzzio IA, Talk AC, Matzel LD. Incremental redistribution of protein kinase C underlies the acquisition curve during in vitro associative conditioning in Hermissenda. Behav Neurosci 1997; 111:739-753.
90. Muzzio IA, Gandhi CC, Manyam U et al. Receptor-stimulated phospholipase A(2) liberates arachidonic acid and regulates neuronal excitability through protein kinase C. J Neurophysiol 2001; 85:1639-1647.
91. Papazian DM, Schwarz TL, Tempel BL et al. Cloning of genomic and complementary DNA from Shaker, a putative potassium channel gene from Drosophila. Science 1987; 237:749-53.
92. Patel AJ, Lazdunski M, Honore E. Lipid and mechano-gated 2P domain K^+ channels. Curr Opin Cell Biol. 2001; 13:422-428.
93. Pongs O, Kecskemethy N, Muller R et al. Shaker encodes a family of putative potassium channel proteins in the nervous system of Drosophila. EMBO J 1988; 7:1087-1096.
94. Poulain C, Ferrus A, Mallart A. Modulation of type A K^+ current in Drosophila larval muscle by internal Ca^{2+}; effects of the overexpression of frequenin. Pflugers Arch 1994; 427:71-79.
95. Power JM, Oh MM, Disterhoft JF. Metrifonate decreases sI_{AHP} in CA1 pyramidal neurons in vitro. J Neurophysiol 2001; 85:319-322.

96. Quinn WG, Harris WA, Benzer S. Conditioned behavior in Drosophila melanogaster. Proc Natl Acad Sci USA 1974; 71:708-712.
97. Quirk MC, Blum KI, Wilson MA. Experience-dependent changes in extracellular spike amplitude may reflect regulation of dendritic action potential back-propagation in rat hippocampal pyramidal cells. J Neurosci 2001; 21:240-248.
98. Renger JJ, Ueda A, Atwood HL et al. Role of cAMP cascade in synaptic stability and plasticity: Ultrastructural and physiological analyses of individual synaptic boutons in Drosophila memory mutants. J Neurosci 2000; 20:3980-3992.
99. Rettig J, Heinemann SH, Wunder F et al. Inactivation properties of voltage-gated K^+ channels altered by presence of β-subunit. Nature 1994; 369:289-294.
100. Rettig J, Wunder F, Stocker M et al. Characterization of a Shaw-related potassium channel family in rat brain. EMBO J 1992; 11:2473-2486.
101. Robertson GA, Warmke JM, Ganetzky B. Potassium currents expressed from Drosophila and mouse eag cDNAs in Xenopus oocytes. Neuropharmacol 1996; 35:841-850.
102. Rogers RF, Matzel LD. G-protein mediated responses to localized serotonin application in an invertebrate photoreceptor. NeuroReport 1995; 13:2161-2165.
103. Saar D, Grossman Y, Barkai E. Reduced after-hyperpolarization in rat piriform cortex pyramidal neurons is associated with increased learning capability during operant conditioning. Eur J Neurosci 1998; 10:1518-1523.
104. Saar D, Grossman Y, Barkai E. Long-lasting cholinergic modulation underlies rule learning in rats. J Neurosci 2001; 21:1385-1392.
105. Sah P. Ca^{2+}-activated K^+ currents in neurones: Types, physiological roles and modulation. Trends Neurosci 1996; 19:150-154.
106. Sah P, Bekkers JM. Apical dendritic location of slow afterhyperpolarization current in hippocampal pyramidal neurons: Implications for the integration of long-term potentiation. J Neurosci 1996; 16:4537-4542.
107. Sakakibara M, Alkon DL, DeLorenzo R et al. Modulation of calcium-mediated inactivation of ionic currents by Ca^{2+}/calmodulin-dependent protein kinase II. Biophys J 1986; 50:319-327.
108. Salkoff L, Butler A, Fawcett G et al. Evolution tunes the excitability of individual neurons. Neuroscience 2001; 103:853-859.
109. Santoro B, Tibbs GR. The HCN gene family: Molecular basis of the hyperpolarization-activated pacemaker channels. Ann N Y Acad Sci 1999; 868:741-64.
110. Schreurs BG, Gusev PA, Tomsic D et al. Intracellular correlates of acquisition and long-term memory of classical conditioning in Purkinje cell dendrites in slices of rabbit cerebellar lobule HVI. J Neurosci 1998; 18:5498-5507.
111. Shah M, Haylett DG. The pharmacology of hSK1 Ca^{2+}-activated K^+ channels expressed in mammalian cell lines. Br J Pharmacol 2000; 129:627-30.
112. Shao LR, Halvorsrud R, Borg-Graham L et al. The role of BK-type Ca^{2+}-dependent K^+ channels in spike broadening during repetitive firing in rat hippocampal pyramidal cells. J Physiol (Lond) 1999; 521:135-146.
113. Sheng M, Tsaur M-L, Jan YN et al. Subcellular segregation of two A-type K^+ channel proteins in rat central neurons. Neuron 1992; 9:271-284.
114. Siegelbaum SA, Camardo JS, Kandel ER. Serotonin and cyclic AMP close single K^+ channels in Aplysia sensory neurones. Nature 1982; 299(5882):413-7.
115. Skoulakis EM, Davis RL. Olfactory learning deficits in mutants for leonardo, a Drosophila gene encoding a 14-3-3 protein. Neuron 1996; 17:931-944.
116. Solomon PR, Vander Schaaf ER, Thompson RF et al. Hippocampus and trace conditioning of the rabbit's classically conditioned nictitating membrane response. Behav Neurosci 1986; 100:729-744.
117. Stocker M, Krause M, Pedarzani P. An apamin-sensitive Ca^{2+}-activated K^+ current in hippocampal pyramidal neurons. Proc Natl Acad Sci USA 1999; 96:4662-4667.
118. Stopfer M, Bhagavan S, Smith BH et al. Impaired odour discrimination on desynchronization of odour-encoding neural assemblies. Nature 1997; 390:70-74.
119. Storm JF Potassium currents in hippocampal pyramidal cells. Prog Brain Res 1990; 83:161-187.
120. Stühmer W, Ruppersberg JP, Schroter KH et al. Molecular basis of the functional diversity of voltage-gated potassium channels in mammalian brain. EMBO J 1989; 8:3235-3244.
121. Sugita S, Baxter DA, Byrne JH. Activators of protein kinase C mimic serotonin-induced modulation of a voltage-dependent potassium current in pleural sensory neurons of Aplysia. J Neurophysiol 1994; 72:1240-1249.
122. Sugita S, Baxter DA, Byrne JH. Modulation of a cAMP/protein kinase A cascade by protein kinase C in sensory neurons of Aplysia. J Neurosci 1997; 17:7237-7244.

123. Sweatt JD. The neuronal MAP kinase cascade: A biochemical signal integration system subserving synaptic plasticity and memory. J Neurochem 2001; 76:1-10.
124. Talk AC, Muzzio IA, Matzel LD. Phospholipases and arachidonic acid contribute independently to sensory transduction and associative neuronal facilitation in Hermissenda type B photoreceptors. Brain Res 1997; 751:196-205.
125. Tanouye MA, Ferrus A. Action potentials in normal and Shaker mutant Drosophila. J Neurogenet 1985; 2:253-257
126. Tempel BL, Papazian DM, Schwarz TL et al. Sequence of a probable potassium channel component encoded at Shaker locus of Drosophila. Science 1987; 237:770-775.
127. Thibault O, Landfield PW. Increase in single L-type calcium channels in hippocampal neurons during aging. Science 1996; 272:1017-1020.
128. Thompson LT, Moyer JR, Disterhoft JF. Transient changes in excitability of rabbit CA3 neurons with a time course appropriate to support memory consolidation. J Neurophysiol 1996; 76:1836-1849.
129. Tsaur M-L, Chou C-C, Shih Y-H et al. Cloning, expression and CNS distribution of $K_v4.3$, an A-type K^+ channel alpha subunit. FEBS Lett 1997; 400:215-220.
130. Tsunoda S, Salkoff L. The major delayed rectifier in both Drosophila neurons and muscle is encoded by Shab. J Neurosci 1995; 15:5209-5221.
131. Tully T, Quinn WG. Classical conditioning and retention in normal and mutant Drosophila melanogaster. J Comp Physiol (A) 1985; 157:263-277.
132. Varga AW, Anderson AE, Adams JP et al. Input-specific immunolocalization of differentially phosphorylated $K_v4.2$ in the mouse brain. Learning Memory 2000; 7:321-332.
133. Vergara C, Latorre R, Marrion NV et al. Calcium-activated potassium channels. Curr Opin Neurobiol 1998; 8:321-329.
134. Woody CD, Gruen E, Birt D. Changes in membrane currents during Pavlovian conditioning of single cortical neurons. Brain Res 1991; 539:76-84.
135. Wright NJ, Zhong Y. Characterization of K^+ currents and the cAMP-dependent modulation in cultured Drosophila mushroom body neurons identified by lacZ expression. J Neurosci 1995; 15:1025-1034.
136. Wu CF, Haugland FN. Voltage clamp analysis of membrane currents in larval muscle fibers of Drosophila: Alteration of potassium currents in Shaker mutants. J Neurosci 1985; 5:2626-2640.
137. Yao WD, Wu CF. Distinct roles of CaMKII and PKA in regulation of firing patterns and K^+ currents in Drosophila neurons. J Neurophysiol 2001; 85:1384-1394.
138. Yeo CH, Hardiman MJ, Moore JW et al. Trace conditioning of the nictitating membrane response in decorticate rabbits. Behav Brain Res 1984; 11:85-88.
139. Yu D, Feng C, Guo A. Altered outward K^+ currents in Drosophila larval neurons of memory mutants rutabaga and amnesiac. J Neurobiol 1999; 40:158-170.
140. Zamanillo D, Sprengel R, Hvalby O et al. Importance of AMPA receptors for hippocampal synaptic plasticity but not for spatial learning. Science 1999; 284:1805-1811.
141. Zhang L, McBain CJ. 1995 Potassium conductances underlying repolarization and after-hyperpolarization in the rat CA1 hippocampal interneurones. J Physiol (Lond) 1995; 488:661-672.
142. Zhao ML, Wu CF. Alterations in frequency coding and activity dependence of excitability in cultured neurons of Drosophila memory mutants. J Neurosci 1997; 17:2187-2199.
143. Zhong Y. Mediation of PACAP-like neuropeptide transmission by coactivation of Ras/Raf and cAMP signal transduction pathways in Drosophila. Nature 1995; 375:588-592.
144. Zhong Y, Pena LA. A novel synaptic transmission mediated by a PACAP-like neuropeptide in Drosophila. Neuron 1995; 14:527-536.
145. Zhong Y, Wu CF. Alteration of four identified K^+ currents in Drosophila muscle by mutations in eag. Science 1991; 252:1562-1564.
146. Zhou Y, Schopperle WM, Murrey H et al. A dynamically regulated 14-3-3, Slob, and Slowpoke potassium channel complex in Drosophila presynaptic nerve terminals. Neuron 1999; 22:809-818.

CHAPTER 2.1

Glutamate Receptors

Gernot Riedel, Jacques Micheau and Bettina Platt

Abstract

Glutamate as a neurotransmitter plays a critical role in multiple processes in the brain from early development to ageing and includes important functions in memory formation. Both ionotropic and metabotropic glutamate receptors are involved in these functions but a close review assessing the contribution of the individual receptor subtypes reveals subtle differences. The longest tradition and extensive work is available for N-methyl-D-aspartate (NMDA) receptors and it seems increasingly clear that NMDA receptor activation is necessary during encoding of new information. The precise function of α-amino-3-hydroxy- 5-methyl- 4-isoxazolepropionic acid (AMPA) receptors proved difficult to assess since block of fast synaptic transmission acts like a local anaesthetic thereby transiently silencing the brain or brain structure. This experimental design, however, is most effective during encoding but also affects recall. Metabotropic glutamate (mGlu) receptors, finally, seem to play an intermediate role during memory consolidation. Evidence in favour of this interpretation is summarised and functional differences between behavioural tasks and brain regions are pointed out.

Introduction

Excitatory neurotransmission in the central nervous system is dominated by one transmitter: glutamate. Due to its now well-established role in many aspects of neuronal communication, plasticity and pathology, it has been the focus of intense research in many laboratories around the world. As a consequence, the amount of literature relevant to glutamate and neuronal function has reached a size that most researchers (especially those at the start of their career) will probably consider overwhelming.

Therefore, the present book chapter aims to give a concise overview of the current knowledge on glutamate receptors and learning and memory. Experimental details however, as well as the vast amount of data obtained by in vitro studies, are only referred to when necessary. To satisfy the more experienced reader, we also provide an extensive review of the literature in tabular form.

Glutamate receptors can be classified into two major classes: ionotropic receptors (iGlu) coupled to cation channels; and metabotropic receptors (mGlu), coupling to intracellular second messenger cascades. Within these two classes, specific receptor types and their subunits have been characterised using molecular, pharmacological, and physiological techniques (Table 1). These properties have been extensively reviewed elsewhere.[58,127,159,172]

Ionotropic receptors to be considered here comprise α-amino-3-hydroxy-5-methyl-4-isoxazoleproprionate (AMPA) and N-methyl-D-aspartate (NMDA) receptors. Kainate receptors only attracted interest more recently[157] and very little is know on their behavioural role. AMPA receptors are the main source of fast excitatory transmission and as such a rather obvious but unspecific keyplayer in learning and memory formation. In contrast, NMDA

From Messengers to Molecules: Memories Are Made of These, edited by Gernot Riedel and Bettina Platt.

Table 1. Glutamate receptor classification

	Ionotropic Receptors			Metabotropic Receptors		
	NMDA	**AMPA**	**Kainate***	**Group I**	**Group II**	**Group III**
Subunits/ Subtypes	NR1 NR2A-D NR3A	GluR1-4 (also called GluRA-GluRD	GluR5-7 KA1&2	1 (a-d), 5 (a&b)	2, 3	4 (a&b), 6, 7 (a&b), 8
Current/ Signalling Cascade	Ca^{2+}, Na^+, K^+	Na^+, K^+, $(Ca^{2+})^1$	Na^+, K^+, $(Ca^{2+})^2$	PLC ↑ → IP3 + DAG → PKC ↑ → Ca_i^{2+}	AC ↓ → cAMP ↓	AC ↓ → cAMP ↓
Agonists	NMDA Ibotenate Quinolinate AMAA	Quinolinate AMPA Domoate Fluorowillar-diine ATPA LY262466 (Kainate)	Kainate Domoate ATPA (R5) Iodowillardiine (R5) Acromelic acid (AMPA)	Quisqualate ACPD DHPG CHPG (5) 3-HPG t-ADA	ACPD DCG-IV L-CCG-I 4-CPG APDC LY354740	AP4
Antagonist	AP5, AP7 MK-801 PCP / TCP Dextromethorphan Ketamine Memantine Ifenprodil (NR2B) CGP37849 CGP39551 CGS19755 LY233536	CNQX, NBQX, DNQX *JST* Barbiturates (e.g. GYKI 52466, GYKI53655) GDEE LY326325 LY215490 SYM 2206 YM90K	CNQX, NBQX, DNQX GAMS AMOA NS-102	MCPG 4-CPG` MPEP (5) LY367385 (1) CPCCOEt AIDA AP3	MCPG MTPG LY341495	(MCPG) MPPG MAP4 MSOP

Continued on next page

Table 1. Continued

	NMDA	AMPA	Kainate*	Group I	Group II	Group III
Modulators	*Positive:* Glycine, D-serine D-cycloserine Polyamines *Negative:* Zn^{2+}, H^{+}, Mg^{2+} 7Cl Kyn HA-966 Arcaine	Anirazetam Cyclothiazide IDRA 21 CX516 1-BCP	Lectins, e.g. Concanvalin A (ConA)			

This table gives an overview over the principle glutamate receptors and lists some of the pharmacological tools used in behavioral studies.
*: kainate receptors have been included for completion despite the lack of behavioral data.
Abbreviations:
Drug names:
ACPD: 1S,3R-1- aminocyclopentane dicarboxylate;
AIDA (=UPF 523): (RS)-1-aminoindan-1,5-dicarboxylic acid;
AMAA: 2-amino-2-(3-hydroxy-5-methyl-4-isoxazolyl)acetic acid;
AMOA: 2-amino-3-(3-(carboxymethoxy)-5-methylisoxazol-4-yl)propionic acid;
AMPA: A-amino-3-hydroxy-5-methyl-4-isoxazoleproprionic acid;
AP3: DL-2-amino-3-phosphonopropionic acid;
AP4: L(+)-2-amino-4-phosphonobutyric acid
AP5: D-2-amino-5-phosphonovaleric acid;
AP7: 2-amino-7-phosphonoheptanoic acid
APDC: (2R,4R)-4-aminopyrrolidine-2,4-dicarboxylic acid;
ATPA: DL-amino-3-hydroxy-5-tertbutyl-4-isoxazolepropionic acid;
1-BCP: 1-(1,3-benzodioxol-5-ylcarbonyl) piperidine;
L-CCG-1: (2S,1′S,2′S)-2-(carboxycyclopropyl)glycine;
CGS 19755: cis-4-phosphonomethyl-2-piperadine carboxylic acid;
CHPG: (R,S)-2-chloro-5-hydroxyphenylglycine;
7Cl Kyn: 7-chlorokynurenic acid;
CNQX: 6-cyano-7-nitroquinoxaline-2,3-dione;
4-CPG: (RS)--ethyl-4-carboxyphenylglycine;
CPP: cis(±)-3-(2-carboxypiperazine-4-yl)propyl-1-phosphonic acid;
CPCCOEt: 7-(Hydroxyimino)cyclopropa[b] chromen-1a-carboxylate ethyl ester;
Cyclothiazide: 6-chloro-3,4-dihydro-3-(2-norbornen-5-yl)-2H-1,2,4-benzothiadiazine-7-sulphonamide-1,1-dioxide;

Continued on next page

Table 1. Continued

CX516: 1-(quinoxalin-6-ylcarbonyl)-piperidine;
DCG-IV: (2S,2′R,3′R)-2-(2′,3′-dicarboxycyclopropyl)glycine;
DHPG: (S)-3,5- dihydroxyphenylglycine;
DNQX: 6,7-dinitroquinoxaline-2,3-dione;
Domoate: [2S-[2a,3b,4b(1Z,3E,5R)]]-2-Carboxy-4-(5-carboxy-1-methyl-1,3-hexadienyl)-3-pyrrolidineacetic acid;
GAMS: γ-D-glutamylaminomethyl sulphonic acid;
GDEE: L-glutamicacid-diethylesther;
GYKI 52466: 1-(4-aminophenyl)4-methyl-7,8-methylenedioxy-5H-2,3-benzodiazepine;
GYKI53655 (= LY300168; LY303070: the active isomer of GYKI53655): 1-(4-aminophenyl)-4-methyl-7,8-methylenedioxy-5H-(3N-methylcarbamate)-2,3-benzodiazepine;
HA966 (±)-3-amino-1-hydroxypyrrolit-2-one;
3-HPG: (RS)-3-hydroxyphenylglycine;
IDRA 21: 7-Chloro-3-methyl-3,4-dihydro-2H-1,2,4-benzothiadiazine S,S-dioxide;
JST: Joro spider toxin;
LY215490: (3SR,4aRS,6RS,8aRS)-6-[2-(1H-tetrazol-5yl)ethyl]decahydroisoquinoline-3-carboxylic acid;
LY262466: DL-2-amino-3-(4-hydroxy-1,2,5-thiadiazol-3-yl)-propanoic acid;
LY326325: (3S,4aR,6R,8aR)-6-[2-(1(2)H-tetrazole-5-yl) ethyl] decahydroisoquinoline-3-carboxylic acid;
LY354740: (2)-aminobicyclo[3.1.0]hexane-2,6-dicarboxylic acid;
LY367385: (S)-(+)--amino-4-carboxy-2-methylbenzeneacetic acid;
MAP4: (S)-2-amino-2-phosphonobutanoic acid;
MCPG: (+)-alpha-methyl-4-carboxyphenylglycine;
Memantine: 1-amino-3,5-dimethyladamantane;
MPEP: 6-methyl-2- (phenylethynyl)-pyridine;
MPPG: (RS)--Methyl-4-phosphonophenylglycine;
MSOP: (RS)-a-Methylserine-O-phosphate;
MTPG: (RS)--Methyl-4-tetrazolylphenylglycine;
MK-801: (+)-5-methyl-10,11-dihydro-5H-dibenzo[a,d]cyclo-hepten-5,10-imine maleate;
NBQX: 2,3-dihydroxy-6-nitro-7-sulfamoylbenzo(F)quinoxaline;
NMDA: N-methyl-D-aspartate;
NPC 12626: 2-amino-4,5-(1,2-cyclohexyl)-7-phosphonoheptanoic acid;
NS-102: 5-nitro-6,7,8,9-tetrahydrobenzo[g]indole-2,3-dione-3-oxime;
PCP: phencyclidine;
SYM 2206: (±)-4-(4-aminophenyl)-1,2-dihydro-1-methyl-2-propylcarbamoyl-6,7-methylenedioxyphthalazine;
tADA: trans-azetidine-2,4-dicarboxylic acid;
TCP: 1-(2-thienyl)-cyclohexyl piperidine.
YM90K: 6-(1H-imidazol-1-yl)-7-nitro-2,3(1H,4H)-quinoxalinedione

Other abbreviations:
PLC (phospholipase C)
PLD (phospholipase D)
cAMP (cyclic aminotrisphosphate)
[1]: Ca^{2+} permeability only for AMPA receptors that do not contain R2
[2]: Ca^{2+} permeability only for GluR6 containing kainate receptors

receptors are undoubtedly the most famous glutamate receptor in the field of learning and memory. The receptor and its associated channel have two properties that render it particularly interesting: firstly, it has a high calcium permeability thus providing access to Ca^{2+}-dependent second messenger cascades relevant to memory formation; secondly, it contains a voltage-gated magnesium block, which requires simultaneous opening of other depolarising conductances to permit activation.[166] Accordingly, coincident activity as it is postulated to occur in Hebbian types of synaptic plasticity can be detected by NMDA receptors. In addition, there is a receptor site for the coagonist glycine and various additional modulatory sites. Of potential interest is the modulatory site within the actual ion channel pore, which provides a use-dependent target for drug development with the potential to modify memory formation (see Table 1).

Metabotropic Glu receptors have been identified in the mid-80s. To date, 3 main classes with different molecular and pharmacological properties have been characterised. Links to different second-messenger cascades (see Table 1) provide access to multiple enzymes, immediate early genes and the production of novel proteins, as well as to the modulation of diverse ion channels (reviewed in refs. 7,48,85,177). Depending on the location of these receptors (e.g., pre or postsynaptic, on GABAergic or glutamatergic neurones) both enhancements and reduction of neuronal excitability is possible. As a general rule, group I mGlu receptors tend to promote excitation while group II/III are more prone to reduce it. In addition, the existence of additional, so far unidentified mGlu receptors such PLD coupled or presynaptic group I mGlu receptors has also been suggested.[87]

In the context of learning and memory, group I mGlu receptors, which couple to phospholipase C and hence to intracellular Ca^{2+} signalling and protein kinase C activation have attracted more interest than group II/III, coupling to adenylate cylclase. Another property of group I mGlu receptors, namely their perisynaptic location,[15,124] is a further reason for the assumed role in memory formation: similar to NMDA receptors, coincident neuronal excitation and subsequent glutamate 'spillover' may be required to activate these receptors.

One obstacle in pharmacological approaches to study glutamate receptor functions in learning and memory research has been the neurotoxic potency of the vast majority of glutamate receptor agonists and even of some of the antagonists. Therefore, other strategies have focussed on glutamate receptor modulators, and, more recently, on genetically modified animals.

Glutamate Receptor Function in Learning and Memory Formation

The starting point of our current interest in glutamate receptor function and memory formation was a publication in 1986 in Nature by Richard Morris from Edinburgh and his co-workers in Irwine (California).[156] They intracerebroventricularly (icv) infused rats with the NMDA-receptor antagonist AP5 through minipumps , and while animals where under drug, they tested the rat's ability to learn a spatial task in an open-field water maze. The milky water in this circular pool prevents the animal from visualising the submerged platform, which is the only means of escape. Compared with vehicle infused animals, AP5-treated subjects were impaired in learning and remembering the location of the submerged platform. Morris interpreted this as a spatial learning deficit and suggested that this might be due to AP5 blocking synaptic plasticity in the hippocampus. This was later confirmed (see Table 2).

Subsequently, numerous other glutamate receptor antagonists and agonists were tested in both spatial and nonspatial learning paradigms and there can be no doubt about the importance of the glutamatergic system in several brain structures during learning and memory formation. For historical reasons, we start of with spatial learning paradigms and NMDA receptors before moving to other behavioral tests.

Spatial Learning

Many thoughtfully developed spatial testing procedures are available to date and the most popular ones are the water maze, the 8-arm radial maze and T- or Y-mazes. These paradigms allow for spatial reference and/or working memory to be assessed and there are also some

Table 2. Effects of glutamate receptor blockade on spatial learning and memory formation

Species	Infusion Time	Infusion Route	Antagonist	Agonist	Effect on Memory	References
			Water Maze			
Rat/ Mouse (Gerbil)	Pretraining	• i.c.v. (minipump), i.HC, i.IC, s.c., i.p. (postnatal d8- d19), i.v.	N: AP5,CPP, MK801, ketamine; 7Cl-kynurenate		Impairment	• 5,11,53,68,76, 76,78,83,125, 140,153,154, 156,160,173, 196,198,242, 254,255,258
			A: NBQX, LY325326			• 69,194,267
			m: MCPG, MAP4			• 90,131,185
				m: ACPD, L-CCG-I, 1S,3S-ACPD, L-AP4		• 90,91,176,244
		• i.p., p.o., i.c.v.	N: NPC 17-742, CGS 19755, CGP 40116, MK801		Impairment due to sensorimotor effects of the drugs	• 3,32,33,82,100, 211,212
			A: CNQX			• 34
		• i.BLA, i.p.	N: CPP, TCP		No effect	• 68,122
		• i.c.v	m: MCPG			• 26
			MAP4 +	m: L-AP4		• 90
		• i. HC (minipump)	MCPG,AIDA			• 188
		• p.o. daily for 2 months	N: Memantine		Facilitation	• 13

continued on next page

Table 2. Continued

Species	Infusion Time	Infusion Route	Antagonist	Agonist	Effect on Memory	References
		• i.p.		*N:* ACPC,DCS, RS 67333 *A:* IDRA 21		• 17,117,178 • 267
Rat / Mouse (Gerbil)	Post-training or Pretesting	• i.c.v., i.IC, i.p., i.HC	*N:* AP5, CPP, MK801 *A:* NBQX, LY325326		No effect	• 54,78,83,140, 153,196,258 • 69,194,267
	(minipump)	• i.c.v., i.HC	*N:* AP5	(working memory)	Impairment	• 225
Mouse (genetically modified and back-crossed to C57-BL/6 strain		• CA1 specific NR1 $^{-/-}$			Impairment	• 240
		• CA1 specific NR1 $^{-/-}$ (inducible)				• 220
		• mGluR1 $^{-/-}$				• 47
		• mGluR5 $^{-/-}$				• 123
		• mGluR2 $^{-/-}$			No effect	• 261
		• mGluR4 $^{-/-}$				• 74
		• NR2B transgenic			Facilitation	• 235,236
		• NR2B substituting NR2C				• 57

continued on next page

Table 2. Continued

Species	Infusion Time	Infusion Route	Antagonist	Agonist	Effect on Memory	References
			Radial Maze			
Rat	Pre-training	• i.p.	N: CGP37849, CGP39551, MK801		Impairment in reference and working memory acquisition	•21,23,27,42,101, 110 ,119,128, 217
		• i.c.v.	m: MAP4	m: (1S,3S)-ACPD, L-AP4		• 90,91
		• i.Nacc core, med. or post. caudate	N: AP5		No effect	• 256
		• i.Nacc shell				• 222
		• i.p.	A: YM90K			• 120
		• i.p.		A: BA-14 BA-74	Facilitation	• 77
Rat / Mouse	Pre-test/ Pre-reversal	• i.c.v. (minipump), • i.p., i.HC, i.LS	N: AP5, PCP, N-allylnormetazocin MK801, CGS19755, CPP, kynurenate		Impairment in reference or working memory or re-mapping	• 25,31,39,52, 99,130,139, 121,126,179, 217,218,258, 259
		• i.p., i.HC	N: MK801		No effect	• 252,256
		• i.Nacc core, med. or post. caudate	N: AP5			• 222

continued on next page

Table 2. Continued

Species	Infusion Time	Infusion Route	Antagonist	Agonist	Effect on Memory	References
			T- and Y-Mazes			
Rat / Mouse	Pre-training	• i.c.v., s.c., i.p.	N: MK801, CPP		Impairment	• 137,138, 144, 94,219,223
				Quinolinic acid		• 148
			m: MCPG, 4-CPG, MPEP			• 8,9,189,191
				m: tADA		• 192
Rat / Mouse	Post-training / Pre-test	• **i.p., i.PC,** i.c.v., i.HC	N: PCP, MK801, NPC, 12626, AP5		Impairment	• 71,79,249-251
			A: CNQX			• 204
			A: NBQX		No effect	• 174
			m: MCPG, MPEP			• 9,191
				m: tADA	Facilitation	• 191
			Other Spatial Tasks			
Rat	Pre-training	• i.p., i.HC, i.AM, i.PC	N: Ketamine, PCP, CPP, AP5, CGS 19755, MK801		Impairment (holeboard, cheese board, open field, 3-panel runway, 3-choice operant task, operant DMTP)	• 102,103,108, 111,164,169-171
			A: DNQX, GYKI52466			• 229
			m: AIDA, 4CPG, MCPG			• 43,44,167, 168,175
			m: MPEP		No effect	• 175

continued on next page

Table 2. Continued

Species	Infusion Time	Infusion Route	Antagonist	Agonist	Effect on Memory	References
Mouse	1 min post-training	• i.Nacc	*N:* AP5, MK801		Impairment (spatial object recognition)	• 1,206,243
	120 min post-training	• i.Nacc	*N:* AP5, MK801		No effect	

Abbreviations: N: NMDA receptor ligand; A: AMPA receptor ligand; m: mGlu receptor ligand; i.AM = intra amygdala, i.BLA = intra basolateral amygdala, i.c.v. = intracerebroventricularly, i.HC = intra hippocampus, i.IC = intra insular cortex, i.LS = intra lateral septum, i.Nacc = intra nucleus accumbens, i.p. = intraperitoneally, i.PC = intra prefrontal cortex, i.v. = intravenously, med. = medial, p.o. = per orally, post. = posterior, s.c. = subcutaneous.

nonspatial variants available for testing sensorimotor effects of the drugs. They are summarised in (Table 2) inclusive of relevant references.

The Open-Field Water Maze

Both apparatus and procedures for conducting experiments with the water maze have been extensively reviewed.[49,230,238] We concentrate on behavioural data and the most influential work is summarized in (Table 2) under the heading 'water maze'. The overall outcome of the survey strongly supports the notion that pretraining inactivation of NMDA receptors impairs learning. In general, this is consistent for all glutamate subtypes, but there are some differences with respect to memory stages. NMDA receptor blockade by competitive (AP5) or noncompetitive (MK-801) antagonism blocks learning with an impairment during the training period. It is thus not surprising that animals do not remember after washout of the drug. Similar results have been reported for the AMPA receptor blocker NBQX. However, these data should be used with caution since blockade of AMPA receptors prevents synaptic transmission of excitatory synapses and this may be equivalent to locally anaesthetising the brain region under investigation. Animals treated with mGlu receptor agonists or antagonists show little, if any, deficit during acquisition, but are impaired during retention testing. This suggests that mGlu receptors play a more prominent role in memory consolidation.

These results have recently been challenged by work indicating that NMDA and AMPA antagonists induce considerable sensorimotor side effects, which would alone explain the learning deficit. However, these studies have used systemic application routes for the drugs. Such general drug effects may be avoided by local micro-infusion of drugs directly into the brain region of interest. Moreover, some antagonists with a relatively weak receptor affinity are sometimes devoid of any effect on learning.

In contrast, pretraining activation of NMDA or AMPA receptors can sometimes enhance memory formation, but the consequences on sensoric and motoric parameters are not well explored. There is a further complication relating to the narrow safety of drugs due to their excitotoxic effects.

Consistent with these data, post-training administration of ionotropic glutamate receptor antagonists via several routes had no effect on spatial reference memory (but see Teather et al[237] for contrasting results). When administered intrahippocampally to rats that have been pretrained in a working memory version of the water maze, however, AP5 caused a delay-dependent deficit confirming a role of hippocampal NMDA receptors in short-term memory. While aged animals with memory impairments have reduced group I mGlu receptor levels in hippocampus,[162] they benefit from post-training infusion of MK-801 and show enhanced long-term memory.[165]

Interesting results partly confirming pharmacological observations have been obtained using the water maze paradigm and genetically modified animals. Deletion of genes encoding for group I mGlu receptors (mGlu1 and 5) or region-specific knockout of subunit NR1 of NMDA receptors impeded spatial reference memory. Both group I mGlu and NMDA receptors mediate a rise in intracellular calcium levels. Over-expression of NR2B or substitution of the NR2C gene with the NR2B gene facilitated water maze learning, possibly due to increased calcium influx through the NMDA receptor. Similar results would be expected for group I mGlu receptor over-expressing animals, but these are not available to date. Group II mGlu receptor knockout mice show no deficit in spatial learning.

The 8-Arm Radial Maze

Numerous publications describe the experimental set-up and design of the radial maze and its major variants.[24,49,118,184] An advantage to the water maze is the fact that radial maze learning enabled distinction between reference and working memory processes within the same experiment. Reference memory refers to task elements that are constant throughout training and testing; working memory refers to elements specific for each trial (i.e., which arms have been visited already). Overwhelming evidence suggests that acquisition of the radial maze is impeded after systemic block of NMDA or mGlu receptors. Sensorimotor side effects of drugs have been reported[259] but could be prevented by reduction of drug doses. Direct infusion of NMDA receptor antagonists into striatal regions has no effect. Contrary to expectation, systemic block of AMPA receptors had no effect on radial maze performance questioning whether doses and bioavalability were sufficient to block neuronal receptors. This needs further clarification, especially in light of the observation, that drugs blocking AMPA receptor desensitisation (for example BA-14) facilitate acquisition learning of the radial maze. Interestingly, radial maze learning is impaired in the presence of group II and group III mGlu receptor agonists, since these receptors seem to play only a minor role in spatial learning in the water maze. The finding is unexpected in light of the fact that group II and group III mGlu receptor agonists induce a decrease of cAMP and thus should facilitate radial maze learning (see Mons and Gulliou in this book). Thus the agonists should facilitate radial maze learning.

Finally, group III mGlu receptor agonists and antagonists blocked radial maze learning, but coapplication of the antagonist MAP-4 with the agonist L-AP4 neutralised the deficits.[90]

Numerous studies have pretrained the animals to asymptote before testing the effects of drugs on both reference and/or working memory. Recall of the reference memory component of the task was normal in animals treated with competitive or noncompetitive NMDA receptor antagonists (but see Marighetto et al[130] for contrasting results). By contrast, working memory was impaired and reversal learning deficits were reported. This latter result may be due to blockade of remapping processes (see Shapiro and colleagues in Table 2) and deficits were reversible by simultaneous treatment with the polyamine site agonist D-cycloserine.[99] D-cycloserine alone facilitated working memory in trained rats[180] and reversed the performance deficits of hippocampally lesioned rats.[215] AP5 infused into the striatum had no effect. Widely unexplored is the role of mGlu receptors in radial maze performance. In one study, animals were trained for several days in a pure working memory paradigm with all arms baited and brain tissue was harvested. Biochemical analysis measuring activity of group I mGlu receptors revealed increased excitability in trained compared to naïve tissue,[161] but it still remains unclear whether the increase in inositide hydrolysis is due to learning of the overall strategy (reference memory) of specific for working memory performance.

Elevated T- and Y-Maze Protocols

Many variations used in the radial maze also apply to T- and Y-mazes so they may be considered to be simple forms of the radial arm maze.[49,184] Both paradigms have strong spatial components (although there are some nonspatial forms as well) and, similar to water and radial maze, block of NMDA, AMPA or mGlu receptors prior to training impeded acquisition of spatial T- and Y-maze tasks (see Table 2 for references). This has been consistently reported for forced choice or spontaneous alternation procedures. Activation of Glu receptors had similar effects. The NMDA receptor antagonist-induced deficits were reversed by selective agonists against the spermine or polyamine binding sites (for example, see refs. 94,137,138,144), and local administration of subtoxic doses of NMDA into the nucleus basalis in rats facilitated performance in a delayed alternation task in experienced animals.[132] By contrast, mGlu receptor antagonists had no effect when administered post-training or pretesting, and mGlu receptor agonists caused marginal enhancement of memory.

An interesting observation throughout is that deficits are particularly striking when the behavioural task includes a delay between different phases, such as matching or nonmatching-to-sample paradigms. In many cases, there was a direct correlation of the performance deficit with the length of delay pointing again at a role of ionotropic Glu receptors in on-line information processing.

Other Spatial Learning Paradigms

Glutamate receptor antagonists have also been tested in less popular spatial learning tasks including holeboard, cheeseboard, 3-panel runway as a working or reference memory task, and operant chambers with strong spatial components. When applied pretraining either systemically or micro-infused into the brain structure of interest, NMDA and AMPA receptor antagonists blocked learning. By contrast, some mGlu receptor antagonists such as AIDA or MCPG (see Table 1) resulted in enhanced within-session performance but impaired long-term memory[43,44] confirming their importance in consolidation processes. MPEP, which is selective for mGlu5 receptors had no effect suggesting that it is not involved in long-term spatial memory formation. Deficits are also apparent in working memory tasks when NMDA or mGlu receptor antagonists are administered pretraining. Impairments due to NMDA receptor blockade were reversed by DCS and spermidine.[168] That NMDA receptor activation is essential during and shortly after acquisition training was also revealed in a spatial object recognition task. In mice, AP5 or MK-801 infusion into the nucleus accumbens impaired memory formation when given 1, but not 120 mins post-training.

Finally, NMDA receptor function was tested in a more natural setting in pigeons homing for their loft. When MK-801 was administered i.p. shortly before the release of the animal, a high percentage of birds never made it to the home loft.[197]

Overall, these results are consistent in suggesting a role of NMDA receptors in spatial learning and on-line processing of trial-specific spatial information. Compelling evidence is available to suggest that the hippocampal NMDA receptor population is of prime importance in this process. Hippocampal mGlu receptors, by contrast, play little if any role in this on-line processing but contribute to the consolidation process required for long-term reference memory formation.

Conditioning of Fear Responses

Several different conditioning protocols using shock-reinforcement to induce fear responses have been used in conjunction with Glu receptor agonists and antagonists. Both lesion and micro-infusion studies support the contention of the amygdala being a central part of the fear-eliciting neural pathway and it is particularly striking how administration of Glu receptor blockers into the basolateral nucleus caused consistent memory impairments (Table 3 and citations therein).

Conditioning to Context and Cue

Operant chamber-like boxes are frequently used for these tests and animals receive mild footshocks shortly after being placed in this novel environment (context). The shock can be in conjunction with a tone or light (cue) or in the absence of any association. Animals react with a freezing behaviour in which they suspend activity and breathing (apart from respiration) and maintain a crouching posture[22] as an index of fear. Few shocks are sufficient to induce lasting memory for days or weeks. Overall, pretraining administration of both NMDA and mGlu receptor antagonists and agonists impaired memory formation of context fear conditioning and in many cases also reduced freezing to the cue, especially when drugs are infused directly into the amygdala. Here, recent work provided strong evidence for the involvement of the NR2 and mGlu5 receptor subtypes in fear conditioning. The observation of Bordi and colleagues[26] that MCPG has no effect on context conditioning possibly needs revision and contrasts with studies using different mGlu receptor knockout variants, all presenting with deficits in contextual fear memory. A similar deficiency was reported for CA1-specific NR1 mutants in trace fear conditioning.[93]

Post-training administration of NMDA receptor antagonists also impaired fear conditioning. In contrast to spatial tasks, this would suggest a role of NMDA receptor-mediated processes in consolidation. A similar function can be postulated for mGlu receptors since recent work has proven that fear conditioning can increase the expression level of mGlu5 receptors in hippocampal subregions in a time-dependent manner during the consolidation period.[40,193] Collectively, prolonged and possibly increased activity of both NMDA and mGlu receptors may contribute to the post-training processing of fear conditioning.

When activated prior to retention testing, group I mGlu receptor agonists caused memory impairments, whereas group II and III agonists enhanced recall. Such results contrast with the

Table 3. Effects of glutamate receptor agonists and antagonists on fear conditioning

Species	Infusion Time	Infusion Route	Antagonist	Agonist	Effect on Memory	References
			Context and Cue Conditioning			
Rat / Mouse	Pre-training	• i.c.v., i.BLA, i.HC, i.p.	N: AP5, MK801, Agmatine		Impairment (context 1d and 28d later)	• 16,62,63,104, 105,115,129,231, 232, 262,266
			m: AIDA, MCPG			• 43,44,73,163
				m: APDC		• 195
			N: Ifenprodil		Impairment (context and cue)	• 199
			m: MPEP			• 200
		• i.c.v., i.BLA, i.HC		N: NMDA		• 266
		• i.c.v.	m: MCPG		No effect	• 26
Rat	Post-training, pre-extinction, pre-retention	• i.BLA, i.p., i.HC	N: AP5, Agmatine		Impairment	• 115,116,231
				m: 3HPG		• 233
			N: AP5		No effect	• 129
				m: L-CCG-I, L-AP4	Enhancement	• 233

Continued on next page

Table 3. Continued

Species	Infusion Time	Infusion Route	Antagonist	Agonist	Effect on Memory	References
Mouse		• mGluR1 K.O.			Impairment (context)	• 4
		• mGluR5 K.O.				• 123
		• mGluR7 K.O.				• 133
Fear Potentiated Startle						
Rat	Pre-training	• i.BLA, i.p.	N: AP5		Impairment	• 38,146,248
			m: MPEP			• 29,64,214
				m: ACPD	Enhancement	• 109
				m: LY3547-40	No effect	• 84,239
Rat	Post-training, pre-test, pre-extinction	• i.BLA, i.AM, i.p.	N: AP5		Impairment	• 61
			A: CNQX			• 107
				m: LY3547-40		• 84,239
			m: MPEP, LY341495		No effect	• 64,239

Abbreviations: N: NMDA receptor ligand; A: AMPA receptor ligand; m: mGlu receptor ligand, i.AM = intra amygdala, i.BLA = intra basolateral amygdala, i.c.v.= intracerebroventricularly, i.HC = intra hippocampal, i.p. = inraperitoneal spatial learning paradigms, for which we have no hard evidence for a role of mGlu receptors in recall. Extinction of fear conditioned responding was NMDA receptor insensitive. Compounds directed against AMPA receptors have not been tested.

Fear Potentiated Startle

In most animals, a loud unsignalled noise can lead to a startle response of different amplitude. When animals are pretrained in a CS-US procedure, say a light indicating the arrival of a footshock, and in phase 2 this light is presented in conjunction with the startling noise, the startle response is significantly potentiated due to light-induced fear. Pretraining administration of NMDA or mGlu5 receptor antagonists blocked fear-potentiated startle and impaired latent inhibition of startle responding.[213] Consistent with this observation, the broad mGlu receptor agonist ACPD enhanced fear-potentiated startle amplitudes while the group II selective agonist LY354740 had no effect, specifically implicating group I mGlu receptors in the formation of fear memories.

This role seems to be reversed for post-training applied mGlu receptor compounds and suggest a differential involvement of the mGlu subtypes in encoding and consolidation of fear conditioning. Similar to classical fear conditioning, microinfusion of NMDA as well as AMPA receptor antagonists post-training or pretest prevented potentiation of the startle amplitude (see Table 3).

Other Fear Conditioning Tasks

Little work has used other fear conditioning procedures. Systemic application of the non-competitive NMDA receptor antagonist MK-801 prior to training blocked conditioned emotional responding,[88] and intra-amygdalar administration of the competitive antagonist AP5 blocked second order fear conditioning.[75] The function of AMPA or mGlu receptors in these paradigms remain elusive.

Avoidance Learning

Avoidance training is widely used and has several advantages compared with spatial learning. Animals have to learn a specific stimulus-response association and this enables better control over the actual learning process. Moreover, the reinforcer used in most studies is a mild footshock. As a consequence, only few trials are necessary to induce long-lasting memory and drug-treatment can therefore selectively target the different stages of memory processing. In inhibitory or passive avoidance paradigms, animals are required to actively inhibit their natural tendency to peck (chick), to step down from a small platform and explore the cage (step-down) or escape from a brightly lid into a dark compartment (step-through) and remain passive in order to avoid aversive reinforcement. By contrast, active avoidance requires animals to escape by jumping over a hurdle into another part of the box (shuttle box) or escape into one arm of the Y-maze. Although seemingly similar in character, pharmacological interference has revealed differences in neuronal mechanisms underlying these paradigms (Table 4).

One-Trial Inhibitory Avoidance in 1 Day-Old Chicks

Presentation of a bead coated with the bitter tasting methyl-anthranylate causes a disgust response in 1 day-old chicks, when they follow their natural tendency to peck. Memory can be recorded minutes, hours or even days later and is expressed by the birds avoiding the presented bead. Systemic or local administration of NMDA or mGlu receptor antagonists pretraining blocked such memory formation. Effects of post-training infusions varied with the time of injection. NMDA receptor activation via D-cycloserine facilitated memory of the avoidance response when injected 1-6 hours, but not immediately after training. Although NMDA receptor antagonism was not detrimental during these time windows, these data suggest that prolonged and exaggerated NMDA receptor activation may be beneficial for memory formation and are thus reminiscent of enhanced spatial learning in mice over-expressing NR2B subtypes. Blocking AMPA receptors by site-directed micro-infusion of quinoxalines into the intermediate hyperstriatum ventrale impaired memory, but were not effective when infused immediately after the training session. Whether this reflects a specific involvement of AMPA receptors remains to be determined since block of AMPA receptors would act like a local anaesthetic completely preventing fast synaptic transmission to occur. Under such circumstances, the data may provide additional support for the fact that the intermediate hyperstriatum ventrale plays a crucial and enduring role in avoidance learning in chicks (for further details and references, see Table 4).

Step-Down Inhibitory Avoidance

Both facilitation of learning and memory formation as well as impairments have been reported for animals trained in step-down passive avoidance paradigms and pretraining exposed to NMDA receptor blockers. This discrepancy remains unresolved but may be due to differences in species used for testing (mouse versus rat) and/or differences in drug doses applied and infusion routes used.

This may also apply to post-training infusions of NMDA receptor antagonists for which variable effects have been obtained. Mondadori and colleagues[150-153] have consistently reported memory enhancements while others found impairments. Preliminary work using AMPA and mGlu receptor antagonists reported impairments when applied post-training or shortly pretest. A prolonged involvement of AMPA receptor function is further supported by the fact that AMPA receptor levels and density in the hippocampus are increased during the consolidation phase.[19,35-37] Metabotropic Glu receptors also play an important role in retrieval of the avoidance memory. The broad spectrum mGlu receptor antagonists MCPG and the selective group I agonist 3-HPG caused impairments in recall while group II and III agonists facilitated memory. Although the exact mechanism has not been revealed, these data strongly indicate a bimodal function of mGlu receptors with group I down-regulating and group II/III enhancing memory formation. More extensive studies are warranted to address this issue.

Step Through Passive Avoidance

Compared with step-down paradigms, step-through inhibitory avoidance has been more popular and results have been more consistent across species and compounds (see Table 4). NMDA as well as antagonists for NMDA, AMPA or mGlu receptors impaired memory formation when administered systemically or intra-cerebrally prior to training and these results have been confirmed in a modified version of step-through passive avoidance using multiple trials. It is consistent with this observation that modulators of NMDA receptors binding to the polyamine or spermine sites facilitate memory formation.

Post-training or pretest administration of both competitive and noncompetitive NMDA receptor antagonists was widely noneffective, but in some cases led to impairments when administered immediately post-training. Similar memory impairments were reported for the mGlu receptor antagonist MCPG. Interesting is also the discrepancy for mGlu receptor agonists such as ACPD. Packard and colleagues[237] reported memory impairments after local intra-striatal infusions while Wiesniewski's group[92,264,265] infused the drug freehand into the ventricle, and thus affected a much greater portion of mGlu receptors in the brain and obtained an enhancement.

Active Avoidance

Glu receptor antagonists have been tested in active avoidance paradigms, mainly using the standard shuttle box design. NMDA receptor antagonists impaired avoidance memory when administered pre or post-training and agonists proved to enhance memory formation when given post-training or even prior to testing. These results are similar for systemic and local treatments and consistent across species.

Table 4. Effects of glutamate receptor agonists and antagonists on inhibitory avoidance learning

Species	Infusion Time	Infusion Route	Antagonist	Agonist	Effect on Memory	References
Inhibitory Avoidance in Birds						
Chick	Pre-training	intracerebral, i.p., i.IMHV (left)	N: AP5, MK801, 7Cl-kynurenate		Impairment	• 30,72,187,226
			m: MCPG			• 89,186
			m: MCPG +	m: ACPD	Reversal of impairment	• 89,186
Chick	Post-training: 0-5 min	IMHV (left)	N: AP5, MK801, 7Cl-kynurenate		No effect	• 30,72,187,226
			A: CNQX, DNQX, NBQX			• 30
	10-30 min			N: DCS		• 228
	1-6 hr		N: AP5, MK801 7Cl-kynurenate			• 30,72,226
				N: DCS	Facilitation	• 228
	1-6 hr		A: CNQX		Impairment	• 187
	10-30 min					• 227,246
	1-6 hr					
Step-Down Inhibitory Avoidance						
Rat / Mouse	Pre-training	• i.p., p.o.	N: CGP 37849, AP7, MK801	Facilitation		• 150-153
		• i.BLA, i.p.	N: AP5, MK801		Impairment	• 18,97,137,138,201-203
		• i.p.	N: MK801 +	N: PRE-084	Attenuation of impairment	• 137,138
Rat / Mouse	Post-training	• i.BLA, i.c.v., i.HC, i.EC	N: AP5, [Ser[1]]-histogranine, arcaine		Impairment	• 65,66,97,136,182, 202,203,207,208
			A: CNQX			• 97
			m: MCPG			• 20

continued on next page

Table 4. Continued

Species	Infusion Time	Infusion Route	Antagonist	Agonist	Effect on Memory	References
		• i.p., p.o., i.AM, i.HC	N: CGP 37849, AP7, MK801		Facilitation	• 150,152,153
				m: ACPD		• 20
				N: Spermidine		• 207,208
			N: Arcaine +	N: Spermidine	Reversal of facilitation	• 207,208
			m: MCPG	m. ACPD		• 20
Rat / Mouse	Pre-test	• i.BLA, i.c.v., i.HC, i.EC, i.PC, i.Nacc	A: CNQX, DNQX		Impairment	• 14,183,234
			m: MCPG			• 14,124
				m: 3HPG		• 233
			A: CNQX	A: AMPA	Reversal of impairment	• 183
			N: AP5, [Ser¹]-histogranine		No effect	• 97,136,202,203
				A: AMPA		• 183
				m: L-CCG-I, L-AP4	Facilitation	• 233
Step through Inhibitory Avoidance						
Rat / Mouse	Pre-training	• i.c.v., i.p., p.o., i.DG,i.ST,i,HCm, i.BLA	N: AP5, PCP, CGP 37849, MK801, CPP, NPC 12626, ketamine, dextromethorphan, 7Cl-kynurenate, dextrorphan; CGP 40116, AP7, N-allylnormetazocine, CGS 19755, riluzole		Impairment	• 50-52,56,80,82,114, 135,141,142,150,152, 158,196,210,221,245
			A: CNQX			• 51
			m: MCPG			• 20
				N: NMDA		• 263
			N: AP5 +	N: NMDA	Reversal of impairment	• 122
			N: MK801 or memantine +	N: NMDA		• 263
			N: CGP 39551 +	N: NMDA	No reversal	• 263
				N: NMDA	No effect	• 122
			A: NBQX			• 147

continued on next page

Table 4. Continued

Species	Infusion Time	Infusion Route	Antagonist	Agonist	Effect on Memory	References
			N: 7-CL-kynurenate		Facilitation	• 247
				N: DCS, HA 966, ACPC		• 112,174,247
Rat / Mouse	Post-training	• i.c.v., i.p., p.o., i.AM, i.ST	N: MK801, AP5, CPP		Impairment	• 41,46,122
			N: AP5, PCP, CGP 40116, CGP 37849, AP7, MK801, CPP		No effect	• 46,50,52,82,150,153,196
				m: ACPD, DHPG	Facilitation	• 92,264,265
Rat	Pre-test	• i.p. , i.AM	N: MK801, AP5, CPP		No effect	• 41,122,196
Rat	Pre-training	• i.AM, i.p.	N: AP5, CPP, MK801; memantine, amantadine CGP37849		Impairment(multi-trial procedure)	• 106,147
			A: CNQX			• 143
Active Avoidance						
Rat / Mouse	Pre-training	• i.HC, i.p.	N: AP5, MK801		Impairment (shuttle box)	• 55
			m: L-AP3		Facilitation (step through)	• 6
Rat / Mouse	Post-training	• i.p., i.HC, i.c.v.	N: γ-LGLA, AP5, CPP		Impairment (Y-maze)	• 45,134,241
			N: MK801, AP5		No effect (shuttle box)	• 55
				N: Milacemid, DCS	Facilitation(shuttle box)	• 70,181
Rat / Mouse	Pre-test	• i.p.		N: Milacemid, DCS	Facilitation(shuttle box)	• 70,181

Abbreviations: N: NMDA receptor ligand; A: AMPA receptor ligand; m: mGlu receptor ligand; i.AM = intra amygdala, i.BLA = intra basolateral amygdala, i.c.v.= intracerebroventricularly, i.DG = intra dentate gyrus, i.EC = intra entorhinal cortex, i.HC = intra hippocampus, i.IMHV = intra intermediate hyperstriatum ventrale, i.Nacc = intra nucleus accumbens, i.p. = intraperitoneally, i.PC = intra prefrontal cortex, i.ST = intrastriatal, p.o. = per orally.

Olfactory and Gustatory Learning Paradigms

Although data using smell and taste-related memory paradigms confirm the importance of the glutamatergic system, interpretation of results need caution due to the limited number of studies, which may only give a preliminary and fragmented picture (Table 5). For olfactory learning such as odour discrimination, odour memory and taste-potentiated odour aversion, pretraining administration of NMDA receptor antagonists via several routes caused memory impairments. Post-training administration of the same drugs into the amygdala was not effective. An interesting observation originally made by Nakanishi's group showed that micro-infusion of mGlu receptor agonists into the olfactory bulb can induce a pregnancy block in mice. This action is due to the activation of group II mGlu receptors.

Systemic and localised infusion of NMDA, AMPA or mGlu receptor antagonists pretraining led to impairments in conditioned taste aversion. In this paradigm, thirsty animals drink a novel substance (saccharose solution) followed by an injection of lithium chloride, which induces malaise. As a result, animals refrain from drinking saccharose solution again despite its sweet taste. Impairment is reflected in animals drinking considerably more than controls. Pretraining administration of the NMDA activator D-cycloserine enhanced memory in rats. Post-training infusion of NMDA or AMPA antagonists into parts of the amygdala or insular cortex also prevented memory formation while the broad mGlu receptor antagonist MCPG had no effect. However, mGlu7 receptor mutant mice were also impaired.

Other Tasks

While a substantial number of studies has investigated the role of NMDA receptors in other behavioural paradigms including open field paradigms, lever-press delayed-matching-to-sample or nonmatching-to-sample as well as classical or trace eye-blink conditioning, the knowledge of AMPA or mGlu receptor function in these tasks is still very fragmented. Overall, drug effects have been very variable, even for the same behavioural test and it would be premature to come to any firm conclusion at present.

Conclusions and Remaining Questions

Collectively, how can these data be summarised? Several themes seem to emerge from the data set presented here.

1) Interpretation of pharmacological experiments using AMPA receptor antagonists are somewhat flawed in their interpretation since this treatment results – at least in the majority of cases – in the blockade of fast synaptic transmission and thus silencing of the whole brain or one brain region in particular. Such a treatment is reminiscent of local anaesthetics such as lidocain or tetrodotoxin and resembles the pharmacological enhancement of inhibition, for instance via activation of γ-amino-butyric acid (GABA) receptor agonists. What may be taken from the AMPA data is thus a more sophisticated way of temporary inactivation or lesion sparing fibres of passage and thus provides data about the time-limited involvement of a particular brain structure in the formation of a particular form of memory. Overall, data suggest that temporary inactivation using AMPA receptor antagonists are most effective when administered prior to training. The fact that learning-induced increases in AMPA-receptor levels have been described for a number of training regimes could reflect the widely observed fact of increased excitability of neurones during the memory consolidation process.

2) The role of NMDA receptors in memory appears to vary depending on the type of memory and the brain regions involved in the specific learning paradigm. From psychopharmacological data, there is strong evidence that NMDA receptor activation is important during learning (encoding) of spatial tasks. Even prolonged NMDA receptor blockade for 7 days post-training was not effective. By contrast, post-training blockade of NMDA receptors caused impairments in fear conditioning, fear-potentiated startle paradigms, inhibitory avoidance or conditioned taste aversion (see Tables 3-5). Memory for negative incentives involves different brain structures compared to spatial learning, i.e., the amygdala as opposed to hippocampus,

Table 5. *Effects of glutamate receptor agonists and antagonists on olfactory and gustatory learning*

Species	Infusion Time	Infusion Route	Antagonist	Agonist	Effect on Memory	References
			Olfactory Tasks			
Rat / Mouse	Pre-training	• i.c.v., i.Olf.bulb; s.c., i.BLA	N: AP5, MK801		Impairment (odor discrimination; odor memory; taste potentiated odor aversion)	• 28,67,81,198,224
Rat	Post-training Pre-test	• i.BLA	N: AP5		No effect(taste potentiated odor aversion)	• 67
Mouse	During pairing	• i.Olf.bulb		m: APDC, DCG-IV	Memory induction (pregnancy block)	• 98,209
			Conditioned Taste Aversion			
Rat	Pre-training	• i.p., i.c.v., i.IC, i.LA, i.BLA	N: Ketamine, MK801, AP5, CPP		Impairment	• 2,59,60,78,145,198, 205,253,257
			A: CNQX			• 260
			m: MCPG			• 260
				N: DCS	Enhancement	• 112
			m: MCPG		No effect	• 60
Rat	Post-training, pre-test	• i.IC, i.LA, i.BLA	N: AP5		Impairment	• 60,78,205
			A: CNQX			• 260
			m: MCPG		No effect	• 260
Mouse		• mGluR7 K.O.			Impairment	• 133

Abbreviations: N: NMDA receptor ligand; A: AMPA receptor ligand; m: mGlu receptor ligand; i.BLA = intra basolateral amygdala, i.c.v. = intracerebroventricularly, i.IC = intra insular cortex, i.Olf.bulb = intra olfactory bulb, i.p. = intraperitoneally, s.c. = subcutaneous.

and competitive and noncompetitive NMDA receptor antagonists are effective when infused directly into the basolateral complex of the amygdala. This is strong evidence for different forms of memory not only involving different parts of the brain, but also engaging different cellular mechanisms. What remains elusive is whether this difference in NMDA receptor function may be due to a difference in subtype expression within the different brain structures. Recent evidence confirms that the NR2 subunit in the basolateral amygdala is very important for fear conditioning and similar work is necessary for other behavioural paradigms. Pharmacological experiments, however, require highly specific drugs, which are not yet available. On the other hand, mutant mice are not yet region- and time-specific, allowing only gross functional assessment of a particular gene.

It appears that differences between competitive and noncompetitive NMDA receptor antagonism are relatively small. Pharmacologically interesting is the existence of various modulatory sites such as polyamine and glycine sites. Blockers acting on these sites caused much less sensory-motor disruptions and agonists have revealed their potency to facilitate memory formation. Transfer to demented humans, however, has met with little therapeutic success.[86,216]

3) The function of mGlu receptors, especially with respect to the different subtypes, seems to be more consistent. Although pretraining infusion of mGlu receptor antagonists is the most effective way of blocking memory formation, acquisition training is not affected by this treatment. This suggests that mGlu receptor activation may take place during learning but plays a more important part in the consolidation processes of memory. Region- and subtype-specific functions are less clear to date owing to the lack of selective compounds, but should attract more attention in the near future. In contrast to agonists acting at the modulatory sites of NMDA receptors, stimulation of mGlu receptors consistently caused memory impairments. This supports the contention that there is a fine balance of endogenous mGlu receptor activity during states of resting and memory formation and its maintenance is crucial for normal learning and memory. Finally, mGlu receptors have recently been implicated in memory retrieval in an inhibitory avoidance task (Table 3). Since very little is known about mGlu receptor function in recall of other paradigms (and forms of memory), this may open a new and intriguing avenue for research with the potential for memory enhancement and clinical implications.

It may appear from the number of publications that we know a great many details about Glu receptor functions in memory formation. What then remains elusive? It will be important in the future to establish the cross talk between the different Glu receptors and develop a more detailed picture about the second messenger systems mediating their actions. The latter can be found in later chapters of this book, but behavioural studies linking receptors with second messengers and enzyme cascades are still sparse. A second and timely topic concerns the role of metaplasticity. This refers to the fact that changes in synaptic plasticity preceding the learning event may have implications for the availability of particular neural mechanisms to learn a second task. Initial reports suggest that such behavioral metaplasticity may render spatial learning NMDA receptor independent, and memory for step-down avoidance is blocked by exposure to a novel environment. Such studies are of importance since they provide information as to the interaction of the different memory systems and memory mechanisms.

References

1. Adriani W, Felici A, Sargolini F et al. N-methyl-D-aspartate and dopamine receptor involvement in the modulation of locomotor activity and memory processes. Exp Brain Res 1998; 123:52-59.
2. Aguado L, San Antonio A, Perez L et al. Effects of the NMDA receptor antagonist ketamine on flavor memory: Conditioned aversion, latent inhibition, and habituation of neophobia. Behav Neural Biol 1994; 61:271-281.
3. Ahlander M, Misane I, Schott PA et al. A behavioral analysis of the spatial learning deficit induced by the NMDA receptor antagonist MK801 (dizocilpine) in the rat. Neuropsychophamacol 1999; 21:414-426.
4. Aiba A, Chen C, Herrup K et al. Reduced hippocampal long-term potentiation and context-specific deficit in associative learning in mGluR1 mutant mice. Cell 1994; 79:365-375.
5. Alessandri B, Battig K, Welzl H. Effects of ketamine on tunnel maze and water maze performance in rat. Behav Neural Biol 1989; 52:196-212.

6. Alvarez EO, Banzan AM. Ventral hippocampal glutamate receptors in the rat: Possible involvement in learning mechanisms of an active avoidance response. J Neural Transm 1999; 106:987-1001.
7. Anwyl R. Metabotropic glutamate receptors: Electrophysiological properties and role in plasticity. Brain Res Rev 1999; 29:83-120.
8. Balschun D, Wetzel W. Inhibition of group I metabotropic glutamate receptors blocks spatial learning in rats. Neurosci Lett 1998; 249:41-44.
9. Balschun D, Wetzel W. Inhibition of mGluR5 blocks hippocampal LTP in vivo and spatial learning in rats. Pharmacol Biochem Behav 2002; 73:375-380.
10. Balschun D, Manahan-Vaughan D, Wagner T et al. A specific role for group I mGluRs in hippocampal LTP and hippocampus-dependent spatial learning. Learn Memory 1999; 6:138-152.
11. Bannerman DM, Butcher SP, Good MA et al. Intraventricular infusion of the NMDA receptor-associated glycine site antagonist 7-chlorokynurenate impairs water maze performance but fails to block hippocampal long-term potentiation in vivo. Neurobiol Learn Memory 1997; 68:252-270.
12. Bannerman DM, Good MA, Butcher SP et al. Distinct components of spatial learning revealed by prior training and NMDA receptor blockade. Nature 1995; 378:182-186.
13. Barnes CA, Danysz W, Parsons CG. Effects of the uncompetitive NMDA receptor antagonist memantine on hippocampal long-term potentiation, short-term exploratory modulation and spatial memory in awake, freely moving rats. Europ J Neurosci 1996; 8:565-571.
14. Barros DM, Izquierdo LA, Souza TME et al. Molecular signalling pathways in the cerebral cortex are required for retrieval of one-trial avoidance learning in rats. Behav Brain Res 2000; 114:183-192.
15. Baude A, Nusser Z, Roberts JD et al. The metabotropic glutamate receptor (mGluR1α) is concentrated at perisynaptic membrane of neuronal subpopulations as detected by immunogold reaction. Neuron 1993; 11:771-787.
16. Bauer EP, Schafe GE, LeDoux JE. NMDA receptors and L-type voltage-gated calcium channels contribute to long-term potentiation and different components of fear memory formation in the lateral amygdala. J Neurosci 2002; 22:5239-5249.
17. Baxter MG, Lanthorn TH, Frick KM et al. D-cycloserine, a novel cognitive enhancer, improves spatial memory in aged rats. Neurobiol Aging 1994; 15:207-213.
18. Benvenga MJ, Spaulding TC. Amnesic effect of the novel anticonvulsant MK-801. Pharmacol Biochem Behav 1988; 30:205-207.
19. Bernabeu R, Cammarota M, Izquierdo I et al. Involvement of hippocampal AMPA glutamate receptor changes and the cAMP/protein kinase A/CREB-P signalling pathway in memory consolidation of an avoidance task in rats. Braz J Med Biol Res 1998; 30:961-965.
20. Bianchin M, Da Silva RC, Schmitz PK et al. Memory of inhibitory avoidance in the rat is regulated by glutamate metabotropic receptors in the hippocampus. Behav Pharmacol 1994; 5:356-359.
21. Bischoff C, Tiedtke PI. Competitive and noncompetitive NMDA receptor antagonists in spatial learning tasks. Europ J Pharmacol 1992; 213:269-273.
22. Blanchard RJ, Blanchard DC. Crouching as an index of fear. J Comp Physiol Psychol 1969; 67:370-375.
23. Boast C, Bartolomeo AC, Morris H et al. 5HT antagonists attenuate MK801-impaired radial arm maze performance in rats. Neurobiol Learn Memory 1999; 71:259-271.
24. Boast CA, Walsh TJ, Bartolomeo AC. The delayed nonmatch-to-sample radial arm maze task: Application to models of Alzheimer's disease. In: Buccafusco JJ, ed. Methods of behavior analysis in neuroscience. Boca Raton, Florida: CRC Press, 2000:167-188.
25. Bolhuis JJ, Reid IC. Effects of intraventricular infusion of the N-methyl-D-aspartate (NMDA) receptor antagonist AP5 on spatial memory of rats in a radial arm maze. Behav Brain Res 1992; 47:151-157.
26. Bordi F, Marcon C, Chiamulera C et al. Effects of metabotropic glutamate receptor antagonist MCPG on spatial and context-specific learning. Neuropharmacol 1996; 35:1557-1565.
27. Bostock E, McKay KE, Fong R et al. Interaction of the opiate and NMDA systems in a spatial learning task. Psychobiol 1992; 20:33-38.
28. Brennan PA. The effects of local inhibition of N-methyl-D-aspartate and AMPA/kainate receptors in the accessory olfactory-bulb on the formation of an olfactory memory in mice. Neurosci 1994; 60:701-708.
29. Brodkin J, Busse C, Sukoff SJ et al. Anxiolytic-like activity of the mGluR5 antagonist MPEP: A comparison with diazepame and buspirone. Pharmacol Biochem Behav 2002; 73:359-366.
30. Burchuladze R, Rose SPR. Memory formation in day-old chicks requires NMDA but not nonNMDA glutamate receptors. Europ J Neurosci 1992; 4:533-538.
31. Butelman ER. A novel NMDA antagonist, MK801, impairs performance in a hippocampal-dependent spatial learning task. Pharmacol Biochem Behav 1989; 34:13-16.

32. Cain DP, Ighanian K, Boon F. Individual and combined manipulation of muscarinic, NMDA, and benzodiazepine receptor activity in the water maze task: Implications for a rat model of Alzheimer dementia. Behav Brain Res 2000; 111:125-137.
33. Cain DP, Saucier D, Boon F. Testing hypotheses of spatial learning: The role of NMDA receptors and NMDA-mediated long-term potentiation. Behav Brain Res 1997; 84:179-193.
34. Cain DP, Saucier D, Hall J et al. Detailed behavioral analysis of water maze acquisition under APV or CNQX: Contribution of sensorimotor disturbances in drug-induced acquisition deficits. Behav Neurosci 1996; 110:86-102.
35. Cammarota M, Bernabeu R, Izquierdo I et al. Reversible changes in hippocampal [^{3}H]-AMPA binding following inhibitory avoidance training in the rat. Neurobiol Learn Memory 1996; 66:85-88.
36. Cammarota M, Bernabeu R, de Stein ML et al. Learning-specific, time-dependent increases in hippocampal Ca2+/calmodulin-dependent protein kinase activity and AMPA GluR1 subunit reactivity. Eur J Neurosci 1998; 10:2669-2676.
37. Cammarota M, Izquierdo I, Wolfman C et al. Inhibitory avoidance training induces rapid and selective changes in [^{3}H]AMPA receptor binding in the rat hippocampal formation. Neurobiol Learn Memory 1995; 64:257-263.
38. Campeau S, Miserendino MJD, Davis M. Intra-amygdala infusion of N-methyl-D-aspartate receptor antagonist AP5 blocks acquisition but not expression of fear-potentiated startle to an auditory conditioned stimulus. Behav Neurosci 1992; 106:569-574.
39. Caramanos Z, Shapiro ML. Spatial memory and N-methyl-D-aspartate receptor antagonists APV and MK-801: Memory impairments depend on familiarity with the environment, drug dose, and training duration. Behav Neurosci 1994; 108:30-43.
40. Casabona G, Platt B, Nicoletti F et al. Learning enhances the expression of mGluR5 receptor protein in rat hippocampal CA1 region. Soc Neurosci Abstr 1997; 23:938.
41. Castellano C, Cestari V, Ciamei A et al. MK-801-induced disruptions of one-trial inhibitory avoidance are potentiated by stress and reversed by naltrexone. Neurobiol Learn Memory 1999; 72:215-229.
42. Chen Z, Zhao Q, Sugimoto Y et al. Effects of histamine on MK-801-induced memory deficits in radial maze performance in rats. Brain Res 1999; 839:186-189.
43. Christoffersen GRJ, Christensen LH, Harrington NR et al. Task-specific enhancement of short-term, but not long-term memory by class I metabotropic glutamate receptor antagonist AIDA in rats. Behav Brain Res 1999; 101:215-226.
44. Christoffersen GRJ, Christensen LH, Hammer P et al. The class I metabotropic glutamate receptor antagonis, AIDA, improves short-term and impairs long-term memory in a spatial task for rats. Neuropharmacol 1999b; 38:817-823.
45. Church AC, Flexner JB, Flexner LB. Complementary memory storage sites in mice: Their development as affected by the competitive NMDA receptor antagonist CPP. Pharmacol Biochem Behav 1995; 52:237-240.
46. Ciamei A, Aversano M, Cestari V et al. Effects of MK-801 and nicotine combinations on memory consolidation in CD1 mice. Psychopharmacol 2001; 154:126-130.
47. Conquet F, Bashir ZI, Davies CH et al. Motor deficit and impairment of synaptic plasticity in mice lacking mGluR1. Nature 1994; 372:237-243.
48. Conn PJ, Pin JP. Pharmacology and functions of metabotropic glutamate receptors. Ann Rev Pharmacol Toxicol 1997; 37:205-237.
49. Crawley JN. What's wrong with my mouse? Behavioral phenotyping of transgenic and knockout mice. New York: Wiley-Liss, 2000.
50. Danysz W. Metaphit fails to antagonize PCP-induced passive avoidance deficit. Pharmacol Biochem Behav 1990; 38:231-233.
51. Danysz W, Wroblewski JT. Amnesic properties of glutamate receptor antagonists. Neurosci Res Comm 1989; 479:9-18.
52. Danysz W, Wroblewski JT, Costa E. Learning impairment in rats by N-methyl-D-aspartate receptor antagonists. Neuropharmacol 1988; 27:653-6.
53. Davis S, Butcher SP, Morris RGM. The NMDA receptor antagonist D-2-amino-5-phosphonopentanoate (D-AP5) impairs spatial learning and LTP in vivo at intracerebral concentrations comparable to those that block LTP in vitro. J Neurosci 1992; 12:21-34.
54. Day M, Morris RGM. Memory consolidation and NMDA receptors: Discrepancy between genetic and pharmacological approaches. Science 2001; 293:755a.
55. Delay ER. Effects of MK-801 on acquisition, retention, and cross-modal transfer of active avoidance behavior in rats. Psychobiol 1996; 24:195-201.
56. DeNoble VJ, Jones KW, Schaeffer CL et al. 3-((±)-2-Carboxy piperazin-4-yl)propyl-1-phophonic acid (CPP) and phencyclidine produce a deficit in passive avoidance retention in rats. Europ J Pharmacol 1990; 175:197-202.

57. Dere E, De Souza-Silva MA, Sedelis M et al. NMDA NR2C-2B subunit substitution in mice improves water-maze learning and object recognition, reduces neophobia and increases acetylcholine in frontal and amygdaloid cortices. 2002. Submitted.
58. Dingledine R, Borges K, Bowie D et al. The glutamate receptor ion channels. Pharmacol Rev 1999; 51:7-61.
59. Escobar ML, Alcocer I, Chao V. The NMDA receptor antagonist CPP impairs conditioned taste aversion and insular cortex long-term potentiation in vivo. Brain Res 1998; 812:246-251.
60. Escobar ML, Alcocer I, Bermudez-Rattoni F. In vivo effects of intracortical administration of NMDA and metabotropic glutamate receptor antagonists on neocortical long-term potentiation and conditioned taste aversion. Behav Brain Res 2002; 129:101-106.
61. Falls WA, Miserendino MJD, Davis M. Extinction of fear-potentiated startle: Blockade by infusion of an NMDA antagonist into the amygdala. J Neurosci 1992; 12:854-863.
62. Fanselow MS, Kim JJ. Acquisition of contextual Pavlovian fear conditioning is blocked by application of an NMDA receptor antagonist DL-2-amino-5phosphonovaleric acid to the basolateral amygdala. Behav Neurosci 1993; 108:210-212.
63. Fanselow MS, Kim JJ, Yipp J et al. Differential effects of the N-methyl-D-aspartate antagonist DL-2-amino-5-phophonovalerate on acquisition of fear of auditory and contextual cues. Behav Neurosci 1994; 108:235-240.
64. Fendt M, Schmid S. Metabotropic glutamate receptors are involved in amygdaloid plasticity. Europ J Neurosci 1999; 15:1535-1541.
65. Ferreira MBC, Da Silva RC, Medina JH et al. Late posttraining memory processing by entorhinal cortex: Involvement of NMDA and GABAergic receptors. Pharmacol Biochem Behav 1992a; 41:767-771.
66. Ferreira MBC, Wolfram C, Walz R et al. NMDA-receptor-dependent, muscimol-sensitive role of the entorhinal cortex in post-training memory processing. Behav Pharmcol 1992b; 3:387-391.
67. Ferry B, Di Scala G. Basolateral amygdala NMDA receptors are selectively involved in the acquisition of taste-potentiated odor aversion in the rat. Behav Neurosci 2000; 114:1005-1010.
68. Filliat P, Blanchet G. Effects of TCP on spatial memory: Comparison with MK-801. Pharmacol Biochem Behav 1995; 51:429-434.
69. Filliat P, Pernot-Marino I, Baubichon D et al. Behavioral effects of NBQX, a competitive antagonist of the AMPA receptor. Pharmacol Biochem Behav 1998; 59:1087-1092.
70. Flood JF, Morley JE, Lanthorn TH. Effects on memory processing by D-cycloserine, an agonist of the NMDA/glycine receptor. Europ J Pharmacol 1992; 221:249-254.
71. Fraser CM, Cooke MJ, Fisher A et al. Interactions between ifenprodil and dizocilpine on mouse behaviour in models of anxiety and working memory. Europ Neuropsychopharmacol 1996; 6:311-316.
72. Freeman FM, Rose SPR. MK-801 blockade of fos and jun expression following passive avoidance training in the chick. Europ J Neurosci 1995; 7:563-569.
73. Frohardt RJ, Guarraci FA, Young SL. Intrahippocampal infusions of a metabotropic glutamate receptor antagonist block the memory of context-specific but not tone-specific conditioned fear. Behav Neurosci 1999; 113:222-227.
74. Gerlai R, Roder JC, Hampson DR. Altered spatial learning and memory in mice lacking the mGluR4 subtype of metabotropic glutamate receptor. Behav Neurosci 1998; 112:525-532.
75. Gewirtz JC, Davis M. Second-order fear conditioning prevented by blocking NMDA receptors in amygdala. Nature 1997; 388:471-474.
76. Gorter JA, de Bruin JPC. Chronic neonatal MK-801 treatment results in an impairment of spatial learning in the adult rat. Brain Res 1992; 580:12-17.
77. Granger R, Stäubli U, Davis M et al. A drug that facilitates glutamatergic transmission reduces exploratory activity and improves performance in a learning-dependent task. Synapse 1993; 15:326-329.
78. Gutierrez H, Gutierrez R, Ramirez-Trejo L et al. Redundnat basal forebrain modulation in taste aversion memory formation. J Neurosci 1999; 19:7661-7669.
79. Handelmann GE, Nevins ME, Mueller LL et al. Milacemide, a glycine prodrug, enhances performance of learning-tasks in normal and amnestic rodents. Pharmacol Biochem Behav 1989; 34:823-828.
80. Harrod SB, Flint RW, Riccio DC. MK-801 induced retrieval, but not acquisition, deficits for passive avoidance conditioning. Pharmacol Biochem Behav 2001; 69:585-593.
81. Hatfield T, Gallagher M. Taste-potentiated odor conditioning: Impairment produced by infusion of an N-methyl-D-aspartate antagonist into basolateral amygdala. Behav Neurosci 1995; 109:663-668.
82. Hauben U, D'Hooge R, Soetens E et al. Effects of oral administration of the competitive N-methyl-D-aspartate antagonist, CGP 40116, on passive avoidance, spatial learning and neuromotor abilities in mice. Brain Res Bull 1999; 48:333-341.
83. Heale V, Harley C. MK-801 and AP5 impair acquisition, but not retention, of the Morris milk maze. Pharmacol Biochem Behav 1990; 36:145-149.

84. Helton DR, Tizzano JP, Monn JA et al. Anxiolytic and side-effect profile of LY354740: A potent, highly selective, orally active agonist for group II metabotropic glutamate receptors. J Pharmacol Exp Ther 1998; 284:651-660.
85. Henley JM, Bruton R, Richmond SA. Metabotropic glutamate receptors. In: Turner AJ, Stephensen FA, eds. Amino acid neurotransmission. London, Portland Press, 1998:113-28.
86. Herting RL. Milacemide and other drugs active at glutamate NMDA receptors as potential treatment for dementia. Ann NY Acad Sci 1991; 640:237-240.
87. Heuss C, Gerber U. G protein-independent signalling by G-protein coupled receptors. Trends Neurosci 2000; 23:469-475.
88. Hoehn-Saric R, McLeod DR, Glowa JR. The effects of NMDA receptor blockade on the acquisition of a conditioned emotional response. Biol Psychiat 1991; 30:170-176.
89. Hoelscher C. Inhibitors of metabotropic glutamate receptors produce amnestic effects in chicks. NeuroReport 1994; 5:1037-1040.
90. Hoelscher C, McGlinchey L, Rowan MJ. L-AP4 (L(+)-2-amino-4-phosphonobutyric acid) induced impairment of spatial learning in the rat is antagonized by MAP4 ((S)-2-amino- 2-methyl-4-phosphonobutanoic acid). Behav Brain Res 1996; 81:69-79.
91. Hoelscher C, McGlinchey L, Anwyl R et al. HFS-induced long-term potentiation and LFS-induced depotentiation in area CA1 of the hippocampus are not good models for learning. Psychopharmacol 1997; 130:174-182.
92. Holy Z, Wiesniewski K. Examination of the influence of 3,5-DHPG on behavioral activity of angiotensin II. Pol J Pharmacol 2001; 53:235-243.
93. Huerta PT, Sun LD, Wilson MA et al. Formation of temporal memory requires NMDA receptors within CA1 pyramidal neurons. Neuron 2000; 25:473-480.
94. Ingram DK, Garofalo P, Spangler EL et al. Reduced density of NMDA receptors and increased sensitivity to dizocilpine-induced learning impairment in aged rats. Brain Res 1992; 580:273-280.
95. Izquierdo LA, Barros DM, Medina JH et al. Novelty enhances retrieval of one-trial avoidance learning in rats 1 or 31 days after training unless the hippocampus is inactivated by different receptor antagonists and enzyme inhibitors. Behav Brain Res 2000; 117:215-220.
96. Izquierdo LA, Barros DM, Ardenghi PG et al. Different hippocampal molecular requirements for short-term and long-term retrieval of one-trial avoidance learning. Behav Brain Res 2000; 111:93-8.
97. Jerusalinsky D, Ferreira MBC, Walz R et al. Amnesia by post-training infusion of glutamate receptor antagonists into the amygdala, hippocampus, and entorhinal cortex. Behav Neural Biol 1992; 58:76-80.
98. Kaba H, Hayashi Y, Higuchi T et al. Induction of an olfactory memory by the activation of metabotropic glutamate receptors. Science 1994; 265:262-264.
99. Kawabe K, Yoshihara T, Ichitani Y et al. Intrahippocampal D-cycloserine improves MK-801-induced memory deficits: Radial arm maze performance in rats. Brain Res 1998; 814:226-230.
100. Keith JR, Galizio M. Acquisition in the Morris swim task is impaired by a benzodiazepine but not an NMDA antagonist: A new procedure for distinguishing acquisition and performance effects. Psychobiol 1997; 25:217-228.
101. Keseberg U, Schmidt WJ. Tetrahydroaminoacridine attenuates dizocilpine-induced behavioral changes. Behav Brain Res 1993; 53:113-118.
102. Kesner RP, Dakis M. Phencyclidine injections into the dorsal hippocampus disrupt long- but not short-term memory within a spatial learning task. Psychopharmacol 1995; 120:203-208.
103. Kesner RP, Dakis M, Bolland BL. Phencyclidine disrupts long- but not short-term memory within a spatial learning task. Psychopharmacol 1993; 111:85-90
104. Kim JJ, DeCola JP, Landeira-Fernandez J et al. N-methyl-D-aspartate receptor antagonist APV blocks acquisition, but not expression of fear conditioning. Behav Neurosci 1991; 105:160-167.
105. Kim JJ, Fanselow MS, DeCola JP et al. Selective impairment of long-term but not short-term conditional fear by the NMDA antagonist APV. Behav Neurosci 1992; 107:1093-1098.
106. Kim M, McGaugh JL. Effects of intra-amygdala injections of NMDA receptor antagonists on acquisition and retention of inhibitory avoidance. Brain Res 1992; 585:35-48.
107. Kim M, Campeu S, Falls WA et al. Infusion of the NonNMDA receptor antagonist CNQX into the amygdala blocks the expression of fear-potentiated startle. Behav Neural Biol 1993; 59:5-8.
108. Kishi A, Ohno M, Watanabe S. Concurrent activation of hippocampal glycine and polyamine binding sites of the N-methyl-D-aspartate receptor synergistically reverses working memory deficits in rats. Neurosci Lett 1998b; 257:131-134.
109. Koch M. Microinjections of the metabotropic glutamate receptor agonist, trans- (±)1-amino-cyclopentane-1,3-dicarboxylate (trans-ACPD) into the amygdala increase the acoustic startle response of rats. Brain Res 1993; 629:176-179.
110. Kretschmer BD, Fink S. Spatial learning deficit after NMDA receptor blockade and state-dependency. Behav Pharmacol 1999; 10:423-428.

111. Lalonde R, Joyal CC. Effects of ketamine and L-glutamic acit diethyl ester on spatial and nonspatial learning in rats. Pharmacol Biochem Behav 1993; 44:539-545
112. Land C, Riccio DC. D-cycloserinea positive modulator of the NMDA receptor, enhances acquisition of a conditioned taste aversion. Psychobiol 1997; 25:210-216.
113. Land C, Riccio DC. D-cycloserine: Effects on long-term retention of a conditioned response and on memory for contextual attributes. Neurobiol Learn Memory 1999; 72:158-168.
114. Lee EHY, Lee CP, Wang HI et al. Hippocampal CRF, NE, and NMDA system interactions in memory processing in the rat. Synapse 1993; 14:144-153.
115. Lee H, Kim JJ. Amygdalar NMDA receptors are critical for new learning in previously fear-conditioned rats. J Neurosci 1998; 18:8444-8454.
116. Lee HJ, Choi JS, Brown TH et al. Amygdalar NMDA receptors are critical for the expression of multiple conditioned fear responses. J Neurosci 2001; 21:4116-4124.
117. Lelong V, Dauphin F, Boulouard M. RS 67333 and D-cycloserine accelerate learning acquisition in the rat. Neuropharmacol 2001; 41:517-522.
118. Levin ED. Use of the radial-arm maze to assess learning and memory in rodents. In: Buccafusco JJ, ed. Methods of behavior analysis in neuroscience. Boca Raton, Florida: CRC Press, 2000:189-200.
119. Levin ED, Bettegowda C, Weaver T et al. Nicotine-dizocilpine interactions and working and reference memory performance of rats in the radial-arm maze. Pharmacol Biochem Behav 1998; 61:335-340.
120. Li H-B, Matsumoto K, Watanabe H. An alpha-amino-3-hydroxy-5-methyl-4-isoxazolepropionic acid (AMPA) antagonist and an N-methyl-D-aspartate (NMDA) channel blocker synergistically impair spatial memory in rats. Biol Pharmac Bull 1998; 211:228-230.
121. Li H-B, Matsumoto K, Yamamoto M et al. NMDA but not AMPA receptor antagonists impair the delay-interposed radial maze performance of rats. Pharmacol Biochem Behav 1997; 58:249-253.
122. Liang KC, Hon W, Davis M. Pre and posttraining infusion of N-methyl-D-aspartate receptor antagonists into the amygdala impair memory in an inhibitory avoidance task. Behav Neurosci 1994; 108:241-253.
123. Lu Y-M, Jia Z, Janus C et al. Mice lacking metabotropic glutamate receptor 5 show impaired learning and reduced CA1 long-term potentiation (LTP) but normal CA3 LTP. J Neurosci 1997; 17:5196-205.
124. Luján R, Nusser Z, Roberts JDB et al. Perisynaptic location of metabotropic glutamate receptors mGluR1 and mGluR5 on dendrites and dendritic spines in the rat hippocampus. Europ J Neurosci 1996; 8:1488-1500.
125. Lukoyanov NV, Paula-Barbosa MM. A single high dose of dizocilpine produces long-lasting impairment of the water maze performance in adult rats. Neurosci Lett 2000; 285:139-142.
126. Lyford GL, Jarrard LE. Effects of the competitive NMDA antagonist CPP on performance of a place and cue radial maze task. Psychobiol 1991; 19:157-160.
127. Madden DR. The structure and function of glutamate receptor ion channels. Nature Rev Neurosci 2002; 3:91-101.
128. Malenfant SA, O'Hearn S, Fleming AS. MK801, an NMDA Antagonist, blocks acquisition of a spatial task but does not block maternal experience effects. Physiol Behav 1991; 49:1129-1137.
129. Maren S, Aharanoov G, Stote DI et al. N-methyl-D-aspartate receptors in the basolateral amygdala are required for both acquisition and expression of conditional fear in rats. Behav Neurosci 1996; 110:1365-1374.
130. Marighetto A, Micheau J, Jaffard R. Effect of intraseptally injected glutamatergic drugs on hippocampal sodium-dependent-high-affinity-choline-uptake in "naive" and "trained" mice. Pharmacol Biochem Behav 1994; 49:689-699.
131. Martin SJ. The role of metabotropic glutamate and NMDA receptors in hippocampal long-term potentiation and spatial learning. Thesis Univ Edinburgh 1998.
132. Mason KI, Mallet PE, Jhamandas K et al. Nucleus basalis injections of N-methyl-D-aspartate enhance memory of rats in the double Y-maze. Brain Res Bull 1999; 48:65-71.
133. Masugi M, Yokoi M, Shigemoto R et al. Metabotropic glutamate receptor subtype 7 ablation causes deficit in fear response and conditioned taste aversion. J Neurosci 1999; 19:955-963.
134. Mathis C, De Barry J, Ungerer A. Memory deficits induced by γ-L-glutamyl-L-aspartate and D-2-amino-5-phosphonovalerate in a Y-maze avoidance task: Relationship to NMDA receptor antagonism. Psychopharmacol 1991; 105:546-552.
135. Mathis C, Paul SM, Crawley JN. The neurosteroid pregnenolone sulfate blocks NMDA antagonist-induced deficits in a passive avoidance memory task. Psychopharmacol 1994; 116:201-206.
136. Maurice T, Pivat A, Lemaire S. The pentadecapeptide [Ser^1]histogranin impairs passive avoidance learning in mice. Europ J Pharmacol 1995; 283:251-254.
137. Maurice T, Su TP, Parish DW et al. PRE084, a sigma selective PCP derivative, attenuates MK-801-induced impairment of learning in mice. Pharmacol Biochem Behav 1994; 49:859-869.

138. Maurice T, Hiramatsu M, Itoh J et al. Behavioral evidence for a modulating role of sigma ligands in memory processes. I. Attenuation of dizocilpine (MK-801)-induced amnesia. Brain Res 1994; 647:44-56.
139. McCann DJ, Rabin RA, Winter JC. Use of the radial maze in studies of phencyclidine and other drugs of abuse. Physiol Behav 1987; 40:805-812.
140. McLamb RL, Williams LR, Nanry KP et al. MK-801 impedes the acquisition of a spatial memory task in rats. Pharmacol Biochem Behav 1990; 37:41-45.
141. Mele A, Catsellano C, Cestari V et al. Interaction between the NMDA competitive antagonist CPP and the dopaminergic system in one-trial inhibitory avoidance in C57BL/6 mice. Neurobiol Learn Memory 1995; 63:143-148.
142. Mele A, Castellano C, Felici A et al. Dopamine-N-methyl-D-aspartate interactions in the modulation of locomotor activity and memory consolidation in mice. Europ J Pharmacol 1996; 308:1-12.
143. Mesches MH, Bianchin M, McGaugh JL. The effects of intra-amygdala infusion of the AMPA receptor antagonist CNQX on the retention performance following aversive training. Neurobiol Learn Memory 1996; 66:324-340.
144. Meyer RC, Knox J, Purwin DA et al. Combined stimulation of the glycine and polyamine sites of the NMDA receptor attenuates NMDA blockade-induced learning deficits in rats in a 14-unit T-maze. Psychopharmacol 1998; 135:290-295.
145. Mickley GA, Schaldach MA, Snyder KJ et al. Ketamine blocks a conditioned taste aversion (CTA) in neonatal rats. Physiol Behav 1998; 64:380-390.
146. Miserendino MJD, Sananes CB, Melia KR et al. Blocking of acquisition but not expression of conditioned fear-potentiated startle by NMDA antagonists in the amygdala. Nature 1990; 345:716-718.
147. Misztal M, Danysz W. Comparison of glutamate antagonists in continuous multiple-trial and single-trial dark avoidance. Behav Pharmcol 1995; 6:550-561.
148. Misztal M, Frankiewicz T, Parsons CG et al. Learning deficits induced by chronic intraventricular infusion of quinolinic acid – protection by MK801 and memantine. Europ J Pharmacol 1996; 296:1-8.
149. Monahan JB, Handelmann GE, Hood WF et al. D-Cycloserine, a positive modulator of the N-methyl-D-aspartate receptor, enhances performance of learning tasks in rats. Pharmacol Biochem Behav 1989; 34:649-653.
150. Mondadori C, Weiskrantz L. NMDA receptor blockers facilitate and impair learning via different machanisms. Behav Neural Biol 1993; 60:205-210.
151. Mondadori C, Borkowski J, Gentsch C. The memory-facilitating effects of the competitive NMDA-receptor antagonist CGP 37849 are steroid sensitive, whereas the memory-impairing effects are not. Psychopharmacol 1996; 124:380-383.
152. Mondadori C, Ortmann R, Petschke F et al. Task-dependent impairment or facilitation of learning by NMDA receptor blockade. In: Alan Liss R, ed. Frontiers in Excitatory Amino Acid Research. 1988:419-26.
153. Mondadori C, Weiskrantz L, Buerki H et al. NMDA receptor antagonists can enhance or impair learning performance in animals. Exp Brain Res 1989; 75:449-456.
154. Morris RGM. Synaptic plasticity and learning: Selective impairment of learning in rats and blockade of long-term potentiation in vivo by the NMDA receptor antagonist AP5. J Neurosci 1989; 9:3040-3057.
155. Morris RGM, Davis M. The role of NMDA receptors in learning and memory. In: Collingridge GL, Watkins JC, eds. The NMDA receptor. Oxford, UK: Oxford University Press, 1994:340-75.
156. Morris RGM, Andersen E, Lynch GS et al. Selective impairment of learning and blockade of long-term potentiation by an N-methyl-D-aspartate receptor antagonist, AP5. Nature 1986; 319:774-776.
157. Mulle C, Sailer A, Perez-Otano I et al. Altered synaptic physiology and reduced susceptibility to kainate-induced seizures in GluR6-deficient mice. Nature 1998; 392:601-605.
158. Murata S, Kawasaki K. Common and uncommon behavioural effects of antagonists for different modulatory sites in the NMDA receptor / channel complex. Europ J Pharmacol 1993; 239:9-15.
159. Nakanishi S, Nakajima Y, Masu M et al. Glutamate receptors: Brain function and signal transduction. Brain Res Rev 1998; 26:230-235.
160. Nemeth H, Varga H, Farkas T et al. Long-term effects of neonatal MK-801 treatment on spatial learning and cortical plasticity in adult rats. Psychopharmacol 2002; 160:1-8.
161. Nicoletti F, Valerio C, Pellegrino C et al. Spatial learning potentiates the stimulation of phosphoinositide hydrolysis by excitatory amino acids in rat hippocampal slices. J Neurochem 1988; 51:725-729.
162. Nicolle MM, Colombo PJ, Gallagher M et al. Metabotropic glutamate receptor-mediated hippocampal phosphoinositide turnover is blunted in spatial learning-impaired aged rats. J Neurosci 1999; 19:9604-9610.

163. Nielsen KS, Macphail EM, Riedel G. Class I mGlu receptor antagonist 1-aminoindan-1,5-dicarboxylic acid blocks contextual but not cue conditioning in rats. Europ J Pharmacol 1997; 326:105-108.
164. Noda A, Noda Y, Kamei H et al. Phencyclidine impairs latent learning in mice: Interaction between glutamatergic systems and sigma (1) receptors. Neuropsychopharmacol 2001; 24:451-460.
165. Norris CM, Foster TC. MK-801 improves retention in aged rats: Implications for altered neural plasticity in aged-related memory deficits. Neurobiol Learn Memory 1999; 71:194-206.
166. Novak L, Bregestovski P, Ascher P et al. Magnesium gates glutamate-activated channels in mouse central neurones. Nature 1994; 307:460-462.
167. Ohno M, Watanabe S. Concurrent blockade of hippocampal metabotropic glutamate and N-methy-D-aspartate receptors disrupts working memory in the rat. Neurosci 1996; 70:303-311.
168. Ohno M, Watanabe S. Enhanced N-methyl-Daspartate function reverses working memory failure induced by blockade of group I metabotropic glutamate receptors in the rat hippocampus. Neurosci Lett 1998; 240:37-40.
169. Ohno M, Yamamoto T, Watanabe S. Effests of intrahippocampal injections of N-methyl-D-aspartate receptor antagonists and scopolamine on working memory and reference memory assessed in rats by a three panel runway task. J Pharmacol Exp Therap 1992; 263:943-950.
170. Ohno M, Yamamoto T, Watanabe S. Amygdaloid NMDA and muscarinic receptors involved in working memory performance of rats. Physiol Behav 1993; 54:993-997.
171. Ohno M, Yoshimatsu A, Kobayashi M et al. β-adrenergic dysfunction exacerbates impairment of working memory induced by hippocampal NMDA receptor blockade in rats. Europ J Pharmacol 1996; 307:21-26.
172. Ozawa S, Kamiya H, Tsuzuki K. Glutamate receptors in the mammalian central nervous system. Prog Neurobiol 1998; 54:581-618.
173. Packard MG, Teather LA. Double dissociation of hippocampal and dorsal-striatal memory systems by posttraining intracerebral injections of 2-amino-5-phosphonopentanoic acid. Behav Neurosci 1997; 111:543-551.
174. Parada-Turska J, Czuczwar SJ, Turski WA. NBQX does not affect learning and memory tasks in mice – a comparison with d-CPPene and ifenprodil. Cogn Brain Res 1992; 1:67-71.
175. Petersen S, Bomme C, Baastrup C et al. Differential effects of mGluR1 and mGluR5 antagonism on spatial learning in rats. Pharmacol Biochem Behav 2002; 73:381-389.
176. Pettit HO, Lutz D, Gutierrez C et al. I.c.v. infusions of ACPD $_{(1S,3R)}$ attenuate learning in a Morris water maze paradigm. Neurosci Lett 1994; 178:43-46.
177. Pin JP, Duvoisin R. The metabotropic glutamate receptors: Structure and functions. Neuropharmacol 1995; 34:1-26.
178. Popik P, Rygielska Z. A partial agonist at strichnine-insensitive glycine sites facilitates spatial learning in aged rats. J Physiol Pharmacol 1999; 50:139-151.
179. Przybyslawski J, Sara SJ. Reconsolidation of memory after its reactivation. Behav Brain Res 1997; 84:241-246.
180. Pussinen R, Sirvio J. Effects of D-cycloserine, a positive modulator of N-methyl-D-aspartate receptors, and ST 587, a putative alpha-1-adrenergic agonist, individually and in combination, on the nondelayed and delayed foraging behaviour of rats assessed in the radial arm maze. J Psychopharmacol 1999; 13:171-179.
181. Quartermain D, Nuygen T, Sheu J et al. Milacemide enhances memory storage and alleviates spontaneous forgetting in mice. Pharmacol Biochem Behav 1991; 39:31-35.
182. Quevedo J, Vianna M, Zanatta MS et al. Involvement of mechanisms dependent on NMDA receptors, nitric oxide and protein kinase A in the hippocampus but not in the caudate nucleus in memory. Behav Pharacol 1997; 8:713-717.
183. Quillfeldt JA, Schmitz PK, Walz R et al. CNQX infused into entorhinal cortex blocks memory expression, and AMPA reverses the effect. Pharmacol Biochem Behav 1994; 48:437-440.
184. Rawlins JNP, Deacon RMJ. Further developments of maze procedures. In: Sahgal A, ed. Behavioural Neuroscience: A practical approach (Vol. I). Oxford, UK: Oxford University Press, 1993:95-106.
185. Richter-Levin G, Errington ML, Maegawa H et al. Activation of metabotropic glutamate receptors is necessary for long-term potentiation in the dentate gyrus and for spatial learning. Neuropharmacol 1994; 33:853-857.
186. Rickard NS, Ng KT. Blockade of metabotropic glutamate receptors prevents long-term memory consolidation. Brain Res Bull 1995; 36:355-359.
187. Rickard NS, Poot AC, Gibbs ME et al. Both nonNMDA and NMDA glutamate receptors are necessary for memory consolidation in the day-old chick. Behav Neural Biol 1994; 62:33-40.
188. Riedel G, Martin SJ, Roloff EvL. Blockade of class I metabotropic glutamate receptors and performance of rats: A watermaze study. 6th Neuropharmacology Conference: Cortical Plasticity 1998:4.20.
189. Riedel G, Wetzel W, Reymann KG. (R,S)-α-methyl-4-carboxyphenylglycine (MCPG) blocks spatial learning in rats and LTP in the dentate gyrus in vivo. Neurosci Lett 1994; 167:141-144.

190. Riedel G, Wetzel W, Reymann KG. Computer-assisted shock-reinforced Y-maze training: A method for studying spatial alternation behaviour. NeuroReport 1994; 5:2061-2064.
191. Riedel G, Wetzel W, Reymann KG. Metabotropic glutamate receptors in spatial and nonspatial learning in rats studied by means of agonist and antagonist application. Learn Memory 1995; 2:243-265.
192. Riedel G, Wetzel W, Kozikowski AP et al. Block of spatial learning by mGluR agonist tADA in rats. Neuropharmacol 1995; 34:559-561.
193. Riedel G, Casabona G, Platt B et al. Fear conditioning-induced time-and subregion-specific increase in expression of mGluR5 receptor protein in rat hippocampus. Neuropharmacol 2000; 39:1943-1951.
194. Riedel G, Micheau J, Lam AGM et al. Reversible neural inactivation reveals hippocampal participation in several memory processes. Nature Neurosci 1999; 2:898-905.
195. Riedel G, Harrington NR, Kozikowski AP et al. Variation of CS salience reveals group II mGluR dependent and independent forms of conditioning in the rat. Neuropharmacol 2002; 43:205-214.
196. Riekkinen M, Stefanski R, Kuitunen J et al. Effects of combined block of α_1-adrenoceptors and NMDA receptors on spatial and passive avoidance behavior in rats. Europ J Pharmacol 1996; 300:9-16.
197. Riters LV, Bingman VP. The NMDA-receptor antagonist MK-801 impairs navigational learning in homing pigeons. Neurobiol. Learn Memory 1994; 62:50-59.
198. Robinson GS, Crooks Jr GB, Shinkman PG et al. Behavioral effects of MK-801 mimic deficits associated with hippocampal damage. Psychobiol 1989; 17:156-164.
199. Rodrigues SM, Schafe GE, LeDoux JE. Intra-amygdala blockade of the NR2B subunit of the NMDA receptor disrupts the acquisition but not expression of fear conditioning. J Neurosci 2001; 21:6889-6896.
200. Rodrigues SM, Bauer EP, Farb CR et al. The group I metabotropic glutamate receptor mGluR5 is required for fear memory formation and long-term potentiation in the lateral amygdala. J Neurosci 2002; 22:5219-5229.
201. Roesler R, Quevedo J, Kuyen CR et al. The NMDA receptor blocker MK-801 inhibits the effect of additional training on retention of an inhibitory avoidance task. Med Sci Res 1997; 25:591-592.
202. Roesler R, Vianna MRM, de-Paris F et al. Infusion of AP5 into the basolateral amygdala impair the formation, but not the expression, of step-down inhibitory avoidance. Braz J Med Biol Res 2000; 33:829-834.
203. Roesler R, Vianna MRM, de-Paris F et al. NMDA receptor antagonism in the basolateral amygdala blocks enhancement of inhibitory avoidance learning in previously trained rats. Behav Brain Res 2000; 112:99-105.
204. Romanides AJ, Duffy P, Kalivas PW. Glutamatergic and dopaminergic afferents to the prefrontal cortex regulate spatial working memory. Neurosci 1999; 92:97-106.
205. Rosenblum K, Berman DE, Hazvi S et al. NMDA receptor and the tyrosine phosphorylation of its 2B subunit in taste learning in the rat insular cortex. J Neurosci 1997; 17:5129-5135.
206. Roullet P, Sargolini F, Oliverio A et al. NMDA and AMPA antagonist infusions into the ventral striatum impair different steps of spatial information processing in a nonassociative task in mice. J Neurosci 2001; 21:2143-2149.
207. Rubin MA, Boemo RL, Jurach A et al. Intrahippocampal spermidine administration improves inhibitory avoidance performance in rats. Behav Pharmacol 2000; 11:57-61.
208. Rubin MA, Stiegemeier JA, Volkweis MA et al. Intra-amygdala spermidine administration improves inhibitory avoidance performance in rats. Europ J Pharmacol 2001; 423:35-39.
209. Rumsey JD, Darby-King A, Harley CW et al. Infusion of the metabotropic receptor antagonist, DCG-IV, into the main olfactory bulb induces olfactory preference learning in rat pups. Develop Brain Res 2001; 128:177-179.
210. Sanger DJ, Joly D. Effects of NMDA receptor antagonists and sigma ligands on the acquisition of conditioned fear in mice. Psychopharmacol 1991; 104:27-34.
211. Saucier D, Cain DP. Spatial learning without NMDA receptors and nonNMDA receptors. Nature 1995; 378:186-189.
212. Saucier D, Hargreaves EL, Boon F et al. Detailed behavioral anlysis of water maze acquisition under systemic NMDA or muscarinic antagonism: Nonspatial pretraining eliminates spatial learning deficits. Behav Neurosci 1996; 110:103-116.
213. Schauz C, Koch M. Blockade of NMDA receptors in the amygdala prevents latent inhibition of fear-conditioning. Learn Memory 2000; 7:393-399.
214. Schulz B, Fendt M, Gasparini F et al. The metabotropic glutamate receptor antagonist 2-methyl-6-(phenylethyl)-pyridine (MPEP) blocks fear conditioning in rats. Neuropharmacol 2001; 41:1-7.
215. Schuster GM, Schmidt WJ. D-cycloserine reverses the working memory impairment of hippocampal-lesioned rats in a spatial learning task. Europ J Pharmacol 1992; 224:97-98.

216. Schwartz BL, Hashtroudi RL, Herting H et al. D-cycloserine enhances implicit memory in Alzheimer patients. Neurology 1994; 46:420-424.
217. Shapiro MI, Carmanos Z. NMDA antagonist MK-801 impairs acquisition but not performance of spatial working and reference memory. Psychobiol 1990; 18:231-243.
218. Shapiro MI, O'Connor C. N-methyl-D-aspartate receptor antagonist MK-801 and spatial mempory representation: Working memory is impaired in an unfamiliar environment but not in a familiar environment. Behav Neurosci 1992; 106:604-612.
219. Shimada A, Spangler EL, London ED et al. Spermidine potentiates dizocilpine-induced impairment of learning performance by rats in a 14-unit T-maze. Europ J Pharmacol 1994; 263:293-300.
220. Shimizu E, Tang YP, Rampon C et al. NMDA receptor-dependent synaptic reinforcement as a crucial process for memory consolidation. Science 2000; 290:1170-1174.
221. Sierocinska J, Nikolaev E, Danysz W et al. Dextrorphan blocks long- but not short-term memory in a passive avoidance task in rats. Europ J Pharmacol 1991; 205:109-111.
222. Smith-Roe SL, Sadeghian K, Kelley AE. Spatial learning and performance in the radial arm maze is impaired after N-methyl-D-aspartate (NMDA) receptor blockade in striatal subregions. Behav Neurosci 1999; 113:703-717.
223. Spangler EL, Breshnahan EL, Garofalo P et al. NMDA receptor channel antagonism by dizocilpine (MK-801) impairs performance of rats in aversively motivated complex maze tasks. Pharmacol Biochem Behav 1991; 40:949-958.
224. Stäubli U, Thibault O, DiLorenzo M et al. Antagonism of NMDA receptors impairs acquisition but not retention of olfactory memory. Behav Neurosci 1989; 103:54-60.
225. Steele RJ, Morris RGM. Delay-dependent impairment of a matching-to-place task with chronic and intrahippocampal infusion of the NMDA-agonist D-AP5. Hippocampus 1999; 9:118-136.
226. Steele RJ, Stewart MG. 7-chlorokynurenate, an antagonist of the glycine binding dite on the NMDA receptor, inhibits memory formation in day-old chicks. Neurobiol Learn Memory 1993; 60:89-92.
227. Steele RJ, Stewart MG. Involvement of AMPA receptors in maintenance of memory for a passive avoidance task in day-old domestic chick (Gallus domesticus). Eur J Neurosci 1995; 7:1297-1304.
228. Steele RJ, Dermon CR, Stewart MG. D-cycloserine causes transient enhancement of memory for a weak aversive stimulus in day-old chicks (Gallus domesticus). Neurobiol Learn Memory 1996; 66:236-240.
229. Stephens DN, Cole BJ. AMPA antagonists differ from NMDA antagonists in their effects on operant DRL and delayed matching to position tasks. Psychopharmacol 1996; 126:249-259.
230. Stewart CW, Morris RGM. The watermaze. In: Sahgal A, ed. Behavioural Neuroscience: A practical approach (Vol. I). Oxford, UK: Oxford University Press, 1993:107-122.
231. Stewart LS, McKay BE. Acquisition deficit and time-dependent retrograde amnesia for contextual fear conditioning in agmatine-treated rats. Behav Pharmacol 2000; 11:93-97.
232. Stiedl O, Birkenfeld K, Palve M et al. Impairment of conditioned contextual fear of C57BL/6J mice by intracerebral injections of the NMDA receptor antagonist APV. Behav Brain Res 2000; 116:157-168.
233. Szapiro G, Barros D, Ardenghi P et al. Facilitation and inhibition of retrieval in two aversive tasks in rats by intrahippocampal infusion of agonists of specific glutamate metabotropic receptor subtypes. Psychopharmacol 2001; 156:397-401.
234. Szapiro G, Izquierdo LA, Alonso M et al. Participation of hippocampal metabotropic glutamate receptors, protein kinase A and mitogen-activated protein kinases in memory retrieval. Neurosci 2000; 99:1-5.
235. Tang YP, Shimizu E, Dube GR et al. Genetic enhancement of learning and memory in mice. Nature 1999; 401:63-69.
236. Tang YP, Wang H, Feng R et al. Differential effects of enrichment on learning and memory formation in NR2B transgenic mice. Neuropharmacol 2001; 41:779-790.
237. Teather LA, Packard MG, Bazan NG. Differential interaction of platelet-activating factor and NMDA receptor function in hippocampal and dorsal striatal memory processes. Neurobiol Learn Memory 2001; 75:310-324.
238. Terry Jr AV. Spatial navigational (water maze) tasks. In: Buccafusco JJ, ed. Methods of behavior analysis in neuroscience. Boca Raton, Florida: CRC Press, 2000:153-166.
239. Tizzano JP, Griffey KI, Schoepp DD. The anxiolytic action of mGluR2/3 receptor agonist, LY354740, in the fear-potentiated startle model in rats is mechanistically distinct from diazepam. Pharmacol Biochem Behav 2002; 73:367-374.
240. Tsien JZ, Huerta PT, Tonegawa S. The essential role of hippocampal CA1 NMDA receptor-dependent synaptic plasticity in spatial memory. Cell 1996; 87:1327-1338.
241. Ungerer A, Mathis C, Melan C et al. The NMDA receptor antagonists, CPP and γ-L-glutamyl-L-aspartate, selectively block post-training improvement of performance in a Y-maze avoidance learning task. Brain Res 1991; 549:59-65.

242. Upchurch M, Wehner JM. Effects of N-methyl-D-aspartate antagonism on spatial learning in mice. Psychopharmacol 1990; 100:209-214.
243. Usiello A, Sargolini F, Roullet P et al. N-methyl-D-aspartate receptors in the nucleus accumbens are involved in detection of spatial novelty in mice. Psychopharmacol 1998; 137:175-183.
244. van der Staay FJ, Antoniecek H, Helpap B et al. Effects of the selective metabotropic glutamate receptor agonist, L-CCG-I, on acquisition of a Morris task by rats. Europ J Pharmacol 1995; 294:361-365.
245. Venable N, Kelly PH. Effects of NMDA receptor antagonists on passive avoidance learning and retrieval in rats and mice. Psychopharmacol 1991; 100:215-221.
246. Venero C, Sandi C. Effects of NMDA and AMPA receptor antagonists on corticosterone facilitation of long-term memory in the chick. Eur J Neurosci 1997; 9:1923-1928.
247. Viu E, Zapata A, Capdevila J et al. $Glycine_B$ receptor antagonists and partial agonists prevent memory deficits in inhibitory avoidance learning. Neurobiol Learn Memory 2000; 74:146-160.
248. Walker DL, Davis M. Involvement of NMDA receptors within the amygdala in short- versus long-term memory for fear conditioning as assessed with fear-potentiated startle. Behav Neurosci 2000; 114:1019-1033.
249. Walker DL, Gold PE. Effects of the novel NMDA antagonist, NPC 12626, on long-term potentiation, learning and memory. Brain Res 1991; 549:213-221.
250. Walker DL, Gold PE. Impairment of spontaneous alternation performance by an NMDA antagonist: Attenuation with nonNMDA treatments. Behav Neural Biol 1992; 58:69-71.
251. Walker DL, Gold PE. Intrahippocampal administration of both the D- and the L-isomers of AP5 disrupts spontaneous alternation behavior and evoked potentials. Behav Neural Biol 1994; 62:159-162.
252. Ward J, Mason SE, Abraham WC. Effects of the NMDA antagonists CPP and MK801 on radial maze performance in rats. Pharmacol Biochem Behav 1989; 35:785-790.
253. Welzl H, Alessandri B, Baettig K. The formation of a new gustatory memory trace is prevented by the noncompetitive NMDA antagonist ketamine. Psychobiol 1990; 18:43-47.
254. Wesierska M, Macias-Gonzales R, Bures J. Differential effect of ketamine on the reference and working memory versions of the Morris water maze task. Behav Neurosci 1990; 104:74-83.
255. Whishaw IQ, Auer RN. Immediate and long-lasting effects of MK-801 on motor activity, spatial navigation in a swimming pool and EEG in the rat. Psychopharmacol 1989; 98:500-507.
256. White AM, Best PJ. The effects of MK-801 on spatial memory and within-session spatial learning. Pharmcol Biochem Behav 1998; 59:613-617.
257. Willner J, Gallagher M, Graham PW et al. N-methyl-D-aspartate antagonist D-APV selectively disrupts taste-potentiated odor aversion learning. Behav Neurosci 1992; 106:315-323.
258. Wilson IA, Puolivali J, Heikkinen T et al. Estrogen and NMDA receptor antagonism: Effects upon reference and working memory. Europ J Pharmacol 1999; 381:93-99.
259. Wozniak DF, Olney JW, Kettinger et al. Behavioral effects of MK-801 in the rat. Psychopharmacol 1990; 101:47-56.
260. Yasoshima Y, Morimoto T, Yamamoto T. Different disruptive effects on the acquisition and expression of conditioned taste aversion by blockades of amygdalar ionoptropic and metabotropic glutammate receptor subtypes in rats. Brain Res 2000; 869:15-24.
261. Yokoi M, Kobayashi K, Manabe T et al. Impairment in hippocampal mossy fibre LTD in mice lacking mGluR2. Science 1996; 273:645-647.
262. Young SL, Bohenek DL, Fanselow MS. NMDA processes mediate anterograde amnesia of contextual fear conditioning induced by hippocampal damage: Immunization against amnesia by context preexposure. Behav Neurosci 1994; 108:19-29.
263. Zajaczkowski W, Frankiewicz T, Parsaons CG et al. Uncompetitive NMDA receptor antagonists attenuate NMDA-induced impairment of passive avoidance learning and LTP. Neuropharmacol 1997; 36:961-971.
264. Zalewska A, Wisniewski K. Behavioural activity of 1S,3R-ACPD, an agonist of metabotropic glutamate receptors. Pol J Pharmacol 1997; 49:239-248.
265. Zalewska-Winska A, Wisniewski K. Behavioural activity of (S)-3,5-DHPG, a selective agonist of group I metabotropic glutamate receptors. Pharmacol Res 2000; 42:239-245.
266. Zhang WN, Bast T, Feldon J. The ventral hippocampus and fear conditioning in rats: Different anterograde amnesias of fear after infusion of N-methyl-D-aspartate or its noncompetitive antagonist MK-801 into the ventral hippocampus. Behav Brain Res 2001; 126:159-174.
267. Zivkovic I, Thompson DM, Bertolino M et al. 7-chloro-3-methyl-3-4-dihydro-2H-1,2,4 benzothiazide S,S-dioxide (IDRA 21): A benzothiadiazine derivative that enhances cognition by attenuating D,L-α-amino2,3-dihydro-5-methyl-3-oxo-4-isoxazolepropionic acid (AMPA) receptor desensitisation. J Pharmacol Exp Therap 1995; 272:300-309.

Chapter 2.2

γ-Amino-Butyric Acid (GABA)

Claudio Castellano, Vincenzo Cestari and Alessandro Ciamei

Abstract

In this chapter studies are considered concerning the actions of $GABA_A$ and $GABA_B$ receptor agonists and antagonists on memory formation in laboratory animals. Peripheral posttraining administrations of GABAergic drugs produce time- and dose-dependent effects on memory in animals tested in a variety of experimental conditions. These effects are not state-dependent, but reflect impairments or enhancements of memory storage processes through influences on central GABAergic receptors. Furthermore, the genetic make-up plays an important role in the modulation of memory processes by $GABA_A$ and $GABA_B$ receptor agonists and antagonists in mice.

Central administrations, as well as lesion studies, show that a number of brain structures, including amygdala, hippocampus, septum and striatum, are involved in the effects of GABAergic agonists and antagonists on memory storage. Finally, GABAergic mechanisms are involved in the effects exerted on memory processes by opioids, benzodiazepines and ethanol.

Introduction

There is extensive evidence showing that memory processes in animals can be influenced by posttraining administration of drugs. In posttraining treatments a drug is administered at some time after the learning or acquisition phase. It is thus possible to eliminate the potential problems of the pretraining administrations, such as alteration by drugs of sensory and motor events involved in learning, focusing on the action of the drugs on memory trace, rather than on performance-related events. In fact, when a drug is administered after training, the animals can be trained and tested while they are not under the direct influence of the drug, and the effects observed can be attributed to influences of the drug on the consolidation of memory, a process which takes place immediately after the training experience. In particular, the studies carried out using this procedure can examine the effects of varying time between the training and the administration of the drugs.[73,74] There is now extensive evidence indicating that retention can be influenced by drugs affecting a variety of neurochemical systems, including catecholaminergic, opioid peptidergic, cholinergic and GABAergic systems.[75,77]

This chapter will deal with the involvement of the GABAergic system in memory formation in laboratory animals. Interaction with other systems in the modulation of memory storage will be also considered.

GABA (γ-aminobutyric acid) is the main inhibitory neurotransmitter in the mammalian nervous system and exerts its effects through specific postsynaptic receptors that influence membrane permeability to chloride ions.[65,96] The GABA receptor is part of a protein complex that also contains receptor sites for benzodiazepines, picrotoxin and barbiturates, associated with a chloride ionophore.[34,93] In the mammalian brain there are two different GABA receptor sites: $GABA_A$ and $GABA_B$.[10] The $GABA_B$ agonist baclofen mimics the effects of GABA on this G-protein coupled receptor, while muscimol is a $GABA_A$ receptor agonist. Picrotoxin is an indirect $GABA_A$ receptor antagonist and blocks GABAergic synaptic transmission by interaction

From Messengers to Molecules: Memories Are Made of These, edited by Gernot Riedel and Bettina Platt.

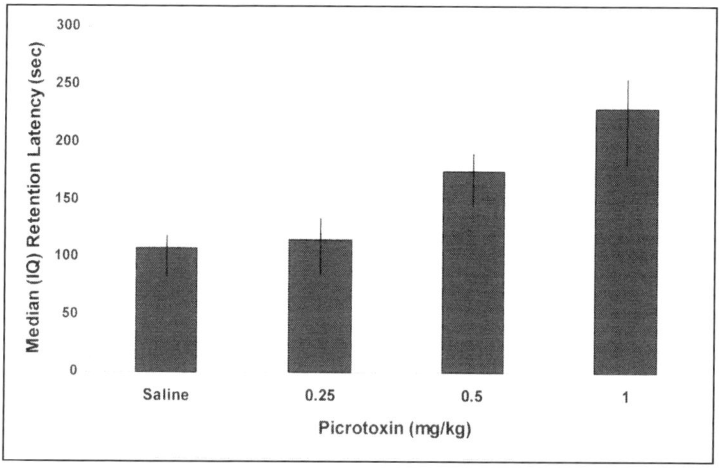

Figure 1. Dose-dependent effects of immediately posttraining i.p. injections of picrotoxin on retention, by mice, of a one-trial step-through inhibitory avoidance task. Each column indicates median response latency (+ interquartile range), in seconds, on the retention test trial (from ref. 24).

occurring at the level of chloride ionophore.[93,94] As concerns bicuculline, electrophysiological and biochemical studies have shown that this drug is a specific $GABA_A$ receptor antagonist with central actions.[35,36,63,117]

GABAergic Drugs and Memory Formation: Peripheral Administrations

GABAergic Antagonists

Picrotoxin

The first study showing a possible involvement of the GABAergic system in memory was carried out by Breen and McGaugh[11] with rats tested in an appetitively-motivated multiple T-maze. Posttraining intraperitoneal (i.p.) administrations of picrotoxin reduced significantly the number of errors made by the animals. This initial finding was confirmed by the observation of other investigators, using aversively-motivated as well as appetitively-motivated tasks. Studies carried out by Bovet et al[9] showed that posttraining i.p. administrations of picrotoxin enhanced the rate of acquisition of a two-way avoidance learning in mice. Furthermore, enhanced learning of rats posttraining injected i.p. with this drug and trained in a Hebb-Williams maze was demonstrated by Garg and Holland.[46]

It must be stressed that, at the time of these studies, the mechanism of action of picrotoxin at the biochemical level was as yet unknown, and only later it was shown that its effects involved activation of the GABAergic neurotransmitter system.

More recently retention improvements have been shown, following posttraining i.p. administration of picrotoxin, in CD1 and CFW mice tested in a one-trial step-through inhibitory avoidance task and in a Y-maze task (Fig. 1).[14,24,28] In these experiments the effects of picrotoxin were dose- and time-dependent. In particular the effects were no more evident when the animals were injected 120 min after training suggesting that they were due to a specific action of the drug on the time-dependent memory consolidation process. Furthermore, picrotoxin treatment did not affect the retention performance of animals unless they received footshock on the training trial. This excludes the possibility that the effects of the drug on retention performance were due to nonspecific influences on response latencies.[72,77] It must

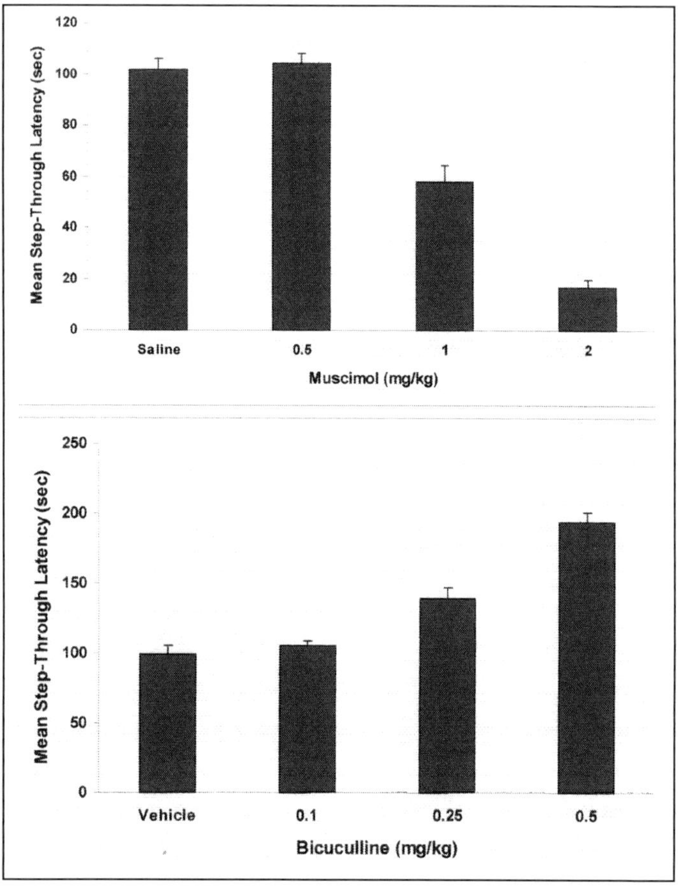

Figure 2. Dose-dependent effects of immediately posttraining i.p. injections of muscimol (upper graph) and bicuculline (lower graph) on retention, by mice, of a one-trial step-through inhibitory avoidance task. Each column indicates mean step-through latency (± SEM), in seconds, on the retention test trial (from ref. 27).

be finally considered that enhancement of latent extinction of conditioned fear, following posttraining i.p. administrations of picrotoxin, has been described in mice tested in a visual discrimination task.[78]

Amnesia, instead of memory improvement, has been reported following posttraining i.p. administration of picrotoxin in rats trained in a one-trial step-down inhibitory avoidance test.[86] As observed by Brioni[12], it is possible that this effect was the consequence of the high shock intensity used in these experiments, and that picrotoxin could induce amnesia at high doses if a high shock is used. In particular, the controls in the Nabeshima and Noda[86] study showed high retention latencies, suggesting that the footshock intensity was high for the animals used in their study. Indeed, some posttraining treatments, such as acetylcholine and epinephrine, enhance retention performance when low footshock intensities are administered in the training, but impair retention when, in the training session, high footshock intensity is used.[48,49,59,60]

Bicuculline

Effects on memory comparable to those exerted by picrotoxin have been found in a number of studies following peripheral (i.p.) bicuculline administration.

Retention enhancement following posttraining bicuculline administrations in rats tested in an active avoidance task has been reported.[115] In further studies CD1 mice were used (Fig. 2).[27] They were injected with the drug immediately or 120 min after training in a one-trial step-through inhibitory avoidance task. As compared with saline injected animals, retention improvement was observed in animals injected immediately after training, while no effect was evident in mice following 120 min posttraining injection. This suggests that the effects exerted by bicuculline were due to a specific action on the time-dependent memory consolidation process. Further, posttraining administrations of bicuculline did not affect retention latencies of unshocked controls, indicating that the effects of the drug on retention performance were not due to nonspecific effects on response latencies.[24] In a further study,[14] carried out with CFW mice, the animals were injected i.p. after training with bicuculline in two aversively-motivated tasks, inhibitory avoidance and Y maze discrimination. Bicuculline methiodide (BMI, a $GABA_A$ receptor antagonist that does not readily cross the blood-brain barrier)-injected groups were also used. In both tasks bicuculline enhanced retention performance of the animals in a dose- and time-dependent way. Further, retention was not affected by the posttraining administration of bicuculline methiodide. These findings suggest the involvement of central GABAergic processes in the effects observed. It must be underlined that other studies have more recently confirmed that bicuculline improves memory consolidation in CD1 mice tested in a one-trial step-through inhibitory avoidance task.[22,25,54] Nabeshima et al[87] found, as in the case of picrotoxin, amnesia also following posttraining i.p. administration of bicuculline in mice tested in an inhibitory avoidance task. As for picrotoxin, this finding could be due to the high footshock intensity used in these experiments (see also above, picrotoxin section).

GABAergic Agonists

Muscimol

There is extensive evidence suggesting that the $GABA_A$ receptor agonist muscimol influences memory processes in animals, after peripheral administration. It has been demonstrated that posttraining i.p. injections of this drug impair memory consolidation in CD1 mice tested in one-trial step-through inhibitory avoidance conditions (Fig. 2).[23,25,26,28,54] In these studies the effects of muscimol were dose- and time-dependent. Furthermore, muscimol treatment did not affect the retention performance of animals unless they received footshock on the training trial, suggesting absence of nonspecific influences by the drug on the response latencies. In a study carried out by Salinas and McGaugh[101] it has been finally demonstrated that muscimol can induce retrograde amnesia for changes in reward magnitude (reward reduction, and reward increase) in male Sprague-Dawley rats.

Baclofen

Findings comparable to those observed with muscimol have been obtained following peripheral (i.p.) administrations of the $GABA_B$ receptor agonist baclofen. In particular, the effects of posttraining i.p. injections of this drug have been studied in two aversively motivated tasks: one-trial step-through inhibitory avoidance and a classical conditioning task (Y maze).[23] The experiments were carried out with CD1 mice. The immediate posttraining administration of baclofen impaired the retention performances of the animals in both experimental conditions. No effect was evident when the drug was injected 120 min after training. These results are in agreement with those of other studies carried out with rats injected posttraining with baclofen and tested in an inhibitory avoidance task.[104]

GABAergic Drugs and Memory: Genotype-Dependent Effects

Studies carried out with the inbred strains of mice C57BL/6 (C57) and DBA/2 (DBA) have shown that the genetic makeup plays an important role in modulating responses to drug administration. As concerns memory processes, strain-dependent effects have been observed, in

C57 and DBA mice, following posttraining injections of a number of drugs, such as opioid agonists and antagonists, corticosterone, nicotine and dopaminergic agents. For a review see reference 20.

Experiments have recently been carried out in which GABAergic agonists and antagonists were administered in these two strains of mice after training in an inhibitory avoidance task.[21] In these experiments immediately posttraining administration of muscimol or baclofen impaired retention in C57 mice, while improving it in the DBA strain. Further, picrotoxin, bicuculline and CGP 35348 (a $GABA_B$ antagonist) dose-dependently improved retention in the former, and impaired it in the latter strain. The fact that the animals of both strain were not different in sensitivity to shock and to light, as well as in memory consolidation, rules out the possibility that the observed strain-dependent effects of GABA receptor agonists and antagonists could be ascribed to differences in these parameters. Further, the effects of the drugs could not be ascribed to nonspecific actions on retention performance, as the latencies during the retention test of those mice that had not received footshock during the training session, were not affected by the posttraining drug administrations. In addition, the effects of all drugs were time-dependent, since they were not observed when the drugs were injected at long intervals (120 min) after training. In summary, these results indicate that the GABA receptors have an opposite role on memory consolidation in C57 and DBA mice. These opposite actions can be tentatively explained on the basis of different strain-dependent distribution of GABA in discrete brain areas. In fact, as compared with DBA mice, higher GABA concentrations have been shown in amygdala, raphe and hippocampus of C57 mice, while lower concentrations of the neurotransmitter are evident in this strain in the olfactory tubercle and the frontal cortex.[31,105] The different distribution of GABA in the brain could be related to a strain-dependent role of the neurotransmitter in the functions of neural circuits or systems in which other neurotransmitters are also involved. Indeed, a role for GABA in the amygdala, on memory consolidation, has been reported.[15,23] Finally, the results of this study confirm that the effects of GABAergic drugs on memory consolidation are not related to GABA receptors subtypes. In fact, within each strain studied, the two GABAergic agonists (A and B) muscimol and baclofen, produced similar effects on retention and the effects of the two $GABA_A$ receptor antagonists and those of the $GABA_B$ receptor antagonists were also similar within each strain. $GABA_A$ and $GABA_B$ receptors are considered to involve different mechanisms in neurotransmission, since activation of the classical $GABA_A$ receptor directly opens Cl^- selective channels, while activation of $GABA_B$ receptor causes an increase in K^+ conductance or a decrease in Ca^{2+} conductance. Activation of $GABA_A$ or $GABA_B$ receptors leads to fast or slow IPSPs, respectively. For a review see ref. 16. In spite of such differences, in this study,[21] selective type A or type B receptor agonists and antagonists produced, within each strain, similar effects on memory consolidation. This might indicate the existence of a similar functional role of the actions of the two receptor types at the synaptic level.

GABAergic Drugs and the State-Dependency Hypothesis

The results of some studies have suggested that the retrograde amnesia which follows some posttraining treatments may be based on state-dependency.[56] According to this hypothesis, the newly acquired informations may be stored in a state induced by the posttraining treatment, and are not subsequently accessible when retention is assessed while the animals are in a normal state. Several studies have reported that, while posttraining administration of β-endorphin or enkephalin impairs retention of newly learned information, the retention of animals injected after training with these, or other hormones (ACTH, epinephrine, that release brain β-endorphin) is comparable to untreated controls if the hormones are administered prior to the retention test. For review see refs. 24-25. These findings lead to the conclusion that, in some circumstances, retention performance may not reflect the degree to which the brain-state at the time of the retention test is congruent with the state which normally occurs or is induced following training. These findings raise the question of whether retention enhancement or impairment produced by posttraining treatments is due to state-dependency rather than to influence on memory storage processes. For example, in studies of retention enhancement resulting from posttraining picrotoxin administrations, the animals are not given treatments shortly prior to the retention test. Thus, a state-dependent interpretation of the retention enhancement produced by a posttraining treatment would require the ad hoc assumption that the state of the animal which normally occurs at the time of retention test is congruent with the state that was induced following the training. As a consequence, the administration of the same retention-enhancing treatment prior to the retention test should decrease the congruence, thus attenuating the retention enhancement induced by the posttraining treatment.[24] This implication was examined by Castellano and McGaugh.[24] The experiments were carried out with CD1 mice tested in a one-trial step-through inhibitory avoidance task.

In a first set of experiments different groups of animals received i.p. injections of saline or picrotoxin immediately after training, and of saline 3 min prior to the retention test, that was carried out 24 h later. Additional groups of animals were injected with saline or picrotoxin (1 mg/kg) 120 min after training, to determine whether the effect on subsequent retention varied with the training-treatment interval. Further, to assess possible aversive effects of picrotoxin injections, another group did not receive footshock but was injected with picrotoxin (1 mg/kg) immediately posttraining and tested 24 h later. Other groups of mice were injected with picrotoxin 30, 10 or 3 min prior to training in order to determine whether the drug affected response latencies at these times following injections. The results of this set of experiments showed that the retention latencies of mice given picrotoxin (0.5 and 1 mg/kg) posttraining were significantly higher than those of saline controls. Moreover, retention was not affected by picrotoxin administration 120 min after training, showing that the effects exerted by the drug were to a specific action on the time-dependent memory consolidation process. Finally, picrotoxin did not affect either training response latencies when administered prior to training or retention latencies of unshocked controls when administered posttraining, indicating that the effects of picrotoxin on retention performance were not due to nonspecific effects on response latencies.

In a second set of experiments mice were trained in the task and given injections of either saline or picrotoxin (0.5 or 1 mg/kg) immediately afterwards. Most of the groups were then given either saline or picrotoxin (0.5 or 1 mg/kg) 3 min prior to the 24 h retention test. Other groups received either saline or picrotoxin (1 mg/kg) 10 or 30 min prior to the 24 h retention test. The results showed that picrotoxin administered prior to the retention test did not affect the retention performance of mice given saline injections posttraining. Further, in all groups given posttraining picrotoxin, the retention latencies of mice given picrotoxin prior to the retention test were comparable to those of mice given saline prior to the test. Thus these findings argue against a state-dependent interpretation of the effects of picrotoxin on memory and are consistent with those of other studies in supporting the view that picrotoxin enhances retention by modulating memory storage processes.

Further studies[25] have shown that the retention-enhancing effect of posttraining administration of bicuculline, and the retention-impairing effects of posttraining administration of muscimol are not state-dependent. In fact, also in this case, the administration of drugs prior to the retention test did not affect the retention latencies of the animals given saline, or bicuculline or muscimol immediately after training.

GABAergic Drugs and Memory Formation: Administrations into Brain Structures

The possibility that GABAergic agonists and antagonists exert their effects on memory as a consequence of specific actions on central GABAergic receptors has been examined by a number of studies.

The first brain structure considered has been the amygdala. It is known that retention can be modulated by posttraining intra-amygdala injections of drugs affecting several neurotrans-

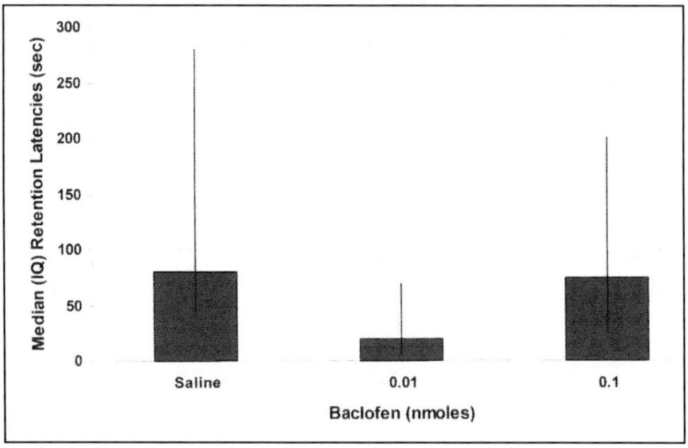

Figure 3. Effect of immediately posttraining intra-amygdala administrations of baclofen on retention of a one-trial step-through inhibitory avoidance task in rats. Data are expressed as the median testing minus training step-through latency (+ interquartile range), in seconds (from ref. 19).

mitters and neuromodulatory systems.[76] Further, GABAergic neurons are widely distributed in this structure and GABAergic cells project mainly within the amygdala.[6,17,67,71,90]

In a study carried out with rats tested in a one-trial step-through inhibitory avoidance task, immediate bilateral intra-amygdala administrations of baclofen impaired retention performance on a 48 h retention test (Fig. 3).[23] Further, bicuculline methiodide (BMI) improved, while muscimol impaired retention (48 h later) in rats tested in a one-trial step-through inhibitory avoidance task and injected into the amygdala with the drugs immediately after training. In particular, the memory enhancing effect of BMI was produced by a dose lower than that necessary to induce convulsions. In a further series of experiments of this study posttraining injections of BMI into the caudate-putamen, a structure which is dorsal to the amygdala, did not affect retention (Figs. 4 and 5).[15] In another study[84] the effects of intra-amygdala infusions of muscimol prior to retention testing was examined. Two sets of experiments were carried out. In the first set, rats were trained in a one-trial step-through inhibitory avoidance task and given bilateral intra-amygdala infusions of vehicle or muscimol, or simultaneous unilateral infusions of each, 5 min before the retention test, 24 h after training. The same procedure was adopted in experiment 2, but two retention measures were taken: initial step-through latency and the number of trials to reach criterion during continuous multiple-trial inhibitory avoidance (CMIA) training. The results showed that infusions of muscimol into the amygdalae prior to the retention test impaired performance in the inhibitory avoidance task. Further, unilateral infusions of muscimol into the right, but not into the left amygdala, prior to the retention test was sufficient to impair retention performance. Although bilateral infusions of muscimol impaired CMIA acquisition, unilateral infusions of the drug into either the right or the left amygdala did not significantly affect the number of trials required to reach criterion during the CMIA acquisition. These results show that the right and the left amygdala are differentially involved in the expression of memory for inhibitory avoidance training and suggest that the different effects observed after unilateral amygdalae infusions of muscimol may depend on the type of task examined.[84] Other experiments, in which muscimol or bicuculline were infused into the amygdala immediately after a reward shift, have shown that GABAergic system is involved in the memory modulation for changes in reward magnitude.[98] Further, facilitation of trace conditioning of odor aversion has been observed following intra-amygdala (basolateral amygdala) bilateral injections of BMI after the presentation of the conditioned stimulus in rats.[43] Long-term memory enhancement in rats tested in a one-trial step-down inhibitory avoidance task and injected immediately after training with picrotoxin in the junction between the central and the

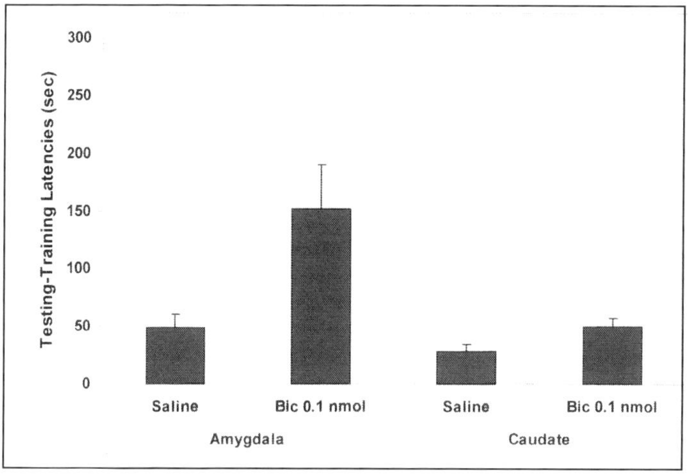

Figure 4. Effect of bicuculline methiodide injections in the amygdala and the caudate-putamen on retention of a one-trial step-through inhibitory avoidance task. The rats with chronic implanted cannulae were trained in the inhibitory avoidance, and were injected immediately after training. Each column represents the mean ± SEM testing minus training latency in seconds (from ref. 15).

basolateral nuclei of amygdala, has also been demonstrated.[7] A series of experiments has been recently carried out in which rats were injected with muscimol into the amygdala immediately after Pavlovian fear conditioning and one-trial inhibitory avoidance tasks.[112] Immediate posttraining infusions of muscimol had no effect on Pavlovian conditioning, but produced a dose-dependent impairing effect in the inhibitory avoidance. However, Pavlovian conditioning was dose-dependently disrupted by pretraining infusions of muscimol. The results of this study indicate that the amygdala plays a critical role in the acquisition of Pavlovian fear conditioning, and is involved in the modulation of memory consolidation of inhibitory avoidance but not of Pavlovian fear conditioning. Further evidence that amygdala is involved in modulating the effects of GABAergic agonists and antagonists on memory comes from lesion studies. Some experiments,[1] carried out with CD1 mice tested in a one-trial step-through inhibitory avoidance task, have shown that bilateral lesions of the amygdala (or the dorsal hippocampus) blocked the memory enhancing effect of posttraining i.p. injections of bicuculline as well as the memory impairing effect of muscimol. In contrast, lesions of the caudate did not influence the retention-modulating effects of posttraining administrations of the two drugs.

Although many studies have demonstrated that systemically administered GABAergic drugs influence memory storage through effects involving the amygdala, it has been shown that learning and memory can be influenced also by injections of GABAergic drugs in brain regions other than this structure. Early studies have for example demonstrated that injections of picrotoxin into the hippocampus[50] and the substantia nigra,[64] and of muscimol into the basal forebrain[90] are followed by retention impairment. It is well known that the processing of spatial information requires an intact hippocampal function, and is sensitive to the disruption of the septal-hippocampal cholinergic pathway. Acquisition improvements of spatial information have been observed in rats after septal lesions, that reduce hippocampal cholinergic activity.[113] Muscimol injections in the medial septum reduce the high affinity choline uptake in the hippocampus. It is interesting to underline that muscimol injections in this structure have been shown to impair, in rats, place navigation in the Morris water maze at the same dose that reduces the high affinity choline uptake in the hippocampus, and without affecting the nonspatial strategies of the animals.[13] In another study,[89] experiments were carried out in which the effects of muscimol administration into the medial septal area prior to training were studied, and

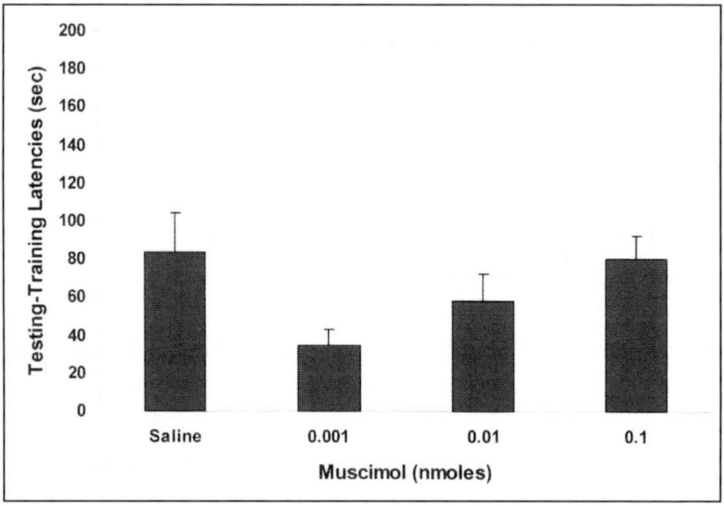

Figure 5. Testing-training latencies to step-through after muscimol injections in the amygdala. Rats with chronic implanted cannulae were trained in a one-trial step-through inhibitory avoidance task and were injected immediately after training. Each column represents the mean ± SEM testing minus training latency in seconds (from ref. 15).

memory was tested at different retention delays in 3 tasks: inhibitory avoidance task, one-trial place learning task and reward alternation task. In all 3 tasks muscimol administration did not impair memory performance at short retention delays, but impaired retention at the longer retention delays. The same authors showed in previous experiments that injections of muscimol into the medial septal area impaired retention performance in an inhibitory avoidance task and in a multiple-trial place learning task, when the injections were carried out pretraining but not posttraining.[13,88] In summary, the findings of Nagahara and McGaugh[89] suggest that the intra-septal muscimol treatment impairs long-term memory by acting on a memory-related process that occurs at the time of training rather than by acting on posttraining consolidation processes. The existence of interactions between the medial septal area and the basolateral amygdala (BLA, which modulates memory encoding in other brain areas including hippocampus) in the processing of memory storage was demonstrated by a recent study.[106] The effects of intra-medial septal infusions of muscimol in rats with BLA lesions were studied. The animals received sham surgery or excitotoxic BLA lesions and were infused with either vehicle or muscimol into the medial septal area 5 min prior to training in a step-through inhibitory avoidance and water maze tasks. The results showed that BLA lesions potentiated the muscimol-induced memory impairment in the inhibitory avoidance task. Further, BLA-lesioned animals given muscimol infusions into the median septal area showed memory impairment also in the water maze task. Taken together, these results give the evidence that BLA interacts with medial septal area in the processing of memory storage.

In a series of experiments the effect of the infusion of muscimol on the role of the entorhinal cortex, amygdala and hippocampus in memory processes was studied[62] in rats bilaterally implanted with cannulae into these structures and trained, after recovery, in a one-trial step-down inhibitory avoidance task. Testing occurred 24 h after training. In particular it has been examined whether the pretraining or pretest intra-entorhinal infusion of muscimol had any influence on the amnesic effect of a posttraining infusion of the same drug into the amygdala and the hippocampus, and on the amnesic effect of cianonitroquinoxaline-dione (CNQX), an antagonist of AMPA glutamatergic receptors, given prior to testing into the amygdala, hippocampus or entorhinal cortex. The results showed that muscimol, infused in the entorhinal cortex 20 min prior to testing, inhibited the amnesic effect of muscimol infused into this area 100 min after training. This result demonstrated that memory-relevant information must be processed by the entorhinal cortex at the time of training in order that this cortex may play a late posttraining role in memory processing. Further: a) pretraining intra-entorhinal muscimol administration did not affect the amnesic effect of the posttraining infusion of muscimol into the amygdala and the hippocampus, or the inhibition of memory expression induced by pretest infusion of CNQX into the amygdala and hippocampus or into the entorhinal cortex; b) pretest intra-entorhinal muscimol administration did not influence the effect of pretest intra-amygdala and intra-hippocampal CNQX administration. According to the authors, these findings indicate that "the cells of the entorhinal cortex that are sensitive to pretraining muscimol are not part of the inputs that lead to posttraining processing by the amygdala and hippocampus or to the intervention of the amygdala, hippocampus and entorhinal cortex in memory expression. Instead, the entorhinal cortex may be an output of the amygdala and hippocampus at the time of memory expression."[62] In a series of experiments,[116] Wistar rats were implanted with cannulae into the CA1 region of the dorsal hippocampus and into the amygdaloid nucleus, the entorhinal cortex and the posterior parietal cortex. The animals were trained in a step-down inhibitory avoidance and received vehicle or muscimol injections 0, 30, 60 or 90 min after training. Retention was measured 24 h after training. The results showed that retention performance was hindered by muscimol administration into both hippocampus and amygdala at 0, but not 30 min posttraining. Furthermore, the drug was amnesic when given into the entorhinal cortex 30, 60 or 90 min after training, or into the parietal cortex only 60 or 90 min after training. The data show a sequential entry in operation, during the posttraining period, of the hippocampus and amygdala, the entorhinal cortex and the posterior parietal cortex in memory processing.

Some experiments have been recently carried out in which the effects of muscimol on memory were studied in rats receiving posttraining intra-hippocampal administrations of muscimol.[110] The results showed that the administration of muscimol into the CA1 region of the dorsal hippocampus impaired long-term memory of habituation to a 5 min exposure to an open field measured 24 h later.

After the study of Brioni et al[15] showing lack of effect following posttraining bicuculline methiodide injections into caudate-putamen of rats tested in an inhibitory avoidance task, some other studies have been carried out in which the effect of posttraining GABAergic antagonists into the striatum have been examined. In a first study[30] the effects of posttraining administrations of picrotoxin and bicuculline were investigated in rats tested in a one-trial step-through inhibitory avoidance task. The results showed that intra-striatal applications of the two GABA antagonists induced dose-dependent and long-lasting impairments of memory consolidation. The discrepancy between these findings and those of Brioni et al[15] might be explained, according to the authors, on the basis of a regional functional heterogeneity of GABA within the structure.[47] It is known that disruption of synaptic activity of some brain structures, including neostriatum and thalamus, is followed by marked deficits in retention of instrumentally conditioned behaviors. Further, when animals in these conditions are given a high number of training trials or high intensities of footshock during training, such disruption is less effective.[38] Thus it has been studied[32] whether, on the basis of the close anatomical and functional relationships between the neostriatum and the substantia nigra, enhanced training with a high level of footshock would prevent the amnesic state induced by picrotoxin and bicuculline infusion into the latter structure. Rats were trained in a step-through inhibitory avoidance task under two footshock levels, 0.2 or 0.4 mA, and were injected posttraining with the GABA antagonists into the substantia nigra and the posterior region of the zona incerta. Retention was measured 24 h later. The results showed amnesia only in the groups injected into the nigra following training under the lower shock intensity. According to the authors, the differences between their findings and previous findings[12,58] showing that the main effect of injections of GABA blockers into amygdala, hippocampus and septum is an impairment of retention in rats tested in the inhibitory avoidance, might be explained on the basis of a dissimilar involvement of limbic and nigrostriatal GABA systems in memory processes, probably due to regional differences in the physiology of GABA. As concerns the lack of effect of GABAergic blockade of the nigra after the high (0.4 mA) intensity of footshock administered during training, it might be related, according to the authors, to central and peripheral events triggered by the increased aversive stimulation. The effects of regional blockade of the striatum on memory consolidation has been examined in some further experiments.[99] In this study rats were trained in a step-through inhibitory avoidance task and received posttraining injections of picrotoxin. Retention test was carried out 24 h later. The data showed that a strong amnesia was produced by picrotoxin injections into the posteroventral and the lateral regions of the striatum, an intermediate degree of impairment was produced by injections into the dorsomedial region, while no retention deficit was evident when the GABA antagonist was injected into the ventromedial part of the anterior striatum. It was thus evident that the retention impairments were higher in the posterior and the lateral striatal regions than in the anterior and medial regions. These results provide strong evidence that the striatal GABAergic activity plays a crucial role in the consolidation of negatively reinforced behaviors. They also demonstrate the existence of a neurochemical heterogeneity within the striatum as concerns memory consolidation and further reflect a differential involvement of limbic and striatonigral GABA in memory processes.

Differential effects of muscimol infusions in different regions of the cingulate cortex on retention have been found.[81] In a series of experiments Wistar rats were bilaterally implanted with cannulae at four different coordinates of the cingulate cortex: 1) the anterior cingulate (AC); 2) the rostral region of the posterior cingulate (RC); 3) the upper portion of the posterior cingulate (UC) and 4) the lower portion of the caudal region of the posterior cingulate (LC). The animals were trained in a step-down inhibitory avoidance task and received infusions of vehicle or muscimol either immediately or 90 or 180 min after training. Muscimol was amnesic when given into any of the three coordinates of the posterior cingulate cortex 90 min after training, and when given into LC immediately posttraining. None of the treatments was effective when given into AC. The results show that the posterior, but not the anterior cingulate cortex regulates memory processing of the inhibitory avoidance task through muscimol-sensitive synapses, relatively late after training.

The involvement of the GABAergic system in the medial precentral prefrontal cortex in memory consolidation has been recently examined.[82] In this study Wistar rats were trained in a step-down inhibitory avoidance learning task and received infusions of muscimol or vehicle into the anterior medial precentral area (Fr2)(CI) or into the junction of Fr1-Fr2 (CII) at different times after training. Muscimol into CI was amnesic when given immediately, 90 or 180 min, but not 270 min after training. When injected into CII muscimol was amnesic when given 90 min, but not 0 or 180 min, after training. The results suggest that the GABAergic system in Fr2 is involved in the consolidation of memory for inhibitory avoidance learning and that timing of involvement of anterior Fr2(CI) is different from that of posterior Fr2(CII).

Interaction with Other Systems

GABA-Opioids Interactions

The involvement of GABAergic mechanisms in the effects of opioids has been demonstrated by a number of studies (for a review see ref. 92). Some findings show that opioid antagonists may act as GABA antagonists. The GABA-induced inhibition of neuronal firing in the olfactory tubercle of rats is antagonized by naloxone; furthermore, this opioid antagonist displaces [^{3}H]GABA from GABA binding sites in rat brain. It has been also shown that picrotoxin- or bicuculline-induced convulsions are enhanced by naloxone.[41,97] Finally, drugs known to facilitate GABA-mediated synaptic inhibition attenuate the rate-decreasing effect of naloxone and picrotoxin on schedule-controlled responding in the pigeon. These observations suggest that this effect of naloxone also is due to antagonism of GABA neurotransmission.[18]

GABAergic mechanisms seem to be also involved in the effects of opioids on memory. In fact some experiments have shown that GABAergic mechanisms are involved in the effects exerted by the opioid antagonists naloxone and naltrexone on memory consolidation in CD1 mice.[23] The experiments have been carried out with animals tested in a one-trial step-through inhibitory avoidance task, and injected i.p. immediately after training with a) naloxone, picrotoxin, bicuculline or muscimol, b) combinations of naloxone (or naltrexone) with each of the GABAergic drugs. Retention was measured 24 h after training. The results showed that: a) the two GABAergic antagonists, picrotoxin or bicuculline, enhanced, while the GABAergic agonist muscimol impaired retention performance; b) the two GABAergic antagonists enhanced, while muscimol attenuated the effects of naloxone and naltrexone on retention. These results suggest that naloxone and naltrexone may influence memory consolidation by interacting with the GABAergic system and by acting as GABAergic antagonists also when their effects on memory are considered. It should be also noted that the effects of opioid antagonists on memory consolidation involve both cholinergic and catecholaminergic mechanisms.[5,45,55,61] Since GABA interacts with cholinergic as well as catecholaminergic systems in the brain,[66] these results suggest that it may also interact with these neurotransmitter systems in the modulation of memory. These findings also suggest that opioid peptides may affect memory storage through GABAergic mechanisms.

The interaction of β-endorphin and GABAergic drugs in the regulation of memory storage has been demonstrated by a further study.[22] Male CD1 mice were trained in a one-trial step-through inhibitory avoidance task and tested 24 h later for retention. Different groups of animals were injected i.p., immediately after training, with β-endorphin, picrotoxin, bicuculline or muscimol, or with combinations of each of the GABAergic drugs with the opioid. The results showed that β-endorphin and muscimol produced a dose-dependent impairment of retention, while the two GABA antagonists enhanced it. Further a low subeffective dose of picrotoxin or bicuculline attenuated the retention-impairing effect of β-endorphin. These findings indicate that β-endorphin influences memory consolidation through an interaction with GABAergic mechanisms. The results of this study can be further considered in the light of previous studies, in which the retention modulating effects of posttraining administrations of drugs affecting opioid peptide and GABAergic systems were attenuated by drugs affecting noradrenergic and cholinergic systems. It has been for example shown that the β-adrenergic receptor blocker propranolol and the muscarinic cholinergic antagonist atropine block the memory-enhancing effects of both naloxone and bicuculline.[52,54,61] Moreover, it has been shown that the adrenergic agonist clenbuterol and the cholinergic agonist oxotremorine attenuate the memory impairing effects of β-endorphin and muscimol.[26,52,55] These data suggest that both GABAergic and opioid influences on memory storage are mediated by influences on noradrenergic and cholinergic systems. It must be underlined that some studies suggest that the noradrenergic influences on memory are mediated by activation of cholinergic neurons.[53] Thus the data obtained concerning the interactions of GABAergic, opioid, adrenergic and cholinergic systems in regulating memory storage suggest that opioid and GABAergic drugs influence memory through a common action on the release of norepinephrine (NE) that, in turn affects acetylcholine (ACh) release (for details on these transmitter systems, see relevant chapters in this book). As concerns the release of NE, it has been shown that opioids inhibit it.[2,85,111] Some studies have also shown that activation of $GABA_A$ receptors induces NE release.[8,103]

Further, other experiments[79] have demonstrated that the memory modulating effects of drugs affecting GABAergic, opioid peptidergic, adrenergic and cholinergic systems are mediated, at least in part, by interactions occurring within the amygdaloid complex. As concerns this point, GABAergic effects on NE release within the amygdala have been recently examined[51] by using in vivo microdialysis and high-performance liquid chromatography (HPLC). In particular the levels of NE within the amygdala were assessed in response to both footshock and GABAergic compounds. It was observed that a 0.55 mA footshock induced a significant increase in NE levels when the microdialysis probe was located within the amygdala. In a further series of experiments rats injected systemically (i.p.) with picrotoxin showed increase in levels of NE within the structure, while systemic (i.p.) injection of muscimol resulted in decreased levels of this neurotransmitter. These data demonstrate that drugs that are capable of modulating memory do so by altering NE levels within the amygdala. These results are supported by previous evidence that amygdala is a critical site for integrating neuromodulatory systems which ultimately influence memory storage through the release of NE within this structure,[80] and by biochemical and pharmacological findings suggesting that treatments that induce the release of NE in the amygdala enhance memory retention whereas treatments that decrease NE release impair retention.[51]

GABA-Benzodiazepines Interactions

It is known that the systemic injections of benzodiazepines induce anterograde amnesia, and that amygdala is involved in the mediation of benzodiazepine-induced memory impairments.[69,109] Some studies have demonstrated that GABAergic mechanisms are involved in the effects of benzodiazepines on memory. GABAergic antagonists block, while GABAergic agonists potentiate, the memory impairing effects of systemically administered benzodiazepines in mice tested in one-trial inhibitory avoidance condition.[87,108] Furthermore, the amnesic effect of intra-amygdala injections of the $GABA_A$ agonist muscimol are blocked by systemic injections of the benzodiazepine antagonist flumazenil.[57] Evidence that regional brain levels of benzodiazepine-like molecules and binding sites changes in the amygdala (as well as in hippocampus and septum) following training on an inhibitory avoidance task provides further support for the view that the amygdaloid complex may be a site at which endogenous benzodiazepines act in modulating memory storage.[114]

The possibility that intra-amygdala administrations of a GABAergic antagonist block the benzodiazepine effect was examined in some studies. In a first study[39] Sprague-Dawley rats were tested in a continuous multiple trial inhibitory avoidance. The animals were implanted bilaterally with cannulae aimed at the amygdala. One week later, ten minutes before training, bicuculline methiodide (BMI) or 10 mM phosphate-buffered saline were infused into the structure. Immediately following the intra-amygdala injections midazolam or saline were injected intraperitoneally. The results showed that intra-amygdala injections of BMI blocked the amnesic effect of systemically administered midazolam on retention. In fact, on a 48 h retention test, the performance of the midazolam-injected animals was significantly poorer than that of the controls, but the retention performance of animals given intra-amygdala injections of bicuculline prior to the systemic injection of midazolam was comparable to that of the controls. These results provide strong support for the view that benzodiazepine induced retention impairment of inhibitory avoidance is mediated by GABAergic activation in the amygdaloid complex. In another study,[40] Sprague-Dawley rats were injected i.p. with either midazolam or vehicle 10 min before training on a multiple-trial inhibitory avoidance task. Immediately after training BMI or vehicle were infused bilaterally into the amygdala. Also in this case the performance of the midazolam-injected animals was significantly poorer than that of controls, on a 48 h retention test. Furthermore, the retention of the midazolam-treated animals given intra-amygdala injections of bicuculline (2.0 pmol / 0.5 _l) was comparable to that of controls. These results provide strong support for the view that posttraining GABAergic activation in the amygdaloid complex mediates benzodiazepines-induced deficits of inhibitory avoidance.

GABA-Ethanol Interactions

GABAergic involvement in the central effects of ethanol has been demonstrated by a number of studies.[37,44,68,70] Acute ethanol administration increases brain GABA content in rats and mice.[29,95,107] Furthermore, in cats, potentiation of the inhibition of cortical neurons by GABA has been shown following ethanol treatment.[91] Finally, ethanol inhibits, through a GABAergic mechanism, the firing of pars reticulata neurons in rats.[83]

On the basis of these findings, Castellano and Pavone[28] investigated the possible involvement of the GABAergic system in ethanol-induced amnesia.[4,27,70] The experiments were carried out with CD1 mice trained in a one-trial step-through inhibitory avoidance task and injected immediately after training with either ethanol or the GABAergic drugs picrotoxin, bicuculline or muscimol alone or in combination with ethanol. The results showed that the retention performance of the animals was impaired by ethanol and muscimol, while it was enhanced by picrotoxin and bicuculline. Furthermore, the ethanol-induced reduction of retention performance was potentiated by the $GABA_A$ agonist, and attenuated by the two $GABA_A$ antagonists. Taken together, the findings confirm the involvement of GABAergic mechanisms in memory consolidation, and demonstrate that this system is involved in the negative effect exerted by ethanol on retention of mice tested in a one-trial inhibitory avoidance task.

Some studies have shown that the major actions of ethanol involve enhancement of the effects of GABA at $GABA_A$ receptors and blockade of the NMDA subtype of excitatory aminoacids (EAA) receptor. Furthermore, tolerance to ethanol results in enhanced EAA neurotransmission and NMDA receptor upregulation, which appear to involve selective increases in NMDA R2B subunit levels and other molecular changes in specific brain loci. (For a review see ref. 42) It has been also shown that excitatory aminoacid antagonists block the cardiovascular and anxiety responses elicited by $GABA_A$ receptor blockade in the basolateral amygdala of rats.[98,102]

There is little cognitive evidence of interaction between NMDA and $GABA_A$ receptors. Some experiments have been carried out, in rats trained on a one-trial step-down inhibitory avoidance, which received immediately after training intra-amygdala, intra-septal or intra-hippocampal injections of drugs. In these experiments picrotoxin counteracted the amnesic effect of the competitive NMDA receptor antagonist AP5.[58] In a recent study,[3] the effect of the noncompetitive NMDA receptor antagonist MK-801 and ethanol combinations on memory consolidation, and the involvement of GABAergic mechanisms in this effect were investigated in CD1 mice injected i.p. with the drugs immediately after training in a one-trial step-through inhibitory avoidance task, and tested for retention 24 h later. The results showed that: a) the retention performances of mice were dose-dependently impaired by immediate posttraining MK-801 (0.2 and 0.3, but not 0.1 mg/kg) and ethanol (1 and 2, but not 0.5 g/kg) administrations; b) an otherwise ineffective dose of MK-801 (0.1 mg/kg) enhanced the deleterious effect exerted by ethanol (1 and 2 g/kg); c) an otherwise ineffective dose of muscimol (0.5 mg/kg) enhanced, while otherwise ineffective doses of picrotoxin (0.25 mg/kg) or bicuculline (0.1 mg/kg) antagonized this effect; d) no effect was observed when the treatments were carried out 120 min after training, suggesting that the effects observed following immediate posttraining administrations were due to influences on the consolidation of memory. From these experiments it is evident that: a) MK-801 enhances ethanol's effects on memory consolidation and b) GABAergic mechanisms are involved in this effect.

Taken together, these results suggest the existence of an interaction between $GABA_A$ and NMDA receptors and ethanol in modulating the consolidation of memory.

References

1. Ammassari-Teule M, Pavone F, Castellano C et al. Amygdala and dorsal hippocampus lesions block the effects of GABAergic drugs on memory storage. Brain Res 1991; 551:104-109.
2. Arbilla S, Langer SZ. Morphine and beta-endorphin inhibit the release of noradrenaline from cerebral cortex but not of dopamine from rat striatum. Nature 1978; 271:559-561.
3. Aversano M, Ciamei A, Cestari V et al. Effects of MK-801 and ethanol combinations on memory consolidation in CD1 mice: Involvement of GABAergic mechanisms. Neurobiol Learn Mem 2002; 77:327-337
4. Bammer G, Chesher GB. An analysis of some effects of ethanol on performance in a passive avoidance task. Psychopharmacol 1982; 77:66-73.
5. Baratti CM, Introini IB, Huygens P. Possible interaction between central cholinergic muscarinic and opioid peptidergic systems during memory consolidation in mice. Behav Neural Biol 1984; 40:155-169.

6. Ben-Ari Y, Kanazawa YI, Zigmond RE. Regional distribution of glutamate decarboxylase and GABA within the amygdaloid complex and stria terminalis system of the rat. J Neurochem 1976; 26:1278-1283.
7. Bianchin M, Mello e Souza T, Medina JH et al. The amygdala is involved in the modulation of long-term memory but not in working or short-term memory. Neurobiol Learn Mem 1999; 71:127-131.
8. Bonanno G, Raiteri M. Release-regulating $GABA_A$ receptors are present on noradrenergic nerve terminals in selective areas of the rat brain. Synapse 1987; 1:254-257.
9. Bovet D, McGaugh JL, Oliverio A. Effects of posttrial administration of drugs on avoidance learning in mice. Life Sci 1966; 5:1309-1315.
10. Bowery NG. Baclofen: 10 years on. Trend Pharmacol Sci 1982; 3:400-403.
11. Breen RA, McGaugh JL. Facilitation of maze learning with posttrial injections of picrotoxin. J Comp Physiol Psychol 1961; 54:498-501.
12. Brioni JD. Role of GABA during multiple consolidation memory. Drug Dev Res 1993; 28:3-27.
13. Brioni JD, Decker MW, Gamboa LP et al. Muscimol injections in the medial septum impair spatial learning. Brain Res 1990; 522:227-234.
14. Brioni JD, McGaugh JL. Post-training administration of GABAergic antagonists enhances retention of aversively motivated tasks. Psychopharmacol 1988; 96:505-510.
15. Brioni JD, Nagahara AH, McGaugh JL. Involvement of the amygdala GABAergic system in the modulation of memory storage. Brain Res 1989; 487:105-112.
16. Cage PW. Activation and modulation of neural K^+ channels by GABA. TINS 1992; 15:46-51.
17. Carlsen J. Immunocytochemical localization of glutamate decarboxylase in the rat basolateral amygdala nucleus, with special reference to GABAergic innervation of amygdalostriatal projection neurons. J Comp Neurol 1988; 273:513-526.
18. Carter RB, Leander JD. Comparison of the effects of naloxone on schedule-controlled responding in the pigeon: Possible GABA-antagonistic effects of naloxone. J Pharm Exp Ther 1984; 230:40-46.
19. Castellano C, Brioni JD, Nagahara AH et al. Post-training systemic and intra-amygdala administration of the $GABA_B$ agonist baclofen impairs retention. Behav Neural Biol 1989; 52:170-179.
20. Castellano C, Cabib S, Puglisi-Allegra S. Psychopharmacology of memory modulation: Evidence for multiple interaction among neurotransmitters and hormones. Behav Brain Res 1996; 77:1-21.
21. Castellano C, Cestari V, Cabib S et al. Strain-dependent effects of post-training GABA receptors agonists and antagonists on memory storage in mice. Psychopharmacology 1993; 111:134-138.
22. Castellano C, Introini-Collison IB, McGaugh JL. Interaction of β-endorphin and GABAergic drugs in the regulation of memory storage. Behav Neural Biol 1993; 60:123-128.
23. Castellano C, Introini-Collison IB, Pavone F et al. Effects of naloxone and naltrexone on memory consolidation in CD1 mice: Involvement of GABAergic mechanisms. Pharmacol Biochem Behav 1989; 32:563-567.
24. Castellano C, McGaugh JL. Retention enhancement with posttraining picrotoxin: Lack of state-dependency. Behav Neural Biol 1989; 51:165-170.
25. Castellano C, McGaugh JL. Effects of post-training bicuculline and muscimol on retention: Lack of state-dependency. Behav Neural Biol 1990; 54:156-164.
26. Castellano C, McGaugh JL. Oxotremorine attenuates retrograde amnesia induced by post-training administration of the GABAergic agonists muscimol and baclofen. Behav Neural Biol 1991; 56:25-31.
27. Castellano C, Pavone F. Naloxone reversible effects of ethanol on passive avoidance behavior in mice. Physiol Psychol 1984; 11:291-295.
28. Castellano C, Pavone F. Effects of ethanol on passive avoidance behavior in the mouse: Involvement of GABAergic mechanisms. Pharmacol Biochem Behav 1988; 29:321-324.
29. Chan AWK. Gamma aminobutyric acid in different strains of mice. Life Sci 1976; 19:597-604.
30. Chàvez ME, Salado-Castillo R, Sanchez-Alvarez M et al. Posttraining injection of GABAergic antagonists into the striatum produces retrograde amnesia. Neurobiol Learn Mem 1995; 63:296-300.
31. Clement J, Simler S, Ciesielski L, Mandel P et al. Age-dependent changes of brain GABA levels, turnover rates and shock-induced aggressive behavior in inbred strains of mice. Pharmacol Biochem Behav 1987; 26:83-88.
32. Cobos-Zapiain GG, Salado-Castillo R, Sanchez-Alvarez M et al. High level of footshock during inhibitory avoidance training prevents amnesia induced by intranigral injection of GABA antagonists. Neurobiol Learn Mem 1996; 65:202-206.
33. Coleman-Mesches K, McGaugh JL. Muscimol injected into the right or left amygdaloid complex differentially affects retention performance following aversively motivated training. Brain Res 1995; 676:183-188.
34. Costa E, Guidotti A. Molecular mechanisms in the receptor action of benzodiazepines. Ann Rev Pharmacol Toxicol 1979; 19:531-545.

35. Curtis DR, Duggan AW, Felix GA et al. GABA, bicuculline and central inhibition. Nature 1970; 226:1222-1224.
36. Curtis DR, Johnson GAR, Game CJA et al. Central action of bicuculline. J Neurochem 1974; 23:605-606.
37. Dar MS, Wools WR. GABA mediation of the central effects of acute and chronic ethanol in mice. Pharmacol Biochem Behav 1985; 22:77-84.
38. Del Guante MA, Rivas-Arancibia S, Quirarte G et al. Over-reinforcement protects against memory deficits induced by muscarinic blockade of the striatum. Bol Est Med Biol Mexico 1990; 38:49-53.
39. Dickinson-Anson H, Coleman K, Mesches MH et al. Bicuculline administered into the amygdala blocks benzodiazepine-induced amnesia. Behav Neural Biol 1993; 60:1-4
40. Dickinson-Anson H, McGaugh JL. Bicuculline administered into the amygdala after training blocks benzodiazepine-induced amnesia. Brain Res 1997; 752:197-202
41. Dingledine R, Iversen LL, Breuker E. Naloxone as a GABA antagonist: Evidence from iontophoretic, receptor binding and convulsant studies. Eur J Pharmacol 1978; 47:19-27.
42. Faingold CL, N'Gouemo P, Riaz A. Ethanol and neurotransmitter interaction. From molecular to integrative effects. Prog Neurobiol 1998; 55:509-535.
43. Ferry B, Di Scala G. Bicuculline administration into basolateral amygdala facilitates trace conditioning of odor aversion in the rat. Neurobiol Learn Mem 1997; 67:80-83.
44. Frye GD, Breese GR. GABAergic modulation of ethanol-induced motor impairment. J Pharmacol Exp Ther 1982; 223:750-755.
45. Gallagher M, Rapp PR, Fanelli RJ. Opiate antagonist facilitation of time-dependent memory processes: Dependence upon intact norepinephrine function. Brain Res 1985; 347:284-290.
46. Garg M, Holland HC. Consolidation and maze learning: The effects of posttrial injections of a stimulant drug (picrotoxin). Psychopharmacologia 1968; 12:96-103.
47. Girault JA, Spampinato U, Savaki HE et al. In vivo release of [^{3}H]gamma-aminobutyric acid in the rat neostriatum. I. Characterization and topographical heterogeneity of the effects of dopaminergic and cholinergic agents. Neurosci 1986; 19:1101-1108.
48. Gold PE, Van Buskirk R. Enhancement and impairment of memory processes with post-trial injections of adrenocorticothropic hormone. Behav Biol 1976; 16:387-400.
49. Gold PE, Van Buskirk R. Posttraining brain norepinephrine concentration: Correlation with retention performance of avoidance training with peripheral epinephrine modulation of memory processing. Behav Biol 1978; 23:509-520.
50. Grecksch G, Matthies H. Differential effects of intrahippocampically or sistemically applied picrotoxin on memory consolidation in rats. Pharmacol Biochem Behav 1981; 14:613-616.
51. Hatfield T, Spanis C, McGaugh JL. Response of amygdalar norepinephrine to footshock and GABAergic drugs using in vivo microdialysis and HPLC. Brain Res 1999; 835:340-345.
52. Introini IB, Baratti CM. The impairment of retention induced by beta-endorphin in mice may be mediated by reduction of central cholinergic activity. Behav Neural Biol 1984; 41:152-163.
53. Introini-Collison IB, Baratti CM. Memory-modulatory effects of centrally acting noradrenergic drugs: Possible involvement of brain cholinergic mechanisms. Behav Neural Biol 1992; 57:248-255.
54. Introini-Collison IB, Castellano C, McGaugh JL. Interaction of GABAergic and ?-noradrenergic drugs in the regulation of memory storage. Behav Neural Biol 1994; 61:150-155.
55. Introini-Collison IB, McGaugh JL. Naloxone and ?-endorphin alter the effects of posttraining epinephrine on retention of an inhibitory avoidance response. Psychopharmacology 1987; 92:229-235.
56. Izquierdo I. Endogenous state dependency: Memory depends on the relation between the neurochemical and hormonal states present after training and at the time of testing. In: Lynch G, McGaugh JL, Weinberger NM, eds. Neurobiology of learning and memory. New York: Guilford, 1984:330-350.
57. Izquierdo I, Da Cunha C, Huang CH et al. Post-training down-regulation of memory consolidation by $GABA_A$ mechanism in the amygdala modulated by endogenous benzodiazepines. Behav Neural Biol 1990; 54:105-109.
58. Izquierdo I, Da Cunha C, Rosat R et al. Neurotransmitter receptors involved in post-training memory processing by the amygdala, medial septum, and hippocampus of the rat. Behav Neural Biol 1992; 58:16-26.
59. Izquierdo I, Diaz RD. Involvement of α-adrenergic receptors in the amnestic and anti-amnestic action of ACTH, β-endorphin and epinephrine. Psychoneuroendocrinol 1984; 9:77-81.
60. Izquierdo I, Diaz RD. Influence on memory of posttraining or pretest injections of ACTH, vasopressin, epinephrine, and ?-endorphin and their interaction with naloxone. Psychoneuroendocrinol 1985; 10:165-172.
61. Izquierdo I, Graudenz M Memory facilitation by naloxone is due to release of dopaminergic and beta-adrenergic systems from tonic inhibition. Psychopharmacology 1980; 67:265-268.

62. Jerusalinsky D, Quillfeldt JA, Walz R et al. Effect of the infusion of the $GABA_A$ receptor agonist muscimol, on the role of the entorhinal cortex, amygdala and hippocampus in memory processes. Behav Neural Biol 1994; 61:132-138.
63. Kardos J, Blasko G, Kerekes P et al. Inhibition of ^{3}H-GABA binding by bicuculline related alkaloids. Biochem Pharmacol 1984; 33:3537-3545.
64. Kim HA, Routtenberg A .Retention disruptions following picrotoxin injections into the substantia nigra. Brain Res 1976; 113:620-625.
65. Krnjevic K, Schwartz S. The action of ?-aminobutyric acid on cortical neurons. Exp Brain Res 1967; 3:320-336.
66. Krogsgaard-Larsen P, Scheel-Kruger J, Kofod H. Pharmacological, biochemical and pharmacological aspects. New York: Academic Press, 1979.
67. Le Gal La Salle G, Paxinos G, Emson P et al. Neurochemical mapping of GABAergic system in the amygdaloid complex and bed nucleus of stria terminalis. Brain Res 1978; 155:397-403.
68. Liljequist S, Engel J. Effects of GABAergic agonists and antagonists on various ethanol-induced behavioral changes. Psychopharmacol 1982; 78:71-75.
69. Lister RG. The amnesic action of benzodiazepines in man. Neurosci Biobehav Rev 1985; 9:87-94.
70. MacInnes JW, Uphouse LL. Effects of alcohol on acquisition and retention of passive avoidance conditioning in different mouse strains. J Comp Physiol Psychol 1973; 84:398-402.
71. McDonald AJ. Immunohistochemical identification of gamma-aminobutyric acid-containing neurons in the rat basolateral amygdala. Neurosci Lett 1985; 53:203-207.
72. McGaugh JL. Time-dependent processes in memory storage. Science 1966; 153:1351-1358.
73. McGaugh JL. Drug facilitation of learning and memory. Ann Rev Pharmacol 1973; 13:229-241.
74. McGaugh JL. Hormonal influences on memory. Ann Rev Psychol 1983; 34:297-323.
75. McGaugh JL. Involvement of hormonal and neuromodulatory systems in the regulation of memory storage. Ann Rev Neurosci 1989a; 12:255-287.
76. McGaugh JL. Modulation of memory storage processes. In: Kelley CM, Stephens BR, eds. Memory: Interdisciplinary approaches New York: Springer-Verlag, 1989b:33-64
77. McGaugh JL. Memory. A century of consolidation. Science 2000; 287:248-251.
78. McGaugh JL, Castellano C, Brioni J. Picrotoxin enhances latent extinction of conditioned fear. Behav Neurosci 1990; 104:264-267.
79. McGaugh JL, Introini-Collison IB, Cahill L et al. Involvement of the amygdala in neuromodulatory influences on memory storage. In: Aggleton J, ed. The amygdala. New York: Wiley-Liss, 1992:201-215
80. McGaugh JL, Introini-Collison IB, Cahill LF et al. Neuromodulatory systems and memory storage. Behav Brain Res 1993; 9:81-90.
81. Mello e Souza T, Roesler R, Madruga M et al. Differential effects of posttraining muscimol and AP5 infusions in different regions of the cingulate cortex on retention for inhibitory avoidance in rats. Neurobiol Learn Mem 1999; 72:118-127.
82. Mello e Souza T, Vianna MR, Rodrigues C et al. Involvement of the medial precentral prefrontal cortex in memory consolidation for inhibitory avoidance in rats. Pharmacol Biochem Behav 2000; 66:615-622.
83. Mereu G, Gessa GL. Low doses of ethanol inhibit the firing of neurons in the substantia nigra, pars reticulata: A GABAergic effect? Brain Res 1985; 360:325-330.
84. Mesches KC, McGaugh JL. Muscimol injected into the right or left amygdaloid complex differentially effects retention performance following aversively motivated training. Brain Res 1995; 676:183-188.
85. Montel H, Starke K, Weber F. Influence of morphine and naloxone on the release of noradrenaline from rat brain cortex slices. Naunyn-Schmied Arch Pharmacol 1974; 283:357-369.
86. Nabeshima T, Noda Y. GABAergic modulation of memory with regard to passive avoidance and conditioned suppression in mice. Psychopharmacology 1988; 94:69-73.
87. Nabeshima T, Tohyama K, Ichihara K et al. Effects of benzodiazepines on passive avoidance response and latent learning in mice: Relationship to benzodiazepines receptors and the cholinergic neuronal system. J Pharm Exp Ther 1990; 255:789-794.
88. Nagahara AH, Brioni JD, McGaugh JL. Effects of intraseptal infusions of muscimol on inhibitory avoidance and spatial learning: Differential effects of pretraining and posttraining administration. Psychobiology 1992; 20:198-204.
89. Nagahara AH, McGaugh JL. Muscimol infused into the medial septal area impairs long-term memory but not short-term memory in inhibitory avoidance, water maze place learning and rewarded alternations tasks. Brain Res 1992; 591:54-61.
90. Nagai T, McGeer PL, McGeer EG. Distribution of GABA-T-intensive neurons in the rat forebrain and midbrain. J Comp Neurol 1983; 218:220-238.

91. Nestoros JN. Ethanol specifically potentiates GABA-mediated neurotransmission in feline cerebral cortex. Science 1980; 209:708-710.
92. Oliverio A, Castellano C, Puglisi-Allegra S. Psychobiology of opioids. Int Rev Neurobiol 1984; 25:277-337.
93. Olsen RW. GABA-benzodiazepine-barbiturate receptor interactions. J Neurochem 1981; 37:1-13.
94. Olsen RW, Ticku MK, Miller T. Dihydropicrotoxin in binding to crayfish muscle sites probably related to gamma-aminobutyric acid receptor-ionophores. Mol Pharmacol 1978; 14:381-390.
95. Rawat AK. Brain levels and turnover rates of presumptive neurotransmitters as influenced by administration and withdrawal of ethanol in mice. J Neurochem 1974; 22:915-922.
96. Roberts E. GABA: The road to neurotransmitter status. In: Olsen RW, Venter JC, eds. Benzodiazepine/GABA receptors and chloride channels: Structural and functional properties. New York: Liss, 1986:1-39
97. Sagratella S, Massotti M. Convulsant and anticonvulsant effects of opioids: Relationship to GABA-mediated transmission. Neuropharmacol 1982; 21:991-1000.
98. Sajdyk TJK, Shekkar A. Excitatory amino acid receptor antagonists block the cardiovascular and anxiety responses elicited by gamma-aminobutyric acid A receptor blockade in the basolateral amygdala of rats. J Pharmacol Exp Ther 1997; 283:969-977.
99. Salado-Castillo R, Diaz del Guante MA, Alvaredo R et al. Effects of regional GABAergic blockade of the striatum on memory consolidation. Neurobiol Learn Mem 1996; 66:102-108.
100. Salinas JA, McGaugh JL. Muscimol induces retrograde amnesia for changes in reward magnitude. Neurobiol Learn Mem 1995; 63:277-285.
101. Salinas JA, McGaugh JL. The amygdala modulates memory for changes in reward magnitude: Involvement of the amygdaloid GABAergic system. Behav Brain Res 1996; 80:87-98.
102. Saltis RP, Cook JC, Gregg AE et al. Interaction of GABA and excitatory amino acids in the basolateral amygdala: Role in cardiovascular regulation. J Neurosci 1997; 17:9367-9374.
103. Schneider D, Schlicker E, Malinowska B et al. Noradrenaline release in the rat vena cava is inhibited by ?-aminobutyric acid via $GABA_B$ receptors but not affected by histamine. Br J Pharmacol 1991; 104:478-482.
104. Schwartzwelder HS, Tilson HA, McLamb WA et al. Baclofen disrupts passive avoidance retention in rats. Psychopharmacology 1987; 92:398-401.
105. Simler S, Puglisi-Allegra S, Mandel P. ?-aminobutyric acid in brain areas of isolated aggressive and nonaggressive inbred strains of mice. Pharmacol Biochem Behav 1982; 16:57-71.
106. Spanis CW, Bianchin MM, Izquierdo I et al. Excitotoxic basolateral amygdala lesions potentiate the memory impairment effect of muscimol injected into the medial septal area. Brain Res 1999; 816:329-336.
107. Sytinsky IA, Guzikov BM, Gomanko MV et al. the gamma-aminobutyric acid (GABA) system in brain during acute and chronic ethanol intoxication. J Neurochem 1975; 25:43-48.
108. Tohyama K, Nabeshima T, Ichihara K et al. Involvement of GABAergic systems in benzodiazepine-induced impairment of passive avoidance learning in mice. Psychopharmacol 1991; 105:22-26.
109. Tomaz C, Dickinson-Anson H, McGaugh JL. Basolateral amygdala lesions block diazepam-induced anterograde amnesia in an inhibitory avoidance task. Proc Natl Acad Sci USA 1992; 89:3615-3619.
110. Vianna MR, Alonso M, Viala H et al. Role of hippocampal signalling pathways in long term memory formation of a nonassociative learning task in the rat. Learn Mem 2000; 7:333-340.
111. Warling LL, Brown SR, Cox BM. Opioid receptor regulation of the release of norepinephrine in brain. Neuropharmacol 1987; 26:987-996.
112. Wilensky AE, Schafe GE, LeDoux JE. The amygdala modulates memory consolidation of fear-motivated inhibitory avoidance learning but not classical fear conditioning. J Neurosci 2000; 20:7059-7066.
113. Winson J. Loss of hippocampal theta rhythm results in partial memory deficit in the rat. Science 1978; 201:160-163.
114. Wolfman C, Da Cunha C, Jerusalinsky D et al. Habituation and inhibitory avoidance training alter brain regional levels of benzodiazepines-like molecules and are affected by intracerebral flumazenil microinjection. Brain Res 1991; 548:74-80.
115. Yonkov DL, Georgiev VP. Memory effects of GABAergic antagonists in rats trained with two active avoidance tasks. Acta Physiol Pharmacol Bulg 1985; 11:44-49.
116. Zanatta MS, Quillfeldt JH, Schaeffer E et al. Involvement of the hippocampus, amygdala, entorhinal cortex and posterior parietal cortex in memory consolidation. Braz J Med Biol Res 1997; 30:235-240.
117. Zukin SR, Young AB, Snyder SH. Gamma-aminobutyric acid binding to receptor sites in the rat central nervous system. Proc Natl Acad Sci USA 1974; 71:4802-4807.

Chapter 2.3

Acetylcholine: I. Muscarinic Receptors

Giancarlo Pepeu and Maria Grazia Giovannini

Abstract

The study of brain muscarinic receptors began more than a century ago, long before the existence of muscarinic receptors was postulated and then demonstrated. However, the effects of drugs acting on these receptors, such as atropine, eserine, pilocarpine and arecoline had been studied much earlier. This chapter is an overview of these studies with the purpose of defining the roles that different subtypes of muscarinic receptors play in the cognitive process. Background information on the anatomy of the brain cholinergic system, the muscarinic receptor subtypes, their transduction mechanisms and their distribution are discussed. The distribution of the receptors is influenced by behavior, age and Alzheimer's disease (AD). Three approaches have been used in order to understand which cognitive processes depend on the activation of muscarinic receptors: i) blockade of the receptors, ii) lesions of the cholinergic pathways and iii) the attempt to correlate the cognitive process with changes in cholinergic neuron activity. From these studies, it may be concluded that muscarinic receptors appear to represent a widespread target system through which acetylcholine, released from the cholinergic network, improves memory performance by augmenting the selectivity of perceptual processing during encoding. Finally, the possibility to correct the cognitive deficits accompanying AD and aging by acting directly or indirectly on muscarinic receptors is examined.

Introduction

It has been known for centuries that the berries, roots and leaves of *Atropa belladonna* and other plants of the Solanaceous family cause disturbances in memory and orientation and, increasing the doses, cause hallucination and delirium.[71] For a long time, the latter effects attracted more attention that the subtle effects on memory.[113] However, the retrograde amnesia following atropine intoxication was noted and reported also in the past. According to Muccioli,[134] patients poisoned with atropine become "idiots". It is interesting that in its "Toxicology" handbook this author already recommended pilocarpine, for its actions on the vagus, as a good antidote for atropine intoxication.

It took a long time to demonstrate that the amnesic effects of atropine and scopolamine, the active ingredient contained in *Atropa belladonna*, *Hyosciamus niger*, and *Datura stramonium*, depended on their blockade of the actions of brain acetylcholine on muscarinic receptors. The first systematic investigation of the effects of belladonna alkaloids on cognition was made by Macht.[114] He reported that after a dose as low as 0.05 mg the rats were unable to solve a familiar maze. No neurochemical mechanism was proposed for this effect, but the author was fully aware that atropine and scopolamine exerted parasympatholytic activity. The discovery of acetylcholine (ACh) in the brain by Chang and Gaddum,[33] the finding that its levels in the brain increased during excitation and decreased during sedation and anaesthesia[63,189] and the observations that anticholinergic agents exerted an amnesic effects[46] all led to the concept that brain ACh may be involved in behavior.[30]

From Messengers to Molecules: Memories Are Made of These, edited by Gernot Riedel and Bettina Platt.

Other steps, however, were necessary before establishing a connection between the brain cholinergic system and its muscarinic receptors with learning and memory.[143] MacIntosh and Obering[115] were the first to demonstrate an outflow of ACh from the cerebral cortex, provided that the superfusing fluid contained an inhibitor of cholinesterase. The cortical cup used in their experiments was the forerunner of the microdialysis technique making it now possible to study ACh release in behaving rats. Besides showing that CNS depressants reduced ACh outflow and CNS stimulants increased it, the cortical cup experiments demonstrated that the stimulation of the reticular formation was accompanied by an increase in ACh release from the cerebral cortex, associated with an electrocortical activation. (For an extensive review of the literature see Pepeu.[147] In the same years, Shute and Lewis[175,176] published the first map of the cholinergic pathways in the rat brain. The map was obtained by staining the cholinesterase contained in the cholinergic neurons and fibres; it was still approximate and lacked many details. However, its general outlay was correct, as it will be shown later, and demonstrated that in the brain a network of cholinergic neurons existed from which ACh was presumably released during their activity.

Another important step towards the demonstration of the neurotransmitter role of ACh in the CNS was the observation that it was possible to stimulate 10 to 20% of neocortical[101] and hippocampal[18] neurons by microiontophoretic application of ACh. In the cortex the neurons were probably the deep pyramidal cells of layer V. The excitation was muscarinic in nature because it could be evoked by other muscarinic agonists besides ACh, and was selectively blocked by atropine.[102] This was the first direct demonstration that muscarinic receptors existed in the CNS, and they were pharmacologically similar to those existing in the parasympathetic system.

The CNS effects of anticholinergic agents were defined as "central anticholinergic syndrome"[112] and emphasis was placed on the amnesic and psychotomimetic effects, and the characteristic EEG modifications. For a long time little attention was devoted to the subtle effects on memory and no attempt was made to define the role of the brain cholinergic system in cognitive processes. In the first congress on the cholinergic mechanisms in the CNS,[84] no paper presented mentioned the word memory. Only Aprison and Hingtgen[6] demonstrated the involvement of a central cholinergic mechanism during drug-induced excitation in avoidance behavior. Nevertheless, even if the association between ascending cholinergic fibers, EEG activation and behavioral arousal had already been discovered, the interest in the role of the cholinergic system in movement control, sleep and wakefulness prevailed over that in the role of memory mechanisms.

Still, in the late fifties and sixties the investigation on the effects of cholinergic drugs on acquisition and retention of conditioned responses had progressively begun using more complex behavioral approaches, as shown by the already large number of references quoted by Longo[112] in his exhaustive review. In the second symposium on cholinergic mechanisms, a review on cholinergic mechanisms and memory was attempted by Moss and Deutsch.[132] Although the authors claimed that "the mechanisms by which atropine might affect behavior are not entirely clear" they concluded that "a full understanding of the effects of cholinergically active drugs may give some insight into certain functions of cholinergic transmission which may be necessary for normal recall of learned behaviors".

An interest in the role of brain ACh in memory was strongly enhanced by the finding that cognitive impairment and loss of forebrain cholinergic neurons are prominent and characteristic landmarks of Alzheimer's disease (AD).[13] Moreover, it was shown that the blockade of muscarinic receptors by scopolamine in young normal subjects induced a cognitive impairment reminiscent of that observed in AD patients.[48] Conversely, it was demonstrated that the activation of muscarinic receptors by ACh, through the increase of its extracellular levels induced by cholinesterase inhibition,[186] or by direct muscarinic agonists[184] may improve the cognitive deficit in AD patients. These findings triggered a large number of studies. In this chapter an attempt will be made to summarize them and present the available information on the role exerted by the central cholinergic neurons, through the activation of muscarinic receptors, on cognitive processes including attention, learning, information storage and recall.

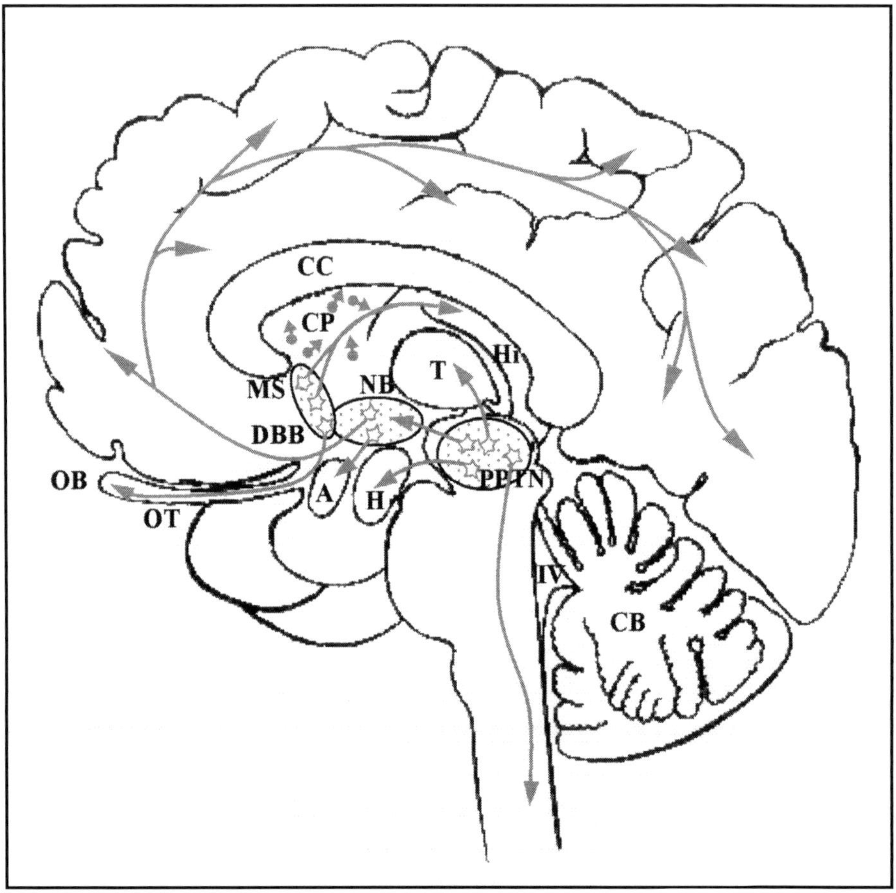

Figure 1. Cholinergic pathways in the human brain. Dotted areas represent the most important nuclei of origin of cholinergic pathways. A: amygdala; CB: cerebellum; CC: corpus callosum; CP: caudate-putamen; DBB: diagonal band of Broca; H: hypothalamus; Hi: hippocampus; MS: medial septum; NB: nucleus basalis of Meynert; OB: olfactory bulb; OT: olfactory tubercle; PPTN: peduncolopontine tegmental nucleus; T: thalamus; IV: fourth ventricle.

Anatomy of Brain Cholinergic Pathways

Muscarinic acetylcholine receptors (mAChRs) are the targets of ACh released from the cholinergic nerve endings and in some cases, are located on cholinergic neurons and nerve endings (autoreceptors). For this reason, a short description of the brain cholinergic system is deemed useful.The central cholinergic system was characterized in the eighties using an anti-choline acetyltransferase (ChAT) antibody[99] and it was divided into 10 relatively well defined populations of cholinergic neurons termed Ch1-Ch10[126] (Fig. 1). Most of these clusters of cholinergic cells are formed by projecting neurons, although there are some well characterized populations of interneurons. The most studied central cholinergic neurons are those found in the basal forebrain because they undergo degeneration in AD. They constitute an aggregate of discontinuous cell islands of large, multipolar cells with extensive dendritic trees. The cholinergic neurons of the basal forebrain, named Ch1-Ch4, give rise to the main cholinergic input to the cortical mantle and hippocampus (MS/DBB projecting primarily to the hippocampus but also to the cortex, and NB, corresponding to the nucleus basalis of Meynert in primates projects diffusely to the cortex and to the amygdala, see (Fig. 1).

Other cholinergic projection neurons include cells extending from the peduncolopontine tegmental nuclei to the floor of the fourth ventricle (Ch5-Ch8). These cells have widespread projections to the forebrain nuclei and to the thalamus, as well as descending projections to the spinal cord. These neurons do not show substantial degeneration in AD.[170]

The striatum contains large aspiny cholinergic interneurons whose dysfunction plays a pathogenetic role in Parkinson's disease. The existence of cortical cholinergic interneurons has been recently confirmed in the rat cerebral cortex by means of immunohistochemical staining for the vesicular ACh transporter.[169] Their role needs to be defined.

Muscarinic Receptors

Subtypes—Transduction Mechanisms

In the mammalian central nervous system (CNS) both nicotinic (see Besheer and Bevins, this book) and mAChRs are present but the density of the muscarinic receptors is much larger.[39]

The first evidence of the existence of more than one muscarinic receptor subtype was given by the work of Riker and Wescoe[156] showing the cardioselective action of gallamine, but it was only at the end of the '70s that the use of the muscarinic receptor antagonist pirenzepine, with higher selectivity for ganglionic than cardiac muscarinic receptors,[79] clearly demonstrated the existence of more than one receptor subtype.

At present, pharmacological, biochemical, immunological, and molecular biological evidence indicates the existence of five mammalian genes (m1-m5) encoding muscarinic receptors (M_1-M_5). The cloning[21,103,149] and expression of these receptor subtypes in cell lines shed light on their function, potential physiological role as well as their signal transduction mechanisms. Recently a gene for a putative sixth muscarinic receptor (m6) was cloned and patented by Millennium Pharmaceuticals Inc. (P1); no details on its pharmacological or potential physiological role are available yet.[51] The five subclasses of mAChRs so far best characterized all have the structural features of the seven-transmembrane helix G-protein-coupled receptor superfamily. Much of the diversity in the structure between the $M_1/M_3/M_5$ sequences compared with the M_2/M_4 sequences resides in the postulated third intracellular loop (i3), responsible for the specificity of coupling to G proteins and which probably determines the quite specific coupling preferences of these two groups.[210] The "odd-numbered" $M_1/M_3/M_5$ mAChRs predominantly couple via the α subunits of the $G_{q/11}$ proteins that activate the enzyme phospholipase Cβ, while the "even-numbered" M_2/M_4 subtypes couple via G_i and G_o α subunits that inhibit adenyl cyclase, as well as to G proteins that directly regulate Ca^{2+} and K^+ channels (for review see ref. 54).

CNS Distribution of the Muscarinic Receptors

While a great diversity of behavioral, physiological and biochemical effects is mediated by mAChRs, the identities of the molecular subtypes responsible for any given neuronal function remain elusive. The complex pharmacology of the mAChR subtypes, together with the lack of drugs with high selectivity has made it difficult to determine the individual roles of m1-m5 receptors in the brain. Identification of the mAChR subtypes in the brain has been accomplished using in situ hybridization to localize their mRNAs,[24,199] or highly selective antibodies to directly localize the proteins;[111] see also Van der Zee and Luiten.[196] All subtypes appear to be present in the brain, although with different densities, localization and relative abundance (Table 1). In the forebrain, the region of interest for AD, the major mAChRs subtypes found are the m1, m2 and m4 proteins. For example, quantitative immunoprecipitation studies showed that in the hippocampus and several areas of human brain neocortex, the m1 receptor accounts for 35-60% whereas the m2 and m4 each accounts for about 15-25% of all binding sites.[59] In contrast, m2 is the most prominent subtype in the basal forebrain and m4 is the most abun-

Table 1. Muscarinic receptors in the central nervous system

	Receptor Subtypes					
	M1	**M2**	**M3**	**M4**	**M5**	**M6**
G-Protein	$G_{q/11}$	$G_{i/o}$	$G_{q/11}$	$G_{i/o}$	$G_{q/11}$	?
Transduction mechanism	↑ IP_3, DAG ↑ $[Ca^{2+}]_i$ ↑ cAMP	↓ cAMP ↑ K^+ currents ↓ Ca^{2+} curre-nts	↑ IP_3, DAG ↑ $[Ca^{2+}]_i$ ↑ cAMP	↓ cAMP ↑ K^+ currents ↓ Ca^{2+} curr-ents	↑ IP_3, DAG ↑ $[Ca^{2+}]_i$ ↑ cAMP	?
Main localization	Neocortex Hippocampus Neostriatum	Moderately abundant	Low levels	Neostriatum Basal forebrain Hippocampus Cortex	Substantia nigra Hippocampus	?
Cellular localization	Pyramidal neurons Striatal spiny neurons	Cholinergic neurons	Neuronal	Striatal spiny neurons	Pyramidal neurons Microglia	?
Synaptic	Post >> pre	Pre >> post	?	Pre and post	?	?

IP_3 inositoltrisphosphate; DAG diacylglicerol; cAMP cyclicAMP; ?, Unknown

dant in caudate-putamen. Immunocytochemistry with specific antibodies has enabled researchers to define a detailed regional and cellular localization of mAChRs in different brain structures. In the medial temporal lobe, the expression of the m1-m4 subtypes shows differences in the regional and laminar patterns.[110] In neocortical areas and hippocampus, the m1 subtype is expressed on all pyramidal neurons, where it is localized in somatodendritic regions, primarily at a postsynaptic level. The pattern of cellular staining for mAChRs in the neocortex is characterized by a clear laminar distribution,[26] with strong immunoreactivity present predominantly in layer 5. Less numerous immunopositive neurons are present in layers 2, 3 and 6.[196] Postsynaptic mAChR subtypes modulate excitatory synaptic transmission in the hippocampus,[76] and an example of this modulation is the enhanced responsiveness of NMDA receptors in area CA1 by activation of M_1 receptors.[116] Double labeling electron microscopic immunocytochemistry has shown that the m1 subunit colocalizes with the NR1 subunit of NMDA receptors in CA1 pyramidal neurons, indicating an appropriate localization for m1 to modulate the activity of the NMDA receptor.[162] The m1 receptor has a similar postsynaptic distribution at excitatory synapses in the striatum[85] and increases the excitability of striatal spiny neurons to the application of NMDA,[28] suggesting that this subunit might play a general role in the modulation of glutamatergic neurotransmission.

The m2 subtype, with its prevalent presynaptic localization in the CNS,[118] is generally believed to act as an autoreceptor inhibiting ACh release. Recently, however, this issue has become controversial. Levey et al.[109] using molecular and immunocytochemical approaches demonstrated that the m2 subtype is present in the basal forebrain not only as a presynaptic cholinergic autoreceptor, but it is also expressed by the remaining population of cells (possibly GABAergic) projecting to the cortex and hippocampus. Also, electron microscopic analysis has shown that in the hippocampus the m2 receptors are present in axons and axon terminals.[109] The m2 receptor is presynaptic in other regions of the brain, including the neocortex,[133] where most of the m2 receptors are located on intrinsic noncholinergic neurons,[109] because complete lesion of the projecting cholinergic neurons almost completely spare the m2 receptors in the terminal field. In the striatum the majority of the m2 receptors act as autoreceptors.[85]

Much less is known about the localization of the other mAChR subtypes in the CNS. The m3 receptor is present throughout the brain,[108] particularly in the basolateral and central amygdala,[108] while the m4 is mostly abundant in the hippocampus and striatum. GABA and glutamate release is inhibited in the basolateral amygdala through m_3-like receptors,[181] suggesting a function of these m3 receptors as heteroreceptors. In the hippocampus the m3 and m4 subtypes are predominantly postsynaptic to the septohippocampal cholinergic terminals.[223] However, the m4 subtype seems to also be presynaptically located on hippocampal associational and commissural projection pathways where it might regulate glutamate release.[110] The distribution profile of the m5 receptor is distinct from the other four subtypes and is enriched in the outer layers of the cortex, specific subfields of the hippocampus, caudate putamen, olfactory tubercle and nucleus accumbens.[155] These studies also demonstrated that the levels of m5 receptor protein expression are apparently higher and more widespread than anticipated from previous in situ hybridization and immunoprecipitation studies.

The mAChRs mediate both excitation and inhibition, depending on the receptor subtype, distribution, subcellular localization, etc. These receptors are found both presynaptically and postsynaptically and, ultimately, their main neuronal effects appear to be mediated through alterations in the properties of ion channels. Excitatory effects result principally from closure of one or more of a number of different K^+ channels,[22] though instances of cation channel openings have also been described,[45,75] while inhibitory effects include opening of K^+ channels and closure of voltage-gated Ca^{2+} channels. The presynaptic auto- and hetero-receptors constitute important feedback loops that control ACh release in an inhibitory manner and represent an important regulatory mechanism for short-term modulation of neurotransmitter release.

Investigations on the physiological role of the various mAChR subtypes are hampered by the lack of selective agonists or antagonists for the specific receptor subtypes. Over the last few years several knockout mice strains have been developed for the M_1,[78] M_2,[69] M_3[119] and M_4 receptors.[70] The use of knockout animals might help to elucidate the physiological functions and pathophysiological implications of each receptor subtype. From these studies it appears that mAChR subtypes are involved in different physiological functions in the CNS, the M_2 receptor being involved in movement, temperature control and nociception,[69] the M_3 in facilitation of food intake,[220] the M_4 in locomotor activity,[70] and the M_5 in water intake, and rewarding brain stimulation.[221] Yamada and coworkers[219] demonstrated that in M_5 receptor knockout mice ACh looses its ability to dilate cerebral arteries and arterioles. Studies on M_1 receptor knockout mice have given contradictory results concerning their role in cognitive mechanisms. Hamilton et al[77] reported that $M_1R^{-/-}$ mutant mice showed defects in LTP induction in hippocampal CA1 neurons and severe impairment in the consolidation of contextual conditioning. On the other hand, it was shown[128] that $M_1R^{-/-}$ mutant mice have a normal working memory, tested with a radial arm maze, a normal spatial memory, tested using a Morris water maze, and a normal freezing levels during contextual fear conditioning. The possibility that compensatory mechanisms occurring during development may help maintaining proper cognitive functions in $M_1R^{-/-}$mice must be considered.

Very recently, release studies with brain slices from M_2 and M_4 receptor single KO mice indicated that autoinhibition of ACh release is mediated primarily by the M_2 receptor in hippocampus and cerebral cortex, but predominantly by the M_4 receptor in the striatum.[224] These results, together with additional receptor localization studies, support the novel concept that autoinhibition of ACh release involves different mAChRs in different regions of the brain.

Alterations in mAChR Expression in Behaviorally-Induced Plasticity

Notwithstanding the consistent trend in distribution pattern of mAChRs in naive animals, the picture of mAChR expression becomes much less static when studied in subjects trained in learning new behavioral paradigms. It has been shown that ACh release is markedly increased in the hippocampus of rats exposed to a novel environment.[65,67,187] In a similar behavioral paradigm investigating spatial orientation, Van der Zee and coworkers[192] showed that in habituated rats mAChRs immunoreactivity increases in cell bodies of pyramidal neurons, but not interneurons, throughout the CA1 region. Changes were also observed in the neocortex, but not in the amygdala.[196] During the habituation period, exploration-associated synaptic changes are likely to occur, and variations in ACh release, accompanied by alterations in mAChRs density might reflect these changes. Similarly, mAChRs immunoreactivity shows a marked increase in pyramidal cells of trace eyeblink conditioned rabbits,[195] an associative learning task in which the hippocampal cholinergic system is critically involved. In rats trained for a passive avoidance conditioned response, increased immunostaining for mAChRs is found in layers 2, 3, and 5 of the prefrontal, cingulate, motor, anterior and posterior sensory cortex and shows a columnar pattern.[194] Since this effect is much less pronounced during recall, the authors concluded that exposure to a new environment and to a painful foot shock, rather than mnemonic processes of the passive avoidance task, are responsible for induction of the changes. Active shock avoidance causes a long lasting and time dependent increase in mAChRs immunoreactivity in the central and basolateral nuclei of the amygdala.[157] Twentyfour hours after training, almost all mAChRs immunoreactivity is lost in the central nucleus, while an increase is present in the basolateral nucleus up to 25 days after the training. The long-lasting, but reversible nature of these changes, indicates that fear conditioning is accompanied by dynamic plasticity of mAChRs immunoreactivity in the amygdaloid complex.

Muscarinic Receptors in Aging and Alzheimer's Disease

Whether or not the mild cognitive deficits associated with aging or the severe memory loss in AD depend, at least partly, on the loss of muscarinic receptors or change in their function is a controversial topic. Age-related loss of muscarinic receptors in rat neocortex and hippocampus is still a matter of debate, with the majority of the studies reporting no changes, and a few reporting a decrease or even an increase in binding sites.[44,196] In a recent PET study using N-[11C]methyl-4-piperidyl benzilate, a decrease in muscarinic cholinergic receptor binding in vivo was observed in several brain areas of aged conscious monkeys.[97] Variation in animal species, rat strain, age of the animal, technique employed and the mAChR subtype studied as well as the poor specificity of the ligands towards individual mAChR subtypes, all contribute to the controversy. For example, Quirion and coworkers[151] showed that memory-impaired, aged Long-Evans rats have higher levels of cortical and hippocampal presynaptic mAChR autoreceptors, determined indirectly through the modulation of ACh release, while other authors, using specific antibodies,[37] did not find any change in the number of mAChRs in the same strain of rats. This latter finding is in line with the observation that cholinergic depletion in the neocortex accomplished by NBM lesion is not accompanied by loss in mAChR immunoreactivity.[194,216] On the other hand, Gill and Gallagher[64] found a significant age-related reduction in M_2 binding sites in the basal forebrain (MS/DBB) and brainstem (PPTN), and the reduction in the basal forebrain was correlated with spatial learning impairment. Both of these ascending cholinergic systems seem to be impaired in the aged rat brain. Since in the MS/DBB complex M_2 receptors are not exclusively located on cholinergic cells,[109] decreased M_2 receptor binding in aged rats may reflect composite alterations affecting cholinergic and non cholinergic (presumably GABAergic) cells. The decreased density in M_2 binding within the basal forebrain in normal aging is consistent with several studies in rodents and primates.[7,137] Taken together all of these findings led to the conclusion that postsynaptic receptors are largely unchanged in the aging brain,[109,118] but their function may be impaired,[58] while there is no consensus regarding the changes in the M_2 receptor subtype.

A reduction in the coupling efficiency to the second messenger system(s), instead of alterations in the number of mAChRs could be the age-related event responsible for cholinergic-related memory dysfunctions. Indeed, it has been shown that the efficiency of hippocampal muscarinic receptors coupling to phosphoinositide (PI) turnover is decreased in cognitively impaired aged rats and this change is highly correlated with the spatial learning index,[37] while there is no difference in the levels of muscarinic receptor proteins between young and aged rats or in rats with impaired spatial learning.

Only recently has information been obtained regarding the molecular subtypes of mAChRs affected by AD. Most studies on muscarinic receptor alterations in AD have concentrated on changes in the pharmacological binding sites in the cortex of aged normal controls and of AD postmortem brains. In AD, the severe loss of cortical cholinergic innervation is accompanied by depletion of m_2 receptors, with a relative stability of m_1 receptors.[118,182] These findings, again, led to the concept that postsynaptic receptors are largely unchanged[58,118] in AD, but may not be functional.[58-60] Quantitative immunoprecipitation followed by radioligand binding demonstrated that m_2 immunoreactivity is decreased, while the m_4 receptor is up-regulated.[59] On the other hand, Mufson and coworkers[135] demonstrated that despite the extensive reduction in cholinergic basal forebrain neurons, the cellular expression of the m_2 receptor is not significantly altered within the basal forebrain of AD patients, suggesting that the reduced levels of the m_2 receptor seen in AD cortex probably reflect changes in other neuronal populations. Some of the m_2-expressing pyramidal neurons may undergo degeneration in AD and thus contribute to the overall loss of m_2 receptors in these patients. Very recently, using muscarinic receptor binding in vivo with [^{11}C]NMBP and PET in healthy volunteers and AD patients,[225] it was demonstrated that muscarinic receptor binding shows an age-related decline, but no evidence of regional changes in AD patients, a finding that is largely in agreement with post-mortem data. At variance with these results, Lai et al[105] in a postmortem investigation found a decrease in M_2 receptor density in the frontal cortex of AD patients with no change in M_1 receptors. However, M_2 receptor density was increased in the frontal cortex of patients with psychotic symptoms compared with those without these symptoms.

The possibility that muscarinic postsynaptic receptors undergo upregulation following degeneration of the cholinergic fibres[141] must be also taken into consideration. It could be speculated that in the AD brain postsynaptic compensatory processes are set forward as a response to a decrease in presynaptic activity. Interestingly,[80,206] it was demonstrated that the mRNA of the m1 subtype increased in the temporal cortex of AD patients, with no changes in m2, m3, and m4 mRNA.[206] A defect in the coupling of muscarinic receptors to GTP-binding protein,[177,207] and/or defective receptor-G protein/phospholipase C coupling[56] in AD brain has also been demonstrated.

Whatever the distribution and dynamics of mAChRs in AD patients, Nitsch and coworkers[140] provided the first and most direct link between muscarinic receptors and AD when they showed that M_1- and M_3-mediated muscarinic stimulation of cortical neurons promotes the processing of the amyloid precursor protein (APP) by the γ-secretase pathway. This pathway splits the APP molecules in the middle of the Aβ domain, therefore precluding the subsequent release of insoluble Aβ.[89] These experiments suggest that loss of mAChRs subtypes during aging and AD may increase the deposition of insoluble Aβ to form plaques.

It must be mentioned that mAChRs are present not only on neurons, but also on glial cells, and are associated with cerebral microvasculature.[196] In contrast to neurons, an increase in the number and intensity of mAChRs-positive astrocytes in cortex and hippocampus of aged rats has been reported.[193] Furthermore, Messamore and coworkers[124] showed, in autopsy brain sections from AD patients, that astrocytes associated with senile plaques possess muscarinic acetylcholine receptors. A brain-specific inflammatory reaction, presumably to the β-amyloid plaques, is an important pathogenetic component of AD (for review see ref. 123). The inflammatory reaction is characterized by activation of microglial cells and astrocytes. The activation of glial cells possessing muscarinic receptors may explain the controversial results on mAChRs expression obtained in aging and AD brain.

Muscarinic Receptors Theta Rhythm and Hippocampal Long Term Potentiation (LTP)

Electrophysiological studies have defined the role of ACh in LTP, an activity-dependent form of synaptic plasticity that is believed to represent the molecular basis of certain forms of learning.[40] The application of ACh or muscarinic agonists to hippocampal slices results in two different actions: i) a rapid, short lasting, selective increase in responsiveness of NMDA receptors in the CA1 area,[116] and ii) a slow-onset facilitation of excitatory postsynaptic potentials (EPSPs) and LTP at Schaffer collaterals-CA1 synapses,[9,19,116] which seems to be mediated by the postsynaptic M_2 receptor subtype[9,174] and is independent of NMDA receptors.[9] This form of LTP involves increased $[Ca^{2+}]_i$ and activation of both serine/threonine and tyrosine kinases. Aged rats, which are cognitively impaired, lack this form of LTP.[171] On the other hand, in the CA3 region of the hippocampus muscarine depresses LTP due to a block of voltage-gated calcium channels.[215]

The hippocampal formation produces a field potential known as rhythmical slow activity (RSA, or theta rhythm) in the electroencephalogram, characterized by a frequency in the range of 3-12 Hz, which is, at least in part, atropine-sensitive.[170] It has been suggested that hippocampal theta rhythm is involved in attention and motivation,[73] and in mechanisms of learning and memory,[87] as well as in the processing of nonspatial working, but not reference memory.[68] An atropine-sensitive theta rhythm has been recorded in immobile, highly aroused animals placed in close proximity to predators,[166] and in rabbits during the presentation of sensory stimuli, but not after habituation.[212] Application of ACh to hippocampal slices induces a theta rhythm of neuronal activity,[88,163] and muscarinic receptor agonists induce theta rhythm, facilitate LTP in vivo and restore learning and memory in a passive avoidance task. This finding demonstrates a possible link between the hippocampal cholinergic system, theta rhythm, LTP in vivo and learning.[92] The presence of theta rhythm might create a permissive environment in the hippocampus for the induction of LTP.[88,138]

Investigations into the intracellular signalling pathways that might be involved indicate that protein kinase A is required for long-lasting LTP induced by stimulation at a theta frequency.[88,138] It has been shown that M_1 and M_4 receptor activation can enhance cAMP, perhaps via $\beta\gamma$ activation of specific adenyl cyclase isozymes[127,185] and muscarinic receptors can also increase levels of intracellular Ca^{2+} from intracellular stores, thus stimulating adenyl cyclase activity in tissues expressing Ca^{2+}-sensitive forms of the enzyme.[36] M_1, M_3 and M_5 receptors indirectly stimulate PKC activity (for review see ref. 54) while the M_2 and M_4 in some cell types can also activate phospholipase C.[188] These events can initiate an overlapping network of signals, including the activation of a mitogen activated protein kinase (p42- p44-MAPK) pathway,[74] which is known to be regulated by muscarinic agonists and antagonists in neuronal tissue.[159] The MAPKs in the CNS are involved, among other actions, in LTP,[52,66,159] fear conditioning,[168] long term spatial memory[172] and associative learning.[8] MAPKs can also be stimulated by growth factors and NMDA receptors, and downstream effectors are transcription factors such as ELK1 or RSK2 (for reviews, see ref. 183 and Selcher et al, this book). The modulation of the MAPK signaling pathway by muscarinic receptors provides support to the concept that ACh exerts its effect on memory through intracellular phosphorylation cascades leading to increased gene transduction and protein synthesis.

Which Cognitive Processes Depend on the Activation of Muscarinic Receptors?

Three experimental approaches have been used for detecting which cognitive process requires activation of muscarinic receptors.

1. Performance of cognitive tasks after blockade of the muscarinic receptors by muscarinic antagonists. This approach has been (and still is) used in humans as well;
2. Performance of cognitive tasks after selective destruction of cholinergic neurons;

3. Measurement of the changes in ACh extracellular levels in discrete brain regions in animal performing cognitive tasks.

The two latter approaches can be used only in animals and do not exclude the possible involvement of nicotinic receptors.

Cognitive Alterations Induced by the Blockade of Muscarinic Receptors

The nonselective muscarinic antagonist scopolamine has been extensively used to study the cognitive alterations induced by the blockade of muscarinic receptors.[112] The interpretation of the pharmacological data is not straightforward, as scopolamine blocks postsynaptic receptors as well as presynaptic autoreceptors, this latter effect being responsible for an increase in ACh.[197]

Scopolamine administered to human subjects impairs rapid information processing[209] and encoding of new memories in humans, primates and rodents (for review see ref. 82). Furthermore, scopolamine impairs the retention of contextual fear conditioning when administered immediately before or immediately post training,[164] as well as spatial learning when administered before training in the spatial version of the Morris water maze.[211] Scopolamine seems to be less disruptive to long term memory storage than to short term memory,[12,68,146] indicating that once information has been consolidated it is no longer susceptible to disruption by the blockade of the muscarinic receptors.

Studies evaluating the involvement of the cholinergic system in memory retrieval have given mixed results. One study showed that scopolamine disrupts retrieval processes in human subjects,[165] while in a different study scopolamine had no effect.[200] Some animal studies also indicate that scopolamine interferes with retrieval of information in rodents,[130] as do NB lesioning.[136]

In both humans and animals the learning deficits caused by muscarinic antagonists are reversed by cholinomimetic agents, such as cholinesterase inhibitors, supporting the role of ACh in learning and memory processes. However, it should be mentioned that rats trained in a Morris water maze under muscarinic antagonism display sensorimotor disturbances that may interfere with their ability to acquire the task, confounding the impairments of learning/memory with motor disturbances.[27]

Cognitive disruption similar to that occurring after scopolamine administration has been observed in animals and humans with all muscarinic receptor antagonists. The memory impairment induced in patients by tricyclic antidepressants such as amitryptiline and analogues or anti-parkinsonian agents such as benztropine may represent a clinically relevant side effect.[1]

Pirenzepine, the only selective M_1 antagonist, does not cross the blood brain barrier. Its intracerebral administration in the rat results in cognitive impairment similar to that induced by scopolamine.[90] Conversely, the administration to rats of the selective M2 antagonists BiBN-99[151] and AFDX 384[198] facilitate the release of ACh from the cholinergic nerve endings by inhibiting the autoreceptor feedback mechanisms and, through this mechanism, restore the object recognition, passive avoidance, and water maze performance impaired by age or by scopolamine in young rats. These findings confirm that an increase in extracellular brain ACh may improve some cognitive processes by acting on M_1 receptors and indicate the blockade of presynaptic muscarinic receptors as a possible target for therapeutic strategies.

Electrophysiological studies, using the selective M_3 antagonist 4-DAMP mustard, demonstrate that M_3 and possibly M_5 receptors, located in the medial septum/diagonal band modulate the impulse flow in the septohippocampal GABAergic pathway.[5]

The muscarinic toxins MT1, MT2 and MT3 from the venom of the snake *Dendroaspis angusticeps*, (Eastern Green Mamba)[3,96] appear to be valuable tools to study receptor pharmacology, physiology and structure/function relationships. MT1 and MT2 toxins are of great interest because they appear to be the first allosteric agonists of muscarinic receptors. It is possible to study the in vivo effects on memory retention of the injection of the toxins into discrete brain regions. MT2 toxin, a highly selective agonist for M_1 receptors of the rabbit vas deferens, when injected into the dorsal hippocampus immediately after an inhibitory avoidance task, improved retention of the task and its effect was antagonized by scopolamine. This finding led the authors to postulate that the M_1 receptor of the dorsal hippocampus is directly involved in positive modulation of memory of this task.[94] On the other hand, MT3 injected into the same region caused amnesia.[95] Since this toxin is a highly selective antagonist for M_4 receptor,[142] acting on the allosteric site of the receptor, the authors suggest that M_4 receptor subtype in the dorsal hippocampus is involved in memory consolidation of a step-down inhibitory avoidance, a task with spatial and strong aversive components.

Cognitive Alteration Induced by Lesions in the Cholinergic Pathways

The prerequisite of lesion studies is the selectivity of the lesion. In order to investigate their role in learning and memory, the lesions must damage only the cholinergic neurons, a condition which is not easily obtained. The first difficulty is represented by the anatomy of the cholinergic system since the basal forebrain cholinergic neurons are not clustered together in a well localized nucleus that can be lesioned easily.[222] For many years the second difficulty was the tool for making the lesion, since a selective cholinergic neurotoxin has been made available only recently. Electrolytic lesions, which destroy both cholinergic and non cholinergic neurons, as well as intermingled neurons and axons, were used in the past (for review see ref. 53,148) before more refined methods were developed. Neurotoxic lesions made by injecting the area with glutamate analogs, such as ibotenic acid, kainate, and quisqualate have been largely used to destroy cholinergic neurons, but they have different degrees of selectivity and effectiveness,[50] and these excitotoxins may damage cholinergic and non cholinergic neurons. A more promising technique has been the development of selective cholinergic neurotoxins. An analog of choline, AF64A, has been used to impair the cholinergic neurons, but it appears more effective on MS than on neocortically projecting cholinergic neurons.[38] An important improvement has been the introduction of the 192IgG-saporin conjugate which permits very selective cholinergic lesions of the basal forebrain cholinergic neurons.[214] Notwithstanding all these limitations, experiments with lesions of the forebrain cholinergic neurons by different means have helped in enlightening the role of ACh in learning and memory.

Destruction of the septo-hippocampal projections, either by kainate or ibotenate, impairs choice accuracy in experimental procedures implying short-term memory[57] and transection at the level of the fimbria-fornix produce deficits in a T-maze performance.[154] According to Casamenti[31] the destruction of the NB obtained by bilateral injection of ibotenic acid results in a complete impairment of passive avoidance conditioned responses lasting at least 6 months. It has been demonstrated that quisqualic acid lesions of the nucleus basalis disrupts working memory evaluated by the object recognition test.[11]

The most selective procedure for the destruction of the cholinergic neurons is the intracerebroventricular or intraparenchymal injection of 192IgG-saporin (see also Jaffard and Marighetto in this book). This immunotoxin acts by coupling the ribosome inactivating toxin saporin to an antibody that recognizes low-affinity NGF (Nerve Growth Factor) receptors (p75), which are located on the cholinergic neurons of the basal forebrain.[150] It may therefore be assumed that the cognitive processes affected by 192IgG-saporin injection depend on the function of cholinergic neurons. Table 2 summarizes the results obtained with this neurotoxin.

Intracerebroventricular injection of 192IgG-saporin results in cell loss in the MS/DBB and NB, whereas cholinergic cells in the adjacent ventral pallidum, septal cholinergic interneurons, and peduncolopontine tegmental nuclei are spared.[83,160] The use of 192IgG-saporin causes up to a 90% reduction in cholinergic presynaptic markers, indicating an almost complete cholinergic deafferentation to the cortex and hippocampus.[160] These changes are accompanied by up to a 35% increase in M_1 receptor density and by a 20% increase in M_2 receptor density in the cortex.[161] Again, increased M_1 receptor density can be regarded as a compensatory mechanism for reduced cholinergic input, while M_2 increase after immunolesion supports the hypothesis that a significant part of M_2 receptors exists as postsynaptic receptors in the cerebral cortex. All the data obtained from behavioral studies performed in rats with i.c.v. injections of 192IgG-saporin are consistent with the finding that only very extensive lesions involving more

Table 2. Behavioral tasks impaired by 192 IgG saporin injections in rodents

Administration Route	Behavioral Test	Effect	References
i.c.v.	Morris water maze	Spatial working Memory	(Leanza et al., 1995) (Waite and Thal, 1996) (Baxter et al., 1995) (Berger-Sweeney et al., 1994) (Walsh et al., 1995)
i.c.v	Radial arm maze	Spatial working Memory	(Wrenn et al., 1999)
i.c.v.	Multiple choice reaction time task	Attentional functions	(Waite et al., 1999)
i.c.v.	Delayed non-matching to position	Operant behavior: working Memory, spatial memory	(Steckler et al., 1995) (McDonald et al., 1997)
MS/DBB and NB	Operant delayed matching -to-position task	Short-term memory	(Leanza et al., 1996) (Torres et al., 1994)
MS/DBB or NB	Morris water maze and radial arm maze	Working memory	(Dornan et al., 1996) (Shen et al., 1996)
MS/DBB or NB	Matching-to-place task	Spatial memory	(Baxter et al., 1995)
Intraseptal	Radial arm maze	Spatial working memory	(Walsh et al., 1996)
Posterior parietal cortex	Conditioned stimulus processing	Attentional functions	(Bucci et al., 1998) (Chiba et al., 1995)
NB	Vigilance task	Sustained attention	(McGaughy et al., 1996)
Intracortical	Sustained attention	Attentional functions	(McGaughy and Sarter, 1998)
MS/DBB	Conditional discrimination	Conditional learning	(Marston et al., 1994)
Amygdala (central nucleus) & SI/NB	Conditioned stimulus processing	Attentional functions	(Han et al., 1999)
NB	Pre-pulse inhibition	Sensorimotor gating	(Ballmaier et al., 2001)
NB	Passive avoidance	Working memory	(Torres et al., 1994)
MS/DBB	Latent inhibition	Attentional functions	(Baxter et al., 1997)
NB	Stimulus discrimination	Operant behavior	(Stoehr et al., 1997)

than 90% of MS/DBB and NB cholinergic neurons result reliably in severely impaired performances (for review see ref. 218). Impairments in water maze acquisition,[107] delayed matching[106] and nonmatching to position task,[120] prepulse inhibition[10] as well as acquisition, but not retention, of an object discrimination[201] have been reported in the rat.

Site directed injection of 192IgG-saporin into distinct nuclei induces cholinergic cell loss in situ and cholinergic hypoactivity in target areas.[208] These lesions are however rather small and circumscribed and did not produce reliable learning and memory deficits in several behavioral tests studied, even after bilateral injections of 192IgG-saporin.[15] Other papers reported impaired retention of a passive avoidance response and of a delayed nonmatching to position task[190] when the toxin was injected into the NB or MS, respectively, and deficit in the acquisition of a working memory task[14] when injected into either or both structures. Intraparenchymal injections of 192IgG-saporin have also been used to study the effects of basal forebrain cholinergic lesions on attentional processing. Most of the studies report disrupted attentional processing in NB- or MS-injected animals,[218] which raises the possibility that the performance deficits in 192IgG-saporin treated rats might be downstream manifestations of disrupted attention.

Cognitive Processes Associated with Changes in Cholinergic Activity

The development of the in vivo microdialysis technique[191] coupled to sensitive analytical systems for the quantification of ACh levels in the nanomolar range has made it feasible to correlate changes in ACh extracellular levels in the brain with cognitive processes in the animal. Using this approach, it has been possible to study the behaviorally-induced activation of the cholinergic system ascending from the NB to the cortex, from the septum to the hippocampus or in other structures, during the performance of behavioral tasks.

A caveat in the interpretation of the results obtained with this approach is that stressor stimuli, such as prolonged handling,[139,158] restraint,[93] and fear,[2] also strongly activate the cholinergic system. Therefore before associating a behavioral response to an increase in ACh release, it is necessary to consider the possible interference of stressors.

Cortical ACh release increased in rats performing a visual attentional task,[41] and during an operant task designed to assess sustained attention and it showed a direct relationship with attentional effort.[86] Similarly, Orsetti and coworkers[144] observed a large increase in cortical and hippocampal ACh release during acquisition of a rewarded operant behavior, but not during its recall. However, a correlation between attentional effort, required by the task difficulties and ACh release has not always been found.[145]

The first exposure to a novel environment causes pronounced behavioral activation.[4,32] Giovannini et al[65] demonstrated that when the rats are moved from the home cage to a novel environment ACh release from the frontal cortex increases during the exploratory activity. Conversely, if the rat is left in the novel environment, exploratory activity and ACh release decrease, presumably as a consequence of habituation. Explorations render the environment familiar[33,67,187] since its features have been memorized. Behavioral habituation, which is usually demonstrated by a decrease in exploratory behavior, provides one of the most elementary forms of learning, both in animals and humans.

The response of an animal to a novel environment includes arousal, attention, anxiety and fear. A novel environment represents a stressful condition and elicits exploratory activity. The increase in ACh release from the frontal cortex and the hippocampus,[67] occurring under such conditions, demonstrates an activation of the forebrain cholinergic neurons.[72] The presence of hippocampal theta rhythm during exploratory activity[213] and attention[73] is further evidence of cholinergic activation since theta rhythm depends on the septohippocampal cholinergic pathway,[170] as already discussed in a previous paragraph. Two questions arise: 1) To which component of the response to a novel environment is the activation of the cholinergic forebrain neurons associated?, and 2) Which cognitive processes depend on the increase in ACh levels in the synaptic cleft?

Motor activity appears to be associated with an increase in ACh release. However, the attempt to establish a direct correlation between motor activity and ACh release has provided contradictory results. While Day et al[42] and Mizuno et al[129] found a correlation, Day and Fibiger,[43] Moore and coworkers,[131] and Thiel and coworkers[187] did not confirm this correlation. Giovannini and coworkers[67] demonstrated a significant correlation between cortical and hippocampal ACh release and motor activity only after habituation, but not during the first exposure to a novel environment. This finding indicates that ACh released during exploration of a novel environment has more than one component: one related to motor activity and the others to attention, arousal, anxiety and fear.

In conclusion, the release experiments in behaving rats support the notion that the forebrain cholinergic system plays a crucial role in the attentional functions. This role is mediated by muscarinic receptors since it is well known that muscarinic antagonists, such as scopolamine, reduce the efficiency of information encoding because they impair the optimal utilization of attentional resources.[49] Further evidence comes from the finding that by enhancing cholinergic activity with the cholinesterase inhibitor donepezil it is possible to reduce inspection time in normal subjects.[91] According to Sarter and colleagues,[167] activated cortical cholinergic inputs enhance cortical sensory and sensory-associational information processing, including filtering of noise and distractors. Furthermore, it may be assumed that optimal operation of information processing results in more efficient memory storage and information retrieval.

Effects of Direct and Indirect Selective Muscarinic Receptor Agonists on Learning and Memory: Therapeutic Implications

The clinical observations that learning and memory are impaired by drugs which block the muscarinic receptors, and pathological conditions characterised by a loss of cholinergic neurons, such as AD, have lead to the attempt to restore or improve the cognitive processes by acting directly or indirectly on the muscarinic receptors. Already in the 19th century both the direct agonist pilocarpine,[134] and the Calabar beans,[100] which contain eserine, were indicated as an antidote for atropine poisoning.

Indirect activation of all subtypes of muscarinic receptors can be obtained by raising the intrasynaptic levels of ACh through the inhibition of acetylcholinesterase, the enzyme specifically devoted to ACh inactivation. Some cholinesterase inhibitors such as tacrine, donepezil, and rivastigmine are currently used in the treatment of the cognitive and behavioral disturbances in AD patients. It is outside the scope of this review to discuss the pharmacological effects of these drugs and their therapeutic results. A recent comprehensive overview was published by Giacobini.[62] The involvement of muscarinic receptors in this therapy is demonstrated by their down regulation during subchronic treatments with cholinesterase inhibitors.[25,153] The extent to which muscarinic receptor down-regulation may affect the therapeutic efficacy of cholinesterase inhibitors has not yet been established.

Indirect activation of muscarinic receptors has been attempted using non selective and selective muscarinic agonists. Arecoline, an alkaloid contained in the Areca nut or Betel nut from the palm tree *Areca catechu*, is a partial agonist with higher M_2 than M_1 receptor affinity.[125] It has a psychostimulant activity well know by the populations of Southwest Asia where the chewing of Betel nuts is a widespread habit. Intravenous administration of arecoline to AD patients resulted in cognitive improvement, particularly in verbal ability.[152] However, arecoline has not been tested extensively for AD therapy.

Several selective agonists for the M1 receptors have been synthesised with the purpose of treating AD.[55] Xanomeline resulted in significant improvement in cognition despite dose-limiting adverse events.[20] The cholinergic side effects include salivation, lacrimation, gastrointestinal disturbances and cardiac rhythm alterations. They are more severe than during treatment with cholinesterase inhibitors and so far they have prevented the therapeutic use of M_1 muscarinic agonists.

Since it has been shown that the administration to rats of the selective M_2 antagonists BiBN-99[151] and AFDX 384[198] facilitate the release of ACh from the cholinergic nerve endings by inhibiting the autoreceptor feedback mechanisms and, through this mechanism, restore cognitive performances impaired by age or by scopolamine in young rats, other more selective M_2 antagonists have been developed for the treatment of AD.[104] For instance, in rodents and non human primates the selective M_2 antagonist SCH57790 was as effective as the ChE inhibitor donepezil in reversing scopolamine induced behavioral deficits.[29] However, no information on the clinical efficacy of these compounds is available yet.

Conclusions

From this overview of about a century of studies on muscarinic receptors, it appears that the widespread target system through which acetylcholine, released from the cholinergic network, improves memory performance by augmenting the selectivity of perceptual processing during encoding.[61] However, it should be remembered that apart from their role in learning and memory, muscarinic receptors are involved in other brain functions, such as motor control, sleep, cardiovascular activity, hormone secretion and pain control.[98]

We still need selective agonists and antagonists through which to improve our understanding of the role the different muscarinic receptor subtypes play. Only with better drugs we may hope to exploit more successfully the muscarinic receptors for therapeutic purposes.

References

1. The Merck Manual of Geriatrics. Whitehouse Station, NJ: Merck Co. Inc., 1995.
2. Acquas E, Wilson C, Fibiger HC. Conditioned and unconditioned stimuli increase frontal cortical and hippocampal acetylcholine release: Effects of novelty, habituation and fear. J Neurosci 1996; 16:3089-3096.
3. Adem A, Asblom A, Johansson G. Toxins from the venom of the green mamba Dendroaspis angusticeps that inhibit the binding of quinuclidinyl benzilate to muscarinic acetylcholine receptors. Biochim Biophys Acta 1988; 968:340-345.
4. Aloisi AM, Casamenti F, Scali C et al. Effects of novelty, pain and stress on hippocampal extracellular acetylcholine levels in male rats. Brain Res 1997; 748:219-226.
5. Alreja M, Wu M, Liu W et al. Muscarinic tone sustains impulse flow in the septohippocampal GABa but not cholinergic pathway: Implication for learning and memory. J Neurosci 2001; 20:8103-8110.
6. Aprison MH, Hingtgen JN. Evidence of a central cholinergic mechanisms functioning during drug induced excitation in avoidance behavior. In: Heilbronn E, Winter A, eds. Drugs and Cholinergic Mechanisms in the CNS. Stockholm: Research Institute of National Defence 1970:543-559.
7. Araujo DM, Lapchak PL, Meaney MJ et al. Effects of aging on nicotinic and muscarinic autoreceptor function in the rat brain: Relationship to presynaptic cholinergic markers and binding sites. J Neurosci 1990; 10:3069-3078.
8. Atkins CM, Selcher JC, Petraitis JJ et al. The MAPK cascade is required for mammalian associative learning. Nat Neurosci 1998; 1:602-609.
9. Auerbach JM, Segal M. A novel cholinergic induction of long-term potentiation in rat hippocampus. J Neurophysiol 1994; 72:2034-2040.
10. Ballmaier M, Casamenti F, Zoli M et al. Selective immunolesioning of cholinergic neurons in nucleus basalis magnocellularis impairs prepulse inhibition of acoustic startle. Neuroscience 2001; 108:299-305.
11. Bartolini L, Casamenti F, Pepeu G. Aniracetam restores object recognition impaired by age, scopolamine and nucleus basalis lesions. Pharmacol Biochem Behav 1996; 53:277-283.
12. Bartolini L, Risaliti R, Pepeu G. Effect of scopolamine and nootropic drugs on rewarded alternation in a T-maze. Pharmacol Biochem Behav 1992; 43:1161-1164.
13. Bartus RT, Dean RL, Beer B et al. The cholinergic hypothesis of geriatric memory dysfunction. Science 1982; 217:408-417.
14. Baxter MG, Bucci DJ, Gorman LK et al. Selective immunotoxic lesions of basal forebrain cholinergic cells: Effects on learning and memory in rats. Behav Neurosci 1995; 109:714-722.
15. Baxter MG, Bucci DJ, Sobel TJ et al. Intact spatial learning following lesions of basal forebrain cholinergic neurons. Neuro Report 1996; 7:1417-1420.
16. Baxter MG, Holland PC, Gallagher M. Disruption of decrements in conditioned stimulus processing by selective removal of hippocampal cholinergic input. J Neurosci 1997; 17:5230-5236.
17. Berger-Sweeney J, Heckers S, Mesulam MM et al. Differential effects on spatial navigation of immunotoxin-induced cholinergic lesions of the medial septal area and nucleus basalis magnocellularis. J Neurosci 1994; 14:4507-4519.
18. Biscoe TJ, Straughan DW. The pharmacology of hippocampal neurones. J Pharm Pharmacol 1965; 17:60-61.
19. Blitzer RD, Gil O, Landau EM, Cholinergic stimulation enhances long-term potentiation in the CA1 region of rat hippocampus. Neurosci Lett 1990; 119:207-210.
20. Bodick NC, Offen WW, Levey AI et al. Effects of xanomeline, a selective muscarinic receptor agonist, on cognitive function and behavioral symptoms in Alzheimer disease. Arch Neurol 1997; 54:465-473.

21. Bonner TI, Buckley NJ, Young AC et al. Identification of a family of muscarinic acetylcholine receptor genes. Science 1987; 237:527-532.
22. Brown DA, Abogadie FC, Allen TG et al. Muscarinic mechanisms in nerve cells. Life Sci 1997; 60:1137-1144.
23. Bucci DJ, Holland PC, Gallagher M. Removal of cholinergic input to rat posterior parietal cortex disrupts incremental processing of conditioned stimuli. J Neurosci 1998; 18:8038-8046.
24. Buckley NJ, Bonner TI, Brann MR. Localization of a family of muscarinic receptor mRNAs in rat brain. J Neurosci 1988; 8:4646-4652.
25. Bushnell PJ, Kelly KL, Ward TR. Repeated inhibition of cholinesterase by chlorpyrifos in rats: Behavioral, neurochemical and pharmacological indices of tolerance. J Pharmacol Exp Ther 1994; 270:15-25.
26. Buwalda B, de Groote L, Van der Zee EA et al. Immunocytochemical demonstration of developmental distribution of muscarinic acetylcholine receptors in rat parietal cortex. Brain Res Dev Brain Res 1995; 84:185-191.
27. Cain DP. Testing the NMDA, long-term potentiation and cholinergic hypotheses of spatial learning. Neurosci Biobehav Rev 1998; 22:181-193.
28. Calabresi P, Centonze D, Gubellini P et al. Endogenous ACh enhances striatal NMDA-responses via M1-like muscarinic receptors and PKC activation. Eur J Neurosci 1998; 10:2887-2895.
29. Carey GJ, Billard W, Binch H et al. SCH 57790, a selective muscarinic M(2) receptor antagonist, releases acetylcholine and produces cognitive enhancement in laboratory animals. Eur J Pharmacol 2001; 431:189-200.
30. Carlton PL. Cholinergic mechanism in the control of behavior by the brain. Psychol Rev 1963; 70:19-39.
31. Casamenti F, Di Patre PL, Bartolini L et al. Unilateral and bilateral nucleus basalis lesions:differences in neurochemical and behavioural recovery. Neuroscience 1988; 24:209-215.
32. Ceccarelli I, Casamenti F, Massafra C et al. Effects of novelty and pain on behavior and hippocampal extracellular ACh levels in male and female rats. Brain Res 1999; 815:169-176.
33. Cerbone A, Sadile AG. Behavioral habituation to spatial novelty: Interference and noninterference studies. Neurosci Biobehav Rev 1994; 18:497-518.
34. Chang HC, Gaddum JH. Choline esters in tissue extracts. J Physiol 1933; 79:255-284.
35. Chiba AA, Bucci DJ, Holland PC et al. Basal forebrain cholinergic lesions disrupt increments but not decrements in conditioned stimulus processing. J Neurosci 1995; 15:7315-7322.
36. Choi EJ, Wong ST, Hinds TR et al. Calcium and muscarinic agonist stimulation of type I adenylylcyclase in whole cells. J Biol Chem 1992; 267:12440-12442.
37. Chouinard ML, Gallagher M, Yasuda RP et al. Hippocampal muscarinic receptor function in spatial learning-impaired aged rats. Neurobiol Aging 1995; 16:955-963.
38. Chrobak JJ, Hanin I, Schmechel DE et al. AF64A-induced working memory impairment: Behavioral, neurochemical and histological correlates. Brain Res 1988; 463:107-117.
39. Clarke PBS. Nicotinic receptors in mammalian brain: localization and relation to cholinergic innervation. Progr Brain Res 1993; 98:77-83.
40. Collingridge GL, Bliss TVP. NMDA receptors-their role in long-term potentiation. TINS 1987; 10:288-293.
41. Dalley JW, McGaughy J, O'Connell MT et al. Distinct changes in cortical acetylcholine and noradrenaline efflux during contingent and noncontingent performance of a visual attentional task. J Neurosci 2001; 21:4908-4914.
42. Day J, Damsma G, Fibiger HC. Cholinergic activity in the rat hippocampus, cortex and striatum correlates with locomotor activity: An in vivo microdialysis study. Pharmacol Biochem Behav 1991; 38:723-729.
43. Day J, Fibiger HC. Dopaminergic regulation of cortical acetylcholine release. Synapse 1992; 12:281-286.
44. Decker MW. The effects of aging on hippocampal and cortical projections of the forebrain cholinergic system. Brain Res 1987; 434:423-438.
45. Delmas P, Niel JP, Gola M. Muscarinic activation of a novel voltage-sensitive inward current in rabbit prevertebral sympathetic neurons. Eur J Neurosci 1996; 8:598-610.
46. Domer FR, Schueler FW. Investigations of the amnesic properties of scopolamine and related compounds. Arch Int Pharmacodyn 1960; 127:449-458.
47. Dornan WA, McCampbell AR, Tinkler GP et al. Comparison of site-specific injections into the basal forebrain on water maze and radial arm maze performance in the male rat after immunolesioning with 192 IgG saporin. Behav Brain Res 1996; 82:93-101.
48. Drachman DA, Leavitt J. Human memory and the cholinergic system: A relationship to aging? Arch Neurol 1974; 30:113-121.

49. Dunne MP, Hartley LR. Scopolamine and the control of attention in humans. Psychopharmacol 1986; 89:94-97.
50. Dunnett SB, Whishaw IQ, Jones GH et al. Behavioural, biochemical and histochemical effects of different neurotoxic amino acids injected into nucleus basalis magnocellularis of rats. Neuroscience 1987; 20:653-669.
51. Eglen RM, Choppin A, Dillon MP et al. Muscarinic receptor ligands and their therapeutic potential. Curr Opin Chem Biol 1999; 3:426-432.
52. English JD, Sweatt JD. A requirement for the mitogen-activated protein kinase cascade in hippocampal long term potentiation. J Biol Chem 1997; 272:19103-19106.
53. Everitt BJ, Robbins TW. Central cholinergic system and cognition. A Rev Psychol 1997; 48:649-684.
54. Felder CC. Muscarinic acetylcholine receptors: Signal transduction through multiple effectors. FASEB J 1995; 9:619-625.
55. Felder CC, Bymaster FP, Ward J et al. Therapeutic opportunities for muscarinic receptors in the central nervous system. J Med Chem 2000; 43:4333-4353.
56. Ferrari-DiLeo G, Flynn DD. Diminished muscarinic receptor-stimulated [3H]-PIP2 hydrolysis in Alzheimer's disease. Life Sci 1993; 53:L439-L444.
57. Flicker C, Dean RL, Watkins DL et al. Behavioral and neurochemical effects following neurotoxic lesions of a major cholinergic input to the cerebral cortex in the rat. Pharmacol Biochem Behav 1983; 18:973-981.
58. Flynn DD, Ferrari-DiLeo G, Levey AI et al. Differential alterations in muscarinic receptor subtypes in Alzheimer's disease: Implications for cholinergic-based therapies. Life Sci 1995; 56:869-876.
59. Flynn DD, Ferrari-DiLeo G, Mash DC et al. Differential regulation of molecular subtypes of muscarinic receptors in Alzheimer's disease. J Neurochem 1995; 64:1888-1891.
60. Flynn DD, Weinstein DA, Mash DC. Loss of high affinity agonist binding to M1 muscarinic receptors in Alzheimer's disease. Implications for the failure of cholinergic replacement therapies. A Neurol 1991; 19:256-261.
61. Furey ML, Pietrini P, Haxby JV. Cholinergic enhancement and increased selectivity of perceptual processing during working memory. Science 2000; 290:2315-2319.
62. Giacobini E. Cholinesterase inhibitors: From the Calabar bean to Alzheimer therapy. In: Giacobini E, ed. Cholinesterases and cholinesterases inhibitors. London: Martin Duniz Ltd., 2000:181-226.
63. Giarman NJ, Pepeu G. Drug-induced changes in brain acetylcholine. Br J Pharmacol 1962; 19:226-234.
64. Gill TM, Gallagher M. Evaluation of muscarinic M2 receptor sites in basal forebrain and brainstem cholinergic systems of behaviorally characterized young and aged Long-Evans rats. Neurobiol Aging 1998; 19:217-225.
65. Giovannini MG, Bartolini L, Kopf SR et al. Acetylcholine release from the frontal cortex during exploratory activity. Brain Res 1998; 784:218-227.
66. Giovannini MG, Blitzer RD, Wong T et al. Mitogen-activated protein kinase regulates early phosphorylation and delayed expression of Ca2+/calmodulin-dependent protein kinase II in long-term potentiation. J Neurosci 2001; 21:7053-7062.
67. Giovannini MG, Rakovska A, Benton RS et al. Effects of novelty and habituation on acetylcholine, GABA and glutamate release from the frontal cortex and hippocampus of freely moving rats. Neuroscience 2001; 106:43-53.
68. Givens B, Olton DS. Local modulation of basal forebrain: Effects on working and reference memory. J Neurosci 1994; 14:3578-3587.
69. Gomeza J, Shannon H, Kostenis E et al. Pronounced pharmacologic deficits in M2 muscarinic acetylcholine receptor knockout mice. Proc Natl Acad Sci USA 1999; 96:1692-1697.
70. Gomeza J, Zhang L, Kostenis E et al. Enhancement of D1 dopamine receptor-mediated locomotor stimulation in M(4) muscarinic acetylcholine receptor knockout mice. Proc Natl Acad Sci USA 1999; 96:10483-10488.
71. Goodman LS, Gilman A. The pharmacological basis of therapeutics. New York: McMillan, 1955.
72. Gray J, McNaughton N. Comparison between the behaviouraleffects of septal and hippocampal lesions: A review. Neurosci Biobehav Rev 1983; 7:119-188.
73. Green JD, Arduini AA. Hippocampal electrical activity in arousal. J Neurophysics 1954; 17:533-557.
74. Gutkind JS. The pathways connecting G protein-coupled receptors to the nucleus through divergent mitogen-activated protein kinase cascades. J Biol Chem 1998; 273:1839-1842.
75. Haj-Dahmane S, Andrade R. Muscarinic activation of a voltage-dependent cation nonselective current in rat association cortex. J Neurosci 1996; 16:3848-3861.
76. Halliwell JV. Physiological mechanisms of cholinergic action in the hippocampus. Prog Brain Res 1990; 84:255-272.

77. Hamilton SE, Hardouin SN, Anagnostaras SG et al. Alteration of cardiovascular and neuronal function in M1 knockout mice. Life Sci 2001; 68:2489-2493.
78. Hamilton SE, Loose MD, Qi M et al. Disruption of the m1 receptor gene ablates muscarinic receptor- dependent M current regulation and seizure activity in mice. Proc Natl Acad Sci USA 1997; 94:13311-13316.
79. Hammer R, Berrie CP, Birdsall NJ et al. Pirenzepine distinguishes between different subclasses of muscarinic receptors. Nature 1980; 283:90-92.
80. Han JS, Holland PC, Gallagher M. Disconnection of the amygdala central nucleus and substantia innominata/nucleus basalis disrupts increments in conditioned stimulus processing in rats. Behav Neurosci 1999; 113:143-151.
81. Harrison PJ, Barton AJ, Najlerahim A et al. Increased muscarinic receptor messenger RNA in Alzheimer's disease temporal cortex demonstrated by in situ hybridization histochemistry. Brain Res Mol Brain Res 1991; 9:15-21.
82. Hasselmo ME, Wyble BP, Wallenstein GV. Encoding and retrieval of episodic memories: Role of cholinergic and GABAergic modulation in the hippocampus. Hippocampus 1996; 6:693-708.
83. Heckers S, Ohtake T, Wiley RG et al. Complete and selective cholinergic denervation of rat neocortex and hippocampus but not amygdala by an immunotoxin against the p75 NGF receptor. J Neurosci 1994; 14:1271-1289.
84. Heilbronn E, Winter A. Drugs and cholinergic mechanisms in the CNS. Stockholm: Research Institute of National Defence, 1970.
85. Hersch SM, Gutekunst CA, Rees HD et al. Distribution of m1-m4 muscarinic receptor proteins in the rat striatum: Light and electron microscopic immunocytochemistry using subtype- specific antibodies. J Neurosci 1994; 14:3351-3363.
86. Himmelheber AM, Sarter M, Bruno JP. Increases in cortical acetylcholine release during sustained attention performance in rats. Brain Res Cogn Brain Res 2000; 9:313-325.
87. Huerta PT, Lisman JE. Heightened synaptic plasticity of hippocampal CA1 neurons during a cholinergically induced rhythmic state. Nature 1993; 364:723-725.
88. Huerta PT, Lisman JE. Bidirectional synaptic plasticity induced by a single burst during cholinergic theta oscillation in CA1 in vitro. Neuron 1995; 15:1053-1063.
89. Hung AY, Haass C, Nitsch RM et al. Activation of protein kinase C inhibits cellular production of the amyloid beta-protein. J Biol Chem 1993; 268:22959-22962.
90. Hunter AJ, Roberts FF. The effect of pirenzepine on spatial learning in the Morris Water Maze. Pharmacol Biochem Behav 1988; 30:519-523.
91. Hutchison CW, Nathan PJ, Mrazek L et al. Cholinergic modulation of speed of early information processing: The effect of donepezil on inspection time. Psychopharmacol 2001; 155:440-442.
92. Iga Y, Arisawa H, Ise M et al. Modulation of rhythmical slow activity, long-term potentiation and memory by muscarinic receptor agonists. Eur J Pharmacol 1996; 308:13-19.
93. Imperato A, Puglisi-Allegra S, Casolini P et al. Changes in brain dopamine and acetylcholine release during and following stress are independent of the pituitary-adrenocortical axis. Brain Res 1991; 538:111-117.
94. Jerusalinsky D, Cervenansky C, Walz R et al. A peptide muscarinic toxin from the Green Mamba venom shows agonist- like action in an inhibitory avoidance learning task. Eur J Pharmacol 1993; 240:103-105.
95. Jerusalinsky D, Kornisiuk E, Alfaro P et al. Muscarinic toxins: Novel pharmacological tools for the muscarinic cholinergic system. Toxicon 2000; 38:747-761.
96. Jolkkonen M, Van Giersbergen PL, Hellman U et al. Muscarinic toxins from the black mamba Dendroaspis polylepis. Eur J Biochem 1995; 234:579-585.
97. Kakiuchi T, Ohba H, Nishiyama S et al. Age-related changes in muscarinic cholinergic receptors in the living brain: A PET study using N-[11C]methyl-4-piperidyl benzilate combined with cerebral blood flow measurement in conscious monkeys. Brain Res 2001; 916:22-31.
98. Karczmar AG. Synaptic, behavioral and toxicological effects of cholinesterase inhibitors in animals and humans. In: Giacobini E, ed. Cholinesterases and cholinesterases inhibitors. London: Martin Duniz Ltd., 2000:157-180.
99. Kimura H, McGeer PL, Peng F et al. Choline acetyltransferase-containing neurons in rodent brain demonstrated by immunohistochemistry. Science 1980; 208:1057-1059.
100. Kleinwächter G. Beobachtungen uber die Wirkung des Calabar-extract gegen Atropinvergiftung. Berl Klin Wschr 1864; 369-371.
101. Krnjevic K, Phillis JW. Sensitivity of cortical neurons to acetylcholine. Experientia 1961; 17:469-474.
102. Krnjevic K, Phillis JW. Pharmacological properties of acetylcholine-sensitive cells in the cerebral cortex. J Physiol 1963; 166:328-350.

103. Kubo T, Fukuda K, Mikami A et al. Cloning, sequencing and expression of complementary DNA encoding the muscarinic acetylcholine receptor. Nature 1986; 323:411-416.
104. Lachowicz JE, Lowe D, Duffy RA et al. SCH 57790: A novel M2 receptor selective antagonist. Life Sci 1999; 64:535-539.
105. Lai MK, Lai OF, Keene J et al. Psychosis of Alzheimer's disease is associated with elevated muscarinic M2 binding in the cortex. Neurology 2001; 57:805-811.
106. Leanza G, Muir J, Nilsson OG et al. Selective immunolesioning of the basal forebrain cholinergic system disrupts short-term memory in rats. Eur J Neurosci 1996; 8:1535-1544.
107. Leanza G, Nilsson OG, Wiley RG et al. Selective lesioning of the basal forebrain cholinergic system by intraventricular 192 IgG-saporin: Behavioural, biochemical and stereological studies in the rat. Eur J Neurosci 1995; 7:329-343.
108. Levey AI, Edmunds SM, Heilman CJ et al. Localization of muscarinic m3 receptor protein and M3 receptor binding in rat brain. Neurosci 1994; 63:207-221.
109. Levey AI, Edmunds SM, Hersch SM et al. Light and electron microscopic study of m2 muscarinic acetylcholine receptor in the basal forebrain of the rat. J Comp Neurol 1995; 351:339-356.
110. Levey AI, Edmunds SM, Koliatsos V et al. Expression of m1-m4 muscarinic acetylcholine receptor proteins in rat hippocampus and regulation by cholinergic innervation. J Neurosci 1995; 15:4077-4092.
111. Levey AI, Kitt CA, Simonds WF et al. Identification and localization of muscarinic acetylcholine receptor proteins in brain with subtype-specific antibodies. J Neurosci 1991; 11:3218-3226.
112. Longo VC. Behavioral and electroencephalographic effects of atropine and related compounds. Pharmacol Rev 1966; 18:965-996.
113. Longo VC. Neuropharmacology and behavior. S. Francisco: Freeman, 1972.
114. Macht DI. A pharmacodynamic analysis of the cerebral effect of atropin, homatropin, scopolamine and related drugs. J Pharmacol Exp Ther 1924; 22:35-48.
115. MacIntosh FC, Oboring PE. Release of acetylcholine from the intact cerebral cortex. XIX. Abstr. Int Physiol Congr 1955; 580-581.
116. Markram H, Segal M. Long-lasting facilitation of excitatory postsynaptic potentials in the rat hippocampus by acetylcholine. J Physiol 1990; 427:381-393.
117. Marston HM, West HL, Wilkinson LS et al. Effects of excitotoxic lesions of the septum and vertical limb nucleus of the diagonal band of Broca on conditional visual discrimination: Relationship between performance and choline acetyltransferase activity in the cingulate cortex. J Neurosci 1994; 14:2009-2019.
118. Mash DC, Flynn DT, Potter LT. Loss of M2 muscarinic receptors in the cerebral cortex in Alzheimer's disease and experimental cholinergic denervation. Science 1985; 228:1115-1117.
119. Matsui M, Motomura D, Karasawa H et al. Multiple functional defects in peripheral autonomic organs in mice lacking muscarinic acetylcholine receptor gene for the M3 subtype. Proc Natl Acad Sci USA 2000; 97:9579-9584.
120. McDonald MP, Wenk GL, Crawley JN. Analysis of galanin and the galanin antagonist M40 on delayed non matching-to-position performance in rats lesioned with the cholinergic immunotoxin 192 IgG-saporin. Behav Neurosci 1997; 111:552-563.
121. McGaughy J, Kaiser T, Sarter M. Behavioral vigilance following infusions of 192 IgG-saporin into the basal forebrain: Selectivity of the behavioral impairment and relation to cortical AChE-positive fiber density. Behav Neurosci 1996; 110:247-265.
122. McGaughy J, Sarter M. Sustained attention performance in rats with intracortical infusions of 192 IgG-saporin-induced cortical cholinergic deafferentation: Effects of physostigmine and FG 7142. Behav Neurosci 1998; 112:1519-1525.
123. McGeer EG, McGeer PL. The importance of inflammatory mechanisms in Alzheimer disease. Exp Gerontol 1998; 33:371-378.
124. Messamore E, Bogdanovich N, Schroder H et al. Astrocytes associated with senile plaques possess muscarinic acetylcholine receptors. Neuro Report 1994; 5:1473-1476.
125. Messer Jr WS, Ellerbrock B, Price M et al. Autoradiographic analyses of agonist binding to muscarinic receptor subtypes. Biochem Pharmacol 1989; 38:837-850.
126. Mesulam MM, Mufson EJ, Vainer BH et al. Central cholinergic pathways in the rat: An overview based on an alternative numenclature (Ch1-Ch6). Neuroscience 1983; 10:1185-1201.
127. Migeon JC, Nathanson NM. Differential regulation of cAMP-mediated gene transcription by m1 and m4 muscarinic acetylcholine receptors. Preferential coupling of m4 receptors to Gi alpha-2. J Biol Chem 1994; 269:9767-9773.
128. Miyakawa T, Yamada M, Duttaroy A et al. Hyperactivity and intact hippocampus-dependent learning in mice lacking the M1 muscarinic acetylcholine receptor. J Neurosci 2001; 21:5239-5250.

129. Mizuno T, Endo Y, Arita J et al. Acetylcholine release in the rat hippocampus as measured by the microdialysis method correlates with motor activity and exhibits a diurnal variation. Neuroscience 1991; 44:607-612.
130. Molinengo L. The action of scopolamine on retrieval and memory storage in rats evaluated in the staircase maze. Behav Neural Biol 1993; 59:18-24.
131. Moore H, Sarter M, Bruno JP. Age-dependent modulation of in vivo cortical acetylcholine release by benzodiazepine receptor ligands. Brain Res 1992; 596:17-29.
132. Moss DE, Deutsch JA. Review of cholinergic mechanisms and memory. In: Waser PG, ed. Cholinergic Mechanisms. New York: Raven Press, 1975:483-492.
133. Mrzljak L, Levey AI, Goldman-Rakic PS. Association of m1 and m2 muscarinic receptor proteins with asymmetric synapses in the primate cerebral cortex: Morphological evidence for cholinergic modulation of excitatory neurotransmission. Proc Natl Acad Sci USA 1993; 90:5194-5198.
134. Muccioli A. Tossicologia Moderna. Roma: Tipografia della Camera dei Deputati. 1895.
135. Mufson EJ, Jaffar S, Levey AI. m2 muscarinic acetylcholine receptor-immunoreactive neurons are not reduced within the nucleus basalis in Alzheimer's disease: Relationship with cholinergic and galaninergic perikarya. J Comp Neurol 1998; 392:313-329.
136. Murai S, Saito H, Masuda Y et al. AF64A disrupts retrieval processes in long-term memory of mice. Neuro Report 1995; 6:349-352.
137. Narang N. In situ determination of M1 and M2 muscarinic receptor binding sites and mRNAs in young and old rat brains. Mech Ageing Dev 1995; 78:221-239.
138. Nguyen PV, Kandel ER. Brief theta-burst stimulation induces a transcription-dependent late phase of LTP requiring cAMP in area CA1 of the mouse hippocampus. Learning Memory 1997; 4:230-243.
139. Nilsson OG, Kalen P, Rosengren E et al. Acetylcholine release in the rat hippocampus as studied by microdialysis is dependent on axonal impulse flow and increases during behavioural activation. Neuroscience 1990; 36:325-338.
140. Nitsch RM, Slack BE, Wurtman RJ et al. Release of Alzheimer amyloid precursor derivatives stimulated by activation of muscarinic acetylcholine receptors. Science 1992; 258:304-307.
141. Nordberg A, Larsson C, Adolfsson R et al. Muscarinic receptor compensation in hippocampus of Alzheimer patients. J Neural Transm 1983; 56:13-19.
142. Olianas MC, Adem A, Karlsson E et al. Rat striatal muscarinic receptors coupled to the inhibition of adenylyl cyclase activity: Potent block by the selective m4 ligand muscarinic toxin 3 (MT3). Br J Pharmacol 1996; 118:283-288.
143. Olivera BM, Gray WR, Zeikus R et al. Peptide neurotoxin from fish-hunting cone snails. Science 1985; 230:1338-1341.
144. Orsetti M, Casamenti F, Pepeu G. Enhanced acetylcholine release in the hippocampus and cortex during acquisition of an operant behavior. Brain Res 1996; 724:89-96.
145. Passetti F, Dalley JW, O'Connell MT et al. Increased acetylcholine release in the rat medial prefrontal cortex during performance of a visual attentional task. Eur J Neurosci 2000; 12:3051-3058.
146. Pazzagli A, Pepeu G. Amnesic properties of scopolamine and brain acetylcholine in the rat. Int J Pharmacol 1964; 4:291-299.
147. Pepeu G. The release of acetylcholine from the rat brain: An approach to the study of central cholinergic mechanism. Progr Neurobiol 1973; 2:257-288.
148. Pepeu G, Casamenti F. Lesioning the nucleus basalis. In: Conn PM, ed. Methods in Neuroscience. San Diego: Academic Press, 1991:139-150.
149. Peralta EG, Ashkenazi A, Winslow JW et al. Distinct primary structures, ligand-binding properties and tissue- specific expression of four human muscarinic acetylcholine receptors. EMBO J 1987; 6:3923-3929.
150. Pizzo DP, Waite JJ, Thal LJ et al. Intraparenchymal infusions of 192 IgG-saporin: Development of a method for selective and discrete lesioning of cholinergic basal forebrain nuclei. J Neurosci Methods 1999; 91:9-19.
151. Quirion R, Wilson A, Rowe W et al. Facilitation of acetylcholine release and cognitive performance by an M2-muscarinic receptor antagonist in aged memory-impaired rats. J Neurosci 1995; 15:1455-1462.
152. Raffaele KC, Asthana S, Berardi A et al. Differential response to the cholinergic agonist arecoline among different cognitive modalities in Alzheimer's disease. Neuropsychopharmacol 1996; 15:163-170.
153. Raiteri M, Marchi M, Paudice P. Adaptation of presynaptic acetylcholine autoreceptors following long- term drug treatment. Eur J Pharmacol 1981; 74:109-110.
154. Rawlins JN, Olton DS. The septo-hippocampal system and cognitive mapping. Behav Brain Res 1982; 5:331-358.

155. Reever CM, Ferrari-DiLeo G, Flynn DD. The M5 (m5) receptor subtype: Fact or fiction? Life Sci 1997; 60:1105-1112.
156. Riker WF, Wescoe WC. The pharmacology of flaxedil with observations on certain analogs. Ann N Y Acad Sci 1951; 54:373-394.
157. Roozendaal B, Van der Zee EA, Hensbroek RA et al. Muscarinic acetylcholine receptor immunoreactivity in the amygdala—II. Fear-induced plasticity. Neurosci 1997; 76:75-83.
158. Rosenblad C, Nilsson OG. Basal forebrain grafts in the rat neocortex restore in vivo acetylcholine release and respond to behavioural activation. Neurosci 1993; 55:353-362.
159. Rosenblum K, Futter M, Jones M et al. ERKI/II regulation by the muscarinic acetylcholine receptors in neurons. J Neurosci 2000; 20:977-985.
160. Rossner S, Schliebs R, Hartig W et al. 192IGG-saporin-induced selective lesion of cholinergic basal forebrain system: Neurochemical effects on cholinergic neurotransmission in rat cerebral cortex and hippocampus. Brain Res Bull 1995; 38:371-381.
161. Rossner S, Schliebs R, Perez-Polo JR et al. Differential changes in cholinergic markers from selected brain regions after specific immunolesion of the rat cholinergic basal forebrain system. J Neurosci Res 1995; 40:31-43.
162. Rouse ST, Marino MJ, Potter LT et al. Muscarinic receptor subtypes involved in hippocampal circuits. Life Sci 1999; 64:501-509.
163. Rowntree CI, Bland BH. An analysis of cholinoceptive neurons in the hippocampal formation by direct microinfusion. Brain Res 1986; 362:98-113.
164. Rudy JW. Scopolamine administered before and after training impairs both contextual and auditory-cue fear conditioning. Neurobiol Learn Mem 1996; 65:73-81.
165. Rusted JM, Warburton DM. Effects of scopolamine on verbal memory; A retrieval or acquisition deficit? Neuropsychobiol 1989; 21:76-83.
166. Sainsbury RS, Heynen A, Montoya CP. Behavioral correlates of hippocampal type 2 theta in the rat. Physiol Behav 1987; 39:513-519.
167. Sarter M, Givens B, Bruno JP. The cognitive neuroscience of sustained attention: Where top-down meets bottom-up. Brain Res Brain Res Rev 2001; 35:146-160.
168. Schafe GE, Atkins CM, Swank MW et al. Activation of ERK/MAP kinase in the amygdala is required for memory consolidation of pavlovian fear conditioning. J Neurosci 2000; 20:8177-8187.
169. Schafer MK, Weihe E, Varoqui H et al. Distribution of the vesicular acetylcholine transporter (VAChT) in the central and peripheral nervous systems of the rat. J Mol Neurosci 1994; 5:1-26.
170. Schliebs R. Basal forebrain cholinergic dysfunction-experimental approaches and the diseased brain. Int J Dev Neurosci 1998; 16:591-593.
171. Segal M, Auerbach JM. Muscarinic receptors involved in hippocampal plasticity. Life Sci 1997; 60:1085-1091.
172. Selcher JC, Atkins CM, Trzaskos JM et al. A necessity for MAP kinase activation in mammalian spatial learning. Learn Mem 1999; 6:478-490.
173. Shen J, Barnes CA, Wenk GL et al. Differential effects of selective immunotoxic lesions of medial septal cholinergic cells on spatial working and reference memory. Behav Neurosci 1996; 110:1181-1186.
174. Shimoshige Y, Maeda T, Kaneko S et al. Involvement of M2 receptor in an enhancement of long-term potentiation by carbachol in Schaffer collateral-CA1 synapses of hippocampal slices. Neurosci Res 1997; 27:175-180.
175. Shute CCD, Lewis PR. Cholinesterase –containing systems of the brain of the rat. Nature 1963; 199:1164-1169.
176. Shute CCD, Lewis PR. Cholinesterase –containing pathways of the hindbrain: Afferent cerebellar and centrifugal cochlear fibres. Nature 1965; 205:242-246.
177. Smith CJ, Perry EK, Perry RH et al Guanine nucleotide modulation of muscarinic cholinergic receptor binding in postmortem human brain—A preliminary study in Alzheimer's disease. Neurosci Lett 1987; 82:227-232.
178. Steckler T, Keith AB, Wiley RG et al. Cholinergic lesions by 192 IgG-saporin and short-term recognition memory: Role of the septohippocampal projection. Neuroscience 1995; 66:101-114.
179. Stewart M, Fox SE. Do septal neurons pace the hippocampal theta rhythm? Trends Neurosci 1990; 13:163-168.
180. Stoehr JD, Mobley SL, Roice D et al. The effects of selective cholinergic basal forebrain lesions and aging upon expectancy in the rat. Neurobiol Learn Mem 1997; 67:214-227.
181. Sugita S, Uchimura N, Jiang ZG et al. Distinct muscarinic receptors inhibit release of gamma-aminobutyric acid and excitatory amino acids in mammalian brain. Proc Natl Acad Sci USA 1991; 88:2608-2611.

182. Svensson AL, Warpman U, Hellstrom-Lindahl E et al. Nicotinic receptors, muscarinic receptors and choline acetyltransferase activity in the temporal cortex of Alzheimer patients with differing apolipoprotein E genotypes. Neurosci Lett 1997; 232:37-40.
183. Sweatt JD. The neuronal MAP kinase cascade: A biochemical signal integration system subserving synaptic plasticity and memory. J Neurochem 2001; 76:1-10.
184. Tariot PN, Cohen RM, Welkowitz JA et al. Multiple-dose arecoline infusions in Alzheimer's disease. Arch Gen Psychiatry 1988; 45:901-905.
185. Taussig R, Quarmby LM, Gilman AG. Regulation of purified type I and type II adenylylcyclases by G protein beta gamma subunits. J Biol Chem 1993; 268:9-12.
186. Thal LJ, Fuld PA, Masur DM et al. Oral physostigmine and lecithin improve memory in Alzheimer disease. Ann Neurol 1983; 13:491-496.
187. Thiel CM, Huston JP, Schwarting RK. Hippocampal acetylcholine and habituation learning. Neurosci 1998; 85:1253-1262.
188. Tietje KM, Goldman PS, Nathanson NM. Cloning and functional analysis of a gene encoding a novel muscarinic acetylcholine receptor expressed in chick heart and brain. J Biol Chem 1990; 265:2828-2834.
189. Tobias JM, Lipton MA, Lepinat AA. Effects of anaesthetics and convulsants on brain acetylcholine content. Proc Soc exp Biol Med 1946; 61:51-54.
190. Torres EM, Perry TA, Blockland A et al. Behavioural, histochemical and biochemical consequences of selective immunolesions in discrete regions of the basal forebrain cholinergic system. Neurosci 1994; 63:95-122.
191. Ungerstedt U. Measurement of neurotransmitter release by intracranial dialysis. In: Marsden CA, ed. Measurement of neurotransmitter release in vivo. New York: John Wiley, 1984:81-105.
192. Van der Zee EA, Compaan JC, Bohus B et al. Alterations in the immunoreactivity for muscarinic acetylcholine receptors and colocalized PKC gamma in mouse hippocampus induced by spatial discrimination learning. Hippocampus 1995; 5:349-362.
193. Van der Zee EA, De Jong GI, Strosberg AD et al. Muscarinic acetylcholine receptor-expression in astrocytes in the cortex of young and aged rats. Glia 1993; 8:42-50.
194. Van der Zee EA, Douma BR, Bohus B et al. Passive avoidance training induces enhanced levels of immunoreactivity for muscarinic acetylcholine receptor and coexpressed PKC gamma and MAP- 2 in rat cortical neurons. Cereb Cortex 1994; 4:376-390.
195. Van der Zee EA, Kronforst-Collins MA, Maizels ET et al. Gamma Isoform-selective changes in PKC immunoreactivity after trace eyeblink conditioning in the rabbit hippocampus. Hippocampus 1997; 7:271-285.
196. Van der Zee EA, Luiten PG. Muscarinic acetylcholine receptors in the hippocampus, neocortex and amygdala: A review of immunocytochemical localization in relation to learning and memory. Prog Neurobiol 1999; 58:409-471.
197. Vannucchi MG, Pepeu G. Muscarinic receptor modulation of acetylcholine release from rat cerebral cortex and hippocampus. Neurosci Lett 1995; 190:53-56.
198. Vannucchi MG, Scali C, Kopf SR et al. Selective muscarinic antagonists differentially affect in vivo acetylcholine release and memory performances of young and aged rats. Neurosci 1997; 79:837-846.
199. Vilaro MT, Mengod G, Palacios JM. Advances and limitations of the molecular neuroanatomy of cholinergic receptors: The example of multiple muscarinic receptors. Prog Brain Res 1993; 98:95-101.
200. Vitiello B, Martin A, Hill J et al. Cognitive and behavioral effects of cholinergic, dopaminergic and serotonergic blockade in humans. Neuropsychopharmacol 1997; 16:15-24.
201. Vnek N, Kromer LF, Wiley RG et al. The basal forebrain cholinergic system and object memory in the rat. Brain Res 1996; 710:265-270.
202. Waite JJ, Thal LJ. Lesions of the cholinergic nuclei in the rat basal forebrain: Excitotoxins vs. an immunotoxin. Life Sci 1996; 58:1947-1953.
203. Waite JJ, Wardlow ML, Power AE. Deficit in selective and divided attention associated with cholinergic basal forebrain immunotoxic lesion produced by 192-saporin; motoric/sensory deficit associated with Purkinje cell immunotoxic lesion produced by OX7-saporin. Neurobiol Learn Mem 1999; 71:325-352.
204. Walsh TJ, Herzog CD, Gandhi C et al. Injection of IgG 192-saporin into the medial septum produces cholinergic hypofunction and dose-dependent working memory deficits. Brain Res 1996; 726:69-79.
205. Walsh TJ, Kelly RM, Dougherty KD et al. Behavioral and neurobiological alterations induced by the immunotoxin 192-IgG-saporin: Cholinergic and noncholinergic effects following i.c.v. injection. Brain Res 1995; 702:233-245.

206. Wang SZ, Zhu SZ, Mash DC et al. Comparison of the concentration of messenger RNA encoding four muscarinic receptor subtypes in control and Alzheimer brains. Brain Res Mol Brain Res 1992; 16:64-70.
207. Warpman U, Alafuzoff I, Nordberg A. Coupling of muscarinic receptors to GTP proteins in postmortem human brain—alterations in Alzheimer's disease. Neurosci Lett 1993; 150:39-43.
208. Wenk GL, Stoehr JD, Quintana G et al. Behavioral, biochemical, histological and electrophysiological effects of 192 IgG-saporin injections into the basal forebrain of rats. J Neurosci 1994; 14:5986-5995.
209. Wesnes K, Warburton DM. Effects of scopolamine and nicotine on human rapid information processing performance. Psychopharmacol 1984; 82:147-150.
210. Wess J. Molecular basis of muscarinic acetylcholine receptor function. Trends Pharmacol Sci 1993; 14:308-313.
211. Whishaw IQ. Dissociating performance and learning deficits on spatial navigation tasks in rats subjected to cholinergic muscarinic blockade. Brain Res Bull 1989; 23:347-358.
212. Whishaw IQ, Dyck R. Comparative potency of tactile, auditory, and visual stimulus repetition in eliciting activated forebrain EEG in the rabbit. Behav Neurosci 1984; 98:333-344.
213. Whishaw IQ, Vanderwolf CH. Hippocampal EEG and behavior: Changes in amplitude and frequency of RSA (theta rhythm) associated with spontaneous and learned movement patterns in rats and cats. Behav Biol 1973; 8:461-484.
214. Wiley RG. Neural lesioning with ribosome-inactivating proteins: Suicide transport and immunolesioning. Trends Neurosci 1992; 15:285-290.
215. Williams S, Johnston D. Muscarinic depression of long-term potentiation in CA3 hippocampal neurons. Science 1988; 242:84-87.
216. Woolf NJ. Cholinoceptive cells in rat cerebral cortex: Somatodendritic immunoreactivity for muscarinic receptor and cytoskeletal proteins. J Chem Neuroanat 1993; 6:375-390.
217. Wrenn CC, Lappi DA, Wiley RG. Threshold relationship between lesion extent of the cholinergic basal forebrain in the rat and working memory impairment in the radial maze. Brain Res 1999; 847:284-298.
218. Wrenn CC, Wiley RG. The behavioral functions of the cholinergic basal forebrain: Lessons from 192 IgG-saporin. Int J Dev Neurosci 1998; 16:595-602.
219. Yamada M, Lamping KG, Duttaroy A et al. Cholinergic dilation of cerebral blood vessels is abolished in M(5) muscarinic acetylcholine receptor knockout mice. Proc Natl Acad Sci USA 2001; 98:14096-14101.
220. Yamada M, Miyakawa T, Duttaroy A et al. Mice lacking the M3 muscarinic acetylcholine receptor are hypophagic and lean. Nature 2001; 410:207-212.
221. Yeomans J, Forster G, Blaha C. M5 muscarinic receptors are needed for slow activation of dopamine neurons and for rewarding brain stimulation. Life Sci 2001; 68:2449-2456.
222. Zaborszky L, Pang K, Somogyi J et al. The basal forebrain corticopetal system revisited. Ann N Y Acad Sci 1999; 877:339-367.
223. Zang Z, Creese I. Differential regulation of expression of rat hippocampal muscarinic receptor subtypes following fimbria-fornix lesion. Biochem Pharmacol 1997; 53:1379-1382.
224. Zhang W, Basile A, Gomeza J et al. Characterization of central inhibitory muscarinic autoreceptors by the useof muscarinic acetylcholine receptor knock-out mice. J Neurosci 2002; 22:1709-1717.
225. Zubieta JK, Koeppe RA, Frey KA et al. Assessment of muscarinic receptor concentrations in aging and Alzheimer disease with [11C]NMPB and PET. Synapse 2001; 39:275-287.

Chapter 2.4

Acetylcholine: II. Nicotinic Receptors

Joyce Besheer and Rick A. Bevins

Abstract

The nicotinic cholinergic system has been widely implicated in mediating learning and/or memory processes in human and nonhuman animals. This chapter highlights various areas of basic research in which stimulation or blockade of nicotinic acetylcholine receptors (nAChRs) has been shown to affect an animal's performance in a variety of tasks thought to measure learning and memory. For example, under certain conditions, stimulation of nAChRs by nicotine (or other nAChRs agonists) can enhance working memory of primates as measured in a delayed matching-to-sample task. Attentional processes are also improved in rats as indexed by a five-choice serial reaction time task. Further, recent research suggests that stimulation of nAChRs by nicotine likely enhances the incentive salience of stimuli. We elaborate on a model by which this enhancement might occur and suggest that the role of this incentive mechanism in relation to learning and memory processes requires more empirical attention. Finally, there appears to be overlap in the processes by which nAChRs affect learning and memory. That is, enhanced incentive salience might be responsible for the increased attentional effects of nicotine, or vice versa. Subsequent research needs to refine the behavioral techniques so as better dissociate, if required, these mechanisms.

Introduction

A survey of the neuropharmacology literature would likely leave even the most critical individual convinced that nicotinic acetylcholine receptors (nAChRs) are involved in learning and/or memory processes in human and nonhuman animals. For example, in humans nAChRs have been implicated in memory and learning difficulties displayed by patients suffering from Alzheimer's disease[66] and attention deficit disorder.[58] In rodent and nonhuman primate models, manipulation of nAChRs can alter performance in such learning tasks as radial-arm maze,[49,52,53,65] Morris water maze,[1,26] T-maze,[6] delayed matching[13] and nonmatching to sample,[38] delayed matching-to-position,[31] 5-choice serial reaction time,[11,41,64] environmental familiarization,[9] passive avoidance,[28,74] signal detection,[14] latent inhibition,[79] "learned helplessness",[36] and context conditioning.[40] In the past decade or so, there have been several thought-provoking reviews on the role of nAChRs in learning, cognition, and memory.[18,56,78] We encourage readers interested in this aspect of nicotinic receptor functioning to seek these other reviews because of the vastness of the relevant literature and the differences in emphasis between reviews—including the present review.

Neuronal nAChRs

Before any detailed discussion of the functional role of nAChRs in learning and memory processes, it may be helpful to provide a brief overview of the main subtypes of neuronal nAChRs and their neuroanatomical localization within the central nervous system (CNS). For a comprehensive review of brain nAChRs we refer the reader to Changeux et al.[18]

From Messengers to Molecules: Memories Are Made of These, edited by Gernot Riedel and Bettina Platt.

Table 1. Distribution of cholinergic neurons in the CNS

Descriptor	CNS Localization	Projection Areas
Ch1	medial septal nucleus	hippocampal complex, limbic cortex
Ch2	ventral nucleus of the diagonal band	hippocampal complex, limbic cortex
Ch3	horizontal limb nucleus of the diagonal band, magnocellular preoptic area	olfactory bulbs, limbic cortex, amygdala
Ch4	nucleus basalis of Meynert	amygdala, neocortex
Ch5, Ch6	pedunculopontine tegmental nucleus, laterodorsal tegmental nucleus, central gray, locus coeruleus	thalamus, substantia nigra (ventral tegmental area and pars compacta), reticular formation, locus coeruleus, cingulate gyrus, subicular cortices, medial prefrontal cortex
Ch7	medial habenula	interpeduncular nucleus
Ch8	parabigeminal nucleus	superior colliculus
Striatal	nucleus accumbens, olfactory tubercle, caudate, putamen, island of Calleja	local interneurons
Hypothalamic	hypothalamus	local interneurons, neocortex

Based on refs. 34, 62,63,98.

Subtypes

Nicotinic acetylcholine receptors are ligand-gated ion channels. This receptor is composed of five polypeptide subunits that form a barrel-like structure around a central ion channel.[21] In contrast to nAChRs located in the periphery which are composed of $\alpha1$, $\beta1$, δ, ε, γ subunits, the standard configuration of the neuronal nAChRs include combinations of α and β subunits. However, $\alpha7$, $\alpha8$, and $\alpha9$ subunits can also form functional nAChRs that consist of a single subunit type.[25] Presently, the subunits $\alpha2$-$\alpha7$ and $\beta2$-$\beta4$ have been identified in the human brain and the distribution of these subunits in the human brain are presently being examined (see ref. 71 for a review of the nAChRs in the human brain). As in other receptor systems, much more work has examined the distribution of nAChR subunits in the rodent brain. In rodent models, the $\alpha3$, $\alpha7$, $\beta2$ subunits, and to a lesser extent the $\alpha4$ subunit are expressed in the hippocampus (see ref. 73). The $\alpha2$, $\alpha4$, $\alpha5$, $\alpha7$, $\beta2$, and $\beta4$ subunits have been identified in the interpeduncular nucleus, and expression of the $\alpha3$, $\alpha4$, $\alpha7$, and $\beta2$ subunits have been reported in the amygdala.[89] For a more detailed description of the distribution of the various subunits see Arneric et al[2] and Shacka and Robinson.[89] Most of the neuronal nAChRs contain $\alpha4\beta2$ or $\alpha7$ subunits[37,90,94,99] and these are the subunits that have been most commonly studied in learning and memory tasks (see later). Although, this chapter focuses on the contribution of central nAChR processes in learning and memory, the role of nAChRs located in the peripheral nervous system should not be ignored (see ref. 89 for discussion of peripheral nAChRs).

Localization

Acetylcholine (ACh) is the endogenous transmitter substance that binds to functional nAChRs. Much of the research localizing these ACh-utilizing (cholinergic) neurons in the central nervous system has employed immunohistochemistry staining for choline acetyltransferase (ChAT); an enzyme required for synthesis of ACh. Table 1 provides a summary of the CNS localization of major clusters of ChAT-containing neurons in mammals. Also, included in (Table 1) are some of the notable brain regions that receive projections from these cholinergic cells. The "Ch" nomenclature allows for simpler designation of diffuse collections of cholinergic neurons.[62,95] For example, Ch3 is located in the basal forebrain and includes neurons in the horizontal limb nucleus of the diagonal band and magnocellular preoptic area of the hypothalamus. We will reference back to this table when discussing the functional importance of nAChRs. For example, cholinergic input from Ch1 may affect learning by modulating $\alpha7^*$ nAChRs in the hippocampus (see later). Because a comprehensive review of the cholinergic system is tangential to our goal, we refer the reader to the following reports for a more detailed discussion of the cholinergic system: Mesulam (refs. 62,63) and Woolf (ref. 98).

Memory

Manipulation of nAChRs has been shown to affect performance in a variety of tasks that assess memory functioning. In nonhuman animals, a majority of this work has been conducted in tasks designed to assess working memory processes. Working memory is defined by Feldman et al[34] as "encoding of task-specific information over short periods of time (e.g., within a single trial or test session)" (p. 272). The delayed matching-to-sample (DMTS) task is a commonly used preparation for assessing working memory. Briefly, in this task an animal is presented with a "sample" stimulus. Following a delay, the animal is presented with the sample stimulus and one or more novel stimuli. During this choice test, the animal is rewarded for choosing the sample stimulus. Thus, working memory processes are recruited to encode the information of the sample stimulus in order to make a correct response during the choice test. That is, working memory allows the animal to discriminate between the sample stimulus and the other stimuli.

The DMTS task has been widely used to examine the role of nAChRs. For example, Buccafusco et al[13] trained mature pig-tailed and rhesus monkeys on the DMTS task using colored lights as the stimuli. After acquisition training, the monkeys were tested at 4 different delays: zero, short, medium, and long. At the zero delay, the choice test occurred immediately following the presentation of the sample stimulus. The other delays were adjusted for each monkey's skill level; on average the short delay was 10.6 sec, the medium delay 39.4 sec, and the long delay 79.4 sec. Presumably, lengthening the delay impairs performance in this task because of the limited capacity of working memory processes.[31] Indeed, performance ranged from 97% correct after the zero delay to 58% correct after the long delay. In order to assess if stimulation of the nicotinic receptors could enhance working memory performance, the animals were administered ABT-418, a nAChR agonist (i.e., cholinergic channel activator). ABT-418 did not affect performance at the zero, short, or medium delays. However, at the long delay, DMTS performance was enhanced by treatment with ABT-418. After a washout period, all animals were tested with nicotine, an agonist at nAChRs with a high affinity for $\alpha4\beta2^*$.[20] Similar to ABT-418, nicotine enhanced performance only at the longest delay.

The inability of nAChR agonists to improve performance in the DMTS task at the shorter delays likely indicates that performance was near optimal levels at the shorter delays, thus making an improvement difficult to observe. However, improvement at the long delay suggests that stimulation of the nAChRs affected memory processes when strained. Recall that at the long delay performance was impaired (58% correct versus 97% correct after no delay). As mentioned earlier, impaired performance in this task after long delays is taken to suggest a disturbance in the capacity of working memory.[31] Thus, one possibility for the enhancement in performance is that the agonists stimulated working memory processes by enhancing the capacity to store the information of the sample stimuli. Given that a major component of working memory tasks involve attention,[12] another tenable and related possibility, is that attentional processes were enhanced such that the neural representation of the sample stimulus was stored/encoded more efficiently. Indeed, stimulation of nAChRs has been shown to enhance attentional processes (see later).

Further, there is a wealth of research examining the role of nAChRs in mediating performance in other tasks that include a working memory component. For example, Levin et al[53-55] has repeatedly found that chronic nicotine treatment enhances performance in a win-shift version of a radial arm maze. In this particular version of the task, the arms of the maze are baited and entries into arms that had been previously visited are scored as errors. The number of errors is taken as a measure of working memory function given that the rat must encode the information about the location of visited arms. Similar to chronic nicotine, AR-R17779, an agonist for the $\alpha7$ subunit, enhances performance in this task.[57] Further, Felix and Levin[35] found that methyllycaconitine (MLA), an antagonist specific for the $\alpha7$ subunit, or dihydro-β-erythroidine (DHBE), an antagonist with affinity for the $\alpha4\beta2^*$ nAChR, infused into the ventral hippocampus impaired performance in the win-shift version of the radial-arm maze. This impairment suggests a role for hippocampal $\alpha4\beta2^*$ and $\alpha7^*$ nAChRs in working memory processes.

Interestingly, the greatest performance improvements resulting from stimulation of nAChRs are commonly reported in tasks that require effortful processing[38,84] or in tasks in which a deficit is produced.[29,56] For example, in the DMTS task described earlier, enhanced performance was not observed until the delay induced a severe deficit in performance. Similar findings have been reported using a novel-object detection task (also referred to as object recognition). In that task, a rat is presented with two identical sample objects. After a delay, the rat is presented with a novel object and one of the previously experienced objects. Rats display a tendency to interact more with the novel object than the familiar object when the delay is 1 h (i.e., delayed nonmatching-to-sample; see refs. 7 and 33). However, Puma et al[77] found that after a 24 h delay rats did not discriminate between the objects (i.e., equal time with the novel and familiar object). Administration of nicotine after exposure to the sample objects reversed this "deficit;" rats spent more time interacting with the novel object during the test that occurred 24 h later. Presumably, stimulation of the nAChRs enhanced the retention (consolidation) of the information about the sample objects during the long delay.[77] Nicotine has also been reported to enhance retention in a passive avoidance task.[74] Interestingly, in that study, mice lacking the $\beta2$ subunit of the nAChR did not show the nicotine-induced enhancement in retention, suggesting a role for this subunit in retention processes (e.g., consolidation, encoding, etc.).

Deficits in performance induced by lesions have also been reversed by nAChR stimulation. For example, Decker et al[29] found that lesions of the septal area, which reduced cholinergic input to the hippocampus (see Table 1), induced an impairment in a spatial discrimination version of a Morris water maze task. Administration of ABT-418 reversed the lesion-induced deficit, but had no effect on intact controls. Interestingly, this drug was subsequently tested by Potter et al[76] in patients with early to moderate Alzheimer's Disease, a disease which is accompanied by memory impairments and degeneration of the cholinergic system. ABT-418 treated patients tested on the Selective Reminding Task had improved recall across a 6 h testing period.

Interestingly, stimulation of nAChRs does not appear to affect performance in tasks that involve reference memory (i.e., use of the same information across trials).[12] For example, administration of SIB-1553A, a $\beta4^*$ nAChR agonist, did not affect performance in mice that were trained to discriminate between a baited and unbaited arm of a T-maze.[12] Similarly, Levin et al[55] found that chronic nicotine treatment did not affect performance in a 16-arm radial maze in which specific arms were repeatedly unbaited. Further, Granon et al[38] reported that antagonism of nAChR by administration of neuronal bungarotoxin (NBT) did not affect performance in a reference memory component of a T-maze task; NBT, however, impaired working memory. Notably, mecamylamine, a noncompetitive nAChR antagonist, infused into the hippocampus did not affect reference memory, but impaired working memory.[67] This result suggests that nAChRs in the hippocampus are involved in mediating working memory, but not reference memory processes.

There clearly exists a wealth of research examining the role of nAChRs in mediating performance in a variety of tasks. Stimulation of nAChRs generally appear to enhance performance in working memory tasks and tasks that involve retention of information across a delay. Further, nAChR processes in working memory and retention (consolidation) mechanisms appear to be especially important in restoring performance induced by degeneration of the cholinergic system, whether the deficit is naturally occurring (i.e., aging) or induced by a lesion. From our reading of the literature, an important factor that may contribute to this restoration includes alterations in attentional processes mediated by nAChRs (see following section). Presumably, increasing attention to specific stimuli would result in better encoding of the information. In turn, this information would be neurally retained more effectively and be more readily available for future use. Further, in a later section (Rewarding/Incentive Effects), we propose another mechanism that may contribute to the enhanced performance observed with nicotinic receptor stimulation.

Attention

The cholinergic system has been implicated in mediating attentional processes in humans and nonhuman animals.[11,50,59,64,85,96] For example, in human studies, smoking a cigarette (e.g., nicotine administration) before presentation of a word list improved the number of words correctly recalled during a later test.[83,96] Specifically, Warburton et al[96] found that more words from the latter part of the list were recalled; this pattern was consistent with an attentional explanation to the extent that attention diminishes towards the latter part of the list. Similarly, Rusted and Eaton-Williams[83] found that nicotine-induced accuracy improvements in word recall was related to the length of the word list. That is, a greater improvement was observed after presentation of a 30-item word list, than a 10-item word list. Nicotine delivered by a transdermal patch has also been shown to enhance performance in a Random Letter Generation task (e.g., participants required to name letters of the alphabet in a random order) and a Stroop test (e.g., participants required to read the ink color of color words), presumably by enhancing attentional processes.[59]

Nonhuman animal studies have also focused on nAChR involvement in attentional processes. The five-choice serial reaction time (5-CSRT) task has been used in rodents to assess the role of various nicotinic agonists and antagonists in attentional processes. Commonly, the apparatus used for this task includes a wall with five distinct holes, each with a light at the rear of the hole. During training, one of the five holes is illuminated for a brief duration. A correct response is registered for nose pokes during the period that the light is illuminated or for a fixed interval after the offset of the light. Daily training sessions usually include about 100 trials (i.e., random light illuminations). Thus, accurate performance in this task involves sustained attention and vigilance throughout the entire session.

Mirza and Stolerman[64] found that increasing the time between each light presentation resulted in a performance decrement, likely because prolonged vigilance was necessary to maintain correct responding. Nicotine administration reversed this deficit suggesting that attentional processes were enhanced. Interestingly, shortening the duration of the stimulus illumination (i.e., weakening the signal strength), decreased correct responding, and increased the latency to make a response.[11] According to the authors, this data pattern indicates that information processing, not necessarily attentional processing, is impaired. Under these conditions, nicotine administration did not enhance correct responding. Together these results suggest that stimulation of nAChRs by nicotine can enhance attentional processes, but may not necessarily affect informational processing (see ref. 92 for a review of the effects of nicotine in the 5-CSRT task). Further, using the short duration light stimulus, Blondel et al[11] replicated the findings of Mirza and Stolerman[64] in that nicotine administration did not affect the number of correct responses. However, the authors did find that nicotine decreased the latency to make a correct response and increased anticipatory responses.

To further assess the role of the specific subunits of the nicotinic receptor that may contribute to attentional processes, Grottick and Higgins[41] assessed various compounds in rats that had failed to meet criterion during 5-CSRT task training. Presumably, attentional/vigilance processes in these rats were slightly impaired given that they had failed to meet the predetermined criterion. Rats were administered nicotine, AR-R 17779 (an $\alpha 7^*$ agonist), or SIB 1765F (an α4β2* agonist). Both nicotine and SIB 1765F improved performance by increasing correct responding and enhancing reaction time. In contrast, AR-R 17779 did not affect correct responding. This data pattern was taken to suggest involvement of the α4β2, but not the α7 subunits, in mediating increased attention/vigilance. Additionally, SIB-1553A, a β4* nAChR agonist, had no effect on correct responding in the 5-CSRT task in aged rats, whereas performance in aged rats was enhanced by nicotine administration.[43] Taken together this work suggests that the β4* nAChR is not involved in attentional processes as measured in a 5-CSRT task.

In sum, performance on tasks designed to measure attention processes can be enhanced by stimulation of nicotinic receptors. Specifically, the α4 and β2 subunits appear to contribute to this enhancing effect. Notably, researchers have begun to examine the feasibility of using nAChR compounds as potential therapeutic agents for attention deficit/hyperactivity disorder (ADHD). Nicotine delivered by a transdermal patch to adults with ADHD showed some effect in aiding attentional processes;[58] however, further research in this area is required.

Rewarding/Incentive Effects

The nAChR agonist nicotine acts on dopaminergic pathways (see Fig. 1) implicated in the rewarding or the incentive-motivational effects of stimuli such as food, play, or copulatory opportunity.[70,88] Notably, nicotine's action in this "incentive/approach system" and the conditioned effect associated with this action is used to explain the acquisition and maintenance of compulsive tobacco use,[44,45,80] and the over 95% relapse rate following abstinence without pharmacotherapy.[16,17] Animal models such as self-administration and intracranial self-stimulation have been employed to elucidate the behavioral and neurobiological processes underlying these effects of nicotine.[15,46,69,81] For example, rodents and nonhuman primates prepared with an intravenous catheter will press more on a lever that produces contiguous intravenous delivery of nicotine.[22,39] The differential increase in responding (self-administration) maintained by nicotine requires normal functioning of the system outline in Figure 1. For example, nicotine self-administration in rats is decreased with bilateral 6-hydroxydopamine lesions of the dopaminergic projections between the ventral tegmental area (VTA) and nucleus accumbens,[23] or by infusions of a nAChR antagonist, DHβE, into the VTA.[24] Further, selective cholinergic lesioning of the pedunculopontine tegmental nucleus, a major cholinergic projection to the VTA, also attenuates nicotine self-administration.[51] Finally, mice lacking the β2 subunit fail to self-administer nicotine.[32]

Additional empirical work has implicated nAChRs located on the cell bodies in the VTA and on the terminals of glutamatergic projections from the prefrontal cortex to the VTA.[27,47,60] These excitatory glutamatergic projections stimulate VTA neurons resulting in dopamine release.[48] Theorists have suggested that this release of dopamine, especially in the nucleus accumbens, is important for various aspects of the rewarding/incentive effects of appetitive stimuli and the conditioned approach effects engendered by these stimuli (de Bruin, this book and refs. 3,10,27). Dopamine release in the nucleus accumbens is increased in vitro and in vivo with nicotine. This prolonged increase in dopamine release appears to be mediated by long-term potentiation of VTA cells containing NMDA receptors—a glutamate receptor selective for the agonist *N*-methyl-D-aspartate.[87] The α7* nAChRs located on the presynaptic terminals of projections from the prefrontal cortex are important for inducing this long-term potentiation in the VTA.[60,88] The α4β2* nAChR located on the cell bodies of VTA neurons quickly desensitize to the presence of nicotine and are unlikely to contribute to the long-term enhancement of dopamine release.[27,75]

Although there is a massive empirical literature studying the functional effects of dopamine release in this system, there is still disagreement as to its role in incentive-related behaviors. The following quote by Dani et al[27] provides a good summary that is consistent with our thinking and will serve as basis for suggesting a broad role of the nAChR-mediated neural plasticity of this system in learning and memory.

"DA [dopamine] concentrations in the NAc [nucleus accumbens] are not a scalar indication of reward. More likely, the DA signal conveys novelty and reward expectation or serves to

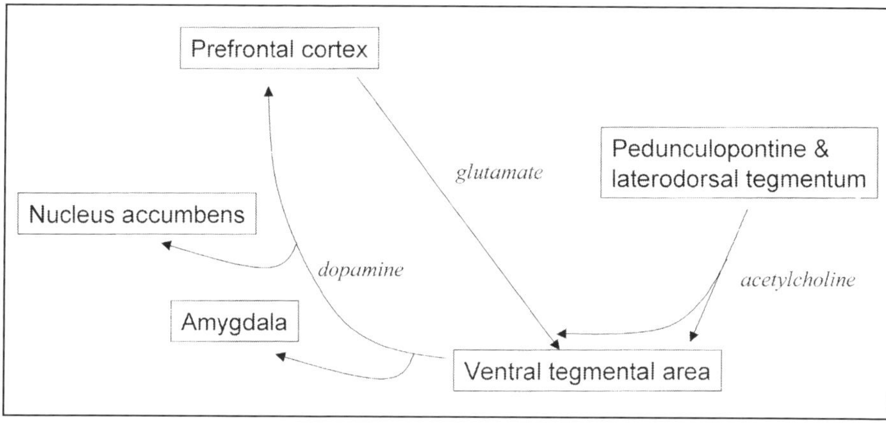

Figure 1. Diagram of the main pathways involved in nicotinic acetylcholine synaptic plasticity of the incentive-motivational (reward) system.

indicate the deviation of the environmental input from the animal's expectations, which were constructed by experience. Thus, DA may participate in the ongoing associative learning of adaptive behaviors as an animal continually updates a construct of environmental salience" (p. 350).

This conceptualization suggests that the dopamine signal that is enhanced by nicotine's action on presynaptic terminals containing α7* nAChRs plays a role in neurally attributing incentive salience to the stimulus input from a continually changing environment. This neural attribution likely occurs through associative learning processes, broadly defined.[3]

Recent research by Caggiula and colleagues[15] provides an important behavioral example of this process. Briefly, rats were trained to self-administer nicotine such that when the response requirement was completed a 1-sec intravenous infusion of nicotine (0.03 mg/kg) was delivered; a 1-sec light co-occurred with the nicotine infusion. Upon establishing stable self-administration behavior, some rats were switched to an extinction phase in which saline replaced nicotine, but the 1-sec cue light still occurred. Although lever press rates decreased with the removal of nicotine, the nicotine-associated light still maintained responding well above controls receiving saline without the cue light. In a separate set of rats, the response-contingent nicotine infusion continued during the extinction phase, but the light signal was removed. Rates of nicotine self-administration also decreased in this group. Notably, the level of responding maintained by nicotine alone was comparable to that maintained by the cue light alone. This result is intriguing if one considers that the rate of behavior maintained by what is conceptualized as the primary reinforcer, nicotine, is similar to that controlled by a cue associated with the effects of nicotine. Caggiula et al[15] concluded that, "nicotine promotes the establishment or magnifies the salience of conditioned reinforcers" (p. 526). We suggest that a plausible mechanism for this enhanced incentive salience is the action of nicotine on the α7* nAChRs located on glutamatergic presynaptic terminals of projections from the prefrontal cortex to the VTA. Of course, this proposal requires empirical attention.

Regardless of the specific neurobiological processes responsible for enhancing the incentive salience of stimuli, this enhancement provides an additional mechanism by which nAChR compounds may broadly affect learning and memory. For example, enhanced attention and/or vigilance (see earlier) may be, at least in part, the result of this process. Stimulus events that occur in the presence of nicotine (or other appropriately selective nAChR agonists) may acquire, or have potentiated, some appetitive property. Presumably this enhanced appetitive quality increases salience and may even require deeper processing given the acquired associations. Indeed, animals in a free-choice situation spend more time in a distinct environment that has been previously paired with appetitive stimuli.[4]

Enhanced/magnified incentive salience of cues may also play a role in the improved acquisition of new tasks observed with some nAChR agonists. For example, we found that acquisition of a T-maze visual discrimination task was faster in chronic nicotine-treated rats than in saline-treated rats.[6] Perhaps nicotine potentiated the incentive effects of the food used to reinforce correct arm choice (i.e., black arm). According to this formulation the incentive salience of the black arm may also be enhanced. That is, any conditioned reinforcing value acquired by the black-arm stimuli repeatedly associated with food may be increased by nicotine. Further, these conditioned effects may be stronger because the appetitive effects of food would also be enhanced (see earlier). Finally, the black-arm stimuli may acquire additional incentive salience by direct association with nicotine. The cumulative increase in the incentive salience of the stimulus events relevant to learning the discrimination thus enhanced acquisition rates relative to saline controls. Interestingly, reversal learning (white-arm now associated with food) was not altered by nicotine pretreatment. Perhaps, the effects of nicotine on the incentive salience require the stimulus events to be relatively novel (i.e., relatively little learning history). Or, perhaps the acquired increase in the incentive salience of black arm cues competed with white arm cues that were now becoming associated with food and nicotine after a long history on nonreinforcement in a manner similar to nonmagnified cues in the saline controls.

As a final note in this section, we found that enhanced acquisition in the T-maze task was predicted by activity in an inescapable novel environment; less reactive rats learned the discrimination faster (see ref. 6 and Fig. 1). Notably, past research on individual differences predicted by reactivity to inescapable novelty has implicated the mesocorticolimbic dopamine system.[82,86] The nAChR-mediated long-term potentiation of the VTA increases dopamine release within this system (see Fig. 1) and suggests a potential process responsible for the predictable difference in T-maze learning produced by nicotine. Of course, the speculations concerning incentive salience and the role of nAChR-mediated long-term potentiation in learning and memory require further research to provide independent evidence for the processes at a neurobiological and behavioral level.

Other Effects

Additional functional effects of nAChRs include alterations in pain, anxiety, appetite, depression, epilepsy, and motoric abilities. Although a comprehensive discussion of these effects is beyond the scope of the present review, their potential influence (direct or indirect) on learning and memory deserves mention. We will use as an example the locomotor effects of centrally located nAChRs. Most of the nonhuman animal research investigating attention, reward, or working memory include controls to assess whether the motoric effects of the nAChR ligand of interest could account for group differences. Such controls are important in that nAChR agonists can alter general locomotor activity.[8,42,61,68,92] Whether the change is locomotor suppression or stimulation depends on such factors as selectivity of ligand, dose, pretreatment or preexposure history, rodent strain, and environmental familiarity. Accordingly, if one is investigating the memory enhancing effects of, say, chronic nicotine, then an index of locomotor stimulation will be important. Arguably, these locomotor stimulant effects could enhance acquisition and performance in certain learning tasks perhaps by producing small decreases in the time between stimulus-outcome or behavior-outcome relations (i.e., improved temporal contiguity) inherent in learning situations.[72,93,97] Along these lines, it is interesting to note that the $\alpha4\beta2^*$ nAChRs[42] and increased dopamine release in the nucleus accumbens[5] appear to be important for the locomotor stimulant effects of nicotine (see section on Reward/Incentive Effects). Likely, future research will begin to more specifically identify the links between the effects we have listed as "Other" and memory. Perhaps the anxiolytic effects of ABT 418[30] or nicotine[19] allow an animal to use neural processing resources released from this decrease in anxiety toward the learning/memory task prescribed by the experimenter.

Closing Remarks

Advances in molecular biology are clearly refining our understanding of the vast structural variation that exists in nAChRs. As these advances continue, so will our understanding of nAChR processes in learning and memory. For example, behavioral geneticists can further develop mutant mice with selective deletions of nAChR subunits. Extensive neurobiological and behavioral assessment of these mice will inform our theoretical models. Also, continued development of selective ligands will allow researchers to dissociate function and receptor subtypes. Finally, from our perspective, we need a better understanding of the psychological constructs measured by current animal models (e.g., radial arm maze, 5-CSRT) and we need further development of new animal models (see Jaffard and Marighetto in this book). Our discussion, for example, of the possibility that working memory models may also be measuring attentional processes highlights this need. If we do not fully understand what the dependent measures are indexing and the factors that affect those measures, we will always be unsure of whether the neurobiological process identified actually reflects the psychological construct (memory, learning, attention, reward, etc.) of interest.

Acknowledgements

We thank D. Taylor, K. Karthigeyan, and M. Palmatier, for their comments on an earlier version of this chapter. R. A. Bevins was supported by United States Public Health Service grant DA11893 while preparing this report; J. Besheer was supported by DA06092. Correspondence concerning this chapter should be addressed to Rick A. Bevins, Department of Psychology, University of Nebraska-Lincoln, Lincoln NE USA 68588-0308, or e-mail rbevins1@unl.edu

References

1. Abdulla FA, Calaminici MR, Stephenson JD et al. Chronic treatments with cholinoceptor drugs influence spatial learning in rats. Psychopharmacol 1993; 111:508-511.
2. Arneric SP, Sullivan JP, Williams M. Neuronal nicotinic acetylcholine receptors: Novel targets for central nervous system therapeutics. In: Bloom FE, Kupfer DJ, eds. Psychopharmacology: The Fourth Generation of Progress. New York: Raven Press Ltd., 1995; 95-110.
3. Balfour DJK, Wright AE, Benwell MEM et al. The putative role of extra-synaptic mesolimbic dopamine in the neurobiology of nicotine dependence. Behav Brain Res 2000; 113:73-83.
4. Bardo MT, Bevins RA. Conditioned place preference: What does it add to our preclinical understanding of drug reward? Psychopharmacol 2000; 153:31-43.
5. Benwell MEM, Balfour DJK. The effects of acute and repeated nicotine treatment on nucleus accumbens dopamine and locomotor activity. Br J Pharmacol 1992; 105:849-856.
6. Besheer J, Bevins RA. Nicotine enhances acquisition of a T-maze visual discrimination: Assessment of individual differences. Behav Pharmacol 2000; 11:613-620.
7. Besheer J, Bevins RA. The role of environmental familiarization in novel-object preference. Behav Processes 2000; 50:19-29.
8. Bevins RA, Besheer J, Pickett KS. Nicotine-conditioned locomotor activity in rats: Dopaminergic and GABAergic influences on conditioned expression. Pharmacol Biochem Behav 2001; 68:135-145.
9. Bevins RA, Koznarova J, Armiger TJ. Environmental familiarization in rats: Differential effects of acute and chronic nicotine. Neurobiol Learn Mem 2001; 75:63-76.
10. Berridge KC, Robinson TE. What is the role of dopamine in reward: Hedonic impact, reward learning or incentive salience. Brain Res Rev 1998; 28:309-369.
11. Blondel A, Sanger DJ, Moser PC. Characterisation of the effects of nicotine in the five-choice serial reaction time task in rats: Antagonist studies. Psychopharmacol 2000; 149:293-305.
12. Bontempi B, Whelan KT, Risbrough VB et al. SIB-1553A, (+/-) - 4- [[2- (1-methyl - 2-pyrrolidinyl) ethyl]thio]phenol hydrochloride, a subtype-selective ligand for nicotinic acetylcholine receptors with putative cognitive-enhancing properties: Effects on working and reference memory performances in aged rodents and nonhuman primates. J Pharmacol Experim Therap 2001; 299:297-306.
13. Buccafusco JJ, Jackson WJ, Terry Jr AV et al. Improvement in performance of delayed matching-to-sample task by monkeys following ABT-418: A novel cholinergic channel activator for memory enhancement. Psychopharmacol 1995; 120:256-266.
14. Bushnell PJ, Oshiro WM, Padnos BK. Detection of visual signals by rats: Effects of chlordiazepoxide and cholinergic and adrenergic drugs on sustained attention. Psychopharmacol 1997; 134:230-241.

15. Caggiula AR, Donny EC, White AR et al. Cue dependency of nicotine self-administration and smoking. Pharmacol Biochem Behav 2001; 70:515-530.
16. Centers for disease control. Smoking cessation during previous year among adults-United States, 1990 and 1991. MMWR 1993; 42:504-507.
17. Centers for disease control. Incidence of initiation of cigarette smoking among US. teens. Tobacco Information and Prevention Source. 1999.
18. Changeux J-P, Bertrand D, Corringer P-J et al. Brain nicotinic receptors: Structure and regulation, role in learning and reinforcement. Brain Res Rev 1998; 26:198-216.
19. Cheeta S, Tucci S, File SE. Antagonism of the anxiolytic effects of nicotine in the dorsal raphé nucleus by di-hydro-β-erythroidine. Pharmacol Biochem Behav 2001; 70:491-496.
20. Connolly J, Boulter J, Heinemann SF. Alpha4-2beta2 and other nicotinic acetylcholine receptor subtypes as targets of psychoactive and addictive drugs. Br J Pharmacol 1992; 105:657-666.
21. Cooper E, Couturier S, Ballivet M. Pentameric structure and subunit stoichiometry of a neuronal acetylcholine receptor. Nature 1991; 350:235-238.
22. Corrigall WA, Coen KM. Nicotine maintains robust self-administration in rats on a limited-access schedule. Psychopharmacol 1989; 99:473-478.
23. Corrigall WA, Franklin KBJ, Coen KM et al. The mesolimbic dopaminergic system is implicated in the reinforcing effects of nicotine. Psychopharmacol 1992; 107:285-289.
24. Corrigall WA, Coen KM, Adamson KL. Self-administrated nicotine activates the mesolimbic dopamine system through the ventral tegmental area. Brain Res 1994; 653:278-284.
25. Couturier S, Bertrand D, Matter JM et al. A neuronal nicotinic acetylcholine receptor subunit (alpha 7) is developmentally regulated and forms a homo-oligomeric channel blocked by alpha-BTX. Neuron 1990; 5:847-856.
26. Curzon P, Brioni JD, Decker MW. Effect of intraventricular injections of dihydro-β-erythroidine (DHβE) on spatial memory in the rat. Brain Res 1996; 714:185-191.
27. Dani JA, Ji D, Zhou F-M. Synaptic plasticity and nicotine addiction. Neuron 2001; 31:349-352.
28. Decker MW, Majchrzak MJ, Arneri SP. Effects of lobeline, a nicotinic receptor agonist, on learning and memory. Pharmacol Biochem Behav 1993; 45:571-576.
29. Decker MW, Curzon P, Brioni JD et al. Effects of ABT-418, a novel cholinergic channel ligand, on place learning in septal-lesioned rats. Eur J Pharmacol 1994; 261:217-222.
30. Decker MW, Brioni JD, Sullivan JP et al. (S)-3-methyl-5-(1-methyl-2-pyrrolidinyl) isoxazole (ABT 418): A novel cholinergic ligand with cognition-enhancing and anxiolytic activities: II. In vivo characterization. J Pharmacol Exper Ther 1994; 270:319-328.
31. Dunnett SB, Martel FL. Proactive interference effects on short-term memory in rats: 1. Basic parameters and drug effects. Behav Neurosci 1990; 104:655-665.
32. Epping-Jordan MP, Picciotto MR, Changeux JP et al. Assessment of nicotinic acetylcholine receptor subunit contributions to nicotine self-administration in mutant mice. Psychopharmacol 1999; 147:25-26.
33. Ennaceur A, Delacour J. A new-one trial test for neurobiological studies of memory in rats. 1: Behavioral data. Behav Brain Res 1988; 31:47-59.
34. Feldman RS, Meyer JS, Quenzer LF. Principles of Neuropsychopharmacology. Sunderland, MA: Sinauer Associates. 1997.
35. Felix R, Levin ED. Nicotinic antagonist administration into the ventral hippocampus and spatial working memory in rats. Neurosci 1997; 81:1009-1077.
36. Ferguson SM, Brodkin JD, Lloyd GK et al. Antidepressant-like effects of the subtype-Selective nicotinic acetylcholine receptor agonist, SIB-1508Y, in the learned helplessness rat model of depression. Psychopharmacol 2000; 152:295-303.
37. Flores CM, Dávila-García MI, Ulrich YM et al. Differential regulation of neuronal nicotinic receptor binding sites following chronic nicotine administration. J Neurochem 1997; 69:2216-2219.
38. Granon S, Poucet B, Thinus-Blanc C et al. Nicotinic and muscarinic receptors in the rat prefrontal cortex: Differential roles in working memory, response selection and effortful processing. Psychopharmacol 1995; 119:139-144.
39. Goldberg SR, Spealman RD, Goldberg DM. Persistent behavior at high rates maintained by intravenous self-administration of nicotine. Science 1981; 214:573-575.
40. Gould TJ, Wehner JM. Nicotine enhancement of contextual fear conditioning. Behav Brain Res 1999; 102:31-39.
41. Grottick AJ, Higgins GA. Effect of subtype selective nicotinic compounds on attention as assessed by the five-choice serial reaction time task. Behav Brain Res 2000; 117:197-208.
42. Grottick AJ, Wyler R, Higgins GA. The $\alpha_4\beta_2$ agonist SIB 1765F, but not the α_7 agonist AR-R 17779, cross-sensitises to the psychostimulant effects of nicotine. Psychopharmacol 2000; 150:233-236.
43. Grottick AJ, Wyler R, Higgins GA. A study of the nicotinic agonist SIB-1553A on locomotion and attention as measured by the five-choice serial reaction time task. Pharmacol Biochem Behav 2001; 70:505-513.

44. Henningfield JE, Schuh LM, Jarvik ME. Pathophysiology of tobacco dependence. In: Bloom FE, Kupfer DJ, eds. Psychopharmacology: The Fourth Generation of Progress. New York: Raven Press, 1995:1715-1729.
45. Henningfield JE, Keenan RM, Clarke PBS. Nicotine. In: Schuster CR, Kuhar MJ, Eds. Pharmacological Aspects of Drug Dependence: Toward an Integrated Neurobehavioral Approach. Heidelberg, Germany: Springer-Verlag, 1996:271-314.
46. Ivanová S, Greenshaw AJ. Nicotine-induced decreases in VTA electrical self-stimulation thresholds: Blockade by haloperidol and mecamylamine but not scopolamine or ondansetron. Psychopharmacol 1997; 134:187-192.
47. Jones S, Sudweeks S, Yakel JL. Nicotinic receptors in the brain: Correlating physiology with function. Trends in Neurosci 1999; 22:555-561.
48. Kalivas PW, Duffy P, Barrow J. Regulation of the mesocorticolimbic dopamine system by glutamic acid receptor subtypes. J Pharmacol Exper Ther 1989; 251:378-387.
49. Kim JS, Levin ED. Nicotinic, muscarinic and dopaminergic actions in the ventral hippocampus and the nucleus accumbens: Effects on spatial working memory in rats. Brain Res 1996; 725:231-240.
50. Krebs SJ, Petros TV, Beckwith BE. Effects of smoking on memory for prose passages. Physiology and Behavior 1994; 56:723-727.
51. Lanca AJ, Adamson KL, Coen KM et al. The pedunculopontine tegmental nucleus and the role of cholinergic neurons in nicotine self-administration in the rat: A correlative neuroanatomical and behavioral study. Neurosci 2000; 96:735-742.
52. Levin ED, McGurk SR, South D et al. Effects of combined muscarinic and nicotinic blockade on choice accuracy in the radial-arm maze. Behav Neural Biol 1989; 51:270-277.
53. Levin ED, Lee C, Rose JE et al. Chronic nicotine and withdrawal effects on radial-arm maze performance in rats. Behav Neural Biol 1990; 53:269-276.
54. Levin ED, Briggs SJ, Christopher NC et al. Persistence of chronic nicotine-induced cognitive functioning. Behav Neur Biol 1992; 58:152-158.
55. Levin ED, Kim P, Meray R. Chronic nicotine working and reference memory effects in the 16-arm radial maze: Interactions with D_1 agonist and antagonist drugs. Psychopharmacol 1996; 127:25-30.
56. Levin ED, Simon BB. Nicotinic acetylcholine involvement in cognitive function in animals. Psychopharmacol 1998; 138:217-230.
57. Levin ED, Bettegowda C, Blosser J et al. AR-R 17779, an $\alpha 7$ nicotine agonist, improves learning and memory in rats. Behav Pharmacol 1999; 10:675-680.
58. Levin ED, Conners CK, Silva D et al. Effects of chronic nicotine and methylphenidate in adults with attention deficit/hyperactivity disorder. Exper Clin Psychopharmacol 2001; 9:83-90.
59. Mancuso G, Warburton DM, Mélen M et al. Selective effects of nicotine on attentional processes. Psychopharmacol 1999; 146:199-204.
60. Mansvelder HD, McGehee DS. Long-term potentiation of excitatory inputs to brain reward areas by nicotine. Neuron 2000; 27:349-357.
61. Menzaghi F, Whelan KT, Risbrough VB et al. Effects of a novel cholinergic ion channel agonist SIB-1765F on locomotor activity in rats. J Pharmacol Exper Ther 1997; 280:384-392.
62. Mesulam M-M. Central cholinergic pathways: Neuroanatomy and some behavioral implications. In: Avoli M, Reader TA, Dykes RW, Gloor P, eds. Neurotransmitters and Cortical Function. New York: Plenum Press,1988:237-260.
63. Mesulam M-M. Structure and function of cholinergic pathways in the cerebral cortex, limbic systems, basal ganglia, and thalamus of the human brain. In: Bloom FE, Kupfer DI, eds. Psychopharmacology: The Fourth Generation of Progress. New York: Raven Press, 1995:135-146.
64. Mirza NR, Stolerman IP. Nicotine enhances sustained attention in the rat under specific task conditions. Psychopharmacol 1998; 138:266-274.
65. Mundy WR, Iwamoto ET. Nicotine impairs acquisition of radial maze performance in rats. Pharmacol Biochem Behav 1988; 30:119-122.
66. Newhouse PA, Potter A, Levin ED. Nicotinic system involvement in Alzheimer's and Parkinson's diseases. Drugs and Aging 1997; 11:206-228.
67. Ohno M, Yamamoto T, Watanabe S. Blockade of hippocampal nicotinic receptors impairs working memory but not reference memory in rats. Pharmacol Biochem Behav 1993; 45:89-93.
68. Palmatier MI, Bevins RA. Examination of SCH-23390, eticlopride, and baclofen on acquisition of nicotine-conditioned hyperactivity in rats. Neuropsychobiol 2002; 45:87-94.
69. Panagis G, Kastellakis A, Spyraki C et al. Effects of methyllycaconitine (MLA), an α_7 nicotinic receptor antagonist, on nicotine- and cocaine-induced potentiation of brain stimulation reward. Psychopharmacol 2000; 149:388-396.
70. Panksepp J. Affective Neuroscience: The Foundations of Human and Animal Emotions. New York: Oxford University Press, 1998.

71. Paterson D, Nordberg A. Neuronal nicotinic receptors in the human brain. Prog Neurobio 2000; 61:75-111.
72. Pavlov IP. Conditioned reflexes. London: Oxford University Press, 1927.
73. Picciotto MR. Common aspects of the action of nicotine and other drugs of abuse. Drug Alcohol Dep 1998; 51:165-172.
74. Picciotto MR, Zoli M, Léna C et al. Abnormal avoidance learning in mice lacking functional high-affinity nicotine receptor in the brain. Nature 1995; 374:65-67.
75. Pidoplichko VI, DeBiasi M, Williams JT et al. Nicotine activates and desensitizes midbrain dopamine neurons. Nature 1997; 390:401-404.
76. Potter A, Corwin J, Lang J et al. Acute effects of the selective cholinergic channel activation (nicotinic agonist) ABT-418 in Alzheimer's disease. Psychopharmacol 1999; 142:334-342.
77. Puma C, Deschaux O, Molimard R et al. Nicotine improves memory in an object recognition task in rats. Eur J Neuropsychopharm 1999; 9:323-327.
78. Rezvani AH, Levin ED. Cognitive effects of nicotine. Biol Psychiatry 2001; 49:258-267.
79. Rochford J, Sen AP, Quirion R. Effect of nicotine and nicotinic receptor agonists on latent inhibition in the rat. J Pharmacol Exp Ther 1996; 277:1267-75.
80. Rose JE. Nicotine addiction and treatment. Ann Rev Med 1996; 47:493-507.
81. Rose JE, Corrigall WA. Nicotine self-administration in animals and humans: Similarities and differences. Psychopharmacol 1997; 130:28-40.
82. Rogué-Pont F, Piazza PV, Kharouby M et al. Higher and longer stress-induced increase in dopamine concentrations in the nucleus accumbens of animals predisposed to amphetamine self-administration. A microdialysis study. Brain Res 1993; 602:169-174.
83. Rusted J, Eaton-Williams P. Distinguishing between attentional and amnestic effects in informational processing: The separate and combined effects of scopolamine and nicotine on verbal free recall. Psychopharmacol 1991; 104:363-366.
84. Rusted JM, Graupner L, Tennant A et al. Effortful processing is a requirement for nicotine-induced improvements in memory. Psychopharmacol 1998; 138:326-368.
85. Rusted JM, Caulfield D, King L et al. Moving out of the laboratory: Does nicotine improve everyday attention? Behav Pharmacol 2000; 11:621-629.
86. Saigusa T, Tuinstra T, Koshikawa N et al. High and low responders to novelty: Effects of a catecholamine synthesis inhibitor on novelty-induced changes in behaviour and release of accumbal dopamine. Neurosci 1999; 88:1153-1163.
87. Schilstrom B, Nomikos GG, Nisell M et al. N-methyl-D-aspartate receptor antagonism in the ventral tegmental area diminishes the systemic nicotine-induced dopamine release in the nucleus accumbens. Neurosci 1998; 82:781-789.
88. Schilstrom B, Svensson HM, Svensson TH et al. Nicotine and food induced dopamine release in the nucleus accumbens of the rat: Putative role of alpha7 nicotine receptors in the ventral tegmental area. Neurosci 1998; 85:1005-1009.
89. Shacka JJ, Robinson SE. Central and peripheral anatomy of nicotinic sites. Med Chem Res 1996:444-464.
90. Shoepfer R, Conroy W, Whiting P et al. Brain α-bungarotoxin binding protein cDNAs and mAbs reveal subtypes of this branch of the ligand-gated ion channel gene superfamily. Neuron 1990; 5:35-48.
91. Stolerman IP, Garcha HS, Mirza NR. Dissociations between the locomotor stimulant and depressant effects of nicotinic agonists in rats. Psychopharmacol 1995; 117:430-437.
92. Stolerman IP, Mirza NR, Hahn B et al. Nicotine in a an animal model of attention. Europ J Pharmacol 2000; 393:147-154.
93. Thorndike EL. Animal intelligence: An experimental study of the associative processes in animals. Psychol Rev 1898; 2:1-109.
94. Wada E, Wada K, Boulter J et al. Distribution of alpha2, alpha3, alpha4, and beta2 neuronal nicotinic receptor subunit mRNAs in the central nervous system: A hybridization histochemical study in the rat. J Comp Neurol 1989; 284:314-335.
95. Wainer BH, Levey AI, Mufson EJ et al. Cholinergic systems in mammalian brain identified with antibodies against choline acetyltransferase. Neurochem Int 1984; 6:163-182.
96. Warburton DM, Rusted JM, Fowler J. A comparison of the attentional and consolidation hypotheses for the facilitation of memory by nicotine. Psychopharmacol 1992; 108:443-447.
97. Wasserman EA, Miller RR. What's elementary about associative learning? Ann Rev Psychol 1997; 48:573-607.
98. Woolf NJ. Cholinergic systems in mammalian brain and spinal cord. Prog Neurobiol 1991; 37:475-524.
99. Zoli M, Le Novère N, Hill JAJ et al. Developmental regulation of nicotinic receptor subunit mRNAs in the rat central and peripheral nervous system. J Neurosci 1995; 15:1912-1939.

Chapter 2.5

Serotonin

Marie-Christine Buhot, Mathieu Wolff and Louis Segu

Abstract

The serotonergic system is widely distributed in the central nervous system and plays a role in many behavioral and physiological processes. However, converging data indicate that serotonin (5-HT) is specifically involved in learning and memory by interacting with major neurotransmitters. Thus, 5-HT modulates acetylcholine and glutamate release in the pathways of first importance for memory functions. The use of global strategies aimed at modifying the 5-HT level allows to estimate the functional implication of 5-HT in several types of memory. Nevertheless, the mechanisms of action are dissected by studying 5-HT receptors. To define the role played by a receptor in memory, one has to consider at least two criteria which are: 1) its linkage to a second messenger, 2) its anatomical, cellular and subcellular locations. According to these criteria and by using both pharmacological approaches and molecular tools such as gene knockout mice, only six types or subtypes among the fourteen 5-HT receptors have presently been demonstrated to be involved in learning and memory. Administration of 5-HT1A/1B and 5-HT3 specific agonists induces impairments in many memory tasks while agonists for 5-HT2A/2C, 5-HT4 and 5-HT6 receptors have generally facilitatory effects. Many of these effects can be reversed by the use of 5-HT receptor specific antagonists. The combination of these pharmacological tools allows to dissect the physiological interaction between different 5-HT receptors and different neurotransmitter systems in various cerebral structures. Such interactions could participate in the control of the signal-to-noise ratio in the information processing. Increasing evidence is consistent with the view that the role of 5-HT in learning and memory becomes more important when the cognitive demand is high. This can be achieved by increasing the complexity of the memory task or by impairing memory abilities, as it is the case in aging. For these reasons, specific 5-HT receptor drugs could prevent memory decline in normal aging or neurodegenerative pathologies such as Alzheimer's disease.

Introduction

Serotonin (5-hydroxytryptamine, 5-HT) presents a wide distribution in the central nervous system (CNS), and thus plays important roles in various behavioral and physiological processes, including activity rythms, food-intake, locomotor activity, behavioral inhibition or emotional states. There is increasing evidence showing that 5-HT is concerned with cognitive functions, especially learning and memory and attentional processes.[24] These functions are not independent from each other or from other behavioral levels. There are indeed some relationships between anxiety and memory, or between learning and behavioral inhibition. It is thus clear that 5-HT may modulate learning and memory by direct or indirect ways. If memory functions are mainly and more directly controlled by other neurotransmitter systems such as the glutamatergic and the cholinergic, it is presently well-established that, by interacting with these systems, 5-HT plays an unneglectible role in memory formation via its various receptors (Fig. 1).

From Messengers to Molecules: Memories Are Made of These, edited by Gernot Riedel and Bettina Platt.

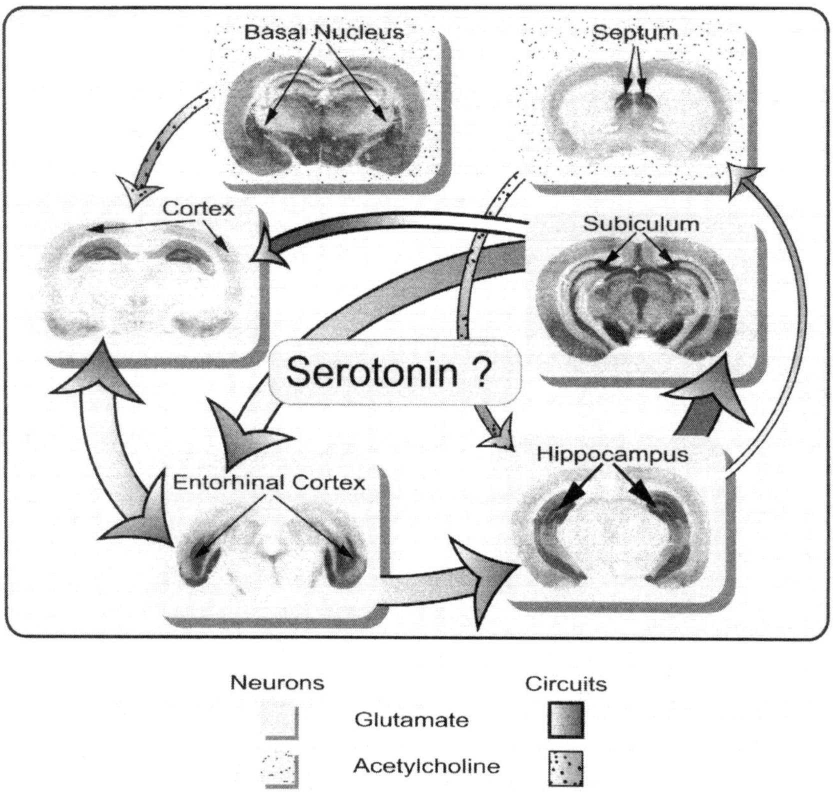

Figure 1. What does serotonin do in the main neuronal circuits underlying memory? The flow of inputs from the environment reaches specialized cortical areas, related information is then transmitted through the entorhinal cortex to the hippocampus where it is processed before being distributed, via the subiculum, to cortical areas. The corticohippocampo-cortical communication allows the consolidation of information into long-term memory. Major anatomical structures underpinning memory processes are interconnected by glutamatergic routes. Acetylcholine, from the nucleus basalis magnocellularis and septum, subserves a cross-talk with these major pathways, hence modulating learning mechanisms. Anatomical structures are illustrated by autoradiography on rat brain sections with specific ligands labeling 5-HT1A ([3H]8-OH-DPAT, for hippocampus, entorhinal cortex and cortex), and 5-HT1B (S-CM-G[125I]TNH2, for subiculum, septum and basal nucleus) receptors.[1]

This chapter thus intends to analyse behavioral and neurobiological data emphasizing the contribution of serotonin in memory, with special reference to the receptor side as the specific multidimentional "target" for serotonin to influence memory systems.

Role of 5-HT in Memory: Global Strategies

Serotonergic projections from the raphe nuclei, where cell bodies are concentrated, are widespread and terminate in brain structures thought to underly memory processes such as the hippocampus, the prefrontal cortex, cortical associative areas, basal ganglia, or thalamus. The projections to the frontal cortex predominantly originate in the dorsal raphe nucleus (DRN), whereas the projections to the hippocampus arise from the median raphe nucleus (MRN). The DRN regulates the excitability of prefrontal cortical neurons,[104] while the MRN plays a direct

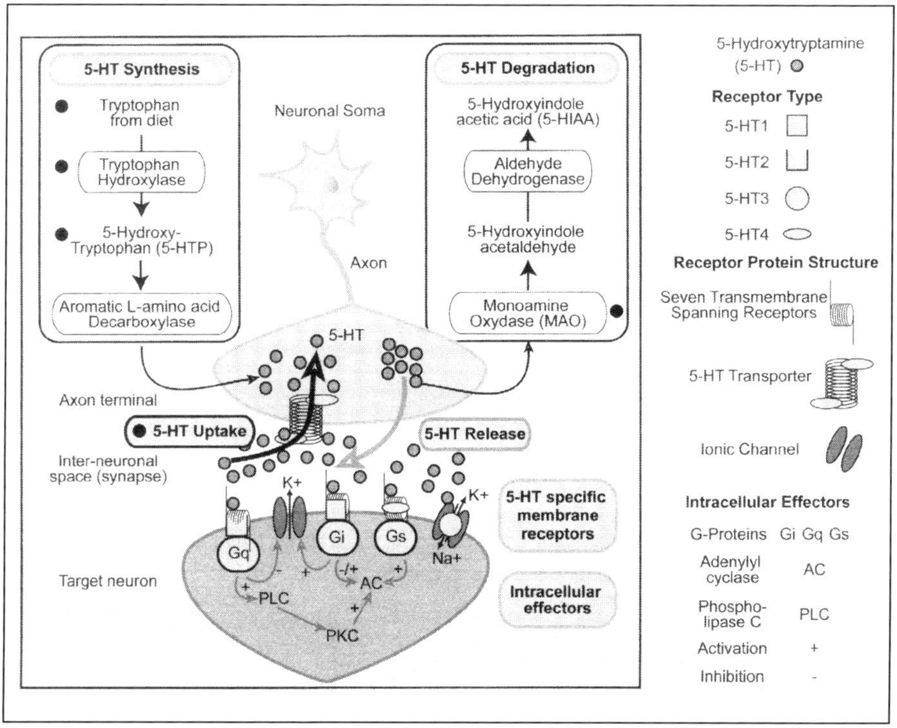

Figure 2. Metabolism and receptor transduction systems of serotonin. Top-half of the figure. 5-HT synthesis is realized in soma and terminals. Tryptophan hydroxylase is specific of 5-HT neurons and is the rate limiting enzyme for 5-HT synthesis. As tryptophan cannot be synthesized by the metabolism, 5-HT synthesis is very sensitive to the presence of this amino acid in the diet. 5-HT degradation takes place in neurons and glial cells. 5-HT uptake is specific for 5-HT neurons and exists in all membrane compartments of the neuron (soma, dendrites, axon terminals). Black dots show the steps which can be experimentally controlled in global strategy approaches to modify the extracellular level of 5-HT. Bottom-half of the figure. Once released in the intercellular space, 5-HT may act on several specific receptor types or subtypes, which are differentiated by their protein structure, and by their intracellular effectors in the target neuron. The multiplicity of effector systems for each receptor type, as well as several common intracellular pathways for different receptor types, increase the diversity of functional effects of the fourteen 5-HT receptor subtypes.

role in the desynchronization of the electroencephalographic activity of the hippocampus,[129] with possible consequences for memory-associated functions of the frontal cortex and the hippocampus, respectively. In the adult rat brain, hippocampal dependent learning increases the number of newborn cells which differentiate into neurons.[50] Depletion of serotonin with 5,7-dihydroxytryptamine (5,7-DHT) reduces neurogenesis, thus serotonin might be a factor stimulating granule cell production.[20]

Nonspecific strategies, which concern much of the earlier studies on the role of serotonin in cognition, used global strategies, by which the effects of increasing or reducing central 5-HT neurotransmission were observed in various learning models (Fig. 2, top-half).

Given the known role of 5-HT in mood disorders such as anxiety, "emotional memory" deserves to be analysed per se in relation to 5-HT. Thus, a change in 5-HT metabolism occurs in the prefrontal cortex, the nucleus accumbens, and the amygdala of rats subjected to the conditioned fear stress (CFS).[65] CFS-induced freezing behavior decreases following treatment with a selective serotonin reuptake inhibitor (SSRI), which is supposed to enhance 5-HT level in the brain.[66] In contrast, global 5-HT lesions induce specific impairments of contextual conditioning in rats, while sparing discrete cue-conditioning.[133] Interestingly, selective lesions of the DRN induce impairments of unconditioned (innate) fear, a response mediated by the DRN-periaqueductal gray (PAG) projection, but have facilitatory effects on learned fear, a response mediated by the DRN-amygdala and DRN-frontal cortex pathways.[51]

Reducing central 5-HT synthesis through L-tryptophan restriction specifically impairs short-term and long-term memory performance in rats and humans.[49,111] Depletion of 5-HT by using the tryptophan hydroxylase inhibitor, parachlorophenylalaline (PCPA), during synaptogenesis decreases synaptic density in the adult rat hippocampus, while inducing spatial learning deficits associated with failure to extinction and, consequently, relearning.[87] In contrast, 5-HT depletion following 5,7-DHT lesioning has been found to facilitate acquisition and performance of various learning tasks in rats such as spatial discrimination,[3] conditional visual discrimination (presumably by reducing proactive interference),[131] and temporal discrimination under certain conditions such as when the task difficulty is increased.[4] Fluoxetine, a SSRI, was found to weaken associative memory in the rat,[97] but aged rats treated with a low dose of the 5-HT precursor (5-hydroxytryptophan, 5-HTP) improved their performance (over controls) in a spatial memory task.[110]

From the above studies inconsistencies emerge about the mechanisms by which 5-HT might be involved in memory functions or dysfunctions. These are attributable, in part, to the application of this global experimental strategy, which modifies the entire serotonergic system and its interactions with other neurotransmitters, such as the cholinergic system.

Serotonergic-Cholinergic Interactions

The serotonergic and cholinergic systems display important functional interactions in learning and memory.[33,115,122] Simultaneous loss of both acetylcholine (ACh) and 5-HT transmission prolongs memory impairment in rats, compared with the effect of the separate loss of cholinergic or serotonergic transmissions.[73,85] Converging data show that 5-HT is involved in the regulation of central cholinergic activity by modulating ACh release in various cerebral structures, e.g., cholinergic pathways from the medial septum/diagonal band of Broca (MS/DBB) to the hippocampus or from the nucleus basalis magnocellularis (NBM) to the cerebral cortex and amygdala.[33,122] Both ACh and 5-HT are crucial for maintaining synapses in the hippocampus and are critically involved in the acquisition of spatial memory.[86,120] Combined disruption of muscarinic and serotonergic functions induces severe deficits in spatial performance in rats,[53] that tacrine, an acetylcholinesterase inhibitor considered to be of high efficacy in the treatment of Alzheimer's disease (AD), is able to alleviate.[54] Intra-hippocampal cografts rich in cholinergic and serotonergic fetal neurons have a selective, beneficial, effect on spatial reference memory impairment induced by extensive lesion of the dorsal septo-hippocampal pathways.[9] Serotonin is able to directly modulate cholinergic septo-hippocampal neurons[94] as well as indirectly by interacting with GABAergic interneurons that synapse on medial septum cholinergic neurons.[43]

5-HT Receptors in Memory Systems

The knowledge of the mechanisms by which 5-HT contributes to learning and memory requires a more consistent and accurate investigation of the role played by specific types or even sub-types of receptors, especially those localized in particular cerebral structures underlying defined cognitive functions. Besides the pharmacological approach which attempts to evaluate the behavioral functions of particular receptors by use of specific drugs, molecular genetic techniques provide useful complementary tools, in particular knockout (KO) mice with a deletion of a single gene coding for a specific receptor, which constitute unique models of selective dysfunctions.

Basic Neurobiological Data

Until now only 5-HT1A, 5-HT1B, 5-HT2, 5-HT3, 5-HT4 and 5-HT6 receptors (among 14 types or subtypes) have been demonstrated to play a role in learning and memory. There are at least two main criteria which have to be taken into account to analyse the potential role played by a receptor in a function such as memory: 1) its linkage to a second messenger, 2) its anatomical, cellular and subcellular locations in the CNS.

5-HT Receptors and Linkage to Second Messenger: Signaling Pathways

Among the fourteen different 5-HT receptor types or subtypes, only 5-HT3 is a transmitter-gated cation channel, all other 5-HT receptors are linked to guanine nucleotide-binding proteins (G proteins) and members of the seven transmembrane (7TM) domain receptors.[15] The class of 5-HT1 receptor is coupled to Gi/o, the 5-HT2 to Gq/11, the 5-HT4/6/7 to Gs (Fig. 2; for reviews see: Hoyer and Martin,[64] Barnes and Sharp[11]). The Gi/o and Gs linkages decrease and increase adenylyl cyclase (AC) activity, respectively, while Gq/11 increases phospholipase C (PLC) activity and can thus also activate adenylyl cyclase through the protein kinase C (PKC) pathway.

Multiplicity of Effector Pathways[84]

A single receptor subtype can be linked to multiple effector pathways within a cell. Native 5-HT1B expressed in OK cells are coupled to both elevation of intracellular calcium and inhibition of adenylyl cyclase.[137] Activation of recombinant h5-HT1B and h5-HT1D in C6 glioma cells produces increases in Ca2+-dependent K+ current.[72] This coupling depends on the neuron type: As an example, 5-HT1A is not coupled to the inhibition of adenylyl cyclase in DRN.[36] Both 5-HT1A and 5-HT1B are linked to Gi and inhibit adenylyl cyclase activity, but in tissues containing the Ca2+-insensitive adenylyl cyclase (e.g., type II), they stimulate the activity of the enzyme. This activation is due to the beta-gamma subunits released from activated proteins Gi.[2] The effect is evidenced in the hippocampus where the 5-HT1A potentiates stimulation of the cyclase promoted by a Gsα-coupled receptor (beta-adrenergic, for example).[5]

Agonist-Directed Differential Signaling of Receptor Stimulus

The agonists may be able to selectively activate a subset of multiple signaling pathways coupled to a single receptor subtype. The 5-HT2A receptor is coupled to PLC-mediated inositol triphosphate (IP3) accumulation and phospholipase A2 (PLA2)-mediated arachidonic acid (AA) release. All 5-HT2A agonists have greater activation capacities for PLA2-AA than for PLC-IP3.[12]

Cross-Talk between 5-HT Receptors

The responsiveness of one receptor system may be regulated by activation of another receptor system, resulting in a "cross-talk" between different receptors. In CHO cells, the h5-HT1B decreases forskolin-stimulated cAMP accumulation and stimulates increases in Ca^{2+}, along with a potentiation of Gq-coupled receptor stimulated second messenger responses (PLC-IP3).[40] The efficacy of 5-HT1A agonists on inhibition of forskolin-stimulated cAMP accumulation is regulated by the activation of phospholipid-coupled receptors.[42]

The interaction between receptors may be due to a physical association. When expressed alone, 5-HT1B or 5-HT1D form monomers and homodimers, while when coexpressed they form heterodimers. This association might exist in several brain areas where both receptors colocalize in the same neuron.[135]

Constitutive Activity and Inverse Agonism

Most agonists (with intrinsic activity) bind with a high affinity to the isomerized and signaling G protein-coupled conformation, and with low affinity to the inactive G protein-uncoupled conformation. Neutral antagonists (with no intrinsic activity) bind to all conformations of the receptor with the same affinity. However, spontaneous activity, not induced by the agonist, has been demonstrated, which may be decreased by inverse agonists (or negative antagonists), that oppose the intrinsic activity of the agonist.

Spontaneous or constitutive activity of the 5-HT1A in transfected HEK-293 cells is attenuated by 5-HT1A partial agonists (buspirone, flesinoxan) sharing an inverse agonist capacity, but not altered by full agonists or antagonists.[2]

The inverse agonist properties of the 5-HT1B antagonist SB224289, observed at cloned receptors, cannot be detected in vivo.[93]

There is an inverse agonism exerted by some 5-HT2A/2C drugs in the nictitating membrane reflex of the rabbit, which suggests the existence of a constitutive activity in vivo.[57] Prolonged treatment with 5-HT2C inverse agonists enhances selectively 5-HT2C-mediated IP3 accumulation, with no receptor upregulation but changes in the expression of G proteins.[13] The in vivo demonstration of a constitutive activity is difficult, but inverse agonists may be useful drugs for the modulation and regulation of 5-HT receptor activity.

Splice Variants

Apart from 5-HT1, all 5-HT receptor genes contain introns. Although all these receptor subtypes respond differentially to the same neurotransmitter, an even higher level of diversity exists because isoforms have been identified for several of these receptor subtypes.[103] Isoforms are either created by RNA editing or by alternative splicing. The variants might exhibit different pharmacology, abundance and anatomical distribution.[69]

Editing of 5-HT2C receptor mRNA results in 11 distinct mRNA species, seven of which are expressed in the rat brain. These different isoforms modify the basal activity of the receptor, and decrease agonist affinity and potency.[25]

The 5-HT4 splice variants differ in the length and composition of their intracellular C terminals after the common splicing site (L358). These variants have a constitutive activity, which is higher in splice variants with short C-terminal sequences than in splice variants with long C-terminal sequences.[35]

Anatomical and Cellular Compartment Locations

Anatomical Distribution

The possibility that 5-HT receptors play a role in memory is based on the evidence that these receptors are present in crucial regions involved in such functions, like the hippocampal formation, the frontal cortex, the striatum, and related structures. Each type or subtype of serotonin receptor, indeed, has a specific regional distribution in the brain.[64] The serotonin transporter is associated to all membrane compartments of 5-HT neurons and thus mimics 5-HT innervation.[138] The 5-HT1A receptor is mainly concentrated in the hippocampus, the septum, the raphe and, to a lesser extent, in cortical areas. The 5-HT1B receptor is found in the substantia nigra, the hippocampus, the dorsal subiculum and, to a lesser extent, in the striatum and cortex. The 5-HT2 receptor is present in a high concentration in cortical regions. The 5-HT3 receptor is widely distributed in the cortex and in all subfields of the hippocampus. The 5-HT4 receptor is mainly localized in the frontal cortex and hippocampus. Lastly, the 5-HT6 receptor, which emerged more recently from the literature as a potential target for cognitive enhancers, is mainly expressed in the nucleus accumbens, striatum, cerebral cortex and hippocampus.

Cellular Location

In neuroanatomical studies, receptors were found to be located: 1) on the cholinergic septo-hippocampal (5-HT1A, 5-HT1B) and NBM-frontal cortex (5-HT2A, 5-HT3, 5-HT4) pathways; 2) on the glutamatergic pyramidal cells in the hippocampus (5-HT1A, 5-HT4, 5-HT6), the subiculum (5-HT1B), the entorhinal and the frontal (5-HT1A, 5-HT2A, 5-HT3) cortices; 3) on GABAergic interneurons (5-HT1A, 5-HT1B, 5-HT2A, 5-HT3) in different regions (Fig. 3).

Subcellular Location

5-HT receptors, like all types of receptors can be found at two main subcellular locations which are somatodendritic and preterminal positions. The somatodendritic receptors modulate the firing rate of the neuron (indirectly controlling the neurotransmitter release in termi-

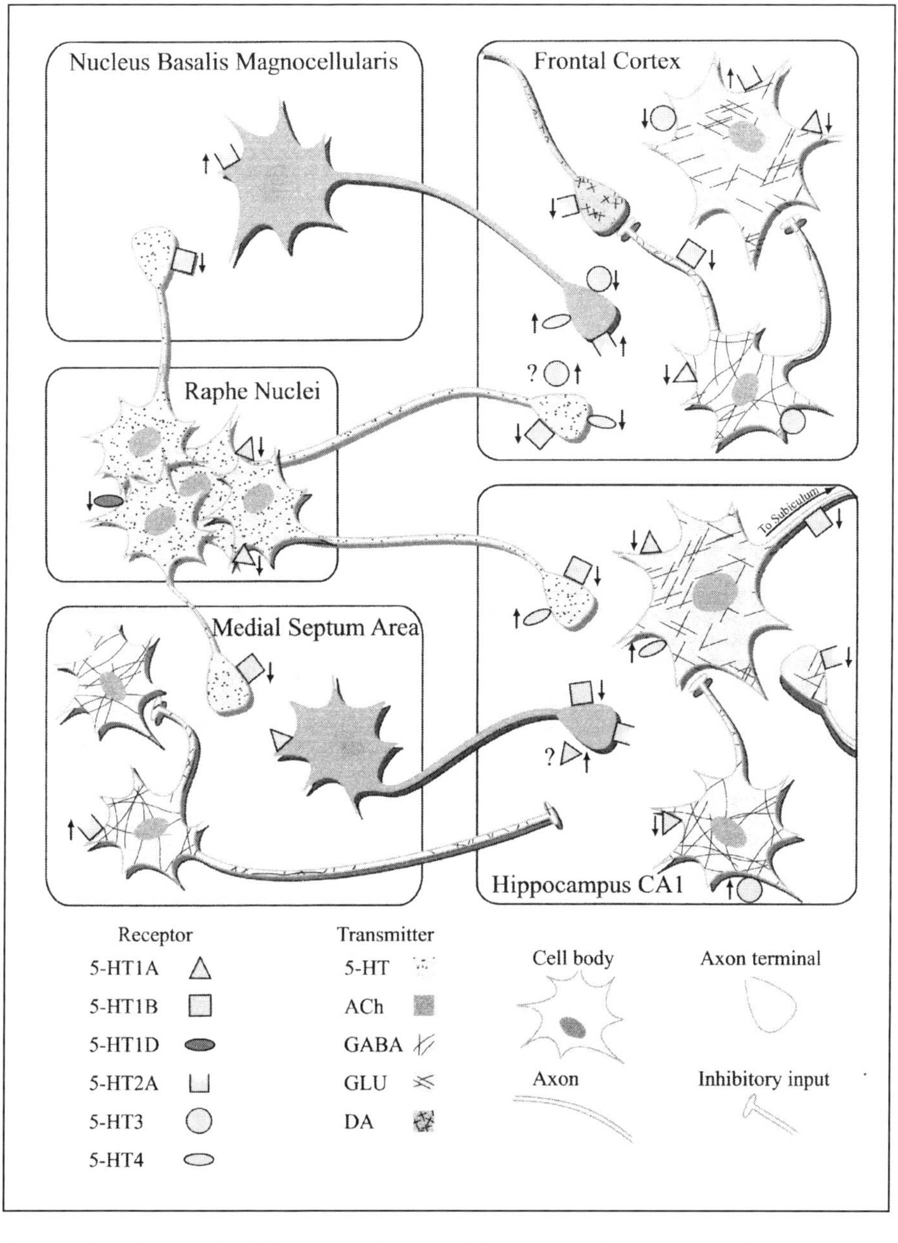

Figure 3. Sites of action of 5-HT receptor subtypes in different neuronal compartments (soma, dendrites, axon terminals) in the septo-hippocampal and basal forebrain-cortical complexes. Schematic representation of data obtained by different authors by using electrophysiology, microscopy and neurotransmitter release methods (see text for references). The functional consequences of the activation of 5-HT receptors are indicated (when known) by an upward arrrow (↑) for an excitatory effect, and a downward (↓) arrow for an inhibitory effect on the target cell (with ? signifying uncertainties). For definitions, please refer to the list of abbreviations.

nals), while preterminal receptors directly control the release of neurotransmitter. Thus, 5-HT receptors can be grouped into three main categories: 1) somatodendritic (5-HT1A, 5-HT3, 5-HT6), 2) preterminal (5-HT1B), 3) both somatodendritic and preterminal (5-HT2, 5-HT4). In addition, receptors can be either auto- or heteroreceptors: Autoreceptors are located on serotonergic neurons whereas heteroreceptors are found on non5-HT-releasing neurons. Autoreceptors have been found in the raphe (5-HT1A, 5-HT1D) and in different projecting areas, such as the septum, hippocampus, the NBM or the entorhinal and frontal cortices (5-HT1B). Heteroreceptors are present in all anatomical structures mentioned above.

Integrating Behavioral and Neurobiological Data

Receptors that Inhibit Adenylyl Cyclase

The 5-HT1A Receptor

The 5-HT1A receptor is characterized by its high concentration in the hippocampus (Fig. 3), as well as in the raphe, where autoreceptors are also likely to exert an indirect influence on cognitive functions. This receptor subtype interacts with other neurotransmitter systems, such as the cholinergic, glutamatergic, GABAergic ones.[52]

8-OH-DPAT, a specific 5-HT1A agonist, dose-dependently impairs the retention of a well-learned maze, as well as acquisition of a novel configuration of the maze.[67] Interestingly, the combination of 8-OH-DPAT and scopolamine, a muscarinic ACh receptor antagonist, even at subthreshold doses, impairs acquisition of the water maze task, but not its retention, in both normal rats and rats with a central 5-HT depletion. This emphasizes the important role of postsynaptic heteroreceptors[112] and the fact that 5-HT1A receptors and muscarinic receptors are located on the same target neurons, i.e., hippocampal pyramidal cells. Both systemic and intra-hippocampal (CA1 region) injections of 8-OH-DPAT impair acquisition of spatial memory tasks in rats, a deficit attributable to the activation of postsynaptic 5-HT1A receptors. Combining both modes of injection with the use of WAY-100135, a 5-HT1A antagonist (with partial agonist properties), Carli and colleagues demonstrated that the activation of hippocampal 5-HT1A receptors selectively impairs spatial but not visual discrimination.[26] In contrast, intra-DRN stimulation of 5-HT1A autoreceptors can compensate spatial learning deficits induced by intra-hippocampal scopolamine injections.[30] WAY-100135 and WAY-100635, another more potent and more selective 5-HT1A antagonist,[44] are equally able to antagonize this deficit.[27,28] Systemic low dose of 8-OH-DPAT prevents intra-hippocampal scopolamine-induced spatial memory impairment, that an intra-DRN injection of WAY-100635 is able to reverse.[32] Similarly, WAY-100635 is able to prevent a cognitive impairment induced by blockade of hippocampal NMDA receptors with MK-801 in a spatial memory task in rats,[31] and in visual and visuospatial discrimination in monkeys.[55]

Stimulation of 5-HT1A receptors has controversial effects on working memory.[60-130] When coadministered intrahippocampally with scopolamine, NAN-190, another 5-HT1A antagonist, is able to reduce scopolamine-induced working memory impairments.[99] There is however no direct reciprocal interaction since physostigmine, a cholinesterase inhibitor generally able to compensate cholinergic dysfunction, fails to compensate the deficit induced by an intra-hippocampal injection of 8-OH-DPAT.[98] WAY-100135 is also able to attenuate the detrimental effect of the blockade of NMDA receptors by MK-801 on working memory measured in rats in a delayed alternation task.[132]

Intra-septal (medial septum) infusion of 8-OH-DPAT in rats impairs spatial learning in a water maze,[14] but facilitates spatial learning in a radial maze in mice.[91] The authors attribute both results, as an alternative hypothesis, to a possible effect of the compound on anxiety. In the same line of research, it was observed that activation of 5-HT1A receptors induces a complex pattern of facilitatory and detrimental effects on learning, which is dependent on the task conditions.[8] The authors have interpreted their results in terms of a particular role played by 5-HT1A receptors in arousal, similar to mild stress. This study represents a particularly interesting example of the complex participation of 5-HT in the balance between emotion and memory mediated by 5-HT1A receptors. Using a behavioral paradigm combining learned helplessness, fear-conditioning and escape learning, Maier et al[78] have emphasized the contribution of 5-HT1A autoreceptors in these behaviors. Systemic and intra-hippocampal administrations of buspirone, a 5-HT1A receptor partial agonist, have detrimental effects on all stages of emotional memory, i.e., acquisition, consolidation and retrieval as studied in rats in different types of avoidance tasks.[75] Intra-amygdala infusion of 8-OH-DPAT (or buspirone) produces a deficit in the retention of avoidance learning,[76] whereas intra-hippocampal infusion of 8-OH-DPAT produces a deficit in the acquisition but not consolidation of fear conditioning.[123] WAY 100635 is able to attenuate or block both of these effects.

Some recent neurobiological data have added further insights to our understanding of the way 5-HT1A receptors may interfere with the septo-hippocampal formation (Fig. 3). Activation of MRN 5-HT1A autoreceptors has been found to produce the hippocampal theta rhythm and a rhythmical firing pattern of MS/DBB neurons, strongly correlating with the theta rhythm.[70] Moreover, 5-HT1A receptors are expressed in a subpopulation of cholinergic neurons belonging to the MS/DBB complex.[68] 5-HT, acting via 5-HT1A receptors, is able on the one hand to decrease CA1 pyramidal cell activity directly, and on the other hand to disinhibit these cells by acting on interneurons, thus controlling the balance between excitation and inhibition in this region.[118] Stimulation of 5-HT1A receptors by BAY x 3702, a high-affinity 5-HT1A agonist, enhances ACh release in the rat cortex and hippocampus, an effect compensated by WAY-100635.[71] This effect may be due to an indirect mechanism or to the contribution of 5-HT7 receptors,[96] which have a good affinity for 5-HT1A agonists.[128]

The main conclusion drawn from these data is that drugs that stimulate 5-HT1A autoreceptors and block hippocampal 5-HT1A receptors may be useful in the symptomatic treatment of human memory disturbances associated with the loss of cholinergic and glutamatergic innervation to the hippocampus.

The results presented here support a consistent role for, especially hippocampal, 5-HT1A receptors in learning and memory, predominantly by way of their relationships with the cholinergic and the glutamatergic systems. In a majority of cases, activation of these receptors compromises learning, whereas their inactivation selectively reduces this detrimental effect and those associated with cholinergic and glutamatergic dysfunction.

More recently, molecular biology provided a generation of specific KO and transgenic mice. The availability of both constitutive and inducible 5-HT1A KO mice will certainly add further critical knowledge regarding the cognitive functions regulated by this receptor subtype.[106-101-58]

The 5-HT1B Receptor

In contrast to the 5-HT1A receptor, the 5-HT1B receptor is predominantly located on axon terminals (Fig. 3). This receptor is present in the hippocampus, the dorsal subiculum and frontal cortex.

We demonstrated that stimulation of hippocampal 5-HT1B receptors, by the specific 5-HT1B receptor agonist CP 93129, impairs the performance of rats trained in a radial arm maze using a procedure that makes it possible to dissociate working versus reference memory errors. This impairment affected the reference memory component of the task more than the working memory component.[23] These results might be explained by the specific cellular and subcellular locations of 5-HT1B receptors in the hippocampal formation[116] (Fig. 3). In particular, stimulating 5-HT1B heteroreceptors located on hippocampal terminals of septal cholinergic neurons induces a decrease in the release of ACh, thus reducing the efficacy of the septo-hippocampal pathway.[34] In addition, stimulating 5-HT1B receptors located on axon terminals of glutamatergic CA1 pyramidal cells[1] induces a decrease in the release of glutamate, thus reducing CA1-subiculum transmission.[16] In this configuration, the 5-HT1B receptors occupy strategic locations for the control of the main inputs and outputs of the hippocampus. In particular, 5-HT, via 5-HT1B receptors, may act as a filter for the transfer of processed information from the hippocampus to cortical areas.

In contrast, stimulation of 5-HT1B receptors has been demonstrated to increase ACh release in the frontal cortex, probably by acting on GABAergic interneurons, thus stimulating cortical pyramidal cells.[37] This is an interesting example of the way a specific receptor is able to influence learning and memory performances in opposite directions, depending on the cerebral structure concerned, in processes which may be either time-dependent (dynamics of the memory trace) or task-dependent (memory systems). For a memory model relying more on the recruitment of the hippocampus (spatial reference memory), we have observed that a specific agonist impairs performance. For a memory model that involves more the recruitment of the frontal cortex (as working memory or sequential processing), one might expect a facilitatory role of 5-HT1B agonists on memory, via an interaction with the cholinergic system.

So far, there was a lack of specific 5-HT1B antagonists enabling to study the functions of this receptor. We thus studied the learning and memory abilities of 5-HT1B KO mice.[117] In an object exploration task, the 5-HT1B KO mice displayed higher exploratory activity than wild type mice,[79] a result consistent with the fact that rats show decreased exploratory activity following specific stimulation (using CP 93129) of hippocampal 5-HT1B receptors.[22] In the Morris water maze, no differences between genotypes was found in the visual cue version of the task, but the 5-HT1B KO mice learned the spatial reference memory task faster than control (wild type) mice and were found to display higher flexibility when confronted with a change in the platform location. Independently, Hamon and colleagues observed that the 5-HT1B KO mice displayed longer periods of paradoxical sleep than wild type mice.[18] Given the positive influence of paradoxical sleep on memory consolidation,[59] this result converges with our findings showing enhanced long term memory performance in 5-HT1B KO mice.[79] Furthermore, recent results from our group indicate that this facilitatory effect of the 5-HT1B gene deletion on memory is accentuated in aged 5-HT1B KO mice.[80]

These results are promising in view of possible therapeutical applications, in particular the use of functionally selective 5-HT1B receptor antagonists with potential use in the treatment of aging-related or AD-associated memory decline.

Receptors that Stimulate Adenylyl Cyclase

The 5-HT4 Receptor

The 5-HT4 receptor is present in the frontal cortex and the hippocampus (Fig. 3). 5-HT4 receptor agonists stimulate adenylyl cyclase, thereby increasing cAMP levels and producing a decreased after-hyperpolarization that may increase neuronal excitability and neurotransmitter release.[41] Stimulation of 5-HT4 receptors by specific agonists selectively facilitates basal ACh release in the frontal cortex of rats, but not in the striatum,[41] and increases the extracellular level of 5-HT in the hippocampus.[48]

These basic modifications in neuronal excitability and/or neurotransmitter release by 5-HT4 ligands in anatomical structures underlying learning and memory suggest an active role for this receptor in these functions. RS67333, a selective 5-HT4 agonist, prevents the performance deficit induced in rats by atropine in the Morris water maze, an effect reversed by the selective 5-HT4 antagonist RS67532.[46] RS17017, another 5-HT4 agonist, enhances dose- and delay-dependently delayed-matching-to-sample performance in aged and young monkeys.[126] Even if the 5-HT4 receptors do not seem to be tonically activated, the preacquisition administration of various 5-HT4 agonists generally improves memory in rats in the Morris water maze,[46] in a social olfactory recognition task assessing working memory,[74] or in an olfactory association learning, assessing long-term memory,[83] whereas the post-training administration impairs consolidation in an autoshaping task in rats.[88] However, post-training administration of 5-HT4 antagonists (SDZ 205557, GR 125487) produces an amnesic effect in the mouse passive avoidance test, that 5-HT4 agonists (BIMU 1, BIMU 8) are able to prevent.[47] More clearly, it was recently observed that selective 5-HT4 agonists, RS 17017 and especially RS 67333, counteract the detrimental effect of the selective 5-HT4 receptor antagonist RS 67532 in the associative olfactory task in rats.[82]

On the whole, these experimental data strongly suggest a beneficial role for 5-HT4 receptor agonists in restoring deficits in learning and memory.

The 5-HT6 Receptor

The 5-HT6 receptor is an emerging target for drug discovery.[19] This receptor, present in the cerebral cortex and the hippocampus is given much attention due to the potential role of some of its antagonists as cognitive enhancers. The 5-HT6 receptor antagonist Ro 04-6790 improves consolidation in operant learning (autoshaping) in rats and is able to reverse the detrimental effect of scopolamine in this task, but not the detrimental effect of dizocilpine.[90] 5-HT6 receptors might be involved in the control of ACh neurotransmission.[17] Ro 04-6790 does not affect the acquisition of the water maze task, but enhances the long-term retention of the platform location.[134] Another 5-HT6 receptor antagonist, SB 271046, also improves retention in the water maze and produces a significant improvement of the performance of aged rats submitted to an operant delayed learning task.[113] This compound was also demonstrated as being able to increase glutamate and aspartate levels in frontal cortex. As it is orally bioavailable, it is currently under investigation in phase I in humans for its potential properties as cognitive enhancer.[92]

Receptors that Stimulate Phospholipase C

The 5-HT2A/2C Receptors

The 5-HT2A and 5-HT2C receptors both stimulate phospholipase C and show a close structural homology. So far, there are no ligands truly selective for each of these receptor subtypes, and few studies are available on the effects of 5-HT2 receptors in learning and memory. They are present in high concentrations in cortical areas and are thought to mediate more attentional than memory processes.[24] Activation of 5-HT2A/2C receptors induces a facilitatory effect on cholinergic release in the rat frontal cortex.[61] MDL100907, a 5-HT2A antagonist, is able to block the excitatory effect of 5-HT on septo-hippocampal neurons, via 5-HT2 receptors, probably located on GABAergic neurons belonging to the MS/DBB complex.[77] MDL100907 is able to selectively abolish the improving effects elicited by DOI, a 5-HT2A/2C receptor agonist, on memory consolidation in rats.[89]

Harvey[56] reviewed different studies dealing with the effects of 5-HT2A/2C receptor agonists and antagonists on associative learning, as assessed by the conditioned avoidance response in rats and the conditioned nictitating membrane response in the rabbit. These two tasks are highly sensitive to specific 5-HT2A/2C drugs. Agonists are consistently observed to enhance learning in both tasks, whereas 5-HT1A receptor agonists either have no effect or retard learning. Interestingly, the author suggests that these drug effects are observed only when the task generates a low level of acquisition. The facilitatory effect of 5-HT2A/2C receptor agonists in learning (see also ref. 98) might, therefore, be particularly efficient in situations in which task difficulty is increased. The 5-HT2C KO mouse exhibits weight gain and a high probability of spontaneous death associated with seizures.[6] This mouse is considered as a robust model for the study of serotonergic mechanisms in epilepsy, which mainly concerns hippocampal dysfunction. The 5-HT2C KO mouse exhibits an abnormal performance in the water maze, as well as a defect in synaptic long-term potentiation restricted to the main input of the hippocampus (perforant path-dentate gyrus).[124] This result elucidates the role of 5-HT2C receptors in neuronal plasticities underlying hippocampal learning and memory functions.

Altogether, these data suggest the use of 5-HT2A/2C agonists as complementary or alternative therapeutic or preventive tools for the treatment of severe memory deficits in humans (e.g., AD patients), and also even in normal aging, since these compounds appear to be beneficial in memory tasks with a high cognitive demand.

A Ligand-Gated Ion Channel

The 5-HT3 Receptor

The 5-HT3 receptor is the only ligand-gated ion channel among the 5-HT receptors.

Inactivation of 5-HT3 receptors has been found to increase the frequency of the hippocampal theta rhythm and the magnitude and duration of long-term potentiation (LTP) in the CA1 of the hippocampus of freely moving rats.[121] These neurophysiological effects have been correlated with improved retention in both spatial and olfactory memory tasks, behaviors which require the integrity of the hippocampus.[121] Systemic administration of a 5-HT3 antagonist (ondansetron) decreases the firing activity of CA1 hippocampal interneurons, with concomitant increases in the firing rate of pyramidal cells.[108] These results indirectly confirm the presence of 5-HT3 receptors on GABAergic interneurons in the rat hippocampus.[95] In the entorhinal cortex, 5-HT3 receptors (probably located on GABAergic interneurons) tonically inhibit the release of ACh.[107]

Initial research on the role played by 5-HT in learning and memory started by demonstrating that 5-HT3 receptor antagonists are able to improve learning and memory or to antagonize the effects of anticholinergic or age-induced memory decline in rodents and primates,[10] which is a more recent discovery for other types or subtypes of 5-HT receptors. The beneficial effect of 5-HT3 antagonists currently receives additional support. Thus, while mCPBG, a 5-HT3 agonist, impairs retention of an associative learning task in rats, ondansetron and tropisetron, 5-HT3 antagonists, improve it.[63] Ondansetron prevents scopolamine-induced impairment of short-term memory retrieval in mice,[114] as well as intra-hippocampal scopolamine induced impairment in rat spatial learning in a water maze.[29] Y-25130, another 5-HT3 antagonist, has a similar compensatory effect on working memory impairment due to cholinergic blockade in rats, but is ineffective to compensate memory impairment due to blockade of glutamatergic transmission.[100]

The cognitive-enhancing property of ondansetron in aged rats is significant and selective in cognitively impaired animals (as determined by a prescreening procedure) when compared with that of a cholinergic agonist.[45] This prescreening procedure might be a particularly sensitive method for detecting those aged subjects who sustain critical memory deficits that can be reversed with drugs. In addition, 5-HT3 antagonists, in contrast to nicotine, have no effect on the acquisition of the water maze task in intact rats, but antagonize both acquisition and retention impairments in rats sustaining a combined lesion of the MS and the NBM.[62] Taken together, these data clearly indicate that the use of 5-HT3 receptor antagonists is an efficient compensatory model for cholinergic dysfunction, maybe better than cholinergic agonists. However, extrapolation to humans is not so clear, as it was found recently that ondansetron is not able to attenuate scopolamine-induced impairment of episodic memory of young healthy volunteers.[21,81] There is, nevertheless, no doubt that the 'cognitive profile' of the 5-HT3 receptor KO mouse will be of great interest.[136]

Current research suggests that there may be subtypes of the 5-HT3 receptor.[38] This hypothesis might explain some functional dissociations, such as the one revealed in a behavioral study demonstrating that ondansetron (on passive avoidance) and tropisetron (on spatial learning) have different efficacies to counteract scopolamine-induced memory impairment.[105] But it must also be kept in mind that different cerebral structures or circuits preferentially underlie the various tasks, recruiting different cognitive operations, that is commonly accepted as memory systems. Hence, Arnsten and colleagues[7] observed mixed drug effects obtained with two 5-HT3 antagonists (ondansetron and SEC-579) in cognitive improvement of aged monkeys submitted to the reversal of a visual discrimination task. Some monkeys either markedly improved or impaired their performance, depending probably on competing controls of cognitive processes between the orbital prefrontal cortex and the inferior temporal cortex. In addition, RS-56812, another 5-HT3 antagonist, was observed to facilitate short-term memory rather than attention in monkeys.[125]

This model of 5-HT3 receptor-mediated control of an inhibitory influence on cholinergic neurons and modulation of hippocampal theta rhythm and LTP may underlie the memory enhancing property of various 5-HT3 antagonists. These compounds may be used in potential therapies aimed at correcting the memory deficits resulting from cholinergic hypofunction.[102]

Conclusions and Perspectives

In the present chapter, we emphasized recent behavioral and neurobiological studies which have increased our understanding of the contribution of 5-HT and its receptors to the mechanisms of memory formation. None of these mechanisms appear to be linked exclusively to the activation (or inactivation) of a single receptor subtype. On the contrary, converging evidence indicates that different subtypes of receptors potentially interact to contribute to a particular function.[127] This is the case of 5-HT1A, 5-HT1B, 5-HT3 and 5-HT4 receptor subtypes in hippocampal functions, and probably also of 5-HT2 and 5-HT3 receptors for those functions depending more critically on cortical control. Interestingly, 5-HT, through some of its receptors, is proposed to play a role both in information processing, by controlling the signal-to-noise ratio, and in the control of behavioral processes, such as behavioral inhibition. It has therefore been suggested that 5-HT plays a selective, and probably crucial, role in situations involving an increased cognitive demand. This has been demonstrated following 5-HT depletion, stimulation of 5-HT2A/2C and 5-HT4 receptors, and inactivation of 5-HT1A and 5-HT3 receptors.

Role of 5-HT in Age- or Alzheimer's Disease-Related Memory Decline

The practical applications of these researches correspond to a crucial need in the population to attempt to delay age-related memory decline or to prevent the behavioral consequences of premature serious neuropathologies such as AD. Learning and memory deficits appearing with aging and AD are positively correlated with the decline in cholinergic neurotransmission.[119] We developed experimental arguments showing that age-related cognitive deficits are associated with a combined dysfunctioning of cholinergic and serotonergic functions,[39,109] and that the functional interaction between these two neurotransmitter systems plays an important role in the maintenance of learning and memory performances.[115,122] Thus, our insight into the mechanisms by which the cross-talk between 5-HT and ACh signals occur, in particular through the mediation of specific types or subtypes of 5-HT receptors, may provide a useful framework for the development of novel drugs to delay or prevent memory decline.

New Tools for Research and Therapeutical Strategies

Through the development of more specific tools, both pharmacology and molecular biology will contribute to a better understanding of the functions of 5-HT receptor subtypes. These tools will include new selective antagonists and agonists, as well as the ability to target changes in genes at specific times, e.g., to bypass the developmental compensatory mechanisms and at specific anatomical sites.[81] However, the behavioral tools will, in all experiments, be decisive. It is thus also crucial to use 'targeted' behavioral models in order to recruit preferentially the functioning of a specific cerebral structure, and thus add further knowledge about memory systems.

References

1. Ait Amara D, Segu L, Buhot MC. Region-specific decrease in 5-HT1A and 5-HT1B binding sites after intra-hippocampal ibotenic acid injections in the rat. Neurosci Lett 2001; 310(1):25-28.
2. Albert PR, Sajedi N, Lemonde S et al. Constitutive G(i2)-dependent activation of adenylyl cyclase type II by the 5-HT1A receptor. Inhibition by anxiolytic partial agonists. J Biol Chem 1999; 274(50):35469-35474.
3. Altman HJ, Normile HJ, Galloway MP et al. Enhanced spatial discrimination learning in rats following 5,7-DHT-induced serotonergic deafferentation of the hippocampus. Brain Res 1990; 518(1-2):61-66.
4. Al-Zahrani SS, Ho MY, Martinez DN et al. Effect of destruction of the 5-hydroxytryptaminergic pathways on the acquisition of temporal discrimination and memory for duration in a delayed conditional discrimination task. Psychopharmacology (Berl) 1996; 123(1):103-110.

5. Andrade R. Enhancement of beta-adrenergic responses by Gi-linked receptors in rat hippocampus. Neuron 1993; 10(1):83-88.
6. Applegate CD, Tecott LH. Global increases in seizure susceptibility in mice lacking 5-HT2C receptors: A behavioral analysis. Exp Neurol 1998; 154(2):522-530.
7. Arnsten AF, Lin CH, Van Dyck CH et al. The effects of 5-HT3 receptor antagonists on cognitive performance in aged monkeys. Neurobiol Aging 1997; 18(1):21-28.
8. Balleine BW, Fletcher N, Dickinson A. Effect of the 5-HT1A agonist, 8-OH-DPAT on instrumental performance in rats. Psychopharmacology (Berl) 1996; 125(1):79-88.
9. Balse E, Lazarus C, Kelche C et al. Intrahippocampal grafts containing cholinergic and serotonergic fetal neurons ameliorate spatial reference but not working memory in rats with fimbria-fornix/cingular bundle lesions. Brain Res Bull 1999; 49(4):263-272.
10. Barnes JM, Costall B, Coughlan J et al. The effects of ondansetron, a 5-HT3 receptor antagonist, on cognition in rodents and primates. Pharmacol Biochem Behav 1990; 35(4):955-962.
11. Barnes NM, Sharp T. A review of central 5-HT receptors and their function. Neuropharmacology 1999; 38(8):1083-1152.
12. Berg KA, Maayani S, Goldfarb J et al. Pleiotropic behavior of 5-HT2A and 5-HT2C receptor agonists. Ann N Y Acad Sci 1998; 861:104-110.
13. Berg KA, Stout BD, Cropper JD et al. Novel actions of inverse agonists on 5-HT2C receptor systems. Mol Pharmacol 1999; 55(5):863-872.
14. Bertrand F, Lehmann O, Lazarus C et al. Intraseptal infusions of 8-OH-DPAT in the rat impairs water-maze performances: Effects on memory or anxiety? Neurosci Lett 2000; 279(1):45-48.
15. Bockaert J, Pin JP. Molecular tinkering of G protein-coupled receptors: An evolutionary success. Embo J 1999; 18(7):1723-1729.
16. Boeijinga PH, Boddeke HW. Activation of 5-HT1B receptors suppresses low but not high frequency synaptic transmission in the rat subicular cortex in vitro. Brain Res 1996; 721(1-2):59-65.
17. Bourson A, Boess FG, Bos M et al. Involvement of 5-HT6 receptors in nigro-striatal function in rodents. Br J Pharmacol 1998; 125(7):1562-1566.
18. Boutrel B, Franc B, Hen R et al. Key role of 5-HT1B receptors in the regulation of paradoxical sleep as evidenced in 5-HT1B knock-out mice. J Neurosci 1999; 19(8):3204-3212.
19. Branchek TA, Blackburn TP. 5-ht6 receptors as emerging targets for drug discovery. Annu Rev Pharmacol Toxicol 2000; 40:319-334.
20. Brezun JM, Daszuta A. Depletion in serotonin decreases neurogenesis in the dentate gyrus and the subventricular zone of adult rats. Neuroscience 1999; 89(4):999-1002.
21. Broocks A, Little JT, Martin A et al. The influence of ondansetron and m-chlorophenylpiperazine on scopolamine-induced cognitive, behavioral, and physiological responses in young healthy controls. Biol Psychiatry 1998; 43(6):408-416.
22. Buhot MC, Naili S. Changes in exploratory activity following stimulation of hippocampal 5-HT1A and 5-HT1B receptors in the rat. Hippocampus 1995; 5(3):198-208.
23. Buhot MC, Patra SK, Naili S. Spatial memory deficits following stimulation of hippocampal 5-HT1B receptors in the rat. Eur J Pharmacol 1995; 285(3):221-228.
24. Buhot MC. Serotonin receptors in cognitive behaviors. Curr Opin Neurobiol 1997; 7(2):243-254.
25. Burns CM, Chu H, Rueter SM et al. Regulation of serotonin-2C receptor G-protein coupling by RNA editing. Nature 1997; 387(6630):303-308.
26. Carli M, Luschi R, Garofalo P et al. 8-OH-DPAT impairs spatial but not visual learning in a water maze by stimulating 5-HT1A receptors in the hippocampus. Behav Brain Res 1995; 67(1):67-74.
27. Carli M, Luschi R, Samanin R. (S)-WAY 100135, a 5-HT1A receptor antagonist, prevents the impairment of spatial learning caused by intrahippocampal scopolamine. Eur J Pharmacol 1995; 283(1-3):133-139.
28. Carli M, Bonalumi P, Samanin R. WAY 100635, a 5-HT1A receptor antagonist, prevents the impairment of spatial learning caused by intrahippocampal administration of scopolamine or 7-chloro-kynurenic acid. Brain Res 1997; 774(1-2):167-174.
29. Carli M, Luschi R, Samanin R. Dose-related impairment of spatial learning by intrahippocampal scopolamine: Antagonism by ondansetron, a 5-HT3 receptor antagonist. Behav Brain Res 1997; 82(2):185-194.
30. Carli M, Bonalumi P, Samanin R. Stimulation of 5-HT1A receptors in the dorsal raphe reverses the impairment of spatial learning caused by intrahippocampal scopolamine in rats. Eur J Neurosci 1998; 10(1):221-230.
31. Carli M, Silva S, Balducci C et al. WAY 100635, a 5-HT1A receptor antagonist, prevents the impairment of spatial learning caused by blockade of hippocampal NMDA receptors. Neuropharmacology 1999; 38(8):1165-1173.
32. Carli M, Balducci C, Samanin R. Low doses of 8-OH-DPAT prevent the impairment of spatial learning caused by intrahippocampal scopolamine through 5-HT(1A) receptors in the dorsal raphe. Br J Pharmacol 2000; 131(2):375-381.

33. Cassel JC, Jeltsch H. Serotonergic modulation of cholinergic function in the central nervous system: Cognitive implications. Neuroscience 1995; 69(1):1-41.
34. Cassel JC, Jeltsch H, Neufang B et al. Downregulation of muscarinic- and 5-HT1B-mediated modulation of [3H]acetylcholine release in hippocampal slices of rats with fimbria-fornix lesions and intrahippocampal grafts of septal origin. Brain Res 1995; 704(2):153-166.
35. Claeysen S, Sebben M, Becamel C et al. Novel brain-specific 5-HT4 receptor splice variants show marked constitutive activity: Role of the C-terminal intracellular domain. Mol Pharmacol 1999; 55(5):910-920.
36. Clarke WP, Yocca FD, Maayani S. Lack of 5-hydroxytryptamine1A-mediated inhibition of adenylyl cyclase in dorsal raphe of male and female rats. J Pharmacol Exp Ther 1996; 277(3):1259-1266.
37. Consolo S, Arnaboldi S, Ramponi S et al. Endogenous serotonin facilitates in vivo acetylcholine release in rat frontal cortex through 5-HT 1B receptors. J Pharmacol Exp Ther 1996; 277(2):823-830.
38. Davies PA, Pistis M, Hanna MC et al. The 5-HT3B subunit is a major determinant of serotonin-receptor function. Nature 1999; 397(6717):359-363.
39. Decker MW, McGaugh JL. The role of interactions between the cholinergic system and other neuromodulatory systems in learning and memory. Synapse 1991; 7(2):151-168.
40. Dickenson JM, Hill SJ. Human 5-HT1B receptor stimulated inositol phospholipid hydrolysis in CHO cells: Synergy with Gq-coupled receptors. Eur J Pharmacol 1998; 348(2-3):279-285.
41. Eglen RM, Wong EH, Dumuis A et al. Central 5-HT4 receptors. Trends Pharmacol Sci 1995; 16(11):391-398.
42. Evans KL, Cropper JD, Berg KA et al. Mechanisms of regulation of agonist efficacy at the 5-HT(1A) receptor by phospholipid-derived signaling components. J Pharmacol Exp Ther 2001; 297(3):1025-1035.
43. Farr SA, Uezu K, Flood JF et al. Septo-hippocampal drug interactions in post-trial memory processing. Brain Res 1999; 847(2):221-230.
44. Fletcher A, Forster EA, Bill DJ et al. Electrophysiological, biochemical, neurohormonal and behavioural studies with WAY-100635, a potent, selective and silent 5-HT1A receptor antagonist. Behav Brain Res 1996; 73(1-2):337-353.
45. Fontana DJ, Daniels SE, Henderson C et al. Ondansetron improves cognitive performance in the Morris water maze spatial navigation task. Psychopharmacology (Berl) 1995; 120(4):409-417.
46. Fontana DJ, Daniels SE, Wong EH et al. The effects of novel, selective 5-hydroxytryptamine (5-HT)4 receptor ligands in rat spatial navigation. Neuropharmacology 1997; 36(4-5):689-696.
47. Galeotti N, Ghelardini C, Bartolini A. Role of 5-HT4 receptors in the mouse passive avoidance test. J Pharmacol Exp Ther 1998; 286(3):1115-1121.
48. Ge J, Barnes NM. 5-HT4 receptor-mediated modulation of 5-HT release in the rat hippocampus in vivo. Br J Pharmacol 1996; 117(7):1475-1480.
49. Gonzalez-Burgos I, Perez-Vega MI, Del Angel-Meza AR et al. Effect of tryptophan restriction on short-term memory. Physiol Behav 1998; 63(2):165-169.
50. Gould E, Beylin A, Tanapat P et al. Learning enhances adult neurogenesis in the hippocampal formation. Nat Neurosci 1999; 2(3):260-265.
51. Graeff FG, Viana MB, Mora PO. Opposed regulation by dorsal raphe nucleus 5-HT pathways of two types of fear in the elevated T-maze. Pharmacol Biochem Behav 1996; 53(1):171-177.
52. Hajos-Korcsok E, Sharp T. 8-OH-DPAT-induced release of hippocampal noradrenaline in vivo: Evidence for a role of both 5-HT1A and dopamine D1 receptors. Eur J Pharmacol 1996; 314(3):285-291.
53. Harder JA, Kelly ME, Cheng CH et al. Combined pCPA and muscarinic antagonist treatment produces a deficit in rat water maze acquisition. Pharmacol Biochem Behav 1996; 55(1):61-65.
54. Harder JA, Kelly ME. The effect of several putative cognition enhancers on a water maze acquisition deficit produced by pCPA + scopolamine combination treatment. Pharmacol Biochem Behav 1997; 56(4):657-661.
55. Harder JA, Ridley RM. The 5-HT1A antagonist, WAY 100 635, alleviates cognitive impairments induced by dizocilpine (MK-801) in monkeys. Neuropharmacology 2000; 39(4):547-552.
56. Harvey JA. Serotonergic regulation of associative learning. Behav Brain Res 1996; 73(1-2):47-50.
57. Harvey JA, Welsh SE, Hood H et al. Effect of 5-HT2 receptor antagonists on a cranial nerve reflex in the rabbit: Evidence for inverse agonism. Psychopharmacology (Berl) 1999; 141(2):162-168.
58. Heisler LK, Chu HM, Brennan TJ et al. Elevated anxiety and antidepressant-like responses in serotonin 5-HT1A receptor mutant mice. Proc Natl Acad Sci USA 1998; 95(25):15049-15054.
59. Hennevin E, Hars B, Maho C et al. Processing of learned information in paradoxical sleep: Relevance for memory. Behav Brain Res 1995; 69(1-2):125-135.
60. Herremans AHJ, Hijzen TH, Olivier B et al. Serotonergic drug effects on a delayed conditional discrimlination task in the rat: Involvement of the 5-HT1A receptor in working memory. J Psychopharmacol 1995; 9:242-250.

61. Hirano H, Day J, Fibiger HC. Serotonergic regulation of acetylcholine release in rat frontal cortex. J Neurochem 1995; 65(3):1139-1145.
62. Hodges H, Sowinski P, Turner JJ et al. Comparison of the effects of the 5-HT3 receptor antagonists WAY-100579 and ondansetron on spatial learning in the water maze in rats with excitotoxic lesions of the forebrain cholinergic projection system. Psychopharmacology (Berl) 1996; 125(2):146-161.
63. Hong E, Meneses A. Systemic injection of p-chloroamphetamine eliminates the effect of the 5-HT3 compounds on learning. Pharmacol Biochem Behav 1996; 53(4):765-769.
64. Hoyer D, Martin G. 5-HT receptor classification and nomenclature: Towards a harmonization with the human genome. Neuropharmacology 1997; 36(4-5):419-428.
65. Inoue T, Tsuchiya K, Koyama T. Regional changes in dopamine and serotonin activation with various intensity of physical and psychological stress in the rat brain. Pharmacol Biochem Behav 1994; 49(4):911-920.
66. Inoue T, Tsuchiya K, Koyama T. Serotonergic activation reduces defensive freezing in the conditioned fear paradigm. Pharmacol Biochem Behav 1996; 53(4):825-831.
67. Kant GJ, Meininger GR, Maughan KR et al. Effects of the serotonin receptor agonists 8-OH-DPAT and TFMPP on learning as assessed using a novel water maze. Pharmacol Biochem Behav 1996; 53(2):385-390.
68. Kia HK, Brisorgueil MJ, Daval G et al. Serotonin1A receptors are expressed by a subpopulation of cholinergic neurons in the rat medial septum and diagonal band of Broca—A double immunocytochemical study. Neuroscience 1996; 74(1):143-154.
69. Kilpatrick GJ, Dautzenberg FM, Martin GR et al. 7TM receptors: The splicing on the cake. Trends Pharmacol Sci 1999; 20(7):294-301.
70. Kinney GG, Kocsis B, Vertes RP. Medial septal unit firing characteristics following injections of 8-OH-DPAT into the median raphe nucleus. Brain Res 1996; 708(1-2):116-122.
71. Koyama T, Nakajima Y, Fujii T et al. Enhancement of cortical and hippocampal cholinergic neurotransmission through 5-HT1A receptor-mediated pathways by BAY x 3702 in freely moving rats. Neurosci Lett 1999; 265(1):33-36.
72. Le Grand B, Panissie A, Pauwels PJ et al. Activation of recombinant h 5-HT1B and h 5-HT1D receptors stably expressed in C6 glioma cells produces increases in Ca2+-dependent K+ current. Naunyn Schmiedebergs Arch Pharmacol 1998; 358(6):608-615.
73. Lehmann O, Jeltsch H, Lehnardt O et al. Combined lesions of cholinergic and serotonergic neurons in the rat brain using 192 IgG-saporin and 5,7-dihydroxytryptamine: Neurochemical and behavioural characterization. Eur J Neurosci 2000; 12(1):67-79.
74. Letty S, Child R, Dumuis A et al. 5-HT4 receptors improve social olfactory memory in the rat. Neuropharmacology 1997; 36(4-5):681-687.
75. Liang KC, Tsui KY, Tyan YM et al. Buspirone impaired acquisition and retention in avoidance tasks: Involvement of the hippocampus. Chin J Physiol 1998; 41(1):33-44.
76. Liang KC. Pre or post-training injection of buspirone impaired retention in the inhibitory avoidance task: Involvement of amygdala 5-HT1A receptors. Eur J Neurosci 1999; 11(5):1491-1500.
77. Liu W, Alreja M. Atypical antipsychotics block the excitatory effects of serotonin in septohippocampal neurons in the rat. Neuroscience 1997; 79(2):369-382.
78. Maier SF, Grahn RE, Watkins LR. 8-OH-DPAT microinjected in the region of the dorsal raphe nucleus blocks and reverses the enhancement of fear conditioning and interference with escape produced by exposure to inescapable shock. Behav Neurosci 1995; 109(3):404-412.
79. Malleret G, Hen R, Guillou JL et al. 5-HT1B receptor knock-out mice exhibit increased exploratory activity and enhanced spatial memory performance in the Morris water maze. J Neurosci 1999; 19(14):6157-6168.
80. Malleret G, Savova M, Hen R et al. 5-HT1B receptor mutation decrease age-related impairement of spatial reference memory performance. Soc Neurosci Abstr 1999; 25:2156.
81. Malleret G, Haditsch U, Genoux D et al. Inducible and reversible enhancement of learning, memory, and long-term potentiation by genetic inhibition of calcineurin. Cell 2001; 104(5):675-686.
82. Marchetti E, Dumuis A, Bockaert J et al. Differential modulation of the 5-HT(4) receptor agonists and antagonist on rat learning and memory. Neuropharmacology 2000; 39(11):2017-2027.
83. Marchetti-Gauthier E, Roman FS, Dumuis A et al. BIMU1 increases associative memory in rats by activating 5-HT4 receptors. Neuropharmacology 1997; 36(4-5):697-706.
84. Marinissen MJ, Gutkind JS. G-protein-coupled receptors and signaling networks: Emerging paradigms. Trends Pharmacol Sci 2001; 22(7):368-376.
85. Markowska AL, Wenk GL. Serotonin influences the behavioral recovery of rats following nucleus basalis lesions. Pharmacol Biochem Behav 1991; 38(4):731-737.
86. Matsukawa M, Ogawa M, Nakadate K et al. Serotonin and acetylcholine are crucial to maintain hippocampal synapses and memory acquisition in rats. Neurosci Lett 1997; 230(1):13-16.

87. Mazer C, Muneyyirci J, Taheny K et al. Serotonin depletion during synaptogenesis leads to decreased synaptic density and learning deficits in the adult rat: A possible model of neurodevelopmental disorders with cognitive deficits. Brain Res 1997; 760(1-2):68-73.
88. Meneses A, Hong E. Effects of 5-HT4 receptor agonists and antagonists in learning. Pharmacol Biochem Behav 1997; 56(3):347-351.
89. Meneses A, Terron JA, Hong E. Effects of the 5-HT receptor antagonists GR127935 (5-HT1B/1D) and MDL100907 (5-HT2A) in the consolidation of learning. Behav Brain Res 1997; 89(1-2):217-223.
90. Meneses A. Effects of the 5-HT(6) receptor antagonist Ro 04-6790 on learning consolidation. Behav Brain Res 2001; 118(1):107-110.
91. Micheau J, Van Marrewijk B. Stimulation of 5-HT1A receptors by systemic or medial septum injection induces anxiogenic-like effects and facilitates acquisition of a spatial discrimination task in mice. Prog Neuropsychopharmacol Biol Psychiatry 1999; 23(6):1113-1133.
92. Miguel-Hidalgo JJ. SB-271046 (SmithKline Beecham). Curr Opin Investig Drugs 2001; 2(1):118-122.
93. Millan MJ, Gobert A, Audinot V et al. Inverse agonists and serotonergic transmission: From recombinant, human serotonin (5-HT)1B receptors to G-protein coupling and function in corticolimbic structures in vivo. Neuropsychopharmacology 1999; 21(2 Suppl):61S-67S.
94. Milner TA, Veznedaroglu E. Serotonin-containing terminals synapse on septohippocampal neurons in the rat. J Neurosci Res 1993; 36(3):260-271.
95. Morales M, Bloom FE. The 5-HT3 receptor is present in different subpopulations of GABAergic neurons in the rat telencephalon. J Neurosci 1997; 17(9):3157-3167.
96. Nakai K, Fujii T, Fujimoto K et al. Effect of WAY-100135 on the hippocampal acetylcholine release potentiated by 8-OH-DPAT, a serotonin1A receptor agonist, in normal and p-chlorophenylalanine-treated rats as measured by in vivo microdialysis. Neurosci Res 1998; 31(1):23-29.
97. Nelson CJ, Jordan WP, Bohan RT. Daily fluoxetine administration impairs avoidance learning in the rat without altering sensory thresholds. Prog Neuropsychopharmacol Biol Psychiatry 1997; 21(6):1043-1057.
98. Ohno M, Kishi A, Watanabe S. Effect of cholinergic activation by physostigmine on working memory failure caused in rats by pharmacological manipulation of hippocampal glutamatergic and 5-HTergic neurotransmission. Neurosci Lett 1996; 217(1):21-24.
99. Ohno M, Watanabe S. Blockade of 5-HT1A receptors compensates loss of hippocampal cholinergic neurotransmission involved in working memory of rats. Brain Res 1996; 736(1-2):180-188.
100. Ohno M, Watanabe S. Differential effects of 5-HT3 receptor antagonism on working memory failure due to deficiency of hippocampal cholinergic and glutamatergic transmission in rats. Brain Res 1997; 762(1-2):211-215.
101. Parks CL, Robinson PS, Sibille E et al. Increased anxiety of mice lacking the serotonin1A receptor. Proc Natl Acad Sci USA 1998; 95(18):10734-10739.
102. Passani MB, Blandina P. Cognitive implications for H3 and 5-HT3 receptor modulation of cortical cholinergic function: A parallel story. Methods Find Exp Clin Pharmacol 1998; 20(8):725-733.
103. Pauwels PJ. Diverse signalling by 5-hydroxytryptamine (5-HT) receptors. Biochem Pharmacol 2000; 60(12):1743-1750.
104. Perez-Vega MI, Feria-Velasco A, Gonzalez-Burgos I. Prefrontocortical serotonin depletion results in plastic changes of prefrontocortical pyramidal neurons, underlying a greater efficiency of short-term memory. Brain Res Bull 2000; 53(3):291-300.
105. Pitsikas N, Borsini F. Different effects of tropisetron and ondansetron in learning and memory paradigms. Pharmacol Biochem Behav 1997; 56(4):571-576.
106. Ramboz S, Oosting R, Amara DA et al. Serotonin receptor 1A knockout: An animal model of anxiety-related disorder. Proc Natl Acad Sci USA 1998; 95(24):14476-14481.
107. Ramirez MJ, Cenarruzabeitia E, Lasheras B et al. Involvement of GABA systems in acetylcholine release induced by 5-HT3 receptor blockade in slices from rat entorhinal cortex. Brain Res 1996; 712(2):274-280.
108. Reznic J, Staubli U. Effects of 5-HT3 receptor antagonism on hippocampal cellular activity in the freely moving rat. J Neurophysiol 1997; 77(1):517-521.
109. Richter-Levin G, Segal M. Age-related cognitive deficits in rats are associated with a combined loss of cholinergic and serotonergic functions. Ann N Y Acad Sci 1993; 695:254-257.
110. Richter-Levin G, Segal M. Serotonin, aging and cognitive functions of the hippocampus. Rev Neurosci 1996; 7(2):103-113.
111. Riedel WJ, Klaassen T, Deutz NE et al. Tryptophan depletion in normal volunteers produces selective impairment in memory consolidation. Psychopharmacology (Berl) 1999; 141(4):362-369.
112. Riekkinen M, Sirvio J, Toivanen T et al. Combined treatment with a 5HT1A receptor agonist and a muscarinic acetylcholine receptor antagonist disrupts water maze navigation behavior. Psychopharmacology (Berl) 1995; 122(2):137-146.

113. Rogers DC, Hagan JJ. 5-HT6 receptor antagonists enhance retention of a water maze task in the rat. Psychopharmacology (Berl) 2001; 158(2):114-119.
114. Roychoudhury M, Kulkarni SK. Effects of ondansetron on short-term memory retrieval in mice. Methods Find Exp Clin Pharmacol 1997; 19(1):43-46.
115. Ruotsalainen S, Miettinen R, MacDonald E et al. The role of the dorsal raphe-serotonergic system and cholinergic receptors in the modulation of working memory. Neurosci Biobehav Rev 1998; 22(1):21-31.
116. Sari Y, Lefevre K, Bancila M et al. Light and electron microscopic immunocytochemical visualization of 5-HT1B receptors in the rat brain. Brain Res 1997; 760(1-2):281-286.
117. Saudou F, Amara DA, Dierich A et al. Enhanced aggressive behavior in mice lacking 5-HT1B receptor. Science 1994; 265(5180):1875-1878.
118. Schmitz D, Empson RM, Heinemann U. Serotonin reduces inhibition via 5-HT1A receptors in area CA1 of rat hippocampal slices in vitro. J Neurosci 1995; 15(11):7217-7225.
119. Sirvio J. Strategies that support declining cholinergic neurotransmission in Alzheimer's disease patients. Gerontology 1999; 45 Suppl 1:3-14.
120. Stancampiano R, Cocco S, Cugusi C et al. Serotonin and acetylcholine relcase response in the rat hippocampus during a spatial memory task. Neuroscience 1999; 89(4):1135-1143.
121. Staubli U, Xu FB. Effects of 5-HT3 receptor antagonism on hippocampal theta rhythm, memory, and LTP induction in the freely moving rat. J Neurosci 1995; 15(3 Pt 2):2445-2452.
122. Steckler T, Sahgal A. The role of serotonergic-cholinergic interactions in the mediation of cognitive behaviour. Behav Brain Res 1995; 67(2):165-199.
123. Stiedl O, Misane I, Spiess J et al. Involvement of the 5-HT1A receptors in classical fear conditioning in C57BL/6J mice. J Neurosci 2000; 20(22):8515-8527.
124. Tecott LH, Logue SF, Wehner JM et al. Perturbed dentate gyrus function in serotonin 5-HT2C receptor mutant mice. Proc Natl Acad Sci USA 1998; 95(25):15026-15031.
125. Terry Jr AV, Buccafusco JJ, Prendergast MA et al. The 5-HT3 receptor antagonist, RS-56812, enhances delayed matching performance in monkeys. Neuroreport 1996; 8(1):49-54.
126. Terry Jr AV, Buccafusco JJ, Jackson WJ et al. Enhanced delayed matching performance in younger and older macaques administered the 5-HT4 receptor agonist, RS 17017. Psychopharmacology (Berl) 1998; 135(4):407-415.
127. Uphouse L. Multiple serotonin receptors: Too many, not enough, or just the right number? Neurosci Biobehav Rev 1997; 21(5):679-698.
128. Vanhoenacker P, Haegeman G, Leysen JE. 5-HT7 receptors: Current knowledge and future prospects. Trends Pharmacol Sci 2000; 21(2):70-77.
129. Vertes RP, Fortin WJ, Crane AM. Projections of the median raphe nucleus in the rat. J Comp Neurol 1999; 407(4):555-582.
130. Warburton EC, Harrison AA, Robbins TW et al. Contrasting effects of systemic and intracerebral infusions of the 5-HT1A receptor agonist 8-OH-DPAT on spatial short-term working memory in rats. Behav Brain Res 1997; 84(1-2):247-258.
131. Ward BO, Wilkinson LS, Robbins TW et al. Forebrain serotonin depletion facilitates the acquisition and performance of a conditional visual discrimination task in rats. Behav Brain Res 1999; 100(1-2):51-65.
132. Wedzony K, Mackowiak M, Zajaczkowski W et al. WAY 100135, an antagonist of 5-HT1A serotonin receptors, attenuates psychotomimetic effects of MK-801. Neuropsychopharmacology 2000; 23(5):547-559.
133. Wilkinson LS, Humby T, Robbins TW et al. Differential effects of forebrain 5-hydroxytryptamine depletions on Pavlovian aversive conditioning to discrete and contextual stimuli in the rat. Eur J Neurosci 1995; 7(10):2042-2052.
134. Woolley ML, Bentley JC, Sleight AJ et al. A role for 5-ht6 receptors in retention of spatial learning in the Morris water maze. Neuropharmacology 2001; 41(2):210-219.
135. Xie Z, Lee SP, O'Dowd BF et al. Serotonin 5-HT1B and 5-HT1D receptors form homodimers when expressed alone and heterodimers when coexpressed. FEBS Lett 1999; 456(1):63-67.
136. Zeitz KP, Guy N, Malmberg AB et al. The 5-HT3 subtype of serotonin receptor contributes to nociceptive processing via a novel subset of myelinated and unmyelinated nociceptors. J Neurosci 2002; 22(3):1010-1019.
137. Zgombick JM, Branchek TA. Native 5-HT1B receptors expressed in OK cells display dual coupling to elevation of intracellular calcium concentrations and inhibition of adenylate cyclase. Naunyn Schmiedebergs Arch Pharmacol 1998; 358(5):503-508.
138. Zhou FC, Tao-Cheng JH, Segu L et al. Serotonin transporters are located on the axons beyond the synaptic junctions: Anatomical and functional evidence. Brain Res 1998; 805(1-2):241-254.

Chapter 2.6

Dopamine

Jan P.C. de Bruin

Abstract

Since the discovery that dopamine occurs in the cerebral cortex and functions as a transmitter a large number of studies has been conducted to examine its precise functions. It was found that administering dopaminergic drugs, either stimulating or inhibiting dopamine receptors, affected various modes of behaviour, including locomotor activity, eating and drinking, reward-related processes, and cognition. This chapter is especially concerned with the role of dopamine receptors in cognitive processes. It focuses on the mesocortical dopaminergic system, which, along with some other cortical areas, innervates the prefrontal cortex (PFC). This system is particularly sensitive to stressful stimuli, which, even when of small magnitude, leads to an elevation of prefrontal dopamine levels. Manipulations of this system have been shown to greatly affect behavioural functions, which depend on the integrity of the prefrontal cortex, such as working memory, attention and behavioural flexibility. To date five subtypes of dopamine receptors have been described (D1-5), all occurring in the brain. Although there are data that all types of dopamine receptors are involved in cognitive processes, available evidence strongly favours dopamine D1 receptors. Either hyper- or hypostimulation of prefrontal D1 receptors leads to dysfunctioning of the prefrontal cortex and behavioural impairments. Models of how D1 receptor actions may impair PFC functions are discussed.

Introduction: Dopamine Receptors in the Brain

The history of dopamine (DA) as a neurotransmitter effective in the central nervous system and affecting cognition-related processes is an interesting one. It is more than four decades ago that DA was recognised as a neurotransmitter present in the brain. During the seventies important progress was made which will be illustrated by four hallmark publications. Thierry and colleagues[60] described in 1973 that DA in the brain was not restricted in its occurrence to the nigrostriatal system and the tubero-infundibular systems, but also occurred "by its own right" (i.e. not as a precursor of noradrenaline) in the cerebral cortex. In subsequent studies, conducted by the same research group,[61] data on functional significance were first described. When subjected to severe stressful stimuli (20 min electric foot shocks) the rate of utilisation of DA was greatly accelerated in the frontal cortex. By contrast, such treatment had no effect on DA utilisation in the dorsal striatum, while a minor effect was found in the ventral striatum (nucleus accumbens). These data are important because they indicate functional differences between three meso-telencephalic dopaminergic systems, the nigrostriatal, the mesolimbic and the mesocortical one.

A few years later evidence was obtained that DA has a great impact on cognitive processes. Two studies, one with primates, the other one with rodents, have provided evidence. In the primate study, Brozoski et al[13] used the technique of local depletion in the prefrontal cortex (PFC) of different neurotransmitters, DA, noradrenaline and serotonine. Using a delay-type of learning task sensitive to damage of the PFC, they showed that depletion of DA in the PFC

From Messengers to Molecules: Memories Are Made of These, edited by Gernot Riedel and Bettina Platt. ©2004 Eurekah.com and Kluwer Academic / Plenum Publishers.

resulted in behavioural impairments quite similar to impairments seen following lesion of that area. Depletion of the other transmitters had no such effects. Moreover, stimulation of the postsynaptic DA receptors with a DA receptor agonist resulted in an amelioration of the behavioural deficiency. Such data can be taken as evidence of the importance of the mesocortical dopaminergic system in behaviour related to PFC functioning.

Comparable findings in rats were obtained by Simon and colleagues.[56] Lesion of the dopaminergic cells in the ventral tegmental area impaired retention of previously learned delayed alternation responses. Thus, although the methodology of this study is different from the one of Brozoski et al,[13] both studies provide evidence for the functional importance of this dopaminergic pathway in delay-type tasks. In addition to these (and other) studies with experimental animals, studies in human patients, suffering from neuropsychiatric diseases such as schizophrenia have indicated the importance of the mesocortical dopaminergic system. Such findings have considerably stimulated research into the basic properties of the dopaminergic innervation of various brain regions, especially frontal cortical ones.

Dopaminergic Systems in the Central Nervous System

The dopaminergic innervation of the brain deviates from that of the other catecholaminergic systems. It is less uniform and more patchy, suggesting localization-related functions. In one of the first reviews on this topic, Moore and Bloom[41] list seven dopaminergic systems in the brain: nigrostriatal, mesocortical, tubero-hypophysial, retinal, incerto-hypothalamic, periventricular and olfactory bulb. The mesocortical system of this list also comprises the previously mentioned mesolimbic one. In view of the scope of this chapter, DA receptors and learning and memory, these two systems will be the focus of interest. Together with the nigrostriatal system they have been described as the meso-telencephalic system[41] with the cells of origin located in the mesencephalon. For the nigrostriatal system the dopaminergic cells are found mainly in the substantia nigra (pars compacta), for the mesocortical and mesolimbic systems mainly in the ventral tegmental area. The latter two systems are distinguished on the basis of differences in their target areas. Their cells of origin in the ventral tegmental area (A10 cell group) are interspersed (see Oades and Halliday[46] for a detailed review). In the field of learning and memory studies the mesolimbic dopaminergic, and especially the mesocortical system clearly have received most of the scientific attention, because the projection areas are limbic ones (e.g., nucleus accumbens and amygdala), and cortical ones (prefrontal, entorhinal, perirhinal cortices), respectively. These areas are known to play a crucial role in various cognitive processes. This interest is also due to the notion that DA has an important role in the pathogenesis of various brain diseases, such as Parkinson disease (nigrostriatal dopaminergic system) and schizophrenia (mesocortical dopaminergic system).

Dopamine Receptor Subtypes

With the progress of various techniques (anatomical, pharmacological, biochemical) in the seventies, it was discovered that there is not just one kind of DA receptor, but that there are more. The first distinction was made between two subtypes, D1 and D2 receptors.[35] The distinction is primarily based on differences in biochemical responses, i.e. stimulation or inhibition of adenylate cyclase (review by Seeman and Grigoriadis[55]). In this latter review it is argued that (at that time) it is necessary to postulate only two types of receptors, D1 and D2. In addition to their opposite effects on adenylate cyclase, they may also mediate different types of behaviour. During the following decade three other DA receptors were described. At present, based on molecular biological studies (gene cloning), five distinct DA receptors, all G-protein-coupled, are recognised. These receptors are clustered in two "families", the "D1-like" receptors (D1 and D5), and the "D2-like" ones (D2, D3 and D4) (for a recent review see ref. 59). The clustering of the various subtypes in these two "families" is based on similarities in their pharmacological profiles, coupling to second messenger systems, nucleotide sequence and genomic organisation.

Using immunocytochemical, mRNA and other techniques the distribution of the various subtypes of DA receptors in the brain has been studied. Although there are discrepancies between studies (perhaps related to the techniques used), and, although there are some differences between rodents and primates, the following, general survey can be given. The distribution of the various subtypes is quite heterogeneous, with some subtypes occurring at high density in one brain region, and being virtually absent in another one. Highest densities of DA receptors are found in the striatal areas (dorsal and ventral striatum) with both D1 and D2 receptors prominent. Both are also found in the frontal cortex, with a higher level of D1 than D2 receptors (e.g., see ref. 25). The distribution of the other subtypes is more region-specific. The highest density of D4 receptors is seen in the frontal cortex, with a low density in other dopamine-innervated areas.[2] D3 receptors are found in the ventral striatum (nucleus accumbens) and in the Islands of Calleja. D5 receptors are found in hippocampus, thalamus, striatum and cortex, but at low densities in all these areas.[59]

Functional Studies Using a Systemic Approach

Concurrent with the distinction between the various subtypes of DA receptors and with the availability of dopaminergic drugs with a high and selective affinity for either one of the DA receptor subtypes, studies became possible to further elucidate the functional properties of the dopaminergic innervation of the brain and to unravel possible functional differences between DA receptor subtypes. There is a large body of studies that have investigated this, using antagonists and/or agonists of these two receptors administered via a peripheral route (systemically). Such studies have revealed functional properties of DA receptors, but generally do not allow a distinction between effects specific for particular brain regions.

The general picture which has emerged from these studies is that central dopaminergic systems are involved in locomotor activity, in eating and drinking behaviour, in reward-related processes, and, last but not least, in cognitive processes. This chapter focusses on cognition-related functions of dopaminergic receptors. However, it is important to briefly review some of the data on other types of behaviour, in which dopaminergic receptors participate. It is clear that the performance in learning and memory paradigms usually require locomotor responses, eating or drinking responses (in the case of food or water restriction), while reward-related processes may also have a great impact on the performance of learning tasks.

Initially, with the development of agonists and antagonists specific for either D1 or D2 receptors, a number of studies have investigated what the functional properties of these two receptors are, and whether their behavioural significance is different. Later studies have addressed the other subtypes of DA receptors.

DA Receptors and Locomotor Activity

One of the first studies in this field[1] examined two DA receptor antagonists, SCH 23390 (highly selective for D1 receptors) and spiperone (highly selective for D2 receptors). Both receptor antagonists fully blocked a supersensitive locomotor response (induced by apomorphine in rats with nucleus accumbens lesions). Using a simple motor task in non-operated rats, both receptor antagonists increased the step-down latency (in a simple bar catalepsy test), and again no difference between the two DA receptor antagonists was detected. The conclusion of these studies is that, notwithstanding distinct differences in their in vitro dopamine binding affinities, no distinction between the two drugs could be established in in vivo tasks.

Another locomotor model to examine properties of dopaminergic drugs is rotational behaviour induced by unilateral striatal lesions. Using such a model Barone et al[8] have examined possible differences between D1 and D2 receptors. Agonists selective for either D1 (SKF 38393) or D2 (LY 171555) receptors were both able to induce ipsilateral turning. Thus, both D1 and D2 receptor systems participate in the regulation of these behaviours.

While both previous studies examined the effect of dopaminergic drugs in animals with a manipulated dopaminergic system (lesions of striatal areas) Sanger[52] has taken a different approach by investigating the effects of dopaminergic drugs in untreated animals, moreover using a locomotor response more directly related to cognition. He found that operant bar pressing (with a FR10 schedule of food reinforcement) was decreased following administration of a D1 receptor antagonist (SCH 23390) in a dose-related manner. Although other studies had shown that typical neuroleptics (e.g., haloperidol, which antagonises both D1 and D2 receptors) also reduced bar pressing in this task, the author argues that the time course is quite different and that, therefore, the actions of D1 and D2 receptors are different.

Using mice as their experimental subjects Tidey and Miczek[62] have examined the effects of D1 and D2 receptor agonists in motor behaviour and schedule controlled responding. While the D2 receptor agonist (quinpirole) decreased both motor activity and schedule-controlled responding, the D1 receptor agonist (SKF 38393) decreased schedule-controlled responding at doses that did not affect motor behaviours. The relative behavioural specificity of the D1 receptor agonist suggests that activation of this receptor alters the temporal patterning of behavior while D2 activation appears to cause a more general suppression of behaviours.

Regarding the effects of manipulation of D1 and D2 dopaminergic receptors on locomotor activity it can be stated that increased dopaminergic transmission increases locomotor activity, while manipulations that decrease dopaminergic transmission decrease locomotor activity. These effects are usually more pronounced for D2 receptors. It is clear that such effects on performance potentially influence any possible effects of manipulations of dopaminergic function on learning and memory formation.

DA Receptors and Reward

DA receptors, for a long time, have been given an important role in reward-related processes, examined e.g., in place conditioning tests and in intracranial self- stimulation paradigms. An example of the former is the study of Hoffman and Beninger[34] using agonists for the D1 (SKF 38393) and D2 (quinpirole) receptors. Although previous studies had little success in differentiating the D1 and D2 receptor subtypes at a behavioural level, the evidence obtained in this study favours the D2 receptor in mediating the reinforcing effects of psychomotor stimulants. This view is also expressed by Beninger and co-workers[9]: agonist studies showed that D2, but not D1 receptor agonists were self-administered, produced place preferences and enhanced responding for conditioned reward.

Data on the role of dopaminergic receptors in intracranial self-stimulation have been collected by Ferrer et al[31] In their paradigms electrodes were localised in the medial PFC, an area of the brain where D1 receptors exceed D2 receptors. While D2 dopaminergic drugs, either receptor agonists or antagonists, failed to affect the rate of self-stimulation, D1 receptor agonists and antagonists both decreased the self-stimulation responses. The conclusion is, that at least in this area of the brain D1, rather than D2, receptors are involved in the rewarding properties of intracranial self-stimulation.

Although this latter study provides evidence for a role of prefrontal DA receptors in reward, it is particularly the mesolimbic dopaminergic system (nucleus accumbens) which is thought to be crucial in mediating reward processes. This topic has recently been reviewed by Spanagel and Weiss.[57]

DA Receptors and Eating and Drinking

Besides their effects on locomotor activity and reward-related processes, dopaminergic drugs are also known to affect eating and drinking. One of the first studies was conducted by Zigmond et al[68] An increased DA release (achieved by d-amphetamine) reduced food intake. This effect was attenuated by DA receptor antagonists (haloperidol or spiroperidol). However at higher doses such DA receptor antagonists also reduced food intake. The interesting conclusion of the authors is that there is an optimal level of dopaminergic activity for the mediation of feeding behaviour and that both increases and decreases from this optimum may disrupt feeding as well as other behaviours. This conclusion is made again in a more specific way, relating D1 receptor activation in the PFC with the performance of working memory (e.g., Desimone,[23] see below). In later studies, when more subtype-selective dopaminergic drugs were available, a distinction between D1 and D2 receptors was made. Clifton and co-workers[17,18] found that especially D2 receptors are involved in food intake, whereas D1 receptors are of greater importance in controlling water intake.

DA Receptors and Cognition

When discussing DA receptors and locomotor activity we already pointed at the role of DA receptors in cognitive performance. This subject has been intensively studied by Goldman-Rakic, Arnsten, and co-workers in the nineties, using both monkeys and rats as experimental subjects. In a study with rhesus monkeys, Arnsten et al[5] have provided compelling evidence that cognitive performance depends on DA D1 receptor mechanisms. In this study both young and aged monkeys were used. In aged primates there is a marked degeneration of the mesocortical dopaminergic system, with a loss of DA in the PFC. Other neurotransmitter systems in this brain area are less affected. Using a working memory task with delayed responding known to be sensitive to prefrontal damage, they found that task performance in young monkeys was impaired following administration of a D1 receptor antagonist (SCH 23390). In aged monkeys, however, this drug had no effect on task performance. By contrast, the D1 receptor agonist (SKF38393) led to an improved performance in the aged monkeys, without affecting performance in the young ones. In the latter group this agonist only improved performance when they had been previously treated with reserpine which leads to a depletion of catecholamines.

Results consistent with these findings were obtained in other studies in which prefrontal dopamine levels had been increased by stressful stimuli such as a loud noise or the odour of a predator, or by an anxiogenic drug which specifically elevates dopamine levels in the PFC. When first exposed to a loud (105-dB) noise stress monkeys performed worse on a spatial working memory task (delayed responding). The deficit in performance was absent at very short delays and also in a visual discrimination task. Pretreatment with an antagonist for the D1 receptor (SCH 233990) prevented the noise-induced impairment in task performance.[6]

In other studies conducted by Murphy and co-workers[44,45] the anxiogenic β-carboline, FG7142, was administered. This drug is known to selectively elevate dopamine turnover in the PFC, without affecting other dopamine terminal fields. Both rats and monkeys were impaired in the performance of a spatial working memory task, while, similar to the Arnsten and Goldman-Rakic[6] study, blockade of D1 receptors prevented the impairment in task performance. Interestingly, in the rats of this study there was a good correlation between performance scores and DA utilisation in the PFC. These data are consistent with those of an earlier study[51] showing that in untreated rats cortical DA concentrations correlated negatively with the number of errors in a T-maze spatial delayed alternation task.

Recently[4] similar procedures were used to examine the possible function of D4 receptors, known to occur at relatively high density in the PFC.[2] In monkeys, pretreated with an antagonist selective for D4 receptors (PNU-101387G), the reduced task performance induced by the anxiogenic β-carboline (FG7142) was reversed. Thus, D4 receptor mechanisms, like D1 ones, also contribute to stress-induced cognitive dysfunction.

Within this line of research the study of Morrow et al[42] is an interesting example of a more naturalistic approach. In their experiments they exposed rats to the odour of a "natural" predator, a fox. Like noise stress or pharmacological stress the fox odour resulted in an increased dopamine metabolism, restricted to the PFC. In a working memory task (delayed non-matching to sample), different from the tasks used in the other studies, it was shown that exposure to the fox odour impaired task performance, but did not suppress overall exploratory behaviour.

Whereas short-lasting stress leads to an increased release of DA in the PFC, resulting in a hyperstimulation of DA receptors and impairing working memory functions, the study of Mizoguchi et al[39] has demonstrated that a long-lasting exposure to stressful stimuli has the opposite effect. There is a decreased level of DA in the PFC and an impairment in a working memory task (T-maze spatial delayed alternation). Interestingly, in this paradigm the administration of a D1 antagonist (SKF 81297) improved working memory performance in a dose- and delay-dependent manner.

The studies reviewed above point to an important role of D1 (and D4) receptor mechanisms in the behavioural tasks examined, while it is suggested that D2 receptors are not involved, perhaps playing only a minor role. However, there is evidence in humans that also DA D2 receptors play a role. Luciana et al[37] have clearly shown that in human volunteers of a relatively young age and not suffering from neurological diseases, spatial working memory was facilitated by D2 receptor activation: the selective D2 receptor agonist, bromocriptine, resulted in improved task performance. However, a later study[43] failed to replicate this finding. Facilitation in the performance of a visuospatial working memory task was seen after administration of a mixed D1-D2 receptor agonist (pergolide), but not after administration of bromocriptine. The discrepancy in findings between these two human studies awaits further experimentation. The findings of the latter study are quite consistent with those of monkey studies indicating a preferential role of (prefrontal) D1 receptors for working memory modulation.

Functional Studies Using a Central Approach

Studies using systemic administration of dopaminergic drugs and examining concomitant behavioural alterations do not directly reveal which brain areas are important for the performance of tasks under investigation. However, sometimes deductions can be made. When the task is sensitive for damage of a particular brain area, and when performance is affected by selective dopaminergic drugs, it can be assumed that the drugs exert their effect in that brain area. This can be exemplified by the study of Packard and White[48] where the beneficial effects of a D2 receptor agonist (LY 171555) on a radial arm maze task were ascribed to the stimulation of D2 receptors in hippocampus and/or nucleus caudatus. Similarly, the effects of selective D1 receptor agonists / antagonists on working memory (reviewed above) were assumed to be due to D1 receptor stimulation / inhibition in the frontal cortex. However, there are other techniques, which may provide such data in a more direct way. The oldest one is local depletion of dopamine, either directed at the cells of origin or targeting dopamine-innervated areas. The other one is based on local, intracerebral administration of dopaminergic drugs to a focal brain area.

Dopaminergic Depletions and Cognitive Impairments

Ever since the availability of drugs that specifically destroy dopaminergic cells and fibres, the technique of local DA depletion has been a powerful tool in examining the contribution of dopaminergic systems in behavioural performance. The most widely used drug in such studies is 6-OHDA (6-hydroxy-dopamine) that will destroy dopaminergic cells and fibres, while leaving postsynaptic terminals intact. Unfortunately, 6-OHDA is not specific for dopamine but will also damage noradrenergic cells and fibres. By using a noradrenergic receptor re-uptake inhibitor, e.g., DMI (desmethylimipramine), administered prior to 6-OHDA, the effects of the latter drug will be more specific causing a degeneration of dopaminergic fibres and cells. Another important issue is that a single administration of 6-OHDA is generally not sufficient to induce large enough dopamine depletion. With dopamine levels at 20-50 % of the original tissue concentration, suspected behavioural impairments might not be detectable. Multiple injections of 6-OHDA therefore may be required. Reduced levels of dopamine are known to result in the development of post-synaptic supersensitivity. This is an explanation for behavioural recovery often witnessed following dopamine depleting brain lesions. Another complicating finding has been that depletion of DA in one area of the brain may result in elevated DA levels in another one. For example, a number of studies[15,24,38] has shown that depletion of DA in the rodent medial PFC leads to elevated DA levels and an increased DA metabolism in the nucleus accumbens.

Notwithstanding these drawbacks, dopaminergic depletions of focal brain areas have contributed to our knowledge of the importance of dopamine for cognitive functioning. The study of Brozoski et al[13] already referred to in the Introduction, is an example of that approach. An example of studies in rodents using this methodology is the one of Bubser and Schmidt[14] also targeting the dopaminergic innervation of the frontal cortex. They found a decreased performance in a spatial delayed alternation task, a working memory task known to depend on the integrity of the PFC. They also described an increased locomotion in their rats with prefrontal DA depletion. This is possibly due to an increased DA receptor activation in the ventral striatum (nucleus accumbens). In one of our own studies,[58] we have also used the technique of reducing the dopaminergic innervation of the PFC. In this study lesions were made in the ventral tegmental area, the site of origin of the dopaminergic cells projecting to the PFC. We found a reduced level of DA in the medial PFC along with an impaired performance of a working memory task, spatial delayed alternation. Thus, our findings are consistent with those of Bubser and Schmidt.[14] However, whereas in the Bubser and Schmidt study some of the findings can be explained by an elevation of DA in the nucleus accumbens, lesioning the VTA[58] will have led to a reduction of DA in that area. The dopaminergic cells of the ventral tegmental area, projecting to PFC and nucleus accumbens are interspersed.[46] Therefore, it is difficult to arrive at a clear conclusion whether the observed behavioural changes are due to DA receptor mechanisms in PFC or nucleus accumbens. An additional disadvantage is that this method does not reveal the role of specific DA receptor subtypes.

Intracerebral Infusion of Dopaminergic Drugs and Cognitive Processes

An important progress in the search for dopaminergic involvement in cognitive processes has been the development of the technique of chronic intracerebral cannulae providing the possibility of infusing dopaminergic drugs into small, well-defined areas of the brain. The areas targeted in this manner have been especially the striatal and prefrontal cortical areas. We will focus on the prefrontal cortical areas.

We will first review studies using local administration of dopaminergic drugs affecting D1 and D2 receptors and examining different cognitive tasks for working memory, attention, and behavioural flexibility. This experimental approach directly targets the prefrontal DA receptors and is expected to further support the indirect evidence provided by the studies reviewed above. The hallmark publication of research in this field was the study of Sawaguchi and Goldman-Rakic.[53] Using a delay response task they examined behavioural changes resulting from local infusions of dopaminergic drugs in the dorsolateral aspect of the PFC in rhesus monkeys. Infusions of a D1 receptor antagonist (SCH2390) resulted in a decreased performance of this working memory task. The impairment was both dose-dependent and delay-dependent, indicating that it was truly working memory that was affected. It is also important to note that performance of a control task, requiring similar sensory and motor functions, was not altered by local administration of the D1 receptor antagonist. Since infusions of a D2 receptor antagonist (raclopride) did not affect the performance of this working memory task, clear evidence was obtained pointing to the D1 receptors in the primate PFC as crucial for working memory mechanisms.

Studies in rats, examining the effects of dopaminergic drugs locally infused in the PFC have not always yielded results which were consistent with those of the primate study of Sawaguchi and Goldman-Rakic.[53] This is surprising and not easily explained in view of the commonalities in function of the PFC of rodents and primates, particularly working memory,[36] and the similarity in dopaminergic innervation of the PFC. One of the first rodent studies using local infusions of dopaminergic drugs was conducted by Brito et al[10] In a working memory task, delayed non-matching to sample, conducted in a T-maze, they found no effects of local infusions of a DA receptor antagonist (sulpiride) in the medial aspect of the PFC; however, a cholinergic receptor antagonist (scopolamine) did interfere with the performance of the delayed non-matching to sample task. The authors conclude that cholinergic, but not dopaminergic mechanisms in the PFC are important for working memory. However, since sulpiride mainly antagonises D2 receptors, a mnemonic role of D1 receptors in the rodent PFC can not be excluded. This topic has been studied by Broersen et al[11,12] The working memory task used was a delayed matching to sample task conducted in operant chambers. The dopaminergic drugs infused in the dorsal[11] or ventral[12] aspect of medial PFC were an agonist (apomorphine) or an antagonist (cis-flupenthixol) affecting both D1 and D2 receptors. Although the drugs affected performance of this task, this did not occur in a delay-dependent manner. Thus no proof was obtained for a role of rodent PFC DA receptors in working memory processes. However, similar to the findings of Brito et al[10] a cholinergic receptor antagonist (scopolamine) interfered with task performance in a delay-dependent manner. Thus, the conclusion was made that in rats cholinergic rather than dopaminergic receptor mechanisms in the PFC are involved in working memory processes. Later studies also failed to show that blockade of DA receptors in the PFC impaired performance in working memory tasks. Using the classical (T-maze) spatial delayed alternation task Romanides et al[50] report that blockade of D1 and/or D2 receptors with SCH 23390 (D1 receptor antagonist) and sulpiride (D2 receptor antagonist) was without effect on working memory performance. Similarly, Zahrt et al[67] described that in a comparable spatial working memory task infusion of the D1 antagonist (SCH 23390) in the medial PFC by itself had no effect on performance. However, in the latter study intraprefrontal infusions were also made with a D1 receptor agonist (SKF 81297). This treatment affected delayed alternation performance and produced a dose-related impairment. Pretreatment with a D1 receptor antagonist reversed the agonist-induced effects. These data suggest that both in primates and rodents prefrontal D1 receptors are involved in working memory processes. Whereas in primates both blockade and stimulation of D1 receptors leads to impaired performance in such tasks, in rodents only "overstimulation" and not blockade of these receptors has been shown to affect task performance.

Seamans et al[54] and Floresco and Phillips[32] have used a spatial (radial arm maze) task with delays between training and testing that far exceed the delay intervals commonly applied in working memory tasks. Intraprefrontal infusions of dopaminergic drugs were given before testing with a training-testing interval of 30 min or 12 h. With an interval of 30 min both a D1 receptor antagonist (SCH 23390) and an agonist (SKF 81297) disrupted memory performance. A D2 receptor antagonist (sulpiride) failed to alter performance. When the long interval (12 h) between training and testing was used the D1 agonist (administered just prior to the testing phase) improved memory performance, an effect opposite to the one seen after a 30 min interval. Apparently, the effect of D1 stimulation depends on the strength of the memory trace. When memory is still good (30 min interval) blockade or stimulation of D1 receptors leads to more errors; when memory has deteriorated (12 h interval) stimulation of D1 receptors improves memory retrieval.

Recent studies have investigated other cognitive properties, besides working memory function, of prefrontal DA receptors, e.g., attentional functions. Granon et al[33] have examined the role of DA (D1 and D2) receptors in the PFC in an attentional task (the five choice serial reaction time task). They used dopaminergic drugs acting on D1 receptors (antagonist, SCH 23390; agonist, SKF 38393) or D2 receptors (antagonist, sulpiride; no agonist). The interesting approach of this study is that individual differences between rats were taken into account in a meaningful way. Based on their scores for task accuracy the rats were divided in a low and high performance group. Antagonising D1 receptors in the PFC impaired performance in the high baseline condition, while in the low baseline condition accuracy was enhanced by the D1 agonist. Thus, both drugs can enhance performance depending on pre-infusion performance scores. The D2 receptor antagonist had no effects on task performance, in any of the groups.

Thus, as a general conclusion of these studies it can be stated that prefrontal D1 receptor mechanisms are important for working memory and attentional processes. Performance levels are of importance for the effects that dopaminergic drugs may exert. This has emanated from studies, which made use of inter-individual differences in performance[33] and from studies in which memory was manipulated by using different intervals between training and testing.[32]

In some of our own studies we have examined prefrontal DA receptor mechanisms in tasks for behavioural flexibility, requiring rats to either select or inhibit lever press responses in an operant chamber.[20] It has been known for quite some time that the capability to respond to changes in task demands in a flexible manner depends on the integrity of the PFC. In humans it has been shown that performance of the Wisconsin Card Sorting Test, a classical test for flexible responding, is impaired with prefrontal cortical damage, whereas the PFC is activated during task performance (see ref. 40 for a recent study). However, most studies relating prefrontal DA receptor mechanisms with cognitive performance have used working memory paradigms. Only recently other functions have been included in this line of research, exemplified by the study of Granon et al[33] In our paradigm for behavioural flexibility rats first learn to press one of two available levers in an operant chamber to obtain a food reward. With a high level of accuracy the rats are subjected to a reversal, i.e. they now have to press the other lever. We first assessed whether reversal learning depends on the PFC. Transient lidocaine-induced inactivation of the PFC impaired reversal learning.[19,20] Subsequently, we have studied the effects of antagonising DA receptors in the PFC. Intraprefrontal infusion of a D1 receptor antagonist (SCH 23390), resulted in impaired reversal learning, comparable with the changes observed following receptor non-specific transient inactivation.[19] Interestingly, we also found that local infusion of an antagonist for D4 receptors (L745,870) also impaired reversal learning.[21] This is, along with the findings of Arnsten et al,[7] one of the first demonstrations of a behavioural significance of this receptor subtype.

Epilogue

In this review we have considered the role of DA receptors in cognitive processes. There are different dopaminergic systems in the mammalian brain and different subtypes of DA receptors. It is especially the mesocortical dopaminergic system innervating the PFC and other cortical areas which is involved in cognitive processes. The actions of DA on these processes are prominently modulated by D1 receptors. Findings from different sources have led to the inverted U-shape model (Desimone[23]). There is an optimal level of D1 receptor stimulation needed to perform tasks, which depend of prefrontal DA. Excessive stimulation resulting from hyperstimulation leads to a deteriorated performance, which can be ameliorated by D1 receptor antagonists. Hypostimulation, a consequence of ageing[22] or chronic stress,[39] also impair performance of PFC-related cognitive tasks, and D1 receptor agonists may then lead to improved performance. The model described by Desimone focuses on prefrontal D1 receptor mechanisms and working memory. Future studies will clarify whether the model can be extended to other types of behaviour related to the PFC, and perhaps to other subtypes of DA receptors (e.g., D4 subtype).

The role of D2 receptors in these modes of behaviour is less obvious. Although most studies, especially those using an intracerebral way of administering dopaminergic drugs have yielded negative findings with respect to cognition-related properties of D2 receptors (note that ref. 26 forms an exception), we have indicated some studies with positive results.[37] With regard to memory formation D4 receptors are the most interesting to date, especially in view of their relative high density in frontal cortical areas.[2] However, most studies of the functional properties of this receptor subtype have been unable to obtain clear functional results (e.g., refs. 16 and 47). Perhaps, when future studies will focus more on functions related with the PFC, a better knowledge of the functional properties of D4 receptors will emerge.

Because of the constraints of this review we have not touched upon possible cognition-related properties of some of the other types of DA receptors. There is recent evidence[49] that conditioned cocaine-seeking behaviour could be inhibited by a dopaminergic drug that selectively acts on D3 receptors as a partial agonist. It has been suggested that this effect can be generalised and that D3 receptor mechanisms would be involved in conditioning processes in general. To our knowledge D5 receptor mechanisms have not yet been linked to cognitive processes, although it has recently been suggested that not only D1 but also D5 receptors are involved in stress-induced impairments of PFC function.[4]

We have also not dealt with other promising approaches, such as the technique of in vivo microdialysis and the use of transgenic animals. In vivo microdialysis offers an important correlative approach. Animals implanted with in vivo microdialysis cannulae can be subjected to a behavioural task, while extracellular samples are taken from a small area of the brain. Sensitive neurochemical methods are available to measure the levels of various catecholamines and their metabolites in such samples. Sampling time (temporal resolution) depends on the neurotransmitter levels of the particular brain region and the sensitivity of the neurochemical detection methods. In the case of DA, levels are commonly assessed with HPLC (high pressure liquid chromatography) and electrochemical detection methods. Sampling time ranges from 5-15 min in prefrontal cortical areas, to one min in striatal areas. Changes in the release of DA during task performance are indicative for the regional involvement of this transmitter in the particular task.[28] Using this technique a number of studies have shown that in a rodent classical conditioning task the conditioning stimulus may be sufficient to elevate prefrontal DA level.[29,30,64] There is one study in primates, which showed that during the performance of a working memory task an increased release of DA was measured in the PFC.[63] Studies in this field have been reviewed in detail by Feenstra.[28]

The technique of in vivo microdialysis can also be used to administer drug to a small area of the brain. This is called "reversed" microdialysis and has the advantage that drug delivery occurs at a slower rate, avoiding the aversive effects of acutely-injected high drug concentrations and of having to manipulate the animal just before subjecting it to the behavioural task.[28] At present, few experimental data are available on this latter approach.

The other promising approach is the use of DA receptor knockouts. At present such knockout mice are available for various DA receptor subtypes. An example of such an approach is the study of El-Ghundi et al[27] which has shown that D1 receptor knockout mice are deficient in a spatial learning task (Morris water maze task). In view of the important role of D1 receptors in cognition it is to be expected that this approach will greatly contribute to our knowledge in this field.

There have been suggestions how DA receptors accomplish their role in cognitive processes. Electrophysiological studies have indicated that dopamine levels in the PFC affect the propagation of signals from apical dendrites to soma in pyramidal cortical cells.[65,66] With insufficient D1 stimulation signals are unfocussed, both spatially and temporally. When there is a hyperstimulation of D1 receptors signals would be oversharpened and not reach the soma. Thus, the PFC would be taken "off-line" during stress, and according to Arnsten[3] ensuring rapid yet reversible loss of prefrontal cortical control over behaviour. Schematic representations, combining behavioural and electrophysiological data, have been given by Zahrt et al[67] and Arnsten.[4] These basic findings have relevance for cognitive functions related to PFC dysfunctioning in humans[67] while evolutionary implications have been hypothesised by Arnsten.[3]

References

1. Amalric M, Koob GF, Creese I et al. "Selective" D-1 and D-2 receptor antagonists fail to differentially alter supersensitive locomotor behavior in the rat. Life Sci 1986; 39:1985-1993.
2. Ariano MA, Wang J, Noblett KL et al. Cellular distribution of the rat D4 dopamine receptor protein in the CNS using anti-receptor antisera. Brain Res 1997; 752:26-34.
3. Arnsten AFT. The biology of being frazzled. Science 1998; 280:1711-1712.
4. Arnsten AFT. Stress impairs prefrontal cortical function in rats and monkeys: Role of dopamine D1 and norepinephrine alpha-1 receptor mechanisms. Prog Brain Res 2000; 126:183-192.
5. Arnsten AFT, Cai JX, Murphy BL et al. Dopamine D1 receptor mechanisms in the cognitive performance of young adult and aged monkeys. Psychopharmacol 1994; 116:143-151.
6. Arnsten AFT, Goldman-Rakic PS. Noise stress impairs prefrontal cortical cognitive function in monkeys: Evidence for a hyperdopaminergic mechanism. Arch Gen Psychiatry 1998; 55:362-368.
7. Arnsten AFT, Murphy BL, Merchant K. The selective dopamine D4 receptor antagonist, PNU-101387G, prevents stress-induced cognitive deficits in monkeys. Neuropsychopharmacol 2000; 23:405-410.
8. Barone P, Davis TA, Braun AR et al. Dopaminergic mechanisms and motor function: Characterization of D-1 and D-2 dopamine receptor interactions. Eur J Pharmacol 1986; 123:109-114.
9. Beninger RJ, Hoffman DC, Mazurski EJ. Receptor subtype-specific dopaminergic agents and conditioned behavior. Neurosci Biobehav Rev 1989; 13:113-122.
10. Brito GNO, Silva SP, Brito LSO. The cholinergic and dopaminergic systems of the prelimbic sector of frontal cortex and memory in the rat. Brazilian J Med Biol Res 1989; 22:1155-1158.

11. Broersen LM, Heinsbroek RPW, De Bruin JPC et al. Effects of local application of dopaminergic drugs into the dorsal part of the medial prefrontal cortex of rats in a delayed matching to position task: Comparison to local cholinergic blockade. Brain Res 1994; 645:113-122.
12. Broersen LM, Heinsbroek RPW, De Bruin JPC et al. The role of the medial prefrontal cortex of rats in short-term memory functioning: Further support for involvement of cholinergic, rather than dopaminergic mechanisms. Brain Res 1995; 674:221-229.
13. Brozoski TJ, Brown RM, Rosvold HE et al. Cognitive deficit caused by regional depletion of dopamine in prefrontal cortex of rhesus monkey. Science 1979; 205:929-932.
14. Bubser M, Schmidt WJ. 6-Hydroxydopamine lesion of the rat prefrontal cortex increases locomotor activity, impairs acquisition of delayed alternation tasks, but does not affect uninterrupted tasks in the radial maze. Behav Brain Res 1990; 37:157-168.
15. Carter CJ, Pycock CJ. Behavioural and biochemical effects of dopamine and noradrenaline depletion within the medial prefrontal cortex of the rat. Brain Res 1980; 192:163-176.
16. Clifford JJ, Waddington JL. Topographically based search for an "ethogram" among a series of novel D4 dopamine receptor agonists and antagonists. Neuropsychopharmacol 2000; 22:538-544.
17. Clifton PG, Rusk IN, Cooper SJ. Stimulation and inhibition of food intake by the selective dopamine D2 agonist, N-0437: A meal pattern analysis. Pharm Biochem Behav 1989; 33:21-26.
18. Clifton PG, Rusk IN, Cooper SJ. Effects of dopamine D1 and dopamine D2 antagonists on the free feeding and drinking patterns of rats. Behav Neurosci 1991; 105:272-281.
19. De Bruin JPC, Feenstra MGP, Broersen LM et al. Response selection and response inhibition: Functional involvement of the prefrontal cortex. Eur J Neurosci 1998; 10(Suppl 10):105.
20. De Bruin JPC, Feenstra MGP, Broersen LM et al. Role of the prefrontal cortex of the rat in learning and decision making: Effects of transient inactivation. Prog Brain Res 2000; 126:103-113.
21. De Bruin JPC, Feenstra MGP, Hoornweg R et al. Dopamine D4 receptors in the medial prefrontal cortex of the rat are involved in behavioural flexibility. Eur J Neurosci 2000; 12(Suppl 11):174.
22. De Keyser J, De Backer J-P. Vauquelin G et al. The effect of aging on the D1 dopamine receptors in human frontal cortex. Brain Res 1990; 528:308-310.
23. Desimone R. Is dopamine a missing link? Nature 1995; 376:549-550.
24. Deutch AY, Clark WA, Roth RH. Prefrontal cortical dopamine depletion enhances the responsiveness of mesolimbic dopamine neurons to stress. Brain Res 1990; 521:311-315.
25. Dewar KM, Reader TA. Distribution of dopamine D1 and D2 receptors in rabbit cortical areas, hippocampus and neostriatum in relation to dopamine contents. Synapse 1989; 4:378-386.
26. Druzin MY, Kurzina NP, Malinina EP et al. The effects of local application of D2 selective dopaminergic drugs into the medial prefrontal cortex of rats in a delayed spatial choice task. Behav Brain Res 2000; 109:99-111.
27. El-Ghundi M, Fletcher PJ, Drago J et al. Spatial learning deficit in dopamine D(1) receptor knockout mice. Eur J Pharmacol 1999; 383:95-106.
28. Feenstra MGP. Dopamine and noradrenaline release in the prefrontal cortex in relation to unconditioned and conditioned stress and reward. Prog Brain Res 2000; 126:133-163.
29. Feenstra MGP, Teske G, Botterblom MHA et al. Dopamine and noradrenaline release in the prefrontal cortex of rats during classical aversive and appetitive conditioning to a contextual stimulus: Interference by novelty effects. Neurosci Lett 1999; 272:179-182.
30. Feenstra MGP, Vogel M, Botterblom MHA et al. Dopamine and noradrenaline efflux in the rat prefrontal cortex after classical aversive conditioning to an auditory cue. Eur J Neurosci 2001; 13:1051-1054.
31. Ferrer JMR, Sanguinetti AM, Vives F et al. Effects of agonists and antagonists of D1 and D2 dopamine receptors on self-stimulation of the medial prefrontal cortex in the rat. Pharmacol Biochem Behav 1983; 19:211-217.
32. Floresco SB, Philips AG. Delay-dependent modulation of memory retrieval by infusion of a dopamine D1 agonist into the rat medial prefrontal cortex. Behav Neurosci 2001; 115:934-939.
33. Granon S, Passetti F, Thomas KL et al. Enhanced and impaired attentional performance after infusion of D1 dopaminergic receptor agents into rat prefrontal cortex. J Neurosci 2000; 20:1208-1215.
34. Hoffman DC, Beninger RJ. Selective D1 and D2 dopamine agonists produce opposing effects in place conditioning but not in conditioned taste aversion learning. Pharmacol Biochem Behav 1988; 31:1-8.
35. Kebabian JW, Calbe DB. Multiple receptors for dopamine. Nature 1979; 277:93-96.
36. Kolb B. Animal models for human PFC-related disorders. Prog Brain Res 1990; 85:501-519.
37. Luciana M, Depue RA, Arbisi P et al. Facilitation of working memory in humans by a D2 dopamine receptor agonist. J Cogn Neurosci 1992; 4:58-68.
38. Mitchell JB, Gratton A. Partial dopamine depletion of the prefrontal cortex leads to enhanced mesolimbic dopamine release elicited by repeated exposure to naturally reinforcing stimuli. J Neurosci 1992; 12:3609-3619.

39. Mizoguchi K, Yuzurihara M, Ishige A et al. Chronic stress induces impairment of spatial working memory because of prefrontal dopaminergic dysfunction. J Neurosci 2000; 20:1568-1574.
40. Monchi O, Petrides M, Peter V et al. Wisconsing card sorting revisited: Distinct neural circuits participating in different stages of the task identified by event-related functional magnetic resonance imaging. J Neurosci 2001; 21:7733-7741.
41. Moore RY, Bloom FY. Central catecholamine neuron systems: Anatomy, and physiology of the dopamine systems. Ann Rev Neurosci 1978; 1:129-169.
42. Morrow BA, Roth RH, Elsworth JD. TMT, a predator odor, elevates mesoprefrontal dopamine metabolic activity and disrupts short-term working memory in the rat. Brain Res Bull 2000; 52:519-523.
43. Müller U, Von Cramon DY, Pollmann S. D1 - versus D2 - receptor modulation of visuospatial working memory in humans. J Neurosci 1998; 18:2720-2728.
44. Murphy BL, Arnsten AFT, Goldman-Rakic PS et al. Increased dopamine turnover in the prefrontal cortex impairs spatial working memory performance in rats and monkeys. Proc Natl Acad Sci USA 1996; 93:1325-1329.
45. Murphy BL, Arnsten AFT, Jentsch JD et al. Dopamine and spatial working memory in rats and monkeys: Pharmacological reversal of stress-induced impairment. J Neurosci 1996; 16:7768-7775.
46. Oades RD, Halliday GM. Ventral tegmental (A10) system: Neurobiology. 1. Anatomy and connectivity. Brain Res Rev 1987; 12:117-165.
47. Oak JN, Oldenhof J, Van Tol HHM. The dopamine D4 receptor: One decade of research. Eur J Pharmacol 2000; 405:303-327.
48. Packard MG, White NM. Memory facilitation produced by dopamine agonists: Role of receptor subtype and mnemonic requirements. Pharmacol Biochem Behav 1989; 33:511-518.
49. Pilla M, Perachon S, Sautel F et al. Selective inhibition of cocaine-seeking behaviour by a partial dopamine D3 receptor agonist. Nature 1999; 400:371-375.
50. Romanides AJ, Duffy P, Kalivas PW. Glutamatergic and dopaminergic afferents to the prefrontal cortex regulate spatial working memory in rats. Neurosci 1999; 92:97-106.
51. Sahakian BJ, Sarna GS, Kantamanemi BD et al. Association between learning and cortical catecholamines in non-drug-treated rats. Psychopharmacol 1985; 86:339-343
52. Sanger DJ. The actions of SCH 23390, A D1 receptor antagonist, on operant and avoidance behavior in rats. Pharmacol Biochem Behav 1987; 26:509-513.
53. Sawaguchi T, Goldman-Rakic PS. D1 dopamine receptors in prefrontal cortex: Involvement in working memory. Science 1991; 251:947-950.
54. Seamans JK, Floresco SB, Phillips AG. D1 receptor modulation of hippocampal-prefrontal cortical circuits integrating spatial memory with executive control functions in the rat. J Neurosci 1998; 18:1613-1621.
55. Seeman P, Grigoriadis D. Dopamine receptors in brain and periphery. Neurochem Int 1987; 10:1-25.
56. Simon H, Scatton B, Le Moal M. Dopaminergic A10 neurones are involved in cognitive functions. Nature 1980; 286:150-151.
57. Spanagel R, Weiss F. The dopamine hypothesis of reward: Past and current status. Trends Neurosci 1999; 22:521-527.
58. Stam CJ, De Bruin JPC, Van Haelst AM et al. The influence of the mesocortical dopaminergic system on activity, food hoarding, social-agonistic behavior and spatial delayed alternation in male rats. Behav Neurosci 1989; 103:24-35.
59. Strange PG. Dopamine receptors. Tocris Reviews 2000; 15:1-4.
60. Thierry A-M, Tassin JP, Blanc G et al. Dopaminergic terminals in the rat cortex. Science 1973; 182:499-451.
61. Thierry A-M, Blanc G, Sobel A et al. Selective activation of the mesocortical DA system by stress. Nature 1976; 263:242-244.
62. Tidey JW, Miczek KA. Effects of SKF 38393 and quinpirole on aggressive, motor and schedule-controlled behaviors in mice. Behav Pharmacol 1992; 3:553-565.
63. Watanabe M, Kodama T, Hikosaka K. Increase of extracellular dopamine in primate prefrontal cortex during a working memory task. J Neurophysiol 1997; 78:2795-2798.
64. Wilkinson LS, Humby T, Killcross AS et al. Dissociations in dopamine release in medial prefrontal cortex and ventral striatum during the acquisition and extinction of classical aversive conditioning in the rat. Eur J Neurosci 1998; 10:1019-1026.
65. Williams GV, Goldman-Rakic PS. Blockade of dopamine D1 receptors enhances memory fields of prefrontal neurons in primate cerebral cortex. Nature 1995; 376:572-575.
66. Yang CR, Seamans JK. Dopamine D1 receptor actions in layers V-VI rat prefrontal cortex neurons: In vitro modulation of dendritic somatic signal integration. J Neurosci 1996; 16:1922-1935.
67. Zahrt J, Taylor JR, Mathew RG et al. Supranormal stimulation of D1 dopamine receptors in the rodent prefrontal cortex impairs spatial working memory performance. J Neurosci 1997; 17:8528-8535.
68. Zigmond MJ, Heffner TG, Stricker EM. The effect of altered dopaminergic activity on food intake in the rat: Evidence for an optimal level of dopaminergic activity for behavior. Prog Neuro-Psychopharmacol 1980; 4:351-362.

CHAPTER 2.7

Adrenaline and Noradrenaline

Marie E. Gibbs and Roger J. Summers

Abstract

Noradrenaline released in the brain can potentially act on any of 9 different receptor subtypes and since activation of the different receptors produces quite different effects both in duration and time-course of memory, noradrenaline can produce very complicated behavioral responses. Using a model of memory derived from work with the chick, we have dissected out the different roles of noradrenaline in memory formation and consolidation and show that its precise role depends on (i) the receptor subtype (ii) the discrete brain location, and (iii) the time at which it is acting relative to the learning experience.

Pharmacology of α- and β-Adrenoceptors in the Central Nervous System

Noradrenaline exerts a wide variety of effects on animal physiology. As a neurotransmitter in the central nervous system it plays an important role in sleep regulation, mood regulation, aggression, the degree of alertness and arousal, eating behaviour as well as modulation of memory storage. The release of noradrenaline within the central nervous system as a result of stimulation—arousal, vigilance, etc, has been regarded as an integral part of memory formation for at least thirty years.[32] Although noradrenaline and adrenaline act via the same adreno-receptors, preipherally released catecholamines do not influence central receptors directly because they do not cross the blood-brain barrier. However, noradrenaline acts more to modulate activity in neuronal pathways rather than as a direct chemical transmitter between cells. The main aim of this chapter is to outline how noradrenaline can modulate memory formation and how it interacts with the different subtypes of adrenoceptor that are found in the brain.

Sources of Noradrenaline in the CNS

Neurones with cell bodies located in the locus coeruleus (loc), subcoeruleus and nuclei of the lateral tegmental system are the major source of noradrenaline in the braïn. Since these neurones branch widely and send both ascending and descending projections to many brain areas and spinal cord; a single neuron can innervate a wide area of the brain. The noradrenergic afferent nerves of the neocortex, hippocampus, amygdala and thalamus run from the loc in the midbrain to the forebrain centres via the dorsal noradrenergic bundle.

Noradrenergic fibres, rather than having discrete synaptic connections, have extensively branched terminals containing multiple varicosities which release the neurotransmitter and this in turn stimulates receptors located on various cell types in the proximity. As well as neurones, the cell types that have receptors for noradrenaline include astrocytes and other glial cells and endothelial cells.

From Messengers to Molecules: Memories Are Made of These, edited by Gernot Riedel and Bettina Platt.

Stimulus for Noradrenaline Release

The neurones of the loc are activated by external and internal stimuli that are novel or significant to the subject. As well as anatomical specificity regarding which areas the fibres project to,[55] there is specificity in terms of the adrenergic receptors in these target areas that respond to noradrenaline. Adrenoceptors are classified into three main subclasses: α_1-, α_2- and β-receptors, each with three subtypes. All are present in brain.

Adrenoceptor Subtypes and Signal Transduction Pathways

The classification of adrenoceptors by functional, receptor binding and molecular cloning techniques has led to the identification of 9 subtypes, α_{1A}/ α_{1B}/α_{1D}; α_{2A}/ α_{2B}/α_{2C} and β_1/ β_2/ β_3-ARs which are classical G-protein-coupled receptors ranging in size from 402 amino acids (β_3) to 560 amino acids (α_{1D}). The adrenoceptors have 7 transmembrane spanning domains, with the third intracellular loop being crucial for coupling to the guanine nucleotide regulatory protein (G-protein). The receptors have several ligand recognition sites located within the transmembrane spanning regions. Intracellularly the adrenoceptors have a C-terminus tail of varying length, depending on the receptor, containing a variety of phosphorylation sites. These phosphorylation sites are largely involved in receptor desensitization. G-proteins couple the receptors to multiple types of intracellular effector proteins and in some cases, couple the receptors directly to certain types of ion channels. G-proteins are heterotrimers composed of single α, β and γ subunits. Combination of an agonist with an adrenoceptor causes coupling with the heterotrimeric G-protein. GDP associated with the Gα subunit exchanges for GTP and the GTP associated Gα dissociates from the $\beta\gamma$ subunits and activates the effector enzymes. Classically adrenoceptors interact with 3 major classes of G-proteins.[71] The α_1-ARs interact with G_q/G_{11} proteins that couple via phospholipase C to phosphoinositol signalling; α_2-ARs couple to G_i and inhibit the activity of adenylate cyclase; whereas β–ARs couple to G_s and activate adenylate cyclase. However there are now many documented cases where particular adrenoceptors demonstrate coupling to multiple G-proteins. The β_3-ARs for example can couple to both Gs and Gi[10] and phosphorylation of β_2-ARs has been shown to switch coupling from G_s to G_i.[17] The activation of effector enzymes results in alterations in intracellular levels of second messengers in target neurons and other cells. Prominent second messengers in the brain include cAMP, cGMP, calcium, the major metabolites of phosphatidyl-inositol (PI) [inositol triphosphate (IP_3) and diacylglycerol (DAG)].[20]

Signal Transduction Pathways of α_1-Adrenoceptors

α_1-ARs are coupled to phospholipase C through G_q, which initiates the hydrolysis of a membrane bound phospholipid to produce two second messengers, diacylglycerol (DAG), which activates PKC, and inositol 1,4,5-triphosphate (IP_3), which acts on its specific intracellular receptor to release Ca^{2+} (Fig. 1). They are selectively stimulated by methoxamine and phenylephrine and selectively blocked by prazosin. α_1-AR subtype selective agonists and antagonists are currently in development.

Signal Transduction Pathways of α_2-Adrenoceptors

α_2-ARs are coupled by G_i to adenyl cyclase or to ion channels. They alter cellular activity either by reducing intracellular levels of cyclic AMP or by modifying activity of Ca^{2+} or K^+ channels (Fig. 2). α_2-ARs are found both pre- and post-synaptically. The presynaptic receptors are regarded as autoreceptors and regulate release of noradrenaline from the terminals. Inhibition of these receptors results in increased neuronal release of noradrenaline.[57,66]

Signal Transduction Pathways of β-Adrenoceptors

β-ARs are coupled by G_s to adenylate cyclase and produce alterations in cellular activity by raising intracellular levels of cyclic AMP (Fig. 3). Pharmacologically, the classical β-ARs have been defined as receptors at which (-)-isoprenaline acts as an agonist and (-)-propranolol as an antagonist. In addition to the classical β_1- and β_2-ARs there is a third subtype, the β_3-AR,

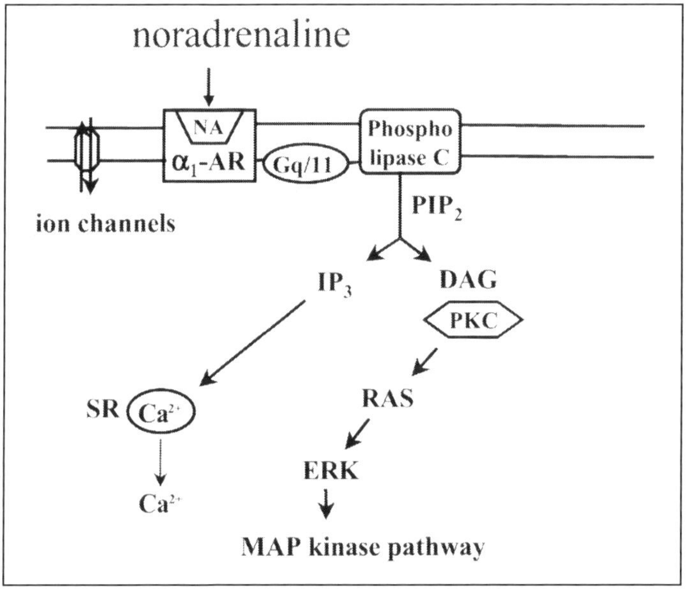

Figure 1. Signal transduction pathways for α_1-adrenocceptors. $G_{q/11}$ couples α_1-adrenenoceptor to PIP_2 (phosphatidyl inositol bisphosphate). The interaction of the G-protein activates phospholipase C-dependent hydrolysis of PIP_2. This generates the second messengers inositol trisphosphate (IP_3) which releases calcium from the intracellular stores of the sarcoplasmic reticulum (SR), and diacylglycerol (DAG), which activates protein kinase C (PKC). This can lead to further interactions with other second messenger dependent protein kinases such as RAS and MAP kinases.

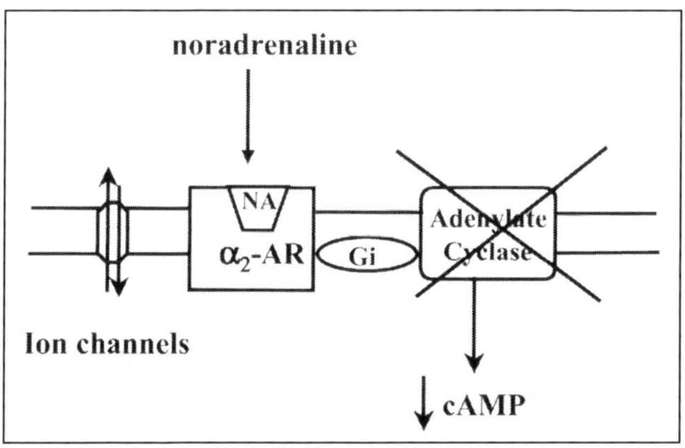

Figure2. Signal transduction pathways for α_2-adrenoceptors. Activation of α_2-adrenoceptors causes G_i to inhibit adenylate cyclase leading to a decrease in cAMP levels.

which has been detected in the brain[72] and is characterized by distinct pharmacology and a short C-terminal region that lacks phosphorylation sites. Unlike the β_1- and β_2-AR, β_3-ARs are resistant to desensitisation involving phosphorylation. There are now highly selective compounds that delineate β_1-, β_2- and β_3-ARs.

During time periods of minutes to hours, phosphorylation of the β_2-AR is closely involved in desensitisation. Agonist occupied receptors are substrates for phosphorylation by G-protein

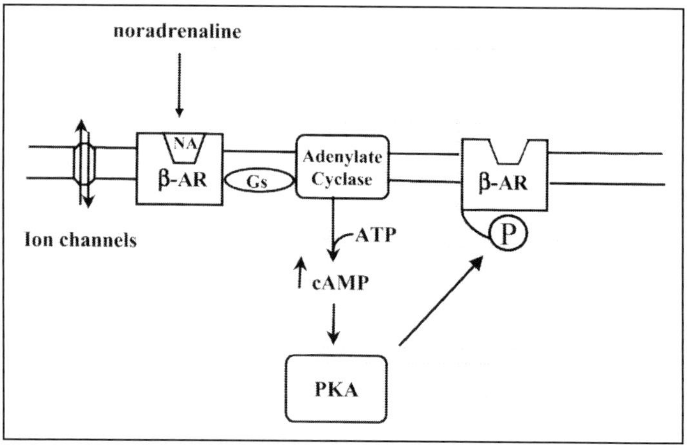

Figure3. Signal transduction pathways for β-adrenoceptors. Activation of β–adrenoceptors couples the receptor to G_s and activates adenylate cyclase increasing cAMP levels within the cell. cAMP activates the kinase PKA leading to phosphorylation and desensitisation of the β-adrenoceptor.

receptor kinases. Phosphorylated receptors uncouple from the signal transduction pathway but are also substrates for β-arrestin, an adaptor protein involved in pinocytosis and subsequent internalisation of receptor complexes. Internalised receptors may undergo proteolysis or be dephosphorylated and re-inserted in the plasma membrane. Receptor phosphorylation by β-AR kinases involves Ser and Thr residues in the carboxy-terminus region. The shorter C-terminus region of the β_3-AR contains few potential phosphorylation sites, which may be the reason that this receptor does not down-regulate in response to prolonged agonist stimulation.

The second messenger systems are also capable of interacting with other kinases such as the CAM kinase and MAP kinase pathways all of which have been implicated in long-term memory formation (see Medina and Cammarota, and also Selcher et al in this book).

Distribution of Adrenoceptor Subtypes in the Brain

Localization of Adrenoceptor Binding

All of the AR subtypes are expressed widely in the brain where they have multiple physiological roles including modulation of cardiovascular responses, sensory input, body temperature and food intake. The distribution of the different adrenoceptor subtypes is not uniform. Within the cortex, α_1 and α_2-ARs are found in different laminae; in the thalamus, α_1-ARs are localized to associational and sensory nuclei, whereas α_2-ARs are localized to periventricular, dorsomedial, midline and intra laminar thalamic nuclei.[53] The amygdala, particularly the central and basolateral-basomedial complex and the CA1 area of the hippocampus in the rat have a high density of α_2-ARs. In contrast, α_1-ARs are found in high density in medial septum, lateral hypothalamus and the basomedial amygdala.[53]

The distribution of β-AR subtypes in brain varies with species, with high densities of localized subtypes found in cortex, hippocampus and striatum.

Localization of mRNA

The presence or lack of mRNA signal in a particular region does not necessarily imply presence or lack of receptor protein since receptors can be transported from cell bodies to distant nerve terminals.[52] Furthermore, it is not always clear whether a signal originates from neuronal or glial cells. Both cell types express a variety of different receptor subtypes.[50,51] Examination of the distribution of the mRNA coding for the three basic AR subtypes enables mapping of those regions in the brain that possess cells which synthesize the distinct receptor subtypes (Table 1).

Table 1. Distribution of adrenoceptor mRNA's in catecholaminergic rat CNS areas involved in memory formation from (adapted from refs. 56,72)

Structure	β_1-	β_2-	β_3-	α_{1A}	α_{1B}	α_{1D}	α_{2A}	α_{2B}	α_{2C}
Cerebral cortex	++	++	+	+++	+++	++	++	-	++
Hippocampus	+-	+	++	++++	-	+++	+-	-	+++
Thalamus	+	+	+	?	+++	+-	+-	+	+
Hypothalamus	-	+	+	?	+	-	++	-	+
Locus coeruleus	-	-	?	++	++	++	+++	-	++
Vl medulla obl	-	-	?	?	++	-	++	-	-
NTS	-	-	?	?	-	-	++	-	-

Localization of α_1-Adrenoceptors

α_{1A}-AR mRNA appears to be widely expressed throughout the CNS, whereas α_{1B}-AR mRNA is heavily concentrated in areas such as the cerebral cortex, thalamus and raphe nuclei. α_{1D}-AR mRNA is found in the cerebral cortex, olfactory bulb, hippocampus, dentate gyrus and reticular thalamic nuclei.

Although all three α_1-AR subtypes are present in rat brain, since subtype-selective agonists or antagonists are only now emerging, it is not yet possible to allocate a particular central effect to particular subtypes.[52,64] However, recent studies using knock out mice suggest that the α_{1B}-AR subtype is involved in the reaction to novelty and in exploration. α_{1B}-AR knockout mice displayed deficits in both passive avoidance and water maze tasks.[38,65]

Localization of α_2-Adrenoceptors

Although presynaptic α_2-ARs are found in the brain, most α_2-ARs are post synaptic.[73] mRNA coding for α_{2A}-ARs is found throughout the brain, especially in the locus coeruleus; α_{2B}-ARs are found only in the thalamus. α_{2C}-ARs have a wide distribution, and expression is particularly intense[45] in the basal ganglia. α_{2A}-ARs appear to be pre- and post synaptically localized in the locus coeruleus.[8,24,39] This may explain why activation of α_2-ARs can both enhance or inhibit memory formation.

Localization of β-Adrenoceptors

Although β-ARs are widely distributed within the CNS, their functional role is still incompletely understood.[33] β_1-ARs are generally associated with forebrain structures such as the cerebral cortex, striatum and hippocampus, while β_2-AR radioligands are found strongly associated with cerebellar membranes.[52] β_1-AR mRNA is found in many areas including the cerebral cortex, thalamic neurones, whilst β_2-AR mRNA is strongest in the olfactory bulb, piriform cortex, hippocampal formation and cerebellar cortex. β_3-AR mRNA was detected in areas associated with memory - cerebral cortex, hippocampus, striatum, and thalamus and also in hypothalmus.[72] The expression of β_3-mRNA in hypothalamic areas is related to feeding and FOS expression also occurs in the hypothalamic areas following icv administration of β_3-AR agonists.[9,72]

Factors Affecting Drug Action at Adrenoceptors

The physiological catecholamines (-)-noradrenaline and (-)-adrenaline activate all of the adrenoceptor subtypes. (-)-Noradrenaline is somewhat more selective for β_1- compared to β_2-ARs. The action of the catecholamines therefore will largely depend on the distribution of receptor subtypes at the site of injection and the rapidity with which these amines are removed by neuronal and extraneuronal uptake processes and enzymic degradation by MAO and COMT (monoamine oxidase and catechol-O-methyl transferase). Results obtained using more selective agonists and antagonists allow a clearer interpretation of results. For most of the receptor subtypes there are a range of selective compounds available with lipophilic or hydrophilic properties and with differing affinities. Lipophilic compounds that penetrate the blood brain barrier can be useful in studies in mammals provided peripheral actions do not interfere (e.g., α_1-AR agonists produce intense vasoconstriction that will prevent or delay absorption following subcutaneous injection).

When agonists or antagonists are administered peripherally, it can be difficult to determine whether the action is mediated peripherally or centrally. Drugs acting centrally when administered peripherally have to be able to cross the blood brain barrier. Unitary action in the brain is unlikely as adrenoceptor subtypes are quite widely spread in different brain areas and activation of receptors in different regions may have opposing actions. Activation of receptors during memory formation can have different effects than when there is no learning, ie processes occurring during memory formation may differentially determine activation. For instance, we have found in the chick that there are different effects depending on the stage of memory processing at the time of administration of the agonist or antagonist. In addition, there may be release of endogenous transmitters within the CNS, including noradrenaline which can confound interpretation of the results.

Memory Studies with Adrenoceptor Agonists and Antagonists in Rats

In the past, studies on the involvement of adrenoceptors in memory formation in rats has focussed on identifying specific brain areas activated by systemically administered agonists and antagonist (Table 2). The search for memory functions related to the release of noradrenaline has concentrated on the amygdala (in particular the baso-lateral amygdala), prefrontal cortex, olfactory cortex and locus coeruleus. There has been very little work involving pharmacological intervention using adrenoceptor agonists and antagonists in the hippocampus. Izquierdo and coworkers[37] have shown differential effects on short- and long-term memory with localized injections of noradrenaline.

Amygdala

There is extensive evidence that β_2-ARs are involved in the modulatory action of noradrenaline in the amygdaloid complex in the rat. The memory enhancing effects of intra-amygdaloid infusion of noradrenaline are blocked by post-training infusion of the β-AR antagonist

Table 2. Commonly used adrenoceptor agonists and antagonists in memory studies with mammals

Subtype	Agonist	Antagonist
α_1-AR	cirazoline phenylephrine methoxamine	prazosin urapidl
α_2-AR	clonidine guanfacine	idazoxan yohimbine
β_1-AR	isoprenaline	atenolol propranolol
β_2-AR	isoprenaline clenbuterol	ICI 118,551 propranolol
β_3-AR	isoprenaline	

propranolol.[41,42] Infusion of the selective β_2-AR agonist, clenbuterol into the amygdala[35] and specifically into the baso-lateral amygdala[21] enhances memory consolidation.

If β_2-ARs are involved in memory consolidation in the amygdala, inhibition of these ARs should produce inhibition of memory. Until recently there was little evidence that systemic injection of propranolol produced memory impairment.[7,54] However, there have been studies showing infusion of propranolol into the amygdala produces a deficit in memory retention.[41] Generally the early reports implicating β_2-ARs in memory modulation demonstrated the ability of propranolol to prevent facilitation either by naloxone or adrenaline[35,42,47] but propranolol by itself did not produce a change in memory. Acute systemic administration of propranolol usually failed to affect memory.[19,69] As Cahill and coworkers[7] point out, variables such as drug dose and release of endogenous noradrenaline induced by the learning situation, may all contribute to the failure to detect inhibition of memory by propranolol, particularly when the drug is administered systemically. Indeed, Schneider et al[62] report enhancement of retention by propranolol in a multiple trial passive avoidance task and points to problems of systemic administration of drugs which can be having effects on a number of different 'circuits' within the brain, or even within the amygdala.[63] However, recently Cahill et al[7] demonstrated impairment of memory in a watermaze task by propranolol.

A series of recent[12,49,75] studies support the notion of noradrenaline release in the amygdala as a consequence of footshock (at levels employed in training paradigms), and as a consequence of systemic injection of adrenaline, naloxone or amphetamine.[21] As well as β_2-AR involvement in the amygdala, there is evidence for the involvement of α_1-ARs in memory modulation in the amygdala.[22,23] The selective α_1-AR antagonist prazosin infused into the basolateral amygdala immediately after inhibitory avoidance training impaired memory tested 48 hours later. Other experiments using prazosin infused into the amygdala, either showed no effect[47] or 'a tendency' to impair.[41] Low doses of the α_1-AR agonist phenylephrine suggested 'a tendency' for impairment of memory. The argument for involvement of α_1-ARs in the amygdala was supported by experiments selectively eliminating the effect of α_2-ARs by adding the antagonist yohimbine to the infusion; and then adding the β_1-AR antagonist atenolol to the phenylephrine and yohimbine mix. The addition of yohimbine reversed the tendency for phenylephrine to inhibit and the addition of atenolol prevented this effect.

In a subsequent experiment,[23] the memory facilitating effect of the β_2-AR agonist clenbuterol was dose-dependently affected by infusion of prazosin (at a dose which inhibited in the earlier experiment), which supported the suggestion that activation of α_1-ARs by noradrenaline serves to modulate the activation of β_2-ARs by noradrenaline. The precise role of these adrenoceptors in the amygdala or basolateral amygdala is not clear. As noradrenaline has the ability to activate all of the adrenoceptors, it is not clear how a meaningful selectivity can be obtained. On the other hand, although small amounts of the drugs are infused into the same area, they may have different lipid solubility and may not remain confined to the same area upon infusion.

The amygdala clearly plays an important role in memory consolidation, and certainly noradrenaline appears to play a role in regulating how the amygdala influences other regions of the brain, eg hippocampus, cortex and striatum which are also implicated in memory storage.[46] It is also probable that other regions of the brain are involved in modulation of information storage.

Hippocampus

Noradrenaline modifies phenomena such as long-term potentiation in the hippocampus. Although there are no pharmacological experiments where manipulation of adrenoceptors in the hippocampus influences memory processing, the presence of adrenoceptors in this area suggests that they may play a role in hippocampal activity associated with memory.

Prefrontal Cortex

Evidence is accumulating from the work of Arnsten and her associates that noradrenaline has an important influence on spatial working memory and attentional functions of the prefrontal cortex. While the cognitive processes of the prefrontal cortex appear to be unaffected by β-AR stimulation, they are impaired by α_1-AR stimulation. Neither systemic administration nor infusion of propranolol into the prefrontal cortex altered working memory performance in monkeys.[2,40] Infusion of isoprenaline into the medial prefrontal cortex had no effect on working memory performance in a T-maze in rats.[1] Stimulation of α-ARs by systemic cirazoline in monkeys[3] or by phenylephrine infused into the prefrontal cortex in rats, impaired working memory,[5] an effect that was blocked by systemic prazosin or co-infusion of the α_1-AR antagonist urapidil respectively. However, infusion of urapidil alone did not enhance working memory. This effect was attributed to α_1-ARs because post-synaptic α_2-AR stimulation in the prefrontal cortex has the opposite effect.

Stimulation of α_2-ARs by systemic injection or by prefrontal cortex infusion of clonidine or guanfacine improved working memory in rats and monkeys.[4] The effects were blocked by co-administration of α_2-AR antagonists such as yohimbine or idazoxan. Yohimbine infused into the prefrontal cortex impaired working memory.

Arnsten suggests that "noradrenaline has a higher affinity for α_2- than for α_1-ARs". Thus α_2-AR mechanisms predominate when basal noradrenaline release is moderate (e.g. normal, attentive, waking) and prefrontal function is optimal, while α_1-AR mechanisms may dominate when higher levels of noradrenaline release occur (eg during stress) contributing to PFC cognitive impairment ".[6]

Olfactory Bulb

In neonatal rat pups, the association of an odour with stimulation of β_2-ARs by the β-AR agonist isoprenaline in the olfactory bulb is sufficient to promote memory consolidation, as is stimulation of the locus coeruleus by the α_2-AR agonist idazoxan.[70] This effect is blocked by systemic injection of propranolol. The acquisition of a response to facial vibrissae stimulation and aversive shock in these rat pups is conditioned to isoprenaline and also blocked by systemic injection of propranolol. These experiments further support the importance of β-ARs in regions other than the amygdala in memory formation.

Locus Coeruleus

Electrical stimulation of the locus coeruleus facilitates memory retrieval in a task when normally there is a considerable decay of memory.[60] This facilitation is blocked by propranolol[19] and is attributed to noradrenaline release stimulated by activation of the locus coeruleus acting on β-ARs. Increased noradrenaline release following blockade of prejunctional α_2-AR by the selective α_2-AR antagonist idazoxan improves memory retention.

β-AR stimulation appears important for a late phase of memory consolidation[59,61] and is also critical for memory reactivation.[56] In agreement with this, blockade of β-AR by propranolol interferes with the reactivation of a learning experience 48 hours after training. Rats trained in an inhibitory avoidance task are not amnestic on first test but on second test after injection of propranolol 5 min after the first test the animals were amnestic tested 24 hours later. The time at which propranolol is effective when injected after first test seems to depend on the nature of the task.

Idazoxan clearly has effects on some aspect of memory processing. Systemic injection before every trial facilitated acquisition of a multi trial visual discrimination test.[60] Whether this effect is due to the effect of idazoxan on visual perception, attention or on memory processing is not clear, as any of these could account for the improved performance. Another experiment is the response to novel and unexpected objects appearing in a familiar environment—rats treated with idazoxan spent more time exploring the novel objects than controls. This suggests that increased release of noradrenaline enhances attention of novel environmental stimuli, which could account for the improvement in the multiple trial task.

These results using an α_2-AR antagonist are in contrast to those reported by Arnsten et al using the α_2-AR agonist clonidine or guanfacine; and indeed to the α_2-AR antagonist yohimbine. As there are both pre- and post -synaptic α_2-ARs in locus coeruleus, the route of injection may determine the outcome on memory processing. One of the problems in equating experiments using agonists and antagonists lies with the route of injection—systemic injection of drugs that cross the blood-brain barrier have the potential to target many regions in the brain; whether the drugs reach all loci at the same time is also an important concern. The nature of the task used may also be important in the interpretation of the drug effects. In the experiments with chicks,[30] noradrenaline affected different stages of memory processing for different reasons, in different locations and at different times relative to the initial acquisition.

Memory Studies with Adrenoceptor Agonists and Antagonists in Chicks

We have systematically investigated the roles of adrenoceptors in the action of noradrenaline in a one trial learning task (discriminated avoidance learning) in the young chick, where the chick discriminates between red and blue beads. The chick at one to two days of age has a number of distinct advantages over rodents in the study of memory, including limited development of the blood brain barrier, the ability to learn about coloured beads in one short trial without any prior knowledge of colored beads. When chicks are readily available from nearby poultry farms (as male rejects from an egg laying strain), they are inexpensive, available in sufficient numbers for large experiments from the same hatch and available soon after sex determination. As they are precocial, they can be used for experiments on the day of arrival and do not need to be housed for more than one day. Birds have a visual system with similar capabilities to humans and are active during the day unlike rodents. The structure of the avian brain, once thought to be very dissimilar to that of mammals, is now considered to have the same components (although perhaps packaged in a different way). For example, there is a close similarity between rats and pigeons in hippocampal functions with respect to spatial memory.[13] The primary visual, somatosensory and motor cortices are also similarly structured.[48]

The one trial peck avoidance task that is used with young chicks in various versions (passive avoidance, or discriminated avoidance) has changed subtly over the years since first introduced by Cherkin and Lee-Teng[11] and Watts and Mark.[74] But the task is still essentially the suppression of pecking at a bead that has the bitter taste of methyl anthranilate. The chick pecks at the bead on the training trial and if it remembers the bitter taste it does not peck at beads of that colour on the retention trial. We now get the chicks to discriminate between a clean red bead (this color was associated with methyl anthranilate on the training trial) and a blue bead and measure memory retention by the ratio of pecks to red and blue on the two successive 10 second retention trials. This gives a memory score for each chick, called the discrimination ratio (DR). If a chick remembers, it pecks less on the red bead and has a DR approaching 1.0 (Fig. 5B); if it forgets, the chick pecks at both beads equally and the DR approaches 0.5. Chicks are trained in groups of up to 20 and memory is measured in separate groups of chicks at prescribed times after training.

The memory time course which results from training with concentrated anthranilate on the bead reveals a model of memory where three stages are delineated by two periods (15 and 55 minutes after training) where the chicks appear to forget (Fig. 4). These two times correspond to the times for short-term, intermediate and long-term memory that have been defined on the basis of susceptibility to different classes of pharmacological drugs.[26,30] These drugs inhibit memory when injected at particular times up to 30 minutes after training and inhibitors for each memory stage have characteristic time windows during which they have to be injected in order to inhibit memory.[26,30] The determination of these times is also important for establishment of the roles of adrenoceptor subtypes in memory (see below, Table 3).

One particular advantage of this task is that the level of reinforcement, ie the concentration of the aversant on the bead, can be altered such that the retention level can be changed from good memory, with a level of 1.0 to poor memory with a retention level of 0.5. With weakly reinforced learning, memory lasts for 30 minutes and then the memory is lost (Fig. 5A). Memory is in a labile form and is not normally kept for longer than 30 minutes unless some event occurs which triggers the consolidation of the labile memory to a permanent form. This can be achieved

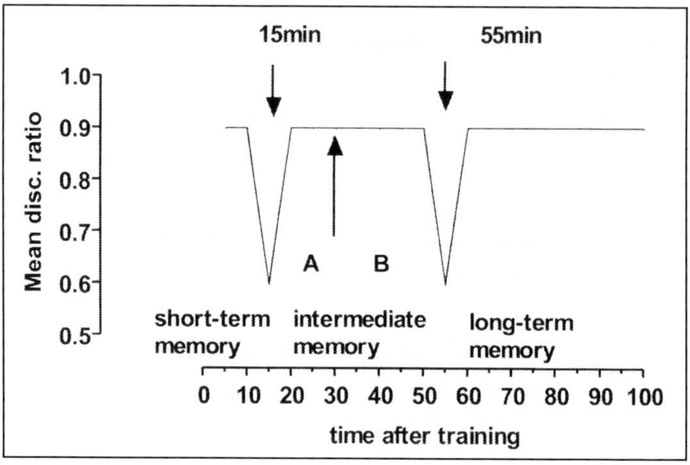

Figure 4. Sequences of memory stages in memory formation in the chick. This model is derived from behavioural and pharmacological experiments where memory is tested in separate groups of chicks at discrete times after training. Memory is measured as the discrimination between red and blue beads, where only the red bead had the aversive taste on training. A discrimination ratio approaching 1.0 is found where chicks avoid the red bead but peck the blue bead; a discrimination ratio approaching 0.5 is found when the chicks forget and peck at both beads equally. Behavioural experiments reveal two times after training when the chicks do not appear to remember. These time points coincide with data from pharmacological experiments, where memory decays after each of the stages following inhibition leaving the previous stage intact (from refs. 26,30). Theoretically, this represents the memory course within any one chick. At 30 minutes some internal event triggers the consolidation of labile memory into permanent storage. This time divides intermediate memory into phase A and phase B, each susceptible to interference by different pharmacological agents. Reprinted from Progress in Neurobiology, 67, 2002, pp. 345-391, Gibbs et al: "Role of adrenoreceptor subtypes in memory consolidation. Figure 1.

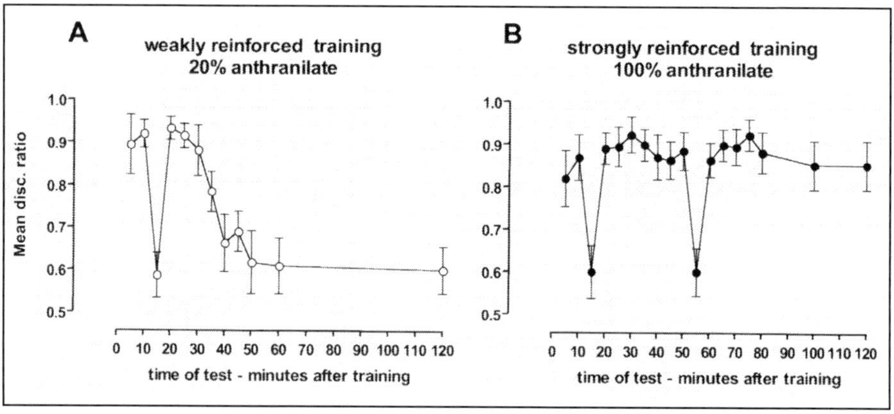

Figure 5. Data from experiments in which the chicks given (A) weakly reinforced training (20% anthranilate) or (B) strongly reinforced training (100% anthranilate). Separate groups of 20 chicks are tested at different times after training. With weakly reinforced training, memory gets weaker in chicks tested after 30-35 minutes; whereas with strongly reinforced training the level of memory retention is still high on the test 120min after training. This memory remains permanently good (at least 72 hours). Reprinted from Progress in Neurobiology, 67, 2002, pp. 345-301. Gibbs et al: Role of adrenoreceptor subtypes in memory consolidation. Figure 2.

Table 3. Adrenoceptor agonists and antagonists used in chick discriminated avoidance learning

Subtype	Agonist	Antagonist	Route	Effect
α_1-AR	Methoxamine		IMHV	inhibit
		Prazosin		enhance
α_2-AR	clonidine		LPO/IMHV	enhance
	oxymetazoline		LPO	enhance
		yohimbine*	LPO	Inhibit/ enhance
		Yohimbine*	subcut	Inhibit/ enhance
β_1-AR	RO363		LPO	enhance
	isoprenaline			enhance
		CGP 20712	LPO/sc	inhibit
		propranolol	LPO/sc	inhibit
		sotalol		inhibit
β_2-AR	isoprenaline		IMHV	enhance
	zinterol		IMHV	Enhance
	clenbuterol			Enhance
	salbutamol			Enhance
	BRL37344†		IMHV	enhance
		Propranolol	IMHV	inhibit
		Sotalol		inhibit
		ICI 118,551		enhance
		timolol		
β_3-AR	isoprenaline		IMHV	enhance
	CL316243		IMHV/sc	enhance
	BRL37344		IMHV	enhance
	CGP12177A		IMHV	enhance
		SR59230A	IMHV/sc	inhibit

*It is clear that yohimbine can enhance or inhibit depending on route of administration and also on the time administered after training.
† High doses.

by injection of noradrenaline or by hormonal or environmental events that may lead to increased endogenous release of noradrenaline. Memory improvement from enhancement of consolidation can be achieved by injection of drugs up to 30 min after training but not later.

Noradrenaline enhances memory of weakly reinforced training (20% anthranilate) when injected subcutaneously[25,30] or when injected into two different brain regions - the intermediate medial hyperstriatum ventrale (IMHV-a multimodal sensory association area - association cortex) and the lobus parolfactorius (LPO- caudate putamen).[30] The dose response relationship at 2 hours after training for intracerebral administration 20 minutes after training is a bell-shaped function, with low doses enhancing and higher doses inhibiting memory.[30] Similar responses are seen in rats with systemic injection of adrenaline[31] or clenbuterol[34] or noradrenaline infused into the amygdala.[43] By pre-administration of selective α- and β-AR antagonists (see Table 2; Fig. 6) we showed that the action of noradrenaline, injected into the IMHV, involves β_2-, β_3-, and α_1-ARs, which are activated at different doses. Low doses of noradrenaline facilitate memory acting via β_3-ARs; higher doses also facilitate memory by acting on β_2-

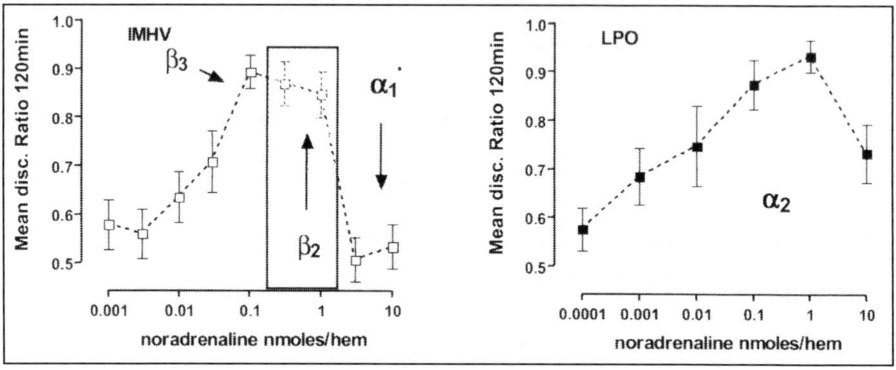

Figure 6. A) Dose response relationship for noradrenaline injected into IMHV 20 minutes after weakly reinforced training. Enhancement of memory consolidation by noradrenaline at doses of 0.1nmoles/hemisphere is attributed to action at β_3-ARs; whereas at 1.0nmole/hemisphere noradrenaline is acting via β_2-ARs. Higher doses of noradrenaline (3-10nmoles/hemisphere) do not enhance weakly reinforced memory and are inhibitory when used with 100% anthranilate. This inhibitory effect is attributed to an action on α_1-ARS. B) Dose response relationship for noradrenaline injected into LPO at 20 minutes after weakly reinforced training. A similar bell-shaped relationship is seen, with the memory enhancing property attributed to β_2-ARs.

ARs; and the highest doses of noradrenaline inhibit memory via α_1-ARs. Pre-administration of the α_2-AR antagonist yohimbine did not affect the response to noradrenaline injected into the IMHV, but did prevent the memory facilitation produced by noradrenaline injected into the LPO. In these experiments, noradrenaline was injected 20 minutes after training, during intermediate memory and prior to the consolidation into long-term memory. Although neither the β_1-AR agonist RO363 nor the selective β_1-AR antagonist CGP20712 influenced memory when given 20 minutes after training, we have evidence that β_1-ARs in the LPO are involved in short-term memory.

Roles for Adrenoceptor Subtypes in the IMHV

α_1-Adrenoceptors

High doses of noradrenaline (10nmol/hemisphere) injected into the IMHV inhibit memory formation in chicks trained with 100% anthranilate, however, high doses of the β-AR agonist isoprenaline do not inhibit.[27] The effect of high doses of noradrenaline is mimicked by injection of the selective α_1-AR agonist methoxamine.[29] Both noradrenaline and methoxamine injected into the IMHV prevent the consolidation of intermediate memory. There are two times of susceptibility to interference by stimulation of α_1-ARs (Fig. 7). Injections made 5 minutes before training or 20 to 25 minutes after training affect memory, but injections at the times in between have no effect. At either time after training, memory was unaffected for 30 minutes, but then rapidly decays (Fig. 8A). Memory does not consolidate.

The specificity of the action of methoxamine on α_1-ARs was demonstrated by pre-administration of the selective α_1-AR antagonist prazosin. Given 5 minutes before methoxamine, prazosin shifted the dose-response curve in parallel to the right, ie. prazosin decreased the ability of methoxamine to inhibit memory.

β_2-Adrenoceptors

The non-selective β-AR agonist isoprenaline enhances memory consolidation in a similar manner to the lower dose range of noradrenaline (0.1 and 3.0 nmol/hemisphere). The action of isoprenaline (1 to 10 nmol) is prevented by pre- administration of the β_1-/β_2-AR antagonist

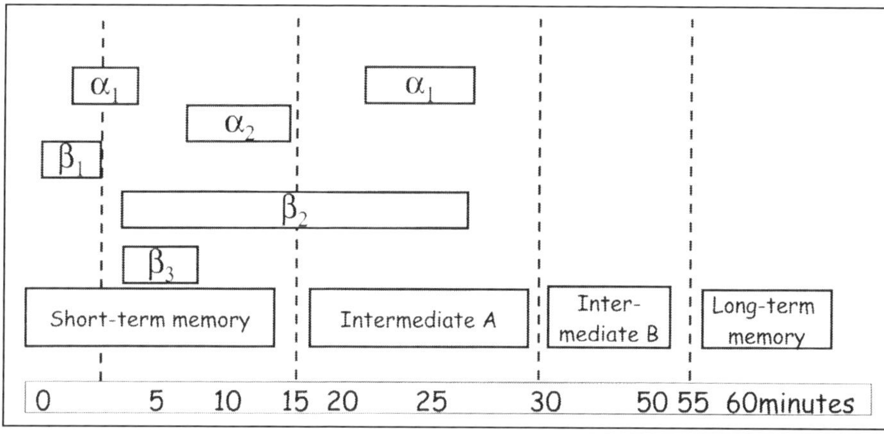

Figure 7. Schematic representation showing times and memory stages at which antagonists to the different subtypes are effective in preventing memory formation.

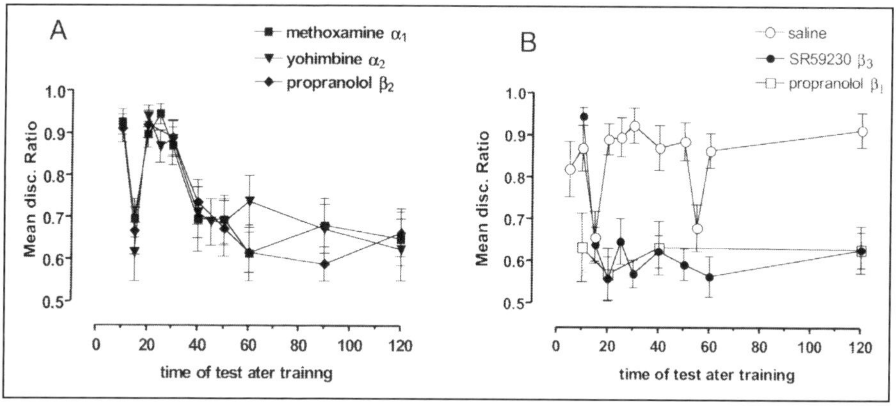

Figure 8. Time of memory decay after injection of different AR subtype antagonists (with 100% anthranilate). A) α_1, α_2 and β_2- AR antagonists do not impair labile memory but memory decays after 30 minutes and there is no consolidation. B) Memory decay after 10 minutes following the β_3-AR antagonist, and from 10 minutes (or earlier) when propranolol's action is attributable to β_1-ARs in the LPO.

propranolol.[27] Although propranolol blocks both β_1- and β_2-ARs, we were not able to demonstrate any effect of β_1-AR agonists or antagonists given 20 minutes after training, suggesting that the action of propranolol on the enhancement of memory by isoprenaline 20 min after training is due to an action at β_2-ARs.

Systemic injection after training of propranolol,[15] sotalol[67,68] and the selective β_2-AR antagonist ICI118551[18] have all been shown to inhibit memory, whereas the β_1-AR antagonist atenolol does not. The selective β_2-AR agonist, salbutamol enhanced weakly reinforced training and facilitated retention in chicks treated with the noradrenergic neurotoxin DSP-4.[14] In order to inhibit memory in IMHV, the intracerebral injections have to be made after training, we found that injections before training into IMHV did not have any effect on memory.[30] However, the injection of propranolol either subcutaneously before or into the LPO immediately after training affects β_1-ARs (Gibbs and Summers, in prep). Propranolol[30] and sotalol[68] can be injected up to 25 minutes after training (Fig. 7). Inhibition of β_2-ARs in the IMHV in

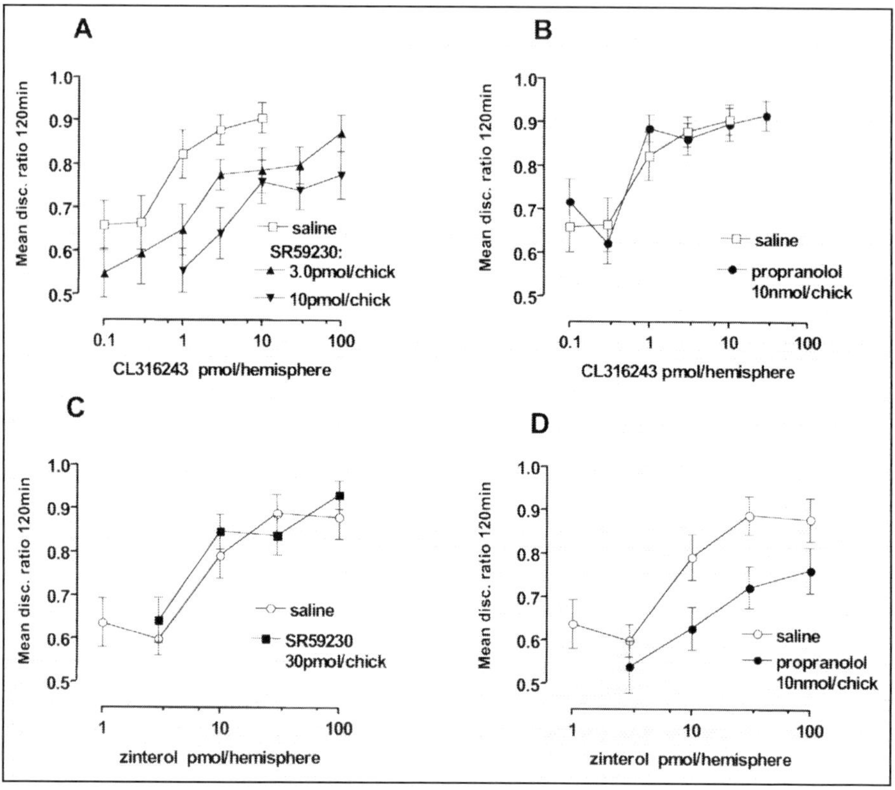

Figure 9. Dose response curves showing specificity of AR subtype action. A. the dose response to CL316243 is shifted in a dose-dependent manner to prior administration of a non amnestic dose of the selective β_3-AR antagonist SR59230 (A); but is not shifted by a non-amnestic dose of propranolol (B). Likewise, the dose response curve to zinterol is not shifted by a non amnestic dose of SR59230, but is shifted by a non amnestic dose of propranolol. (From ref. 27). Reprinted from Neuroscience, 95, 2000, pp. 913-922. Gibbs et al: Separate roles for β_2- and β_3-adrenoceptors in memory consolidation. Figure 3.

chicks given strongly reinforced training prevents consolidation and memory loss occurs after 30 minutes (Fig. 8A). This discrepancy in results between subcutaneous injection and injection into particular brain regions points out the importance of full investigations of the actions of drugs which may act at several receptor subtypes.

The selective β_2-AR agonist, zinterol will enhance the consolidation of weakly reinforced learning when injected at 0 to 25 minutes after training and when given immediately after training, increases the duration of both short-term and intermediate memory.[27] The specificity of action of zinterol was demonstrated by experiments in which we pre-administered (5 minutes after training with 20% anthranilate) non-amnestic doses of either propranolol or the selective β_3-AR antagonist SR59230 (Table 3). The dose-response curve for zinterol was shifted to the right by propranolol but not by SR59230 (Fig. 9C and D).

β_3-Adrenoceptors

The action of the selective β_3-AR agonist CL316243, shows specificity for the β_3-AR in memory functioning. The dose response curve to CL316243 injected into the IMHV at 20 minutes after weakly reinforced learning shows dose dependent consolidation of memory. Prior administration of a non-amnestic dose of the selective β_3-AR antagonist SR59230 shifts the dose response curve to the right in a dose dependent manner (Fig. 9A). Propranolol does not shift the dose response curve to CL316243 (Fig. 9B). Activation of both the β_3- and the β_2-AR are able to promote consolidation of memory. It appears that the action of low doses of noradrenaline and isoprenaline is on β_3-ARs as the selective antagonist SR59230 inhibits the action of 0.1 nmol/hemisphere of noradrenaline and 0.03nmol/hemisphere of isoprenaline. This difference in the effective doses is in accord with the differences in affinity of the two catecholamines for the receptor. We hypothesize that endogenously released noradrenaline acts firstly on β_3-ARs and then β_2-ARs are recruited. The level of endogenous noradrenaline may be related to the level of reinforcement or in the case of aversive learning, to the level of stress invoked by the learning situation.

The selective β_3-AR agonist CL316243 also enhances memory consolidation.[27] Injection immediately after training results in a memory time course that appears to be the same as that seen with strongly reinforced training. Injection up to 20 minutes after weakly reinforced training is effective in promoting consolidation. The time window during which strongly reinforced learning can be inhibited by SR59230 is limited to injection 5 minutes after training (Fig. 7) and memory loss occurs after 10 minutes following training. The limited action of this antagonist was puzzling, but may be explained by the added complication of endogenous release of noradrenaline which will occur in many learning situations. In support of this interpretation, injection of SR59230 to chicks given weakly reinforced training was effective with injections at any time during the first part of intermediate memory (ITMA), i.e. from 10 to 25 minutes after training. To obtain this result, chicks had to be tested at 30 minutes after training. The endogenous noradrenaline, released with strongly reinforced training, may not act on β_3-ARs when inhibited by the antagonist, but can still act on β_2-ARs. In the weakly reinforced situation there should be less endogenous noradrenaline release.

We have been able to show enhancement of consolidation with two other β_3-AR agonists—BRL37344 and CGP12177. However, BRL37344 appears to have two actions. At low doses (100pmol/hemisphere) it acts via β_3-AR, whereas at higher doses (1 nmol / hemisphere) it acts on β_2-ARs.[28]

Roles for Adrenoceptor Subtypes in the LPO

β_1-Adrenoceptors

As mentioned above, β_1-ARs are involved in the LPO at the time of acquisition, during short-term memory, but do not appear to be involved in the IMHV at any of the times tested. The antagonist CGP20712 inhibits strongly reinforced training when injected into the LPO up to 2.5 minutes after training, and the selective agonist RO363 enhances memory when given up to 2.5 minutes after training (Gibbs and Summers, unpublished observation). Whether these agonists and antagonists act by enhancing consolidation or by some other action on short-term memory is not yet clear. It is possible that the β_1-ARs are involved with the effects of arousal or attention in the basal ganglia at the time of acquisition and short-term memory.

α_2-Adrenoceptors

α_2-ARs seem to be involved in memory consolidation in the LPO rather than in the IMHV. Pre-administration of the selective α_2-AR antagonist yohimbine prevents enhancement of memory formation by noradrenaline when injected into the LPO (Fig. 6B), but has no effect on noradrenaline injected into the IMHV.[30]

Injection of the selective α_2-AR agonist clonidine into the IMHV produces an effect, but at a higher dose and at more restricted times than when injected into LPO where injection is effective over the full time of labile memory. However, clonidine is lipophilic and crosses membranes readily, which makes it less than ideal for determination of sites of action. Experiments were therefore conducted with a hydrophilic α_2-AR agonist, oxymetazoline which was found to have no effect on memory when injected into the IMHV, but is as effective as clonidine when given into the LPO. Likewise, yohimbine was not effective in the IMHV. Memory loss

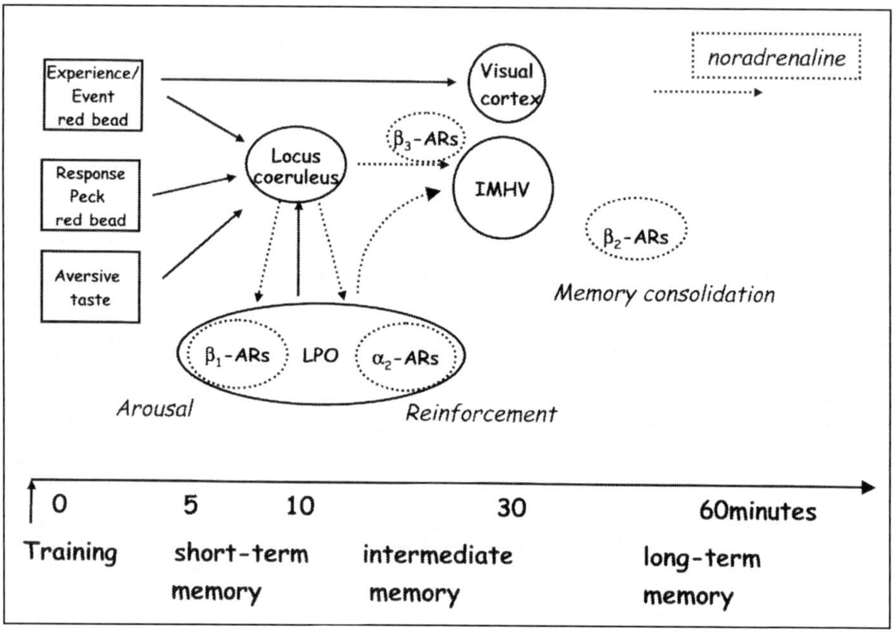

Figure 10. A model of noradrenergic modulation of memory formation in the consolidation of memory in the chick. Noradrenergic input from the locus coeruleus activates β_1-ARs in LPO (basal ganglia) at the time of or shortly after training. β_3-AR activation in IMHV (association cortex) occurs before β_2-AR activation in the IMHV, with the latter being dependent on α_2-AR activation in LPO. It is likely that there could be other interactions between noradrenaline and adrenoceptors in other parts of the brain not yet explored.

after yohimbine occurs after 30 minutes (Fig. 8A). However, yohimbine is only effective in inhibiting consolidation when injected 10 or 15 minutes after training.

The danger of basing findings solely on drug administration by systemic injection is exemplified in our recent finding that yohimbine has a biphasic effect. Given subcutaneously 10 or 15 minutes after training, it inhibits memory consolidation, whereas given by the same route 2.5 or 25 minutes after training, it enhances consolidation. This does not occur following injections into either LPO or IMHV at these times. The may indicate an action of yohimbine in the locus coeruleus, and be related to the contradictory findings mentioned earlier.

Summary

By examining the response to administration of selective adrenoceptor subtype agonists and antagonists we have been able to map the time-course of noradrenaline involvement in memory acquisition and consolidation in chicks. By varying the route of administration, based on the response to subcutaneous injection, we have located different areas in the brain which, as well as being involved in memory,[16] are involved in differential adrenoceptor activation at different times after training (Fig. 10).

Our research demonstrates the importance of systematic investigation to determine the time in the memory processing sequence when drugs are effective, the dose-response relationships for each drug and the site of action in the brain in order to conclude that a particular receptor is or is not involved in memory formation. Even when these factors are taken into account, the effects described are most likely still an oversimplification of the events in the brain that are influenced by noradrenaline in the acquisition and consolidation of a memory, since we have not touched on the possible involvement of noradrenaline in areas like the amygdala (archistriatum), hippocampus and prefrontal cortex (caudal neostriatum) in the avian brain. But it does emphasize the caution needed when interpreting results from experiments using a single dose, a single route of injection and a single time of administration.

Input into different primary sensory brain regions and into various multimodal association areas will occur at the same time and cellular processing of the information will probably occur in more than one (if not many) of these areas. The potential is there for different influences or modulators employing many different neurotransmitter systems

As Lipp and Wolfer[44] conclude - the reticular formation (and noradrenaline) has the potential to integrate and coordinate activity in many different brain regions during the acquisition and consolidation of memory. Other neurotransmitter and hormonal influences will also play a part in memory formation, some of which may involve noradrenaline (e.g., see ref. 58).

In the rodent, memory research is now focussing on hippocampus, amygdala, and the prefrontal cortex, all of which are influenced by noradrenergic input. It is likely that these areas all have a different response to noradrenaline dependent on the subtype, and relative distribution of adrenoceptors. When information is available on the action of selective adrenoceptor agonists and antagonists administered to all these brain areas, a clearer picture of the interrelationships may emerge. The role of noradrenaline in modulating memory formation may be more complex than that of other neurotransmitters.

References

1. Arnsten AFT. Through the looking glass: Differential noradrenergic modulation of prefrontal cortical function. Neural Plasticity 2000; 7:133-146.
2. Arnsten AF, Goldman-Rakic PS. Alpha 2-adrenergic mechanisms in prefrontal cortex associated with cognitive decline in aged nonhuman primates. Science 1985; 230:1273-1276.
3. Arnsten AFT, Jentsch JD. The alpha-1 adrenergic agonist, cirazoline, impairs spatial working memory performance in aged monkeys. Pharmacol Biochem Behav 1997; 58:55-59.
4. Arnsten AF, Cai JX, Goldman-Rakic PS. The alpha-2 adrenergic agonist guanfacine improves memory in aged monkeys without sedative or hypotensive side effects: Evidence for alpha-2 receptor subtypes. J Neurosci 1988; 8:4287-4298.
5. Arnsten AFT, Mathew R, Ubriani R et al. α-1 noradrenergic receptor stimulation impairs prefrontal cortical cognitive function. Biol Psychiatry 1999; 45:26-31.
6. Arnsten AFT, Steere JC, Jentsch DJ et al. Noradrenergic influences on preforntal cortical cognitive function: Opposing actions at postjunctional α_1 versus α_2-adrenergic receptors. Adv Pharmacol 1998; 42:764-767.
7. Cahill L, Pham CA, Setlow B. Impaired memory consolidation in rats produced with β-adrenergic blockade. Neurobiol Learn Mem 2000; 74:259-266.
8. Callado LF, Stamford JA. α_{2A}- but not $\alpha_{2B-/C}$ -adrenoceptors modulate noradrenaline release in rat locus coeruleus: Voltammetric data. Europ J Pharmacol 1999; 366:35-39.
9. Castillo-Melendez M, McKinley MJ, Summers RJ. Intracerebroventricular administration of the β_3-adrenoceptor agonist CL 316243 causes Fos immunoreactivity in discrete regions of rat hypothalamus. Neurosci Lett 2000; 290:161-164.
10. Chaudhry A, MacKenzie RG, Georgic LM et al. Differential interaction of beta 1- and beta 3-adrenergic receptors with G_I in rat adipocytes. Cell Signalling 1994; 6:457-465.
11. Cherkin A, Lee-Teng E. Interruption by halothane of memory consolidation. Federation Proc 1965; 24:328.
12. Clayton EC, Williams CL. Adrenergic activation of the nucleus tractus solitarius potentiates amygdala norepinephrine release and enhances retention performance in emotionally arousing and spatial memory tasks. Behav Brain Res 2000; 112:151-158.
13. Colombo M, Broadbent N. Is the avian hippocampus a functional homologue of the mammalian hippocampus? Neurosci Biobehav Rev 2000; 24:465-484.
14. Crowe SF, Shaw S. Salbutamol overcomes the effect of the noradrenergic neurotoxin DSP-4 on memory function in the day-old chick. Behav Pharmacol 1997; 8:216-222.
15. Crowe SF, Ng KT, Gibbs ME. Possible noradrenergic involvement in training stimulus intensity. Pharmacol Biochem Behav 1991; 39:717-722.
16. Csillag A. Striato-telencephalic and striato-tegmental circuits: Relevance to learning in domestic chicks. Behav Brain Res 1999; 98:227-236.
17. Daaka Y, Luttrell LM, Lefkowitz RJ. Switching of the coupling of the β_2-adrenergic receptor to different G proteins by protein kinase A. Nature 1997; 390:88-91.

18. Davies DC, Payne JM. Amnesia of a passive avoidance task due to the β_2-adrenoceptor antagonist ICI 118,551. Pharmacol Biochem Behav 1989; 32:187-190.
19. Devauges V, Sara SJ. Memory retrieval enhancement by locus coeruleus stimulation: Evidence for mediation by β-receptors. Behav Brain Res 1991; 43:93-97.
20. Duman RS, Nestler EJ. Signal transduction pathways for catecholamine receptors. In: Bloom FE, Kupfer DJ, eds. Psychopharmacology: The Fourth Generation of Progress. New York: Raven Press, 1995:303-320.
21. Ferry B, McGaugh JL. Clenbuterol administration into the basolateral amygdala post-training enhances retention in an inhibitory avoidance task. Neurobiol Learn Mem 1999; 72:8-12.
22. Ferry B, Roozendaal B, McGaugh JL. Basolateral amygdala noradrenergic influences on memory storage are mediated by an interaction between β- and α-adrenoceptors. J Neurosci 1999; 19:5119-5123.
23. Ferry B, Roozendaal B, McGaugh JL. Involvement of α_1-adrenoceptors in the basolateral amygdala in modulation of memory storage. Europ J Pharmacol 1999; 372:9-16.
24. French N. α_2-Adrenoceptors and I_2 sites in the mammalian central nervous system. Pharmacol Therap 1995; 68:175-208.
25. Gibbs ME. Modulation of cycloheximide-resistant memory by sympathomimetic agents. Pharmacol Biochem Behav 1976; 4:703-707.
26. Gibbs ME, Ng KT. Psychobiology of memory: Towards a model of memory formation. Biobehav Rev 1977; 1:113-136.
27. Gibbs ME, Summers RJ. Separate roles for β_2- and β_3-adrenoceptors in memory consolidation. Neurosci 2000; 95:913-922.
28. Gibbs ME, Summers RJ. Enhancement of memory consolidation in chicks by β_3-adrenoceptor agonists. Europ J Pharmacol 2000; 413:235-240.
29. Gibbs ME, Summers RJ. Stimulation of α_1-adrenoceptors inhibits memory consolidation in the chick. Europ J Neurosci 2001; 14:1369-1376.
30. Gibbs ME, Summers RJ. Role of adrenoceptor subtypes in memory consolidation. Prog Neurobiol 2002; 67:345-391.
31. Gold PE, van Buskirk R. Posttraining brain norepinephrine concentrations: Correlation with retention performance of avoidance training and with peripheral epinephrine modulation of memory processing. Behav Biol 1978; 23:509-520.
32. Harley CW. A role for norepinephrine in arousal, emotion and learning?: Limbic modulation by norepinephrine and the kety hypothesis. Prog Neuro-Psychopharmacol Biol Psychiat 1987; 11:419-458.
33. Hieble JP, Ruffolo RR. Functions mediated by β-adrenoceptor activation. In: Ruffolo RR, ed. β-Adrenoceptors: Molecular Biology, Biochemistry and Pharmacology. Basel: Karger, 1991:173-209.
34. Introini-Collison I, Baratti CM. Opioid peptiderguc systems modulate the activity of β-adrenergic mechanisms during memory consolidation processes. Behav Neural Biol 1986; 46:227-241.
35. Introini-Collison IB, McGaugh JL. Involvement of the amygdala in the memory-enhancing effects of clenbuterol. Psychopharmacol 1991; 104:541-544.
36. Introini-Collison IB, Nagahara AH, McGaugh JL. Memory enhancement with intra-amygdala post-training naloxone is blocked by concurrent administration of propranolol. Brain Res 1989; 476:94-101.
37. Izquierdo I, Medina JH, Izquierdo LA et al. Short- and long-term memory are differentially regulated by monoaminergic systems in the rat brain. Neurobiol Learn Mem 1998; 69:219-224.
38. Knauber J, Muller WE. Decreased exploratory activity and impaired passive avoidance behaviour in mice deficient for the α_{1b}-adrenoceptor. European Neuropsychopharmacol 2000; 10:423-427.
39. Lee A, Rosin DL, van Bockstaele EJ. α_{2A}-adrenergic receptors in the rat nucleus locus coeruleus: Subcellular localization in catecholaminergic dendrites, astrocytes and presynaptic axon terminals. Brain Res 1998; 795:157-169.
40. Li BM, Mei ZT. Delayed-response deficit induced by local injection of the alpha 2- adrenergic antagonist yohimbine into the dorsolateral prefrontal cortex in young adult monkeys. Behav Neural Biol 1994; 62:134-139.
41. Liang KC, Chen LL, Huang T-E. The role of amygdala norepinephrine in memory formation: Involvement in the memory enhanceing effect of peripheral epinephrine. Chin J Physiol 1995; 38:81-91.
42. Liang KC, Juler RG, McGaugh JL. Modulating effects of posttraining epinephrine on memory: Involvement of the amygdala noradrenergic system. Brain Res 1986; 368:125-133.
43. Liang KC, McGaugh JL, Yao H-Y. Involvement of amygdala pathways in the influence of post-training intra-amygdala norepinephrine and peripheral epinephrine on memory storage. Brain Res 1990; 508:225-233.
44. Lipp HP, Wolfer DP. Genetically modified mice and cognition. Curr Opin Neurobiol 1998; 8:272-280.
45. MacDonald E, Scheinin M. Distribution and pharmacology of alpha(2)-adrenoceptors in the central nervous system. J Physiol Pharmacol 1995; 46:241-258.
46. McGaugh JL. Memory- a century of consolidation. Science 2000; 287:248-251.

47. McGaugh JL, Introini-Collison IB, Nagahara AH. Memory-enhancing effects of posttraining naloxone: Involvement of β-noradrenergic influences in the amygdaloid complex. Brain Res 1988; 446:37-49.
48. Medina L, Reiner A. Do birds possess homologues of mammalian primary visual, somatosensory and motor cortices? TINS 2000; 23:1-12.
49. Men D, McCarty R, Gold PE. Enhanced release of norepinephrine in rat hippocampus during spontaneous alternation tests. Neurobiol Learn Mem 1999; 72:289-300.
50. Muyderman H, Hansson E, Nilsson M. Adrenoceptor-induced changes of intracellular K^+ and Ca^{2+} ind astrocytes and neurons in rat cortical primary cultures. Neurosci Lett 1997; 238:33-36.
51. Muyderman H, Sinclair J, Jardemark K et al. Activation of β-adrenoceptors opens calcium-activated potassium channels in astroglial cells. Neurochem Int 2001; 38:269-276.
52. Nicholas AP, Hokfelt T, Pieribone VA. The distribution and significance of CNS adrenoceptors examined with in situ hybridisation. TiPS 1996; 17:245-255.
53. Nichols AJ, Ruffolo RR. Functions mediated by α-adrenoceptors. In: Ruffolo RRJ, ed. αAdrenoceptors: Molecular Biology, Biochemistry and Pharmacology. Basel: Karger, 1991:115-179.
54. Nielson KA, Czech DA, Laubmeier KK. Chronic administration of propranolol impairs inhibitory avoidance retention in mice. Neurobiol Learn Mem 1999; 71:248-257.
55. Paspalas CD, Papadopoulos GC. Ultrastructural relationsips between noradrenergic nerve fibres and non-neuronal elements in the rat cerebral cortex. Glia 1996; 17:133-146.
56. Przbyslawski J, Roullet P, Sara SJ. Attenuation of emotional and nonemotional memories after their reactivation: Role of β adrenergic receptors. J Neurosci 1999; 19:6623-6628.
57. Raiteri M. Presynaptic autoreceptors. J Neurochem 2001; 78:673-675.
58. Roozendaal B, Quirarte GL, McGaugh JL. Glucocorticoids interact with the basolateral amygdala β-adrenoceptor-cAMP/PKA system in influencing memory consolidation. Europ J Neurosci 2002; 15:553-560.
59. Roullet P, Sara SJ. Consolidation of memory after its reactivation: Involvement of β-noradrenergic receptors in the late phase. Neural Plasticity 1998; 6:63-68.
60. Sara SJ, Devauges V. Idazoxan, an α2 antagonist, facilitates memory retrieval in the rat. Behav Neural Biol 1989; 51:401-411.
61. Sara SJ, Roullet P, Przbyslawski J. Consolidation of memory for odor-reward association: β-adrenergic receptor involvement in the late phase. Learn Mem 1999; 6:88-96.
62. Schneider AM, Koven NS, Lombardo KA et al. Beta-adrenergic receptor blockade by propranolol enhances retention in a multitrial passive-avoidance procedure. Behav Neurosci 2000; 114:1256-1260.
63. Simson PE, Naylor JC, Gibson B et al. Dose-sensitive excitation and inhibition of spontaneous amygdala activity by propranolol. Pharmacol Biochem Behav 2001; 69:85-92.
64. Sirvio J, MacDonald E. Central alpha1-adrenoceptors: Their role in the modulation of attention and memory formation. Pharmacol Therap 1999; 83:49-65.
65. Spreng M, Cotecchia S, Schenk F. A behavioral study of alpha-1b adrenergic receptor knockout mice: Increased reaction to novelty and selectively reduced learning capacities. Neurobiol Learn Mem 2001; 75:214-229.
66. Starke K. Presyntaptic autoreceptors in the third decade: Focus on α_2-adrenoceptors. J Neurochem 2001; 78:685-693.
67. Stephenson RM. Monoamine systems and memory formation. In: Andrew RJ, ed. Neural and Behavioural Plasticity: The Use of the Domestic Chick as a Model. New York: Oxford University Press 1991:419-439.
68. Stephenson RM, Andrew RJ. Amnesia due to β-antagonists in a passive avoidance task in the chick. Pharmacol Biochem Behav 1981; 15:597-604.
69. Sternberg DB, Isaacs KR, Gold PE et al. Epinephrine facilitation of appetitive learning: Attenuation with adrenergic receptor antagonists. Behav Neural Biol 1985; 44:447-453.
70. Sullivan RM, Stackenwalt G, Nasr F et al. Association of an odor with activation of olfactory bulb noradrenergic β-receptors or locus coeruleus stimulation is sufficient to produce learned approach responses to that odor in neonatal rats. Behav Neurosci 2000; 114:957-962.
71. Summers RJ, McMartin LR. Adrenoceptors and their second messenger systems. J Neurochem 1993; 60:10-23.
72. Summers RJ, Papaioannou M, Harris S et al. Expression of β_3-adrenoceptors in rat brain. Brit J Pharmacol 1995; 116:2547-2548.
73. U'Prichard DC. Biochemical characteristics and regulation of brain alpha 2-adrenoceptors. Ann N Y Acad Sci 1980; 430:55-75.
74. Watts ME, Mark RF. Separate actions of ouabain and cycloheximide on memory. Brain Res 1971; 25:420-423.
75. Williams CL, Men D, Clayton EC et al. Norepinephrine release in the amygdala after systemic injection of epinephrine or escapable footshock: Contribution of the nucleus of the solitary tract. Behav Neurosci 1998; 112:1414-1422.

CHAPTER 2.8

Histamine

Rüdiger U. Hasenöhrl and Joseph P. Huston

Abstract

Histamine is an important but largely neglected modulator in the central nervous system. Histaminergic neurons are located exclusively in the posterior hypothalamus, the tuberomammillary nucleus (TM), from where they project to almost all brain regions, with ventral areas (hypothalamus, basal forebrain, amygdala) receiving a particularly strong innervation. Here we summarize behavioural data based on TM-lesions as well as on electrophysiological, neurochemical and pharmacological studies related to histamine agonists and antagonists. The outcome of these studies provides evidence that the brain histamine system is a) involved in neural plasticity and functional recovery following unilateral damage of the brain and b) may subserve inhibitory functions in the control of reward and learning processes.

Introduction

In recent years, evidence has accrued from neurochemical, electrophysiological and pharmacological studies that histamine functions as a central neurotransmitter and/or neuromodulator.[43,106,120] The characteristic distribution of histamine and the presence of specific histamine receptive sites in the brain[121] underlined such a possibility and incited further investigation to unravel the role of this biogenic amine in brain function (see ref. 14 for review). Much attention has been focused on the effects of histamine or its agonists and antagonists on various (electro)physiological parameters and behaviours.[41,51,150] The existence of a histaminergic neuronal system was, however, disputed until immunocytochemical studies revealed the existence of histaminergic neurons in the brain,[98,128,145] which allowed a more precise anatomical and neurochemical analysis of this neuronal system beyond the pharmacological approach. This chapter will start with a short description of some of the properties of the neuronal histamine system, its receptors and its chemoarchitecture in the brain; then we present a summary of experiments, which investigated a possible involvement of the TM histamine system in neural plasticity, reinforcement and memory functions.

The Histaminergic Neuron System

The presence of histamine can be demonstrated in two major pools—in neurons as well as in mast cells, which, however, are relatively scarce in the brain.[35,128] Histamine is synthesized in a single step from L-histidine by the enzyme histidine decarboxylase (HDC). Up to now, no high-affinity uptake system for histamine has been reported and termination of its action in brain appears to require catabolism to telemethylhistamine, which is further metabolised by MAO.[120] Inhibition of histamine synthesis can be achieved by alpha-fluoromethylhistidine, an irreversible inhibitor of HDC,[70] which is frequently used as a research tool with which to investigate the functional role of neuronal histamine.[94]

Histaminergic neurons are exclusively located in the posterior hypothalamus, specifically in the tuberomammillary (TM) nucleus (Fig. 1). Fibres arising from the TM constitute two ascending pathways: one laterally, via the medial forebrain bundle, and the other periventricularly.

From Messengers to Molecules: Memories Are Made of These, edited by Gernot Riedel and Bettina Platt.

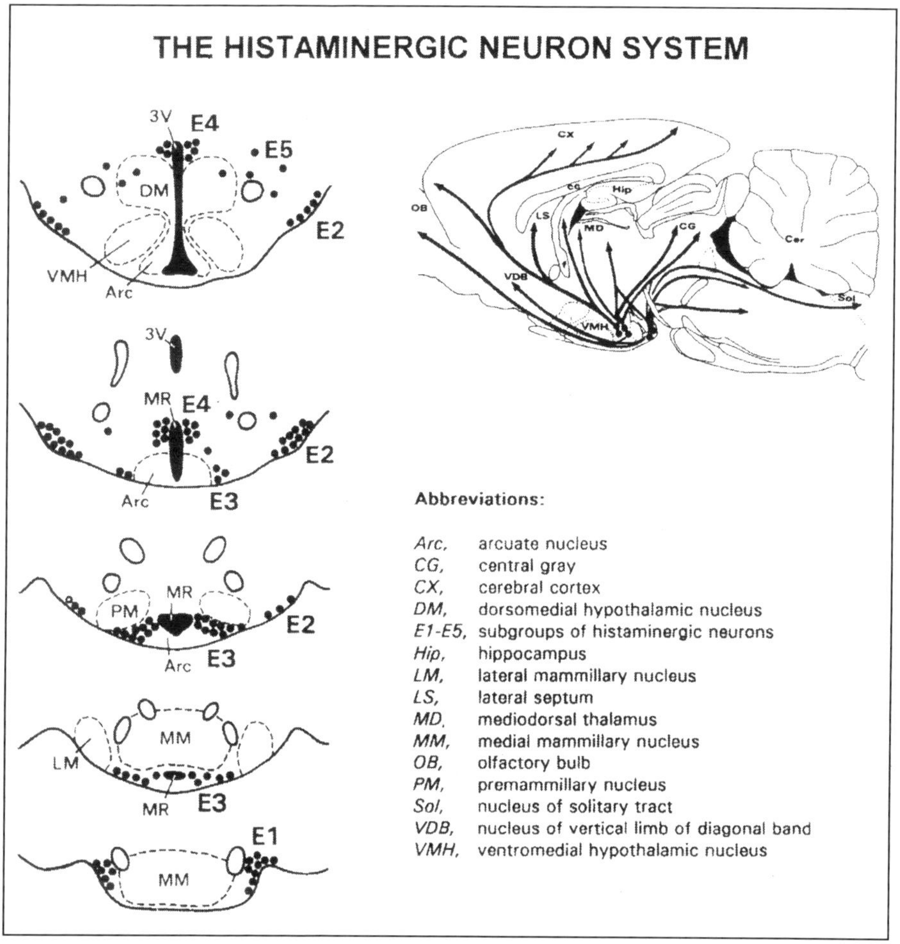

Figure 1. Left: A series of schematic drawings of frontal sections through the posterior part of the hypothalamus of the rat, illustrating the topographical localization of subgroups E1-E5 of histaminergic neurons. (Modified after ref. 142). Right: Schematic diagram of histaminergic pathways in the rat brain. (Modified after ref. 120)

These two pathways combine in the diagonal band of Broca to project to many telencephalic areas. The projections from the TM to the various brain regions are bilateral with ipsilateral predominance and, curiously, no differences have been reported so far in the projection sites of the cell bodies of the respective subgroups, E1 to E5.[58] In most brain areas histamine is released from varicosities, mostly at non-synaptic sites, indicating modulatory functions similar to those found for other biogenic amines.[142]

The diverse actions of the histaminergic neuronal system appear to be mediated by at least three classes of receptors, namely H_1, H_2 and H_3, which differ in pharmacology, localization and the intracellular response they mediate.[50,75,121] The H_1 receptor was initially defined in functional assays and by the design of potent antagonists, the so-called 'antihistamines'. The widespread distribution of H_1 receptive sites in areas involved in arousal, such as thalamus, cortex and cholinergic cell groups in tegmentum and basal forebrain, possibly accounts for the sedative properties of most H_1 antagonistic compounds. High densities of H_1 receptors are also found in hypothalamus, septum, nucleus accumbens and in several hippocampal areas.[11,97] It is interesting to note that several drugs used in the treatment of psychiatric disorders, such as tricyclic antidepressants and neuroleptics, also have significant H_1 receptor blocking ability.[50] The finding that most of the peripheral actions of histamine cannot be blocked by classical antihistamines led to the proposal of an additional class of histamine receptors.[6] This second subtype was validated pharmacologically by Black and co-workers[9] and designated the H_2 receptor. Like the histamine H_1 receptor, the H_2 type has a widespread distribution in brain and spinal cord with high densities in basal ganglia, hippocampus and amygdala; unlike the H_1 receptors, H_2 receptive sites are present in low density in septal areas, hypothalamic and thalamic nuclei.[103] Furthermore, H_1 and H_2 receptors show partial overlap in several brain regions, including hippocampus, nucleus coeruleus, ventral tegmental area and substantia nigra, where the receptors can interact in a synergistic manner. A third histamine receptor subtype, H_3, has received much interest in recent years, as this receptor was initially detected as an autoreceptor mediating feedback inhibition of histamine synthesis and release.[4,5] High concentrations of H_3 receptors are found in neostriatum, nucleus accumbens and in cingulate/infralimbic cortices,[104] whereas its density is relatively low in hypothalamus, which contains most of the histaminergic axons and perikarya in the brain. This distribution pattern suggests that the majority of H_3 receptors are not autoreceptors. Actually, H_3-receptors can also function as heteroreceptors, modulating the activity of other monoaminergic,[118] glutamatergic[12] and peptidergic systems.[81] Recently, a fourth histamine receptor has been cloned and characterised.[91] This H_4 receptor is primarily found on intestinal tissue and immune active cells and, thus, differs markedly from the H_3 receptor, whose expression seems to be restricted to brain.[77,78]

In spite of many different suggestions mainly derived from observations of responses to locally applied histamine or related compounds, only few physiological roles of the histaminergic neuronal system are relatively well documented.[14,45,56] Recent research emphasis has been placed on the possible role of neuronal histamine in the control of the waking state[86,122] and circadian rhythmicity,[15,139] in autonomic[37] and neuroendocrine processes,[68] in the regulation of seizure susceptibility,[60,154,155] in motivated behaviours like feeding and drinking,[72,89,138] in affective processes such as fear/anxiety,[59,114] in neuropsychiatric disorders,[28,90,107] in learning and memory processes[96,102] and in the control of reward or reinforcement.[49,108] Furthermore, histamine was found to promote survival of developing hippocampal tissue[23] and to alleviate neuronal damage subsequent to experimental brain lesion,[1,32] and may, therefore, be important for processes related to neurogenesis and neuronal functional recovery.

The Role of the Tuberomammillary Nucleus Projection System in Neural Plasticity and Functional Recovery

Swanson[135] was one of the first to report who reported extrahypothalamic projections of the TM. Specific crossed and uncrossed projections from the TM to the caudate putamen were then described by Watanabe et al[145] with immunohistochemical methods and by Steinbusch et al[129] with retrograde tracing by fluorescent dyes. We confirmed these findings by using the horseradish peroxidase (HRP) tracing technique, and extended them by showing neuroplastic changes in tuberomammillary-striatal projections in relation to recovery from behavioural asymmetries induced by hemivibrissotomy[147] and unilateral 6-hydroxydopamine (6-OHDA) lesion of the substantia nigra (SN). Hemivibrissotomy, which stands for the removal of the tactile hairs (vibrissae) on one side of the rat's snout, induces a transient asymmetry in the side of the face used to scan the wall while traversing the edge of an open field (i.e., 'thigmotactic scanning'; see Fig. 2A) from which rats recover over time.[55] Time-related to these behavioural changes we found an increase in 'strength' (i.e., in structure and/or activity) in the uncrossed and crossed projections from the TM to the caudate nucleus. Rats examined four to twenty days after unilateral clipping of vibrissae had more HRP-labelled cells in the crossed and uncrossed projections from the TM nuclei to the caudate nucleus on the side of intact vibrissae (i.e., to the hemisphere deprived of vibrissal sensory input) compared to projections to the caudate nucleus on the side of vibrissae removal (Fig. 2B). The neuronal asymmetries in the

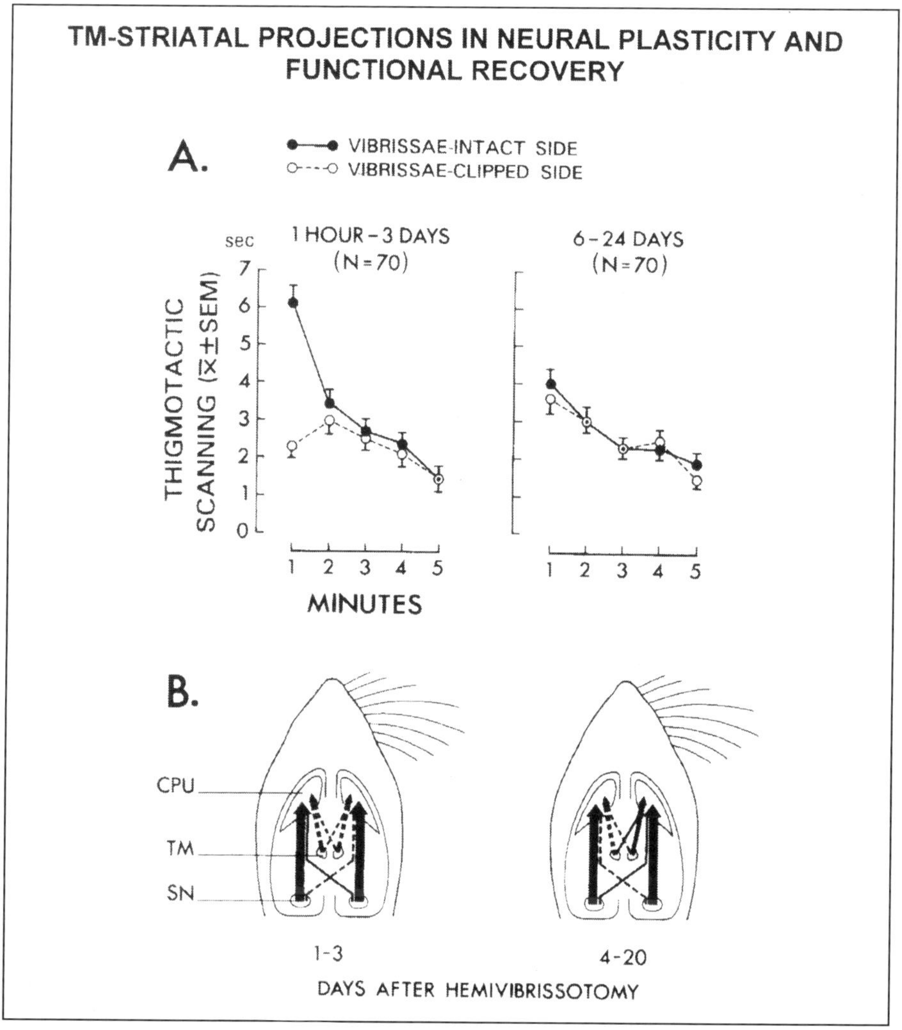

Figure 2. A) Duration (mean±SEM, in seconds) of thigmotactic scanning with the vibrissae-intact side (continuous line) and the vibrissae-clipped side (broken line) through a 5-min test session. Rats tested between 1 h and 3 days after hemivibrissotomy exhibited a strong asymmetry in scanning during the first minute of testing, as they scanned more with the vibrissae-intact side (left). This behavioural asymmetry was absent in rats tested 6 days, or later, after clipping the vibrissae (right). B) Corresponding neuronal changes in the tuberomammillary-striatal projections 1-3 days (left) and 4-20 days after hemivibrissotomy (right) in comparison with the findings in the nigro-striatal projections. (Data from ref. 55 and 147)

TM-striatal projections were in the same direction as the asymmetries previously found in nigro-striatal projections after hemivibrissotomy.[131,132] However, unlike in the nigro-striatal pathway, apparent neuronal asymmetries in the tuberomammillary-striatal projections were only evident during the period when the rats had recovered from the behavioural asymmetry. Given the coincidence of changes in striatal afferents from the SN and from the TM, both being correlated in time with recovery from behavioural asymmetries after hemivibrissotomy, it is conceivable that an interaction between histamine and dopamine (DA) could play a role in the control of compensatory processes and recovery of function. In line with this suggestion, nigrostriatal DA denervation was found to induce a marked up-regulation of H_3-receptors in the striatum, which was reduced by dopamine D_1 receptor stimulation.[112] Furthermore, partial destruction of the TM resulted in increased DA and serotonin levels in neostriatum,[79] corroborating the notion that histamine has an inhibitory impact on striatal monoamine activity under physiological conditions.[117] Taken together, these data provided first evidence for a role of the tuberomammillary-striatal system in behavioural plasticity subsequent to unilateral removal of the vibrissae, in concert with the nigro-striatal system.

Based on these anatomical findings we were interested in a behavioural correlate of a lesion in the TM region. Therefore, we investigated the influence of a unilateral direct current (DC) lesion in the TM region on thigmotactic scanning behaviour. Destruction of the TM region was found to produce more thigmotactic wall scanning behaviour with the vibrissae contralateral to the lesion; the histamine precursor histidine reversed the effects of the TM lesion, suggesting that histamine is involved in this effect.[148] In contrast, a unilateral 6-OHDA lesion of the SN produced more wall scanning behaviour with the vibrissae ipsilateral to the lesion.[130] The finding that lesions in the SN and TM have opposite effects on scanning behaviour suggests that the projections (perhaps to the striatum) could represent a reciprocally acting regulatory system in terms of sensorimotor processes, possibly involving DA and histamine. In accordance with the idea of a reciprocal relationship between the TM system (histaminergic) and the SN (dopaminergic) is the finding that functional recovery from a unilateral 6-OHDA lesion of the SN was associated with an enhancement of the nigro-striatal projections,[87] whereas in rats that failed to recover from the nigral lesion, an enhancement of the TM-striatal projections (based on the extent of HRP-labelling in TM and SN after HRP injection into the caudate-putamen region) was observed.[88] These findings suggest that the increase in HRP-labelling seen in the tuberomammillary-caudate projections indicates an enhancement of histaminergic activity, which, in turn may be related to the lack of recovery from a unilateral SN lesion, and to the increase in asymmetry that develops over time in such animals.

The Role of the Histaminergic Neuronal System in the Control of Reinforcement

A number of pharmacological studies have examined the role of histamine in reinforcement processes. For example, the self-administration of histamine and histamine-blocking compounds has been evaluated.[7,113] The injection of histamine and histamine antagonists was also studied in combination with rewarding brain stimulation[22,110,140,146] and in drug-discrimination tests.[36] Their effects were assessed on operant behaviour[8,84,136] and in conditioned place preference tasks, either alone[82,134] or in combination with stimulants[80] and opioids.[66,133] The results of these experiments provided evidence that histamine agonists may have aversive properties, whereas histamine antagonists, particularly those blocking the H_1-receptor, can exert reinforcing as well as reward potentiating effects.

The TM nucleus itself has largely been neglected in the search for the neural mechanisms underlying reinforcement. Some studies, in which the hypothalamic region was mapped for reinforcing properties of electrical stimulation reported negative or ambivalent stimulation effects in the posterior hypothalamus, the region where the TM is located.[93] Given the evidence for a role of the TM projections in neural plasticity and functional recovery and the proposed reciprocal relation between histaminergic and dopaminergic mechanisms (see above), a series of experiments was performed to examine the possible involvement of the TM in the brain's reinforcement system. In the first experiment, the effects of an electrolytic lesion in the rostroventral part of the TM (E2-region) on lateral hypothalamic self-stimulation were assessed.[143] From the second day post-lesion, the response rate gradually increased in TM-lesioned animals and peaked on day thirteen in the ipsilateral hemisphere only. Although there was no further increase over subsequent days, response rates remained elevated during the following seven weekly tests (Fig. 3A). Since electrolytic lesions lead to general tissue damage, it was not possible to pinpoint with certainty that the TM neurons were responsible for the observed

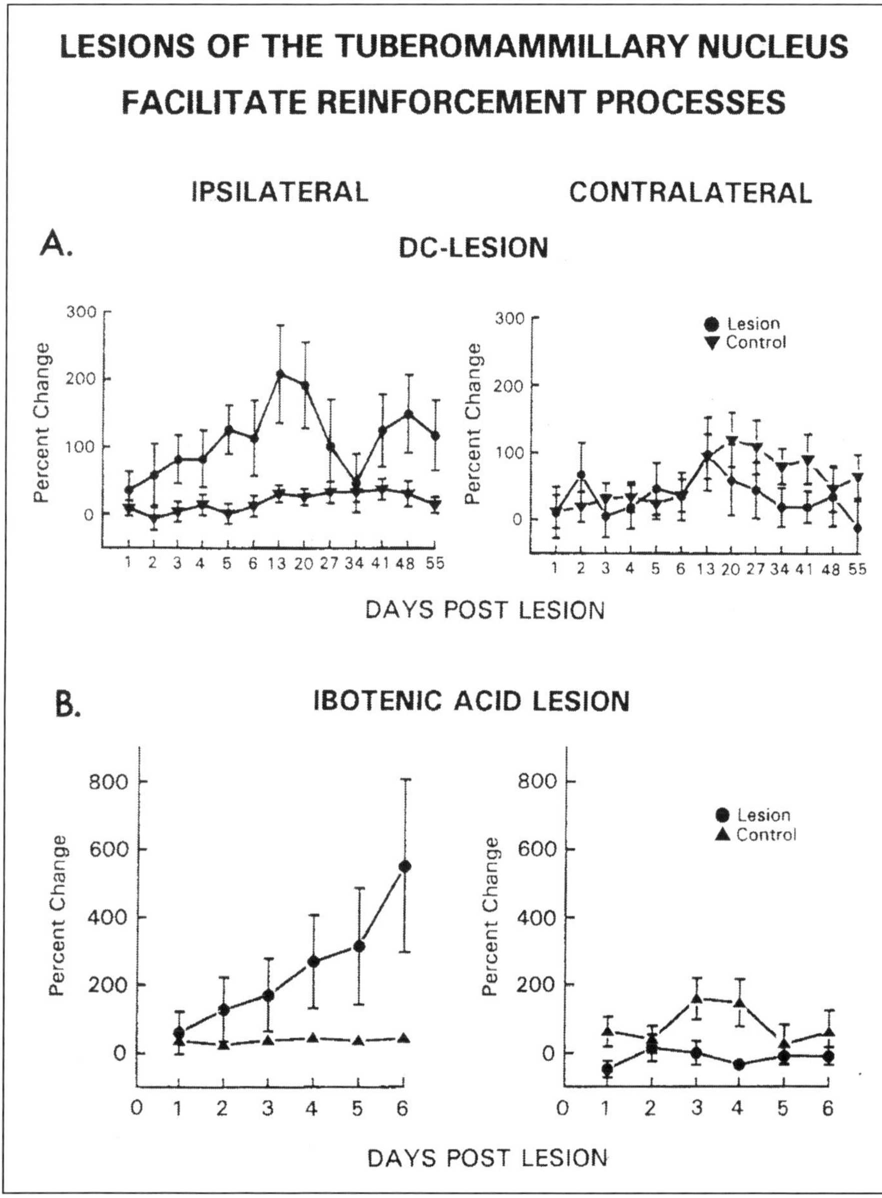

Figure 3. Lateral hypothalamic self-stimulation ipsi- (left) and contralateral (right) to the side of a unilateral electrolytic (A) or ibotenic acid lesion (B) in the region of the tuberomammillary nucleus. Rats were implanted bilaterally with stimulating electrodes in the lateral hypothalamus and unilaterally with one lesion electrode/injection cannula in the TM area. Following three days of baseline testing, one half of the animals were given an electrolytic or excitotoxic TM lesion. Response rates are expressed as mean (±SEM) percentage of corresponding baseline values. (Data from refs. 143 and 144)

'disinhibition' of reinforcement in this experiment. Thus, another study was performed to determine, whether the observed increase in response rate was due to the destruction of intrinsic TM neurons or to the destruction of passing fibres.[144] Therefore, lateral hypothalamic self-stimulation was examined following a unilateral TM-lesion with ibotenic acid, which destroys cell bodies but can spare fibres of passage.[16,119] Figure 3B shows increasing response rates obtained from the hemisphere ipsilateral to the excitotoxic TM lesion. As in the previous experiment, there were no changes in rate when the animals stimulated themselves in the lateral hypothalamus contralateral to the lesion, and therefore, an interpretation of the rate increase in terms of an unspecific enhancement of general arousal can be ruled out.[76] Since the response curves revealed in both studies were very similar, it can be concluded that the destruction of TM-intrinsic neurons was critical for the effects, rather than the denervation of remote structures induced by damaged fibres of passage. Furthermore, it is important to note that in both experiments facilitation of self-stimulation only occurred after destruction of the E2- but not of the E1-subgroup of the TM. This dissociation can be considered as the first indication for a functional specificity of a cell population within the TM, which until now has been defined on anatomical grounds only.

To establish whether the observed effects following TM lesions involve a histaminergic component, pharmacological studies were performed to investigate the effects of different histamine-receptor blocking drugs in the nucleus accumbens (NAcc) and the nucleus basalis of the ventral pallidum.[108,157] Both structures are known to play an important role in reinforcement-related processes[53,71] receive specific histaminergic input[99,129] and contain all three histamine receptor subtypes.[121] In order to assess the effects of histamine receptor blockade on reinforcement, the administration of the histamine antagonists was either combined with lateral-hypothalamic self-stimulation, or their effects were examined with the corral version of the conditioned place preference paradigm.[46] In the NAcc, the administration of the H_1-blocking drug chlorpheniramine produced a lateralised increase of hypothalamic self-stimulation and was effective in inducing a conditioned corral preference, indicative of a positively reinforcing action. Furthermore, the effects of chlorpheniramine were found to be restricted to the caudal part of the NAcc, since injection of the H_1-antagonist into the rostral NAcc did not affect the behaviour in either paradigm.[157] In the ventral pallidum, chlorpheniramine as well as the H_2-antagonist ranitidine were tested for possible reinforcing effects by the use of the corral method.[108] The results are summarized in (Fig. 4): A single intrabasalis injection of chlorpheniramine increased the sojourn time in the corral previously paired with the drug treatment in a dose-dependant manner, indicative of a reinforcing action of the H_1-antagonist. In contrast, the H_2-antagonist ranitidine did not significantly influence the preference behaviour within the entire dose range tested.

Taken together, the outcome of this series of lesion studies and pharmacological experiments suggest that the TM and its histaminergic projections (specifically the histaminergic efferents to basal forebrain) exert inhibitory effects on reinforcement under normal conditions. Reducing histaminergic activity either by a partial destruction of TM-intrinsic histamine neurons or by inhibiting histaminergic transmission at H_1-receptive sites apparently results in a disinhibition of reinforcement. The described inhibitory function of the TM in the control of intracranial self-stimulation and the effects of histaminergic agonists and antagonists on various measures of reinforcement stand in sharp contrast to the effects of DA on reinforcement. It is widely accepted that DA agonists facilitate and DA antagonists inhibit brain stimulation reward.[151] Thus, DA seems to influence the brain's reinforcement system in a way, which is again reciprocal to histamine. The brain's reinforcement mechanism or mechanisms can be considered as being activated in a tonic fashion by DA and histamine, with the further promoting, and the latter diminishing reinforcement, i.e., the effectiveness of a reinforcing stimulus to increase the probability of recurrence of a preceding operant behavior, as also evidenced by changes in the organism's degree of 'preference for' that stimulation or for place cues that have been associated with such reinforcing stimulation.

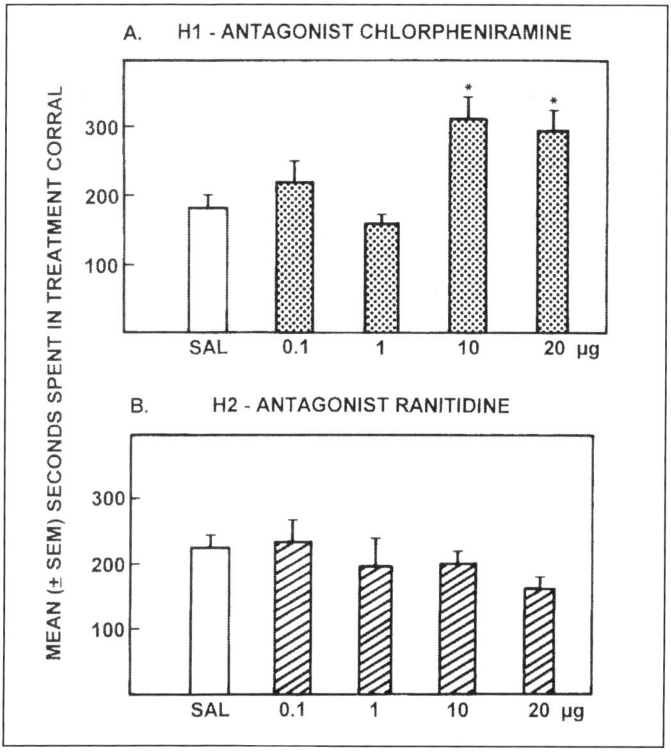

Figure 4. Mean (+SEM) time in seconds spent in the treatment corral during test for conditioned corral preference. The corral apparatus was a circular open field, which could be divided into 4 quadrants (corrals) of equal size, identical floor and wall texture, and identical colour. Spatial orientation inside the apparatus was provided by external cues located in the surroundings. During the conditioning session, the animals received a single injection of different doses of (A) the H_1-antagonist chlorpheniramine, (B) the H_2-antagonist ranitidine, or vehicle (SAL; 0.5 µl) into the ventral pallidum, after which they were confined for 15 min to one quadrant of the apparatus (treatment corral). During test for conditioned corral preference, the undrugged animals were again placed into the corral for 15 min and the time spent in each quadrant was scored (open corral). *$p<0.05$, significantly different from vehicle controls; Mann-Whitney U-test, two-tailed. (Data from ref. 108)

The Role of the Histaminergic Neuronal System in the Control of Learning and Mnemonic Processes

The role of the histaminergic system in learning and memory has been generally investigated pharmacologically, with contradictory results.[95,101] For example, histamine was reported to improve inhibitory and active avoidance conditioning,[24,65] whereas administration of H_1-antagonists disrupted learning in an active avoidance task.[61,64] Both histamine and acetylcholine reversed the impairing effects of H_1 receptor antagonist injection,[61] suggesting an interaction between the central histaminergic and cholinergic system in learning. Thioperamide, a histamine H_3-antagonist, was found to improve the retention performance of adult[105] and senescence-accelerated mice,[85] whereas H_3-agonists such as imetit or (R)-alpha-methylhistamine produced learning disruption.[10] Furthermore, histamine was reported to improve memory retrieval in old and hippocampus-lesioned rats.[62,63] On the contrary, histamine has been shown to reduce active avoidance responding, an effect mediated via the H_1-receptive site,[137] and long-term depletion of neuronal histamine by alpha-fluoromethylhistidine proved to be effective in facilitating active avoidance[3] and radial maze learning (ref. 113, but see ref. 20). Microinjection of histamine into the dentate area and subiculum complex was reported to adversely affect active avoidance conditioning via histamine H_1-receptive sites.[3] Furthermore, the H_3-receptor agonist (R)-alpha-methylhistamine was found to improve navigation performance in the Morris water maze task.[111,124]

The reasons for these discrepant findings require clarification; however, the main problem with the reliability of the data may lie in the effectiveness/specificity of the histaminergic drugs tested, rather than, for example, in the use of different learning tasks, modes of injection and variation in the time of injection in relation to the learning trials, etc. Moreover, the exact functions of the histamine receptor subtypes remain to be determined. For example, mutant mice lacking the H_1-receptor showed reduced aggressive and exploratory behaviours but no apparent change in learning capacities.[152,153] Furthermore, although functionally characterized as an inhibitory autoreceptor, the histamine H_3-receptor is not restricted to presynaptic elements of the histaminergic neurons, but can also function as a heteroreceptor modulating the activity of several other transmitter system.[115,116]

Lesion Studies

Our strategy in investigating the role of the histaminergic neuronal system in learning and mnemonic processes was guided by a theory of reinforcement,[54] which proposes that reinforcers 'strengthen' behaviour by preventing memory traces from fading out and therefore leading to learning (consolidation). Based on this theory, one aim of the present studies was to examine possible effects of lesions in the TM region on learning and memory processes in adult rats. Given the parallelism between reinforcing and memory-promoting effects of manipulations of the brain, it was hypothesized that lesion of the TM region could have a facilitatory effect on learning and mnemonic processing in addition to its facilitatory effect on reinforcement processes. Furthermore, it was asked whether TM lesions might exert a beneficial action on the performance of aged rats, which are considered an animal model for learning and memory disturbances related to aging and nervous system disorders like Alzheimer's disease.[33]

In the first series of experiments,[30,67] adult and aged rats with a bilateral electrolytic lesion in the TM region were tested along with sham-lesioned controls in a set of learning tasks, which differed in terms of complexity and reward contingencies (habituation of exploratory activity, inhibitory avoidance retention, discrimination learning). An improvement was found in every test applied, indicating that TM lesions can generally enhance learning and memory capacities independent of the special demands of a given task. Moreover, age-related learning deficits were strongly diminished by the lesion (Fig. 5A-C). The fact that habituation learning was improved is important. For one, this test of memory, as assessed by behavioural habituation, does not involve application of conventional reinforcers, such as food or escape from or avoidance of aversive stimulation, and secondly, the TM histaminergic system is thought to play a role in stress, perception of pain, and thermoregulation,[51,95] which are important factors in aversive conditioning, but not for habituation. This makes an interpretation of the performance enhancement following TM lesion simply in terms of an interaction between the lesion and physiological processes induced by a punishing/aversive stimulus, unlikely. Based on these initial findings, the objectives of a follow-up study were two-fold.[30] In order to determine whether the facilitation of learning and memory was due to the destruction of intrinsic TM neurons, adult and aged rats with bilateral ibotenic acid lesions of the TM region were tested along with vehicle-injected controls in the Morris water maze, in which old rats display marked performance deficits.[47] In addition, the number of histamine cells was determined at the site of the neurotoxic lesion by immunohistochemistry using specific antibodies against the amine.[128] The main finding of this study was that adult and aged rats with neurotoxic lesions of the TM showed accelerated navigation performance in the course of place learning in the maze and an improved ability to locate the platform site during a spatial probe trial (Fig. 6). Inspection of the site of the ibotenic acid microinjection in the TM region revealed a marked decrease of

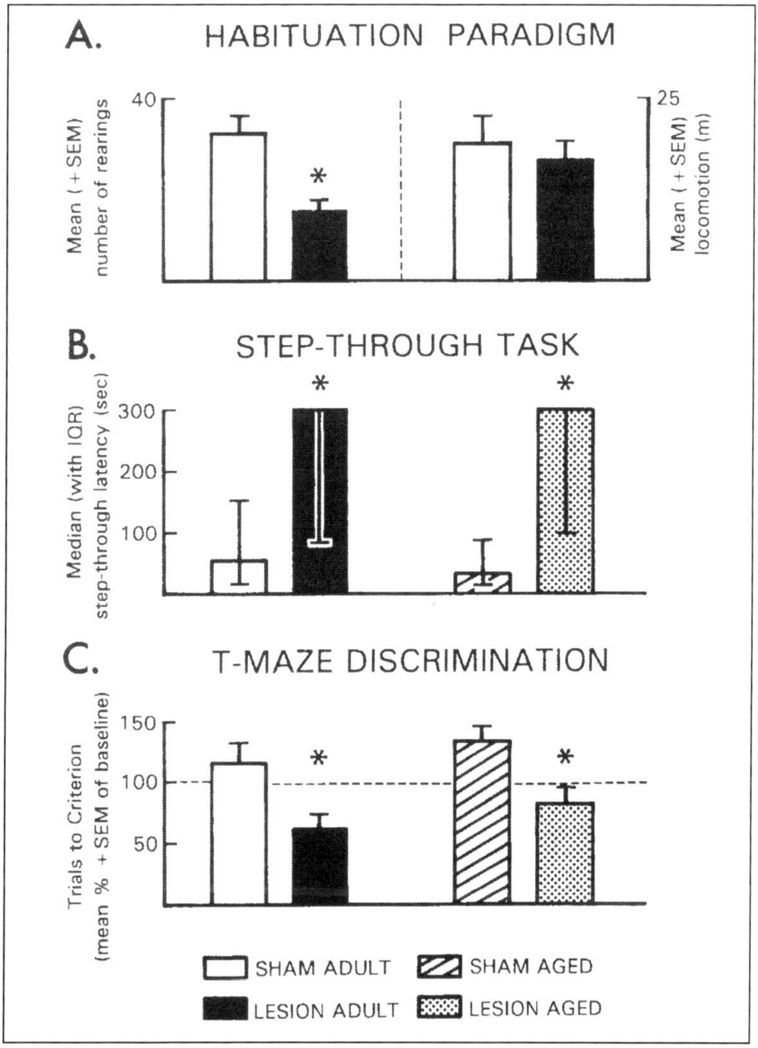

Figure 5. The effects of a bilateral DC lesion in the TM region on performance of adult (3-month-old) and aged (31-month-old) rats in different learning tasks. (A) Habituation paradigm: Mean (+SEM) number of rearing (left) and mean (+SEM) distance travelled in the open field (right) during test for habituation learning. Habituation was measured in an open field by recording the number of rearing and square crossings during 5 min of free exploration in one baseline and one test session with a baseline-test interval of 7 days (aged rats had to be disqualified from this task because baseline activity was too low to determine possible effects of the TM lesion on habituation). (B) Step-through task: Median (with interquartile range) step-through latency revealed in the retention test. Immediately after the rat had entered the dark compartment in the third familiarization trial, a foot shock was applied (training). Retention of the step-through response was measured 24 hours after shock administration with a 300 s cut-off. (C) T-maze discrimination: During one baseline trial, the rats were trained to swim to the end of the left or right arm of the maze in order to escape. After a training-test interval of 7 days, the animals were retrained on the same task to the same criterion. The figure depicts the number of trials required to reach the criterion of 5 successive correct choices in the T-maze during the retention test. Trials to criterion during retention test are expressed as mean (+SEM) percentage of corresponding baseline values (= 100%, dashed line). *$p<0.05$ vs. sham-lesioned controls, indicative of enhanced learning (Data from refs. 30 and 67)

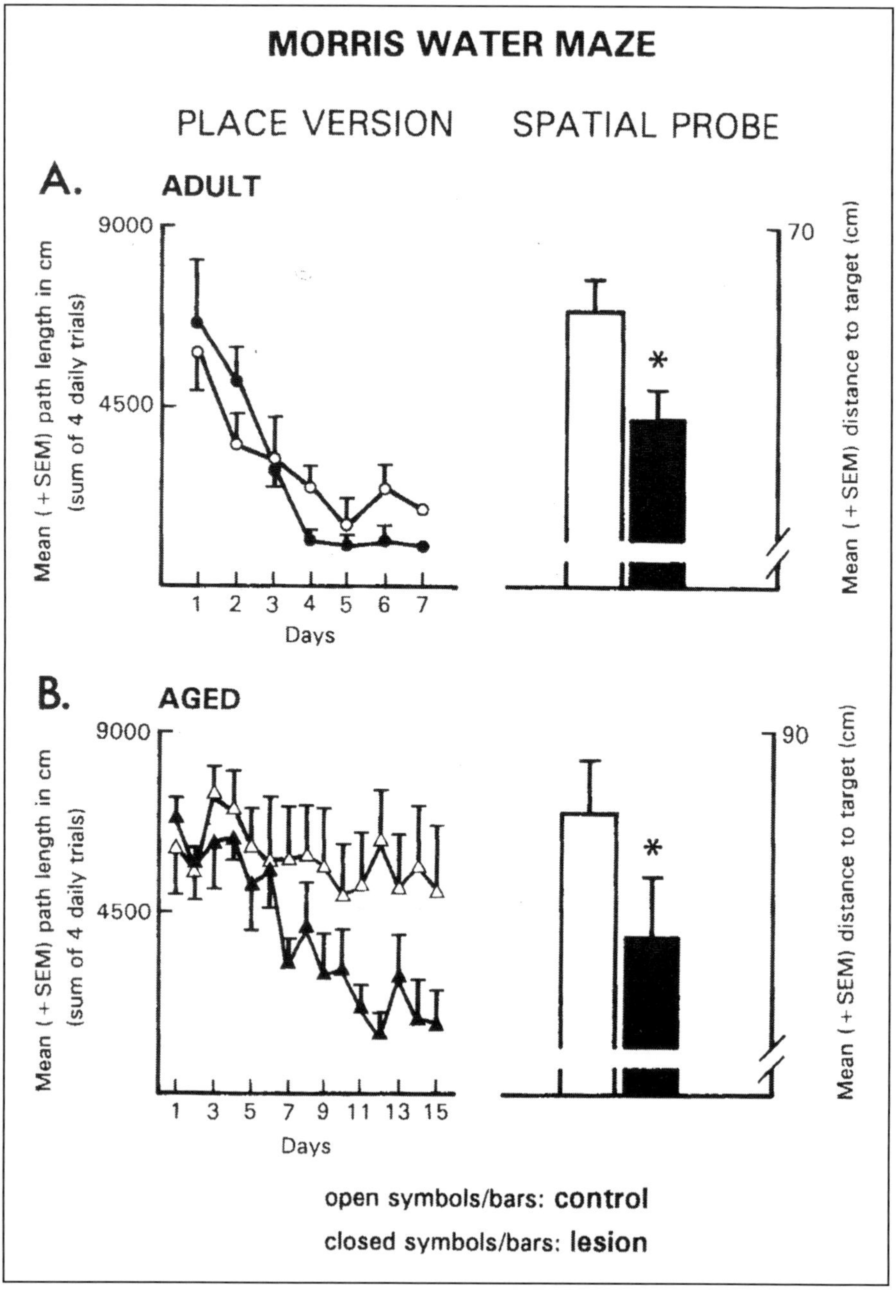

Figure 6. The effects of a bilateral ibotenic acid lesion in the TM-region on the navigation performance of (A) adult (3-month-old) and (B) aged (28 to 31-month-old) rats in the Morris water maze. Left: Mean (±SEM) path length to find the hidden platform in the place version of the maze. Right: Mean (+SEM) distance to target during a spatial probe trial without platform. $^{*}p<0.05$ vs. vehicle-injected controls. (Data from ref. 30)

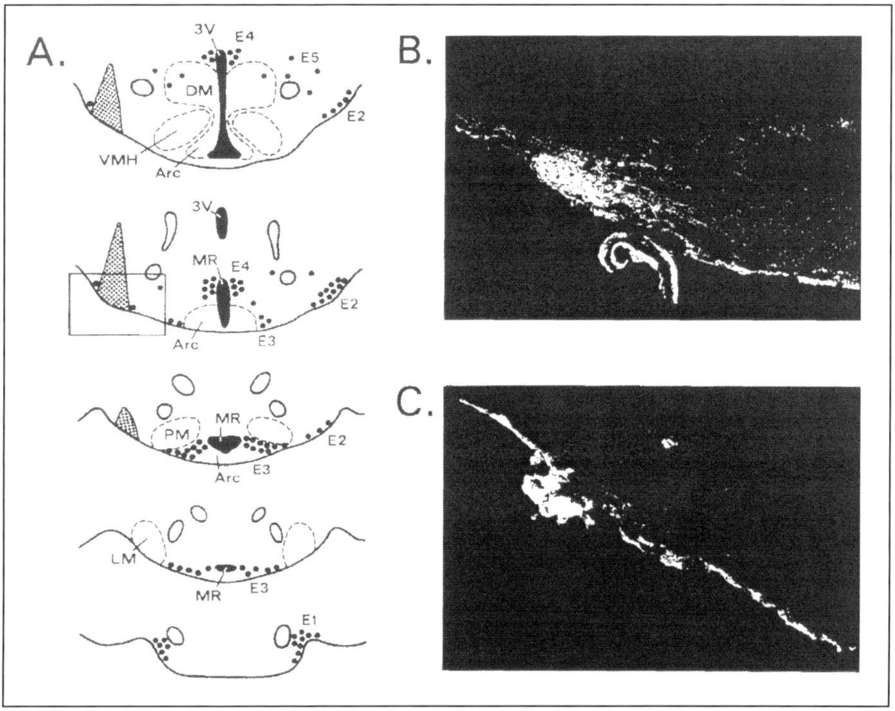

Figure 7. Effects of ibotenic acid injections into the E2-subregion of the TM on histamine-positive neurons of the TM. (A) Schematic drawings of frontal sections through the posterior hypothalamus illustrating the location of subgroups of histaminergic neurons (E1-E5) in the TM region (Modified after ref. 142). The dotted area indicates the extent of a representative lesion induced by ibotenic acid, based on cresyl violet staining. Box indicates the location from where the photomicrographs were taken. (B) and (C): Photomicrographs of coronal sections showing histamine immunoreactive neurons and fibres in the E2-subgroup of the TM two weeks after injection of (B) vehicle or (C) ibotenic acid; magn. x 200. (from ref. 30.)

histamine-staining neurons mainly in the rostral part of the TM (Fig. 7), suggesting that the facilitatory effects on maze navigation observed after TM lesion might be related to reduced histaminergic activity produced by a partial destruction of TM-intrinsic histaminergic cells. It is important to note that the facilitatory effects on learning and memory parameters produced by irreversible TM lesions could be mimicked by a transient inactivation of this brain region with the short-acting local anaesthetic lidocaine.[31] This suggests, that the beneficial effects on reinforcement and memory observed after permanent TM lesion were not a function of long-term compensatory processes of the brain, but, instead, a direct result of an inhibition of (histaminergic) TM activity induced by the lesion.

Pharmacological Approach

Congruent with the outcome of the lesion studies are the results from our pharmacological experiments dealing with the effects of histamine antagonists on different aspects of learning, which are also suggestive for an inhibitory action of the biogenic amine. Thus, we found that the histamine H_1 receptor antagonist chlorpheniramine, but not the H_2 antagonist ranitidine, can exert memory-promoting effects when administered into nucleus accumbens or ventral pallidum.[49,109] The peripheral injection of chlorpheniramine improved appetitive learning in goldfish[125] and the compound ameliorated learning deficits in behaviourally impaired old rats

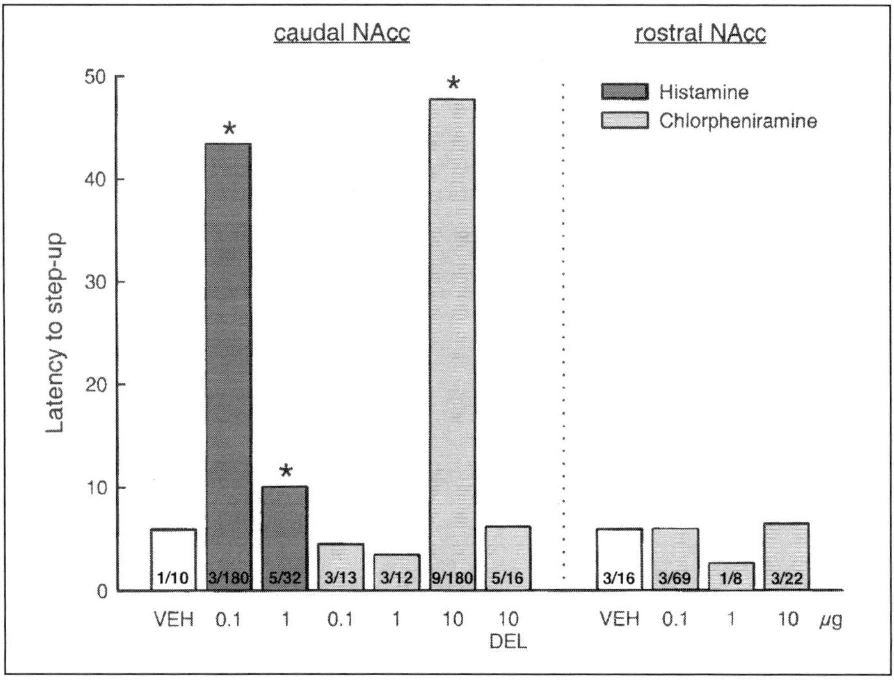

Figure 8. Effect of intra-accumbens histamine and chlorpheniramine injection on the performance of the uphill avoidance task, which involves punishment of a high-probability turning response on a tilted platform.[127] Immediately after the learning trial, that is, after a shock was administered upon performing the response, different doses of chlorpheniramine and histamine were injected unilaterally into caudal or rostral NAcc. Retention is expressed as median latency (s) to step-up, measured 24 h after shock administration with a 180 s cut-off. Numbers at the bottom of the histograms indicate 25./75. percentiles. Controls included vehicle-injected (VEH; 0.5 μl) rats and a group administered 10 μg chlorpheniramine 5 h after training (DEL). *$p<0.05$ vs. vehicle controls, indicative of enhanced learning. Note: The failure of the delayed post-trial injection of the antagonist to influence learning indicates that the compound influenced learning by modulating early memory storage processes, rather than by acting on performance variables during retrieval of the task or by influencing memory consolidation going on 5 h after learning or later. (from ref. 49)

when administered systemically[29] or into the ventricles.[48] Furthermore, neurochemical experiments with in vivo microdialysis in anaesthetized rats revealed that peripheral administration of chlorpheniramine can produce increases in extracellular levels of acetylcholine in the cortex and dopamine in nucleus accumbens,[25,26] confirming that the memory enhancing and reinforcing effects of H_1 antagonists involve dopaminergic[134] and cholinergic mechanisms.[124] However, histaminergic antagonists, particularly those acting at H_1 receptors, have been seriously questioned with regard to their selectivity, since they can bind also to receptors other than histaminergic ones.[73,92,123] Hence, we compared the H_1 antagonist chlorpheniramine with histamine in their effects on learning following injection into different subregions of the NAcc. With respect to the proposed inhibitory function of histamine in reward-related processes (see above), we expected that accumbal injection of the biogenic amine should produce a behavioural profile distinct from or even opposite to that obtained after intra-NAcc administration of the H_1 receptor antagonist; that is, histamine injection should be non-rewarding or even interfere negatively with reward and memory processes.[49] However, incongruent with this premise intraaccumbens injection of both chlorpheniramine as well as histamine produced a conditioned place preference and facilitated negative reinforcement learning when administered post-trial (Fig. 8). These effects were evident only when drug infusion was performed into the caudal-shell but not into the rostral subregion of the NAcc, providing further evidence for behavioural relevance of the known histaminergic innervation of this brain region with a functional subdivision on its rostrocaudal axis.[157] Interestingly, it was shown that locally administered histamine can increase DA levels in the NAcc, whereby the histamine-induced DA release could be blocked by peripheral administration of the H_1-antagonist pyrilamine, which itself was found to decrease DA levels after local intraacumbens injection.[34] These findings suggest that histamine and H_1 antagonists could be operative via quite different neurochemical mechanisms within the NAcc, which, however, can produce a quite similar behavioural profile. Thus, it seems that the promnestic and reinforcing effects of chlorpheniramine involve pharmacodynamic aspects beyond its antagonistic activity at H_1-receptive sites. Furthermore, these data imply that the behavioural changes observed after manipulations of the TM-histamine system may not necessarily be related to disturbances specific to histaminergic neurotransmission.

Tuberomammillary Modulation of Hippocampal Signal Transfer

The hippocampus is thought to play an important role in memory formation[27] and in reward-related processes.[156] The hippocampus receives histaminergic fibres through both a ventral and a dorsal route[57] and contains all three histamine receptor subtypes.[5,39] Furthermore, a number of electrophysiological studies have demonstrated that TM histamine projections are involved in the subcortical modulation of neuronal excitability and synaptic plasticity in the hippocampal circuitry.[13,41,44] Given the evidence for an inhibitory role of the TM in reinforcement and mnemonic processes and the functional link between TM and hippocampus, we[149] gauged whether activation of the TM could modulate evoked field potentials in the dentate gyrus, frequently used to study electrophysiological correlates of learning.[17,38] Therefore, paired-pulses of electrical stimulation were delivered to the perforant path (PP) and evoked field potentials (fEPSPs) were recorded in the dentate gyrus (DG) in freely moving rats. Before activating the PP, the TM was triggered by electrical stimulation when the rat explored an unfamiliar environment [Type I, 'theta' behaviour, including walking, sniffing and rearing according to Vanderwolf[141]] or when the animals showed Type II, 'non-theta' behaviour, including grooming, awake immobility and slow-wave sleep. The results indicate that activation of the histaminergic TM nucleus in the freely moving rat differentially affected the efficacy of afferent transmission to the hippocampus, depending on the behavioural state of the animal. Prestimulation of the TM was found to modulate neuronal transmission in the PP during learning-related exploratory behavior, but not during 'non-theta' related behaviours, including awake immobility (Fig. 9). During exploration both the conditioning as well as the test response of the dentate fEPSPs decreased with increasing TM train stimulation intensities, whereas the population-spikes were unchanged. Similar excitability changes in the PP-dentate area were previously observed following glutamate microinjections into the TM in vivo (unpublished results) and in hippocampal slices exposed to high concentrations of histamine.[40,42] Taken together, these results indicate that the TM and the hippocampus may comprise a common system involved in the inhibition of the brain's reinforcement system and suggest that the TM projection system exerts its inhibitory action on associative functioning by interfering negatively with the signal transfer across the PP-granule cell synapses of the dentate gyrus. Congruent with this hypothesis, it was recently found that histamine and certain H_2 antagonists can inhibit high frequency oscillations ('ripples') in hippocampus CA1 subfield,[69] that are known to be associated with processes related to memory formation and certain behavioural states such as slow wave sleep;[21] application of histamine H_1 receptor antagonists had the opposite effect and facilitated the occurrence of ripples. Furthermore, behavioural studies revealed that a lesion of the hippocampus can amplify rewarding hypothalamic stimulation[156] and microinjection of histamine into the dentate area was reported to adversely affect active avoidance conditioning.[3]

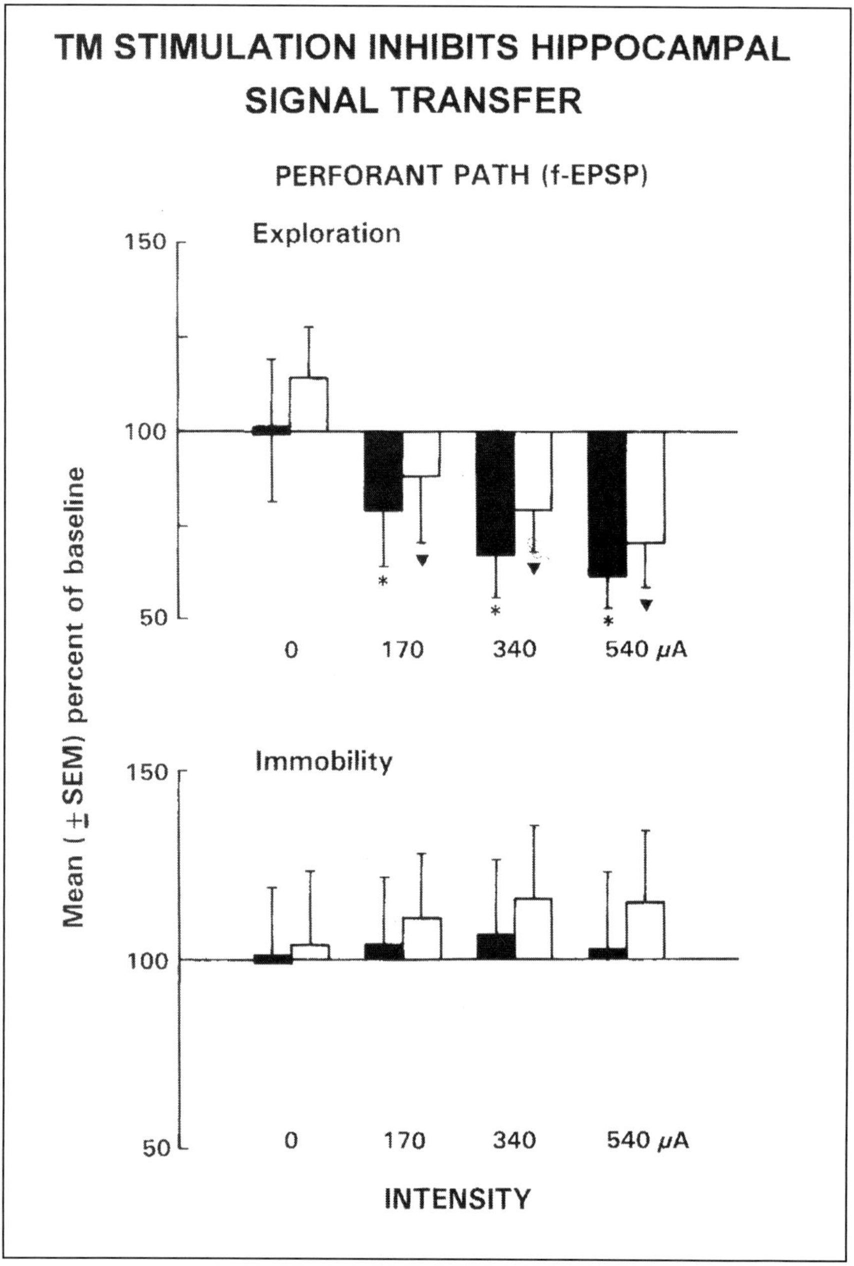

Figure 9. Normalized fEPSP slopes recorded in the dentate area evoked by paired-pulse stimulation (ISI= 30 ms) of the perforant path 50 ms following train stimulation in the TM with three different current intensities (0= no train, baseline) during exploration and awake immobility. The values given are mean (±SEM) percentage of the respective baseline condition (=100%). Black bars: Response on the conditioning (first) pulse (RC). White bars: Response on the test (second) pulse (RT). The Wilcoxon test for related samples performed on raw data was used to test for within group differences; *$p<0.05$ vs. conditioning pulse (RC0); tp< 0.05 vs. test pulse (RT0). (from ref. 149)

Conclusions

Results from this laboratory suggest that TM histamine projections are involved in behavioural asymmetries and in subsequent behavioural recovery after hemivibrissotomy and unilateral 6-OHDA lesions of the substantia nigra. Furthermore, our findings indicate that the histaminergic neuronal system (histamine fibres arising from E2-subgroup) may function as an inhibitory neurochemical substrate in the control of reinforcement and mnemonic processes. Both amplification of rewarding hypothalamic stimulation as well as facilitation of mnemonic processes were demonstrated following destruction of the TM. On the other hand, electrical or chemical stimulation of the TM was found to negatively interfere with the signal transfer in the hippocampus during learning-related behaviours. Moreover, administration of the histamine H_1-receptor antagonist chlorpheniramine, but not the H_2-receptor antagonist ranitidine, was found to exert reinforcing effects and to promote learning in projection areas of the TM known to be crucial for reward and memory, namely, the ventral pallidum and the NAcc. However, the finding that histamine itself can have beneficial effects on learning and reward-related functions do not implicitly support this view and suggest that the behavioural effects observed after destruction of the TM might involve neurochemical processes other than a lesion-induced downregulation of histaminergic activity. Possible mechanisms that might account for the behavioural effects could involve TM-lesion produced alterations in diverse neurochemical systems that are colocalized and functionally linked to histamine such as GABA, glutamate, adenosine and certain neuropeptides.[2,74,126] Thus, it remains to be determined which endogenous processes are related to the inhibitory control of TM neurons, thereby affecting processes of learning and memory. This can only be achieved through knowledge of the distinct and opposite modulatory actions that the TM-histamine system might exert by activating different receptor subtypes on different neuronal systems involved in reinforcement and learning processes.

Nevertheless, our results are the first to focus on an inhibitory element in the neural system underlying the reinforcement process ('stamping-in'). Up to now, such an inhibitory substrate has been largely ignored or neglected in the attempt to characterize the neural basis of the reinforcement system.[52] Furthermore, we found that lesions of the TM or blockade of certain histamine receptors generally induced changes in behavioural parameters, which were opposite to those known to occur after destruction or pharmacological manipulations of the substantia nigra.[19,55] Such an antagonism was evident for turning and thigmotactic scanning, lateral-hypothalamic self-stimulation, place conditioning and mnemonic functioning. The evidence that the TM, the substantia nigra, and their transmitters DA and histamine can act in a reciprocal fashion with regard to the behaviours investigated so far, may be indicative of a functional link between the tuberomammillary-striatal and the nigrostriatal system.

Finally, another aspect should be pointed out. The lesions in the TM region and the application of the H_1-blocking drug chlorpheniramine not only improved learning in adult rats but also ameliorated performance deficits of aged rats, which are proposed to be an animal model for Alzheimer's disease.[33] This finding is interesting in the light of recent studies, showing increased levels of histamine in aged rats[83] as well as in Alzheimer's disease patients with mental deterioration (ref. 18; but see ref. 100). Based on these findings, histamine antagonists, particularly those acting at the H_1-receptive site, or H_3-agonists could be considered in terms of their possible therapeutic and/or protective role in Alzheimer patients, and also in patients suffering from other neuropathies, such as Parkinson's disease.

Acknowledgments

This work was supported by grants from the German National Science Foundation to J.P.H. and R.U.H. The experiments were carried out in accordance with the German Law on the Protection of Animals and were approved by the state authority.

References

1. Adachi N, Oishi R, Itano Y et al. Aggravation of ischemic neuronal damage in the rat hippocampus by impairment of histaminergic neurotransmission. Brain Res 1993; 602:165-168.
2. Airaksinen MS, Alanen S, Szabat E et al. Multiple neurotransmitters in the tuberomammillary nucleus: Comparison of rat, mouse, and guinea pig. J Comp Neurol 1992; 323:103-116.
3. Alvarez EO, Banzan AM. Hippocampus and learning. Possible role of histamine receptors. Medicina (B-Aires) 1996; 56:155-160.
4. Arrang JM, Garbarg M, Schwartz JC. Autoinhibition of histamine synthesis mediated by presynaptic H_3 receptors. Neurosci 1987; 23:149-157.
5. Arrang JM, Garbarg M, Lancelot JC et al. Highly potent and selective ligands for histamine H_3-receptors. Nature 1987; 327:117-123.
6. Ash AS, Schild HO. Receptors mediating some actions of histamine. Br J Pharmacol 1966; 27:427-439.
7. Beardsley PM, Balster RL. The intravenous self-administration of antihistamines by rhesus monkeys. Drug Alcohol Depend 1992; 30:117-126.
8. Bergman J, Spealman RD. Some behavioral effects of histamine H_1 antagonists in squirrel monkeys. J Pharmacol Exp Ther 1986; 239:104-110.
9. Black JW, Duncan WA, Durant CJ et al. Definition and antagonism of histamine H2-receptors. Nature 1972; 236:385-390.
10. Blandina P, Giorgetti M, Bartolini L et al. Inhibition of cortical acetylcholine release and cognitive performance by histamine H_3 receptor activation in rats. Br J Pharmacol 1996; 119:1656-1664.
11. Bouthenet ML, Ruat M, Sales N et al. A detailed mapping of histamine H_1-receptors in guinea-pig central nervous system established by autoradiography with [125I]iodobolpyramine. Neurosci 1988; 26:553-600.
12. Brown RE, Reymann KG. Histamine H_3 receptor-mediated depression of synaptic transmission in the dentate gyrus of the rat in vitro. J Physiol 1996; 496(Pt1):175-184.
13. Brown RE, Fedorov NB, Haas HL et al. Histaminergic modulation of synaptic plasticity in area CA1 of rat hippocampal slices. Neuropharmacol 1995; 34:181-190.
14. Brown RE, Stevens DR, Haas HL. The physiology of brain histamine. Prog Neurobiol 2001; 63:637-672.
15. Burns TA, Huston JP, Sherman RL et al. Circadian variation of brain histamine in goldfish. Paper Presented at the 10th European Congress of Ichthyology. Prague, Sept 3-7, 2001.
16. Buscher W, Schugens M, Wagner U et al. Interhemispheric relationship between lateral hypothalamic self-stimulation and the region of the nucleus tegmenti pedunculo-pontinus. Brain Res 1989; 487:321-334.
17. Buzsaki G, Grastyan E, Czopf J et al. Changes in neuronal transmission in the rat hippocampus during behavior. Brain Res 1981; 225:235-247.
18. Cacabelos R, Yamatodani A, Niigawa H et al. Brain histamine in Alzheimer's disease. Meth Find Exp Clin Pharmacol 1989; 11:353-360.
19. Carey RJ. Unilateral 6-hydroxydopamine lesions of dopamine neurons produce bilateral self-stimulation deficits. Behav Brain Res 1982; 6:101-114.
20. Chen Z, Sugimoto Y, Kamei C. Effects of intracerebroventricular injection of alpha-fluoromethylhistidine on radial maze performance in rats. Pharmacol Biochem Behav 1999; 64:513-518.
21. Chrobak JJ, Lorincz A, Buzsaki G. Physiological patterns in the hippocampo-entorhinal cortex system. Hippocampus 2000; 10:457-465.
22. Cohn CK, Ball GG, Hirsch J. Histamine: Effect on self-stimulation. Science 1973; 180:757-758.
23. Curtis M, Bergman H, Price ML et al. Hypothalamic tissue stimulates hippocampal pyramidal neuron survival during development: Evidence from intraocular double transplants. Hippocampus 1995; 5:584-594.
24. De Almeida MA, Izquierdo I. Intracerebroventricular histamine, but not 48/80, causes posttraining memory facilitation in the rat. Arch Int Pharmacodyn Ther 1988; 291:202-207.
25. Dringenberg HC, De Souza-Silva MA, Rossmüller J et al. Histamine H_1 receptor antagonists produce increases in extracellular acetylcholine in rat frontal cortex and hippocampus. J Neurochem 1998; 70:1750-1758.
26. Dringenberg HC, De Souza-Silva MA, Schwarting RKW et al. Increased levels of extracellular dopamine in neostriatum and nucleus accumbens after histamine H_1 receptor blockade. Naunyn-Schmiedebergs Arch Pharmacol 1998; 358:423-429.
27. Eichenbaum H, Cohen NJ, Otto T et al. Memory representation in the hippocampus: Functional domain and functional organization. In: Squire LR, Weinberger NM, Lynch G, McGaugh JL, eds. Memory: Organization and Locus of Change. New York: Oxford UP, 1991:163-204.

28. Fernandez-Novoa L, Cacabelos R. Histamine function in brain disorders. Behav Brain Res 2001; 124:213-233.
29. Frisch C, Hasenöhrl RU, Huston JP. The histamine H_1-antagonist chlorpheniramine facilitates learning in aged rats. Neurosci Lett 1997; 229:89-92.
30. Frisch C, Hasenöhrl RU, Haas HL et al. Facilitation of learning after lesions of the tuberomammillary nucleus region in adult and aged rats. Exp Brain Res 1998; 118:447-456.
31. Frisch C, Hasenöhrl RU, Huston JP. Memory improvement by post-trial injection of lidocaine into the tuberomammillary nucleus, the source of neuronal histamine. Neurobiol Learn Mem 1999; 72:69-77.
32. Fujitani T, Adachi N, Nagaro T et al. Histaminergic H_2 action protects hippocampal CA1 neurons by prolonging the onset of the anoxic depolarization in gerbils. J Neurochem 1996; 67:2613-2615.
33. Gallagher M, Nicolle MM. Animal models of normal aging: Relationship between cognitive decline and markers in hippocampal circuitry. Behav Brain Res 1993; 57:155-162.
34. Galosi R, Lenard L, Knoche A et al. Dopaminergic effects of histamine administration in the nucleus accumbens and the impact of H_1-receptor blockade. Neuropharmacol 2001; 40:624-633.
35. Garbarg M, Barbin G, Bischoff S et al. Dual localization of histamine in an ascending neuronal pathway and in non-neuronal cells evidenced by lesions in the lateral hypothalamic area. Brain Res 1976; 106:333-348.
36. Gauvin DV, Carl KL, Briscoe RJ et al. Cross-generalization between a cocaine cue and two antihistamines. Eur J Pharmacol 1995; 294:281-288.
37. Göthert M, Garbarg M, Hey JA et al. New aspects of the role of histamine in cardiovascular function: Identification, characterization, and potential pathophysiological importance of H_3 receptors. Can J Physiol Pharmacol 1995; 73:558-564.
38. Green EJ, McNaughton BL, Barnes CA. Exploration-dependent modulation of evoked responses in fascia dentata: Dissociation of motor, EEG, and sensory factors and evidence for a synaptic efficacy change. J Neurosci 1990; 10:1455-1471.
39. Green JP. Histamine receptors in brain. In: Iverson LL, Iverson SD, Snyder SH, eds. Handbook of Psychopharmacology. New York: Plenum Press, 1983:17:384-420.
40. Greene RW, Haas HL. Effects of histamine on dentate granule cells in vitro. Neurosci 1990; 34:299-303.
41. Haas HL. Electrophysiology of histamine receptors. In: Schwartz JC, Haas HL, eds. The histamine receptor. New York: Wiley-Liss, 1992:161-177.
42. Haas HL, Greene RW. Effects of histamine on hippocampal pyramidal cells of the rat in vitro. Exp Brain Res 1986; 62:123-130.
43. Haas HL, Reiner PB, Greene RW. Histaminergic and histaminoceptive neurons: Electrophysiological studies in vertebrates. In: Watanabe T, Wada H, eds. Histaminergic neurons: Morphology and Function. Boca Raton: CRC Press, 1991:195-208.
44. Haas HL, Sergueeva OA, Vorobjev VS et al. Subcortical modulation of synaptic plasticity in the hippocampus. Behav Brain Res 1995; 66:41-44.
45. Hasenöhrl RU, Huston JP. Neuronal histamine: Physiology, behavior and brain disorders. Special Issue of Behav Brain Res 2001; 124:103-250.
46. Hasenöhrl RU, Oitzl MS, Huston JP. Conditioned place preference in the corral: A procedure for measuring reinforcing properties of drugs. J Neurosci Meth 1989; 30:141-146.
47. Hasenöhrl RU, Frisch C, Nikolaus S et al. Chronic administration of neurokinin SP improves maze performance in aged Rattus norvegicus. Behav Neural Biol 1994; 62:110-120.
48. Hasenöhrl RU, Weth K, Huston JP. Intraventricular infusion of the histamine H_1 receptor antagonist chlorpheniramine improves maze performance and has anxiolytic-like effects in aged hybrid Fischer 344xBrown Norway rats. Exp Brain Res 1999; 128:435-440.
49. Hasenöhrl RU, Kuhlen A, Frisch C et al. Comparison of intra-accumbens injection of histamine with histamine H_1-receptor antagonist chlorpheniramine in effects on reinforcement and memory parameters. Behav Brain Res 2001; 124:203-211.
50. Hill SJ. Histamine receptors in the mammalian central nervous system: Biochemical studies. Prog Med Chem 1987; 24:29-84.
51. Hough LB. Cellular localization and possible functions for brain histamine: Recent progress. Prog Neurobiol 1988; 30:469-505.
52. Huston JP. Searching for the neural mechanism of reinforcement (of "stamping-in"). In: Hoebel BG, Novin D, eds. The Neural Basis of Feeding and Reward. New York: Academic Press, 1982:75-83
53. Huston JP, Hasenöhrl RU. The role of neuropeptides in learning: Focus on the neurokinin substance P. Behav Brain Res 1995; 66:117-127.

54. Huston JP, Mueller CC, Mondadori C. Memory facilitation by posttrial hypothalamic stimulation and other reinforcers: A central theory of reinforcement. Biobehav Rev 1977; 1:143-150.
55. Huston JP, Steiner H, Weiler HT et al. The basal ganglia-orofacial system: Studies on neurobehavioral plasticity and sensory-motor tuning. Neurosci Biobehav Rev 1990; 14:433-446.
56. Huston JP, Wagner U, Hasenöhrl RU. The tuberomammillary nucleus projections in the control of learning, memory and reinforcement processes: Evidence for an inhibitory role. Behav Brain Res 1997; 83:97-105.
57. Inagaki N, Yamatodani A, Ando-Yamamoto M et al. Organization of histaminergic fibers in the rat brain. J Comp Neurol 1988; 273:283-300.
58. Inagaki N, Toda K, Taniuchi I et al. An analysis of histaminergic efferents of the tuberomammillary nucleus to the medial preoptic area and inferior colliculus of the rat. Exp Brain Res 1990; 80:374-380.
59. Ito C. The role of brain histamine in acute and chronic stresses. Biomed Pharmacother 2000; 54:263-267.
60. Kamei C. Involvement of central histamine in amygdaloid kindled seizures in rats. Behav Brain Res 2001; 124:243-250.
61. Kamei C, Tasaka K. Participation of histamine in the step-through active avoidance response and its inhibition by H_1-blockers. Jpn J Pharmacol 1991; 57:473-482.
62. Kamei C, Tasaka K. Effect of intracerebroventricular injection of histamine on memory impairment induced by hippocampal lesions in rats. Jpn J Pharmacol 1992; 58[Suppl.1]:55.
63. Kamei C, Tasaka K. Effect of histamine on memory retrieval in old rats. Biol Pharm Bull 1993; 16:128-132.
64. Kamei C, Chung YH, Tasaka K. Influence of certain H_1-blockers on the step-through active avoidance response in rats. Psychopharmacol 1990; 102:312-318.
65. Kamei C, Okumura Y, Tasaka K. Influence of histamine depletion on learning and memory recollection in rats. Psychopharmacol 1993; 111:376-382.
66. Kamei J, Morita K, Ohsawa M et al. Effects of epinastine on the antitussive and rewarding effects of dihydrocodeine in mice. Methods Find Exp Clin Pharmacol 1999; 21:663-668.
67. Klapdor K, Hasenöhrl RU, Huston JP. Facilitation of learning in adult and aged rats following bilateral lesions of the tuberomammillary nucleus region. Behav Brain Res 1994; 61:113-116.
68. Knigge U, Kjaer A, Larsen PJ et al. Effect of histamine on gene expression and release of proopiomelanocortin-derived peptides from the anterior and intermediate pituitary lobes in conscious male rats. Neuroendocrinol 1995; 62:319-325.
69. Knoche A, Yokoyama H, Ponomarenko A et al. High-frequency oscillation in the hippocampus of the behaving rat and its modulation by the histaminergic system. Hippocampus 2003; 13:273-280.
70. Kollonitsch J, Perkins LM, Patchett AA et al. Selective inhibitors of biosynthesis of aminergic neurotransmitters. Nature 1978; 274:906-908.
71. Koob GF, Bloom FE. Cellular and molecular mechanisms of drug dependence. Science 1988; 242:715-723.
72. Kraly FS, Tribuzio RA, Keefe ME et al. Endogenous histamine contributes to drinking initiated without postprandial challenges to fluid homeostasis in rats. Physiol Behav 1995; 58:1137-1143.
73. Kubo N, Shirakawa O, Kuno T et al. Antimuscarinic effects of antihistamines: Quantitative evaluation by receptor-binding assay. Jpn J Pharmacol 1987; 43:277-282.
74. Lantos TA, Gorcs TJ, Palkovits M. Immunohistochemical mapping of neuropeptides in the premamillary region of the hypothalamus in rats. Brain Res Rev 1995; 20:209-249.
75. Leurs R, Smit MJ, Timmerman H. Molecular pharmacological aspects of histamine receptors. Pharmacol Ther 1995; 66:413-463.
76. Liebman JM. Discriminating between reward and performance: A critical review of intracranial self-stimulation methodology. Neurosci Biobehav Rev 1983; 7:45-72.
77. Liu C, Ma X, Jiang X et al. Cloning and pharmacological characterization of a fourth histamine receptor (H_4) expressed in bone marrow. Mol Pharmacol 2001; 59:420-426.
78. Lovenberg TW, Roland BL, Wilson SJ et al. Cloning and functional expression of the human histamine H_3 receptor. Mol Pharmacol 1999; 55:1101-1107.
79. Maisonnette S, Huston JP, Brandao M et al. Behavioral asymmetries and neurochemical changes after unilateral lesions of tuberomammillary nucleus or substantia nigra. Exp Brain Res 1998; 120:273-282.
80. Masukawa Y, Suzuki T, Misawa M. Differential modification of the rewarding effects of methamphetamine and cocaine by opioids and antihistamines. Psychopharmacol 1993; 111:139-143.
81. Matsubara T, Moskowitz MA, Huang Z. UK-14,304, R(-)-alpha-methyl-histamine and SMS 201-995 block plasma protein leakage within dura mater by prejunctional mechanisms. Eur J Pharmacol 1992; 224:145-150.

82. Mattioli R, Nelson CA, Huston JP et al. Conditioned place-preference analysis in the goldfish with the H_1 histamine antagonist chlorpheniramine. Brain Res Bull 1998; 45:41-44.
83. Mazurkiewicz-Kwilecki IM, Prell GD. Age-related changes in brain histamine. Agents Actions 1984; 14:554-557.
84. McKearney JW. Stimulant actions of histamine H_1 antagonists on operant behavior in the squirrel monkey. Psychopharmacol 1982; 77:156-158.
85. Meguro K, Yanai K, Sakai N et al. Effects of thioperamide, a histamine H_3 antagonist, on the step-through passive avoidance response and histidine decarboxylase activity in senescence-accelerated mice. Pharmacol Biochem Behav 1995; 50:321-325.
86. Monti JM, Jantos H, Leschke C. The selective histamine H_1-receptor agonist 2-(3- trifluoromethylphenyl) histamine increases waking in the rat. Eur Neuropsychopharmacol 1994; 4:459-462.
87. Morgan S, Nomikos G, Huston. Behavioral analysis of asymmetries induced by unilateral 6-OHDA injections into the substantia nigra. Behav Neural Biol 1993; 60:241-250.
88. Morgan S, Baker D, Huston JP. Relationship between behavioral recovery from unilateral 6-OHDA lesion of the substantia nigra and changes in the tuberomammillary-striatal projection as measured by HRP labeling. Restor Neurol Neurosci 1998; 12:1-9.
89. Morimoto T, Yamamoto Y, Yamatodani A. Brain histamine and feeding behavior. Behav Brain Res 2001; 124:145-150.
90. Nakamura S, Ohnishi K, Nishimura M et al. Large neurons in the tuberomammillary nucleus in patients with Parkinson's disease and multiple system atrophy. Neurology 1996; 46:1693-1696.
91. Oda T, Morikawa N, Saito Y et al. Molecular cloning and characterization of a novel type of histamine receptor preferentially expressed in leukocytes. J Biol Chem 2000; 275:36781-36786.
92. Oishi R, Shishido S, Yamori M et al. Comparison of the effects of eleven histamine H_1-receptor antagonists on monoamine turnover in the mouse brain. Naunyn Schmiedebergs Arch Pharmacol 1994; 349:140-144.
93. Olds ME, Olds J. Approach-avoidance analysis of rat diencephalon. J Comp Neurol 1963; 120:259-295.
94. Onodera K, Yamatodani A, Watanabe T. Effects of alpha-fluoromethylhistidine on locomotor activity, brain histamine and catecholamine contents in rats. Methods Find Exp Clin Pharmacol 1992; 14:97-105.
95. Onodera K, Yamatodani A, Watanabe T et al. Neuropharmacology of the histaminergic neuron system in the brain and its relationship with behavioral disorders. Prog Neurobiol 1994; 42:685-702.
96. Orsetti M, Ghi P, Di Carlo G. Histamine H_3-receptor antagonism improves memory retention and reverses the cognitive deficit induced by scopolamine in a two-trial place recognition task. Behav Brain Res 2000; 1124:235-242.
97. Palacios JM, Wamsley JK, Kuhar MJ. The distribution of histamine H_1-receptors in the rat brain: An autoradiographic study. Neurosci 1981; 6:15-37.
98. Panula P, Yang HY, Costa E. Histamine-containing neurons in the rat hypothalamus. Proc Natl Acad Sci USA 1984; 81:2572-2576.
99. Panula P, Pirvola U, Auvinen S et al. Histamine-immunoreactive nerve fibers in the rat brain. Neurosci 1989; 28:585-610.
100. Panula P, Rinne J, Kuokkanen K et al. Neuronal histamine deficit in Alzheimer's disease. Neurosci 1998; 82:993-997.
101. Passani MB, Bacciottini L, Mannaioni PF et al. Central histaminergic system and cognition. Neurosci Biobehav Rev 2000; 24:107-113.
102. Philippu A, Prast H. Importance of histamine in modulatory processes, locomotion and memory. Behav Brain Res 2001; 124:151-159.
103. Pollard H, Bouthenet ML. Autoradiographic visualization of the three histamine receptor subtypes in the brain. In: Schwartz JC, Haas HL, eds. The Histamine Receptor. New York: Wiley Liss, 1992; 179-192.
104. Pollard H, Moreau J, Arrang JM et al. A detailed autoradiographic mapping of histamine H_3 receptors in rat brain areas. Neurosci 1993; 52:169-189.
105. Prast H, Argyriou A, Philippu A. Histaminergic neurons facilitate social memory in rats. Brain Res 1996; 734:316-318.
106. Prell GD, Green JP. Histamine as a neuroregulator. Annu Rev Neurosci 1986; 9:209-254.
107. Prell GD, Green JP, Kaufmann CA et al. Histamine metabolites in cerebrospinal fluid of patients with chronic schizophrenia: Their relationships to levels of other aminergic transmitters and ratings of symptoms. Schizophr Res 1995; 14:93-104.
108. Privou C, Knoche A, Hasenöhrl RU et al. The H_1 and H_2-blockers chlorpheniramine and ranitidine applied to the nucleus basalis magnocellularis region modulate anxiety and reinforcement related processes. Neuropharmacol 1998; 37:1019-1032.

109. Privou C, Li J-S, Hasenöhrl RU et al. Enhanced learning by post-trial injection of H_1- but not H_2-histaminergic antagonists into the nucleus basalis magnocellularis region. Neurobiol Learn Mem 1999; 71:308-324.
110. Rassnick S, Kornetsky C. L-Histidine attenuates the effects of pentazocine on rewarding brain-stimulation. Life Sci 1991; 48:1729-1736.
111. Rubio S, Begega A, Santin LJ et al. Improvement of spatial memory by (R)-alpha-methylhistamine, a histamine H_3 receptor agonist, on the Morris water maze in rat. Behav Brain Res 2001; 124:177-181.
112. Ryu JH, Yanai K, Zhao X-L et al. The effect of dopamine D_1 receptor stimulation on the up-regulation of histamine H_3-receptors following destruction of the ascending dopaminergic neurones. Br J Pharmacol 1996; 118:585-592.
113. Sakai N, Sakurai E, Sakurai E et al. Depletion of brain histamine induced by alpha-fluoromethylhistidine enhances radial maze performance in rats with modulation of brain amino acid levels. Life Sci 1998; 62:989-994.
114. Sannerud CA, Kaminski BJ, Griffiths RR et al. Maintenance of H-sub-1 antagonists self-injection in baboons. Exp Clin Psychopharmacol 1995; 3:26-32.
115. Santos NR, Huston JP, Brandao ML. Escape behavior under tonic inhibitory control of histamine H_2-receptor mediated mechanisms in the midbrain tectum. Behav Brain Res 2001; 124:167-175.
116. Schlicker E, Betz R, Göthert M. Histamine H_3 receptor-mediated inhibition of serotonin release in the rat brain cortex. Naunyn-Schmiedebergs Arch Pharmacol 1988; 337:588-590.
117. Schlicker E, Behling A, Lummen G et al. Mutual interaction of histamine H_3-receptors and alpha 2-adrenoceptors on noradrenergic terminals in mouse and rat brain cortex. Naunyn-Schmiedebergs Arch Pharmacol 1992; 345:639-646.
118. Schlicker E, Fink K, Detzner M et al. Histamine inhibits dopamine release in the mouse striatum via presynaptic H_3 receptors. J Neural Transm Gen Sect 1993; 93:1-10.
119. Schlicker E, Malinowska B, Kathmann M et al. Modulation of neurotransmitter release via histamine H_3 heteroreceptors. Fundam Clin Pharmacol 1994; 8:128-137.
120. Schwarcz R, Hökfelt T, Fuxe K et al. Ibotenic acid-induced neuronal degeneration: A morphological and neurochemical study. Exp Brain Res 1979; 37:199-216.
121. Schwartz JC, Arrang JM, Garbarg M et al. Histaminergic transmission in the mammalian brain. Physiol Rev 1991; 71:1-51.
122. Schwartz JC, Arrang JM, Bouthenet ML et al. Histamine receptors in brain. In: Uvnäs B, ed. Histamine and Histamine Antagonists. Berlin: Springer, 1991:191-242.
123. Servos P, Barke KE, Hough LB et al. Histamine does not play an essential role in electrocortical activation during waking behavior. Brain Res 1994; 636:98-102.
124. Shishido S, Oishi R, Saeki K. In vivo effects of some histamine H_1-receptor antagonists on monoamine metabolism in the mouse brain. Naunyn Schmiedebergs Arch Pharmacol 1991; 343:185-189.
125. Smith CP, Hunter AJ, Bennett GW. Effects of (R)-alpha-methylhistamine and scopolamine on spatial learning in the rat assessed using a water maze. Psychopharmacol 1994; 114:651-656.
126. Spieler RE, Nelson CA, Huston JP et al. Post-trial administration of H_1 histamine receptor blocker improves appetitive reversal learning and memory in goldfish, Carassius auratus. Neurosci Lett 1999; 277:5-8.
127. Staines WA, Daddona PE, Nagy JI. The organization and hypothalamic projections of the tuberomammillary nucleus in the rat: An immunohistochemical study of adenosine deaminase-positive neurons and fibers. Neurosci 1987; 23:571-596.
128. Stäubli U, Huston JP. Up-hill avoidance: A new passive-avoidance task. Physiol Behav 1979; 22:775-776.
129. Steinbusch HWM, Mulder AH. Immunohistochemical localization of histamine in neurons and mast cells in the rat brain. In: Björklund A, Hökfeld T, Kuhar MJ, eds. Handbook of chemical neuroanatomy. Classical Transmitters and Transmitter Receptors in the CNS (Part II). Amsterdam: Elsevier, 1984:3:126-140.
130. Steinbusch HWM, Sauren Y, Groenewegen H et al. Histaminergic projections from the premammillary and posterior hypothalamic region to the caudate-putamen complex in the rat. Brain Res 1986; 368:389-393.
131. Steiner H, Bonatz AE, Huston JP et al. Lateralized wall-facing versus turning as measures of behavioral asymmetries and recovery of function after injection of 6-hydroxydopamine into the substantia nigra. Exp Neurol 1988; 99:556-566.
132. Steiner H, Weiler HT, Morgan S et al. Asymmetries in crossed and uncrossed nigrostriatal projections dependent on duration of unilateral removal of vibrissae in rats. Exp Brain Res 1989; 77:421-424.

133. Steiner H, Weiler HT, Morgan S et al. Time-dependent neuroplasticity in mesostriatal projections after unilateral removal of vibrissae in the adult rat: Compartment-specific effects on horseradish peroxidase transport and cell size. Neurosci 1992; 47:793-806.
134. Suzuki T, Takamori K, Misawa M et al Effects of the histaminergic system on the morphine-induced conditioned place preference in mice. Brain Res 1995; 675:195-202.
135. Suzuki T, Mori T, Tsuji M et al. Evaluation of the histamine H_1-antagonist-induced place preference in rats. Jpn J Pharmacol 1999; 81:332-338.
136. Swanson LW. The projections of the ventral tegmental area and adjacent regions: A combined fluorescent retrograde tracer and immunofluorescence study in the rat. Brain Res Bull 1982; 9:321-353.
137. Takada K, Barrett JE, Allen MS et al. Punishment of schedule-controlled behavior with beta-carboline injections: Antagonism and comparisons with other compounds. J Pharmacol Exp Ther 1992; 261:138-145.
138. Tasaka K, Kamei C, Akahori H et al. The effects of histamine and some related compounds on conditioned avoidance response in rats. Life Sci 1985; 37:2005-2014.
139. Tuomisto L. Regulation of feeding behavior, with special reference to histamine. J Physiol Pharmacol 1994; 45:469-477.
140. Tuomisto L, Lozeva V, Valjakka A et al. Modifying effects of histamine on circadian rhythms and neuronal excitability. Behav Brain Res 2001; 124:129-135.
141. Unterwald EM, Kucharsky LT, Williams JE et al. Tripelennamine: Enhancement of brain-stimulation reward. Life Sci 1984; 34:149-153.
142. Vanderwolf CH. Hippocampal electrical activity and voluntary movement in the rat. Electroencephalogr Clin Neurophysiol 1969; 26:407-418.
143. Wada H, Inagaki N, Yamatodani A et al. Is the histaminergic neuron system a regulatory center for whole-brain activity? Trends Neurosci 1991; 14:415-418.
144. Wagner U, Weiler HT, Huston JP. Amplification of rewarding hypothalamic stimulation following a unilateral lesion in the region of the tuberomammillary nucleus. Neurosci 1993; 52:927-932.
145. Wagner U, Segura-Torres P, Weiler T et al. The tuberomammillary nucleus region as a reinforcement inhibiting substrate: Facilitation of ipsihypothalamic self-stimulation by unilateral ibotenic acid lesions. Brain Res 1993; 613:269-274.
146. Watanabe T, Taguchi Y, Shiosaka S et al. Distribution of the histaminergic neuron system in the central nervous system of rats: A fluorescent immunohistochemical analysis with histidine decarboxylase as a marker. Brain Res 1984; 295:13-25.
147. Wauquier A, Niemegeers CJ. Effects of chlorpheniramine, pyrilamine and astemizole on intracranial self-stimulation in rats. Eur J Pharmacol 1981; 72:245-248.
148. Weiler HT, Steiner H, Huston JP. Plasticity in crossed and uncrossed tuberomammillary-striatal projections in relation to recovery from behavioral asymmetries induced by hemivibrissotomy. Neurosci 1990; 37:463-469.
149. Weiler HT, Wagner U, Huston JP. Unilateral lesion in the tuberomammillary nucleus region: Behavioral asymmetries and effects of histamine precursor. Behav Brain Res 1992; 49:167-173.
150. Weiler HT, Hasenöhrl RU, Van Landeghem AAL et al. Differential modulation of hippocampal signal transfer by tuberomammillary nucleus stimulation in freely moving rats dependent on behavioral state. Synapse 1998; 28:294-301.
151. White JM, Rumbold GR. Behavioural effects of histamine and its antagonists: A review. Psychopharmacol 1988; 95:1-14.
152. Wise RA, Rompre PP. Brain dopamine and reward. Annu Rev Psychol 1989; 40:191-225.
153. Yanai K, Son LZ, Endou M et al. Targeting disruption of histamine H_1 receptors in mice: Behavioral and neurochemical characterization. Life Sci 1998; 62:1607-1610.
154. Yanai K, Son LZ, Endou M et al. Behavioural characterization and amounts of brain monoamines and their metabolites in mice lacking histamine H_1 receptors. Neurosci 1998; 87:479-487.
155. Yokoyama H, Onodera K, Iinuma K et al. 2-Thiazolylethylamine, a selective histamine H_1 agonist, decreases seizure susceptibility in mice. Pharmacol Biochem Behav 1994; 47:503-507.
156. Yokoyama H, Onodera K, Maeyama K et al. Clobenpropit (VUF-9153), a new histamine H_3 receptor antagonist, inhibits electrically induced convulsions in mice. Eur J Pharmacol 1994; 260:23-28.
157. Zimmermann P, Wagner U, Krauth J et al. Unilateral lesion of dorsal hippocampus enhances reinforcing lateral hypothalamic stimulation in the contralateral hemisphere. Brain Res Bull 1997; 44:265-271.
158. Zimmermann P, Privou C, Huston JP. Differential sensitivity of the caudal and rostral nucleus accumbens to the rewarding effects of a H1-histaminergic receptor blocker as measured with place-preference and self-stimulation behavior. Neurosci 1999; 94:93-103.

CHAPTER 2.9

Adenosine and Purines

Trevor W. Stone, M-R. Nikbakht and E. Martin O'Kane

Abstract

Adenosine can act on four subtypes of receptor, of which the A_1 and A_{2A} subtypes have received the most attention experimentally. The A_1 receptors are primarily inhibitory by depressing transmitter release or causing hyperpolarisation, while the A_{2A} receptors often cause overall excitation by direct depolarisation or the facilitation of transmitter release. Activation of these receptors can also modulate neuronal sensitivity to classical transmitters by altering receptor function, especially of acetylcholine and glutamate receptors, two of the transmitters most closely involved in processess of learning and memory. Both the A_1 and A_{2A} receptors have been shown to modulate synaptic plasticity in areas such as the hippocampus, although the relationship between these effects and the influence on individual classical transmitters remains unclear at present. Adenine nucleotides are also known to be active at receptors in the brain, and some forms of long-term potentiation may be in part attributable to the local release of ATP. Together, the purine nucleosides and nucleotides represent strong candidates for major physiological regulators of the cellular processes underlying neuronal excitability and synaptic plasticity.

Origin of Adenosine in the Extracellular Fluid

Adenosine is normally present in the extracellular fluid at a concentration of around 1μM or less.[13,28,29,137,185,264] The origin of this adenosine remains unclear with some authors supporting the view that the nucleoside is transported out of cells by bi-directional membrane transporters when the intracellular level of free adenosine exceeds a threshold level while others argue that enzymes such as adenosine deaminase and adenosine kinase maintain intracellular free adenosine at a low concentration, and that extracellular nucleoside is primarily the consequence of metabolism of ATP which has been released from cells as a neurotransmitter, cotransmitter or trophic factor, for example. Release can be stimulated by cellular depolarisation produced by transmitters such as glutamate and acetylcholine.[44,185]

Adenosine Receptors

To date four types of adenosine receptor have been cloned, namely adenosine A_1, A_{2A}, A_{2B}, and A_3 receptors.[83,149,151,266] Adenosine A_1 and A_2 receptors occur widely distributed throughout the CNS, with the heaviest density of A_1 receptors in the hippocampus and of 'classical' A_{2A} receptors in the striatum and limbic areas such as the nucleus accumbens and olfactory tubercle.[117,138,165] A_1 receptors in the hippocampus have been localised to granule cell bodies and dendrites and to pyramidal neurons, but do also occur on glial cells.

There is, however, uncertainty as to whether the A_{2A} receptors found throughout the CNS are homogeneous. Molecular biology has revealed only a single population of sites, but there are significant pharmacological differences, especially in the binding affinities of 2-[4-(2-carboxyethyl)-phenylethylamino]-5'N-ethyl-carboxamido-adenosine (CGS 21680) and 4-(2-[7-amino-2-{2-furyl}{1,2,4}-triazolo{2,3-a}-(1,3,5}triazin-5-yl-amino]ethyl)phenol (ZM

From Messengers to Molecules: Memories Are Made of These, edited by Gernot Riedel and Bettina Platt.

241385), which suggest functional differences, probably attributable to other cellular or membrane components linked to, and modifying responsiveness of, the receptors themselves. There is certainly a high affinity binding site for the prototypical (striatal) A_{2A} receptor agonist CGS 21680, but the detailed characterisation of CGS 21680 binding sites in the hippocampus reveals that they are not identical in their properties and pharmacology to those present (in much greater abundance) in the striatum.[49,118,119,143]

Adenosine and Learning

Adenosine analogues exhibit a range of behavioural effects (see ref. 227), which include sedation,[14,45,74,218,220] anticonvulsant activity,[15,65,74] anti-nocisponsive effects,[3,110,190,259] inhibition of aggression[181] and suppression of operant responding.[34,35]

Surprisingly, however, relatively little attention has been paid to purine modulation of memory processes. Several studies have indicated that adenosine analogues can suppress aspects of learning such as the acquisition of conditioned reflexes[254,255] and conditioned avoidance responding.[153,172,262] Depressant effects on working memory[177] and specifically tests of spatial memory[247] have also been reported. These generally inhibitory actions have formed the basis of current interest in the potential use of xanthine derivatives as cognition enhancers.[203,221,233]

Adenosine Receptor Subtypes and Learning

Few of these studies have been designed specifically to clarify the relative importance of the different adenosine receptor subtypes. Hooper et al[111] addressed this question using one of the simplest tests of memory function—spontaneous alternation in a Y-maze. Tests involving spontaneous alternation became widely used after the classical studies of Dennis,[59] Douglas and Isaacson[63] and Anisman,[10] and are based on the tendency of rodents to enter that arm of a Y-maze least recently visited. Alternation scores by definition are significantly greater than 0.5, the proportion of alternations expected if the animal was selecting arms purely by chance. Some authors have attempted to interpret spontaneous alternation behaviour in terms of habituation to the most recently explored arm of the Y-maze,[10,97,132] but it is now generally accepted that spontaneous alternation behaviour reflects spatial working memory[19,184;200,231,239,248] as originally proposed by Dennis.[59]

This system was used to examine the effects of purine receptor ligands with some selectivity for acting at A_1 and A_2 adenosine receptors. The A_1 receptor selective agonist N6-cyclopentyladenosine (CPA) did not change spontaneous alternation behaviour alone, but it prevented the decrease of spontaneous alternation scores produced by scopolamine. The A_1 receptor selective antagonist 1,3-dipropyl-8-cyclopentylxanthine (DPCPX) prevented this scopolamine reversal by CPA although it had no effect when administered alone. The nonselective adenosine receptor antagonist 8-(p-sulphophenyl)theophylline (8PST), which does not cross the blood-brain barrier, had no effect upon alternation behaviour or arm entries. The A_2 receptor selective agonist (N6-[2-(3,5-dimethoxyphenyl)-2-(2-methylphenyl)ethyl]adenosine (DPMA), and the A_2 receptor selective antagonist 1,3-dimethyl-1-propargylxanthine (DMPX) had no effect on alternation behaviour alone and did not modify the effect of scopolamine.

These results contrast with previous studies using more complex experimental paradigms. Normile and Barraco,[172] for example, observed that CPA attenuated retention in a passive avoidance test. Winsky and Harvey[254] reported that R-phenylisopropyl-adenosine (R-PIA) reduced the acquisition of a conditioned avoidance response and similar results were claimed by Martin et al.[153] Conversely, acute administration of an A_1 receptor antagonist has been claimed to facilitate learning.[203] The explanation for the ability of these groups to find effects of the purines tested alone, and the results of Hooper et al[111] is not clear, although different behavioural tests were used in each case, and the adenosine receptor ligands were also different. In the case of R-PIA, only very low doses can be used, less than 1 mg/kg, if depression of overall motor activity is to be avoided. In the work of Normile and Barraco[172] the doses of CPA found to be effective were over 0.5 mg/kg - doses over ten-fold greater than the doses used by Hooper et al.[111] Similar results were obtained by Zarrindast and Bijan[262] who only obtained effects on passive avoidance learning at R-PIA doses of 0.125 mg/kg or above. In the study by Martin et al[153] the ED_{50} doses of R-PIA and CPA were found to be 10 mg/kg and 1.5 mg/kg respectively, doses far in excess of those used by Hooper et al.[111] While it remains possible that inhibition of learning does occur at these high doses, the work of Hooper et al[111] clearly indicates that at low doses, a reversal of scopolamine-induced memory deficits can be obtained.

When purine receptor ligands were combined with scopolamine, however, it was clear that A_1 but not A_2 receptor activation could modify working memory deficits induced by scopolamine. Perhaps even more significantly, however, the fact that neither the A_1 receptor selective antagonist DPCPX[25,144] nor the A_2 receptor antagonist DMPX had any effects themselves upon spontaneous alternation behaviour or arm entries, and did not modify scopolamine's elimination of spontaneous alternation behaviour suggests that activation of A_1 or A_2 receptors by endogenous adenosine is not normally involved in spatial working memory.

Of great interest is the later finding that blockade of A_{2A} receptors by DMPX could reverse the detriment to learning caused by the NMDA receptor channel blocker dizocilpine,[86] possibly implying that different neural mechanisms and/or pathways were involved in the disruption of learning produced by scopolamine and dizocilpine. A facilitatory effect of a more selective A_{2A} receptor antagonist—7-(2-phenylethyl)-5-amino-2 -(2-furyl)-pyrazolo- [4,3e] -1,2,4-triazolo [1,5-c]-pyrimidine (SCH 58261)—was also reported by Kopf et al[133] using an inhibitory avoidance test.

A potential confounding factor in studies with purines is the influence of locomotor depression. Importantly, however, locomotor activity was unchanged by CPA at any of the doses used in the analysis of alternation behaviour. Furthermore, it would be expected that a decrease in total entries, which would imply an increased time between successive entries, might allow greater time for forgetting the previous arm, thus hindering spontaneous alternation behaviour. Anisman[10] has investigated specifically the relationship between arm entries and spontaneous alternation behaviour using three strains of mice with different degrees of locomotor activity. Despite this difference, all strains showed the same level of spontaneous alternation behaviour. In the same study, it was noted that scopolamine increased arm entries in two strains but not the third, whereas it eliminated spontaneous alternation behaviour in all mice. In addition, Drew et al[66] observed no correlation between arm entries and spontaneous alternation behaviour. Nevertheless, the motor effects of modulating purine receptor function may contribute to some instances of apparent changes of learning behaviour. Blockade of A_2 receptors, for example, increases motor activity[234] and still needs to be excluded as a factor in the memory-enhancing effects of A_2 antagonists.

Cellular Actions of Adenosine

Given this evidence for adenosine receptor modulation of learning, what are the cellular processes which may underlie the behavioural effect? This section will be approached by dealing firstly with studies of the effects of purines and their antagonists on synaptic plasticity, and then considering in more detail the various sites and mechanisms of action of adenosine and its analogues which could underlie the change of plasticity.

Once in the extracellular space adenosine is able to act on its receptors to modulate neuronal activity in the central nervous system by a variety of actions including inhibition of the release of neurotransmitters such as glutamate,[33,38,78,191,201] acetylcholine,[51,125,135,154,225] dopamine,[160,263] serotonin[82] and noradrenaline,[120,130] by acting at the A_1 receptor. The A_2 receptors on the other hand tend to increase the release of some of these transmitters[40,126,225] including the all-important glutamate.[42,176,189] Depression of release is directed largely against excitatory transmitters, with little influence on the release of inhibitory transmitters.[109,136,261]

While there is general agreement that A_1 receptors depress, and A_{2A} receptors increase, the release of acetylcholine,[51,125,135,154,225] studies of GABA release have proved more controversial. It has been reported, for example, that A_{2A} receptors can increase[41,155] or decrease[125,134,166] GABA release. It is almost certain that methodological differences, especially the use of radiolabelled versus endogenous material, account for some of these differences, but clarification would be valuable.

The inhibitory effects of adenosine are often mediated by an inhibition of calcium influx into synaptic terminals[5,6,87,205,258] or by a decreased availability of calcium within the terminals to the active site for release.[214] A_1 receptors have been shown to inhibit presynaptic ω–conotoxin sensitive calcium channels,[260] while several authors have shown inhibitory actions on N, P and Q-type channels.[6,98,244] In dissociated hippocampal neurons, Mogul et al[164] described the inhibition by A_1 receptors of N-type currents, while A2B receptors appeared to increase P-type channels. Synaptic currents can still be inhibited by A_1 receptors in the presence of calcium channel blockers, suggesting that there is an additional component to the presynaptic activity of adenosine which is independent of calcium movement.[204]

In postsynaptic cell somata, adenosine can alter neuronal polarisation, A_1 receptors often inducing hyperpolarisation, while A_2 receptors often cause depolarisation.[7,71,141,238] The hyperpolarisations have usually been ascribed to the opening of potassium channels in the hippocampus and elsewhere[99,100,182,196,241] or of chloride channels.[150] The channels on hippocampal pyramidal neurons are sensitive to blockade by barium[21] which not only prevents the direct hyperpolarisation by adenosine, but also prevents the changes by adenosine of spike activation threshold and EPSP / spike coupling.

There is some evidence that adenosine can act partly via ATP-sensitive potassium channels. For example, tolbutamide and glibenclamide, blockers of those channels, can reduce the postsynaptic actions of adenosine, including the changes of EPSP / spike coupling, at concentrations which do not alter the presynaptic actions.[112,174]

Distinguishing the site of action of adenosine—presynaptic or postsynaptic—is difficult to achieve in the mammalian hippocampus in view of the difficulty of recording directly from synaptic terminals. The use of paired-pulse stimuli is far from ideal, but does provide a window on that distinction which lends a different view from simple measures of spike or postsynaptic potential size, or the demands of quantal analysis. Using the paired-pulse approach Higgins and Stone[108] attempted to examine the effects of adenosine specifically on the presynaptic terminals, as described below.

Adenosine and Synaptic Plasticity

The activation of A_1 receptors was shown to suppress the induction of LTP, provided that adenosine (the agonist used in those experiments) was applied within one minute of the inducing tetanus; there was no effect of adenosine if applied 5 minutes after stimulation.[11] A_1 receptors may even respond to endogenous levels of adenosine sufficiently to regulate the degree of LTP and LTD induced by electrical stimulation, since antagonists at these receptors increase the amplitude of both these phenomena.[55,56] However, while Fuji et al[91-93] confirmed the ability of adenosine to restrain the size of LTP, they reported that the presence of the nonselective adenosine receptor antagonist 8-cyclopentyl-theophylline (CPT) also decreased the size of a subsequent depotentiation, implying that in their system, endogenous adenosine was contributing to, or facilitating, the extent of depotentiation. Whether this is simply a species difference between rats and guinea-pigs would be interesting to establish.

There are several reports that adenosine, released spontaneously or as the result of neuronal activation, can participate in the physiological regulation of synaptic transmission. Thus, even low frequency stimulation of hippocampal axons can apparently release enough adenosine to inhibit synaptic transmission.[163] Given the numerous physiological factors which can in turn modulate adenosine levels extracellularly, or can modify the results of activating adenosine receptors, this modulatory role of adenosine could play a pivotal role in many aspects of hippocampal function, including those related to learning and memory. A recently described example of this has been reported by Huang et al.[113] This group studied the depotentiation of hippocampal potentials following the enhancement by LTP. When a period of low frequency stimulation was initiated within a few minutes of the initial LTP, stable depotentiation was obtained which could be mimicked by bath application of adenosine, and was prevented by the A_1 receptor antagonist DPCPX.

Interestingly, the LTP obtained by stimulation in the presence of an A_1 receptor blocker was dependent on the activation of NMDA receptors, whereas LTD induced under similar conditions was not.[54] The intriguing result was that of a dissociation between the effects of adenosine receptor activation (A_1 or A_2) on synaptic potentials and on EPSP-spike coupling[91,92] providing some of the clearest evidence for nonparallel changes of presynaptic and postsynaptic actions of adenosine which in turn imply a complex, state- and environment-sensitive modulation by adenosine of synaptic plasticity.

A_2 receptors also do appear to contribute to classical, NMDA-dependent LTP, since agonists increase[56] and most importantly antagonists reduce the amplitude of tetanus-induced potentiated potentials[91,122,123,212] Antagonists were only effective when applied within a relatively short time window after an inducing tetanus; application after 45 minutes, for example, failed to modify the potentiated potential size, suggesting that the A_2 receptors are more important for the induction of LTP than for its maintenance.[122] The A_{2A} antagonist produced a substantially greater facilitation of depotentiation when studied using evoked excitatory postsynaptic potentials compared with postsynaptic population spikes, suggesting that the effect is preferentially expressed presynaptically rather than postsynaptically.[94]

The A_{2A} receptor population in the hippocampus was reported not to greatly influence electrophysiological activity, possibly because these receptors show low affinity for agonists.[147] This lack of effect of the A_{2A} receptor agonist CGS 21680 was later confirmed by Kessey and Mogul,[122] although less selective agonists could increase synaptic potentials, while antagonists reduced them, leading to the proposal that it is the A2B population which can most readily modulate transmission. However, Sebastiao and Ribeiro[209] used concentrations of CGS 21680 which they believed to be more selective for A_{2A} receptors, and without the complicating activation of A_1 receptors which was noted by Lupica et al.[147] This, and later work from the same laboratory[50] demonstrated that A_{2A} receptors could enhance transmission, an effect which probably stems in part from the ability of A_2 receptor agonists to increase presynaptic calcium conductances.[98,164] The similar enhancement of transmission recorded by Li and Henry[141] was accompanied by a slowly developing depolarisation which was responsible for a post-inhibitory excitatory action of adenosine. Paradoxically, however, the absence of any change of paired-pulse facilitation in response to A_2 receptor activation would seem to exclude a presynaptic site of action.[122] While this paradox has not yet been fully resolved, part of the explanation is that A_{2A} receptors can facilitate postsynaptic responses to AMPA, allowing the emergence of an NMDA receptor-independent form of LTP.[122]

Paired-Pulse Inhibition

Endogenous adenosine may play a significant part in other aspects of synaptic transmission in addition to LTP and LTD. The phenomenon of paired-pulse inhibition is believed to reflect the depletion of presynaptic stores of transmitter and any decrease of that inhibition should indicate a specifically presynaptic inhibitory site of action of an agent. Higgins and Stone[108] concluded that adenosine probably contributed to that fraction of paired-pulse inhibition which was not blocked by bicuculline and was not therefore mediated by GABA release from local interneurones. In the same study it was revealed that adenosine contributed also to the inhibition produced by twin stimuli separated by only 30ms, implying that a rapid release of adenosine might allow this substance to function as a classical neurotransmitter. The reduction of paired-pulse inhibition produced by CPT in the presence of bicuculline was only partly reversible, raising the possibility that endogenous adenosine itself may play a role in long-term changes of neuronal excitability. In terms of understanding fully the relationship between adenosine receptors and learning, it would be valuable to have a clearer view than is available at present on the relative magnitudes of the various actions of adenosine on hippocampal transmission and, in particular, whether all those actions are optimally expressed under the same or different environmental conditions existing under varying physiological and pathological conditions.

Several groups have previously reported an inhibitory effect of adenosine A_1 receptors on population excitatory postsynaptic potentials (popEPSP), population spikes (PS) and the relationship between the two i.e., EPSP-spike (E-S) coupling in the CA1 area of rat hippocampus.[173] The popEPSP gives primarily a measure of membrane potential changes generated by excitatory synapses on the apical dendrites of CA1 pyramidal neurones. The population spike reflects the summated firing of CA1 pyramidal neurones[8] and gives a measure of the excitability of the postsynaptic neurone. EPSP-spike coupling gives an indication of the ability of a given level of synaptic depolarisation to induce the postsynaptic cell to fire an action potential. Intracellularly, changes in excitability can be measured as a change in EPSP slope or as a change in the firing probability of the cell.[2,9,206,237]

Interactions between Adenosine and Cholinergic Neurotransmission

Central cholinergic systems have been widely implicated in learning and memory processes (see ref. 101 for an excellent review; and also the relevant chapters in this book). The postnatal time course of development of cholinergic neurones parallels closely the development of spontaneous alternation behaviour.[76] As more recently observed by Dunbar et al,[68] specific cholinergic markers such as choline acetyltransferase, in areas of the brain believed to be associated with learning such as the hippocampus, correlate with spatial learning ability. The selective block of hippocampal muscarinic M1 receptors has also been shown to impair working memory in rats.[178]

Consistent with this view, cholinergic antagonists have been canvassed as a means of inducing a pharmacological model of the memory disturbance encountered in Alzheimer's and other degenerative disorders.[64,222,223,240] The alkaloid scopolamine has been shown to impair learning in a variety of paradigms and in a range of species including humans[64,222,240] and accordingly reduces alternation scores in rodents and other species.[19,76,116,178,200,226,231,235,240,248]

Cellular Mechanisms of Adenosine / Acetylcholine Interactions

Both adenosine[148] and acetylcholine[192,210] act on presynaptic receptors to regulate glutamate release from synaptic terminals, including those of the CA1 Schaffer collateral and commissural axons. Again, paired-pulse inhibition was used as a sensitive indicator of presynaptic terminal function.[10,29,58,159,252] Nikbakht and Stone[169] demonstrated that, using this protocol, both adenosine, acting at A_1 receptors and oxotremorine-M acting at M2 receptors[24,105] were able to depress transmitter release at short interpulse intervals (10ms), and facilitate release at longer intervals (20 and 50ms) as shown by others.[72]

There is long-standing evidence that the activation of adenosine receptors can suppress responses to muscarinic receptors[43] and similar data have been collected from experiments on sensory ganglia[36] as well as the hippocampus.[23] A more recent study sought to establish whether the interactions between adenosine and muscarinic receptors were apparent specifically on presynaptic terminals.[169] The presynaptic inhibitory effects of A_1 and M2 receptors are occlusive: the combination of agonists at these sites has a less than additive effect upon transmitter release from CA1 terminals (Fig. 1). This suggests that they are acting via a common mechanism. Previous work has suggested that the suppression of transmitter release is mediated by a reduction of calcium influx or calcium availability to the release process.[214] The blockade of calcium channels by adenosine and muscarinic receptors exhibits occlusion[193] and might, therefore, underlie their occlusive interaction on transmitter release. However, several groups have reported that presynaptic cholinomimetic effects in the hippocampus are not mediated by a suppression of calcium channels,[202] so that potassium conductances may be more relevant. Raising extracellular potassium levels or adding 4-aminopyridine to block potassium channels suppressed the responses to both CPA and oxotremorine-M, suggesting that both receptor types are operating by increasing potassium conductance in the axon terminals. These channels may, therefore, represent a common site of action. A similar convergence was reported by McCormick and Williamson[156] on postsynaptic sites. These effects could be secondary to the

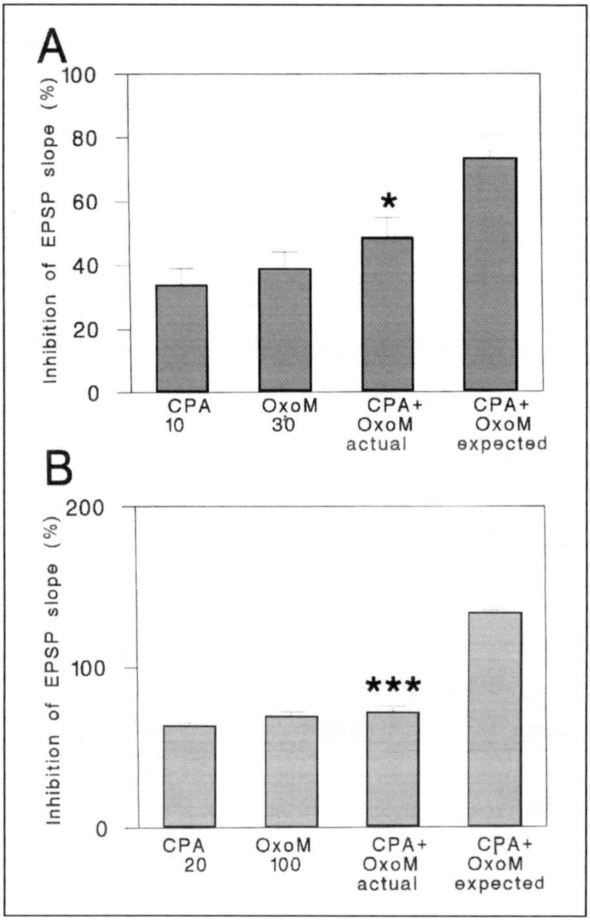

Figure 1. Histograms summarising the effect on the EPSP slope in rat hippocampal slices of two combinations of CPA and oxotremorine-M. In both cases, the effects of CPA and oxotremorine-M produce comparable degrees of inhibition of the response, but their combined addition produces an effect which is not significantly greater than either alone. Both were able to produce 100% inhibition at sufficiently high concentrations. The final column indicates the predicted effect if responses to the two agents had been additive. The actual combined response (ACT) was significantly different from the predicted additive response ($p < 0.001$, $n = 5$ for (A), $n = 3$ for (B)) even when the addition in B was limited to the theoretical maximum of 100%. (Reproduced with permission from ref. 169).

reported effects on calcium conductances, or the calcium changes could be secondary to the changes of potassium movements altering the polarisation state of the presynaptic terminals.

Adenosine and Acetylcholine Release

Acetylcholine is one of the transmitters whose release is modulated by the activation of adenosine receptors. However, while several groups have demonstrated an overall inhibitory effect of adenosine A_1 receptors or stimulation by A_{2A} receptors in the CNS,[50,51,125,135] Cunha et al[48,51] uncovered regional variations within the hippocampus such that release was modulated only by inhibitory A_1 receptors in the CA1 area, whereas it was inhibited by A_1 and increased by A_{2A} receptors in CA3. If subtle differences such as these occur in other areas of brain, and if different regions of hippocampus are involved in different aspects of learning, then dissecting out the relative roles of A_1 and A_2 receptors in learning will not be simple.

The effects of purines on acetylcholine release in the hippocampus in vivo need to be examined however, since Materi et al[154] have demonstrated recently that A_1 receptors suppress evoked but not spontaneous release of acetylcholine from the rat neocortex and, more surprisingly, that A_{2A} receptors did not modify spontaneous or evoked release. It is, therefore, important to establish whether the effects of A_{2A} receptors seen in vitro do not represent an experimental artefact and that they do also occur in the intact, conscious animal, in which the release of acetylcholine is known to be a critical factor in wakefulness. Some differences between neocortex and hippocampus may in fact reflect differences in the source of extracellular adenosine (adenosine efflux or nucleotide metabolism, see ref. 47) and differences in the accessibility of adenosine to A_1 and A_{2A} receptors which they could produce.[47]

Interactions between Purines and Glutamate Receptors

The activation of glutamate receptor subtypes is now known to be important for several aspects of long-term plastic changes including LTP and LTD, and there is now evidence for a variety of ways in which the activation of adenosine receptors can modify the presence or actions of glutamate.

Adenosine and Glutamate Release

As in the case of acetylcholine, there is evidence for a dual modulation by purines of glutamate release in the brain, A_1 receptors inhibiting and A_2 receptors increasing release.[176,189,216]

Adenosine and Glutamate Receptor Interactions

A close relationship may exist between the presence of adenosine receptors and the extent to which NMDA receptors can participate in plastic changes of neurotransmission. Klishin et al[128] noticed that in the presence of an increased ratio of extracellular calcium to magnesium in hippocampal slices, blockade of A_1 adenosine receptors induced a long-lasting increase in the NMDA receptor-mediated component of excitatory postsynaptic currents relative to the nonNMDA component. The authors proposed that a proportion of NMDA receptors may normally be functionally masked by A_1 receptors, and it is these which are made available to the transmission process after A_1 blockade. This would certainly account for the facilitation of learning reported by some groups using A_1 receptor antagonists (e.g., ref. 203), and could be highly relevant to physiological learning (in the absence of pharmacological agents) if the A_1 receptors are inactivated by other transmitters or receptors. The conclusion that there may be a population of 'latent' NMDA receptors suppressed under resting conditions by endogenous levels of adenosine was supported by the demonstration that, after blocking all NMDA receptor function with the channel blocker dizocilpine, an NMDA receptor-mediated component of transmission could be restored by perfusing slices with 8-cyclopentyltheophylline. One explanation of this finding is that removing the influence of endogenous adenosine had again revealed a population of NMDA receptors which had not previously contributed to glutamate sensitivity and which had therefore escaped blockade by the use-dependent agent dizocilpine.[129]

In dissociated hippocampal pyramidal neurons, de Mendonca et al[57] found that A_1 receptor activation would suppress ionic conductances induced by NMDA. This raises the possibility that intense stimulation of neurons, whether pathologically by hypoxia-ischaemia or physiologically during memory formation, might lead to a degree of NMDA receptor activation which is limited by local increases in adenosine concentration. Certainly, Mitchell et al[163] have concluded that adenosine can be released by quite low levels of hippocampal fibre stimulation, reaching local levels high enough to inhibit further transmitter release. This report requires reexamination, however, to assess whether the cells studied were exhibiting a homogeneous response to A_1 receptors, since it has been reported that on a sub-population of striatal neurons

A_1 receptors do not modify NMDA receptor activation, whereas both A_{2A} receptors and A_3 receptors were able to inhibit NMDA-induced currents.[171,256]

As long ago as 1988, it was reported that the presynaptic inhibitory effects of adenosine on glutamate release in the hippocampal CA1 region were dependent on the presence of magnesium, since removal of this ion from the superfusing medium prevented responses to adenosine.[16] This change was later shown to be reproduced by superfusing N-methyl-D-aspartate (NMDA), and prevented by including blockers of the NMDA-sensitive receptors (such as dizocilpine or 2-amino-5-phosphono-pentanoic acid) before the removal of magnesium.[17] This result suggested that activation of NMDA receptors was involved in the suppression of adenosine sensitivity. Also consistent with this view was the weaker ability of adenosine receptor activation to suppress neuronal firing induced by microiontophoretically applied NMDA compared with firing induced by acetylcholine or quisqualate.[18] Although interpreted as consistent with a postsynaptic locus for the interaction between NMDA and adenosine receptors, it is difficult to be certain of the site of action of agents applied by microiontophoresis,[228] and attempts to do so by, for example, lowering extracellular calcium, complicate interpretation by modifying neuronal excitability and receptor function.

There remains a major question as to the site of the adenosine / NMDA interaction—presynaptic or postsynaptic. The interaction has therefore been reexamined using the paired-pulse paradigm, which is widely accepted as providing a more accurate indication of presynaptic events than the study of population spikes and postsynaptic potentials.[106,107,252,258] Paired-pulse inhibition at interpulse intervals of around 10 ms reflects the depletion of transmitter from presynaptic stores,[26,106,252] and is reduced by agents or procedures which decrease transmitter release. Paired-pulse facilitation, on the other hand, at longer interpulse intervals, results from the residual intraterminal calcium which increases transmitter release.[58,107,127,258] There is already ample evidence for the existence of presynaptic glutamate receptors[85] and especially presynaptic NMDA receptors[27,83,121,152,180] on terminals in the hippocampus and other regions of CNS.

Data showed that NMDA receptor activation suppresses the inhibitory effects of adenosine on transmitter release assessed using paired-pulse interactions both with population spikes and population EPSPs. This interaction occurs at levels of NMDA receptor activation which are not themselves sufficient to alter paired-pulse inhibition and strongly suggests that the primary site of the interaction is presynaptic. The fact that the interaction can also be observed in the presence of bicuculline suggests that the receptors involved are likely to be located on the main terminals of the Schaffer collateral fibres, and not on inhibitory interneurones. In addition, the suppression of adenosine sensitivity can be produced by methods other than the direct activation of NMDA receptors. Thus, the induction of LTP, which is known to involve the activation of NMDA receptors by synaptically released glutamate, or the application of exogenous glycine which can enhance the activation of NMDA receptors[162] and induce or facilitate LTP in regions such as the hippocampus[213] and superior colliculus[1,188,249] also reduced adenosine responses. Responses to baclofen were unaffected.

One explanation for some of the earlier data of Bartrup and Stone[17] was proposed by Smith and Dunwiddie,[217] who argued that the effects of magnesium removal could simply reflect the altered balance between calcium and magnesium in determining the amount of transmitter release and thus account for the loss of sensitivity to adenosine. However, the finding that application of NMDA itself mimicked the effects of low magnesium, while NMDA antagonists prevented it, indicates that this cannot represent the whole explanation and that amino acid receptors probably contribute to the phenomenon. Of course, it is still possible that the activation of NMDA receptors changes sensitivity to adenosine by way of an alteration of intracellular calcium levels or availability to the transduction mechanism.

The inhibitory effect of adenosine on population spikes, and the decrease of paired-pulse inhibition assessed using either population spikes or population excitatory postsynaptic potentials, were suppressed by performing the experiments in magnesium-free medium, or by superfusion of the slices with NMDA at a concentration (4 μM) which did not itself affect potential size. The suppressant effect of NMDA was prevented by 2-amino-5-phosphonopentanoic acid. All these interactions were still seen in the presence of bicuculline methobromide, 30 μM. Neither α-amino-3-hydroxy-5-methyl-4-isoxazole-propionic acid (AMPA) nor kainate produced a suppression of adenosine responses. The presence of NMDA did not modify the effects of baclofen on population potentials or paired-pulse inhibition. Activating NMDA receptors by the induction of LTP or by superfusion with glycine also reduced significantly the effects of adenosine on population spikes and paired-pulse interactions. Increasing population potential size by a mechanism which did not involve the activation of NMDA receptors (increasing stimulus strength) did not change sensitivity to adenosine. When adenosine receptor-selective agonists were tested, it was found that NMDA did not modify the inhibitory effect of the A_1 receptor agonist N^6-cyclopentyladenosine (CPA), but did enhance the excitatory effect of the A_{2A} receptor agonist 2-[p- (2-carboxyethyl)phenylethylamino] -5'-N-ethylcarboxamidoadenosine (CGS 21680). The combined response to NMDA and CGS21680 was prevented by the A_{2A} receptor selective antagonist 4- (2- 7-amino-2-(2-furyl)[1,2,4] triazolo [2,3a][1,3,5]triazin-5-ylamino]ethyl)phenol (ZM 241385). It was concluded that NMDA receptor activation can suppress neuronal sensitivity to adenosine by acting at presynaptic sites, and that this interaction results from an increase in the excitatory action of A_{2A} receptors, rather than a depression of A_1 receptor function.[170]

This result appears particularly surprising in view of demonstrations that the activation of A_{2A} receptors can suppress neuronal responses to NMDA in slices and patch-clamp experiments[57,171,256] (Fig. 2). It should be emphasised, however, that the interactions described in the present study involved a concentration of NMDA which was not active when tested alone. It therefore seems that the simultaneous activation of A_{2A} and NMDA receptors at low (subthreshold) concentrations produce an increase of glutamate release and neuronal excitability, whereas their combined activation at higher concentrations - which are themselves depolarising - results in antagonism.

Overall, therefore, NMDA receptor activation seems able to modify selectively the presynaptic responses to activation of A_{2A} adenosine receptors, leading to the masking of adenosine's inhibitory activity on transmitter release. The physiological significance of this is potentially interesting. Craig and White[44] have proposed that adenosine A_1 receptors present a barrier to the actions of NMDA receptors which must be overcome if the full effects of NMDA receptor activation are to be observed in phenomena such as LTP. The present work suggests that part of the mechanism of overcoming this barrier may be that, under conditions in which the amount of adenosine released by neurons and glia is greatly increased so that the relatively low affinity A_{2A} receptors are activated, the inhibitory A_1 receptors effects are overcome. Such a sequence provides at least one rationale for the otherwise curious coexistence of inhibitory A_1 and facilitatory A_{2A} receptors on the same population of glutamatergic terminals, and is consistent with earlier proposals that A_{2A} receptor activation can suppress responses mediated by A_1 receptors.[50,137,173]

By affecting NMDA receptors adenosine may have a fundamental role to play in controlling the dynamics of neuronal interactions. The nonspecific blockade of adenosine receptors has been claimed to block what may be a crucial role for adenosine of preventing dendritic spikes generated by NMDA receptors.[142] Such a blockade releases the tendency, described by many authors in the presence of adenosine antagonists and with specific experimental conditions (e.g., ref. 32) for neurons to generate spontaneous bursts of action potentials.

Other Receptor Interactions

In addition to the potential interactions between purine receptors and acetylcholine and glutamate receptors—two of the neurotransmitters most clearly and consistently related to learning and memory—there is also evidence for interaction between adenosine receptors themselves, between adenosine and dopamine receptors, and between adenosine and peptide receptors.

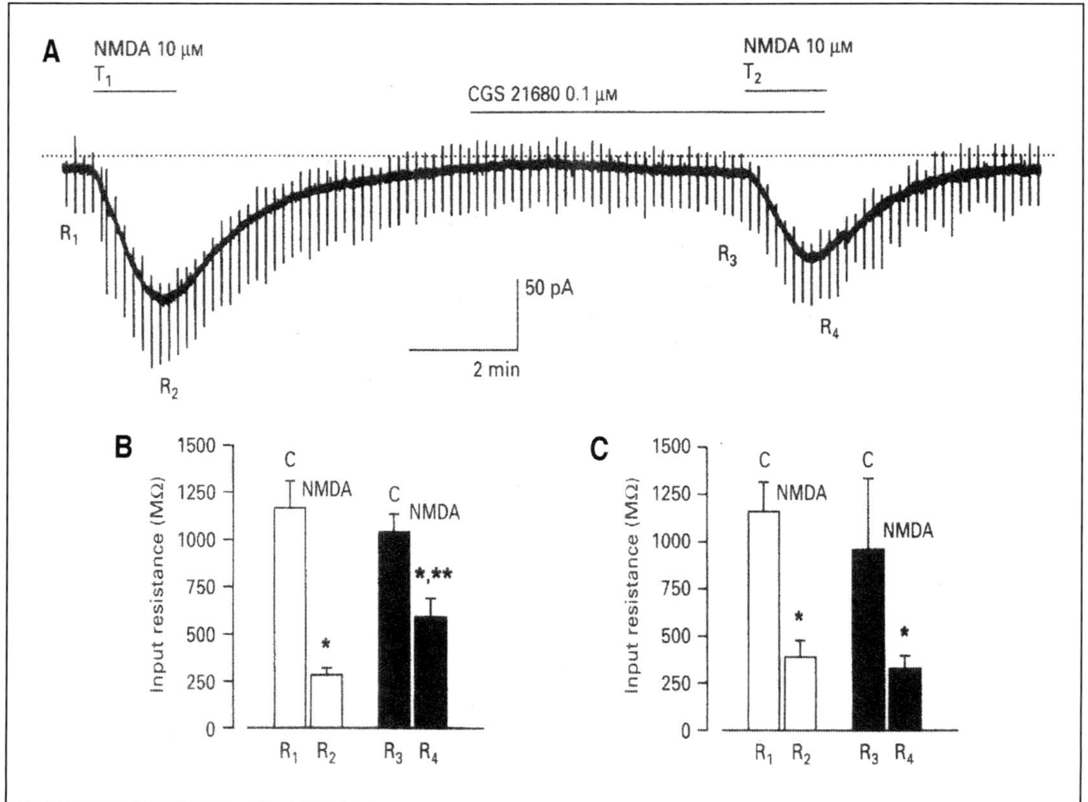

Figure 2. CGS21680 inhibited the conductance of NMDA receptor channels. The holding potential was –80mV in this and all subsequent experiments. (A) Experimental procedure to assess possible effects of CGS21680 on the conductance of NMDA-activated channels. The input resistance of striatal medium spiny neurons was monitored by applying hyperpolarising voltage steps, 10 to 20mV in amplitude, and 100ms in duration every 10 seconds. The input resistance was measured 4 times: immediately before the first application of NMDA (10μM; R1 before T1), during the maximum response to NMDA (10μM; R2 during T1), immediately before the second challenge with NMDA (10μM) in the presence of CGS21680 (0.1μM; R3 before T2), and during the maximum response to NMDA (10μM) in the presence of CGS21680 (0.1μM; R4 during T2). (B) Input resistance values in 8 neurons sensitive to CGS21680. Means of 3 current responses at R1, R2, R3, and R4 respectively were obtained according to the scheme in (A), either in the absence (open columns) or in the presence of CGS21680 (0.1μM; solid columns). *P < 0.001 significant differences from the respective controls (C) in the absence of NMDA (R2 compared with R1, and R4 compared with R3, respectively); **P < 0.001, significant difference between NMDA alone and NMDA plus CGS21680 (R2 and R4). (C) Input resistance in 4 medium spiny neurons which did not respond to CGS21680. Here, the NMDA (10μM)-evoked increase in membrane conductance was uninfluenced by CGS21680 0.1μM; compare R2 with R4). *P0.001, significant differences from the respective controls in the absence of NMDA (10μM). (Reproduced with permission from ref. 171).

Adenosine Receptor Interactions

The activation of both A_{2A}[50,137,173] and A_3 receptors[69] has been shown to suppress the activation of A_1 receptors. There is evidence that these interactions may involve the enhanced desensitisation of the A_1 receptors.[62,69]

One form of interaction between A_{2A} and A_1 receptors is illustrated in Figs. 3 and 4 Activation of adenosine A_1 receptors with the specific agonist CPA caused a greater inhibition of PS amplitudes than popEPSP slopes indicating that the resulting E-S dissociation is due mainly to postsynaptic effects of A_1 receptor activation.[173] When adenosine A_{2A} receptors are coactivated with A_1 receptors, using CGS 21680 along with CPA, the postsynaptic inhibitory effects of A_1 receptor activation are significantly attenuated showing that cross-talk exists between the two types of receptor. The attenuation of A_1 receptor-mediated inhibitory responses by adenosine A_{2A} receptor activation is not due to a functional antagonism between excitatory versus inhibitory effects of the two receptor types, since activation of A_{2A} receptors by CGS21680 does not cause a significant degree of excitation.[173]

Cunha et al[50] have previously shown an attenuation of adenosine A_1 receptor responses on PS amplitude by A_{2A} receptor activation in the rat hippocampus, while Dixon et al[62] reported a desensitisation of adenosine A_1 receptors by A_{2A} receptors in the rat striatum, an effect mediated by protein kinase C. A reduction of A_1 receptor binding can also be demonstrated in the presence of A_{2A} receptor agonists, and this is also mediated via protein kinase C.[146] In the hippocampus, however, we found no evidence that the A_1 receptor effects could be modified by inhibitors of protein kinases (A or C). Previous investigators have also demonstrated a lack of relationship between cAMP levels and the electrophysiological effects of adenosine[70,73,195,257] except in forskolin treated hippocampal slices.[88] There is growing evidence for an interaction between adenosine and nitric oxide systems.[61,157,186] However, neither the competitive nitric oxide synthase inhibitor L-nitroarginine methylester (L-NAME) nor the brain specific inhibitor 7-nitroindazole (7NI) showed any effects on the inhibition obtained upon addition of CPA, suggesting that nitric oxide does not play a significant role in the inhibition seen with A_1 activation.

In contrast, blockade of potassium channels with barium attenuated the postsynaptic actions of adenosine A_1 receptor activation. It is well established that, postsynaptically, adenosine increases potassium conductance,[179,210,241] and it has been shown that barium will selectively block the postsynaptic hyperpolarising effects of adenosine.[4,96,103,238] The E-S dissociation caused by adenosine A_1 receptor activation studied extracellularly, and the directly measured effect on spike threshold recorded intracellularly, are also prevented by barium. The possibility exists that a similar suppression of a potassium current may be the mechanism by which adenosine A_{2A} receptor activation causes inhibition of A_1 receptor-mediated changes of spike threshold. Since barium can block several potassium currents,[37,102] it is not clear which of these might be involved in the A_1/A_{2A} receptor interaction. It is unlikely that the IA current is involved, however, as Pan et al[182] have shown that, whereas barium blocks the postsynaptic hyperpolarisation induced by adenosine, it does not prevent adenosine activation of the A-current.

Adenosine and Dopamine

There are several reports of dopamine receptors exerting a modulatory influence on synaptic plasticity, and there has been much interest in the receptor-receptor interaction between D2 dopamine and A_{2A} adenosine receptors[79,80,83] and between D1 and A_1 receptors.[81] It is not yet clear whether D2 receptors are involved in the physiological regulation of learning and memory, although dopamine itself does suppress spontaneous alternation[253] and D2 dopamine agonists reverse scopolamine-induced memory deficits.[253] It is possible, therefore, that this represents another site at which endogenous purines could interact to modify learning in a psychologically state-dependent manner.

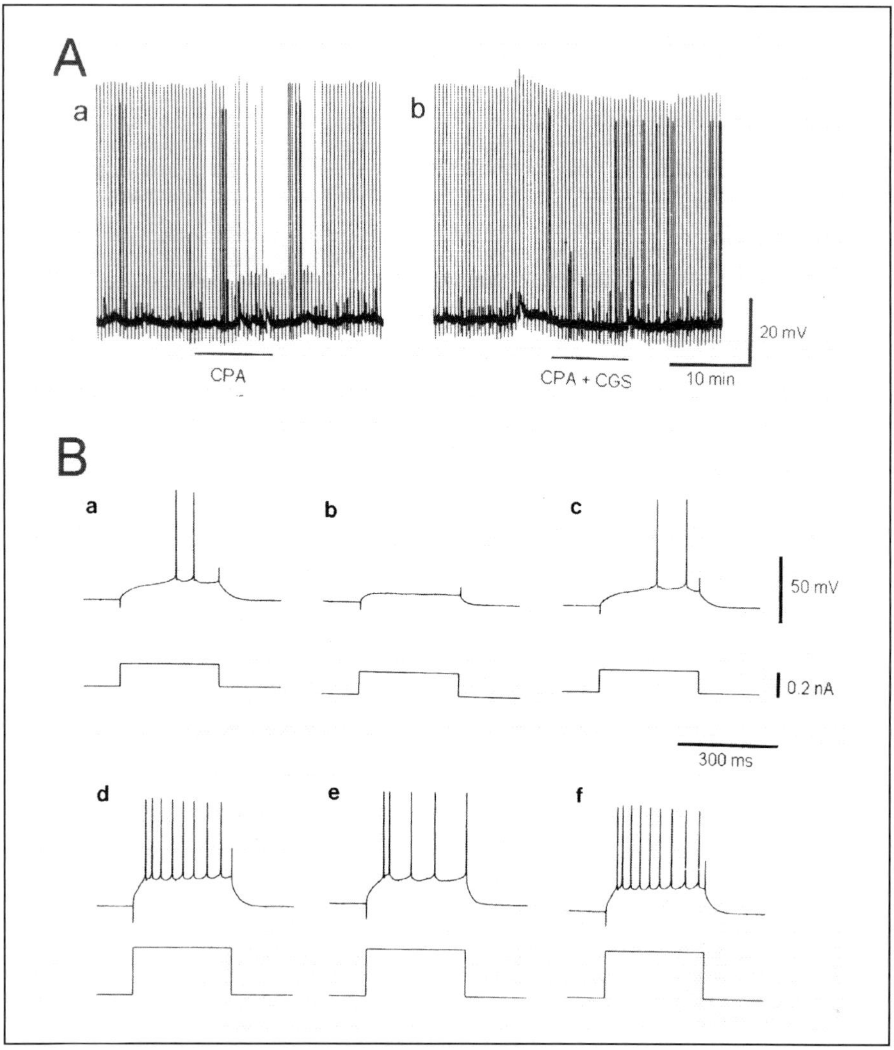

Figure 3. Intracellular recordings from pyramidal neurons in the hippocampus. Record A (a) illustrates the membrane potential of a neurone which is stimulated by a depolarising current pulse just sufficient to induce the production of an action potential on most occasions (0.2 nA delivered every 30s). During the period indicated by the bar below the record, N6-cyclopentyladenosine (CPA) was perfused at 100nM, and causes failure of action potential initiation with no accompanying change of membrane potential. In record A(b), the cell is superfused with CPA 100nM plus CGS 21680 at 30nM. The latter compound was perfused for 15 min before the addition of CPA. The elevation of spike threshold is now blocked and there is some evidence of increased synaptic activity and spontaneous action potentials, with little overall change of membrane potential. Records in B show this effect on a more expanded time scale from a different cell. B(a) and (d) represent responses of the cell to pulses of 0.2 and 0.4nA in the control state. Record (b) shows the failure of spike initiation and (e) a reduced number of spikes produced during superfusion with CPA 100nM. The latter record (e) also shows an increase in the degree of after-hyperpolarisation which probably contributed to the reduced spike number. Records (c) and (f) show the prevention of the CPA effect when coperfused with CGS21680 at 30nM. Calibrations: 20mV and 10 min in A; 50mV, 0.2nA and 300ms for B. (Reproduced with permission from ref. 175).

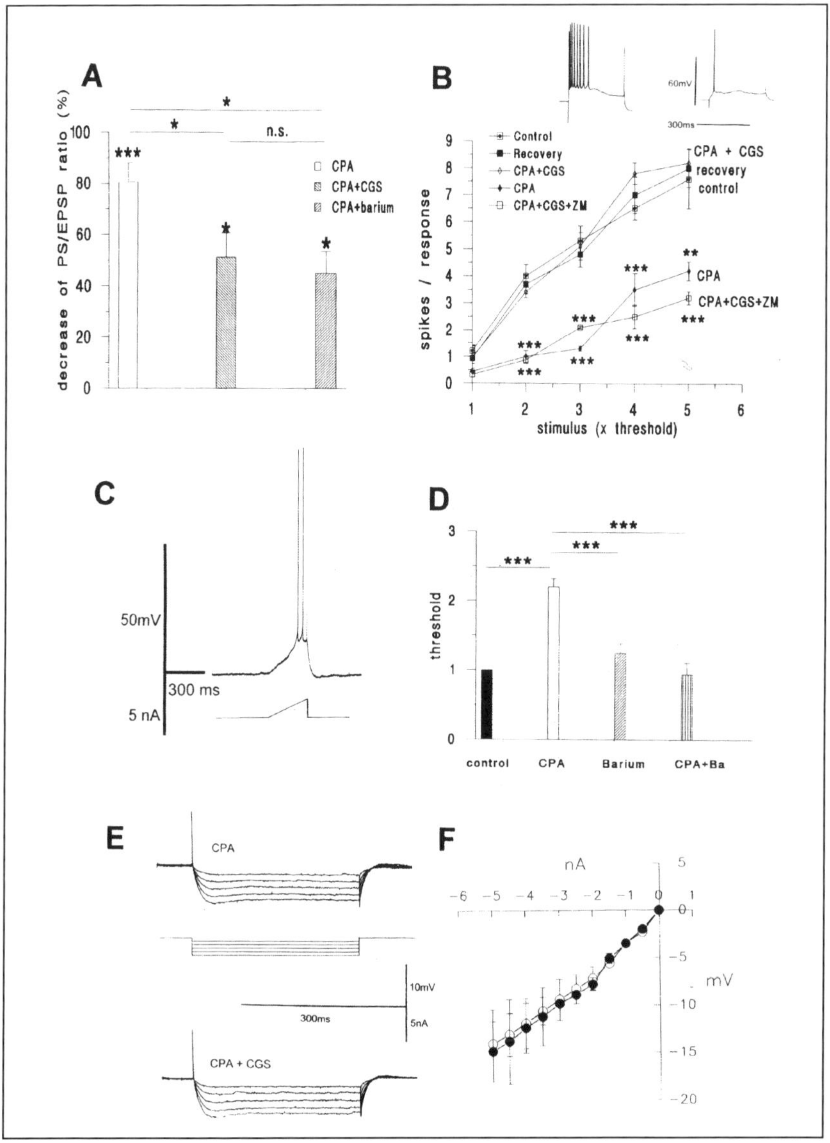

Figure 4. (A) Effect of N6-cyclopentyladenosine (CPA) on the population spikes (PS) and population excitatory postsynaptic potentials (popEPSP) ratio in rat hippocampal slices. While 50nM CPA decreased this by 80.1 ± 6.41% compared with control values, barium or CGS21680 greatly reduced this effect. (B) summarises the number of spikes evoked by intracellular pulses of varying amplitude and shows the depressant effect of CPA, the block of this effect by CGS21680, and the ability of ZM241385 to prevent the effect of CGS21680.(mean ± s.e.mean for n = 5). The insets show representative records at threshold and 5 x threshold for a typical cell. (C)illustrates the use of a depolarising current ramp to determine the threshold for spike initiation, and (D) summarises the results showing the elevation of threshold by CPA, the non-significant increase of threshold by barium and the blockade by barium of the CPA effect. (E) shows representative records of membrane voltage in response to hyperpolarising current pulses in the presence of CPA alone or in CPA plus CGS21680, and (F) summarises the pooled data for CPA (open circles) and CPA with CGS21680 (closed circles) indicating the absence of any changes at the concentrations used here.

Adenosine and Peptides

There is growing evidence for an ability of adenosine receptors to modify responses to certain neuropeptide hormones and neurotransmitters. The targets studied to date include calcitonin gene-related peptide (CGRP) and vasoactive intestinal peptide (VIP). Correia-de-Sa and Ribeiro[39] demonstrated that A_{2A} receptors could facilitate the actions of CGRP at motor nerve terminals, and more recently have shown complex interactions in the hippocampus. A_1 receptors exert a restricting effect on responses to CGRP in the hippocampus, so that effects of the peptide can only be observed when the influence of endogenous adenosine is removed by a suitable antagonist.[208] Conversely, A_{2A} receptors enhance the response to CGRP. The studies on purine/peptide interactions to date have been well reviewed by Ribeiro.[197]

The Effects of Ageing on Adenosine Receptors

Intriguingly for any consideration of purines and learning, Corsi et al[42] have reported that A_{2A} receptors increase spontaneous glutamate release only in young, not old, rats. Although this work was performed in the striatum (in vivo), could a similar change in the hippocampus help to account for the declining memory so often associated with ageing? Interestingly, 8-cyclopentyltheophylline (CPT) has been shown to increase acetylcholine release from hippocampal slices only from relatively young (4 and 12 month) rats, but not from 24-month rats.[224] The change was apparently attributable to a reduced A_1 receptor density rather than affinity. The authors noted that in the older animals, there was a parallel increase in the level of extracellular adenosine, with the possibility that this has induced a down-regulation of A_1 receptors and thus reduced A_1 sensitivity. This in turn could mean that the physiological regulation of synaptic plasticity by adenosine is less effective with ageing.

The decline in the density, though not the affinity, of A_1 receptors in the hippocampus and other regions of brain has also been noted by Cunha et al[46,48] in old animals (24 months) compared with young ones (6 weeks). Conversely, there is an increase in the density of A_{2A} receptors in the hippocampus[46,48] and cortex.[145] These binding studies were supported by electrophysiological data showing a parallel decrease in the efficacy of CPA to decrease neurotransmission by activating A_1 receptors[208] (Fig. 5). However, CPX generated a greater increase of potential size, implying that there was a greater concentration of functionally active A_1 receptors despite the reduced apparent density. The authors attempted to explain this seeming paradox by suggesting that the reduced number of receptors is accompanied by an increased relative activation by endogenous adenosine. This explanation would certainly be consistent with the numerous reports of an increased level of extracellular adenosine in the brain of aged rats, but is difficult to reconcile with the studies of acetylcholine release.[224]

At the junior end of the ageing spectrum, the presynaptic inhibitory effects of A_1 receptors are poorly developed shortly after birth and become increasingly apparent only over the first few weeks of life. This is probably related to the similar time course of development of those adenosine-sensitive processes relevant to transmitter release.[67]

Trophic Functions of Nucleosides

Long-term memory formation is usually assumed to involve some form of permanent or semi-permanent structural change in cells, whether neurons, glia or both. It is therefore pertinent to any discussion of learning to note that a number of purines and pyrimidines have been shown to have marked trophic effects on cells, altering neuronal growth or viability and glial proliferation. While this review is concentrated on the events surrounding the initial establishment of a memory trace, these long-term changes are clearly important and, as more is learnt about their cellular mechanisms and the nature of the receptors involved, these could form new targets for future generation drugs intended to reverse memory and cognitive decline. The trophic actions of purines and related compounds have been the subject of excellent and detailed review by Rathbone et al.[194]

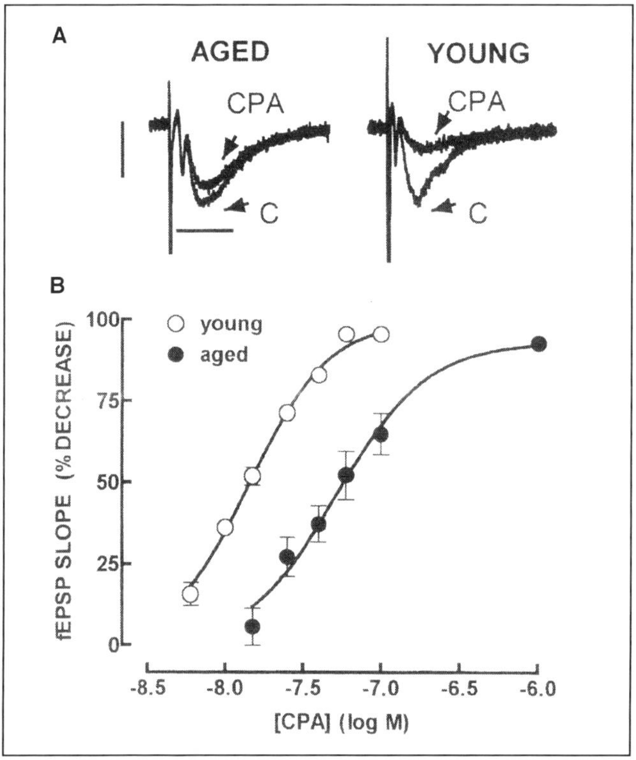

Figure 5. Comparison between the effects of the adenosine A_1 receptor agonist CPA on field excitatory postsynaptic potentials (fEPSPs) recorded from the CA1 area of hippocampal slices taken from young adult (6 weeks) and and aged (24 months) rats. (a) shows trace recordings of averaged fEPSPs obtained in one experiment with an aged rat (left), and in another experiment with a young rat (right); in each panel the fEPSP obtained in the same slice in control conditions (C) and 30-34 mins after the application of CPA (40nM) are superimposed; calibration bars 500μV, 10ms. (b) shows the log-concentration response curves for the inhibitory effects of CPA on the slope of fEPSPs in aged and young adult rats; on the ordinate 0% corresponds to the fEPSP slope before CPA application (0.42 ± 0.06 mV/ms in young and 0.42 ± 0.10 mV/ms in aged rats) and 100% represents the complete inhibition of fEPSPs. The data for each curve were obtained from 4-5 experiments, except for saturating concentrations of CPA (60-100nM) in young animals, which represent results from 2 experiments; the s.e.mean are shown when they exceed the symbols in size. (Reproduced with permission from ref. 208)

Nucleotides and Synaptic Plasticity

ATP has been shown to produce excitation of neurones in several regions of the central nervous system[89,104,232,242] and to modulate membrane potassium[114,167,198,243] or Ca^{2+} conductances.[31,53,131] In addition, it is now recognised that ATP can function as a fast excitatory neurotransmitter in the locus coeruleus,[75,168] peripheral ganglia[95,215] and between cultured neurones[77] often with a pharmacology suggestive of a P2 purinoceptor rather than an indirect effect such as ion chelation or metabolic modification.

Binding and molecular biology data suggest the presence of $P2_{X3}$, $P2_{X4}$ and $P2_{X6}$ receptor subunits and their messenger RNAs in the hippocampus.[12,22,124,139,161,211,219,236] Homomeric assemblies of $P2_{X4}$ subunits respond poorly to αβ-methyleneATP and are relatively insensitive to suramin. Combinations of $P2_{X4}$ and $P2_{X6}$ subunits, however, have been shown recently to be sensitive to the agonist effects of αβ-methyleneATP and blockade by suramin.[139]

Despite this evidence, several studies have failed to detect any consistent functional responses to adenine nucleotides on neuronal networks and synaptic transmission in the mammalian hippocampus which cannot be explained by metabolism to adenosine.[52,73,140,187,229,230] Only depressant effects were noted when ATP was applied by microiontophoresis to single neurones which were spontaneously active or excited by glutamate.[60,229] Furthermore, ATP and derivatives depressed evoked potentials[52,73,230] even when analogues were used which were resistant to hydrolysis and had selective actions on $P2_X$ and $P2_Y$ receptors respectively.[230] Cunha et al[52] have recently performed a careful analysis suggesting that ATP must first be metabolised by ecto-nucleotidases, and that it was the adenosine (or perhaps adenosine 5'monophosphate—ref. 1999) generated, acting at P1 purinoceptors, that caused inhibition of hippocampal synaptic transmission. On the other hand, von Kugelgen et al[245,246] have argued that nucleotides can act directly on P2 receptors to modulate transmitter release. The relative importance of nucleotides acting at P2 receptors, nucleotides acting at P1 receptors, and prior metabolism to nucleosides, has been discussed by Ross et al.[199]

There have also been reports of functional responses to $P2_Y$ receptor stimulation. Zhang et al,[265] for example, indicated that a series of ATP analogues promoted the release of dopamine in the rat striatum, with an order of potency suggesting the mediation of this effect via $P2_Y$ receptors. Most relevant to questions of learning are interactions with glutamate, and Mendoza-Fernandez et al[158] have recently reported that activation of $P2_Y$ receptors can suppress glutamate release from hippocampal pyramidal neurons. If reproducible, this would represent one of the first indications that $P2_Y$ receptors could be involved in synaptic plasticity.

ATP and Synaptic Plasticity

On the other hand, there are recent reports of ATP contributions to synaptic transmission[169,183] as well as reports of long-term potentiation being induced in the hippocampus in response to ATP.[30,90,175,250] The superfusion of rat hippocampal slices with ATP induces an initial depression of evoked potential size which is followed by a rebound facilitation which is not reproduced by αβ-methyleneATP, βγ-methyleneATP, or the dinucleotide P1,P6-diadenosine hexaphosphate (Fig. 6). The post-ATP facilitation can be prevented by the adenosine A_1 receptor antagonists or adenosine deaminase. The adenosine A_{2A} receptor antagonist 8-(chlorostyryl)-caffeine did not affect the inhibition but prevented the post-ATP facilitation. The NMDA receptor antagonist 2-amino-5-phosphonopentanoic acid prevented the establishment of post-ATP facilitation. Suramin prevented the induction but not the maintenance phase of the post-ATP facilitation. The repeated induction of post-ATP facilitation by bursts of electrical stimulation designed to saturate the normal mechanisms of long-term potentiation prevented the induction of post-ATP facilitation. However, repeated applications of ATP to achieve saturation of its receptor did not prevent the subsequent induction of electrically-evoked long-term potentiation. It was concluded that ATP could induce a form of synaptic facilitation which resembles only partially that induced by electrical stimulation and which may require the simultaneous activation of P1 and P2 receptors.

The results suggested that the depressant action of ATP is due to its conversion to adenosine, but that the subsequent enhancement of potential size requires the activation of both P1 and P2-receptors. Finding the precise balance between the activation of these sites to produce long-term potentiation experimentally and confirm this idea may be difficult.

ATP has been shown to elicit the release of glutamate[115] and such an effect could account for the production of the long-lasting facilitation. The involvement of a common mechanism is supported by the finding that the saturation of electrically-induced long-term potentiation prevents the establishment of ATP-induced long-term facilitation. This in turn raises the question of whether electrically-induced long-term potentiation might involve the activation of ATP receptors. Complicating this issue, however, the prior application of ATP did not prevent electrically-induced long-term potentiation, implying a significant difference at some point along the signalling pathways employed by the two procedures. The existence of differences is

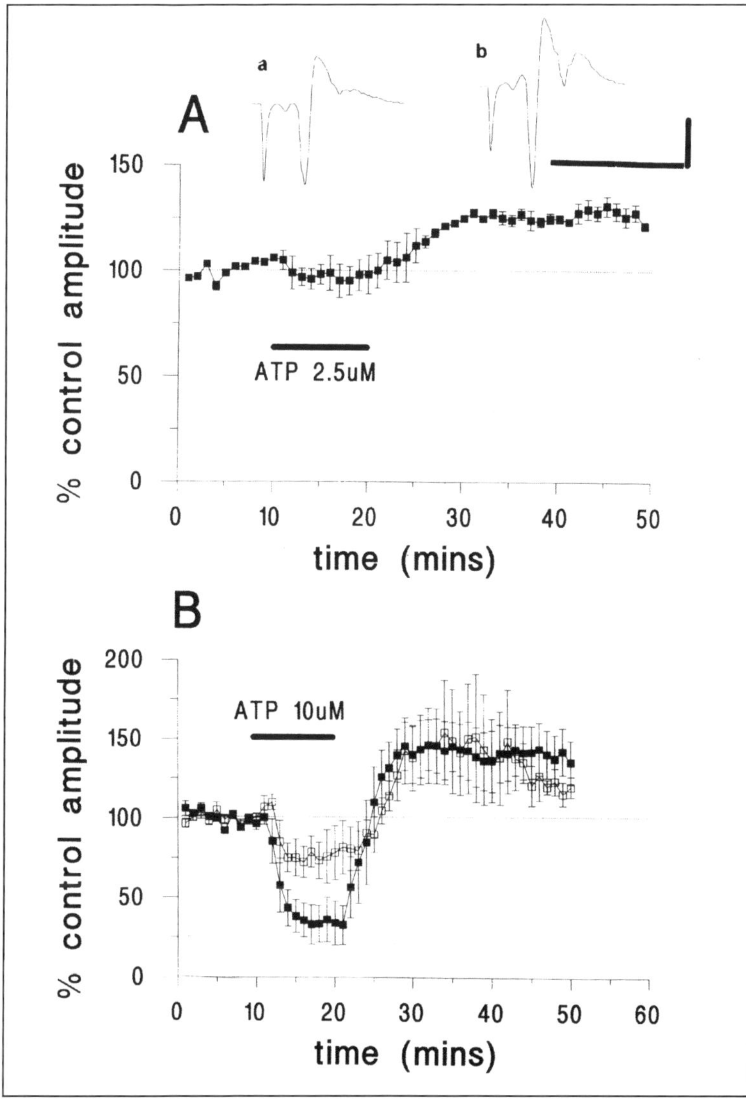

Figure 6. Plots showing the effect of ATP on the evoked potentials in rat hippocampal slices. The plots show the size of the potentials expressed as a percentage of the initial amplitude. In A, ATP at 2.5μM produces no effect during its presence, but generates a facilitation of potential size on washout. The insets show sample records obtained (a) immediately before the addition of ATP and (a) 30 min after ending the ATP perfusion. Plot B illustrates the effect of superfusion of ATP 10μM, which produces an initial depression of population spike and (filled squares; n = 5) and population EPSP (open squares; n = 3), both of which recover to an increased size on washout, although there is a subsequent decline in EPSP slope towards baseline values. Symbols indicate the mean ± s.e.m. Calibration bars in A: 1mV and 20ms. (Reproduced with permission from ref. 173).

supported by the finding that E-S dissociation is not associated with ATP-induced long-term facilitation, whereas an enhancement of E-S coupling always accompanies electrically-evoked long-term potentiation.[20]

It has been suggested that ATP-induced long-term facilitation is due to ecto-protein kinase enzymes using ATP as a substrate, on the basis that stable analogues such as βγ-methyleneATP, αβ-methyleneATP and βγ-imidoATP do not mimic the facilitation induced by ATP[251] and that an inhibitor of ecto-protein kinase, K-252b, prevents the establishment of long-term facilitation by ATP.[90] The accumulated evidence suggests that ATP itself is needed to activate long-term facilitation by a mechanism which may involve protein phosphorylation, but that P2 receptors are also implicated. The latter statement is supported by the use of both suramin and PPADS as receptor blockers. Suramin is known to exhibit a range of other actions, most particularly that of inhibiting ecto-nucleotidases,[207,267,268] an action which could prevent the use of ATP for phosphorylation and thus be consistent with a metabolic, rather than a receptor-mediated involvement in long-term potentiation. Of course, there is no reason why both these mechanisms should be exclusive, and a situation could be envisaged in which the activation of P2 receptors is required at the same time, or within a defined temporal window around the use of ATP in a phosphorylation process. Interestingly, the involvement of ecto-protein kinase in electrically-induced long-term potentiation has been proposed by Chen et al[30] with the demonstration that LTP can be prevented by a monoclonal antibody to the catalytic domain of protein kinase C, increasing the possibility that ATP may contribute to electrically-induced long-term potentiation.

Summary

It is clear that both adenosine and ATP can modulate neuronal activity at a variety of sites and by a range of mechanisms which involve both direct effects as well as the modulation of the release and receptor sensitivity to other neurotransmitters. Taken together with the evidence for the ability of adenosine receptor agonists and antagonists to affect learning at the behavioural level, it seems likely that at least some of these cellular sites of action are relevant to the physiological regulation of learning. Perhaps even more importantly, there is the possibility that targeting purine receptors will provide new agents for the pharmacological enhancement of learning in the elderly population, and those with disorders of cognitive function.

References

1. Abe K, Xie F, Watanabe Y et al. Glycine facilitates induction of long-term potentiation of evoked potentials in rat hippocampus. Neurosci Lett 1990; 117:87-92.
2. Abraham WC, Gustafsson B, Wigstrom H. Long-term potentiation involves enhanced synaptic excitation relative to synaptic inhibition in guinea-pig hippocampus. J Physiol (London) 1987; 394:367-380.
3. Ahlijanian MK, Takemori AE. Effects of R-PIA and caffeine on nociception and morphine-produced analgesia, tolerance and dependence in mice. Europ J Pharmacol 1985; 112:171-179.
4. Akhondzadeh S, Stone T W. Interactions between adenosine and GABA receptors on hippocampal neurones. Brain Res 1994; 665:229-236.
5. Ambrosio AF, Malva JO, Carvallo AP et al. Modulation of calcium channels by activation of adenosine A_1 receptors in rat striatal glutamatergic nerve terminals. Neurosci Lett 1996; 220:163-166.
6. Ambrosio AF, Malva JO, Carvallo AP et al. Inhibition of N-, P/Q- an dother types of calcium channels in rat hippocampal nerve terminals by the adenosine A_1 receptor. Europ J Pharmacol 1997; 340:301-310.
7. Ameri A, Jurna I. Adenosine A_1 and nonA1 receptors: Intracellular analysis of the actions of adenosine agonists and antagonists in rat hippocampal neurones. Brain Res 1991; 546:69-78.
8. Andersen P, Bliss TVP, Skrede KK. Unit analysis of the hippocampal population spike. Exp Brain Res 1971; 13:208-221.
9. Andersen P, Sundberg SH, Sveen O et al. Possible mechanisms for long-lasting potentiation of synaptic transmission in hippocampal slices from guinea-pigs. J Physiol 1980; 302:463-482.
10. Anisman H. Dissociation of disinhibitory effects of scopolamine: Strain and task factors. Pharmacol Biochem Behav 1975; 3:613-618.
11. Arai A, Kessler M, Lynch G. The effects of adenosine on the development of LTP. Neurosci Lett 1990; 119:41-44.
12. Balcar VJ, Li Y, Killinger S et al. Autoradiography of P2x ATP receptors in the rat brain, Brit J Pharmacol 1995; 115:302-306.

13. Ballarin M, Herrera-Marschitz M, Casas M et al. Striatal adenosine levels measured in vivo by microdialysis in rats with unilateral dopamine denervation. Neurosci Lett 1987; 83:338-44.
14. Barraco RA, Coffin VL, Altman H J et al. Central effects of adenosine analogs on locomotor activity in mice and antagonism of caffeine. Brain Res 1983; 272:392-395.
15. Barraco RA, Swanson TH, Phillis JW et al. Anticonvulsant effects of adenosine analogues on amygdala-kindled seizures in rats. Neurosci Lett 1984; 46:317-322.
16. Bartrup JT, Stone TW. Interactions of adenosine and magnesium on rat hippocampal slices. Brain Res 1988; 463:374-379.
17. Bartrup JT, Stone TW. Activation of NMDA receptor-couples channels suppresses the inhibitory action of adenosine on hippocampal slices. Brain Res 1990; 530:330-334.
18. Bartrup JT, Addae JI, Stone TW. Depression of purine induced inhibition during NMDA receptor mediated activation of hippocampal pyramidal cells - an iontophoretic study. Brain Res 1991; 564:323-327.
19. Beninger RJ, Jhamandas K, Boegman RJ et al. Effect of scopolamine and unilateral lesions of the forebrain on T-maze spatial discrimination and alternation in rats. Pharmacol Biochem Behav 1986; 24:1353-1360.
20. Bernard C, Wheal HV. Expression of EPSP/spike potentiation following low frequency and tetanic stimulation in CA1 area of the rat hippocampus. J Neurosci 1995; 15:6542-6551.
21. Birnstiel S, Gerber U, Greene RW. Adenosine-mediated synaptic inhibition—partial blockade by barium does not prevent anti-epileptiform activity. Synapse 1992; 11:191-196.
22. Bo X, Burnstock G. Distribution of $\alpha\beta$-methylene ATP binding sites in rat brain and spinal cord. Neuro Report 1994; 5:1601-1604.
23. Brooks PA, Stone TW. Purine modulation of cholinomimetic responses in the rat hippocampal slice. Brain Res 1988; 458:106-114.
24. Brundege JM, Dunwiddie TV. Modulation of excitatory synaptic transmission by adenosine released from single hippocampal neurons. J Neurosci 1996; 16:5603-5612.
25. Bruns RF, Fergus JH, Badger EW et al. Binding of the A_1-selective adenosine antagonist 8-cyclopentyl-1,3-dipropylxanthine to rat brain membranes. Arch Pharmacol 1987; 335:59-63.
26. Burke JP, Hablitz JJ. Presynaptic depression of synaptic transmission mediated by activation of metabotropic glutamate receptors in rat neocortex. J Neurosci 1994; 14:5120-5130.
27. Cai N, Kiss B, Erdo L. Heterogeneity of NMDA receptors regulating the release of dopamine and acetylcholine from striatal slices. J Neurochem 1991; 57:2148-2151.
28. Carswell HV, Graham DI, Stone TW. Kainate-evoked release of adenosine from the hippocampus of the anaesthetised rat: Possible involvement of free radicals. J Neurochem 1997; 68:240-247.
29. Chen Y, Graham DI, Stone TW. Release of endogenous adenosine and its metabolites by the activation of NMDA receptors in the rat hippocampus in vivo. Brit J Pharmacol 1992; 106:632-638.
30. Chen W, Wieraszko A, Hogan MV et al. Surface protein phosphorylation by ecto-protein kinase is required for the maintenance of hippocampal long-term potentiation. Proc Natl Acad Sci USA 1996; 93:8688-8693.
31. Chen ZP, Levy A, Lightman SL. Activation of specific ATP receptors induces a rapid increase of intracellular Ca2+ in rat hypothalamic neurones. Brain Res 1994; 641:249-256.
32. Chesi AJR, Stone TW. Alkylxanthine-induced epileptiform activity in the rat hippocampal slice. Exp Brain Res 1997; 113:303-310.
33. Clark M, Dar MS. Release of endogenous glutamate from rat cerebellar synaptosmes: Interactions with adenosine and ethanol. Life Sci 1989; 44:1625-1635.
34. Coffin VL, Spealman RD. Modulation of the behavioural effects of clordiazepoxide by methylxanthines and analogues of adenosine in squirrel monkeys. J Pharmacol Exp Ther 1985; 235:724-728.
35. Coffin VL, Spealman RD. Behavioural and cardiovascular effects of analogues of adenosine in cynomolgus monkeys. J Pharmacol Exp Ther 1987; 241:76-83.
36. Connolly GP, Stone TW. Adenosine selectively depresses muscarinic compared with nonmuscarinic receptor mediated depolarisation of the rat superior cervical ganglion. Gen Pharmacol 1995; 26:865-873.
37. Cook NS. The pharmacology of potassium channels and their therapeutic potential. Trends Pharmacol Sci 1988; 9:21-28.
38. Corradetti R, Le Conte F, Moroni F et al. Adenosine decreases aspartate and glutamate release from rat hippocampal slices. Europ J Pharmacol 1984; 140:19-26.
39. Correia-de-Sa P, Ribeiro JA. Potentiation by tonic A_{2A} adenosine receptor activation of CGRP-facilitated [3H]acetylcholine release from rat motor nerve endings. Brit J Pharmacol 1994; 111:582-588.
40. Correia-de-Sa P, Sebastiao AM, Ribeiro JA. Inhibitory and excitatory effects of adenosine receptor agonists on evoked transmitter release from phrenic nerve endings of the rat. Brit J Pharmacol 1991; 103:1614-1620.
41. Corsi C, Melani A, Bianchi L et al. Effect of adenosine A_{2A} stimulation on GABA release from the striatum of young and aged rats in vivo. NeuroReport 1999; 10:3933-3937.

42. Corsi C, Melani A, Bianchi L et al. Striatal A_{2A} adenosine receptors differentially regulate spontaneous anad K+-evoked glutamate release in vivo in young and aged rats. NeuroReport 1999; 10:687-691.
43. Cox RT, Walker RJ. The inhibition of an acetylcholine-induced current in cell F1 of Helix aspersa by adenosine. Brit J Pharmacol 1985; 86:449p.
44. Craig CG, White TD. Low level NMDA receptor activation provides a purinergic threshold against further NMDA-mediated neurotransmission in the cortex. J Pharmacol Exp Therap 1992; 260:1278-1284.
45. Crawley JN, Patel J, Marangos PJ. Behavioural characterization of two long lasting adenosine analogues: Sedative properties and interaction with diazepam. Life Sci 1981; 29:2623-2630.
46. Cunha RA, Constantino MD, Sebastiao AM et al. Modification of A_1 and A_{2A} adenosine receptor binding in aged striatum, hippocampus and cortex of the rat. NeuroReport 1995; 6:1583-1588.
47. Cunha RA, Correia-de-Sa P, Sebastiao AM et al. Preferential activation of excitatory adenosine receptors at rat hippocampal and neuromuscular synapses by adenosine formed from released adenine nucleotides. Brit J Pharmacol 1996; 119:253-260.
48. Cunha RA, Johansson B, Fredholm BB et al. Adenosine A_{2A} receptors stimulate acetylcholine release from nerve terminals of the rat hippocampus. Neurosci Lett 1995; 196:41-44.
49. Cunha RA, Johansson B, Constantino MD et al. Evidence for high affinity binding sites for the adenosine A_{2A} receptor agonist [3H]CGS21680 in the rat hippocampus and cerebral cortex that are different from striatal A_{2A} receptors. Naunyn-Schmiedeberg's Ach Pharmacol 1996; 353:261-271.
50. Cunha RA, Johansson B, Van Der Ploeg I et al. Evidence for functionally important adenosine A_{2A} receptors in the rat hippocampus. Brain Res 1994; 649:208-216.
51. Cunha RA, Milusheva E, Vizi ES et al. Excitatory and inhibitory effects of A_1 and A_{2A} adenosine receptor activation on the electrically evoked [3H]acetylcholine release from different areas of the rat hippocampus. J Neurochem 1994; 63:207-214.
52. Cunha RA, Sebastião AM, Ribeiro JA. Inhibition by ATP of hippocampal synaptic transmission requires localized extracellular catabolism by ecto-nucleotidases into adenosine and channeling to adenosine A_1 receptors. J Neurosci 1998; 18:1987-1995.
53. Dave S, Mogul DJ. ATP receptor activation potentiates a voltage-dependent calcium channel in hippocampal neurones. Brain Res 1996; 715:208-216.
54. De Mendonca A, Ribeiro JA. LTP observed upon blockade of adenosine A_1 receptors in rat hippocampus is NMDA receptor-dependent. Neurosci Lett 2000; 291:81-84.
55. De Mendonca A, Almeida A, Bashir ZI et al. Endogenous adenosine attenuates long-term depression and depotentiation in the CA1 region of the rat hippocampus. Neuropharmacol 1997; 36:161-167.
56. De Mendonca A, Ribeiro JA. Endogenous adenosine modulates LTP in the hippocampus. Neurosci 1994; 62:385-390.
57. De Mendonca A, Sebastiao AM, Ribeiro JA. Inhibition of NMDA receptor-mediated currents in isolated rat hippocampal neurones by adenosine A_1 receptor activation. NeuroReport 1995; 6:1097-1100.
58. Debanne D, Guerineau NC, Gahwiler B et al. Paired-pulse facilitation and depression at unitary synapses in rat hippocampus: Quantal fluctuation affects subsequent release. J Physiol 1996; 491:163-176.
59. Dennis WJ. Spontaneous alternation in rats as an indicator of the persistence of stimulus traces. J Comp Psychol 1939; 28:305-312.
60. Di Cori S, Henry JL. Effects of ATP and AMP on hippocampal neurones of the rat in vitro. Brain Res Bull 1984; 13:199-201.
61. Dirnagl U, Niwa K, Lindauer U et al. Coupling of cerebral blood flow to neuronal activation: Role of adenosine and nitric oxide. Am J Physiol 1994; 267:H296-H301.
62. Dixon AK, Widdowson L, Richardson PJ. Desensitisation of the A_1 receptor by the A_{2A} receptor in rat striatum. J Neurochem 1997; 69:315-321.
63. Douglas RJ, Isaacson RL. Spontaneous alternation and scopolamine. Psychonom Sci 1966; 4:283-284.
64. Drachman DA. Memory and cognitive function in man: Does the cholinergic system have a specific role? Neurology 1977; 27:783-790.
65. Dragunow M. Purinergic mechanisms in epilepsy. Progr Neurobiol 1988; 31:85-108.
66. Drew WG, Miller LL, Baugh EL. Effects of THC, LSD-25 and scopolamine on continuous, spontaneous alternation in the Y-maze. Psychopharmacol 1973; 32:171-182.
67. Dumas TC, Foster TC. Late developmental changes in the ability of adenosine A_1 receptors to regulate synaptic transmission in the hippocampus. Develop Brain Res 1998; 105:137-139.
68. Dunbar GL, Rylett RJ, Schmidt BM et al. Hippocampal choline acetyltransferase activity correlates with spatial learning in aged rats. Brain Res 1993; 604:266-272.
69. Dunwiddie TV, Diao L, Kim HO et al. Activation of hippocampal adenosine A_3 receptors produces a desensitisation of A_1 receptor-mediated responses in rat hippocampus J Neurosci 1997; 17:607-614.

70. Dunwiddie TV, Fredholm BB. Adenosine receptors mediating inhibitory electrophysiological responses in rat hippocampus are different from receptors mediating cyclic AMP accumulation. Naunyn-Schmiedeberg's Arch Pharmacol 1984; 326:294-301.
71. Dunwiddie TV, Fredholm BB. Adenosine A_1 receptors inhibit adenylate cyclase and neurotransmitter release and hyperpolarise pyramidal neurons in rat hippocampus. J Pharmacol Exp Therap 1989; 249:31-37.
72. Dunwiddie TV, Haas HL. Adenosine increases synaptic facilitation in the in vitro rat hippocampus, evidence for presynaptic site of action. J Physiol 1985; 373:47-62.
73. Dunwiddie TV, Hoffer BJ. Adenine nucleotides and synaptic transmission in the invitro rat hippocampus. Br J Pharmacol 1980; 69:59-68.
74. Dunwiddie TV, Worth T. Sedative and anticonvulsant effects of adenosine analogs in mouse and rat. J Pharmacol Exp Therap 1982; 220:70-76.
75. Edwards FA, Gibb AJ, Colquoun D. ATP-receptor-mediated synaptic currents in the central nervous system. Nature 1992; 359:144-147.
76. Egger GJ, Livesey PJ, Dawson RG. Ontogenetic aspects of central cholinergic involvement in spontaneous alternation behaviour. Develop Psychobiol 1973; 6:289-299.
77. Evans RJ, Derkach V, Surprenant A. ATP mediates fast synaptic transmission in mammalian neurones. Nature 1992; 357:503-505.
78. Fastbom J, Fredholm BB. Inhibition of (3H) glutamate release from rat hippocampal slices by L-PIA. Acta Physiol Scand 1985; 125:121-123.
79. Ferre S. Adenosine-dopamine interactions in the ventral striatum. Psychopharmacol 1997; 133:107-120.
80. Ferre S, Von Euler G, Johansson J et al. Stimulation of high-affinity adenosine-A_2 receptors decreases the affinity of dopamine D2 receptors in rat striatal membranes. Proc Natl Acad Sci USA 1991; 88:7238-7241.
81. Ferre S, Rimondini R, Popoli P et al. Stimulation of adenosine A_1 receptors attenuates dopamine D1 receptor-mediated increases of NGF1A, c-fos and jun-B mRNA levels in the dopamine-denervated striatum and dopamine D1 receptor-mediated turning behaviour. Europ J Neurosci 1999; 11:3884-3892.
82. Feuerstein TJ, Hertling G, Jackisch R. Modulation of hippocampal serotonin release by endogenous adenosine. Europ J Pharmacol 1985; 107:233-242.
83. Fink JS, Weaver DR, Rivkees SA et al. Molecular cloning of the rat A_2 adenosine receptor: Selective coexpression with D_2 dopamine receptors in rat striatum. Mol Brain Res 1992; 14:186-195.
84. Fink K, Bonisch H, Gothert M. Presynaptic NMDA receptors stimulate noradrenaline release in the cerebral cortex. Europ J Pharmacol 1990; 185:115-117.
85. Forsythe ID, Clements JD. Presynaptic glutamate receptors depress excitatory monosynaptic transmission between mouse hippocampal neurones. J Physiol 1990; 429:1-16.
86. Fraser CM, Fisher A, Cooke MJ et al. Purine modulation of dizocilpine effects on spontaneous alternation. Psychopharmacol 1997; 130:334-342.
87. Fredholm BB, Dunwiddie TV. How does adenosine inhibits transmitter release. Trends Pharmacol Sci 1988; 9:130-134.
88. Fredholm BB, Jonzon B, Lindström K. Adenosine receptor mediated increases and decreases in cyclic AMP in hippocampal slices treated with forskolin. Acta Physiol Scand 1983; 117:461-463.
89. Frohlich R, Boehm S, Illes P. Pharmacological characterisation of P2 purinoceptor types in rat locus coeruleus neurones, Europ J Pharmacol 1996; 315:255-261.
90. Fuji S, Kato H, Furuse H et al. The mechanism of ATP-induced long-term potentiation involves extracellular phosphorylation of membrane proteins in guinea-pig hippocampal CA1 neurons. Neurosci Lett 1995; 187:130-132.
91. Fuji S, Kato H, Ito KI et al. Effects of A_1 and A_2 adenosine receptor antagonists on the induction and reversal of LTP in guinea-pig hippocampal slice of CA1 neurons. Cell Molec Neurobiol 2000; 20:331-350.
92. Fuji S, Kuroda Y, Ito K et al. Effects of adenosine receptors on the synaptic and EPSP-spike components of LTP and depotentiation in the guinea-pig hippocampus. J Physiol 1999; 521:451-466.
93. Fuji S, Sekino Y, Kuroda Y et al. 8-cyclopentyltheophylline an adenosine A_1 receptor antagonist inhibits the reversal of long-term potentiation in hippocampal CA1 neurons. Europ J Pharmacol 1997; 331:9-14.
94. Fuji S, Wakizaka A, Sekino Y et al. Adenosine A_2 receptor antagonist facilitates the reversal of LTP (depotentiation) of evoked postsynaptic potentials but inhibits that of population spikes in hippocampal CA1 neurons. Neurosci Lett 1992; 148:148-150.
95. Galligan JJ, Bertrand PP. ATP mediates fast synaptic potentials in enteric neurones. J Neurosci 1994; 14:7563-7571.
96. Gerber U, Green RW, Haas HL et al. Characterization of inhibition mediated by adenosine in the hppocampus of the rat in vitro. J Physiol 1989; 417:567-578.

97. Glanzer M. Stimulus satiation: An explanation of spontaneous alternation and related phenomena. Psychol Rev 1953; 60:257-268.
98. Goncalves ML, Cunha RA, Ribeiro JA. Adenosine $A(_{2A})$ receptors facilitate 45Ca2+ uptake through class A calcium channels in rat hippocampal CA3 but not CA1 synaptosomes. Neurosci Lett 1997; 238:73-77.
99. Greene RW, Haas H L. Adenosine actions on CA_1 pyramidal neurones in rat hippocampal slices. J Physiol 1985; 366:119-127.
100. Greene RW, Haas HL. The electrophysiology of adenosine in the mammalian CNS. Progr Neurobiol 1991; 36:329-341.
101. Grette Lydon R. Cholinergic neurones and memory: And overview and historical perspective. In Stone TW, ed. CNS Neurotransmitters and Neuromodulators, Acetylcholine. Boca Raton: CRC Press, 1994:1:Chapter 14.
102. Guatteo E, Federici M, Siniscalchi A et al. Whole cell patch-clamp recordings of rat midbrain dopaminergic neurons isolate a sulphonylurea- and ATP-sensitive component of potassium currents activated by hypoxia. J Neurophysiol 1998; 79:1239-1245.
103. Haas HL, Greene RW. Adenosine enhances afterhyperpolarization and accommodation in hippocampal pyramidal cells. Pflugers Archiv - Europ J Physiol 1984; 402:244-247.
104. Harms L, Finta EP, Tschopl M et al. Depolarisation of rat locus coeruleus neurones by ATP. Neurosci 1992; 48:941-952.
105. Harvey J, Lacey MG. A postsynaptic interaction between dopamine D1 and NMDA receptors promotes presynaptic inhibition in the rat nucleus accumbens via adenosine release. J Neurosci 1997; 17:5271-5280.
106. Hashimoto K, Kano M. Presynaptic origin of paired-pulse depression at climbing fibres-Purkinje cell synapses in the rat cerebellum. J Physiol 1998; 506:391-405.
107. Hess G, Kuhnt U, Voronin LL. Quantal analysis of paired-pulse facilitation in guinea-pig hippocampal slices. Neurosci Lett 1987; 77:187-192.
108. Higgins MJ, Stone TW. The contribution of adenosine to paired-pulse inhibition in the normal and disinhibited hippocampal slice. Europ J Pharmacol 1996; 317:215-223.
109. Hollins C, Stone TW. Adenosine inhibition of GABA release from slices of rat cerebral cortex. Brit J Pharmacol 1980; 69:107-112.
110. Holmgren M, Hednar T, Nordberg G et al. Anti-nociceptive effects in the rat of an adenosine analogue N6-phenylisopropyladenosine. J Pharm Pharmacol 1983; 35:679.
111. Hooper N, Fraser C, Stone TW. Effects of purine analogues on spontaneous alternation in mice. Psychopharmacol 1996; 123:250-257.
112. Hosseinzadeh H, Stone TW. Tolbutamide blocks postsynaptic but not presynaptic effects of adenosine on hippocampal CA1 neurons. J Neural Transm 1998; 105:161-172.
113. Huang CC, Liang YC, Hsu. A role for extracellular adenosine in time-dependent reversal of LTP by low-frequency stimulation at hippocampal CA1 synapses. J Neurosci 1999; 19:9728-9738.
114. Ikeuchi Y, Nishizaki T. ATP-evoked potassium currents in rat striatal neurones are mediated by a P2 purinergic receptor, Neurosci Lett 1995; 19089-92.
115. Inoue K, Nakazawa K, Fujimori K et al. Extracellular ATP-evoked glutamate release in cultured hippocampal neurons. Neurosci Lett 1992; 134:215-218.
116. Itoh J, Ukai M, Kameyama T. Dynorphin A(1-13) markedly improves scopolamine-induced impairment of spontaneous alternation performance in mice. Europ J Pharmacol 1993; 236:341-345.
117. Jarvis MF, Williams M. Direct autoradiographic localisation of adenosine A_2 receptors in the brain using the A_2-selective agonist, [^{3}H]CGS 21680. Europ J Pharmacol 1989; 168:243-246.
118. Johansson B, Fredhom BB. Further characterisation of the binding of the adenosine receptor agonist [3H]CGS21680 to rat brain using autoradiography. Neuropharmacol 1995; 34:393-403.
119. Johansson B, Georgiev V, Parkinson FE et al. The binding of the adenosine A_2-selective agonist [3H]CGS 21680 to rat cortex differs from its binding to rat striatum. Europ J Pharmacol 1993; 247:103-110.
120. Jonzon B, Fredholm BB. Adenosine receptor mediated inhibition of noradrenaline release form slices of rat hippocampus. Life Sci 1984; 35:1971-1979.
121. Kato K, Zorumski CF. Modulation of long-term potentiation in the hippocampus by NMDA-mediated presynaptic inhibition. Neurosci 1999; 92:1261-1272.
122. Kessey K, Mogul DJ. NMDA-independent LTP by adenosine A_2 receptor-mediated postsynaptic AMPA potentiation in hippocampus. J Neurophysiol 1997; 78:1965-1972.
123. Kessey K, Trommer BL, Overstreet LS et al. A role for adenosine A_2 receptors in the induction of LTP in the CA1 region of rat hippocampus. Brain Res 1997; 756:184-190.
124. Kidd EJ, Grahames CBA, Simon J et al. Localisation of P2X purinoceptor transcripts in the rat nervous system. Molec Pharmacol 1995; 48:569-573.
125. Kirk IP, Richardson PJ. Adenosine A_{2A} receptor-mediated modulation of striatal [3H]GABA and [3H]acetylcholine release. J Neurochem 1994; 62:960-966.

126. Kirkpatrick KA, Richardson PJ. Adenosine receptor mediated modulation of acetylcholine release from rat striatal synaptosomes. Brit J Pharmacol 1993; 110:949-954.
127. Kleschevnikov A M, Sokolov M V, Kuhnt V et al. Changes in paired-pulse facilitation correlate with induction of long-term potentiation in area CA1 of rat hippocampal slices. Neurosci 1997; 76:829-843.
128. Klishin A, Lozovaya N, Krishtal O. A_1 adenosine receptors differentially regulate the NMDA and nonNMDA receptor-mediated components of hippocampal excitatory postsynaptic current in a Ca/Mg-dependent manner. Neurosci 1995; 65:947-953.
129. Klishin A, Tsintsadze T, Lozovaya N et al. Latent NMDA receptors in the recurrent excitatory pathway between hippocampal CA1 pyramidal neurons: Calcium-dependent activation by blocking adenosine A_1 receptors. Proc Natl Acad Sci USA 1995; 92:12431-12435.
130. Koch H, von Kugelgen I, Starke K. P2-receptor mediated inhibition of noradrenaline release in the rat hippocampus. Naunyn.-Schmied Arch Pharmacol 1997; 355:707-715.
131. Koizumi S, Inoue K. Inhibition by ATP of calcium oscillations in rat cultured hippocampal neurones, Brit J Pharmacol 1997; 122:51-58.
132. Kokkinidis L, Anisman H. Interaction between cholinergic and catecholaminergic agents in a spontaneous alternation task. Psychopharmacol 1976; 48:261-265.
133. Kopf SR, Melani A, Pedata F et al. Adenosine and memory storage: Effect of A_1 and A_2 receptor antagonists. Psychopharmacol 1999; 146:214-219.
134. Kurokawa M, Kirk IP, Kirkpatrick KA et al. Inhibition by KF17837 of adenosine A_{2A} receptor-mediated modulation of striatal GABA and acetylcholine release. Brit J Pharmacol 1994; 113:43-48.
135. Kurokawa M, Koga K, Kase H et al. Adenosine A_{2A} receptor- mediated modulation of striatal acetylcholine release in vivo. J Neurochem 1996; 66:1882-1888.
136. Lambert NA, Teyler TJ. Adenosine depresses excitatory but not fast inhibitory transmission in area CA1 of the rat hippocampus. Neurosci Lett 1991; 122:50-52.
137. Latini S, Bordoni F, Pedata F et al. Extracellular adenosine concentrations during in vitro ischaemia in rat hippocampal slices. Br J Pharmacol 1999; 127:729-739.
138. Latini S, Pazzagli M, Pepeu G et al. A_2 adenosine receptors: Their presence and neuromodulatory role in the central nervous system. Gen Pharmacol 1996; 27:925-933.
139. Le KT, Villeneuve P, Ramjaun AR et al. Sensory presynaptic and widespread somatodendritic immunolocalization of central ionotropic P2X ATP receptors. Neurosci 1998; 83:177-190.
140. Lee KS, Schubert P, Emmert H et al. Effect of adenosine versus adenine nucleotides on evoked potentials in a rat hippocampal slice preparation. Neurosci Lett 1981; 23:309-314.
141. Li H, Henry JL. Adenosine A_2 receptor mediation of preand postsynaptic excitatory effects of adenosine in rat hippocampal slices. Europ J Pharmacol 1998; 347:173-182.
142. Li H, Henry JL. Adenosine receptor blockade reveals NMDA receptor- and voltage-sensitive dendritic spikes in rat hippocampal CA1 pyramidal cells in vitro. Neurosci 2000; 100:21-31.
143. Lindstrom K, Ongini E, Fredholm BB. The selective A_{2A} receptor antagonist SCH 58261 discriminates between two different binding sites for [3H]CGS 21680 in the rat striatum. Arch Pharmacol 1996; 354:539-541.
144. Lohse MJ, Klotz KN, Lindenborn-Fotinos J et al. 8-Cyclopentyl-1,3-dipropyl-xanthine (DPCPX) - a selective high affinity antagonist radioligand for A_1 adenosine receptors. Arch Pharmacol 1987; 336:204-210.
145. Lopes LV, Cunha RA, Ribeiro JA. Increase in the number, G-protein coupling and efficiency of facilitatory adenosine A_{2A} receptors in the limbic cortex, but not striatum of aged rats. J Neurochem 1999; 73:1733-1738.
146. Lopes LV, Cunha RA, Ribeiro JA. Crosstalk between A_1 and A_{2A} adenosine receptors in the hipocampus and cortex of young adult and aged rats. J Neurophysiol 1999; 82:3196-3203.
147. Lupica CR, Cass WA, Zahniser NR et al. Effects of the selective adenosine agonist CGS21680 on in vitro electrophysiology, cAMP formation and dopamine release in rat hippocampus and striatum. J Pharmacol Exp Ther 1990; 252:1134-1141.
148. Lupica CR, Proctor WR, Dunwiddie TV. Presynaptic inhibition of excitatory synaptic transmission by adenosine in rat hippocampus: Analysis of unitary EPSP variance measured by whole cell recording. J Neurosci 1992; 12:3753-3764.
149. Maenhaut C, Van Sande J, Libert F et al. RDC8 codes for an adenosine A_2 receptor with physiological constitutive activity. Biochem Biophys Res Comm 1990; 173:1169-1178.
150. Mager R, Ferroni S, Schubert P. Adenosine modulates a voltage-dependent chloride conductance in cultured hippocampal neurons. Brain Res 1990; 532:58-62.
151. Mahan LC, Mcvittie LD, Smyk-Randall EM et al. Cloning and expression of an A_1 adenosine receptor from rat brain. Mol Pharmacol 1991; 40:1-7.
152. Martin D, Bustos GA, Bowe MA et al. Autoreceptor regulation of glutamate and aspartate release from slices of the hippocampal CA1 area. J Neurochem 1991; 56:1647-1655.

153. Martin GE, Rossi DJ, Jarvis MF. Adenosine agonists reduce conditioned avoidance responding in the rat. Pharmacol Biochem Behav 1993; 45:951-958.
154. Materi LM, Rasmusson DD, Semba K. Inhibition of synaptically evoked cortical acetylcholine release by adenosine: An in vivo microdialysis study in the rat. Neurosci 2000; 97:219-226.
155. Mayfield RD, Larson G, Orona RA et al. Opposing actions of adenosine A(2a) and dopamine D2 receptor activation on GABA release in the basal ganglia: Evidence for an A(2a)/D2 receptor interaction in globus pallidus. Synapse 1996; 22:132-136.
156. McCormick DA, Williamson A. Convergence and divergence of neurotransmitter action in human cerebral cortex. Proc Natl Acad Sci USA 1989; 86:8098-8102.
157. McKie LD, Bass BL, Dunkin BJ et al. Nitric oxide mediates the blood flow response to intravenous adenosine in the rabbit. Circulatory Shock 1994; 43:103-106.
158. Mendoza-Fernandez V, Andrew RD, Barajas-Lopez C. ATP inhibits glutamate synaptic release by acting at P2Y receptors in pyramidal neurons of hippocampal slices. J Pharmacol Exp Therap 2000; 293:172-179.
159. Mennerick S, Zorumski CF. Paired-pulse modulation of fast excitatory synaptic currents in microculture of rat hippocampal neurons. J Physiol 1995; 488:85-101.
160. Michaelis ML, Michaelis EK, Myers SL. Adenosine modulation of synaptosomal dopamine release. Life Sci 1979; 24:2083-2092.
161. Michel AD, Humphrey PPA. Distribution and characterisation of [3H]αβ-methyleneATP binding sites on the rat. Arch Pharmacol 1993; 348:608-617.
162. Minota S, Miyazaki T, Wang MY et al. Glycine potentiates NMDA responses in rat hippocampal CA1 neurons. Neurosci Lett 1989; 100:237-240.
163. Mitchell JB, Lupica CR, Dunwiddie TV. Activity-dependent release of endogenous adenosine modulates synaptic responses in the rat hippocampus. J Neurosci 1993; 13:3439-3447.
164. Mogul DJ, Adams ME, Fox AP. Differential activation of adenosine receptors decreases N-type but potentiates P-type Ca2+ current in hippocampal CA3 neurons. Neuron 1993; 10:327-334.
165. Moreau JL, Huber G. Central A_{2A} receptors: An overview. Brain Res Rev 1999; 31:65-82.
166. Mori A, Shindou T, Ichimura M et al. The role of adenosine A_{2A} receptors in regulating GABAergic synaptic transmission in striatal medium spiny neurons. J Neurosci 1996; 16:605-611.
167. Nakazawa K, Inoue K, Inoue K. ATP reduces voltage-activated potassium current in cultured rat hippocampal neurones. Europ J Physiol 1994; 429:143-145.
168. Nieber K, Polechen W, Illes P. Role of ATP in fast excitatory synaptic potentials in locus coeruleus neurones of the rat. Brit J Pharmacol 1997; 122:423-430.
169. Nikbakht M-R, Stone TW. Occlusive responses to adenosine A_1 receptor and muscarinic M2 receptor activation on hippocampal presynaptic terminals. Brain Res 1999; 829:193-196.
170. Nikbakht M-R, Stone TW. Suppression of presynaptic responses to adenosine by activation of NMDA receptors. Europ J Pharmacol 2001; 427:13-25.
171. Norenberg W, Wirkner K, Illes P. Effect of adenosine and some of its structural analogues om the conductance of NMDA receptor channels in a subset of rat neostriatal neurones. Brit J Pharmacol 1997; 122:71-80.
172. Normile HJ, Barraco RA. N6-cyclopentyladenosine impairs passive avoidance retention by selective action at A_1 receptors. Brain Res Bull 1991; 27:101-104.
173. O'Kane EM, Stone TW. Interaction between adenosine A_1 and A_2 receptor-mediated responses in the rat hippocampus in vitro. Eur J Pharmacol 1998; 362:17-25.
174. O'Kane EM, Stone TW. Adenosine A_1 receptor-mediated changes of EPSP-spike coupling are prevented by barium. Brit Neurosci Assoc Abst 1999; 15:128[53.04].
175. O'Kane EM, Stone TW. Characterisation of ATP-induced long-term facilitation of synaptic transmission in rat hippocampal slices. Europ J Pharmacol 2000; 409:159-166.
176. O'Regan MH, Simpson RE, Perkins LM et al. The selective A_2 adenosine receptor agonist CGS21680 enhances excitatory amino acid release from the ischaemic rat cerebral cortex. Neurosci Lett 1992; 138:169-172.
177. Ohno M, Watanabe S. Working memory failure by stimulation of hippocampal adenosine A_1 receptors in rats. NeuroReport 1996; 25:3013-3016.
178. Ohno,M. Yamamoto T, Watanabe S. Blockade of hippocampal M1 muscarinic receptors impairs working memory performance of rats. Brain Res 1994; 650:260-266.
179. Okada Y, Ozawa S. Inhibitory action of adenosine on synaptic transmission in the hippocampus of the guina-pig invitro. Europ J Pharmacol 1980; 68:483-492.
180. Overton P, Clark D. NMDA increases the excitability of nigrostriatal dopamine terminals. Europ J Pharmacol 1991; 201:117-120.
181. Palmour RM, Lipowski CJ, Simon CK et al. Adenosine analogues inhibit fighting in isolated male mice. Life Sci 1989; 44:1293-1301.
182. Pan WJ, Osmanovic SS, Shefner SA. Characterization of the adenosine A_1 receptor-activated potassium current in rat locus coeruleus neurons. J Pharmacol Exp Therap 1995; 273:537-544.

183. Pankratov Y, Lalo U, Castro E et al. ATP receptor-mediated component in the excitatory synaptic transmission in the hippocampus. Prog Brain Res 1999; 120:237-249.
184. Parada-Turska J, Turski WA. Excitatory amino acid antagonists and memory: Effect of drugs acting at NMDA receptors in learning and memory tasks. Neuropharmacol 1990; 29:1111-1116.
185. Pazzagli M, Corsi C, Latini S et al. In vivo regulation of extracellular adenosine levels in the cerebral cortex by NMDA and muscarinic receptors. Europ J Pharmacol 1994; 254:277-282.
186. Peralta C, Hotter G, Closa D et al. Protective effect of preconditioning on the injury associated to hepatic ischaemia-reperfusion in the rat: Role of nitric oxide and adenosine. Hepatology 1997; 25:934-937.
187. Piper AS, Hollingsworth M. ATP and methylene ATP produce relaxation of guinea-pig isolated trachealis muscle via actions of P1 purinoceptors. Eur J Pharmacol 1996; 307:183-189.
188. Platt B, Bate JR, Roloff EVL et al. Glycine induces a novel form of long-term potentiation in the superficial layers of the superior colliculus. Brit J Pharmacol 1998; 125:293-300.
189. Popoli P, Betto P, Reggio R et al. Adenosine A_{2A} receptor stimulation enhances striatal extracellular glutamate levels in rats. Europ J Pharmacol 1995; 287:215-217.
190. Post C. Anti-nociceptive effects in mice after intrathecal injections of NECA. Neurosci Lett 1984; 51:325-3.
191. Prince D A, Stevens C F. Adenosine decreases transmitter release at central synapses. Proc Natl Acad Sci USA 1992; 89:8586-8590.
192. Psarropoulou C, Beaucher J, Harnois C. Comparison of the effects of M1 and M2 muscarinic receptor activation in the absence of GABAergic inhibition in immature rat hippocampal CA3 area. Develop Brain Res 1998; 107:285-290.
193. Qian J, Saggau P. Presynaptic inhibition of synaptic transmission in the rat hippocampus by activation of muscarinic receptors: Involvement of presynaptic calcium influx. Brit J Pharmacol 1997; 122:511-519.
194. Rathbone MP, Middlemiss PJ, Gysbers JW et al. Trophic effects of purines in neurons and glial cells. Progr Neurobiol 1999; 59:663-690.
195. Reddington M, Schubert P. Parallel investigations of the effect of adenosine on evoked potentials and cyclic AMP accumulation in hippocampus slices of the rat. Neurosci Lett 1979; 14:37-42.
196. Regenold JT, Illes P. Inhibitory adenosine A_1-receptors on rat locus coeruleus neurones. An intracellular electrophysiological study. Naun Schmied Arch Pharmacol 1990; 341:225-231.
197. Ribeiro JA. Adenosine A_{2A} receptor interactions with receptors for other neurotransmitters and neuromodulators. Europ J Pharmacol 1999; 375:101-113.
198. Robertson SJ, Rae MG, Rowan EG et al. Characterisation of a P2x-purinoceptor in cultured neurones of the rat dorsal root ganglia. Brit J Pharmacol 1996; 118:951-956.
199. Ross FM, Brodie MJ, Stone TW. AMP as a mediator of ATP effects at P1 purinoceptors. Brit J Pharmacol 1998; 124:818-824.
200. Sarter M, Bodewitz G, Stephens DN. Attenuation of scopolamine induced impairment of spontaneous alternation behaviour by antagonist but not inverse agonist and agonist b-carbolines. Psychopharmacol 1988; 94:491-495.
201. Scanziani M, Capogna M, Gähwiler B H et al. Presynaptic inhibition of miniature excitatory synaptic currents by baclofen and adenosine in the hippocampus. Neuron 1992; 9:919-927.
202. Scanziani M, Gahwiler BH, Thompson SM. Presynaptic inhibition of excitatory synaptic transmission by muscarinic and metabotropic glutamate receptor activation in the hippocampus : Are calcium channels involved? Neuropharmacol 1995; 34:1549-1557.
203. Schingnitz G, Kufner-Muhl U, Ensinger H et al. Selective A_1-antagonists for treatment of cognitive deficits. Nucleosides & Nucleotides 1991; 10:1067-1076.
204. Scholz KP, Miller RJ. Inhibition of quantal transmitter release in the absence of calcium influx by a G protein-linked adenosine receptor at hippocampal synapses. Neuron 1991; 8:1139-1150.
205. Schubert P. Physiological modulation by adenosine: Selective blockade of A_1 receptors with DPCPX enhances stimulus train-evoked neuronal calcium influx in rat hippocampal slices. Brain Res 1988; 458:162-165.
206. Schwartzkroin PA, Webster K. Long-lasting facilitation of a synaptic potential following tetanization in the in vitro hippocampal slice. Brain Res 1975; 89:107-119.
207. Schwarzbaum PJ, Frischmann ME, Krumschnabel G et al. Functional role of ecto-ATPase activity in goldfish hepatocytes. Am J Physiol 1998; 274:R1031-R1038.
208. Sebastiao AM, Cunha RA, de Mendonca A et al. Modification of adenosine modulation of synaptic transmission in the hippocampus of aged rats. Brit J Pharmacol 2000; 131:1629-1634.
209. Sebastiao AM, Ribeiro JA. Evidence for the presence of excitatory A_2 adenosine receptors in the rat hippocampus. Neurosci Lett 1992; 138:41-44.
210. Segal M. Intracellular analysis of a postsynaptic action of adenosine in the rat hippocampus. Europ J Pharmacol 1982; 79:193-199.

211. Seguela P, Haghighi A, Soghomonian JJ et al. A novel neuronal P2x ATP receptor ion channel with widespread distribution in the brain. J Neurosci 1996; 16:448-455.
212. Sekino Y, Ito KI, Miyaka H et al. An adenosine A_2 antagonist inhibits induction of LTP of evoked synaptic potentials but not of the population spike in hippocampal CA1 neurons. Biochem Biophys Res Comm 1991; 181:1010-1014.
213. Shahi K, Baudry M. Glycine induced changes in synaptic efficacy in hippocampal slices involve changes in AMPA receptors. Brain Res 1993; 627:261-266.
214. Silinsky EM. Inhibition of transmitter release by adenosine: Are calcium currents depressed or are the intracellular effects of calcium impaired? Trends Pharmacol Sci 1986; 7:180-185.
215. Silinsky EM, Gerzanich V, Vanner SM. ATP mediates excitatory synaptic transmission in mammalian neurones, Brit J Pharmacol 1992; 106:762-763.
216. Simpson RE, O'Regan MH, Perkins LM et al. Excitatory transmitter amino acid release from the ischaemic rat cerebral cortex: Effects of adenosine receptor agonists and antagonists. J Neurochem 1992; 58:1683-1690.
217. Smith DA, Dunwiddie TV. Effects of bivalent cations on adenosine sensitivity in the rat hippocampal slice. Brain Res 1993; 617:61-68.
218. Snyder SH, Katins JS, Annau Z et al. Adenosine receptors and the actions of the methylxanthines. Proc Natl Acad Sci USA 1981; 78:3260-3264.
219. Soto F, Garcia-Guzman M, Gomez-Hernandez JM et al. P2X4: An ATP-activated ionotropic receptor cloned from rat brain, Proc Natl Acad Sci USA 1996; 93:3684-3688.
220. Spealman RD, Coffin VL. Behavioural effects of adenosine analogues in squirrel monkeys: Relation to adenosine A_2 receptors. Psychopharmacol 1986; 90:419-421.
221. Spedding M, Williams M. Developments in purine and pyrimidine receptor-based therapeutics. Drug Develop Res 1996; 39:436-441.
222. Spencer DG, Lal H. Effects of anticholinergic drugs on learning and memory. Drug Develop Res 1983; 3:489-502.
223. Spencer DG, Pontecorvo MJ, Heise GA. Central cholinergic inolvement in working memory: Effects of scopolamine on continuous nonmatching and discrimination performance in the rat. Behav Neurosci 1985; 99:1049-1065.
224. Sperlagh B, Zsilla G, Baranyi M et al. Age-dependent changes of presynaptic neuromodulation via A_1 adenosine receptors in rat hippocampal slices. Intern J Develop Neurosci 1997; 15:739-747.
225. Spignoli G, Pedata F, Pepeu G. A_1 and A_2 adenosine receptors modulate acetylcholine release from brain slices. Europ J Pharmacol 1984; 97:341-342.
226. Squire LR. Effect of pretrial and post-trial administration of cholinergic and anticholinergic drugs on spontaneous alternation. J Comp Physiol Psychol 1969; 74:41-45.
227. Stone TW, Simmonds HA. Purines: Basic and Clinical Aspects. Dordrecht: Kluwer Press, 1991.
228. Stone TW. Microiontophoresis and Pressure Ejection. Chichester: Wiley, 1985.
229. Stone TW, Perkins MN. Adenine dinucleotide effects on cortical neurones. Brain Res 1981; 229:241-245.
230. Stone TW, Cusack N. Purinoceptors in hippocampal pathways. Br J Pharmacol 1989; 97:631-635.
231. Stone WS, Rudd RJ, Gold PE. Glucose attenuation of scopolamine and age induced deficits in spontaneous alternation behaviour and regional brain [3H]-2-deoxyglucose uptake in mice. Psychobiol 1992; 20:270-279.
232. Sun MK, Wahlstedt C, Reis DJ. Action of externally applied ATP in rat reticulospinal vasomotor neurones. Europ J Pharmacol 1992; 224:93-96.
233. Suzuki F, Shimada J, Shiozaki S et al. Adenosine A_1 antagonists. 3. Structureactivity relationships on amelioration against scopolamine- or N6-((R)-phenylisopropyl)-adenosine-induced cognitive disturbance. J Med Chem 1993; 36:2508-2518.
234. Svenningsson P, Hall H, Sedvall G et al. Distribution of adenosine receptors in the postmortem human brain: An extended autoradiographic study. Synapse 1997; 27:322-335.
235. Swonger AK, Rech RH. Serotonergic and cholinergic involvement in habituation of activity and spontaneous alternation of rats in a Y maze. J Comp Physiol Psychol 1972; 81:509-522.
236. Tanaka J, Murate M, Wang CZ et al. Cellular distribution of the P2X4 ATP receptor mRNA in the brain and nonneuronal organs of rats. Arch. Histol Cytol 1996; 59:485-490.
237. Taube JS, Schwartzkroin PA. Mechanisms of long-term potentiation: EPSP/spike dissociation, intradendritic recordings, and glutamate sensitivity. J Neurosci 1988; 81632-1644.
238. Thompson SM, Haas HL, Gähwiler BH. Comparison of the actions of adenosine at pre and postsynaptic receptors in the rat hippocampus invitro. J Physiol 1992; 451:347-363.
239. Tobe A, Egawa M, Nagai R. Effect of MCI-2016 on the scopolamine induced deficit of spontaneous alternation behaviour in rats. Japan J Pharmacol 1983; 33:775-784.
240. Troster AI, Beatty WW, Staton RD et al. Effect of scopolamine on anterograde and remote memory in humans. Psychobiol 1989; 17:12-18.

241. Trussel LD, Jackson MB. Adenosine activated potassium conductance in cultured striatal neurones. Proc Natl Acad Sci USA 1985; 82:4857-4861.
242. Tschopl M, Harms L, Norenberg W et al. Excitatory effects of ATP on rat locus coeruleus neurones. Europ J Pharmacol 1992; 213:71-77.
243. Ueno S, Harata N, Inoue K et al. ATP-gated current in dissociated rat nucleus solitarii neurones. J Neurophysiol 1992; 68:778-785.
244. Umemiya M, Berger AJ. Activation of adenosine A_1 and A_2 receptors differentially modulates calcium channels and glycinergic synaptic transmission in rat brainstem. Neuron 1994; 13:1439-1446.
245. Von Kugelgen I, Spath L, Starke K. Stable adenine nucleotides inhibit [3H]-noradrenaline release in rabbit brain cortex slices by direct action at presynaptic adenosine A_1- receptors. Naunyn Schmied Arch Pharmacol 1992; 346:187-196.
246. Von Kugelgen I, Spath L, Starke K. Evidence for P2-purinoceptor-mediated inhibition of noradrenaline release in rat brain cortex. Br J Pharmacol 1994; 113:815-822.
247. Von Lubitz DKJE, Paul IA, Bartus RT et al. Effects of chronic administration of adenosine A_1 receptor agonist and antagonist on spatial learning and memory. Europ J Pharmacol 1993; 16:271-280.
248. Warburton DM, Heise GA. Effects of scopolamine on spatial double alternation in rats. J Comp Physiol Psychol 1972; 81:523-532.
249. Watanabe Y, Saito H, Abe K. Effects of glycine and structurally related amino acids on generation of long-term potentiation in rat hippocampal slices. Europ J Pharmacol 1992; 223:179-184.
250. Wieraszko A, Seyfried TN. ATP-induced synaptic potentiation in hippocampal slices. Brain Res 1989; 491:356-359.
251. Wieraszko A, Ehrlich YH. On the role of extracellular ATP in the induction of long-term potentiation in the hippocampus. J Neurochem 1994; 63:1731-1738.
252. Wilcox KS, Dichter MA. Paired-pulse depression in cultured hippocampal neurons is due to a presynaptic mechanism independent of GABA-B autoreceptor activation. J Neurosci 1994; 14:1775-1788.
253. Willig F, Van de Velde D, Laurent J et al. The Roman strains of rats as a psychogenetic tool for pharmacological investigation of working memory: Example with RU 41656. Psychopharmacol 1992; 107:415-424.
254. Winsky L, Harvey JA. Retardation of associative learning in the rabbit by an adenosine analogue as measured by classical conditioning of the nictitating membrane response. J Neurosci 1986; 6:2684.
255. Winsky L, Harvey JA. Effects of L-PIA, caffeine, theophylline and rolipram on the acquisition of conditioned avoidance response in rats. J Pharmacol Exp Therap 1987; 241:223-229.
256. Wirkner K, Assmann H, Koles L et al. Inhibition by adenosine A_{2A} receptors of NMDA but not AMPA currents in rat neostriatal neurons. Brit J Pharmacol 2000; 130:259-269.
257. Worley PF, Baraban JM, McCarren M et al. Cholinergic phosphatidylinositol modulation of inhibitory, G protein-linked, neurotransmitter actions: Electrophysiological studies in rat hippocampus. Proc Natl Acad Sci USA 1987; 84:3467-3471.
258. Wu LG, Saggau P. Presynaptic calcium is increased during normal synaptic transmission and paired-pulse facilitation but not long-term potentiation in area CA1 of hippocampus. J Neurosci 1994; 14:645-654.
259. Yarbrough GG, McGuffin-Clineschmidt TC. In vivo behavioural assessment of CNS purinergic receptors. Europ J Pharmacol 1981; 76:137-144.
260. Yawo H, Chuhma N. Preferential inhibition of omega-conotoxin sensitive presynaptic calcium channels by adenosine autoreceptors. Nature 1993; 365:256-258.
261. Yoon KW, Rothman SM. Adenosine inhibits excitatory but not inhibitory synaptic transmission in the hippocampus. J Neurosci 1991; 11:1375-1380.
262. Zarrindast MR, Bijan S. Effects of adenosine receptor agonists and antagonists on acquisition of passive avoidance learning. Europ J Pharmacol 1994; 256:233-239.
263. Zetterstrom T, Fillenz M. Adenosine agonists can both inhibit and enhance in vivo striatal dopamine release. Europ J Pharmacol 1990; 180:137-143.
264. Zetterstrom T, Vernet L, Ungerstedt U et al. Purine levels in the intact brain, studied with an implanted perfused hollow fibre. Neurosci Lett 1982; 29:111-115.
265. Zhang YX, Yamashita H, Ohshita T et al. ATP increases extracellular dopamine level through stimulation of P2Y purinoceptors in the rat striatum. Brain Res 1995; 691:205-212.
266. Zhou QY, Li C, Olah ME et al. Molecular cloning and characterization of an adenosine receptor: The A_3 adenosine receptor. Proc Natl Acad Sci USA 1992; 89:7432-7436.
267. Ziganshin AU, Ziganshina LE, King BF et al. Effects of P2-purinoceptor antagonists on degradation of adenine nucleotides by ecto-nucleotidases in folliculated oocytes of Xenopus laevis. Biochem Pharmacol 1996; 51:897-901.
268. Zimmerman H. Biochemistry, localisation and functional roles of ecto-nucleotidases in the nervous system. Progr Neurobiol 1996; 49:589-618.

Chapter 3.1

Cannabinoids

Lianne Robinson, Bettina Platt and Gernot Riedel

Abstract

Despite its long tradition in human psychopharmacology, animal studies on the effect of marijuana and its constituents in memory formation are relatively recent. They have been aided by both the development of synthetic cannabinoid drugs and the identification of specific receptors located in both central nervous and peripheral tissue. On one hand, cannabinoid receptor agonists induced memory deficits, especially in the short-term, but not in the long-term domain. Antagonists, on the other hand, efficiently reversed these deficits but had little effect when given alone. This suggested: 1) cannabinoid receptors exist in memory-relevant structures and can potentially modulate memory, but 2) there is no endocannabinoid tone in these brain areas and 3) endocannabinoids are not released during learning or they play no major part in memory processing. During the last 2 years, however, the generation of cannabinoid receptor knockout mice coupled with increased sophistication of learning experiments and pharmacological tools revealed compelling evidence for an active role of the endocannabinoid system in forgetting and extinction of memories.

Introduction

The resin made from the flowers and leaves of the hemp plant *Cannabis sativa* is more commonly known as cannabis or marijuana. It contains approximately 60 terpenophenolic compounds which are referred to as plant cannabinoids, with the primary psychoactive constituent being Δ^9-tetrahydrocannabinol (Δ^9THC).[32] Marijuana has been used for hundreds of years all over the world for both recreational and medicinal purposes. However, there has been an increase in recreational marijuana utilisation in the west since the 1960's and it is now the most widely used illicit and the third most common recreational drug after alcohol and tobacco (see ref. 47 for review). Marijuana consumption is most common in under 30's with almost 50% of 18 year olds having tried it at some point. A number of adverse effects have been reported following marijuana use including nausea, sickness, vomiting, dizziness and headaches. Unpleasant cognitive side effects can also be experienced such as paranoia, depression, fear, anxiety and hallucinations. Users also report a number of positive effects including relaxation, calming and relief from stress. Cannabinoids have been studied for both their actions as drugs of abuse and also for potential therapeutic applications.

Cannabinoid Receptors

At least two cannabinoid receptors have been cloned and characterised in mammalian tissue so far, namely 'CB_1' (see ref. 68) and 'CB_2' (see ref. 76). Both are G protein coupled membrane receptors, coupled to $G_{i/o}$ proteins, inhibiting adenylate cyclase activity and stimulating mitogen activated protein kinase. CB_1 receptors also couple to ion channels via $G_{i/o}$ proteins (see ref. 86 for review). A-type potassium channels are activated, while D-type potassium or N and P/Q type voltage-dependent calcium channels in hippocampal neurones are inhibited.[44,45,89]

From Messengers to Molecules: Memories Are Made of These, edited by Gernot Riedel and Bettina Platt.

The distribution of the two cannabinoid receptors is different: CB_1 receptors are found throughout the central nervous system and also in peripheral tissues including immune cells, reproductive tissues, the heart and lungs.[85] In the brain, they are predominantly expressed presynaptically and regulate the release of various neurotransmitters including gamma-amino-butyric-acid (GABA), acetylcholine, dopamine, norepinephrine, glutamate, cholecystokinin and serotonin.[86] CB_1 receptor mRNA and immunoreactivity[25] is particularly evident throughout the brain, in areas including the cerebral cortex, basal ganglia, hippocampus, lateral caudate-putamen, substantia nigra pars reticulata and globus pallidus.[84,85] By contrast, CB_2 receptors are mainly expressed in the periphery, tonsils, spleen and immune cells including B-cells and natural killer cells. CB_2 receptors are suggested to be responsible for the potential immunosuppressant and anti-inflammatory effects of cannabinoids.[2]

Another nonCB_1 and nonCB_2 receptor has been proposed in the brain based on evidence from CB_1 knockout mice.[5,21] This receptor, sometimes referred to as CB_3, is also G protein-coupled and sensitive to agonists like WIN55,212-2 and anandamide but not CP55,940 or Δ^9THC (see below for pharmacology, and ref. 102 for review). Its activity is also blocked by the CB_1 receptor antagonist SR141716A suggesting that actions reported with this antagonist may not always be due to CB_1 antagonism.

Cannabinoid Receptor Ligands

Endogenous cannabinoid receptor ligands (endocannabinoids) such as anandamide and 2-arachidonoyl glycerol have been isolated[20,70] and are able to bind and activate the cloned cannabinoid receptors. A common property of these endocannabinoids, sometimes referred to as eicosanoids, is that they are chemical derivatives of the polyunsaturated fatty acid, arachidonic acid. They serve as neurotransmitters/neuromodulators, are not stored in vesicles but instead are synthesised on demand and released from neurones as a result of depolarisation and Ca^+ influx.[23,87] They activate presynaptic receptors and are then removed from the extracellular space by a membrane transport process and hydrolysed by the microsomal enzyme, fatty acid amide hydrolase, with anandamide being hydrolysed to arachidonic acid and ethanolamine.

R(+)Methanandamide is a synthetic analogue of anandamide more resistant to hydrolysis than anandamide. Other synthetic cannabinoid receptor agonists can be categorised into 3 chemical drug groups. 1) Classical 2) Nonclassical and 3) aminoalkylindoles. Classical cannabinoids include plant-derived compounds such as cannabinol and cannabidiol, and their synthetic analogues. The best known is Δ^9THC, however other cannabinoids in this group include the synthetic analogue 11-hydroxy Δ^8-dimethylheptyl (HU210). Nonclassical cannabinoids consist of bicyclic and tricyclic analogues of Δ^9THC but lack the pyran ring; CP55,940 is perhaps the most widely used. The aminoalkylindoles are structurally different from the classical and nonclassical cannabinoids, resulting in binding differences. WIN55,212-2 is a member of this class of cannabinoids.

Receptor agonists differ in their affinities and efficacies for the cannabinoid receptors.[85] Δ^9THC binds equally to CB_1 and CB_2 in the nanomolar range, acting as a partial agonist for the CB_1 receptor, similar to anandamide. Cannabinol and cannabidiol have lower affinities and efficacy for CB_1 receptors than Δ^9THC. In contrast, HU210, WIN55,212-2 and CP55,940 have greater affinities for the CB_1 receptor. For WIN 55,212-2, an inactive isomer, WIN55,212-3, exists which can serve as a convenient control.

Several cannabinoid receptor antagonists have also been synthesised, e.g., SR141716A is a well-studied and extremely potent CB_1 selective antagonist (see refs. 85 and 86 for review). SR141716A is able to reverse or prevent the CB_1 receptor agonist-mediated effects in vivo and in vitro.[6,9,59,84,101] If administered alone at high doses it may behave like an inverse agonist, producing responses opposite to CB_1 agonists.[57,92] SR141716A appears to be more potent against noneicosanoid cannabinoid agonists. Structural analogues including AM281 and AM251 have been developed, with AM281 being slightly less potent than SR141716A. Similar to

SR141716A, AM281 attenuates the effects of cannabinoid receptor agonists and when administered alone can behave like an inverse agonist.[30]

Cannabinoid Receptors Modulate Memory Formation

Most of the animal research summarised below has been guided by work on human subjects, for which disruption of short-term memory is a widely acknowledged effect of marijuana or Δ^9THC.[11-13,24,29,72,95] Direct effects on memory storage, however, have not been found.[3,24] Interestingly, marijuana increased cerebral blood flow predominantly in paralimbic regions of the frontal lobes and the cerebellum, but reduced blood flow in the temporal lobe.[82] Such hypoactivity may be the neural basis of cognitive alterations. It is also in line with observations that marijuana addicts have reduced amplitudes in P300, an event related potential reflecting attentional resource allocation and active working memory[54] and that monkeys treated with Δ^9THC chronically have predominantly slow wave EEGs (1-2Hz) in hippocampus, amygdala and septum.[96]

Human data have subsequently been confirmed in work on a variety of other species including monkeys[1,4,26,31,78,94,103] and rodents (Table 1 and see also below). Due to refinement of pharmacological and physiological techniques, however, a much more complex framework has emerged for various animal species and there is now accumulation of evidence to suggest a modulatory role of cannabinoid receptors in several memory processes, such as encoding, consolidation, and even forgetting.

Spatial Learning

Water Maze

Training in an open water-filled tank containing a submerged platform is a popular learning paradigm tackling spatial and thus hippocampal-dependent memory.[75] Animals learn to find the submerged platform in the opaque water in relation to distal cues and reduce their latency to swim and climb onto it over days. Despite its long tradition as a behavioural test, reports on the effects of cannabinoids are relatively recent. The initial report by Ferrari and colleagues[28] revealed evidence for HU210-induced dose-dependent spatial acquisition deficits in rats for a reference memory paradigm in which the platform location was constant throughout training. A visible platform test showed no differences between the groups excluding sensory perception as a contributing factor. More recently, three reports extended this finding to Δ^9THC in rats and mice[14,73,101] and found that once spatial memory is acquired, consolidation and recall is not sensitive to cannabinoid treatment unless drug doses caused considerable motor side effects. When exposed to a working memory paradigm, in which the location of the platform was changed on a daily basis, Δ^9THC-treated mice were impaired in finding the platform despite extensive pretraining over weeks. Consequently, Varvel and coworkers,[61,101] claimed that spatial working memory in mice is more sensitive to cannabinoid treatment. Since mice trained in the reference memory paradigm were completely overtrained, however, this claim needs qualification since animals were only tested for recall and recall seems to be cannabinoid insensitive. In the working memory task, animals had to learn the platform location every day and this acquisition was not surprisingly, disrupted by Δ^9THC. A similar impairment of acquisition of reference memory has now been reported by Da Silva and Takahashi.[14] Finally, the working memory paradigm of Varvel et al,[101] which was also used more recently for the testing of CB_1 null mutants,[100] may not be an allocentric (guided by distal cues) task due to the fact that the same release site was used in each session. This makes it more egocentric (animals learn to swim to the left or right according to their own location independent of the environment) than the authors may accept and, as a consequence, a hippocampally-independent task.[51] At any rate, the CB_1 knock out mice showed no deficits in acquisition of a reference memory task with constant platform location, but were impaired in reversal learning suggesting a deficit in task flexibility.[100] $CB_1^{-/-}$ mice were not different from wildtype littermates in a

Table 1. Effects of cannabinoid treatment on memory formation

Animal	Behavioural Task	Drug (admin. & dose)	Effect on Acquisition/Learning	Effect on Consolidation	Effect on Retrieval	Other Observations	Refs.
			Spatial Learning				
Rat	Water maze	Pre-training					28
(Wistar)	Reference memory	HU210 (25-100 mg/kg i.p.)	IMPAIRMENT (dose-dependent)				
	Visible platform		No effect				
Mouse	Water maze	Post-training					101
(C57BL/6)	Reference memory	Δ^9THC 3 – 100 mg/kg i.p. (+SR141716A 3 mg/kg i.p.)			IMPAIRMENT (100 mg/kg) REVERSAL	due to side effects of the high dose	
	Working memory	Δ^9THC 0.3-10 mg/kg i.p. (+SR141716A 1-10 mg/kg i.p.)	IMPAIRMENT (3-10 mg/kg) REVERSAL (3-10 mg/kg)				
Rat	Water maze	Post-training					73
(Wistar)	Reference memory	Δ^9THC (6-10mg/kg)			No effect		

Continued on next page

Table 1. Continued

Animal	Behavioural Task	Drug (admin. & dose)	Effect on Acquisition/Learning	Effect on Consolidation	Effect on Retrieval	Other Observations	Refs.
Mouse	Water maze	Pre-training					14
(male albino)	Reference memory	Δ^9THC 4 - 8mg/kg i.p. (+SR141716A 1mg/kg i.p.)	IMPAIRMENT (8 mg/kg) REVERSAL				
		Δ^9THC 8 mg/kg i.p.	IMPAIRMENT				
		Post-training		No effect			
		Pre-test			No effect		
	Working memory	Δ^9THC 8 mg/kg i.p.	IMPAIRMENT				
Mouse	Water maze						100
CB_1 $^{-/-}$	Reference memory		No effect				
	Reversal learning		IMPAIRMENT				
	Working memory		IMPAIRMENT				
Rat	8-arm RM	Pre-training					97
(Wistar)	(working memory)	Δ^9THC (20 mg/kg i.p.: chronic for 3-6 months, then 1 month drug free)	IMPAIRMENT				

Continued on next page

Table 1. Continued

Animal	Behavioural Task	Drug (admin. & dose)	Effect on Acquisition/Learning	Effect on Consolidation	Effect on Retrieval	Other Observations	Refs.
Rat	8-arm RM	Pre-testing					77
(Wistar)	(working memory + delay 5s and 1h)	Acute: Δ^9THC (1.25 mg/kg i.p.)		IMPAIRMENT (post-delay)			
		Chronic: Δ^9THC (5 mg/kg i.p.for 90 days)		IMPAIRMENT (post-delay)			
Rat (Wistar)	8-arm RM (working memory + 5s delay)	Pre-testing				No changes in serotonin metabolism in cortex, hippocampus etc.	74
		Δ^9THC (5mg/kg p.o.)	IMPAIRMENT (pre and post delay)				
Rat	8-arm RM	Pre-testing					60
(Sprague-Dawley)	(working memory)	Δ^9THC (1-5.6 mg/kg i.p.)	IMPAIRMENT (dose-dependent)				
		CP-55,940 (0.125-0.25 mg/kg i.p.)	IMPAIRMENT (dose-dependent)				
		WIN-55,212-2 (1-10 mg/kg i.p.)	IMPAIRMENT (dose-dependent)				
		CP-55,940 (5-10 µg/rat i.HC.)	IMPAIRMENT (dose-dependent)				

Continued on next page

Table 1. Continued

Animal	Behavioural Task	Drug (admin. & dose)	Effect on Acquisition/Learning	Effect on Consolidation	Effect on Retrieval	Other Observations	Refs.
Rat	8-arm RM	Pre-testing					59
(Wistar)	(working memory + 5-30s delay)	Δ^9THC (2-6mg/kg i.p.)	IMPAIRMENT (dose-dependent, delay-independent)				
		Δ^9THC (5mg/kg i.p.) + SR141716A	REVERSAL				
		Δ^9THC (5mg/kg i.p.) + Physostigmine	IMPAIRMENT NO reversal				
Rat	8-arm RM	Pre-testing					57
(Sprague-Dawley)	(working memory + delay 1-24hrs)	SR141716A (3mg/kg i.p.)	IMPROVEMENT (delay-dependent)				
Rat	8-arm RM	Pre-testing				Noradrenaline in hippocampus reduced	42
		Δ^9THC	IMPAIRMENT				
Rat	8-arm RM	Pre-testing					73
(Wistar)	(Working memory)	Δ^9THC(2-6mg/kg)	IMPAIRMENT (6mg/kg)				
		+SR141716A (0.01-0.1mg/kg)	REVERSAL				
	(Working + reference memory)	Δ^9THC(2-6mg/kg i.p.)	IMPAIRMENT (working memory; not reference memory)				

Continued on next page

Table 1. Continued

Animal	Behavioural Task	Drug (admin. & dose)	Effect on Acquisition/Learning	Effect on Consolidation	Effect on Retrieval	Other Observations	Refs.
Rat (Sprague-Dawley)	T-maze, delayed spatial alternation	Pre-testing					
		Δ^9THC (5mg/kg i.p.)	IMPAIRMENT			Activation of dopamine and noradrenaline in prefrontal cortex	53
		Δ^9THC (5mg/kg i.p.) for 7 days +SR141716A	REVERSAL			Depression of hippocampal acetylcholine release	80
	T-maze, black-white discrimination	Δ^9THC (5mg/kg i.p.)	No effect				53
	T-maze, visual form discrimination	Δ^9THC (5-10 mg/kg i.p.)	No effect				73
Rat (Sprague-Dawley)	2-lever DMTP	Pre-test				CA1 complex spike cells have reduced firing in sample phase	43
		Δ^9THC (0.75-2 mg/kg i.p.)	IMPAIRMENT (delay dependent; dose-dependent >0.75mg/kg)				
Rat (Wistar)	2-lever DMTP	Chronic (during testing)					17
		Δ^9THC (10 mg/kg i.p.)	IMPAIRMENT (adaptation after 35 days)				

Continued on next page

Table 1. Continued

Animal	Behavioural Task	Drug (admin. & dose)	Effect on Acquisition/Learning	Effect on Consolidation	Effect on Retrieval	Other Observations	Refs.
Rat (Long-Evans)	2-lever DNMTP	Pre-test				CA1 and CA3 complex spike cells have reduced firing in sample phase	34,35, 36
		WIN-55,212-2 (0.25-0.75 mg/kg i.p.)	IMPAIRMENT (dose-dependent)				
		+ SR141716A (1.5mg/kg i.p.)	REVERSAL				
		SR141716A (1.5mg/kg i.p.)	No effect				
Rat	2-lever DNMTP	Pre-test					41
(Sprague-Dawley)		WIN 55,212-2 (0.5-2mg/kg i.p.)	IMPAIRMENT (2 mg/kg)				
Rat (Wistar)	3-lever DMTP *(delay 3s)*	Pre-test					71
		Δ^9THC (10mg/kg i.p.)	IMPAIRMENT				
Rat (Wistar)	2-lever DNMTP *(delay 4-16s)*	Pre-test					63
		Δ^9THC (0.5-4 mg/kg i.p)	IMPAIRMENT (dose-dependent)				
		Anandamide (0.25-2mg/kg i.p.)	IMPAIRMENT (dose-dependent)				

Continued on next page

Table 1. Continued

Animal	Behavioural Task	Drug (admin. & dose)	Effect on Acquisition/Learning	Effect on Consolidation	Effect on Retrieval	Other Observations	Refs.
Rat (Wistar)	2-lever DNMTP *(delay 4-16s)*	Pre-test					64
		Δ^9THC (2 or 4 mg/kg i.p)	IMPAIRMENT				
		+ SR141716A	partial REVERSAL				
		Anandamide (2mg/kg i.p.)	IMPAIRMENT				
		+SR141716A	REVERSAL				
		SR141716A (0.05-2 mg/kg i.p.)	No effect				
			Fear Conditioning				
Mice C57Bl/6	Auditory fear conditioning	Pre-training SR141716A	No effect				66
		Pre-extinction SR141716A	IMPAIRMENT				
and $CB_1^{-/-}$			Normal acquisition, IMPAIRMENT in extinction				
Rat (Sprague-Dawley)	Acoustic startle + prepulse inhibition of acoustic startle	Pre-training					65
		CP-55,940 (0.01-0.1 mg/kg i.p.)			IMPAIRMENT		
		+ SR141716A (10mg/kg i.p.)			REVERSAL		
		SR141716A (1-10 mg/kg i.p.)			No effect		

Continued on next page

Table 1. Continued

Animal	Behavioural Task	Drug (admin. & dose)	Effect on Acquisition/Learning	Effect on Consolidation	Effect on Retrieval	Other Observations	Refs.
Rat (Wistar)	Pre-pulse inhibition of fear potentiated startle	Pre-testing					93
		WIN55,212-2 1.2 mg/kg i.p.			IMPAIRMENT		
			Avoidance Learning				
Mice - CD-1	Step-through passive avoidance	Post-training (0 – 120 min)				Effects antagonized by dopamine D1 and D2 antagonists	7
		Anandamide (1.5-6mg/kg i.p.)		IMPAIRMENT (for injections at 0 min post-training)			
Mice - C57Bl/6 - DBA/2	Step-through passive avoidance	Post-training (0-120 min)				Effects antagonized by naltrexone (opioide antagonist)	8
		Anandamide (1.5-6mg/kg i.p.)		IMPAIRMENT in DBA/2 IMPROVEMENT in C57Bl/6			
Rat (Wistar)	Step-through passive avoidance	Post-training (30s)					91
		Anandamide (3.6nmol/5µl i.c.v.)		IMPAIRMENT (24h, but not 15 min later)		More slow-wave sleep and REM	

Continued on next page

Table 1. Continued

Animal	Behavioural Task	Drug (admin. & dose)	Effect on Acquisition/Learning	Effect on Consolidation	Effect on Retrieval	Other Observations	Refs.
		Arachidonic acid (3.6nmol/5μl i.c.v.)		IMPAIRMENT (15 min and 24h later)		Less slow-wave sleep	
Rat (Wistar)	Step-through passive avoidance	Pre-training,	IMPAIRMENT				73
		Post-training		IMPAIRMENT			
		Pre-test			IMPAIRMENT		
		Δ^9THC (10mg/kg i.p.)					
Rats (albino)	Active avoidance	Cannabidiol (3.5 mg/kg i.p.)	IMPAIRMENT				48
Rat (Wistar)	Active avoidance (shuttle box)	Chronic (3 months, then 1.5 month drug free)					98
		Δ^9THC (20 mg/kg i.p.)	IMPROVEMENT				
$CB_1^{-/-}$ mice	Active avoidance		IMPROVEMENT				67
			Olfactory and Gustatory Paradigms				
Rat (Wistar)	Social recognition task	Post-training				Partial reversal by scopolamine (0.06mg/kg i.p.)	99
		SR141716A (0.03-3mg/kg s.c.)		IMPROVEMENT (0-5min, but not longer delays)			

Continued on next page

Table 1. Continued

Animal	Behavioural Task	Drug (admin. & dose)	Effect on Acquisition/Learning	Effect on Consolidation	Effect on Retrieval	Other Observations	Refs.
Rat (Wistar)	Social recognition task	Pre-training					93
		WIN55,212-2 0.6 or 1.2 mg/kg i.p.			IMPAIRMENT		
			Other Tasks				
Rat (Wistar)	Object recognition task	Pre-testing					93
		WIN55,212-2 0.6 or1.2 mg/kg i.p.			IMPAIRMENT		
Rat (alkohol preferring)	Object recognition task (15 min interval)	Pre-testing					9
		Δ^9-THC					
		0.2-5 mg/kg i.p.			No effect		
		10 mg/kg i.p.			IMPAIRMENT		
		Δ^9-THC + SR141716A (10 + 1 mg/kg i.p.)			REVERSAL		
CB1 -/- mice	Object recognition task				IMPROVEMENT (24 and 48 hr post-acquisition)		62,90

Continued on next page

Table 1. Continued

Animal	Behavioural Task	Drug (admin. & dose)	Effect on Acquisition/Learning	Effect on Consolidation	Effect on Retrieval	Other Observations	Refs.
Rat (Long – Evans)	Repeated acquisition procedure with light-related keypressing	Pre-test					6
		Cannabidiol (3.2-100mg/kg i.p.)	No effect				
		Anandamide (0.01-18mg/kg i.p.)	No effect				
		Δ^9THC (3.2-18mg/kg i.p.)	IMPAIRMENT				
		+ SR141716A (1mg/kg i.p.)	REVERSAL				
		R-methanandamide (1-18mg/kg i.p.)	IMPAIRMENT				
		+ SR141716A (1mg/kg i.p.)	REVERSAL				
		SR141617A (1-32 mg/kg i.p.)	No effect				

i.c.v. = intracerebroventricularly; i.p. = intraperitoneally; p.o. = per orally; s.c. = subcutaneous working memory version of the task, and were also insensitive to Δ^9THC, WIN55,212-2 or methanandamide treatment, which disrupted working memory in wild-type littermates in a SR141716A-sensitive manner.

Radial Arm Maze

In this paradigm, animals are kept on 80-85% of their free-feeding body weights and have to retrieve food placed at the distal end of 8 arms radiating from an octagonal central platform. Although widely used for cannabis research, the 8-arm radial maze is not ideal to assess the effects of synthetic or endogenous cannabinoids on memory formation. This has two main reasons. First, cannabis or Δ^9THC as well as other synthetic analogues depress locomotor activity (see refs. 18,19 and 91 for review), which can obviously influence latencies, and a prolongation of the task may also have confounding consequences on attention. Second, Δ^9THC is well known to increase appetite thereby affecting the motivation of already hungry animals to perform in this food rewarded task.[22,69]

Nevertheless, numerous reports suggest that cannabinoids impair performance in the 8-arm radial maze, especially when short-term memory is tested. This has been originally reported in chronic experiments with Δ^9THC administered for 3 or 6 months,[97] and corroborated later for acute infusions of Δ^9THC,[42,58,60,73,74,77] acute systemic administration of the full synthetic CB_1 receptor agonists WIN55,212-2 and CP55,940 (see ref. 60) or local infusion of CP55,940 directly into the hippocampus.[60]

In most cases, animals were trained to criterion performance in a working memory task with all 8 arms baited. Systemic treatment with cannabinoids increased the number of working memory errors (reentries of previously baited arms) (see refs. 73,74,97) with low doses not affecting the amount of time required to complete the 4 visits.[42] Short-term memory can be tested by introducing a short delay between visits to arms 1-4 and 5-8. Cannabinoids also disrupt performance after a 5s,[74] 30s,[42] or 1h delay period,[77] but reports did not distinguish between revisits to arms 1-4 or 5-8. However, post-delay performance was prolonged in the Δ^9THC group,[42,77] suggesting a dissociation between impaired short-term memory and reduced overall activity. Mishima and coworkers[73] recently also distinguished between working and reference memory errors using a protocol with 4 baited and 4 nonbaited arms. Δ^9THC (6mg/kg) significantly impaired working memory but not reference memory. However, inspection of their data suggests that reference memory was also impaired, warranting a replication of this study.

Cannabinoid effects were due to CB_1 receptor activation since the selective antagonist SR141716A reversed deficits induced by Δ^9THC.[58,73] If no or very short delays were employed between arm visits 4 and 5, SR141716A alone had no effects.[56,57,73] Prolonging this delay, however, revealed a memory enhancement in the SR141716A group[57] suggesting that the ability to deal with increased task difficulty may be aided by CB_1 receptor blockade.

T-and Y Maze Procedures

Spontaneous alternation in the T-maze or Y-maze is another popular spatial paradigm, in which animals alternate between the arms. Similar to the 8-arm radial maze, this task has the disadvantage of using food reward in most cases. Δ^9THC administered systemically prior to training reduced the alternation score and diminished acetylcholine release in the hippocampus.[79] When administered chronically (twice per day 5mg/kg i.p.) for up to one week, the alternation impairment persisted.[80] More interestingly, both acetylcholine depression and alternation impairment was fully reversible upon SR141716A treatment, suggesting that *no* tolerance developed after chronic 5 day Δ^9THC exposure.

Delayed alternation is another possible training protocol in which animals are rewarded for choosing any goal box in trial 1. After an inter-trial interval spent in the start box, animals are rewarded in the next trial only when entering the opposite arm (nonmatch). Animals trained to criterion (80% correct responses) were impaired after Δ^9THC treatment and had reduced monoamine turnover in their prefrontal cortex.[53] They were, however, not impaired in brightness discrimination[52] or visual discrimination of forms procedures[73] using the same apparatus.

Delayed-Match-to-Position

Several studies have tested for short-term memory in delayed-match-to-position (DMTP) or delayed-match-to-sample (DMTS) tasks. They are usually performed in a conditioning chamber and animals learn to press a lever during the sample phase and press the same (match) or opposite (nonmatch) lever during the choice phase. These tasks have a spatial component (right or left). In addition, task difficulty can be modified by increasing the delay between the sample and the choice phase with rats coming down to chance levels at delays of about 30-40 seconds.[16] The group of Deadwyler has extensively used this task and determined a) the hippocampal involvement during performance,[38] and b) the electrophysiological activity of CA3 and CA1 ensembles of neurones during the different phases of the task[15,16,33,37,39] using multielectrode recording techniques. In brief, principle neurones distinguish between the sample, delay and match phase. In a series of elegant studies, this group and others have provided compelling evidence for a modulatory role of cannabinoids in delayed-match-to-sample performance. Systemic administration of Δ^9THC prior to testing led to dose- and delay-dependent performance deficits suggesting compromised short-term memory.[43] In addition, the hippocampal firing during the sample phase was greatly diminished[43] leading to ensemble miscodes increasing the probability for the occurrence of errors especially at long, but not very short delays. A follow-up study further established that animals develop behavioural tolerance to Δ^9THC after about 35 days of exposure followed by a short withdrawal period of 2 days.[17] Behavioural sensitization, however, develops within 4 days of repeated treatment.[71] The same results were obtained for delayed-nonmatch-to-position protocols,[34,63] WIN55,212-2,[35,36,41] or anandamide[63] applications, and deficits are reversed by the CB_1 receptor antagonist SR141716A.[35,36,64] Interestingly, SR141716A alone had no effect.[36,64] While Hampson and Deadwyler have made a strong case for a role of CB_1 receptors in encoding, Han and Robinson[40] recently challenged this view and showed that cannabinoids (WIN55,212-2 and Δ^9THC) can shorten time estimation in the rat providing a nonmemory-related explanation for the behavioural deficit observed in DMTP/DNMTP procedures. Interestingly, SR141716A prolonged modal response times providing compelling evidence for endocannabinoid participation in time estimation.

Conditioning of Fear

Auditory fear conditioning is a standard procedure used in animal research (for review, see ref. 10). Animals are placed in a small chamber and a tone is presented which coterminates with a mild footshock. This induces a typical crouching posture and immobility, and the length of this freezing response is probed upon reexposure to the chamber and presentation of the tone. Both animals exposed to the CB_1 receptor antagonist SR141716A or CB_1 knockout mice show no difficulty in acquiring this conditioning. When given a number of extinction trials, in which no footshock is presented, wildtype mice reduce the amount of freezing over 6 days. By contrast, both SR141716A treated mice or knockouts did not extinguish the freezing response suggesting that the endocannabinoid system plays a major role in the forgetting and extinction of aversive memories.[66] This reexposure to the tone (but not shock) during extinction also induced the release of endocannabinoids such as anandamide and 2-AG in the basolateral amygdala, and this not only confirms the importance of the amygdala in fear conditioning, but also that on-demand release of endocannabinoids controls their extinction.[66]

A second paradigm, the acoustic startle response, is based on a naturally occurring reaction to loud noise. If coupled with a 20ms prepulse (pure tone) delivered 30-500ms prior to the startle stimulus the startle reaction is considerably inhibited, termed prepulse inhibition. Prepulse inhibition is a measure of sensory-motor gating and involves a multitude of brain stem areas and transmitter systems.[55] Rats injected with WIN55,212-2 show less prepulse inhibition compared to controls, corroborating similar findings using CP55,940 and suggesting a deficit in sensorimotor integration.[65] The effect of CP55,940 was fully reversed by SR141716A, but the antagonist had no effect when given alone.[65] Given the normal startle reaction even under drug treatment, these data confirm a specific role of CB_1 receptors in sensory-motor gating, the involvement of the endocannabinoid system remains elusive.

Avoidance Tasks

Inhibitory Avoidance

A number of studies have assessed the role of cannabinoid receptors in step-through inhibitory avoidance tasks. Animals placed in the light part of a box tend to escape into darkness. This is punished by a mild footshock and animals readily learn to inhibit this natural escape tendency. Mice and rats treated with anandamide immediately, but not several hours post-training were impaired in retention performance.[7,8] This is due to an effect of anandamide on memory consolidation since injections 2 hours post-training had no effect[7,8] and were strain-specific with C57Bl/6 mice showing memory facilitation.[8] A more detailed time course with systemic pretraining, post-training and pretest injection of Δ^9THC in rats revealed memory deficits independent of injection time.[73] This suggests that cannabinoid receptors may play important roles in encoding, consolidation and retention processes of emotional memories, but the reasons for the discrepancies between these data remain to be established

Active Avoidance

Few reports are available on active avoidance procedures, in which the animals have to escape in a predetermined time window in order to avoid a mild footshock. In rats cannabidiol, a weak CB_1 receptor agonist, given i.p. resulted in an acquisition impairment, but had no effect on consolidation of the conditioned responding.[48] In agreement with the idea that the effect is mediated via CB_1 receptors, CB_1 null mutants showed an increase in active avoidance responses[67] consistent with memory enhancement. Enhanced acquisition of the active avoidance paradigm was reported for rats that had been chronically treated with Δ^9THC for 3 months, but were left untreated for 30 or 118 days before exposure to the shuttle box. Previously drug-treated animals outperformed controls during initial training and were faster to reach asymptotic levels.[98] Such a result is reminiscent of hippocampal lesions, which also facilitated active avoidance learning[46] suggesting that systemically administered Δ^9THC may have acted via hippocampal cannabinoid receptors and have thereby compromised hippocampal function.

Olfactory and Gustatory Memory Paradigms

There are various behavioural models to assess olfactory memory, for example using a social recognition task. Adult animals meet a juvenile conspecific and explore him/her via anogenital sniffing for say 5 mins. Reexposure to the same (familiar) juvenile after minutes tol hour will lead to reduced anogenital sniffing reflecting recognition memory. If exposed to a different juvenile, however, anogenital sniffing will be high. Social recognition memory is affected by the CB_1 receptor agonist WIN55,212-2 which reduced short-term memory (30 mins) in a dose-dependent manner without affecting anogenital exploration per se.[93] This observation is in line with the original work by Terranova and coworkers[99] who used subcutaneous injection of the CB_1 receptor antagonist SR141716A at various time points after the first presentation of the juvenile. When injected immediately or 5 minutes post-presentation of the juvenile, SR141716A in doses of 0.1-3 mg/kg enhanced social recognition memory tested after a retention interval of 120 min. This memory enhancement was effective in both aged rats and aged mice, and was antagonised by scopolamine suggesting a tight interaction with the cholinergic system. The work raises at least two remaining questions: 1) Was the effect due to SR141716A antagonism at CB_1 receptors or due to inverse agonism? 2) What memory processes are affected by SR141716A? Speculation about the latter suggests that CB_1 receptor activation during the retention interval supports memory decline or forgetting in agreement with recent work on emotional memory.[66] It is highly questionable, however, whether SR141716A is 'active upon consolidation' (see ref. 99) of a memory which may not be consolidated at all.

Other Paradigms

Another test for short-term memory, but with a different psychological quality, is object recognition.[27] In this task, animals are exposed to an environment containing an object A, and are reexposed to object A and another novel object B after a predetermined time interval. Given that rodents have a natural tendency to evaluate novelty over familiarity, exploration of object B is measured and, if higher in comparison to A, indicates memory for object A. Brain structures involved in object recognition include entorhinal and perirhinal cortex. Rats have been tested in object recognition after i.p. injection of Δ^9THC (see ref. 9) or WIN55,212-2.[93] While 0.2 –5 mg/kg Δ^9THC was not effective, 10 mg/kg reduced the amount of exploration of object B upon reexposure suggesting a memory deficit. The deficit was reversed by simultaneous administration of SR141716A (1mg/kg) suggesting selective CB_1-mediated action of this effect.[9] However, SR141716A alone was not tested, which leaves the possibility that endogenous cannabinoid release during object recognition might limit the length of the recognition memory. This idea is supported by work that utilised CB_1 null mutants.[62,90] Animals showed normal exploratory activity to object A but memory is enhanced when tested against object B 24 or 48 hours later. While one needs to be cautious with the interpretation of such data, they predict that a similar effect should be observed in normal animals under SR141716A. It then remains to be established whether this is due to endocannabinoid release during encoding or during the consolidation period.

A different but highly complex learning protocol was used by Brodkin and Moerschbacher.[6] For up to 14 weeks, animals were trained to respond to a sequence of lights by pressing appropriate keys in a modified conditioning box. Once asymptotic performance criteria were met, drugs like cannabidiol and anandamide were injected, but had no effect on performance. By contrast, Δ^9THC and R-methanandamide, a more stable anandamide analogue impaired performance in a dose-related manner. This impairment was reversed by SR141716A, but the antagonist had no effect on its own.[6]

Conclusions

In summary, the aforementioned behavioural data seem to provide strong evidence for a role of cannabinoids in learning and formation of different forms of memory. However, it could be argued that because systemic infusion of Δ^9THC or other synthetic cannabinoid receptor agonists has numerous unspecific effects, hard proof is difficult to obtain and many studies may be flawed by the overshadowing effects of CB_1 activation-induced changes in activity or reward rather than learning.

While this may explain a false negative due to increased appetite under cannabinoid treatment, the picture is somewhat complicated by the observation of Δ^9THC-induced suppression of ambulations when exploring a novel open-field.[49,50,81] This may be due to anxiogenic properties of cannabinoids and could generate higher levels of emotionality and neophobia reflected in a degree of immobility.[83] Reduced activity may lead to the impression, for instance in inhibitory avoidance tasks, of memory enhancement owing to longer latencies to step down or step through. However, reduced activity does not necessarily imply learning or memory impairments and it is therefore important to monitor carefully parameters like reaction times or swim speed etc., which enable to discern overall motor performance and learning.

Finally, cannabinoids affect signal detection. For instance, Presburger and Robinson[88] recently showed that Δ^9THC significantly and selectively disrupted signal detection accuracy with light as a signal in rats at a stimulus duration of 100, but not 300 or 1000 ms. This result is revealing and has important implications for stimulus presentations during behavioural learning. The results differ from those obtained for muscarinic and glutamatergic antagonists, which did not produce deficits in visual attention.[88]

Despite all these problems with systemic administration of cannabinoids there is now strong evidence for a role of CB_1 receptors in memory formation. For delay-dependent short-term memory tasks, CB_1 receptors may be able to modulate the encoding processes. By contrast, CB_1 receptors may play a role in consolidation and even recall in memory formation of avoidance tasks. These effects are likely to be mediated by different CB_1 receptor populations located in different brain regions and a better understanding of their function requires more localised administration of selective CB_1 agonists and antagonists. This would also reveal more detailed information about the role of the endocannabinoid system in memory processing.

References

1. Aiger TG. Delta-9-tetrahydrocannabinol impairs visual recognition memory but not discrimination learning in rhesus monkeys. Psychopharmacol 1988; 95:507-511.
2. Ameri A. The effects of cannabinoids on the brain. Prog Neurobiol 1999; 58:315-348.
3. Belmore SM, Miller LL. Levels of processing and acute effects of marijuana on memory. Pharmacol Biochem Behav 1980; 13:199-203.
4. Branch MN, Dearing ME, Lee DM. Acute and chronic effects of Δ^9-tetrahydrocannabinol on complex behavior in squirrel monkeys. Psychopharmacol 1980; 71: 247-256.
5. Breivogel CS, Griffin G, Di Marzo V et al. Evidence for a new G protein-coupled cannabinoid receptor in the mouse brain. Mol Pharmacol 2001; 60:155-163.
6. Brodkin J, Moerschbacher JM. SR141716A antagonizes the disruptive effects of cannabinoid ligands on learning in rats. J Pharmacol Exp Ther 1997; 282:1526-1532.
7. Castellano C, Cabib S, Palmisano A et al. The effects of anandamide on memory consolidation in mice involve both D1 and D2 dopamine receptors. Behav Pharmacol 1997; 8:707-712.
8. Castellano C, Ventura R, Caibb S et al. Strain-dependent effects of anandamide on memory consolidation in mice are antagonized by naltrexone. Behav Pharmacol 1999; 10:453-457.
9. Ciccocioppo R, Antonelli L, Biondini M et al. Memory impairment following combined exposure to Δ^9-tetrahydrocannabinol and ethanol in rats. Europ J Pharmacol 2002; 449:245-252.
10. Crawley JN. What is Wrong with my Mouse? New York: Wiley-Liss, 2000.
11. Darley CF, Tinklenberg JR, Roth WT et al. Influence of marihuana on storage and retrieval processes in memory. Mem Cogn 1973a;1:196-200.
12. Darley CF, Tinklenberg JR, Roth WT et al. Marihuana and retrieval of short-term memory. Psychopharmacologia 1973b; 29:231-238.
13. Darley CF, Tinklenberg JR, Roth WT et al. The nature of storage deficits and state-dependent retrieval under marihuana. Psychopharmacol 1974; 37:139-149.
14. Da Silva GE, Takahashi RN. SR141716A prevents Δ^9-tetrahydrocannabinol-induced spatial learning deficit in a Morris-type water maze in mice. Prog Neuro-Psychopharmacol Biol Psychiat 2002; 26:321-325.
15. Deadwyler SA, Hampson RE. Anatomic model of hippocampal encoding of spatial information. Hippocampus 1999; 9:397-412.
16. Deadwyler SA, Bunn T, Hampson RE. Hippocampal ensemble activity during spatial delayed-nonmatch-to-sample performance in rats. J Neurosci 1996; 16:354-372.
17. Deadwyler SA, Heyser CJ, Hampson RE. Complete adaptation to the memory disruptive effects of delta-9-THC following 35 days of exposure. Neurosci Res Comm 1995; 17:9-18.
18. DeSanty KP, Dar MS. Cannabinoid-induced motor incoordination through the cerebellar CB_1 receptor in mice. Pharmacol Biochem Behav 2001; 69:251-259.
19. DeSanty KP, Dar MS. Involvement of the cerebellar adenosine A1 receptor in cannabinoid-induced motor incoordination in the acute and tolerant state in mice. Brain Res 2001; 905:178-187.
20. Devane WA, Hanus L, Breuer A et al. Isolation and structure of a brain constituent that binds to the cannabinoid receptor. Science 1992; 258:1946-1949.
21. Di Marzo V, Breivogel CS, Tao Q et al. Levels, metabolism, and pharmacological activity of anandamide in CB(1) cannabinoid receptor knockout mice: Evidence for nonCB(1), nonCB(2) receptor mediated actions of anandamide in mouse brain. J Neurochem 2000; 75:2434-2444.
22. Di Marzo V, Goparaju SK, Wang L et al. Leptin-regulated endocannabinoids are involved in maintaining food intake. Nature 2001; 410:822-825.
23. Di Marzo V, Melck D, Bisogno T et al. Endocannabinoids: Endogenous cannabinoid receptor ligands with neuromodulatory action. Trends Neurosci 1998; 21:521-528.

24. Drew WG, Miller LL. Cannabis: Neural mechanisms and behavior – A theoretical review. Pharmacology 1974; 11:12-32.
25. Egertova M, Elphick MR. Localisation of cannabinoid receptors in the rat brain using antibodies to the intracellular C-terminal tail of CB. J Comp Neurol 2000; 422:159-171.
26. Evans EB, Wenger GR. Effects of drugs of abuse on acquisition of behavioral chains in squirrel monkeys. Psychopharmacol 1992; 107:55-60.
27. Ennaceur A, Delacour J. A new one-trial test for neurobiological studies of memory in rats: I Behavioral data. Behav Brain Res 1988; 31:47-59.
28. Ferrari F, Ottani A, Vivoli R et al. Learning impairment produced in rats by the cannabinoid agonist HU210 in a water-maze task. Pharmacol Biochem Behav 1999; 64:555-561.
29. Fletcher JM, Page JB, Francis DJ et al. Cognitive correlates of long-term cannabis use in Costa Rican men. Arch Gen Psychiatry 1996; 53:1051-1057.
30. Gifford AN, Tang Y, Gatley SJ et al. Effect of the cannabinoid receptor SPECT agent, AM281, on hippocampal acetylcholine release from rat brain slices. Neurosci Lett 1997; 238:84-86.
31. Gluck JP, Ferraro DP, Marriott RG. Retardation of discrimination reversal by delta-9-tetrahydrocannabinol in monkeys. Pharmacol Biochem Behav 1973; 1:605-608.
32. Goani Y, Mechoulam R. Isolation, structure and partial synthesis of an active constituent of hashish. J Am Chem Soc 1964; 86:1646-1647.
33. Hampson RE, Deadwyler SA. Ensemble codes involving hippocampal neurons are at risk during delayed performance tests. Proc Natl Acad Sci USA 1996; 93:13487-13493.
34. Hampson RE, Deadwyler SA. Role of cannabinoid receptors in memory storage. Neurobiol Disease 1998; 5:474-482.
35. Hampson RE, Deadwyler SA. Cannabinoids, hippocampal function and memory. Life Sci 1999; 65:715-723.
36. Hampson RE, Deadwyler SA. Cannabinoids reveal the necessity of hippocampal neural encoding for short-term memory in rats. J Neurosci 2000; 20:8932-8942.
37. Hampson RE, Heyser CJ, Deawyler SA. Hippocampal cell firing correlates of delayed-match-to-sample performance in the rat. Behav Neurosci 1993; 107:715-739.
38. Hampson RE, Jarrard LE, Deadwyler SA. Effects of ibotenate hippocampal destruction on delayed matching and nonmatching to sample behavior in rats. J Neurosci 1999; 19:1492-1507.
39. Hampson RE, Simeral JD, Deadwyler SA. Distribution of spatial and nonspatial information in dorsal hippocampus. Nature 1999; 402:610-614.
40. Han CJ, Robinson JK. Cannabinoid modulation of time estimation in the rat. Behav Neurosci 2001; 115:243-246.
41. Han CJ, Pierre-Louis J, Scheff A et al. A performance-dependent adjustment of the retention interval in the delayed nonmatching-to-position paradigm differentiates effects of amnestic drugs in rats. Europ J Pharmacol 2000; 403:87-93.
42. Hernandez-Tristan R, Arevalo C, Canals S et al. The effects of acute treatment with Delta(9)-THC on exploratory behaviour and memory in the rat. J Physiol Biochem 2000; 56:17-24.
43. Heyser CJ, Hampson RE, Deadwyler SA. Effects of delta-9-tetrahydrocannabinol on delayed match to sample performance in rats. Alterations in short-term memory associated with changes in task specific firing of hippocampal cells. J Pharmacol Exp Ther 1993; 264:294-307.
44. Howlett AC. Cannabinoid compounds and signal transduction mechanisms. In: Pertwee RG, ed. Cannabinoid Receptors: Molecular Biology and Pharmacology. London: Academic Press, 1995:167-204.
45. Howlett AC. The CB_1 cannabinoid receptor in the brain. Neurobiol Disease 1998; 5:405-416.
46. Isaacson RL, Douglas RJ, Moore, RY. The effect of radical hippocampal ablation on acquisition of avoidance response. J Comp Physiol Psychol 1961; 54:625-628.
47. Iversen LL. The Science of Marijuana. New York: Oxford Univ Press, 2000.
48. Izquierdo I, Nasello AG. Effects of cannabidiol and diphenylhydantoin on the hippocampus and learning. Psychopharmacologia 1973; 31:167-175.
49. Jaerbe TUC, Hiltunen AJ. Cannabimimetic activity of cannabinol in rats and pigeons. Neuropharmacol 1987; 26:219-228.
50. Jaerbe TUC, Andrzejewski ME, DiPatrizio NV. Interaction between the CB1 receptor agonist Δ^9-THC and the CB1 receptor antagonist SR-141716 in rats. Open field revisited. Pharmacol Biochem Behav 2002; 73:911-919.
51. Jarrard LE. On the role of the hippocampus in learning and memory in the rat. Behav Neural Biol 1993; 60:9-26.
52. Jentsch JD, Andrusiak E, Tran A et al. HA966 prevents increases in prefrontal dopamine turnover and impairments in spatial working memory induced by psychomimetic drugs. FASEB J 1996; 20:A716.

53. Jentsch JD, Andrusiak E, Tran A et al. Δ^9-tetrahydrocannabinol increases prefrontal cortical catecholaminergic utilization and impairs spatial working memory in the rat: Blockade of dopaminergic effects by HA966. Neuropsychopharmacology 1997; 16:426-432.
54. Johnson JP, Muhleman D, MacMurray J et al. Association between the cannabinoid receptor gene (CNR1) and the P300 event-related potential. Molec Psychiat 1997; 2:169-171.
55. Koch M. The neurobiology of startle. Prog Neurobiol 1999; 59:107-128.
56. Lichtman AH. The novel cannabinoid antagonist SR141716A does not enhance working memory in rats. Faseb J 1996; 10:4109-4109.
57. Lichtman AH. SR141716A enhances spatial memory assessed in a radial-arm maze task in rats. Europ J Pharmacol 2000; 404:175-179.
58. Lichtman AH, Martin BR. Δ^9-tetrahydrocannabinol impairs spatial memory through cannabinoid receptor mechanism. Psychopharmacol 1996; 126:125-131.
59. Lichtman AH, Martin BR. The selective cannabinoid antagonist SR 141716A blocks cannabinoid-induced antinociception in rats. Pharmacol Biochem Behav 1997; 57:7-12.
60. Lichtman AH, Dimen KR, Martin BR. Systemic or intrahippocampal cannabinoid administration impairs spatial memory in rats. Psychopharmacol 1995; 119:282-290.
61. Lichtman AH, Varvel SA, Martin BR. Endocannabinoids in cognition and dependence. Prostaglandins Leukotrines Essential Fatty Acids 2002; 66:269-285.
62. Maccarrone M, Valverde O, Barbaccia ML et al. Age-related changes of anandamide metabolism in CB_1 cannabinoid receptor knockout mice: Correlation with behaviour. Europ J Neurosci 2002; 15:1178-1186.
63. Mallet PE, Beninger RJ. The endogenous cannabinoid receptor agonist impairs memory in rats. Behav Pharmacology 1996; 7:276-284.
64. Mallet PE, Beninger RJ. The cannabinoid CB_1 receptor antagonist SR141716A attenuates the memory impairment produced by Δ^9-tetrahydrocannabinol or anandamide. Psychopharmacology 1998; 140:11-19.
65. Mansbach RS, Rovetti CC, Winston EN et al. Effects of cannabinoid receptor antagonist SR141716A on the behavior of pigeons and rats. Psychopharmacol 1996; 124:315-322.
66. Marsicano G, Wotjak CT, Azad SC et al. The endogenous cannabinoid system controls extinction of aversive memories. Nature 2002; 418:530-534.
67. Martin M, Ledent C, Parmentier et al. Involvement of CB1 cannabinoid receptors in emotional behaviour. Psychopharmacol 2002; 159:379-387.
68. Matsuda LA, Lolait SJ, Brownstein MJ et al. Structure of a cannabinoid receptor and functional expression of the cloned cDNA. Nature 1990; 346:561-564.
69. Mechoulam R, Fride E. A hunger for cannabinoids. Nature 2001; 410:763-765.
70. Mechoulam R, Ben-Shabat S, Hanus L et al. Identification of an endogenous 2-monoglyceride, present in canine gut, that binds to cannabinoid receptors. Biochem Pharmacol 1995; 50:83-90.
71. Miyamoto A, Yamamoto T, Watanabe S. Effect of repeated administration of Δ^9-tetrahydrocannabinol on delayed matching-to-sample performance in rats. Neurosci Lett 1995; 201:139-142.
72. Miller LL, Branconnier RJ. Cannabis: Effects on memory and the cholinergic limbic system. Psychol Bull 1983; 93:441-456.
73. Mishima K, Egashira N, Hirosawa N et al. Characteristics of learning and memory impairment induced by Δ^9-tetrahydrocannabinol in rats. Jpn J Pharmacol 2001; 87:297-308.
74. Molina-Holgado F, Gonzalez MI, Leret ML. Effect of Δ^9-tetrahydocannabinol on short-term memory in the rat. Physiol Behav 1995; 57:177-179.
75. Morris RGM. Developments of a water-maze procedure for studying spatial learning in that rat. J Neurosci Methods 1984; 11:47-60.
76. Munro S, Thomas KL, Abu-Shaar M. Molecular characterization of a peripheral receptor for cannabinoids. Nature 1993; 365:61-65.
77. Nakamura EM, da Silva EA, Concilio GV et al. Reversible effects of acute and long-term administration of Δ-9-tetrahydrocannabinol (THC) on memory in the rat. Drug Alcohol Dep 1991; 28:167-175.
78. Nakamura-Palacios EM, Winsauer PJ, Moerschbacher JM. Effects of the cannabinoid ligand SR141716A alone or in combination with Δ^9-tetrahydrocannabinol or scopolamine on learning in squirrel monkeys. Behav Pharmacol 2000; 11:377-389.
79. Nava F, Carta G, Battesi AM et al. D_2 dopamine receptors enable Δ^9tetrahydrocannabinol induced memory impairment and reduction of hippocampal extracellular acetylcholine concentration. Brit J Pharmacol 2000; 130:1201-1210.

80. Nava F, Carta G, Colombo G et al. Effects of chronic Δ^9tetrahydrocannabinol treatment on hippocampal extracellular acetylcholine concentration and alternation performance in the T-maze. Neuropharmacol 2001; 41:392-399.
81. Navarro M, Fernandez Ruiz JJ, de Miguel R et al. Motor disturbances induced by an acute dose of Δ^9-tetrahydrocannabinol: Possible involvement of nigrostriatal dopaminergic alterations. Pharmacol Biochem Behav 1993; 45:291-298.
82. O'Leary DS, Block RI, Koeppel JA et al. Effects of smoking marijuana on brain perfusion and cognition. Neuropsychopharmacol 2002; 26:802-816.
83. Onaivi ES, Green MR, Martin BR. Pharmacological characterisation of cannabinoids in the elevated plus-maze. J Pharmacol Exp Ther 1990; 253:1002-1009.
84. Pertwee RG. Pharmacology of cannabinoid CB1 and CB2 receptors. Pharmacol Therapeut 1997; 74:129-180.
85. Pertwee RG. Pharmacology of cannabinoid receptor ligands. Curr Med Chem 1999; 6:635-664.
86. Pertwee RG, Ross RA. Cannabinoid receptors and their ligands. Prostag Leukotr Ess Fatty Acids 2002; 66:101-121.
87. Piomelli D, Beltramo M, Giuffrida A et al. Endogenous cannabinoid signaling. Neurobiol Disease 1998; 5:462-473.
88. Presburger G, Robinson JK. Spatial signal detection in rats is differentially disrupted by Δ-9-tetrahydrocannabinol, scopolamine and MK-801. Behav Brain Res 1999; 99:27-34.
89. Porter AC, Felder CC. The endocannabinoid nervous system: Unique opportunities for therapeutic intervention. Pharmacol Therap 2001; 90:45-60.
90. Reibaud M, Obinu MC, Ledent C et al. Enhancement of memory in cannabinoid CB_1 receptor knock-out mice. Europ J Pharmacol 1999; 379:R1-R2.
91. Rodriguez de Fonseca F, Del Arco I, Martin-Calderon JL et al. Role of the endogenous cannabinoid system in the regulation of motor activity. Neurobiol Disease 1998; 5:483-501.
92. Rowland NE, Mukherjee M, Robertson K. Effects of the cannabinoid receptor antagonist SR 141716A, alone and in combination with dexfenfluramine or naloxone, on food intake in rats. Psychopharmacol 2001; 159:111-116.
93. Schneider M, Koch M. The cannabinoid agonist WIN 55,212-2 reduces sensorimotor gating and recognition memory in rats. Behav Pharmacol 2002; 13:29-37.
94. Schulze GE, McMillan DE, Bailey JR et al. Acute effects of delta-9-tetrahydrocannabinol in rhesus monkeys as measured by performance in a battery of complex operant tests. J Pharmacol Exp Ther 1988; 245:178-186.
95. Schwartz RH, Gruenewald PJ, Klitzner M et al. Short-term memory impairments in cannabis-dependent adolescents. Am J Dis Chil 1989; 143:1214-1219.
96. Stadnicki SW, Schaeppi U, Rosenkrantz H et al. Crude marihuana extract: EEG and behavioral effects of chronic oral administration in rhesus monkeys. Psychopharmacologia 1974; 37:225-233.
97. Stiglick A, Kalant H. Learning impairment in the radial-arm maze following prolonged cannabis treatment in rats. Psychopharmacol 1982; 77:117-123.
98. Stiglick A, Llewellyn ME, Kalant H. Residual effects of prolonged cannabis treatment on shuttle-box avoidance in the rat. Psychopharmacol 1984; 84:476-479.
99. Terranova J-P, Storme J-J, Lafon N et al. Improvement of memory in rodents by the selective CB1 cannabinoid receptor antagonist, SR141716. Psychopharmacol 1996; 126:165-172.
100. Varvel SA, Lichtman AH. Evaluation of CB_1 receptor knockout mice in the Morris water maze. J Pharmacol Exp Ther 2002; 301:915-924.
101. Varvel SA, Hamm RJ, Martin BR et al. Differential effects of Δ^9THC on spatial reference and working memory in mice. Psychopharmacol 2001; 157:142-150.
102. Wilson RI, Nicoll RA. Endocannabinoid signalling in the brain. Science 2002; 296:678-682.
103. Winsauer PJ, Lambert P, Moerschbacher JM. Cannabinoid ligands and their effects on learning and performance in rhesus monkeys. Behav Pharmacol 1999; 10:497-511.

Chapter 3.2

Opioids

Makoto Ukai, Ken Kanematsu, Tsutomu Kameyama and Takayoshi Mamiya

Abstract

To date, numerous publications have reported on the effects of opioids in learning and memory, and much research has been conducted on novel opioid receptor ligands as anti-amnesic drugs. The involvement of the opioid system including opioid peptides and their receptors has been increasingly revealed by anatomical, electrophysiological and biochemical approaches. Here, we review the relationship between opioid systems and learning and memory, and discuss several of our novel findings.

Introduction

More than twenty years have passed since the implication of opioids in learning and memory was revealed. At that time, the effects of nonselective opioid receptor ligands (e.g., morphine, naloxone, etc.) were studied on learning and memory functions from the standpoint of behavioral pharmacology. Then, the influence of opioid receptor ligands on memory processes (acquisition, consolidation, retention and retrieval) was investigated in detail. Furthermore, anatomical, electrophysiological and biochemical studies have been conducted using not only whole animals but also brain slices. In particular, the localization of opioid peptides, the alternation of long-term potentiation (LTP) as well as long-term depression (LTD), known indices of synaptic efficiency, and the quantitative changes of neurotransmitter release caused by opioid receptor ligands in the hippocampus suggested a link between opioids and memory function.

In 1997, endogenous μ-opioid receptor ligands (endomorphins-1 and 2) have finally been isolated and purified from mammalian brain, leading to the supposition that the opioid system may contribute to learning and memory functions, and that opioid agents could be useful drugs in the therapy of memory disturbances.

To date, numerous publications have reported on the effects of opioids in learning and memory, and much research has been conducted on novel opioid receptor ligands as anti-amnesic drugs since the initial report by Castellano that morphine impaired learning and memory.[3] In particular, the involvement of the opioid system including opioid peptides and their receptors in learning and memory has been increasingly revealed by anatomical, electrophysiological and biochemical approaches.

Distribution of Opioid Peptides and Their Receptors in the Hippocampus

Opioid Peptides

The immunohistochemical localization of proenkephalin-derived peptides in the hippocampus has indicated their presence in the somatic fibers of specific neural populations. In the

From Messengers to Molecules: Memories Are Made of These, edited by Gernot Riedel and Bettina Platt.

rat dentate gyrus, enkephalins were observed in fibers of the lateral perforant path and the temporoammonic tract[4,11,23] (Table 1), and both fiber tracts originate in the entorhinal cortex. Additionally, some proenkephalin-derived peptide immunoreactivity was found in the mossy fiber projection, mainly consisting of granule cell axons projecting to the proximal dendrites of CA3 pyramidal cells.[4,23]

In contrast, prodynorphin-derived peptides exist in a more restricted neuronal population within the hippocampus than proenkephalin-derived peptides (Table 1). Dynorphins are produced primarily by granule cells, and the immunoreactivity is densest in the mossy fiber pathway.[4,10,23,24] Dynorphin is also detected in the dentate molecular layer in guinea pigs. Electron microscopy revealed that 75% of the dynorphin-containing dense core vesicles in the molecular layer exist in dendrites and most likely in granule cells.[10]

Receptors

There are important differences in the pattern of opioid receptor expression among species. In rats, μ-opioid receptor mRNA has been demonstrated in neurons scattered in both granule cell and pyramidal cell layer, but also in stratum oriens and radiatum.[2,20] Similarly, receptor autoradiography, using [^{3}H]DAMGO, has revealed a high density of μ-opioid binding sites in the stratum lacunosum moleculare and the pyramidal cell layer. The binding sites are moderately dense in CA3 stratum radiatum but absent from stratum lucidum.[24] In the dentate gyrus, receptors seem to be concentrated in the layers adjacent to the granule cell layers. More recently, differential distribution of endomorphin-1 and endomorphin-2-like immunoreactivities in rodents has been reported.[21]

μ-Opioid receptor-like immunoreactivity was found in nonpyramidal neurons that are likely to be GABAergic interneurons.[1] Punctate labeling was also observed in the surrounding dentate granule and pyramidal cells, consistent with localization at presynaptic terminals. Furthermore, it was reported that μ-opioid receptors are functionally coupled to this channel.[28]

Compared to μ-opioid receptor expression, δ-opioid receptor expression is more widespread. In the rat, δ-opioid receptor mRNA has been demonstrated in neurons within the pyramidal and granule cell body layers.[20] Moreover, [^{3}H]DADLE autoradiography has shown that the distribution of δ-opioid receptor binding sites is generally similar to that of μ-opioid receptor binding sites. However, δ-opioid receptor binding sites are more abundant, particularly in the dentate molecular layer.[24] Using antisera raised against the cytoplasmic tail of the mouse δ-opioid receptor, it was demonstrated that the δ-opioid receptor is localized on GABAergic terminals surrounding the somata of a subpopulation of cells in the stratum radiatum, stratum pyramidale, and stratum oriens.[1] These cell bodies include both pyramidal and nonpyramidal cells. Similarly, in the dentate gyrus, immunoreactive GABAergic terminals surround the cell bodies of nongranule cells in stratum moleculare and the hilus. The cells associated with δ-opioid receptor immunoreactive terminals include both GABAergic and nonGABAergic cells.[1]

On the other hand, κ-opioid receptor distribution in hippocampus also shows marked differences among species, with much higher levels in the guinea pig than in the rat. Moreover, the κ_2 receptor with low affinity for dynorphins and U-69593 seems to be the main κ subtype expressed in rat hippocampus[34,38] but the κ_1 subtype, which is defined by its high affinity for the synthetic agonist U-69593, is also expressed in guinea pig hippocampus.

An autoradiographic study showed that κ-opioid receptors are present in greatest abundance in the pyramidal and granule cell layers and areas adjacent to these layers, as well as in CA3 stratum lucidum in the rat.[24] κ-Opioid receptor specific antibodies revealed additional receptors in the inner molecular layer of pyramidal cells in guinea pigs[10] and in the middle molecular layer of the ventral dentate gyrus in rats.[23] It is interesting to note that the pattern of κ-opioid receptor expression in guinea pigs, as opposed to rats, resembles the pattern seen in the human brain.[27]

Table 1. Distribution of opioid peptides in hippocampal region

Regions	Cells/Pathway	Peptides	References
CA1/2			
Stratum lacunosum-moleculare	Lateral temporoammonic tract	Enkephalins	J Comp Neurol, 198, 335-350 (1981).
All layers	Interneurons	Enkephalins	J Comp Neurol, 198, 335-350 (1981).
CA3			
Stratum lucidum	Mossy fiber	Dynorphins Enkephalins	Brain Res, 331, 366-370 (1985); J Neurosci, 5, 808-816 (1985); Proc Natl Acad Sci USA, 80, 589-593 (1985).
Stratum lacunosum-moleculare	Lateral temporoammonic tract	Enkephalins	J Comp Neurol, 198, 335-350 (1981).
All layers	Interneurons	Enkephalins	J Comp Neurol, 198, 335-350 (1981).
Dentate Gyrus			
Outer molecular layer	Lateral perforant path	Enkephalins	J Comp Neurol, 198, 335-350 (1981).
All layers	Interneurons	Enkephalins	J Comp Neurol, 198, 335-350 (1981).
Hilus	Mossy fiber collaterals	Enkephalins Dynorphins	Brain Res, 331, 366-370 (1985); J Neurosci, 5, 808-816 (1985); Proc Natl Acad Sci USA, 80, 589-593 (1985).
Molecular layer	Granule cell dendrites	Dynorphins	J Neurosci, 14, 3736-3750 (1994).

Effects of Opioid Receptor Ligands on Long-Term Potentiation in Hippocampal Regions (Table 2)

LTP is characterized by the enhancement of synaptic responses that are induced by a particular event, such as high-frequency stimulation, and lasts for hours or days. It is widely regarded as a cellular substrate for learning and memory processes. The mechanisms of LTP are beyond the scope of this review and have been thoroughly reviewed elsewhere.[19,25] Here, we will focus on the roles of opioids in LTP of the hippocampal mossy fiber.

There have been many demonstrations of μ- and δ-opioid receptor modulation of LTP at the mossy fiber-CA3 pyramidal cell synapse. It was reported that local application of naloxone blocked LTP of CA3 field EPSPs evoked by mossy fiber stimulation in the rat. In contrast, naloxone had no effect on CA3 LTP evoked by stimulation of commissural afferents. The conclusion drawn from those studies was that the high-frequency tetanus used to induce LTP also evoked the release of opioid peptides from mossy fibers, and these opioids were necessary for LTP induction. The specific receptor type responsible for the above effects was later determined as the μ-opioid receptor. The μ-opioid receptor antagonist CTOP also blocked mossy fiber LTP, and both the δ-opioid receptor antagonist naltrindole and the κ-opioid receptor antagonist nor-binartorphimine (norBNI) were without any effects.[9] The effects of CTOP were restricted to mossy fiber potentiation, because commissural fiber LTP was not blocked. CTOP also attenuated post-tetanic potentiation, which is a marked facilitation observed in the first few seconds to minutes after mossy fiber tetanus. Post-tetanic potentiation is thought to reflect increased glutamate release as a result of accumulated calcium in the presynaptic terminal. Further information regarding μ-opioid receptor mediated modulation of LTP was gained by pairing local application of DAMGO with mossy fiber stimulation. A brief mossy fiber tetanus, which had been shown to be insufficient for inducing LTP, was delivered following application of a low dose of DAMGO, which alone did not induce LTP. When the two events were paired, LTP was induced, suggesting that the μ-opioid receptor agonist lowered the thresholds of the induction of mossy fiber LTP.[7,8]

The effects of κ-opioid receptor mediated modulation of LTP are quite different from those of μ-opioid receptor mediated modulation. In the guinea pig hippocampus, the κ-opioid receptor antagonist norbinaltorphimine facilitated LTP, that is, a tetanus paradigm that failed to induce LTP in control experiments was sufficient to induce LTP when applied while κ-opioid receptors were blocked. Thus, the tetanic stimulation released endogenous dynorphins from mossy fiber terminals, which functioned to increase the threshold for LTP induction. The inhibitory effects of endogenous dynorphins could be overcome by increasing the number of pulses in the tetanus. When this longer tetanus was delivered in the presence of naloxone (at a concentration that blocks κ-opioid receptors), the magnitude of LTP was not significantly changed. This means that κ-opioid receptors do not modulate mossy fiber LTP.[9] Furthermore, the inhibition of mossy fiber transmission by endogenously released dynorphins is masked when LTP is induced simultaneously. However, the processes of potentiation do not prevent the processes of dynorphin-mediated inhibition, because a second tetanus delivered 30 min after an LTP tetanus elicits a transient, norBNI-sensitive inhibition.[29] In addition to raising the threshold for mossy fiber LTP, dynorphins released by the tetanic stimulation also serve to inhibit excitatory transmission at parallel, unpotentiated mossy fiber synapses.[37] This heterosynaptic depression is transient and naloxone-sensitive. Thus, the tetanized synapses are relatively unaffected by dynorphins because they are simultaneously potentiated, while neighboring synapses are depressed. It has been reasoned that the heterosynaptic depression effectively limits the LTP to the tetanized pathway, which may be important for learning and memory functions ascribed to the hippocampus.

Table 2. Effects of various opioid receptor ligands on the long-term potentiation in the hippocampal regions

Recording	Stimulation				References
μ-Opioid receptor					
CA1 pyramidal cells	Schaffer lateral	naloxone	antagonist	±	Brain Res, 280, 127-138 (1983).
CA3 pyramidal cells	Mossy fiber	naloxone	antagonist	⇩	Brain Res Bull, 27, 219-223 (1991). Brain Res, 821, 286-293 (1999).
CA3 pyramidal cells	Commissural fiber	naloxone	antagonist	±	Brain Res Bull, 27, 219-223 (1991).
Granule cells	Perforant path	naloxone	antagonist	⇩	Brain Res, 567, 42-50 (1991).
CA3 pyramidal cells	Mossy fiber	CTOP	antagonist	⇩	J Pharmacol Exp Ther, 263, 725-733 (1992).
CA3 pyramidal cells	Commissural fiber	CTOP	antagonist	±	J Pharmacol Exp Ther, 263, 725-733 (1992).
CA3 pyramidal cells	Mossy fiber	DAMGO	agonist	⇧	Brain Res, 821, 286-293 (1999).
δ-Opioid receptor					
Granule cells	Perforant path	ICI 174864	antagonist	⇩	Brain Res, 567, 42-50 (1991).
κ-Opioid receptor					
CA1 pyramidal cells	Mossy fiber	Dynorphin A	agonist	⇩	Nature, 362, 423-427 (1993)
Granule cells	Perforant path	Dynorphin	agonist	⇩	Nature, 363, 451-454 (1993)
Granule cells	Perforant path	U69593	agonist	⇩	J Neurosci, 14, 4740-4747 (1994). J Neurosci, 20, 4379-4388 (2000)

⇧: facilitation, ⇩:inhibition, ±:no change

Effects of Opioid Receptor Ligands on Learning and Memory in Hippocampal Regions

One role of the hippocampus in spatial memory processes is that hippocampal lesions induce the impairment of spatial performance.[30] In a behavioral test, naloxone was found to induce an enhancement of spatial memory.[12] The impairment induced by dentate granule cell stimulation was blocked by pretreatment with naloxone, indicating that endogenous hippocampal opioid systems are implicated in the memory deficit.[5] Since granule cells contain both enkephalin and dynorphins, it remains to be determined which opioid system mediates the amnesic effects of dentate granule cell stimulation. The presence of dynorphin A and dynorphin A (1-8) in hippocampus resulted in the impaired performance of learning, however, their effects were blocked by naloxone pretreatment.[22] On the other hand, aged rats with memory deficit have elevated levels of dynorphin peptides compared with young healthy rats.[18] The finding that dynorphins released from granule cells following intense activity impair spatial memory is consistent with the inhibition of hippocampal LTP by dynorphins.[36] These findings indicate that endogenous opioid systems are involved in the negative modulation of learning and memory.

Effects of Opioid Receptor Ligands on Learning and Memory Tasks (Table 3)

It is well established that peripherally injected opioid receptor agonists and antagonists modulate learning and memory in a wide variety of tasks.[26] On the whole, agonists tend to inhibit learning and memory and antagonists facilitate them, with some exceptions. Relatively little is known, however, about the behavioural significance of opioids in hippocampus. Gallagher et al[12] found that systemically administered opioid receptor antagonists facilitate water-maze learning, which is highly sensitive to hippocampal damage. It was later determined that β-endorphin and naloxone can affect hippocampal function by disrupting modulatory inputs from the medial septal area (cholinergic/GABAergic, see Pepeu and Giovannini and Castellano et al, this book) and the locus coeruleus (noradrenergic, see Gibbs and Summers, this book). Furthermore, aged rats with poor spatial learning ability have elevated hippocampal levels of dynorphin and preprodynorphin mRNA, while aged cohorts with normal spatial learning ability do not.[18] It was shown that applying a 10-sec sinusoidal electrical stimulation to mossy fibers after acquisition of a radial arm maze task impaired subsequent retention, and this effect was blocked by systemic administration of naloxone.[5] The study implicated the mossy fiber–CA3 region in spatial memory, but it was not determined whether the effects of naloxone were due to a direct action on the hippocampus, and whether the effects of electrical stimulation on CA3 cells were monitored. It was also found that the rapid learning to self-administer dynorphin A into the CA3 region is blocked by the coadministration of naloxone. It is suggested that the CA3 region plays an important role in opioid dependence, which may be related to clinical observations that drug craving and compulsive drug seeking are aroused by memory of past drug reinforcement. Moreover, some genetically inbred strains of mice with abnormal exploratory behavior and learning ability were found to have aberrant mossy fiber terminal fields and altered dynorphin levels.[6] Further characterization of these strains may provide insight into normal mossy fiber function.[35] Finally, studies combining behavioral electrophysiology with focal manipulation of opioidergic transmission will be very useful for clarifying the contribution of opioids to hippocampal function.

Ameliorating Effects of Opioid Receptor Ligands on Models of Learning and Memory Impairment (Table 4)

Here, we have focussed on the effects of μ- and κ-opioid receptor agonists on animal models of learning and memory impairment. In particular, not only endomorphin-1, but also dynorphin A and the synthetic agent U-50488H ameliorated the impairment of learning and

Table 3. Effects of opioid receptor ligands on the various learning and memory tasks

	Tasks	Effects	References
μ-Receptors			
Agonists			
Morphine	Water Y-maze	⇩	Psychopharmacologia, 42, 235-242 (1975).
Morphine	Passive response	⇩	Psychopharmacology, 66, 199-203 (1979).
Levorphanol	Passive avoidance response	⇩	Life Sci, 23, 1973-1978 (1978).
Endomorphins	Y-maze	⇩	Eur J Pharmacol, 395, 211-215 (2000).
Endomorphins	Passive avoidance response	⇩	Eur J Pharmacol, 421, 115-119 (2001).
β-Endorphin	Passive response	⇩	Psychopharmacology, 69, 111-115 (1980). Neurosci Lett, 19, 197-201 (1980).
γ-Endorphin	Passive response	⇩	Neurosci Lett, 19, 197-201 (1980).
DAMGO	Y-maze	⇩	J Pharmacol Exp Ther, 269, 15-21 (1994).
TAPA	Y-maze	⇩	Gen Pharmacol, 29, 453-456 (1997).
TAPA	Passive avoidance response	⇩	Eur J Pharmacol, 239, 237-240 (1993). Eur J Pharmacol, 287, 245-249 (1995).
Antagonists			
Naloxone	Passive avoidance response	⇧	Life Sci, 23, 1973-1978 (1978).
Naloxone	Passive response	⇧	Psychopharmacology, 66, 199-203 (1979).
Naloxone	Rearing response	⇧	Psychopharmacology, 67, 265-268 (1980).
Naloxone	8-arm radial maze	⇧	Science, 221, 975-976 (1983).
Naloxone	8-arm radial maze	⇧	Behav Neural Biol, 44, 374-385 (1985).
Diprenorphine	8-arm radial maze	⇧	Science, 221, 975-976 (1983).
Naltrexone	8-arm radial maze	⇧	Behav Neural Biol, 44, 374-385 (1985).
δ-Receptors			
Agonists			
DPLPE	Passive avoidance response	⇩	Eur J Pharmacol, 338, 1-6 (1997).
[D-Ala^2] Deltorphin	Passive avoidance response	⇩	Eur J Pharmacol, 338, 1-6 (1997).
DPLPE	Passive avoidance response	⇩	Behav Neurosci, 103, 429-437 (1989).
[leu^5] Enkephalin	Passive avoidance response	⇩	Behav Neurosci, 103, 429-437 (1989).
Antagonists			
ICI 174864	Passive avoidance response	⇧	Pharmacol Biochem Behav, 52, 683-687 (1995).
κ-Receptors			
Agonists			
Dynorphin A(1-8)	Radial maze	⇩	Pharmacol Biochem Behav, 35, 429-432 (1990).
Dynorphin A(1-13)	Peck avoidance learning	⇩	Psychopharmacology, 108, 235-240 (1992).
Dynorphin A(1-13)	Aversive & appetitive learning	⇩	Peptides, 14, 1165-1170 (1993).
Dynorphin B	Water maze	⇩	Neuroscience, 85, 375-382 (1998).
U-50488H	Peck avoidance learning	⇩	Psychopharmacology, 108, 235-240 (1992).
Tifluadom	Passive avoidance response	⇩	Behav Brain Res, 15, 177-181 (1985).
Bremazocine	Passive avoidance response	⇩	Arch Int Pharmacodyn Ther, 283, 199-208 (1986).
Dynorphin A(1-17)	Passive avoidance response	⇧	Life Sci, 47, 1453-1462 (1990).
U-50488H	Peck avoidance learning	⇧	Psychopharmacology, 108, 235-240 (1992).
U-69593	Y-maze	⇧	Brain Res, 856, 259-280 (2000).
Antagonists			
MR-1452	Passive avoidance response	⇧	Arch Int Pharmacodyn Ther, 283, 199-208 (1986).
norBNI	Y-maze	⇩	Behav Brain Res, 109, 229-241 (2000).

⇧: enhancement, ⇩: impairment

Table 4. Attenuating effects of κ-opioid receptor ligands on the impaired models of learning and memory

	Tasks	Impaired Model	References
κ-Opioid receptor ligands			
Dynorphin A(1-13)	Y-maze Passive avoidance response Elevated plus-maze	ischemia	Eur J Pharmacol, 234, 9-15 (1993).
Dynorphin A(1-13)	Passive avoidance response	basal forebrain lesion	Brain Res, 625, 355-356 (1993).
Dynorphin A(1-13)	Y-maze	scopolamine	Eur J Pharmacol, 236, 341-345 (1993).
Dynorphin A(1-13)	Passive avoidance response	scopolamine	Eur J Pharmacol, 274, 89-93 (1995).
Dynorphin A(1-13)	Y-maze	pirenzepine	Eur J Pharmacol, 281, 173-178 (1995).
Dynorphin A(1-13)	Y-maze	DAMGO	J Pharmacol Exp Ther, 269, 15-21(1994).
Dynorphin A(1-13)	Passive avoidance response	cycloheximide	Eur J Pharmacol, 313, 11-15 (1996).
Dynorphin A(1-13)	Elevated plus-maze	galanin	Neuropharmacology, 33, 1167-1169 (1994).
Dynorphin A(1-13)	Passive avoidance response	galanin	Hum Psychopharmacol Clin Exp, 12, 243-248 (1997).
Dynorphin A(1-13)	Passive avoidance response	carbon monoxide	Eur J Pharmacol, 282, 185-191 (1995).
Dynorphin A(1-13)	Y-maze	carbon monoxide	Pharmacol Biochem Behav, 56, 73-79 (1997).
Dynorphin A(2-13)	Y-maze Passive avoidance response	carbon monoxide	Brain Res, 859, 303-310 (2000).
Dynorphin A(2-13)	Y-maze Passive avoidance response	β-amyloid	Neuroreport , 11, 431-435 (2000).
U-50488H	Y-maze Passive avoidance response	carbon monoxide	Eur J Pharmacol, 315, 119-125 (1996).
U-50488H	Passive avoidance response	carbachol	Neurosci Lett, 236, 45-48 (1997).
U-50488H	Passive avoidance response	carbachol	Br J Pharmacol, 123, 920-926 (1998).
U-50488H	Passive avoidance response	scopolamine	J Pharmacol Exp Ther, 284, 858-867 (1998).
U-50488H	Y-maze	pirenzepine	Eur J Pharmacol, 281, 173-178 (1995).
U-50488H	Passive avoidance response	mecamylamine	J Pharmacol Exp Ther, 284, 858-867 (1998).
U-50488H	Passive avoidance response	dizocilpine	J Pharmacol Exp Ther, 284, 858-867 (1998).
U-50488H	3-way runway	ischemia	Eur J Pharmacol, 193, 357-361 (1991).
U-50488H	Y-maze Passive avoidance response Elevated plus-maze	ischemia	Brain Res, 619, 223-228 (1993).

memory induced by scopolamine and pirenzepine (muscarinic receptor antagonists).[13,14,31-33] The finding that opioid receptor agonists attenuate learning and memory impairments is controversial, because many reports have shown that opioid receptor agonists can produce memory deficits (see Table 3). It appears that opioid receptor agonists improve memory impairments by acting directly on neurotransmitters rather than on the cellular mechanisms of LTP. For example, the increase in acetylcholine release induced by scopolamine was suggested to be the result of a positive feedback after the blockade of pre and/or postsynaptic muscarinic receptors. Although U-50488H attenuated the increase in acetylcholine release by muscarinic receptor antagonists, dynorphin A (1-13) did not affect this increase. We speculate that dynorphin A (1-13) may postsynaptically activate cholinergic systems because κ-opioid receptor agonists inhibit the spontaneous release of acetylcholine.[13] Furthermore, we examined the effects of dynorphin A and U-50488H on the impairment of learning and memory induced by carbon monoxide exposure and transient ischemia in mice.[15-17] κ-Opioid receptor agonists attenuated the impairment in models of delayed amnesia accompanied by neuronal death. In contrast, although the exact mechanism underlying the effects of endomorphin-1 remains elusive, it is likely that the opioid peptide improved short-term memory disturbances resulting from cholinergic dysfunction via μ_1-opioid receptors.[31] These findings suggest that κ- and μ-opioid receptor agonists act as neuromodulators and are useful drugs for the improvement of learning and memory impairments.

Future Expectation

Here we have briefly reviewed the relationship between the opioid system and the learning and memory function in the hippocampus. Recently, not only conventional pharmacological techniques but also molecular biological approaches have been employed in the detailed investigation of opioid systems (endogenous opioid and its receptor function). Thus, we believe that the significance of opioid systems in the learning and memory will be clarified fully in the near future.

References

1. Bausch SB, Patterson TA, Appleyard SM et al. Immunocytochemical localization of delta opioid receptors in mouse brain. J Chem Neuroanat 1995; 8:175-189.
2. Bunzow JR, Zhang G, Bouvier C et al. Characterization and distribution of a cloned rat mu-opioid receptor. J Neurochem 1995; 64:14-24.
3. Castellano C. Effects of morphine and heroin on discrimination learning and consolidation in mice. Psychopharmacologia 1975; 42:235-242.
4. Chavkin C, Shoemaker WJ, McGinty JF et al. Characterization of the prodynorphin and proenkephalin neuropeptide systems in rat hippocampus. J Neurosci 1985; 5:808-816.
5. Collier TJ, Routtenberg A. Selective impairment of declarative memory following stimulation of dentate gyrus granule cells: A naloxone-sensitive effect. Brain Res 1984; 310:384-387.
6. Crusio WE, Schwegler H, Brust I et al. Genetic selection for novelty-induced rearing behavior in mice produces changes in hippocampal mossy fiber distributions. J Neurogenet 1989; 5:87-93.
7. Derrick BE, Martinez Jr JL. Frequency-dependent associative long-term potentiation at the hippocampal mossy fiber-CA3 synapse. Proc Natl Acad Sci USA 1994; 91:10290-10294.
8. Derrick BE, Martinez Jr JL. Opioid receptor activation is one factor underlying the frequency dependence of mossy fiber LTP induction. J Neurosci 1994; 14:4359-4367.
9. Derrick BE, Rodriguez SB, Lieberman DN et al. Mu opioid receptors are associated with the induction of hippocampal mossy fiber long-term potentiation. J Pharmacol Exp Ther 1992; 263:725-733.
10. Drake CT, Terman GW, Simmons ML et al. Dynorphin opioids present in dentate granule cells may function as retrograde inhibitory neurotransmitters. J Neurosci 1994; 14:3736-3750.
11. Gall C, Brecha N, Karten HJ et al. Localization of enkephalin-like immunoreactivity to identified axonal and neuronal populations of the rat hippocampus. J Comp Neurol 1981; 198:335-50.
12. Gallagher M, King RA, Young NB. Opiate antagonists improve spatial memory. Science 1983; 221:975-976.
13. Hiramatsu M, Murasawa H, Mori H et al. Reversion of muscarinic autoreceptor agonist-induced acetylcholine decrease and learning impairment by dynorphin A (1-13), an endogenous kappa-opioid receptor agonist. Br J Pharmacol 1998; 123:920-926.

14. Hiramatsu M, Murasawa H, Nabeshima T et al. Effects of U-50,488H on scopolamine-, mecamylamine- and dizocilpine-induced learning and memory impairment in rats. J Pharmacol Exp Ther 1998; 284:858-867.
15. Hiramatsu M, Sasaki M, Nabeshima T et al. Effects of dynorphin A (1-13) on carbon monoxide-induced delayed amnesia in mice. Pharmacol Biochem Behav 1997; 156:73-79.
16. Itoh J, Ukai M, Kameyama T. U-50,488H, a kappa-opioid receptor agonist, markedly prevents memory dysfunctions induced by transient cerebral ischemia in mice. Brain Res 1993; 619:223-228.
17. Itoh J, Ukai M, Kameyama T. Dynorphin A-(1-13) potently prevents memory dysfunctions induced by transient cerebral ischemia in mice. Eur J Pharmacol 1993; 234:9-15.
18. Jiang HK, Owyang VV, Hong JS et al. Elevated dynorphin in the hippocampal formation of aged rats: Relation to cognitive impairment on a spatial learning task. Proc Natl Acad Sci USA 1989; 86:2948-2951.
19. Madison DV, Schuman EM. LTP, post or pre? A look at the evidence for the locus of long-term potentiation. New Biol 1991; 3:549-557.
20. Mansour A, Fox CA, Burke S et al. Mu, delta, and kappa opioid receptor mRNA expression in the rat CNS: An in situ hybridization study. J Comp Neurol 1994; 350:412-438.
21. Martin-Schild S, Gerall AA, Kastin AJ et al. Differential distribution of endomorphin 1 and endomorphin 2-like immunoreactivities in the CNS of the rodent. J Comp Neurol 1999; 405, 450-471.
22. McDaniel KL, Mundy WR, Tilson HA. Microinjection of dynorphin into the hippocampus impairs spatial learning in rats. Pharmacol Biochem Behav 1990; 35:429-35.
23. McGinty JF, Henriksen SJ, Goldstein A et al. Dynorphin is contained within hippocampal mossy fibers: Immunochemical alterations after kainic acid administration and colchicine-induced neurotoxicity. Proc Natl Acad Sci USA 1983; 80:589-593.
24. McLean S, Rothman RB, Jacobson AE et al. Distribution of opiate receptor subtypes and enkephalin and dynorphin immunoreactivity in the hippocampus of squirrel, guinea pig, rat, and hamster. J Comp Neurol 1987; 255:497-510.
25. Nicoll RA, Kauer JA, Malenka RC. The current excitement in long-term potentiation. Neuron 1988; 1:97-103.
26. Olson GA, Olson RD, Kastin AJ. Endogenous opiates: Peptides 1990; 11:1277-304.
27. Quirion R, Pilapil C, Magnan J. Localization of kappa opioid receptor binding sites in human forebrain using [^{3}H]U69,593: Comparison with [^{3}H]bremazocine. Cell Mol Neurobiol 1987; 7:303-307.
28. Simmons ML, Chavkin C. κ-Opioid receptor activation of a dendrotoxin-sensitive potassium channel mediates presynaptic inhibition of mossy fiber neurotransmitter release. Mol Pharmacol 1996; 50:80-85.
29. Simmons ML, Terman GW, Gibbs SM et al. L-type calcium channels mediate dynorphin neuropeptide release from dendrites but not axons of hippocampal granule cells. Neuron 1995; 14:1265-1272.
30. Sinnamon HM, Freniere S, Kootz J. Rat hippocampus and memory for places of changing significance. J Comp Physiol Psychol 1978; 92:142-155.
31. Ukai M, Katoh T, Mamiya T. Endomorphin-1 improves scopolamine-induced impairment of short-term memory via μ_1-opioid receptor in mice. NeuroReport 2002; 12:3723-3727.
32. Ukai M, Watanabe Y, Kameyama T. Effects of endomorphins-1 and -2, endogenous μ-opioid receptor agonists, on spontaneous alternation performance in mice. Eur J Pharmacol 2000; 395:211-215.
33. Ukai M, Watanabe Y, Kameyama T. Endomorphins 1 and 2, endogenous μ-opioid receptor agonists, impair passive avoidance learning in mice. Eur J Pharmacol 2001; 421:115-119.
34. Unterwald EM, Knapp C, Zukin RS. Neuroanatomical localization of kappa 1 and kappa 2 opioid receptors in rat and guinea pig brain. Brain Res 1991; 562:57-65.
35. van Abeelen JH, Janssens CJ, Crusio WE et al. Y-chromosomal effects on discrimination learning and hippocampal asymmetry in mice. Behav Genet 1989; 19:543-549.
36. Wagner JJ, Terman GW, Chavkin C. Endogenous dynorphins inhibit excitatory neurotransmission and block LTP induction in the hippocampus. Nature 1993; 363:451-454.
37. Weisskopf MG, Zalutsky RA, Nicoll RA. The opioid peptide dynorphin mediates heterosynaptic depression of hippocampal mossy fibre synapses and modulates long-term potentiation. Nature 1993; 362:423-427.
38. Zukin RS, Eghbali M, Olive D et al. Characterization and visualization of rat and guinea pig brain kappa opioid receptors: Evidence for kappa 1 and kappa 2 opioid receptors. Proc Natl Acad Sci USA 1988; 85:4061-4065.

Chapter 3.3

Neuropeptides

David De Wied and Gábor L. Kovács

Introduction

Neuropeptides are peptides involved in nervous system function. They are synthesized in cells in large precursor proteins, and generally several biologically active peptides are contained in the same precursor molecule. Various biochemical processes control the quantities of neuropeptides, as well as the nature of their biological activity, through size, form and derivatization of the end product. In this way neuropeptides with different, opposite, and often more selective properties are formed from the same precursor.

The generation of neuropeptides is a cell/specific phenomenon. They are co-localized with classical neurotransmitters and released when the system is stimulated. Neuropeptides may act as neurotransmitters, but in most instances modulate neuronal activity in conjunction with the neurotransmitter, with which they are co-localized. In recent years, receptors have been found for many of the neuropeptides in the brain.

The effects on learning and memory processes by various neuropeptide families have been reviewed earlier.[65] This manuscript mainly—although not exclusively—summarizes findings of the literature, published after 1995.

Posterior Pituitary Peptides (Vasopressin, Oxytocin)

Vasopressin exerts a long-term facilitating effect on learning and memory processes, which can be clearly demonstrated in aversive-conditioning. The influence of vasopressin is time-dependent, i.e., the effectiveness of neuropeptide treatment depends on the time interval between the learning or retention trial and the treatment. Vasopressin and related peptides facilitate consolidation as well as retrieval processes. These neuropeptides also prevent and reverse retrograde amnesia induced by various amnesic treatments, which is another measure to determine effects on retrieval processes.[8,27,31]

The effect of vasopressin on learning and memory processes in non-aversive (food-rewarded, sexually motivated etc.) tasks has been controversial for a long time.[20,37] Paban et al[92] using a non-aversive visual discrimination task observed a tendency to improve performance following intrahippocampal injection of vasopressin in male Balb/c mice at different stages of the learning process. Engelmann et al[38] discussed the effects of vasopressin and also oxytocin on behavior. They concluded that the results found in aversively motivated behavior is often different from those obtained in non-aversive behavior and also depend on the area of the brain under study.

The capacity of male rats to remember familiar conspecifics is called social recognition. It is a form of short-term memory. It measures the amount of time an adult male rat spends to investigates a juvenile. Recognition lasts 30 min and has disappeared after 120 min. Social recognition is also facilitated by vasopressin and attenuated by oxytocin.[97] However, low doses of oxytocin microinjected into the preoptic area facilitates social recognition.[98] Dluzen et al[35] infused vasopressin and oxytocin bilaterally into the olfactory bulb in male rats and found that both peptides preserved recognition at 120 min. Antagonists of the two neuropeptides had no

From Messengers to Molecules: Memories Are Made of These, edited by Gernot Riedel and Bettina Platt.

effect. Thus the olfactory bulb represents an important central nervous system target site where these neuropeptides can act to preserve social recognition responses. In contrast to social recognition, object recognition reflects a form of habituation, which is not under the control of vasopressin.[40]

Learning and memory processes are disturbed in animals deficient in vasopressin [e.g., in the homozygous variant of the Brattleboro strain of rats (HODI);[8]]. Disturbances can be observed in various single and multiple learning paradigms and these disturbances can be normalized by vasopressin and some behaviorally active (non-endocrine) fragments, such as Des-Gly^9-Arg^8 -LVP.[8]

In contrast to vasopressin, oxytocin facilitates extinction of conditioned avoidance behavior. The performance in a one-trial passive avoidance paradigm is also time-dependent following systemic, or central administration. In general, it can be stated that vasopressin and oxytocin exert opposite effects on fear-motivated avoidance behavior and hence the hypothesis has been put forward that oxytocin is an amnesic neuropeptide.[27,65]

Evidence for a role of endogenous vasopressin and oxytocin stems from results of studies with central (intracerebroventricular, i.c.v.) injections of specific antisera against neurohypophyseal neuropeptides. I.c.v. administration of anti-vasopressin serum induced severe disturbances in active and passive avoidance behavior. Anti-oxytocin serum, on the other hand, resulted in an improved performance. Time gradient studies with anti-vasopressin serum or anti-oxytocin serum also pointed to a modulatory role of endogenous vasopressin and oxytocin on consolidation as well as retrieval processes.[27,65]

Several attempts have been made to determine the sites of action of vasopressin and oxytocin on learning and memory processes. Microinjections of small amounts of vasopressin in the dentate gyrus, in the dorsal septal nuclei, or in the dorsal raphe nuclei improved passive avoidance behavior, when administered after the learning trial. Microinjection of vasopressin into various limbic areas also improved passive avoidance behavior of rats that had been made amnesic by pentylenetetrazol. Of various brain structures tested, the ventral hippocampus appeared to be the most sensitive area for vasopressin to improve passive avoidance behavior in rats.[65]

Alescio-Lautier et al[1] also found that the hippocampal structure is involved in the memory effect of vasopressin. Using a Go-No Go visual discrimination task in mice, these authors have shown that both parts of the hippocampus mediate the effect of endogenous or exogenous vasopressin on memory processes, but the ventral part is more sensitive. They showed that micro-injection of vasopressin into the ventral hippocampus, generates different behavioral effects, depending on whether treatment is performed at the beginning or in the middle of the learning process, suggesting that the mnemonic context is an important factor for understanding the effect of vasopressin on memory in the ventral hippocampus.[92]

Paban et al[94] used a polyclonal antibody to determine Fos and Fos-like proteins. In non-conditioned mice, vasopressin evoked a time-dependent increase in Fos and Fos-like protein expression in the dentate gyrus, CA1 and CA3 hippocampal fields, lateral septum, bed nucleus of the stria terminalis, and basolateral and central amygdaloid nuclei. In contrast, in conditioned mice, an increase in the level of Fos expression was detected only in the dentate gyrus, ventral CA3 hippocampal field, and the lateral septum. Thus, the pattern observed after post-training injection of vasopressin was not the same as that evoked by vasopressin alone.

Endogenous vasopressin and oxytocin in the dorsal septum or in the ventral hippocampus is of particular importance for learning and memory processes. Engelmann et al[39] found evidence that the two nonapeptides are not only secreted from the neurohypophysis into the general circulation, but—probably upon some specific stimuli, and largely independently of their peripheral release—are also released intracerebrally (e.g., into septum). The experiments provide additional evidence for an involvement of endogenous vasopressin and oxytocin in the regulation of learning and memory processes. Metzger et al[73] determined whether the injection of vasopressin or vasopressin antisera into the ventral hippocampus has an effect on retrieval and relearning of a visual discrimination task in mice. Pretest microinjection of vasopressin into the ventral hippocampus alleviated forgetting observed after a prolonged interval of 24 days between the acquisition of information and its retrieval. Conversely, immunoneutralization of endogenous vasopressin in the ventral hippocampus by microinjection of vasopressin antisera resulted in a drastic impairment of retrieval and relearning.

Similarly to vasopressin, oxytocin has also been found to exert locus-specific effects on learning and memory.[65] Oxytocin attenuates memory consolidation upon microinjection into the hippocampal dentate gyrus or the midbrain dorsal raphe nucleus. The central amygdaloid nuclei did not respond to oxytocin, although this region receives a relatively dense oxytocinergic innervation. It is possible, however, that endogenous oxytocin in the amygdala is involved in retrieval rather than consolidation processes. Bilateral injections of oxytocin in the dorsal hippocampus, in rats, impaired acquisition and accelerated extinction of conditioned avoidance behavior in a shuttle-box.[155] The data suggested that the attenuating effect of oxytocin on acquisition of shuttle-box avoidance behavior is, at least partly, mediated by the hippocampus.

Vasopressin treatment may also counteract behavioral effects, induced by other neuropeptides. In a study of Izquierdo et al,[61] retrograde amnesia was induced in rats trained in a step-down inhibitory avoidance task, by an i.p injection of β-endorphin or by the administration of an electroconvulsive shock. Pretest i.p. injections of vasopressin (or ACTH) reversed the amnesia.

Based on electrophysiological findings, one might conclude that a neurotransmitter-like effect is associated with vasopressin in limbic brain structures. Vasopressin applied iontophoretically to neurons in the lateral septal area or ventral hippocampus in vivo, excites approximately 30 percent of the neurons tested. The remaining 70 percent were not excited by vasopressin. The response of these neurons to glutamate but not to acetylcholine was markedly potentiated.[64] This suggests that the more important influence of vasopressin is on neuromodulation. In this respect it is of interest that vasopressin is capable of modulating long-term potentiation, which is believed to be an electrophysiological basis of memory processes.[133] Winnicka and Wisniewski[144-146] found that the positive influence of vasopressin on memory processes is mediated by excitatory amino acids, since it was abolished by NMDA receptor antagonists. These authors also reported that bilateral transections of glutamatergic temporo-entorhinal connections attenuated the facilitating effect of vasopressin on retrieval. Vasopressin may also modulate various other neurotransmitter systems in the central nervous system. It enhanced noradrenaline turnover in the hypothalamus, thalamus, and medulla oblongata but not in the septum, preoptic area, hippocampus and amygdala in rats. Microinjection of vasopressin into the dentate gyrus enhanced, while injection into the dorsal septal area decreased, noradrenaline turnover. Destruction of the coerulo-telencephalic noradrenergic system with 6-OHDA prevented the effect of vasopressin on passive avoidance behavior following post-learning administration (consolidation), but not when the peptide was injected prior to the retention test (retrieval). Thus, the coerulo-telencephalic noradrenaline system may mediate the effects of vasopressin on memory consolidation.[65]

Diaz Brinton et al[32] summarized the cellular mechanisms of the action of vasopressin on hippocampal cells. They suggest that receptors for vasopressin are present in both neurons and glial cells. In the periphery, vasopressin is a potent mitogen in some proliferative cell types, which also suggests a possible association between vasopressin receptor activation and the proliferative capacity of astrocytes. These authors therefore investigated whether vasopressin would induce the expression of the immediate early response gene, NGFI-A, which is associated with initiation of mitogenesis. Cultured hippocampal glial cells were exposed to vasopressin or a selective vasopressin V1 receptor agonist and in situ hybridization for NGFI-A mRNA was conducted. Results of these experiments demonstrated that vasopressin induced a highly significant dose-dependent increase in the number of cells expressing NGFI-A. Jachimowicz et al[62] tested the hypothesis that nitric oxide (NO), which functions as a novel type of inter-cellular messenger in the central nervous system, participated in the facilitating effect of arginine vasopressin on learning and memory. Their results, however, indicate that central action of vasopressin is probably independent of NO activity in the brain.

Born et al[9] reviewed human research on the effect of vasopressin on memory processes. Although the human studies yielded less consistent results than those in rats, they indicate that vasopressin is able to improve declarative memory formation, which is the type of memory essentially relying on hippocampal function. The effect appears to center on the encoding process for memory. The regulation of voluntary selective attention and arousal do not appear to be primary targets of vasopressin effects in humans. Mediation of effects by peripheral changes could be excluded since the central effects were observed in studies using intranasal vasopressin administration which provides a direct access to the brain.

Vasopressin and oxytocin are converted to highly selective memory molecules in the brain[13] as [Cyt^6]AVP-(4-9/5-9) and [Cyt^6]AVP-(4-8/4-9). These are more effective than the parent nonapeptide, in terms that a lower amount of these peptide fragments facilitated passive avoidance behavior in all brain regions investigated. Following microinjections into the ventral hippocampus, [Cyt^6]AVP-(4-8/5-8) were more effective in a post-learning than in a pre-retention treatment schedule. [Cyt^6]AVP-(4-9/5-9), on the other hand, was more effective when injected shortly before the retention trial. It was suggested, therefore, that active fragments of vasopressin selectively influence different phases of information processing. This is in agreement with the findings using systemic administration.[24] Analyzing the effects of AVP-(4-8), Du et al[36] found a putative receptor-mediated signaling pathway involving second messenger IP3, immediately-early gene c-fos transcription and protein kinase PKC, CaMKII and MAPK. These authors also described a peptide-induced enhancement of some crucial functional proteins such as calmodulin, nerve growth factor (NGF) and brain-derived nerve growth factor (BDNF). Tanabe et al[125] described the effects of a newly synthesized cationized AVP-(4-9) analogue on learning and memory in rats, using the passive avoidance test. The cationized peptide fragment was highly effective after a s.c. treatment route, most likely because it more easily penetrates the blood-brain-barrier.

NC-1900 is a stable peptide analog of AVP-(4-9), with a five-fold longer half-life than that of AVP-(4-9) itself. Sato et al[112] investigated the mechanism of action of NC-1900 on learning and/or memory impairment in passive avoidance task and on damage of cultured cerebro-cortical neurocytes induced by glutamic acid. NC-1900 ameliorated impairments of learning and/or memory induced by intracisternal injection of glutamic acid and impairments induced by intracisternal NMDA, AMPA-antagonist CNQX and by the $mGlu_1$ receptor agonist 3,5-dihydroxyphenylglycine, but not by the kainate receptor agonist domoic acid nor MK-801 in mice. NC-1900 also ameliorated the cell damage. These results suggest that NC-1900 may serve as a remedy in patients with certain brain disorders induced by excess glutamic acid.

Ponomareva et al[95] reported studies on the effects of intranasal administration of a new analog of arginine-vasopressin fragment AVP-(6-9), i.e., D-MPRG, on the learning ability of rats with positive and negative reinforcement. The peptide improved learning, i.e., accelerated the acquisition of a conditioned active avoidance behavior both when given before or immediately after training sessions. The peptide had a greater effect when animals were trained with negative reinforcement. Analysis of the results suggests that the action of D-MPRG is mainly on perception processes, i.e., extraction of the conditioned stimulus from the environmental surroundings and evaluation and enhancement of its biological significance. In addition, the peptide prevented extinction of an acquired habit and improved consolidation, though this effect was weaker than its effect on perception. In recent experiments of Dietrich and Allen,[33,34] $AVP_{4\text{-}9}$ enhanced radial arm maze performance. AVP-(4-9) treated animals showed enhancement in performance as well as increases in the rate of learning, indicating that they learned the task faster. The overall memory enhancement was due to improved working as well as reference memory. Rats with NMDA lesions in the hippocampus showed a marked deficit in working memory that was not ameliorated by AVP-(4-9); however, the improvement in reference memory produced by the compound was as large as in healthy animals. It is concluded that the vasopressin fragment has a more general effect on memory and that its site of action includes but is not limited to the hippocampus.

Which receptor is involved in the effects of the neurohypophyseal hormones on learning and memory processes? Kidney vasopressin V_2 receptors which mediate the antidiuretic effect, activate adenylate cyclase. Vascular vasopressin V_1 receptors are coupled to phospholipase C to generate inositol-1,4,5- triphosphate (IP3) and to increase intracellular Ca^{2+} concentrations. The liver membrane or hepatocyte receptors are also coupled to IP3 and Ca^{2+}. The same holds for the pituitary vasopressin receptors. These show marked differences in affinity as well as in cellular and endocrine responsiveness to a number of vasopressin analogues and this receptor is regarded as a subtype of the vasopressin V_1 receptor, namely the vasopressin V_{1b} receptor, as opposed to the liver and vascular V_{1a} receptor. The oxytocin receptor is also coupled to IP3 and Ca^{2+} and mediates the contraction of ovine myometrial cells and myoepithelial cells of the bovine mammary gland.[129] Vasopressin and oxytocin receptors have also been found in the brain. Receptors in the brain, selective for vasopressin, are presumably of the vasopressin V_{1a} receptor type. There is also evidence for central vasopressin V_{1b} receptor and circumstantial evidence for the existence of central vasopressin V_2 receptors. In the hippocampus both oxytocin and vasopressin V_1 receptors have been detected. Expression of mRNA was found for the vasopressin V_{1a} receptor in the cortex.[27,115]

As mentioned above, the ventral hippocampus is the area most sensitive to the effect of the neurohypophyseal hormones on avoidance behavior. The facilitating effect of vasopressin and the attenuating effects of oxytocin on passive avoidance behavior are blocked by a vasopressin V_1, a vasopressin V_2, and by an oxytocin receptor antagonist. Since this oxytocin receptor antagonist does not have a high affinity for vasopressin V_{1a} receptors, while the vasopressin receptor antagonists possess relative high affinity for oxytocin sites, these results suggest that the memory effects of the neurohypophyseal hormones may be mediated by the oxytocin receptor in the ventral hippocampus. Vasopressin and related peptides may act on this non-selective receptor as agonists while oxytocin and related peptides, which have opposite effects, may act as "inverse" agonists.[27] Paban et al[93] have shown that the enhancing effect of vasopressin, when microinjected into the ventral hippocampus of mice, is antagonized by pretreatment with the vasopressin V_1 receptor antagonist $d(CH_2)^5$Tyr(Me)-vasopressin. The vasopressin V_2 receptor antagonist microinjected into the ventral hippocampus did not alter the enhancing effect of vasopressin on retrieval and relearning. In contrast, the oxytocin receptor antagonist blocked the vasopressin-enhancing effect on retention processes. Thus these observations on mice partly confirm the previous conclusions of the Utrecht group on rats, that both vasopressin V_1 receptors and oxytocin receptors seem to be involved in the enhancing effect of vasopressin on memory retention.

In a recent experiment, Nakayama et al[82] pharmacologically characterized the putative binding site and mechanism of intracellular signaling of AVP-(4-9). Radioligand binding assay showed that AVP-(4-9) could detect specific binding sites in the rat hippocampus membrane preparations, and the binding site was specifically displaced by AVP-(4-9). AVP-(4-9) caused the $[Ca^{2+}]$(i) increase via release from intracellular calcium store as well as influx from extracellular calcium. AVP-(4-9) could not detect vasopressin V_{1a}, V_{1b} and V_2 receptors. For the first time, this study provides evidence to show that AVP(4-9) activates an IP3/$[Ca^{2+}]_{(i)}$ pathway and intracellular calcium concentrations through a novel type of receptor in rat hippocampus, which might be potentially important for the mnemonic effect of these selective neuropeptides.

Tanabe et al[126] conducted studies to clarify the mechanisms by which AVP-(4-9) affects memory processes. AVP-(4-9) enhanced the basal and the high-potassium-evoked acetylcholine release from rat hippocampal slices. A vasopressin V_1-selective antagonist (but not a vasopressin V_2-selective antagonist) inhibited AVP-(4-9)-stimulated basal acetylcholine release. AVP-(4-9) also facilitated the passive-avoidance response of scopolamine-induced memory-deficient mice. These findings suggest that AVP-(4-9) stimulates acetylcholine release via vasopressin V_1-like receptors. The results also suggest that the mechanism of action of AVP-(4-9) on learning and memory is mediated by the cholinergic system in the brain.

ACTH/MSH and Opioid Peptides

ACTH/MSH neuropeptides facilitate the deficient acquisition of shuttle-box avoidance behavior of hypophysectomized rats, delay extinction of shuttle box avoidance behavior and pole-jumping avoidance behavior, and facilitate passive avoidance behavior of intact rats.[25] Classical endocrine activity of ACTH/MSH neuropeptides can be clearly dissociated from behavioral effects. A great number of structure-activity studies were performed using active and passive avoidance behavior as the behavioral bioassay. The main conclusion of these studies was that ACTH-(4-7) was the smallest peptide to be fully active. γ-2-MSH, which differs from α-MSH in various amino acid residues, attenuates acquisition and facilitates extinction of active avoidance behavior and attenuates passive avoidance behavior. Structure-activity studies also revealed a number of highly selective ACTH-(4-9) analogs. ORG 2766 with 3 modifications in the ACTH-(4-9) sequence is markedly more potent in active and passive avoidance behavior, while its intrinsic endocrine effects are negligible.[50]

Different hypotheses have been offered to explain the influence of ACTH/MSH neuropeptides. One hypothesis suggested that ACTH/MSH neuropeptides increase the motivational value of the consolidating stimulus, thereby modifying the input and the external mechanisms of learning and memory processes. This effect may be caused by a selective arousal in limbic midbrain structures. This hypothesis would easily explain the physiological significance of ACTH/MSH peptides in aversive (stressful) behavioral situations. ACTH also influences attention and concentration. This hypothesis offers an explanation for the putative physiological effects of ACTH/MSH peptides in non-aversive learning. The electrophysiological finding that ACTH and related peptides increase the sensitivity (mean and peak frequency of theta-activity following stimulation of the mesencephalic reticular formation) of the hippocampus in rats is evidence for a selective arousal effect of these neuropeptides in limbic-midbrain structures. As outlined above, these effects might have important consequences for learning and memory processes, especially for the retrieval of stored information.[19,25] The posterior thalamic region seems to be an essential structure for the effect of ACTH-related peptides on avoidance behavior. This is also indicated by the fact that bilateral lesioning of the parafascicular nuclei inhibit the effect of ACTH-(1-10) on extinction of pole-jumping avoidance behavior.[7]

It has been suggested that ACTH/MSH peptides improve avoidance behavior via an increase in the turnover of noradrenaline in the brain. Many reports indeed showed an increase in catecholamine and serotonin turnover or content in different brain regions after the treatment with ACTH or behaviorally active ACTH fragments.[136] It has been reported that the behaviorally highly active ACTH-(4-9)-fragment (ORG 2766) and related peptide fragments affect acetylcholine turnover e.g., in the hippocampus and the frontal cortex of the rat. Behaviorally active ACTH fragments were found to antagonize glutamate binding.[57] Horvath et al[54] studied the effect of ORG 2766 applied in early postnatal life when brain structures and neuronal pathways are still developing. The aim was to see whether such treatment during development would result in permanent changes in adult behavioral performance. Animals treated with ORG 2766 during early postnatal life learned faster in the spatial Morris water-maze. The treatment had a positive effect on performance during the acquisition phase of the learning task, while memory retrieval was not affected. Learning in a non-spatial active avoidance task was not affected by postnatal $ACTH_{4-9}$ treatment. In addition, there were no differences in the open field test, the defensive burying test, the elevated plus maze and the conditioned fear test. The latter supports the conclusion that the differences in water-maze performance was due to a difference in learning speed, rather than a difference in anxiety or behavioral stress reactivity.

Also in humans, neuropeptides of the ACTH/MSH family are modulators of cognitive function. Their neurobehavioral activity is principally encoded in the 4-10 fragment of the ACTH/MSH molecule; in humans, it has been shown to pertain primarily to functions of attentive stimulus/response processing.[119] The effects support the view of a de-focusing action of ACTH during selective attention that could serve to improve the organism's adaptation to stress stimuli.[77] These results could provide the basis for developing a new, specific, and "soft" neuropharmacology.[41]

Data on the role of β-endorphin in learning and memory processes are rather ambiguous. β-Endorphin has been found to delay the extinction of pole-jumping active avoidance behavior and to facilitate the retention of passive avoidance behavior.[26] Flood et al[46] found that in mice, partially trained to avoid footshock in a T-maze, both intra-amygdaloid and intraventricular injections of β-endorphin resulted in amnesia. Izquierdo[60] suggested, that a difference in the neurohumoral state of an animal after the learning trial and the retention trial results in poor retrieval. According to this hypothesis, β-endorphin does not affect consolidation but merely influences retrieval processes. In accordance with this hypothesis Netto and Maltchik[83] found that a single injection of β-endorphin prior to the retention test enhances retrieval. Although these results are consistent with an interpretation of an anti-amnesic action of endorphins, especially on retrieval processes, other hypotheses regarding changes in arousal, fear-motivation, or response to stress were not explored. In contrast, others found postlearning facilitating effects of enkephalins or β-endorphin on passive avoidance behavior, and a dose-dependent dual effect of β-endorphin administered prior to the retention test.[26] It has been shown in these experiments that the effect of β-endorphin on performance in learning situations is largely dependent on the dose of the neuropeptide. Smaller doses of β-endorphin facilitate passive avoidance behavior, while higher doses have attenuating effects. This dose-dependent dual effect might be related to the fact that β-endorphin affects learning and memory processes (or 'second order' physiological processes closely associated to learning and memory) by more than one neuronal, or neurochemical mechanism. In that respect, it is of interest that β-endorphin can be converted in the brain to γ-endorphin (β-endorphin-[1-17]), α-endorphin (β-endorphin-[1-16]), and smaller fragments. α- and γ-endorphin exert opposite effects on the performance in active and passive avoidance tasks. The differential effect of β- and γ-endorphin on avoidance behavior has been replicated by other groups of investigators as well, showing, in addition, opposite effects of the two endorphins in a lever press response for food.[26] It has been suggested[24] that β-endorphin possesses amphetamine-like, while γ-endorphin (and various active fragments thereof) possesses neuroleptic-like, activities. Effects of these peptides on the performance of animals in a learning situation might thus also be secondary to these amphetamine- and neuroleptic-like effects and do not necessarily suggest (but also do not exclude) an involvement of β-endorphin and related neuropeptides in mechanisms of learning and memory.

Long-term potentiation in the mossy fibre pathway to the CA3 region in the hippocampus, an animal model of memory acquisition, is modulated by dynorphin peptides. Sandin et al[108] investigated the possible role of hippocampal dynorphin in spatial learning in the Morris water task. After microinjection into the CA3 region of the dorsal hippocampus, dynorphin B was found to impair spatial learning. The synthetic κ_1-selective opiate receptor antagonist nor-binaltorphimine also injected into the hippocampus fully blocked the acquisition impairment caused by dynorphin B, while nor-binaltorphimine alone did not affect learning performance. These findings suggest that dynorphin peptides could play a modulatory role in hippocampal plasticity by acting on hippocampal kappa-receptors and thereby impair spatial learning.

The recently discovered endogenous mu-selective opioid peptide, endomorphin-2, and the endogenous δ-selective opioid peptide, Leu-enkephalin, were tested for their ability to affect spatial learning in the Morris water task.[109] It was found that microinjection of endomorphin-2 into the CA3 region of the rat hippocampus significantly impaired spatial learning. Leu-enkephalin did not have an effect on spatial learning. Neither peptide had affected motor performance as measured by swim speed. The results indicate that mu-receptors in the CA3 region of the rat hippocampus are more relevant than δ-receptors for spatial learning. Orphanin/nociceptin, a 17-amino-acid peptide, is an endogenous peptide. Its receptor is similar to δ- and κ-opioid receptors (approximately 65% homology). It has been reported that i.c.v. injection of orphanin/nociceptin can antagonize morphine analgesia, whereas i.c.v. antibody injection against the peptide can reverse morphine tolerance. Nocistatin is a recently characterized neuropeptide possessing an antagonistic effect on orphanin/nociceptin.[91] Since orphanin/nociceptin and nocistatin are derived from the same preprohormone, the processing in the CNS may play an important role in determining the effectiveness of morphine analgesia. Sandin et al[107] investigated the possible role of hippocampal orphanin/nociceptin in spatial learning and in spontaneous locomotion. Male rats were trained in the Morris water task after microinjection of nociceptin/orphanin into the CA3 region of the dorsal hippocampus. Nociceptin/orphanin was found to severely impair spatial learning without interfering with swimming performance. Intrahippocampal injection of nociceptin/orphanin markedly decreased exploratory locomotor activity including vertical movements (rearing). The data suggest that nociceptin/orphanin is a potent modulator of synaptic plasticity within the hippocampus.

Noda et al[87] demonstrate that nociceptin-knockout mice show greater learning ability in the water maze task, an enhanced latent learning in the water finding task, better memory in the passive avoidance task, and further, larger long-term potentiation in the hippocampal CA1 region than wild-type mice. Nociceptin itself impaires passive avoidance behavior in wild-type mice. Thus, the nociceptin system seems to play negative roles in learning and memory.

Hypophyseotropic Peptides (CRF, Somatostatin)

Corticotropin-releasing factor (CRF) is the principal activator of the pituitary-adrenocortical system. However, CRF containing neurons were found outside the endocrine hypothalamus in brain structures of primary importance for learning and memory processes, e.g., in the cortex, amygdala, thalamus, locus coeruleus, brainstem, etc. CRF affects behavioral processes related to learning and memory. The hypothesis has been put forward that CRF primarily exerts anxiogenic effects and has arousal properties. Anxiety has profound effects on learning and memory processes.[19,113]

Croiset et al[19] reviewed the effects of CRF on learning and memory processes. Most of the information has been derived from studies on avoidance behavior in rats. Aversive stimuli are associated with the release of stress hormones and neuropeptides. Neuropeptides not only affect attention, motivation, concentration and arousal or vigilance, but also anxiety and fear. In this way, they participate in learning and memory processes. Furthermore, neuropeptides as CRF and vasopressin modulate the release of stress hormones such as epinephrine. In turn, systemic catecholamines enhance memory consolidation. CRF and vasopressin are co-localized in neurons from the nucleus paraventricularis, which project to nuclei in the brainstem involved in autonomic regulation. Both CRF and vasopressin have effects in the same direction on behavior, learning and memory processes and stress responses (e.g., release of catecholamines and ACTH). These neuropeptides may act synergistically or in a concerted action aimed at learning to adapt to environmental demands.

Nijsen et al[84] investigated the role of the endogenous corticotropin-releasing hormone (CRF) system in the regulation of heart rate, PQ interval (a measure of vagal activity), gross activity and release of ACTH, noradrenaline and adrenaline into the blood during conditioned fear in freely moving rats. I.c.v. infusion of α-helical CRF-(9-41), a non-selective CRF receptor antagonist, under resting conditions had no significant effect on gross activity, heart rate and PQ interval, indicating that at this dose it was devoid of agonist effects. Conditioned-fear rats showed freezing behavior, associated with an increase in heart rate, PQ interval, noradrenaline and adrenaline. This indicates that the cardiac effects were the result of co-activation of the sympathetic and parasympathetic nervous system. The i.c.v. pre-treatment of rats with α-helical CRF significantly reduced the conditioned-fear-induced tachycardiac and ACTH response, and enhanced the increase in PQ interval, without affecting the noradrenaline and adrenaline response. These results suggest that endogenous CRF reduces the vagal response to conditioned-fear stress in rats.

Diamant and De Wied[30] found that the fragment CRF-(34-41) given i.c.v. was as active as CRF-(1-41) in attenuating passive avoidance behavior, CRF-(28-41) had a minor effect and CRF-(1-8) was without effect. CRF-(34-41) did not possess ACTH-releasing effects as the parent hormone. The dissociation between endocrine, autonomic and behavioural activities suggested the existence of different CRF receptors.

Different CRF receptors have been cloned. The stimulation of the HPA-axis by CRF is mediated by CRF_1 receptors, which bind CRF in a highly preferential manner. A CRF_2 receptor is widely distributed in subcortical areas which project to autonomic brain stem areas and therefore may be involved in autonomic regulation. A splice variant of the CRF_2 receptor has been identified and designated as CRF_{2b} receptor. CRF_1 and CRF_{2a} receptors have different tissue distributions. CRF_{2a} receptor was not detected in the neocortex and cerebral cortex in contrast to CRF_1 receptor.[118] It was almost undetected in the pituitary lobes. CRF_{2a} receptor mRNA is decreased in response to food and maternal deprivation in rats. A second mammalian CRF-related neuropeptide urocortin has been found[135] which binds with high affinity to both CRF receptors. The limited overlap, between the distribution of CRF and urocortin in the rat suggests that these two peptides have distinct physiological roles. Urocortin is less potent in generating acute anxiety effects and generalized behavioural activation but more potent in suppressing appetite. This suggests that urocortin mediates some stress related effects attributed originally to CRF by serving as a ligand for the CRF_{2a} receptor. Recently a new member of the CRF-related neuropeptides was cloned, urocortin II. It has no appreciable activity on CRF_1 receptors. Transcripts encoding urocortin II are expressed in discrete regions of the rodent CNS including the PVN and the nucleus arcuatus and in the brain stem (locus coeruleus).[103] Central administration of this peptide indicates that it is involved in autonomic and appetitive control (suppression of night-time feeding), but not in general behavioural activation.

Radulovic et al[100] demonstrated a differential modulation of learning and anxiety by CRF through CRF_1 receptor and CRF_2 receptor. As learning paradigm, context- and tone-dependent fear conditioning of the mouse was used. Injection of CRF into the dorsal hippocampus before training enhanced learning through CRF_1 receptor as demonstrated by the finding that this effect was prevented by the local injection of the unselective CRF receptor antagonist astressin, but not by the CRF_2 receptor specific antagonist antisauvagine-30 (anti-Svg-30). In contrast, injection of CRF into the lateral intermediate septum impaired learning through CRF_2 receptors, as demonstrated by the ability of antisauvagine-30 to block this effect. When antisauvagine-30 was injected alone into the lateral intermediate septum, learning was enhanced. Such tonic control of learning was not observed when astressin or antisauvagine-30 were injected into the dorsal hippocampus. Injection of CRF after the training into the dorsal hippocampus and the lateral intermediate septum also enhanced and impaired learning, respectively. It was suggested that CRF acts on memory consolidation and that the observed effects reflects changes of associative learning and not arousal, attention, or motivation.

Wu et al[152] demonstrated that the local injection of an antisense oligonucleotide against CRF into the hippocampus significantly impaired the retention performance of an inhibitory avoidance task in rats. Experiments performed by Liebsch et al,[69] in which antisense oligodeoxynucleotides corresponding to either the rat CRF_1 or CRF_2 receptor mRNA were infused chronically via osmotic minipumps into the lateral ventricle. The rats were subjected to social defeat and immediately afterwards tested on the elevated plus-maze. CRF_1 receptor antisense oligodeoxynucleotide infusion was found to exert an anxiolytic-like effect, whereas CRF_2 receptor antisense oligodeoxynucleotide infusion had no effect on defeat-induced anxiety-related behavior. In contrast, the CRF_2 receptor antisense oligodeoxynucleotide increased immobility in a forced swim test. No influence of either oligodeoxynucleotides was found on general locomotor activity in an open field or on short-term memory performance in a social discrimination test. The results support the hypothesis that the two CRF receptor subtypes selectively mediate differential effects of endogenous CRF or CRF-related peptides at the brain level with the CRF_1 receptor contributing predominantly to emotional behavior and the CRF_2 receptor being involved in the regulation of stress coping behavior.

Chen et al[18] studied the role of CRF in the locus coeruleus, which gives rise to ascending noradrenergic neurons of the coeruleo-telencephalic tract and which has been implicated in attention and behavioral arousal. Microinjections of CRF into the locus coeruleus significantly improved retention performance. Intra-hippocampal destruction of catecholaminergic neurons by 6-OHDA antagonized the memory-enhancing effect of CRF in the locus coeruleus. This finding suggests that the dorsal noradrenergic pathway is involved in the effects of CRF on memory processes. This is of interest, since vasopressin also needs this pathway for its effect on memory consolidation. Since the locus coeruleus is regarded as an anatomical substrate for anxiety, CRF may enhance memory processes through its anxiogenic actions. It may well be that the effects of vasopressin on arousal are also mediated through this pathway. Clear morphological evidence on the co-localization of these two peptides in the locus coeruleus is, however, missing.

Wang et al[138] have found that CRF, injected into the dentate gyrus of hippocampus produced a dose-dependent and long-lasting enhancement in synaptic efficacy of these neurons, as measured by an increase in the amplitude and slope of population excitatory postsynaptic potentials, as well as the amplitude of population spikes. This effect of CRF was completely blocked by pretreatment with a cAMP inhibitor, and partially blocked by a NMDA receptor inhibitor. These results suggest that CRF-induced potentiation simulates the late phase of tetanization-induced long-term potentiation. cAMP seems to be the messenger mediating this effect. Ma et al[70] investigated the effects of CRF injected into the dentate gyrus of the hippocampus on brain-derived neurotrophic factor (BDNF) mRNA expression and studied whether NMDA receptors mediate the effects of CRF on BDNF mRNA expression in the dentate gyrus. Since both CRF and BDNF gene expressions are involved in memory processing in rats, these authors further investigated whether CRF facilitates memory retention through enhanced BDNF mRNA expression in the hippocampus. Effect of direct BDNF injection into the dentate gyrus on retention performance in rats was also assessed. Results indicated that CRF produced a dose-dependent increase in BDNF mRNA level and consistently improved retention performance in rats in an inhibitory avoidance learning task. BDNF antisense oligonucleotide treatment, at a concentration, which did not affect retention performance alone, blocked the memory-enhancing effect of CRF. However, acute and chronic BDNF injection into the dentate gyrus did not improve memory performance in rats. These results suggest that at least one of the mechanisms responsible for the memory-facilitating effect of CRF is mediated through enhanced BDNF mRNA expression in the hippocampus. Astressin, a novel unselective CRF receptor antagonist, has been found to be particularly potent in inhibiting the HPA axis. Astressin also significantly reverses the anxiogenic-like response induced by social stress.[121]

Somatostatin is highly concentrated in the extra-hypothalamic areas of the brain, including the frontal and parietal cortex and the hippocampus. At these locations somatostatin may play a fundamental role in the modulation of cognitive functions. Indeed, somatostatin was found to affect behavioral processes related to learning and memory.[15,137] Following i.c.v. administration, the neuropeptide inhibits extinction of an active avoidance response and attenuates retrograde amnesia induced by electroconvulsive shock in rats.

Somatostatin increased the turnover of acetylcholine in the hippocampus, brainstem and the diencephalon of rats. A facilitated release of cortical and hippocampal serotonin and noradrenaline was also observed in in vivo and in vitro experiments following somatostatin administration.[65] Schettini[114] found that the activation of somatostatin receptors inhibited adenylate cyclase and reduced intracellular Ca^{2+} levels in the brain. The peptide caused hypopolarization of hippocampal and cortical cells by inducing outward K^+ currents. Florio et al[47] found a significant reduction of pre-prosomatostatin mRNA levels in aged animals in the frontal and the parietal cortex, but not in the hypothalamus. These results demonstrate that age-related alterations in somatostatin gene expression occur in the rat. This suggests that such alterations may participate in the behavioral and cognitive impairments that occur during aging.

Sanchez-Alavez et al[106] studied the effect of cortistatin a recently discovered neuropeptide related to somatostatin, named after its predominantly cortical expression and ability to depress cortical activity, on learning and memory. Cortistatin-14 shares 11 of the 14 amino acids with somatostatin-14, yet their nucleotide sequences and chromosomal localization clearly indicate that they are products of separate genes. Now cloned from human, mouse and rat sources, cortistatin binds to all 5 cloned somatostatin receptors and shares many pharmacological and functional properties with somatostatin including the depression of neural activity. However corticostatin also has properties distinct from somatostatin including induction of slow-wave sleep, apparently by antagonism of the excitatory effects of acetylcholine on the cortex and reduction of locomotor activity.[120] Its mRNA is related to γ-amino-butyric-acid (GABA)-containing cells in the cerebral cortex and hippocampus. Cortistatin modulates the electrophysiology of the hippocampus and cerebral cortex of rats; hence it may modulate mnemonic processes. Results showed that the administration of either cortistatin or somatostatin into the hippocampal CA1 area deteriorates memory consolidation in a dose –dependent manner and facilitates extinction of learned behavior. Cortistatin is more potent in this respect than somatostatin. As far as the mechanism of action is concerned however, cortistatin increases cAMP, while somatostatin has the opposite effect.

Brain-Gut Peptides (CCK, Neuropeptide Y, Galanin)

Cholecystokinin (CCK) is a peptide originally discovered in the gastorintestinal tract but also found in high density in the mammalian brain. The C-terminal sulphated octapeptide fragment of cholecystokinin (CCK-8) constitutes one of the major neuropeptides in the brain; CCK-8 has been shown to be involved in numerous physiological functions such as feeding behavior, central respiratory control and cardiovascular tonus, vigilance states, memory processes, nociception, emotional and motivational responses. CCK-8 interacts with nanomolar affinities with two different receptors designated cholecystokinin receptor Type A (CCK_1 receptor) and Type B (CCK_2 receptor).[86] The functional role of CCK and its binding sites in the brain and periphery has been investigated thanks to the development of potent and selective CCK receptor antagonists and agonist. The physiological and pathological implications of CCK_2 receptor have been demonstrated in CCK_2 receptor deficient mice obtained by gene targeting (Nagata et al, 1996).

Peptides related to CCK have been detected in the brain. The predominant form is the C-terminal octapeptide (CCK-8). Pathways of CCK-8 have been demonstrated in the cerebral cortex, as well as in subcortical structures. In the cerebral cortex CCK is present in very high concentrations. CCK has been shown to co-exist with dopamine in several dopamine-containing neurons.

In relation to learning and memory processes, early findings have shown that peripheral injections of CCK-8 impaired acquisition and facilitated extinction of active avoidance behavior. Fekete et al[42] measured the acquisition of shuttle-box avoidance behavior, extinction of bench-jumping active avoidance behavior, food-motivated conditioned approach behavior, and one-trial learning passive avoidance behavior. These authors found that following peripheral administration, both the sulfated and non-sulfated octapeptide, the C-terminal tetra-, penta-, hexa-, and heptapeptides were almost equally active on extinction of active avoidance behavior and on passive avoidance behavior. CCK-8 was found to impair acquisition in a shuttle-box (two way) avoidance paradigm. In contrast, in a passive avoidance learning paradigm CCK-8 improved retention (lengthened avoidance latency), when the neuropeptide was injected either after the single learning trial[42] or prior to the retention test.[134] Other findings also argue for a significant role of CCK in information processing,[58] since CCK receptor agonists and antagonists have repeatedly been demonstrated to improve and impair, respectively, learning and memory functions.[46,74] The effect of subcutaneously injected caerulein (a nonselective CCK receptor agonist) on memory impairment induced by protein kinase C inhibitors was examined in rats.[124] Intracerebroventricular injection of protein kinase C inhibitors caused marked memory impairment in a one-trial passive avoidance test and in a Morris water maze. When rats were pretreated with caerulein before the training trials, the CCK receptor agonist offered protection. Itoh et al[59] studied the effect of subcutaneously administered caerulein on amnesia induced by protein synthesis inhibitors in passive and active avoidance behavior and in the Morris water maze test. The amnesic effect of the protein synthesis inhibitors was abolished by combined administration with caerulein.

Harro and Oreland[51] studied the effect of CCK receptor agonists and antagonists on the ability to acquire an appetitively motivated task and to influence spatial memory. Studies have been carried out in which endogenous CCK was blocked in the posterior cingulate cortex of mice using a local injection of CCK-8 antiserum;[74] and memory effects were tested using visual discrimination conditioning. Injection of CCK-8 antiserum 10-15 min before each session produced substantial learning impairment in the discrimination task. But when injections were stopped, animals began to learn the task normally, showing that the CCK antiserum effect was reversible. When the antiserum was administered at the same dose before a single test session 14 days after the end of the initial training, the retention was also affected. These results show that cingulate CCK can affect retrieval processes. Mice, partially trained to avoid foot shock in a T-maze, showed enhanced retention relative to vehicle-injected mice when treated peripherally with CCK-8. Both intra-amygdaloid and intraventricular injections of β-endorphin resulted in amnesia. The effects of CCK-8 showed a differential ability to block amnesia induced by β-endorphin. This data suggests that the memory enhancement produced by peripherally administered CCK-8 involves the amygdala and that CCK-8 interacts with opioid amnesic mechanisms within the amygdala to alter memory processing.[46]

The involvement in memory processes of the neuropeptide CCK through interaction with CCK_2 receptors was studied in a recent experiment of Sebret et al[117] A two-trial recognition memory task was used. The positive effect of a selective CCK_2 agonist, BC 264, i.p. administered in the retrieval phase of the task, was also observed after injection into the dorsal subiculum/CA1 of the hippocampus but not into the caudate/putamen nucleus or into the prefrontal cortex of rats. The CCK_2 antagonist L-365,260 injected into this region of the hippocampus abolished the effect of BC 264 injected i.p. Furthermore, L-365,260 injected into the hippocampus suppressed recognition of the novel arm normally found in the controls, when it was injected before the acquisition or the retrieval phase of the task. In addition, an increase in extracellular levels of CCK-like immunoreactivity in the hippocampus of rats during the acquisition and retention phase of the task was observed. CCK_2 receptor-deficient mice also have an impairment in performance in this memory task. Together, these results support a physiological role of the CCK-ergic system through interaction with CCK_2 receptors in the hippocampus to improve performance of rodents in spatial recognition. Nomoto et al[88] investigated the behavior of Otsuka Long-Evans Tokushima Fatty (OLETF) rats, which are lacking CCK_1 receptors and which are hyperphagic, obese, and diabetic. OLETF rats were performing poorly in an elevated eight-arm radial maze, where the sequence of arms entered and the time spent there were recorded. The number of errors was significantly higher, and that of the correct choices was significantly lower in OLETF rats compared to the controls. The LTP of the population spike amplitude, measured following stimulation of the perforant path to activate comissural fibers projecting to the dentate gyrus of the hippocampus, was also significantly lower in the OLETF than in the control rats. From these observations, one might conclude that learning and memory functions are impaired in the OLETF rats due to the absence of CCK_1 receptors.

Winnicka and Wisniewski[143] studied the involvement of dopaminergic projection to the hippocampus on the effect of CCK-8US and caerulein on memory in male rats. CCK-8US and cerulein were given s.c., immediately after a single learning trial in a passive avoidance situation, after bilateral 6-OHDA lesions to the dentate gyrus of the hippocampus. Bilateral 6-OHDA lesions to the hippocampus significantly attenuated the facilitating effect of CCK-8US and caerulein on retention of passive avoidance behavior. These results may indicate that the facilitating effect of CCK-8US and cerulein on memory is, in part, mediated by dopaminergic projections from the ventral tegmental area to the dentate gyrus of the hippocampus. In a similar experiment[143] the involvement of dopaminergic projections to the central amygdala were also analyzed. Bilateral 6-OHDA lesions of the central amygdala totally abolished the facilitating effect of CCK-8US and cerulein on retention of passive avoidance behavior. These results may indicate that the effect of CCK-8US and cerulein on memory (at least if motivated aversively) is mediated by dopaminergic projections from the ventral tegmental area to the central amygdala.

Huston et al[56] examined the effects of the CCK fragments Boc-CCK-4 and CCK-8s on memory, reinforcement and anxiety following unilateral injection into the central nucleus of the amygdala. A one-trial uphill avoidance task was used. Post-trial injection of Boc-CCK-4 or CCK-8s was found to improve the retention performance in a narrow dose range. The hypermnestic effects of Boc-CCK-4 and CCK-8s were no longer evident when injection was performed 5 h after the learning trial. Authors used the elevated plus-maze to gauge anxiogenic properties of intra-amygdala injections and a circular open field for a single conditioning trial in one of the four restricted quadrants to measure aversive effects of the two peptides. No indication for anxiogenic or aversive influences were found.

Although it is known that CCK-4 triggers panic attacks, the specific involvement of peripheral or central CCK receptors in various adaptive processes such as emotion, memory and anxiety has yet to be demonstrated. Ladurelle et al[67] investigated the biochemical and pharmacological effects resulting from the administration of BC264, the highly potent and selective CCK_2 receptor agonist able to cross the blood-brain barrier. Very low doses of BC264 increased exploration of animals submitted to an unknown territory but were devoid of anxiogenic properties in the elevated plus maze. BC264 increased locomotion and rearings of rats newly placed in an open field and improved their spontaneous alternation in a Y-maze. The use of vagotomized animals showed that the increased alternation induced by BC264 did not require an intact vagus nerve, but the locomotor activation did. These behavioral effects are prevented by the prior i.p. administration of the CCK_2 receptor antagonist L-365,260 but not by the CCK_1 receptor antagonist L-364,718. These effects depend on dopaminergic systems, since they were blocked by D1 or D2 antagonists. In addition, bilateral perfusion in freely moving rats with BC264 at pharmacologically active doses was found to increase the extracellular levels of dopamine, DOPAC and HVA in the anterior part of the nucleus accumbens. These results show that activation of CCK_2 receptors by BC264 does not produce anxiogenic-like effects but appears to improve motivation and attention, whereas other CCK_2 agonists such as BocCCK4 induce anxiogenic responses.

Neuropeptide Y (NPY) is an amidated 36 amino acid peptide with a wide distribution in the central and peripheral nervous system. NPY is highly concentrated in the hippocampus and the amygdala.[80] Cholinergic interactions of NPY in the neocortex have been reported.[99] Of particular interest are the findings on the potential influence of NPY transmission in memory and cognition. Post-training i.c.v. administration of NPY to mice resulted in improved retention, when mice were retested 7 days later.[80] When testing the performance of mice in a T-maze active avoidance task, i.c.v. administered NPY had no effect on acquisition but improved retention. Peripheral administration had no effect. The effect of NPY on memory retention was time-dependent. When NPY was administered immediately prior to the retention test, enhanced recall was observed. As NPY did not alter acquisition, this enhanced recall most probably reflects enhanced retrieval of previously stored memories. NPY was found to reverse retrograde amnesia induced by scopolamine treatment and by protein synthesis inhibitors.[80] NPY improved retention when injected into the rostral portion of the hippocampus and in the septum. Conversely it impaired retention when injected into the caudal portion of the hippocampus and in the amygdala. Injections of NPY into the caudate nucleus, thalamus, or into cortical sites above the rostral hippocampus were without effect.[44,80] The physiological role of NPY on T-maze avoidance was studied following local microinjections of NPY antibodies into various brain structures. NPY antibodies caused amnesia when injected into the rostral hippocampus and septum and were found to facilitate the behavior when administered into the caudal hippocampus or the amygdala.[45]

Part of the activity of NPY to modify learning and memory processes is likely to reside in the C-terminal part of the molecule, since the C-terminal peptide fragment, NPY-(20-36), was as active as the whole molecule. A shorter C-terminal fragment, NPY-(26-36), was ineffective.[44] Two distinct subtypes of NPY receptors have been found, a postsynaptic (Y_1), for which effects could only be obtained with the complete NPY molecule, and a presynaptic (Y_2) receptor, for which effects could be elicited by long C-terminal fragments as well as the whole molecule. Taken together, it is likely that Y_1 receptors mediate the effects of NPY on food intake, while Y_2 receptors are responsible for the effects of NPY and NPY fragments on learning and memory processes.[44] The latter effect, which is localized in the hippocampus, most probably is the result of an inhibition of the release of GABA from the basket cells[80] and explains why NPY facilitates the firing of glutamate-containing pyramidal cells.

In a recent experiment, Thorsell et al[130] found that exogenous NPY reduces experimental anxiety in a wide range of animal models. The generation of an NPY-transgenic rat has provided a unique model to examine the role of endogenous NPY in control of stress and anxiety-related behaviors. Locomotor activity and baseline behavior on the elevated plus maze were normal in transgenic subjects. Two robust phenotypic traits were observed. In one trait transgenic subjects showed a markedly attenuated sensitivity to behavioral consequences of stress, as they were insensitive to the normal anxiogenic-like effect of restraint stress on the elevated plus maze and displayed absent fear suppression of behavior in a punished drinking test. In the other trait a selective impairment of spatial memory acquisition was found in the Morris water maze. Control experiments suggest these traits to be independent. These phenotypic traits were accompanied by an overexpression of prepro-NPY mRNA and NPY peptide and decreased NPY-Y1 binding within the hippocampus, a brain structure implicated both in memory processing and stress responses. These data support and extend a previously postulated anti-stress action of NPY and provide novel evidence for a role of NPY in learning and memory.

Galanin, a 29 amino-acid neuropeptide, affects diverse processes throughout the nervous system and coexists with several "classical" neurotransmitters, including norepinephrine, serotonin, and acetylcholine.[104] Galanin coexists with acetylcholine in neurons of the medial septum, diagonal band, and nucleus basalis of Meynert. The cholinergic forebrain neurons appear to play a significant role in learning and memory, as suggested by a severe loss of these neurons in Alzheimer's disease. In the ventral hippocampus, galanin inhibits the release of acetylcholine and inhibits carbachol-stimulated phosphatidylinositol hydrolysis. Galanin impairs choice accuracy in learning and memory paradigms in rats.[104] Malin et al[71] investigated whether galanin, administered i.c.v. immediately after the learning trial, might interfere with a one-trial discriminative reward learning task. Galanin infused rats showed significantly less retention. Administered before the retention trial, galanin had no effect, suggesting that galanin may interfere with memory consolidation rather than memory retrieval or task performance.

To test the possibility that galanin acts on the cell bodies of medial septal neurons,[49] two measures of septohippocampal function were assessed following intra-septal microinfusion of galanin or two of its synthetic fragments: galanin-(1-16) and galanin-(21-29). The behavioral measure was choice accuracy in a memory task in a T-maze. The electrophysiological measure was hippocampal theta activity recorded from the dentate hilus. Both the galanin fragment, galanin-(1-16), and the complete peptide, galanin-(1-29), decreased choice accuracy and hippocampal theta activity in a dose-dependent fashion. Sensorimotor performance was unaffected by the neuropeptide. These findings demonstrate that galanin impairs memory when administered directly into the medial septal area and suggest that galanin inhibits medial septal activity. The involvement of endogenous galanin in learning has also been demonstrated by the use of a high-affinity galanin receptor antagonist M35 [galanin- (1-13)-bradykinin-(2-9)amide]. I.c.v. administration of M35 facilitated acquisition of spatial learning in the Morris swim maze without an increase in swim speed. Thus, M35 shortened escape latency, reduced the number of failures to reach the platform, and shortened the path length to reach the hidden platform. M35 also tended to enhance retention performance seven days after the last training session.[89] Age-related alterations in cue-training and place-training tasks were evaluated[21] and compared to alterations in galanin-like immunoreactive neurons in the medial septal area of the rat. The majority of aged male rats exhibited impaired performance in a Morris water maze, as compared to young rats. In addition, there was a significant loss of galanin-like immunoreactive cells in the medial septal-diagonal band complex, and a marginal loss of septo-hippocampal galanin positive neurons in aged rats. Schott et al[116] investigated the role of galanin in hippocampally mediated functions such as spatial learning and memory. Galanin was infused via bilateral chronic cannulae into different areas of the hippocampal formation, which are characterized by different galanin receptor subtypes and also by different galanin innervation patterns. The effects of galanin on spatial learning were examined in the Morris swim maze. Infusions of galanin into both the dorsal and ventral dentate gyrus, which mainly contain galanin GAL_2 receptor mRNA and a high degree of galanin-noradrenaline coexistence, significantly retarded spatial acquisition without affecting swim speed or performance in the visible platform test. This spatial learning deficit was fully blocked by pretreatment with the non-selective galanin receptor antagonist M35.

Analysis of retention performance suggested that the major effect of intrahippocampal galanin is mediated via a specific disruption of acquisition mechanisms of importance for performance in the trial. Galanin infused into the ventral CA1 (a mainly galanin GAL_1 receptor mRNA expressing region) or into anterior, ventral CA3 regions did not produce any deficits in spatial learning compared to control animals. These results suggest that galanin mediates its action on spatial learning mainly through the galanin GAL_2 receptor subtype in areas where most of the galanin is present in noradrenergic terminals. A possible role for the galanin GAL_1 receptor subtype in cognition in the dorsal and ventral hippocampus remains to be defined. The results suggest a differential functional role for galanin and galanin receptor subtypes within subregions of the hippocampal formation. Ogren et al[90] suggested that galanin is a potent modulator of basal acetylcholine release in the rat forebrain. These effects appear to be related to the activation of galanin GAL_1 (ventral hippocampus) and galanin GAL_2 (dorsal hippocampus) receptors, respectively. Thus, galanin perfused through a microdialysis probe decreased basal acetylcholine release in the ventral hippocampus, while it enhanced acetylcholine release in the dorsal hippocampus. This finding indicates that galanin may act via different mechanisms within subsystems of the hippocampus. This hypothesis has received support from studies in the Morris maze. Galanin infused into the ventral hippocampus impaired while infusion into the dorsal hippocampus tended to facilitate spatial learning. However, the authors suggest that the effects of galanin on acetylcholine release and on spatial learning, which are due to activation of GALreceptors, may be indirectly mediated via noradrenaline or 5-HT transmission as Galanin is also a potent inhibitor of mesencephalic 5-HT neurotransmission in vivo. Misane et al[76] provided evidence that galanin can modulate brain serotoninergic (5-HT) neurotransmission in vivo and, particularly, 5-HT_{1A} receptor-mediated transmission. Galanin (given i.c.v.) dose-dependently attenuated the impairment of passive avoidance retention induced by the selective 5-HT_{1A} receptor agonist 8-hydroxy-2-(di-n-propyloamino)tetralin (8-OH-DPAT) when injected prior to training. This impairment appears to be mainly related to activation of 5-HT_{1A} receptors in the CNS. Wrenn and Crawley[148] recently reviewed the effects of galanin on learning and memory processes and hypothesized that galanin is localized in brain pathways involved in both cognition and affect. Galanin may inhibit learning and memory by inhibiting neurotransmitter release and neuronal firing rate. Two signal transduction mechanisms through which galanin exerts its inhibitory actions are through inhibition of phosphatidyl inositol hydrolysis and adenylate cyclase.

Substance P

Substance P is present in various brain (including limbic) nuclei, where the neurons are intimately associated or colocalized with neurons containing classical neurotransmitters, e.g., acetylcholine in the basal forebrain nucleus or dopamine in the striatum.[2] Substance P is considered to be a putative transmitter substance in sensory nerves exerting a slow excitatory influence. Substance P, given i.p., disrupted learning to turn off an aversive stimulus that was conditioned to an acoustic stimulus.[55] In a hexagonal maze, which measures activity, exploratory efficiency, habituation, and perimeter walking, injection of substance P facilitated perimeter walking only.[81] In a radial maze, substance P produced facilitation of long-term and short-term memory without affecting activity. When the effect of pre- and post-learning injections of substance P was tested on performance in the radial maze configuration, only pretrial injections facilitated performance with respect to measures of efficiency and short- and long-term memory. Virtually no effect was seen with post-learning injections. In the early studies, post-learning injections of substance P into the substantia nigra and into the amygdala was found to disrupt passive avoidance learning and thus resulted in amnesia.[55] In contrast, microinjections of substance P into the lateral hypothalamus or the medial septal nucleus improved avoidance learning.[122] Differential functions and individual sensitivity of various anatomical brain sites may not be the only reason for the opposite effects of substance P on learning and memory processes. Following unilateral injection into the nucleus accumbens, the intact peptide (substance P-[1-11]) and the C-terminal fragment substance P-(7-11) disrupted, whereas the N-terminal fragment substance P-(1-7) facilitated, passive avoidance behavior.[48] It has been concluded that—similarly to various other neuropeptides—substance P may require processing by enzymatic cleavage to activate moieties which modulate avoidance behavior.[48] It thus seems that substance P can modulate avoidance learning and facilitate or inhibit performance depending on the site of injection and the formation of biologically active fragments.

Tomaz and Nogueira[131] found that peripheral post-training substance P administration in rats enhances memory in a dose-and time-dependent way. The effect of substance P on retention was observed across tasks with different response requirements and in the absence of explicit punishment. The memory-enhancing effects are long-lasting, until 21 days post-training, and are mediated, at least in part, via interactions with the endogenous opioid system. The mnemotropic effects of peripherally administered substance P are sensitive to the functional integrity of the vagus, suggesting that the vagus nerve may be one pathway by which systemic substance P influences memory storage processes in the brain. Furthermore, the data indicated that these effects seemed to be encoded by different substance P sequences, the N-terminal substance P1-7, but not the C-terminal hepta- and hexapeptide sequences being responsible for the memory-promoting effects, thus confirming earlier results by Gaffori et al[48] Data of Santangelo et al[110] suggest that substance P-like in mammals—can facilitate memory in goldfish in an inhibitory avoidance test.

Recently, the receptor for substance P (the tachykinin NK_1 receptor), has been proposed as possible target for new antidepressant and anxiolytic therapies.[111] Localized administration of substance P in the central nervous system may produce anxiogenic or anxiolytic responses, depending on the animal species and site of injection. Conflicting results have been obtained from the use of tachykinin NK_1-receptor antagonists, and issues of drug access and species specificity have further clouded the roles of substance P in stress-related behaviors. Central tachykinin NK_1 receptors are thought to modulate aversion, whereas the periaqueductal gray matter (PAG) is a common pathway for the integration of fear behaviors. Mongeau et al[78] determined whether injection of an NK_1 agonist (GR73632) into subregions of the PAG would alter fear-related behaviors. Behavioral inactivity was increased by GR73632 injected into the caudodorsal PAG or the dorsal raphe. Flight behavior induced by stimulation of the dorsal PAG or by a foot shock was decreased after injection of GR73632 into the dorsal PAG. Rats that had 6 pairings of a tone with a foot shock after injection of GR73632 into the dorsal PAG, displayed more freezing behavior than controls at the beginning of the session. It is concluded that tachykinin NK_1 receptors in the dorsal PAG modulate the unconditional but not the mnemonic aspects of fear behaviors.

Ukai et al[132] investigated the effects of intracerebroventricular injection of substance P on scopolamine-induced impairment of spontaneous alternation performance in the mouse. The neuropeptide alone did not influence either spontaneous alternation performance or total arm entries. Scopolamine impaired spontaneous alternation performance was accompanied by an increment in total arm entries. In contrast, substance P significantly attenuated the scopolamine-induced impairment of spontaneous alternation performance. The effects of substance P on scopolamine-induced impairment of spontaneous alternation performance were almost completely reversed by pretreatment with WIN 62577, a tachykinin NK_1 receptor antagonist. These results suggest that substance P affects scopolamine-induced impairment of spontaneous alternation performance through the mediation of tachykinin NK_1 receptors. In a recent review, Hasenohrl et al[52] showed that substance P can have memory-promoting, reinforcing and anxiolytic-like effects when administered systemically or centrally into the nucleus basalis of the ventral pallidum. These effects seem to be mediated via the substance P preferring tachykinin NK_1 receptor and differentially related to N- versus C-terminal fragments of the undecapeptide. Substance P injection into the ventral pallidum can lead to increases of acetylcholine in the frontal cortex and of dopamine in the nucleus accumbens, suggesting that the hypermnestic, positively reinforcing and anxiolytic effects observed upon basal forebrain injection of substance P are mediated by activation of the nucleus accumbens-ventral pallidum circuitry. Furthermore, substance P and certain substance P fragments may not only be considered to have beneficial behavioral effects in normal animals, but can also prevent lesion-induced functional deficits and improve the speed of recovery. This indicates that substance P agonists might also have a neuroprotective capacity in parallel with recovery-promoting actions. Based on these findings, one might suppose that substance P- like vasopressin- has time-dependent facilitatory effect on learning and memory processes.

Natriuretic Peptides, Angiotensin

The natriuretic peptide family is composed of at least three ligands: atrial natriuretic peptide (ANP), brain natriuretic peptide (BNP) and C-type natriuretic peptide (CNP). ANP and BNP regulate body fluid homeostasis and blood pressure.[22,72] CNP acts mainly as a vasodilator with little natriuretic activity, and is found principally in the central nervous system and endothelial cells.[123] The natriuretic peptides have a widespread distribution in the brain. They possess a variety of CNS functions, involving body fluid homeostasis and cardio-vascular function.

Two types of active receptors for the natriuretic peptides have been identified as $ANPA_A$ and $ANPA_B$ receptor. CNP is a selective ligand for the $ANPA_B$ receptor, whereas ANP possesses the highest selectivity for the $ANPA_A$ receptor. BNP activates guanylate cyclase through both $ANPA_A$ and $ANPA_B$ receptor. The third natriuretic peptide Type C receptor, $ANPA_C$ receptor, is known to have its major role in the clearance of natriuretic peptides from the circulation.

Electrophysiological studies have demonstrated that ANP is a potent inhibitor of vasopressin neurons to prevent vasopressin secretion.[154]

Data suggest that ANP might influence learning and memory processes. Bidzseranova et al[5] investigated the effects of rat atrial natriuretic peptide (ANP-[1-28]) on passive avoidance behavior in rats following administration into a lateral ventricle immediately after the learning trial. ANP-(1-28) dose-dependently facilitated passive avoidance behavior. When injected before the learning trial, ANP had the same effect. When, however, the peptide was given shortly before the retention trial, there was no effect. The data suggest that ANP-(1-28) facilitates acquisition and the consolidation of passive avoidance behavior. Electroconvulsive shock-induced partial retrograde amnesia could also be prevented by i.c.v. administered ANP.[5] Structure activity studies revealed that the active moiety of ANP resides in the sequence ANP-(15-23).[6] The same authors, in addition, showed that i.c.v. injection of an ANP antiserum attenuated passive avoidance behavior when administered immediately after the learning trial. It also facilitated extinction of an active avoidance response. The results suggest that endogenous ANP is involved in the modulation of learning and memory processes. According to the same authors, BNP has similar effects as ANP on avoidance behavior.

Jahn et al[63] found evidence that CNP exerts effects opposite to those of ANP, on the performance in the elevated plus maze. Low CNP doses did not significantly facilitate the behavior of rats in the plus maze. At higher doses (0.5 – 5 ug i.c.v.) CNP had distinct anxiogenic properties. Authors concluded that CNP may have anxiogenic, while ANP has anxiolytic-like properties after i.p., i.c.v. and intra-amygdala infusion in rats. The data suggest opposite effects of CNP and ANP on anxiety-related behavior and neuroendocrine regulation in rats, which appear to be mediated via different receptor occupation and brain regions. The hypothesis that the anxiogenic effect of CNP is mediated by CRF, which possesses anxiogenic effects, was investigated by using an antagonist of CRF α-helical CRF at both CRF receptors. The anxiogenic effect of CNP was entirely blocked by α-helical CRF.

Telegdy et al[128] studied the action of CNP on passive avoidance learning in rats. The involvement of transmitters was investigated by pre-treating the animals with different neurotransmitter receptor blockers. CNP administered into the lateral brain ventricle caused a dose-dependent facilitation of learning and consolidation of passive avoidance learning, but was ineffective on retrieval. Pretreatment of the animals with atropine, haloperidol or the nitric oxide synthase inhibitor nitro-L-arginine abolished the action of CNP. Phenoxybenzamine, naloxone, bicuculline, propranolol and methysergide were ineffective. The results suggest that CNP improves learning and consolidation in a passive avoidance paradigm, but is ineffective on retrieval processes. In the action of CNP, dopamine, acetylcholine and nitric oxide could be the mediating transmitters.

Abundant evidence indicates that angiotensin II (Ang II) can influence central nervous system activity. Effects on blood pressure, thirst, salt appetite, and release of such pituitary hormones as vasopressin, oxytocin, ACTH, and LHRH have been reported.[149] Angiotensin immunoreactive neurons have been visualized in the brain. Among the effector peptides of the brain renin angiotensin system (RAS), Ang II and Ang III [(Ang-(2-8)], have the same affinity for the two pharmacologically well-defined receptors: type 1 (angiotensin AT_1 receptor) and Type II (angiotensin AT_2 receptor). In rodents two angiotensin AT receptor subtypes, angiotensin AT_{1A} receptor and angiotensin AT_{1B} receptor, have been isolated. Angiotensin AT_{1A} receptor and angiotensin AT_2 receptor mRNA are predominantly expressed in the brain and angiotensin AT_{1B} receptor in the pituitary. Limited overlap was found between expression of angiotensin AT_{1A} receptor and angiotensin AT_2 receptor mRNAs in the brain. The neural expression of angiotensin AT_{1a} and angiotensin AT_2 receptors was demonstrated in the subfornical organ, the hypothalamus, and the lateral septum. Angiotensin AT_{1A} receptor expression was localized in CRF- but not in AVP-containing neurons. These findings point to a central role of angiotensin in cardiovascular regulation, water metabolism, pituitary function and behavior. Other fragments of Ang I have been found in the course of years. One of these is the fragment Ang-(1-7). It is the most pleiotropic of the metabolites found because it exerts effects that may be identical as well as opposite to those of Ang II.[43] It activates antihypertensive mechanisms as it stimulates the synthesis and release of vasodilator prostaglandins, augments metabolic actions of bradykinin and facilitates the release of nitric oxide. Ang-(1-7) may therefore act as a negative feed back hormone of the pressor and trophic actions of Ang II. Effects on learning and memory have not been reported.

Behavioral effects of Ang II include effects on exploratory and stereotype behavior as well as on learning and memory processes. A single subcutaneous injection of Ang II failed to modify extinction of active avoidance behavior in rats.[23] However, administered i.c.v. it facilitated retention of a food-motivated T-maze task, shuttle box avoidance training, and passive avoidance behavior.[11] Chalas and Conway[17] found no evidence for involvement of Ang II in spatial learning in a water maze. Neither the ACE inhibitor ceramapril nor ramipril altered the increase in path length in the maze produced by scopolamine. Administration of the substrate, renin that leads to Ang II formation, did not alter water maze performance over 5 days of training. The angiotensin receptor antagonist, losartan, has been shown to improve basal and scopolamine-impaired performance in a habituation task and reverse the inhibition in long-term potentiation produced by diazepam. However, neither losartan, nor ramipril reversed diazepam-impaired acquisition of the spatial memory task over 5 days of training. Winnicka[140] showed that the facilitory effect of Ang II on passive avoidance retention is mediated by dopamine projections to the central amygdala as 6-OHDA lesions abolished the effect of Ang II. Braszko et al[12] found that the ACE inhibitor trandolapril given orally either acutely or chronically, attenuated acquisition of active avoidance behavior. The treatment however did not affect consolidation and retention of passive avoidance behavior, object recognition and locomotor activity in an open field. These authors conclude that physiological levels of Ang II may be required for effective learning. Baranowska et al[3] reported that i.c.v. Ang II facilitated acquisition but did not affect extinction of the response. Others found similar effects after i.c.v. administration of Ang II.[149]

Belcheva et al[4] investigated the effects Ang II, micro-injected into the CA1 hippocampal area of male Sprague Dawley rats, on learning and memory in a shuttle box. Bilateral micro-injections of Ang II improved learning, i.e., increased the number of avoidances during the two training days. Interestingly, Ang II facilitated learning and memory, only when micro-injected into the left CA1 hippocampal area.

Several reports have been published on negative effects of Ang II on learning and memory. Raghavendra et al[101] showed that immobilization stress and i.c.v. injection of Ang II in mice and rats produced an increase in tail-flick latency. Similarly, immobilization stress and post training i.c.v. injection of Ang II impaired retention in the plus maze and in the passive avoidance step-down test. Both these responses were reversed by prior treatment with the Ang I receptor antagonist losartan and an Ang II receptor antagonist. Naloxon attenuated immobilization-induced analgesia and -retention impairment but not that of Ang II. Nikolova et al[85] also showed that ACE inhibitors improved learning and memory in active and passive avoidance behavior. Raghavendra et al[102] further studied the potential nootropic effects of captopril and losartan. Postlearning administration of captopril but not losartan improved learning in the second trial of the acquisition test. However, both drugs were equally effective when administered prior to the training. The retention enhancing effect of the antagonists were reversed by post training administration of L-NAME, dizocilpine or scopolamine.

Sakagawa et al[105] have shown that angiotensin AT_2 receptor deficient mice displayed anxiety-like behavior in comparison with wild-type mice. In the passive avoidance task, no differences were found between wild-type mice and angiotensin AT_2 receptor deficient mice. In contrast, the pain threshold was significantly lower in angiotensin AT_2 receptor deficient mice as compared with findings in wild-type mice. It thus was concluded that the angiotensin AT_2 receptor does not influence learning behavior. Since angiotensin AT_2 receptor deficient mice also have increased sensitivity to pain and decreased levels of brain β-endorphin, angiotensin AT_2 receptors may mediate pain threshold.

Smaller fragments of Ang I as Ang-(3-7) and Ang-(3-8), the latter known as Ang IV, which lack most of the physiological effects of the parent hormones,[150] have been found to improve memory. Binding sites also known as AT 4 receptors are widely distributed in the brain with high densities in the hippocampus, neocortex and motor nuclei in guinea pigs and monkeys. Binding sites have been found also in post mortem human brains with the use of iodinated Norleucine 1-Ang IV (Nle(1)-AngIV), with a higher affinity than the naturally occurring compound. It did not bind to Ang II receptors. High densities were found throughout the cerebral cortex, the claustrum, chorioid plexus, hippocampus and pontine nucleus.[16]

Structure-activity studies showed that Ang-(3-7) was as active, as Ang II in acquisition of active and retention of passive avoidance behavior. Holy et al[53] found that 3-7(4)Phe Ang-(3-7) and Ang-(3-7) as well as Ang II, stimulated the rate of acquisition of conditioned avoidance behavior. These peptides had no effect on activity in an open field nor on the retention of a spatial task in the Morris maze. In the passive avoidance test Ang II had a more pronounced effect on retention than the two fragments. All peptides potentiated apomorphine and amphetamine induced stereotypy. Braszko[10] also reported that Ang II as well as Ang-(3-7) improved retention of passive avoidance behavior. These effects were abolished by the selective AT1 losartan or AT2 receptor antagonist CGP 42112 A. These antagonists also abolished Ang II as well as Ang-(3-7)-induced enhancement of apomorphine stereotypy. Winnicka and associates showed that destruction of the dopamine projections to the central amygdala and the CA4 field of the hippocampus, abolished respectively diminished the effect of i.c.v. administered Ang-(3-7) given 15 min. before the retention test. The concomitant increase in spontane-

Table 1. Amino acid sequences of neuropeptides (human) modulating learning and memory processes

Vasopressin	Cys-Tyr-Phe-Gln-Asn-Cys-Pro-Arg-Gly-NH_2
Oxytocin	Cys-Tyr-Ile-Gln-Asn-Cys-Pro-Leu-Gly-NH_2
ACTH	Ser-Tyr-Ser-Met-Glu-His-Phe-Arg-Trp-Gly-Lys-Pro-Val-Gly-Lys-Lys-Arg-Arg-Pro-Val-Lys-Val-Tyr-Pro-Asn-Gly-Ala-Glu-Asp-Glu-Ser-Ala-Glu-Ala-Phe-Pro-Leu-Glu-Phe
α-MSH	N-Acetyl-Ser-Tyr-Ser-Met-Glu-His-Phe-Arg-Trp-Gly-Lys-Pro-Val-NH_2
β-Endorphin	Tyr-Gly-Gly-Phe-Met-Thr-Ser-Glu-Lys-Ser-Gln-Thr-Pro-Leu-Val-Thr-Leu-Phe-Lys-Asn-Ala-Ile-Ile-Lys-Asn-Ala-His-Lys-Lys-Gly-Gln
γ-Endorphin	Tyr-Gly-Gly-Phe-Met-Thr-Ser-Glu-Lys-Ser-Gln-Thr-Pro-Leu-Val-Thr-Leu
α-Endorphin	Tyr-Gly-Gly-Phe-Met-Thr-Ser-Glu-Lys-Ser-Gln-Thr-Pro-Leu-Val-Thr
Endomorphin-2	Tyr-Pro-Phe-Phe-NH_2
Astressin	D-Phe-His-Leu-Leu-Arg-Glu-Val-Leu-Glu-Nle-Ala-Arg-Ala-Glu-Gln-L
Orphanin/nociceptin	Phe-Gly-Gly-Phe-Thr-Gly-Ala-Arg-Lys-Ser-Ala-Arg-Lys-Leu-Ala-Asn-Gln
Nocistatin	Thr-Glu-Pro-Gly-Leu-Glu-Glu-Val-Gly-Glu-Ile-Glu-Gln-Lys-Gln-Leu-Gln
CRF	Ser-Glu-Glu-Pro-Pro-Ile-Ser-Leu-Asp-Leu-Thr-Phe-His-Leu-Leu-Arg-Glu-Val-Leu-Glu-Met-Ala-Arg-Ala-Glu-Gln-Leu-Ala-Gln-Gln-Ala-His-Ser-Asn-Arg-Lys-Leu-Met-Glu-Ile-Ile-NH_2
Somatostatin	Ala-Gly-Cys-Lys-Asn-Phe-Phe-Trp-Lys-Thr-Phe-Thr-Ser-Cys
Cortistatin	Pro-Cys-Lys-Asn-Phe-Phe-Trp-Lys-Thr-Phe-Ser-Ser-Cys-Lys
CCK-8	Asp-Tyr-Met-Gly-Trp-Met-Asp-Phe-NH_2
Coerulein	pGlu-Gln-Asp-Tyr(SO_3H)-Thr-Gly-Trp-Met-Asp-Phe-NH_2
NPY	Tyr-Pro-Ser-Lys-Pro-Asp-Asn-Pro-Gly-Glu-Asp-Ala-Pro-Ala-Glu-Asp-Met-Ala-Arg-Tyr-Tyr-Ser-Ala-Leu-Arg-His-Tyr-Ile-Asn-Leu-Ile-Thr-Arg-Gln-Arg-Tyr-NH_2
Galanin	Gly-Trp-Thr-Leu-Asn-Ser-Ala-Gly-Tyr-Leu-Leu-Gly-Pro-His-Ala-Val-Gly-Asn-His-Arg-Ser-Phe-Ser-Asp-Lys-Asn-Gly-Leu-Thr-Ser
Substance P	Arg-Pro-Lys-Pro-Gln-Gln-Phe-Phe-Gly-Leu-Met-NH_2
ANP	Ser-Leu-Arg-Arg-Ser-Ser-Cys-Phe-Gly-Gly-Arg-Met-Asp-Arg-Ile-Gly-Ala-Gln-Ser-Gly-Leu-Gly-Cys-Asn-Ser-Phe-Arg-Tyr
BNP	Ser-Pro-Lys-Met-Val-Gln-Gly-Ser-Gly-Cys-Phe-Gly-Arg-Lys-Met-Asp-Arg-Ile-Ser-Ser-Ser-Ser-Gly-leu-Gly-Cys-Lys-Val-Leu-Arg-Arg-His
CNP	Gly-Leu-Ser-Lys-Gly-Cys-Phe-Gly-Leu-Lys-Leu-Asp-Arg-Ile-Gly-Ser-Met-Ser-Gly-Leu-Gly-Cys
Angiotensin II	Asp-Arg-Val-Tyr-Ile-His-Pro-Phe

ous locomotor activity in rats lesioned to the central amygdala and a decrease in rats lesioned to the hippocampus were unlikely to interfere with the cognitive effect of the peptide.[147] Similar lesions to the nucleus accumbens and septi lateralis showed that lesions to the nucleus accumbens but not to the nucleus septi lateralis abolished Ang II and Ang-(3-7) induced facilitation of retrieval of object recognition.[142] These effects are also abolished by NMDA receptor antagonists. Bilateral destruction of glutaminergic temporo-entorhinal connections abolished the facilitatory effect of both angiotensins on recall of passive avoidance behavior.[143] Winnicka[141] also studied the influence of Ang II and AII (3-7) given i.c.v., in rats in which the dopamine projections to the nucleus accumbens and to the nucleus septi lateralis were lesioned with 6-OHDA. Angiotensin II and its 3-7 fragment significantly improved passive avoidance retention in sham-operated rats. Bilateral 6-OHDA lesions to the dopaminergic projections to the nucleus accumbens but not to the nucleus septi lateralis blocked these effects. Bilateral lesions of the dopaminergic projections to the central amygdala also abolished, and to the CA4 field of the hippocampus significantly diminished, the effect of Ang-(3-7). These results suggest that

Table 2. Shortest active amino acid sequences of neuropeptides modulating learning and memory processes

Vasopressin	AVP_{4-8}
Oxytocin	OXT_{4-8}
ACTH	$ACTH/MSH_{4-7}$
α-MSH	$ACTH/MSH_{4-7}$
β-Endorphin	Complete sequence
γ-Endorphin	β-$Endorphin_{6-17}$
α-Endorphin	β-$Endorphin_{6-16}$
Endomorphin-2	Complete sequence
Astressin	Complete sequence (?)
Orphanin/nociceptin	Complete sequence (?)
Nocistatin	Complete sequence (?)
CRF	CRF_{34-41}
Somatostatin	Complete sequence (?)
Cortistatin	Complete sequence (?)
CCK	CCK_{30-33} (CCK4)
Coerulein	Complete sequence (?)
NPY	NPY_{20-36}
Galanin	$Galanin_{1-16}$, $Galanin_{21-29}$
Substance P	Substance P_{1-7}, Substance P_{7-11}
ANP	ANP_{15-23}
BNP	Complete sequence (?)
CNP	Complete sequence (?)
Angiotensin II	$Angiotensin_{3-7}$, $Angiotensin_{3-8}$

the anatomical substrate of facilitating retrieval of information activity of AII(3-7) is closely related to the dopaminergic projection from the ventral tegmental area and substantia nigra to the central amygdala and to the hippocampus.[147]

Delorenzi et al[29] in the crab Chasmagnathus found that Ang IV enhances long-term memory stronger than Ang II. The effect was dose dependent and salarasin reversible. Kramar et al[66] investigated the effects of two Ang IV analogs, Nle(1)-AngIV (an AT_4 receptor agonist) and Nle 1-Leual 3-AngIV (an AT_4 receptor antagonist), on long-term potentiation (LTP). Excitatory postsynaptic field potentials were recorded from the CA1 stratum radiatum following stimulation of the Schaffer collateral pathway. Activation of AT_4 receptors by Nle(1)-AngIV enhanced synaptic transmission and increased LTP. Paired stimulation before and during infusion of Nle(1)-AngIV indicated no change in paired-pulse facilitation as a result of AT_4 receptor activation suggesting that the underlying mechanism(s) responsible for Nle(1)-AngIV-induced increase in synaptic transmission and LTP is likely a postsynaptic event. These results extended previous findings from behavioral data in that AT_4 receptor agonists and antagonists are capable of activating, and inhibiting, learning and memory pathways in the hippocampus, and suggest that the AT_4 receptor subtype is involved in synaptic plasticity.

Ang II impairs learning and memory when administered directly or released into the hippocampal dentate gyrus and inhibits LTP in medial perforant path-dentate granule cell synapses, Wayner et al[139] studied the effect of Ang IV on LTP in the same synapses. It significantly enhanced LTP and the enhancement was both dose and time dependent. An inverted U-type dose related effect was observed, when administered before the first tetanus. A complex time related effect was observed with a maximum at 5 min., a return to normal LTP at 30 min and a minimum below normal at 90 min and return to normal again at 120 min. The enhancement could be prevented by pre-treatment with divalinal, an Ang IV antagonist without any effect on normal LTP. The agonists Nle(1)-AngIV was less effective than the parent compound but exhibited the same time dependent effects. Both produced a significant suppression of LTP at 90 min that remains to be explained. However the inhibition was dose dependent and blocked by divalinal. This compound did not affect Ang II induced inhibition of LTP. AT_4 receptors and cholinergic neurons are closely associated in regions involved in cognitive processing, such as the hippocampus and neocortex. Lee et al[68] therefore postulated that AT_4 receptors affect cognitive processing by modulating cholinergic neurotransmission. Ang IV potentiated depolarization-induced [(3)H]ACh release from the rat hippocampus. Potentiation of release was attenuated by the angiotensin AT_4 receptor antagonist, divalinal. Ang IV-induced potentiation was not affected by angiotensin AT_1 and angiotensin AT_2 receptor antagonists. These results indicate that stimulation of AT_4 receptors can potentiate depolarization-induced release of ACh from hippocampal slices and suggest that potentiation of cholinergic transmission may be a mechanism by which angiotensin AT_4 receptor ligands enhance cognition. Tchekalarova et al[127] found that Ang IV exerted a dose dependent (inverted U) improvement of passive avoidance retention. Adenosine A1 is involved in this effect as theophiline enhanced and the selective antagonist cyclopentyladenosine attenuated Ang IV-induced memory effect. Wright et al[151] investigated the role of angiotensin AT_4 receptor in the acquisition of this spatial learning task. Chronic i.c.v. infusion of an angiotensin AT_4 receptor agonist Nle(1)-AngIV via osmotic pump facilitated the rate of acquisition to solve this task, whereas treatment with an angiotensin AT_4 receptor antagonist significantly interfered with the acquisition of successful search strategies. Animals with bilateral knife cuts of the perforant path, a major afferent hippocampal fiber bundle originating in the entorhinal cortex, displayed deficits in solving this task. This performance deficit could be reversed by acute i.c.v infusion of a second angiotensin AT_4 receptor agonist (Norleucinal). These results suggest that the brain Ang IV – angiotensin AT_4 receptor system plays a role in the formation of spatial search strategies and memories.

Amyloid Peptides

Yamaguchi and Kawashima[153] investigated the neurotoxicity of amyloid-β-(25-35) protein, which is thought to be the active site of amyloid-β, a 42 amino acid peptide chain. A single i.c.v. injection of amyloid-β-(25-35) induced a marked decrease in latency in step-through passive avoidance task, impaired radial-arm maze performance, and induced a decrease in choline-acetyltransferase activity in the medial septum, cortex and hippocampus, but not in the basal forebrain of rats. The reverse sequence of amyloid-β-(25-35) was without harmful effects on passive avoidance performance. These results suggest that learning and cognitive disturbance induced by i.c.v. injection of amyloid-β-(25-35) is associated with a dysfunction of cholinergic activity in the brain. Vaccinations with amyloid- peptide can dramatically reduce amyloid deposition in a transgenic mouse model of Alzheimer's disease. To determine if the vaccinations had deleterious or beneficial functional consequences, Morgan et al[79] tested eight months of amyloid-β vaccination in a different transgenic model for Alzheimer's disease in which mice develop learning deficits as amyloid accumulates. Vaccination with amyloid-β protects transgenic mice from the learning and age-related memory deficits that normally occur in this mouse model for Alzheimer's disease. During testing for potential deleterious effects of the vaccine, all mice performed superbly on the radial-arm water-maze test of working memory. Later, at an age when untreated transgenic mice show memory deficits, the amyloid-β vaccinated transgenic mice showed cognitive performance superior to that of the control transgenic mice and, ultimately, performed as well as non-transgenic mice.

The amyloid-β/A4 protein precursor has been implicated in age-associated plastic changes at synapses that might contribute to memory loss in Alzheimer's disease. As amyloid-β/A4 has previously been reported to have multiple functions during normal development, Mileusnic et al[75] employed a one-trial passive avoidance task in day-old chicks to study its role in the process of memory formation. Administration of anti-amyloid-β/A4 antibodies, injected 30 min pretraining, prevented memory for a one-trial passive avoidance task in day-old chicks without effects on general behavior or initial acquisition. Amnesia was apparent at 30 min post-training and lasted for at least 24 h. The same result was obtained by down-regulation of amyloid-β/A4 expression by amyloid-β/A4-antisense, injected 8-12 h pretraining. However, injections of anti-amyloid-β/A4 antibodies or amyloid-β/A4 antisense at later post-training time did not cause amnesia for the task. Unlike antibodies and antisense, injection of the amyloid-β/A4-328-332 pentapeptide, in either orientation, 30 min pretraining, was able to rescue the memory and prevented antisense-induced amnesia. The post-training time within which the antibody- and antisense-induced amnesia, and within which the amyloid-β/A4 peptides prevent amnesia, correspond to that during which memory formation is vulnerable to disruption of the putative signal transduction functions of amyloid β/A4. These results suggest that amyloid-β/A4 is required during an early phase of memory formation, and the memory enhancing effect is localized within a pentamer sequence of the growth-promoting domain of amyloid-β/A4.

Conclusions

Animals, as well as human beings, acquire new information about their environment by learning and subsequent retention of that information. The brain interacts with the internal and external environment through axon discharges and synaptic transmission, and it follows that the substrates of memory are triggered by and act upon these physiological events. The integrated activity of primary physiological and behavioral processes is a necessary condition for the occurrence of memory.

The strategy for studying the biology of learning and memory is based on the belief that information is stored as changes in neuronal interactions in the brain. If learning and memory processes involve the formation of new synaptic contacts or modification of existing ones, then these modifications are likely to require changes in the quantity, turnover, metabolism, release, or receptor-mediated events of specific biochemical mechanisms. Although the precise nature of these changes is not yet understood, a good deal is known about the morphology, physiology, and biochemistry of neurons and about the ways in which neurons can change the way they communicate with other neurons. A major discovery of the past three decades has been that within the cascade of processes involved in learning and memory, neuropeptides play a significant part.

Memory encoding is the result of the formation of specific spatiotemporal patterns of activation of neuronal networks. Neuropeptides (peptidergic neurons) may either be part of these networks or may modulate the activity of these networks. Many of the neuropeptides known to facilitate learning and memory processes are also present in limbic or cortical structures. The areas involved, i.e., limbic-midbrain areas, in particular, are innervated by neuropeptide systems or are characterized by the presence of neuropeptides and neuropeptide receptors. This suggests the participation of these compounds in the activity of these areas. Here, neuropeptides also affect biochemical and electrophysiological processes intimately involved in the formation of memory (long-term potentiation, neuronal excitability of the hippocampus, modification of the responses of the neurons to glutamate, functions of NMDA receptors, phosphoinositide metabolism, the expression of immediate early genes, etc.).

None of the neuropeptides discussed above is specific in the sense that their only effect would be on information processing. Almost all have well-characterized, wide-spread endocrine activities either on the pituitary gland or in the periphery. Release of 'endocrine' neuropeptides may occur in response to specific stimuli (e.g., vasopressin release to thirst, oxytocin release to suckling, CCK release to hunger and satiety, etc.). In some situations (anxiety, fear, etc.) the release of these neuropeptides occurs in the brain exclusively.[39]

One of the most important discoveries is the principle that classical endocrine activities and central nervous activities of the same neuropeptides can be dissociated; thus, potent behavioral activities may reside in smaller, more selective peptides, which are devoid of endocrine activity.[24] This conclusion was originally based on experiments with vasopressin, oxytocin, ACTH/MSH and the endorphins. In agreement with this hypothesis, however, are various observations based on molecular and neurobiochemical studies. Neuropeptides are formed following gene expression in nerve cells, and produced in large precursor molecules which through a series of processes express the genetic information into biologically active peptides. Specific binding sites in the brain, receptor-coupled biochemical events, gene activation have been described for many of these peptide fragments. This principle has now been widely accepted and described for many other peptides affecting learning and memory processes (e.g., angiotensin, natriuretic peptides, CCK, substance P, NPY, etc.). This concept gave an explanation for the multitude of behavioral effects of neuropeptides involved in various brain structures and in different learning situations.

As the human genome with 'only approximately 30.000 genes expresses proteins with multiple information, it seems that peptide- and proteohormones of the endocrine system express multiple neurohormones. A typical example of this is the POMC molecule. POMC is synthesised in and released from the corticotrophs in the adenohypophysis, the melanotrophs in the neurointermediate lobe and in hypothalamic neurons. In the adenohypophysis it is a precursor for β-lipotropin (β-LPH) and ACTH. β-LPH in turn is a precursor for γ-LPH, β-endorphin (βE) and β-MSH. ACTH released from the pituitary melanotrophs is the precursor for α-MSH and ACTH-(1-16), corticotrophin-like intermediate lobe peptide (CLIP), while the N-terminal part of POMC, pro-γ-MSH, is the precursor for the various γ-MSH's. These peptides are precursor molecules for neuropeptides of the second generation. ACTH-1-16 and α-MSH generate fragments as those found following incubation in synaptosomal membranes that could well be responsible for the nervous system effects of the melanocortins.[13]

Another aspect of specificity is whether neuropeptides are selectively involved in cognitive processes of memory consolidation and retrieval, or whether they also have secondary effects through 'second order' events which modulate the input to information processing, i.e., processes such as perception, motivation, emotionality, attention, or arousal.

Memory processes, the retrieval of memory in particular, must be intimately related to perception, attention, and stimulus selection. Time- and dose-dependent post-learning effects, modification of neuronal excitability, the presence of these neuropeptides, their genetic apparatus and their receptors in brain structures critically involved in information are the most important criteria for a putative physiological involvement of a neuropeptide in cognitive processes. The greater effectiveness following intracerebral administration into these brain structures than following systemic treatment, and disturbances of cognitive processes following immunoneutralization, receptor blockade, anti-sense treatment or those of the peptide-deficient knock-out animals are essential arguments as well. The hypothesis has been put forward that some neuropeptides do not directly affect information processing per se, but modify it through motivation (motivational states are inferred mechanisms postulated to explain the intensity and direction of a variety of complex behaviors), attention, or arousal. For example, ACTH/MSH peptides are thought to act via sympathetic activation on motivation and attention, while CRF is thought to activate arousal mechanisms in autonomic nuclei of the brainstem. The hypothesis that neuronal peptides are often not released under basal conditions, but become released as auxiliary messengers in synaptic signaling upon activation, is in complete agreement with their putative role in biochemical events of information processing, arousal, attention, or motivation.

A certain degree of specificity may also appear at the level of the interaction of neuropeptides with classical transmitters or other peptidergic pathways in the brain. It is widely accepted that most neuropeptides modulate the ongoing neuronal activity of other transmitters. The physiological importance of these interactions is related to the fact that neuronal peptides have been found to be co-localized with one or more transmitters. It has been suggested that a neuron releases the same combination of transmitters at all terminals. It might be that co-existing transmitters and peptides interact in a synergistic as well as in an antagonistic manner, whereby neuropeptides are functioning as auxiliary messengers. This is in accordance with the finding, that the presence or absence of neuropeptides never induce 'all or none' type effects in information processing. The result of different peptidergic inputs that modulate neurotransmission might be part of synaptic plasticity.

Despite intensive experimentation and theorizing, the discussion of the role of neuropeptides in learning and memory processes makes no claim to completeness. One of the conclusions is that the mammalian brain does not possess a single neuropeptidergic mechanism that could account for the modulation of learning or memory processes. On the contrary, a symphony of neuropeptides of different chemical nature, localization and origin seem to act in concert with each other and with classical transmitters, and in some instances one of them may become more effective or even specific in a particular behavioral situation. What makes the role of some of these neuropeptides very attractive, is that they may contribute to plastic changes in the connectivity of neurons whose relationships are being reconstructed during learning and memory formation.

References

1. Alescio-Lautier B, Paban V, Soumireu-Mourat B. Neuromodulation of memory in the hippocampus by vasopressin. Eur J Pharmacol 2000; 405:63-72.
2. Bannon MJ, Haverstick DM, Shibata K et al. Preprotachykinin gene expression in the forebrain: regulation by dopamine. Ann N Y Acad Sci 1991; 632:31-37.
3. Baranowska D, Braszko JJ, Wisniewski K. Effect of angiotensin II and vasopressin on acquisition and extinction of conditioned avoidance in rats. Psychopharmacol1983; 81:247-251.
4. Belcheva I, Ternianov A, Georgiev V. Lateralized learning and memory effects of angiotensin II microinjected into the rat CA1 hippocampal area. Peptides 2000; 21:407-411.
5. Bidzseranova A, Gueron J, Balaspiri L et al. Intracerebroventricularly administered atrial natriuretric peptide (ANP) antiserum attenuates fear-motivated learning behavior in rats. Peptides 1992; 13:957-960.
6. Bidzseranova A, Gueron J, Toth G et al. Structure-activity studies on the effects of atrial natriuretic peptide, brain natriuretic peptide and their analogs on fear-motivated learning behavior in rats. Neuropeptides 1992; 23:61-65.
7. Bohus B, De Wied D. Failure of α-MSH to delay extinction of conditoned avoidance behavior in rats with lesions in the parafscicular nuclei of the thalamus. Physiol Behav 1967; 2:221-223.
8. Bohus B, De Wied D. The vasopressin deficient Brattleboro rats: a natural knockout model used in the search for CNS effects of vasopressin. Prog Brain Res 1998; 119:555-573.
9. Born J, Pietrowsky R, Fehm HL. Neuropsychological effects of vasopressin in healthy humans. Prog Brain Res 1998; 119: 619-643.
10. Braszko JJ. The contribution of AT1 and AT2 angiotensin receptors to its cognitive effects. Acta Neurobiol Exp (Warsz) 1996; 56:49-54.
11. Braszko JJ, Lasienko J, Kupryszewski G et al. Behavioral effects of angiotensin II and angiotensin II-(4-8)-pentapeptide in rats. Physiol Behav 1988; 44:327-332.
12. Braszko JJ, Paslawski L, Karwowska-Polecka W et al. Trandolapril attenuates acquisition of conditioned avoidance in rats. Pol J Pharmacol 2000; 52:195-201.
13. Burbach JP, Bohus B, Kovacs GL et al. Oxytocin is a precursor of potent behaviourally active neuropeptides. Eur J Pharmacol 1983;94:125-131.
14. Burbach JPH, Lebouille JLM, Wang X-C. Metabolic conversion of pro-opiomelanocrtin peptides by brain peptidases: ACTH, α-MSH and β-endorphin. In: De Wied D, Ferrari W, eds. Central Actions of ACTH and Related Peptides. Padova, Liviana Press, 1986; 53-68.
15. Cacabelos R, Niigawa H, Rodriguez-Arnao MD et al. Influence of somatostatin and growth hormone-releasing factor on behavior. Clinical and therapeutic implications in neuropsychiatric disorders. Horm Res 1988; 29:129-132.
16. Chai SY, Bastias MA, Clune EF et al. Distribution of angiotensin IV binding sites (AT4 receptor) in the human forebrain, midbrain and pons as visualised by in vitro receptor autoradiography. J Chem Neuroanat 2000; 20:339-348.
17. Chalas A, Conway EL. No evidence for involvement of angiotensin II in spatial learning in water maze in rats. Behav Brain Res 1996; 81:199-205.
18. Chen MF, Chiu TH, Lee EH. Noradrenergic mediation of the memory-enhancing effect of corticotropin-releasing factor in the locus coeruleus of rats. Psychoneuroendocrinol 1992; 17:113-124.
19. Croiset G, Nijsen MJ, Kamphuis PJ. Role of corticotropin-releasing factor, vasopressin and the autonomic nervous system in learning and memory. Eur J Pharmacol 2000; 405:225-234.
20. Dantzer R, Bluthe RM. Vasopressin and behavior: from memory to olfaction. Regul Pept 1993; 45:121-125.
21. De Bilbao F, Senut MC, Sweeney J et al. Comparative study in aged rats of behavioral deficiencies and morphometrical modifications of galanin containing neurons in the medial septal area. C R Acad Sci III 1991; 312:503-509.

22. De Bold AJ, Flynn TG. Cardionatrin I—a novel heart peptide with potent diuretic and natriuretic properties. Life Sci 1983; 33:297-302.
23. De Wied D. Long term effect of vasopressin on the maintenance of a conditioned avoidance response in rats. Nature 1971; 232:58-60.
24. De Wied D. The neuropeptide concept. Prog Brain Res 1987; 72:93-108.
25. De Wied D, Croiset G. Stress modulation of learning and memory processes. In: Jasmin G, Proschek L, eds. Stress Revisited. 2. Systemic Effects of Stress. Karger, Basel, 1991; 167-199.
26. De Wied D, Croiset G. Behavioral effects of neuropeptdies related to the stress hormones. Monography Russian Academy of Medical Sciences. Russian Academy of Medical Sciences, Moscow, 2002; in press.
27. De Wied D, Diamant M, Fodor M. Central nervous system effects of the neurohypophyseal hormones and related peptides. Front Neuroendocrinol 1993; 14:251-302.
28. De Wied D, Gaffori O, Burbach JP et al. Structure activity relationship studies with C-terminal fragments of vasopressin and oxytocin on avoidance behaviors of rats. J Pharmacol Exp Ther 1987; 241:268-274.
29. Delorenzi A, Locatelli F, Romano A et al. Angiotensin II (3-8) induces long-term memory improvement in the crab Chasmagnathus. Neurosci Lett 1997; 226:143-146.
30. Diamant M, De Wied D. Structure-related effects of CRF and CRF-derived peptides: dissociation of behavioral, endocrine and autonomic activity. Neuroendocrinology 1993; 57:1071-1081.
31. Diaz Brinton R. Vasopressin in the mammalian brain: the neurobiology of a mnemonic peptide. Prog Brain Res 1998; 119:177-99.:177-199.
32. Diaz Brinton R, Yamazaki R, Gonzalez CM et al. Vasopressin-induction of the immediate early gene, NGFI-A, in cultured hippocampal glial cells. Mol Brain Res 1998; 57:73-85.
33. Dietrich A, Allen JD. Vasopressin and memory. I. The vasopressin analogue AVP4-9 enhances working memory as well as reference memory in the radial arm maze. Behav Brain Res 1997; 87:195-200.
34. Dietrich A, Allen JD. Vasopressin and memory. II. Lesions to the hippocampus block the memory enhancing effects of AVP4-9 in the radial maze. Behav Brain Res 1997; 87:201-208.
35. Dluzen DE, Muraoka S, Engelmann M et al. The effects of infusion of arginine vasopressin, oxytocin, or their antagonists into the olfactory bulb upon social recognition responses in male rats. Peptides 1998; 19:999-1005.
36. Du YC, Yan QW, Qiao LY. Function and molecular basis of action of vasopressin 4-8 and its analogues in rat brain. Prog Brain Res 1998; 119:163-75.
37. Engelmann M, Bures J, Landgraf R. Vasopressin administration via microdialysis into the septum interferes with the acquisition of spatial memory in rats. Neurosci Lett 1992; 142:69-72.
38. Engelmann M, Wotjak CT, Neumann I et al. Behavioral consequences of intracerebral vasopressin and oxytocin: focus on learning and memory. Neurosci Biobehav Rev 1996; 20:341-358.
39. Engelmann M, Wotjak CT, Ebner K et al. Behavioural impact of intraseptally released vasopressin and oxytocin in rats. Exp Physiol 2000; 85 Spec No:125S-130S.
40. Everts HG, Koolhaas JM. Lateral septal vasopressin in rats: role in social and object recognition? Brain Res 1997; 760:1-7.
41. Fehm HL, Perras B, Smolnik R et al. Manipulating neuropeptidergic pathways in humans: a novel approach to neuropharmacology? Eur J Pharmacol 2000; 405:43-54.
42. Fekete M, Szipöcs I, Halmai L et al. Cholecystokinin and the central nervous system. Front Horm Res 1987; 15:175-251.
43. Ferrario CM. Angiotension-(1-7) and antihypertensive mechanisms. J Nephrol 1998; 11:278-283.
44. Flood JF, Morley JE. Dissociation of the effects of neuropeptide Y on feeding and memory: evidence for pre- and postsynaptic mediation. Peptides 1989; 10:963-966.
45. Flood JF, Baker ML, Hernandez EN et al. Modulation of memory processing by neuropeptide Y varies with brain injection site. Brain Res 1989; 503:73-82.
46. Flood JF, Garland JS, Morley JE. Evidence that cholecystokinin-enhanced retention is mediated by changes in opioid activity in the amygdala. Brain Res 1992; 585:94-104.
47. Florio T, Ventra C, Postiglione A et al. Age-related alterations of somatostatin gene expression in different rat brain areas. Brain Res 1991; 557:64-68.
48. Gaffori O, Stewart JM, De Wied D. Influence of substance P and fragments on passive avoidance behavior. Experientia 1984; 40:89-91.
49. Givens BS, Olton DS, Crawley JN. Galanin in the medial septal area impairs working memory. Brain Res 1992; 582:71-77.
50. Greven HM, De Wied D. Influence of peptides structurally related to ACTH and MSH on active avoidance behaviour in rats. A structure-activity relationship study. Front Horm Res 1977; 4: 140-152.
51. Harro J, Oreland L. Cholecystokinin receptors and memory: a radial maze study. Pharmacol Biochem Behav 1993; 44:509-517.

52. Hasenohrl RU, Souza-Silva MA, Nikolaus S et al. Substance P and its role in neural mechanisms governing learning, anxiety and functional recovery. Neuropeptides 2000; 34:272-280.
53. Holy Z, Wisniewski K, Jachimowicz A et al. Behavioural activity of angiotensin II (3-7)4Phe—analogue of natural fragment 3-7 of angiotensin II. Rocz Akad Med Bialymst 1996; 41:191-201.
54. Horvath KM, Meerlo P, Felszeghy K et al. Early postnatal treatment with ACTH4-9 analog ORG 2766 improves adult spatial learning but does not affect behavioural stress reactivity. Behav Brain Res 1999; 106:181-188.
55. Huston JP, Staubli U. Substance P and its effects on learning and memory. In: Martinez JL, Jensen RA, Messing RB, Rigter H, McGaugh JL, eds. Endogenous Peptides and Learning and Memory Processes. New York, Academic Press, 1981; 521-540.
56. Huston JP, Schildein S, Gerhardt P et al. Modulation of memory, reinforcement and anxiety parameters by intra-amygdala injection of cholecystokinin-fragments Boc-CCK-4 and CCK-8s. Peptides 1998; 19:27-37.
57. Ito M, Yu O, Chiu TH. Interactions of ACTH4-10 and ACTH1-24 with L-[3H]glutamate binding sites and GABA/benzodiazepine/picrotoxin receptor complexes in vitro. Brain Dev 1988; 10:106-109.
58. Itoh S, Lal H. Influences of cholecystokinin and analogues on memory processes. Drug Develop Res 1990; 21:257-276.
59. Itoh S, Takashima A, Maeda Y. Protective effect of cerulein on memory impairment induced by protein synthesis inhibitors in rats. Peptides 1992; 13:1007-1012.
60. Izquierdo I. Endogenous state dependency: memory depend on the relation between the neurohumoral and hormonal states present after training and at the time of testing. In: Lynch G, McGaugh JL, Weinberger NM, eds. Neurobiology of Learning and Memory. New York, Guilford, 1984; 333-350.
61. Izquierdo LA, Schroder N, Ardenghi P et al. Systemic administration of ACTH or vasopressin reverses the amnestic effect of posttraining beta-endorphin or electroconvulsive shock but not that of intrahippocampal infusion of protein kinase inhibitors. Neurobiol Learn Mem 1997; 68:197-202.
62. Jachimowicz A, Holy Z, Wisniewski K. The participation of nitric oxide in the facilitator effect of arginine vasopressin on memory. Acta Neurobiol Exp (Warsz) 1998; 58:37-45.
63. Jahn H, Montkowski A, Knaudt K et al. Alpha-helical-corticotropin-releasing hormone reverses anxiogenic effects of C-type natriuretic peptide in rats. Brain Res 2001; 893:21-28.
64. Joels M, Urban IJ. Arginine-vasopressin enhances the responses of lateral septal neurons in the rat to excitatory amino acids and fimbria-fornix stimuli. Brain Res 1984; 311:201-209.
65. Kovacs GL, De Wied D. Peptidergic modulation of learning and memory processes. Pharmacol Rev 1994; 46:269-291.
66. Kramar EA, Armstrong DL, Ikeda S et al. The effects of angiotensin IV analogs on long-term potentiation within the CA1 region of the hippocampus in vitro. Brain Res 2001; 897:114-121.
67. Ladurelle N, Keller G, Blommaert A et al. The CCK-B agonist, BC264, increases dopamine in the nucleus accumbens and facilitates motivation and attention after intraperitoneal injection in rats. Eur J Neurosci 1997; 9:1804-1814.
68. Lee J, Chai SY, Mendelsohn FA et al. Potentiation of cholinergic transmission in the rat hippocampus by angiotensin IV and LVV-hemorphin-7. Neuropharmacol 2001; 40:618-623.
69. Liebsch G, Landgraf R, Engelmann M et al. Differential behavioural effects of chronic infusion of CRH 1 and CRH 2 receptor antisense oligonucleotides into the rat brain. J Psychiatr Res 1999; 33:153-163.
70. Ma YL, Chen KY, Wei CL et al. Corticotropin-releasing factor enhances brain-derived neurotrophic factor gene expression to facilitate memory retention in rats. Chin J Physiol 1999; 42:73-81.
71. Malin DH, Novy BJ, Lett-Brown AE et al. Galanin attenuates retention of one-trial reward learning. Life Sci 1992; 50:939-944.
72. McGregor A, Richards M, Espiner E et al. Brain natriuretic peptide administered to man: actions and metabolism. J Clin Endocrinol Metab 1990; 70:1103-1107.
73. Metzger D, Alescio-Lautier B, Bosler O et al. Effect of changes in the intrahippocampal vasopressin on memory retrieval and relearning. Behav Neural Biol 1993; 59:29-48.
74. Meziane H, Devigne C, Tramu G et al. Effects of anti-CCK-8 antiserum on acquisition and retrieval by mice in an appetitive task. Peptides 1993; 14:67-73.
75. Mileusnic R, Lancashire CL, Johnston AN et al. APP is required during an early phase of memory formation. Eur J Neurosci 2000; 12:4487-4495.
76. Misane I, Razani H, Wang FH et al. Intraventricular galanin modulates a 5-HT1A receptor-mediated behavioural response in the rat. Eur J Neurosci 1998; 10:1230-1240.
77. Molle M, Albrecht C, Marshall L et al. Adrenocorticotropin widens the focus of attention in humans. A nonliner electroencephalographic analysis. Psychosom Med 1997; 59:497-502.

78. Mongeau R, De Oca BM, Fanselow MS et al. Differential effects of neurokinin-1 receptor activation in subregions of the periaqueductal gray matter on conditional and unconditional fear behaviors in rats. Behav Neurosci 1998; 112:1125-1135.
79. Morgan D, Diamond DM, Gottschall PE et al. A beta peptide vaccination prevents memory loss in an animal model of Alzheimer's disease. Nature 2000; 408:982-985.
80. Morley JE, Flood JF. Neuropeptide Y and memory processing. Ann N Y Acad Sci 1990; 611:226-231.
81. Nagel JA, Welzl H, Battig K et al. Facilitation of tunnel maze performance by systemic injection of the neurokinin substance P. Peptides 1993; 14:85-95.
82. Nakayama Y, Takano Y, Shimohigashi Y et al. Pharmacological characterization of a novel AVP(4-9) binding site in rat hippocampus. Brain Res 2000; 858:416-423.
83. Netto CA, Maltchik M. Beta-endorphin enhancement of retrieval in a three-session paradigm. Behav Neural Biol 1990; 54:97-104.
84. Nijsen MJ, Croiset G, Diamant M et al. Endogenous corticotropin-releasing hormone inhibits conditioned-fear-induced vagal activation in the rat. Eur J Pharmacol 2000; 389:89-98.
85. Nikolova JG, Getova DP, Nikolov FP. Effects of ACE-inhibitors on learning and memory processes in rats. Folia Med (Plovdiv) 2000; 42:47-51.
86. Noble F, Roques BP. CCK-B receptor: chemistry, molecular biology, biochemistry and pharmacology. Prog Neurobiol 1999; 58:349-379.
87. Noda Y, Mamiya T, Manabe T et al. Role of nociceptin systems in learning and memory. Peptides 2000; 21:1063-1069.
88. Nomoto S, Miyake M, Ohta M et al. Impaired learning and memory in OLETF rats without cholecystokinin (CCK)-A receptor. Physiol Behav 1999; 66:869-872.
89. Ogren SO, Hokfelt T, Kask K et al. Evidence for a role of the neuropeptide galanin in spatial learning. Neurosci 1992;51:1-5.
90. Ogren SO, Schott PA, Kehr J et al. Modulation of acetylcholine and serotonin transmission by galanin. Relationship to spatial and aversive learning. Ann N Y Acad Sci 1998; 863:342-363.
91. Okuda-Ashitaka E, Ito S. Nocistatin: a novel neuropeptide encoded by the gene for the nociceptin/orphanin FQ precursor. Peptides 2000; 21:1101-1109.
92. Paban V, Alescio-Lautier B, Devigne C et al. Effects of arginine8-vasopressin administered at different times in the learning of an appetitive visual discriminative task in mice. Behav Brain Res 1997; 87:149-157.
93. Paban V, Alescio-Lautier B, Devigne C et al. The behavioral effect of vasopressin in the ventral hippocampus is antagonized by an oxytocin receptor antagonist. Eur J Pharmacol 1998; 361:165-173.
94. Paban V, Alescio-Lautier B, Devigne C et al. Fos protein expression induced by intracerebroventricular injection of vasopressin in unconditioned and conditioned mice. Brain Res 1999; 825:115-131.
95. Ponomareva NS, Voskresenskaya OG, Kamenskii AA et al. Improvements in the selective perception and training of rats using an original analog of the C-terminal fragment of vasopressin. Neurosci Behav Physiol 2000; 30:225-229.
96. Popik P, Vetulani J. Opposite action of oxytocin and its peptide antagonists on social memory in rats. Neuropeptides 1991; 18:23-27.
97. Popik P, van Ree JM. () Neurohypophyseal peptides and social recognition in rats. Prog Brain Res 1998; 119: 415-436.
98. Popik P, Wolterink G, De Brabander H et al. Neuropeptides related to [Arg8]vasopressin facilitates social recognition in rats. Physiol Behav 1991; 49:1031-1035.
99. Poulakos JJ, Meyer EM, Pruysers CR et al. Neuropeptide Y-cholinergic interactions in neocortical tissue. Ann N Y Acad Sci 1990; 611: 219-225.
100. Radulovic J, Ruhmann A, Liepold T et al. Modulation of learning and anxiety by corticotropin-releasing factor (CRF) and stress: differential roles of CRF receptors 1 and 2. J Neurosci 1999; 19:5016-5025.
101. Raghavendra V, Chopra K, Kulkarni SK. Brain renin angiotensin system (RAS) in stress-induced analgesia and impaired retention. Peptides 1999; 20:335-342.
102. Raghavendra V, Chopra K, Kulkarni SK. Comparative studies on the memory-enhancing actions of captopril and losartan in mice using inhibitory shock avoidance paradigm. Neuropeptides 2001; 35:65-69.
103. Reyes TM, Lewis K, Perrin MH et al. Urocortin II: A member of the corticotropin-releasing factor (CRF) neuropeptide family that is selectively bound by type 2 CRF receptors. Proc Natl Acad Sci USA 2001; 98:2843-2848.
104. Robinson JK, Crawley JN. The role of galanin in cholinergically-mediated memory processes. Prog Neuropsychopharmacol Biol Psychiatry 1993; 17:71-85.

105. Sakagawa T, Okuyama S, Kawashima N et al. Pain threshold, learning and formation of brain edema in mice lacking the angiotensin II type 2 receptor. Life Sci 2000; 67:2577-2585.
106. Sanchez-Alavez M, Gomez-Chavarin M, Navarro L et al. Cortistatin modulates memory processes in rats. Brain Res 2000; 858:78-83.
107. Sandin J, Georgieva J, Schott PA et al. Nociceptin/orphanin FQ microinjected into hippocampus impairs spatial learning in rats. Eur J Neurosci 1997;9:194-197.
108. Sandin J, Nylander I, Georgieva J et al. Hippocampal dynorphin B injections impair spatial learning in rats: a kappa-opioid receptor-mediated effect. Neurosci 1998; 85:375-382.
109. Sandin J, Ogren SO, Terenius L. Endomorphin-2 but not Leu-enkephalin modulates spatial learning when microinjected in the CA3 region of the rat hippocampus. NeuroReport 2000; 11:3659-3662.
110. Santangelo EM, Morato S, Mattioli R. Facilitatory effect of substance P on learning and memory in the inhibitory avoidance test for goldfish. Neurosci Lett 2001; 303:137-139.
111. Santarelli L, Gobbi G, Debs PC et al. Genetic and pharmacological disruption of neurokinin 1 receptor function decreases anxiety-related behaviors and increases serotonergic function. Proc Natl Acad Sci USA 2001; 98:1912-1917.
112. Sato T, Tanaka K, Ohnishi Y et al. The ameliorating effects of a novel NC-1900 on impairments of learning/memory caused by glutamic acid. Nippon Yakurigaku Zasshi 114 Suppl 1999; 1:198P-203P.
113. Sawchenko PE, Imaki T, Potter E et al. The functional neuroanatomy of corticotropin-releasing factor. Ciba Found Symp 1993; 172:5-21; discussion 21-9.:5-21.
114. Schettini G. Brain somatostatin: receptor-coupled transducing mechanisms and role in cognitive functions. Pharmacol Res 1991; 23:203-215.
115. Schoots O, Hernando F, Knoers NV et al. Vasopressin receptors: structural functional relationships and role in neural and endocrine regulation. Results Probl Cell Differ 1999; 26: 107-133.
116. Schott PA, Hokfelt T, Ogren SO. Galanin and spatial learning in the rat. Evidence for a differential role for galanin in subregions of the hippocampal formation. Neuropharmacol 2000; 39:1386-1403.
117. Sebret A, Lena I, Crete D et al. Rat hippocampal neurons are critically involved in physiological improvement of memory processes induced by cholecystokinin-B receptor stimulation. J Neurosci 1999; 19:7230-7237.
118. Smagin GN, Dunn AJ. The role of CRF receptor subtypes in stress-induced behavioural responses. Eur J Pharmacol 2000; 405:199-206.
119. Smolnik R, Perras B, Molle M et al. Event-related brain potentials and working memory function in healthy humans after single-dose and prolonged intranasal administration of adrenocorticotropin 4-10 and desacetyl-alpha-melanocyte stimulating hormone. J Clin Psychopharmacol 2000; 20:445-454.
120. Spier AD, de Lecea L. Cortistatin: a member of the somatostatin neuropeptide family with distinct physiological functions. Brain Res Rev 2000; 33:228-241.
121. Spina MG, Basso AM, Zorrilla EP et al. Behavioral effects of central administration of the novel CRF antagonist astressin in rats. Neuropsychopharmacol 2000; 22:230-239.
122. Staubli U, Huston JP. Facilitation of learning by post-trial injection of substance P into the medial septal nucleus. Behav Brain Res 1980; 1:245-255.
123. Suga S, Nakao K, Hosoda K et al. Receptor selectivity of natriuretic peptide family, atrial natriuretic peptide, brain natriuretic peptide, and C-type natriuretic peptide. Endocrinology 1992; 130:229-239.
124. Takashima A, Yokota T, Maeda Y et al. Pretreatment with caerulein protects against memory impairment induced by protein kinase C inhibitors in the rat. Peptides 1991; 12:699-703.
125. Tanabe S, Shishido Y, Furushiro M et al. Facilitation of passive avoidance response by newly synthesized cationized arginine vasopressin fragment 4-9 in rats. Pharmacol Biochem Behav 1997; 57:251-256.
126. Tanabe S, Shishido Y, Nakayama Y et al. Effects of arginine-vasopressin fragment 4-9 on rodent cholinergic systems. Pharmacol Biochem Behav 1999; 63:549-553.
127. Tchekalarova J, Kambourova T, Georgiev V. Interaction between angiotensin IV and adenosine A(1) receptor related drugs in passive avoidance conditioning in rats. Behav Brain Res 2001; 123:113-116.
128. Telegdy G, Kokavszky K, Nyerges A. Action of C-type natriuretic peptide (CNP) on passive avoidance learning in rats: involvement of transmitters. Eur J Neurosci 1999; 11:3302-3306.
129. Thibonnier M, Coles P, Thibonnier A et al. The basic and clinical pharmacology of nonpeptide vasopressin receptor antagonists. Annu Rev Pharmacol Toxicol 2001; 41:175-202.
130. Thorsell A, Michalkiewicz M, Dumont Y et al.) Behavioral insensitivity to restraint stress, absent fear suppression of behavior and impaired spatial learning in transgenic rats with hippocampal neuropeptide Y overexpression. Proc Natl Acad Sci USA 2000; 97:12852-12857.

131. Tomaz C, Nogueira PJ. Facilitation of memory by peripheral administration of substance P. Behav Brain Res 1997; 83:143-145.
132. Ukai M, Shinkai N, Ohashi K et al. Substance P markedly ameliorates scopolamine-induced impairment of spontaneous alternation performance in the mouse. Brain Res 1995; 673:335-338.
133. Urban IJA. Effects of vasopressin and related peptides on neurons of the rat lteral septum and ventral hippocampus. Progr Brain Res 1998; 119: 285-310.
134. Van Ree JM, Gaffori O, De Wied D. In rats, the behavioral profile of CCK-8 related peptides resembles that of antipsychotic agents. Eur J Pharmacol 1983; 93:63-78.
135. Vaughan J, Donaldson C, Bittencourt J et al. Urocortin, a mammalian neuropeptide related to fish urotensin I and to corticotropin-releasing factor. Nature 1995; 378:287-292.
136. Versteeg DHG. Interactions of peptides related to ACTH, MSH and β-LPH with neurotransmitters in the brain. Pharmacol Ther 1986; 11:535-557.
137. Vécsei L. Behavioral, pharmacological and neurochemical studies of somatostatin. Lund, 1989.
138. Wang HL, Wayner MJ, Chai CY et al. Corticotrophin-releasing factor produces a long-lasting enhancement of synaptic efficacy in the hippocampus. Eur J Neurosci 1998; 10:3428-3437.
139. Wayner MJ, Armstrong DL, Phelix CF et al. Angiotensin IV enhances LTP in rat dentate gyrus in vivo. Peptides 2001; 22:1403-1414.

Chapter 3.4

Nerve Growth Factors and Neurotrophins

Catherine Brandner

Abstract

Cell division, cell death, and cell differentiation are hallmarks of embryogenesis. Such processes are supported by neurotrophins that have the capacity of regulating not only developmental processes, but also neuronal survival, morphological adaptation, and neural plasticity.

Nerve growth factor (NGF) is a prototype of neurotrophins capable of influencing survival and differentiation of neuronal cells during development. It has been shown that the cholinergic neurons of the basal forebrain critically depend on NGF for differentiation and survival. Exogenous administrations of NGF induced up or down regulation of cholinergic enzymatic activity that in turn altered the number of muscarinic receptors. Apart from its trophic function through activation of tyrosine kinase A receptors, NGF also can be neurotoxic through activation of $p75^{NTR}$ receptors.

Such dual and opposite effect suggests that exogenous NGF supplementation could alter the exact maturation of the cholinergic system either in a positive or in a negative way, depending on the period of the treatment. This was confirmed with NGF treatment during postnatal week two generating an adult-like spatial learning capacity despite animals being less than 5 weeks old. This superiority was maintained into adulthood. By contrast, NGF treatment during postnatal week one impaired spatial learning and hindered development into adult-like efficiency. These results reveal a developmentally crucial period for spatial learning mechanisms with a critical modulatory role of NGF.

Introduction

Among growth factors, neurotrophins appear to play a critical role particularly in neurite outgrowth and terminal arborization. In addition to their classical role in neuronal differentiation and survival, neurotrophins have been strongly implicated in axon pathfinding.[63] The notion that growth factors can guide growing axons to their targets was introduced more than 20 years ago.[51] This assumption was confirmed by culture experiments showing that nerve growth factor (NGF) induces a chemotactic response of sensory neurons.[31] The growing tip of the axons, the growth cones expresses growth factor receptors. The neurotrophic influence seems to depend on the receptor-mediated uptake and on the retrograde axonal transport toward the soma of the responsive neuron.[44,69] In vivo and in vitro experiments led to the hypothesis that growth factors promote the development of innervation. As will be revealed later, the effects of NGF on the development of cholinergic neurons support this idea. However, the concentration of NGF required to produce this effect appears to be higher than endogenous levels of NGF. To explain this paradox, it has been postulated that most of the developing neurons die during embryogenesis, because of their insufficient ability to compete for the limited amount of a trophic factor[42,45,24] and undergo synaptic plasticity.[66]

From Messengers to Molecules: Memories Are Made of These, edited by Gernot Riedel and Bettina Platt.

Neurotrophin Expression and Regulation of Neurogenesis during Development

The five closely related factors nerve growth factor (NGF), brain-derived neurotrophic factor (BDNF), neurotrophin-3 (NT3), -4 (NT4) and -5 (NT5) constitute the neurotrophin family. In the developing rat nervous system, the distribution of NT3, BDNF and NGF transcripts display a simultaneous increase in their expression between the 11th and 12th embryonic day and are widely distributed at embryonic days 12 and 13. This timing approximately corresponds with the developmental onset of neurogenesis (for example, see ref. 4). Despite the simultaneous gene expression of neurotrophins, the levels differ greatly at early embryonic stages. Nt-3 is the most highly expressed in the embryo while BDNF is expressed lowest. During development, NT-3 expression appears to follow proliferation, migration and differentiation of neurons and decreases within CNS regions as they mature. BDNF expression in the newborn rat is most prominent in CNS regions in which neurogenesis has already occurred and increases with maturation. Finally, NGF expression varies locally during development, but these variations do not follow a consistent pattern.[47] This absence of specificity would suggest a more general action of NGF on neurogenesis.

Embryogenesis, characterized by continuous cell division, death and differentiation is supported by neurotrophins, which regulate developmental processes, neuronal survival, morphology, and neural plasticity. The synaptic targets of the cells that enable neurotrophins neuronal survival. A partial or complete deletion of targets results in reduced innervation of neurons and reduced numbers of surviving neurons.[35] However, it seems that the target neurons may not be the only source of trophic support for neuronal survival.[58] Directional guidance of the growth cones appears to depend on second messengers, particularly cAMP, and the growth cone behavior seems to be regulated by the sum of second-messenger signals generated by several guidance cues.[50] Data have shown that an abrupt change in levels of guidance molecules are necessary for steering axons to an intermediate point or to a synaptic target.[51,72] This seems to indicate that a uniform pattern of guidance molecules prevents growth cones to extract guidance information and to enter their targets correctly. It appears that once secreted, immediate binding to cofactors could spatially restrict the actions of growth factors. For example, NGF and BDNF are immediately catabolized after secretion by the cell surface.[5] As NGF expression presents variations during development, it could also play a particularly important role in functional neuronal connections.

Neurotrophin Receptors

Actions of NGF depend on specific receptors. Their activation can lead to a wide range of responses, and these responses seem to depend on the activation of distinct second-messenger pathways. Growth factors bind to different tyrosine kinase members, NGF binds to trkA, BDNF to trkB, NT3 to trkB and trkC and NT4 to trkB, while the p75^{NTR} low affinity receptor binds NGF, BDNF and NT3-4 (Fig. 1).

The p75^{NTR} has no catalytic intracellular tyrosine kinase domain, but it is capable of mediating the neurotrophin signals. The ligand binding of p75^{NTR} increases the high-affinity trkA binding sites and enhances trkA autophosphorylation and selectivity for neurotrophin ligands. The trk-independent pathway of p75^{NTR} increases intracellular ceramide levels and further NFκB transcription factor[17] and JNK kinase.[18] Conversely, trkA activation can inhibit p75^{NTR}-mediated signaling, but the mechanism of this inhibition remains unclear.[40]

Nerve Growth Factor and the Basal Forebrain Cholinergic System

NGF is the most widely studied and characterized polypeptide growth factor capable of influencing survival and differentiation of neural cells during development.[44,45] Although, this prototype neurotrophic factor is well known to regulate the survival of neuronal populations, its function in the control of nerve growth remains unclear. Investigations on newborn and

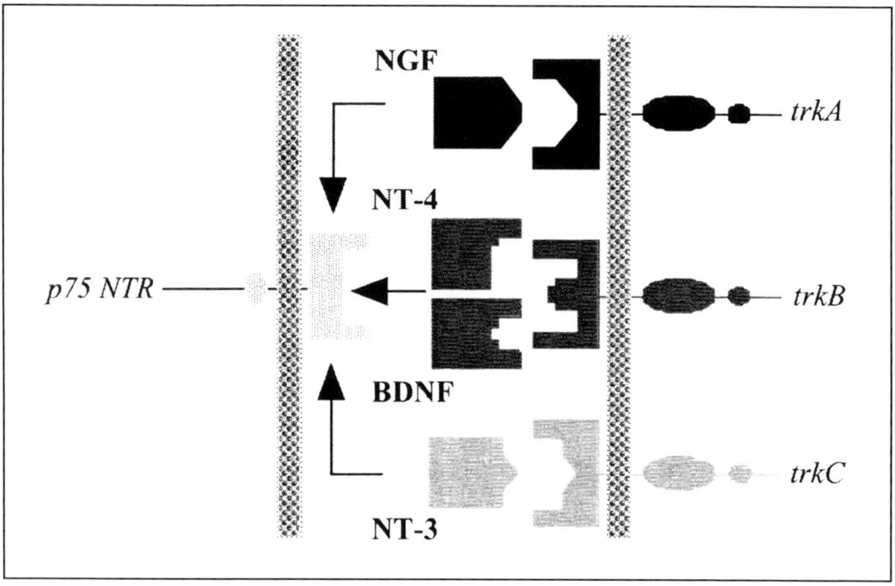

Figure 1. Schematic representation of neurotrophin binding showing that NGF binds to trkA, BDNF and NT4 to trkB, NT3 to trkC while the $p75^{NTR}$ low affinity receptor can bind NGF, BDNF and NT3-4.

adult rats have shown that exogenous NGF affects at least two parameters of the basal forebrain and striatal cholinergic neurons:

- It induces a selective and prominent increase of choline acetyltransferase (ChAT) activity.[29,55,56,39,25]
- It enlarges the size of the cholinergic neurons.

Enzyme activity is a classical measure for evaluating the maturation of neuronal tissue. The development of enzyme activity in the terminal fields of the cholinergic forebrain system takes place during the first four postnatal weeks.[65] During ontogeny, the basal forebrain cholinergic neurons depend on NGF for their differentiation and the expression of neurotransmitter phenotype.[42,37] Fiber terminals originating in the septal complex are present within the hippocampus formation by at least fetal day 20. Septal terminals are diffusely distributed initially and segregate to their mature position during the second postnatal week.[55] Cholinergic enzyme activity increases between birth and PN 5 in the hippocampus and frontal cortex. It reaches a peak value by PN 30 except in striatum, which achieved maximal activity at PN 60. This increase in activity is transient, and a major decrease is observed between PN 30 and 60.[66]

The cholinergic neurons of the basal forebrain depend on NGF for their differentiation and survival, and the expression of neurotransmitter phenotype.[77,43,37] Exogenous administration of NGF in neonatal rats produces an up and down-regulation of muscarinic cholinergic receptors in the cerebral cortex that could be correlated with concomitant changes in ChAT activity.[21] In neonatal mice, a single intracerebroventricular injection of NGF enhanced reactivity to the muscarinic blocker scopolamine suggesting an acceleration of cholinergic maturation.[2,16] This result supports the notion of the trophic action of NGF for these neurons since NGF antibody administration produced a decrease of ChAT activity in the hippocampus, septal area, cortex, and striatum of rat pups.[71] On the other hand, NGF can kill neurons during normal development by activating the $p75^{NTR}$ receptor, and this apoptotic effect through $p75^{NTR}$ receptor is not shared by the other neurotrophins. For example, it has been shown that $p75^{NTR}$ positive, trkA-negative cholinergic neurons in the basal forebrain of mice are normally eliminated within two weeks after birth.[68] During apoptosis, the absence of trkA expression appears to be a common denominator of NGF-induced cell death in the CNS. The expression of trkA often comes after that of p75^{NTR,} suggesting that in the intact organism, NGF-induced toxicity may be limited to early stages of development.[24] More generally, these effects indicate that the occurrence of endogenous NGF in the CNS is physiologically relevant for regulating the function of forebrain cholinergic neurons.

Taken together, these results suggest that exogenous NGF supplementation during development could either promote cholinergic maturation through trophic actions or damage the functionality of the system by an apoptotic effect. This dual and opposite role seems to rely on the maturation of the central cholinergic system depending on the presence of different NGF receptors. Thus, we could hypothesize that high levels of NGF in the first postnatal week are susceptible to induce behavioral disturbances through preferential activation of p75^{NTR}. On the other hand, exogenous NGF given during the second postnatal week is likely to accelerate cholinergic maturation and thus enhanced cognitive abilities. Establishment of NGF actions in the developing nervous system cannot be achieved in the absence of behavioral data.

Behavioral Studies of NGF Administrations

A large body of data support the hypothesis that normal spatial learning and memory processes depend on cholinergic function in the hippocampus and cortex. Maturation of spatial behavior, like learning and memory capacities in general appear relatively late in development. Data from experimental neuropsychology, comparative anatomy and field research show that behavioral adaptation requiring accurate spatial memory are most often mediated by permanent or transitory changes in the functional configuration of the hippocampus and cortex. Functional activity of the hippocampus and cortex rely on cholinergic input from the basal forebrain. These structures require 4-8 weeks to develop and NGF appears to play an important role in their maturation. Icv NGF injections during the first postnatal week produce a reduction of both ChAT and AChE activities in hippocampus. Neurochemical changes are also detected in 120-day old rats and are accompanied by an increase in the density of muscarinic receptors in the cerebral cortex.[72] The same treatment given during the second postnatal week induced a decrease in the muscarinic receptor number that return to control values shortly after treatment has ceased. Such up and down regulation of muscarinic receptors is associated with concomitant changes in ChAT activity[72,21] and might reflect the selection of cholinergic terminals. These results suggest the presence of critical periods during postnatal development withNGF injections having opposite effects on the maturation of the central cholinergic system. This also depends on the presence of different NGF receptors. We could thus expect that exogenous NGF administrations during these critical periods could differentially affect development and maintenance of cognitive abilities like spatial learning.

Does Early Icv NFG Injections Alter the Development of Spatial Abilities in Immature Rats?

In rats, navigation tasks like the Morris navigation task allow the study of spatial learning and memory processes. In the classical procedure ("place only"), animals learn to find a hidden platform on the basis of distant landmarks in the environment. The relational properties of the surrounding cues, no one of which is necessary, direct the movements towards a goal. This behavior is considered as a "place response". In contrast, "cue responses" are movements guided by a specific cue. In the Morris task for example, this behavior is observed when the platform is made visible. For normal adult rats with intact spatial abilities, no single landmark is necessary for place discrimination in the Morris navigation task. The addition of a conspicuous cue signaling the presence of the hidden platform induces the development of straight swim paths and the removal of the cue does not alter the memory of the goal position. In immature rats, the goal seeking response critically depends on cue presentation.[63,64,8] This effect, corresponding to an overshadowing of the distant cues by the more proximal one, is dependent on central cholinergic function.[9] Thus, training in the presence of a salient cue indicating a goal location is a task particularly sensitive to modifications in memory processing. In these studies, immature rats (28 control, 28 icv NGF injected: 14 in days 2 and 3 and 14 in days 12 and 13) were trained in both a 'place only' and a 'place & cue' version of the Morris navigation task starting at PD22 (see refs. 10, 11). This task was chosen owing to previous experiments[8] in which cholinergic manipulations has particularly severe behavioral consequences in rats applying a mixed learning strategy that combined cue response with place response (Fig. 2A).

During the fourth postnatal week, adult-like spatial learning abilities emerge in normal rats,[62] but the full repertoire of spatial strategies is yet to develop.[8] Thus, training of immature rats showed a progressive reduction of escape latency in both place only and place & cue version with animals expressing a bias toward the training quadrant (Fig. 2B and C). A tendency towards better spatial performance was observed in rats injected with NGF on postnatal days 12 and 13, but this was only apparent in the place & cue condition. In rats treated on postnatal days 2 and 3, the spatial abilities were not clearly altered by the treatment. Further overtraining (stabilisation) in the "place only" condition revealed an improvement of escape efficacy in the 12/13 day NGF group (Fig. 2A) which was similarly observed in the place & cue version (Fig. 2C). It is interesting to note a decrease in efficiency of rats NGF-treated on days 2 and 3 and trained in the place & cue, but not the place only, version.

Probe trials, during which the platform and the suspended cue were removed, were used to measure spatial memory. In line with training data, NGF treatment increased the time spent in the training quadrant during probe trial 1. Data are summarized in (Fig. 3) A more restrictive measure of accuracy (annulus crossings), however, revealed that only spatial memory of rats treated with NGF on days 12 and 13 was enhanced (Fig. 3A and B).

Following "place & cue" training, the expected overshadowing of distant landmarks was observed in both control and NGF 2/3 days treated rats. However, spatial memory of rats treated on days 12 and 13 was not affected by this training procedure (Fig. 3B). We continued to train animals with the platform at a new location (Fig. 2A, phase 2). This procedure exaggerated the superior performance of the NGF 12/13-day group, especially in the place only version. Such a difference is surprising given that the cue hanging above the new position was expected to exert a powerful attraction. This effect, however, was not observed. Following learning of the new location, the improvement of the spatial accuracy observed in rats treated with NGF on days 12- and 13 was maintained during another probe trial. This was independent of the training condition. NGF rats treated on days 2 and 3 and trained in a "place only" condition displayed a decrease of spatial memory following training to the new location (Fig. 3C and D).

In the present work, immature rats showed a progressive reduction of escape time in both place only and place & cue conditions and they expressed a bias toward the training quadrant, but their efficacy was limited. If trained without the hanging cue, the immature subjects showed rapid learning of the new spatial position, as indicated by the time spent in the new training quadrant during the second probe trial. As expected also, we measured a significant overshadowing effect of the presence of the cue upon the performance of immature rats. This appeared as a lack of bias toward the most recently trained position when escape has been facilitated by the presence of a cue hanging above the platform during training. Indeed, these rats showed a rapid adaptation of escape to the new position while they were allowed vision of the cue. However, following removal of this cue, they gave no indication that they had memorized the position of the platform relative to the distant room cues. Immature subjects seem to pay less attention to distant room cues when trained with a salient cue associated with the invisible target. This effect, could be due to the relative importance of the proximal cue that prevents an allocentric use of the more distant landmarks.[8,63] In comparison, the performance of rats treated on days 12 and 13 was comparable to what can be expected from normal adult rats in these conditions (see ref. 9): rapid acquisition of escape, accurate memory of the spatial position, efficient use of the salient cue to learn about the position of the platform relative to distant environmental cues and rapid learning of a new escape position.

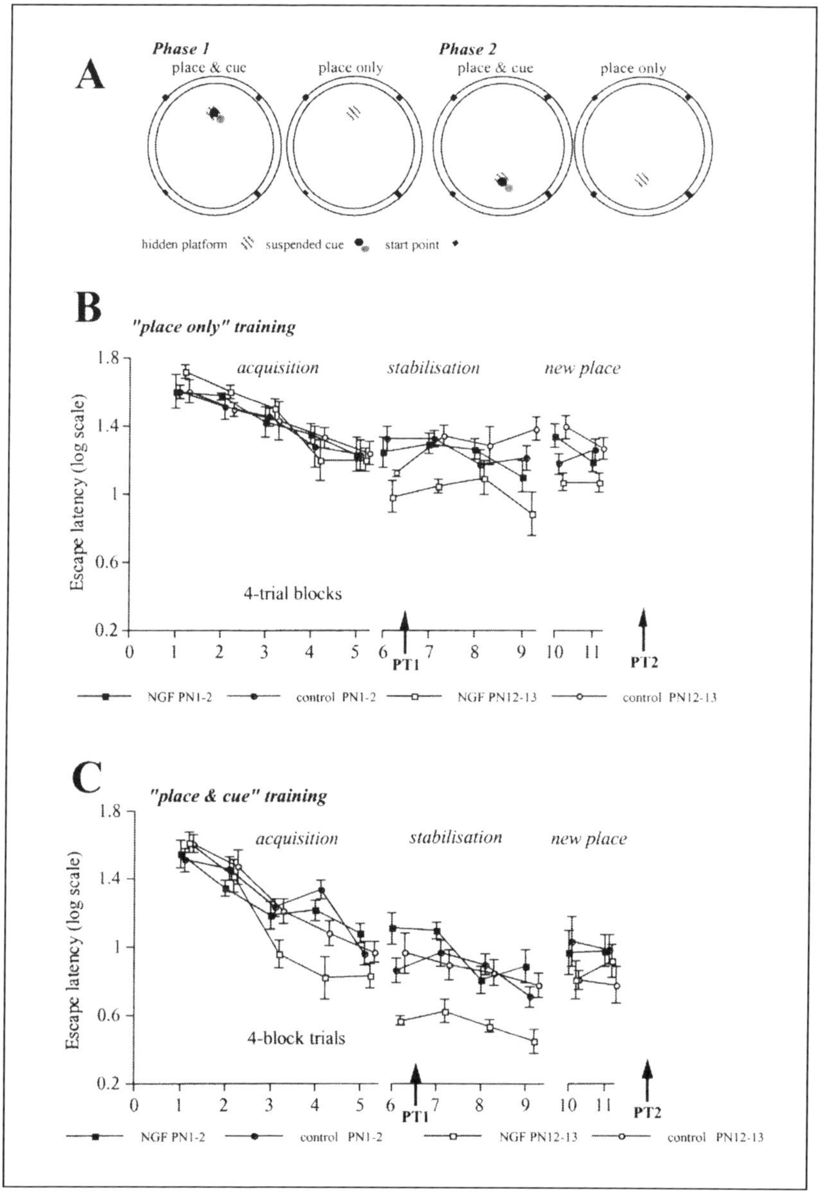

Figure 2. A) Schematic representation of the apparatus used for the Morris navigation task. Positions of the platform and, if provided, hanging cues are indicated. B) Mean (±sem) escape latencies (logarithmic scale) during training in the Morris navigation task by control (N=28) and NGF treated, 22 day-old rats trained in the absence of the suspended cue ("place only" condition). Acquisition, blocks 1-5, corresponds to the 20 first trials. Stabilization, blocks 6-9, corresponds to the asymptote of the escape latency. New place, block 10 and 11, the location of the hidden platform was changed. C) Mean (±sem) escape latencies (logarithmic scale) during training in the Morris navigation task by control and NGF treated, 22 day-old rats trained in the "place & cue" condition. Acquisition, blocks 1-5, corresponds to the 20 first trials. Stabilization, blocks 6-9, corresponds to the asymptote of the escape latency. New place, block 10 and 11, the location of the hidden platform was changed. PT1/2=probe trail 1/2; PN1,2=postnatal week 1,2.

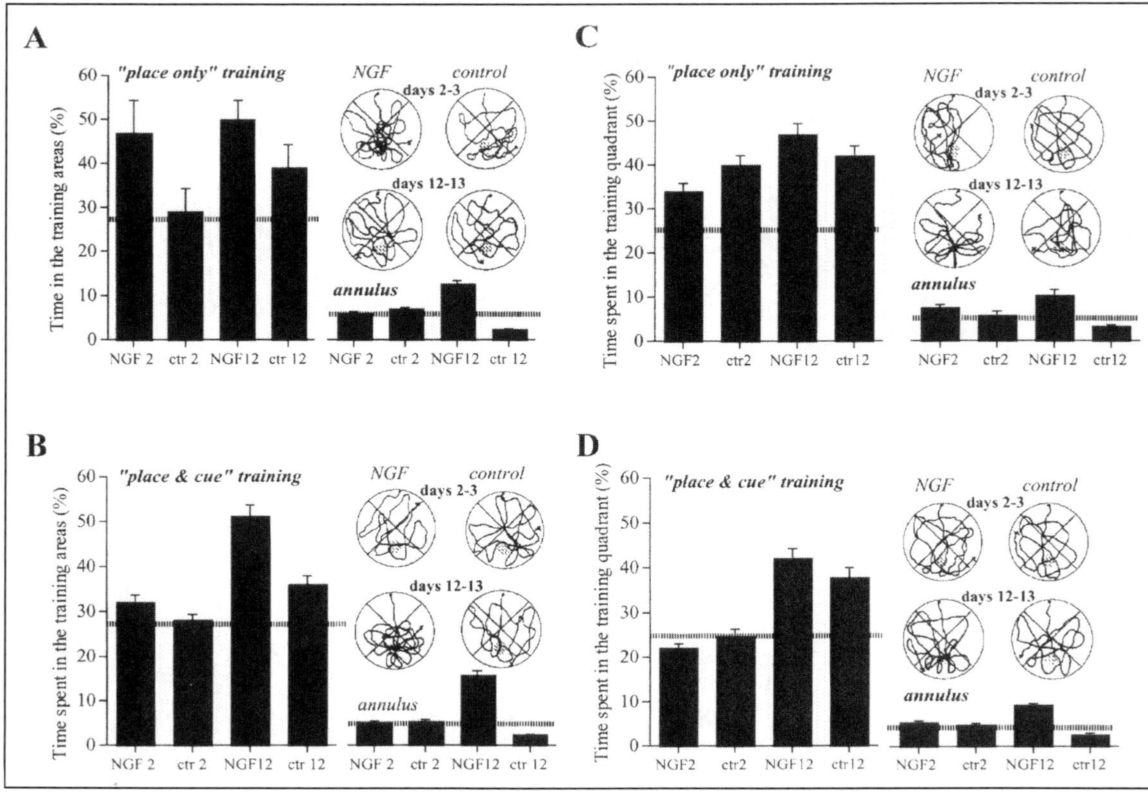

Figure 3. Following training, a probe trial (PT1) was given between block 6 and 7 of the stabilization phase, following trial 24. A) Mean (±sem) of the percentage of time spent in the training quadrant of the pool during a 60-second probe trial after training in the place only condition. Swim paths taken by representative rats are also given. Annulus crossings were measured as percentage of time spent in the platform area (Ø14-cm) of the training quadrant. B) Probe trial data for the place & cue trained groups. Mean (±sem). C) Probe trial 2 (PT2) was given after reversal training. Data from place only group. Mean (± sem). D) Probe trial 2 data of animal trained in the «place and cue» condition. Mean (± sem).

Are These Effects Maintained in Adulthood?

To assess long-term effects of NGF treatment, some of the rats treated on days 12 and 13 were tested in an 8-arm radial maze at the age of six months. All rats treated on days 2 and 3 were retrained in the Morris navigation task following the same procedure at the age of two months.

Rats Treated on Days 12 and 13

A general improvement on place learning ability that was observed in immature rats treated with NGF on days 12 and 13 was maintained in adulthood. The long-term effect of NGF treatment, assessed at six months in the radial maze task, was demonstrated by an early reduction of the errors in the NGF treated rats. A secondary effect of treatment was displayed by sex comparison. During the free choice phase, female control rats made a higher number of reentries while NGF treated female rats performed like male rats. Although the forced choice phase confirmed the efficiency of the NGF treated rats, the effects of treatment and sex were no more consistent as if such procedure elevated attentional processes in control rats (Fig. 4).

Rats Treated on Days 2 and 3

The effect of NGF injections on days 2 and 3 upon adult spatial performance was more extensive and appeared as a general impairment in spatial learning and memory abilities. Retrained at the age of two months, control rats showed equally efficient capacities in a cued and a noncued training condition. In contrast, NGF treated rats showed a decrease in escape efficacy particularly marked in a cued condition (Fig. 5A).

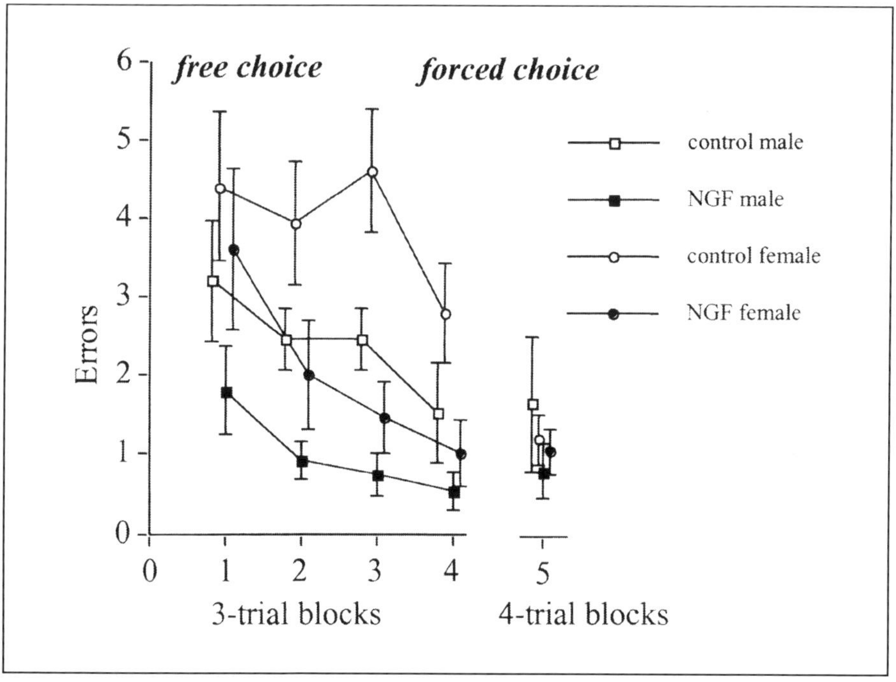

Figure 4. Mean number of errors (±sem) per block in the free choice and forced choice acquisition phases of the radial maze testing applied to 6 month-old control and NGF rats treated on days 12-13. MALE-FEMALE comparison.

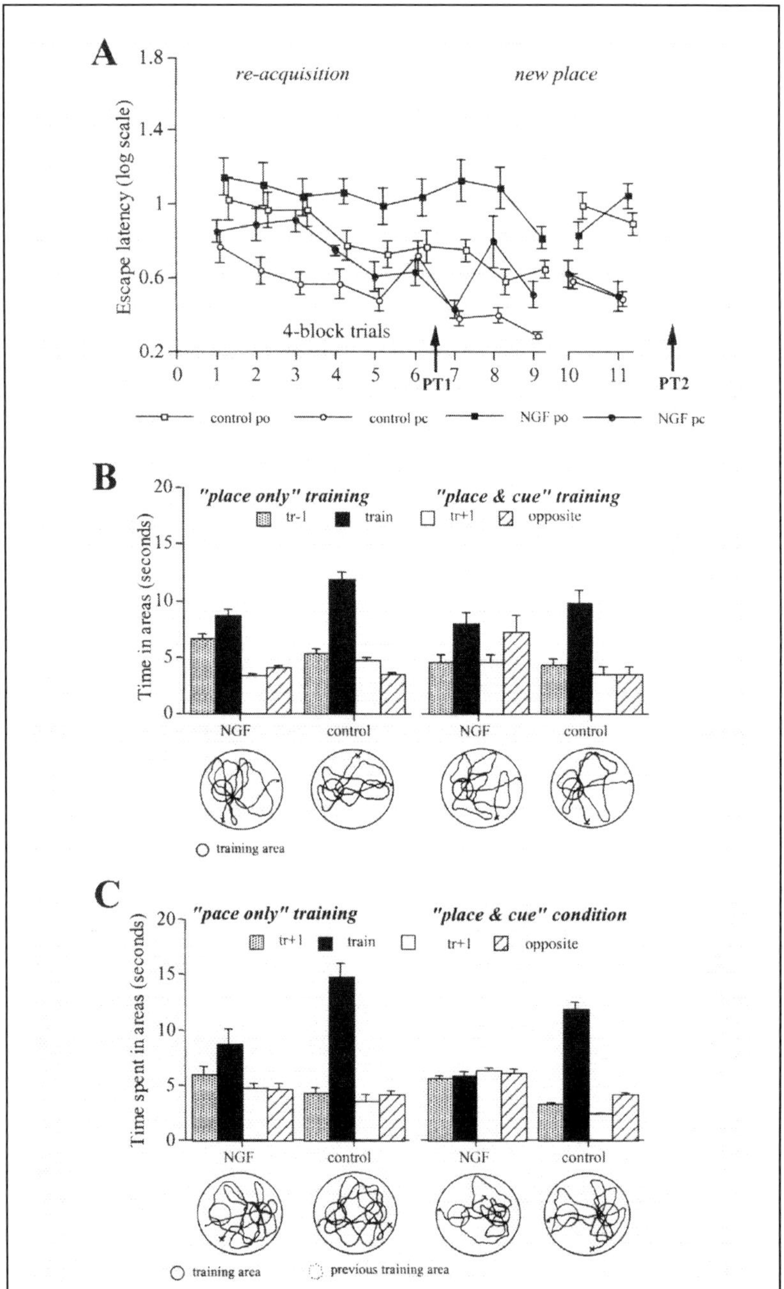

Figure 5. A) Mean (±sem) escape latencies (logarithmic scale) during retraining in the Morris navigation task in the "place & cue" and the "place only training" conditions by NGF and control two month-old adult rats. (For details, see Fig. 2B). B) Mean (±sem) of probe trial 1 administered after 24 training trials. (For details, see Fig. 3). (quadrant labeling: tr+1 = adjacent right ; train = training ; tr-1 = adjacent left ; opposite = opposite). C) Mean (± sem) of probe trial 2 after reversal training. (For details, see Fig. 3).

Compared with control animals, the spatial bias for the training quadrant measured during the first probe trial was consistently reduced in treated rats. This effect was independent of training condition (Fig. 5B).

Reversal learning distinguished the two learning procedures with a place & cue reversal being learnt more readily (Fig. 5A). Subsequent probe trials confirmed that NGF 2/3-day postnatally treated animals did not remember the new platform compared with controls (Fig. 5C).

In general, the treated rats appeared significantly impaired in all aspects of the place learning task when adults. They had longer escape latencies and weaker biases toward the training sector following acquisition in both training conditions. The treated rats showed reduced capacities to learn a new position. This later deficit was particularly obvious following training with a cued platform since they showed nearly no bias toward this position during the probe trial. In contrast, control rats showed flexible and accurate behavior. These results suggest that the treated rats were especially sensitive to the effect of a salient cue overshadowing the more subtle distal cues.

Discussion

NGF appears to regulate specifically the postnatal maturation of the central cholinergic nervous system (for a review, see ref. 15). Cholinergic projections to the hippocampus are essential for normal learning and memory capacities[43] (see also Jaffard and Marighetto, and Pepeu and Giovannini in this book). Neurotrophic factors contribute substantially to many of the neuronal changes in the brain (for review, see refs. 6, 44, 67). For example, BDNF appears to modulate transmission and plasticity in the hippocampus during development,[30] can enhance synaptic transmission in the adult hippocampus, and increases BDNF and NT-3 mRNA in the CA1 region of hippocampal slices have been shown following long-term potentiation (LTP).[60] In vivo, recent studies have shown that spatial and contextual learning are related to the expression of BDNF mRNA in the hippocampus.[31,34,42] The development of LTP in the hippocampus is influenced by the activity of septohippocampal cholinergic fibers[13,28] generating the theta rhythm.[3,54,70] An optimal tuning of the cholinergic system is indispensable for efficient spatial learning, but there are diverse interpretations as to the function of cholinergic activation for solving spatial tasks. Cholinergic blockade with muscarinic antagonists impairs various components of spatial abilities (see Pepeu and Giovannini in this book) such as the sensitivity to distant cues,[33,74,76] the organisation of exploratory responses[75,12] or the development of appropriate behavioural strategies necessary for the acquisition of movement sequences under distal cue guidance.[74] Since cholinergic blockade was most efficient when administered before training (see refs. 33, 1) this treatment might interfere with the initial storage of information, or with the process by which ongoing information is integrated before the selection of an appropriate behavioural strategy. Along this line, experiments have confirmed that cholinergic dysfunction does affect the attention to environmental stimuli.[58,14] This suggests an involvement of cholinergic transmission in attentional processes and in the selection of an appropriate strategy as well, which does not necessarily preclude a participation in memory processes.

Icv injections or infusion of NGF in young adult rats have been shown to prevent retrograde neuronal death, to promote recovery after damage to the septohippocampal pathway and to improve retention of a spatial memory task in impaired aged rats (Markowska et al., 1996).[36,22,61] Likewise, NGF injections appear to compensate for deficits induced by septo-hippocampal lesions or ageing.[26,27] In particular, NGF could modulate both the number and appearance of basal forebrain cholinergic neurons of cognitively impaired aged rats[23] and increase the number and size of cholinergic synaptic elements.[19]

Trophic action of NGF is known to prevent neuronal death. This property seems to rely on trkA receptor activation that regulates neuronal function like synaptic plasticity.[38] During development, however, NGF binding to low affinity $p75^{NTR}$ receptors appears to induce neuronal death.[24] Van der Zee and collegues (1996) have shown that $p75^{NTR}$ receptor mediates apoptosis of approximately 25% of the cholinergic basal forebrain neurons in mice between postnatal day 6 and 15, but only in cholinergic neurons that lack trkA receptors. In adulthood, by contrast, recovery after injury could be mediated by the p75^{NTR} low-affinity neurotrophin receptor. Van der Zee et al[69] now provide evidence that NGF infusion after fimbria fornix transection did not induce a reversal of choline ChAT expression in adult p75^{NTR} deficient mice.

Opposite to expectations derived from binding studies, postnatal NGF administration resulted in cognitive enhancement, which was particularly obvious in rats trained in the presence of a salient cue. Efficient spatial representation requires attention to each of the different cues, despite large inequalities in salience. This may be due to differences in size, contrasts with the background, or varying distances from the pool and the platform. Different treatments inducing cholinergic system modifications in rats indicate that either impairment or enhancement was more consistent when rats were trained in the presence of a salient local cue.[9,7]

It is known that a salient local cue facilitates escape. It is assumed that a cued task does not require spatial memory per se, but rather an association between the cue and the goal for the development of a guidance strategy. It has also been demonstrated that lesions of the striatum affect the cued task while lesions of the fornix reduce performance in place tasks.[20] In most cued tasks, the platform is visible, so that a rapid escape can be based on the single rule of approaching the conspicuous platform. In our task, however, the cue does not precisely coincide with the platform and offers only a partial support to landing. When the rat is in the proximity of the target, the cue is positioned above its head and thus might appear less salient. Such a cued task requires to chain at least two different strategies, i.e., a cue guidance combined with a memory of the platform position of the relative more distant room cues. As discussed by Jaffard and Meunier,[39] an optimal behaviour might require an active process "that readily upsets the imbalance between competing memory systems." This suggests that the elaboration and the use of a spatial representation might require a temporary memory of the local cues' salience, and one of the main functions of the cholinergic system might be the modulation of attentional processes by the balancing of the relative importance of the various components of the environment.

Finally, NGF seems to have a dual role that consists in preventing or inducing neuronal cholinergic death during development. This early regulation of the cholinergic system appears to be critical for the development of normal spatial capacity. The effects of early exogenous NGF administrations depend on the maturational state of the neuronal tissue. Given during periods crucial for development, NGF will preferentially induce p75^{NTR} receptor expression that could lead to cholinergic cell segregation and produce spatial impairments depending on the modification of attentional processes.

References

1. Aigner TG, Walker DL, Mishkin M. Comparison of the effects of scopolamine administered before and after acquisition in a test of visual recognition memory in monkeys. Behavl Neural Biol 1991; 55:61-67.
2. Alleva E, Bignami G. Development of mouse activity, stimulus reactivity, habituation, and response to amphetamine and scopolamine. Physiol and Behav 1985; 34:519-523.
3. Alonso A, Gaztelu JM, Bruno W et al. Cross-correlation anlysis of septohippocampal neurons during theta-rhythm. Brain Res 1987; 413:135-146.
4. Altman J, Bayer SA. The development of the rat spinal cord. In Advances in Anatomy Embryology and Cell Biology 85. New York: Springer Verlag, 1984.
5. Blochl A, Thoenen H. Characterization of nerve growth factor (NGF) release from hippocampal neurons: Evidence for a constitutive and an unconventional sodium-dependent regulated pathway. Eur J Neurosci 1995; 7:1220-8.
6. Bothwell M. Fonctional interactions of neurotrophins and neurotrophin receptors. Ann Rev Neurosci 1995; 18:223-253.
7. Brandner C. Perinatal choline treatment helps to disregard a salient but misleading cue for navigation in naive adult rats. Brain Res 2002; 928:85-95.
8. Brandner C. Effets promnésiants de traitements stimulant le développement et le maintien du système cholinergique chez le rat. PhD Thesis, Lausanne University 1999.

9. Brandner C, Schenk F. Septal lesions impair the acquisition of a cued place navigation task: Attentional or memory deficit? Neurobiol Learn Mem 1998; 69:106-125.
10. Brandner C, Vantini G, Schenk F. Enhanced visuospatial memory following intracerebroventricular administration of nerve growth factor. Neurobiol Learni Mem 2000a; 73:49-67.
11. Brandner C, Vantini G, Schenk F. Postnatal intracerebroventricular administrations of NGF alter spatial memory in adult hood. Behav Brain Res 2000b; 111:165-173.
12. Buhot M-C, Soffie M, Poucet B. Scopolamine affects the cognitiveprocesses involved in selective object exploration more than locomotor activity. Psychobiol 1989; 17:40-417.
13. Buszaki G. Two-stage model of memory trace formation: A role for «noisy» brain states. Neurosci 1992; 3:551-570.
14. Callahan MJ, Kinsora JJ, Harbaugh RE et al. Continuous infusion of scopolamine impairs sustained attention of rhesus monkeys. Neurobiol Aging 1993; 14:147-151.
15. Calamandrei G, Alleva E. Neuronal growth factors, neurotrophins and memory deficiency. Behav Brain Res 1995; 66:129-132.
16. Calamandrei G, Volanzano A, Alleva E. NGF and cholinergic control of behavior: Anticipation and enhancement of scopolamine effects in neonatal mice. Dev Brain Res 1991; 61:237-241.
17. Carter BD, Kaltschmidt C, Kaltschmidt B et al. Selective activation of NF-kappa B by nerve growth factor through the neurotrophin receptor p75. Science 1996; 272:542-5.
18. Casaccia-Bonnefil P, Carter BD, Dobrowsky RT et al. Death of oligodendrocytes mediated by the interaction of nerve growth factor with its receptor p75. Nature 1996; 383:716-9.
19. Cuello AC. Trophic factor therapy in the adult CNS: Remodeling of injured basalo-cortical neurons. Prog Brain res 1994; 100:213-221.
20. Devan BD, Goad EH, Petri HL. Dissociation of hippocampal and striatal contributions to spatial navigation in the water maze. Neurobiol Learn Mem 1996; 6:305-323.
21. Eva C, Fusco M, Brusa R et al. Intracerebroventricular administration of nerve growth factor affects muscarinic cholinergic receptors in the cerebral cortex of neonatal rats. Neurochem Inter 1994; 24:57-65.
22. Fisher W, Wictorin K, Bjorklund LR et al. Amelioration of cholinergic neuron atrophy and spatial memory impairment in aged rats by nerve growth factor. Nature 1987; 329:65-68.
23. Fisher W, Bjorklund LR, Chen K et al. NGF improves spatial memory in aged rodents as a function of age. Journal of Neuroscience 1991; 11:1889-1906.
24. Frade jm, Barde ya. Nerve growth factor: Two receptors, multiple functions. Bioessays 1998; 20:137-45.
25. Fusco M, Vantini G, Cavicchioli L et al. Effect on NGF on the adult intact and lesioned septohippocampal cholinergic system. Society for Neuroscience Abstracts 1988; 500:18.
26. Gage FH, Armstrong DM, Williams LR et al. Morphological response of axotomized septal neurons to nerve growth factor. J Comp Neurol 1988; 269:147-155.
27. Gage FH, Chen KS, Buszaki G et al. Experimental approaches to age-related cognitive impairments. Neurobiol Aging 1988; 9(5-6):645-655.
28. Garcia R, Jaffard R. The hippocampo-septal projection in mice: Long-term potentiation in the lateral septum. Neuro Report 1992; 3:193-196.
29. Gnahn H, Hefti F, Heumann R et al. NGF-induced increase of choline acetyltransferase (ChAT) in the neonatal rat forebrain: Evidence for a physiological role of NGF in brain? Dev Brain Res 1983; 9:45-52.
30. Gottschalk W, Pozzo-Miller LD, Figurov A et al. Presynaptic modulation of synaptic transmission of synaptic transmission and plasticity by brain-derived neurotrophic factor in the developing hippocampus. J Neurosci 1998; 18:6830-6839.
31. Gomez-Pinilla F, So V, Kesslak JP. Spatial learning induces neurotrophin receptor and synapsin I in the hippocampus. Brain Res 2001; 904:13-19.
32. Gundersen RW, Barrett JN. Neuronal chemotaxis: Chick dorsal-root axons turn toward high concentrations of nerve growth factor. Science 1979; 206:1079-80.
33. Hagan JJ, Tweedie F, Morris RGM. Lack of task specificity and absence of posttraining effects of atropine on learning. Behav Neurosci 1986; 100:483-493.
34. Hall J, Thomas KL, Everitt BJ. Rapid and selective induction of BDNF expression in the hippocampus during contextual learning. Nature Neurosci 2000; 3:533-535.
35. Hamburger V, Oppenheim RW. Naturally-occuring neuronal death in vertebrates. Neurosci Comments 1982; 1:38-55.
36. Hefti F. Nerve growth factor promotes survival of septal cholinergic neurons after fimbrial transections. Journal of Neuroscience 1986; 6:2155-2162.
37. Hefti F, Lapchak, PA. Pharmacology of nerve growth factor in the brain. Adv Pharmacol 1993; 24:239-273.

38. Holtzman DM, Li Y, Parada LF et al. P140trk mRNA marks NGF responsive forebrain neurons: Evidence that trk gene expression is induced by NGF. Neuron 1992; 9:465-478.
39. Jaffard R, Meunier M. Role of the hippocampal formation in learning and memory. Hippocampus 1993; 3:203-218.
40. Johnston MV, Rutkowski LJ, Wainer BH et al. NGF effects on developing forebrain cholinergic neurons are regionally specific. Neurochem Res 1987; 12:985-994.
41. Kaplan DR, Miller FD. Signal transduction by the neurotrophin receptors. Curr Opin Cell Biol 1997; 9:213-221.
42. Kesslak JP, So V, Choi J et al. Learning upregulates brain-derived neurotrophic factor messenger ribonucleic acid: A mechanism to facilitate encoding and circuit maintenance? Behav Neurosci 1998; 112:1012-1019.
43. Knipper M, Yamamura HI. Cholinergic mechanism and trophic factors. A preliminary study. In Treatment of Age-Related Cognitive Dysfunction: Pharmacological and Clinical Evaluation. In: Racagni G, Mendlewicz J. eds. International academy of biomedical drug research. Basel, Krager 1992; 2:19-34.
44. Korsching S. The neurotrophic factor concept: A reexamination. J Neurosci 1993; 13:2739-48.
45. Levi-Montalchini R, Calissano P. Nerve growth factor as a paradigm for other polypeptide growth factors. Trends neurosci 1986; 9:473-477.
46. Levi-Montalchini R. The nerve growth factor 35 years later. Science 1987; 237:1154-1162.
47. Lewin GR, Barde YA. Physiology of the neurotrophins. Ann Rev Neurosci 1996; 19:289-317.
48. Maisonpierre PC, Belluscio L, Friedman B et al. NT-3, BDNF, and NGF in the developing rat nervous system: parallel as well as reciprocal patterns of expression. Neuron 1990; 5:501-509.
49. Markowska AL, Koliatsos VE, Brecker SJ et al. Human nerve growth factor improves spatial memory in aged but not in young rats. J Neurosci 1994; 14:4815-4824.
50. Mc Donald RJ, White NM. A triple dissociation of memory systems: Hippocampus, amygdala, and dorsal striatum. Behav Neurosci 1993; 107:3-22.
51. McFarlane S, Holt CE. Growth factors: A role in guiding axons? Trends Cell Biol 1997; 7:424-430.
52. McFarlane S, Mcneill L, Holt CE. FGF signaling and target recognition in the developing Xenopus visual system. Neuron 1995; 15:1017-28.
53. Mnesini Chen, MG Chen JS, Levi-Montalcini R. Sympathetic nerve fibers ingrowth in the central nervous system of neonatal rodent upon intracerebral NGF injections. Archives Italiennes de Biologie 1978; 116:53-84.
54. Miller CL, Bickford PC, Wiser AK et al. Long-term potentiation disrupts auditory gating in the rat hippocampus. J Neurosci 1995; 15(8):5820-30.
55. Milner TA, Loy R, Amaral DG. An anatomical study of the development of the septo-hippocampal projection in the rat. Dev Brain Res 1983; 8:343-371.
56. Mobley WC, Rutkowski JL, Tennekoon GI et al. Choline acetytransferase in striatum of neonatal rats increased by nerve growth factor. Science 1985; 229:284-287.
57. Mobley WC, Rutkowski JL, Tennekoon GI et al. Nerve growth factor increases choline acetyltransferase activity in developing basal forebrain neurons. Mol Brain Res 1986; 1:53-62.
58. Muir JL, Dunnett SB, Robbins TW et al. Attentional functions of the forebrain cholinergic systems: Effects of intraventricular hemicholinium, physostigmine, basel forebrain lesions and intra cortical grafts on a multiple-choice serial reaction time task. Exp Brain Res 1992; 89:611-622.
59. Oppenheim RW. Celle death during development of the nervous system. Ann Rev Neurosci 1991; 14:453-501.
60. Patterson SL, Grover LM, Schwartzkroin PA et al. Neurotrophin expression in rat hippocampal slices: A stimulus paradigm inducing LTP in CA1 evokes increases in BDNF and NT-3 mRNAs. Neuron 1992; 9:1081-1088.
61. Pelleymounter MA, Cullen MJ, Baker MB et al. The effects of intrahippocampal BDNF and NGF on spatial learning in aged Long Evans rats. Mol Chem Neuropathol 1996; 29(2-3):211-226.
62. Rudy JW, Stalder-Morris S, Albert P. Ontogeny of spatial navigation behaviors in the rat: Dissociation of "proximal" and "distal" cue-based behaviors. Behav Neurosci 1987; 101:62-73.
63. Schenk F. Development of place navigation in rats from weaning to puberty. Behav Neural Biol 1985; 43:69-85.
64. Schenk F, Brandner C. Indirect effects of peri- and postnatal choline treatment on place-learning abilities in rat. Psychobiol 1995; 23 (4):302-313.
65. Song HJ, Poo mM M. Signal transduction underlying growth cone guidance by diffusible factors. Curr Opin Neurobiol 1999; 9(3):355-63.
66. Thal LJ, Gilbertson E, armstrong dm et al. Development of the basal forebrain cholinergic system: Phenotype expression prior to target innervation. Neurobiol Aging 1991; 13:67-72.
67. Thoenen H. Neurotrophins and neuronal plasticity. Science 1995; 270:593-598.

68. Van der Zee CE, Ross GM, Riopelle RJ et al. Survival of cholinergic forebrain neurons in developping p75 NGF deficient mice. Science 1996; 274:1729-1732.
69. Van der Zee CE, Ross EM, Hagg T. Delayed NGF infusion fails to reverse axotomy-induced degeneration of basal forebrain cholinergic neurons in adult p75 LNTR-deficient mice. Neurosci in press.
70. Vanderwolf CH, Kramis R, Gillespie LA et al. Hippocampal rhythmical slow activity and neocortical low voltage fast activity: Relations to behaviour. In: Isaacson RL and Pribram KH, eds. The Hippocampus, Neurophysiology and behaviour. 1975:2:101-128.
71. Vantini G, Schiavo N, Dimartino A et al. Evidence for a physiological role of nerve growth factor in the central nervous system of neonatal rats. Neuron 1989; 3:267-273.
72. Vantini G, Tria MA, Schiavo N et al. Intracerebroventicular (icv) administartion of NGF to newborn rats ultimately induces a permanent deficit of cholinergic markers in the hippocampus (The NGF pardox). Society for Neuroscience Abstract16: 1990; 155.
73. Wang HU, Anderson JD. Eph family transmembrane ligands can mediate repulsive guidance of trunk neural crest migration and motor axon outgrowth. Neuron 1997; 18:383-396.
74. Whishaw IQ, Petrie BF. Cholinergic blockade in the rat impairs strategy selection but not learning and retention of nonspatial visual dicrimination problems in a swimming pool. Behav Neurosci 1988; 120:662-677.
75. Whishaw IQ, Tomie JA. Cholinergic receptor blockade produces impairments in a sensorimotor subsystem for place navigation in the rat: Evidence from sensory, motor, and acquisition tests in a swimming pool. Behav Neurosci 1987; 101:603-616.
76. Whishaw IQ. Dissociating performance and learning deficits on spatial navigation tasks in rats subjected to cholinergic muscarinic blockade. Brain Res Bull 1989; 23:347-358.
77. Whittemore SR, Seiger A. The expression, localisation and functional significance of nerve growth factor in the central nervous system. Brain Res Revi 1987; 12:438-464.

Chapter 3.5

Eph Receptors and Their Ephrin Ligands in Neural Plasticity

Robert Gerlai

Abstract

Eph receptor tyrosine kinases are largely known for their involvement in brain development. But, as these receptors are also expressed in the adult, their possible role in the mature nervous system has begun to be explored. Emerging evidence for the involvement of Eph receptors in synaptic plasticity, learning and memory is discussed in this chapter. It is forecast that the actions of Eph receptors in the adult brain will attract significant attention, and research into their roles will have relevance for the human clinic, particularly in the area of CNS disorders associated with abnormal neural plasticity and memory loss.

Introduction

Tyrosine kinases including the receptors of neurotrophic factors such as NGF (Nerve Growth Factor), BDNF (Brain Derived Neurotrophic Factor) and neurotrophins NT3 and NT/4-5, have enjoyed considerable attention because of their newly discovered roles in neural plasticity and learning and memory. These proteins that were previously thought to function exclusively in brain development, now are known to be key players in synaptic processes thought to underlie LTP and memory formation.[62-67] Some argue now that the development of the brain and the development of the memory trace are not fundamentally different in terms of underlying molecular mechanisms.

The focus of the present review is the newest and largest receptor tyrosine kinase family, the Eph tyrosine kinases. Interestingly, the history of research into the function of these kinases is fairly similar to that of the "traditional" nerurotrophic factors and their receptors. The initial functional characterization of Eph tyrosine kinase receptors was also focussed on brain development. Eph receptors were found to mediate the establishment of topographic connections and migration of neuronal cells during ontogenesis. The fact that Eph receptors are expressed in the adult brain escaped attention for several years after the discovery of these receptors. Recently, however, such expression has been clearly demonstrated and the question regarding the possible role these receptors may play in the adult central nervous system has been raised. Here I review the emerging evidence for Eph receptor involvement in neural plasticity, and argue that the actions of Eph kinases in the adult brain will attract much attention and will become a prolific research area, perhaps even more so than in the case of neurotrophins and their tyrosine kinase receptors.

The Promiscuous Family of Eph Receptors

Eph receptors form the largest family of tyrosine kinase receptors with highly conserved amino acid sequence and perhaps function across vertebrate species (for most recent reviews

From Messengers to Molecules: Memories Are Made of These, edited by Gernot Riedel and Bettina Platt.

see refs. 1, 2). Their ligands, the ephrins, are also a highly abundant class of molecules.[1,2] Two main classes of Eph receptors are differentiated, A and B. This classification is based on the homology of the extracellular domains of the receptors and on their ligand preference.[3,4] EphA receptors bind ephrinA ligands and EphB receptors bind ephrinB ligands. The ephrin ligands, similarly to their receptors, are characterized by higher sequence homology within a class. The A and B classes of ephrins are also different in the way these ligands are attached to the cell membrane. EphrinA ligands are glycosylphosphatidylinositol (GPI) anchored. EphrinB ligands, however, span the cell membrane as they possess a transmembrane and a cytoplasmic domain. Importantly, the ephrin ligand must be membrane bound in order for it to activate its receptor. Soluble ephrin extracellular domains are inhibitory as they bind to the Eph receptors but are unable to initiate dimerization and autophosphorylation of the receptor. Artificial aggregation of soluble ligands mimics the endogenous physiological conformation of the ligands and can be used to activate the Eph receptor.[5] In summary, under physiological conditions receptor-ligand interaction requires cell-cell contact.[6]

The majority of studies investigating the function of Eph receptors has been largely limited to exploring the developmental role of these receptors.[7] Interestingly, however, recently both the receptors and their ligands were found to be expressed in the mature mammalian brain (see e.g., ref. 9 and references therein). This has raised the intriguing possibility that Eph receptors have a role beyond development. Here the first pieces of evidence supporting a role for Eph kinases in the adult nervous system is reviewed. The discussion will be focused on the involvement of Eph receptors in synaptic plasticity and learning and memory. The possible mechanisms of their action will also be outlined.

Eph Receptors Are in the Right Places and at the Right Time

The expression of Eph receptors has been thoroughly investigated in the developing brain. It has been found to be complex, temporally controlled, and tissue specific. Recently, however, continued expression in the adult CNS has been demonstrated by in situ hybridization and immunohistochemical analysis. For example, a strong signal for EphA5, a member of the Eph tyrosine kinase family, was found in all hippocampal neuronal fields, in the cortex, and in the amygdala of the adult rat brain.[8] The results were confirmed in two inbred strains of mice (C57BL/6 and DBA/2) by in situ hybridization.[8] Strong EphA5 mRNA expression was observed in the hippocampus, and a milder but still clearly detectable message was seen in the cortex, the amygdala, the thalamus and the hypothalamus.[9] The presence of EphA5 protein was also revealed.[9] It was found in hippocampal tissue in a phosphorylated form, which implies that the Eph kinase was present in an activated form in the adult mouse brain. EphrinA5, a ligand of the EphA5 receptor, was not detected by in situ hybridization in mice.[9] Nevertheless, a more sensitive technique, quantitative real time RT-PCR demonstrated the presence of mRNA of this and other ephrin ligands including ephrinA2.[9] Other studies using immunostaining revealed the presence of EphA3 and EphA4 receptors and the ephrinA2 ligand in both the adult rat and mouse brains[10,11] Clearly, these findings imply a possible functional role for the Eph receptors and their ligands in the adult brain.

The mere presence of these receptors and their ligands in adult brain tissue does not allow one to speculate what role these molecules may play there. However, analysis of their microstructural localization may offer some clues. Eph receptors and ephrinB ligands were found to co-localize with PDZ binding proteins in subcellular fractions (crude synaptosomes, and pre- and post-synaptic membranes) of adult rat cortex, indicating that these molecules may be present at synapses in vivo.[12] Moreover, immunohistochemical double labeling for synaptophysin and for Eph receptors or ephrinB ligands has confirmed synaptic localization of these proteins in hippocampal neuronal cultures.[12] Based on these observations a potential role for Eph kinases in the physiology of the synapse has been suggested,[12] an idea that has gained considerable support by the results of in vivo and ex vivo analyses of the function of Eph receptors.

Eph Receptors: "New" Players in the Adult Brain

Perhaps the first indication that Eph receptors may function in the adult brain came from a study in which kainate induced excitotoxicity and its effects on Eph gene expression were studied.[13] Kainate injection was found to induce the expression of Eph tyrosine kinases, namely EphA4, EphB2 and EphA5. Quantification of the expression levels of these receptors showed significant temporal changes. The results suggested that Eph receptors/ligands might function in neuronal pathfinding after sprouting subsequent to neuronal denervation in the adult, potentially implicating these receptors in such human brain diseases as epilepsy or spinal cord injury.[14] For instance, upon spinal cord injury EphB3 was found to be overexpressed in a rat model of contusive spinal cord trauma suggesting that EphB3 may contribute to the unfavorable environment for axonal regeneration.[68] In another study, ephrinA5 was found to be involved in selective inhibition of spinal cord neurite outgrowth and cell survival[14] again suggesting that Eph receptors significantly impair regeneration after injury in the adult CNS. Another interesting recent finding relevant for adult brain injury and repair concerns the expression of EphB1-3 and EphA4 receptors and their ephrinB ligands in the subventricular zone (SVZ) of the lateral ventricles in the adult mammalian brain.[69] SVZ, the largest remaining germinal zone of the adult brain contains neuroblast cells migrating rostrally to the olfactory bulb. The Eph receptors were demonstrated to mediate the migration and proliferation of these cells[69] raising the intriguing possibility that modulation of Eph receptor function may allow one to develop therapeutic applications by influencing neurogenesis in the adult brain. Finally, in a recent study, investigators using a kindling model found that activation or deactivation of Eph receptors can alter the development of behavioral seizures and change both the extent and the pattern of mossy fiber sprouting.[70] In summary, it appears that Eph receptors are involved in processes following injury to the adult brain. But what do they do in the normal brain?

Function of Eph Receptors in the Normal Brain: Role in Plasticity and Memory

The above question has been difficult to address because of the scarcity of good molecular tools with which one can manipulate Eph function. Specific pharmacological agents are not available for Eph tyrosine kinases. Antisense oligonucleotide knock down approaches have not been attempted. Gene targeting, although successfully employed with a number of Eph receptors and their ligands, has had limited use for the analysis of adult neural function because disruption of a single gene encoding a particular receptor or ligand could be compensated for by the presence of sister molecules. That is, functional redundancy made it difficult for the investigators to analyze the disruption of single members of this large protein family. Another complication in these studies is that these receptors and ligands are involved in CNS development. Thus if their disruption by gene targeting is not compensated for, the effects almost certainly will manifest as significant developmental abnormalities which would make the analysis of their adult neural function complicated. Perhaps, an inducible and cell type restricted knock out approach could adequately address the confounding effects of developmental alterations. But such an approach has not been attempted for these kinases. Furthermore, because of the high redundancy in the Eph family (overlapping expression and high homology between sister receptors or ligands), significant compensation may be expected if a single gene encoding one Eph receptor or ephrin ligand is mutated[15] thus double, triple, quadruple, etc. knock outs may be needed. Ultimately, creating all permutations of absence vs. presence of the normal form of certain members of this family may be required, clearly a daunting task that could take decades of experimentation. To solve the above problems an alternative molecular tool, the immunoadhesins[16] was utilized.

The immunoadhesins (Fig. 1) employed in the functional analysis of EphA receptors[8,9] were comprised of the ligand-binding domain of the EphA5 receptor (EphA5-IgG) or the receptor-binding domain of the ephrin-A5 ligand (ephrinA5-IgG). These immunoadhesins had opposing effects. EphA5-IgG scavenged the endogenous ligand and acted as an antagonist,

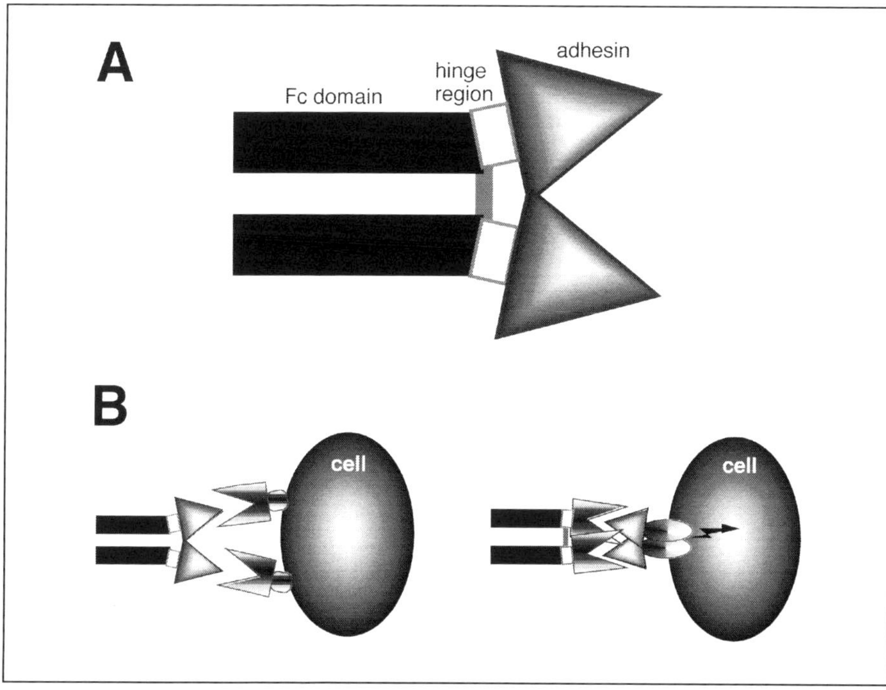

Figure 1. Immunoadhesins in the functional characterization of Eph receptors. Immunoadhesins (A) are genetically engineered proteins that consist of the Fc portion of an IgG molecule attached to a cell-surface protein (for review see 16). Immunoadhesins are disulfide-linked homodimers structurally similar to antibodies. They contain an adhesin region derived from a receptor or cell-surface ligand (triangles), the hinge region (white rectangles) and the Fc portion (black rectangles). Immunoadhesins bind to their target (B) with high affinity and specificity because the binding capacity of their adhesin domain is identical to that of the receptor or ligand of interest. For example, the receptor immunoadhesin EphA5–IgG (panel B left side) binds to ephrinA ligands anchored to the cell surface. By scavenging the ligands, it acts as a competitive antagonist of EphA function. The ligand immunoadhesin ephrinA5–IgG (panel B right side) Fc domain, black; receptor-binding domain of ligand attached to the Fc, "claw" shape) binds to EphA receptors (triangle and elliptic shape) and elicits receptor dimerization, which leads to receptor activation and intracellular signaling (but see below).
It is important to stress that these immunoadhesins recognize the ligand or the receptor on the basis of the high-affinity ligand-receptor interaction.[16,17] Immunoadhesins therefore may obviate the lack of EphA selective pharmacological agents and, as a result of the unaltered binding sites, immunoadhesins are capable of binding all the relevant proteins that the endogenous Eph receptor and the ephrins would bind. As Eph receptors are promiscuous and interact with several ephrin ligands,[3] immunoadhesins allow the manipulation of all functionally relevant ligands and receptors without the confounding effects of compensation by related molecules, as occurs in gene targeting experiments.[15,59,58]
Several caveats must also be mentioned, however. First, the ability of immunoadhesins to act as agonists may depend on the experimental conditions and the particular target receptor the immunoadhesin is supposed to bind. Eliciting receptor dimerization may require cross linking several immunoadhesins, i.e., the creation of immunoadhesin multimers.[16] Second, even the monomer is large enough not to be able to cross the blood brain barrier. Thus the in vivo delivery of the immunoadhesin requires time consuming, delicate, and invasive stereotaxic brain surgery. Third, the immunoadhesin solution may contain endotoxin, a bacterial lipoprotein-polysaccharide complex that may have significant toxic effects in the brain. Fourth, the immunoadhesin, as a foreign protein, may elicit an immune response. Despite these caveats that can complicate the interpretation of immunoadhesin effects, immunoadhesins have been successfully used in the functional analysis of neurotrophic factors and their tyrosine kinase receptors as well as ephrins and their Eph receptors (for a recent review and methods see refs. 60, 61). Figure modified from ref. 60.

whereas ephrinA5-IgG worked as an EphA agonist by dimerizing and initiating the autophosphorylation cycle of the receptor.[6,17]

Acute administration of EphA5-IgG, the EphA antagonist, resulted in EphA receptor deactivation leading to a significant impairment in long-term potentiation (LTP) in rat hippocampal slices.[8] Conversely, the agonist immunoadhesin, ephrinA5-IgG, led to synaptic potentiation resembling LTP.[8] These results provided the first direct evidence demonstrating that Eph tyrosine kinases participate in synaptic plasticity in vitro.

The question whether similar effects may be seen in vivo has also been addressed.[9,18] In these studies, the synaptoplastic and behavioral effects of in vivo chronic (7 day long) bilateral intrahippocampal immunoadhesin infusion were investigated. Although the induction of LTP was found normal in hippocampal slices of C57BL/6 mice previously infused with EphA5-IgG, the potentiated response was shown to decay faster when compared to control slices. The synaptoplastic changes correlated with behavioral alterations. Mice that received bilateral intrahippocampal infusion of EphA5-IgG for a week exhibited impaired T-maze spontaneous alternation (Figs. 2 and 3) as well as disrupted context-dependent fear conditioning performance (Figs. 4 and 5.), behavioral aberrations indicative of hippocampal abnormalities.[19,20,21] Thus, inhibition of EphA activity impaired neuronal plasticity, which manifested both in electrophysiological as well as behavioral tests. A potential concern could be that the impairment was due to non-specific effects but perhaps general impairment of health or brain function. However, the effects of ephrinA5-IgG induced Eph activation could not be explained by a non-specific action of this immunoadhesin. When infused into the hippocampus of DBA/2

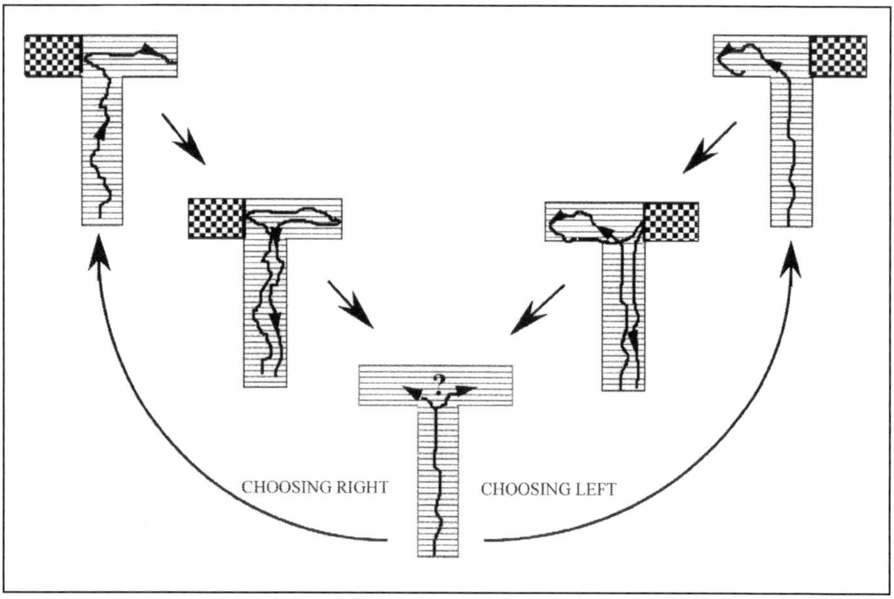

Figure 2. The T-maze Continuous Alternation Task (T-CAT). Mice are allowed to alternate between the left and right arms of the T-maze throughout a 15-trial session. Once they have entered a particular arm, a guillotine door is lowered to block entry to the opposite arm (checkered area). The door is removed only after the mice have returned to the start arm, allowing a new alternation trial to be started. Alternation rate is calculated as the ratio between alternating choices and total number of choices (50%, random choice; 100%, alternation at every trial; 0%, no alternation). Time to complete 15 choices is recorded. In addition, several motor and posture patterns are also measured (not shown).

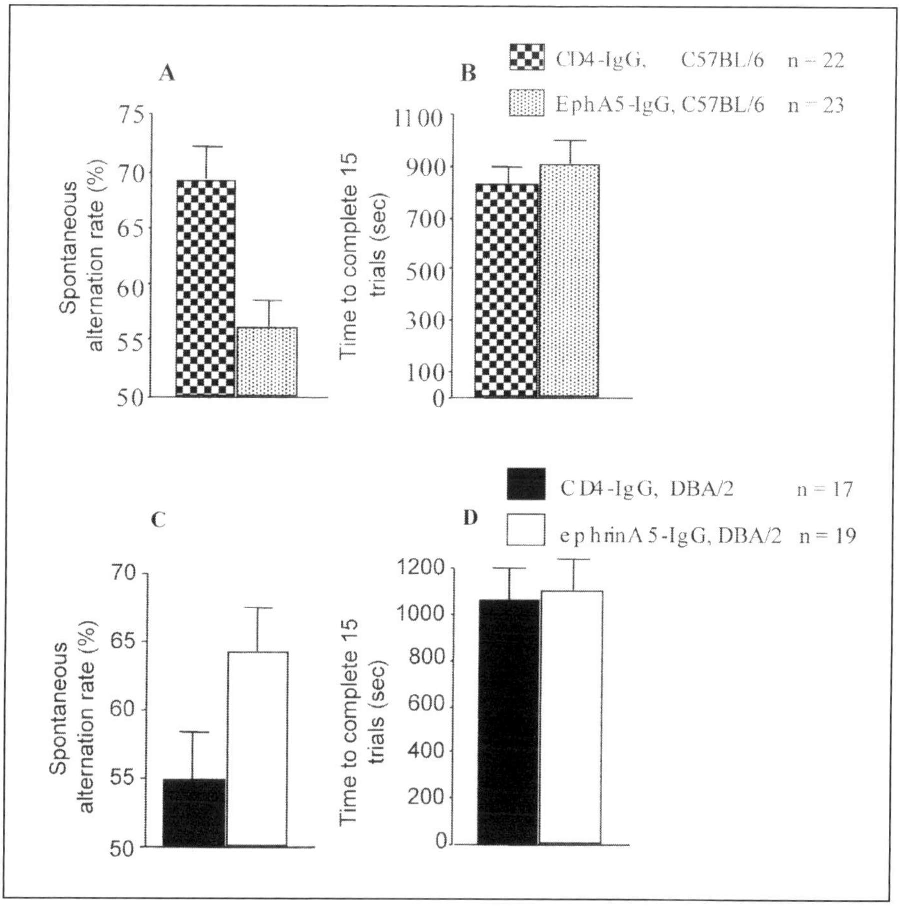

Figure 3. EphA receptors mediate spontaneous alternation performance in the T-maze. Infusion of EphA5-IgG impairs alternation performance in C57BL/6 mice (A) while ephrinA5-IgG improves alternation performance in DBA/2 mice (C) in the T-maze spontaneous alternation task. The changes are not related to task completion time (B, D) indicating unaltered motor performance or motivation. Mean ± standard error are shown. Sample sizes (n) are also indicated.

mice, a strain with impaired hippocampal function,[21,22,23,24] ephrinA5-IgG led to significantly improved LTP and this improvement correlated with superior performance in both the T-maze alternation task and the context dependent fear conditioning test as compared to control. These results were replicated in another strain (C57BL/6) of mice with the use of modified stimulation and testing protocols[9] suggesting that the findings are robust and not unique to a particular inbred mouse strain. Lastly, the involvement of Eph receptors in consolidation of memory has also been demonstrated[18] in a ketamine anesthesia induced retrograde amnesia model. In this work, ephrinA5-IgG, infused after ketamine induced disruption of memory consolidation, significantly improved cognitive performance in a hippocampus dependent manner (Fig. 6). In conclusion, the electrophysiological and behavioral observations obtained support a role for Eph receptors in neural plasticity in the adult mammalian brain.

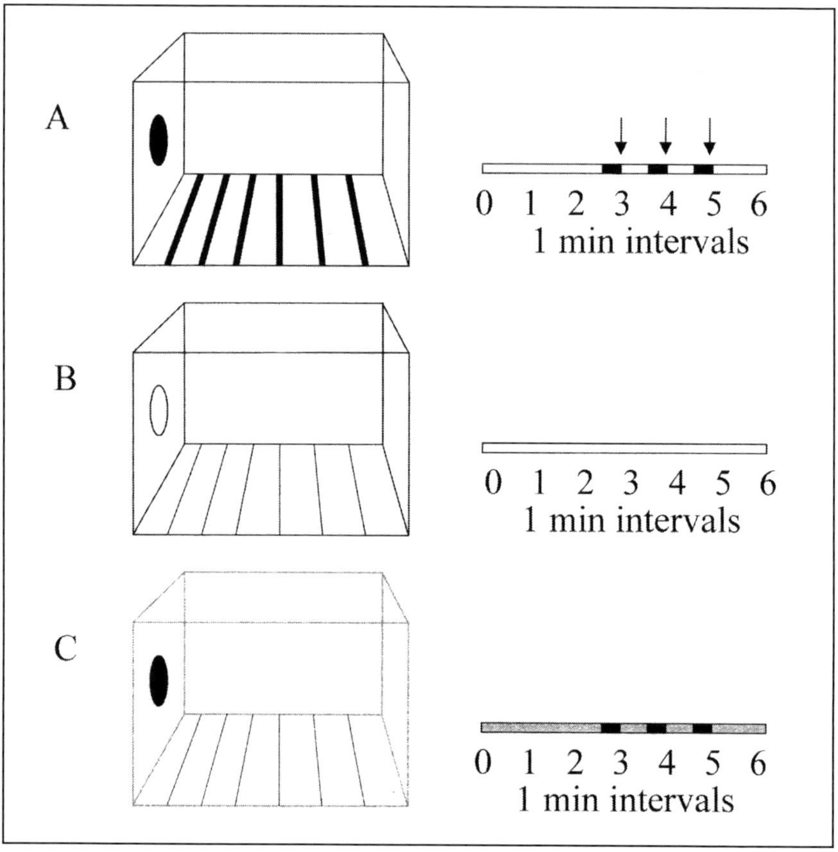

Figure 4. The fear conditioning paradigm. The paradigm has three phases: a training phase (A), a context dependent test (B), and a cue dependent test (C). For training, mice receive 3 electric foot shocks (1 sec, 0.7 mA, indicated by the thick black bars on the bottom of the cage) each preceded by an 80 dB, 2900 Hz, 20 sec long tone cue (indicated by the black filled circle on the wall). The context test is performed in the training chamber but no shock (thin bars) or tone (empty circle) is delivered. The cue test is carried out in another chamber identical in size but different in visual, olfactory, and tactile cues from those of the training chamber. Tone signals identical to the one used in training are given (black filled circle) but no shock (thin bars) is delivered. Behavior is video-recorded and later quantified using event recording computer programs. Behavior elements correlated with fear, primarily freezing, are measured. The timing of stimulus delivery in each phase of the paradigm is also shown: solid black bars represent the tone, the arrows the shock, and the gray shading the different context.

Mechanisms Mediating Eph Action: The First Working Hypotheses

Admittedly, the potential neurobiological mechanisms underlying the observed behavioral and electrophysiological effects are speculative at this point. The findings obtained so far, however, have led to the emergence of working hypotheses that may be tested in future mechanistic studies. The recent observation showing that Eph receptors and ephrinB ligands contain PDZ recognition motifs and are bound and clustered by PDZ proteins at pre- and postsynaptic sites of neuronal synapses in vitro suggests that Eph receptors are properly positioned to mediate synaptic plasticity.[12,25] Moreover, as Eph receptor and ephrin ligand binding interaction requires cell-cell contact (both the ligand and the receptor are membrane bound), Eph receptor mediated signaling can be achieved in a highly localized manner, a crucial prerequisite in the

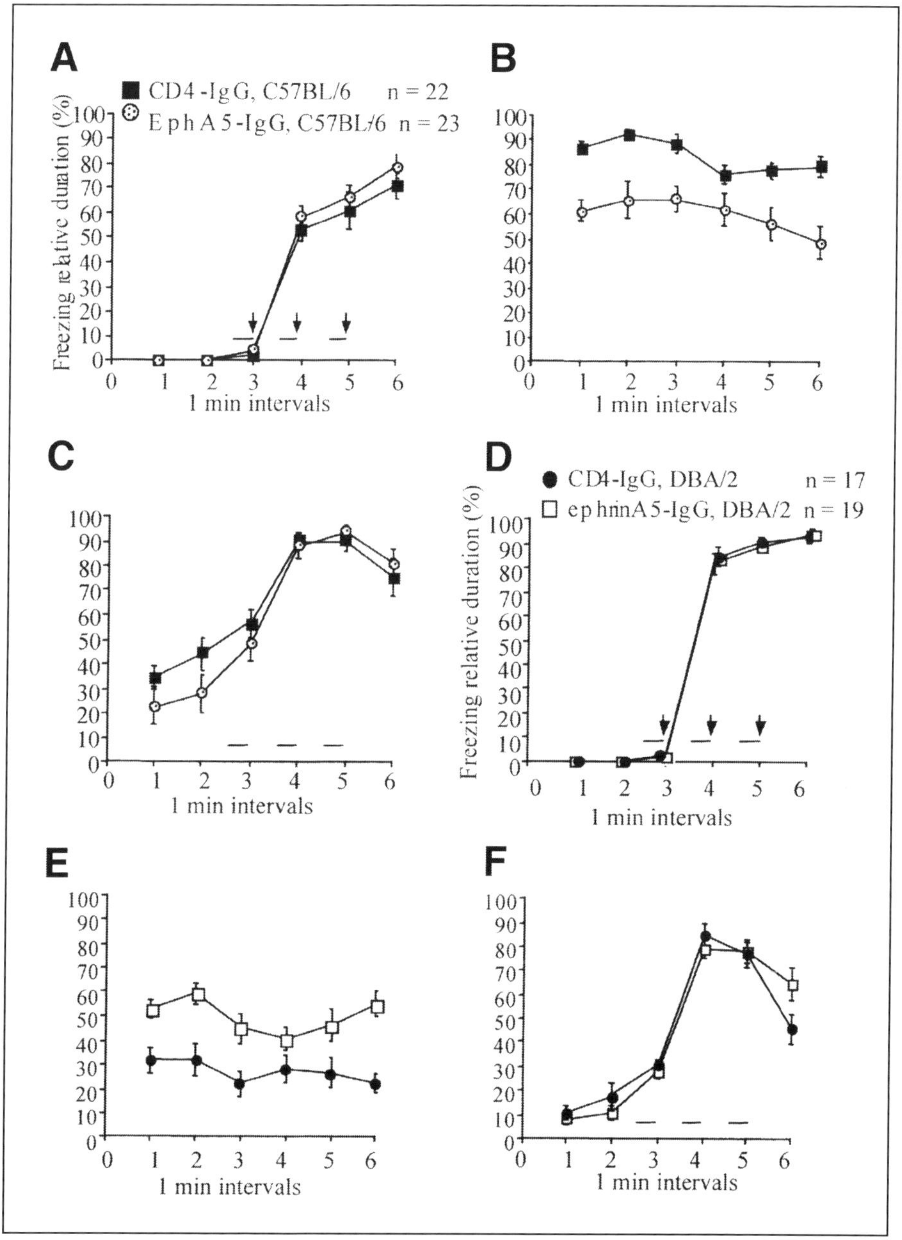

Figure 5. EphA receptors mediate cognitive performance in a context dependent manner in fear conditioning. The performance of EphA5-IgG infused C57BL/6 mice was significantly impaired compared to control (CD1-IgG infused mice) in the context test (B) but not in other phases of the paradigm (A training, C cue test). The performance of ephrinA5-IgG infused DBA/2 mice after fear-conditioning was significantly improved (increased freezing) compared to the control animals in a context-dependent manner (D training, E context test, F cue test). Note that both the context and the cued tests were carried out 24 hours after the fear conditioning. Mean ± standard error are shown. Sample sizes (n) are also indicated. Thin solid lines represent the delivery of tone and the arrows the shocks. (Modified from ref. 9)

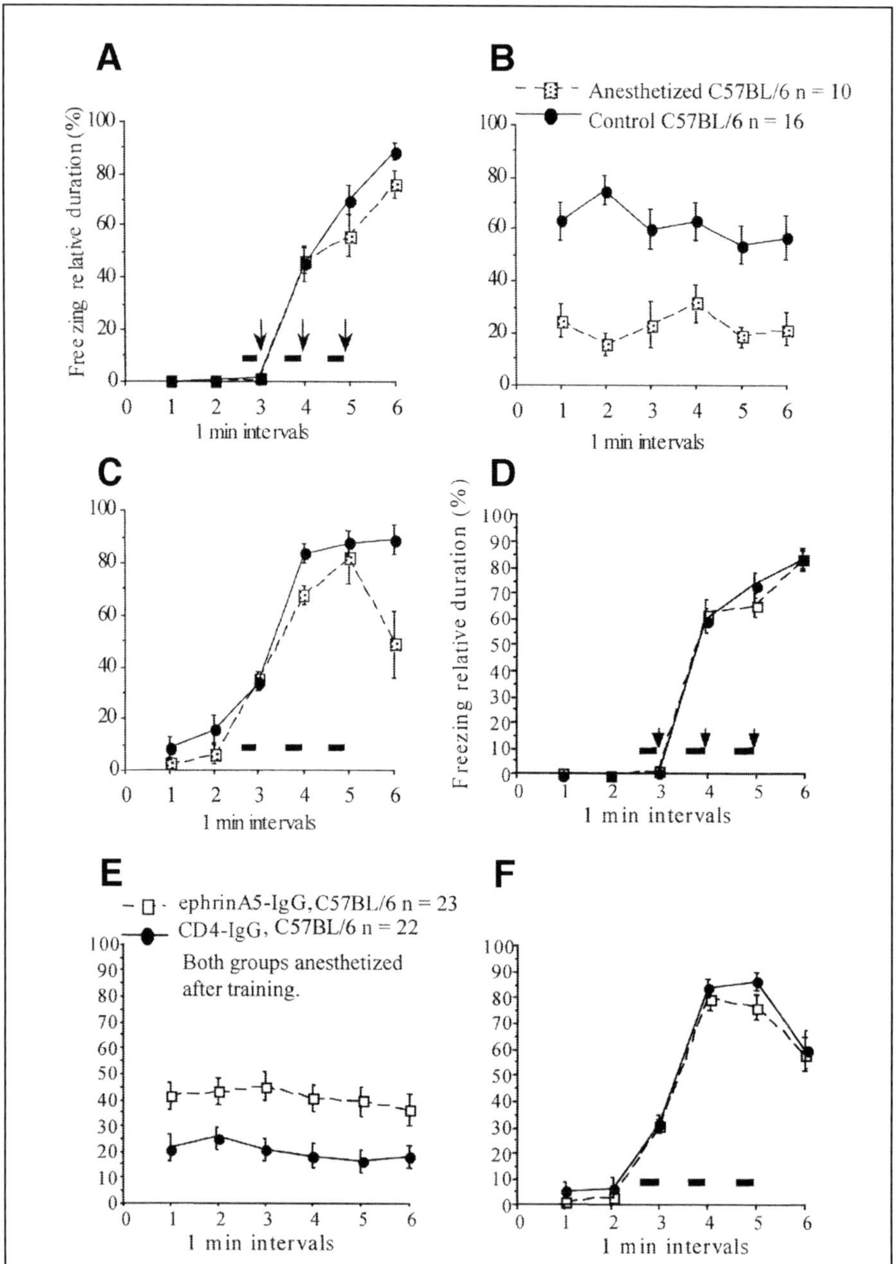

Figure 6. EphA receptors are involved in consolidation of memory. The performance of C57BL/6 mice were significantly disrupted by surgical anesthesia (ketamine) delivered 90 min after completion of training (A). The retrograde amnesia is robust in the context test (B), and almost completely absent in the cue test (C). EphrinA5-IgG infusion significantly ameliorates surgical anesthesia induced retrograde amnesia (D training, E context test, F cue test) in C57BL/6 mice. Mean ± standard error are shown. Sample sizes (n) are also indicated. Thin solid lines represent the delivery of tone and the arrows the shocks. (Modified from).[18]

activation/deactivation of single synapses essential for proper stimulus processing. Eph receptors may interact with a number of proteins through their PDZ binding domains that mediate cytoskeletal processes[12] and thus potentially affect a range of subcellular mechanisms influencing synaptic transmission and/or plasticity. Such mechanisms may include, for example, the trafficking and docking of presynaptic vesicles,[26] the clustering of neurotransmitter receptors, e.g., AMPA-R and NMDA-R,[27] and the formation of "perforated" synapses associated with LTP[28,29,30] and perhaps with memory formation. Interestingly, a member of the Eph family, the EphA5 receptor, has been shown to mediate actin polymerization, and its activation by administration of ephrinA5-IgG leads to actin depolymerization and axonal growth cone collapse in neuronal cell cultures and cortical explants.[6] Depolymerization of actin, a component of the scaffolding of thc synapse, may allow the synapse to undergo plastic structural modification. Indeed, actin has been found to be a crucial component of the cytoskeleton present in presynaptic as well as postsynaptic terminals[31,32,33] and has been shown to be associated with structural changes underlying synaptic plasticity[34,31,35,32] affecting both presynaptic and postsynapric mechanisms including paired pulse facilitation, and LTP.[36] Remarkably, it has been demonstrated that application of the EphA agonist ephrinA5-IgG, which destabilizes actin filaments[6] improves LTP. Therefore, the assumption that EphA receptor activation mobilizes the synapse by destabilizing actin filaments thus allowing the synapse to undergo structural modifications necessary for plastic changes to take place is not far fetched. Perhaps this hypothesis may be tested by detailed electron- or confocal microscopy analyses coupled with electrophysiological manipulation and monitoring of the synapse.

The possibility that Eph receptors play roles in cytostructural processes is consistent with the changes that were observed in the expression of the *tubulin* and *MAP2* (microtubule associated protein 2) genes in response to EphA5-IgG or ephrinA5-IgG treatment.[9] Tubulin and MAP2 were overexpressed as a result of EphA receptor inactivation and were underexpressed due to receptor activation in the adult mouse hippocampus. First, these findings are compatible with the known arresting effects of ephrinA ligands on axonal and dendritic growth during CNS development.[17,6,15] Second, they are also consistent with the suggested cytostructural role of the Eph receptors in neural plasticity: removal of the structural components tubulin and MAP2 may be a prerequisite of plastic changes of the synapse. In the adult brain, where major developmental alterations do not take place, transcriptional regulation of tubulin, and perhaps other genes of cytoskeletal proteins, may subserve the development of new or altered synaptic connections, i.e., neural plasticity as previously assumed.[37,38,39]

Although the above hypotheses are plausible, they are not the only possible ones. Eph receptors may also influence synaptic mechanisms via mediating adhesion processes. For example, phosphorylation of L1, a transmembrane adhesion molecule, was demonstrated following EphB2 activation,[40] and disruption of L1 function by anti-L1 antibody application was shown to impair synaptic plasticity.[41] EphA receptor induced signaling via ephrinA ligands (e.g., ephrinA5) should also be mentioned here as it was shown to increase the attachment of neuronal cells to the extracellular matrix,[42] a process that may influence synaptic plasticity.[43]

Furthermore, Eph receptors contain a cytoplasmic sequence motif, YEPD, that mediates binding src non-receptor tyrosine kinases, including src and fyn.[44] fyn is involved in the phosphorylation of NMDA-R,[45] a key player in LTP,[46] and fyn null mutant mice exhibit impaired spatial learning and blunted hippocampal LTP.[47] src also modulates NMDA-R function[48] and plays a crucial role in LTP.[49] LTP, and NMDA-R itself, has been implicated in acquisition and consolidation of memory .[50,46,51,52,19,53,54,55] Thus, src kinase mediated synaptic plasticity may be a potential substrate of Eph action. Lastly, EphB receptors have been shown to directly interact with NMDA receptors, a process that may influence synapse formation and function.[56]

Involvement of Eph receptors in adult neural plasticity implies that Eph receptor function must be modulated in a precise time and location specific manner. At this point, however, it is unclear how this is achieved. Ephrin ligands, compared to their receptors, are expressed at low levels in the adult brain[9] implying that perhaps a considerable proportion of Eph receptors is not activated under basal conditions. It is plausible that localized induction of expression of the ligands is the primary process that leads to receptor activation at the appropriate synaptic sites, however, this has not been investigated. Perhaps sensitive single cell PCR techniques or expression profiling using gene arrays will be able to address this question. It is also possible that proper clustering of the GPI anchored membrane bound or transmembrane ephrin ligands underlies receptor activation, as at least two ligand molecules need to be in close proximity to induce receptor dimerization and initiate the autophosphorylation process.[16] Although no direct evidence has been obtained to confirm the validity of this suggestion, ephrinA5 ligands have been found in specialized membrane rafts, called caveolae, which perhaps facilitate clustering of EphA receptors[42] and eprhinB ligands.[12] Activity dependent induction of EphA and EphB receptors (e.g., EphA4, EphA5, EphB2) at the mRNA level has been demonstrated in the hippocampus[13] suggesting that transcriptional regulation of the receptors may be possible. Alternatively, or additionally, modulation of Eph receptor signaling may be achieved through the tyrosine phosphorylation sites identified at the juxtamembrane, SAP, and kinase domains of the Eph receptor (reviewed in refs. 1, 2). But again, the molecular components involved in such processes are not well understood. Similarly, the downstream elements of Eph signalling are not yet elucidated. Nevertheless, based on the binding domains identified on the Eph receptor, downstream molecular interactions could involve numerous signaling pathways acting through src family cytoplasmic tyrosine kinases, the RasGAP pathway, the LMW-PTP phophotyrosine phosphatase, PI3 kinase, the Grb2, Grb10 and SLAP adaptor proteins, and several PDZ domain containing proteins including GRIP (reviewed in refs. 1, 2). Finally, signal transduction via ephrin ligands must also be mentioned. EphrinB ligands possess a cytoplasmic domain and have been clearly shown to transduce signals (reviewed in ref. 57) and ephrinA ligands (ephrinA5), as already mentioned, may also be involved in signal transduction (for review see refs. 1, 2).

Concluding Remarks

The molecular cascade of events in which Eph receptors are involved, including both the upstream and downstream elements, are far from understood. The potential neurobiological mechanisms associated with Eph action are also highly speculative. Nevertheless, the gross anatomical localization of Eph receptors and ephrin ligands in the adult brain, and the localization of some of these proteins at the synapse, suggest that this receptor system is involved not only in development of the brain but also in adult neural function. This conclusion is now supported by the findings demonstrating that significant changes occur in synaptic plasticity following acute or chronic modulation of Eph function in hippocampal slices and that significant changes are also observed in learning and memory after chronic modulation of Eph function in vivo.

This is a promising start by all means, but much needs to be done before the exact role of Eph receptors in adult neural function can be understood. Characterization of the signaling pathways upstream and downstream of the Eph receptor will be a complex task given the multitude of potential molecular interactions in which these receptors and ligands participate. It is also not clear whether different members of the Eph receptor tyrosine kinase family have spatially and/or temporally distinct roles in the adult brain. Inducible and cell type restricted gene targeting or the use of immunoadhesins and perhaps novel small molecules, specific pharmacological tools to be developed for particular Eph receptors, will advance our understanding of the actions of the Eph receptors. Ultimately, these techniques will enable us to address the intriguing question whether the development of our brain and the development of our memories share common molecular mechanisms.

References

1. Wilkinson DG. Multiple roles of Eph receptors and ephrins in neural development. Nature Rev Neurosci 2001; 2.
2. Mellitzer G, Xu Q, Wilkinson DG. Control of cell behaviour by signalling through Eph receptors and ephrins. Curr Opin Neurobiol 2000; 10:400-408.
3. Orioli D, Klein R. The Eph receptor family: Axonal guidance by contact repulsion. Trends Genet 1997; 13:354-359.
4. Eph Nomenclature. Committee Unified nomenclature for Eph family receptors and their ligands, the ephrins. Cell 1997; 90:403-404.
5. Davis S, Gale NW, Aldrich TH et al. Ligands for EPH-related receptor tyrosine kinases that require membrane attachment or clustering for activity. Science 1994; 266:816-819.
6. Meima L, Kljavin IJ, Shih A et al. AL-1-induced growth cone collapse of rat cortical neurons is correlated with REK7 expression and rearrangement of the actin cytoskeleton. Eur J Neurosci 1997; 9:177-188.
7. Frisén J, Holmberg J, Barbacid M. Ephrins and their Eph receptors: Multitalented directors of embryonic development. The EMBO Journal 1999; 18:5159-5165.
8. Gao W-Q, Shinsky N, Armanini MP et al. Regulation of hippocampal synaptic plasticity by the tyrosine kinase receptor, REK7/EphA5, and its ligand, AL-1/Ephrin-A5. Mol Cell Neurosci 1998; 11:247-259.
9. Gerlai R, Shinsky N, Shih A et al. Regulation of learning by EphA receptors: A protein targeting study. J Neurosci 1999; 19:9538-9549.
10. Mecteau M, Moers S, Doucet G. Distribution of EphA3 And EphA4 Receptors in the Neonatal and Adult Mouse and Rat Brain. Soc Neurosci New Orleans Abstr # 2000; 218:8.
11. Mamou BC, DesGroseillers L, Chazal G et al. Distribution of EphA5 and Ephrin-A2 in the Brain of Newborn and Adult Mouse. Soc Neurosci New Orleans Abstr # 2000; 218:9.
12. Torres R, Firestein BL, Dong H et al. PDZ proteins bind, cluster, and synaptically colocalize with Eph receptors and their ephrin ligands. Neuron 1998; 21:1453-1463.
13. Moreno-Flores MT, Wandosell F. Up-regulation of Eph tyrosine kinase receptors after excitotoxic injury in adult hippocampus. Neurosci 1999; 91:193-201.
14. Yue Y, Su J, Cerretti DP et al. Selective inhibition of spinal cord neurite outgrowth and cell survival by the Eph family ligand ephrin-A5. J Neurosci 1999; 19:10026-10035.
15. Frisén J, Yates PA, McLaughlin T et al. Ephrin-A5 (AL-1/RAGS) is essential for proper retinal axon guidance and topographic mapping in the mammalian visual system. Neuron 1998; 20:235-243.
16. Chamow SM, Ashkenazi A. Immunoadhesins: Principles and applications. Trends Biotech 1996; 14:52-60.
17. Winslow JW, Moran P, Valverde J et al. Cloning of AL-1, a ligand for an eph-related tyrosine kinase receptor involved in axon bundle formation. Neuron 1995; 14:973-981.
18. Gerlai R, McNamara A. Anesthesia induced retrograde amnesia is ameliorated by ephrinA5-IgG in mice: Evidence for Eph receptor tyrosine kinase involvement in mammalian memory. Behav Brain Res 2000; 108:133-143.
19. Kim JJ, Fanselow MS. Modality-specific retrograde amnesia of fear. Science 1992; 256:675-677.
20. Gerlai R. A new continuous alternation task in T-maze detects hippocampal dysfunction in mice: A strain comparison and lesion study. Behav Brain Res 1998a; 95:91-101.
21. Gerlai R. Contextual Learning and Cue Association in Fear Conditioning in Mice: A Strain Comparison and a Lesion Study. Behav Brain Res 1998b; 95:191-203.
22. Crusio WE, Bertholet JY, Schwegler H. No correlations between spatial and non-spatial reference memory in a T-maze task and hippocampal mossy fibre distribution in the mouse. Behav Brain Res 1990; 41:251-259.
23. Matsuyama S, Namgung U, Routtenberg A. Long-term potentiation persistence greater in C57BL/6 than DBA/2 mice: Predicted on basis of protein kinase C levels and learning performance. Brain Res 1997; 763:27-130.
24. Nguyen PV, Duffy SN, Young JZ. Differential maintenance and frequency-dependent tuning of LTP at hippocampal synapses of specific strains of inbred mice. Journal of Neurophysiology 2000; 84:2484-2493.
25. Hsueh Y-P, Sheng M. Eph receptors, ephrins, and PDZs gather in neuronal synapses. Neuron 1998; 21:1227-1229.
26. Calakos N, Scheller RH. Synaptic vesicle biogenesis, docking, and fusion: A molecular description. Physiol Rev 1996; 76:1-29.

27. Allison DW, Gelfand VI, Craig AM. Role of actin in anchoring postsynaptic receptors in cultured hippocampal neurons: Differential attachment of NMDA versus AMPA receptors. J Neurosci 1998; 18:2423-2436.
28. Edwards FA. LTP - a structural model to explain the inconsistencies. Trends Neurosci 1995; 18:250-255.
29. Geinisman Y, deToledo-Morrell L, Morrell F et al. Structural synaptic correlate of long-term potentiation: Formation of axospinous synapses with multiple, completely partitioned transmission zones. Hippocampus 1993; 3:435-446.
30. Geinisman Y, deToledo-Morrell L, Morrell F. Induction of long-term potentiation is associated with an increase in the number of axospinous synapses with segmented postsynaptic densities. Brain Res 1991; 566:77-88.
31. Fifkova E, Delay RJ. Cytoplasmic actin in neuronal processes as a possible mediator of synaptic plasticity. J Cell Biol 1982; 95:345-350.
32. Landis DM, Reese TS. Cytoplasmic organization in cerebellar dendritic spines. J Cell Biol 1983; 97:1169-1178.
33. Cohen RS, Chung SK, Pfaff DW. Immunocytochemical localization of actin in dendritic spines of the cerebral cortex using colloidal gold as a probe. Cell Mol Neurobiol 1985; 5:271-284.
34. Hayashi K, Ishikawa R, Ye LH et al. Modulatory role of drebrin on the cytoskeleton within dendritic spines in the rat cerebral cortex. J Neurosci 1996; 16:7161-7170.
35. Westrum LE, Jones DH, Gray EG et al. Microtubules, dendritic spines and spine appratuses. Cell Tissue Res 1980; 208:171-181.
36. Kim CH, Lisman JE. A role of actin filament in synaptic transmission and long-term potentiation. J Neurosci 1999; 19:4314-4324.
37. Mileusnic R, Rose SPR, Tillson P. Passive avoidance learning results in region-specific changes in concentration of and incorporation into colchicine-binding proteins in the chick forebrain. J Neurochem 1980; 34:1007-1015
38. Rose PRS. Synaptic Plasticity, Learning and Memory. In: Baudry M, Thompson RF, Davis JL, eds. Synaptic Plasticity. Cambridge, USA: MIT Press, 1993:209-229.
39. Scholey A, Bullock S, Rose SPR. Passive avoidance learning in the young chick results in time- and locus-specific elevations of alpha-tubulin immunoreactivity. Neurochem Int 1992; 21:343-350.
40. Zisch AH et al. Tyrosine phosphorylation of L1 family adhesion molecules: Implication of the Eph kinase Cek5. J Neurosci Res 1997; 47:655-665.
41. Luth A, Laurent JP, Figurov A et al. Hippocampal long-term potentiation and neural cell adhesion molecules L1 and NCAM. Nature 1944; 372:777-779.
42. Davy A, Gale NW, Murray EW et al. Compartmentalized signaling by GPI-anchored ephrin-A5 requires the fyn tyrosine kinase to regulate cellular adhesion. Genes Dev 1999; 13:3125-3135.
43. Murase S, Schuman EM. The role of cell adhesion molecules in synaptic plasticity and memory Current Opinion in Cell Biology. 1999; 11:549-553.
44. Ellis C, Kasmi F, Ganju P et al. A juxtamembrane autophosphorylation site in the Eph fanmily receptor tyrosine kinase, Sek, mediates high affinity interaction with p59fyn. Oncogene 1996; 12:1727-1736.
45. Grant SGN. Analysis of NMDA receptor mediated synaptic plasticity using gene targeting: Roles of Fyn and FAK non-receptor tyrosine kinases. Journal of Physiology 1996; 90:337-338.
46. Morris RGM, Anderson E, Lynch GS et al. Selective impairment of learning and blockade of long-term potentiation by a N-methyl-D-aspartate receptor antagonist, AP5. Nature 1986; 319:774-776.
47. Grant SGN, O'Dell TJ, Karl Kaet al. Impaired Long-Term Potentiation, Spatial Learning, and Hippocampal Development in fyn Mutant Mice. Science 1992; 258:1903-1910.
48. Yu X-M, Askalan R, Keil GJ et al. NMDA channel regulation by channel-associated protein tyrosine kinase Src. Science 1997; 275:674-678.
49. Lu YM, Roder JC, Davidow J et al. Src activation in the induction of long-term potentiation in CA1 hippocampal neurons. Science 1998; 279:1363-1367.
50. Bliss TVP, Collingridge GL. A synaptic model of memory: Long-term potentiation in the hippocampus. Nature 1993; 361:34-39.
51. Flood JF, Baker ML, Davis JL. Modulation of memory processing by glutamic acid receptor agonists and antagonists. Brain Research 1990; 521:197-202.
52. Izquierdo I, Medina J. Role of amygdala, hippocampus and entorhinal cortex in memory consolidation and expression. Brazilian J Med Biol Res 1984; 26:573-589.
53. Kim M, McGaugh JL. Effects of intra-amygdala injections of NMDA receptor antagonists on acquisition and retention of inhibitory avoidance. Brain Research 1992; 585:35-48.

54. Ungerer A, Mathis C, Melan C et al. The NMDA receptor antagonists, CPP and g-L-glutamyl-L-aspartate, selectively block post-training improvement of peformance in a Y-maze aviodance learning task. Brain Research 1991; 549:59-65.
55. Shimizu E, Tang Y-P, Rampon C et al. NMDA Receptor-Dependent Synaptic Reinforcement as a Crucial Process for Memory Consolidation. Science 2000; 290:1170-1174.
56. Dalva M, Takasu MA, Lin MZ et al. EphB receptors interact with NMDA receptors and regulate excitatory synapse formation. Cell 2000; 103:945-956.
57. Brückner K, Klein R. Signaling by Eph receptors and their ephrin ligands. Curr Opin Neurobiol 1998; 8:375-382.
58. Gerlai R. Gene targeting studies of mammalian behavior: Is it the mutation or the background genotype? Trends Neurosci. 1996a; 19:177-181.
59. Gerlai R. Gene targeting in Neuroscience: The systemic approach. Trends Neurosci 1996b; 19:188-189.
60. Gerlai R. Protein targeting: Altering receptor kinase function in the brain. Trends Neurosci 2000; 23:236-239.
61. Gerlai R, Cairns B, Van Bruggen N et al. Protein targeting in the analysis of learning and memory: A potential alternative approach to gene targeting Experimental Brain Research 1998a; 123:24-35.
62. Thoenen H. Neurotrophins and neuronal plasticity. Science 1995; 270:593-598.
63. Schuman E. Growth factors sculpt the synapse. Science 1997; 275:1277-1278.
64. McAllister AK, Katz LC, Lo DC. Neurotrophins and synaptic plasticity. Ann Rev Neurosci 1999; 22:295-318.
65. Lu B, Chow A. Neurotrophins and hippocampal transmission and plasticity. J Neurosci Res 1999; 58:76-87.
66. Black IB. Trophic regulation of synaptic plasticity. J Neurobiol 1999; 41:108-118.
67. Kang H et al Neurotrophins and time: Different roles of TrkB signaling in hippocampal long-term potentiation. Neuron 1997; 19:653-664.
68. Miranda JD, White LA, Marcillo AE et al. Induction of Eph B3 after spinal cord injury. Experimental Neurology 1999; 156:218-222.
69. Conover JC, Doetsch F, Garcia-Verdugo J-M et al. Disruption of Eph/ephrin signaling affects migration and proliferation in the adult subventricular zone. Nat Neurosci 2000; 3:1091-1097.
70. Xu B, Li S, Brown A et al. Regulation of Epileptogenesis and Kindling-Induced Mossy Fiber Sprouting by the Axon Guidance Molecules EphA Receptor and EphrinA Ligand. Mol Cell Neurosci 2003; 24:984-999.

Chapter 3.6

Corticosteroids

Carmen Sandi

Abstract

Glucocorticoid hormones, released from the adrenal glands, easily access the brain where they can affect neural structure and function through the binding to two types of intracellular receptors, the mineralocorticoid receptor (MR) and the glucocorticoid receptor (GR). Secretion of these steroids is activated by exposure to stressful situations, and growing evidence indicates that they can interact with the neurobiological mechanisms subserving memory formation. After a brief description of these hormones and their receptors, we will review the scientific literature questioning whether glucocorticoid release, during the processing of certain types of information, could play a role on the neural processes involved in long-term memory formation. Emphasis will be made on findings that have shown a differential role of the two corticosteroid receptors on cognitive function, with MRs involved in behavioural reactivity to novel situations, and GRs in the consolidation of the newly acquired information. Which could be the mediating mechanisms involved in glucocorticoid actions is one of the key questions to be addressed when dealing with the capacity of these hormones to modulate memory storage. Recent evidence suggesting that glucocorticoids could induce their memory effects, at least partially, by regulating expression and function of synaptic proteins (in particular, cell adhesion molecules) will be presented. Finally, the behavioural and neural outcomes induced by chronic exposure to hypercortisolemic situations—a field that has received increasing attention over the past decade—will be reviewed.

Glucocorticoid Hormones and Receptors

Glucocorticoids and the Hypothalamic-Pituitary-Adrenal (HPA) Axis

Glucocorticoids are a major subgroup of steroid hormones, which are produced by the adrenal cortex under the regulatory influence of the adrenocorticotropin hormone (ACTH). They are important elements of the hypothalamic-pituitary-adrenal (HPA) axis, a neuroendocrine circuit critically involved in the response to stress and emotions and in the maintenance of homeostasis in the organism (see Fig. 1). The parvocellular neurons of the paraventricular nucleus (PVN) in the hypothalamus are the endpoint that integrates inputs from different neurotransmitter systems throughout the brain, including influences from the prefrontal cortex, hippocampus, amygdala, and septum. These neurons secrete corticotrophin-releasing hormone (CRH) and arginine vasopressin (AVP), which stimulate the pituitary to release ACTH (for discussion of the roles of these peptides, see de Wied and Kovac, this book).

Therefore, glucocorticoids (cortisol being the major naturally occurring glucocorticoid in humans, and corticosterone in several other animal species, including rats, mice and chicks) are the final products of the HPA system which, under basal conditions, shows a pulsatile and circadian secretion of the different hormones involved. Under exposure to physical or psychological stress, the brain structures involved in the regulation of this circuit stimulate the PVN, which then triggers the chain of endocrine responses on the different components of the axis.

From Messengers to Molecules: Memories Are Made of These, edited by Gernot Riedel and Bettina Platt.

Figure 1. Schematic representation of the hypothalamus-pituitary-adrenocortical (HPA) axis. Corticotropin Releasing Hormone (CRH) is released from the hypothalamus and activates the pituitary, where it activates the secretion of Adrenocorticotropin (ACTH). When ACTH stimulates the adrenal cortex, glucocorticoid secretion is activated. Circulating glucocorticoids can then inhibit HPA axis, by inhibiting ACTH secretion at the pituitary and CRH secretion at the hypothalamus. In addition, glucocorticoids can get access to the brain and also inhibiting HPA axis activation through their binding to specific mineralocorticoid (MR) and glucocorticoid (GR) receptors in different brain areas (particularly the hippocampus and the frontal cortex).

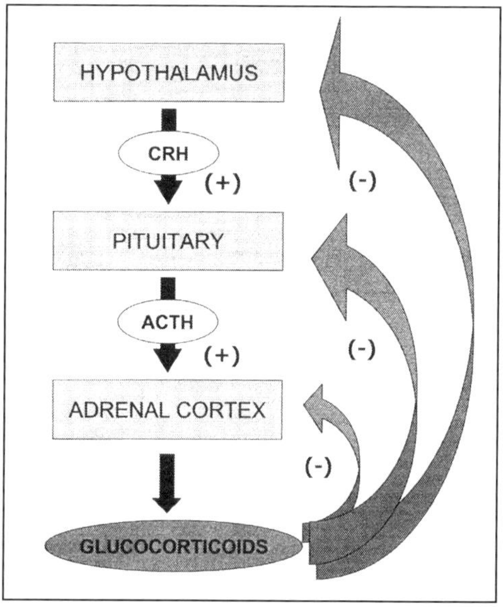

As important as this stress response is for adaptation and survival, as important is the termination of its activation, since sustained exposure to elevated levels of these hormones (particularly glucocorticoids) is well known to be highly deleterious – and potentially lethal- for the organism. Thus, glucocorticoids play a key role in the termination of their own release by inhibiting the secretion of ACTH and CRH at the level of the pituitary and hypothalamus, respectively, and also by interacting with other brain structures, among which the hippocampus plays a very important role.[30]

In addition to displaying a wide number of actions at different levels of the organism – including the regulation of glucose levels, blood pressure, and the immune response-, due to their lipophilic nature, glucocorticoids can readily enter the brain, where they affect neural function and behaviour mainly by interacting with two types of intracellular receptors: the glucocorticoid receptor (GR) and the mineralocorticoid receptor (MR). Although increasing evidence indicates that glucocorticoids can also exert rapid, non-genomic actions by interacting with different proteins of the cell membrane (including neurotransmitter receptors and putative non-genomic receptors for glucocorticoids),[37] this mechanism of action is still poorly understood. Here, we will focus on the classic corticosteroid genomic actions through the intracellular corticosteroid receptors.

Corticosteroid Receptors in the Brain

The intracellular corticosteroid receptors belong to the superfamily of nuclear hormone receptors. These are part of a cytoplasmic multiprotein complex, which, in addition to a receptor and several other molecules, involves heat shock proteins (hsp). When a corticosteroid hormone binds to a receptor, a conformational change is induced in the receptor molecule, which then leads to a cascade of events, including the dissociation of the receptor from the hsp complex, and the translocation of the receptor-ligand complex to the nucleus, where it can modulate gene transcription. Depending on a number of factors (such as the cellular context or specific physiological conditions), gene expression can either be activated or repressed, either through a direct interaction of the ligand-activated MR or GR with specific DNA sequences, or by the interaction of the activated receptor with other transcription factors, such as the activating protein (AP-1), the nuclear factor kB (NFKB), or the cAMP-response element-binding protein[12] (CREB, see Frankland and Josselyn, this book) (see Fig. 2).

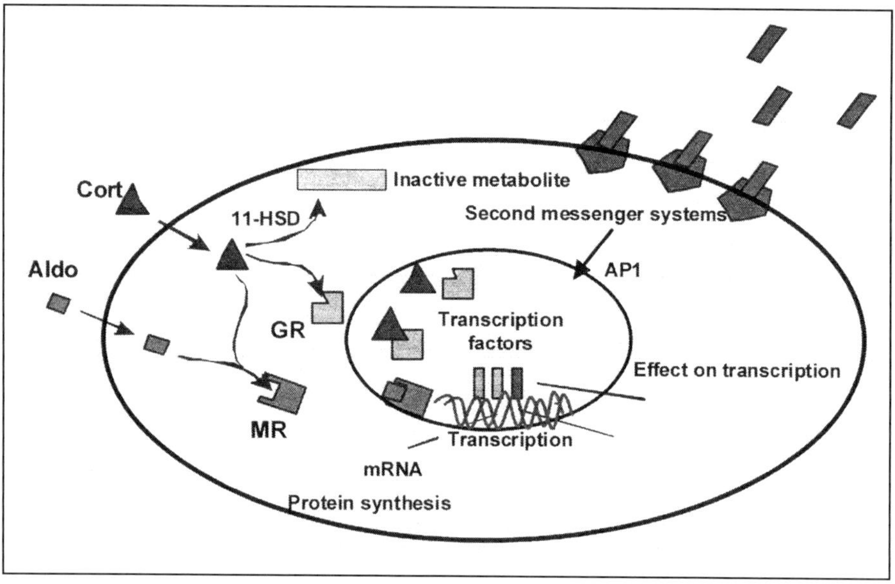

Figure 2. Genomic mechanisms of action of mineralocorticoid (MR) and glucocorticoid (GR) receptors. Due to their lipohilic nature, corticosteroids can easily cross cell membranes. When an agonist ligand (i.e., aldosterone or corticosterone) binds to the intracellular corticosteroid receptors, a chain of events is triggered, including the translocation of the hormone-receptor complex to the nucleus. There, they can either directly or indirectly (through the interaction with other transcription factors, such as AP1) modulate gene transcription, eventually facilitating or inhibiting the synthesis of specific proteins. Cort: corticosterone; Aldo: aldosterone; GR: glucocorticoid receptor; MR: mineralocorticoid receptor; 11-HSD: 11beta-Hydroxysteroid dehydrogenase.

MRs and GRs differ in their affinity to bind different ligands. In particular, affinity of MRs to bind corticosterone is approximately 10-fold higher than the affinity of GRs. As a consequence, at physiological conditions when corticosteroid levels are low (i.e., at the circadian trough at rest), whereas MRs are largely occupied (around 70-80%), GRs only show a low occupancy (around 10%). However, under situations of enhanced corticosteroid levels (i.e., under stressful circumstances), activation of GRs is considerably increased.

Both receptor types also differ in their respective distribution throughout the brain. Although they are co-localized in a number of brain structures involved in emotion and cognition, such as hippocampus, septum and amygdala, the GR has a much wider distribution in the brain, with its highest expression being observed in brain areas involved in the regulatory feedback of the HPA axis, including the pituitary, paraventricular nucleus of the hypothalamus, and the hippocampus. In the context of the present chapter, it is important to emphasize that a particularly high density of both MRs and GRs is found in hippocampal neurons.[24,25]

Hippocampal MRs have been implicated in the control of the inhibitory tone that the hippocampus exerts on the HPA axis,[52] as well as in the maintenance of neuronal excitability in the CA1 subfield.[26] Importantly, the expression of MRs in the hippocampus has been found to be rapidly upregulated by acute stress exposure, an effect which seems to potentiate the inhibitory tonus of these receptors on the activity of the HPA axis.[52] In contrast, GR action appears to be regulated by the hormone level. Thus, when corticosteroid levels are increased, their activation of GRs (in addition to MRs) has been associated with a facilitation of HPA activation and reduced neuronal activity in the hippocampus. A balance in corticosteroid actions mediated via MRs and GRs has been proposed to be critical for the control of homeostasis.[12]

Role of Glucocorticoids on Memory Consolidation

The idea that stress hormones, released during training experiences, can modulate the storage of information, was proposed after the observation that emotionally arousing experiences generally lead to stronger memories than more ordinary events.[20,22] A wide body of data has lead to recognise that peripheral catecholamines (adrenaline and noradrenaline), secreted as part of the stress response by the activation of the sympathetic nervous system, display important modulatory actions on learning and memory processes[21] (see also, Gibbs and Summers, this book). In addition, intensive research, involving a great variety of experimental approaches, has also indicated a key role of glucocorticoids in the storage of information.

The interest in glucocorticoid actions on cognitive processes is multiple. First, although glucocorticoids are peripheral hormones, their lipophilic nature allows them to readily cross the blood-brain barrier and to get access to the brain. Second, the high density of corticosteroid receptors expressed in brain areas involved in learning and memory, such as the hippocampus, septum, cerebral cortex and amygdala, denotes their key location to affect cognitive processing. Third, considerable evidence has shown that protein synthesis is required for long-term memory storage (Stork and Welzel, and also Mileusnic in this book for review). Given that the classic mechanism of corticosteroid action is to modulate gene transcription (with immediate effects on the synthesis of a number of proteins), this functional regulation might have important consequences both on the structural and functional characteristics of the nervous system, including the neurobiological processes involved in memory formation. Furthermore, it is nowadays well established that glucocorticoids affect numerous cellular and molecular events in brain cells,[12,38] the main substrate of behaviour and cognition.

In order to question whether these hormones could actually affect cognitive function, different approaches have been used to investigate the role of glucocorticoids on memory formation:

Manipulation of the Degree of Stress Involved in the Training Task

Some studies have evaluated to what extent the strength of a long-term memory could be related to the degree of stress involved in the training situation. One way to assess this question is to manipulate the intensity of the stressor used as the unconditioned stimulus (US) in a particular task, and to subsequently evaluate whether any relationship can be observed between posttraining corticosterone levels and the degree of memory displayed by the animals.

In training tasks in which the US is a footshock, it is the intensity of the shock which is generally varied. Thus, experiments performed in the contextual fear conditioning task, involving groups of rats that received different shock intensities (0.2, 0.4 and 1 mA), observed a direct relationship between the stressor intensity experienced at training and the level of freezing displayed by rats at the testing session. Besides, posttraining corticosterone levels showed a positive correlation with the strength at which fear conditioning is established into a long-term memory.[8] However, in the passive avoidance task, it has been reported that very high shock intensities, instead of resulting in potentiated memory, might have the opposite, inhibitory, effect on memory formation, an effect which seems to resemble the amnestic phenomenon associated to the experience of traumatic situations.

In the water maze task, a similar phenomenon has been described by manipulating the temperature of the pool water during the acquisition phase. Rats learning the task at a water temperature of 19 °C showed a greater retention of the platform location on the second day of training than rats trained at 25 °C. Again, a relationship was found between the strength of memory and corticosterone levels displayed by rats after the first training session, with rats trained on the experimental conditions that led to a stronger and longer-lasting memory (i.e., at 19 °C) showing the highest circulating hormone levels.[68]

Therefore, these studies suggest the existence of a correlational relationship between corticosterone secretion during training and the strength at which long-term memory is established. The following experimental approaches are complementary to this one, and were designed to evaluate the possible existence of a causal relationship between these two phenomena.

Inhibition of Glucocorticoid Secretion

A more direct way to question whether training-induced glucocorticoid secretion might play a role in memory formation is to interfere, around the time of training, either with hormone secretion, or with its neural action, and then to evaluate whether that might have any impact on later retention of the task.

Inhibition of hormone secretion can be accomplished through either surgical adrenalectomy or by injecting inhibitors of glucocorticoid synthesis, such as metyrapone or aminogluthetimide. In adrenalectomized animals, there is a total absence of circulating corticosterone levels. Training rats under such conditions has proved to impair memory formation for a number of tasks, including contextual fear conditioning[51] and water maze learning.[48,58]

Corticosteroid synthesis inhibitors induce a partial chemical adrenalectomy, causing a dose-dependent reduction of plasma corticosterone levels. Pretraining injection of glucocorticoid synthesis inhibitors has also been reported to interfere, in a dose-dependent manner, with the strength and duration of newly formed memories. Among other tasks, this effect has again been reported for the water maze[59] and contextual fear conditioning.[7] Interestingly, the effect in the fear conditioning paradigm, in addition to being dose-related was also dependent on the stressor intensity used during training. Whereas systemic administration of a metyrapone dose of 100 mg/kg impaired memory in animals trained at either 0.4 or 1.0 mA shock intensity, a lower dose of 50 mg/kg was only effective to decrease memory in the 0.4 mA condition, suggesting an interaction of the drug dose effectiveness with the endogenous corticosterone levels induced by the training experience.

Therefore, these experiments further supported the idea that training-induced corticosterone release plays an important role in the mechanisms that determine the strength of memories.

Inhibition of Glucocorticoid Action

Given that corticosterone is released from the adrenal glands, the question, then, is whether its effects are mediated via specific brain receptors or attributable to a peripheral action.

Several studies have addressed this question by administering selective MR or GR antagonists directly into the brain. In the water maze task, intracerebral injection of GR antagonists, either before or immediately after the first training session, decreased long-term retention of the platform location.[48,56] However, administration of a MR antagonist, although it slightly changed rats' searching pattern at training, failed to affect subsequent retention of the task. As for the contextual fear conditioning task, the intracerebroventricular (icv) pretraining injection of a GR (RU-38486), but not a MR (RU-28316), antagonist diminished subsequent retention of conditioned freezing in rats trained at an intermediate shock condition (0.4 mA), but failed to affect retention levels in rats trained at a high shock intensity (1 mA).[6] This could be either due to the fact that neural processes that mediate memory for drastic experiences might be under the influence of several physiological systems operating in a co-ordinated and redundant manner and, therefore, manipulating only one physiological system (such as MRs or GRs) might be insufficient to interfere with such a memory. However, the fact that we also found that a pretraining systemic injection of metyrapone (100 mg/kg) was effective to diminish the level of fear conditioning[7] (see above), suggests that the lack of effect observed with the icv administration of the GR antagonist (100 ng) might have also been due to the use of a dose of the compound not high enough to antagonise all the relevant receptors, even though it should be noted that the same dose has been shown to be sufficient to interfere with a number of behavioural processes in other studies.[48,28]

From these findings and data obtained in other learning tasks and animal species (see Fig. 3), quite different roles have been proposed for each receptor type on cognitive processes.[11,53,62] Thus, activation of MRs seems to be essential for sensory integration and response selection. However, GR activation has been more directly implicated in the mechanisms of memory consolidation.

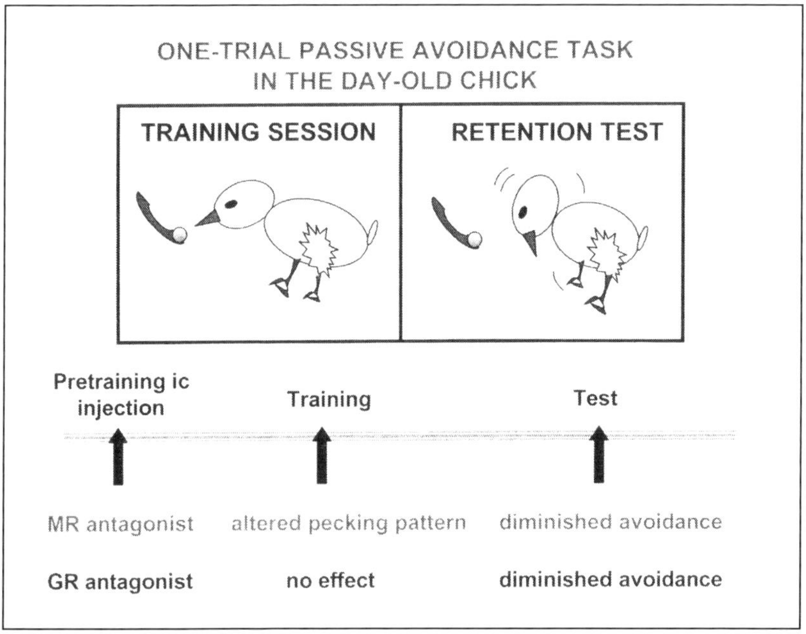

Figure 3. Schematic representation of the one-trial passive avoidance task in the day-old chick and the effects of the administration of mineralocorticoid (MR) and glucocorticoid (GR) receptor antagonists on memory for this task. On the training session, due to their innate tendency to peak salient objects in their environment, chicks peck at a bright bead which has been coated in an aversive substance. As a consequence, they display a disgust response and, when subsequently submitted to a retention test, they generally avoid pecking at a similar dry bead. Therefore, avoidance of the bead at test is an index of memory for the task. When chicks are intracerebrally (ic) injected with a MR antagonist prior to training, in addition to a reduction on avoidance at retention, they show an altered reactivity response during the training situation. However, when a GR antagonist is administered, only retrieval for the task is affected. Therefore, MRs are suggested to influence response activation at training, whereas GRs have been proposed to participate in the memory-facilitating mechanisms induced by glucocorticoids in the consolidation period. Based on data taken from refs. 66 and 67.

More recently, genetic tools developed to interfere with GR function have also provided further support for the role of GRs in memory formation as suggested by psychopharmacological studies. Thus, memory was impaired in experiments in which antisense oligonucleotides that prevented the synthesis of GR were injected in rats[28] and in 'knock out' mice of the genes encoding for the GR.[49]

Potentiation of Glucocorticoid Action

Another way to explore whether posttraining glucocorticoid levels could have an impact on the consolidation of newly acquired information is to assess whether the strength of a memory could be potentiated by increasing corticosterone levels during the posttraining period. This idea has been validated under experimental conditions inducing low to moderate levels of both learning and corticosterone release. Posttraining injections (systemic or central) of either corticosterone or synthetic corticosteroid agonists have proved to facilitate subsequent retention for a number of tasks, including passive avoidance,[18,54] brightness discrimination[42] contextual fear conditioning,[6] and water maze learning.[68]

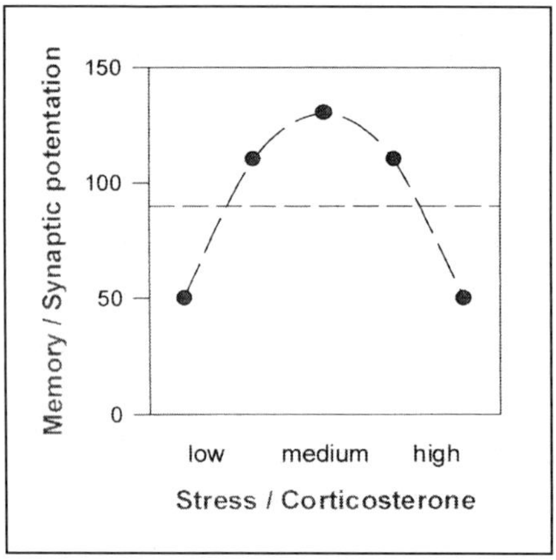

Figure 4. Memory or synaptic strength and corticosteroid activation. A wide body of data has suggested that conditions in which either very low (as found, for example, in adrenalectomized animals) or very high (as it happens, for example, under strong stressful circumstances) circulating levels of glucocorticoids impair memory formation or synaptic potentiation (such as LTP). However, both memory storage and synaptic strength seem to be facilitated under mild stressful conditions involving glucocorticoid levels able to partially activate GRs, in addition to most MRs.

However, a different picture can be observed when animals display high circulating levels of corticosteroids during the posttraining period, either due to the injection of high doses of the steroid, or to the interaction of moderate dose injections with considerably stressful training situations. Thus, in one-day old chicks, the same corticosterone dose that facilitated memory for weak training conditions on the passive avoidance task, impaired memory in chicks trained in a strong learning version of the task.[64] Similarly, whereas posttraining corticosterone injections enhanced consolidation processes in rats trained under water maze conditions involving intermediate stress levels (i.e., warm temperature),[68] administration of dexamethasone impaired memory under more stressful conditions of the same task.[58]

Therefore, although there is considerable evidence to propose that endogenously released glucocorticoids potentiate memory consolidation processes by acting through brain GRs during the posttraining period, the whole range of modulatory actions induced by these steroids on cognitive processes requires to be described by an inverted-U shape relationship (Fig. 4). Thus, whereas intermediate glucocorticoid levels leading to the partial activation of GRs (in addition to MRs occupancy) facilitate consolidation, impaired retention is generally observed in animals trained under conditions of either absence (lack of MR and GR occupancy) or excess (high occupancy of MRs and GRs) of circulating corticosteroid hormone.

It is interesting to note that electrophysiological studies have shown similar biphasic modulatory actions of glucocorticoids on synaptic plasticity. Expression of hippocampal long-term potentiation (LTP)—a long-lasting increase of synaptic efficacy induced by high-frequency electrical stimulation which appears to be relevant to memory—is dependent upon glucocorticoid levels. Whereas adrenalectomy and elevated glucocorticoid levels interfere with synaptic potentiation, both in vivo and in vitro,[13,50] intermediate glucocorticoid levels leading to partial occupation of GRs (in addition to MRs) lead to optimal expression of this phenomenon.[76] Therefore, it seems that whereas MR's activation facilitates LTP, GR-induced effects depend upon their degree of occupation. Thus, while facilitating effects would follow their weak or partial occupancy, their extensive occupation would inhibit the induction of this type of synaptic plasticity. Interestingly, inverted-U shape relationships have also been described for a number of cellular effecs of glucocorticoids, including calcium influx and responsiveness to transmitters.[12] In addition, it is interesting to mention that GR activation has been implicated in the facilitation of long-term depression (LTD), a physiological phenomenon opposite to LTP that consists on a weakening of synaptic efficacy.[50,81]

Neural Mechanisms Involved in Glucocorticoid Actions on Memory Consolidation

Increasing attention is being devoted to the study of the neurobiological mechanisms by which glucocorticoids affect the processes of memory consolidation. Here, we will address this question at two different levels of analysis, by firstly dealing with the brain structures that have been implicated in glucocorticoid actions and, then, reviewing the cellular and molecular mechanisms that might mediate their effects.

Brain Regions Implicated in Glucocorticoid Effects on Memory Consolidation

The hippocampus is one of the brain structures implicated in the facilitating effects of GR activation on consolidation. Whereas intra-hippocampal administration of corticosterone or synthetic GR agonists enhanced memory consolidation for different tasks, infusions of a GR antagonist induced the opposite, impairing, effect on the storage of spatial orientation learning in the water maze.[42,57] Although glucocorticoids could induce these facilitating effects by direct binding to hippocampal GRs, a regulatory influence of the amygdala has been proposed to be required for glucocorticoids to influence memory consolidation. In particular, the basolateral nucleus of the amygdala (BLA) has been implicated in the glucocorticoid-induced memory consolidation processes involving the hippocampus (Fig. 5).

In fact, studies involving a combination of amygdala lesions and systemic injections of glucocorticoids have led to the view that the BLA, but not the adjacent central nucleus of the amygdala (CEA), is critically involved in the memory-enhancing effects of posttraining glucocorticoids.[54,55] This critical participation of the BLA in mediating glucocorticoid effects in memory consolidation involve binding of corticosterone to GRs in this brain area, as shown in experiments in which (i) local administration of a selective GR agonist enhanced retention for the passive avoidance task when administered into the BLA, but was ineffective when infused into the CEA; (ii) administration of a GR antagonist into the BLA interfered with memory for spatial orientation in the water maze.[56]

In addition, the integrity of the amygdala β-adrenergic system seems to be required for these facilitating actions of glucocorticoid on cognition.[53]

Cellular and Molecular Mechanisms Involved on Glucocorticoid Effects on Memory Consolidation

Protein synthesis mechanisms have been reported to be required for the transfer of information into a long-term memory storage in a variety of learning tasks. Given that glucocorticoids can regulate (either facilitating or inhibiting) the synthesis of a large number of proteins, including several which have been critically implicated in neural plasticity[10,14] the possible involvement of a protein regulatory action on the modulatory actions of glucocorticoids on cognition is receiving increasing attention.[62]

Using the one-trial passive avoidance task in day-old chicks, a set of experiments was performed to evaluate whether the memory-facilitating effect that induces corticosterone in a weak training version of this learning model is dependent upon protein synthesis.[65] By injecting the protein synthesis inhibitor anisomicyn at different times with regards to training, it was found that this inhibitor was effective to reverse the facilitatory effect of the steroid on retention when injected up to 4-5 h after training, but not at later time points. Therefore, these results suggested that receptor-mediated changes in gene expression were involved in the facilitating effect of corticosterone on consolidation. One important question to address from these findings was, therefore, which type/s of proteins could be relevant in this context.

Evidence obtained in the same chick learning model suggested that a family of fucosylated glycoproteins could play a major role, not only in the mechanisms of memory formation, as previously described,[60] but also in the memory enhancement induced by glucocorticoids. Firstly,

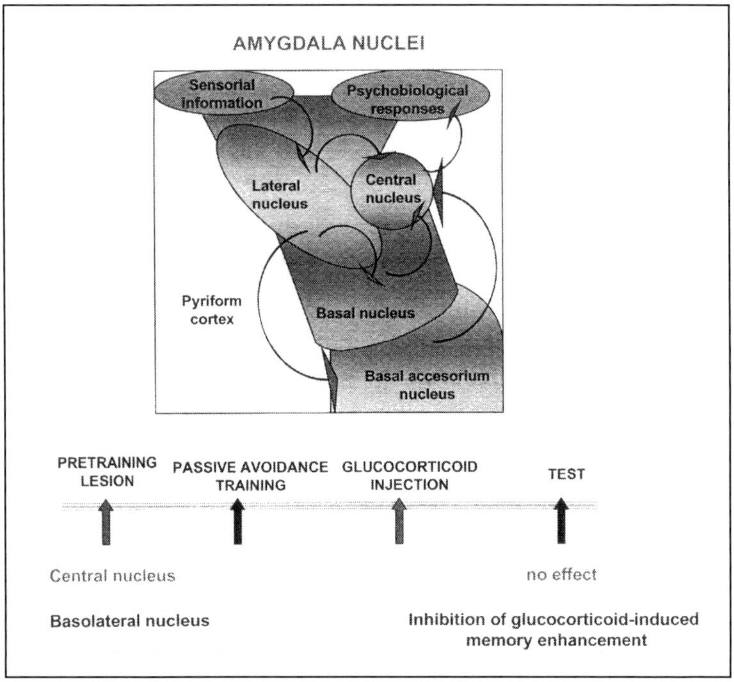

Figure 5. Role of amygdala nuclei in the memory-facilitating effect of glucocorticoids. Cartoon showing neural circuits involved in the communication among different amygdala nuclei. When the central nucleus is lesioned in rats before being exposed to a training session in the passive avoidance task, the enhancement of memory induced by a posttraining glucocorticoid injection is preserved. However, when it is the basolateral nucleus which is lesioned, the facilitation of memory induced by glucocorticoids is prevented. Therefore, the basolateral nucleus has been proposed to play a critical role on the memory-facilitating effects of glucocorticoids. Based on data taken from ref. 55.

biochemical experiments showed that the intracerebral injection of a corticosterone dose that facilitated the storage of the avoidance response, enhanced protein fucosylation in a brain region largely implicated in learning and memory in the chick.[70] Subsequent psychopharmacological studies showed that the administration of the fucosyl-glycoprotein synthesis inhibitor, 2-deoxygalactose (2-DG), prevented the facilitating effect of corticosterone on retention when injected 5.5 – 7.5 h posttraining, but was ineffective if injected at earlier or later time points,[65] therefore implicating glycoprotein fucosylation in the cognitive actions of the steroid. Interestingly, studies performed in rats also showed that corticosterone administration [at a dose that induces circulating stress levels of this steroid and facilitates consolidation of water maze learning (see above)] resulted in decreased glycoprotein expression in the hippocampus when evaluated 3 h post-injection, an effect which was suggested to be related to synaptic restructuring mechanisms.[79]

Among the different synaptic membrane glycoproteins, the cell adhesion molecules (CAMs) of the immunoglobulin superfamily (including the neural CAM –NCAM- and L1) have received particular attention in the search for the cellular and molecular mechanism of memory (see Fig. 6 and Regan, this book). CAMs are cell surface macromolecules that participate in target recognition and synapse stabilisation.[46] They have been largely implicated in cell-cell interactions during development of the nervous system[16] and in activity-dependent synaptic plasticity in adulthood,[17,46] including the synaptic changes underlying learning and memory

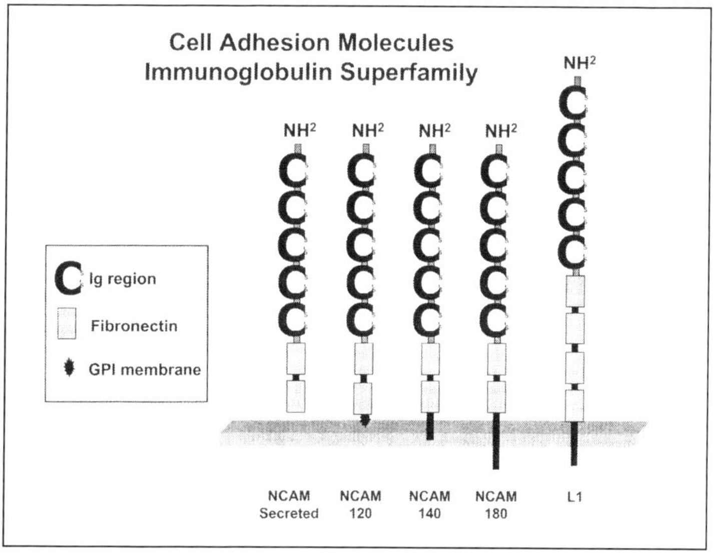

Figure 6. Schematic representation of the cell adhesion molecules of the immunoglobulin superfamily, NCAM and L1. These molecules are formed by five immunoglobulin domains and several fibronectin domains. NCAM is expressed in several isoforms which differ in their molecular weights and mode of attachment to the cytoplasmic membrane.

processes.[75] Furthermore, the post-translational modification of NCAM that consists in the addition of α-2,8-linked polysialic acid (PSA) homopolymers to its fifth immunoglobulin-like domain, by attenuating interactions mediated by NCAM and other related molecules,[61] provides another mechanism for structural plasticity. In fact, PSA-NCAM has also been implicated in memory formation[15,19,47] and synaptic plasticity.[2,45]

Interestingly, these CAMs seem to be implicated in corticosteroid actions in cognitive processes. Thus, corticosterone facilitation of memory formation in the day-old chick was shown to be inhibited by intracerebral administration of NCAM antibodies 5.5 h posttraining.[70] In rats, an acute corticosterone injection, although not affecting NCAM levels in the hippocampus, significantly enhanced NCAM expression in frontal cortical areas when evaluated 8 and 24 h post-injection.[63]

Interestingly, different effects were found when rats were submitted to a training experience involving different stressor intensities (and, as noted above, inducing different posttraining corticosterone levels), as shown for the contextual fear conditioning paradigm. As opposed to the lack of effect induced by the single injection of corticosterone in the hippocampus, rats trained in this fear conditioning task (at either 0.2, 0.4, or 1 mA shock intensity) showed a marked regulation of hippocampal CAMs which was dependent upon time and stressor intensity.[41] At 12 h post-training, conditioned animals displayed reduced NCAM, but increased L1, expression. The group trained at the highest shock intensity (1 mA) also presented decreased PSA-NCAM expression. However, at 24 h posttraining, the 1 mA group exhibited increased NCAM and L1 expression, but decreased expression of PSA-NCAM levels. The pattern of CAMs expression found in the 1 mA group (which is the one that shows higher posttraining corticosterone levels and develops the stronger and longer lasting levels of fear conditioning) supports the view that, after a first phase of synaptic de-adherence during consolidation, NCAM and L1 might participate in the stabilization of selected synapses underlying the establishment of long-term memory for contextual fear conditioning. They also suggest that glucocorticoids might play a role in the observed regulation of CAMs.

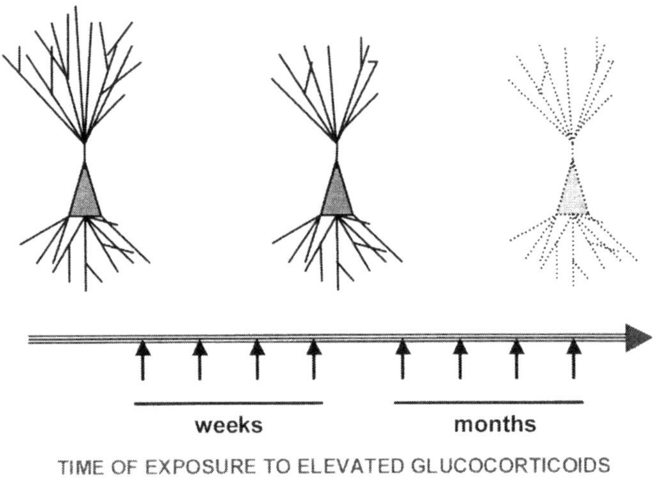

Figure 7. Schematic representation of the effect of sustained exposure to elevated levels of glucocorticoids on hippocampal CA3 neurons. Animal studies have shown that pyramidal CA3 neurons undergo a time-dependent neurodegeneration process when expossed to excessive glucocorticoids or to sustained stress. When these conditions are mantained for 3-4 weeks, a reversible atrophy of apical dendrites is observed. However, if they last for longer time periods or occur in association with other neural challenges, an irreversible cell death could happen. Based on references cited by ref. 38.

In addition to glycoprotein regulation, evidence suggests that glucocorticoids could also exert their effects on learning and memory processes by their reported modulatory actions on other cellular and molecular factors implicated in cell function and survival, such as growth and neurotrophic factors.[4,74]

Furthermore, glucocorticoids can increase glutamate concentrations in the hippocampus as well as in other brain regions,[31,44] and given that glutamate has been largely involved in the mechanisms of memory formation (see Riedel et al., this book), a regulatory action of corticosterone through glutamatergic mechanisms has also been proposed.[78]

Effects of Chronic Exposure to Elevated Glucocorticoid Levels on Cognitive and Neural Function

Although the physiological responses to stress trigger a chain of reactions in the body to promote adaptation to the changing circumstances, it is certainly critical for the system to have an efficient mechanism to restrain these defense reactions to stress. However, when this machinery is damaged (which could happen, for example, when the activation of stress systems is either excessive or maintained chronically), the individual become more prone to develop different psychological and psychiatric disturbances. Conversely, such disturbances are generally associated with an inability to develop adaptive responses under challenging circumstances which, in turn, can potentiate the stress responses of the individual.

Neural Consequences of Chronic Glucocorticoid Exposure

Chronic exposure to either exogenous or stress-induced endogenous glucocorticoids has been associated with deficits in learning, memory and retrieval. These effects have mainly been related to the finding that chronic exposure to high levels of glucocorticoids can lead to atrophy of the hippocampus, both in animals and humans.[38,33] In animals, vulnerability of hip-

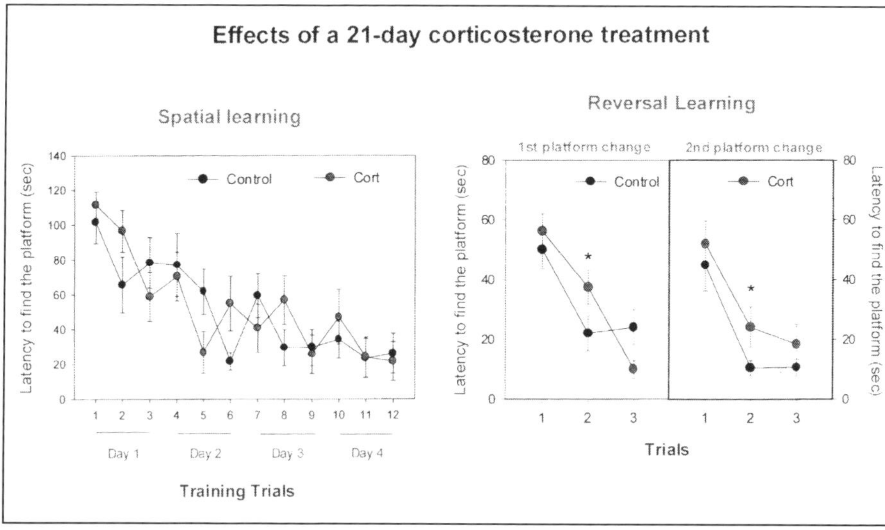

Figure 8. Effects of chronic exposure to high glucocorticoid levels on spatial orientation learning and reversal learning. The results show that animals submitted to a corticosterone (Cort) treatment for 21 days did not differ from controls in their spatial learning abilities in the Morris water maze. However, when they were subsequently trained to find the platform in other locations of the pool, a consistent deficit was observed in corticosteroid-treated animals, which indicates a detrimental effect of the treatment on reversal learning abilities.

pocampal neurons to a number of different insults was shown to be increased under chronic hypercortisolemia.[72] Besides, although there has been some controversy as to whether prolonged exposure to stress or to high glucocorticoids levels could eventually result in neuronal loss, there is a general agreement that these treatments can result in more subtle structural changes in hippocampal neurons and, particularly, in the CA3 subregion. Thus, an atrophy of apical dendrites of CA3 pyramidal neurons –the site of mossy fiber projection- has been consistently found after 3-4 weeks of chronic stress or corticosterone treatments, both in rats[35,77,80] and primitive primates.[36] In addition, atrophy has also been found in granule and CA1 pyramidal cells[77,80] (see Fig. 7).

This glucocorticoid-induced neural damage has attracted considerable attention for the critical role of the hippocampus in cognitive function and evidence which supports, in humans, the same type of interactions between chronic glucocorticoid exposure and hippocampal morphology.[34,71] In addition, high levels of glucocorticoids have been hypothesised to accelerate brain ageing,[29,73] which is frequently associated with learning and memory impairments.[40] However, it should be noted that besides these changes in hippocampal structure, the possibility that other brain areas can be functionally altered by chronic stress and/or glucocorticoid administration has been suggested by several neurochemical and biochemical studies.[23,43,63,79]

Cognitive Consequences of Chronic Glucocorticoid Exposure

Although the initial reports suggested that sustained exposure to stress or high glucocorticoid levels would result in cognitive impairments, the current picture suggest that the outcome depends on a number of factors, including the duration of the temporal exposure to these treatments, the type of demands involved in the cognitive tasks and individual differences in the vulnerability to develop alterations in hippocampal structure and function. Thus, whereas exposure to these treatments for around 3 months or longer seems to impair the acquisition of spatial learning in a variety of mazes,[3,9,39] shorter exposure (for 3 weeks) of rats to stress or corticosterone regimes (which have been shown to induce reversible atrophy of pyramidal CA3 neurons as described above), hardly affected this type of learning.[1,3,32] However, a cognitive deficit for these 3-week treatments is detected when rats are evaluated in tasks that require to develop a plastic strategy, such as in reversal learning paradigms (Fig. 8).

A surprising effect has been reported for the effect of a 3-week restraint stress regime on another hippocampus-dependent paradigm, the contextual fear conditioning task. Given that the hippocampus has been implicated to play a significant role in this learning task,[27] a decreased conditioning was expected as the outcome of chronic stress. However, chronically stressed rats developed enhanced conditioning, not only when trained at an intermediate shock condition (0.4 mA),[5] but also when trained at a high shock intensity (1 mA), an experimental condition that, by itself, leads to considerably high levels of conditioned freezing.[69] Since morphological experiments showed that the potentiation of fear conditioning was also observed in stressed rats treated with daily injections of tianeptine (a tricyclic antidepressant that facilitates serotonin reuptake and prevents the development of hippocampal atrophy), the hippocampal alterations induced by chronic stress do not seem to be involved in the conditioning enhancement. It is quite possible that other brain regions, known to be neurochemically affected by this type of chronic regimes (particularly the amygdala, but also the prefrontal cortex or other cortical areas), are implicated in the observed potentiation of fear conditioning.

Therefore, chronic exposure to high glucocorticoid levels predispose individuals to develop enhanced fear conditioning responses, impaired acquisition of spatial learning and reduced behavioural flexibility, a variety of behavioural alterations that accompany and/or underlie many psychopathological disorders. However, it is also important to mention that recent findings indicate that individuals differ in their vulnerability to develop stress-induced cognitive alterations, a phenomenon that in rats seems to be related to the behavioural trait of reactivity to novelty, as well as to the type of demand involved in the stressful situation. Thus, whereas rats characterised by a hightened locomotor response (HR) in a novel environment show learning impairments after exposure to a social stress regime, the low reactive rats (LR) appear to be more affected when submitted to a restraint stress regime (Touyarot, Venero and Sandi, unpublished observations).

Conclusion

Evidence available to date indicates that glucocorticoids can exert profound effects on cognitive and neural function. However, these effects are varied and complex and, therefore, it is not possible to simplify their outcome. In general terms, the evidence discussed favors the hypothesis that GR activation induced by corticosterone released during the processing of information can contribute to the strength of newly formed memories. However, although we cannot conclude that exposure to chronic glucocorticoid elevations impairs performance in all tasks that require the integrity of the hippocampus, the behavioural alterations observed after chronic stress treatments suggest that their constellation of effects could be related to a number of psychopathological disorders. The recent observations of individual differences in the vulnerability to this type of disturbances after chronic stress supports the interest to develop future investigations addressed to find possible behavioural and physiological markers efficient to predict vulnerability of particular individuals to show cognitive alterations after sustained exposure to stressful situations.

References

1. Bardgett ME, Newcomer JW, Taylor GT. The effect of chronic corticosterone on memory performance in the platform maze task. Physiol Behav 1996; 59:1111-1115.
2. Becker CG, Artola A, Gerardy-Schahn R et al. The polysialic acid modification of the neural cell adhesion molecule is involved in spatial learning and hippocampal long-term potentiation. J Neurosci Res 1996; 45:143-152.
3. Bodnoff SR, Humphreys AG, Lehman JC et al. Enduring effects of chronic corticosterone treatment on spatial learning, synaptic plasticity, and hippocampal neuropathology in young and mid-aged rats. J Neurosci 1995; 15:61-69.

4. Chao HM, McEwen BS. Glucocorticoids and the expression of mRNA for neurotrophins, their receptors and GAP-43 in the rat hippocampus. Mol Brain Res 1994; 26:271-276.
5. Conrad CD, Magariños AM, LeDoux JE et al. Repeated restraint stress facilitates fear conditioning independently of causing hippocampal CA3 dendritic atrophy. Behav Neurosci 1999; 113:902-913.
6. Cordero MI, Sandi C. A role for brain glucocorticoid receptors in contextual fear conditioning: dependence upon training intensity. Brain Res 1998; 786:11-17.
7. Cordero MI, Kruyt ND, Merino JJ et al. Glucocorticoid involvement on memory formation in a rat model for traumatic memory. Stress 2002; 5:73-79.
8. Cordero MI, Merino JJ, Sandi C. Correlational relationship between shock intensity and corticosterone secretion on the establishement and subsequent expression of contextual fear conditioning. Behav Neurosci 1998; 111:885-891.
9. Dachir S, Kadar T, Robinson B et al. Cognitive deficits induced in young rats by long-term corticosterone administration. Behav Neural Biol 1993; 60:103-109.
10. Datson NA, van Der Perk J, de Kloet ER et al. Identification of corticosteroid-responsive genes in rat hippocampus using serial analysis of gene expression. Eur J Neurosci 2001; 14:675-89.
11. de Kloet ER, Oitzl MS, Joëls M. Stress and cognition: are corticosteroids good or bad guys? Trends Neurosci 1999; 22:422-426.
12. de Kloet ER, Vreugdenhil E, Oitzl MS et al. Brain corticosteroid receptor balance in health and disease. Endocr Rev 1998; 19:269-301.
13. Diamond DM, Bennet MC, Fleshner M et al. Inverted-U relationship between the level of peripheral corticosterone and the magnitude of hippocampal primed burst potentiation. Hippocampus 1992; 2:421-430.
14. Dokas LA, Schlatter LK, Barr CS. Corticosteroid-induced proteins in brain. Ann NY Acad Sci 1994; 746:157-163.
15. Doyle E, Nolan PM, Bell R et al. Hippocampal NCAM180 transiently increases sialylation during the acquisition and consolidation of a passive avoidance response in the adult rat. J Neurosci Res 1992; 31:513-523.
16. Edelman GM, Jones FS. Gene regulation of cell adhesion: a key step in neural morphogenesis. Brain Res Rev 1998; 26:337-352.
17. Fields RD, Itoh K. Neural cell adhesion molecules in activity-dependent development and synaptic plasticity. Trends Neurosci 1996; 19:473-480.
18. Flood JF, Vidal D, Bennet EL et al. Memory facilitating and anti-amnestic effects of corticosteroids. Pharmacol Biochem Behav 1978; 8:81-87.
19. Fox GB, O'Connell AW, Murphy KJ et al. Memory consolidation induces a transient and time-dependent increase in the frequency of neural cell adhesion molecule polysialylated cells in the adult rat hippocampus. J Neurochem 1995; 65:2796-2799.
20. Gold PE, Buskirk R. Facilitation of time-dependent memory processes with posttrial amygdala stimulation: effect on memory varies with footshock level. Brain Res 1975; 89:509-513.
21. Gold PE, McCarty RC. Stress regulation of memory processes. Role of peripheral catecholamines and glucose. In: Friedman MJ, Charney DS, Deutch AY, eds. Neurobiological and Clinical Consequences of Stress: From Normal Adaptation to PTSD. Philadelphia: Lippincott-Raven Publishers, 1995:151-162.
22. Gold PE, McGaugh JL. A single-trace, two process view of memory storage processes. In: Deutsch D, Deutsch JA, eds. Short-Term Memory. New York: Academic Press, 1975:355-378.
23. Grootendorst J, Oitzl MS, Dalm S et al. Stress alleviates reduced expression of cell adhesion molecules NCAM, L1 and deficits in learning and corticosterone regulation of apolipoprotein E knockout mice. Eur J Neurosci 2001; 14:1505-1514.
24. Joëls M, de Kloet ER. Control of neuronal excitability by corticosteroid hormones. Trends Neurosci 1992; 15:25-30.
25. Joëls M, de Kloet ER. Mineralocorticoid and glucocorticoid receptors in the brain. Implications of ion permeability and transmitter systems. Prog Neurobiol 1994; 43:1-36.
26. Joëls M, de Kloet ER. Effects of glucocorticoids and norepinephrine in the excitability in the hippocampus. Science 1989; 245:1502-1505.
27. Kim JJ, Fanselow M. Modality-specific retrograde amnesia of fear. Science 1992; 256:675-676.
28. Korte SM, de Kloet ER, Buwalda B et al. Antisense to the glucocorticoid receptor in hippocampal dentate gyrus reduces immobility in forced swim test. Eur J Pharmacol 1996; 301:19-25.
29. Landfield PW, Baskin E, Pitler T. Brain aging correlates: retardation by hormonal-pharmacological treatments. Science 1981; 214:581-584.
30. López JF, Akil H, Watson SJ. Role of biological and psychological factors in early development and their impact on adult life. Neural circuits mediating stress. Biol Psychiatry 1999; 46:1461-1471.

31. Lowy MT, Wittenberg L, Novotney S. Adrenalectomy attenuates stress-induced elevations in extracellular glutamate concentrations in the hippocampus. J Neurochem 1993; 61:1957-1960.
32. Luine VN, Spencer RL, McEwen BS. Effects of chronic corticosterone ingestion on spatial memory performance and hippocampal serotonergic function. Brain Res 1993; 616:65-70.
33. Lupien SJ, Lepage M. Stress, memory, and the hippocampus: can't live with it, can't live without it. Behav. Brain Res 2001; 127:137-158.
34. Lupien SJ, de Leon M, de Santi S et al. Cortisol levels during human aging predict hippocampal atrophy and memory deficits. Nature Neurosci 1998; 1:69-73.
35. Magariños AM, McEwen BS. Stress-induced atrophy of apical dendrites of hippocampal CA3c neurons: involvement of glucocorticoid secretion and excitatory amino acid receptors. Neuroscience 1995; 69:89-98.
36. Magariños AM, McEwen BS, Flugge G et al. Chronic psychosocial stress causes apical dendritic atrophy of hippocampal CA3 pyramidal neurons in subordinate tree shrews. J Neurosci 1996; 16: 3534-3540.
37. Makara G, Haller J. Non-genomic effects of glucocorticoids in the neural system. Evidence, mechanisms and implications. Progr Neurobiol 2001; 65:367-390.
38. McEwen BS. Stress and hippocampal plasticity. Annu Rev Neurosci 1999; 22:105-122.
39. McLay RN, Freeman SM, Zadina JE. Chronic corticosterone impairs memory performance in the Barnes maze. Physiol Behav 1998; 63:933-937.
40. McLay RN, Freeman SM, Harlan RE et al. Aging in the hippocampus: Interrelated actions of neurotrophins and glucocorticoids. Neurosci Biobehav Rev 1997; 1:615-629.
41. Merino JJ, Cordero MI, Sandi C. Hippocampal regulation of the cell adhesion molecules NCAM and L1 by contextual fear conditioning is dependent upon time and stressor intensity. Eur J Neurosci 2000; 12:3283-3290.
42. Micheau J, Destrade C, Soumireu-Mourat B. Time-dependent effects of posttraining intrahippocampal injections of corticosterone on retention of appetitive learning tasks in mice. Eur J Pharmacol 1985; 106:39-46.
43. Mizoguchi K, Yuzurihara M, Ishige A et al. Chronic stress induces impairment of spatial working memory because of prefrontal dopaminergic dysfunction. J Neurosci 2000; 20:1568-1574.
44. Moghaddam B, Molinao ML, Stein-Behrens BA et al. Glucocorticoids mediate the stress-induced extracellular accumulation of glutamate. Brain Res 1994; 655:251-254.
45. Muller D, Wang C, Skibo G et al. PSA-NCAM is required for activity-induced synaptic plasticity. Neuron 1996; 17:413-422.
46. Murase S, Schuman, EM. The role of cell adhesion molecules in synaptic plasticity and memory. Curr Op Cell Biol 1999; 11:549-553.
47. Murphy KJ, O'Connell AW, Regan CM. Repetitive and transient increases in hippocampal neural cell adhesion molecule polysialylation state following multi-trial spatial training. J Neurochem 1996; 67:1268-1274.
48. Oitzl M, de Kloet ER. Selective corticosteroid antagonists modulate specific aspects of spatial orientation learning. Behav Neurosci 1992; 106:62-71.
49. Oitzl MS, de Kloet ER, Joëls M et al. Spatial learning deficits in mice with a targeted glucocorticoid receptor gene disruption. Eur J Neurosci 1998; 9:2284-2296.
50. Pavlides C, Watanabe Y, Magariños AM et al. Oppossing roles of type I and type II adrenal steroid receptors in hippocampal long-term potentiation. Neuroscience 1995; 68:387-394.
51. Pugh CR, Tremblay D, Fleshner M et al. A selective role for corticosterone in contextual fear conditioning. Behav Neurosci 1997; 111:503-511.
52. Reul JMHM, Gesing A, Droste S et al. The brain mineralocorticoid receptor: greedy for ligand, mysterious in function. Eur J Pharmacol 2000; 405:235-249.
53. Roozendaal B. Glucocorticoids and the regulation of memory consolidation. Psychoneuroendocrinol 2000; 25:213-238.
54. Roozendaal B, McGaugh JL. Amygdaloid nuclei lesions differentially affect glucocorticoid-induced memory enhancement in an inhibitory avoidance task. Neurobiol Learn Mem 1996; 65:1-8.
55. Roozendaal B, McGaugh JL. The memory-modulatory effect of glucocorticoids depend on an intact stria terminalis. Brain Res 1996; 709:243-250.
56. Roozendaal B, McGaugh JL. Glucocorticoid receptor agonist and antagonist administration into the basolateral but not central amygdala modulates memory storage. Neurobiol Learn Mem 1997; 67:176-179.
57. Roozendaal B, McGaugh JL. Basolateral amygdala lesions block the memory-enhancing effect of glucocorticoid administration in the dorsal hippocampus of rats. Eur J Neurosci 1997; 9:76-83.
58. Roozendaal B, Bohus B, McGaugh JL. Dose-dependent suppression of adreno-cortical activity with metyrapone: effects on emotion and learning. Psychoneuroendocrinol 1996; 21:681-693.

59. Roozendaal B, Cahill L, McGaugh JL. Interaction of amotionally activated neuromodulatory systems in regulating memory storage. In: Ishikawa K, McGaugh JL, Sakata H, eds. Brain Processes and Memory. Amsterdam: Elsevier, 1996:39-54.
60. Rose SPR. How chicks make memories: the cellular cascade from c-fos to dendritic remodelling. Trends Neurosci 1991; 14:390-397.
61. Rougon G. Structure, metabolism and cell biology of polysialic acid. Eur J Cell Biol 1993; 61:197-207.
62. Sandi C. Role and mechanisms of action of glucocorticoids in memory formation. Neural Plasticity 1998; 6:39-49.
63. Sandi C, Loscertales M. Opposite effects on NCAM expression in the rat frontal cortex induced by acute vs. chronic corticosterone treatments. Brain Res 1999; 828:127-134.
64. Sandi C, Rose SPR. Training-dependent biphasic effects of corticosterone in memory formation for a passive avoidance learning task in chicks. Psychopharmacology 1997; 133:152-160.
65. Sandi C, Rose SPR. Protein synthesis- and fucosylation-dependent mechanisms in corticosterone facilitation of long-term memory in the chick. Behav Neurosci 1997; 111:1098-1104.
66. Sandi C, Rose SPR. Corticosteroid receptor antagonists are amnestic for passive avoidance learning in day-old chicks. Eur J Neurosci 1994; 6:1292-1297.
67. Sandi, C, Rose SPR. Corticosterone enhances long-term retention in one-day old chicks trained in a weak passive avoidance learning paradigm. Brain Res 1994; 647:106-112.
68. Sandi C, Loscertales M, Guaza C. Experience-dependent facilitating effect of corticosterone on spatial memory formation in the water maze. Eur J Neurosci 1997; 9:637-642.
69. Sandi C, Merino JJ, Cordero MI et al. Effects of chronic stress on contextual fear conditioning and the hippocampal expression of the neural cell adhesion molecule, its polysialylation, and L1. Neuroscience 2001; 102:329-339.
70. Sandi C, Rose SPR, Mileusnic R et al. Corticosterone facilitates long-term memory formation via enhanced glycoprotein synthesis. Neuroscience 1995; 69:1087-1093.
71. Sapolsky RM. Stress, glucocorticoids, and damage to the nervous system: The current state of confusion. Stress 1996; 1:1-29.
72. Sapolsky R. Stress, glucocorticoids and their adverse neurological effects: Relevance to aging. Exp Gerontol 1999; 3:721-732.
73. Sapolsky RM, Krey LC, McEwen BS. Prolonged glucocorticoid exposure reduces hippocampal neuron number: Implications for aging. J Neurosci 1985; 5:1221-1226.
74. Schaaf MJM, Sibug RM, Duurland R et al. Corticosterone effects on BDNF mRNA expression in the rat hippocampus during Morris Water Maze training. Stress 1999; 3:173-183
75. Schachner M. Neural recognition molecules and synaptic plasticity. Curr Op Cell Biol 1997; 9:627-634.
76. Seidenbecher T, Reymann KG, Balschun D. A post-tetanic time window for the reinforcement of long-term potentiation by appetitive and aversive stimuli. Proc Natl Acad Sci USA 1997; 94:1494-1499.
77. Sousa N, Lukoyanov NV, Madeira MD et al. Reorganization of the morphology of hippocampal neurites and synapses after stress-induced damage correlates with behavioral improvement. Neuroscience 2000; 97:253-266.
78. Venero C, Sandi C. Effects of NMDA and AMPA receptor antagonists on corticosterone facilitation of long-term memory in the chick. Eur J Neurosci 1997; 9:1923-1928.
79. Venero C, Guaza C, Sandi C. Regional and temporal modulation of brain glycoprotein synthesis by corticosterone. NeuroReport 1996; 7:2819-2822.
80. Wooley C, Gould E, McEwen BS. Exposure to excess glucocorticoids alters dendritic morphology of adult hippocampal pyramidal neurons. Brain Res 1990; 531:225-231.
81. Xu L, Anwyl R, Rowan MR. Spatial exploration induces a persistent reversal of long-term potentiation in rat hippocampus. Nature 1998; 394:497-500.

CHAPTER 4.1

Adenylyl Cyclases

Nicole Mons and Jean-Louis Guillou

Abstract

Although a number of signal transduction pathways have been implicated in short- and long-term adaptative changes in neuronal plasticity and memory formation, there is increasing evidence that cross-talk between the cAMP- and Ca^{2+}-regulatory pathways may play a pivotal role in learning and memory processes. The fact that adenylyl cyclases (AC), in both invertebrates and mammals, are potentially subject to a wide range of influences has given rise to the notion that they can act as molecular coincidence detectors which are able of yielding a unique integrated response when simultaneously exposed to multiple stimuli. In this review, we discuss the role of AC in the molecular mechanisms underlying the induction and/or expression of memory in various organisms that perform different behavioral tasks, ranging from studies of implicit memory for the acquisition of fear, such as behavioral sensitization in *Aplysia* or classical conditioning in *Drosophila,* to explicit forms of long-term memory (LTM) and synaptic plasticity in the rodent brain.

Introduction

Intracellular adenosine 3',5'-cyclic monophosphate (cAMP) is generated from ATP in a reaction catalyzed by adenylyl cyclase (AC) in response to a variety of extracellular signals, such as hormones, neurotransmitters, and other regulatory molecules, via the activation of specific receptors. The cAMP then propagates the hormone signal either by stimulating cAMP-dependent protein kinase (PKA) or directly by inducing protein-protein interactions independently of any phosphorylation (for a review see ref. 74). The subsequent activation of PKA controls multiple cell functions, including metabolism, cell growth, differentiation, ion channel activity, synaptic transmission, gene transcription and memory formation (for a review see refs. 61,118). A major surprise to emerge from the cloning and expression of the mammalian AC family is that most, if not all, ACs are potentially subject to dynamic control by multiple regulatory influences and that distinct coincident signals can be translated into a unique integrated response. In addition to their capability to respond via either Gi_{α} or Gs_{α} subunits, the activity of particular AC can also be regulated either directly or indirectly by a variety of signals, including cytosolic calcium ions ($[Ca^{2+}]_i$), protein kinase C (PKC) and $\beta\gamma$ subunits of G proteins (for a review see refs. 37,79,131). The fact that ACs are subject to this wide range of influences has given rise to the notion that they can act as «coincident signal detectors», which are capable of yielding a unique response when simultaneously exposed to multiple regulatory cues. In addition to interacting with various signalling pathways, increasing evidence indicates that the various ACs show distinct cellular distributions and subcellular compartmentation. Indeed, the role and specificity of the cAMP/PKA-signaling pathway may be critical in the regulation of synaptic functions by virtue of this restricted synaptic compartmentation. In particular, the selective targeting of both AC and PKA to discrete subcellular localizations via interaction with specific anchoring proteins, in juxtaposition with other Ca^{2+}-regulated signaling molecules (mainly

From Messengers to Molecules: Memories Are Made of These, edited by Gernot Riedel and Bettina Platt.

kinases and phosphatases) optimizes the reception and propagation of the signal carried by cAMP, such as those required for the establishment of learning and memory.

In this review, we will first outline the essential evidence implicating the cAMP/AC/PKA pathway in short- and long-term sensitization of the siphon and gill withdrawal reflex (GSWR) in *Aplysia*. We will examine the associative learning defects in the fruit fly, *Drosophila melanogaster*, caused by mutational perturbations of the cAMP cascade, focusing on the genes *rutabaga* (rut) and *dunce* (dnc), which encode a Ca^{2+}/calmodulin (CaM)-responsive AC and a cAMP-specific phosphodiesterase, respectively. We will then briefly survey current evidence that mammalian AC isoforms are uniquely regulated by a variety of influences and are spatially organized for integrating coincident cellular signals and thus, modulate local regulatory components subserving early and late memory processes. We will then proceed to outline recent data implicating mammalian Ca^{2+}/CaM-sensitive and -insensitive ACs as molecular coincidence detectors subserving synaptic function and memory formation. Particular attention will be given on the recent genetic studies demonstrating that Ca^{2+}/CaM-stimulated ACs may have a crucial role in the hippocampus-dependent LTP and memory.

Adenylyl Cyclases and Memory Formation in Invertebrates

Molecular Mechanisms Underlying Memory in Aplysia californica

In the marine snail *Aplysia californica*, the role of the cAMP-dependent signaling pathway in short- and long-term memory (LTM) comes from studies on sensitization of the GSWR, which is a nonassociative form of learning.[55] A weak stimulus (conditioned stimulus, CS) to the siphon leads to the animal's defensive response that includes the GSWR. [27,29,116] The amplitude and duration of defensive withdrawal reflexes become enhanced when stimulation of the siphon is coupled to strong noxious stimulus (unconditioned stimulus, US), which is usually a shock to the tail. Whereas one single stimulation to the tail produces short-term sensitization (few minutes to hours), repeated spaced stimulations produce a long-term sensitization that lasts from days to weeks.[35,116] The GSWR is controlled by sensory (SN) and motor (MN) neurons[27] and cellular analyses of the SN-MN synapses demonstrated that the site of induction and expression for sensitization is the presynaptic SN. Both short- and long-term facilitation induced by sensitizing stimuli (US presented alone or unpaired with CS) activate serotonergic (5-hydroxytryptamine; 5-HT) receptors and lead to increased cAMP levels, activation of PKA and thus, modulation of membrane channels and other effector proteins that contribute to enhanced transmitter release (for a review see refs 23,82). Five spaced 5-HT pulses can cause long-term facilitation by inducing a prolonged activation of PKA and translocation of its catalytic units into the nucleus of SN where it activates transcription factors belonging to the cAMP-response element binding protein (CREB) family.[11,39] Manipulation of the signaling cascades in the presynaptic SN, such as intracellular injection of cAMP, induces long-term changes which can be blocked by anisomycin, an inhibitor of protein synthesis.[28,112,117] Thus, it is likely that the transient 5HT-induced elevation of cAMP can lead to long-term facilitation in *Aplysia*. Anisomycin is not effective, however, when applied 12-15 hr after the cAMP injection, suggesting that transient cAMP elevation induced a signaling cascade of enduring process, such as protein synthesis whose products continue to be synthesized for several hours after cAMP levels have returned to baseline.[104] All these findings support the hypothesis that the specific temporal activation of the cAMP cascade, dependent on distinct stimulation parameters, may be critical for the induction of long-lasting neuronal and behavioral changes in *Aplysia*.

Although direct evidence is lacking, cellular studies suggested that a dually-regulated AC serves as a coincidence detector for detection of US-CS contingencies (with 5-HT release and Ca^{2+} influx, respectively).[2,4,75] By injecting a peptide inhibitor of PKA into the SN, Bao et al[10] revealed that activation of the cAMP cascade is crucial for both associative and nonassociative facilitations. In contrast, associative, but not nonassociative, facilitation of SN-MN synapse is attenuated by either by presynaptic injection of Ca^{2+} chelators or a postsynaptic injection of an N-methyl-D-aspartate (NMDA)-receptor antagonist or a strong postsynaptic hyperpolarization, suggesting that associative facilitation requires activation of a Ca^{2+}-sensitive AC in the presynaptic SN and coincident elevation in $[Ca^{2+}]_i$ in both post- and presynaptic SN-MN. It was proposed that a postsynaptic site of detection involving Ca^{2+} influx through NMDA receptor-gated channels might serve for presynaptic glutamate release and postsynaptic depolarization to initiate induction of associative plasticity.[10,82] The resulting rise in postsynaptic Ca^{2+} might induce the release of a retrograde signal which acts presynaptically by activation of Ca^{2+}/CaM-stimulated AC. The *Aplysia* AC has not been cloned yet but it is clearly distinct from mammalian types 1 and 8 which do not require sequential application of Ca^{2+}/CaM and Gs_{α} to be synergistically stimulated.[38] In addition to activation of the cAMP/PKA cascade, 5HT acting on different receptor subtypes can also activate other kinases, including PKC[23,122] and the mitogen-activated protein kinase (MAPK).[89] Recent studies indicated that prolonged activation of PKC is involved in the long-term facilitatory actions of 5-HT that are mediated primarily by the cAMP/PKA cascade, suggesting that AC activity can be modulated via cross-talk between different signal transduction pathways in the *Aplysia* SN.[12-121]

The Drosophila System

The cAMP signaling cascade has a crucial role in LTM of associative olfactory learning in which the fruit fly *Drosophila melanogaster* is presented with two novel odors, and then trained to avoid a particular odor by pairing that odor with an electric shock.[138] Repeated, temporally-spaced training trials induced a stable, long-lasting memory that requires protein synthesis.[137] This memory can be dissectable into a medium-term memory (lasting few hours) which requires activation of PKC activity and a LTM (over 1 day) which requires a PKA- and nitric oxide-dependent processes.[45,91] In the mushroom bodies, which mediate olfactory learning, multiple conditioning trials induced temporal dynamics of PKA activation which depend both on the sequence of CS (which triggers odor-specific Ca^{2+}-mediated process) and US stimulation and also on the number of conditioning trials.[46] Mutational analyses of associative learning behavior have identified genes that are required for olfactory associative memory and their molecular characterization indicating that they all affect, albeit in different ways, the cAMP signaling cascade.[44,45,47,84,137] Gene disruptions of G-protein α subunit (*dGs*$_{\alpha}$), AC (*Rutabaga*), cAMP phosphodiesterase (*dunce*), catalytic (*DCO*) and regulatory subunits (*dPKA-RI*) of cAMP-dependent protein kinase (PKA) and cAMP-response element binding (CREB) (*dCREB2*) impair olfactory learning and/or memory formation in flies (for a review see refs. 111,136). Interestingly, both *rutabaga* and *dunce* are severely affected in initial memory acquisition and subsequent consolidation whereas relatively intact learning scores immediately after training are observed in *dPKA-RI*, *DCO* and *dCREB2* mutations. A neuronal model involving the cAMP cascade has been proposed for olfactory associative learning.[40,83] In this model, the *rutabaga* AC acts as a molecular coincidence integrator of associative learning cues responding synergistically to activated Gs_{α} and Ca^{2+} signals.[40] Interestingly, *rutabaga* AC shows high similarity to mammalian Ca^{2+}/CaM-stimulable AC isoforms (types 1 and 8).[26,157] It has been proposed that integration of sensory inputs from olfactory cues (increased $[Ca^{2+}]$i) and footshock (activation of *dGs*$_{\alpha}$) in mushroom body neurons may lead to activation of AC and produce a synergistic increase of cAMP levels which then, may act as the primary mediator of downstream events that are responsible for long-term functional and structural changes. Zars et al[157] have recently reported that a cell type-specific gene targeting the *rutabaga* gene in Kenyon cells (the primary afferents of which convey olfactory inputs via the antenno-glomerular tract) restores olfactory learning, and indicates that mushroom bodies are a critical locus for the signal-integrating properties of *rutabaga* AC.

A Specific Role for Mammalian Adenylyl Cyclases in Learning and Memory Processes: Heterogeneity of Mammalian Adenylyl Cyclases

Since the original cloning of the first mammalian AC isoform by Kuprinski et al[80] nine isoforms have now been identified and characterized in brain, revealing variable sensitivities to regulators such as G proteins, Ca^{2+}, CaM and protein kinases.[37,74,79,131] Hydropathy analysis predicted that all isoforms are large (1080-1248 amino acids) polypeptides consisting of a short and variable cytoplasmic N-terminal region, followed by a double six-transmembrane spanning motif (M1 and M2) and two 40 kDa cytoplasmic domains (C1 and C2). Whereas the transmembrane domains are not highly conserved among ACs, two subregions of the cytosolic domains (termed C1a and C2a) are well conserved within a particular AC isoform, they also share homology with the cytoplasmic domains of Drosophila rutabaga AC, bacterial and yeast AC and even with the catalytic domains of membrane-bound guanylyl cyclases, suggesting that both eukaryotic and prokaryotic AC share the same ancestral origin.[123,126] These homologies among the C1a and C2a domains from the same or different mammalian ACs suggest that the cytosolic domains constitute the site for cAMP synthesis. Indeed, molecular studies showed that a soluble chimeric construct consisting of C1a from type 1 and C2a from type 2 contains all of the catalytic apparatus of the wild type AC and is responsive to Fsk, Gs_{α}-and $G_{\beta\gamma}$ subunits.[126]

Based on their similarities in sequences and their distinct regulation by Ca^{2+} and G-protein signaling pathways, mammalian AC isoforms have been divided into distinct subfamilies as (1) Ca^{2+}-stimulated AC types 1, 8 and 3 (types 1 and 8 act as coincidence detectors for positive cross-talk between Ca^{2+}/CaM and Gs_{α} whereas stimulation of type 3 by Ca^{2+}/CaM is strictly conditional and requires concomitant activation by Gs_{α} or forskolin (Fsk)); (2) Ca^{2+}-inhibited ACs (types 5 and 6); (3) Ca^{2+}-insensitive ACs (types 2, 4 and 7 are insensitive to $[Ca^{2+}{}_{i}]$, but stimulated by Gs_{α} and $\beta\gamma$ under coincidental activation by Gs and Gi) and (4) Ca^{2+}/ CaM-dependent protein phosphatase (calcineurin)-inhibited type 9 (for a review see refs. 37,38,123).

Diversity in the Regulation of Mammalian Adenylyl Cyclases by G Proteins Ca^{2+} Signals and Phosphorylation

In light of their varied and complex modes of regulation by G-proteins, kinases (PKA, PKC, MAPK and CaMkinase), phosphatases (calcineurin), Ca^{2+} and Ca^{2+}/CaM, mammalian ACs have been proposed to serve as critical « coincidence » detectors i.e., they could respond synergistically to multiple signals that arrive from independent transductional pathways to efficiently increase cAMP production[6,20,95] (see Fig. 1). All of the ACs are regulated in type-specific patterns, and their mechanisms of regulation are often highly synergistic or conditional.

Regulation by G-proteins

Although the different isoforms differ greatly in their pattern of regulation, all ACs share the capacity to be stimulated by the plant diterpene Fsk and Gs_{α} *in* vitro (except type 9). However, the Ca^{2+}/CaM-stimulated isoforms (types 1 and 8) are insensitive to Gs in vivo.[67,145] The different mammalian ACs exhibit different susceptibilities for activation by Fsk, Gs_{α} or both. Coincident stimulation by both Fsk and Gs_{α} results in synergistic, non competitive, stimulation of enzymatic activity for Ca^{2+}-insensitive ACs (types 2, 4, 7) and Ca^{2+}-inhibited type 5 whereas the two activators act independently for type 1.[124,125,130]

Although stimulation through Gs_{α} is the principal mechanism whereby ACs are activated, the activity of certain isoforms is also regulated by the family of Gi-related proteins (Gi, Go, Gz) that can be activated by diverse hormones and neurotransmitters (i.e., adenosine, epinephrine and cannabinoids). Inhibition of catalytic activity by Gi_{α} is selective and variable degrees

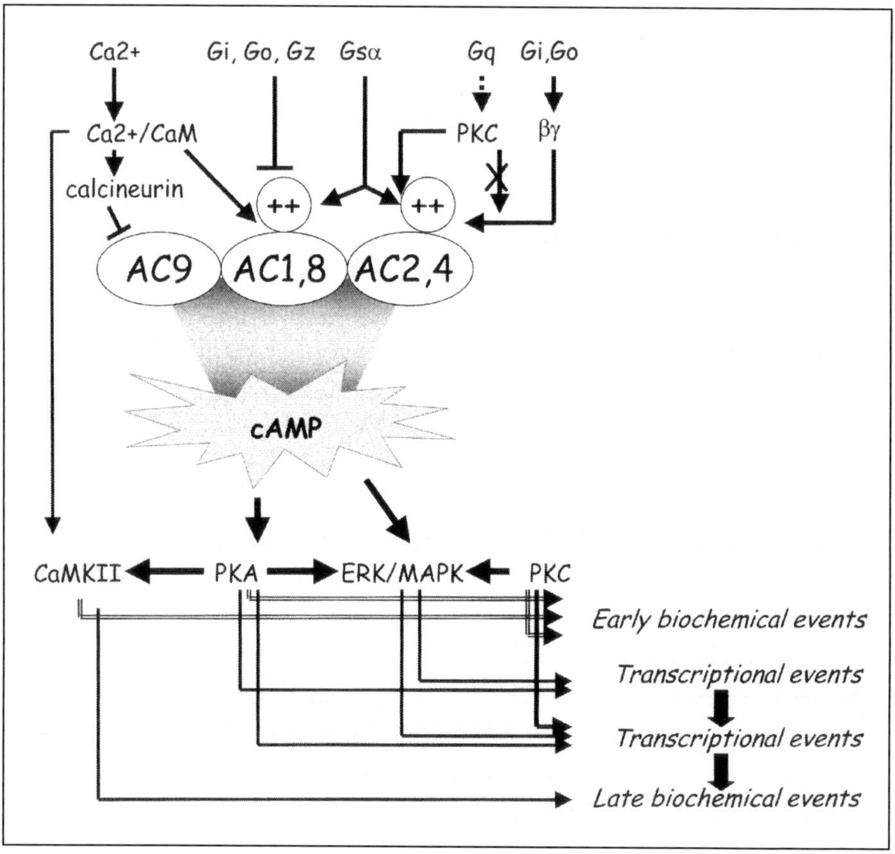

Figure 1. Complex regulatory patterns of hippocampal AC by various G protein subunits, Ca^{2+}/CaM, kinase (PKC), phosphatase (calcineurin). The different Ca2+-sensitive and insensitive AC act as molecular coincidence detectors i.e., they could respond to multiple signals that arrive from independent pathways to efficiently increase cAMP level and activate PKA activity. The AC/cAMP/PKA pathway, in addition to participate to early biochemical events, also interacts with other kinases (CaMKII, ERK/MAPK) to regulate transcriptional and translational events required for the establishment of late biochemical events. Stimulatory signals are shown as arrows and inhibitory signals as plungers. Abbreviations are described in the text. (adapted from ref. 74).

of inhibition have been reported.[129] The ability of AC to integrate multiple regulatory inputs from the α and the βγ subunits released from Gi is isoform-specific.[127,130] Reconstitution or transfection studies demonstrated that activated Gi selectively inhibits types 3, 5 and 6.[43,133] whereas types 1, 2, 7, 8 and 9 are less sensitive to Gi_{α}.[109,129,130] The inhibition is noncompetitive with Gs_{α}, arguing that Gs_{α} and Gi_{α} bind to separate nonoverlapping sites on the AC protein. Type 1 is slightly inhibited by Gi_{α}, but in the presence of βγ subunits released from hormonal activation of Gi, the CaM (or Fsk or Gs_{α})-stimulated activity of type 1 is inhibited by $G_{\beta\gamma}$ subunits. Interestingly, Nielsen et al[103] have shown that type 1, but not type 8, is inhibited by activation of Gi-coupled receptors in vivo.

Coincidence regulation has also been proposed for Ca^{2+}-insensitive isoforms (types 2, 4 and 7) which are only weakly inhibited by activated Giα, but are synergistically stimulated by βγ subunits in the presence of Gs_{α}.[130,131] In addition to in vitro regulation of βγ subunits, types 2 and 4 also act as coincidence detectors of paired Gi and Gs inputs with βγ potentiation in vivo.[52,86,140] The cotransfection of HEK-293 cells with the Gi-coupled serotoninergic receptor (5-HT1A or 5-HT1B), type 2 and Gi$_\alpha$ greatly stimulates AC activity and this activation is blocked by pertussis toxin and a G$_{\beta\gamma}$ antagonist.[4] Similarly, activation of the 5-HT1A receptor in tissues in which type 2 is highly expressed (e.g., hippocampus) potentiates actions of Gs-coupled receptors (e.g., β-adrenergic receptor in CA1 neurons) by Gβγ-mediated activation of type 2 ACs.[5,24]

Regulation by Ca^{2+}

Stimulation by Ca^{2+}. Profound physiological significance derives from the regulation of mammalian AC by Ca^{2+}, which provides a means of integrating the activities of the two crucial cAMP- and Ca^{2+}-regulated signalling pathways.[38] Submicromolar concentrations of Ca^{2+} elicit a prominent stimulation of type 1 and 8 ACs, in the presence of CaM.[25,27,80] In vitro stimulation of type 3 by Ca^{2+}/CaM requires low micromolar $[Ca^{2+}]_i$ and is seen only in the presence of activated G$_\alpha$s or Fsk.[33]

Inhibition by Ca^{2+}. All AC activities are inhibited by high, nonphysiological submillimolar levels of $[Ca^{2+}]_i$, possibly by competition with magnesium which is required for catalysis.[59] Submicromolar $[Ca^{2+}]_i$ directly inhibits the activity of types 5 and 6, independently of CaM.[38] This inhibition by $[Ca^{2+}]_i$ is additive to that elicited by receptors acting via Gi$_\alpha$.[130] Interestingly, both types 5 and 6 are weakly expressed in regions associated with learning and memory, including the hippocampus and cortex, suggesting that a direct inhibitory control of AC by Ca^{2+} is not critical for memory processes.

Inhibition by Ca^{2+}/Calcineurin. Calcineurin-dependent dephosphorylation represents another mode of regulating cAMP production by which Ca^{2+} signals may exert an indirect negative control on AC. In HEK-293 or COS7 cells transfected with AC9 or in AtT20 cells that express predominantly endogenous AC9, the inhibition of cAMP synthesis by a rise in $[Ca^{2+}]_i$ is alleviated by specific inhibitors of calcineurin (FK506 or cyclosporin A).[7]

In vivo regulation of AC by Ca^{2+}. When Ca^{2+}-sensitive ACs are directly regulated by changes in $[Ca^{2+}]_i$, studies in non excitable cells demonstrated that the positive or negative regulation of AC activity is strictly dependent on capacitative Ca^{2+} entry (CCE), activated secondary to the emptying of intracellular Ca^{2+} pool.[49,50] In contrast, Ca^{2+} release from internal stores or non specific Ca^{2+} entry via ionophore is unable to regulate Ca^{2+}-sensitive ACs.[32] In neuronal cells in which the CCE plays a modest role, both CCE and prominent voltage-gated Ca^{2+} entry appear equally efficacious at regulating Ca^{2+}-sensitive ACs, indicating that Ca^{2+}-sensitive AC is closer to the CCE channel than the voltage gated Ca^{2+} channel.[48]

Regulation by Protein Kinases

Serine/threonine phosphorylation of specific isoforms of ACs by protein kinases (PKC, PKA, CaMK) is a very important regulatory mechanism allowing a direct and efficient control of cAMP production within the cell (see also relevant chapters in this book).

PKC. The PKC-mediated phosphorylation of AC isoforms positively regulates types 1-5 and 7 but inhibits type 6.[69] Intriguingly, the G$_{\beta\gamma}$ potentiation of the Gs$_\alpha$-stimulated activities for types 2 and 4 is abolished by the PKC-mediated phosphorylation, indicating that PKC can exert an inhibitory effect on activated Ca^{2+}-insensitive types 2 and 4.[128] PKC synergistically increases the activity of type 2 evoked by Gs$_\alpha$ or G$_{\beta\gamma}$ whereas it inhibits Gs$_\alpha$-activated activity of type 4.[71,86,154] These findings strongly suggest that activation of PKC pathway greatly reduces the ability of type 2 to integrate coincident signals from Gi$_\alpha$- and Gs$_\alpha$-coupled receptors. Thus, the role of type 2 (or type 4) to mediate cross-modulation of synaptic plasticity between Gi$_\alpha$ and Gs$_\alpha$-coupled receptors in hippocampal neurons might be affected upon activation of PKC.[4,5]

PKA. Both Fsk- and Gs$_\alpha$-stimulation of Ca^{2+}-inhibitable types 5 and 6 are inhibited by PKA-mediated phosphorylation,[30,70] suggesting that both types 5 and 6 are under feedback inhibition by cAMP cascade. This effect is isoform-specific since types 1 and 2 are not susceptible to PKA-mediated loss of Gs$_\alpha$ stimulation.[30]

CaMkinase. The best example of rapid desensitization of AC by CaM kinase phosphorylation is provided by the negative effect exerts by CaMKII on type 3 in olfactory signaling.[144,147] Wayman et al[146]also reported that CaMKIV functions as a negative feedback regulator of Ca^{2+}-stimulation of type 1 activity, without affecting basal and Fsk-stimulated activity in vivo. Since type 1, but not type 8, is subject to inhibition by both CaM kinase and Gi-coupled receptors, it is suggested that the two Ca^{2+}-stimulated ACs may have very distinct regulatory properties and thus, the presence of both types 1 and 8 in a particular neuron is not redundant.

Potential Targets of cAMP

cAMP-binding proteins. In addition to PKA activation, cAMP also regulates the activity of specific guanine nucleotide exchange factors (cAMP-GEF). Two genes have been identified for cAMP-GEF also called Epac (exchange protein directly activated by cAMP). They exhibit both a cAMP-binding site and a domain that is homologous to domains GEF for Ras and Ras-like GTPase (Rap1).[41,42,77] Recent studies reveal complex regulation of Rap1 by cAMP including PKA-independent activation and PKA-dependent negative feedback regulation.[139] As one Epac isoform (Epac 2) is strongly expressed in restricted brain areas, including the hippocampus (mainly CA3 and DG), the cortex and the cerebellum,[77] a PKA-independent activation of Rap1 by Epac 2 may provide a direct mechanism for cAMP to activate the Rap1-MAPK/ERK cascade and thus, to stimulate the gene transcription in a PKA-independent manner. Furthermore, the restricted expression of Epac 2 could contribute to region- and cell type-specific cAMP-mediated neuronal functions.

Cyclic nucleotide gated ion channels (CNGC). As the CNGC conduct Ca^{2+} entry under the control of cAMP and cGMP,[153] Fagan et al [51] proposed that they could also participate in the Ca^{2+} feedback regulation of Ca^{2+}-sensitive AC, independently of voltage-operated Ca^{2+} channels and Ca^{2+} stores. Regulation of AC by Ca^{2+}-dependent CNGC modulation is particularly important in the context of short-term adaptation and desensitization in olfactory cilia, because Ca^{2+} transients present in the olfactory cilia following cAMP-mediated gating of CNGC inhibits the activity of AC3 via phosphorylation by CaMKII and also via a down-regulation of CNGC affinity to cAMP.[159]

The Specific Distribution and Expression Levels of Mammalian Adenylyl Cyclases in Brain

Although all AC isoforms are present in the brain, the various ACs are distributed in quite distinct patterns throughout the different regions. In situ hybridization studies showed that (1) only four AC isoforms are highly expressed in the brain (e.g., types 1, 2, 5 and 9); (2) many brain areas express multiple AC isoforms and (3) Ca^{2+}-sensitive ACs are expressed in specific regions (e.g., type 3 in olfactory cilia, type 5 in basal ganglia; type 1 in areas implicated in memory formation)whereas others are widely distributed (e.g., types 2, 6, 7, 9) (for a review see refs. 61,93,94).

In the hippocampus, at least six AC isoforms (types 1,2,4,7,8,9) are expressed in the CA1-CA3 pyramidal layers and the dentate gyrus (DG). The pattern of expression of type 1 in the hippocampus provides a good example of cell-type specific expression of an individual AC isoform.[97,152] Type 1 is expressed predominantly in the CA1-CA2 fields and the DG whereas it is barely expressed above background in CA3 field. Compared to type 1, the level of expression of type 8 in the hippocampus is weaker[25,98] Since most forms of hippocampal LTP require increased $[Ca^{2+}]i$ which markedly elevates cAMP levels,[76,87,102] the presence of types 1 and 8 in hippocampal subfields strongly suggests that Ca^{2+}-mediated increased cAMP level depends upon these two Ca^{2+}-stimulated ACs. In addition to types 1 and 8, high levels of mRNA encoding for Ca^{2+}-insensitive, PKC-stimulated type 2 and Ca^{2+}/calcineurin-inhibited type 9 are also expressed in all hippocampal subfields. Specific isoform-antibodies against types 2 and 9 have been developed to examine the distribution of the protein in the brain. Labeling for type 2 is found in the dendritic subfields of the CA1-CA3 pyramidal and the granular cells and type

2 colocalizes with the dendritic marker (MAP2), suggesting that type 2 plays an important role for the generation of the cAMP signal in the postsynaptic compartment.[9] Type 9 also appears implicated in postsynaptic mechanisms underlying synaptic plasticity since it is also present in the dendritic fields in both hippocampus and neocortex and it colocalizes with calcineurin in synaptic structures of most cerebral neurons.[8,119]

Ultrastructural analysis using anti-AC antibodies that recognize a domain common to all mammalian AC confirmed that AC immunoreactivity is highly distributed near postsynaptic densities in dendritic spines of hippocampal CA1 region.[96] Dendritic spines are areas of high concentrations of Ca^{2+} channels and pumps,[100,156] as well as PKA and CaMKII,[78,85] ACs may thus be precisely where they are most efficacious in the integration and propagation of Ca^{2+} signals. We might expect that cAMP would need to diffuse only a short distance before activating the anchored PKA, thereby greatly facilitating the local downstream phosphorylation steps that are responsible for short-term modifications.

Adenylyl Cyclase and Long-Term Potentiation

LTP is a robust and persistent modification of synaptic transmission in response to transient stimuli and is thought to be a candidate cellular mechanism for mediating some forms of explicit hippocampus-dependent memory. LTP requires stimulation of NMDA receptors, postsynaptic depolarization and Ca^{2+} influx into the postsynaptic cell in the Schaffer collateral/commissural synapses in area CA1 and the perforant path/DG synapses[31,102] whereas LTP in the mossy fibers is initiated presynaptically through voltage-sensitive Ca^{2+} channels.[63,102,148]

In contrast to the general agreement that the late phase of LTP (L-LTP) requires activation of AC and cAMP-dependent PKA, the issue of whether early phase of LTP (E-LTP) depends on a rise in cAMP level is not clear. Several pharmacological and genetic studies showed that interfering with the cAMP signal does not affect E-LTP.[1,53,64,105,149] However, recent studies demonstrated that inhibition of the cAMP/PKA pathway indeed decreases E-LTP.[19,106,155] Blitzer et al[18,19] proposed a postsynaptic mechanism by which the cAMP pathway may act as keeping the «gate open» for the induction of LTP by controlling the activity of protein phosphatases, such as calcineurin (see Fig. 2). They proposed that the gating mechanism comprises two opposite PKA and calcineurin pathways, which converge on the regulatory protein inhibitor-1 (I-1), a specific blocker of protein phosphatase-1 (PP1). The cAMP pathway, through activation of I-1 and inhibition of PP1, enables the autophosphorylation of CaMKII to occur and thereby, enhances CaMKII activity.[18] As calcineurin could mediate the decrease in synaptic strength through dephosphorylation of I-1 and thus, activation of PP1,[99] the interactions between the two cAMP and Ca^{2+} signals at this point may play a key role in the modulation of LTP.[18,19,22,134] As shown by Raman et al[108] in cultured hippocampal CA1 neurons, inhibition of PKA prevented recovery of NMDA receptors from calcineurin-mediated dephosphorylation induced by synaptic activity whereas elevation of PKA activity by Fsk, cAMP analogs or β-adrenergic receptor agonists can antagonize the effects of calcineurin. Moreover, Malleret et al[88] showed that the enhancement of E-LTP in area CA1 after altering calcineurin activity could be prevented by blocking PKA. Taken together, the findings suggest that a PKA/ calcineurin gate represents a major activator/suppressor mechanism for regulating E-LTP. Blitzer et al[19] proposed that the direct mechanism for coupling increases when Ca^{2+} influx leading to rises in cAMP levels and this might be through activation of types 1 and 8. Interestingly, type 9 which is under inhibitory control by calcineurin, is inhibited by the same range of $[Ca^{2+}]_i$ that stimulates type 1.[7] Thus, it is possible that cAMP generated by type 9 also provides a critical link in the balance between phosphorylation/dephosphorylation cascades that controls LTP.

In addition to the Ca^{2+} signal, cAMP-induced synaptic plasticity can also be modulated by neurotransmitter receptors acting on Gs_α, Gi_α or $\beta\gamma$ subunits of G proteins. Thus, by acting as a molecular coincidence detector to integrate signals from PKC- and Gs/Gi-protein-regulated pathways, it is possible that the cAMP cascade arising from activation of Ca^{2+}-insensitive type 2 also participates in the molecular events that trigger LTP (See Fig. 1). In particular, electro-

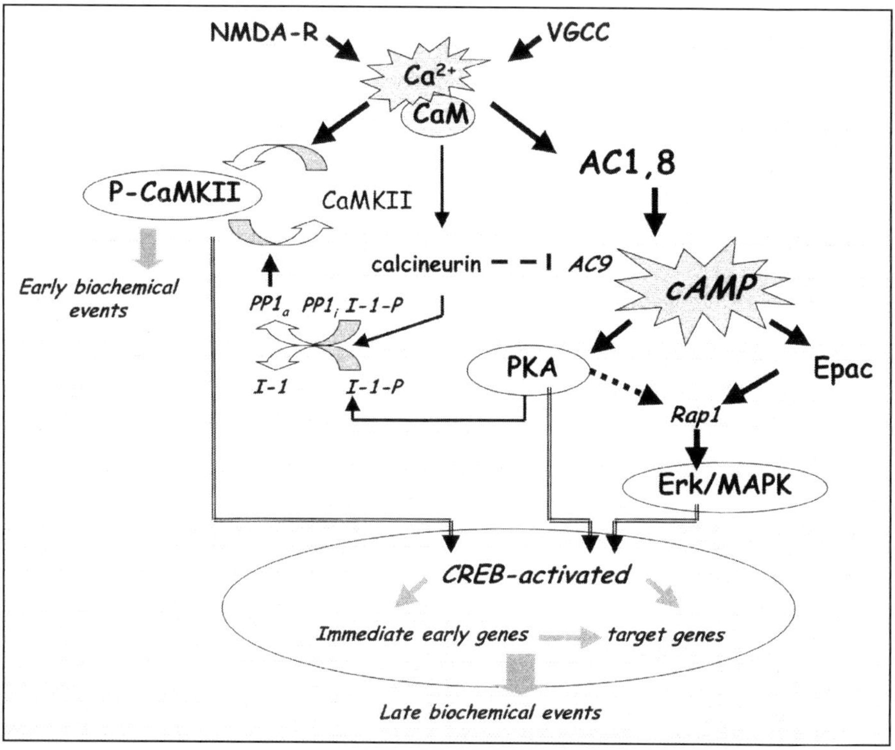

Figure 2. Postulated interactions between Ca^{2+}/CaM-stimulated types 1 and 8 and Ca^{2+}-regulated pathways in the early and late biochemical events underlying LTP and memory formation. Increased $[Ca^{2+}]$i arising from NMDA-R or VGCC induces elevation of intracellular cAMP via activation of type 1 or type 8. The resulting activation of the cAMP/PKA pathway, through phosphorylation of I-1 and inhibition of PP1, acts as keeping the «gate open» for Ca^{2+}-dependent biochemical events by inhibiting calcineurin and thus, maintaining CaMKII activity. Abbreviations are described in the text. (adapted from refs. 18,19,74).

physiological studies reported that, in hippocampal CA1 neurons, agonist stimulation of Gi-coupled 5-HT1A, GABA-B and α-adrenergic receptors leads to liberation of βγ complex and potentiates Gs_{α}-mediated actions of β-adrenergic receptor via activation of type 2 AC (or type 4).[4,5]

Studies using pharmacological inhibitors or genetic manipulation have implicated the cAMP cascade in the late phase of LTP (L-LTP) in all hippocampal pathways.[54,62,66,73,148,150] There is increasing evidence that cross-talk between the Ca^{2+}, cAMP and mitogen-activated protein kinase (MAPK) pathways is critical for the stimulation of CREB and thus, the expression of genes required for the formation of LTP and LTM (see Fig. 2).[68,89] In the hippocampus, a rise in intracellular cAMP activates the Erk/MAPK cascade, much as it does in lateral amygdala, and coactivation of the cAMP and MAPK pathways by Ca^{2+} is essential for phosphorylation of CREB and L-LTP formation.[62,65,110] In this context, the induction of arg3.1/arc mRNA in primary culture hippocampal neurons is strictly dependent on the coactivation of PKA and Erk/MAPK pathways.[143] In neuronal cells, the effect of cAMP has been proposed to involve the sequential activation of Ca^{2+}/CaM-sensitive ACs (types 1,8) and the phosphorylation and activation by PKA of Rap-1, then the coupling of Rap-1 to B-Raf results in the activation of ERK/MAPK pathway.[56,115,142] Although PKA plays a crucial role in the activation of CREB, activation of Rap1 by cAMP-GEFII may also provide another mechanism by which cAMP can stimulate the Erk/MAPK pathway and thus, can induce gene transcription in a PKA-independent manner.[42,77]

To investigate the role of Ca^{2+}/CaM-stimulated ACs in LTP, mice lacking either type 1 or type 8 (AC1 or AC8KO) or both ACs (DKO) were analyzed for several forms of LTP.[120,141,150] Surprinsingly, LTP at the Schaffer collateral/CA1 pyramidal cell synapse was not affected in the KO mice whereas it was impaired in the DKO mice.[150,151] Moreover, hippocampal Ca^{2+}-stimulated AC activity was partially reduced in KO mice whereas response to Ca^{2+} was totally abolished in DKO. These observations suggest that the two Ca^{2+}-stimulated AC1 and AC8 can, at least in part, substitute to each other for cAMP production in hippocampal CA1 region. In contrast to hippocampal CA1 LTP, AC1KO mice exhibit impaired mossy fiber/CA3 and cerebellar parallel fiber L-LTP, suggesting that presynaptic forms of LTP strictly depend upon AC1.[120,141] In addition, since administration of Fsk (a nonselective stimulator of ACs) to DKO mice in hippocampal CA1, (or to AC1KO in mossy fiber) can restore L-LTP, it thus appears that postsynaptic activation of hippocampal ACs, other than types 1 and 8, could also modulate L-LTP.

Are Ca^{2+}-Stimulated Adenylyl Cyclases Critical for Memory

Behavioural studies have provided evidence that AC activity is critical for learning and memory functions in mammals. A first study in our laboratory reported that AC activity was altered in mouse hippocampus following learning tasks. After acquisition of a spatial discrimination task performed in a 8-arm radial maze (a hippocampus-dependent task), Fsk-stimulated AC activity was down-regulated in the hippocampus and negatively correlated with the response accuracy attained by the subjects.[57] In contrast, AC activity was increased following acquisition of a bar-pressing task, which is an hippocampal-independent task.[72] Arguments based on phylogenetic adaptation supported our proposal that these opposite learning-induced alterations of AC activity might reflect an interaction between two (or more) competing memory systems at the hippocampal level, in which ACs might have a critical role. Meanwhile, Wu et al[151] reported that AC1KO could acquire normally, as compared to controls, a task where they are required to find a hidden platform in the standard water maze task. Moreover, AC1KO did not keep searching the quadrant where the platform had been previously located. This observation was interpreted as a spatial memory deficit although no argument excluded the possibility that these animals might be more flexible (i.e., search for the platform elsewhere). Whatever the case, the deficit was marginal and could be explained by the fact that 50-60 % of the Ca^{2+}-stimulated AC activity remained in the hippocampus of AC1KO, suggesting that up-regulation of AC8 might have compensated the absence of AC1 function. To test this hypothesis, behavioural responses of AC8KO, AC1KO and DKO mice were analyzed.[150] The results showed that the single mutants had normal LTM for contextual and passive avoidance learning whereas the DKO mice displayed a lower inhibitory response than controls after 30 minutes, but not 5 minutes, following acquisition of a single trial step-through passive avoidance paradigm. Also, DKO mice expressed a lower level of conditioned-fear when exposed, after 8 days (but not 24 h), to the context in which they had previously received an electric shock. Thus, it was hypothetized that hippocampal Ca^{2+}-stimulated AC activity may be required for LTM, but not for short-term memory. This conclusion is in agreement with the idea that a cAMP cascade in the hippocampus is involved in the late, but not the early, phase of a memory consolidation process occurring after inhibitory learning in rats. Bernabeu et al[13,14] showed that rats submitted to step-down passive avoidance learning displayed a time-dependent increase in hippocampal cAMP levels with a peak at 3-6 hr after training. This was supported by findings that intrahippocampal infusion of 8-Br-cAMP (a stable analogue of cAMP) or Fsk enhanced memory retrieval when given 3 or 6 hr (but not earlier than 3 hr) after the acquisition.[13,14,15,16,17] Moreover, activation of dopamine D1, β-noradrenergic or 5-HT1A receptors also modulates cAMP levels at 3-6 hr after training, and an increase in cAMP level is coincident in time with increases in PKA activity, and in phosphorylated CREB and c-fos immunoreactivities in the hippocampus after training. As emphasised by Wong et al[150] the memory deficits of the DKO lacking AC1 and AC8 resembled those previously described in CREB deficient mutants in fear-conditioning experiments.[21] They hypothesized that Ca^{2+} activation of type 1 and type 8 ACs play a crucial role in LTM because they can generate the critical cAMP signal required for Ca^{2+} stimulation of the CREB/CRE-mediated transcription (see Fig. 2). The use of similar fear-conditioning methods in both studies supported this conclusion. However, the interpretations of these experiments have relied on the assumption that this task is sensitive to hippocampal lesions in mice. Several years later, authors of the study of the CREB mutants reported behavioural findings, which were crucial for the interpretation of transgenic experiments with the widely used fear-conditioning paradigms. They demonstrated that hippocampal-lesioned mice are impaired in spatial versions of the Morris water maze task but can show contextual fear conditioning[34] suggesting that, at least in some conditions (such as those used in the DKO study), the hippocampus may not be necessary for task acquisition. A second issue to consider is that AC8KO mice do not show normal increases in behavioural markers of anxiety when subjected to repeated stress, such as repetitive testing in the plus-maze or restraint preceding plus-maze testing, suggesting a role for type 8 in the modulation of anxiety.[114] This observation is of significance because anxiolytic-like effects could have interfered with the estimation of retention performance of the DKO mice in tasks such as passive avoidance or fear-conditioning.

All these recent results gained from genetic strategies strengthened the hypothesis for a role of type 1 and/or type 8 in memory formation, which initially, was based only on brain locations and functional considerations related to their regulatory properties (see above). However, the conclusions remain still elusive and controversial. Considering that selective pharmacological tools are not available yet, further characterizations of the behavioural phenotypes of these genetically modified animals appear indispensable and should help to detail what is the exact nature of the memory processes in which the Ca^{2+}-stimulated ACs have a role.

Among the pharmacological strategies, inhibitors of PKA activity have been commonly used to inhibit the cAMP signaling cascade and were shown to impair memory performance in a variety of tasks (including spatial learning) in correlation with impaired LTP in the hippocampus (for a review see ref. 92 and Vianna and Izquierdo in this book). Conversely, stimulation of PKA activity was used to demonstrate a role of PKA in the maintenance of LTP.

Pharmacological approaches supporting the view that an elevation in cAMP in the hippocampus is important for memory are based on the following data obtained using the passive avoidance paradigm: 1) Post-trial injections of Fsk or 8Br-cAMP in the hippocampus improved memory retrieval in the step-down passive avoidance[13] and (2) in DKO mice, unilateral administration of Fsk to the CA1 subfield immediately before training was shown to restore LTM of passive avoidance.[150] Recent studies in our laboratory have shown that increased hippocampal cAMP levels produced by local infusions of Fsk improve memory in a similar kind of task but impair spatial learning in water-maze tasks (unpublished data). The latter result is not isolated since Taylor et al[132] also reported that injection of Sp-cAMP into the prefrontal cortex impair working memory in a delayed alternation task performed in a T-maze, suggesting that activation of PKA activity produces deleterious effects in spatial memory tasks. These findings greatly contrast with an extensive body of literature indicating that enhancement of the PKA pathway improves memory formation. Indeed, increased cAMP levels can oppositely alter mechanisms subserving different memory systems, suggesting mechanisms leading to "cognitive enhancement" are not universal (see ref. 132 for further discussion).

Adenylyl Cyclases Up or Down Depending on Task Demands

Even though Ca^{2+}-stimulated AC might have a crucial role in the memory function of the hippocampus, these AC isoforms probably constitute only one part of a complex molecular system in which, interactions between diverse sources of cAMP (including from Ca^{2+}-insensitive isoforms), would optimise the hippocampal functioning depending on the learning situation. Since the insight of Tolman[135] that animals can learn about a particular experience in more than one way, it is now widely accepted that there exist multiple forms of memory and that the underlying neural substrates are distributed throughout the brain.[113] An important implica-

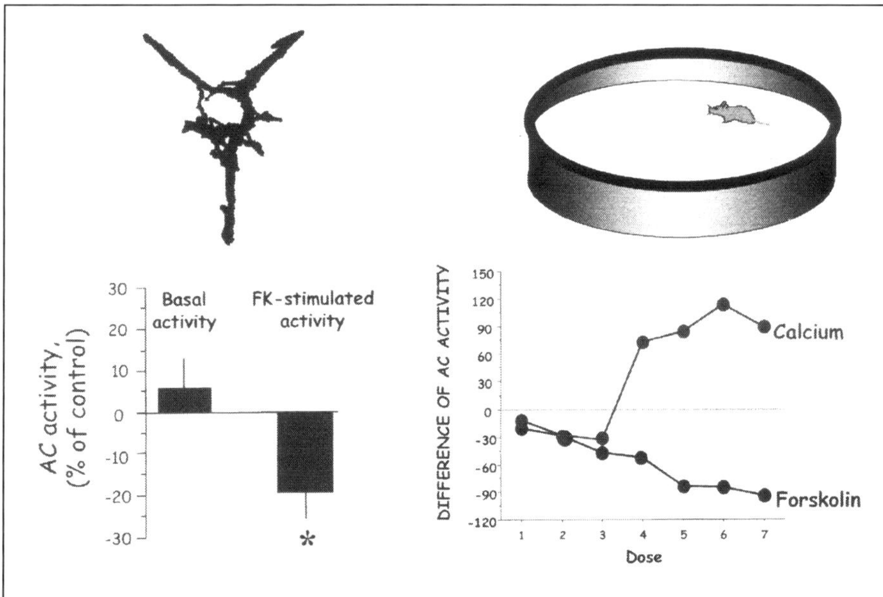

Figure 3. Opposite regulations of Fsk-stimulated and Ca^{2+}-stimulated AC activity occurs following spatial learning in the hippocampus. (A), In an 8-arms radial maze, mice were trained to discriminate 3 arms which were constantly baited. The top of the figure shows a representative track recorded at the end of learning and illustrates searching patterns occurring selectively into the 3 baited arms of the maze. The graph bellow shows changes in hippocampal AC activity in mice who had learned this task as compared to naive animals (controls). Fsk-stimulated AC activity was reduced after learning. (B) summarizes results obtained in mice who learned to locate a hidden platform in a circular water maze. In the hippocampus, in response to stimulation by Fsk, the AC activity was dose-dependently reduced after learning whereas, in sharp contrast, the AC responses were increased as function of the Ca^{2+} concentration.

tion of this notion is that these different memory systems interact synergistically or competitively to produce behaviour.[90] One consequence of this is that an animal may use different strategies in order to deal with a learning situation. Moreover, recent data have shown that hippocampal lesions facilitate the use of alternative learning strategies[80-107] that are normally overridden by hippocampal-dependent memory processing. Jaffard and Meunier[72] have reported data showing neurochemical or electrophysiological alterations in the hippocampus following the acquisition of tasks, which are not dependent on the hippocampal formation. Further, more neurobiological changes can be opposite to those observed following acquisition of hippocampal-dependent tasks and furthermore, one pharmacological treatment (like a lesion) can produce differential memory effects (no effect, facilitation or impairment) as a function of task demands.[36] In the context of these findings, opposite alterations in hippocampal AC activity following acquisition of hippocampal-dependent or hippocampal-independent learning have been reported.[57,58,60] Increased Fsk-stimulated AC activity was observed after acquisition of a bar-pressing task (hippocampal-independent task) whereas a decrease occurred after acquisition of place learning in an 8-arm radial maze (see Fig. 3). Moreover, we showed that cysteamine-induced depletion of somatostatin produced an increase in AC activity in the hippocampus and improved acquisition of the bar-pressing task whereas place learning was impaired.[58] Changes in AC activity were also studied following spatial learning in the water maze. Again, responses to Fsk were dose-dependently decreased in the hippocampus. However, in sharp contrast, responses to Ca^{2+} were enhanced. In other words, nonselective stimulation of hippocampal ACs was reduced whereas selective stimulation by Ca^{2+} was selectively increased.[60]

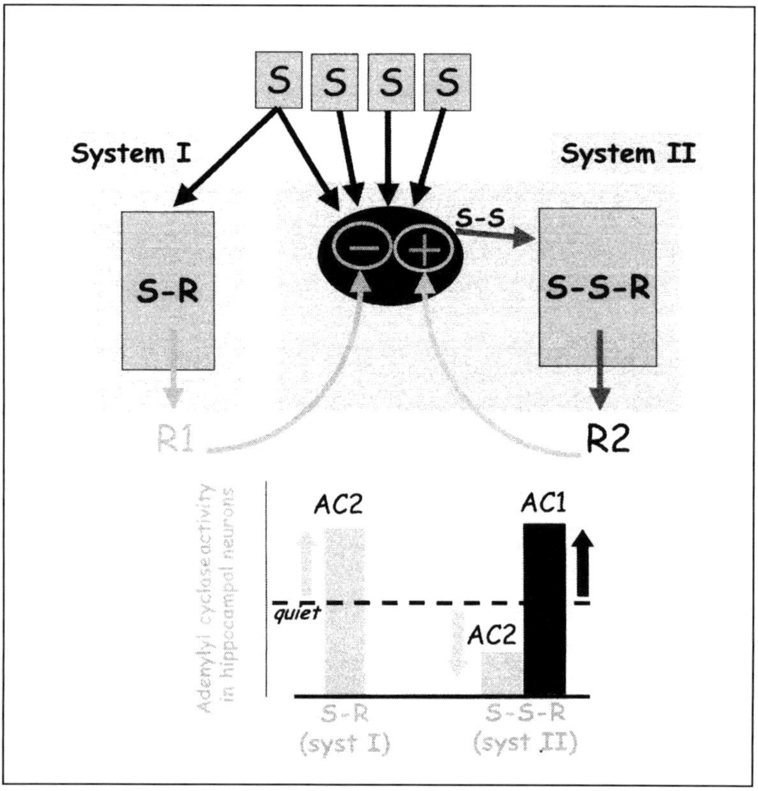

Figure 4. This model proposes the existence of two memory systems, a system I (coding for stimulus-response associations) and a second memory system (system II) coding for stimulus-stimulus associations, i.e., relational associations). When a novel learning situation occurs, both memory systems are *a priori* activated by incoming stimuli (S), process and emitted responses. In simple learning conditions, the responses emitted by system 1 (R1) can be sufficiently adapted to deal with the problem. In this condition, the inhibition (negative feedback) of the hippocampal functioning, blocking nonuseful information processing (SS-R), would speed-up acquisition. In contrast, when learning conditions are complex and required relational encoding between the stimuli, the adaptation of the responses emitted by system II (R2) would trigger a positive feedback to strengthen the hippocampus functioning. The bi-directional regulations of hippocampal AC activity, as observed following learning in tasks respectively involving each of these two kinds of information processing, might reflect a modulation of the hippocampal functioning. Further, we hypothesized that signals involving type 2 might be involved in this regulatory process. Beyond this, Ca^{2+}-sensitive ACs (types 1, 8) could be necessary to compute specific hippocampal functions such as the establishment of relational representations and/or spatial mapping.

This suggests that an up-regulation of Ca^{2+}-stimulated ACs associated with a down-regulation of other AC isoforms might be as critical for spatial learning (see Fig. 3). Because the Ca^{2+}-insensitive type 2 AC is highly expressed within the hippocampal subfields, it was hypothesized that cAMP-signaling occurring at synapses expressing type 2 AC could also modulate hippocampus functioning as a function of the task demand. Based on widely accepted memory theories, which postulated the existence of at least two memory systems,[113] a model was proposed to explain why bi-directional regulations might be relevant.[72,95] This model confers a modulatory role on hippocampal functioning to signaling pathways involving the Ca^{2+}-insensitive type 2 AC, (see Fig. 4). As a function of the task demand, activation of type 2 AC would block the information processing in hippocampus. Conversely, decreased cAMP levels at synapses involving type 2 would be permissive for such a function to occur. In agreement with the extensive literature on the role of a Ca^{2+}-stimulated ACs in memory formation, this model also integrates the idea that type 1 (and/or type 8) could be a critical component of an information processing system underlying the establishment of relational representation or spatial mapping.

Summary and Conclusions

Studies over the past few years have firmly established that members of the AC family play a key role in the complex intracellular network underlying synaptic plasticity and memory formation. As ACs are regulated by diverse extracellular stimuli through multiple signaling cascades, they could act as coincidence detectors to generate a unique cAMP response which then makes cross-talk with other signalling pathways to enable specific cellular functions. Overall, more detailed insight into the targeting of the different mammalian ACs in the neuronal compartments and the identification of complex mechanisms by which cAMP regulates other signaling systems, such as the Rap1-ERK pathway, as well as the knowledge of specific crosstalk between ACs and other cellular components, will be critical for a richer understanding of how the different ACs participate in the regulation of synaptic efficacy and memory formation.

References

1. Abel T, Nguyen PV, Barad M et al. Genetic demonstration of a role for PKA in the late phase of LTP and in hippocampus-based long-term memory. Cell 1997; 88:615-626.
2. Abrams TW, Castellucci VF, Camardo JS et al. Two endogenous neuropeptides modulate the gill and siphon withdrawal reflex in Aplysia by presynaptic facilitation involving cAMP-dependent closure of a serotonin-sensitive potassium channel. Proc Natl Acad Sci USA 1984; 81:7956-7960.
3. Abrams TW, Karl KA, Kandel ER. Biochemical studies of stimulus convergence during classical conditioning in Aplysia: dual regulation of adenylate cyclase by Ca^{2+}/calmodulin and transmitter. J Neurosci 1991; 11:2655-2666
4. Albert PR, Sajedi N, Lemonde S et al. Constitutive G(i2)-dependent activation of adenylyl cyclase type II by the 5-HT1A receptor. Inhibition by anxiolytic partial agonists. J Biol Chem 1999; 274:35469-35474.
5. Andrade R. Enhancement of β-adrenergic receptors by Gi linked receptors in rat hippocampus. Neuron 1993; 10:83-88
6. Anholt RR. Signal transduction in the nervous system: Adenylyl cyclases as molecular coincidence detectors. Trends Neurosci 1994; 17:37-41.
7. Antoni FA, Palkovits M, Simpson J et al. Ca2+/calcineurin-inhibited adenylyl cyclase, highly abundant in forebrain regions, is important for learning and memory. J Neurosci 1998; 18:9650-9661.
8. Antoni FA, Simpson J, Sosunov A. Calcineurin regulated adenylyl cyclase in the brain. In: Gold B, Fisher G, Hergegen T, eds. Immunophilins in the Brain-FKBP-Ligands: Novel Strategies treatments of Neurodegenerative disorders. Barcelona: Prous Sci, 2000.
9. Baker LP, Nielsen MD, Impey S et al. Regulation and immunohistochemical localization of betagamma-stimulated adenylyl cyclases in mouse hippocampus. J Neurosci 1999; 19:180-192.
10. Bao JX, Kandel ER, Hawkins RD. Involvement of presynaptic and postsynaptic mechanisms in a cellular analog of classical conditioning at Aplysia sensory-motor neuron synapses in isolated cell culture. J Neurosci 1998; 18:458-466.
11. Bartsch D, Casadio A, Karl KA et al. CREB1 encodes a nuclear activator, a repressor, and a cytoplasmic modulator that form a regulatory unit critical for long-term facilitation. Cell 1998; 95:211-223.
12. Baxter DA, Canavier CC, Clark Jr JW et al. Computational model of the serotonergic modulation of sensory neurons in Aplysia. J Neurophysiol 1999; 82:2914-2935.
13. Bernabeu R, Bevilaqua L, Ardenghi P et al. Involvement of hippocampal cAMP/cAMP-dependent protein kinase signaling pathways in a late memory consolidation phase of aversively motivated learning in rats. Proc Natl Acad Sci USA 1997; 94:7041-7046
14. Bernabeu R, Cammarota M, Izquierdo I et al. Involvement of hippocampal AMPA glutamate receptor changes and the cAMP/protein kinase A/CREB-P signalling pathway in memory consolidation of an avoidance task in rats. Braz J Med Biol Res 1997; 30:961-965.
15. Bernabeu R, Schmitz P, Faillace MP et al. Hippocampal cGMP and cAMP are differentially involved in memory processing of inhibitory avoidance learning. Neuro Report 1996; 31:585-5888.
16. Bevilaqua L, Ardenghi P, Schroder N et al. Drugs acting upon the cyclic adenosine monophosphate/protein kinase A signalling pathway modulate memory consolidation when given late after training into rat hippocampus but not amygdala. Behav Pharmacol 1997; 8:331-338.

17. Bevilaqua L, Ardenghi P, Schroder N et al. Agents that affect cAMP levels or protein kinase A activity modulate memory consolidation when injected into rat hippocampus but not amygdala. Braz J Med Biol Res 1997; 30:967-970.
18. Blitzer RD, Connor JH, Brown GP et al. Gating of CaMKII by cAMP-regulated protein phosphatase activity during LTP. Science 1998; 280:1940-1942.
19. Blitzer RD, Wong T, Nouranifar R et al. Postsynaptic cAMP pathway gates early LTP in hippocampal CA1 region. Neuron 1995; 15:1403-1414.
20. Bourne HR, Nicoll R. Molecular machines integrate coincident synaptic signals. Cell 72/ Neuron 1993; 10:65-75.
21. Bourtchuladze R, Frenguelli B, Blendy J et al. Deficient long-term memory in mice with a targeted mutation of the cAMP-responsive element-binding protein. Cell 1994; 79:59-68.
22. Brown GP, Blitzer RD, Connor JH et al. Long-term potentiation induced by theta frequency stimulation is regulated by a protein phosphatase-1-operated gate. J Neurosci 2000; 20:7880-7887.
23. Byrne JH, Kandel ER. Presynaptic facilitation revisited: state and time dependence. J Neurosci 1996; 16:425-435.
24. Cadogan AK, Kendall DA, Marsden CA. Serotonin 5-HT1A receptor activation increases cyclic AMP formation in the rat hippocampus in vivo. J Neurochem 1994; 62:1816-1821.
25. Cali JJ, Zwaagstra JC, Mons N et al. Type VIII adenylyl cyclase. A Ca^{2+}/calmodulin-stimulated enzyme expressed in discrete regions of rat brain. J Biol Chem 1994; 269:12190-12195.
26. Cann MJ, Levin LR. Genetic characterization of adenylyl cyclase function. Adv Second Messenger Phosphoprotein Res 1998; 32:121-135.
27. Carew TJ, Hawkins RD, Kandel ER. Differential classical conditioning of a defensive withdrawal reflex in Aplysia californica. Science 1983; 219:397-400.
28. Castellucci VF, Blumenfeld H, Goelet P et al. Inhibitor of protein synthesis blocks long-term behavioral sensitization in the isolated gill-withdrawal reflex of Aplysia. J Neurosci 1989; 20:1-9.
29. Castellucci VF , Pinsker H , Kupfermann I et al. Neuronal mechanisms of habituation and dishabituation of the gill-withdrawal reflex in Aplysia. Science 1970; 167:1745-1748.
30. Chen Y, Harry A, Li J et al. Adenylyl cyclase 6 is selectively regulated by protein kinase A phosphorylation in a region involved in Galphas stimulation. Proc Natl Acad Sci U S A 1997; 94:14100-14104.
31. Chetkovich DM, Gray R, Johnston D et al. N-methyl-D-aspartate receptor activation increases cAMP levels and voltage-gated Ca^{2+} channel activity in area CA1 of hippocampus. Proc Natl Acad Sci USA 1991; 88:6467-6471.
32. Chiono M, Mahey R, Tate G et al. Capacitative Ca2+ entry exclusively inhibits cAMP synthesis in C6-2B glioma cells. Evidence that physiologically evoked Ca2+ entry regulates Ca(2+)-inhibitable adenylyl cyclase in nonexcitable cells. J Biol Chem 1995; 270:1149-1155.
33. Choi EJ, Xia Z, Storm DR. Stimulation of the type III olfactory adenylyl cyclase by calcium and calmodulin. Biochemistry 1992; 31:6492-6498.
34. Cho YH, Friedman E, Silva AJ. Ibotenate lesions of the hippocampus impair spatial learning but not contextual fear conditioning in mice. Behavioural Brain Research 1999; 98:77-87.
35. Cleary LJ, Lee WL, Byrne JH. Cellular correlates of long-term sensitization in Aplysia. J Neurosci 1998; 18:5988-5998.
36. Cohen NJ, Eichenbaum H. Memory, amnesia and the hippocampal system. Bradford Book, The MIT Press.
37. Cooper DMF, Karpen JW, Fagan K et al. Ca^{2+}-sensitive adenylyl cyclases. Adv Second Messenger Phosphoprotein Res 1998; 32:23-53.
38. Cooper DMF, Mons N, Karpen JW. Adenylyl cyclases and the interaction between calcium and cAMP signalling. Nature 1995; 374:421-424.
39. Dash PK, Hochner B, Kandel ER. Injection of the cAMP-responsive element into the nucleus of Aplysia sensory neurons blocks long-term facilitation. Nature 1990; 345:718-721
40. Davis RL, Cherry J, Dauwalder B et al. The cyclic AMP system and Drosophila learning. Mol Cell Biochem 1995; 149-150:271-278.
41. De Rooij J, Rehmann H, van Triest M et al. Mechanism of regulation of the Epac family of cAMP-dependent RapGEFs. J Biol Chem 2000; 275:20829-20836.
42. De Rooij J, Zwartkruis FJ, Verheijen MH et al. Epac is a Rap1 guanine-nucleotide-exchange factor directly activated by cyclic AMP. Nature 1998; 396:474-477.
43. Dessauer CW, Tesmer JJ, Sprang SR et al. Identification of a Gi alpha binding site on type V adenylyl cyclase. J Biol Chem 1998; 273:25831-25839.
44. DeZazzo J, Tully T. Dissection of memory formation: from behavioral pharmacology to molecular genetics. Trends Neurosci 1995; 18:212-218.
45. Dubnau J, Tully T. Gene discovery in Drosophila: new insights for learning and memory. Annu Rev Neurosci 1998; 21:407-444.

46. Dubnau J, Tully T. Functional anatomy: from molecule to memory. Curr Biol 2001; 11:R240-243
47. Dudai Y, Jan Y-N, Byers D et al. Dunce, a mutant of Drosophila deficient in learning. Proc Natl Acad Sci USA 1976; 73:1684-1688.
48. Fagan KA, Graf RA, Tolman S et al. Regulation of a Ca2+-sensitive adenylyl cyclase in an excitable cell. Role of voltage-gated versus capacitative Ca2+ entry. J Biol Chem 2000; 275:40187-40194.
49. Fagan KA, Mahey R, Cooper DMF. Functional colocalization of transfected Ca(2+)-stimulable adenylyl cyclases with capacitative Ca^{2+} entry sites. J Biol Chem 1996; 271: 12438-12244.
50. Fagan KA, Mons N, Cooper DMF. Dependence of the Ca^{2+}-inhibitable adenylyl cyclase of C6-2B glioma cells on capacitative Ca^{2+} entry. J Biol Chem 1998; 273:9297-9305.
51. Fagan KA, Rich TC, Tolman S et al. Adenovirus-mediated expression of an olfactory cyclic nucleotide-gated channel regulates the endogenous Ca^{2+}-inhibitable adenylyl cyclase in C6-2B glioma cells. J Biol Chem 1999; 274:12445-12453.
52. Federman AD, Conklin BR, Schrader KA et al. Hormonal stimulation of adenylyl cyclase through Gi-protein beta gamma subunits. Nature 1992; 356:159-161.
53. Frey U, Huang YY, Kandel ER. Effects on cAMP simulate a late phase of LTP in hippocampal CA1 neurons. Science 1993; 260:1661-1664.
54. Frey U, Krug M, Reymann KG et al. Anisomycin, an inhibitor of protein synthesis, blocks late phases of LTP phenomena in the hippocampal CA1 region in vitro. Brain Res 1988; 452:57-65.
55. Frost WN, Kandel ER. Structure of the network mediating siphon-elicited siphon withdrawal in Aplysia. J Neurophysiol 1995; 73:2413-2427.
56. Grewal SS, Fass DM, Yao H et al. Calcium and cAMP signals differentially regulate cAMP-responsive element-binding protein function via a Rap1-extracellular signal-regulated kinase pathway. J Biol Chem 2000; 275:34433-34441.
57. Guillou JL, Micheau J, Jaffard R. The hippocampus as a site of interaction between memory systems. Comptes Rendus de la Sociëté de Biologie (Paris) 1994; 188:459-467.
58. Guillou JL, Micheau J, Jaffard R. The opposite effects of cysteamine on the acquisition of two different tasks in mice are associated with bidirectional testing-induced changes in hippocampal adenylyl cyclase activity. Behav Neurosci 1998; 4:900-908.
59. Guillou JL, Nakata H, Cooper DMF. Inhibition by calcium of mammalian adenylyl cyclases. J Biol Chem 1999; 274:35539-35545.
60. Guillou JL, Rose GM, Cooper DMF. Differential activation of adenylyl cyclases by spatial and procedural learning. J Neurosci 1999; 19:6183-6190.
61. Hanoune J, Defer N. Regulation and role of adenylyl cyclase isoforms. Annu Rev Pharmacol Toxicol 2001; 41:145-174.
62. Huang YY, Kandel ER. Modulation of both the early and the late phase of mossy fiber LTP by the activation of beta-adrenergic receptors. Neuron 1996; 16:611-617.
63. Huang YY, Kandel ER, Varshavsky L et al. A genetic test of the effects of mutations in PKA on mossy fiber LTP and its relation to spatial and contextual learning. Cell 1995; 83:1211-1222.
64. Huang YY, Li XC, Kandel ER. cAMP contributes to mossy fiber LTP by initiating both a covently mediated early phase and macromolecular synthesis-dependent late phase. Cell 1994; 79:69-79
65. Huang YY, Martin KC, Kandel ER. Both protein kinase A and mitogen-activated protein kinase are required in the amygdala for the macromolecular synthesis-dependent late phase of long-term potentiation. J Neurosci 2000; 20:6317-6325.
66. Huang YY, Nguyen PV, Abel T et al. Long-lasting forms of synaptic potentiation in the mammalian hippocampus. Learn Mem 1996; 3:74-85.
67. Impey S, Wayman G, Wu Z et al. Type I adenylyl cyclase functions as a coincidence detector for control of cyclic AMP response element-mediated transcription: synergistic regulation of transcription by Ca^{2+} and isoproterenol. Mol Cell Biol 1994; 14:8272-8281.
68. Impey S, Obrietan K, Wong ST et al. Cross talk between ERK and PKA is required for Ca2+ stimulation of CREB-dependent transcription and ERK nuclear translocation. Neuron 1998; 21:869-883.
69. Ishikawa Y. Regulation of cAMP signaling by phosphorylation. Adv Second Messenger Phosphoprotein Res 1998; 32:99-120.
70. Iwami G, Kawabe J, Ebina T et al. Regulation of adenylyl cyclase by protein kinase A. J Biol Chem 1995; 270:12481-1244.
71. Jacobowitz O, Chen J, Premont RT et al. Stimulation of specific types of Gs-stimulated adenylyl cyclases by phorbol ester treatment. J Biol Chem 1993; 268:3829-3832.
72. Jaffard R, Meunier M. Role of the hippocampal formation in learning and memory. Hippocampus 1993; 3:203-208.
73. Johnston D, Williams S, Jaffe D et al. NMDA-receptor-independent long-term potentiation. Annu Rev Physiol 1992; 54:489-505.

74. Jordan JD, Landau EM, Iyengar R. Signaling networks: the origins of cellular multitasking. Cell 2000; 103:193-200.
75. Kandel ER, Abrams T, Bernier L et al. Classical conditioning and sensitization share aspects of the same molecular cascade in Aplysia. Cold Spring Harb Symp Quant Biol 1983; 48:821-830.
76. Kauer JA, Malenka RC, Nicoll RA. NMDA application potentiates synaptic transmission in the hippocampus. Nature 1988; 334:250-252.
77. Kawasaki H, Springett GM, Mochizuki N et al. A family of cAMP-binding proteins that directly activate Rap1. Science 1998; 282:2275-2279.
78. Klauck TM, Scott JD. The postsynaptic density: A subcellular anchor for signal transduction enzymes. Cell Signal 1995; 7:747-757.
79. Krupinski J, Cali JJ. Molecular diversity of the adenylyl cyclases. In Advances Second Messenger Phosphoprotein Res 1998; 32:53-79.
80. Krupinski J, Coussen F, Bakalyar HA et al. Adenylyl cyclase amino acid sequence: possible channel- or transporter-like structure. Science 1989; 244:1558-1564.
81. Kubie JL, Sutherland RJ, Muller RU. Hippocampal lesions produce a temporally graded retrograde amnesia on a dry version of the Morris swimming task. Psychobiology 1999; 27:313-330.
82. Lechner HA, Byrne JH. New perspectives on classical conditioning: a synthesis of Hebbian and nonHebbian mechanisms. Neuron 1998; 20:355-358.
83. Levin LR, Han PL, Hwang PM et al. The Drosophila learning and memory gene Rutabaga encodes a Ca^{2+}/calmodulin-responsive adenylyl cyclase. Cell 1992; 68:479-489.
84. Livingstone MS. Genetic dissection of Drosophila adenylate cyclase. Proc Natl Acad Sci USA 1985; 82:5992-5996.
85. Ludvig N, Burmeister V, Jobe PC et al. Electron microscopic immunocytochemical evidence that the calmodulin-dependent cyclic nucleotide phosphodiesterase is localized predominantly at postsynaptic sites in tat brain. Neuroscience 1991; 44:491-500.
86. Lustig KD, Conklin BR, Herzmark P et al. Type II adenylylcyclase integrates coincident signals from Gs, Gi, and Gq. J Biol Chem 1993; 268:13900-13905.
87. Malenka RC, Kauer JA, Zucker RS et al. Postsynaptic calcium is sufficient for potentiation of hippocampal synaptic transmission. Science 1988; 242:81-84.
88. Malleret G, Haditsch U, Genoux D et al. Inducible and reversible enhancement of learning, memory, and long-term potentiation by genetic inhibition of calcineurin. Cell 2001; 104:675-686.
89. Martin KC, Michael D, Rose JC et al. MAP kinase translocates into the nucleus of the presynaptic cell and is required for long-term facilitation in Aplysia. Neuron 1997; 18:899-912.
90. McDonald RM, Ergis AM, Winocur G. Functional dissociation of brain regions in learning and memory: evidence for multiple systems. In: Fuster J, Jelicic M, eds. Unitary versus multiple systems account of memory. London: Oxford University Press, 1999:66-103.
91. Menzel R, Muller U. Learning from a fly's memory. Nature 2001; 411:433-434.
92. Micheau J, Riedel G. Protein kinases: which one is the memory molecule? Cell Mol Life Sci 1999; 55:534-548.
93. Mons N. Analysis of adenylyl cyclase subspecies gene expression in brain by in situ hybridization histochemistry. In: Roche P, ed. G Proteins. San Diego: Academic Press, 1996:19-34.
94. Mons N, Cooper DMF. Adenylate cyclases: critical foci in neuronal signaling. Trends Neurosci 1995; 18:536-542.
95. Mons N, Guillou JL, Jaffard R. The role of Ca^{2+}/calmodulin-stimulable adenylyl cyclases as molecular coincidence detectors in memory formation. Cell Mol Life Sci 1999; 55:525-533.
96. Mons N, Harry A, Dubourg P et al. Immunohistochemical localization of adenylyl cyclase in rat brain indicates a highly selective concentration at synapses. Proc Natl Acad Sci USA 1995; 92:8473-8477.
97. Mons N, Yoshimura M, Cooper DMF. Discrete expression of Ca^{2+}/calmodulin-sensitive and Ca^{2+}-insensitive adenylyl cyclases in rat brain. Synapse 1993; 14:51-59.
98. Muglia LM, Schaefer ML, Vogt SK et al. The 5'-flanking region of the mouse adenylyl cyclase type VIII gene imparts tissue-specific expression in transgenic mice. J Neurosci 1999; 19:2051-2058.
99. Mulkey RM, Endo S, Shenolikar S et al. Involvement of a calcineurin/inhibitor-1 phosphatase cascade in hippocampal long-term depression. Nature 1994; 369:486-488.
100. Muller W, Connor JA. Dendritic spines as individual neuronal compartments for synaptic Ca^{2+} responses. Nature 1991; 354:73-76.
101. Nicoll RA, Kauer JA, Malenka RC. The current excitement in long-term potentiation. Neuron 1988; 1:97-103.
102. Nicoll RA, Malenka RC. Contrasting properties of two forms of long-term potentiation in the hippocampus. Nature 1995; 377:115-118.

103. Nielsen MD, Chan GC, Poser SW et al. Differential regulation of type I and type VIII Ca2+-stimulated adenylyl cyclases by Gi-coupled receptors in vivo. J Biol Chem 1996; 271:33308-33316.
104. O'Leary FA, Byrne JH, Cleary LJ. Long-term structural remodeling in Aplysia sensory neurons requires de novo protein synthesis during a critical time period. J Neurosci 1995; 15:3519-3525.
105. Otmakhova NA, Lisman JE. D1/D5 dopamine receptor activation increases the magnitude of early long-term potentiation at CA1 hippocampal synapses. J Neurosci 1996; 16:7478-7486.
106. Otmakhova NA, Otmakhov N, Mortenson LH et al. Inhibition of the cAMP pathway decreases early long-term potentiation at CA1 hippocampal synapses. J Neurosci 2000; 20:4446-4451.
107. Pearce JM, Roberts ADL, Good M. Hippocampal lesions disrupt navigation based on cognitive maps but not heading vectors. Nature 1998; 396:75-77.
108. Raman IM, Tong G, Jahr CE. B-adrenergic regulation of synaptic NMDA receptors by cAMP-dependent protein kinase. Neuron 1996; 16:415-421.
109. Rhee MH, Bayewitch M, Avidor-Reiss T et al. Cannabinoid receptor activation differentially regulates the various adenylyl cyclase isozymes. J Neurochem 1998; 71:1525-1534.
110. Roberson ED, English JD, Adams JP et al. The mitogen-activated protein kinase cascade couples PKA and PKC to cAMP response element binding protein phosphorylation in area CA1 of hippocampus. J Neurosci 1999; 19:4337-4348.
111. Roman G, Davis RL. Molecular biology and anatomy of Drosophila olfactory associative learning. Bioessays 2001; 23:571-581.
112. Schacher S, Castellucci VF, Kandel ER. cAMP evokes long-term facilitation in Aplysia sensory neurons that requires new protein synthesis. Science 1988; 240:1667-1669.
113. Schacter DL, Tulving E. What are the memory systems of 1994? In: Schacter DL, Tulving E, eds. Memory systems. Cambridge: The MIT Press, 1994:1-38.
114. Schaefer ML, Wong ST, Wozniak DF et al. Altered stress-induced anxiety in adenylyl cyclase type VIII-deficient mice. J Neurosci 2000; 20:4809-4820.
115. Schmitt JM, Stork PJ. beta 2-adrenergic receptor activates extracellular signal-regulated kinases (ERKs) via the small G protein rap1 and the serine/threonine kinase B-Raf. J Biol Chem 2000; 275:25342-25350.
116. Scholz KP, Byrne JH. Long-term sensitization in Aplysia: biophysical correlates in tail sensory neurons. Science 1987; 235:685-687.
117. Scholz KP, Byrne JH. Intracellular injection of cAMP induces long-term reduction of neuronal K+ currents. Science 1988; 240:1664-1666.
118. Skalhegg BS, Tasken K. Specificity in the cAMP/PKA signaling pathway. Differential expression, regulation, and subcellular localization of subunits of PKA. Front Biosci 2000; 5:D678-693.
119. Sosunov SA, Kemaikin SP, Kurnikova IA et al. Expression of adenylyl cyclase type ix and calcineurin in synapses of the central nervous system. Bull Exp Biol Med 2001; 131:172-175.
120. Storm DR, Hansel C, Hacker B et al. Impaired cerebellar long-term potentiation in type 1 adenylyl cyclase mutant mice. Neuron 1998; 20:1199-1210.
121. Sugita S, Baxter DA, Byrne JH. Modulation of a cAMP/protein kinase A cascade by protein kinase C in sensory neurons of Aplysia. J Neurosci 1997; 17:7237-7244.
122. Sugita S, Goldsmith JR, Baxter DA et al. Involvement of protein kinase C in serotonin-induced spike broadening and synaptic facilitation in sensorimotor connections of Aplysia. J Neurophysiol 1992; 68:643-651.
123. Sunahara RK, Beuve A, Tesmer JJG et al. Exchange of substrate and inhibitor specificities between adenylyl and guanylyl cyclases. J Biol Chem 1998; 273:16332-16338.
124. Sunahara RK, Dessauer CW, Gilman AG. Complexity and diversity of mammalian adenylyl cyclases. Ann Rev Pharmacol Toxicol 1996; 36:461-480.
125. Sutkowski EM, Tang WJ, Broome CW et al. Regulation of forskolin interactions with type I, II, V, and VI adenylyl cyclases by Gs alpha. Biochemistry 1994; 33:12852-12859.
126. Tang WJ, Hurley JH. Catalytic mechanism and regulation of mammalian adenylyl cyclases. Mol Pharmacol 1998; 54:231-240.
127. Tang WJ, Krupinski J, Gilman AG. Expression and characterization of calmodulin-activated (type I) adenylylcyclase. J Biol Chem 1991; 266:8595-8603.
128. Taussig R, Gilman AG. Mammalian membrane-bound adenylyl cyclases. J Biol Chem 1995; 270:1-4.
129. Taussig R, Iniguez-Lluhi JA, Gilman AG. Inhibition of adenylyl cyclase by Gi alpha. Science 1993; 261:218-221.
130. Taussig R, Tang WJ, Hepler JR et al. Distinct patterns of bidirectional regulation of mammalian adenylyl cyclases. J Biol Chem 1994; 269:6093-6100.
131. Taussig R, Zimmermann G. Type-specific regulation of mammalian adenylyl cyclases by G protein pathways. Adv Second Messenger Phosphoprotein Res 1998; 32:81-98.

132. Taylor JR, Birnbaum S, Ubriani R et al. Activation of cAMP-dependent protein kinase A in prefrontal cortex impairs working memory performance. J Neurosci 1999; 19:RC23.
133. Tesmer JJ, Sprang SR. The structure, catalytic mechanism and regulation of adenylyl cyclase. Curr Opin Struct Biol 1998; 8:713-719.
134. Thomas MJ, Moody TD, Makhinson M et al. Activity-dependent beta-adrenergic modulation of low frequency stimulation induced LTP in the hippocampal CA1 region. Neuron 1996; 17:475-482
135. Tolman ED. Cognitive maps in rats and men. Psychol Rev 1948; 55:189-208.
136. Tully T, Bolwig G, Christensen J et al. A return to genetic dissection of memory in Drosophila. Cold Spring Harb Symp Quant Biol 1996; 61:207-218.
137. Tully T, Préat T, Boynton SC et al. Genetic dissection of consolidated memory in Drosophila. Cell 1994; 79:35-47.
138. Tully T, Quinn WG. Classical conditioning and retention in normal and mutant Drosophila melanogaster. J Comp Physiol 1985; 157:263-277.
139. Tsygankova OM, Saavedra A, Rebhun JF et al. Coordinated regulation of Rap1 and thyroid differentiation by cyclic AMP and protein kinase A. Mol Cell Biol 2001; 21:1921-1929
140. Tsu RC, Wong YH. Gi-mediated stimulation of type II adenylyl cyclase is augmented by Gq-coupled receptor activation and phorbol ester treatment. J Neurosci 1996; 16:1317-1323.
141. Villacres EC, Wong ST, Chavkin C et al. Type I adenylyl cyclase mutant mice have impaired mossy fiber long term potentiation. J Neurosci 1998; 18:3186-3194.
142. Vossler MR, Yao H, York RD et al. cAMP activates MAP kinase and Elk-1 through a B-Raf- and Rap1-dependent pathway. Cell 1997; 89:73-82.
143. Waltereit R, Dammermann B, Wulff P et al. Arg3.1/Arc mRNA induction by Ca^{2+} and cAMP requires protein kinase A and mitogen-activated protein kinase/extracellular regulated kinase activation.J Neurosci 2001; 21:5484-5493.
144. Wayman GA, Impey S, Storm DR. Ca^{2+} inhibition of type III adenylyl cyclase in vivo. J Biol Chem 1995; 270:21480-21486.
145. Wayman GA, Impey S, Wu Z et al. Synergistic activation of the type I adenylyl cyclase by Ca^{2+} and Gs-coupled receptors in vivo. J Biol Chem 1994; 269:25400-24405.
146. Wayman GA, Wei J, Wong S et al. Regulation of type I adenylyl cyclase by calmodulin kinase IV in vivo. Mol Cell Biol 1996; 16:6075-6082.
147. Wei J, Wayman G, Storm DR.Phosphorylation and inhibition of type III adenylyl cyclase by calmodulin-dependent protein kinase II in vivo. J Biol Chem 1996; 271:24231-24235.
148. Weisskopf MG, Castillo PE, Zalutzky RA et al. Mediation of hippocampal mossy fiber long-term potentiation by cyclic AMP. Science 1994; 23:1878-1882.
149. Winder DG, Mansuy IM, Osman M et al. Genetic and pharmacological evidence for a novel, intermediate phase of long-term potentiation suppressed by calcineurin. Cell 1998; 92:25-37.
150. Wong ST, Athos J, Figueroa XA et al. Calcium-stimulated adenylyl cyclase activity is critical for hippocampus-dependent long-term memory and late phase LTP. Neuron 1999; 23:787-798.
151. Wu ZL, Thomas SA, Villacres EC et al. Altered behavior and long-term potentiation in type 1 adenylyl cyclase in mice. Proc Natl Acad Sci USA 1995; 92:220-224.
152. Xia ZG, Refsdal CD, Merchant KM et al. Distribution of messenger-RNA for the calmodulin-sensitive adenylyl cyclase in rat brain: expression in areas associated with learning and memory. Neuron 1991; 6:431-443.
153. Yau KW. Cyclic nucleotide-gated channels: an expanding new family of ion channels. Proc Natl Acad Sci U S A 1994; 91:3481-3483.
154. Yoshimura M, Cooper DMF. Type-specific stimulation of adenylyl cyclases by protein kinase C. J Biol Chem 1993; 268:4604-4607.
155. Yu TP, McKinney S, Lester HA et al. Gamma-aminobutyric acid type A receptors modulate cAMP-mediated long-term potentiation and long-term depression at monosynaptic CA3-CA1 synapses. Proc Natl Acad Sci U S A 2001; 98:5264-5269.
156. Yuste R, Denk W. Dendritic spines as basic functional units of neuronal integration. Nature 1995; 375:682-684.
157. Zars T, Wolf R, Davis R et al. Tissue-specific expression of a type I adenylyl cyclase rescues the rutabaga mutant memory defect: in search of the engram. Learn Mem 2000; 7:18-31.
158. Zimmermann G, Taussig R. Protein kinase C alters the responsiveness of adenylyl cyclases to G protein alpha and betagamma subunits. J Biol Chem 1996; 271:27161-27166.
159. Zufall F, Leinders-Zufall T. The cellular and molecular basis of odor adaptation. Chem Senses 2000; 25:473-481.

CHAPTER 4.2

Phospholipases and Oxidases

Christian Hölscher

Abstract

Memory formation is dependent on a series of biochemical cascades that alter synaptic transmission and neuronal activity. Phospholipases are key enzymes in these cascades that produce second messengers, which interact with a host of target systems, such as transmitter uptake systems, transmitter release, and intracellular calcium stores. One of the main second messengers is arachidonic acid, which also acts as a substrate for lipoxygenases and cyclooxygenases. These enzymes metabolise arachidonic acid to second messengers such as prostaglandins and leukotrienes. All of the transmitters have been shown to be of importance in the induction of synaptic plasticity and in learning and memory formation in rodents. In learning tasks such as a passive avoidance task in day-old chicks, inhibitors of phospholipases prevented the consolidation of memory from 1 h after training onwards, while inhibitors of cyclooxygenases blocked memory consolidation from 2 h onwards. These results show that messengers synthetised by phospholipases and oxidases are most likely part of a serial messenger cascade that underlies memory formation. Such a cascade could enable the system to filter information and enable forgetting before complete consolidation of long-term memory.

Introduction

As has been described in detail in the previous chapters, memory formation is a process that largely depends on neuronal metabolic mechanisms. Neurotransmitters are passed on between neurons and activate specific receptors on the postsynaptic site, and the information has to be transmitted beyond the neuronal membrane. Some receptors are linked to ion channels that promote influx of Ca^{2+} (a second messenger; see the chapter on Ca^{2+}channels), other receptors are linked to G-proteins that are located on the other side of the membrane inside the cell. These in turn activate enzymes that release second messengers or modulate ion channels. An important family of enzymes that generates second messenger is the group of phospholipases. The activity of the phospholipases induces the release of a range of second messengers. These messengers then can be metabolised further by downstream enzymes, such as the cyclooxygenases. All of these messengers interact with a multitude of target systems: ion channels, receptors, intracellular calcium stores, transmitter release systems, cytoskeleton modifying systems, immediate early gene activation, and more. This chapter will describe some of the known types of phospholipases and their role in the complex network of cellular activities that underlies memory formation.

From Messengers to Molecules: Memories Are Made of These, edited by Gernot Riedel and Bettina Platt.

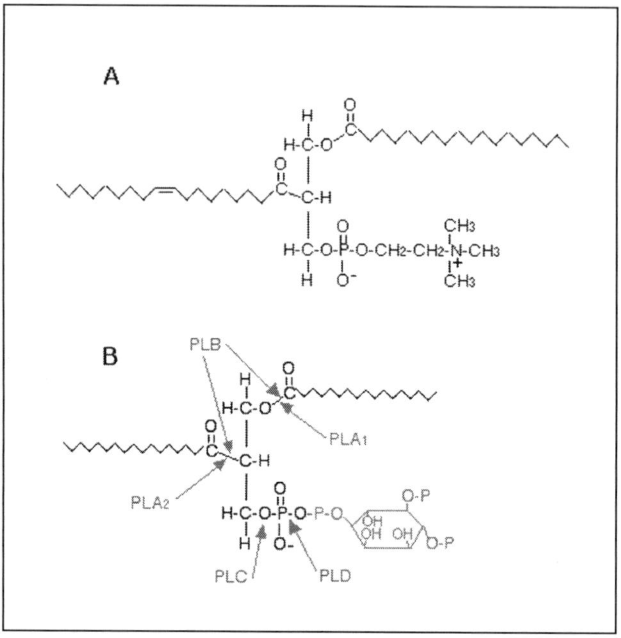

Figure 1. A) Shown is the structure of phosphatidylcholine, a phospholipid. Note the saturated fatty acid residue at the C1 position (top) and the unsaturated fatty acid residue at the C2 position (left). The lower right is a choline residue linked to the glycerol-3-phosphate frame. Triacylglycerates do not contain such groups, instead they have 3 fatty acid residues. B) Phospholipases are defined by their enzymatic actions. Shown is the molecule phosphoinositol bisphosphate, which contains inositol-1,4,5-triphosphate as a residue (lower right molecule). This residue is an important second messenger that modulates intracellular Ca^{2+} concentrations (see text for details). The locations where the phospholipases splice the molecule are shown. For details see Karlson.[57]

Phospholipases

What Are Phospholipases?

Phospholipases are enzymes that degrade phospholipids. Phospholipids are glycerolphosphate groups with two fatty acid residues and a functional group (see Fig. 1A). They are lipophilic and part of cell membranes. Depending on the pathway of catabolism one differentiates between several types of phospholipases. Phospholipase type A1 (PLA_1) degrades phospholipids by cutting off a fatty acid residue at a defined site (see Fig. 1B for details on phospholipase metabolism). Two molecules are produced that both act as second messengers. A different group is called phospholipase type A2 (PLA_2) which separates the substrate at a different site (see Fig. 1B). Furthermore, phospholipase type B, C, and D (PLB-PLD) are known. PLC metabolises phosphoinositol bisphosphate (PIP_2) to diacylglycerates (DAG) and inositoltriphosphate (IP_3). Both DAG and IP_3 are important second messengers. PLD metabolises phospholipids at a different site than PLC (see Fig. 1B)(for further details on phospholipases see refs. 16,36,57,99). The best described phospholipases are PLA_2 and PLC and they appear to play the most important roles in neuronal metabolism.

It is important to state that phospholipases are found in all cell types. They are a part of the basic biochemical machinery that is required for cell metabolism. However, neuronal isoforms of phospholipases are known that have particular properties and play specific roles in neuronal communication. These isoforms and their neuron-specific roles will be described in this chapter.

Phospholipases are mostly located near the membranes close to their substrates. However, a cytosolic Ca^{2+}-dependent isoform of PLA_2 exists. Similar to γ-protein kinase C (PKCγ), PKA_2 binds to the cell membrane when activated by Ca^{2+}. This isoform is neuron-specific and appears to play an important role in neuronal communication and in the induction of synaptic plasticity (long-term potentiation of synaptic transmission, LTP).[70,113]

Phospholipases produce two molecules when degrading phospholipids. One is a functional group (such as IP_3, or arachidonic acid), the other is diacylglycerate (DAG). All of those molecules act as a second messengers. DAG can activate PKC (see Nogues et al in this book, or ref. 15 for review). The released DAG can be further degraded by a DAG lipase to release more fatty acids.[29,73]

The type of fatty acids that is released by phospholipase activity depends on the kind of phospholipids that are present in the membranes. In non-neuronal cells (eg. in adipose cells), the types of fatty acids are quite diverse and can be saturated, unsaturated, or poly-unsaturated. In neurons, however, the concentration of the poly-unsaturated fatty acid arachidonic acid (ArA) is relatively high in cell membranes, and the percentage of ArA that is released after stimulation of neuronal activity and transmission is very high compared to other fatty acids, such as oleic acid or linoleic acid.[9,15,19]

The second messenger ArA has a number of important biological properties. It evokes Ca^{2+} release from intracellular stores,[62,91] increases glutamate release,[69] modulates ion channels,[107] and inhibits uptake of glutamate in neurons and astrocytes.[107,114]

How Are Phospholipases Activated?

Phospholipases are activated by several mechanisms. Ca^{2+} -sensitive PLA types are activated after neuronal activity which opens voltage-dependent Ca^{2+} channels, or the Ca^{2+} channel that is associated with the *N*-methyl-*D*-aspartate sensitive glutamate receptor (NMDA receptor). A second mechanism is the direct activation via metabotropic receptors that are linked to a G-protein. G-proteins act as interfaces between metabotropic receptors and intracellular target molecules, such as second messenger generating systems[4] or membrane-bound channels.[92] One family of metabotropic receptors that plays important roles in neuronal communication is the metabotropic glutamate receptor family (mGluR)[10,54] (see also Riedel et al in this book).

As has been described in the section 'Glutamate receptors', there are several subtypes of mGluRs, divided into three main groups. Group I (mGluR1 and 5) is coupled to a PLC via G proteins and modulate the synthesis of inositol-1,4,5-triphosphate.[80,94] These mGluRs can also be linked to PLD,[24,81] but most likely only transiently during development.[62]

Other metabotropic receptors that are linked to PLC are the acetylcholine receptors. The release of ArA via PLC can be triggered by carbachol, an acetylcholine agonist.[101] In primary cortical cultures from mice lacking the muscarinergic type 1 acetylcholine receptor, agonist-stimulated phosphoinositide hydrolysis was reduced by more than 60% compared to cultures from wild type mice.[39] Release of ArA can also be induced by activation of a type 2 serotonin receptor.[31]

It is of interest to note in this context that beta-amyloid fragments, a class of polypeptides that accumulates in the brains of Alzheimer's disease, inhibited the cytosolic PLC in the presence of increased Ca^{2+} concentrations.[102] In another study, it was found that acetylcholine receptors are uncoupled from PLC by beta-amyloid fragments and the production of inositol phosphates was compromised.[60]

Phospholipase activity is under strict control. Piomelli and Greengard[83]studied PLA_2 activity in in vitro enzyme assays and found that it is modulated by casein kinase II, CaMkinase II, and protein kinase A.

Arachidonic Acid (ArA), a Second Messenger

ArA is one of the main messengers produced by phospholipase activity.[73] Unlike other second messengers, the eicosanoids such as ArA and metabolites of ArA are able to leave the cell in which they are generated and act as first messengers on neighbouring cells. Therefore, a role as an intercellular messenger is plausible.[84]

The conditions for release and the neurophysiological effects of ArA have been investigated in numerous experiments. Evidence for a role of ArA in neuronal transmission in particular will be discussed here.

Release of ArA

As mentioned above, ArA is released by metabolising phospholipids by PLA_2 or PLC activity.[4] In striatal neurons, ArA was released by coactivation of the α-amino-3-hydroxy-5-methyl-4-isoxazole-propionic acid sensitive glutamate receptor (AMPA receptor, an ionotropic glutamate receptor subtype) and mGluRs and voltage gated Ca^{2+} channels.[10,95] In primary cultures of striatal neurons, this release was sensitive to quinacrine, an inhibitor of PLA_2, that also blocked LTP in this preparation. An inhibitor of diacylglycerol (DAG) lipase, another source for ArA release,[29,73] had no effect on LTP formation at early time points.[29]

In the hippocampal slice of the rat, NMDA receptor activation caused release of ArA and oleic acid. This release was inhibited by inactivation of PLA_2, or by NMDA receptor blockade with MK-801 (dizocilpine).[82]

Other studies showed that release of ArA following activation of the NMDA receptor is due to the activation of a Ca^{2+}-dependent PLA_2. For example, the NMDA- stimulated ArA release in primary cultures of rat hippocampal neurons was inhibited by the NMDA receptor channel blockers Mg^{2+}, 5-amino phosphonovalerate (AP5), or the PLA_2 and lipoxygenase inhibitor Nordihydroguaiaretic acid (NDGA).[92,103]

These results suggest that there is a chain of events that starts with the activation of NMDA receptors, an increase of intracellular Ca^{2+} levels, and the activation of Ca^{2+}-dependent phospholipases that induce ArA release.

In a different study, it was observed that the fatty acids ArA, 12-hydroxy-6,8,11,14-eicosatetraenoic acid (12-HETE), and 12-hydroperoxyeicosatetraenoic acid (12-HPETE) were released after LTP induction in the rat in vivo. However, the concentrations of 12-HETE release declined already after 1 h to baseline,[8] while ArA levels remained high. In a follow up study, after the induction of LTP in the hippocampal slice, ArA concentration in the postsynaptic membrane fraction increased, due to PLA_2 activity in the first minutes after induction. 45 min and 3 h later, PLC activity had been responsible for the release.[19] This shows that PLC plays a role in the LTP-induced increase of ArA, but the release is later than PLA_2 dependent ArA release. In synaptoneurosomes of rat cortex neurons, 30% of arachidonic acid release was inhibited by neomycin, an inhibitor of PLC, and 60% by quinacrine, an inhibitor of PLA_2. The effect was additive when both inhibitors were given.

Time Course of Release

After the induction of LTP in the hippocampal slice, ArA concentration in the postsynaptic membrane fraction increased. This increase had a defined time course. Release of ArA that was due to PLA_2 activity, which took place in the first 2.5 minutes after induction, but went down shortly thereafter. The PLC and DAG lipase-linked pathway of ArA release had a different time course. ArA concentrations were increased as late as 45 min and 3 h after stimulation.[19,20] This later wave of ArA release can hardly be due to the time that PLC or DAG need to release ArA. The data might show a second step needed for LTP consolidation by prolonging duration of a second messenger signal.

Targets of ArA

ArA has a number of effects in neurons, from channel modulation to reuptake inhibition. All these different activities seem to work towards changing the basal state of the neuron to a state of increased excitability.

In the rat hippocampal slice, ArA can directly mobilise intracellular Ca^{2+} independent of IP_3,[68,91] in rat hippocampal synaptosomes, ArA and metabolites of ArA[67-69] were able to stimulate hydrolysis of PIP_2[68,69] and release of glutamate.[69] ArA inhibits uptake of glutamate in neurons and astrocytes in cultures of rat cortex tissue. The uptake mechanisms in neurons are 20 times more sensitive to modulation than the ones in astrocytes.[114] In a patch-clamp study, ArA strongly inhibited glutamate uptake in glial cells.[6]

Cis-fatty acids such as ArA can activate protein kinase C (PKC), an enzyme involved in key processes of LTP formation, as well as in memory formation (see Nogues et al in this book). As described before, PLC turns PIP_2 over into IP_3 and DAG. Lester et al[65] and Kato et al[58] showed that ArA and DAG act synergistically to activate PKC in vitro and in vivo. In another study, ArA increased B-50 phosphorylation,[93] a presynaptic protein which is a PKC substrate and which is associated with transmitter release.[27,93] ArA is rapidly cleared from the cytosol, which is important for the deactivation of the messenger signal after the transmitter-releasing stimulus has stopped.[73]

ArA and Metabolites of ArA As Transmitters and 'Retrograde Messengers' in Synaptic Plasticity

Apart from playing a role as intracellular messengers, a further role that these messengers might play is that of a feedback signal to the presynaptic site (Fig. 2). Since changes can be observed at the presynaptic site after induction of LTP, it has been speculated that a feedback signal from the postsynaptic site must exist to relay the information that the presynaptic neuron has successfully activated the postsynaptic neuron.[109]

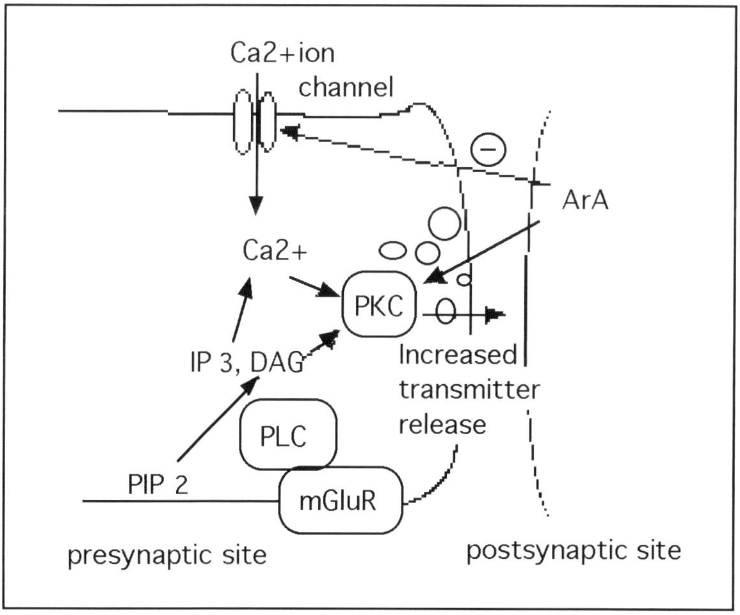

Figure 2. Proposed model of increase or decrease of transmitter release after the activation of phospholipases and the postsynaptic release of ArA. Presence of ArA reduces presynaptic Ca^{2+} influx which will in turn reduce transmitter release. If ArA is present along with DAG and Ca^{2+}, PKC will be activated which causes increases in transmitter release. The DAG could be released by a PLC which is linked to a presynaptic mGluR. The IP_3 released by the PLC increases intracellular Ca^{2+} levels via internal store depletion. See text for details.

In sensory neurons of the sea snail *Aplysia*, where synaptic plasticity appears to be expressed mostly at the presynaptic site, extensive studies of the activity of ArA in neuronal communication have been conducted. Metabolites of ArA are released by sensory neurons in response to inhibitory transmitters and directly target a class of K^+ channel, increasing the probability of their opening. This causes hyperpolarisation and shortening of action potentials. In vertebrate neurons, other types of K^+ channels have been found to be sensitive to ArA and to other polyunsaturated fatty acids.[107]

In the motor end plate of *Xenopus*, an ArA metabolite has been found to play a role as a retrograde messenger. In the muscle, a G-protein dependent release of ArA into the membrane was observed. Injecting a non-hydrolysable GTP analogue into the muscle to activate the G-protein increased spontaneous firing rates of the innervating neurons. Since the muscle is the postsynaptic site and the motor neuron the presynaptic site, retrograde communication must have taken place. Analysis of the ArA metabolites in the neurons identified 5-HPETE as a prime candidate.[40]

Prostaglandins (PG) were shown to be excreted after activation of neurons that use noradrenaline (NA) or acetylcholine (ACh) as transmitters. The prostaglandin PGE seems to have a negative effect on NA release. Indomethacin, a cyclooxygenase (COX) inhibitor which prevents PG formation, increased NA release. In intestines, PG increased ACh evoked muscle contraction, and indomethacin produced a decrease of contractions in the gut. While PG played a negative feedback role in the NA system, it activated the ACh system in the peripheral nervous system in a positive feedback manner[41]

These ArA and ArA metabolite -related changes in synaptic transmission suggest that these messengers indeed travel between neurons and act as some kind of retrograde messenger.

How to Ensure Selectivity of ArA Messenger Activity

If ArA and metabolites would change synaptic activity in an indiscriminate manner, any ArA molecules that diffuse to other neurons than to the target neuron would create a chaotic situation by indiscriminately upregulating neuronal transmission. There is evidence that the target neurons require more than one signal for modulation to prevent a non-selective change in synaptic activity. ArA can modulate the release of the neurotransmitter glutamate. Ca^{2+}-dependent glutamate transmitter release was found to be inhibited by ArA in a PKC independent fashion in cerebrocortical synaptosomes.[42,44] This effect seemed to be due to the reduction of Ca^{2+} entry into the presynaptic site that was caused by ArA.[43] Yet, if the preparation was incubated with ArA and low concentrations of phorbol esters simultaneously, PKC was activated, and glutamate release potentiated.[73] The same result was obtained when incubating the preparation with ArA and a synthetic DAG analogue. The increase of glutamate release was Ca^{2+} dependent.[115] In the presence of ArA, only very low concentrations of phorbol esters were needed to activate PKC and to increase glutamate release. The authors suggest that ArA serves as a retrograde messenger which only increases transmitter release if it coincides with a second signal such as Ca^{2+} entry into the presynapse or diacylglycerol formation.[45,46] This mechanism would ensure that only previously active neurons are upregulated in their transmitter release, as these neurons have increased intracellular Ca^{2+} concentrations when the ArA signal arrives. There is further evidence for such a mechanism. A PKC substrate which is associated with transmitter release is the presynaptic protein B-50.[27] Perfusion with high concentrations of ArA increased B-50 phosphorylation in a synaptoneurosome preparation. Addition of Ca^{2+} to the medium facilitated this, presumably because activation of the Ca^{2+} -sensitive PKC required only low amounts of ArA to increase its activity.[93]

This 'simultaneity detection system' is not unique. Similar modes of operation have been suggested for other molecules, such as PKC,[30,44] the NMDA receptor,[21] or nitric oxide (NO).[50] See also Bourne and Nicoll[13] for a discussion on coincidence detecting systems in the nervous system. Since ArA will diffuse into neighbouring cells and synapses which are not activated at the same time as the neuron that releases ArA, a potentiation of that inactive neuron would not be sensible. Instead, the inhibiting effect could suppress neurons that are not firing at the same time as the active neurons which causes ArA release. If the neighbouring neuron is not active, or if this activity is not in synchrony with the ArA releasing neuron, it is most probably not working in cooperation with the active neuron. Suppression of such 'non-cooperating' neurons thereby keeps spontaneous firing 'noise' and interference down. If ArA reaches a synapse that recently has been active, Ca^{2+} levels will be high due to activation of voltage dependent channels, and diacylglycerol will be available, for instance via activation of presynaptic glutamate metabotropic receptors that are linked to a PLC.[23,44,45]

Contradictory results observed in the hippocampal slice preparation could be explained by this model. Perfusion of the slice preparation with the mGluR agonist 1S,3R-1-amino-cyclopentyl-1,3-dicarboxylic acid (1S-3R-ACPD) produced LTP in area CA1.[11] O'Mara et al[77] and Collins and Davies,[22] however, did not find a potentiation of transmission after perfusion with 1S-3R-ACPD in the CA1 area. Instead, a depression (LTD) was observed. Only when ArA was perfused along with 1S-3R-ACPD, a potentiation developed over 30 min. ArA alone also created a slight depression of field excitatory postsynaptic potentials (EPSPs).[22] A similar result was published by Zhang and Dorman[115] who found that KCl-induced depolarisation increased glutamate release when ArA and a diacylglycerol analogue were added. This process was Ca^{2+} dependent.

The Role of ArA in LTP Formation

In various experiments, ArA and metabolites of ArA were shown to play a role in the induction of LTP. In in vivo studies of the dentate gyrus of the rat, NDGA blocked the synaptic component of LTP and the associated increase in release of glutamate. LTP produced a sustained increase of ArA release that was blocked by NDGA.[68] After the induction of LTP in the hippocampal slice, ArA concentration in the postsynaptic membrane fraction increased. The release was PLA_2 and PLC dependent.[19] The specific PLA_2 inhibitor bromophenacyl bromide caused a large reduction in the magnitude of LTP in the CA1 field of the hippocampal slice.[70]

In vivo blockade of PLA_2 by quinacrine in the hippocampus of the rat inhibited oleate release and LTP formation, the effect was reversible by application of oleate.[66] In the CA1 region of rat slices, ArA produced LTP and LTD, which was inhibited by NDGA[28,76] or AP5.[76] ArA together with 1-oleyl-2-acetyl glycerol (a DAG analogue) induced LTP in the CA1 region of guinea pig slices in low Mg^{2+} concentration, this was blocked by the phospholipase inhibitors neomycin and 2-nitro-4-carboxylphenyl-N,N-diphenylcarbamate (NCDC).[59] ArA induced activity-dependent LTP in the hippocampus of the rat in vitro and in vivo. ArA itself was not capable of inducing synaptic plastic effects. However, weak tetanic stimulation that did not potentiate synaptic transmission by itself was required to induce LTP in the slice.[8,76,108]

It seems to follow from these results that ArA is of importance in LTP induction. However, the comparatively slow effect of ArA in LTP induction made Williams et al[108] suggest that ArA only plays a role as a 'slower' retrograde messenger, while faster messengers such as NO precede the ArA signal (see also Fig. 3).

Learning Experiments: Evidence for the Role of Phospholipase Activity in Memory Formation

Inhibiting the activity of phospholipases in learning experiments have shown that these enzymes are of importance for memory formation or consolidation. In a one -trial passive avoidance task of the chick, animals learn to associate the unattractive taste of methylanthranilate and the colour of a bead. The animals only need one trial to form the association. This task offers the advantage of being able to 'time' the steps of memory formation.

In one study, bilateral intracerebral injections of the PLA_2 and lipoxygenase inhibitor nordihydroguaiaretic acid (NDGA) or the PLA_2 inhibitor aristolochic acid (AST) were made into the intermediate medial hyperstriatum ventrale (IMHV), an area that is of importance for the formation of memories formed by learning this task. Pre-training injections of either inhibitor produced lasting amnesia for the avoidance response. The onset of amnesia

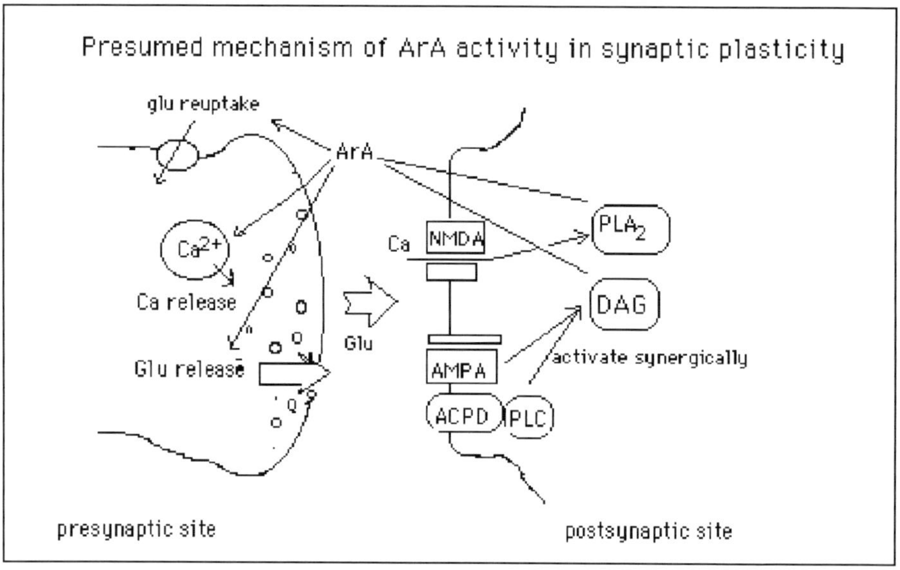

Figure 3. Schematic graph depicting the activity of the presumed retrograde messenger ArA. After NMDA receptor activation (in the cortex, or hippocampus), or synchronous AMPA and mGluR activation (in the striatum), ArA is released via PLA_2 (NMDA receptor dependent) or PLC (mGluR dependent) activation. ArA causes intracellular Ca^{2+} release and increases transmitter release, while transmitter re-uptake is inhibited. Additionally, membrane fluidity is changed by the increased percentage of unsaturated fatty acids, which modulates receptors activity. This enhances synaptic plasticity.

for both inhibitors NDGA and AST was at 1.25hr post-training (see Figs 4 and 5). Injection of drugs post-training had no effect on retention.[53] The results support the theory that ArA release is a necessary step in the relatively early events mediating the synaptic plasticity associated with memory formation. Other phospholipase inhibitors have produced similar results. For example, the PLA_2 inhibitor bromoenol lactone impaired spatial memory formation in mice.[32]

In further support of the notion that the release of ArA plays an important role in memory formation, a study of young and aged rats showed that aged rats with learning impairments had lower concentrations of ArA in their brain tissue, while saturated fatty acids were increased.[106]

Chicks were given bilateral intracerebral injections of NDGA (5µl of a 4mM solution) or saline 30min before training and tested at the stated times post-training (n=13-18 per group; *=p<0.05). For details see Hölscher and Rose.[53]

Chicks were given bilateral injections of 5 µl of a 4 mM AST solution or saline 30 min before training, testing was after 1, 1.25 hr or 24 hr subsequently. (n=15-18 per group; *=p<0.05). For details see Hölscher and Rose.[53]

A Different Second Messenger Released by PLA_2: Platelet-Activated Factor (PAF)

As an example for another cellular messenger that is released by PLA_2 and that plays important roles in memory formation is the platelet-activating factor (PAF; 1-O-alkyl-2-acyl-sn-3-phosphocholine). Some results of importance are that the induction of LTP can be blocked by a PAF receptor antagonists in area CA1 of the rat hippocampus,[2] in the dentate gyrus,[59] and in other areas of the brain.[37] Furthermore, PAF was found to inhibit ionotropic GABA receptor activity[17] and to increase glutamate transmitter release, perhaps as a retrograde synaptic messenger.[63] A mouse strain lacking the PAF receptor also showed impaired LTP in some areas of the hippocampus.[18] PAF furthermore couples synaptic events with gene expression by

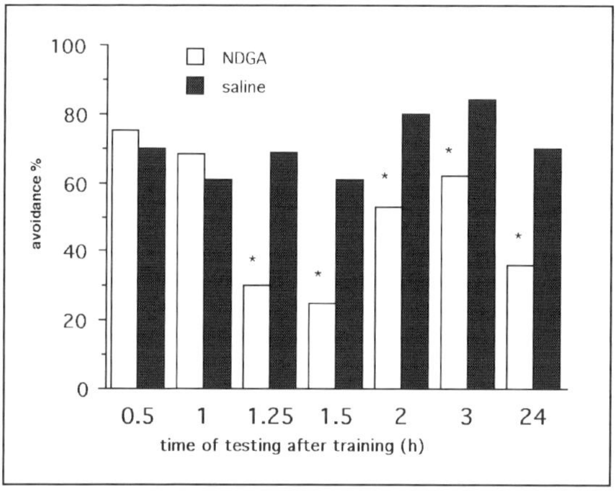

Figure 4. Effect of the phospholipase inhibitor NDGA on retention of a one-trial passive avoidance task when injected pre-training and tested post-training. Animals were trained to avoid to peck a bead dipped in an unpleasant substance. Shown are the percentages of animals pecking or avoiding the bead that they were trained on. Avoidance indicates memory retention, while pecking suggests that the animals had forgotten the task.

stimulating a FOS/JUN/AP-1 transcriptional signalling system, as well as transcription of COX-2 (inducible prostaglandin synthase, see below).

Most interestingly in the context of this chapter is the indication that PAF enhances memory formation if infused into the hippocampus or other learning-related brain areas.[56,104]

Oxygenases That Are of Importance in Memory Formation

ArA serves as a substrate for cyclooxygenases (COX) and lipoxygenases, leading to products such as prostaglandins and leukotrienes (see Fig. 8), which are cellular messengers themselves.[23,57,73] In the following sections, a brief description of their role will be given to cast some light on other parts of the biochemical cascades that form the basis of memory formation.

Lipoxygenases

ArA metabolism via the lipoxygenase pathway leads to formation of a family of messengers, such as leukotrienes. The first step in leukotriene biosynthesis is catalyzed by the enzyme lipoxygenase and results in the formation of hydroperoxy-6,8,11,14-eicosatetraenoic acid (5-HPETE) in the case of 5-lipoxygenase, which may be converted enzymatically (via the action of a peroxidase) or nonenzymatically to the corresponding hydroxy acid, 5-HETE (see McMillan et al, 72 for a summary).

A Ca^{2+}-dependent lipoxygenase has been described.[72] This isoform is activated by increased intracellular Ca^{2+}levels, as are observed after activation of neurons, due to opening of voltage dependent Ca^{2+} channels, opening of NMDA receptor associated Ca^{2+} channels, or due to release of Ca^{2+} from intracellular stores. Activation of neurons therefore could activate or prime the lipoxygenase through Ca^{2+} influx, as it is the case with many other Ca^{2+} dependent enzymes. In the canine brain, a 12-lipoxygenase was found to be widely distributed: it was localised in hippocampus, cortex, and basal ganglia.[75] Lipoxygenases and their products (e.g., 15-HETE) also have been found in the chick cerebrum and cerebellum.[38]

As described earlier, extensive studies of the activity of ArA metabolites in neuronal communication have been conducted. 12-lipoxygenase derivatives of ArA are released by sensory neurons in response to inhibitory transmitters and directly target a class of K^+ channel, increasing the probability of their opening. This causes hyperpolarisation and shortening of action potentials. In vertebrate neurons, other types of K^+ channels have been found to be

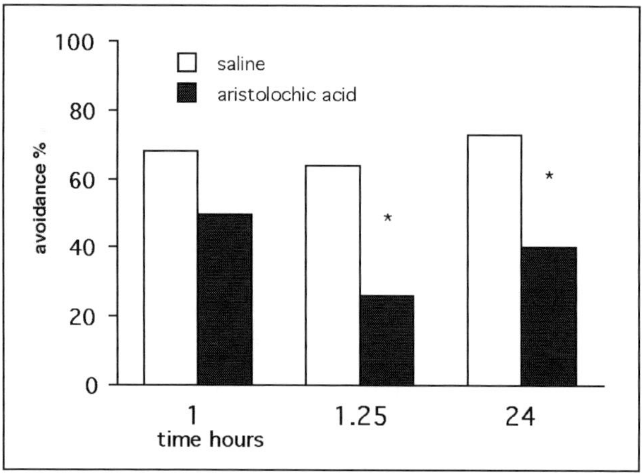

Figure 5. Effect of the phospholipase inhibitor aristolochic acid (AST) on retention when injected pre-training (4mM) and tested post-training.

sensitive to ArA, to lipoxygenase products, and to other polyunsaturated fatty acids.[107] Piomelli et al[84] isolated lipoxygenase metabolites of ArA, 5-HETE and 12-HETE, from *Aplysia* nervous tissue, as well as cyclooxygenase products, prostaglandins such as PGE_2 and $PGF_{2\alpha}$.

In a behavioural study, the lipoxygenase product 1-oleoyl-2-docosahexaenoyl-sn-glycero-3-phosphorylcholine (ODHPC), a phosphatidylcholine, enhanced discriminatory shock avoidance learning in rats. Also, ODHPC enhanced LTP of population spikes in the CA1 region.[55] These results indicate that lipoxygenase products are involved in the induction of synaptic plasticity and memory formation.

Cyclooxygenases

What Are Cyclooxygenases?

Cyclooxygenases (COX) convert ArA into a variety of metabolites, mainly into prostacyclins, prostaglandins, or thromboxanes[57] (see also Fig. 7). The enzymes oxidise two double bonds with two oxygen molecules and form a two-ring system (PGG_2) in an epoxy-reaction. This metabolite is very unstable and can be converted to any of the three main groups of COX products.

COXs are found in all tissues. Enzyme concentrations vary greatly depending on tissue type.[99] At least two isoforms exist. COX-1 is a constitutive enzyme, the enzyme is always present and the level of expression is regulated in a steady state mode. The enzyme is present in neurons, macrophages, fibroblasts, and endothelial cells.

COX-2 is an inducible form of cyclooxygenase. It is induced by pro-inflammatory agents including interleukin-1β and lipopolysaccharides. Anti-inflammatory steroids such as dexamethasone inhibit the induction of COX-2 but do not affect levels of COX-1. Additionally, COX-2 synthesis can be induced by a number of other molecules of which cAMP, interleukin-1, leukotrienes, and ArA are noteworthy. The expression of the enzyme is furthermore modulated by PKC, phorbol esters and DAG-induced prostaglandin synthesis. COX-2 is predominantly expressed in endothelial cells, fibroblasts, and macrophages, and above all, in neurons.[1,14,111] In an in vivo dialysis study in the hippocampus, an increase in the release of prostaglandins after stimulation of the NMDA receptor was observed.[64] The time course of COX-2 induction is surprisingly constant, the time measured after COX had been induced by various agents was always around 2 h.[26]

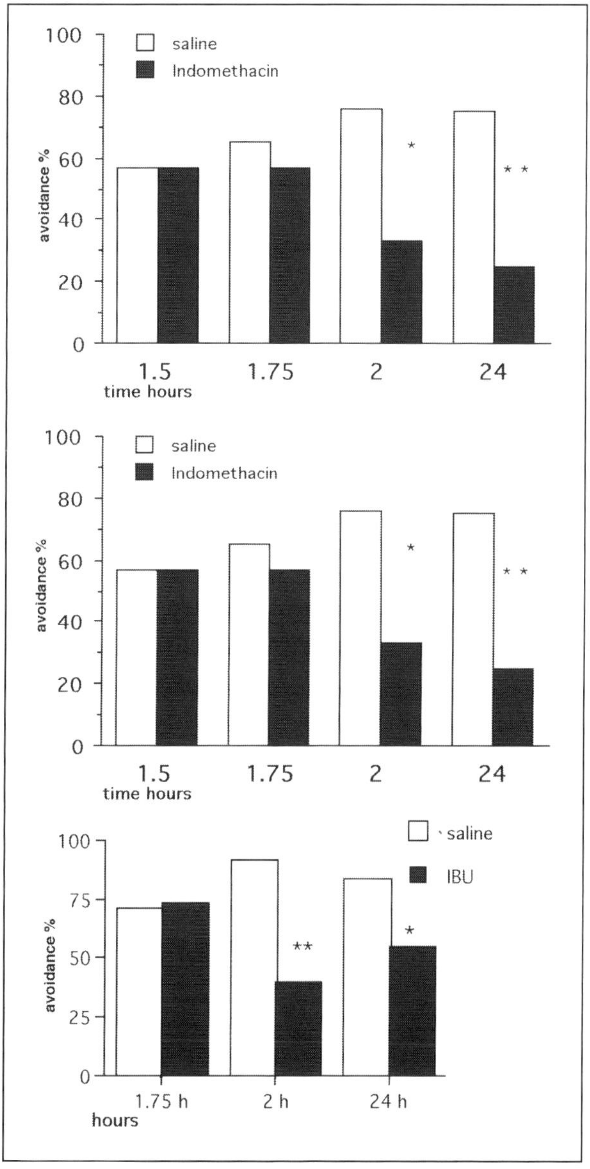

Figure 6. Effect of inhibitors of cyclooxygenase inhibitors on retention of a one-trial passive avoidance task when injected pre-training and tested post-training.

Learning Experiments: Evidence for the Role of COX-2 Activity in Memory Formation

To test if arachidonic acid is metabolised to other messengers by COXs, the effect of inhibitors of these enzymes were tested in a one-trial passive avoidance task in the chick. The cyclooxygenase inhibitors Indomethacin, Naproxen, and Ibuprofen caused amnestic effects at all concentrations tested when injected intracerebrally (i.c.) before training. The onset of amnestic effects was always 2 h after training, independent of drug type, concentration, and injection time before training (see Fig. 6, and ref. 47). In a second study, the injection of the selective COX-2 inhibitor, SC58125 (see ref. 97) or dexamethasone before training showed amnestic effects for training on a one-trial passive avoidance task at 2 h but not 1 h after training.[48]

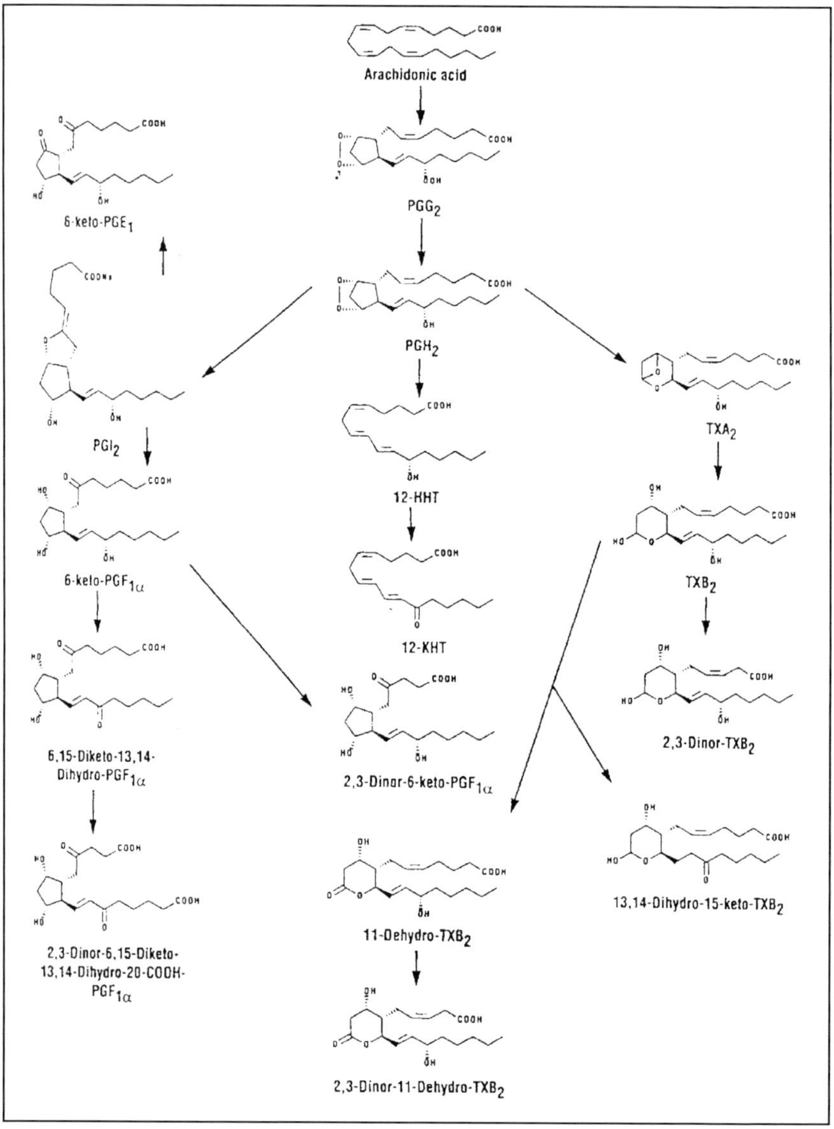

Figure 7. Scheme of metabolism of ArA via the cyclooxygenase pathway II to prostaglandins (PG) or prostacyclins (PGI_2) and thromboxanes (TX) as an example of the diversity of these pathways. Seven major pathways are known, three lipoxygenase (5-, 12, and 15-lipoxygenase), two cyclooxygenase, and two P-450-cytochrome oxidase pathways.[36,99]

A follow up study analysed the release of COX products (prostaglandins) from brain tissue using the ELISA technique. The release of cyclooxygenase products into the extracellular fluid was measured at 1, 2, and 3 h post-training. An increase of prostaglandin production was see after 2 and 3 h, but not after 1 h. A cyclooxygenase inhibitor, ibuprofen, inhibited the training-dependent increase of cyclooxygenase products 2 h and 3 h after learning when injected pre-training, as did dexamethasone which prevents cyclooxygenase induction. The selective COX-2 inhibitor, SC58125 had the same effect.[97]

The delay of release 2 h after training suggests that the drugs prevent induction of COX-2, which takes around 2 h.[26,74] The results indicate that COX-2 products play a role in memory consolidation in the chick when learning this task. They also suggest that COX-2 induction plays a role in memory consolidation several hours after the learning experience (see Fig. 7, and refs. 47,48).

These studies of the role of COX in memory formation in chicks have been corroborated by similar studies in rats, suggesting that the molecular mechanisms are similar in different species. For example, it was shown that the effects of the Cox inhibitor ibuprofen impaired spatial learning in rats as well as the development of LTP in vivo.[78] In another study it was shown that conditioning of animals in a lever-pressing task was dependent on COX activity. Interesting enough, this investigation concluded that the main effect of the cannabinoid (CB1) receptor agonist tetrahydrocannabinol on behaviour is mediated through COX activity as it was blocked by COX antagonists such as diclofenac or indomethacin.[112]

In transgenic mice that overexpressed COX-2 in neurons, memory impairments were observed.[1] This finding suggests that an uncontrolled release or an unphysiologically high level of COX products interferes with mechanisms necessary for normal memory formation.

Chicks were given bilateral intracerebral injections of the cyclooxygenase inhibitors indomethacin, naproxen, ibuprofen, or saline 30min before training and were tested at the stated times subsequently (n=13-18 per group; *=$p<0.05$, **=$p<0.01$). For details see ref. 47.

Cooperation of ArA and Metabolites of ArA As Messengers in Neuronal Systems

The question arises why several neuronal messenger systems are in operation that appear to serve similar functions. One argument in favour of parallel messenger systems is that the system is very stable through redundancy. In fact, biological systems tend to make use of redundancy when processes are concerned that are essential for life. A different possibility is that only a subset of neurons use a particular type of messenger while others use a different type to avoid cross talk. This appears to be the case for the neurotransmitter nitric oxide synthase (NOS), which is produced in the different areas of the hippocampus of the rat at different quantities[50]. A different line of arguments comes from a theoretical approach. Neuronal signalling networks were constructed with experimentally obtained constants and analysed by computational methods to understand their role in complex biological processes. These networks exhibited emergent properties such as integration of signals across multiple time scales, generation of distinct outputs depending on input strength and duration, and self-sustaining feedback loops. Feedback can result in bistable behaviour with discrete steady-state activities, well-defined input thresholds for transition between states and prolonged signal output, and signal modulation in response to transient stimuli. These properties of signaling networks raise the possibility that information for "learned behaviour" of biological systems may be stored within intracellular biochemical reactions that comprise signalling pathways.[7] Hence, a multitude of biochemical inputs can be more than just the sum of its parts and can produce surprising effects and produce novel qualities.

Evidence for the parallel use of different signal systems has been collected in many investigations in several species. In the marine mollusc *Aplysia*, Piomelli et al[83] identified the lipoxygenase pathway products 5-HETE and 12-HETE as well as COX products such as the prostaglandins PGE_2 and $PGF_2\alpha$ as neurotransmitters. In the hippocampal slice of the rat, 12-HPETE and 12-HETE release was increased after LTP induction.[68,9] In postganglionic neurons, both messengers NO and ArA modulate calcium currents. In one study, the effect of the NO generating drug nitroprusside was abolished by the NOS inhibitor N^G-nitro-L-arginine methyl ester (NAME), while the effect of ArA was unchanged by NAME,[61] showing that both systems are independent from each other. Finally, in the sensory pain-pathway, NO[86] and COX products[3] act as messengers, they are co-located with dopamine, Histamine and neuropeptides.

These results show us that several signaling systems act simultaneously in neuronal communication. So what does this tell us about the different function of these transmitter system? Perhaps we have to look at other parameters. If we go back to the one-trial passive avoidance task (PAT) of the chick we can have a look at the timing of memory formation and consolidation.

The Timing of Memory Formation

Discrete Time Windows of Messenger Activity

Studies with the NOS inhibitors nitro-*L*-arginine (L-NARG) and 7-nitro indazole[49,51,52] produced amnesia for the one-trial passive avoidance task in the chick. The interesting observation here was that injection of the nitric oxide synthase inhibitors pre-training resulted in amnesia for the task after 15 minutes of training.

As shown earlier, inhibitors of phospholipases are effective from 1 h onwards after training,[53] the time point when release of arachidonic acid into the extracellular fluid was the highest.20

Injection of cyclooxygenase inhibitors before training produced amnestic effects from 2 h onwards after training.[47] The learning-related increase of release of prostaglandins observed in the saline group followed the same time course, i.e., a large increase of release after 2 h post-training compared to 1 h post-training values.[48] The delay of 2 h can be explained by the time the induction process of COX-2 takes, as measured in different cell types. As mentioned before, the time course takes about 2 h.[26]

It appears that nitric oxide, arachidonic acid, and arachidonic acid metabolites act together as messengers, lined-up in a linear cascade. Nitric oxide is an uncharged molecule that is released quickly and that diffuses across cell membranes without much resistance.[33,100] In contrast, lipids are released rather slowly and tend to 'stick' to membranes for a longer time.[36,99] Therefore, they are better suited as longer-lasting messengers. Arachidonic acid is a molecule that is fairly unstable due to its double bonds, it is not only metabolised by oxidases but by oxygen radicals or other free radicals.[36] Hence, ArA is not a good messenger for longer time durations. Oxidase products, however, have a longer life span but take longer to be synthetised.[26] These properties could explain the observed time windows of memory formation. Hence, induction of key enzymes in neurons appear to be of importance for memory consolidation and synaptic plastic processes. This has been shown before in experiments that analysed the time-course of development of LTP in the hippocampus. Several enzymes such as Ca^{2+}/calmodulin-dependent kinase II, and γ-PKC were found to be induced in the course of LTP consolidation.[105]

Figure 8 summarises the three different time courses of amnesia development after drug injection. One has to postulate that each messenger has a peak of production and is present only in low concentration before and after this peak. Otherwise, a compensation of effects due to loss of one messenger by other messengers should occur, and no amnesia would develop. This might well be the case in other areas of the chick brain or in brains of other species.

Defined Steps in Memory Formation

Defined time windows for activity of drugs in learning-tasks have been observed before. Rosenzweig et al[90] as well as Gibbs[34] discriminate between different steps of memory formation in the chick. They differentiate between three phases: short term memory (0-15 min), intermediate memory (15-55 min), and long-term memory (>55 min). Different drugs can interrupt one or several phases of this cascade, and produce amnestic effects at the end of each phase, which has not been interrupted. This interruption is independent of the concentration of drugs, or the time point of injection before training (see also ref. 90). Rose and collaborators found another time point that appears to be intrinsic in memory consolidation in the chick. The second wave of glycoprotein synthesis that was first described by Pohle et al[85] in the rat and was later found

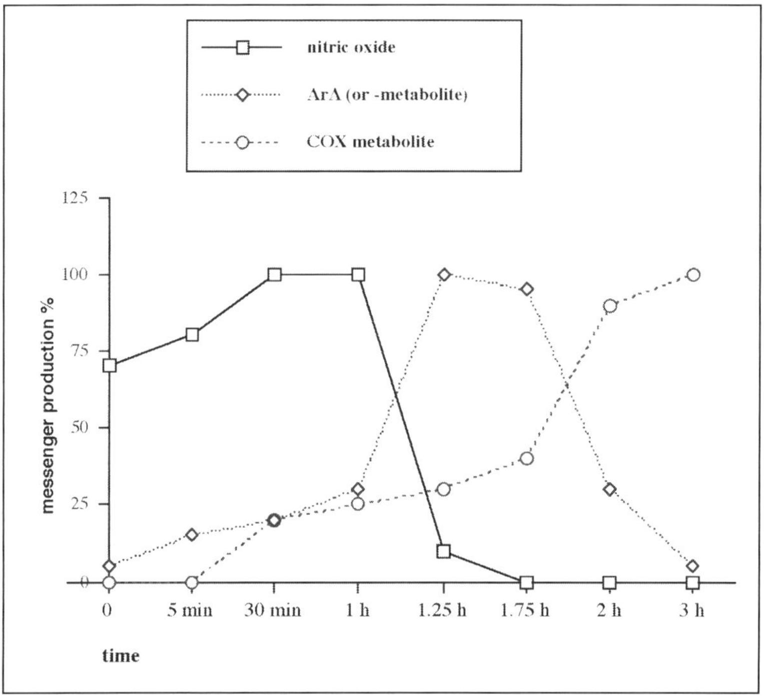

Figure 8. Scheme of possible production levels of neuronal messengers over time. This scheme is based on the data presented in this chapter. A first wave of NO is proposed since inhibition of NO synthase produced amnestic effects after 30 min. After NO production starts to decrease, ArA is released in greater quantity with a rapid increase after about 1 h and decrease after about 1.75 h. ArA is most likely metabolised in this time period, and the release of newly formed ArA is inhibited, since an amnestic effect becomes unmasked after 1.75 h when injecting COX inhibitors. The third wave, which consists of cyclooxygenase metabolites, starts after 1.75 h and continues for an unknown time period. For details see Hölscher and Rose;[53] Hölscher.[47,48]

in the chick.[15,96] This second wave of expression is found 6 h to 8 h post-training. Interestingly enough, an increase of spontaneous neuronal bursting in that time period had been observed[35] (see refs. 88,89 for a review).

Defined phases in the metabolic cascade of memory formation in mammals are not unknown either. In memory formation of rats[79] or humans,[5] a division between short-term, long-term, and working memory has been suggested. McGaugh[71] noted as early as 1968 that there are defined time windows for sensitivity of drugs injected in rats tested in passive avoidance training tasks. Pohle et al[85] observed two defined time windows of [^{3}H]fucose incorporation in the area CA1 and CA3 after training of rats in a discrimination task. Since fucose is incorporated into glycoproteins, a second phase of memory-related glycoprotein synthesis might be expressed in hippocampal neurons. Regan[87] reported that intraventricular infusion of antibodies to neuronal cell adhesion molecules (N-CAM) disrupted consolidation of a passive avoidance response in the rat when administered between 6 and 8 h post-training.

A Potential Role for Defined Time Windows of Messenger Systems in Memory Formation

Why do distinct time windows for metabolic processes in memory formation exist? Each step in the biochemical cascade continues only for a limited amount of time, and after that, the activity of the receptor/enzyme is not essential for establishing the memory trace any more. Then, inhibiting the activation of the specific step in the chain of events cannot prevent memory consolidation. But what role would these defined time windows play in memory formation?

Clearly, it is of importance to filter information before it is stored in long-term memory. To be able to filter memory input, several steps of consolidation are required. Initial activation by NO could be a form of short-term memory, or a priming step, for further consolidation that needs input of a different quality or quantity to be maintained past this stage. The PLA_2 linked ArA release might be the next step in this mechanism, which has to be followed by the 3rd step, PLC activation, to ensure intermediate memory formation. A fourth step could be the metabolism of ArA by COXs or lipoxygenases to messengers with longer half-lives that would ensure a long lasting signal. Memory consolidation could be terminated at any of these steps. If the conditions for long-term memory formation are not met, this signal could be interrupted and the memory trace would be retained only for a limited amount of time. Clearly, memory can be retained for different lengths of time. Not all information has to be kept, or should be kept, in storage for the whole lifetime. If the activation of the NMDA receptor and the subsequent activation of protein kinases, and the synthesis of glycoproteins, always resulted in long-term memory formation, the phenomenon of forgetting would be hard to explain. In a weak-stimulus passive avoidance task of the day old chick as employed by Bourne et al[12] or by Crowe et al[25], amnesia is observed after about 5 hours, before a second wave of glycoprotein synthesis occurs,[12,96] or the main part of post-training neuronal bursting is observed.[35] A termination of memory formation at this step is possible. In other words, even if the synthesis of glycoproteins, the production of immediate early genes, the upregulation of receptor sensitivity via protein kinase activity already has happened, the mechanism can still be stopped, and memory formation halts. All these different biochemical steps appear to act independently from each other, but all are required in order to establish a long-term change in synaptic efficacy. Only then, blocking the synthesis of a retrograde messenger as late as 2 h post-training could evoke amnesia. The results presented in this chapter therefore not only illuminate the process of learning and memory formation but also potential mechanisms of forgetting.

References

1. Andreasson KI, Savonenko A, Vidensky S et al Age-dependent cognitive deficits and neuronal apoptosis in cyclooxygenase-2 transgenic mice. J Neurosci 2001; 21:8198-8209.
2. Arai A, Lynch G. Antagonist of the platelet -activating factor receptor block long-term potentiation in hippocampal slices. Eur J Neurosci 1992; 4:411-419.
3. Arrigo-Reina R, Chiechio S, Catti T. Increased analgesic potenty of morphine following repeated anaphylactoid reaction by compound 48/80 involves cyclooxygenase products of arachidonic acid. Agents Actions 1993; 38, Suppl C PT:C269-C271.
4. Axelrod J, Burch RM, Jelsema CL. Receptor-mediated activation of phospholipase A_2 via GTP-binding proteins: arachidonic acid and its metabolites as second messengers. Trends Neurosci 1988; 11:117-123.
5. Baddeley AD, Hitch GJ. Working Memory. New York: Oxford University Press 1988.
6. Barbour B, Szatkowski M, Ingledew N et al. Arachidonic acid induces a prolonged inhibition of glutamate uptake into glial cells. Nature 1989; 342:918-920.
7. Bhalla US, Iyengar R. Emergent properties of networks of biological signaling pathways. Science 1999; 283:381-387.
8. Bliss TVP, Errington ML, Lunch MA et al. Presynaptic mechanisms in hippocampal long-term potentiation. Cold Spring Harbor Symposia on Quantitative Biology 1990; 55:119-129.
9. Bliss TVP, Lynch MA. Long-term potentiation of synaptic transmission in the hippocampus: Properties and mechanisms. In: Long-Term Potentiation: From Biophysics to Behavior. Alan R. Liss, Inc. 1988: 3-72.
10. Bockaert J, Pin J, Fagni L. Metabotropic glutamate receptors: an original family of G protein-coupled receptors. Fundam Clin Pharmacol 1993;7:473-485.
11. Bortolotto ZA, Collingridge GL. Activation of glutamate metabotropic receptors induces long-term potentiation. Eur J Pharmacol 1992; 214:297-298.
12. Bourne RC, Davies DC, Stewart DC et al. Cerebral glycoprotein synthesis and long-term memory formation in the chick (Gallus domesticus) following passive avoidance training depends on the nature of the aversive stimulus. Eur J Neurosci 1990; 3:243-248.

13. Bourne HR, Nicoll R. Molecular machines integrate coincident synaptic signals. Neuron 1993; 10 (Suppl.):65-76.
14. Breder CD, Dewitt D, Kraig RP. Characterization of inducible cyclooxygenase in rat brain. J Comp Neurol 1995; 355:296-315.
15. Bullock S, Rose SP, Zamani R. Characterisation and regional localisation of pre- and postsynaptic glycoproteins of the chick forebrain showing changed fucose incorporation following passive avoidance training. J Neurochem 1992; 58:2145-2154.
16. Byrne JH, Zwartjes R, Homayouni R et al. Roles of second messenger pathways in neuronal plasticity and in learning and memory. Adv Sec Mess Phosphoprot Res 1993; 27:47-108.
17. Chen C, Bazan NG. Platelet-activating factor inhibits ionotropic GABA receptor activity in cultured hippocampal neurons. NeuroReport 1999; 10:3831-3835.
18. Chen C, Magee JC, Marcheselli V et al. Attenuated LTP in hippocampal dentate gyrus neurons of mice deficient in the PAF receptor. J Neurophysiol 2001; 85:384-390.
19. Clements MP, Bliss TVP, Lynch MA. Increase in arachidonic acid concentration in a postsynaptic membrane fraction following the induction of LTP in the dentate gyrus. Neurosci 1991; 45:379-389.
20. Clements MP, Rose SPR. Time-dependent increase in release of arachidonic acid following passive avoidance training in the day-old chick. J Neurochem 1996; 67:1317-1323.
21. Collingridge GL, Bliss TVP. NMDA receptors—their role in long-term potentiation. Trends Neurosci 1987; 10:288-293.
22. Collins DR. Davies SN. Co-administration of (1S, 3R)-1-aminocyclopentate-1,3-dicarboxylic acid and arachidonic acid potentiates synaptic transmission in rat hippocampal slices. Eur J Pharmacol 1993; 240:325-326.
23. Conn PJ, Desai MA. Pharmacology and physiology of metabotropic glutamate receptors in mammalian central nervous system. Drug Dev Res 1991; 24:207-229.
24. Conn PJ, Pin JP. Pharmacology and functions of metabotropic glutamate receptors. Ann Rev Pharmacol Toxicol 1997; 205:205-237.
25. Crowe SF, Ng KT, Gibbs ME. Memory formation processes in weakly reinforced learning. Pharmacol Biochem Behav 1989; 33:881-887.
26. De Witt DL. Prostagandin endoperoxide synthase: regulation of enzyme expression. Biochim et Biophys Acta 1991; 1083:121-134.
27. Dekker LV, DeGraan PNE, Versteeg DHG et al Phosphorylaltion of B-50 (GAP43) is correlated with neurotransmitter release in rat hippocampal slices. J Neurochem 1989; 52:24-30.
28. Drapeau C, Pellerin L, Wolfe LS et al. Long-term changes of synaptic transmission induced by arachidonic acid in the CA1 subfield of the rat hippocampus. Neurosci Lett 1990; 115:286-292.
29. Dumuis A, Pin JP, Oomagari K et al. Arachidonic acid released from striatal neurons by joint stimulation of ionotropic and metabotropic quisqualate receptors. Nature 1990; 347:182-184.
30. Farley J, Auerbach S. Protein kinase C activation induces conductance changes in Hermissenda photoreceptors like those seen in associative learning. Nature 1986; 319:220-223.
31. Felder CC, Kanterman RY, MA AL et al, Serotonin stimulates PLA_2 and the release of arachidonic acid in hippocampal neurons by a type 2 serotonine receptor that is independent of inositolphospholipid hydrolysis. Proc Natl Acad Sci USA 1990; 87:2187-2191.
32. Fujita S, Ikegaya Y, Nishiyama N et al. Ca2+-independent phospholipase A2 inhibitor impairs spatial memory of mice. Jpn J Pharmacol 2000; 83:277-278.
33. Gally JA, Montague RP, Reeke GN et al. The NO hypothesis: Possible effects of a short-lived, rapidly diffusible signal in the development and function of the nervous system. Proc Natl Acad Sci USA, 1990; 87:3547-3551.
34. Gibbs ME, Ng KT, Crowe S. Hormones and the timing of phases of memory formation. In: Andrews RJ, ed. Neural and behavioural plasticity in the chick. Oxford: Oxford Univ. Press, 1991:441-455.
35. Gigg J, Patterson TA, Rose SPR. Training-induced increases in neuronal activity recorded from the forebrain of the day-old chick are time dependent. Neurosci 1993; 56:771-776.
36. Graber R, Sumida C, Nunez EA. Fatty acids and cell signal transduction. J Lipid Med Cell Signal 1994; 9:91-116.
37. Grassi S, Francescangeli E, Goracci G. Platelet-activating factor and group I metabotropic glutamate receptors interact for full development and maintenance of long-term potentiation in the rat medial vestibular nuclei. Neurosci 1990; 94:549-559.
38. Greenberg SS, Xie J, Wang Y. Ethanol relaxes pulmonary artery by release of prostaglandin and nitric oxide. Alcohol 1993; 10:21-29.
39. Hamilton SE, Nathanson NM. The M{sub1} receptor is required for muscarinic activation of MAP kinase in murine cerebral cortical neurons. J Biol Chem 2001; 276:122-129.

40. Harish OE, Poo M. Retrograde modulation at developing neuromuscular synapses: Involvement of a G protein and arachidonic acid cascade. Neuron 1992; 9:1201-1209.
41. Hedqvist P. Basic mechanisms of prostaglandin action on autonomic neurotransmission. Ann Rev Pharmacol Toxicol 1977; 17:259-279.
42. Herrero I, Miras-Portugal MT, Sánchez-Prieto J. Inhibition of glutamate release by arachidonic acid in rat cerebrocortical synaptosomes. J Neurochem 1991; 57:718-721.
43. Herrero I, Castro E, Miras-Portugal MT et al. Glutamate exocytosis evoked by 4-aminopyridine is inhibited by free fatty acids released from rat cerebrocortical synaptosomes. Neurosci Lett 1991; 126:41-44.
44. Herrero I, Miras-Portugal MT, Sánchez-Prieto J. Activation of protein kinase C by phorbol esters and arachidonic acid required for the optimal potentiation of glutamate exocytosis. J Neurochem 1992; 59:1574-1577.
45. Herrero I, Miras-Portugal MT, Sánchez-Prieto J. PKC-independent inhibition of glutamate exocytosis by arachidonic acid in rat cerebrocortical synaptosomes. FEBBS Lett 1992; 296:317-319.
46. Herrero I, Vázquez E, Miras-Portugal MT et al. A decrease in $[Ca^{2+}]c$ but not in cAMP mediates L-AP4 inhibition of glutamate release: PKC mediated suppression of this inhibitory pathway. Eur J Neurosci 1996; 8:700-709.
47. Hölscher C. Inhibitors of cyclooxygenases but not lipoxygenases have amnestic effects in the chick. Eur J Neurosci 1995; 7:1360-1365.
48. Hölscher C. Release of prostaglandins during memory consolidation in the chick. Eur J Pharmacol 1995; 294:253-259.
49. Hölscher C. 7-Nitro indazole, a neuron specific nitric oxide synthase inhibitor, produces amnesia in the chick. Learn Mem 1995; 1:9-11.
50. Hölscher C. Nitric oxide, the enigmatic neuronal messenger: Its role in synaptic plasticity. Trends Neurosci 1997; 20:298-303.
51. Hölscher C, Rose SPR. An inhibitor of nitric oxide synthesis prevents memory formation in the chick. Neurosci Lett 1992; 145:165-167.
52. Hölscher C, Rose SPR. Inhibiting synthesis of the putative retrograde messenger nitric oxide results in amnesia in a passive avoidance task in the chick. Brain Res 1993; 619:189-194.
53. Hölscher C, Rose SPR. Inhibitors of phospholipase A_2 produce amnesia for a passive avoidance task in the chick. Behav Neural Biol 1994; 61:225-232.
54. Hölscher C, Gigg J, O'Mara S. Metabotropic glutamate receptor activation and blockade. consequences for long-term potentiation, learning and neurotoxicity. Neurosci Biobeh Rev 1999; 23:399-410.
55. Izaki Y, Hashimoto M, Arita J. Enhancement by 1-oleoyl-2-docosahexaenoyl phosphatidylcholine of long-term potentiation in the rat hippocampal CA1 region. Neurosci Lett 1999; 260:146-148.
56. Izquierdo I, Fin C, Schmitz PK et al. Memory enhancement by intrahippocampal, intraamygdala, or intraentorhinal infusion of platelet-activating factor measured in an inhibitory avoidance task. Proc Natl Acad Sci USA 1995; 92:5047-5051.
57. Karlson P. Kurzes Lehrbuch der Biochemie. Stuttgart: Georg Thieme Verlag, 1988.
58. Kato K, Uruno K, Saito K et al. Both arachidonic acid and 1-oleoyl-2-acetyl glycerol in low magnesium solution induce long-term potentiation in hippocampal CA1 neurons in vitro. Brain Res 1991; 563:94-100.
59. Kato K, Zorumski CF. Platelet-activating factor as a potential retrograde messenger. J Lipid Mediat Cell Signal 1996; 14:341-348.
60. Kelly JF, Furukawa K, Barger SW et al. Amyloid beta-peptide disrupts carbachol-induced muscarinic cholinergic signal transduction in cortical neurons. Proc Natl Acad Sci USA 1996; 93:6753-6758.
61. Khurana G, Bennett MR. Nitric oxide and arachidonic acid modulation of calcium currents in postganglionic neurones of avian cultured ciliary ganglia. Br J Pharmacol 1993; 109:480-484.
62. Klein J, Iovino M, Vakil M et al. Ontogenetic and pharmacological studies on metabotropic glutamate receptors coupled to phospholipase D activation. Neuropharmacol 1997;36:305-311.
63. Kornecki E, Wieraszko A, Chan J et al. Platelet activating factor (PAF) in memory formation: role as a retrograde messenger in long-term potentiation. J Lipid Mediat Cell Signal 1996; 14:115-126.
64. Lazarewicz JW, Salinska E. N-methyl-D-aspartate-evoked release of cyclo-oxygenase products in rabbit hippocampus: An in vivo microdialysis study. J Neurosci Res 1995; 40:660-666.
65. Lester DS, Colling CC, Etcheberrigaray R et al. Arachidonic acid and diacylglycerol act synergistically to activate protein kinase C in vitro and in vivo. Biochem Biophys Res Commun 1991; 179:1522-1528.
66. Linden DJ, Sheu F-S, Murakami K et al. Enhancement of long-term potentiation by cis-unsaturated fatty acid: relation to protein kinase C and phospholipase A2. J Neurosci 1987; 7:3783-3792.

67. Lynch MA, Bliss TVP. LTP of synaptic transmission in the hippocampus: effect of calmodulin and oleoyl-acetyl-glycerol (OAG) on release of ^{3}H glutamate. Neurosci Lett 1986; 65:171-176.
68. Lynch MA, Errington ML, Bliss TVP. Nordihydroguaiaretic acid blocks the synaptic component of long-term potentiation and the associated increases in release of glutamate and arachidonate: an in vivo study in the dentate gyrus of the rat. Neurosci 1989; 30:693-701.
69. Lynch MA, Voss VL. Arachidonic acid increases inositol phospholipid metabolism and glutamate release in synaptosomes prepared from hippocampal tissue. J Neurochem 1990; 55:215-221.
70. Massicotte G, Olivier MW, Lynch G et al. Effect of bromophenylacylbromide, a phospholipase A_2 inhibitor, on the induction and maintenance of LTP in hippocampal slices. Brain Res 1990; 537:49-53.
71. McGaugh JL. A multi-trace view of memory storage processes. In: Bovert R, ed. Attuali orentamenti della ricerca sull'apprendimento e la memoria. Rome: Academia Nazional dei Lincei, Quaderno, 1968:13-24.
72. McMillan RM, Masters DJ, Sterling WW et al. Biosynthesis of Leukotriene B_4 in human leukocytes. In: Bailey M, ed. Prostaglandins, Leukotriens, and Lipoxins. New York: Plenum Press 1985:chapter 62.
73. Naor Z. Is arachidonic acid a second messenger in signal transduction? Mol Cell Endocrinol 1991; 80:C181-C186.
74. Newman SP, Flower RJ, Croxtall JD. Dexamethasone suppression of IL-1β-induced cyclooxygenase 2 expression is not mediated by lipocortin-1 in A549 cells. Biochem Biophys Res Commun 1994; 202:931-939.
75. Nishiyama M, Watanabe T, Ueda N et al. Arachidonate 12-lipoxygenase is located in neurons, glial cells, and endothelial cells of the canine brain. J Histochem Cytochem 1993; 41:111-117.
76. O'Dell TJ, Hawkins RD, Kandel ER et al. Tests of the roles of two diffusible substances in long-term potentiation: evidence for nitric oxide as a possible early retrograde messenger. Proc Natl Acad Sci USA 1991; 88:11285-11289.
77. O'Mara S, Rowan MJ, Anwyl R. Metabotropic glutamate receptor-induced homosynaptic long-term depression and depotentiation in the dentate gyrus of the rat hippocampus in vitro. Neuropharmacol 1995; 34:983-989.
78. O'Mara S, Shaw K, Commins S. Broad-spectrum cyclooxygenase inhibition reversibly blocks long-term potentiation and spatial learning via a brain-derived neurotrophic factor-dependent mechanism. Soc Neurosci Meeting, San Diego, 2001; abstract 190.20.
79. Olton DS, Becker JT, Handelmann GE. Hippocampus, space and memory. Behav Brain Sci 1979; 2:313-365.
80. Palmer E, Nangel-Taylor K, Krause JD et al. Changes in excitatory amino acid modulation of phosphoinositide metabolism during development. Dev Brain Res 1990; 51:132-134.
81. Pellegrini-Giampietro DE, Torregrossa SA, Moroni F. Pharmacological characterization of metabotropic glutamate receptors coupled to phospholipase D in the rat hippocampus. Br J Pharmacol 1996;118:1035-1043.
82. Pellerin L, Wolfe S. Release of arachidonic acid by NMDA-receptor activation in the rat hippocampus. Neurochem Res 1991; 16:983-989.
83. Piomelli D, Greengard P. Bidirectional control of PLA_2 acitvity by Ca2+/Calmodulin-dependent protein kinase II, cAMP- dependent protein kinase, and casein kinase II. Proc Natl Acad Sci USA 1991; 88:6770-6774.
84. Piomelli D, Volterra A, Dale N et al. Lipoxygenase metabolites of arachidonic acid as second messengers for presynaptic inhibition of Aplysia sensory cells. Nature 1987; 328:38-43.
85. Pohle WM, Ruthrich H-L, Popov N et al. Fucose incorporation into rat hippocampus structures after acquisition of a brightness discrimination. Acta Biol Med Germ 1979; 38:53-63.
86. Przewlocki R, Machelska H, Przewlocka B. Inhibition of nitric oxide synthase enhances morphine antinociception in the rat spinal cord. Life Sci 1993; 53:PL1-PL5.
87. Regan CM. Regulation of neural cell adhesion molecule (NCAM) sialylation state. Int J Biochem 1991; 23:513-523.
88. Rose SPR. How chicks make memories: the cellular cascade from c-fos to dendritic remodelling. Trends Neurosci 1991; 14:390-397.
89. Rose SPR. On chicks and Rosetta stones. In: Butters S, ed. Neuropsychology of memory. New York: Guilford Press 1992:547-556.
90. Rosenzweig MR, Bennett EL, Martinez JL et al. Studying stages of memory formation with chicks. In: Butters S, ed. Neuropsychology of memory. New York: Guilford Press 1992:533-546.
91. Rotman EI, Brostrom MA, Brostrom CO. Inhibition of protein synthesis in intact mammalian cells by arachidonic acid. Biochem J 1992; 282:487-494.

92. Sanfeliu C, Hunt A, Patel AJ. Exposure to N-methyl-D-aspartate increases release of arachidonic acid in primary cultures of rat hippocampal neurons and not in astrocytes. Brain Res 1990; 526:241-248.
93. Schaechter JD, Benowitz LI. Activation of protein kinase C by arachidonic acid selectively enhances the phosphorylation of GAP-43 in nerve terminal membranes. J Neurosci 1993; 13:4361-4371.
94. Schoepp DD, Johnson BG. Inhibition of excitatory amino acid-stimulated phosphoinositide hydrolysis in the neonatal rat hippocampus by 2-amino-3-phosphonopropionate. J Neurochem 1989; 53:1865-1868.
95. Schoepp DD, Johnson BG, Smith ECR et al. Stereoselectivity and mode of inhibition of phosphoinositide-coupled excitatory amino acid receptors by 2-amino-3-phosphonopropionic acid. Mol Pharmacol 1990; 38:222-228.
96. Scholey AB, Rose SPR, Zamani MR et al. A role for the neural cell adhesion molecule in a late, consolidating phase of glycoprotein synthesis six hours following passive avoidance training of the young chick. Neurosci 1993; 55:449-509.
97. Seibert K, Zhang Y, Leahy K et al. Pharmacological and biochemical demonstration of the role of cyclooxygenase 2 in inflammation and pain. Proc Natl Acad Sci USA 1994; 91:12013-12017.
98. Sharon D, Vorobiov D, Dascal N. Positive and negative coupling of the metabotropic glutamate receptors to a G protein-activated K+ channel, GIRK, in Xenopus oocytes. J Gen Physiol 1997; 109:477-490.
99. Shen T-Y, Winter CA. Chemical and biological studies on indomethacin, sulindac and their analogs. Adv Drug Res 1977; 12:89-245.
100. Snyder SH. Nitric Oxide: More jobs for that molecule. Nature 1994; 372:504-505.
101. Strosznajder J, Samochocki M. Ca^{2+}-independent, Ca^{2+}-dependent, and carbachol-mediated arachidonic acid release from rat brain cortex membrane. J Neurochem 1991; 57:1198-1206.
102. Strosznajder JB, Zambrzycka A, Kacprzak MD et al. Amyloid beta peptide 25-35 modulates hydrolysis of phosphoinositides by membrane phospholipase(s) C of adult brain cortex. J Mol Neurosci 1999; 12:101-109.
103. Tapia-Arancibia LT, Rage F, Récasens M et al. NMDA receptor activation stimulates phospholipase A_2 and somatostatin release from rat cortical neurons in primary cultures. Eur J Pharamacol 1992; 225:253-262.
104. Teather LA, Packard MG, Bazan NG. Effects of posttraining intrahippocampal injections of platelet-activating factor and PAF antagonists on memory. Neurobiol Learn Mem 1998; 70:349-363.
105. Thomas KL, Laroche S, Errington ML et al. Spatial and temporal changes in signal transduction pathways during LTP. Neuron 1994; 13:737-746.
106. Ulmann L, Mimouni V, Roux S et al. Brain and hippocampus fatty acid composition in phospholipid classes of aged-relative cognitive deficit rats. Prostaglandins Leukot Essent Fatty Acids 2001; 64:189-195.
107. Volterra A, Trotti D, Cassutti P et al. A role for the arachidonic acid cascade in fast synaptic modulation: Ion channels and transmitter uptake systems as target proteins. Eur J Pharmacol 1992; 217:225-227.
108. Williams JH, Bliss TVP. An in vitro study of the effect of lipoxygenase and cyclo-oxygenase inhibitors of arachidonic acid on the induction and maintenance of long-term potentiation in the hippocampus. Neurosci Lett 1989; 107:301-306.
109. Williams JH, Errington ML, Li Y-G et al. The search for retrograde messengers in long-term potentiation. The Neurosciences 1993; 5:149-158.
110. Williams JH, Errington ML, Lynch MA et al. Arachidonic acid induces a long-term activity-dependent enhancement of synaptic transmission in the hippocampus. Nature 1989; 341:739-742.
111. Yamagata K, Andreasson KI, Kaufmann WE et al. Expression of a mitogen-inducible cyclooxygenase in brain neurons: regulation by synaptic activity and glucocorticoids. Neuron 1993; 11:371-386.
112. Yamaguchi T, Shoyama Y, Watanabe S et al. Behavioral suppression induced by cannabinoids is due to activation of the arachidonic acid cascade in rats. Brain Res 2001; 889:149-154.
113. Yoshihara Y, Yamaji M, Kawasaki M. Ontogeny of cytosolic phospholipase A_2 activity in rat brain. Biochem Biophys Res Commun 1992; 185:350-355.
114. Yu ACH, Chan PH, Fishman RA. Effects of arachidonic acid on glutamate and γ-aminobutyric acid uptake in primary cultures of rat cerebral cortical astrocytes and neurons. J Neurochem 1986; 47:1181-1189.
115. Zhang L, Dorman RV. Synergistic potentiation of glutamate release by arachidonic acid and oleoyl-acetyl-glycerol. Brain Res Bull 1993; 32:437-441.

CHAPTER 4.3

Protein Kinase A

Monica R.M. Vianna and Ivan Izquierdo

Abstract

Memories are believed to rely upon enduring morphologic and functional changes at synapses activated by learning events. Experiments carried out in the past two decades have indicated that several cellular mechanisms need to be activated in order for the synaptic changes to take place. Among these general cellular mechanisms, enzymatic cascades including the cAMP-dependent protein kinase (protein kinase A, PKA) signaling pathway in CA1 region of the hippocampus have been demonstrated to be crucial to memory processing.

The importance of the PKA pathway to memory formation is indicated by its unique profile of activation following learning experiences: PKA has two peaks of activity during long-term memory consolidation period, the first within the first few minutes after training, and the second in a protracted way, beginning 2-3 h after the experience, after most enzymatic cascades have ceased their contribution.

The coincident increase of nuclear phosphorylated form of the cAMP-responsive element binding protein (CREB) transcription factor at these specific periods, together with memory sensitiveness to inhibitors of gene transcription and protein synthesis during PKA active periods suggest this signaling pathway may contribute actively to the synthesis of new proteins, a crucial event for long-term memory (LTM) establishment. Simultaneously to its involvement in LTM formation, the PKA pathway in the hippocampus is critical in the first hour after training for the establishment of short-term memory (STM) and has contributed to the demonstration of STM and LTM independence.

Recently it has also been shown that PKA contributes crucially to memory retrieval and extinction, probably involving distinct mechanisms of activation among the variety of events that have been shown to influence PKA activity.

Introduction

Memories are considered since Ramón y Cajal[60] to rely upon enduring morphologic and functional changes at synapses activated by learning processes (see also Geinisman et al, this book). Experiments carried out in the past two decades have indicated that several enzymatic cascades need to be activated in order for the synaptic changes to take place. It is clear that the enzymatic signaling pathways do not carry the mnemonic information themselves, but act instead as amplifying systems without which the protein synthesis dependent synaptic changes inherent to memory would not be correctly or sufficiently activated.

Some of these enzymatic cascades act directly at synapses in order to enhance transmitter release or receptor function in the CA1 region of the hippocampus: the Ca^{2+}/calmodulin-dependent kinases (Medina and Cammarota, this volume), protein kinase C (see Noguès, this volume, and ref. 66) and perhaps tyrosine kinases (Gerlai, this volume; and also ref. 58). These mechanisms may be viewed as modulators of the input to the hippocampus; i.e., the connec-

From Messengers to Molecules: Memories Are Made of These, edited by Gernot Riedel and Bettina Platt.

tion between fibers afferent to CA1 and the postsynaptic membrane of CA1 pyramidal cells. Other enzymatic cascades boost mechanisms triggered by second messengers in order to activate general cellular cascades and, particularly, gene transcription and the resulting protein synthesis. These mechanisms are also activated in CA1 pyramidal cells, but in view of their nature may affect mainly the interactions of these cells with their output connections. Among these general cellular mechanisms, the cAMP-dependent protein kinase (protein kinase A, PKA) signaling pathway in CA1 stands out for several reasons.

First and foremost, it is the only signaling pathway in the hippocampus that is activated twice after training experiences that produce memories: the first time briefly, within the first few minutes after training, and the second time in a protracted way, beginning 2-3 h after the experience (Fig. 1A).[6,71] The second peak of PKA activity occurs long after other major cascades involved in memory have ceased their function (see chapters by Noguès and Cammarota and Medina in this book, and also refs. 6,7 and 84).

Second, the two peaks of increased PKA activity are coincident in time both with an increase of nuclear phosphorylated form of the cAMP-responsive element binding protein (CREB) transcription factor (Fig. 1B) (see refs. 6, 69 and Frankland and Josselyn, this book) and with periods in which memory formation is peculiarly sensitive to inhibitors of gene transcription[42] and protein synthesis.[47,48,59] PKA-mediated phosphorylation of $CREB_1$, a constitutive transcription factor that is essential in CA1 for the maintenance both of LTM[6,30] and LTP,[33] is indeed a marker of LTM processing;[75] its absence is a marker of amnesia.[69] A large variety of plastic events in other nervous tissues has also been shown to depend on PKA and CREB phosphorylation.[4,13] Thus, differently from the other kinases that are activated during or shortly after acquisition and remain active during a limited period following the learning experience, PKA seems more likely to be responsible for the mediation of the late and long-lasting cellular modifications thought to underlie long-term memory (LTM) storage.

Third, the PKA pathway in the hippocampus is critical in the first hour after training for the establishment of short-term memory (STM).[71,73] As will be seen in the next section, STM is a separate form of memory that lasts about 3-6 h and runs parallel to the consolidation phase of long-term memory (LTM).[36,41]

Finally, the PKA signaling pathway interacts strongly and at several points with the PKC and MAPK pathways,[53] both of which are critical for memory processing in the hippocampus (see chapters by Noguès et al and Selcher et al in this book). The first peak of PKA activity may correlate with the need for MAPK or PKC activity in the hippocampus for the establishment of STM,[44,71] and the second peak of PKA may correlate with the requirement for MAPK activity 3 h after training in order for long-term memory to become consolidated.[77]

Short- and Long-Term Memory

LTMs are not immediately established in their definitive form.[50,51] This process takes hours and requires a sequence of molecular events that are believed to result in structural and functional long-lasting modifications at synapses of brain areas involved in memory storage.[29,39,46] As was postulated years ago,[50] a parallel STM system is in charge during the hours that it takes for LTM to become effectively consolidated.[35-37,40,41.71-73]

The findings that led to the discovery that STM pertains to separate system parallel to the first hours of LTM consolidation involved the demonstration that a variety of treatments can block STM while leaving LTM intact for the same task in a given individual. This was first shown for short- and long-term facilitation in *Aplysia* (see ref. 21) and then for one-trial inhibitory (passive) avoidance learning in rats.[35-37,40,41]

The method of using localized and timed infusions of drugs with specific molecular actions in order to extricate STM from LTM is the only procedure that can effectively separate the two memory types.[36,37,40,41] The learning-associated metabolic changes that are seen in the CA1 region of the hippocampus and elsewhere in the first 3-6 hours that follow acquisition,[12,39,58] could in principle underlie one or another form of memory. Only the specific inhibition of one

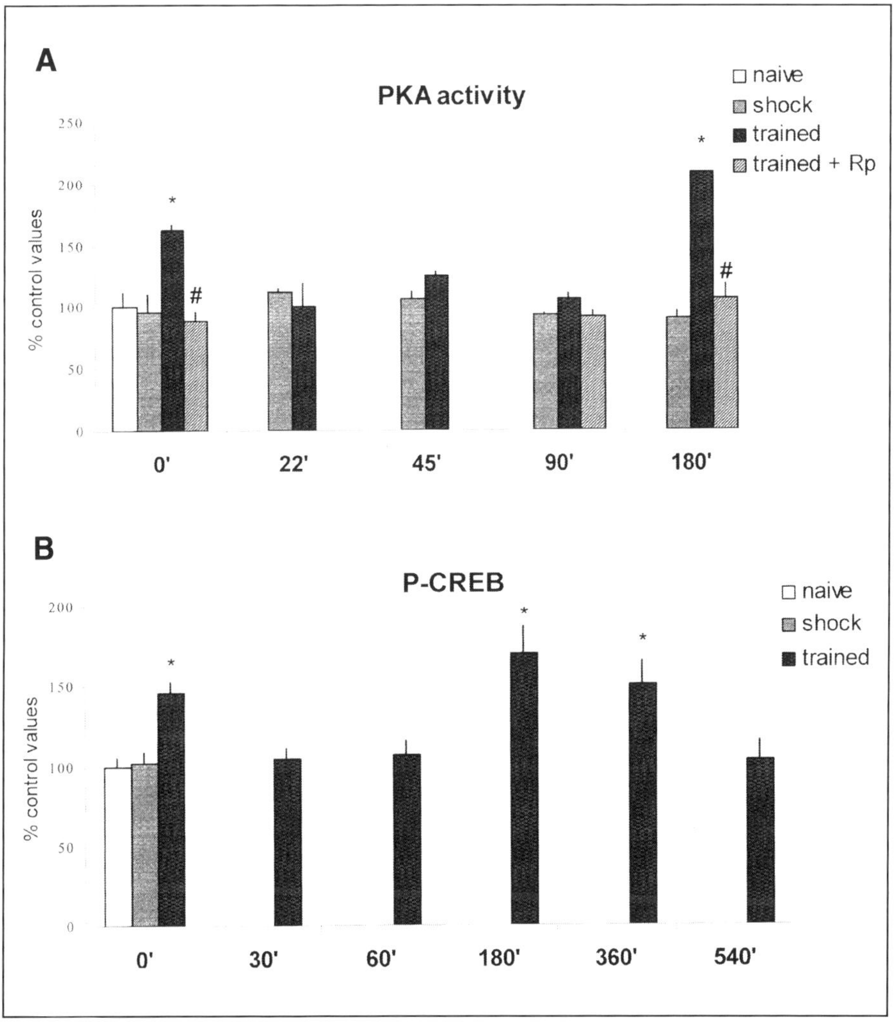

Figure 1. A) PKA activity measured in the CA1 region of 2mm-wide slices taken from the infused sites of the rat hippocampus. Data are expressed as as percentage of saline treated naive control values (means $\pm$ SEM pmol/min/mg protein) N = 2-4 per group. The white column indicates values from naive saline treated animals. The grey columns indicate values from saline treated animals submitted to shock of identical intensity (0.4 mA) to that used in inhibitory avoidance training. The black columns show values for trained animals at the indicated period following training session. The striped columns show the data from animals trained and submitted to Rp-cAMPs (0.1 mg) infusion in that region 0 min after training. * indicate significant differences from all other groups at $p < 0.001$ in a Duncan test. # indicates significant differences from trained animals at that time period. Results taken from ref. 71.
B) Quantitative densitometric analysis of immunocytochemistry of P-CREB (in CA1 region of the dorsal hippocampus of naive, shocked and trained in the inhibitory avoidance task rats. Data are expressed as percentage of saline treated naive control values (means ± SE of relative OD per mm^2). N= 5. The white column indicates values from naive animals. The grey columns indicate values from animals submitted to shock of identical intensity (0.4 mA) to that used in inhibitory avoidance training. The black columns show values for trained animals at the indicated period following training session. * indicate $p< 0.001$ in comparison with naive control values in Newman-Keuls test after ANOVA. Results taken from refs. 6 and 39.

or more of these changes at different times in the post-training period can determine whether they pertain to STM or to LTM formation.[36,41,43]

The hippocampal events that underlie the processing of STM and LTM begin with the activation of glutamate NMDA receptors in the CA1 region of the hippocampus. The changes include activation of the PKA and MAPK cascades and an increase of nuclear P-$CREB_1$ and c-fos protein levels.[6,12,39] They are not seen in animals treated with posttraining intra-hippocampal infusions of the NMDA receptor blocker D-aminophosphonopentanoic acid (AP5), which also causes amnesia.[12]

One-Trial Avoidance

The task of choice for the study of variables that affect or are affected by memory formation has been for many years a form of contextual fear called one-trial inhibitory avoidance.[39,74] This task is learned in a few seconds and therefore permits a clear-cut dichotomy between effects on acquisition and effects on the post-acquisition period in which STMs and LTMs are formed.[27] In addition, this task offers a clear-cut definition between acquisition, consolidation, retrieval and extinction.[74] This is not the case with multi-trial tasks, such as the diverse varieties of spatial learning; in these tasks, acquisition, consolidation and retrieval are distributed over several sessions and are impossible to extricate one from the other in each session. Further, many experiments carried out over the past decade have shown that one-trial avoidance depends mainly on the functional integrity of specific molecular mechanisms in the CA1 region of the hippocampus, and in several of its connections, namely, the entorhinal, posterior parietal and cingulate cortex,[2] the basolateral amygdala and the medial septum.[38,39]

Many of the experiments to be commented here used one-trial step-down inhibitory (passive) avoidance. For this task, rats are placed on a platform (CS) and receive a footshock (US) when they step down from it onto a grid. Animals learn to remain longer on the platform than they do on the training session (CR). The name "avoidance" should not obscure the fact that this is not really an instrumental task: it is acquired through a single CS(context)-US (footshock) pairing, which implies an impossibility for the animal to actually use the CR (to stay in a safe platform or compartment) as an instrument to avoid the US. When the animals are tested, they are exposed to the CS alone, which is actually the method of choice for initiating extinction.[74] In fact, one-trial avoidance is extinguished by repeated testing at 24 h intervals.[74]

The cAMP/PKA Signaling Pathway

The second-messenger cyclic 3'-5'adenosine monophosphate (cAMP) regulates several important pre- and post-synaptic events, mainly by the activation of the cAMP-dependent protein kinase (PKA) and subsequent targets. The intracellular levels of cAMP are directly regulated by cAMP-phosphodiesterase, a membrane bound enzyme that breaks down cAMP molecules, and by adenylyl cyclase (AC), the enzyme that produces cAMP from ATP molecules. Several types of adenylyl cyclase have been identified and their activation constitute an important regulatory step in the cascade.[14,19] ACs are targets of dopaminergic D1, β-noradrenergic and serotonergic 1A receptor coupled G proteins that either can have stimulatory or inhibitory effect upon cAMP production. Also, AC can be directly sensitive to intracellular raises in calcium-calmodulin levels resulting from the activation of NMDA receptors, voltage-dependent calcium channels, or intracellular Ca^{2+} release.[12,55,78]

PKA consists of a tetrameric holoenzyme composed by two regulatory subunits constitutively linked to two catalytic subunits. cAMP binds to the regulatory subunits, inducing a conformational change that results in the release of catalytic subunits. Once separated, catalytic subunits can phosphorylate Ser/Tre residues on its substrates either in the synapse and its periphery, or, when translocated to the soma, in the nucleus.[24] Drugs acting specifically upon catalytic and regulatory PKA subunits have been extensively used to characterize the physiological role of PKA.[1,6,71,73]

Different PKA isoforms of both regulatory and catalytic subunits have been identified, resulting either from expression of different genes or from alternative mRNA splicing of the same precursor.[24] This diversity permits several combinations between them. Several holoenzyme subtypes have been characterized, with specific catalytic dynamics, substrate affinity and cellular location. PKAs also vary in relation to the subcellular compartment in which they are more prevalent, and this seems to be related to specific anchoring proteins that bind PKA in resting conditions.[20,22]

Among PKA neural substrates, the constitutively expressed regulatory transcription factor CREB is a prominent candidate to mediate PKA mechanisms in LTM storage and has been proposed to act as molecular switch from short- to long-lasting synaptic modifications.[6,13,23,33,69] When catalytic subunits of PKA translocate to the nucleus they phosphorylate CREB on[133] Ser, activating the protein and directly linking cAMP transduction pathways to gene expression and protein synthesis.[13,67,79] There are several molecular forms of CREB; the one most widely believed to participate in memory consolidation is $CREB_1$ (see ref. 67 and Frankland and Josselyn, this book).

There is no direct evidence whatsoever as to what proteins are specifically synthetized through the activation of the cAMP/PKA/P-CREB pathway, with the possible exception of c-fos.[12] However, very solid and abundant evidence suggests that: 1) many of these are new proteins; 2) some are cell adhesion molecules (Regan, this volume), and 3) the action of these mediates changes in synaptic ultrastructure (Geinisman et al, this volume) and synaptic organization. Thus, the cAMP/PKA/CREB signaling pathway is crucial for the regulation of the synaptic events that are at the core of memory formation. It does NOT carry the information, but enables it to be carried.

PKA Involvement in Long-Term Memory Formation

The first inklings of involvement of the cAMP/PKA/P-CREB pathway in the maintenance of CA1 LTP[49] and in LTM storage appeared several years ago.[17] Subsequently, the participation of this pathway in memory formation was characterized in different plastic processes in several species, from facilitation in *Aplysia*[17] to odor conditioning in *Drosophila*,[70,79] spatial learning in the mouse,[10,11,30] and aversive learning in the chick[65,81] and the rat.[6,69,71,72] The various experiments were performed using both genetically modified animals that were unable to express CREB or PKA correctly,[10,11,79] and PKA inhibitors,[1,6,71,73] cAMP analogues,[1] or anti-sense CREB[30] infused at different times after the original training The studies on transgenic or knockout animals were useful to establish the need of PKA or CREB for memory formation, and the pharmacologic experiments revealed both this and the precise timing of the intervention of cAMP, PKA or CREB in the process.[39]

As mentioned above, the cAMP/PKA/P-CREB pathway is activated twice after inhibitory avoidance training: briefly within the first few minutes after acquisition, and again 2-6 h later (Fig. 1A).[6,12,69,71-73] The two posttraining peaks of PKA activity in the CA1 region of the rat hippocampus are accompanied by increased levels of P-CREB (Fig. 1B)[6,69] and are necessary for LTM formation.[71] Also, they are coincident in time with the two phases in which memory of the one-trial task is sensitive to the infusion into CA1 of inhibitors of transcription[42] or of protein synthesis.[59] Inhibition of AC or PKA at the time of either peak blocks LTM formation.[1,6,8]

The second peak of post-training PKA activity depends on the first: if this is abolished by Rp-cAMPs given into CA1, the second peak of PKA is not seen.[71] Further, the second peak of post-training PKA activity depends on the prior activation of glutamate NMDA receptors at the time of training: if these are blocked by AP5 given post-training, the increase of intranuclear PKA activity that takes place 2 h later is not seen.[12]

The reliance of memory formation on a double wave of metabolic activity in the hippocampus was first described by Matthies and his collaborators in the '80s,[28,47,48] and confirmed by many others using various forms of aversive conditioning in the rat[6,10,39] and the chick.[65,81]

A peak of increased PKA activity is seen in the entorhinal cortex but not in the parietal cortex 3 h after training in the one-trial task.[58] The hippocampus is interconnected through the entorhinal cortex to several other regions of the cortex.[34]

The first posttraining peak of PKA activity occurs without any detectable concomitant change in cellular cAMP levels.[6] It must, then, result from a quick activation of the enzyme somehow triggered by glutamate or by noradrenergic receptor activation (see below). The early PKA peak could also result from cross-talk with the concomitant activation of other enzymatic systems, such as CaMKII (Cammarota and Medina, this volume), PKC or Src.[53,80] In the hippocampus, cAMP levels increase slowly 60 min after inhibitory avoidance training, and attain a peak at 180-360 min. The maximum rise in cAMP levels is supposed to trigger PKA activation: it correlates with the second PKA peak of activity.[6] The cAMP increase cannot be attributed to changes in cAMP-specific phosphodiesterase, and might be consequence of enhanced adenylylcyclase activity.[6]

Various neurotransmitter systems associated with alertness, anxiety, emotion or mood affect PKA activity indirectly, through actions on G-protein coupled receptors that regulate AC. Dopaminergic D1 and β-noradrenergic receptors enhance, and serotonin 1A (5HT1A) receptors inhibit adenylyl cyclase activity (see Buhot et al, de Bruin, and also Gibbs and Summers in this book), and thereby alter cellular cAMP levels. Forskolin stimulates adenylyl cyclase; 8-Br-cAMP mimicks the effects of cAMP, including that upon the regulatory subunits of PKA. It was found that the infusion of the D1 agonist, SKF38393, of norepinephrine, of the 5HT1A receptor antagonist, NAN-190, of forskolin or of 8-Br-cAMP enhances LTM when given posttraining into CA1, the entorhinal cortex or the posterior parietal cortex; in contrast, infusions of the PKA inhibitor, KT5720, of the D1 antagonist. SCH23390, of the β-blocker timolol, or of the 5HT1A agonist, 8-HO-DPAT hinders LTM formation.[1,6] The effect of these substances on memory is probably related to the well-known fluctuations of memory processes that occur in relation to mood, anxiety levels or emotion. The effect of these substances on memory varies with the time after training at which they are given, and with the brain structure into which they are infused. The time-windows of the effectiveness of each drug may or may not correlate with the occurrence of PKA activity peaks. Thus, at the immediate posttraining period, norepinephrine enhances and KT5720 inhibits memory consolidation when given into CA1; but only 3 or 6 h later all the drugs become effective as mentioned when given into this structure.[1,8] In contrast, SKF38393, SCH23390, norepinephrine, timolol, 8-HO-DPAT, NAN-190, KT5720, forskolin and 8-Br-cAMP were effective when given into the entorhinal cortex 0, 3 or 6 h after training, or into the parietal cortex 3 or 6 h after training.[1] Obviously, these studies point to the need of AC/PKA activity at precise moments of the post-training period, regardless of whether this activity is at a peak or not.

It is interesting to note here that despite the similarity in nature and time-course between the second peak of PKA activation in memory formation[6] and in the involvement of PKA in CA1 LTP,[33] there are some significant differences. First, the existence of an early peak of PKA activity in LTP has not been clearly demonstrated and whether this may be at all necessary for the occurrence of the second peak, as is the case in memory formation.[71] Second, the second, late peak of PKA activity and P-CREB levels that follows training and is necessary for LTM is modulated by dopaminergic D1, β-noradrenergic and 5HT1A receptors in CA1[1,6] and by the muscarinic cholinergic input coming from the medial septum.[69] The late CREB-dependent phase of LTP in CA1 is apparently only modulated by D1 receptors in CA1;[31] it is instead modulated by β-noradrenergic receptors only in CA3.[32] The differences may of course be due to the fact that the LTP work was carried out in tissue slices, in which modulatory input is absent.

As mentioned, both PKA peaks correlate with an increase in nuclear $CREB_1$ phosphorylation at^{133} Ser in CA1, and with a sensitivity of memory to inhibitors of transcription or of protein synthesis. Antisense CREB infused into CA1 blocks the persistence of spatial LTM beyond 4h (Frankland and Josselyn, this book and also ref. 30). Inhibitors of transcription[42] or of protein synthesis[10,59] given at the time of training or 3 h later but not in the period in between also block LTM. There is evidence that the key proteins syntetized 3-5 or more h after training for the construction of long-lasting memories involve glycoproteins related to cell adhesion (Regan, this volume) promoting morphological changes at the synapses involved in each particular learning experience (see Geinisman et al, this book).

The intracellular mediators of glutamatergic, monoaminergic or cholinergic transmission to protein synthesis stimulation remain unclear. The hypothesis that there must be a relation between the receptors and PKA involvement is substantiated by the following findings: a) intact NMDA receptors are needed for the second peak of PKA activity;[12] b) D1, β, 5HT1A and cholinergic receptors in the hippocampus modulate PKA and produce the changes in memory formation that would be predicted from their biochemical effects; c) PKA is the only kinase described to follow the double wave activation pattern that coincides with protein synthesis requirement during LTM consolidation, and this depends on the early participation of glutamate AMPA, NMDA and metabotropic receptors (mGluRs);[39] d) PKA activates CREB at the times in which changes in gene activation and protein synthesis are essential for memory formation: around the time of training and again 2-6 h later.[59,74]

The role of mGluRs in the activation of PKA should be investigated. Such a role is to be predicted from their physiological action.[62] Intact mGluRs in rat CA1 are necessary for memory formation in the first few minutes posttraining,[8a] as they have shown to be for the establishment of LTP.[9,64] The participation of class I metabotropic receptor and specifically mGluR5 in memory formation has been recently ascertained.[15,63] Moreover, their different contribution to short- and long-term memories suggest a distinct contribution to short- and long-term memory (see Riedel et al, this book).

The metabolic intracellular scenarios of each period of PKA activation, ranging from neurotransmitter actions and their consequences on second messengers to the cross-talk between PKA and other signaling pathways[53] are different, and this will have to be taken into account when a fully descriptive formal hypothesis on the cellular processes necessary for memory formation is established. The formulation of such a hypothesis is still several experiments away from current knowledge.

PKA Involvement in Short-Term Memory Formation

PKA is separately involved in STM and in LTM. In fact, this separation contributed to demonstrate the dichotomy between STM and LTM.[71,73]

When infused into CA1 immediately after training, canceling the first peak of hippocampal PKA activity,[71] competitive inhibitors of the catalytic (KT5720) and the regulatory (Rp-cAMP) subunits of PKA cause amnesia for STM and LTM. (Fig. 2A).[41,71,73] In contrast, the stimulant of the regulatory subunit, Sp-cAMP, enhanced retention of both memory types (Fig. 2B).[71] Therefore in the immediate posttraining period, PKA is obviously necessary for the formation of both STM and LTM.

However, the infusion into CA1 of KT5720 or Rp-cAMPs from 22 to 90 min after training blocks STM completely but has no effect on LTM (Fig. 3A); Sp-cAMPs given 22-90 min posttraining selectively enhances STM.[71,73] When given 170-180 min post-training, these drugs affect LTM again (Fig. 3A), while having no influence on the retrieval of STM (see refs. 41, 71, 73). Stimulators of PKA subserve the same time profile than inhibitors influencing LTM (Fig. 3B). Therefore, PKA is necessary for STM formation during the first hour or so after training, and it is necessary for LTM formation only at the time of its two peaks.[71] The PKA substrates(s) involved in its role in STM are not known; clearly, they do not include P-CREB, which re-

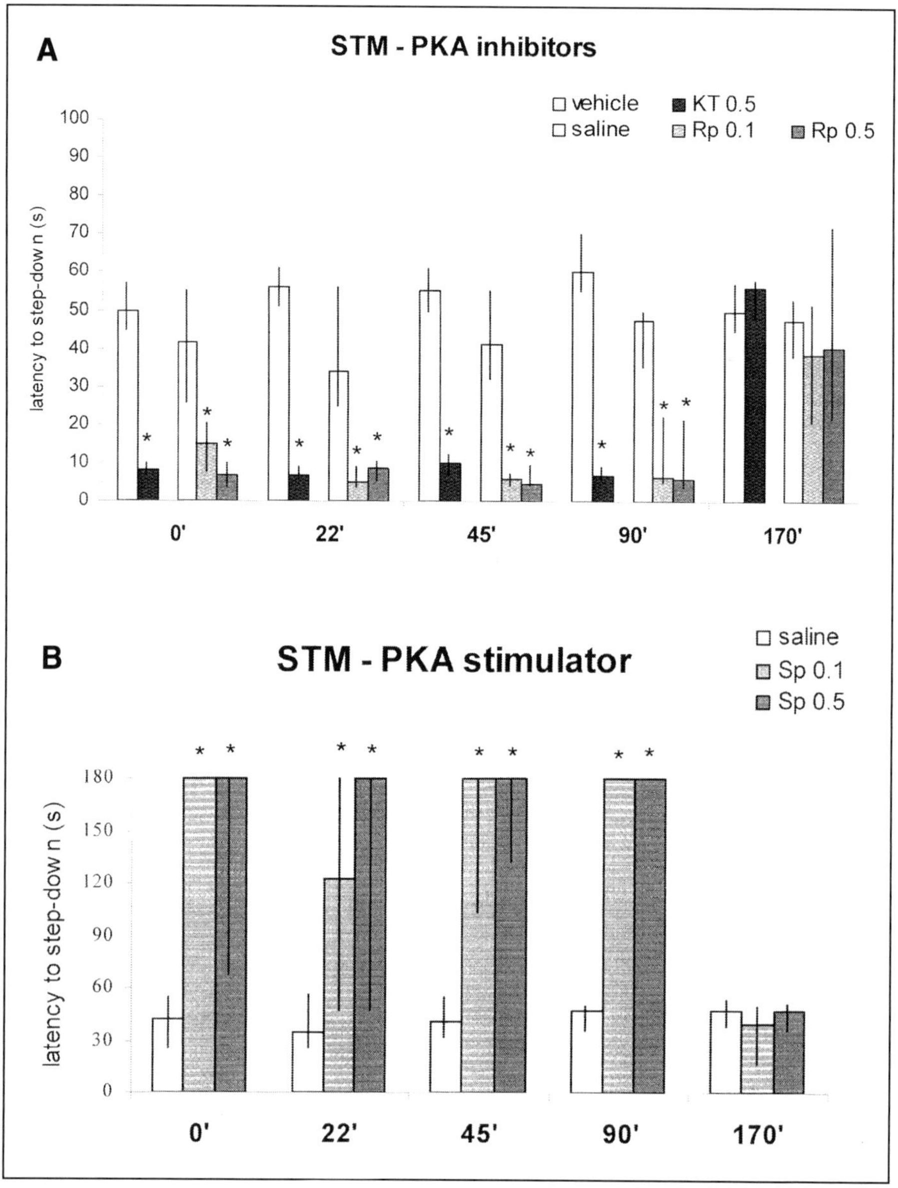

Figure 2. Short-Term Memory. A) Data are expressed as median (interquartile range) latency to step down from the platform on STM test sessions. N = 8-10 per group. A STM was measured at 180 min from training in animals that received intrahippocampal infusions of vehicle (20% dimethylsulfoxide), the PKA catalytic subunit inhibitor KT5720 (0.5mg) (KT 0.5), saline or the PKA regulatory inhibitor Rp-cAMPs at two doses (0.1 and 0.5 mg) (Rp 0.1 and Rp 0.5, respectively). Infusions were given 0, 22, 45, 90 or 170 min after inhibitory avoidance training, * indicate significant differences from controls at $p < 0.01$ level in Mann-Whitney u tests, two-tailed. B) STM was measured at 180 min from training in animals that received intrahippocampal infusions of saline or the PKA regulatory stimulator Sp-cAMPs at two doses (0.1 and 0.5 mg) (Sp 0.1 and Sp 0.5, respectively). Infusions were given 0, 22, 45, 90 or 170 min after inhibitory avoidance training. * indicate significant differences from controls at $p < 0.01$ level in Mann-Whitney U tests, two-tailed.

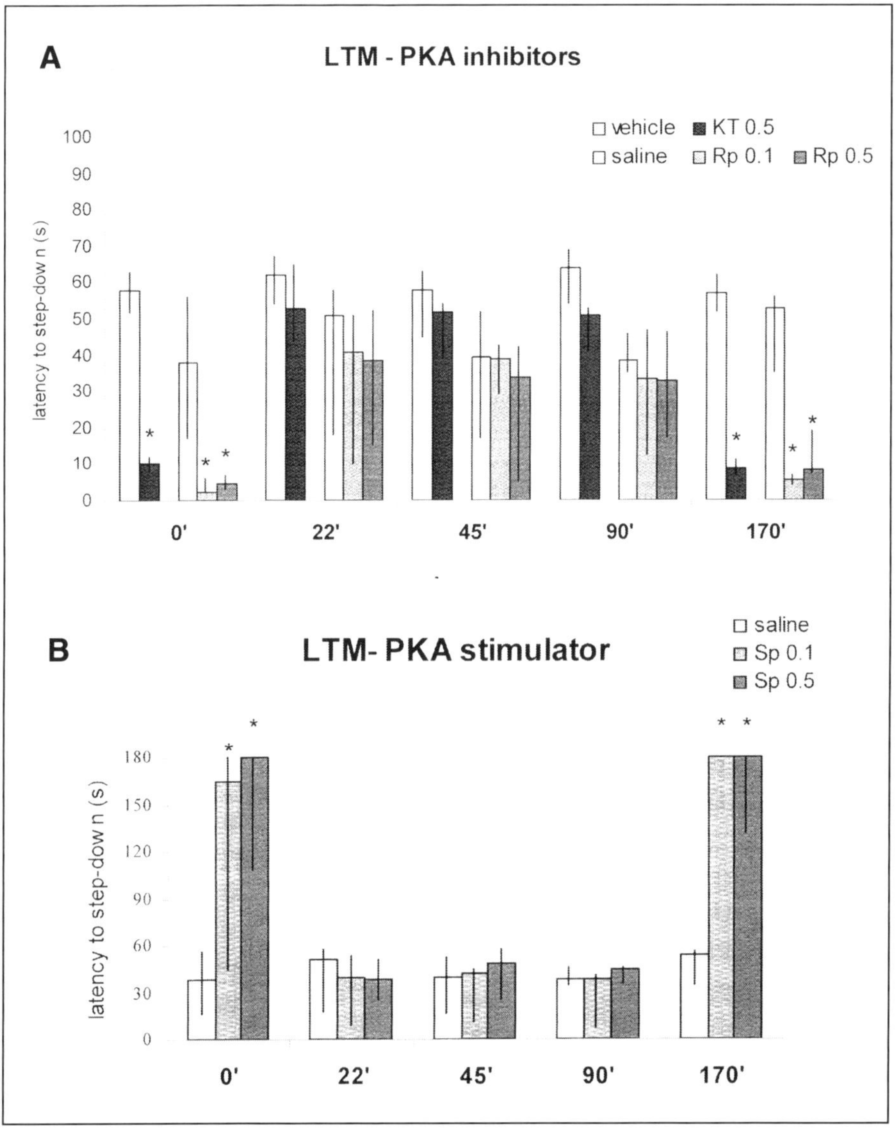

Figure 3. Long-Term Memory. A) Data are expressed as median (interquartile range) latency to step down from the platform on LTM test sessions. N = 8-11 per group. A LTM was measured 24h after training session in animals that received intrahippocampal infusions of vehicle (20% dimethylsulfoxide), the PKA catalytic subunit inhibitor KT5720 (0.5mg) (KT 0.5), saline or or the PKA regulatory inhibitor Rp-cAMPs at two doses (0.1 and 0.5 mg) (Rp 0.1 and Rp 0.5, respectively). Infusions were given 0, 22, 45, 90 or 170 min after inhibitory avoidance training, * indicate significant differences from controls at $p < 0.01$ level in Mann-Whitney u tests, two-tailed. B) Long-Term Memory. Data are expressed as median (interquartile range) latency to step down from the platform on LTM test sessions. N = 8-11 per group. LTM was measured 24h after training session in animals that received intrahippocampal infusions of saline or the PKA regulatory stimulator Sp-cAMPs at two doses (0.1 and 0.5 mg) (Sp 0.1 and Sp 0.5, respectively). Infusions were given 0, 22, 45, 90 or 170 min after inhibitory avoidance training. * indicate significant differences from controls at $p < 0.01$ level in Mann-Whitney u tests, two-tailed.

mains at basal levels in the period between 5 and 180 min post-training.[8,69] In principle, the PKA substrates needed for its influence on STM may include receptors, enzymes or cytoskeletal synaptic components whose function can be temporary modulated by phosphorylation.[23]

STM is also modulated by dopamine D1, β-noradrenergic and 5HT1A receptors in CA1 and in the entorhinal cortex. As mentioned, these receptors regulate PKA indirectly through influences on AC. However, this modulation is quite different in both brain structures, and it is also different from that observed for LTM.[37] Thus, overall, there is a very strong monoaminergic modulation of memory processes by pathways and receptors involved in the perception of and reaction to changes in alertness, mood, emotion or anxiety levels, but the final outcome of this modulation in terms of cognitive events is difficult to predict. Monoaminergic receptors in the CA1 region and in the entorhinal cortex acting simultaneously may have similar, different or even opposite effects on STM and LTM depending on the degree to which each of these receptors is activated.[41,42] In animals, including humans, this will probably depend on the particular mood, degree of alertness or emotional state of the subjects, and these, as is known, vary subtly from minute to minute in daily life.

PKA Involvement in Memory Retrieval

For reasons that escape us, most research on the molecular mechanisms in memory has centered on memory formation.[38,39,46] Until 2 years ago there were very few experiments on the mechanisms of memory retrieval. This is so in spite of the obvious fact that memories can only be measured indirectly, through retrieval.

Again, most research on this topic centered on the one-trial step-down avoidance task. Retrieval of this task measured 3 h after training (STM retrieval) is blocked by the infusion into CA1 of the glutamate AMPA receptor antagonists, CNQX or DNQX, or by the generic mGluR receptor blocker, MCPG; it is unaffected by NMDA antagonists or by inhibitors of PKA or MAPK.[44] Therefore, STM retrieval appears to depend only on the integrity of the regular glutamatergic transmission through AMPA and metabotropic receptors in CA1. The relation between glutamate receptor activation, particularly mGluRs, and PKA in retrieval deserves to be studied as well, as has been pointed out above in connection with consolidation. Actually, more is known about the role of mGluRs in retrieval than about that in consolidation and its down-stream effectors. In the hippocampus, together with AMPA glutamatergic receptors, mGluRs are crucial for retrieval and its involvement is suggested to be responsible for PKA and MAPK signaling pathways activation.[68]

In contrast, retrieval of the one-trial avoidance task measured 24 h or 31 days after training depends on mGluRs and on the PKA and MAPK cascade: it is blocked by the infusion into CA1 of MCPG (or, at the 24 h interval, CNQX or DNQX), or of the inhibitor of PKA, Rp-cAMPs, or of the MAPK inhibitor PD098059.[2,43,68] Retrieval measured 1 or 31 days after training is enhanced by the pre-test infusion into CA1 of the PKA activator, Sp-cAMPs, which underscores a key role of PKA in retention test performance.[2] AP5 given into CA1 has no effect on retrieval, and DNQX given into this structure at the time of testing 31 days after training also has no effect.[2,43]

Very importantly, all these substances have similar effects to those observed in CA1, when given into the entorhinal, posterior parietal or anterior cingulate cortex prior to the test session, with two exceptions: AP5 also blocked retrieval when given into the parietal or cingulate cortex, and DNQX was ineffective when given into the cingulate cortex.[2] This shows that the retrieval of a task as deceivingly simple as one-trial avoidance requires similar and simultaneous metabolic activity in many regions of the brain.[42]

The basolateral amygdala, which plays a major role in the consolidation of various forms of fear conditioning, including one-trial avoidance[38] also participates in retrieval. Among the molecular systems studied (see above), only DNQX was able to block retrieval when infused into the basolateral amygdala prior to testing.[2] This does not detract from the role of that structure on retrieval. It merely shows that this role is metabolically simpler than that of the cortex. The basolateral amygdala is a site of action of glucocorticoids in the modulation of retrieval[18] and there is evidence for a role of it in the modulation of the emotional content of memories both at the time of consolidation and at the time of retrieval.[38]

The involvement of PKA in retrieval occurs without any detectable change of the activity of the enzyme in CA1. This stands in contrast to the MAPK pathway enzymes, p42 and p44, which are increased following the test session (ref. 68 and see also Selcher et al in this book).

The need for regular ongoing PKA activity in CA1, entorhinal, parietal and cingulate cortex in order for retrieval to take place is underlined by the fact that the infusion of D1 or β receptor agonists or of a 5HT1A antagonist in all these structures prior to testing enhances retrieval, whereas that of D1 or β receptor antagonists or of a 5HT1A antagonist depresses retrieval of the one-trial task.[3] Again, this modulation by the pathways and receptors involved in emotion, mood, alertness or anxiety occurs simultaneously in all the brain structures mentioned.

PKA Involvement in Extinction

Retrieval of the one-trial avoidance task, or of most fear conditioning procedures for that matter, is usually carried out without the unconditioned stimulus, i.e., the footshock(s).[42,74] This is precisely the necessary condition for extinction to take place.[57] Recent evidence indicates that extinction of fear-motivated tasks requires the integrity of the hippocampus.[16] Extinction involves a new learning of opposite sign to the original learning: animals learn a CS-no shock contingency instead of the previously acquired CS-US contingency.

We have observed that extinction of the one-trial avoidance task really begins in the first test session. If further test sessions are repeated at 24 h intervals, retrieval becomes gradually diminished.[74] A variety of treatments given into CA1 either before or after the first test sessions hinders extinction. Among these, the most relevant are the transcription blocker, DRB,[42] the protein synthesis inhibitor, anisomycin,[74] the NMDA receptor antagonist, AP5, the CaMKII inhibitor, KN-62, the MAPK inhibitor, PD098059, and the PKA inhibitor, Rp-cAMPs.[42,74]

These pharmacological findings indicate that extinction is indeed a new learning, requiring transcription and protein synthesis (see also ref. 5) as much as the original learning does, as well as a key role of NMDA receptors, CaMKII, PKA and MAPK. The main difference between memory formation of aversive learning and its extinction is that in the former all these molecular processes act in a sequential way, whereas in extinction they appear to act simultaneously at the time of the first retrieval test.[42]

Summary

PKA plays a pivotal role in the consolidation, retrieval and extinction of memories. The cAMP/PKA/P-CREB pathway is involved twice in the consolidation of LTM: first at the time of training, and then again 2-6 h later. PKA is involved in STM formation during at least the first hour after training. The role of PKA in memory formation is not restricted to the hippocampus: PKA is also necessary in the entorhinal and posterior parietal cortex, where it can be up- or down-regulated by receptors involved in anxiety or mood. It is believed that the protein synthesis that mediates the effective laying down of memory traces through changes at the synaptic level is specifically activated by the PKA signaling pathway.

Further, PKA activity is necessary for retrieval in CA1, entorhinal cortex, posterior parietal and anterior cingulate cortex. Extinction is normally initiated by the first retrieval test after a training experience. Extinction also requires intact on-going PKA activity in the CA1 region.

Cross-talk between the PKA pathway and others may occur and it may be crucial for memory processes.[53] It remains to be studied and analysed in detail.

References

1. Ardenghi P, Barros D, Izquierdo LA et al. Late and prolonged post-training memory modulation in entorhinal and parietal cortex by drugs acting on the cAMP/protein kinase A signalling pathway. Behav Pharmacol 1997; 8:745-751.
2. Barros DM, Izquierdo LA, Mello e Souza T et al. Molecular signaling pathways in the cerebral cortex are required for retrieval of one-trial avoidance learning in rats. Behav Brain Res 2000; 114:183-192.
3. Barros DM, Mello e Souza T, De David T et al. Simultaneous modulation of retrieval by dopaminergic D_1, b-noradrenergic, serotoninergic 1A and cholinergic muscarinic receptors in cortical structures of the rat. Behav Brain Res 2001; 124:1-7.
4. Bartsch D, Ghirardi M, Skehel PA et al. Aplysia CREB2 represses long-term facilitation: Relief of repression converts transient facilitation into long-term functional and structural change. Cell 1995; 83:979-992.
5. Berman DE, Dudai Y. Memory extinction, learning anew and learning the new: Dissociations in the molecular machinery of learning in the cortex. Science 2001; 291:2417-2419.
6. Bernabeu R, Bevilaqua L, Ardenghi P et al. Involvement of hippocampal D1/D5 receptor-cAMP signaling pathways in a late memory consolidation phase of an aversively-motivated task in rats. Proc Natl Acad Sci USA 1997; 94:7041-7046.
7. Bernabeu R, Schmitz P, Faillace MP et al. Hippocampal cGMP and cAMP are differentially involved in memory processing of an inhibitory avoidance learning. Neuro Report 1996; 7:585-588.
8. Bevilaqua L, Ardenghi P, Schröder N et al. Drugs acting upon the protein kinase A/CREB pathway modulate memory consolidation when given late after training into rat hippocampus but not amygdala. Behav Pharmacol 1997; 8:331-338.
9. Bortolotto ZA, Fitzjohn SM, Collingridge GL. Roles of metabotropic glutamate receptors in LTP and LTD in the hippocampus. Curr Opin Neurobiol 1999; 9:299-304.
10. Bourchuladze R, Abel T, Berman N et al. Differential training procedures recruit either one or two critical periods for contextual memory consolidation. Learn Mem 1998; 5:465-474.
11. Bourtchouladze R, Freguelli B, Blendy J et al. Deficient long-term memory in mice with a targeted mutation of the cAMP-responsive element binding protein. Cell 1994; 79:59-68.
12. Cammarota M, Bevilaqua LR, Ardenghi P et al. Learning-associated activation of nuclear MAPK, CREB and Elk-1, along with Fos production, in the rat hippocampus after a one-trial avoidance learning: Abolition by NMDA receptor blockade. Mol Brain Res 2000; 76:36-46.
13. Carew TJ. Molecular enhancement of memory formation. Neuron 1996; 16:5-8.
14. Choi EJ, Xia Z, Villacres EC et al. The regulatory diversity of the mammalian adenylyl cyclases. Curr Opin Cell Biol 1993; 5:269-73.
15. Christoffersen GR, Christensen LH, Harrington NR et al. Task-specific enhancement of short-term, but not long-term, memory by class I metabotropic glutamate receptor antagonist 1-aminoindan-1,5-dicarboxylic acid in rats. Behav Brain Res 1999; 101:215-226.
16. Corcoran KA, Maren S. Hippocampal inactivation disrupts contextual retrieval of fear memory after extinction. J Neurosci 2001; 21:1720-1726.
17. Dash PK, Hochner B, Kandel ER. Injection of the cAMP-responsive element into the nucleus of Aplysia sensory neurons blocks long-term facilitation. Nature 1990; 345:718-21.
18. de Quervain DJK, Roozendaal B, McGaugh JL. Stress and glucocorticoids impair retrieval of long-term memory. Nature 1998; 394:787-790.
19. Edelhoff S, Villacres EC, Storm D et al. Mapping adenylyl cyclase genes type I, II, III, IV, V and VI in mouse. Mamm Genome 1995; 6:111-113.
20. Edwards AS, Scott JD. A-kinase anchoring proteins: Protein kinase A and beyond. Curr Opin Cell Biol 2000; 12:217-221.
21. Emptage NJ, Carew TJ. Long-term synaptic facilitation in the absence of short-term facilitation in Aplysia neurons. Science 1993; 262:253-256.
22. Felciello A, Gottesman MX, Avedimento EV. The biological functions of A-kinase anchor proteins. J Mol Biol 2001; 308:99-114.
23. Ferrer I, Blanco R, Rivera R et al. CREB-1 and CREB-2 immunoreactivity in the rat brain. Brain Res 1996; 712:159-164.
24 Francis SH, Corbin JD. Structure and function of cyclic nucleotide-dependent protein kinases. Annu Rev Physiol 1994; 56:237-272.
25. Geinisman Y. Synaptogenesis and structural remodelling. In: Riedel G, Platt B, eds. Memories are made of these. Austin: RG Landes Co., 2002: in press.
26. Gerlai RT. Ephrins. In: Riedel G, Platt B, eds. Memories are made of these. Austin: RG Landes Co., 2002: in press.

27. Gold PE. The use of avoidance training in studies of modulation of memory storage. Behav Neural Biol 1986; 46:87-98.
28. Grecksch G, Matthies H. Two sensitiv periods for the amnestic effect of anisomycin. Pharmacol Biochem Behav 1979; 12:663-665.
29. Greenough WT. Morphological and molecular studies of synaptic memory mechanisms: Links to the fragile X mental retardation syndrome. In: Gold PE, Greenough WT, eds. Memory Consolidation: Essays in honor of James L McGaugh – A time to remember. Washington: American Psychological Association, 2001.
30. Guzowski JF, McGaugh JL. Antisense oligodeoxynucleotide-mediated disruption of hippocampal cAMP response element binding protein levels impairs consolidation of memory for water maze training. Proc Natl Acad Sci USA 1997; 94:2693-2698.
31. Huang YY, Kandel ER. D1/D5 receptor agonists induce a protein synthesis-dependent late potentiation in the CA1 region of the hippocampus. Proc Natl Acad Sci USA 1995; 92:2446-2450.
32. Huang Y-Y, Kandel ER. Modulation of both the early and the late phase of mossy fiber LTP by the activation of beta-adrenergic receptors. Neuron 1996; 16:611-617.
33. Huang Y-Y, Li X-C, Kandel ER. cAMP contributes to mossy fiber LTP by initiating both a covalently mediated early phase and a macromolecular synthesis-dependent late phase. Cell 1994; 79:69-79.
34. Hyman BT, Van Hoesen GW, Damasio AR. Memory-related neural systems in Alzheimer's disease: An anatomic study. Neurology 1990; 40:1721-1730.
35. Izquierdo I, Ardenghi PG, Barros DM et al. Consolidation of short- and long-term memory. In: Gold PE, Greenough WT, eds. Memory Consolidation. Washington American Psychological Association, 2001:79-112.
36. Izquierdo I, Barros DM, Mello e Souza T et al. Mechanisms for memory types differ. Nature 1998; 293:635-636.
37. Izquierdo I, Izquierdo LA, Barros DM et al. Differential involvement of cortical receptor mechanisms in working, short- and long-term memory. Behav Pharmacol 1998; 9:421-427.
38. Izquierdo I, McGaugh JL. Behavioural pharmacology and its contribution to the molecular basis of memory consolidation. Behav Pharmacol 2000; 11:517-534.
39. Izquierdo I, Medina JH. Memory formation: The sequence of biochemical events in the hippocampus and its connection to activity in other brain structures. Neurobiol Learn Mem 1997; 68:285-316.
40. Izquierdo I, Medina JH, Izquierdo LA et al. Short- and long-term memory are differentially regulated by monoaminergic systems in the rat brain. Neurobiol Learn Mem 1998; 69:219-224.
41. Izquierdo I, Medina JH, Vianna MRM et al. Separate mechanisms for short- and long-term memory. Behav Brain Res 1999; 103:1-11.
42. Izquierdo I, Vianna MRM, Izquierdo LA et al. Memory retrieval and its lasting consequences In: Palomo T, Archer T, Beninger RJ, Borrell J, Murray R eds. Neurodevelopmental liabilities in brain disease states. Madrid: Fundacion Cerebro y Mente. 2002.
43. Izquierdo LA, Barros DM, Ardenghi PG et al. Different hippocampal molecular requirements for short- and long-term retrieval of one-trial avoidance learning. Behav Brain Res 2000; 111:93-98.
44. Izquierdo LA, Barros DM, Medina JH et al. Novelty enhances retrieval of one-trial avoidance learning in rats 1 or 31 days after training unless the hippocampus is inactivated by different receptor antagonists and enzyme inhibitors. Behav Brain Res 2000; 117:215-220.
45. Izquierdo LA, Vianna M, Barros DM et al. Short- and long-term memory are differentially affected by metabolic inhibitors given into hippocampus and entorhinal cortex. Neurobiol Learn Mem 2000; 73:141-149.
46. Kandel ER, Squire LR. Neuroscience: Breaking down scientific barriers to the study of brain and mind. Science 2000; 290:1113-20.
47. Matthies H. Neurobiological aspects of learning and memory. Ann Rev Psychol 1989; 40:381-409.
48. Matthies H. In search of the cellular mechanism of learning. Prog Neurobiol 1989; 32:277-349.
49. Matthies H, Reymann KG. Protein kinase A inhibitors prevent the maintenance of hippocam pal long-term potentiation. Neuro Report 1993; 4:712-714.
50. McGaugh JL. Time-dependent processes in memory storage. Science 1966; 153:1351-1359.
51. McGaugh JL. Memory: A century of consolidation. Science 2000; 287:248-251.
52. Medina JH, Cammarota M. Ca^{+2}/calmodulin-dependent kinase II. In: Riedel G, Platt B, eds. Memories are made of these. Austin: RG Landes Co., 2002:in press.
53. Micheau J, Riedel G. Protein kinases: Which one is the memory molecule? Cell Molec Life Sci 1999; 55:534-548.
54. Mileusnic R. Protein synthesis: II new proteins. In: Riedel G, Platt B, eds. Memories are made of these. Austin: RG Landes Co., 2002:in press.

55. Mons N, Guillou JL, Jaffard R. The role of Ca^{+2}/calmodulin-stimulable adenylyl ciclase as molecular coincidence detectors in memory formation. Cell Mol Life Sci 1999; 55:525-533.
56. Nógues X, Pascale A, Micheau J et al. Protein kinase C. In: Riedel G, Platt B, eds. Memories are made of these. Austin: RG Landes Co., 2002: in press.
57. Pavlov IP. Conditioned Reflexes. New York: Dover, 1956.
58. Pereira P, Ardenghi P, de Souza MM et al. Effects of infusions of the tyrosine kinase inhibitor radicicol into the hippocampus on short- and long-term memory of the inhibitory avoidance task. Behav Pharmacol 2001; 12:299-302.
59. Quevedo J, Vianna MRM, Roesler R et al. Two time-windows of anisomycin-induced amnesia for inhibitory avoidance training in rats: Protection from amnesia by pretraining but not pre-exposure to the apparatus. Learn Mem 1999; 6:600-607.
60. Ramón y Cajal S. Neue Darstellung vom histologischen Bau des Zentralnervösen System. Archiv Anat Physiol (Anat) 1893; 319-428.
61. Regan C. Cell adhesion molecules. In: Riedel G, Platt B, eds. Memories are made of these. Austin: RG Landes Co., 2002:in press.
62. Riedel G. Function of metabotropic glutamate receptors in learning and memory. Trends Neurosci 1996; 19:219-224.
63. Riedel G, Casabona G, Platt B et al. Fear conditioning-induced time- and subregion-specific increase in expression of mGlu5 receptor protein in rat hippocampus. Neuropharmacol 2000; 39:1943-1951.
64. Riedel G, Reymann KG. Metabotropic glutamate receptors in hippocampal long-term potentiation and learning and memory. Acta Physiol Scand 1996; 15:1-19.
65. Rose SPR. Time-dependent processes in memory formation revisited. In: Gold PE, Greenough WT, eds. Memory Consolidation. Washington: American Psychological Association, 2001: 113-128.
66. Routtenberg A. It's about time. In: Gold PE, Greenough WT, eds. Memory Consolidation. Washington: American Psychological Association, 2001: 113-128.
67. Silva AJ, Kogan JH, Frankland PW et al. CREB and memory. Ann Rev Neurosci 1998; 21:127-148.
68. Szapiro G, Izquierdo LA, Alonso M et al. Participation of hippocampal metabotropic receptors, protein kinase A and mitogen-activated protein kinases in memory retrieval. Neurosci. 2000; 99:1-5.
69. Taubenfeld SM, Wiig KA, Bear MF et al. A molecular correlate of memory and amnesia in the hippocampus. Nature Neurosci 1999; 2:309-310.
70. Tully T. Discovery of genes involved with learning and memory: An experimental synthesis of Hirschian and Benzerian perspectives. Proc Natl Acad Sci USA 1996; 93:13460-13467.
71. Vianna MR, Izquierdo LA, Barros DM et al. Differential role of hippocampal cAMP-dependent protein kinase in short- and long-term memory. Neurochem Res 2000; 25:621-626.
72. Vianna MRM, Barros DM, Silva T et al. Pharmacological demonstration of the differential involvement of protein kinase C isoforms in short- and long-term memory formation and retrieval in rats. Psychopharmacol 2000; 150:77-84.
73. Vianna MRM, Izquierdo LA, Barros DM et al. Intrahippocampal infusion of an inhibitor of protein kinase A separates short- from long-term memory. Behav Pharmacol 1999; 10:223-227.
74. Vianna MRM, Szapiro G, McGaugh JL et al. Retrieval of memory for fear-motivated training initiates extinction requiring protein synthesis in the rat hippocampus. Proc Natl Acad Sci USA 2001; 98:12251-12254.
75. Viola H, Furman M, Izquierdo LA et al. Phosphorylated cAMP response element-binding protein as a molecular marker of memory processing: Effect of novelty. J Neurosci 2000; 20: 1-5.
76. Walz R, Roesler R, Barros DM et al. Effects of post-training infusions of a mitogen-activated protein kinase kinase inhibitor into the hippocampus or entorhinal cortex on short- and long-term retention of inhibitory avoidance. Behav Pharmacol 1999; 10:723-730.
77. Walz R, Roesler R, Quevedo J et al. Time-dependent impairment of inhibitory avoidance retention in rats by posttraining infusion of a mitogen-activated protein kinase kinase inhibitor into cortical and limbic structures. Neurobiol Learn Mem 2000; 73:11-20.
78. Wong ST, Athos J, Figueroa XA et al. Calcium-stimulated adenylyl cyclase activity is critical for hippocampal dependent long-term memory and late phase LTP. Neuron 1999; 23:787-798.
79. Yin JCP, Tully T. CREB and the formation of long-term memory. Curr Opin Neurobiol 1996; 6:264-268.
80. Zhao W, Cavallaro S, Gusev P et al. Nonreceptor tyrosine protein kinase pp60c-src in spatial learning: Synapse-specific changes in its gene expression, tyrosine phosphorylation, and protein-protein interactions. Proc Natl Acad Sci USA 2000; 97:8098-8103.
81. Zhao WQ, Polya GM, Wang BH et al. Inhibitors of cAMP-dependent protein kinase impair long-term memory formation in day-old chicks. Neurobiol Learn Mem 1995; 64:106-18.

Chapter 4.4

Protein Kinase C

Xavier Noguès, Alessia Pascale, Jacques Micheau and Fiorenzo Battaini

Abstract

This chapter reviews the involvement of PKC in cognition and in the brain pathologies affecting cognition. PKC is a family of enzymes. Its activation process is described in the first part. In the second part, we analyse the role of PKC in synaptic plasticity, a neuronal property which may be required for the acquisition of information. Both electrophysiological data and behavioural pharmacology suggest the involvement of this enzyme in metaplasticity. Then, the experiments directly addressing the involvement of PKC in cognition at various phylogenetic levels are presented according to the psychological question addressed. Finally, we show that the brain pathologies causing cognitive deficits are characterised by PKC abnormalities and the putative therapeutic ways mediated by PKC are presented.

Introduction

The involvement of protein kinase C (PKC) in the biological bases of cognition is now widely accepted. This family of enzymes was discovered in 1977,[59] and the first paper showing its role in learning processes was published 10 years later.[74] Since then, the number of publications addressing the issue of its involvement in various aspects of cognition has increased (see ref. 116, 176 for a review). Paradoxically, although these publications are consistent in recognizing the basic involvement of PKC in cognitive processes, both the mode of activation induced by learning and the precise locus of involvement of this enzyme remain unclear.

An interesting issue concerning the neurobiology of cognition is the role of this enzyme in the molecular mechanisms of neuronal and synaptic plasticity. According to most of the present theories on the biological bases of cognition, the property of neuronal plasticity appears to be a sine qua non for an organism to be able to perform cognitive processing. Basically, four types of theories, which have proposed plausible biological devices of active information processing can be distinguished. The local hypothesis supposes that the activity of one neuron may represent one information (see ref. 9 for a detailed description of this view and ref. 141 for a formal description of computational processes involved). The hypothesis of distributed static representation is quite more recent. In this hypothesis, one information is represented by the coactivation of a defined set of cells. Such a view was proposed by Hopfield,[53] McNaughton and Morris[96] or Schneider and Detweiler.[149] A hypothesis of distributed dynamic representation has been proposed by Kohonen.[67] Finally, Freeman and Skarda[42] proposed a hypothesis emphasising the absence of biological representations of information. All of these hypotheses have been supported by the demonstration that each proposed device was able to perform an active cognitive processing of information. All of these hypotheses require properties of neuronal plasticity.

The aim of the present chapter is to present an up-to-date view of the topic, both by presenting the data demonstrating the involvement of PKC in cognition and by raising issues which remain to be elucidated. In its first part, the PKC family is briefly presented through its

From Messengers to Molecules: Memories Are Made of These, edited by Gernot Riedel and Bettina Platt.

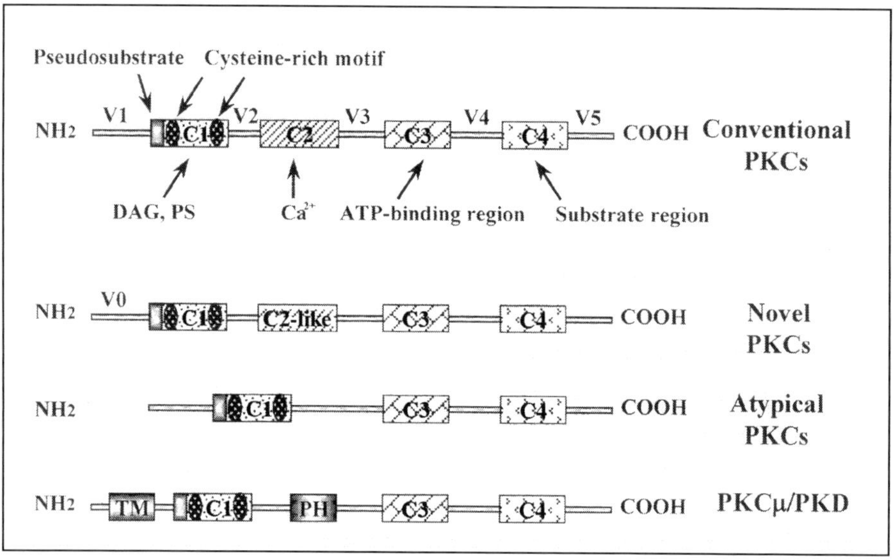

Figure 1. PKCs structure. DAG: diacyglycerol; PS: phosphatidylserine; TM: transmembrane domain; PH: pleckstrin homology domain; C1-C4: conserved regions, V1-V5: variable regions; V0: variable region only present in novel PKCs.

molecular structure and its mode of activation. The second part of the chapter shows the putative roles of PKC in the biochemical processes providing the neurons with the properties required to generate cognitive abilities. In the third part, we describe the set of experiments directly assessing the involvement of PKC in cognitive processes and attempting to identify in which cognitive processes PKC is involved. Finally, in the fourth and fifth parts, we present the current knowledge of the involvement of PKC in various pathologies affecting cognitive function and the putative therapeutic pathways mediated by interventions on PKC.

Protein Kinase C: Who Is It?

PKC: A Family of Phosphorylating Enzymes

Extracellular signals (neurotransmitters, hormones, growth factors) are decisive for cell-cell communication. Their interaction with specific cellular receptors triggers a cascade of events that allows the transduction of the message inside the single cell. In this context, a strategic role is played by protein kinases through phosphorylation of specific substrates. The PKC family includes 12 different isoenzymes classified as calcium-dependent (conventional PKCs: α, βI, βII and γ) and calcium-independent isoforms (novel PKCs: δ, ε, η, θ and μ; atypical PKCs: ζ, ι and λ) according to their calcium sensitivity.[111]

Conventional and novel PKCs require diacylglycerol (DAG) and phosphatidylserine (PS) for full activation, while atypical isoenzymes are DAG-insensitive (for a review see ref. 61). Novel PKCs appear to be selectively stimulated by DAG generated via the phospholipase D pathway that is not associated with a calcium increase.[46]

PKCs are characterized by a tissue-specific expression, e.g., while a, d and ζ isoforms are ubiquitous, PKCγ is selectively expressed in the CNS, PKCη is mainly found in lung and skin and PKC θ is highly present in skeletal muscle.[111,124]

The primary amminoacidic PKC sequence comprises conserved domains (C1-C4) separated by variable regions (V1-V5; cf. Fig. 1). The C1 domain contains the pseudosubstrate region responsible for keeping the enzyme in the inactive form (folded conformation) when the physiological activators (e.g., calcium, DAG and PS) are absent. A tandem cysteine-rich motif (C1a and C1b) follows the pseudosubstrate sequence and represents the DAG/phorbol ester binding site. Mutational studies have shown that C1a and C1b are not equivalent. In the atypical PKCs, there is only a single cysteine-rich motif present that seems more similar to C1a, whose function is still unclear.[100] The C2 domain constitutes the recognition site for the calcium ion and is lacking in calcium-independent isoforms, although a C2-like sequence has been described in novel PKCs. The C2-like region does not contain carboxyl residues and is consequently unable to bind calcium.[109] PS also binds to C2, but several studies have indicated the contribution of both C1 and V1 in mediating this interaction.[84] The cysteine-rich motif and C2 regions do not appear to be the only portions involved in the phospholipid interaction. In fact, the participation of the pseudosubstrate through its basic residues has been described.[139]

The C3 domain represents the ATP binding motif and is highly conserved in all human PKC isoenzymes. The substrate binding site responsible for substrate specificity is localized in the C4 region (reviewed in ref. 51).

Protein kinase μ (human) and its murine homolog (PKD) seem to form a PKC related-kinase class. They contain an N-terminal transmembrane domain and lack the pseudosubstrate region. PKCμ/PKD is closest to novel PKCs in terms of stimulation conditions, since it is insensitive to calcium ion and requires DAG and PS for activation. PKCμ/PKD is also characterized by the presence of a PH (pleckstrin homology) region and does not catalyze significant phosphorylation of typical PKC substrates including histone and myelin basic protein. This unique pattern of substrate recognition and its Golgi localization suggest the possibility that different effects from those related to novel PKCs are coupled to this class of kinase.[139]

Newly synthesized PKC associates with the membrane where it is probably phosphorylated by the 3-phosphoinositide-dependent kinase (PDK)-1 on the exposed activation loop. This phosphorylation correctly aligns residues for catalysis, allowing subsequent autophosphorylation at two key positions in the carboxy terminus. The phosphorylation at the first of these sites brings the PKC into a catalytically competent conformation, while the second one accounts for the release of the enzyme into the cytosol.[31,110] This suggests a role of phosphorylation in directing the subcellular localization of PKCs and a possible modulation exerted by dephosphorylation processes.[110]

Involvement of Anchoring Proteins in PKC Activation

Short-term stimulation of PKCs seems to be involved in signal transduction pathways leading to short-term events, while prolonged activation may be associated with long-term processes like proliferation and differentiation.

The various PKC isoforms have been found in different subcellular compartments including the cytosol, cell membranes, cytoskeleton and nucleus in both inactive and active conformations. In the inactive state (folded conformation), the pseudosubstrate region binds the catalytic domain. Following the interaction with the specific cofactors, there is a decrease in the affinity of the pseudosubstrate for the catalytic site that can thus exert its phosphorylating function, while PKC shifts to the open, active conformation (reviewed in ref. 128). Moreover, the localization in specific subcellular sites may account for the different functions associated with the individual isoenzymes, despite substrate specificity or sensitivity to activators.

Extracellular signals not only lead to activation of PKCs through the production of cofactors, but also induce the redistribution of the individual kinase C isoenzymes from one compartment to another, an event known as translocation.[69]

Studies on trypsin sensitivity have indicated the importance of targeting proteins in this translocation mechanism.[102] A primary function played by targeting proteins is to localize the individual isoforms in the appropriate position to respond to distinct activating signals. Secondly, these anchoring proteins bring PKCs close to the specific substrates.[62] In this context, Receptors for Activated C Kinase (RACKs) have a crucial importance. RACKs bind with PKCs at a site distinct from the substrate or the catalytic region, suggesting a direct protein-protein interaction. These anchoring proteins appear isoform-specific and seem to bind and shuttle the individual PKCs to a different compartment only when they are in the active conformation.[138] It has been suggested that another group of proteins may interact with PKCs in their inactive state. In analogy with RACKs, these proteins have been named RICKs (Receptors for Inactivated C Kinase) and they are probably involved in localizing the various isoenzymes in subcellular sites distinguished from those related to RACKs.[103] An intriguing recent hypothesis arising from studies on astrocytes proposes that PKCs may translocate through actin filaments from one RACK pool to another localized in a distinct compartment, and that RACKs are mainly acceptors and not shuttle proteins for PKCs in different functional states.[130]

The coordination of cellular signals often involves phosphorylation/dephosphorylation processes requiring the control of multiple protein kinases and phosphatases. Protein phosphatase 2A (PP2A) can dephosphorylate and thus inactivate PKC; moreover, inhibition of PP2A prevents PKC down-regulation.[48] A fascinating possibility to be investigated is that kinases, phosphatases and specific substrates could be brought to the same subcellular site through anchoring proteins.[155] Targeting proteins might then ensure the appropriate phosphorylation/dephosphorylation level of specific substrates and guarantee a more adequate response to cellular dynamism.

PKC in Synaptic Plasticity

Synaptic plasticity is a physiological phenomenon whereby specific patterns of neural activity give rise to changes in synaptic efficacy and neural excitability that long outlast the events that trigger them. The phenomenon called long-term potentiation (LTP) is a remarkable example of synaptic plasticity, which has quickly become an attractive model as a biological substrate of learning and memory formation (for a comprehensive review, see ref. 91). LTP, first described by Bliss and Lømo[16] at perforant path-granule cell synapses of the rabbit, has since been observed in different brain areas such as the amygdala and the neocortex. The molecular mechanisms that underlie LTP have been extensively investigated and early studies have pointed to the prominent role played by phosphorylation reactions.[19,83,144] Therefore, among the protein kinases implicated in these processes, PKC has been the main focus of attention.

Early Evidence for PKC Involvement in LTP

A first set of experiments performed in the dorsal hippocampus by the Routtenberg group demonstrated a selective increase in the in vitro phosphorylation of protein F1 (also called B-50, GAP 43, neuromodulin) in animals shortly sacrificed after LTP initiation in the dentate gyrus. This increase in F1 phosphorylation persisted at least three days and was positively correlated with the change in enhancement of both the population spike and the EPSP.[83,145] As evidence indicated that protein F1 was a substrate for protein kinase C, it was assumed by the same group that LTP would induce an increase in PKC activity. Consequently, they showed that high-frequency stimulation of the perforant path in the dorsal hippocampus resulted in a translocation of PKC activity from the cytosol to the membrane compartment. However, this activation did not occur immediately as the changes were detectable only one hour after the termination of the LTP procedure.[3,145] This time-dependent redistribution of PKC activity was the first evidence suggesting a role of PKC in the persistence of the change in synaptic efficacy but not in its initiation.

In Which LTP Phase Is PKC Involved?

It is generally assumed that the mechanisms underlying LTP are divided into two distinct phases: the induction phase of short duration (about 30 s), during which a rise in postsynaptic intracellular concentration of Ca^{2+} triggers the enhancement of the synaptic response; and the maintenance phase, during which cascades of events bring about a long-lasting and selective increase in synaptic efficacy. A great deal of work has gone into elucidating the participation of PKC in both phases. Pharmacological evidence from experiments using both PKC inhibitors and PKC activating phorbol esters support the contention that PKC activation is necessary for the maintenance of LTP but not its induction.[82,137] However, the broad spectrum of activity of these compounds may have led to some confusing effects due to functional cross-talk between Ca^{2+}/calmodulin and PKC pathways that may occur during the induction of LTP.[182] Moreover, findings using more specific tools to assess PKC activity indicate that activation of PKC contributes to both the induction and the maintenance of hippocampal LTP.[66,183] Other reports have more or less restricted the involvement of PKC to the induction phase of the LTP, most likely by modulating the activity of the NMDA receptor.[14,105,118] Although there is still much discussion on the role of PKC in the induction and maintenance of LTP (see also "Recent Findings: The Role of PKC in Synaptic plasticity, Revisited", below), another current debate concerns the synaptic locus of PKC requirement for LTP.

Is PKC Requirement for LTP Presynaptic, Postsynaptic or Both?

There has been much controversy as to whether LTP in various fields of the hippocampus is mediated by presynaptic or postsynaptic mechanisms. Although this debate continues, recent observations point to a predominantly postsynaptic site for LTP expression, at least in the Schaffer-collateral-commissural synapses in region CA1.[156] In the context of this debate, the locus of PKC activity necessary for LTP has been largely investigated. Based on pharmacological experiments examining PKC activator and inhibitor effects on LTP, it has been proposed that established LTP is expressed through a mechanism localized in the presynaptic terminal.[87,88] Other pharmacological studies showed evidence that PKC activation is necessary for LTP persistence at both presynaptic and postsynaptic sites, but with a different time window.[56] This assumption is strengthened by findings demonstrating that proteins involved in synaptic plasticity like the presynaptic associated protein F1/GAP 43 and the postsynaptic protein neurogranin (also called RC3, BICKS and p17) possess PKC phosphorylation sites.[131] However, although it is not an exclusive hypothesis, more evidence is in favor of a postsynaptic target for PKC involvement in LTP maintenance (see ref. 156). For example, the injection of antibodies against the postsynaptic protein neurogranin in hippocampal slices was shown to prevent LTP.[39] Moreover, postsynaptic injection of $PKC_{19\text{-}31}$, a pseudo-substrate inhibitor of PKC, blocked Ca^{2+}/calmodulin-induced potentiation in CA1 neurons of a hippocampal slice.[182,183]

In conclusion, to paraphrase Huang and collaborators[56] it is attractive to propose that different protein kinase C subtypes differentially localized to presynaptic or postsynaptic elements may sequentially participate in LTP generation and expression.

PKC Isozymes: Who Is Doing What?

While behavioural experiments have highlighted the prominent role played by the γ isoform of PKC in the cellular mechanisms of memory formation (see "Evidence for the Involvement of PKC in Cognitive Processes"), electrophysiological investigations analyzing the biochemical mechanisms of LTP have unravelled a more striking diversity in the possible actions of the different PKC isoforms. Experiments with hippocampal LTP induction in vivo seem to converge to emphasize changes in PKC γ immunoreactivity[5,6,171] or in the expression of PKC γ mRNA[98,99] after LTP. Indeed, LTP induced in the dentate gyrus synapses produced an early but transient PKC translocation restricted to the γ isoform.[5,6] Following this initial activation, other isoforms like the PKC α/β may, through a different molecular cascade, be transiently activated during a later post-tetanic phase.[6] However, long-term increases of only γ PKC mRNA were detected 24 hr after LTP induction.[171] The prominent role played by this PKC isoform in synaptic plasticity has been confirmed by studies with transgenic mice lacking PKC γ that showed diminished LTP.[1] However, these mice exhibited only slight memory impairment, suggesting the participation of other kinases. In vitro studies appear to paint a more complex picture of the respective involvement of the different PKC isoforms in the cellular mechanisms underlying LTP. First, LTP, which occurs in many pathways of the brain, is differentially sensitive to the pattern of stimulation required for its induction. It is now well established that LTP in mossy fiber-CA3 pathway differs in many respects from LTP in the Schaffer collateral/commissural (SC)-CA1 pathway. For instance, the latter depends on NMDA receptors while the former does not (see ref. 91). PKC involvement may vary according to the form of LTP that is concerned. LTP in mossy fiber-CA3 pathway was associated with a detectable translocation of the presynaptically enriched PKC α and ε isozymes, whereas no changes were observed in γ and β isoform distribution.[157] In partial agreement with these findings, LTP induction in the CA1 in hippocampal slices was shown to initiate a 15-minute delayed translocation to membrane of the PKC α/β isoform.[79,160] This PKC activation was thought to reflect the redistribution of the predominantly presynaptic PKC α that was accompanied by an increase in the presynaptic protein F1/GAP-43 phosphorylation.[79] However, these results differ from a study by Sacktor et al. in which a transient immediate translocation of the PKC isozymes α, βI, βII, γ, δ, ζ and η into the membrane compartment was reported after a tetanus to SC-fibres.[147] After this rapid and very transient activation of PKC isozymes only a proteolytic activation of PKC ζ was observed to generate a persistently activated PKC. Moreover, the increase in PKM ζ was shown to be correlated with the degree of EPSP potentiation.[125] Sacktor's group has extended the key role played by PKM ζ in synaptic plasticity by showing a decrease in this constitutively active kinase C isoform in long-term depression (LTD) maintenance. It was thus concluded that a bidirectional regulation of PKC, more precisely PKC ζ, may participate in the molecular mechanisms of LTP and LTD.[54] In addition, the participation of PKC isoforms may have a regional specificity. It has recently been reported that the deletion of the PKC β gene, which is predominantly expressed in the neocortex, in area CA1 of the hippocampus and in the basolateral nucleus of the amygdala, results in defects in both cue and contextual versions of fear conditioning. However, PKC β knock-out mice showed normal hippocampal synaptic transmission and LTP, indicating a critical role for the β isoform of PKC in learning-related signal transduction mechanisms that take place in the basolateral nucleus of the amygdala.[185]

Calcium-dependent and -independent isoforms of PKC are potentially involved in molecular mechanisms of LTP. However, their respective participation appears to depend on their cellular and subcellular localization and on the compartmentalization of PKC-substrate interactions. Another aspect requiring our attention concerns the stimulation history of the synapse.

Recent Findings: The Role of PKC in Synaptic Plasticity Revisited

New compounds that more selectively target PKC have led to the reconsideration of the role of kinase in synaptic plasticity. Potent and highly selective inhibitors of PKC failed to prevent the induction of LTP in the CA1 region of adult rat hippocampal slices, suggesting that PKC was unlikely to be directly involved in LTP induction or expression.[18] In contrast, PKC appears to play a role in a form of metaplasticity that regulates the induction of LTP.[18] Metaplasticity refers to the plasticity of synaptic plasticity; it is a change in the ability to induce subsequent synaptic plasticity, such as LTP or LTD.[2] Interestingly, this view had also been suggested by using a pharmacological and behavioural approach.[113,115] Bortolotto and collaborators[17] have previously demonstrated that activation of metabotropic glutamate receptors (mGluR) can facilitate the subsequent induction of LTP. It has been proposed that the stimulation of mGluR activates a molecular switch whose setting requires the activation of PKC.[18] More generally, recent experimental data support the idea that PKC may play a critical role in modulating bidirectional changes in synaptic strength. Depending on the previous synaptic history, PKC activation was shown to affect the threshold for eliciting LTD or for suppressing LTP induction.[161] The molecular mechanisms by which PKC may modulate these forms of metaplasticity are still unknown. However, this modulation may not occur directly by phosphorylation of the NMDA receptor, but rather by that of associated targeting, anchoring or signalling protein(s).[76,191]

The whole picture is rather more complicated because PKC does not function in isolation but rather in a complex cross-talk within the intracellular network.[101] A growing body of evi-

Table 1. Studies on the involvement of hippocampal PKC in the classical conditioning of the nictitating membrane in the rabbit

PKC Assessment	Result	Ref.
PdBu binding in slices	Increase in the CA1 after 3 sessions of training. No significant changes in the CA3.	120
PdBu binding in slices	Increase in the stratum oriens, decrease in the stratum pyramidale relative to 1 session of training.	120
PdBu binding in slices	Increase in the CA3 stratum oriens after 1 session of training. No significant changes in the CA1, CA3 stratum pyramidale, CA3 stratum radiatum, and DG.	148
PKC βII and γ immunoreactivity.	Increase in PKCγ in the CA1 and CA3. No changes in PKC α, β I and II. Positive correlation between the ability to learn and PKCg immunoreactivity in the CA1 and CA3.	175
In vitro immunoblotting assay	No translocation. No changes in the amount of PKC.	175
PKC activity	Translocation from the cytosol to the membranes (3 sessions)	8
PKC activity	Potentiation in the ability to be activated in the synaptosomal membrane fraction in the CA1, CA2, CA3 after 3 days of training.	163
In situ hybridization	No significant changes in α,β, γ and \epsiion PKC mRNAs.	27

dence is showing that several signaling pathways interact to refine the molecular mechanisms mediating short- and long-term forms of synaptic plasticity.[138,164,182]

Evidence for the Involvement of PKC in Cognitive Processes

Habituation Sensitization and Classical Conditioning in Invertebrates

Together with some elementary forms of classical conditioning, habituation and sensitization have been thoroughly studied in *Aplysia* and *Hermissenda.* In such animals, these types of learning can easily be explained by the properties and the organization of a few well identified neurons.

Basically, the involvement of PKC has been demonstrated by measuring PKC activity during behavioral sensitization of the defensive reflex in *Aplysia* (increase in the membrane fraction[147] and classical conditioning in *Hermissenda* (decrease in the cytosolic fraction[106]). The amount of [^{3}H]-PdBu binding also increases following training.[97] PdBu (Phorbol-12,13-dibutyrate) is a PKC activator which can be used at low concentrations as a specific radioligand for quantitative autoradiography, either in situ or in vitro. It is then considered as an index of the total amount of conventional and novel PKCs. The functional aspect of these changes has been demonstrated by showing that the administration of various PKC inhibitors impairs learning[28,38,94,106] and that both the intracellular injection of PKC and the extracellular administration of the PKC activator PdBu mimic the electrophysiological changes induced by learning.[37]

Finally, several studies on *Drosophila* mutants exhibiting low learning abilities show that they have a low or nonexistent expression of PKC.[24,49] These data suggest a wide ubiquity of the involvement of PKC in learning and memory processes in the animal kingdom.

Classical Conditioning

Basically, most of the studies on classical conditioning have used [^{3}H]-PdBu binding to study the changes in the hippocampus following nictitating membrane conditioning (see Table 1). Most of them demonstrate changes induced by training though these changes do not occur

Table 2. Involvement of PKC in emotional response conditioning

Task	PKC Assessment	Structure	Result	Ref.
Sequence of electric shocks; re-exposure to the context	PdBu binding on slices	Hippocampus (dorsal): CA1, CA3, DG	No effects of stress. Increase in the CA1 induced by re-exposure to the context 5 days later	154
Sequence of electric shocks; re-exposure to the context	PdBu binding on slices	Amygdala: basolateral nucleus; Thalamus: dorsomedial nucleus	Increase up to 24 h after stress. Increase after re-exposure 5 days later	154
Sequence of electric shocks; re-exposure to the context	PdBu binding on slices	Somato-sensory cortex	No effects of stress. No effect of re-exposure to the context 5 days later	154
US: electric shocks; CS: context + tone; R: freezing	H7 injections	Amygdala: basolateral and central nucleus	No effects in central nucleus; decreased freezing both in context and auditory conditioning in basolateral nucleus	45
US: electric shocks; CS: context + tone; R: freezing	PKCβ deficient	Whole brain	Induced amnesia both for context and tone	185
US: electric shocks; CS: context; R: freezing	PKCγ deficient	Whole brain	Mutants freeze less than wild type mice when re-exposed to the context.	1

Studies have all been conducted on rats excepted for the studies using transgenic animals (mice). CS: conditioned stimulus; US: unconditioned stimulus; R: response.

systematically in the same hippocampal fields. However, the results are difficult to conciliate into a unifying model. Studies on PKC activity measured in vitro also failed to give a homogeneous set of results. The first[8] displayed an obvious translocation of PKC activity, i.e., a significant decrease in the cytosol associated with an increase in the membrane-bound fraction (relative to the total amount of PKC activity). In another study, though PKC activity did not change, its ability to be activated was potentiated in the synaptosomal fraction.[163]

In situ immunostaining shows the involvement of PKCγ in these changes.[175] Interestingly, this increase in PKCγ immunoreactivity and changes in PKC activity are not due to a neosynthesis of the enzyme but rather to changes in its molecular conformation, given that conditioning may not change the mRNA level of various PKC isozymes.[27]

Classical conditioning of the emotional response in rodents (see Table 2) has also provided some interesting results. A correlation approach using [^{3}H]-PdBu binding shows the involvement of hippocampal PKC in the memory component of a task, whereas in the amygdala, PKC is involved both in stress and response to the context.[154]

To verify the functional role of these changes, both PKC inhibitor injection and transgenic mice have been used. PKC inhibition in the amygdala induces amnesia.[45] Obviously, both PKC β and γ transgenic mice show a deficit (respectively, refs. 185 and 1), but this is less pronounced for PKC γ. It might be interesting to use methods such as antisense injections or inducible gene knock-out's in order to palliate the limitations of conventional transgenic animals.

Table 3. Data showing the involvement of PKC in passive avoidance of electric shocks

PKC Assessment	Structure	Result	Ref.
PdBu binding in slices	Hippocampus	Increase in the CA1, CA2, CA3, and DG	15
PdBu binding in slices	Amygdala	Increase	15
PdBu binding in slices	Cortex : Frontal, parietal, entorhina	Increase	15
PdBu binding in slices	Striatum	No effect	15
γPKC immunoreact.	Cortex : motor, prefrontal, cingular, sensorial	Strong increase in PKCg immunoreactivity	174
cPKC immunoreact. in synaptic plasma membrane	Hippocampus	PKC α: no changes. PKCβ I: increase. PKC βII: increase; cumulative with shock. PKC γ: n.s. increase induced by shock	127
PKC activity	Hippocampus	Increase in synaptosomal membrane	21
H7, melittin	Intra-cerebro-ventricular	Impairment	169
H7 and staurosporin	Hippocampus	Impairment	188
Staurosporin or CGP-41231 injections	Dorsal hippocampus	Impairment of retention when injected up to 120 mn	63
Go6976. Inhibits: α and βI PKC	Hippocampus (dorsal): CA1	Induces amnesia when injected up to 110 mn after acquisition	127
Go6976 (inhibits α and βI PKCs) and Go7874 (all PKCs)	Dorsal CA1	Post-acquisition injections: Impairment of short-term and long-term memories Pre-retrieval injections: Impairment when test occurs 24h after acquisition but not 3h.	177
PMB and NPC15437	Amygdala	Post-acquisition injections: Impaired memory	178
PMB and NPC15437	Caudate nucleus	No effects	178
PMA	Systemic	Prevent scopolamine induced amnesia	74
PMA	Hippocampus	Improvement of retention 24h after learning.	188
PKCβ deficient	Whole brain	No effects	185

All experiments have been conducted on rats excepted for Refs. 74 and 185 which have been performed on mice.

Passive Avoidance

As shown in (Table 3), passive avoidance of electric shocks induces an increase in [^{3}H]-PdBu binding in several cerebral structures.[15] This increase may be related to the increase in PKCγ immunoreactivity[174] or in PKC βI and βII.[127] Pharmacologic evidence also supports the involvement of PKC and shows that its activation may be necessary for 1 to 2 hours after training.

Other studies have used passive avoidance of food intake (either conditioned taste aversion or conditioned visual aversion; see Table 4). Basically, most of these studies use a pharmacologic approach and support the previously described findings. Interestingly, several of them have been conducted on chicks, thus contributing to show the wide involvement of PKC in memory in the animal kingdom.

Table 4. Involvement of PKC in passive avoidance of food intake

Species	PKC Assessment	Structure	Result	Ref.
Chicks	In vitro immunologic assay	IMHV	Unilateral increase in the particulate/ soluble ratio of α and β PKCs	20
Chicks	PMB and melittin	Forebrain	Impaired memory test performed 3h after training	4
Chicks	H7 and melittin	Forebrain	Impaired memory test performed 3h after training	20
Chicks	Melittin and PMA	Forebrain	Melittin impaired memory. PMA prevented melittin induced amnesia, improved memory after a weakly reinforced training, but impaired memory after a strongly reinforced learning	190
Chicks	13 different kinase inhibitors	IMHV	H7 impaired memory	150
Chicks	Chelerythrine	IMHV	Impaired memory tested 24h after training	151
Rats	PKC activity	Parabrachial nucleus	Increase in the cytosol 24h and 48h but not 12h or 5 days after acquisition. No changes in particulate fraction.	70
Rats	PMB and H7	Gustatory cortex	Impaired learning when injected between CS and US. PMB impaired retention also when injected 30 mn after US	189
Rats	PMB and H7	Amygdala	Impaired learning when injected between CS and US	189
Rats	PMB	Thalamus: ventral posteromedial nucleus	No effects	189
Rats	Chelerythrine	Parabrachial nucleus	Impaired learning	189

IMHV: intermediate medial hyperstriatum ventrale; CS: conditioned stimulus; US: unconditioned stimulus

Operant Conditioning

Few studies have addressed the issue of the involvement of PKC in operant conditioning. All have been conducted by using a pharmacological approach. Consistently, they support the involvement of PKC in this kind of learning (Table 5).[92,162]

Perceptual Discrimination Learning and Perceptual Memory

The studies addressing the issue of Involvement of PKC in perceptual memory are presented in the (Table 6) Basically, it is still widely thought that PKC is involved in a wide panel of tasks. However, these five studies contain two kinds of intriguing results. The first is the learning-induced lateralization of staining in the piriform cortex. Although this result does not call into question the involvement of PKC in learning and memory, it provides interesting insight into rodent neuropsychology, given that studies supporting a lateralization of brain functions in rodents are rare. The second is the discrepancy between two sets of data obtained from the same team.[44,121] The first showed changes in the hippocampus after cue learning in the water maze, whereas the second failed to show this phenomenon though the tasks were quite similar. The authors explained this discrepancy by the difference in task components. As

Table 5. Data showing the involvement of PKC in operant conditioning

Biehavior	Task	PKC Assessment	Structure	Result	Ref.
Active avoidance	Action: must start within 5s. Reward: avoiding an electric shock	NPC15437	Systemic	Impaired the tempora component of the task	92
Active avoidance	Action: must choose the right side. Reward: avoiding an electric shock	NPC15437	Systemic	No effects on the spatial component	92
Skinner box	Action: press a lever Reward: piece of food	GF109203X	Intra-cerebro-ventricular	Impaired retention	162

All three experiments have been conducted on mice. NPC15437 and GF109203X are PKC inhibitors.

Table 6. Evidences of the involvement of PKC in perceptual memory

Behavior	Task	PKC Assessment	Structure	Result	Ref.
Visual memory	Cue learning in the water maze	PdBu binding on slices	Hippocampus	Decrease in the CA3. No significant changes in the CA1 and DG	121
Visual memory	Cue learning in the water maze	PdBu binding on slices	Occipital and cingulate cortex, caudate, putamen, habenula	No significant changes	121
Olfactory memory	Olfactory discrimination	PdBu binding on slices	Piriform cortex	Lateralisation induced by learning	122
Olfactory, auditory and visual memory	Olfactory, auditory and visual discrimination	PdBu binding on slices	Hippocampus	Increase in the CA1 and CA3	122
Visual memory	Cue learning in the water maze	PdBu binding on slices	Cingulate, motor, somato-sensory and striate cortices, hippocampus, globus pallidus, medial geniculate and superior colliculus	No significant changes	44
Visual / spatial memory	Learn a visual pattern associated with cocaine	H7 and chelerythrine	Intra-crerbro-ventricular	Impaired choice when administered 10min but not 30min after learning	23
Visual / spatial memory	Guidance strategy in the water maze	PKCγ deficient	Whole brain	No differences with normal mice	1

All three experiments have been conducted in rats.

shown in this paper and in other reviews on PKC and memory, PKC appears to be involved in a wide range of cognitive processes. Although it cannot yet be formally excluded, this view thus appears unlikely. Such changes in the results of experiment replications have already been found and quite different explanations have been proposed.[114,116]

Cognitive Mapping and Spatial Discrimination

Given the putative role of the hippocampus in cognitive mapping since O'Keefe and Nadel's theory[119] the involvement of hippocampal PKC in spatial learning has widely been investigated.

Place learning in the water maze induced a decrease in [^{3}H]-PdBu binding in the CA3[121] or both in the CA3 and CA1.[44] These results are in obvious opposition with studies using immuno-histochemistry which showed increases in PKCγ immunoreactivity in most of the hippocampal areas.[13,34,173] Spatial discrimination learning decreases cytosolic PKC activity[112] although long-term exposure to an enriched environment increases cytosolic PKC activity.[134]

The functional role of hippocampal PKC in cognitive mapping has been demonstrated by studies showing that intra-hippocampal PKC inhibition impairs learning whereas its acute pharmacological activation improves long-term retention.[113] It is interesting to underline that, although slowed down, learning is possible even when PKC is inhibited. This result supports the concept that PKC is more involved in the regulation of the synaptic learning steps (metaplasticity) rather than in the induction of long-term plasticity itself. Basically, the injected drugs reach a wide variety of synapses, and it is unlikely that the treatment specifically modifies the learning relevant synapses. Moreover, as shown by electrophysiological studies, the pharmacological activation or inhibition of PKC respectively increases or decreases the efficiency of synapses over a period of several hours. As suggested earlier,[115] it is thus probable that the activation of PKC in a whole set of synapses may nonspecifically and transiently increase synaptic efficiency. Then, the response probabilities of synapses specifically activated by learning will be enhanced, thereby inducing a long-term synapse-specific plasticity (see Fig. 2).

Finally, a negative correlation has been found between PKC activity and the ability to learn spatial discrimination tasks both in cytosolic and membrane-bound fractions,[112,114] and a positive correlation was found between basal cytosolic PKC activity and the ability to use cognitive maps (Table 7).[129]

Therefore, it is obvious that spatial discrimination affects hippocampal PKC. However, given that contradictions appear when several methods of PKC assessment are used, cognitive mapping might induce changes in PKC conformation in the hippocampus, leading to an increase in its immunoreactivity. Its proteolysis or limited proteolysis which may follow its activation or participate in it may lead to a decrease in both PdBu binding and calcium- and phospholipid-dependent cytosolic PKC activity (measured post-mortem in vitro).

Table 8 shows studies concerning the involvement of PKC in cognitive mapping in other regions of the brain or nonspecifically in the whole brain.

Motor Activity, Anxiety, Stress

Motor activity, anxiety and stress are cognitive components, which are not specifically but are almost systematically related to learning and memory. In nonhuman learning tasks, they are almost always associated with the experimentally measured learning performance. The study of their association with PKC shows several contradictory results. The simple exploration of a maze induces an increase in hippocampal PKCγ immunoreactivity. However, it is difficult to know whether this increase is due to the motor activity associated with exploration, to the latent cognitive mapping naturally associated with exploration,[172] or to the stress induced by the new environment. The latter view is supported by the study of Fordyce and Wehner[40] which shows that physical activity increases PKC activity in a cytosolic fraction (loosely bound fraction). Motor activity has also been positively correlated with in vitro PdBu binding.[159] Although these results are puzzling, the more consistent reports tend to show that the modifi-

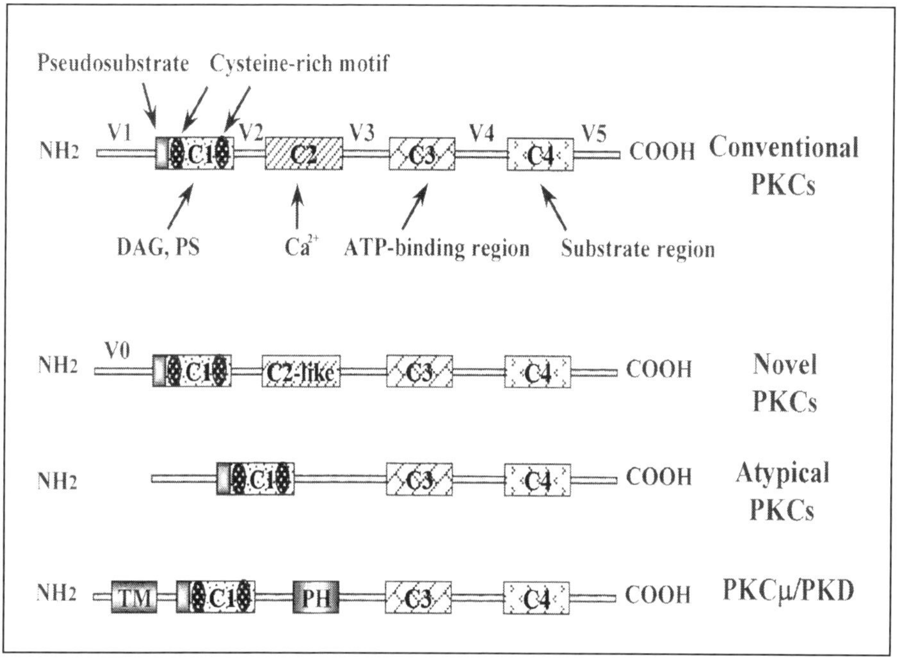

Figure 2. An Illustration of How PKC May Control the "Learning Step" of Synapses (Metaplasticity) During Learning.This figure shows that more stimulations are required in a system where PKC is poorly activated (upper pathway) than in a system where it is strongly activated (lowest pathway) in order to reach the same level of long-term synaptic changes. Pharmacological data have shown that the administration of PKC activators transiently increases synaptic transmission probability. This may make it possible to facilitate the specific potentiation of the relevant synapses for a long period and thus may speed up learning. PKC inhibitors prevent artificially induced LTP but do not affect conventional synaptic transmission. This view may help to understand why basal stores of PKC are correlated with the ability to learn, and why PKC inhibitors slow down learning but do not prevent it whereas PKC activators facilitate learning. (Fig. adapted from ref. 113).

cations in PKC induced by motor activity are only nonfunctional correlates of behavior. Indeed, except for PMA (phorbol myristate acetate) which may have large side effects, pharmacological interventions on PKC generally do not have any effect on motor activity.

More interestingly, stress appears to induce increases in hippocampal PKCγ immunoreactivity[71] whereas PKC inhibitors may induce a mild anxiogenic effect.[177]

PKC and Neuronal Pathologies Impairing Cognition

The involvement of PKC in disease states became apparent 20 years ago when it was discovered that PKC was the receptor for tumor-promoting phorbol esters.[22] Since then PKC has been demonstrated to be involved in a variety of other pathologies, in addition to tumor cell growth, such as cardiovascular disease, diabetes, immune and infectious diseases.[43] PKC controls not only proliferation but also cell survival, and recent studies have attempted to define the role of specific PKC isoforms in regulating apoptotic pathways. While different PKC isoforms are anti-apoptotic, depending on the cell system investigated, PKC δ in particular, is emerging as a general pro-apoptotic intermediate in a number of cells,[31] including neurons.[68] PKC is ubiquitous and most of the PKC isoforms are expressed in brain tissues, where the highest levels of kinase activity are present.[166] In addition to the control of neuronal plasticity discussed above, PKC modifies a variety of other neuronal conditions. For instance, regenerating

Table 7. Data showing the involvement of hippocampal PKC in cognitive mapping

Species	Task	PKC Assessment	Result	Ref.
Rats	Place learning in the water maze	PdBu binding on slices	Decrease in the CA3. No significant changes in the CA1 and DG	121
Rats	Place learning in the water maze	PdBu binding on slices	Decrease in the CA3 and DG. Not significant decrease in the CA1	44
Rats	Hole-board maze; learning food location	γPKC immunoblotts in synaptic membrane	Anterior hippocampus: no changes. Posterior hippocampus: translocation from the cytosol to the membrane after 4 days of training; not after 11 days	34
Mouse	Place learning in the hole-board maze.	γPKC immunoreact.	Strong increase in PKCγ in the CA1, CA3 and DG	173
Mouse	Hole-board maze; exploration	γPKC immunoreact.	Slight increase in PKCγ in the CA1, CA3 and DG	173
Rats	Place learning in the hole-board maze.	PKC immunoreact.	Increase in PKCγ in the CA1 after 9 sessions. Strong increase in PKCγ in the CA1, CA3 and DG after 17 sessions. No changes in PKC α, β I and II.	
Rats	Place learning in the water maze	α, βII and γPKC immunoreact.	Positive correlation only between performance and particulate γPKC, and performance and soluble bII PKC. In old animals: only significant negative correlation between performance and soluble γPKC	25
Mouse	Place learning in the water maze	PKC activity	Negative correlation between performance and PKC activity, both in cytosol and particulate fractions. Analyses performed on genetically different groups of mice, not on individual data.	186
Mouse	Place learning in the radial maze	Soluble PKC activity	No correlation between performance and PKC activity, either on calcium dependant or independent PKCs.	129
Mouse	Spatial mapping, radial maze	Soluble PKC activity	Positive correlation between performance and calcium-dependant PKC activity. No correlation between performance and calcium-independent PKC activity.	129
Rats	6 or 12 days of exposure to an enriched environment	PKC activity	Increase in cytosolic PKC activity after 12 days of exposure. No changes in the particulate fraction. No effects of a 6 days exposure.	134
Mouse	Place learning in the radial maze	PKC activity	Decrease in cytosolic fraction; persists at least 24h. No effect in membrane fraction	112
Mouse	Exploration, radial maze	PKC activity	No effect on PKC activity	112
Mouse	Place learning in the radial maze	PKC activity	Negative correlation between performance and particulate PKC activity	112

continued on next page

Table 7. Continued

Species	Task	PKC Assessment	Result	Ref.
Mouse	Place learning in the radial maze	PKC activity	Negative correlation between performance and particulate PKC activity	112
Mouse	Place learning in the radial maze	PKC activity	Negative correlation of learning ability both with particulate and soluble PKC activity. No correlation with the level of training	114
Mouse	Place learning in the, radial maze	Intra-hippocampal PMB injections	Impaired retention (between 24h delayed sessions). No effects on working memory (intra-session learning)	115
Mouse	Place learning in the radial maze	Intra-hippocampal PMB injections	Improved long term retention (16 days)	115
Mouse	Place learning in the radial maze	Intra-hippocampal OAG injections	Improved long term retention (16 days)	115

nerve fibers, a "neuronal growth process", involves the activation and change of localization of various PKC isoforms.[65]

We will mainly consider here the clinical studies in which changes in central nervous system (CNS) PKC activity and isoform levels have been related to chronic and acute neurodegenerative pathologies involving memory loss. Other CNS-related conditions in which PKC may play a role will also be discussed.

Chronic Neurodegeneration

Alzheimer's Disease (AD)

One of the initial clinical features of AD is an impairment in short-term memory; as the disease progresses other cognitive processes are lost. Pathological characteristics of the disease are senile plaques, neurofibrillary tangles and neuronal loss involving cholinergic pathways.[75] Since M1 and M3 acethylcholine receptors are coupled to PKC activation, this enzymatic system has been studied extensively in this pathology. In AD, brain PKC β and ε are reported to be deficient in the temporal cortex in terms of levels and activities.[93] For the calcium-dependent isoform, the changes are due to modified isoform degradation because PKC β mRNA levels are not modified. Crude extracts from hippocampus, temporal and frontal cortex have a decreased PKC activity[179] while such activities are unchanged in all brain areas of AD patients when the PKC enzyme is purified from endogenous modulators, thus suggesting the possibility of increased endogenous inhibitors of PKC activity in AD brain.[78]

[^{3}H]-PdBu binding studies suggest the possible decline in PKC in parallel with neurofibrillary staging in the enthorhinal cortex, subiculum and hippocampus, on the one hand, and on the other, with amyloid staging in the subiculum.[73] For a more comprehensive review on PKC isoforms and activity in AD, see ref. 64. A different approach has been taken investigating the response of PKC to activation.[179] By incubating AD brain slices in vitro with phorbol esters, a decreased translocation of the enzyme was observed in cortex and hippocampus. Following cholinergic stimuli, AD brain also shows an impairment in coupling, such as interaction of muscarinic M1 receptors with G proteins.[75] This deficit in signal transduction (receptors, G proteins, PKC activation) may contribute to the loss of memory processes and to the limited therapeutic responses to drugs activating the cholinergic system in AD patients. The significance of an impaired PKC activation in AD brain is further underlined by data analyzing

Table 8. Involvement of non-hippocampal PKC in cognitive mapping

Task	PKC Assessment	Structure	Result	Ref
Place learning in the water maze	Immunoreact.	Neostriatum	No correlations with performance (α, βII and γ PKC in soluble and particulate fractions)	25
Place learning in the water maze	PdBu binding on slices	Occipital and cingulate cortex, caudate, putamen, habenula	No significant changes.	121
Place learning in the water maze	PdBu binding on slices	Caudate nucleus	Decrease.	44
Place learning in the water maze	PdBu binding on slices	Cingulate, motor, somato-sensory and striate cortices, globus pallidus, medial geniculate and superior colliculus	No significant changes.	44
Exploration of the passive avoidance apparatus	PKC γ immunoreact.	Cortex: prefrontal, cingular, motor, sensoria	Slight increase in PKCγ immunoreactivity	174
Place learning in the water maze	H7, melittin	Intra-cerebro-ventricular	Impairment	169
Place learning in the water maze	PdBu injections	Intra-cerebro-ventricular	Improvement of learning and retention.	133
Place learning in the water maze	PKCγ deficient	Whole brain	Mild impairment and obviously a change in the problem solving strategy.	1

All experiments have been conducted on rats except for ref. 1 which have been conducted on transgenic mice.

RACK proteins. These "receptors" for PKCs are also important regulators of PKC activation in brain tissues.[10] In AD frontal cortex, there is a deficit in RACK1 levels. The specificity of this effect is indicated by the observation that PKCβII, which specifically binds RACK1[140] is not changed in the same tissues,[11] suggesting that the dissociation in levels of expression in the same pathological samples is a specific rather than a general effect of AD on PKC signaling. These data suggest that in addition to a lipid environment,[143] other changes in the protein environment may impair PKC activation and function in AD brain. Moreover, for RACK1 there may be a selective area effect because the levels of this protein are unchanged in the temporal cortex of AD patients.[153] In addition to a deficit in muscarinic signal transduction, a deficit in PKC activation in AD may lead to an increased amyloidogenic processing of the amyloid precursor protein (APP). This aspect has received much attention in recent years through the use of peripheral cellular models (reviewed in Refs. 36, 128 and136). However, the PKC-induced α-secretase APP processing in brain tissues may be independent from β-amyloid production, as observed in human primary neuronal cultures[80] and in animals with a constitutive over-activation of brain PKC.[142]

Basal Ganglia Diseases

In Parkinson's (PD) and Huntington (HD) disease patients, cognitive functions deteriorate with the progression of the illness. Studies analyzing PKC levels in these neurodegenerative diseases have utilized radiolabeled phorbol ester binding in striatal tissues from patients with different degrees of mental deterioration.[165] PKC levels were reduced in HD and in PD patients with dementia but not in those with preserved cognitive functions. The immunoassay of the PKC isoforms involved showed a decrease in βII isoform in HD putamen and an increase in α isoform. The data may indicate a correlation among PKC levels, neuronal loss and cognitive function. (PKC βII is expressed in GABAergic striatal neurons specifically degenerating in HD). The increase in PKC α could be viewed as an indication of reactive gliosis accompanying the neurodegeneration.[166]

Amyotrophic Lateral Sclerosis (ALS)

ALS is a chronic disease in which abnormalities in various systems (excitatory aminoacids, calcium channels, superoxide dismutase) induce CNS degeneration. The disorder is due to cell death in motor neurons of the ventral horn of the spinal cord and in cortical neurons providing their afferent input, without compromising memory systems.[30] In human spinal cord autopsies, PKC activity is increased in both cytosolic and membrane fractions and this effect is related to an increase in the calcium-dependent immunoreactive isoforms.[77] A tendency to increased PKC activity in motor cortex was observed, while in the visual cortex no changes in PKC were detected. These data are in accordance with the observation that cognitive functions are maintained in this pathology, and that the spinal cord-restricted overactivation of PKC could contribute to the motor neuron degeneration. It is interesting to recall that in addition to having an anti-glutamatergic effect, riluzole, the only drug used to control ALS, also has an inhibitory action on PKC.[117]

Acute Neurodegeneration

Cerebral Ischemia

Anoxia/ischemia is a devastating condition for both cardiac and brain tissues. According to studies in animal and cell cultures, an increase in intracellular calcium activates protein kinases, including PKC, thereby favoring interactions with second messengers/activators and translocation to specific regions of the plasma membrane where kinases can then be deactivated by proteolytic cleavage.[32] According to the time point investigated, PKC can be either activated or down-regulated/degraded. A study in 10 patients who died after ischemic stroke (between 1 and 52 days) investigated PKC activity and isoform expression in infarcted, penumbra and contralateral unaffected tissues.[72] Ischemic penumbra showed an increase in PKC γ, while in infarcted tissues there was an increase in PKC β isoforms. The α isoform was unchanged. PKC activity changes did not match PKC isoform expression. A cascade of PKC isoform activation could thus be associated with the progression of the ischemic damage.

Brain Trauma

Experimental data have indicated that brain trauma attenuates spatial learning in rats.[41] Preclinical data have studied PKC involvement in traumatic brain injury. PKC activation was present in regions undergoing neuronal degeneration; the α and β PKC isoforms were involved in such an effect loss.[126] In another study, fluid percussion brain injury activated a chain of events linking endothelin 1 production, PKC activation, oxygen radical formation, and impaired ATP-sensitive potassium channel function, ultimately leading to vascular tone contraction.[7] Studies in humans have shown that free radical scavengers improve the outcome of severe head injury.[104] It is possible to postulate that modulation of PKC could be an additional and/or different strategy to control the damage.

Affective Disorders

Bipolar

The antimanic effect of lithium also has memory deficits as a side effect. Its action has been related to interference with inositol phospholipid turnover (and thus with DAG production) and more recently to interference with the brain PKC system.[81] In preclinical studies, chronic lithium treatment down-regulates PKC α and ε in the hippocampus.[89] In cerebral autopsies from bipolar patients, the PKC translocation in frontal slices is significantly augmented in comparison to control tissues.[180] Concerning the PKC isoforms involved, the γ and ζ were increased in membrane structures. In addition, studies in living patients have demonstrated that this increased PKC translocation can be observed in platelets (treated with serotonin, thrombin and phorbol esters) of bipolar patients during the manic state of the illness. No such changes have been reported in major depression and in schizophrenia. Moreover, the mania-related increase in PKC activity (under unstimulated conditions) decreased as a consequence of lithium treatment and of the improvement in related symptoms.[181] It has also been postulated that changes in brain PKC are important in the pathophysiology of the illness.[47]

Unipolar and Obsessive Compulsive Disorder

In autopsies from suicides with diagnosed depression, the levels of PKC (assessed as phorbol ester binding) were higher in soluble fractions from the frontal cortex and in the particulate fraction from the hippocampus only in drug-free patients, while the antidepressant-treated patients did not show such changes.[26] In obsessive compulsive patients, platelet PKC is overactivated as assessed by the decrease in serotonin uptake.[90] No data have been reported on the isoforms of PKC involved in the pathology-related changes.

Pharmacological Modulation of PKC: The Goal of Isoenzyme Selectivity

The identification of the conditions in which PKC is specifically involved has suffered until recently from the lack of isoform-selective activators/inhibitors. Phorbol esters do not have absolute specificity for PKC[139] and the "classical" PKC inhibitors can interfere with other kinase systems (for a review see ref. 51).

From what has been presented so far, it is clear that specific PKC isoforms are involved in various brain pathologies as well as in other systems. Targeting a compound to a specific isoform is an important condition for obtaining specificity of action and, hopefully, limited side effects when the compound is tested preclinically and clinically. PKC isoform-selective compounds are now available and their effects in pathologies are under investigation.

PKC Inhibitors

Staurosporine Derivatives

A derivative of staurosporine, LY333531, shows 10-fold selectivity for PKCβ over other isotypes.[60] This compound improves retinal and renal complications in streptozotocin diabetic rats (associated with a specific up-regulation of PKCβII in vascular tissues). LY333531 is under clinical testing in diabetic patients and in CNS diseases associated with HIV-1[43] and also as a potentiator of cytotoxic cancer therapies.[170] It has no signs of general toxicity, probably due to its high isoform specificity.

Peptides

Other PKC isoform-selective inhibitors of peptidic nature have been produced by the Mochly-Rosen group who started with the working hypothesis that in the sequence of each PKC isoform there must be a RACK-selective recognition site that is exposed only after that PKC is "activated".[29] These peptides have been tested and their specificity of PKC interaction demonstrated utilizing both cardiac and neuronal cells.[103] Isoform-selective peptides specifically inhibit PKC βI, βII, δ, ε, η and θ.[158] Using these compounds, PKC ε has been demonstrated to be specifically activated in cardiac preconditioning (see below and in neuronal growth[58]). For their use in animals, their degradation must be controlled and the manner in which they cross the membrane barriers must be evidenced.

Antisense Oligodeoxynucleotides (ODN)

Another approach to isoenzyme selectivity is to use ODN to block the expression of a particular PKC isoform. ISIS 3521 is a 20-mer phosphorothyolate oligo that specifically inhibits PKC α. It has also been tested in the clinical setting as an anticancer compound.[95] Like LY33353, ISIS3521 has been found to be well tolerated in preclinical and clinical studies.[43] Other PKC isoform-specific ODNs are under study (see ref. 184 for a review).

PKC Activators

The control of disease states may alternatively require the activation of PKC. For instance, while PKC is generally overexpressed in tumor cells, it is underexpressed in various colon cancers.[52] PKC activators are natural products like phorbol and ingenol esters, teleocidins, aplysiatoxin and briostatins.[85] Since these compounds are not rapidly metabolised like the physiological activator diacylglycerol, persistent PKC activation may lead to enzyme down-regulation. Briostatin is in clinical trials for cancer and malignant melanoma.[43] RACK-derived peptides stimulate PKC β and ε specifically.[86] These peptides can be delivered to isolated or cultured cells or expressed in transgenic mice to investigate the role of different PKC isoforms. Using these tools an experimental setting in which PKC activation plays an important role has been discovered. Brief periods of ischemia can protect brain and cardiac tissues from a subsequent more severe ischemic insult, a phenomenon termed ischemic preconditioning. The activation of PKC is the first step in a cascade of events leading to tissue protection in heart tissues. Cardioprotection has been observed in all animal species so far tested and may be operant in humans as well.[107] The process has been extensively investigated in cardiac myocytes and the use of PKC isoform selective peptide activators/inhibitors has demonstrated the relevant role played by the activation of PKC ε in preconditioning protection.[33] Such protection may depend on inhibition of calcium influx through L-type calcium channels.[55] In this model PKC δ activation has the opposite effect, i.e., preconditioning is inhibited.[33] These studies suggest a new approach for improving cardiac survival after ischemic insult. In cultured brain cells and slices, ischemic preconditioning does not require PKC activation.[135,167] However, this may be applicable to in vitro models because in animals, down-regulation of PKC γ may provide the neuroprotection of preconditioning.[152]

Conclusions

The actual concept on the biological bases of information storage and processing is based on Hebb's[50] and Rosenblatt's[141] theories, which demonstrated that neuronal assemblies are able to store and process information if they have some properties of plasticity and a suitable organization in their connections. The property of plasticity was discovered in the nervous tissue in 1973. The identification of the molecular bases underlying this phenomenon then became one of the main goals for neurochemistry. Protein kinase C was one of the first intracellular molecules to be identified as having a key role in this property.

Therefore, given its involvement in neuronal plasticity and particularly in synaptic plasticity, its involvement in information storage was also hypothesized. Interestingly, experiments directly devoted to verifying this hypothesis supported this view at each phylogenetic level tested.

Whatever its concentration, PKC is almost ubiquitous in the central nervous system. In the hippocampus, it appears to be involved in a wide variety of tasks. Conversely, its activation in the other structures seems to be more task-specific. This may call into question the concept presently in vogue in animal neuropsychology. Basically, it emphasizes that the hippocampus may be involved only in certain kind of tasks but not in others.[35,119,123,146]

According to the involvement of PKC in the cellular mechanisms of synaptic plasticity, it is to be expected that PKC activation is mainly involved in information storage. Therefore, the changes in PKC induced by factors such as motor activity and stress may be surprising. As shown in (Table 9), the effects of pharmacologic intervention on PKC on motor activity are quite controversial and the most specific PKC inhibitors have no effects. The relationships with stress and anxiety are more obvious, and it can be speculated that some changes in PKC induced by motor activity are due to the stress induced by the exploration of a new environment. Accordingly, a putative concept is that the stress promoted by the exploration of new environments induces changes both in motor activity and PKC activations. This might explain the correlations between PKC activity and motor activity. In this way, this activated PKC could contribute to learn the rules governing new environments, and thus to learn tasks. Feed-back loops could then control the level and the duration of this activation. Though plausible, further experiments will be needed to test this theory. Indeed, although the connexionist concept is attractive, alternative views emphasizing quite different roles for neuronal plasticity in information storage have been proposed.[42]

The second issue which remains to be clarified concerns the differences in the results obtained either with the same biochemical method on the same task, with the same biochemical method on different kinds of tasks, or on the same task but with different biochemical methods. Though the interactionist approach can account for the differences in results obtained in the first case,[116] the solution to the two latter issues may be found once better knowledge in PKC biophysical and activation processes is obtained. In the first section, we have briefly presented an up-to-date view of PKC activation. Since the discovery of PKC, the model of PKC activation has radically changed. First, it was thought that PKC activation consisted in a translocation from the cytosol to the membrane.[69] Then, three-stage[12] and five-stage models[57] were proposed. As shown in the first part of this article, the present model is again quite different.

Accordingly, it is probable that better knowledge in PKC biochemistry will lead to breakthroughs in neuropsychology by showing that a brain structure, which is activated during a task, may have various qualitative forms of activation.

As described in the two last sections, most of the brain pathologies affecting cognitive abilities are associated with PKC changes. Though PKC changes are probably a consequence and not a cause of the pathology in most cases, they likely participate actively in the cognitive impairment observed. Therefore, PKC is considered as a putative target for therapeutical research. We have shown how different isoforms may play quite opposing roles in controlling a function. However, it is also possible that more than one isoform controls a certain function, in which case a broader spectrum of PKC selectivity would be needed. Another possibility is the use of PKC compounds interacting with other drugs to improve certain pathologies. Given the prominent role of PKC in neuronal function, a variety of other nervous system pathologies may reveal changes in selected isoform levels and/or activation. From the preclinical studies (with both models of pathologies or with transgenic mice that overexpress or have one isoform deleted in general, or in a tissue-specific manner), it may be hypothesized that particular forms of memory, alcohol withdrawal phenomena, nociception, heart failure and tumor suppression could benefit from selected PKC isoform interactions.[31,108,184] The use of PKC-specific compounds will be fundamental to establish specific roles and possibly new therapeutic approaches for a number of other pathologies. In this way, better knowledge of the activation mode of the various PKC isoforms and of their role in cell physiology will help in targeting pharmacological agents.

Finally, the recent discovery that PKC may be more involved in the regulation of the "learning step" (metaplasticity) of the synapse rather than directly in synaptic plasticity is an exciting research prospect. In 1962, Rosenblatt showed that networks of artificial neurons (the perceptrons) were able to perform some cognitive tasks if they had some specific properties.

Table 9. Relations of PKC with motor activity, anxiety and stress

Behavior	Species	Task	PKC Assessment	Structure	Result	Ref.
Motor activity	Rats	Exploration, working memory, radialmaze	In vitro PdBu	Cortex binding	Positive correlation between the number of visited arms and particulate binding. No correlation with the soluble fraction.	159
Motor activity	Rats	Exploration, working memory, radial maze	In vitro PdBu binding	Hippo-campus	Positive correlation between the number of visited arms and particulate binding. Negative correlation between the number of visited arms and calcium-dependant soluble binding.	159
Motor activity	Mouse	Treadmill	PKC activity	Hippo-campus	Acute physical activity: increase in membrane associated PKC activity. Chronic physical activity: decrease in membrane associated PKC activity. No effects on cytosolic PKC activity.	40
Loco-motor activity	Rats	Open-field	Go6976 (inhibits α and βI PKCs) and Go7874 (all PKCs)	Dorsal CA1	No effects	177
Motor activity	Rats	Measure of spontaneous motor activity	H7 injections	Intra-cerebro-ventricular	Decrease cocaine induced hyperactivity; no effects on non treated animals	23
Motor activity	Mouse	Open-field	GF109203X	Intra-cerebro-ventricular	No effects	162
Motor activity	Rats	Measure of spontaneous motor activity	PMA	Intra-cerebro-ventricular	Induces hypoactivity	132
Motor activity	Mutant mice	Open-field and rotarod	PKC β deficient	Whole brain	No effects	185
Psycho-social stress	Rats	Prolonged con-frontation to dominant animals	PKC γ immunoreact.	Hippo-campus	Increase in PKCγ in the CA1 and CA3. Decreases the PKCγ increase induced by learning spatial tasks in posterior hippocampus.	71
Anxiety	Rats	Elevated plus-maze	Go6976 (in-hibits α and βI PKCs) and Go 7874 (all PKCs)	Dorsal CA1	Mild anxiogenic effect	177

Like synaptic potentiation and synaptic depression, the learning step is one of these properties. It is now widely used to improve the cognitive abilities displayed by more recent artificial neural networks. If confirmed, the involvement of PKC in the regulation of the synaptic learning step may fill in one of the remaining blanks in neuroscience knowledge regarding a connexionist theory of the cognitive abilities of the biological systems. Moreover, it may help to explain and predict some of the effects of PKC activation and inhibition on behavioural tasks.

References

1. Abeliovich A, Paylor R, Chen C et al. PKC γ mutant mice exhibit mild deficits in spatial and contextual learning. Cell 1993; 75(7):1263-71.
2. Abraham WC, Bear MF. Metaplasticity: The plasticity of synaptic plasticity. Trends Neurosci 1996; 19:126-130.
3. Akers RF, Lovinger DM, Colley PA et al. Translocation of protein kinase C activity may mediate hippocampal long-term potentiation. Science 1986; 231:587-9.
4. Ali SM, Bullock S, Rose SPR. Protein kinase C inhibitors prevent long-term memory formation in the one-day-old chick. Neurosci Res Com 1988; 3(3):133-140.
5. Angenstein F, Riedel G, Reymann KG et al. Hippocampal long-term potentiation in vivo induces translocation of protein kinase C γ. NeuroReport 1994; 5:381-384.
6. Angenstein F, Riedel G, Reymann KG et al. Transient translocation of protein kinase C γ in hippocampal long-term potentiation depends on activation of metabotropic glutamate receptors. Neuroscience 1999; 93:1289-1295.
7. Armstead WA. Superoxide generation links protein kinase C activation to impaired ATP-sensitive K+ channel function after brain injury. Stroke 1999; 30:153-159.
8. Bank B, DeWeer A, Kuzirian AM et al. Classical conditioning induces long-term translocation of protein kinase C in rabbit hippocampal CA1 cells. Proc Natl Acad Sci USA 1988; 85:1988-1992.
9. Barlow HB The role of single neurons in the psychology of perception. Quart J Exp Psychol 1985; 37A:121-145.
10. Battaini F, Pascale A, Paoletti R et al. The role of anchoring protein RACK1 in PKC activation in the aging rat brain. Trends Neurosci 1997; 20:410-415.
11. Battaini F, Pascale A, Lucchi L et al. Protein kinase C anchoring deficit in postmortem brains of Alzheimer's disease patients. Exp Neurol 1999; 159:559-564.
12. Bazzi MD, Nelsestuen GL. Constitutive activity of membrane-inserted protein kinase C. Biochem Biophys Res Commun 1988; 152(1):336-43.
13. Beldhuis HJA, Everts HGJ, Van der Zee EA et al. Amygdala kindling-induced seizures selectively impair spatial memory. 1. Behavioral characteristics and effects on hippocampal neuronal protein kinase C isoforms. Hippocampus 1992; 2(4):397-410.
14. Ben-Ari Y, Aniksztejn L, Bregestovski P Protein kinase C modulation of NMDA currents: An important link for LTP induction. Trends Neurosci 1992; 15:333-339.
15. Bernabeu R, Izquierdo I, Cammarota M et al. Learning-specific, time-dependent increase in [^{3}H]phorbol dibutyrate binding to protein kinase C in selected regions of the rat brain. Brain Res 1995; 685(1-2):163-8.
16. Bliss TVP, Lømo T Long-lasting potentiation of synaptic transmission in the dentate area of the anaesthetized rabbit following stimulation of the perforant path. J Physiol 1973; 232:331-356.
17. Bortolotto ZA, Bashir ZI, Davies CH et al. A molecular switch activated by metabotropic glutamate receptors regulates induction of long-term potentiation. Nature 1994; 368:740-743.
18. Bortolotto ZA, Collingridge GL A role for protein kinase C in a form of metaplasticity that regulates the induction of long-term potentiation at CA1 synapses of the adult rat hippocampus. Eur J Neurosci 2000; 12:4055-4062.
19. Browning M, Dunwiddie M, Bennett W et al. Synaptic phosphoproteins specific changes after repetitive stimulation of the hippocampal slice. Science 1979; 203:60-62.
20. Burchuladze R, Potter J, Rose SP Memory formation in the chick depends on membrane-bound protein kinase C. Brain Res 1990; 535(1):131-8.
21. Cammarota M, Paratcha G, Levi de Stein M et al. B-50/GAP-43 phosphorylation and PKC activity are increased in rat hippocampal synaptosomal membranes after an inhibitory avoidance training. Neurochem Res 1997; 22(4):499-505.
22. Castagna M, Takai Y, Kaibuchi K et al. Direct activation of calcium-activated phospholipid-dependent protein kinase by tumor promoting phorbol esters. J Biol Chem 1982; 257:7847-7851.

23. Cervo L, Mukherjee S, Bertaglia A et al. Protein kinases A and C are involved in the mechanisms underlying consolidation of cocaine place conditioning. Brain Res 1997; 775(1-2):30-6.
24. Choi KW, Smith RF, Buratowski RM et al. Deficient protein kinase C activity in turnip, a Drosophila learning mutant. J Biol Chem 1991; 266(24):15999-16006.
25. Colombo PJ, Wetsel WC, Gallagher M. Spatial memory is related to subcellular concentrations of calcium-dependent protein kinase C isoform in young and aged rats. Proc Natl Acad Sci USA 1997; 94:14195-14199.
26. Coull MA, Lowther S, Katona CL et al. Altered brain protein kinase C in depression a post-mortem study. Eur Neuropsychopharmacol 2000; 10:283-288.
27. Craig AM, Olds JL, Schreurs BG et al. Quantitative distribution of protein kinase C α, β, γ, and ε mRNAs in the hippocampus of control and nictitating membrane conditioned rabbits. Brain Res Mol Brain Res 1993; 19(4):269-76.
28. Crow T, Forrester J. Down-regulation of protein kinase C and kinase inhibitors dissociate short- and long-term enhancement produced by one trial conditioning of Hermissenda. J Neurophysiol 1993; 69(2):636-641.
29. Csukai M, Mochly-Rosen D. Pharmacological modulation of protein kinase C isozymes: The role of RACKs and subcellular localization. Pharmacol Res 1999; 39:253-259.
30. Demaerschalj BM, Strong MJ. Amyotrophic lateral sclerosis. Curr Treat Options in Neurol 2000 2:13-22.
31. Dempsey EC, Newton AC, Mochly-Rosen D et al. Protein kinase C isozymes and the regulation of diverse cell responses. Am J Physiol Lung Cell Mol Physiol 2000; 3:L429-38.
32. Domanska-Janik K. Protein serine/threonine kinase (PKA, PKC, CaMKII) involved in ischemic brain pathology. Acta Neurobiol Exp 1996; 56:579-585.
33. Dorn GW, Souroujon MC, Liron T et al. Sustained in vivo cardiac protection by a rationally designed peptide that causes protein kinase C translocation. Proc Natl Acad Sci USA 1999; 96:12798-12803.
34. Douma BR, Van der Zee EA, Luiten PG. Translocation of protein kinase Cγ occurs during the early phase of acquisition of food rewarded spatial learning. Behav Neurosci 1998; 112(3):496-501.
35. Eichenbaum H, Otto T, Cohen NJ. The hippocampus - What does it do? Behav Neural Biol 1992; 57:2-36.
36. Bhagavan RS. Ionic and signal transduction alterations in Alzheimer's disease: Relevance of studies in peripheral cells. Mol Neurobiol 2000; 20:93-109.
37. Farley J, Auerbach S. Protein kinase C activation induces conductance changes in Hermissenda photoreceptors like those seen in associative learning. Nature 1986; 319:220-223.
38. Farley J, Schuman E. Protein kinase C inhibitors prevent induction and continued expression of cell memory in Hermissenda type B photoreceptors. Proc Natl Acad Sci USA 1991; 88:2016-2020.
39. Fedorov NB, Pasinelli P, Oestreicher AB et al. Antibodies to postsynaptic PKC substrate neurogranin prevent long-term potentiation in hippocampal CA1 neurons. Eur J Neurosci 1995; 7:819-822.
40. Fordyce DE, Wehner JM. Effects of aging on spatial learning and hippocampal protein kinase C in mice. Neurobiol Aging 1993; 14(4):309-17.
41. Fox GB, Fan L, LeVasseur RA. Effect of traumatic brain injury on mouse spatial and nonspatial lerning in the Barres circular maze. J Neurotrauma 1998; 15:1037-1046.
42. Freeman WJ, Skarda C. Spatial EEG patterns, nonlinear dynamics and perception: The neo-sherringtonian view. Brain Res Rev 1985; 10:147-175.
43. Goekjian PG, Jirousek MR. Protein kinase C in the treatment of disease: Signal transduction pathways, inhibitors, and agents in development. Curr Med Chem 1999; 6:877-903.
44. Golski S, Olds JL, Mishkin M et al. Protein kinase C in the hippocampus is altered by spatial but not cued discriminations: A component task analysis. Brain Res 1995; 676(1):53-62.
45. Goosens KA, Holt W, Maren S. A role for amygdaloid PKA and PKC in the acquisition of long-term conditional fear memories in rats. Behav Brain Res 2000; 114(1-2):145-52.
46. Ha KS, Exton JH. Activation of actin polymerization by phosphatidic acid derived from phosphatidylcholine in IIC9 fibroblasts. J Cell Biol 1993; 6(Pt 2):1789-96.
47. Hahn CG, Friedman E. Abnormalities in protein kinase C signalling and the pathophysiology of bipolar disorders. Bipolar Disord 1999; 1:81-86.
48. Hansra G, Bornancin F, Whelan R et al. 12-O-tetradecanoylphorbol-13-acetate-induced dephosphorylation of protein kinase C alpha correlates with the presence of a membrane-associated protein phosphatase 2A heterotrimer. J Biol Chem 1996; 271(51):32785-8.
49. Hardie RC, Peretz A, Suss-Toby E et al. Protein kinase C is required for light adaptation in Drosophila photoreceptors. Nature 1993; 363:634-637.
50. Hebb DO. The organization of behavior. Wiley, New York: 1949.

51. Hofmann J. The potential for isoenzyme-selective modulation of protein kinase C. FASEB J 1997; 11:649-669.
52. Hofmann J. Modulation of protein kinase C in antitumor treatment. Rev Physiol Biochem Pharmacol 2001; 142:1-96.
53. Hopfield JJ. Neural networks and physical systems with emergent collective computational abilities. Proc Natl Acad Sci USA 1982; 79:2554-2558.
54. Hrabetova S, Sacktor TC. Bidirectional regulation of protein kinase M ζ in the maintenance of long-term potentiation and long-term depression. J Neurosci 1996; 16:5324-5333.
55. Hu K, Mochly-Rosen D, Boutjdir M. Evidence for functional role of εPKC isozyme in the regulation of cardiac calcium channels. Am J Physiol Heart Circ Physiol 2000; 279:H2658-H2664.
56. Huang YY, Colley PA, Routtenberg A. Postsynaptic then presynaptic protein kinase C activity may be necessary for long-term potentiation. Neuroscience 1992; 49:819-827.
57. Huang KP, Huang FL. How is protein kinase C activated in CNS. Neurochem Int 1993; 22(5):417-33.
58. Hundle B, McMahon T, Dadgar J. An inhibitory fragment derived from εPKC prevents enhancement of NGF responses by ethanol and phorbol esters. J Biol Chem 1997; 272:15028-15035.
59. Inoue M, Kishimoto A, Takai Y et al. Studies on a cyclic nucleotide-independent protein kinase and its proenzyme in mammalian tissues. II. Proenzyme and its activation by calcium-dependent protease from rat brain. J Biol Chem 1977; 252(21):7610-7616.
60. Ishii H, Jirousek MR, Koya D et al. Amelioration of vascular dysfunctions in diabetic rats by an oral PKC β inhibitor. Science 1996; 272:728-731.
61. Jaken S. Protein kinase C isozymes and substrates. Current Opinion in Cell Biol 1998; 8:168-173.
62. Jaken S, Parker PJ. Protein kinase C binding partners. Bioessays 2000; 3:245-54.
63. Jerusalinsky D, Quillfeldt JA, Walz R et al. Post-training intrahippocampal infusion of protein kinase C inhibitors causes amnesia in rats. Behav Neural Biol 1994; 61(2):107-9.
64. Jin LW, Saito T. Changes in protein kinases in brain aging and Alzheimer's disease. Drugs & Aging 1995; 6:136-149.
65. Kawano S, Okajima S, Mizguchi et al. Immunochemical distribution of calcium-independent. PKC subtypes (δ, ε and ζ) in regenerating axonal growth cones of rat peripheral nerve. Neuroscience 1997; 81:263-273.
66. Klann E, Chen SJ, Sweatt JD. Mechanism of protein kinase C activation during the induction and maintenance of long-term potentiation probed using a selective peptide substrate. Proc Natl Acad Sci USA 1993; 90:8337-8341.
67. Kohonen T. Self-organization and associative memory. New York: Springer-Verlag, 1984.
68. Koponen S, Goldsteins G, Keinannen R et al. Induction of PKC δ subspecies in neurons and microglia after transient global brain ischemia. J Cereb Blood Flow Metab 2000; 20:93-102.
69. Kraft AS, Anderson WB. Phorbol esters increase the amount of calcium, phospholipid-dependent protein kinase associated with plasma membrane. Nature 1983; 301:621-623.
70. Krivanek J. Conditioned taste aversion and protein kinase C in the parabrachial nucleus of rats. Neurobiol Learn Mem 1996; 65(2):154-62.
71. Krugers HJ, Douma BR, Andringa G et al. Exposure to chronic psychosocial stress and corticosterone in the rat: Effects on spatial discrimination learning and hippocampal protein kinase Cγ immunoreactivity. Hippocampus 1997; 7(4):427-36.
72. Krupinski J, Slevin MA, Kumar P et al. Protein kinase C expression and activity in the human brain after ischaemic stroke. Acta Neurobiol Exp 1998; 58:13-21.
73. Kurumatani T, Fastbom J, Bonkale WL et al. Loss of inositol 1,4,5 trisphosphate receptor sites and decreased PKC levels correlate with staging of Alzheimer's disease neurofibrillary pathology. Brain Res 1998; 796:209-221.
74. Laborit H, Zerbib R. Phorbol-Ester Antagonize scopolamine-induced amnesia of a passive avoidance. Res Comm Psychol Psychiat Behav 1987; 12:105-117.
75. Ladner CJ, Lee JM. Pharmacological drug treatment of Alzheimer's disease: The cholinergic hypothesis revisited. J Neuropathol Exp Neurol 1998; 57:719-731.
76. Lan JY, Skeberdis VA, Jover T et al. Protein kinase C modulates NMDA receptor trafficking and gating. Nature Neurosci 2001; 4:382-390.
77. Lanius RA, Paddon HB, Mezei M et al. A role for amplified PKC activity in the pathogenesis of amyotrophic lateral sclerosis. J Neurochem. 1995; 65:927-930.
78. Lanius RA, Wagey R, Sahl B et al. Protein kinase C activity and protein levels in Alzheimer's disease. Brain Res 1997; 764:75-80.
79. Leahy JC, Luo Y, Kent CS et al. Demonstration of presynaptic protein kinase C activation following long-term potentiation in rat hippocampal slices. Neuroscience 1993; 52:563-574.

80. LeBlanc AC, Koutroumanis M, Goodyer CG. Protein kinase C activation increases release of secreted amyloid precursor protein without decreasing Abeta production in human primary neuron cultures. J Neurosci 1998; 18:2907-2913.
81. Lenox RH, Hahn CG. Overview of the mechanisms of action of lithium in the brain: Fifthy-year update. J Clin Psychiatry 2000; 61S:5-15.
82. Linden DJ, Routtenberg A. The role of protein kinase C in long-term potentiation: A testable model. Brain Res Rev 1989; 14:279-296.
83. Lovinger DM, Akers RF, Nelson RB et al. A selective increase in phosphorylation of protein F1, a protein kinase C substrate, directly related to three day growth of long term synaptic enhancement. Brain Res 1985; 343:137-143.
84. Luo JH, Kahn S, O'Driscoll K et al. The regulatory domain of protein kinase C β 1 contains phosphatidylserine- and phorbol ester-dependent calcium binding activity. J Biol Chem 1993; 268:3715-3719.
85. Ma D. Recent advances in the discovery of protein kinase C modulators based on the structures of natural protein kinase C activators. Curr Med Chem 2001; 8:191-202.
86. Mackay K, Mochly-Rosen D. Localization, anchoring and functions of PKC isozymes in the heart. J Mol Cell Cardiol 2001; 33:1301-1307.
87. Malenka RC, Madison DV, Nicoll RA. Potentiation of synaptic transmission in the hippocampus by phorbol esters. Nature 1986; 321:175-177.
88. Malinow R, Schulman H, Tsien RW. Inhibition of postsynaptic PKC or CaMKII blocks induction but not expression of LTP. Science, 1989; 245:862-866.
89. Manji HK, Lenox RH. PKC signalling in the brain: Molecular transduction of mood stabilization in the treatment of manic-depressive illness. Biol Psychiatry 1999; 46:1328-1351.
90. Marazziti D, Masala I, Rossi A et al. Increased inhibitory activity of protein kinase C on the serotonin transporter in OCD. Neuropsychobiology 2000; 41:171-177.
91. Martin SJ, Grimwood PD, Morris RGM. Synaptic plasticity and memory: An evaluation of the hypothesis. Annu Rev Neurosci 2000; 23:649-711.
92. Mathis C, Lehmann J, Ungerer A. The selective protein kinase C inhibitor, NPC 15437, induces specific deficits in memory retention in mice. Eur J Pharmacol 1992; 220(1):107-10.
93. Matsushima H, Shimohama S, Chachin M et al. Calcium-dependent and independent protein kinase C changes in the brains of patients with Alzheimer's disease. J Neurochem 1996; 67:317-323.
94. Matzel LD, Lederhendler II, Alkon DL. Regulation of short-term associative memory by calcium-dependent protein kinase. J Neurosci 1990; 10(7):2300-2307.
95. McKay RA, Miraglia LJ, Cummins LL et al. Characterization of a potent and specific class of antisense oligonucleotide inhibitor of human protein kinase C-α expression. J Biol Chem 1999; 274:1715-1722.
96. McNaughton BL, Morris RGM. Hippocampal synaptic enhancement and information storage within a distributed memory system. Trends Neurosci 1987; 10:408-415.
97. McPhie DL, Matzel LD, Olds JL et al. Cell specificity of molecular changes during memory storage. J Neurochem 1993; 60(2):646-651.
98. Meberg PJ, Barnes CA, McNaughton BL et al. Protein kinase C and F1/GAP-43 gene expression in hippocampus inversely related to synaptic enhancement lasting three days. Proc Natl Acad Sci USA 1993; 90:12050-12054.
99. Meberg PJ, Valcourt EG, Routtenberg A. Protein F1/GAP-43 gene expression patterns in hippocampus are altered 1-2 h after LTP. Mol Brain Res 1995; 34:343-346.
100. Mellor H and Parker PJ. The extended protein kinase C superfamily. Biochem J 1998; 332:281-292.
101. Micheau J, Riedel G. Protein kinases: Which one is the memory molecule? Cell Mol Life Sci 1999; 55:534-548.
102. Mochly-Rosen D, Khaner H, Lopez J. Identification of intracellular receptor proteins for activated protein kinase C. Proc Natl Acad Sci USA 1991; 88:3997-4000.
103. Mochly-Rosen D, Gordon AS. Anchoring proteins for protein kinase C: A means for isozyme reactivity. FASEB J 1998; 12:35-42.
104. Muizelaar JP, Marmaroe A, Young HF et al. Improving the outcome of severe head injury with the oxygen radical scavenger polyethylene glycol-conjugated superoxide dismutase: A phase II trial. J Neurosurg 1993; 78:375-382.
105. Muller D, Buchs PA, Dunant Y et al. Protein kinase C activity is not responsible for the expression of long-term potentiation in hippocampus. Proc Natl Acad Sci USA 1990; 87:4073-4077.
106. Muzzio IA, Talk AC, Matzel LD. Incremental redistribution of protein kinase C underlies the acquisition curve during in vitro associative conditioning in Hermissenda. Behav Neurosci 1997; 111(4):739-753.

107. Nakano A, Cohen MV, Downey JM. Ischemic preconditioning: From basic mechanisms to clinical applications. Pharmacol Ther 2000; 86:263-275.
108. Naruse K, King GL. Protein kinase C and myocardial biology and function. Circ Res 2000; 86:1104-1107.
109. Newton AC. Protein kinase C: Structure, function, and regulation. J Biol Chem 1995; 270:28495-28498.
110. Newton AC. Regulation of protein kinase C. Curr Opin Cell Biol 1997; 2:161-7.
111. Nishizuka Y. The molecular heterogeneity of protein kinase C and its implication in cellular regulation. Nature 1988; 334:661-665.
112. Noguès X, Micheau J, Jaffard R. Protein kinase C activity in the hippocampus following spatial learning tasks in mice. Hippocampus 1994; 4(1):71-7.
113. Noguès X. Implication de la kinase C dans la mémoire de l'espace chez le rongeur: Approches corrélatives et interventionnistes. Doctoral thesis Talence Université Bordeaux I 1994.
114. Noguès X, Micheau J, Jaffard R. Correlations between hippocampal protein kinase C activity and learning abilities in a spatial reference memory task. Psychobiol 1996a; 24(1):1-8.
115. Noguès X, Jaffard R, Micheau J. Investigations on the role of hippocampal protein kinase C on memory processes pharmacological approach. Behav Brain Res 1996b; 75(1-2):139-46.
116. Noguès X. Protein kinase C, learning and memory: A circular determinism between physiology and behaviour. Prog Neuro-Psychopharmacol & Biol Psychiat 1997; 21:507-529.
117. Noh KM, Hwang JY, Shin HC, Koh JY. A novel neuroprotective mechanism of riluzole: Direct inhibition of protein kinase C. Neurobiol. Disease 2000; 7:375-383.
118. O'Connor JJ, Rowan MJ, Anwyl R. Tetanically induced LTP involves a similar increase in AMPA and NMDA receptor components of the excitatory postsynaptic current: Investigations of the involvement of mGlu receptors J Neurosci 1995; 15:2013-2020.
119. O'Keefe J, Nadel L. The hippocampus as a cognitive map. Behav Brain Sci 1979; 2:487-533.
120. Olds JL, Anderson ML, McPhie DL et al. Imaging of memory-specific changes in the distribution of protein kinase C in the hippocampus. Science. 1989; 245:866-9.
121. Olds JL, Golski S, McPhie DL et al. Discrimination learning alters the distribution of protein kinase C in the hippocampus of rats. J Neurosci 1990; 10(11):3707-13.
122. Olds JL, Bhalla US, McPhie DL et al. Lateralization of membrane-associated protein kinase C in rat piriform cortex: Specific to operant training cues in the olfactory modality. Behav Brain Res 1994; 61(1):37-46.
123. Olton DS, Becker J, Handelmann G. Hippocampus, space and memory. Behav Brain Sci 1979; 2:313-365.
124. Osada S, Mizuno K, Saido T et al. A new member of protein kinase C family nPKC θ predominantly expressed in skeletal-muscle. Mol Cell Biol 1992; 12:3930-3938.
125. Osten P, Valsamis L, Harris A et al. Protein synthesis-dependent formation of protein kinase M ζ in long-term potentiation. J Neurosci 1996; 16:2444-2451.
126. Padmaperuma B, Mark R, Dhillon HS et al. Alterations in brain protein kinase C after experimental brain injury. Brain Res 1996; 714:19-26.
127. Paratcha G, Furman M, Bevilaqua L et al. Involvement of hippocampal PKCβI isoform in the early phase of memory formation of an inhibitory avoidance learning. Brain Res 2000; 855(2):199-205.
128. Pascale A, Govoni S, Battaini F. Age-related alteration of PKC, a key enzyme in memory processes: Physiological and pathological examples. Mol Neurobiol 1998a; 16(1):49-62.
129. Pascale A, Noguès X, Marighetto A et al. Cytosolic hippocampal PKC and aging: Correlation with discrimination performance. Neuroreport 1998b; 9:725-729.
130. Pascale A, Grimaldi M, Russel JT et al.Translocation of Protein Kinase C-βII from cytosolic to membrane-associated receptor for activated C kinase 1 (RACK-1) requires actin stress fibers integrity (manuscript in preparation).
131. Pasinelli P, Ramakers GMJ, Urban IJA et al. Long-term potentiation and synaptic protein phosphorylation. Behav. Brain Res 1995; 66:53-59.
132. Patishi Y, Bersudsky Y, Belmaker RH. Phorbol ester intracerebroventricularly induces a behavioral hypoactivity that is not affected by chronic or acute lithium. Eur Neuropsychopharm 1996; 6:39-41.
133. Paylor R, Rudy JW, Wehner JM. Acute phorbol ester treatment improves spatial learning performance in rats. Behav Brain Res 1991; 45(2):189-93.
134. Paylor R, Morrison SK, Rudy JW et al. Brief exposure to an enriched environment improves performance on the Morris water task and increases hippocampal cytosolic protein kinase C activity in young rats. Behav Brain Res 1992; 52(1):49-59.
135. Perez-Pinzon MA, Born JG. Rapid preconditioning neuroprotection following anoxia in hippocampal slices: Role of the K+ ATP channel and protein kinase C. Neuroscience 1999; 89:453-459.

136. Racchi M, Govoni S. Rationalizing a pharmacological approach on the amyloid precursor protein metabolism. Trends Pharmacol Sci 1999; 20:418-423.
137. Reymann KG. Mechanisms underlying synaptic long-term potentiation in the hippocampus: Focus on postsynaptic glutamate receptors and protein kinases. Funct Neurol Suppl 1993; 5:8:7-32.
138. Roberson ED, English JD, Adams JP et al. The mitogen-activated protein kinase cascade couples PKA and PKC to cAMP response element binding protein phosphorylation in area CA1 of hippocampus. J Neurosci 1999; 19:4337-4348.
139. Ron D, Kazanietz MG. New insights into the regulation of protein kinase C and novel phorbol ester receptors. FASEB J 1999; 13:1658-76.
140. Ron D, Jiang Z, Yao L et al. Coordinated movement of RACK1 with activated βII PKC. J Biol Chem 1999; 274(38):27039-27046.
141. Rosenblatt F. Principles of neurodynamics. Washington: Spartan Books, 1962.
142. Rossner S, Beck M, Stahl T et al. Constitutitve overactivation of protein kinase C in guinea pig brain inceaseses α-secretory APP processing without decreasing β-amyloid generation. Eur J Neurosci 2000; 12:3191-3200.
143. Roth GS, Joseph JA, Preston Mason R. Membrane alterations as causes of impaired signal transduction in Alzheimer's disease and aging. Trends Pharmacol Sci 1995; 13:203-206.
144. Routtenberg A. Phosphoprotein regulation of memory formation: Enhancement and control of synaptic plasticity by protein kinase C and protein F1. Ann NY Acad Sci 1985; 444:203-209.
145. Routtenberg A. Synaptic plasticity and protein kinase C. Prog Brain Res 1986; 69:211-234.
146. Rudy JW, Sutherland RJ. Configural association theory and the hippocampal formation: An appraisal and reconfiguration. Hippocampus 1995; 5(5):375-89.
147. Sacktor TC, Schwartz JH. Sensitizing stimuli cause translocation of protein kinase C in Aplysia sensory neurons. Proc Natl Acad Sci USA 1990; 87:2036-2039.
148. Scharenberg AM, Olds JL, Schreurs BG et al. Protein kinase C redistribution within CA3 stratum oriens during acquisition of nictitating membrane conditioning in the rabbit. Proc Natl Acad Sci USA 1991; 88:6637-6641.
149. Schneider W, Detweiler M. A connexionist/control architecture for working memory. The Psychol. of Learning and Memory 1987; 21:53-119.
150. Serrano PA, Beniston DS, Oxonian MG et al. Differential effects of protein kinase inhibitors and activators on memory formation in the 2-day-old chick. Behav Neural Biol 1994; 61(1):60-72.
151. Serrano PA, Rodriguez WA, Pope B et al. Protein kinase C inhibitor chelerythrine disrupts memory formation in chicks. Behav Neurosci 1995; 109(2):278-284.
152. Shamloo M, Wieloch T. Rapid decline in PKCγ levels in the synaptosomal fraction of rat hippocampus after ischemic preconditioning. Neuroreport 1999; 10:931-935.
153. Shimohama S, Kamiya S, Taniguchi T et al. Intracellular receptors for activated C-kinase in the postmortem human brain: No alteration in Alzheimer's disease. Alzheimer Disease and Associated Disorders 1998; 12:384-386.
154. Shors TJ, Elkabes S, Selcher JC et al. Stress persistently increases NMDA receptor-mediated binding of [^{3}H]PdBu (a marker for protein kinase C) in the amygdala, and reexposure to the stressful context reactivates the increase. Brain Res 1997; 750:293-300.
155. Sim AT, Scott JD. Targeting of PKA, PKC and protein phosphatases to cellular microdomains. Cell Calcium 1999; 5:209-17.
156. Soderling TR, Derkach VA. Postsynaptic protein phosphorylation and LTP. Trends Neurosci 2000; 23:75-80.
157. Son H, Madelian V, Carpenter DO. The translocation and the involvement of protein kinase C in mossy fiber-CA3 long-term potentiation in hippocampus of the rat brain. Brain Res 1996; 739:282-292.
158. Souroujon M, Mochly-Rosen D. Peptide modulators of protein-protein interactions in intracellular signaling. Nat Biotechnol 1998; 16:919-924.
159. Spieler K, Schoch P, Martin JR et al. Environmental stimulation promotes changes in the distribution of phorbol ester receptors. Pharmacol Biochem Behav 1993; 46(3):553-60.
160. Staak S, Behnisch T, Angenstein F. Hippocampal long-term potentiaion: Transient increase but not persistent translocation of protein kinase C isozymes α et β. Brain Res 1995; 682:55-62.
161. Stanton PK. Transient protein kinase C activation primes long-term depression and suppresses long-term potentiation of synaptic transmission in hippocampus. Proc Natl Acad Sci USA 1995; 92:1724-1728.
162. Stemmelin J, Mathis C, Ungerer A. GF109203X, a selective inhibitor of protein kinase C, impairs retention performance in an operant task. Neuroreport 1999; 10(13):2805-9.
163. Sunayashiki-Kusuzaki K, Lester DS, Schreurs BG et al. Associative learning potentiates protein kinase C activation in synaptosomes of the rabbit hippocampus. Proc Natl Acad Sci USA 1993; 90(9):4286-9.

164. Sweatt JD. The neuronal MAP kinase cascade: A biochemical signal integration system subserving synaptic plasticity and memory. J Neurochem 2001; 76:1-10.
165. Tanaka C, Nishino N, Hashimoto T et al. Second messenger systems in brains of patients with Parkinson's or Huntington's disease. Adv Neurol 1993; 60:175-180.
166. Tanaka C, Nishizuka Y. The protein kinase C family for neuronal signaling. Annu Rev Neurosci 1994; 17:551-567.
167. Tauskela JS, Chakravarthy BR, Murray CL et al. Evidence from cultured rat cortical neurons of differences in the mechanisms of ischemic preconditioning of brain and heart Brain Res 1999; 827:143-151.
168. Takai Y, Kishimoto A, Inoue M et al. Studies on a cyclic nucleotide-independent protein kinase and its proenzyme in mammalian tissues. I. Purification and characterization of an active enzyme from bovine cerebellum. J Biol Chem 1977; 252(21):7603-9.
169. Takashima A, Yokota T, Maeda Y et al. Pretreatment with caerulein protects against memory impairment induced by protein kinase C inhibitors in the rat. Peptides 1991; 12(4):699-703.
170. Teicher BA, Alvarez E, Mendelsohn LG et al. Enzymatic rationale and preclinical support for a potent PKC β inhibitor in cancer therapy. Adv Enzyme Regul 1999; 39:313-327.
171. Thomas KL, Laroche S, Errington ML et al. Spatial and temporal changes in signal transduction pathways during LTP. Neuron 1994; 13(3):737-45.
172. Tolman EC. Cognitive maps in rats and men. Psychol Rev 1948; 55:189-208.
173. Van der Zee EA, Compaan JC, de Boer M et al. Changes in PKC γ immunoreactivity in mouse hippocampus induced by spatial discrimination learning. J Neurosci 1992; 12(12):4808-15.
174. Van der Zee EA, Douma BRK, Bohus B et al. Passive avoidance training induces enhanced levels of immunoreactivity for muscarinic acetylcholine receptor and coexpressed PKCγ and MAP-2 in rat cortical neurons. Cereb Cortex 1994; 4:376-390.
175. Van der Zee EA, Kronforst-Collins MA, Maizels ET et al. γ Isoform-selective changes in PKC immunoreactivity after trace eyeblink conditioning in the rabbit hippocampus. Hippocampus 1997; 7(3):271-85.
176. Van der Zee EA, Douma BRK. Historical review of research on protein kinase C in learning and memory. Prog Neuro-Psychopharmacol & Biol Psychiat 1997; 21:379-406.
177. Vianna MR, Barros DM, Silva T et al. Pharmacological demonstration of the differential involvement of protein kinase C isoforms in short- and long-term memory formation and retrieval of one-trial avoidance in rats. Psychopharmacology 2000; 150(1):77-84.
178. Walker DL, Gold PE. Intra-amygdala kinase inhibitors disrupt retention of a learned avoidance response in rats. Neurosci Lett 1994; 176(2):255-8.
179. Wang HY, Pisano MR, Friedman E. Attenuated protein kinase C activity and translocation in Alzheimer's disease brain. Neurobiol Aging 1994; 15:293-298.
180. Wang HY, Friedman E. Enhanced protein kinase C activity and translocation in bipolar affective disorder brains. Biol Psychiatry 1996; 40:568-575.
181. Wang HY, Markowitz P, Levinson D et al. Increased membrane-associated protein kinase C activity and translocation in blood platelets from bipolar affective disorder patients. J Psychiatr Res 1999; 33:171-179.
182. Wang JH, Kelly PT. Postsynaptic injection of Ca^{2+}/CaM induces synaptic potentiation requiring CaMKII and PKC activity. Neuron 1995; 15:443-452.
183. Wang JH, Kelly PT. The balance between postsynaptic Ca^{2+}-dependent protein kinase and phosphatase activities controlling synaptic strength. Learn & Mem 1996; 3:170-181.
184. Way KJ, Chou E, King GL. Identification of PKC-isoform-specific biological actions using pharmacological approaches. Trends Pharmacol Sci 2000; 5:181-187.
185. Weeber EJ, Atkins CM, Selcher JC et al. A role for the β isoform of protein kinase C in fear conditioning. J Neurosci 2000; 20(16):5906-5914.
186. Wehner JM, Sleight S, Upchurch M. Hippocampal protein kinase C activity is reduced in poor spatial learners. Brain Res 1990; 523(2):181-7.
188. Yang HC, Lee EH. Protein kinase C activation facilitates memory retention in rats. Chin J Physiol 1993; 36(2):115-123.
189. Yasoshima Y, Yamamoto T. Rat gustatory memory requires protein kinase C activity in the amygdala and cortical gustatory area. Neuroreport 1997; 8(6):1363-1367.
190. Zhao WQ, Sedman GL, Gibbs ME et al. Effect of PKC inhibitors and activators on memory. Behav Brain Res 1994; 60(2):151-160.
191. Zheng X, Zhang L, Wang AP et al. Protein kinase C potentiation of N-methyl-D-aspartate receptor activity is not mediated by phosphorylation of N-methyl-D-aspartate receptor subunits. Proc Natl Acad Sci USA 1999; 96:15262-15267.

CHAPTER 4.5

CaMKinase II

Martín Cammarota and Jorge H. Medina

Abstract

The modification of synaptic properties by means of protein phosphorylation has been, for long, recognized as a core and unifying principle in the study of the molecular mechanisms underlying the formation and storage of new memories. The intrinsically transient nature of this posttranslational modification, at first glance a property not compatible with the protracted durability of the mnemonic trace, has encouraged the search for protein kinases able to act as "memory devices" capable of auto-preserving its own enzymatic activity even long after the stimuli that promoted its initial onset disappeared. The unique ability of CaMKII to respond to a rise in the intracellular levels of calcium with an autophosphorylation step that both, converts the enzyme into a calcium independent form and promotes its translocation to the post synaptic density where it can phosphorylate plasticity-related substrates, has made this kinase the paradigmatic example of such a "memory device". In this chapter we review the molecular properties of CaMKII and discuss recent experimental findings that extend our knowledge about its participation in plastic processes.

Introduction

Ca^{2+}/calmodulin (CaM)-dependent protein kinase II (CaMKII) is an ubiquitous and broad-specificity Ser/Thr protein kinase which plays a central role in synaptic plasticity and learning and memory.[56,57] This fact does not come as a surprise since CaMKII has demonstrated to fulfil a key role mediating neuronal responses to intracellular calcium oscillations such as transcriptional and translational regulation, receptor and channel function and neurotransmitter synthesis and release.[10]

CaMKII isoenzymes are encoded by at least 4 genes, which are selectively expressed in different tissues.[10] The highly homologous α and β isoforms are primarily found in the central nervous system (CNS),[8,73] from where CaMKII can be purified as a multisubunit enzymatic complex of 8-12 α and β subunits (450-650 kDa). Although these subunits are present in an approximate ratio of 3:1, it is important to keep in mind that this ratio represents just the average composition of what could be a heterogeneous array of holoenzymes formed by various combinations of α and β subunits. In fact, some authors have suggested that forebrain CaMKII consists mainly of homomultimers of these two subunits. This is a particularly important observation, since it has been shown that in hippocampal pyramidal cells the mRNA coding for αCaMKII is found at high concentration in dendrites while the distribution of the β subunit messenger seems to be restricted to the neuronal soma, suggesting that αCaMKII mRNA is locally translated into dendritic-localised polysomes and that the holoenzymes synthetised there are primarily composed by α-subunits.[46,65,77,95]

It has been recognized early that the assembling as a holoenzyme is an essential modulatory step in CaMKII behaviour. Regulation of the enzyme activation implies an auto-inhibitory pseudo-substrate domain which, in the basal state, occupies the kinase's catalytic site and inhibits the binding of substrates (for the representation of CaMKII functional domains, see Fig. 1).

From Messengers to Molecules: Memories Are Made of These, edited by Gernot Riedel and Bettina Platt.

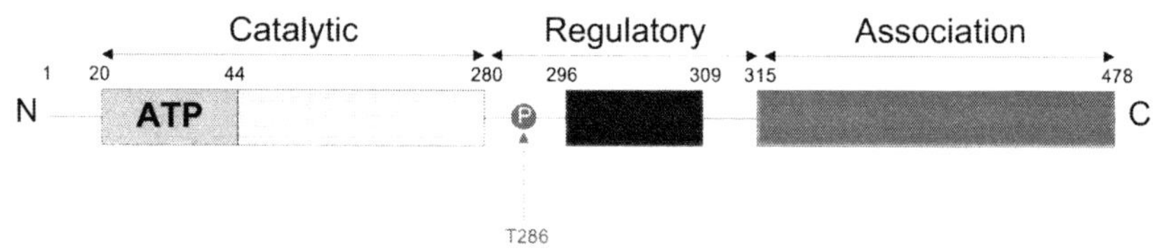

Figure 1. Schematic representation of αCaMKII functional domains. Rat αCaMKII is a polypeptide 478 aminoacids long. Its N-terminal end (residues 1-280) contains the kinase catalytic domain, including both the Mg^{2+}/ATP binding site and the substrate-recognition site. Aminoacids 281-314 define the so-called regulatory domain which can be further separated into two overlapping regions: the autoinhibitory domain and the CaM binding domain. The C-terminal extreme, denominated association domain (residues 315-478) participates in the assembly of the holoenzyme.

Ca^{2+}/CaM activates CaMKII by interacting with a target sequence that overlaps the kinase catalytic domain, altering its conformation and releasing the blockade. The activated kinase does not only phosphorylate exogenous substrates but it is also (and importantly) able to display a prominent autophosphorylation activity targeting the Thr 286 residue within the pseudo-substrate domain. Thr 286 autophosphorylation is an intersubunit/intraholoenzyme phenomena that requires Ca^{2+}/CaM binding to two different and adjacent subunits, one acting as kinase and the other serving as substrate. This autophosphorylation decreases the CaM dissociation rate by more than three orders of magnitude and causes the autophosphorylated enzyme to remain active even after CaM has dissociated from it. Trapping of CaM maintains it bound to CaMKII for longer times and promotes the likelihood of more subunits to enter the trapped state, increasing in that way the number of fully activated kinase molecules.[10,20,33,34] Considering that in certain brain regions[23] and neuronal organelles, such as the hippocampal post-synaptic densities (PSD),[50] CaMKII represents a considerable percentage of the total protein content and could be in molar excess with respect to CaM, the conversion of CaMKII into a high-affinity Ca^{2+}/CaM-binding form can contribute to the subcellular redistribution of CaM during periods of high frequency Ca^{2+} oscillations, modulating the availability of CaM and regulating calmodulin-dependent pathways whose activation could have a detrimental effect on Thr 286 phosphorylation levels. Thus, rapid Thr 286 autophosphorylation produces the appearance of an autonomous, constitutively-active form of CaMKII which catalyses additional regulatory autophosphorylation on different sites and that is able to translate a transient elevation in the intracellular Ca^{2+} concentration into a persistent and Ca^{2+}-independent up-regulation of kinase activity.

CaMKII: Synaptic Plasticity and Memory Processing

Over the last decades a multidisciplinary effort has been made in the search of the molecular and cellular mechanisms responsible for learning and memory. Most of these studies have utilised as a working hypothesis that postulated by Donald Hebb in 1949 which proposes that a given synapse would be strengthened when the pre and postsynaptic neurons were coincidentally activated. Today it is an almost universally accepted hypothesis that the storage of new information under the form of long term memories requires the occurrence of activity-dependent plastic mechanisms, such as those involved in long-term potentiation (LTP).[13,39,64,81,82,89] In the CA1 region of the hippocampus, LTP is triggered by changes in the frequency of intradendritic Ca^{2+} oscillations largely mediated through the NMDA type of glutamatergic receptor (rNMDA). This fact points to the necessity of molecular devices able to sense and properly interpret variations in postsynaptic spine Ca^{2+} produced by the activation of the rNMDA. The demonstration that, at least in vitro, CaMKII is capable to decipher the message coded in the amplitude and duration of individual Ca^{2+} spikes and translates it into distinct amounts of autonomous activity, strongly suggests that this kinase could be such a molecular adaptor.[20]

Several evidences indicate that CaMKII has a prominent participation in the early phase of LTP and during the consolidation of different types of memories. In the CA1 region of the hippocampus, the induction of rNMDA-dependent LTP produces a rapid increase in both CaMKII autonomous activity and Thr 286 autophosphorylation.[25,26] Postsynaptic injection of Ca^{2+}/CaM induces synaptic potentiation requiring CaMKII activity[111] and inhibition of postsynaptic CaMKII abolishes the induction of LTP.[76] Tetanic stimulation increases dendritic synthesis and accumulation of CaMKII in hippocampal neurons[77] and the injection of a truncated, constitutively active form of the kinase into CA1 pyramidal neurons results in an increase in the size of the excitatory postsynaptic currents and occludes subsequent LTP.[58] Furthermore, targeted disruption of the αCaMKII gene impairs LTP.[37]

Long-term memories can be classified into associative and nonassociative depending on the mechanisms required for its formation. Associative memories are based on the acquisition of a predictive link between a specific event and a stimulus. Nonassociative memories are acquired when repeated or continuous exposure to a novel stimulus changes behavioral responses to it. In mammals, some forms of associative and nonassociative memories involve the participation of the hippocampal formation.[40,68,106,123]

The use of transgenic technology together with both pharmacological and biochemical approaches have unequivocally demonstrated the requirement of CaMKII activation in the formation of memories for different associative training paradigms in a variety of animal models. Mutant flies, expressing a specific CaMKII inhibitor peptide under the transcriptional control of an inducible promoter, present serious learning deficits.[32] In mice, expression of a constitutively activated Ca^{2+}-independent form of CaMKII results in deficiencies in spatial learning and fear conditioning.[67] Homozygous αCaMKII mutant mice exhibit impaired spatial learning[90] and recently,[24] it has been shown that although mice which are heterozygous for a CaMKII null mutation show normal memory retention for contextual fear and water maze task 1-3 days after training, these animals are amnesic for the mentioned paradigms when tested 10-50 days post-training.

In the rat, one-trial step-down avoidance learning is associated with a rapid and specific increase in hippocampal CaMKII activity[14] and, pharmacological evidences indicate that its participation is restricted to a narrow posttraining time window and involved, mainly, with events related with the consolidation of long-term memories. For long-term retention of the avoidance response measured 24 hs posttraining, the intrahippocampal infusion of a CaMKII blocker causes full retrograde amnesia only when given immediately after the acquisition session; 30 min posttraining, CaMKII inhibition produces just partial amnesia and, interestingly, no effect at all in long term memory formation is observed when the drug is administered 2 or 3 hrs posttraining[115] or when short term avoidance memory is measured 90-120 min after training.[42] Passive avoidance,[122] but not imprinting,[93] also induces CaMKII phosphorylation in the chick intermediate medial hyperstriatum ventrale (IMHV) and lobus parolfactorius (LPO) and Tan and Liang have reported an increase in amygdalar CaMKII activity after training rats in a step-through inhibitory avoidance learning task.[103] The same group has published evidence indicating that the Morris water maze training task induces CaMKII activation in the rat hippocampus and that the retention of the memory associated with this paradigm is positively correlated with the levels of hippocampal CaMKII autonomous activity.[102]

Much less is known about the participation of CaMKII in the formation of nonassociative memories, though, using one of the most elementary nonassociative learning tasks, the behavioural habituation to a novel environment, we have found that blockade of hippocampal CaMKII at two different time periods, one around training and the other 3 hrs after, impairs spatial habituation.[41,109] Interestingly, inhibition of other hippocampal signalling pathways such as PKA, ERK1/2 and PKC, which are crucially involved in many hippocampal-dependent associative tasks (see chapters by Nugues et al, Vianna and Izquierdo, and Selcher et al in this book), has no effect on long-term memory of habituation.

In contrast to memory acquisition and consolidation, the information about the molecular mechanisms of memory retrieval, and particularly the role of CaMKII in this process, is surprisingly scarce and fragmentary. Using the one-trial inhibitory avoidance task, we found that pretest inhibition of CA1 CaMKII with doses of KN-62 that fully block memory formation when given immediately after training,[115] does not alter retention test performance.[100] It is noteworthy that blockade of rNMDA prior to the test session also produces no effect on memory retrieval, indicating that some hippocampal mechanisms which are crucial for encoding and/or consolidation (rNMDA and CaMKII), are apparently not involved in retrieval (see ref. 70, 100 and below).

Is hippocampal CaMKII involved in the core mechanisms of memory processing? The answer to this question seems to be "yes". Several pharmacological interventions enabled a distinction between core and modulatory mechanisms in memory formation (Fig. 2). For example, memory deficits produced by treatments that act upon modulatory systems, such as electroconvulsive shock or systemic β-endorphin, can be attenuated by treatment with different drugs and hormones. Conversely, amnesia produced by inhibition of hippocampal CaMKII cannot be reverted by the subsequent administration of neither PKA activators into the CA1 region of the hippocampus nor by systemic ACTH or vasopressin given at the time of retrieval, three well-known treatments that enhance memory expression.[6,43]

CaMKII is able to translocate to the PSD in an activity-dependent manner and it is believed that in this subsynaptic specialisation is where CaMKII plays its more important "plastic" role. CaMKII phosphorylates several PSD-associated substrates, like, among others, PSD-95 (a member of the membrane-associated guanylate kinase family of proteins which constitutes a core component of the PSD),[119] the GluR1 subunit of the AMPA receptor (rAMPA)[62] and the NR2A/B subunits of the rNMDA,[16] altering the properties of some of them in a way that is thought to contribute to the establishment of LTP (see below). In the PSD, CaMKII is able to associate with different binding partners, including densin-180 (a transmembrane glycoprotein member of the LAP [leucine-rich repeat (LRR) and PDZ (PSD-95, Dlg, Z0-1)] family of proteins implicated in the organisation and subcellular sorting of multimolecular complexes),[98,110] F-actin[3] and, fundamentally, with the rNMDA. Binding of αCaMKII to the NR2B subunit has been reported to require Thr 286 autophosphorylation[96] and several studies have indicated that dephosphorylation of CaMKII promotes its release from the PSD, an event that is mainly mediated by protein phosphatase 1 (PP1).[97,120] At this respect, a recent and exciting work from Waxham's group, presents experimental evidence backing the intriguing hypothesis postulating that CaMKII could have an intrinsic ADP-dependent, phosphatase inhibitor-independent autodephosphorylating activity which targets the Thr 286 site and abolishes its Ca^{2+}/CaM autonomous activity.[52]

Interestingly, studies using green fluorescent protein-tagged CaMKII have demonstrated that NMDA-stimulated translocation of this kinase to the PSD, although stimulated by Ca^{2+}/CaM and indirectly prolonged by autophosphorylation, does not require this posttranslational modification.[86] In a recent report, Schulman's group has elegantly solved these apparent discrepancies.[7] The authors showed that NR2B contains two different CaMKII binding regions: an autophosphorylation-dependent binding site (NR2B-P) and a Ca^{2+}/CaM-regulated interacting motif (NR2B-C), mapping within residues 839-1120 and 1120-1482 of NR2B, respectively. Binding of CaMKII to NR2B-C increases its affinity for CaM similarly to what happens after autophosphorylation. This effect is mimicked by a short sequence contained within residues 1289-1310 of NR2B-C, which is homologous to the segment surrounding Thr 286 in the autoinhibitory domain of αCaMKII. The interaction of CaMKII with this region

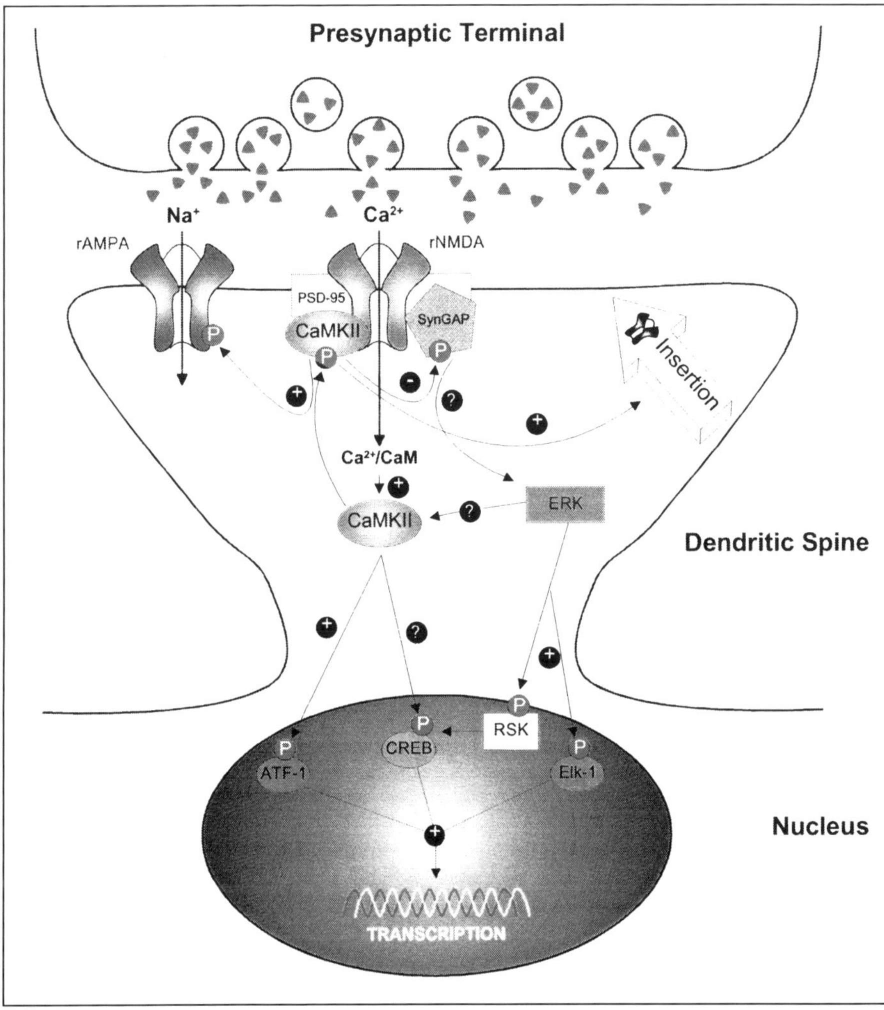

Figure 2. Postsynaptic effects of CaMKII activation during plastic events. As a consequence of glutamate-enhanced exocytosis from the presynapse, the postsynaptic terminal is strongly depolarized through the activation of AMPA receptors (rAMPA). This results in the removal of the Mg^{2+} block from the NMDA receptor (rNMDA), allowing an increase in the Ca^{2+} influx into the postsynaptic neuron. The elevation of the intradendritic Ca^{2+} levels induces, in turn, the formation of Ca^{2+}/calmodulin (Ca^{2+}/CaM) complexes and the Thr 286 autophosphorylation of CaMKII, which translocates to the post-synaptic density where it interacts, among others, with the rNMDA. Once there, CaMKII phosphorylates the GluR1 subunit of the rAMPA at Ser 831, increasing the conductance of the receptor-associated channel. Different sources indicate that the activation of this kinase is a necessary step in the cascade of events that conclude with the insertion of rAMPA into formerly silent synapses. CaMKII also promotes the rNMDA-dependent activation of the ERK cascade through the inhibitory phosphorylation of the Ras-GTPase activator, p135 SynGap (SynGaP), contributing to the ERK-mediated gene expression. Although controversial, there exist some evidence indicating that CaMKII could directly up-regulate transcription through the phosphorylation of the transcription factors ATF-1 and CREB. See the text for further details.

of NR2B-C promotes the phosphorylation of the peptide and increases the kinase's autonomous activity in a Thr 286 autophosphorylation-independent way, generating a kinase activity that does not require Ca^{2+}/CaM and that cannot be reverted by the action of protein phosphatases. Moreover, the mentioned interaction blocks the inhibitory secondary autophosphorylation of CaMKII at Thr 305/306 that is normally initiated when CaM dissociates from the autonomous enzyme. In this way, and in contrast to what happens with the interaction with other binding proteins, the association of CaMKII with the rNMDA may occur following stimuli that produces only a small increment in the intradendritic levels of Ca^{2+} and that are not able to induce a large increase in Thr 286 autophosphorylation. Binding of CaMKII to the phosphorylation-independent site on NR2B might, in turn, facilitates Thr 286 phosphorylation of other subunits in the translocated holoenzyme and hence, further increase the strength of the interaction, promoting the association of the autophosphorylated subunits with the NR2B phosphorylation-dependent binding site.

The association of CaMKII with the rNMDA can also be attained through its interaction with the NR2A subunit. The group of Di Luca at the University of Milan has identified a sequence present in the cytosolic C-terminus of NR2A (residues 1412-1419) that is able to specifically bind CaMKII.[27] This sequence contains a PKC phosphorylation site (Ser 1416) and the authors showed that the phosphorylation of this residue decreases αCaMKII binding to the rNMDA without affecting the total amount of CaMKII present in a triton-insoluble subcellular fraction from rat hippocampus. Moreover, PSD-95, which directly interacts with the rNMDA and induces clustering of channel proteins into the PSD, competes with CaMKII for binding to the NMDA receptor complex.[28] At this regard, and using rNMDA subunit-deficient mice, it has been recently shown that the activation of CaMKII that occurs in the lateral/basolateral amygdala after retrieval of an auditory fear conditioning memory, involves the NR2A subunit, suggesting the existence of a functional association between this subunit and the up-regulation of the CaMKII cascade that follows the expression of a mnemonic trace.[70]

The results mentioned above, together with those describing that the synaptic clustering of αCaMKII is totally dependent on an intact actin cytoskeleton,[3] indicate that the localisation of CaMKII close to the postsynaptic receptor machinery is a highly and finely tuned process that requires the concerted participation of several kinases, phosphatases and scaffold proteins. The fact that CaMKII is able to phosphorylate several of its binding partners indicates that the enzyme could be capable to modulate its own rate of association, modifying some of the properties of its interacting proteins. This hypothesis is further supported by findings showing that CaMKII phosphorylation of PSD proteins inhibits its own translocation.[120]

Downstream Effectors of the CaMKII Cascade

What are the consequences of CaMKII translocation? How does the activation of this kinase contribute to the establishment of LTP and in which way is this contribution related to the formation of new memories? CaMKII phosphorylates several PSD-associated proteins, including PSD-95, α and β tubulin, the GTPase dynamin, the type IV intermediate filament protein α-internexin and cAMP phosphodiesterase,[120] although the functional consequences of most of these phosphorylations are not known. Despite this fact, there exists a considerable amount of evidence indicating that one of the principal events mediated by CaMKII during the early phase of LTP is related with the upregulation of AMPA receptors. As mentioned above, CaMKII phosphorylates the GluR1 subunit of the rAMPA at Ser 831 and it has been shown that this phosphorylation potentiates AMPA-mediated currents,[4,58,69,104] presumably through a mechanism that involves the stabilisation of the receptor into a high conductance state.[21] rNMDA stimulation promotes the Ca^{2+} and CaMKII-dependent phosphorylation of the rAMPA[104] and, moreover, induction of LTP is associated with a delayed enhancement in AMPA-mediated responses which is accompanied by the CaMKII-dependent phosphorylation of GluR1 at Ser 831.[5] Interestingly, inhibitory avoidance learning in the rat is also associated with an increase in hippocampal GluR1 phosphorylation, possibly through a mechanism involving CaMKII activation.[14]

Besides the direct effect that phosphorylation of the GluR1 subunit might have on the electrophysiological properties of the rAMPA, CaMKII could mediated some other responses that are known to involve plastic changes of this receptor. A high proportion of synapses in the CA1 area of the hippocampus transmit with rNMDA but not with rAMPA, indicating that they are nonfunctional at normal resting potentials. Surprisingly, these same synapses acquire AMPA-type responses following LTP induction.[54] These findings, together with earlier results showing that both glycine-induced LTP in hippocampal slices[72,84] and in vivo induced LTP in the rat hippocampus[63,107] produce an increase in the number of ^{3}H-AMPA binding sites, fuelled the appearance of the "silent synapse" hypothesis (for a recent review see ref. 61). This hypothesis postulates that LTP produces the conversion of synapses lacking rAMPA responses into fully functional AMPAergic terminals, maybe by means of the insertion of previously extrasynaptically localised rAMPA. This idea received further support when it was shown that induction of LTP causes the rNMDA-dependent redistribution of green fluorescent protein-tagged GluR1 from intracellular pools into dendritic spines.[87] Although this process has been reported to be independent of GluR1 Ser 831 phosphorylation, it can be mimicked by increasing the activation state of CaMKII[35] and quite recently it has been shown that spontaneous activation of rNMDA in hippocampal neuronal cultures provokes the rapid recruitment of rAMPA into morphological silent synapses (i.e., synapses that contains rNMDA but not rAMPA), an event accompanied by the translocation of CaMKII into those synapses and the phosphorylation of GluR1 at Ser 831.[55] These results are in agreement with both, recent findings showing that KN-62 blocks the NMDA-induced increase in GluR1 and GluR2/3 associated with hippocampal synaptic plasma membranes (SPM)[12] and those indicating that incubation of hippocampal SPM under conditions suitable for CaMKII activation and autophosphorylation promotes a CaMKII-dependent increase in PSD-associated ^{3}H-AMPA binding sites,[14] suggesting the participation of CaMKII in this insertion mechanism and providing a plausible explanation to the observed increase in hippocampal ^{3}H-AMPA binding sites that follows inhibitory avoidance learning in the rat.[15]

A role for CaMKII in Ca^{2+} dendrite membrane trafficking has been suggested.[59] There is experimental evidence for the formation of new synapses (or the remodelling of existing ones) after the induction of LTP[1,71] as well as during the consolidation phase of both an avoidance training and the water maze learning task (see ref. 74,75 and Geinisman et al, this book). Tetanic stimulation promotes a rNMDA-dependent increase in the number of dendritic spines surrounding the area of stimulation, which is blocked by KN-93, a specific CaMKII inhibitor.[22,60,108] Moreover, in Drosophila, expression of a constitutively active form of CaMKII results in the phosphorylation of the disc large protein (DLG), a homologue of the mammalian PSD family of proteins, causing rearrangement of the synaptic structure.[53] These findings suggest that long-lasting CaMKII up-regulation, as it occurs during activity-dependent synaptic plasticity, could help in the synaptic remodelling of the PSD scaffold.

Together with CaMKII, PKA and PKC, the intracellular signalling pathway mediated through the activation of the mitogen-activated protein kinases ERK1/2 is one the best characterised in relation to plastic mechanisms (for recent reviews see ref. 99 and Selcher et al in this book). Up-regulation of ERK1/2 is classically attained through the sequential activation of the GTPase Ras, Raf-1 and the ERK kinase MEK and this pathway can modulate the phosphorylation state of the transcription factors ElK-1 and CREB.[19] A couple of years ago, and almost simultaneously, Huganir's and Kennedy's groups at the John Hopkins University and California Institute of Technology, respectively, reported the existence of a novel, synaptically-localized Ras-GTPase activating protein named p135 SynGAP. p135 SynGAP colocalizes with PSD-95, the rNMDA and the synapse associated protein SAP-102 and it is almost exclusively present in hippocampal neurons. p135 SynGAP stimulates Ras GTPase activity, suggesting that it is negatively coupled to the activation of the ERK pathway.[18,51] Interestingly, p135 SynGAP is a substrate of CaMKII which inhibits its Ras-GTPase activating activity. Moreover, it has been shown that in cortical neurons, inhibition of CaMKII partially blocks the NMDA-induced phosphorylation of ERK1/2.[45] These findings encourage the tempting hypothesis that, by means of blocking the inactivation of GTP-bound Ras, CaMKII could directly couple rNMDA stimulation with the up-regulation of the ERK pathway and hence, indirectly contribute to the activation of the inducible transcription factors that is needed for the establishment of late LTP and the formation of long-lasting memories (see ref. 99 and also Selcher et al and Frankland and Josselyn in this book]. In addition, there exists evidence suggesting that, in turn, ERK1/2 could also control CaMKII up-regulation during LTP. When LTP is induced into the CA1 area of the rat hippocampus using a protocol that involves theta-pulse stimulation and activation of β adrenergic receptors, it produces an ERK-dependent potentiation which is accompanied by a transient, colocalized and rNMDA-dependent increase in ERK and CaMKII phosphorylation. These early increases are followed by a delayed, actinomycin-D/anisomycin-dependent augmentation of CaMKII protein levels. Both, CaMKII phosphorylation and expression are blocked by preincubation of the hippocampal slices in the presence of the MEK inhibitor PD98059, indicating that ERK1/2 are likely to participate as an up-stream factor in the events that regulate CaMKII function during neuronal plasticity.[29]

As mentioned above CaMKII subunits are encoded by a family of 4 related genes: α,β,γ and δ and it has been described that alternative splicing can generate isoforms containing a nuclear localization signal (NLS) homologous to that present in the SV40 large T antigen.[11,94] As demonstrated by immunostaining of brain sections[11] and despite the large size of the holoenzyme (400-600 kDa), it seems that CaMKII does have access to the neuronal nucleus where it could directly regulate the activity of different transcription factors and hence gene expression. Overexpression of a nuclear-localised isoform of the δ CaMKII subunit promotes BDNF transcription in NG108-15 cells[101] and it has been shown that, in hippocampal neurons, Ca^{2+}/calmodulin-dependent kinases intensify the activity and expression of the CCAAT enhancer element-binding protein β (C/EBPβ),[121] a transcription factor involved in the switch from short to long-term facilitation in Aplysia[2] and in the consolidation of hippocampal dependent memories.[105]

As well as PKA and the ERK-activated kinase $_{p90}$RSK, CaMKII can phosphorylate CREB at Ser 133 [99] but it is not able to promote the activity of this transcription factor, maybe due to the phosphorylation of a secondary site at Ser 142 that prevents CREB dimerization and binding to the CREB-binding protein (CBP).[117] Despite this fact, there are several lines of evidence suggesting the involvement of CaMKII in the activation of CREcontaining genes. At this respect, early studies have shown that CaMKII is able to directly phosphorylate the activating transcription factor 1 (ATF-1; a member of the ATF/CREB family of transcription factors) on Ser 63, suggesting that in that way it could mediate transactivation of Ca^{2+}/cAMP responsive genes.[88] (See Fig. 2 for a schematic diagram of the postsynaptic effects of CaMKII activation during plastic events).

CaMKIV: A New (and Important) Player in the Plasticity Team

Translocation of CaMKII into the nucleus is thought to be regulated by the phosphorylation of a Ser residue adjacent to the NLS sequence in the nuclear-targeted isoforms. The finding that both CaMKI and CaMKIV are able to phosphorylate that site and hinder the accessibility of CaMKII to the nucleus shows the existence of complicated interrelation mechanisms among the different members of the Ca^{2+}/CaM-dependent kinases family.[36] As it happens with CaMKII, both CaMKI and CaMKIV are able to bind Ca^{2+}/CaM complexes and this interaction activates these enzymes through a mechanism that involves displacement of an autoinhibitory domain from the substrate and Mg^{2+}/ATP binding sites.[91] CaMKI and IV exist as monomeric proteins[78] and to attain full activity they require phosphorylation of a Thr residue present in its activation loop. The phosphorylation of these sites (Thr 177 in CaMKI and Thr 196 in CaMKIV) is carried out by another Ca^{2+}/CaM dependent kinase (Ca^{2+}/CaM dependent kinase kinase or CaMKK) and to occur it needs binding of Ca^{2+}/CaM to both CaMKK and its substrates.[92] CaMKI is enriched in neuronal processes and synapses[60] but, in contrast, CaMKIV and the β isoform of CaMKK are enriched in the nucleus.[44,92] This subcellular distribution suggests a role for CaMKIV in the onset of nuclear responses to synaptic stimulation and may be related with the observed effect of CaMKIV-signalling inhibition in the protein synthesis-dependent phase of different plastic events. It has been early shown that CaMKIV regulates CREB-dependent gene expression[66] and that it is able to phosphorylate CREB and the transcriptional coactivator CBP.[9,17] KCl-mediated depolarisation and rNMDA stimulation increase CaMKIV activity in hippocampal neurons[48] and disruption of CaMKIV expression blocks KCl-induced CREB phosphorylation.[9] CaMKIV mutant mice present a decrease in both basal and experimentally-induced levels of phosphoCREB together with impaired LTP.[38] Recently, Eishichi Miyamoto and his group at Kumamoto University, have shown that induction of LTP in the CA1 region of the rat hippocampus is accompanied by a transient (back to control levels within 30 minutes) increase in CaMKIV activity which is associated with an enhanced expression of the immediate early gene c-fos and a rapid CaMK-dependent/PKA-independent activation of CREB.[49] These findings, together with those showing that the rapid (0-10 min) CREB phosphorylation observed after synaptic stimulation is mainly due to the activation of Ca^{2+}/CaM-dependent kinases,[116] suggest that the early up-regulation of CREB phosphorylation observed after LTP induction and memory formation[40,105] could be caused by an initial increase in CaMKIV activity followed by the occurrence of late ERK or PKA-dominated events. The use of transgenic mice in which the expression of a dominant-negative mutant form of CaMKIV is restricted to the postnatal forebrain has partially confirmed this hypothesis. These animals show intact synaptic functionality as well as normal early LTP but, in contrast, activity-dependent c-Fos expression and CREB phosphorylation, together with late LTP and consolidation of long term memory, are impaired.[47]

The existence of cross-talk mechanisms between CaMKIV and other signalling cascades have also been reported. For example, PKA phosphorylates CaMKK, inhibiting its activity and abolishing CaMKIV activation.[112] In turn, CaMKIV can phosphorylate and inhibit adenylate cyclase I and III,[113,114] modulating cAMP levels and PKA activation. The prevalence of one or the other of these reciprocal control mechanisms seems to depend on the sequence in which the stimulation agents are presented. The rise in cAMP levels elicited by forskolin blocks the subsequent Ca^{2+}-dependent increase in CaMKK and CaMKIV activities; conversely, if ionomicin-induced Ca^{2+} mobilisation precedes forskolin stimulation, then CaMKK inhibition is greatly reduced, maybe because Ca^{2+}/CaM binding to CaMKK reduces the accessibility of PKA to its phosphorylation site.[112] This auto-regulatory mechanism may have important consequences in the activation state of several downstream pathways, including those involving ERK. PKA activity is required for both depolarisation and NGF-induced ERK activation in PC12 cells[79,118] and it has been suggested that the pathway linking these two kinases is likely to require PKA-mediated phosphorylation of Rap-1, a Ras-related GTP binding protein able to interact with B-Raf and promote ERK phosphorylation.[31] Interestingly, it has been shown that both calcium and cAMP-induced increases in CREB functionality need activation of the Rap-1/ERK pathway[30] and that CaMKIV is able to phosphorylate Rap-1 "in vitro", potentially strengthening its interaction with B-Raf and hence, inducing the ERK cascade.[83]

Concluding Remarks

Although extensive research has linked CaMKII activity and, by extension, the phosphorylation of its downstream substrates to synaptic plasticity, only a small body of evidence bridges the gap between this protein kinase and behaviour. In general, the existing information is based on pharmacological data and, with a few exceptions, one have to admit that very little is known about the role that CaMKII substrates actually play in learning events. Taking into account this caveat, and though the involvement of CaMKII in memory retrieval is still controversial, the current knowledge points to a central role of this enzyme in the acquisition of new information and/or in the early stages of memory formation of associative and nonassociative tasks. Hopefully, the availability of new molecular and biochemical tools, in particular the use of phospho-specific antibodies together with penetratins and antisense technologies will allow us to fully understand how, when and where CaMKII and its "cousin" CaMKIV, a "newcomer" in this scenario that deserves much attention, participate in the mnemonic processes.

References

1. Agnihotri N, Lopez-Garcia J, Hawkins R et al. Morphological changes associated with long-term potentiation. Histol Histopathol 1998; 13:1155-1162.
2. Alberini C, Ghirardi M, Metz R et al. C/EBP is an immediate-early gene required for the consolidation of long-term facilitation in Aplysia. Cell 76:1099-1114.
3. Allison D, Chervin A, Gelfand V et al. Postsynaptic scaffolds of excitatory and inhibitory synapses in hippocampal neurons: Maintenance of core components independent of actin filaments and microtubules. J Neurosci 2000; 20:4545-4554.
4. Barria A, Derkach V, Soderling T. Identification of the Ca^{2+}/calmodulin-dependent protein kinase II regulatory phosphorylation site in the alpha-amino-3-hydroxyl-5-methyl-4-isoxazole-propionate-type glutamate receptor. J Biol Chem 1997; 272:32727-32730.
5. Barria A, Muller D, Derkach V et al. Regulatory phosphorylation of AMPA-type glutamate receptors by CaM-KII during long-term potentiation. Science 1997; 276:2042-2045.
6. Barros D, Izquierdo L, Sant'Anna M et al. Stimulators of the cAMP cascade reverse amnesia induced by intra-amygdala but not intrahippocampal KN-62 administration. Neurobiol Learn Mem 1999; 71:94-103.
7. Bayer K, De Koninck P, Leonard A et al. Interaction with the NMDA receptor locks CaMKII in an active conformation. Nature 2001; 411:801-805.
8. Bennet M, Erondu N, Kennedy M. Purification and characterization of a calmodulin-dependent protein kinase that is highly concentrated in brain. J Biol Chem 1983; 258:12735-12744.
9. Bito H, Deisseroth K, Tsien R. CREB phosphorylation and dephosphorylation: A Ca^{2+} and stimulus duration-dependent switch for hippocampal gene expression. Cell 1996; 87:1203-1214.
10. Braun A, Schulman H. The multifunctional Ca^{2+}/calmodulin-dependent protein kinase: Form to function. Annu Rev Physiol 1995; 57:417-452.
11. Brocke L, Srinivasan M, Schulman H. Developmental and regional expression of multifunctional Ca^{2+}/calmodulin-dependent protein kinase isoforms in rat brain. J Neurosci 1995; 15:6797-6808.
12. Broutman G, Baudry M. Involvement of the secretory pathway for AMPA receptors in NMDA-induced potentiation in hippocampus. J Neurosci 2001; 21:27-34.
13. Brun V, Ytterbo K, Morris RGM et al. Retrograde amnesia for spatial memory induced by NMDA receptor-mediated long-term potentiation. J Neurosci 2001; 21:356-362.
14. Cammarota M, Bernabeu R, Levi De Stein M et al. Learning-specific, time-dependent increases in hippocampal Ca^{2+}/calmodulin-dependent protein kinase II activity and AMPA GluR1 subunit immunoreactivity. Eur J Neurosci 1998; 10:2669-2676.
15. Cammarota M, Izquierdo I, Wolfman C et al. Inhibitory avoidance training induces rapid and selective changes in ^{3}H-AMPA receptor binding in the rat hippocampal formation. Neurobiol Learn Mem 1995; 64:257-264.
16. Caputi A, Gardoni F, Cimino M et al. CaMKII-dependent phosphorylation of NR2A and NR2B is decreased in animals characterized by hippocampal damage and impaired LTP. Eur J Neurosci 1999; 11:141-148.
17. Chawla S, Hardingham G, Quinn D et al. CBP: A signal-regulated transcriptional coactivator controlled by nuclear calcium and CaM kinase IV. Science 1998; 281:1505-1509.
18. Chen H, Rojas-Soto M, Oguni A et al. A synaptic Ras-GTPase activating protein (p135 SynGAP) inhibited by CaM kinase II. Neuron 1998; 20:895-904.
19. Davis S, Vanhoutte P, Pages C et al. The MAPK/ERK cascade targets both Elk-1 and cAMP response element-binding protein to control long-term potentiation-dependent gene expression in the dentate gyrus in vivo. J Neurosci 2000; 20:4563-4572.
20. De Koninck P, Schulman H. Sensitivity of CaM kinase II to the frequency of Ca^{2+} oscillations. Science 1998; 279:227-230.
21. Derkach V, Barria A, Soderling T. Ca^{2+}/calmodulin-kinase II enhances channel conductance of alpha-amino-3-hydroxy-5-methyl-4-isoxazolepropionate type glutamate receptors. Proc Natl Acad Sci USA 1999; 96:3269-3274.
22. Engert F, Bonhoeffer T. Dendritic spine changes associated with hippocampal long-term synaptic plasticity. Nature 1999; 399:66-70.

23. Erondu N, Kennedy M. Regional distribution of type II Ca^{2+}/calmodulin-dependent protein kinase in rat brain. J Neurosci 1985; 13:1010-1021.
24. Frankland P, O'Brien C, Ohno M et al. Alpha-CaMKII-dependent plasticity in the cortex is required for permanent memory. Nature 2001; 411:309-313.
25. Fukunaga K, Muller D, Miyamoto E. Increased phosphorylation of Ca^{2+}/calmodulin-dependent protein kinase II and its endogenous substrates in the induction of long-term potentiation. J Biol Chem 1995; 270:6119-6124.
26. Fukunaga K, Stoppini L, Miyamoto E et al. Long term potentiation is associated with an increased activity of Ca^{2+}/calmodulin-dependent protein kinase II. J Biol Chem 1993; 268:7863-7867.
27. Gardoni F, Bellone C, Cattabeni F et al. Protein kinase C activation modulates alpha-calmodulin kinase II binding to NR2A subunit of N-methyl-D-aspartate receptor complex. J Biol Chem 2001; 276:7609-7613.
28. Gardoni F, Schrama L, Kamal A et al. Hippocampal synaptic plasticity involves competition between Ca^{2+}/calmodulin-dependent. protein kinase II and postsynaptic density 95 for binding to the NR2A subunit of the NMDA receptor. J Neurosci 2001; 21:1501-1509.
29. Giovannini M, Blitzer R, Wong T et al. Mitogen-activated protein kinase regulates early phosphorylation and delayed expression of Ca^{2+}/calmodulin-dependent protein kinase II in long-term potentiation. J Neurosci 2001; 21:7053-7062.
30. Grewal S, Fass D, Yao H et al. Calcium and cAMP signals differentially regulate cAMP-responsive element-binding protein function via a Rap1-extracellular signal-regulated kinase pathway. J Biol Chem 2000; 275:34433-34441.
31. Grewal S, Horgan A, York R et al. Neuronal calcium activates a Rap1 and B-Raf signaling pathway via the cyclic adenosine monophosphate-dependent protein kinase. J Biol Chem 2000; 275:3722-3728.
32. Griffith L, Verselis L, Aitken K et al. Inhibition of calcium/calmodulin-dependent protein kinase in Drosophila disrupts behavioral plasticity. Neuron 1993; 10:501-509.
33. Hanson P, Schulman H. Neuronal Ca^{2+}/calmodulin-dependent protein kinases. Ann Rev Biochem 1992; 61:559-601.
34. Hanson P, Meyer T, Stryer L et al. Dual role of calmodulin in autophosphorylation of multifunctional CaM kinase may underlie decoding of calcium signals. Neuron 1994; 12:943-956.
35. Hayashi Y, Shi S, Esteban J et al. Driving AMPA receptors into synapses by LTP and CaMKII: Requirement for GluR1 and PDZ domain interaction. Science 2000; 287:2262-2267.
36. Heist E, Srinivasan M, Schulman H. () Phosphorylation at the nuclear localization signal of Ca^{2+}/calmodulin-dependent protein kinase II blocks its nuclear targeting. J Biol Chem 1998; 273:19763-19771.
37. Hinds H, Tonegawa S, Malinow R. CA1 long-term potentiation is diminished but present in hippocampal slices from αCaMKII mutant mice. Learn Mem 1998; 5:344-354.
38. Ho N, Liauw J, Blaeser F et al. Impaired synaptic plasticity and cAMP response element-binding protein activation in Ca^{2+}/calmodulin-dependent protein kinase type IV/Gr-deficient mice. J Neurosci 2000; 20:6459-6472.
39. Holscher C. Long-term potentiation: A good model for learning and memory? Prog. Neuro-Psychopharmacol. Biol Psychiatry 1997; 21:47-68.
40. Izquierdo I, Medina J. Memory formation: The sequence of biochemical events in the hippocampus and its connection to activity in other brain structures. Neurobiol Learn Mem 1997; 68:285-316.
41. Izquierdo I, Da Cunha C, Rosat R et al. Neurotransmitter receptors involved in memory processing by the amygdala, septum and hippocampus of rats. Behav Neural Biol 1992; 58:16-25.
42. Izquierdo I, Medina J, Vianna M et al. Separate mechanisms for short- and long-term memory. Behav Brain Res 1999; 103:1-11
43. Izquierdo L, Schroder N, Ardenghi P et al. Systemic administration of ACTH or vasopressin reverses the amnestic effect of posttraining beta-endorphin or electroconvulsive shock but not that of intrahippocampal infusion of protein kinase inhibitors. Neurobiol Learn Mem 1997; 68:197-202.
44. Jensen K, Ohmstede C, Fisher R et al. Acquisition and loss of a neuronal Ca^{2+}/calmodulin-dependent protein kinase during neuronal differentiation. Proc Natl Acad Sci USA 1991; 88:4050-4053.
45. Judson Chandler L, Sutton G, Dorairaj N et al. N-methyl D-aspartate receptor mediated bidirectional control of extracellular-signal regulated kinase activity in cortical neuronal cultures. J Biol Chem 2001; 276:2627-2636.
46. Kanaseki T, Ikeuchi Y, Sugiura H et al. Structural features of Ca^{2+}/calmodulin-dependent protein kinase II revealed by electron microscopy. J Cell Biol 1991; 115:1049-1060.
47. Kang H, Sun L, Atkins C et al. An important role of neural activity-dependent. CaMKIV signaling in the consolidation of long-term memory. Cell 2001; 106:771-783.

48. Kasahara J, Fukunaga K, Miyamoto E. Activation of Ca^{2+}/calmodulin-dependent protein kinase IV in cultured rat hippocampal neurons. J Neurosci Res 2000; 59:594-600.
49. Kasahara J, Fukunaga K, Miyamoto E. Activation of calcium/calmodulin-dependent protein kinase IV in long term potentiation in the rat hippocampal CA1 region. J Biol Chem 2001; 276:24044-24050.
50. Kennedy M. The postsynaptic density at glutamatergic synapses. Trends Neurosci 1997; 20:264-268.
51. Kim J, Liao D, Lau L et al. SynGAP: A synaptic RasGAP that associates with the PSD-95/SAP90 protein family. Neuron 1998; 20:683-691.
52. Kim S, Hudmon A, Volmer A et al. CaM-kinase II dephosphorylates Thr(286) by a reversal of the autophosphorylation reaction. Biochem Biophys Res Commun 2001; 282:773-780.
53. Koh Y, Popova E, Thomas U et al. Regulation of DLG localization at synapses by CaMKII-dependent phosphorylation. Cell 1999; 98:353-363.
54. Liao D, Hessler N, Malinow R. Activation of postsynaptically silent synapses during pairing-induced LTP in CA1 region of hippocampal slice. Nature 1995; 375:400-404.
55. Liao D, Scannevin R, Huganir R. Activation of silent synapses by rapid activity-dependent synaptic recruitment of AMPA receptors. J Neurosci 2001; 21:6008-6017.
56. Lisman J. The CaMKII hypothesis for storage of synaptic memory. Trends Neurosci 1994; 17:406-412.
57. Lisman J, Goldring M. Feasibility of long-term storage of graded information by the Ca^{2+}/ calmodulin-dependent protein kinase molecules of the postsynaptic densities. Proc Natl Acad Sci USA 1988; 85:5320-5324.
58. Lledo P, Hjelmstad G, Mukherfi S et al. Ca^{2+}/calmodulin-dependent protein kinase II and long-term potentiation enhance synaptic transmission by the same mechanism. Proc Natl Acad Sci USA 1995; 92:11175-11179.
59. Maletic-Savatic M, Koothan T, Malinow R. Calcium-evoked dendritic exocytosis in cultured hippocampal neurons. Part II: Mediation by calcium/calmodulin-dependent protein kinase II. J Neurosci 1998; 18:6814-6821.
60. Maletic-Savatic M, Malinow R, Svoboda K. Rapid dendritic morphogenesis in CA1 hippocampal dendrites induced by synaptic activity. Science 1999; 283:1923-1927.
61. Malinow R, Mainen Z, Hayashi Y. LTP mechanisms: From silence to four-lane traffic. Curr Opin Neurobiol 2000; 10:352-357.
62. Mammen A, Kameyama K, Roche K et al. Phosphorylation of the alpha-amino-3-hydroxy-5-methylisoxazole-4-propionic acid receptor GluR1 subunit by calcium/calmodulin-dependent kinase II. J Biol Chem 1997; 272:32528-32533.
63. Maren S, Tocco G, Standley S et al. Postsynaptic factors in the expression of long-term potentiation (LTP): Increased glutamate receptor binding following LTP induction in vivo. Proc Natl Acad Sci USA 1993; 90:9654-9658.
64. Martin S, Grimwood P, Morris RGM. () Synaptic plasticity and memory: An evaluation of the hypothesis. Annu Rev Neurosci 2000; 23:649-711.
65. Martone M, Pollock J, Jones Y et al. Ultrastructural localization of dendritic messenger RNA in adult brain hippocampus. J Neurosci 1996; 16:7437-7446.
66. Matthews R, Guthrie C, Wailes L et al. Calcium/calmodulin-dependent protein kinase types II and IV differentially regulate CREB-dependent gene expression. Mol Cell Biol 1994; 14:6107-6116.
67. Mayford M, Bach M, Huang Y et al. Control of memory formation through regulated expression of a CaMKII transgene. Science 1996; 274:1678-1683.
68. McGaugh J. Memory—A century of consolidation. Science 2000; 287:248-251.
69. McGlade-McCulloh E, Yamamoto H, Tan S et al. Phosphorylation and regulation of glutamate receptors by calcium/calmodulin-dependent protein kinase II. Nature 1993; 362:640-642.
70. Moriya T, Kouzu Y, Shibata S et al. Close linkage between calcium/calmodulin kinase II alpha/beta and NMDA-2A receptors in the lateral amygdala and significance for retrieval of auditory fear conditioning. Eur J Neurosci 2000; 12:3307-3314.
71. Muller D, Toni N, Buchs P. Spine changes associated with long-term potentiation. Hippocampus 2000; 10:596-604.
72. Musleh W, Bi X, Tocco G et al. Glycine-induced long-term potentiation is associated with structural and functional modifications of alpha-amino-3-hydroxyl-5-methyl-4-isoxazolepropionic acid receptors. Proc Natl Acad Sci USA 1997; 94:9451-9456.
73. Nairn A, Hemmings H, Greengard P. Protein kinases in the brain. Annu Rev Biochem 1985; 54:931-976.
74. O'Malley A, O'Connell C, Regan C. Ultrastructural analysis reveals avoidance conditioning to induce a transient increase in hippocampal dentate spine density in the 6 hour post-training period of consolidation. Neurosci 1998; 87:607-613.

75. O'Malley A, O'Connell C, Murphy K et al. Transient spine density increases in the mid-molecular layer of hippocampal dentate gyrus accompany consolidation of a spatial learning task in the rodent. Neurosci 2000; 99:229-232.
76. Otmakhov N, Griffith L, Lisman J. Postsynaptic inhibitors of Ca^{2+}/calmodulin-dependent protein kinase type II block induction but not maintenance of pairing induced long-term potentiation. J Neurosci 1997; 17:5357-5365.
77. Ouyang Y, Rosenstein A, Kreiman G et al. Tetanic stimulation leads to increased accumulation of Ca^{2+}/calmodulin-dependent protein kinase II via dendritic protein synthesis in hippocampal neurons. J Neurosci 1999; 19:7823-7833.
78. Park I, Soderling T. Activation of Ca^{2+}/calmodulin dependent protein kinase IV by CaM-kinase kinase in Jurkat T lymphocytes. J Biol Chem 1995; 270:19320-19324.
79. Park J, Park J, Bae K et al. Protein kinase A activity is required for depolarization-induced proline-rich tyrosine kinase 2 and mitogen-activated protein kinase activation in PC12 cells. Neurosci Lett 2000; 290:25-28.
80. Picciotto M, Zoli M, Bertuzzi G et al. Immunochemical localization of calcium/calmodulin-dependent protein kinase I. Synapse 1995; 20:75-84.
81. Richter-Levin G, Yaniv D. Is LTP in the hippocampus a useful model for learning-related alterations in gene expression? Rev Neurosci 2001; 12:289-296.
82. Sacchetti B, Lorenzini C, Baldi E et al. Long-lasting hippocampal potentiation and contextual memory consolidation. Eur J Neurosci 2001; 13:2291-2298.
83. Sahyoun N, McDonald O, Farrell F et al. Phosphorylation of a Ras-related GTP-binding protein, Rap-1b, by a neuronal Ca^{2+}/calmodulin-dependent protein kinase, CaM kinase G. Proc Natl Acad Sci USA 1991; 88:2643-2647.
84. Shahi K, Baudry M. Glycine-induced changes in synaptic efficacy in hippocampal slices involve changes in AMPA receptors. Brain Res 1993; 627:261-266.
85. Shaywitz A, Greenberg M. CREB: A stimulus-induced transcription factor activated by a diverse array of extracellular signals. Annu Rev Biochem 1999; 68:821-861.
86. Shen K, Meyer T. Dynamic control of CaMKII translocation and localization in hippocampal neurons by NMDA receptor stimulation. Science 1999; 284:162-166.
87. Shi S, Hayashi Y, Petralia R et al. Rapid spine delivery and redistribution of AMPA receptors after synaptic NMDA receptor activation. Science 1999; 284:1811-1816.
88. Shimomura A, Ogawa Y, Kitani T et al. Calmodulin-dependent protein kinase II potentiates transcriptional activation through activating transcription factor 1 but not cAMP response element-binding protein. J Biol Chem 1996; 271:17957-17960.
89. Shors T, Matzel L. Long-term potentiation: What's learning got to do with it? Behav Brain Sci 1997; 20:597-614.
90. Silva A, Paylor R, Wehner J et al. Impaired spatial learning in alpha-calcium-calmodulin kinase II mutant mice. Science 1992; 257:206-211.
91. Soderling T. Structure and regulation of calcium/calmodulin-dependent protein kinases II and IV. Biochim Biophys Acta 1996; 1297:131-138.
92. Soderling T. The Ca^{2+}-calmodulin-dependent protein kinase cascade. Trends Biochem Sci 1999; 24:232-236.
93. Solomonia R, Kiguradze T, McCabe B et al. Neural cell adhesion molecules, CaM kinase II and long-term memory in the chick. Neuro Report 2000; 11:3139-3143.
94. Srinivasan M, Edman C, Schulman H. Alternative splicing introduces a nuclear localization signal that targets multifunctional CaM kinase to the nucleus. J Cell Biol 1994; 126:839-852.
95. Steward O. mRNA localization in neurons: A multipurpose mechanism? Neuron 1997; 18:9-12.
96. Strack S, Colbran R. Autophosphorylation-dependent targeting of calcium/ calmodulin-dependent protein kinase II by the NR2B subunit of the N-methyl- D-aspartate receptor. J Biol Chem 1998; 273:20689-20692.
97. Strack S, Barban M, Wadzinski B et al. Differential inactivation of postsynaptic density-associated and soluble Ca^{2+}/calmodulin-dependent protein kinase II by protein phosphatases 1 and 2A. J Neurochem 1997; 68:2119-2128.
98. Strack S, Robison A, Bass M et al. Association of calcium/calmodulin-dependent kinase II with developmentally regulated splice variants of the postsynaptic density protein densin-180. J Biol Chem 2000; 275:25061-25064.
99. Sweatt J. The neuronal MAP kinase cascade: A biochemical signal integration system subserving synaptic plasticity and memory. J Neurochem 2001; 76:1-10.
100. Szapiro G, Izquierdo L, Alonso M et al. Participation of hippocampal metabotropic glutamate receptors, protein kinase A and mitogen-activated protein kinases in memory retrieval. Neurosci 2000; 99:1-5.

101. Takeuchi Y, Yamamoto H, Miyakawa T et al. Increase of brain-derived neurotrophic factor gene expression in NG108-15 cells by the nuclear isoforms of Ca^{2+}/ calmodulin-dependent protein kinase II. J Neurochem 2000; 74:1913-1922.
102. Tan S, Liang K. Spatial learning alters hippocampal calcium/calmodulin-dependent protein kinase II activity in rats. Brain Res 1996; 711:234-240.
103. Tan S, Liang K. Inhibitory avoidance learning alters the amygdala calcium/calmodulin-dependent protein kinase II activity in rats. Brain Res 1997; 748:227-233.
104. Tan S, Wenthold R, Soderling T. Phosphorylation of AMPA-type glutamate receptors by calcium/ calmodulin-dependent protein kinase II and protein kinase C in cultured hippocampal neurons. J Neurosci 1994; 14:1123-1129.
105. Taubenfeld S, Milekic M, Monti B et al. The consolidation of new but not reactivated memory requires hippocampal C/EBPbeta. Nat Neurosci 2001; 4:813-818.
106. Thiel C, Huston J, Schwarting R. Hippocampal acetylcholine and habituation learning. Neurosci 1998; 85:1253-1262.
107. Tocco G, Maren S, Shors T et al. Long-term potentiation is associated with increased ^{3}H-AMPA binding in rat hippocampus. Brain Res 1992; 573:228-234.
108. Toni N, Buchs P, Nikonenko I et al. LTP promotes formation of multiple spine synapses between a single axon terminal and a dendrite. Nature 1999; 402:421-425.
109. Vianna M, Alonso M, Viola H et al. Role of hippocampal signaling pathways in long-term memory formation of a nonassociative learning task in the rat. Learn Mem 2000; 7:333-340.
110. Walikonis R, Oguni A, Khorosheva E et al. Densin-180 forms a ternary complex with the (alpha)-subunit of Ca^{2+}/calmodulin-dependent protein kinase II and α-actinin. J Neurosci 2001; 21:423-4333.
111. Wang J, Kelly P. Postsynaptic injection of Ca^{2+}/calmodulin induces synaptic potentiation requiring CaMKII and PKC activity. Neuron 1995; 15:443-452.
112. Wayman G, Tokumitsu H, Soderling T. Inhibitory cross-talk by cAMP kinase on the calmodulin-dependent protein kinase cascade. J Biol Chem 1997; 272:16073-16076.
113. Wayman G, Wei J, Wong S et al. Regulation of type I adenylyl cyclase by calmodulin kinase IV in vivo. Mol Cell Biol 1996; 16:6075-6082.
114. Wei J, Wayman G, Storm D. Phosphorylation and inhibition of type III adenylyl cyclase by calmodulin-dependent protein kinase II in vivo. J Biol Chem 1996; 271:24231-24235.
115. Wolfman C, Fin C, Dias M et al. Intrahippocampal or intraamygdala infusion of KN-62, a specific inhibitor of calcium/calmodulin-dependent protein kinase II, causes retrograde amnesia in the rat. Behav Neural Biol 1994; 61:203-205.
116. Wu G, Deisseroth K, Tsien R. Activity-dependent CREB phosphorylation: Convergence of a fast, sensitive calmodulin kinase pathway and a slow, less sensitive mitogen-activated protein kinase pathway. Proc Natl Acad Sci USA 2001; 98:2808-2813.
117. Wu X, McMurray C. Calmodulin kinase II attenuation of gene transcription by preventing cAMP response element-binding protein (CREB) dimerization and binding to the CREB-binding protein. J Biol Chem 2001; 276:1735-1741.
118. Yao H, York R, Misra-Press A et al. The cyclic adenosine monophosphate-dependent protein kinase (PKA) is required for the sustained activation of mitogen-activated kinases and gene expression by nerve growth factor. J Biol Chem 1998; 273:8240-8247.
119. Yoshimura Y, Aoi C, Yamauchi T. Investigation of protein substrates of Ca^{2+}/calmodulin-dependent protein kinase II translocated to the postsynaptic density. Brain Res Mol Brain Res 2000; 81:118-128.
120. Yoshimura Y, Sogawa Y, Yamauchi T. Protein phosphatase 1 is involved in the dissociation of Ca^{2+}/calmodulin-dependent protein kinase II from postsynaptic densities. FEBS Lett 1999; 446:239-242.
121. Yukawa K, Tanaka T, Tsuji S et al. Expressions of CCAAT/Enhancer-binding proteins beta and delta and their activities are intensified by cAMP signaling as well as Ca^{2+}/calmodulin kinases activation in hippocampal neurons. J Biol Chem 1998; 273:31345-31351.
122. Zhao W, Lawen A, Ng K. Changes in phosphorylation of Ca^{2+}/calmodulin-dependent protein kinase II (CaMKII) in processing of short-term and long-term memories after passive avoidance learning. J Neurosci Res 1999; 55:557-568.
123. Zhu X, McCabe B, Aggleton J et al. Differential activation of the rat hippocampus and perirhinal cortex by novel visual stimuli and a novel environment. Neurosci Lett 1997; 229:141-143.

Chapter 4.6

MAP Kinases

Joel C. Selcher, Edwin J. Weeber and J. David Sweatt

Abstract

In neurons the extracellular-signal-regulated kinases (ERKs) are emerging as important regulators of neuronal function. This chapter will discuss the great progress that has been made in identifying the essential components of the signal transduction pathways that lead to ERK activation and several critical downstream effectors. In addition, we will discuss the experimental basis supporting the concept that the ERK pathway interacts with synaptic mechanisms involved in synaptic plasticity and learning and memory processes. In particular, the necessity of the ERK signal transduction cascade for long-term potentiation in mammalian hippocampus and certain spatial and associative forms of hippocampus-dependent learning. Further discussions will include current studies focused on the role of the ERK signal transduction cascade on neuronal excitability and gene expression, implicating the ERKs as molecular signal integrators and coincidence detectors.

Introduction and Background

Perhaps the most daunting scientific challenge of the new millennium will be deciphering the cellular mechanisms governing cognitive processes in the brain. Although once probably considered as far-fetched an endeavor as say sequencing of the human genome, efforts to understand behavior at the molecular level are beginning to progress. In particular, the cellular mechanisms underlying learning and memory are of considerable interest. We possess the remarkable ability to take in an experience, store it for a seemingly limitless amount of time, and then recall it in a form that is often as vivid and evocative as the initial experience being remembered. Because learned information often results in modifications to measurable behaviors, it is also perhaps the most amenable of the cognitive processes to experimentation in lower animals such as rodents. Although still in a relatively nascent stage, intense biochemical and molecular investigation into how organisms encode and store information is currently underway. In this chapter we will focus on recent efforts to understand the role of the MAP kinase superfamily of signal transduction cascades in mammalian learning and memory.

The ERK/MAPK Cascade

As described in several preceding chapters, recent advances have given us a much more detailed understanding of the signal transduction mechanisms subserving learning in the intact animal and one fact that has become clear is that protein kinases play a critical role in these processes. Most recently, the Mitogen-Activated Protein Kinase (MAPK) superfamily of signaling cascades has achieved some notoreity as a player in learning and memory. The MAPK superfamily includes three subfamilies: the ERK (extracellular signal-regulated kinase) family, the p38 MAPK family, and the jun kinase (JNK) family. Each subcategory of the superfamily has a common motif, a characteristic core cascade of three kinases. The first kinase in each cascade is a so-called MAP kinase kinase kinase (MAPKKK, e.g., Raf-1 and B-Raf in the ERK cascade, see Fig. 1) which activates the second, a MAP kinase kinase (MAPKK, e.g., MEK in

From Messengers to Molecules: Memories Are Made of These, edited by Gernot Riedel and Bettina Platt.

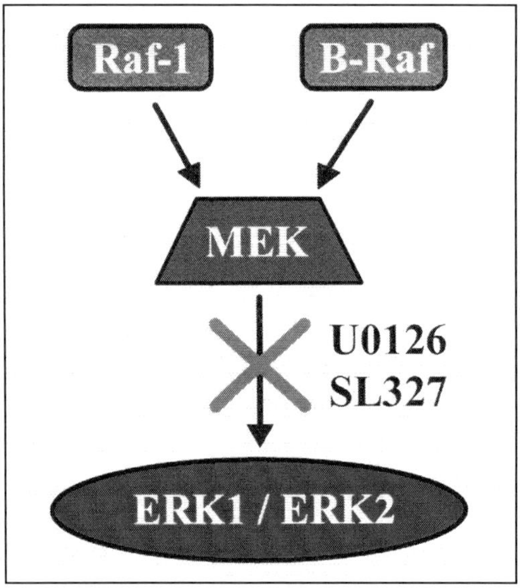

Figure 1. ERK activation and inhibition. Schematic of the ERK MAPK cascade. The classic MAP kinase cascade consists of at least 3 protein kinases beginning with the serine/threonine kinases, Raf-1 and B-Raf, which phosphorylate the MAP kinase kinase MEK. MEK, in turn, is a dedicated, dual-specificity kinase that phosphorylates and thereby activates both the ERK1 and ERK2 isoforms of MAP kinase. These isoforms are the only known effectors of MEK, and MEK is the only known direct activator of the ERK MAPKs. Also depicted are the MEK inhibitors U0126 and SL327. SL327 is capable of crossing the blood-brain barrier and is therefore amenable to behavior studies.

the ERK cascade), by serine/threonine phosphorylation. MAPKKs are dual specificity kinases which in turn activate a MAP kinase (e.g., ERK1, ERK2) by phosphorylating threonine and tyrosine moieties (reviewed in ref. 44).

In addition to serving as a target for growth factor/tyrosine kinases in general, ERK serves as an important point of convergence for the PKC and PKA pathways, both of which have been shown to play critical roles in synaptic plasticity and learning (for review see refs. 1, 2, 19, 102 and also Nogues et al and Vianna and Izquierdo in this book). For example, PKC regulates ERK activity through an interaction with either Ras or Raf-1 leading to activation of MEK and consequently the ERKs. Interestingly, a family of phorbol ester-binding Ras/Rap guanine nucleotide exchange factors (GEFs) was recently discovered that allows the second messenger diacylglycerol (DAG) to achieve ERK activation independent of PKC activation.[35]

In another example of cross talk between two kinase systems, PKA can also couple, both negatively and positively to the ERK cascade. In some cell types, PKA can attenuate ERK activity through inhibition of the Ras/Raf-1 pathway. In an important breakthrough, however, Stork and coworkers discovered that cAMP could also be positively coupled to ERK activation in neurons by signaling through the Ras homologue, Rap1 to activate the B-Raf pathway.[100] Like Raf-1, B-Raf is a serine/threonine kinase that can activate MEK and therefore ERK. In addition, a cAMP-responsive GEF was recently discovered that also leads to ERK activation, independent of PKA.[58]

Simplifying what is a rather complex network of upstream activators is the fact that ERK activity is exclusively regulated by MEK. Dual phosphorylation by MEK has been shown to be both necessary and sufficient for ERK activation. This is a convenient feature of the ERK system as it allows for monitoring of ERK activation using commercially available phospho-specific antibodies recognizing phosphorylation at Thr202 and Tyr204 (see ref. 81 for example). In addition to easing detection of ERK activation, this attribute also has been capitalized upon to create three pharmacologic tools used to investigate the ERK cascade experimentally: the MEK inhibitors PD098059,[5] U0126,[34] and SL327.[8] By inhibiting MEK, these agents effectively block ERK activation and lend themselves well to studies in vitro. Of these inhibitors, SL327 is particularly important for behavioral studies due to its ability to cross the blood-brain barrier and achieve effective concentrations in the CNS when administered systemically.[8]

Targets of the ERK Cascade

We will focus on two prominent targets of ERK, regulation of gene expression and potassium channel function. We focus on these because they are the two neuronal targets for which the most information is available. Other intriguing potential targets include the protein synthesis machinery, synapsin, and the cytoskeleton—see section 8 of this book.

Because the formation of long-term memory is thought to depend on gene expression and protein synthesis,[30,41] an attractive mechanism by which ERK activation could contribute to long-term memory formation involves ERK's well-documented role in gene expression.[88] ERK has been shown to be an important regulator of a number of transcription factors including c-Jun, Elk-1 and CREB. Members of the CREB family respond to increases in intracellular calcium and cAMP by activating the transcription of genes containing an upstream CRE sequence. A wide variety of studies have implicated CREB-dependent transcription in long-term synaptic plasticity and long-term memory in both invertebrate and vertebrate systems (for review, see Frankland and Josselyn in this book). Injection of CRE containing nucleotides into Aplysia cultured neurons selectively blocked long-term, but not short-term, facilitation of neurotransmitter release.[29] Both phosphorylation of CREB and CRE mediated transcription have been shown to increase during the induction of LTP.[32,50] In addition, genetic manipulations of CREB in *Drosophila* and mice lead to deficits in the formation and retention of long-term memory.[20,42,109] It is currently unclear what upstream mechanisms lead to learning-related activation of CREB.

ERK has recently been observed to regulate CREB phosphorylation in the hippocampus, most likely using the protein kinase RSK2 as an intermediary.[50] RSK2, a downstream effector of ERK, phosphorylates CREB at Ser_{133} thereby activating it.[107] In addition, ERK appears to mediate the phosphorylation of CREB by other kinase systems such as PKA[52,81] and PKC[81] in the hippocampus. CREB activation produced by both phorbol esters and forskolin requires activation of ERK, as MEK inhibitors block this activation of CREB in area CA1 of hippocampal slices. Gaining a better understanding of the regulation of gene expression by the ERK cascade should provide important insight into the mechanisms underlying information storage at both the synaptic and behavioral levels.

Potassium Channel Modulation

As described by Vernon and Giese in this book, potassium (K^+) channels serve as the principal regulators of membrane excitability in the central nervous system. These channels flux K^+ currents that can influence the likelihood of spike generation in a variety of ways. Certain K^+ currents, such as the transient A-type and inward rectifier currents, contribute to the establishment of the resting membrane potential (V_m). Transient K^+ currents also act to raise the threshold for action potential initiation from dendritic depolarizations. Many other K^+ currents participate in membrane repolarization following the firing of an action potential, while others control the frequency at which repetitive firing occurs.[54] Alterations in the properties of these channels (e.g., voltage-dependence, activation or inactivation profiles, distribution) have profound effects on membrane excitability. For example, depolarizing V_m or slowing the rate of repolarization could lead to generation of one or multiple spikes in response to stimulation that normally would not produce an action potential.

Potassium channels can also regulate membrane excitability via their influence over the size and shape of action potentials as they back-propagate into the dendrites. At one time, dendrites were seen as little more than cables passively carrying excitatory post-synaptic potentials (EPSPs) from the synapse to the soma. In this classic view, the influence of an individual EPSP was dependent on its location within the neuron, as signals from more distant synapses were attenuated to a greater extent than were proximal inputs. EPSPs originating in various regions of the dendritic tree were "summed" in the axon hillock near the soma, and if the total depolarization reached a critical threshold level, an action potential was generated. Action potentials then propagated forward (orthodromically) along axons toward connections with postsynaptic neurons. However, the concept of a passive role for dendrites in synaptic integration has recently been discredited with the discovery of voltage-gated ion channels in the dendrites and the demonstration of action potentials spreading backward (antidromically) into the dendritic tree. These back-propagating action potentials are important feedback mechanisms as they allow communication between the soma and an activated synapse in the dendrites.

Back-propagating action potentials appear substantially different when recorded in the dendrites compared to those recorded in the soma. For one, action potentials are significantly wider with a much slower rate of repolarization in distal dendrites than in the cell body. The amplitude of back-propagating action potentials has also been found to decrease progressively with distance from the soma. These two characteristics of dendritic depolarizations stem from differences in the electrical properties of dendrites based on nonuniform distributions of two types of potassium channels: Ca^{2+}-gated K^+ channels and voltage-gated, transient K^+ channels.[47,79]

The decreasing amplitude of action potentials as they back-propagate into the dendrites results from an increasing density of transient A-type K^+ channels with distance from the soma. These channels activate at potentials near the resting membrane potential and inactivate rapidly. These biophysical properties allow K^+ channels to function in the dampening of dendritic depolarizations. The high density of these channels in the distal dendrites prevents the initiation of dendritic action potentials, reduces the amplitude of back-propagating action potentials and restricts the size of EPSPs.[47]

A-type K^+ channels can be modulated in a number of ways. Due to their rapid inactivation, subthreshold depolarization following synaptic activity decreases the current subserved by these channels. EPSPs with appropriate timing and amplitude can release the dampening effect of these K^+ channels and amplify backpropagating action potentials. With the proper timing, this pairing can also induce robust LTP, suggesting a role for these postsynaptic action potentials in Hebbian forms of synaptic plasticity.[64] Similar to synaptic depolarization, protein kinase cascades have also been shown to modulate transient K^+ currents presumably through phosphorylation of K^+ channels.[46] In the dendrites of CA1 pyramidal neurons, activators of both PKA and PKC have been shown to decrease the voltage-dependent activation of these currents leading to amplification of back-propagating action potentials.

Kv4.2 As an Effector for ERK

Recent research suggests that ERK modifies transient A-type potassium currents. Johnston and colleagues have demonstrated that MEK inhibitor application leads to downregulation of A-type K^+ currents in CA1 hippocampal dendrites. Both PD098059 and U0126 produce a hyperpolarizing shift in the activation curve for this current and block the enhancement in back-propagating action potential amplitude observed in response to activators of PKA and PKC (Yuan, et al, submitted). These findings suggest a role for ERK in the modulation of dendritic A-type potassium currents.

While the identity of the K^+ channel subunits responsible for this transient A-type K^+ current is currently unknown, evidence points to the *Shal*-type Kv4.2 channels as the leading candidate to mediate these currents. One of two transient K^+ channels found in the hippocampus, Kv4.2 is abundantly expressed in somatic and dendritic regions of hippocampal pyramidal neurons in area CA1.[89] At the ultrastructural level, these channels are predominately localized on the postsynaptic membrane associated with presynaptic terminals.[6]

Examination of the Kv4.2 amino acid sequence has revealed consensus sites on the C-terminal domain suitable for phosphorylation by ERK. Using phospho-specific antibodies generated against the channel at putative ERK phosphorylation sites, Kv4.2 appears to be an excellent substrate for ERK in the hippocampus.[4] Using this antibody, immunohistochemical studies have exposed a fascinating pattern of input-specific labeling in the hippocampus, with high levels of staining in stratum radiatum and sparse labeling in stratum pyramidale.[99] In addition, recent studies have demonstrated ERK-dependent phosphorylation of Kv4.2 in hippocampal area CA1 following activation of ERK by PKA, PKC and β-adrenergic receptor stimulation (Adams et al,

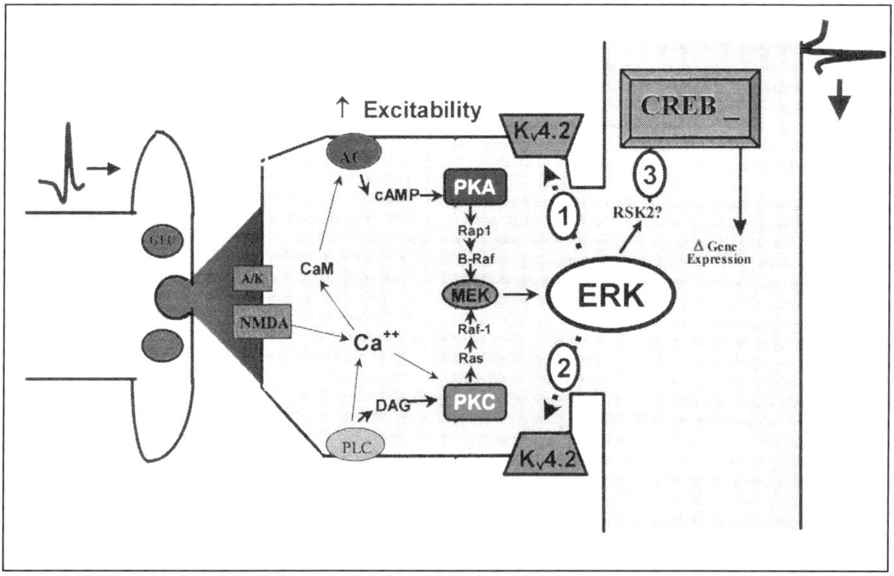

Figure 2. Working model. This model, as discussed in the text, highlights three potential sites of action for ERK in LTP and learning and memory. To summarize:

- ERK may phosphorylate Kv4.2, thereby downregulating K^+ currents and essentially amplifying individual EPSPs. The larger EPSPs signaling to the soma could cause action potential generation.
- Action potentials initiated in the axon back-propagate into the dendrites. Again, ERK phosphorylation of Kv4.2 could relieve the dampening effect that K^+ currents have on these backpropagating action potentials. A larger depolarization in the dendrite resulting from the larger action potential could remove the Mg^{++} block from NMDA receptor allowing for calcium influx.
- The calcium entry via NMDA receptor-mediated channels could activate ERK via one of the pathways shown above. A possible third site of action could be phosphorylation of the transcription factor CREB leading to changes in gene expression.

unpublished observations). Taken together, these studies identify phosphorylation of Kv4.2 by ERK as an attractive potential mechanism underlying modulation of A-type K+ currents, and by this mechanism regulate the triggering of synaptic plasticity and memory formation (Fig. 2).

Hippocampal Involvement in Learning

In the early 1950s, numerous patients with intractable epileptic seizures underwent experimental surgeries involving removal of one or both temporal lobes. The surgeries appeared at first glance to be successful, as removal had the desired beneficial effect of ameliorating the seizures. However, in patients who had unknown temporal lesions in the opposite brain hemisphere or who underwent bilateral temporal ablations, the surgery had tragic results. One such patient, H.M., displayed severe anterograde amnesia following his surgery.[87] Although memories formed prior to the surgery were relatively intact, H.M. seemed incapable of forming new long-term memories. For example, if given a three-digit number to memorize, H.M. could remember the number as long as he was able to continuously rehearse. If distracted even momentarily, H.M. would lose not only the memory of the number but also the memory of having been given the task in the first place. Thus, for H.M., the impairment was somewhere in the transition from short-term memory into long-term memory. This suggested to researchers that the hippocampus was a critical structure in the formation of new memories.

Results from the extensive testing of amnesic patients like H.M. and from animal lesion studies have contributed greatly to our understanding of the neuroanatomical distribution of memory systems.[36] Having identified the hippocampus as a brain region likely to be involved in the formation of long-term memories, the next major task was in determining how plastic changes in the hippocampus are related to behavioral changes and what mechanisms underlie these plastic changes. This endeavor seemingly required a model system in the mammalian brain analogous to the reductionist approach used in early studies in the marine mollusc *Aplysia*. Hippocampal long-term potentiation has been proposed as one such model.

Long-Term Potentiation

Hebbian theory stated that the strength of a synaptic input would increase if it repeatedly caused the postsynaptic neuron to fire action potentials. By 1973, Hebb's theory proved prophetic with experimental evidence demonstrating activity-dependent changes in synaptic strength. In that year, Timothy Bliss and Terje Lømo showed that repetitive high-frequency activation of the perforant path evoked a long-lasting enhancement of synaptic strength in the dentate gyrus of the rabbit hippocampus.[16] The discovery of this phenomenon, known since as long-term potentiation (LTP), captivated learning researchers and has led to generations of intense study. It has since been found to exist in in vitro preparations of hippocampus slices and in many other brain regions. A complementary phenomenon known as long-term depression (LTD) has also been identified in some of these synaptic pathways.[53,97] In LTD, low-frequency stimulation yields long-lasting decreases of synaptic strength.

Due to the fact that information storage is believed to occur at the synaptic level, alterations in synaptic strength have garnered considerable attention in the field of learning and memory. LTP shares many attributes with behavioral learning and has been proposed as a memory mechanism due to the similarity of many of these characteristics including its durability, specificity, cooperativity and associativity. Like learning, LTP can be defined as a long-lasting change in output in response to a transient input. The persistence of this effect has been demonstrated to extend many hours in vitro and several weeks in vivo. Forms of LTP are specific in that the change in efficacy is restricted to tetanized pathways; even inputs convergent on the same dendritic region of neurons are not potentiated.[7] In general, LTP requires cooperative interaction of afferent fibers, which in essence means there is an intensity threshold. This threshold allows strong inputs to induce LTP, whereas weaker inputs cannot.[70] Through LTP's associative property, however, a weak input can induce potentiation, provided a strong convergent input is activated at the same time.[13] This particular feature of LTP has stimulated a great deal of interest as it has been likened to classical conditioning with weak and strong inputs corresponding to conditioned and unconditioned stimuli, respectively. LTP has been demonstrated at synapses throughout the nervous system but is most heavily studied in the limbic system, primarily the hippocampus.

Deciphering the specific molecular mechanisms underlying LTP is a rather daunting and complicated task. At this point, a comprehensive analysis of every molecule implicated in LTP has become a near impossibility due to the sheer volume of studies. With the wide variety of agents that have been found to modulate LTP, it is difficult to attempt to put them all in one model. Another complication has been the lack of consistency in findings from one lab to another. For these reasons, we will outline just a few of the more pertinent concepts underlying LTP concentrating on those that are most widely accepted.

The primary triggering mechanism of LTP involves a rise in postsynaptic calcium. Preventing the rise in Ca^{2+} with postsynaptic injection of Ca^{2+} chelators blocks the induction of LTP in hippocampal neurons,[63] while release of Ca^{2+} sequestered by caging compounds leads to enhanced synaptic strength that mimics LTP.[65] This rise in Ca^{2+} can be accomplished a number of ways, but the predominant mechanism is influx of extracellular Ca^{2+} through NMDA receptors.[26] Other mechanisms for raising postsynaptic calcium include influx through voltage-sensitive Ca^{2+} channels and release of Ca^{2+} from intracellular stores.

Once triggered, two different types of LTP mechanisms are employed and must be distinguished. Induction mechanisms have been defined as transient biochemical events that serve to initiate the formation of LTP. These processes engender the persisting changes, or expression mechanisms, that directly support synaptic potentiation.[80,96] To date, most LTP studies have focused on induction mechanisms at the expense of expression mechanisms.

Another theme common to many discussions of LTP involves determining the locus of the alterations that sustain enhanced synaptic strength. With the triggering signal firmly established as a rise in postsynaptic Ca^{2+}, it is not surprising that induction mechanisms are also believed to reside in the postsynaptic cell. Calcium-dependent protein kinases are currently the leading candidate molecules to subserve these induction processes. However, what maintains this enhanced efficacy: presynaptic changes in neurotransmitter release or postsynaptic changes in receptor responsivity? Many scientific wars regarding the locus of the expression mechanism have been waged across the synaptic cleft. Most of the data from LTP in hippocampal area CA1 appear to favor enhanced AMPA receptor-mediated postsynaptic currents.[57,61] If, however, the expression of LTP is wholly or even partially due to presynaptic events, the release of a retrograde messenger is necessary to account for the jump from a postsynaptic induction site to a presynaptic expression site. Suggested retrograde signal molecules include nitric oxide,[77] carbon monoxide,[95] arachadonic acid,[103] and platelet-activating factor.[33]

LTP As a Model for Learning and Memory

Countless hours and dollars of scientific resources have been spent in an exhaustive effort to better understand the considerable complexities of synaptic plasticity, yet one overriding and critical question remains: does LTP equal memory?[94] In other words, does LTP have any natural physiological relevance to behavior? Conclusive answers to these questions are most certainly beyond the reach of this chapter, but we do want to evaluate some of the currently available data (reviewed in refs. 10, 48, 68, 90, and 94).

Most of the data linking activity-dependent alterations in synaptic strength (namely, LTP and LTD) and experience-induced behavioral modifications come courtesy of correlational studies. In one of the first examples of these studies, aging rats displayed impairment in the rate of acquisition of a spatial memory task involving navigation around a circular platform task. The decline in behavioral performance was shown to correlate with statistical significance to a corresponding decline in the persistence of LTP in these rats.[9,11] Saturation studies have also provided a link between LTP and learning. The premise of these experiments was as follows: if synaptic potentiation was necessary for the formation of new memories, then eliminating the capacity for further potentiation by inducing saturating levels of LTP should impair learning. For example, Castro et al[23] tetanized the perforant pathway for over 30 days producing persistent LTP in the dentate gyrus and then trained animals in the Morris water maze task. The persistently tetanized animals showed spatial learning impairments[23] that disappeared with decay of the dentate gyrus LTP. More recent studies, however, have cast doubt on these observations.[12]

Since these initial experiments, another form of correlational study has become increasingly prevalent. In these experiments, manipulations that block the induction of hippocampal LTP are shown to produce commensurate effects on a particular learning paradigm. For example, many studies have been undertaken to assess the effect of pharmacological blockade of NMDA receptors on LTP and hippocampus-dependent learning. LTP induced at a number of synapses in the hippocampus has been shown to depend on activation of the NMDA receptor.[26,45] Following this observation, numerous laboratories have demonstrated impairments in a variety of hippocampus-dependent learning tasks including spatial learning, T-maze alternation, and contextual fear conditioning.[39,68,75] To provide a more specific example, Morris[75] impaired spatial learning in the water maze with intra-hippocampal infusions of AP5 at a dose sufficient to block LTP.

Results from these experiments were not received without skepticism, as standard caveats of drug experiments apply, especially with regards to the behavioral experiments. For one, the locus of the drug's effect comes into question. With intraperitoneal, intraventricular and even more direct infusions into the hippocampus, drug diffusion to other brain areas remained a possibility. If this was the case in the NMDA antagonist studies, for instance, the performance deficits seen in the behavioral tests may represent nonspecific effects on NMDA receptor-dependent sensory or motor processes rather than learning effects. Although baseline hippocampal synaptic transmission is independent of the NMDA receptor, NMDA antagonists show effects on normal transmission in other brain regions, such as the thalamus.[68] An animal that cannot process sensory information normally is unlikely to be able to store it properly.

The introduction of gene-targeting techniques has provided a means to avoid these specificity concerns (at least at one level) and has supplied researchers with a new format for inhibitor studies comparing mechanisms underlying LTP and learning. In the first of this new breed of study, Silva et al[91,92] generated mice lacking the gene for the alpha subunit of calcium/calmodulin-dependent protein kinase II (CaMKII). Upon electrophysiological and behavioral characterization, these mice demonstrated impaired hippocampal LTP[92] correlated with substantial spatial learning deficits.[91] In the behavior experiments, the mice lacking α-CaMKII exhibited slower escape latencies and poor performance in probe trials in the hidden platform of the Morris water maze.

These genetic techniques, however, were also burdened by an inherent caveat. The absence of a gene product during development could theoretically produce nonspecific effects. In fact, some genes proved so important developmentally that their ablation prevented the survival of mutant mice. Standard knockout of the R1 subunit of the NMDA receptor resulted in lethality soon after birth,[24] thus seriously limiting its utility. Other genetic manipulations produce anatomical defects that can also complicate interpretation of these knockout studies. For example, the deletion of the tyrosine kinase fyn in a study by Grant et al[43] resulted in impairments in LTP and learning. However, it was later shown that this genetic manipulation interferes with normal myelination and resulted in gross anatomic abnormalities in the hippocampus.[108] Additionally, mutant mice often have altered behavior independent of learning. As with nonspecific drug effects, aberrant behavior produced by the absence of a specific gene can often confound processes related to the performance of the measured response in behavioral learning tasks, thereby complicating interpretation of results. As mentioned above, α-CaMKII knockout mice displayed slower escape latencies in the Morris water maze compared to wildtype littermates.[91] One puzzling aspect of this deficit was the fact that it existed during the first training trials before any learning could have occurred in either set of mice. This finding was suggestive of a nonspecific effect on sensory or motor abilities as opposed to a definite impairment in spatial learning abilities.

In response to these criticisms, gene-targeting approaches have now been fine-tuned to the point that expression patterns of mutated genes can be confined both temporally and regionally. The CreloxP technique produces gene knockouts restricted to specific brain subregions. This regional specificity is precise enough to essentially limit expression to area CA1 of the hippocampus.[98] A further refinement of this technology allows for temporal control on transgene expression and thus minimizes developmental concerns.[66] For example, Mansuy et al[66] demonstrated that by combining the tetracycline-controlled activator (rtTA) system with the α-CaMKII promotor, expression of an active form of the calcium-dependent phosphatase calcineurin could be induced specifically in forebrain structures by administration of doxycycline. Mice overexpressing calcineurin showed impaired LTP and spatial learning. One additional advantage of this system is that it allows for the dissociation of a protein's role in distinct phases of learning (acquisition, consolidation, retrieval) based on the timing of the transgene induction.

To this point, we have focused on positive correlations between LTP and learning. However, a number of studies have failed to show this association. Targeted deletion of the AMPA receptor subunit GluR-A resulted in impairment of LTP in area CA1 of the hippocampus despite normal glutamatergic synaptic transmission.[110] Despite the lack of hippocampal LTP in these mice, they learned normally in the Morris water maze task. Mice generated lacking the γ-isoform of PKC showed similar results, although later experiments found that normal LTP could strangely be rescued in these mice by preceding LTP-inducing tetanization with a bout of low-frequency simulation.[2,3] In another example of a dissociation of LTP and learning, Meiri et al[71] utilized a unique technique for investigating the role of the presynaptic A-type potassium channel Kv1.4 in these phenomena. Intraventricular microinjection of antisense oligodeoxyribonucleotides (ODN) to the Kv1.4 gene resulted in reductions in Kv1.4 mRNA and protein levels, eliminated LTP yet had no effect on spatial learning in the rat.[71] In a third example, mice transgenic for mutant PSD-95, a protein purported to function in NMDA receptor localization, displayed remarkably heightened levels of LTP with a range of stimulation parameters.[73] This enhanced LTP was surprisingly accompanied by impaired performance in the Morris water maze. Compared to the prior examples, however, this result was less disheartening to proponents of the "LTP = learning" hypothesis due to the lack of normal bi-directional control of synaptic plasticity in the PSD-95 mutant mice that might lead to the observed learning impairment.

So, the results are certainly ambiguous, but thus far, we have concentrated on studies involving the hippocampus. However, defining the relationship between LTP and learning will be very difficult to do in the hippocampus, based on the complexity of the behaviors mediated by this structure. Resolution of this debate may require work in brain regions where the circuitry involved in a learning task is better understood, such as in the amygdala for fear conditioning or the cerebellum for eyeblink conditioning.

Indeed, recent work on auditory fear conditioning in the amygdala provides perhaps the most compelling data to date arguing for the use of LTP-like mechanisms during learning. In fear conditioning, a neutral auditory conditioned stimulus (CS) is paired with an aversive footshock (unconditioned stimulus; US). Information from the CS and the US are believed to converge in the lateral amygdala. Auditory information that represents the CS is carried from the thalamus to the amygdala and LTP has been demonstrated at the synapse between the auditory thalamus and the lateral amygdala (LA). Interestingly, artificial induction of LTP at this synapse leads to enhanced responses in the lateral amygdala to natural auditory stimuli.[82] This demonstrates that the amygdala can utilize the mechanism of LTP in the processing of auditory stimuli. Enhanced field potentials in response to the CS in the lateral amygdala have also been observed following fear conditioning.[83] This potentiation only accompanied learning; unpaired presentation of the CS and US did not lead to learning nor did it result in the thalamo-LA LTP. In a complementary finding, synaptic responses of LA neurons were consistently enhanced in in vitro slices prepared from fear conditioned compared to control rats.[69] These studies are very exciting as they suggest that fear conditioning induces a form of LTP, lending credibility to the hypothesis that LTP underlies learning.

ERK in Hippocampal Synaptic Plasticity

Considerable evidence already exists implicating ERK in various forms of synaptic plasticity in a wide variety of systems. Long-term facilitation of the sensory-motor neuron synapse in *Aplysia* causes translocation of ERK into the nucleus of presynaptic neurons. Furthermore, inhibition of ERK by either anti-MAPK antibodies or PD098059 blocks long-term, but not short-term, facilitation of the sensory-motor synapse.[67,72] ERK has also been shown to be activated during an in vitro Pavlovian conditioning paradigm in *Hermissenda*, and this activation is blocked by pretreatment with PD098059.[28] Activation of the ERK isoforms of MAPK have been demonstrated to be necessary for the induction of NMDA receptor-dependent LTP in area CA1 of the rat hippocampus (see Fig. 3),[8,37,38,51,52,104,105] NMDA receptor-independent LTP in area CA1,[56] LTP in the dentate gyrus,[27] LTP in vivo,[31,84] and LTP of the amygdalar inputs into the insular cortex.[55] In addition, long-term depression (LTD) in the rat hippocampus is associated with long-lasting decreases in ERK immunoreactivity.[76]

Surprisingly, inhibiting ERK activation produced differential effects, depending on the species, on hippocampal long-term potentiation induced with a high-frequency stimulation (HFS) paradigm consisting of two trains of 1-sec, 100-Hz tetani (Fig. 3). In vehicle-treated hippoc-

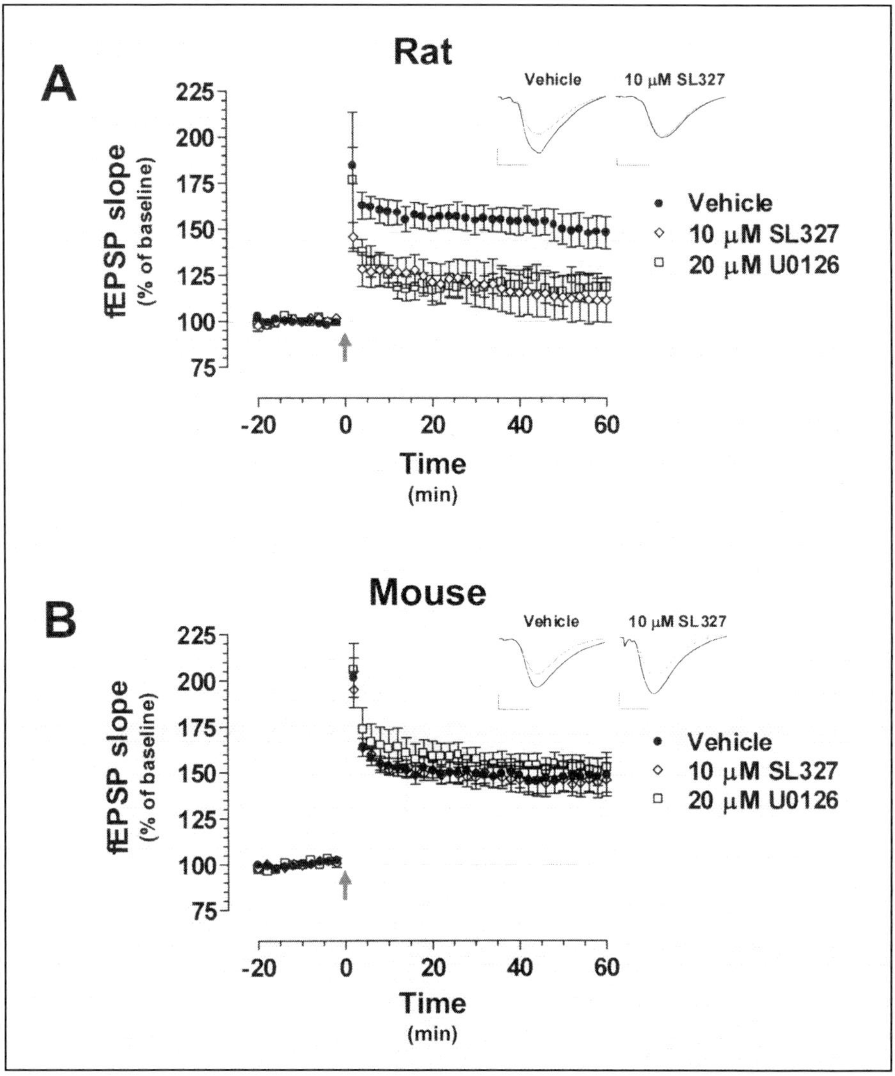

Figure 3. HFS-LTP requires ERK activation in rat but not mouse hippocampus. A) LTP induced with a pair of 100-Hz tetani in rat hippocampal slices in the presence of either vehicle (0.1% DMSO; n = 10 slices), 10 μM SL327 (n = 7), or 20 μM U0126 (n = 5). *Inset*, representative traces from vehicle- and SL327-treated rat slices before (gray) and after (black) tetanic stimulation. Scale bars are 0.5 mV by 8 msec. B) Mouse hippocampal slices exposed to the same LTP-induction paradign in the presence of either vehicle (n = 19), 10 μM SL327 (n = 11), or 20 μM U0126 (n = 10). *Inset*, representative traces from vehicle- and SL327-treated mouse slices before (gray) and after (black) tetanization. Scale bars are 1 mV by 8 msec.

ampal slices from rats, HFS resulted in a significant and stable increase in the initial slope of the EPSP. Consistent with results obtained previously in a number of studies, this enhancement in synaptic strength was blocked by application of either 10 μM SL327 or 20 μM U0126.[8,38,105] In the mouse hippocampus, however, MEK inhibitors had no effect on LTP induced with this high-frequency tetanic stimulation. These findings suggest that LTP induced with a pair of 100-Hz tetani requires ERK activation in rats but not in mice. This would then suggest the novel idea that these two closely related species do not employ identical molecular mechanisms for the induction of this particular form of synaptic plasticity.

Mice are notoriously variable depending on their particular genetic background strain. To ensure that our observation is due to a species difference as opposed to a quirk of a particular mouse strain, the LTP experiments were conducted with hippocampi from three different mouse strains: the widely used C57BL/6J strain, another inbred strain 129S1/SvImJ, and an outbred strain CD-1. The CD-1 mice serve as a simple control for their outbred counterpart, the Sprague-Dawley rat. MEK inhibitors failed to impair LTP in hippocampal slices from all three of these mouse strains, suggesting that this is not merely a strain or inbreeding effect (data not shown).

Theta-Like LTP Induction Paradigms Require ERK Activation

The behavioral relevance of traditional high-frequency stimulation paradigms such as the one described above (2 trains of 1-sec, 100-Hz tetanic stimulation) is highly questionable considering this amount of activity would rarely if ever occur in a normal brain. However, LTP in the hippocampus can also be induced with naturalistic patterns of stimulation that emulate the firing pattern of hippocampal pyramidal neurons. This natural brain rhythm called the theta rhythm occurs as a rhythmic oscillation in hippocampal activity in the 4- to 12-Hz range when exploring the space of a novel environment.[15]

Short bursts of 100-Hz stimulation delivered at 200 msec intervals (to fall within the 4-12 Hz theta range) can very effectively induce LTP lasting weeks in vivo.[59,93] In anesthetized rats, LTP is induced preferentially with short trains of pulses applied at the peak of theta rhythm.[15] Pulses applied at the trough of the theta rhythm had no effect or produce depression. In CA1 hippocampal slices, the theta rhythm can be elicited by pharmacologic stimulation of cholinergic receptors. As in the whole animal, stimulus trains timed to the peak but not the trough of this sinusoidal rhythm resulted in reliable LTP induction.[49]

The effects of MEK inhibition on LTP induction paradigms that mimick this theta rhythm have been tested. Unlike the results obtained with LTP induced by high-frequency tetani, patterned stimulation in the theta frequency required ERK activation in the mouse. MEK inhibitor-treated slices showed impaired (but not completely blocked) LTP elicited by "theta frequency stimulation" consisting of stimulation at 5-Hz for 30 sec.[101,104]

A Necessity for ERK Activation for Mammalian Learning

As described in section two of this book, two paradigms that have been used extensively in studying learning are the conditioned fear task[60] and the Morris water task.[74] In fear conditioning, animals learn to associate neutral stimuli with a foot shock. In the spatial learning version of the Morris water task, animals learn to utilize various distal visual cues to navigate through a pool of water to locate a hidden escape platform. Lesions of the hippocampus has previously been shown to impair learning in variants of both of these tasks in mice.[62] Evidence from studies of hippocampal synaptic plasticity and of the behaving animal suggests that activation of protein kinases may contribute to the formation of these types of memories, however, the underlying signal transduction mechanisms remain largely unknown.[1,2,3,8,85,91,102]

In early studies, we found that the ERK MAPK cascade is required for fear conditioning in the rat,[8] as has also been observed subsequently by Schafe et al.[85] In the study by Atkins et al,[8] hippocampal ERK activation increased 1 hour after training with a fear conditioning protocol. This increased ERK activation was prevented by injection of the NMDA antagonist MK801, a drug that blocks both LTP and fear conditioning.[18,25] The NMDA receptor-mediated activation of ERK was required for learning, as inhibiting ERK activation by intraperitoneal administration of SL327 blocked both contextual and cue learning in the rat.

Due to the emergence of mice as the standard genetic and behavioral model, we examined the role of ERK in fear learning in mice (Fig. 4). In these studies, SL327 injected systemically crossed the blood-brain barrier at concentrations sufficient to inhibit basal levels of ERK acti-

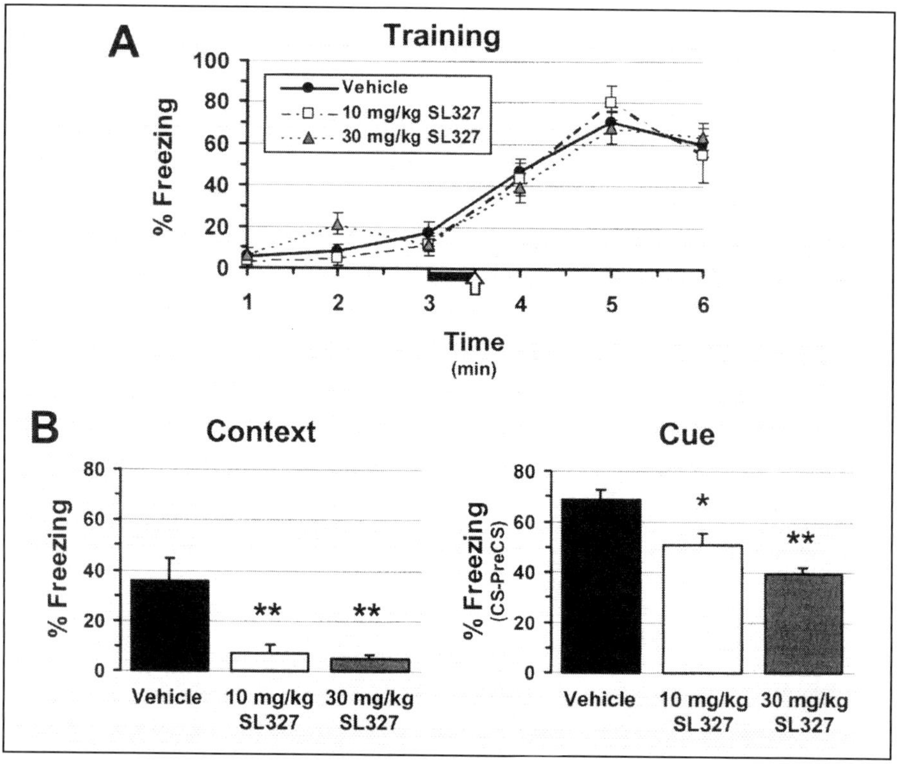

Figure 4. SL327 blocks contextual fear conditioning and attenuates cue learning following 1 CS-US pairing. A) Freezing responses during the training phase are shown. A tone (solid bar) was paired with a foot shock (⇧) between minutes 3 and 4. Baseline behavior (before presentation of the tone) and shock response (after the foot shock) were similar for all groups. B) Mice given SL327 (10 mg/kg, n = 5 or 30 mg/kg, n = 7) demonstrated significant reductions in freezing to the context 24 hours after receiving one pairing of tone and shock as compared to animals injected with vehicle (n = 6), *left*. Administration of 10 or 30 mg/kg SL327 was also sufficient to significantly attenuate freezing to the cue (CS - PreCS), *right*. There was no difference between the PreCS values for any of the groups. **represents $p < 0.01$, *represents $p < 0.05$.

vation, and could effectively eliminate contextual learning in mice as it did in rats.[8] SL327 administration also significantly attenuated cue learning, although these drug-treated mice did display considerable freezing in response to the white noise following its pairing with a foot shock. More intense training paradigms consisting of more than a single CS-US pairing rescued the deficit observed in cue learning.

Of course, a number of other laboratories also investigated the ERK cascade in associative learning in a wide variety of systems. ERK is activated in response to a novel taste, and infusion of PD098059 into the insular cortex of rats attenuates conditioned taste aversion.[14] Genetic manipulations of upstream members of the cascade that leads to ERK MAPK activation also engender learning impairments. For example, the *Drosophila* mutant *leonardo* which lacks 14-3-3, a protein important in the activation of Raf-1 (MAPKKK) by Ras, shows learning-related impairments.[22] In addition, mice lacking Ras-GRF, a guanine-nucleotide exchange factor that induces Ras activation, display impairments in cue fear conditioning and abnormal amygdalar LTP.[21] However, these mice show normal hippocampal function and no significant deficits in spatial learning.

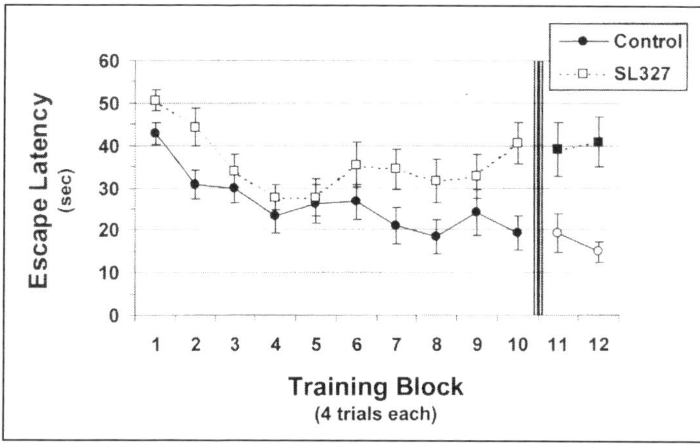

Figure 5. SL327 impairs performance during training on the hidden platform version of the Morris water maze. Average escape latency during training on the hidden platform task. Performance for mice injected with vehicle (n = 13) improved over the course of the training. Mice treated with 30 mg/kg SL327 (n = 11) took significantly longer to locate the escape platform. The dark vertical line between blocks 10 and 11 represents the drug switch on day 6. On that day, SL327-trained mice (■) received DMSO (n = 9), while vehicle-trained animals (O) received SL327 (n = 11).

Spatial Learning Requires ERK

Administration of SL327 to mice during training in the Morris water maze produced significant spatial learning deficits. SL327-treated mice took significantly longer to locate the escape platform during training compared to vehicle-treated controls (Fig. 5). Furthermore, vehicle-treated control mice performed significantly better than drug-treated animals in probe trials conducted following training suggesting that SL327 impairs spatial learning performance (Figs. 6 and 7). These results demonstrate a necessity for ERK activation in spatial learning.

These findings are consistent with those reported by Dash and colleagues[29] who described an increase in ERK phosphorylation in pyramidal cells of the CA1/CA2 subfield of the dorsal hippocampus in response to behavioral training in the Morris water maze.[17] This group also demonstrated a necessity for ERK activation for spatial learning in rats, although they only saw behavioral impairments on the first retention trial 48 hours after training following PD098059 infusions into the hippocampus. These differences seen in the two studies could be due to different training or testing paradigms, species of test subjects, or differences in the efficacy and time course of MEK inhibition by SL327 versus PD098059. One particularly intriguing possibility to explain the different observations is the locus of the drug effect. While Dash and colleagues[29] selectively infused a MEK inhibitor into the hippocampus, our intraperitoneal administration of SL327 inhibits MEK throughout the central nervous system. Thus, the observed differences might simply be accounted for by involvement of ERK activation in extra-hippocampal areas during spatial learning.

When administered post-training, SL327 produced no impairments in water maze performance, demonstrating that inhibition of ERK activation had virtually no effect on the performance of animals that had already learned the water maze task (Figs. 5 and 8). This suggests that ERK activation is required for the formation of memory, but is unnecessary for ongoing maintenance of memory. These results are consistent with the LTP studies involving ERK performed by English et al,[38] in which application of PD098059 before tetanization was shown to block the induction of LTP whereas PD098059 application 30 min after tetanization had no effect on the expression of established LTP.

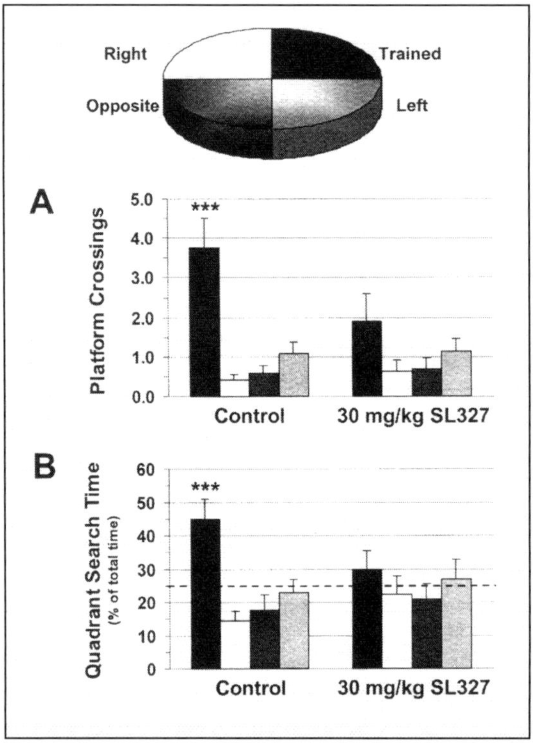

Figure 6. Administration of SL327 significantly impaired spatial learning in the Morris water maze. A) During the probe trials on days 4 and 5, mice treated with vehicle crossed the area where the platform had been in the trained quadrant significantly more frequently than a corresponding area in the alternate quadrants. SL327-treated mice did not exhibit the same selectivity in their search. B) Control mice spent significantly more time searching in the trained quadrant than in any of the alternate quadrants. However, mice injected with 30 mg/kg SL327 did not spend more time searching in the trained quadrant. The dotted line represents chance (25% of the time in each quadrant). ***significantly larger ($p < 0.01$) than all three of the other quadrants.

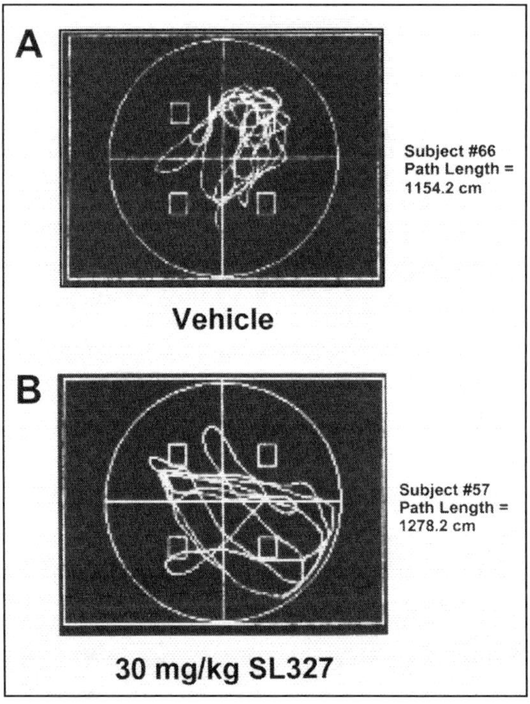

Figure 7. Mice treated with SL327 failed to employ a spatial search strategy during probe trials. A) Representative probe trial of a vehicle-treated mouse. The swim path trace shown here provides an excellent example of a selective search. This particular subject was trained with the platform located in the northeast quadrant. During the probe trial, this mouse spent 56% of the time in the correct quadrant and crosses the exact area where the platform had been 9 times. B) Representative probe trial of an SL327-treated mouse. This trace does not represent a selective search. This mouse was trained with the platform in the northwest quadrant, but during the probe trial, the subject crossed this patform area only once and spent 32% of the time in this quadrant (versus 34% in the opposite quadrant).

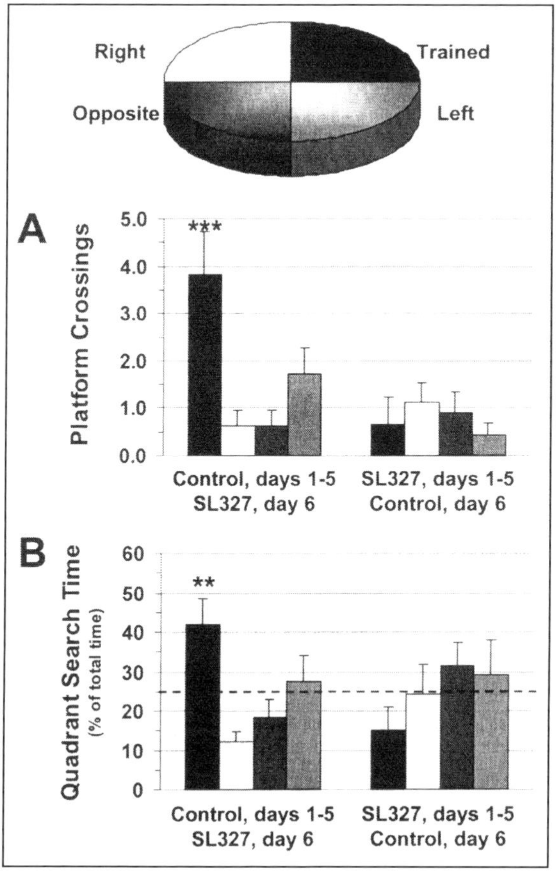

Figure 8. SL327 had little effect on mice that had previously learned the Morris water maze task. After administration of SL327 on day 6 (drug switch), the vehicle-trained mice still showed significantly more platform crossings (A) and spent more time searching (B) in the trained quadrant compared to the other three quadrants. SL327-trained mice who were administered vehicle on day 6 did not display a selective search strategy. ***significantly larger ($p < 0.05$) than all three of the other quadrants. **significantly larger ($p < 0.05$) than right and opposite quadrants, but not quite significantly different than the left quadrant ($p = 0.06$).

Dissociation of Contextual Fear Conditioning and HFS-LTP

In addition to suggesting a species difference in the molecular mechanisms underlying synaptic plasticity, these results highlight an apparent disconnect between the mouse LTP data and the mouse contextual learning data (compare Fig. 3 with Fig. 4). Although 100-Hz LTP in the mouse hippocampus is unaffected by MEK inhibition, administration of SL327 has been shown to impair contextual fear conditioning, a learning paradigm dependent on the integrity of the hippocampus.[40,78] Based on the lack of correspondence between these results, LTP induced with a pair of 100-Hz tetani does not appear to be a viable in vitro physiological model for contextual fear conditioning in the mouse.

We and others have observed that other forms of synaptic plasticity present in the mouse hippocampus are ERK-dependent and could therefore reconcile murine LTP and contextual fear learning. For example, in a recent study, Winder and colleagues demonstrated that LTP induced by theta frequency stimulation (5 Hz for 30 sec) is blocked by application of U0126 in hippocampal slices prepared from C57BL/6 mice.[104] It should also be noted that similar to the present study this group observed no effect of U0126 application on LTP induced by a single 100-Hz train in these mice. Thus, the theta-frequency stimulation used in Winder et al[104] may represent a more physiologically relevant LTP-induction paradigm in the mouse and may serve as a better model for hippocampus-dependent learning in this species.

Specific Contributions of ERK Isoforms to LTP and Learning

Taken together, these findings described above build a convincing argument for a requirement of the ERK isoforms of MAPK in the molecular events that underlie information storage at both the synaptic and behavioral levels. However, these inhibitor studies do not address the specific contribution of one versus the other ERK isoforms, i.e., ERK1 versus ERK2. To this end, ERK1 knockout mice were tested in various behavioral and physiological paradigms in order to elucidate the role of this particular protein in sensory, motor, and learning systems.

Emotional learning was assayed in ERK1-deficient mice using a standard fear-conditioning paradigm. As in the pharmacological studies described above, mice were placed in a novel environment or context and were exposed to two pairings of an acoustic cue and mild footshock. Learning was assessed 24 hours after training by measuring freezing behavior, a behavioral index of fear, in response to representation of either the context (the training cage) or representation of the auditory cue within an entirely novel environment. Long-term retention of these contextual and cue memories were also conducted two weeks after training.

Mice lacking the ERK1 isoform of MAPK displayed levels of conditioned fear similar to wildtype control mice (data not shown). We observed no difference in freezing levels during the training phase, suggesting an identical acquisition of conditioned fear in the two sets of mice. Compared to controls, ERK1 knockout mice displayed normal freezing behavior in response to representation of the context and to the delivery of the cue within a new context. Longer-term fear retention in the two groups also appeared similar, as fear associated with both the context and the cue was intact in both the mutant and wildtype mice when tested two weeks after training.

One important caveat to this study, as with other studies involving knockout animals, involves developmental compensation due to the mutation. The fact that the gene and its product were missing throughout development means that the knockout affects every ERK1-dependent function during development. No obvious compensatory changes were seen in the basal levels of ERK2 or in stimulated levels of phosphorylated ERK2 in the hippocampi of knockout mice. Therefore, behavioral and physiological characterization of these animals should provide an accurate assessment of the role of ERK1 in mouse learning.

Similarly, mice deficient in ERK1 also showed no impairments in tests of hippocampal physiology. ERK1 knockout mice displayed normal synaptic transmission, short-term plasticity, and long-term plasticity as tested with three different LTP induction paradigms. Both high-frequency (HFS) and theta burst (TBS) stimulation paradigms produced significant LTP in ERK null mice that was indistinguishable from controls.

So, what insight do these findings provide regarding the role of the ERK2 isoform of MAPK in synaptic plasticity and learning? One obvious explanation for the lack of a functional effect would suggest that ERK1 and ERK2 play redundant roles. In such a scenario, ERK2 can compensate for the loss of ERK1 in these knockout mice, thereby preventing the detection of a learning impairment. A second explanation would allow for the possibility that ERK2 plays a predominant role in the plastic changes accompanying learning, and its selective activation is necessary for learning to occur.

The fact that the two ERK isoforms are coordinately regulated in most in vitro is consistent with the first idea. ERK1 and ERK2 are both activated solely by MEK1 and MEK2, share very similar substrate profiles, and display a high degree of sequence homology.[88] Thus, the absence of an overt physiological phenotype in the ERK1 knockout mice and the finding that the animals are behaviorally similar to wild-type littermates would support the idea that the ERK isoforms have significant functional redundancy.

Nevertheless, based on a number of recent findings suggesting that ERK2 may be more selectively involved in mammalian learning, we favor the second hypothesis. Relative to ERK1, for example, the ERK2 isoform of MAPK shows higher basal levels of activation in the hippoc-

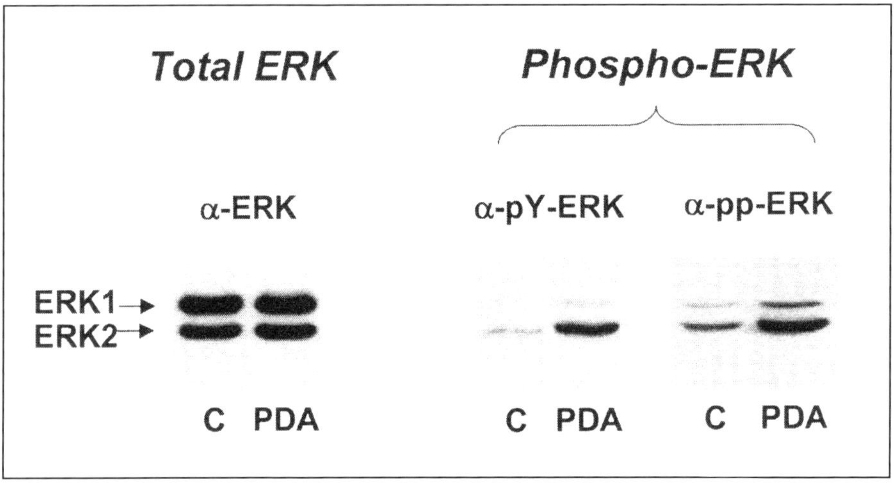

Figure 9. Selective activation of ERK2 in the hippocampus. Although their protein expression levels are similar (*left*), we often see selective activation of the ERK2 (p42) isoform but not the ERK1 (p44) isoform of MAPK. This preferential activation of ERK2 over ERK1 seems to be particularly prevalent in the hippocampus. The Western blots shown above (done by J.P. Adams) indicate a selective activation of ERK2 following exposure to the phorbol ester PDA (*right*). Although the ERK1 band is enhanced following PDA application, it still makes up a very small percentage of the activated ERK, as measured by phospho-specific antibodies. Deciphering the relative contributions of these two isoforms is one of the goals of the present study.

ampus as assessed using phospho-selective antibodies (demonstrated in Fig. 9). It has also been demonstrated that hippocampal ERK2 displays a high degree of responsiveness to a variety of signal transduction pathways critical to synaptic plasticity and learning.[76,81,86] More interestingly perhaps, selective activation of ERK2 has been previously demonstrated following induction of a physiological model for learning. In these experiments, phospho-ERK2 but not phospho-ERK1 levels were significantly increased in area CA1 of the rat hippocampus one hour after delivery of LTP-inducing tetanic stimulation.[37] This preferential activation of the ERK2 isoform of MAPK in the same region of the rat hippocampus was also observed in experimental animals one hour after exposure to contextual fear conditioning.[8]

More direct testing of the specific role of ERK2 is obviously needed to distinguish between the two alternatives proposed above. At this point many attempts have made to develop mice that specifically lacked the ERK2 isoform by a number of laboratories. Unfortunately, all efforts to this point have failed. Evidence from these unsuccessful experiments suggests that ERK2 is necessary for normal development, as deletion of this gene produces lethality in embryonic stages. It should nonetheless be noted that the absence of one isoform of ERK proves lethal while the absence of the other yields no discernible effect. These findings are quite suggestive that there may be significant differences in the functions subserved by these two isoforms, especially considering that protein expression levels for both isoforms are roughly the same in normal mice. Fortunately, conditional and inducible knockout strategies offer some hope that the developmental effects of ERK2 targeting can be overcome. Generation of a viable ERK2 mutant mouse will allow a direct assessment of ERK2's contribution to MAPK functioning in the hippocampus.

Biochemical Attributes That Make ERK Suited for Memory Formation

In this final section we would like to briefly touch on some particular attributes of the ERK signaling cascade that we feel make this system particularly well suited for playing a role in learning and memory. We will highlight the fact that the ERK cascade is a highly amplified system, that the cascade can operate as a signal integrator, and that, as with most kinases, that ERK is a pluripotent enzyme capable of coordinately regulating diverse downstream targets.

Signal Amplification

One attribute of the MAPK cascades that has been commented on extensively in the literature is their enormous capacity for signal amplification. This capacity arises out of the serial linkage of three enzymes: the MAPKKK, MAPKK, and the MAPK (e.g., ERK). Serial coupling of three catalysts confers a theoretical potential for signal amplification that seemingly is far beyond anything the cell might ever require. However, the hallmark of highly amplified systems is that they can operate in an essentially all-or-none fashion, conferring a capacity for generating a biochemical step function for triggering cellular events. We speculate that this capacity makes the ERK cascade ideally suited as an upstream trigger for establishing a memory trace – wherein what is called for is that a single, transient stimulus can be capable of establishing robust and lasting change.

Signal Integration

As we commented upon in the first section of this chapter, ERK is a downstream target of a wide variety of signal transduction systems. Serving as a convergence point for a variety of signal transduction systems, including dopaminergic and β-adrenergic signaling through PKA and metabotropic glutamatergic and cholinergic signaling through PKC,[81] ERK is a particularly attractive molecule for the integration of a variety of biochemical signals. Synergistic effects on ERK activation have been demonstrated following coactivation of inputs merging onto this kinase cascade.[101] This capacity is coincidence detection at the biochemical level, wherein the simultaneous presence of two signals achieves a unique effect. As has been widely discussed in the context of the NMDA receptor, this capacity allows for powerful information processing at the cellular level.

Temporal Integration

Tsien and colleagues have recently described a striking example of temporal integration at the level of this molecule.[106] In these experiments, the duration of ERK activation was critically dependent on the temporal pattern of stimulation. Multiple spaced stimuli (in this case, K^+-induced membrane depolarizations) gave rise to sustained increases in ERK phosphorylation, while single or "massed" stimulus delivery resulted in transient ERK activation. Interestingly, this sustained activation of ERK by spaced stimuli is reminiscent of numerous studies showing that memory formation in certain behavioral tasks also requires or is enhanced by spaced training.

Response Coordination

ERK, being an enzyme with many downstream targets, is capable of eliciting a coordinated cellular response at a wide variety of levels. This is not a unique attribute of ERK, of course, as most protein kinases share this capacity, and it has been held for many years that kinases evolved in part in order to be able to recruit a coordinated multicomponent response in regulating the metabolism of cells. In the context of ERK's role in neurons it is interesting that ERK appears to be capable of regulating ion channels, gene expression, the protein synthesis machinery, neurotransmitter release, and the cytoskeleton. The neuron may capitalize upon this attribute in order to use a single system for simultaneously triggering both short-term and long-term changes in neuronal function.

Summary

The MAPK superfamily of signal transduction cascades has been wieldy implicated in regulating cell growth, differentiation, and stress responses. However, recent studies have highlighted a prominent new role for these cascades in the central nervous system. Many studies now document roles for MAPK cascades in LTP at a wide variety of mammalian central synapses, and new studies indicate a role for both the ERK and p38 cascades in LTD in the hippocampus and cerebellum. In vivo studies have demonstrated a requirement for the extracellular signal-regulated kinase (ERK) cascade in associative learning, spatial learning and novel taste learning. A new report also has demonstrated a role for p38 in cerebellum-dependent eye-blink conditioning. In this chapter we have overviewed this wide variety of studies implicating MAPK cascades as general mediators of plasticity and memory formation in the mature nervous system. We also have speculated about likely intracellular targets of these cascades and discussed attributes of these signal transduction systems that may make them particularly well suited for involvement in plastic phenomena in the CNS. It will be interesting in the future to track the increasing understanding of the roles for these important signal transduction cascades in memory in particular and cognitive processing in general.

Acknowledgements

Some parts of the text and figures for this chapter were adapted, with modifications, from the following two publications:

Selcher J, Atkins CM, Trzaskos JM et al. A necessity for MAP kinase activation in mammalian spatial learning. Learning and Memory 1999; 6:478-490.

Selcher JD, Nekrasova T, Paylor R et al. Mice lacking the ERK1 isoform of MAPK kinase are unimpaired in emotional learning. Learning and Memory 2001; 8:11-19.

References

1. Abel T, Nguyen PV, Barad M et al. Genetic demonstration of a role for PKA in the late phase of LTP and in hippocampus-based long-term memory. Cell 1997; 88:615-626.
2. Abeliovich A, Chen C, Goda Y et al. Modified hippocampal long-term potentiation in PKCg-mutant mice. Cell 1993; 75:1253-1262.
3. Abeliovich A, Paylor R, Chen C et al. PKCg mutant mice exhibit mild deficits in spatial and contextual learning. Cell 1993; 75:1263-1271.
4. Adams JP, Anderson AE, Varga AW et al. The A-type potassium channel Kv4.2 is a substrate for the mitogen-activated protein kinase ERK. J Neurochem 2000; 75:2277-2287.
5. Alessi DR, Cuenda A, Cohen P et al. PD 098059 is a specific inhibitor of the activation of mitogen-activated protein kinase kinase in vitro and in vivo. J Biol Chem 1995; 270:27489-27494.
6. Alonso G, Widmer H. Clustering of KV4.2 potassium channels in postsynaptic membrane of rat supraoptic neurons: An ultrastructural study. Neurosci 1997; 77:617-621.
7. Andersen P, Sundberg SH, Sveen O et al. Specific long-lasting potentiation of synaptic transmission in hippocampal slices. Nature 1977; 266:736-737.
8. Atkins CM, Selcher JC, Petraitis JJ et al. The MAPK cascade is required for mammalian associative learning. Nat Neurosci 1998; 1:602-609.
9. Barnes CA. Memory deficits associated with senescence: A neurophysiological and behavioral study in the rat. J Comp Physiol Psychol 1979; 93:74-104.
10. Barnes CA. Involvement of LTP in memory: Are we 'searching under the street light'? Neuron 1995; 15:751-754.
11. Barnes CA, McNaughton BL. An age comparison of the rates of acquisition and forgetting of spatial information in relation to long-term enhancement of hippocampal synapses. Behav Neurosci 1985; 99:1040-1048.
12. Barnes CA, Jung MW, McNaughton BL et al. LTP saturation and spatial learning disruption: Effects of task variables and saturation levels. J Neurosci 1994; 14:5793-5806.
13. Barrionuevo G, Brown TH. Associative long-term potentiation in hippocampal slices. Proc Natl Acad Sci USA 1983; 80:7347-7351.
14. Berman DE, Hazvi S, Rosenblum K et al. Specific and differential activation of mitogen-activated protein kinase cascades by unfamiliar taste in the insular cortex of the behaving rat. J Neurosci 1998; 18:10037-10044.

15. Bland BH, Seto MG, Sinclair BR et al. The pharmacology of hippocampal theta cells: Evidence that the sensory processing correlate is cholinergic. Brain Res 1984; 299:121-131.
16. Bliss TVP, Lomo T. Long-lasting potentiation of synaptic transmission in the dentate area of the anaesthetized rabbit following stimulation of the perforant path. J Physiol 1973; 232:331-356.
17. Blum S, Moore AN, Adams F et al. A mitogen-activated protein kinase cascade in the CA1/CA2 subfield of the dorsal hippocampus is essential for long-term spatial memory. J Neurosci 1999; 19:3535-3544.
18. Bordi F, Marcon C, Chiamulera C et al. Effects of the metabotropic glutamate receptor antagonist MCPG on spatial and context-specific learning. Neuropharmacol 1996; 35:1557-1565.
19. Bourtchouladze R, Abel T, Berman N et al. Different training procedures recruit either one or two critical periods for contextual memory consolidation, each of which requires protein synthesis and PKA. Learn Mem 1998; 5:365-374.
20. Bourtchouladze R, Frenguelli B, Blendy J et al. Deficient long-term memory in mice with a targeted mutation of the cAMP-responsive element-binding protein. Cell 1994; 79:59-68.
21. Brambilla R, Gnesutta N, Minichiello L et al. A role for the Ras signalling pathway in synaptic transmission and long-term memory. Nature 1997; 390:281-286.
22. Broadie K, Rushton E, Skoulakis EM et al. Leonardo, a Drosophila 14-3-3 protein involved in learning, regulates presynaptic function. Neuron 1997; 19:391-402.
23. Castro CA, Silbert LH, McNaughton BL et al. Recovery of spatial learning deficits after decay of electrically induced synaptic enhancement in the hippocampus. Nature 1989; 342:545-548.
24. Chen C, Tonegawa S. Molecular genetic analysis of synaptic plasticity, activity-dependent neural development, learning, and memory in the mammalian brain. Ann Rev Neurosci 1997; 20:157-84.
25. Coan EJ, Saywood W, Collingridge GL. MK-801 blocks NMDA receptor-mediated synaptic transmission and long term potentiation in rat hippocampal slices. Neurosci Lett 1987; 80:111-114.
26. Collingridge GL, Kehl SJ, McLennan H. Excitatory amino acids in synaptic transmission in the Schaffer collateral-commissural pathway of the rat hippocampus. J Physiol 1983; 334:33-46.
27. Coogan AN, O'Leary DM, O'Connor JJ. P42/44 MAP kinase inhibitor PD98059 attenuates multiple forms of synaptic plasticity in rat dentate gyrus in vitro. J Neurophysiol 1999; 81:103-110.
28. Crow T, Xue-Bian J-J, Siddiqi V et al. Phosphorylation of mitogen-activated protein kinase by one-trial and multi-trial classical conditioning. J Neurosci 1998; 18:3480-3487.
29. Dash PK, Hochner B, Kandel ER. Injection of the cAMP-responsive element into the nucleus of Aplysia sensory neurons blocks long-term facilitation. Nature 1990; 345:718-721.
30. Davis HP, Squire LR. Protein synthesis and memory: A review. Psychol Bull 1984; 96:518-559.
31. Davis S, Vanhoutte P, Pages C et al. The MAPK/ERK cascade targets both Elk-1 and cAMP response element- binding protein to control long-term potentiation-dependent gene expression in the dentate gyrus in vivo. J Neurosci 2000; 20:4563-4572.
32. Deisseroth K, Bito H, Tsien RW. Signaling from synapse to nucleus: Postsynaptic CREB phosphorylation during multiple forms of hippocampal synaptic plasticity. Neuron 1996; 16:89-101.
33. Del Cerro S, Arai A, Lynch G. Inhibition of long-term potentiation by an antagonist of platelet-activating factor receptors. Behav Neural Biol 1990; 54:213-217.
34. DeSilva DR, Jones EA, Favata MF et al. Inhibition of mitogen-activated protein kinase kinase blocks T cell proliferation but does not induce or prevent anergy. J Immunol 1998; 160:4175-4181.
35. Ebinu JO, Bottorff DA, Chan EY et al. RasGRP, a Ras guanyl nucleotide- releasing protein with calcium- and diacylglycerol-binding motifs. Science 1998; 280:1082-1086.
36. Eichenbaum H. How does the brain organize memories? Science 1997; 277:330-332.
37. English JD, Sweatt JD. Activation of p42 mitogen-activated protein kinase in hippocampal long term potentiation. J Biol Chem 1996; 271:24329-24332.
38. English JD, Sweatt JD. A requirement for the mitogen-activated protein kinase cascade in hippocampal long-term potentiation. J Biol Chem 1997; 272:19103-19106.
39. Fanselow MS, Kim JJ, Yipp J et al. Differential effects of the N-methyl-D-aspartate antagonist DL-2-amino-5-phosphonovalerate on acquisition of fear of auditory and contextual cues. Behav Neurosci 1994; 108:235-40.
40. Frankland PW, Cestari V, Filipkowski RK et al. The dorsal hippocampus is essential for context discrimination but not for contextual conditioning. Behav Neurosci 1998; 112:863-874.
41. Frey U, Krug M, Reymann KG et al. Anisomycin, an inhibitor of protein synthesis, block late phases of LTP phenomena in the hippocampal CA1 region in vitro. Brain Res 1988; 452:57-65.
42. Gass P, Wolfer DP, Balschun D et al. Deficits in memory tasks of mice with CREB mutations depend on gene dosage. Learn Mem 1998; 5:274-288.
43. Grant SGN, O'Dell TJ, Karl KA et al. Impaired long-term potentiation, spatial learning, and hippocampal development in fyn mutant mice. Science 1992; 258:1903-1910.

44. Grewal SS, York RD, Stork PJ. Extracellular-signal-regulated kinase signalling in neurons. Curr Opin Neurobiol 1999; 9:544-553.
45. Harris EW, Ganong AH, Cotman CW. Long-term potentiation in the hippocampus involves activation of N-methyl-D-aspartate receptors. Brain Res 1984; 323:132-137.
46. Hoffman DA, Johnston D. Downregulation of transient K+ channels in dendrites of hippocampal CA1 pyramidal neurons by activation of PKA and PKC. J Neurosci 1998; 18:3521-3528.
47. Hoffman DA, Magee JC, Colbert CM et al. K^+ channel regulation of signal propagation in dendrites of hippocampal pyramidal neurons. Nature 1997; 387:869-875.
48. Holscher C. Synaptic plasticity and learning and memory: LTP and beyond. J Neurosci Res 1999; 58:62-75.
49. Huerta PT, Lisman JE. Bidirectional synaptic plasticity induced by a single burst during cholinergic theta oscillation in CA1 in vitro. Neuron 1995; 15:1053-1063.
50. Impey S, Mark M, Villacres EC et al. Induction of CRE-mediated gene expression by stimuli that generate long- lasting LTP in area CA1 of the hippocampus. Neuron 1996; 16:973-982.
51. Impey S, Obrietan K, Wong ST et al. Cross talk between ERK and PKA is required for Ca2+ stimulation of CREB- dependent transcription and ERK nuclear translocation. Neuron 1998; 21:869-883.
52. Impey S, Smith DM, Obrietan K et al. Stimulation of the cAMP response element (CRE)-mediated transcription during contextual learning. Nature Neurosci 1998; 1:595-601.
53. Ito M, Sakurai M, Tongroach P. Climbing fibre induced depression of both mossy fibre responsiveness and glutamate sensitivity of cerebellar Purkinje cells. J Physiol 1982; 324:113-34.
54. Johnston D, Wu SM-S. Foundations of cellular neurophysiology. Cambridge: MIT Press, 1997.
55. Jones MW, French PJ, Bliss TV et al. Molecular mechanisms of long-term potentiation in the insular cortex in vivo. J Neurosci 1999; 19:RC36.
56. Kanterewicz BI, Urban NN, McMahon DB et al. The extracellular signal-regulated kinase cascade is required for NMDA receptor-independent LTP in area CA1 but not area CA3 of the hippocampus. J Neurosci 2000; 20:3057-3066.
57. Kauer JA, Malenka RC, Nicoll RA. A persistent postsynaptic modification mediates long-term potentiation in the hippocampus. Neuron 1988; 1:911-917.
58. Kawasaki H, Springett GM, Toki S et al. A Rap guanine nucleotide exchange factor enriched highly in the basal ganglia. Proc Natl Acad Sci USA 1998; 95:13278-13283.
59. Larson J, Wong D, Lynch G. Patterned stimulation at the theta frequency is optimal for the induction of hippocampal long-term potentiation. Brain Res 1986; 368:347-50.
60. LeDoux JE. Emotion: Clues from the brain. Ann Rev Psychol 1995; 46:209-235.
61. Liao D, Hessler NA, Malinow R. Activation of postsynaptically silent synapses during pairing-induced LTP in CA1 region of hippocampal slice. Nature 1995; 375:400-404.
62. Logue SF, Paylor R, Wehner JM. Hippocampal lesions cause learning deficits in inbred mice in the Morris water maze and conditioned-fear task. Behav Neurosci 1997; 111:104-113.
63. Lynch G, Larson J, Kelso S et al. Intracellular injections of EGTA block induction of hippocampal long-term potentiation. Nature 1983; 305:719-721.
64. Magee JC, Johnston D. A synaptically controlled, associative signal for Hebbian plasticity in hippocampal neurons. Science 1997; 275:209-213.
65. Malenka RC, Kauer JA, Zucker RS et al. Postsynaptic calcium is sufficient for potentiation of hippocampal synaptic transmission. Science 1988; 242:81-84.
66. Mansuy IM, Winder DG, Moallem TM et al. Inducible and reversible gene expression with the rtTA system for the study of memory. Neuron 1998; 21:257-265.
67. Martin KC, Michael D, Rose JC et al. MAP kinase translocates into the nucleus of the presynaptic cell and is required for long-term facilitation in Aplysia. Neuron 1997; 18:899-912.
68. Martin SJ, Grimwood PD, Morris RGM. Synaptic plasticity and memory: An evaluation of the hypothesis. Annu Rev Neurosci 2000; 23:649-711.
69. McKernan MG, Shinnick-Gallagher P. Fear conditioning induces a lasting potentiation of synaptic currents in vitro. Nature 1997; 390:607-611.
70. McNaughton BL, Douglas RM, Goddard GV. Synaptic enhancement in fascia dentata: Cooperativity among coactive afferents. Brain Res 1978; 157:277-293.
71. Meiri N, Sun MK, Segal Z et al. Memory and long-term potentiation (LTP) dissociated: Normal spatial memory despite CA1 LTP elimination with Kv1.4 antisense. Proc Natl Acad Sci USA 1998; 95:15037-15042.
72. Michael D, Martin KC, Seger R et al. Repeated pulses of serotonin required for long-term facilitation activate mitogen-activated protein kinase in sensory neurons of Aplysia. Proc Natl Acad Sci USA 1998; 95:1864-1869.

73. Migaud M, Charlesworth P, Dempster M et al. Enhanced long-term potentiation and impaired learning in mice with mutant postsynaptic density-95 protein. Nature 1998; 396:433-439.
74. Morris RGM. Spatial localisation does not depend on the presence of local cues. Learn Motivation 1981; 12:239-260.
75. Morris RGM. Synaptic plasticity and learning: Selective impairment of learning rats and blockade of long-term potentiation in vivo by the N-methyl-D-aspartate receptor antagonist AP5. J Neurosci 1989; 9:3040-3057.
76. Norman ED, Thiels E, Barrionuevo G et al. Long-term depression in the hippocampus in vivo is associated with protein phosphatase-dependent alterations in extracellular signal- regulated kinase. J Neurochem 2000; 74:192-198.
77. O'Dell TJ, Hawkins RD, Kandel ER et al. Tests of the roles of two diffusible substances in long-term potentiation: Evidence for nitric oxide as a possible early retrograde messenger. Proc Natl Acad Sci USA 1991; 88:11285-11289.
78. Phillips RG, LeDoux JE. Differential contribution of amygdala and hippocampus to cued and contextual fear conditioning. Behav Neurosci 1992; 106:274-285.
79. Poolos NP, Johnston D. Calcium-activated potassium conductances contribute to action potential repolarization at the soma but not the dendrites of hippocampal CA1 pyramidal neurons. J Neurosci 1999; 19:5205-5212.
80. Roberson ED, English JD, Sweatt JD. A biochemist's view of long-term potentiation. Learn Mem 1996; 3:1-24.
81. Roberson ED, English JD, Adams JP et al. The mitogen-activated protein kinase cascade couples PKA and PKC to cAMP response element binding protein phosphorylation in area CA1 of hippocampus. J Neurosci 1999; 19:4337-4348.
82. Rogan MT, LeDoux JE. LTP is accompanied by commensurate enhancement of auditory-evoked responses in a fear conditioning circuit. Neuron 1995; 15:127-136.
83. Rogan MT, Staubli UV, LeDoux JE. Fear conditioning induces associative long-term potentiation in the amygdala. Nature 1997; 390:604-607.
84. Rosenblum K, Futter M, Jones M et al. ERKI/II regulation by the muscarinic acetylcholine receptors in neurons. J Neurosci 2000; 20:977-985.
85. Schafe GE, Nadel NV, Sullivan GM et al. Memory consolidation for contextual and auditory fear conditioning is dependent on protein synthesis, PKA, and MAP kinase. Learn Mem 1999; 6:97-110.
86. Schmid RS, Graff RD, Schaller MD et al. NCAM stimulates the Ras-MAPK pathway and CREB phosphorylation in neuronal cells. J Neurobiol 1999; 38:542-558.
87. Scoville WB, Milner B. Loss of recent memory after bilateral hippocampal lesions. J Neurolog Neurosurg Psychiat 1957; 20:11-21.
88. Seger R, Krebs EG. The MAPK signaling cascade. Faseb J 1995; 9:726-735.
89. Sheng M, Tsaur M-L, Jan YN et al. Subcellular segregation of two A-type K^+ channel proteins in rat central neurons. Neuron 1992; 9:271-284.
90. Shors TJ, Matzel LD. Long-term potentiation: What's learning got to do with it? Behav Brain Sci 1997; 20:597-614; discussion 615-655.
91. Silva AJ, Paylor R, Wehner JM et al. Impaired spatial learning in a-calcium-calmodulin kinase II mutant mice. Science 1992; 257:206-211.
92. Silva AJ, Stevens CF, Tonegawa S et al. Deficient hippocampal long-term potentiation in alpha-calcium-calmodulin kinase II mutant mice. Science 1992; 257:201-206.
93. Staubli U, Lynch G. Stable hippocampal long-term potentiation elicited by 'theta' pattern stimulation. Brain Res 1987; 435:227-234.
94. Stevens CF. A million dollar question: Does LTP = memory? Neuron 1998; 20:1-2.
95. Stevens CF, Wang Y. Reversal of long-term potentiation by inhibitors of heme oxygenase. Nature 1993; 364:147-149.
96. Sweatt JD. Toward a molecular explanation for long-term potentiation. Learn Mem 1999; 6:399-416.
97. Thiels E, Xie X, Yeckel MF et al. NMDA receptor-dependent LTD in different subfields of hippocampus in vivo and in vitro. Hippocampus 1996; 6:43-51.
98. Tsien JZ, Chen DF, Gerber D et al. Subregion- and cell type-restricted gene knockout in mouse brain. Cell 1996; 87:1317-1326.
99. Varga AW, Anderson AE, Adams JP et al. Input-specific immunolocalization of differentially phosphorylated Kv4.2 in the mouse brain. Learn Mem 2000; 7:321-332.
100. Vossler MR, Yao H, York RD et al. cAMP activates MAP Kinase and Elk-1 through a B-Raf- and Rap-1-dependent pathway. Cell 1997; 89:73-82.
101. Watabe AM, Zaki PA, O'Dell TJ. Coactivation of beta-adrenergic and cholinergic receptors enhances the induction of long-term potentiation and synergistically activates mitogen-activated protein kinase in the hippocampal CA1 region. J Neurosci 2000; 20:5924-5931.

102. Weeber EJ, Atkins CM, Selcher JC et al. A role for the beta isoform of protein kinase C in fear conditioning. J Neurosci 2000; 20:5906-5914.
103. Williams JH, Errington ML, Lynch MA et al. Arachidonic acid induces a long-term activity-dependent enhancement of synaptic transmission in the hippocampus. Nature 1989; 341:739-742.
104. Winder DG, Martin KC, Muzzio IA et al. ERK plays a regulatory role in induction of LTP by theta frequency stimulation and its modulation by beta-adrenergic receptors. Neuron 1999; 24:715-726.
105. Wu SP, Lu KT, Chang WC et al. Involvement of mitogen-activated protein kinase in hippocampal long- term potentiation. J Biomed Sci 1999; 6:409-417.
106. Wu GY, Deisseroth K, Tsien RW. Spaced stimuli stabilize MAPK pathway activation and its effects on dendritic morphology. Nat Neurosci 2001; 4:151-158.
107. Xing J, Ginty DD, Greenberg ME. Coupling of the RAS-MAPK pathway to gene activation by RSK2, a growth factor-regulated CREB kinase. Science 1996; 273:959-963.
108. Yagi T, Shigetani Y, Okado N et al. Regional localization of Fyn in adult brain; studies with mice in which fyn gene was replaced by lacZ. Oncogene 1993; 8:3343-3351.
109. Yin JC, Wallach JS, Del Vecchio M et al. Induction of a dominant negative CREB transgene specifically blocks long-term memory in Drosophila. Cell 1994; 79:49-58.
110. Zamanillo D, Sprengel R, Hvalby O et al. Importance of AMPA receptors for hippocampal synaptic plasticity but not for spatial learning. Science 1999; 284:1805-1181.

Chapter 4.7

Phosphatases

Pauleen C. Bennett and Kim T. Ng

Abstract

Evidence implicating protein serine/threonine phosphorylation in memory formation and related brain processes continues to accumulate at a rapid rate. Several protein kinases have been demonstrated to play critical roles in models of synaptic plasticity, such as long-term potentiation, and also to be necessary for memory formation to proceed in numerous vertebrate and invertebrate species, across a range of learning tasks. Research into equivalent roles for protein phosphatases began more recently and, although preliminary data are promising, evidence implicating these enzymes in memory formation remains less compelling. In this chapter theoretical reasons for hypothesising that phosphatases will be critically involved in biological information storage processes are discussed. Evidence implicating phosphatases in neurophysiological models of synaptic plasticity is then reviewed, as is available evidence directly implicating phosphatases in memory formation. Our research group has produced relatively convincing evidence that each of three classes of serine/threonine phosphatases are essential for memory formation in day-old chicks trained on an established single-trial, passive-avoidance, learning task. Our pharmaco-behavioural data are broadly commensurate with more recent studies using gene technologies in rodent learning models, which predict an especially important role in memory modulation for the calcium-dependent phosphatase, PP2B (also called calcineurin). The temporal specificity of the results obtained thus far, and the impressive similarities across studies from vastly different species and models of synaptic plasticity, is consistent with the view that memory formation may rest upon a complex sequence of more-or-less transient events involving post-translational modifications to existing proteins. Kinases and phosphatases are likely to be equally important in fine-tuning critical events in the intra-cellular milieu.

Introduction

Almost ten years ago, in a review of the contribution made by *Aplysia californica* to our understanding of brain plasticity and memory formation, it was claimed that "*the study of cellular mechanisms underlying memory is being reduced to problems of activation of second-messenger systems, modulation of membrane channels and other cellular processes by phosphorylation, and regulation of gene expression and protein synthesis*".[27] No truer words have been spoken and in the last ten years, thousands of papers relevant to these issues have been published. While many questions remain unanswered, the literature is beginning to provide a coherent account of information storage processes in the brain. In this chapter, reasons for focusing on phosphorylation and, in particular, dephosphorylation are briefly reviewed. The contribution made to information storage processes by one class of enzymes, protein serine/threonine phosphatases, is then described.

From Messengers to Molecules: Memories Are Made of These, edited by Gernot Riedel and Bettina Platt.

Protein Phosphorylation: What Is It and Why Is It Significant?

The cells that make up biological organisms comprise numerous components. Many are proteins, the sequence of which is encoded genetically. Proteins carry small, localized, electrostatic charges that force each molecule to adopt a three-dimensional configuration specific for that protein. This configuration determines its function. Several mechanisms have evolved by which the configuration and function of proteins can be altered. Some are irreversible, resulting in permanent structural and functional change. Others involve temporary reactions catalysed by opposing enzyme classes. These permit reversible changes in cell functioning and potentially operate as 'on/off' or 'more/less' switches. Protein phosphorylation and dephosphorylation are the most common reversible post-translational modifications. They are used ubiquitously in biological systems to alter cell functioning in response to signals arising from the extracellular environment.

Phosphorylative regulation involves a complex series of events (Fig. 1). A protein is phosphorylated when the terminal phosphate is transferred from adenosine triphosphate (ATP) to a hydroxyl group within the protein. This requires, first, that ATP be complexed to a divalent cation, usually magnesium (Mg^{2+}). A second requirement is the involvement of an enzyme, a protein kinase, which catalyses the phosphate transfer. The protein is dephosphorylated when the phosphate molecule is removed via hydrolysis, a reaction catalysed by enzymes known as protein phosphatases. As phosphate molecules are negatively charged, their addition to, or deletion from, a protein alters the distribution of electrostatic forces between that section of the protein and other sections. This can induce a change in the protein's three-dimensional configuration and modify its functional state. Phosphorylative changes are typically rapid and reversible, persisting only until a further enzyme reverses the initial change. They can be maintained, however, extending and/or amplifying a response to the signal initially responsible for enzyme activation.

The brain is a rich source of kinases, phosphatases and substrate proteins, many of which are unique to neural tissue,[204] and phosphorylation is implicated in many brain processes and mediates numerous nerve cell responses (reviewed in refs. 68 and 194). Indeed, the number of brain proteins regulated by phosphorylation appears to increase at the same rate as new proteins are described and it is difficult to locate a well-characterized protein not regulated either directly or indirectly by this process. Most of the events that are fundamental to normal operations in the brain, and to synaptic transmission in particular, are regulated by phosphorylation. For example, it seems likely that all voltage-gated ion channels may be subject to some form of phosphorylative regulation (reviewed in ref. 135). In some cases phosphorylation plays a mediatory role, being an obligatory step in the opening or closing of a channel. More commonly, phosphorylation is modulatory, altering channel sensitivity to factors responsible for opening or closing the channel, altering the time course of channel opening, and so forth. Neurotransmitter synthesis typically also proceeds only following phosphorylation of rate-limiting enzymes, and several Ca^{2+}-dependent kinases and phosphatases act to both mediate and modulate the process of neurotransmitter release.[160] Postsynaptic effector molecules strongly implicated in important functional changes in the brain are similarly regulated by phosphorylation. These include many ligand-gated ion channels and G-protein-coupled receptors (reviewed in ref. 101). The consequences of phosphorylation vary with each particular system but include desensitisation, internalisation and activation.

While changes in acute pre- and post-synaptic events may underlie permanent information storage, most theorists argue that they are more likely to maintain *transient* changes in synaptic efficacy, with other, more stable, mechanisms being responsible for permanent storage.[216] Consistent with this, formation of permanent memories depends on synthesis of new proteins[208] and changes in gene expression.[45] Phosphorylation and dephosphorylation are intimately involved in initiating transcription and translation, with over 20 critical proteins being regulated by these processes.[212] The same processes also regulate the fibrillar cytoskeletal proteins that interact to dynamically determine the shape of a cell at any given time.[150] Phosphorylative

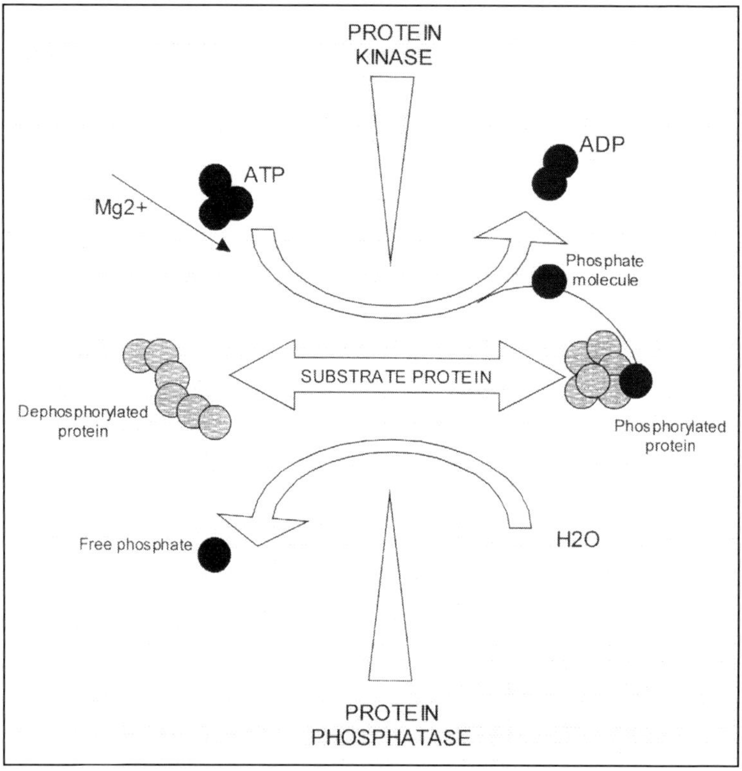

Figure 1. Schematic representation of the processes of protein phosphorylation and dephosphorylation.

regulation of one cytoskeletal element, microtubule-associated protein-2 (MAP2), has been proposed to play a particularly important role in memory formation.[97] MAP2 phosphorylation may act as a trigger for the incorporation of new cytoskeletal proteins, perhaps 'tagging' recently activated sites to allow specificity of plastic changes. This may, in turn, lead to rearrangement and enlargement of critical synaptic areas.[216]

Finally, numerous enzymes responsible for regulating all aspects of brain functioning are controlled, either directly or indirectly, by phosphorylation and dephosphorylation. Indeed, many kinases and phosphatases, in addition to being regulated by other means, are themselves regulated by phosphorylation.[131] Others are subject to regulation by specific proteins, some of which are inhibitory and some of which target the enzymes to specific cellular locations.[207] Many of these proteins are, in turn, regulated by phosphorylation and dephosphorylation. Complex sequences of phosphorylative events often act either competitively, additively or synergistically to bring about extremely subtle, rapid, and precise changes in cell functioning. These may be transient or persist indefinitely according to the requirements of the organism.[204]

Enzymes That Regulate Phosphorylation in the Brain

Several amino acids are capable of accepting phosphate molecules. Under physiological conditions, however, phosphate binding generally occurs on the hydroxyl group of either serine (Ser) or threonine (Thr) residues or of tyrosine residues. Some degree of overlap has been identified[103] but the enzymes catalysing these two types of reactions are generally distinct. Kinases and phosphatases responsible for the phosphorylation of Ser/Thr residues have been strongly implicated both in the regulation of brain processes relevant to memory formation[151,194] and in the regulation of memory formation itself.[130,169] Kinases that phosphorylate tyrosine residues have also been implicated in memory formation[152,211] but less attention has been paid to phosphatases that dephosphorylate tyrosine residues. In this chapter, we are concerned primarily with evidence implicating Ser/Thr phosphatases in memory formation.

Protein Serine/Threonine Phosphatases in the Brain

Phosphatases have not been as intensively studied as kinases and, even fifteen years ago, no firm conclusions could be drawn regarding how they were regulated or about their physiological roles.[141] This was due mainly to a mistaken assumption that phosphatases, by generally turning things 'off', act merely to restore homeostasis. Turning things 'on' via kinase activity was thought to be the primary means of effecting change. Consistent with this view, phosphatases were not typically observed to be directly responsive to known second-messenger systems. They were also found to have broad and overlapping substrate specificities in vitro. They were assumed, therefore, to be constitutively active, non-discriminating enzymes, responsible for keeping proteins in a dephosphorylated state in 'resting' cells.

It is now evident that the activity of most phosphatases is highly regulated through complex and often indirect routes.[154,207] Phosphatases have central roles in the regulation of many cellular processes and phosphatase activity is often necessary to initiate important cellular events, not just to terminate events initiated by kinases.[22,213] Two alternative classification systems were originally proposed. One of these used a nomenclature based on the enzymes' differing requirements for specific cations.[129] The other, more successful, system was developed initially to distinguish among enzymes involved in the regulation of metabolism in skeletal muscle and liver.[83] It is widely applicable to other biological systems, however, and is now used almost universally.

Under this classificatory scheme phosphatases are categorised on the basis of whether they preferentially dephosphorylate the α or β subunit of one specific protein, phosphorylase kinase, and on the basis of their sensitivity to nanomolar concentrations of two endogenous inhibitor proteins, identified by Huang and Glinsmann[77] and now referred to as phosphatase inhibitor-1 (INH-1) and phosphatase inhibitor-2 (INH-2). Type 1 phosphatases (PP1s) preferentially dephosphorylate the β-subunit of phosphorylase kinase and are inhibited by the two inhibitor proteins. Type 2 phosphatases (PP2) preferentially dephosphorylate the α-subunit of phosphorylase kinase and are relatively insensitive to INH-1 and INH-2. Type 2 phosphatases are further divided into three groups on the basis of their requirement for divalent cations. PP2A enzymes do not have a requirement for divalent cations. PP2B enzymes (also called calcineurin) are dependent on Ca^{2+} and PP2C enzymes on Mg^{2+}.

The classification system proposed by Ingebritsen and Cohen[83] is quite robust. The four types of phosphatase activity can be distinguished in all mammalian tissues and even in cells from primitive organisms. With few exceptions, the distinctions drawn among the categories apply across species. Various other criteria, including sensitivity towards various exogenous inhibitory agents, can now be used to distinguish among the four classes of enzymes.[171] In addition, studies of the primary sequences of each phosphatase group have confirmed that unique sequences account for the four types of activity. PP1 and PP2A are among the most highly conserved enzymes known, sharing significant sequence homology with each other and with PP2B.[84] PP2C is unrelated to the other phosphatases and has derived from a distinct gene family.[36]

Despite general support for the above classification system, recent developments have challenged both its specificity and generality.[35] Additional levels of classification have been introduced to account for reports that different isoforms of each type of phosphatase are the products of distinct genes. Although highly conserved, the isoforms differ in distribution and may have distinctive properties.[34] Additional notation has also been required to account for the fact that PP1 and PP2A typically exist as high molecular weight complexes; a common catalytic subunit being associated with a range of non-catalytic proteins.[171] A less tractable difficulty comes from the isolation of enzymes which do not fit recognized categories and others which cannot be distinguished using established criteria.[35] Such difficulties might eventually force revision of the current nomenclature, but Table 1 summarizes the structure, distribution, regulation, and substrate specificity of the main Ser/Thr phosphatase classes recognised at this time. This information is expanded in refs. 33, 154, 209 and in the papers listed in Table 1.

Why Phosphorylation Might Be Critical for Memory Formation

The demonstrated involvement of phosphorylative regulation in almost every facet of cell functioning provides justification for examining the contributions made by this process to memory formation. One of the difficulties inherent with research in this area, however, lies in distinguishing brain processes that play a direct role in retaining information from those that play a less direct, but equally critical, role, perhaps being necessary for general cell functioning, without which memory cannot proceed. Phosphorylation might be expected to participate in several 'housekeeping' functions that indirectly support memory formation. When one considers several assumptions that underlie memory research, however, it is clear that phosphorylation and dephosphorylation are also suited to several roles that may be directly involved in the establishment of memory.

The first of these assumptions is that memory formation consists of multiple stages, a concept that is not without controversy. Many years ago debate raged over whether memory stages formed in parallel or were sequentially dependent.[66,126] Sequentially-dependent models prevailed, but have recently been challenged within many memory paradigms .[46,88,125] Whether memory stages form in parallel or in sequence is yet to be fully resolved. That two or more biochemically- and temporally-distinct memory stages exist, however, is unquestioned.[88] One or more labile short-term memory stages are assumed to precede the expression of permanent memories in all existing animal models, and these must be supported by the cellular processes underlying memory formation.

A second assumption is that not all learned experiences are retained. Kety[94] argued that it makes evolutionary sense to conserve memory capacity by consolidating only information that is 'biologically significant'. At some point between a learning experience and the relatively permanent changes presumed to underlie long-term memory, therefore, some mechanism must appraise the significance of each experience. The 'biological significance' of an experience depends at least partially on the level of arousal associated with it.[127] In vertebrates, this may be mediated, at least partially, by the neurohormone noradrenaline.[143] Some of the evidence supporting this claim is provided by Gibbs et al in this book. The important point at this stage is that, during consolidation, the developing memory trace must be represented in such a way that it remains vulnerable to modulating factors.

In 1897, Sherrington[172] coined the term 'synapse' to describe connecting points between neurons. He also noted that these areas were likely to be involved in learning, arguing that:

"*Shut off from all opportunities of reproducing itself and adding to its number by mitosis or otherwise, the nerve cell directs its pent-up energy towards amplifying its connections with its fellows, in response to the events which stir it up. Hence, it is capable of an education unknown to other tissues.*" (Sherrington, 1897/1948).

Hebb[72] more formally proposed that information may initially be represented in the brain as a pattern of electrical activity, and that the 'strength' of synapses linking simultaneously active pre- and post-synaptic neurons may be permanently enhanced. Theoretical and empirical advances have supported Hebb's description of associative memories. Only changes at the level of synapses appear able to account for the amount of information stored by biological organisms. In addition, computer models have demonstrated that simple neural network systems, which store information in the form of weighted modifications to 'synaptic' connections, are capable of learning feats which resemble those exhibited by biological organisms.[163] Most importantly, it has been demonstrated unequivocally that synapses contain mechanisms able to

Table 1. Properties of the four serine/threonine phosphatase classes that predominate in nervous tissue

Enzyme (Alternative Nomenclature)	Classificatory Properties	Known Forms of Regulation	Distribution in Brain	Specific Reviews
PP1 (ATP- and magnesium-dependent protein phosphatase)	1. Preferentially dephosphorylates the β-subunit of phosphorylase kinase. 2. Sensitive to INH-1 and INH-2 3. Sensitive to Okadaic acid ($IC_{50} < 10$ nM) 4. Does not require divalent cations for activation 5. Broad specificity	1. Targeted to substrates and sites by subunits which are regulated by phosphorylation 2. Inhibited by INH-1, DARPP-32 and PKA-R, all activated by PKA and inactivated by PP2B. 3. Inhibited by INH-2, regulated by a complex sequence of phosphorylation events 4. Affected by a deinhibitor protein, regulated by phosphorylation 5. Phosphorylated by a cyclin-dependent kinase 6. Specific inhibitors regulate PP1 in cell nuclei	Widely distributed in cytosolic, synaptosolic, synaptic membrane and synaptic junction fractions	(Bollen and Stalmans, 1992) (Watanabe, 2001)
PP2A (Polycation-stimulated protein phosphatase)	1. Preferentially dephosphorylates the α-subunit of phosphorylase kinase. 2. Insensitive to INH-1 and INH-2 3. Sensitive to Okadaic acid ($IC_{50} < 0.1$ nM) 4. Does not require divalent cations for activation 5. Broad specificity	1. Distribution of different isoforms is regulated at the level of transcription 2. Sensitive to polycations, polyamines and phosphoproteins 3. Phosphorylated by PKA and engages in auto-dephosphorylation 4. Inhibited by G-substrate, activated by PKG and inactivated by PP2B 5. Exists as a complex of a catalytic subunit and one ormore noncatalytic subunits which may act to regulate or localize activity	Extremely concentrated in brain as compared to other tissues. Appears to be predominantly cytosolic.	(Sontag, 2001) (Millward et al., 1999)

continued on next page

Table 1. Properties of the four serine/threonine phosphatase classes that predominate in nervous tissue

Enzyme (Alternative Nomenclature)	Classificatory Properties	Known Forms of Regulation	Distribution in Brain	Specific Reviews
PP2B (calcineurin) (Calcium-dependent protein phosphatase)	1. Preferentially dephosphorylates the α-subunit of phosphorylase kinase. 2. Insensitive to INH-1 and INH-2 3. Partially sensitive to Okadaic acid ($IC_{50} < 10$ mM) 4. Dependent on Ca^{2+} for activation 5. Restricted range of substrates	1. Extremely sensitive to Ca^{2+}/CaM 2. Requires metal cofactors 3. Targeted to sites by myristoylation 4. Phosphorylation by dephosphorylated form of CaMK-II 5. Constitutively active form can be generated by limited proteolysis, mediated by calpain	Found in both cytosolic and particulate fraction. Comprises up to 1% of total brain protein	(Yakel, 1997) (Guerini, 1997) (Klee et al., 1998) (Sugiura et al., 2001) (Hemenway and Heitman, 1999)
PP2C (Magnesium-dependent protein phosphatase)	1. Preferentially dephosphorylates the α-subunit of phosphorylase kinase. 2. Insensitive to INH-1 and INH-2 3. Insensitive to Okadaic acid 4. Dependent on Mg^{2+} for activation 5. Broad specificity	1. Dependent on Mg^{2+}	Higher levels in brain than in other tissues, mainly in cytosolic fractions. Not concentrated in nerve terminals.	

fully account for transient and permanent changes in their responses to neurochemical signals. Research into mechanisms underlying synaptic plasticity currently consumes a good portion of the neuroscience community.

These assumptions, that memory formation consists of multiple stages, that a mechanism exists to selectively initiate permanent consolidation only of those experiences that are biologically significant, and that memories reside in modifications to synaptic efficacy, dictate that research into memory formation concentrates on brain processes able to meet these requirements. Processes regulated by kinases and phosphatases are particularly suited to such roles and exhibit several appropriate characteristics, including:

A. a capacity for rapid responsivity and reversibility;
B. a capacity for specificity and precision;
C. a capacity for signal integration and
D. a capacity to prolong changes and induce permanent functional alterations

Capacity for Rapid Responsivity and Reversibility

Kinases and phosphatases respond directly or indirectly to second-messenger activation and typically involve changes to existing proteins. Because substrate proteins include ion channels and receptors, phosphorylative changes can instigate immediate change in the electrical or chemical sensitivity of synapses. Phosphorylation cascades can act, therefore, to rapidly facilitate or inhibit the activity of individual synapses and might even 'store' information for a short period, allowing the cell to respond more rapidly or forcefully to further stimulation. In addition, the changes produced by phosphorylation are often transient, with substrates returning to their basal state as enzyme activity levels subside. This is attractive because, if information is initially represented as reversible changes in the phosphorylation state of specific proteins, modulatory factors could directly strengthen or weaken the existing representation by stimulating or inhibiting kinase or phosphatase activity. Whenever enzyme activity is not appropriately reinforced, the neurons involved could simply return to a basal state, ready to respond to further stimulation.

Capacity for Specificity and Precision

Although most kinases and phosphatases have broad and overlapping substrate specificities in vitro, they each regulate highly specific responses in vivo. This is partly because of the localized distribution of enzymes and/or substrates,[82] and partly because enzymes, although activated by the same mechanisms, may be differentially sensitive to certain activating events.[38,62] The latter is most readily demonstrated by reference to the prolific second-messenger, Ca^{2+}. In addition to direct regulation of ion channels, Ca^{2+} activates many types of cytosolic receptor molecules. These include various isoforms of protein kinase C (PKC), activated by Ca^{2+} in conjunction with 1,2-diacylglycerol (DAG), and a group of proteases called calpains. Ca^{2+} also binds to a number of binding proteins, the most prominent of which is calmodulin (CaM). The Ca^{2+}/CaM complex regulates a variety of enzymes and receptors, including several kinases, PP2B, several types of adenylate cyclase, a number of phosphodiesterases, inositol 1,4,5-tri phosphate, and nitric oxide synthetase.[203] The different effects are regulated by temporal factors and by phosphorylation-mediated changes in the binding of CaM to different proteins.[164] Increases in Ca^{2+} are also often confined to specific intracellular regions so that the response reflects the spatial organization of effector molecules.[32;155] In addition, affinity for Ca^{2+}/CaM differs widely among target proteins, so that a small influx may result in an effect directly opposed to that induced by a larger influx.[96] The differential sensitivity of specific enzymes to Ca^{2+} appears to be a critical factor in the regulation of synaptic plasticity and is discussed later (see also Sun and Alkon, this book).

Capacity for Signal Integration

Kinases and phosphatases often participate in networks of cross-communication between different messenger systems, mediated by proteins that include receptors, second messengers, kinases, phosphatases, regulatory proteins and so forth. Complex interrelationships, whereby a single enzyme is regulated by an interplay of many factors, allow signal convergence and amplification.[170] Some substrates have multiple sites that can be acted upon by different kinases and phosphatases and these events may occur independently and have opposing or synergistic effects. In others, an initial phosphorylation event alters substrate conformation such that it becomes more or less susceptible to further phosphorylation events. Multiple phosphorylations can generate a variety of protein forms in response to different extracellular events. Alternatively, they may ensure that a cell must experience a precise sequence of events before triggering a particular response. Phosphorylation and dephosphorylation processes can, therefore, integrate information arising from different sources, perhaps allowing a developing memory trace to respond not only to the presence of information itself, but also to modulatory factors that signal whether or not the information is to be retained.

PP1 is particularly well suited to an integrative role, being regulated by a variety of events. One is regulation by various inhibitory proteins. The most common of these inhibitors is INH-1; others include the 'Dopamine and cyclic AMP-Regulated Phospho-Protein' (DARPP-32), and a specific type of regulatory subunit, R-II, which usually forms part of an isoform of a protein kinase A (PKA) holoenzyme. All of these proteins become potent inhibitors of PP1 following their phosphorylation by PKA, activated in response to cyclic adenosine monophosphate (cAMP). They are inactivated following dephosphorylation by PP2B, activated in response to Ca^{2+}.[207] This may permit the existence of regulatory cascades such as those depicted in Figure 2. The physiological significance of cascades such as these is currently being explored.

Capacity to Prolong Changes and Induce Permanent Functional Alterations

While phosphorylation and dephosphorylation events are, in many cases, quite transient, the activity of most kinases and some phosphatases can be prolonged via several means including, in particular, autophosphorylation, through which an activated kinase phosphorylates itself as well as other substrates. Autophosphorylated kinases continue to phosphorylate substrates following degradation of the message responsible for their activation. Provided phosphatase activity is constrained, therefore, they could maintain information while structural modifications, proposed by many to be necessary to produce permanent memory, occur. The idea that kinase autophosphorylation may act to store information has been comprehensively explored by Lisman, who argued in 1985 that a bistable autophosphorylating kinase could store information indefinitely.[104]

According to Lisman's initial model, a switching mechanism could be formed from a kinase capable of autophosphorylation and a phosphatase able to dephosphorylate the kinase. In an initial state the kinase is dephosphorylated and incoming information triggers autophosphorylation. If only a small proportion of the available kinase molecules are activated, they are rapidly dephosphorylated, causing loss of the 'biologically unimportant' information. If the stimulus is sufficiently strong to activate many molecules, however, the limited amount of phosphatase available cannot counteract the autophosphorylation. All kinase molecules are eventually activated and the activity is maintained indefinitely. It results, therefore, in permanent phosphorylation of critical substrates and a stable record of the information.

This model represents only one way in which phosphorylation and dephosphorylation could contribute to prolonged information storage. Another popular, not necessarily mutually exclusive, model holds that permanent memory formation depends on changes in gene expression and protein synthesis. Any change to gene expression or protein synthesis is likely to depend heavily on phosphorylation since both transcription and translation are regulated at many points by changes in the phosphorylation state of critical substrate proteins.[212]

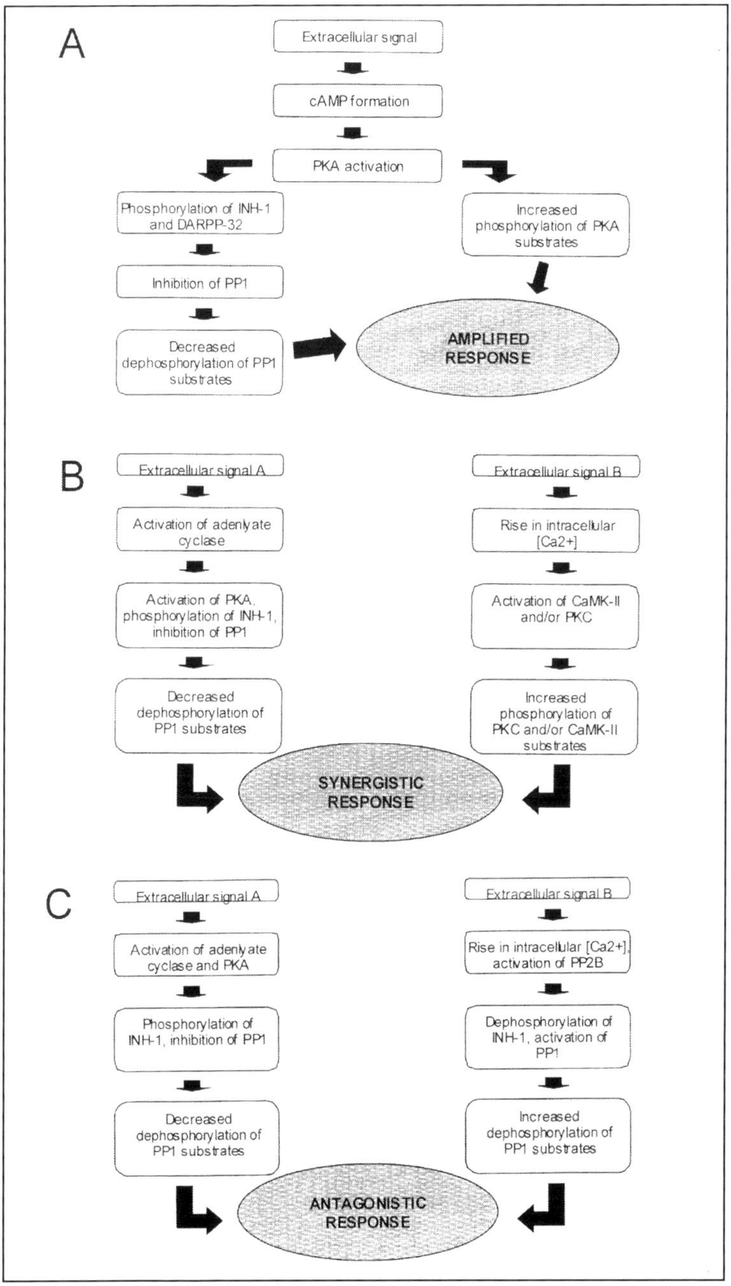

Figure 2. A) Signal amplification-a cAMP signal stimulates PKA, resulting in increased phosphorylation of PKA substrates including INH-1 and DARPP-32. This leads to inhibition of PP1, decreasing dephosphorylation of PKA substrates so that the response to PKA is amplified. B) Signal synergism-phosphorylation of INH-1 in response to activation of PKA results in inhibition of PP1. If Ca^{2+}-dependent systems are also active, this synergistically increases phosphorylation of substrates for these systems. C) Signal antagonism-dephosphorylation of INH-1 by PP2B means that extracellular signals that act through Ca^{2+}-driven systems can antagonise the actions of PKA.

Phosphorylation in Information Storage Processes

A strong theoretical rationale exists, therefore, for investigating the involvement of phosphorylation and dephosphorylation in memory formation. No other regulatory mechanisms exhibit the potential to contribute simultaneously to so many aspects of information storage, including:

- initial registration of incoming information
- the decision to retain or reject information
- integration of various types of information
- maintenance of relatively labile stages of memory preceding the formation of a permanent engram
- permanent, possibly transcription-dependent, changes in synaptic functioning

Given this rationale, it is not surprising that numerous studies have examined whether kinases are directly involved in information storage. Roles for phosphatases have been considered only more recently, but evidence is rapidly accumulating which suggests that these enzymes may have roles as central as those attributed to kinases in synaptic plasticity[213] and in memory formation itself.[157] The issue is particularly pressing given the potential importance of bidirectionality in information storage processes. If memory formation depends, as is widely believed, on changes in synaptic efficacy, mechanisms probably exist by which such efficacy can be both increased and decreased.

Early Empirical Evidence of a Role for Phosphorylation in Memory Formation

It was shown almost thirty years ago that a training experience increases incorporation of radioactively labelled phosphate into nuclear proteins in rodents.[115,116] Synaptosome-enriched fractions from trained mice showed greater incorporation of labelled phosphate than did fractions from untrained mice.[64,153] In rats, increased phosphate incorporation was localized to the hippocampus and caudate nucleus.[178] Four synaptic proteins exhibited altered rates of phosphate incorporation following active-avoidance training.[112,113] Only some synaptic proteins phosphorylated in vitro showed altered phosphorylation following training (reviewed in ref. 162).

After this promising beginning, subsequent studies, attempting to examine the nature and function of phosphorylation changes following training, initially encountered substantial difficulties. These arose in part from the complexity of the systems regulating phosphorylation and dephosphorylation events, which make it difficult to interpret changes observed in vitro.[159] They also arose in part from the complexity of the average vertebrate central nervous system, where billions of neurons render it impossible at present to accurately track the flow of information from sensory organs to putative storage sites, much less observe the molecular processes responsible.[167] Additionally, researchers have tended to conduct isolated studies using different species, tasks and methodologies. This initially worked against the systematic development of a comprehensive knowledge base in the area of vertebrate memory formation.

Due to these difficulties, much of the information implicating phosphorylation in memory comes from studies in which researchers have utilized a 'simple' systems approach,[49] involving learning in invertebrate species or synaptic plasticity in dissociated neurons and tissue slices. Of central importance in the latter have been investigations into the phenomena of long term potentiation (LTP) and long term depression (LTD), both of which depend critically on phosphorylation and dephosphorylation events.[86,194]

Long Term Potentiation: A Model for Increasing Synaptic Efficacy

Long-term potentiation (LTP) is a phenomenon observed when brief trains of high-frequency stimulation (HFS) are delivered to monosynaptic excitatory pathways in hippocampal neurons, resulting in a persistent increase in synaptic efficacy. First reported in 1973,[18,19] this phenomenon is consistent with the proposal that neurons may store information as changes in synaptic efficacy and has been extensively studied as a putative memory-related brain process. While tissue from various brain areas exhibits LTP, it is most often studied in tissue slices containing one of two excitatory hippocampal synaptic connections: the perforant path to dentate granule cell synapse and the Schaffer collateral/commissural (CA3 pyramidal cell axon) to CA1 pyramidal cell synapse. The mammalian hippocampus has a highly organized structure amenable to the types of studies required to investigate LTP and has been strongly implicated in several kinds of memory processing.[87,179,193] The nature of any relationship between LTP and memory remains controversial.[11,124] There is arguably some overlap between the two processes, however, and many treatments which disrupt or facilitate LTP similarly affect memory formation (discussed in refs. 86 and 158).

Hippocampal LTP is characterized by a requirement for simultaneous presynaptic activity and postsynaptic depolarisation, involving the neurotransmitter glutamate. During HFS glutamate is released in large quantities at many synapses. The flow of ions through non N-methyl-D-aspartate (NMDA)-linked glutamate channels depolarises the postsynaptic neuron sufficiently to dislodge the Mg^{2+} blocking NMDA glutamate receptors (GluRs), with the subsequent local influx of Ca^{2+} being a critical trigger for the induction of a major form of LTP.[119] Strong evidence implicates NMDA-GluRs in some kinds of learning.[23,39,102,156] Like memory, LTP may consist of two or more temporally-distinct stages, beginning with Ca^{2+} influx and ending with a protein synthesis-dependent change in synaptic structure which occurs many hours later.[76,183]

Understanding LTP depends on identifying the processes responsible for transforming a rise in intracellular Ca^{2+} into lasting synaptic modifications. It is therefore of interest that, in addition to numerous other effector molecules, Ca^{2+} activates many kinases and phosphatases. In conjunction with DAG, Ca^{2+} activates several isoforms of PKC.[142] In conjunction with CaM, it is responsible for activation of the prolific calcium/calmodulin-dependent kinase – type II (CaMK-II)[107] and PP2B,[213] the latter of which may then dephosphorylate PP1 inhibitors, activating PP1.[207] Through its effects on Ca^{2+}/CaM-regulated forms of adenylate cyclase, Ca^{2+} may also contribute to activation or deactivation of PKA,[3,132] which may directly oppose the effects of PP2B on PP1 inhibitors (see Fig 2). Finally, through stimulation of Ca^{2+}/CaM-dependent forms of nitric oxide synthase, it promotes activation of protein kinase G (PKG).[166] Numerous studies have investigated roles for kinases in LTP, and each of the major Ser/Thr kinase groups has been implicated in different parts of this cellular process (see reviews in ref. 177 and 183). CaMK-II appears to play a particularly critical role.[107] Expression of LTP may depend on increased α-amino-3-hydroxy-5-methyl-4-isoxazoleproprionic acid (AMPA) receptor activity and/or AMPA receptor endocytosis, both of which may occur at least partially through phosphorylation of the AMPA receptor subunit, GluR1, mediated by CaMK-II .[8,119]

Protein Phosphatases in LTP

The fact that numerous kinase-mediated events contribute to LTP justifies examination of whether phosphatases may be similarly involved, and several studies have now addressed this issue (reviewed in ref. 213). NMDA GluR subunits are reported to be closely associated with PP2A[29] and decreased PP2A activity, possibly associated with increased phosphorylation of a non-catalytic PP2A subunit, has been reported in hippocampal slices following induction of LTP.[57] There is also evidence to suggest that PP1 may be targeted to NMDA GluRs by the receptor associated protein, yotiao, which also binds PKA.[210] In this study it was found that anchored PP1 was constitutively active, limiting basal channel activity, and that activation of PKA or inhibition of PP1 was required to enhance NMDA receptor currents. Hence, PP1 activity may modulate NMDA GluR activity. Accordingly, inhibition of PP1 and PP2A typically enhances normal transmission in rat hippocampal slices,[55,73] although it also increases responses of cultured neurons to agonists for nonNMDA-GluRs.[206] Inhibition of PP1 indirectly, by activation of PKA and subsequent phosphorylation of INH-1, facilitates LTP induction.[20,24] PP1 activation, via genetic suppression of INH-1, affects certain forms of LTP in some hippocampal regions.[2] These data suggest that these phosphatases, particularly PP1, normally constrain synaptic activity.

Application of NMDA to adult rat striatal slices and hippocampal slices results in rapid dephosphorylation of two PP2B substrates, the PP1 inhibitor DARPP-32[70] and the cytoskeletal protein MAP2.[69,133] This may indicate that occupancy of NMDA-glutamate receptors results in an immediate increase in PP2B activity. That this activation is particularly rapid is indicated by the observation that dephosphorylation of MAP2 precedes phosphorylation events. PP2B is more sensitive to Ca^{2+} than most Ca^{2+}-dependent kinases[96] and a dephosphorylation event is also reported to either precede[67] or co-occur[47] with phosphorylation events during depolarisation-induced neurotransmitter release.

In visual cortex from young rats, inhibition of PP2B *facilitates* induction of LTP without affecting basal transmission, an effect that appears to be postsynaptic.[56] Inhibition of PP2B synthesis or activity by chronic administration of antisense oligonucleotides against the catalytic subunits of PP2B similarly *facilitates* LTP induction in the hippocampus of anaesthetised rats.[81] In a more recent study a truncated form of PP2B, under the control of a promoter molecule, was overexpressed in the forebrain of mice,[214] resulting in a 75% increase in Ca^{2+}-dependent phosphatase activity in the hippocampus. When hippocampal slices from these mice were exposed to HFS they failed to demonstrated 'normal' levels of LTP but, instead, exhibited what was termed an intermediate form of LTP, since it decayed to baseline levels within three hours. Consistent with the proposal that overexpression of PP2B may alter PP2B/PKA dynamics in favour of PP2B, the capacity for LTP was reinstated if PKA was activated. It was concluded that PP2B may act as an inhibitory constraint that acts to regulate the synaptic induction of longer forms of LTP by PKA.[214]

In hippocampal tissue, however, investigations have also found that PP2B inhibition *disrupts* NMDA-GluR-mediated LTP, with no effect on low frequency stimulation-induced depotentiation of previously potentiated synapses.[108,110] Another found that inhibition of PP2B *prevented* expression of LTP beyond 30 minutes.[205] A third reported that PP2B activation is *essential* for the depression of a gamma-aminobutyric acid $(GABA)_A$-mediated inhibitory postsynaptic potential that normally accompanies an increased excitatory post synaptic potential during LTP.[109] These contradictory results characterize much of the literature concerned with the molecular processes underlying LTP but suggest that a simplistic model, in which phosphatases merely reverse the effects of kinases, may not account for the involvement of phosphatases in either LTP or general synaptic transmission. Given that LTP is generally accepted by the scientific community as a valid model of cellular processes relevant to memory formation, possible roles for phosphatases require additional investigation.

Long-Term Depression: A Model for Decreasing Synaptic Efficacy

Initial reports that synaptic efficacy could be depressed in response to specific stimulation protocols were welcomed as providing a necessary adjunct to LTP as a model for biological information storage processes.[114,118] The relevance of LTD to memory formation was difficult to establish,[123] however, and it was not until the late 1990's that LTD was demonstrated to occur in vivo in adult animals.[74,180,192] Since then, progress in understanding the mechanisms underlying various forms of LTD has been impressive. Some forms of LTD are postulated to be critically dependent on phosphatase activity, making this process of great interest in the present context. In the interests of brevity only hippocampal LTD will be considered. For a review of processes underlying cerebellar LTD see ref. 85. For a review of other forms of LTD, including hippocampal LTD, see refs. 12 and 93.

Mechanisms underlying the induction, maintenance and expression of hippocampal LTD remain unclear, primarily because they appear to vary depending on such things as the induction protocol employed and the age and strain of the animal.[93,100] A critical variable for post-synaptically induced, activity-dependent LTD may be a modest level of postsynaptic activation during induction.[12] This may result in a localized but perhaps more moderate influx of Ca^{2+} than is required for LTP.[41,218] Because Ca^{2+} activates a range of enzymes it is difficult to predict the consequences of Ca^{2+} influx. It has been argued, however, that PP2B has a higher affinity for Ca^{2+}/CaM than most other enzymes, and that CaMK-II and Ca^{2+}/CaM-dependent forms of adenylate cyclase, in particular, require higher levels of Ca^{2+} than does PP2B.[198] These claims have led to the development of a widely cited bi-directional model of synaptic plasticity, building on the theoretical model first proposed by John Lisman.[104]

Lisman's Bi-Directional Model of Synaptic Plasticity

According to Lisman's first attempt to apply his bi-directional model of synaptic plasticity to specific neural substrates (see ref. 105; most recently updated in ref. 107), information storage in synapses may depend on prolonged autophosphorylation of CaMK-II. CaMK-II is a prolific multisubunit kinase known to engage in intermolecular autophosphorylation following stimulation by Ca^{2+}/CaM.[78] This typically results in translocation of the enzyme to postsynaptic density sites and prolongs its activity until it is dephosphorylated or autophosphorylated at another site. It is not clear which enzymes are responsible for dephosphorylation of CaMK-II. PP2B is relatively ineffective in vitro,[92] while PP1,[44] PP2A[131] and PP2C[56] all readily dephosphorylate CaMK-II. In vivo the situation is more complex, and it seems that the location of the enzyme determines its sensitivity to different phosphatases. Cytoplasmic CaMK-II is most readily dephosphorylated by PP2A.[181] Postsynaptic density associated CaMK-II differs, however, in being completely insensitive to PP2A, so that PP1 predominately dephosphorylates the enzyme.[220] This may be because PP1 and CaMK-II are bound tightly into postsynaptic density positions by relevant scaffolding proteins.[117,207]

Lisman proposes that a moderate increase in Ca^{2+} may preferentially activate PP2B. Because PP2B is relatively inactive against CaMK-II, it is argued that the primary effect of this activation may be dephosphorylation of INH-1. This is followed by activation of PP1 and dephosphorylation of CaMK-II,[44] which leads either directly or indirectly, via reduced phosphorylation of critical substrates such as NMDA[195] and AMPA[197] GluRs, to a 'reduction' in the strength of information stored at the synapse. Conversely, a massive increase in Ca^{2+} is postulated to activate CaMK-II, increasing the number of molecules undergoing autophosphorylation and stimulating phosphorylation of relevant substrates. Simultaneously, stimulation of adenylate cyclase activates PKA, leading to phosphorylation of INH-1, inhibition of PP1, and a reduction in dephosphorylation of CaMK-II and other substrates. This is argued to represent an increase in the strength of stored information.

Protein Phosphatases in LTD

Lisman's model has received a great deal of theoretical and empirical support (reviewed in refs. 106 and 222). In the interests of brevity, only evidence examining the involvement of phosphatases in relevant forms of synaptic plasticity is considered here. During LTD in the adult hippocampus in vivo, there is a transient (< one hour) increase in PP1 activity and a more persistent (> one hour) increase in PP2A activity.[191] LTD in the CA1 region can be blocked by bath-application of phosphatase inhibitors, with the agents used acting primarily against PP2A and PP1.[138] In this study phosphatase inhibition, after establishment of LTD, also reversed LTD, and loading of postsynaptic neurons with a membrane-impermeable phosphatase inhibitor blocked LTD induction. These results suggest that postsynaptic phosphatase activity may be required for induction of LTD and that continued phosphatase activity may be necessary for its maintenance. Mulkey et al.[138] speculated that PP1 was more likely to play a critical role than PP2A simply because it is concentrated in isolated synaptic junctions, known to dephosphorylate several important synaptic proteins, including CaMK-II, and could potentially be activated indirectly by Ca^{2+} acting via PP2B. To test this hypothesis, inhibitors of PP2B were bath-applied to hippocampal slices or injected directly into postsynaptic neurons. Each protocol blocked LTD induction, leaving LTP unaffected. To further examine whether

PP2B may regulate LTD induction via INH-1, thiophosphorylated INH-1, unable to be dephosphorylated by PP2B, was injected into postsynaptic neurons. This blocked LTD, an effect not observed when dephosphorylated INH-1 was applied. cAMP analogues also prevented LTD induction, presumably by activating PKA, keeping INH-1 in a phosphorylated state and, hence, preventing activation of PP1 by PP2B.[137]

These early results suggested that hippocampal LTD induction may require a postsynaptic phosphatase cascade beginning with stimulation of PP2B by Ca^{2+}/CaM and culminating with increased PP1 activation. Reports from other sources soon confirmed that phosphatase inhibition disrupts LTD in hippocampal tissue[139,217] and in tissue from other brain areas[95,196] in vitro, and in hippocampal brain regions in vivo.[145,190] The picture is now confused, however, by the identification of forms of hippocampal LTD that appear not to require phosphatase activity[148] or that require activation of PP1, but not PP2B.[165] Hodgkiss and Kelly[75] initially argued that only de novo LTD is blocked by PP2B inhibition. These authors found that PP2B inhibition had no effect on LTD induction in slices that had previously undergone depression, although LTD in naïve slices was reduced. More recently, it has been reported that LTD induced by bath application of NMDA[89,134] or by activation of metabotropic GluRs[148] is not sensitive to the phosphatase inhibition that disrupts synaptically-induced LTD.

Synaptic activation of NMDA GluRs in cultured hippocampal neurons results in a redistribution of PP1 to synaptic locations.[134] Peptides that inhibit the binding of PP1 to proteins that target the enzyme to these locations prevent initiation of LTD and, in cells where LTD is already induced, result in an increase in synaptic strength. These data indicate that PP1 activity may contribute to the maintenance of LTD for at least 30 minutes following induction.[134] Inhibition of phosphatases also prevents low frequency stimulation from depotentiating previously potentiated synapses,[146] and renders it more likely to result in LTP. Such complexity is not easily accommodated within existing models and awaits verification. In addition, very little attention has been paid to the possibility that different forms of PP2B may have unique roles in specific aspects of synaptic plasticity. Each PP2B isoform has a distinct distribution in the brain, perhaps indicative of distinct roles.[187] Consistent with this possibility, it has been reported that mice selectively lacking a brain-enriched PP2B isoform, Aa, are significantly deficient in depotentiation but exhibit normal NMDA-GluR-mediated LTD and LTP.[230] Such evidence suggests that roles for different PP2B isoforms require further investigation.

Also awaiting further elaboration are downstream targets for phosphatase cascades other than INH-1 and CaMK-II. Potential targets are plentiful (see review in ref. 213) and include AMPA receptors. At least one form of hippocampal LTD appears to depend on AMPA GluR internalisation[111] and AMPA GluRs have been shown to downregulate following dephosphorylation.[99] A recent study has shown that both PKA and PP2B may be anchored to GluR1, a critical AMPA GluR phosphorylation site, by a common anchoring protein, and that this anchoring may confer a Ca^{2+}/PP2B-mediated downregulation on GluR1 currents.[189] It is of interest, then, that a recent report examining regulation of AMPA GluR endocytosis in response to activation of NMDA receptors found that inhibitors of PP2B were able to block this effect.[13] Inhibitors of PP1 failed to do so, potentially indicating that PP2B has effects on AMPA GluR functioning that are independent of its role in activating PP1. Issues such as this need to be addressed by future research.

Phosphatase Involvement in Invertebrate Memory Models

Few vertebrate models support the systematic exploration of cellular events underlying memory formation. These events have been probed, however, using invertebrates such as the marine molluscs *Aplysia californica* and *Hermissenda crassicornis* and the fruit fly *Drosophila melanogaster*. Information storage in each of these species has been found to depend on phosphorylation, with *Aplysia* and *Drosophila* serving below as exemplars of two approaches to understanding information storage processes. A recent study using *Hermissenda* has also implicated phosphatases in some aspects of memory formation in this species.[140] For a general discussion of the utility of studying invertebrate learning, see reference 98.

Protein Phosphatases in *Aplysia* Learning and Memory

Aplysia has been used to study various forms of learning (reviewed in ref. 27). Sensitisation of the gill-and-siphon-withdrawal-reflex has been most extensively studied and provides evidence of a role for PKA-mediated phosphorylation in non-associative memory processes. PKC is also indirectly implicated in learning in *Aplysia*[26] but the exact role for this enzyme is less clear and it is not discussed here.

Sensitisation occurs when *Aplysia* learns to strengthen a reflex withdrawal response to previously neutral stimuli following presentation of a noxious stimulus at another site.[90] This involves a presynaptic increase in neurotransmitter release at monosynaptic connections between sensory and motor neurons. The neurotransmitter is serotonin (5-HT), which increases the activity of adenylate cyclase in the membrane of sensory cells. This increases the concentration of cAMP and, subsequently, activates PKA. PKA transiently phosphorylates many proteins, one of which is either a 5-HT-sensitive K^+ channel, the S-type K^+ channel, or a protein that regulates this channel, and another of which is a voltage-gated K^+ channel.[184] Phosphorylation closes these channels, reducing K^+ currents that repolarise the cell following an action potential. This broadens subsequent action potentials so that more Ca^{2+} enters the presynaptic terminal via voltage-gated Ca^{2+}channels and more neurotransmitter is released. Thus, PKA plays a critical role in the transient strengthening of synaptic efficacy that underlies short-term sensitisation (STS).

The same enzyme, PKA, may also be involved in more persistent responses. Sensitisation of the gill-withdrawal response in *Aplysia*, following a single presentation of a noxious stimulus, lasts for only several minutes. Four presentations produce sensitisation for up to a day, however, and 64 presentations produce sensitisation for several weeks. This long-term sensitisation (LTS) resembles STS in that it involves enhanced 5-HT release, modulation of S-type K^+ channels, depression of K^+ currents, and phosphorylation of the same proteins phosphorylated following STS.[184] LTS is uniquely dependent, however, on protein synthesis and gene expression.[10] Evidence suggests that cAMP-responsive element binding protein (CREB)-like transcriptional activators may be required for expression of LTS[43] and that PKA-mediated phosphorylation may be responsible for initiating these prolonged cellular events.[42]

Extracts from *Aplysia* contain phosphatase activities corresponding to the four major groups,[52] with PP1 accounting for over 75% of membrane-associated activity.[51] It is established that modulatory changes induced by 5-HT or cAMP can be opposed by a second transmitter, FMRFamide. The molecular consequences of FMRFamide have not been determined but, because the actions of 5-HT and FMRFamide converge at the S-type K^+ channel, it is anticipated that this transmitter may act via PKA inhibition or phosphatase activation. Administration of FMRFamide decreases phosphorylation in *Aplysia* neurons and reverses the increase in phosphorylation which normally coincides with 5-HT administration.[185]

Ichinose and Byrne[79] prepared voltage-clamped sensory neurons from *Aplysia* and examined steady-state membrane currents following administration of either a phosphatase inhibitor or purified phosphatases. They also examined whether these substances could modulate 5-HT- and cAMP-induced inward currents and FMRFamide-induced outward currents. Phosphatase inhibition in preparations treated with 5-HT or cAMP enlarged and prolonged inward currents, presumably either by facilitating phosphorylation necessary to induce inward currents or by preventing dephosphorylation that normally returns the channels to a basal state. By contrast, injection of phosphatases led to the opening of K^+ channels and modulation of their susceptibility to closure by 5-HT. Phosphatase inhibition in preparations treated with FMRFamide reduced outward currents in previously unstimulated cells, while administration of purified phosphatases mimicked aspects of the response to FMRFamide. Phosphatase inhibition produced an even larger reduction in outward currents in FMRFamide-treated cells that had previously been treated with 5-HT, supporting the hypothesis that FMRFamide may directly antagonize 5-HT by opening closed channels, perhaps stimulating dephosphorylation of PKA-phosphorylated channel components.

These results indicate that phosphatases influence the magnitude and time course of cAMP-dependent responses in *Aplysia* and are consistent with the view that FMRFamide may act via stimulation of phosphatase activity. This has not yet been implicated in behavioural change but is interesting given the lack of information concerning endogenous regulators of phosphatase activity, perhaps indicating that these enzymes are responsive to as yet undiscovered second-messenger systems. The actual phosphatase involved in the response has not been identified, although preliminary evidence suggests that PP1 may be responsible for dephosphorylating proteins involved in the regulation of K^+ currents in *Aplysia* neurons, with PP2A playing a minor role.[40]

Protein Phosphatases in Drosophila *Learning and Memory*

Drosophila demonstrates numerous forms of learning. Most studies, however, use an olfactory shock-avoidance conditioning procedure in which flies are placed en masse in a setting in which they are sequentially exposed to two odours, one of which is paired with shock. To test for conditioned avoidance the flies are later placed in a T-shaped apparatus and the two odours delivered in air currents, one from each arm of the T. Avoidance is measured as the percentage of flies migrating away from the odour previously paired with shock. Approximately 90% of wild-type flies avoid the conditioned stimulus for up to one hour post-training, with retention remaining significant for 24 hours.[200] As with other species, additional training trials increase the number of flies who retain memory for the training event over longer periods.

Five temporally distinct stages of memory formation have been identified in *Drosophila*[199] and, quite remarkably, almost every mutant gene which produces deficits in learning or memory encodes products involved in second-messenger systems and phosphorylative cascades. Most evidence specifically implicates cAMP-dependent processes in memory formation,[60,174,219] although roles for other kinases have also been identified.[31,91] Importantly, it appears that steady state levels of cAMP may be less critical than an unidentified aspect of cAMP dynamics.[229] Several attempts have been made to model the dependence of *Drosophila* learning on PKA dynamics, with most models holding that prolonged activity of a specific type of PKA may maintain a labile memory trace while structural changes are initiated.[5] Prolonged activation of PKA is attributed to either autophosphorylation or Ca^{2+}-dependent proteolytic degradation of dissociated regulatory subunits, which is argued to make them more sensitive to cAMP.

A role for phosphatases in memory formation in *Drosophila* seems likely. If phosphatase activity is not constrained, autophosphorylated or dissociated PKA regulatory subunits would be dephosphorylated, resulting in reassociation of the regulatory and catalytic subunits and loss of stored information.[25] Conversely, a low level of phosphatase activity may mean that any transient cAMP signal results in prolonged activation of PKA, making it impossible to selectively store information. *Drosophila* expresses PP1, PP2A and PP2B, each enzyme being similar to its mammalian counterpart.[149] Despite the potential importance of phosphatases in memory formation in *Drosophila* they have received little attention. One study has shown that a mutant *Drosophila* strain carrying the Su-var(3)6^{01} mutation, known to affect the structural gene of PP1, is impaired in performance on various learning tasks.[6] The mutants did not demonstrate deficiencies in their sensory abilities, strengthening claims that the Su-var(3)6^{01} mutation may selectively impair memory formation. The Su-var(3)6^{01} mutation does impair other cellular processes, however, making it necessary to confirm, through other means, that PP1 is necessary for memory formation in this species.

Phosphorylation in Vertebrate Memory Models

Many studies have shown that inhibition of each major kinase group is sufficient to prevent vertebrates from learning and/or retaining some kinds of information, as is genetically induced depletion of certain kinases (reviewed in refs. 130 and 169). Others have reported that kinase stimulation facilitates learning, while still more have identified brain proteins that show altered rates of phosphorylation following training. While there is a wealth of evidence implicating each major kinase group in memory formation, only a few studies have examined roles for phosphatases. These are described below.

Many vertebrate training paradigms require repeated exposure to a learning situation. This makes it impossible to define exactly when learning occurs and may hinder the systematic exploration of dynamic events that occur early in the consolidation process. For this reason, a great deal of valuable information has been gained from studies using the day-old chick and a single-trial passive-avoidance learning (PAL) task. This species and variants of the task have been widely used to explore events which occur soon after learning[53,144,161] and the paradigm is particularly useful for pharmaco-behavioural investigations, which can be carried out rapidly and with sufficient animals to provide adequate statistical power. A similar model exists in rats and has been used extensively to pharmacologically explore events underlying memory formation in this species.[86] Similarities between critical cellular processes across the two models encourage generalization but, thus far, roles for protein phosphatases in memory formation have only been investigated within the chick model.

Kinases and Phosphatases in Chick Memory Formation

For the past three decades Ng, Gibbs, and their colleagues have investigated memory formation in groups of day-old-chicks trained using a single-trial PAL task in which chicks learn to avoid a red coloured bead which has been coated during a single, ten second training trial, with an aversive chemical, methylanthranilate (reviewed in ref. 144). By testing groups of chicks at precise times relative to learning, it has been possible to document the temporal characteristics of the memory trace that develops after the training experience. In untreated chicks trained on the PAL task retention levels, indicated by selective avoidance of the red bead during a retention test, remain high for at least two weeks post-training, although two transient 'dips' in retention are observed, one at approximately 15 minutes post-training and the other at about 55 minutes post-training. These dips, possibly analogous to the 'Kamin' effects identified in rodents, coincide with the emergence of amnestic effects following administration of a number of inhibitors. They are believed, therefore, to reflect transition points between different stages of memory formation. This has supported the development of a temporally precise, three-stage model, although studies within a similar chick paradigm have identified what appear to be even later stages of memory formation.[161] If the aversive training stimulus is diluted to 10 or 15%, high retention levels are maintained for only thirty minutes or so post-training. This weakly reinforced version of the task provides a useful tool with which to investigate treatments expected to facilitate memory formation.

A number of studies have addressed the involvement of phosphorylation in memory formation in the chick and strong evidence implicates each of the four major multifunctional Ser/Thr kinases in this process. The temporal specificity of the model has permitted demonstration of the fact that each enzyme group appears necessary for quite specific stages of the memory formation process. From relevant pharmaco-behavioural studies CaMK-II appears to be required in both hemispheres of the chick brain for early stages of memory formation, although this requirement may occur slightly later in the left hemisphere than in the right hemisphere.[223,225] PKC and PKA appear necessary only in the left hemisphere; PKC prior to 25-30 minutes post-training[228] and PKA prior to 60 minutes post-training.[227] PKG inhibition also impairs memory, with bilateral inhibition causing a prolonged but transient period of memory loss, with onset from approximately 100 minutes post-training.[48]

Preliminary studies measuring activity levels of the various kinases and changes in substrate phosphorylation[1,226] are generally consistent with the behavioural data. Importantly, however, these studies indicate that periods of kinase activity may be transient, suggesting that dephosphorylation may also take place following a learning experience. It has also been found that the effects of PKC inhibition and activation are dependent on the level of reinforcement associated with the task.[228] Pharmacological activation of PKC following weakly-reinforced training facilitates long term memory formation. Conversely, activation of PKC following strongly-reinforced training causes memory impairment. This may indicate that too much kinase activity is as detrimental to memory as too little, and again heralds significant roles for the phosphatases likely to be required in order to curtail kinase activity at critical time points. It also underscores the efficacy of continuing to investigate brain processes relevant to memory formation within the framework provided by a temporally precise model, derived from a single-trial learning task within which the strength of reinforcement of a training trial can be systematically manipulated.[161]

The involvement of three Ser/Thr phosphatase classes in memory formation has been studied by our research group and, in recent years, evidence implicating all three enzyme classes in memory formation in the chick has been published.[14,16,17,224] Although interpretation of the data is made difficult by the poor specificity of available pharmacological inhibitors, a review of these studies is useful in delineating some of the issues that will require further investigation in the future. Pharmacological approaches are acknowledged to be particularly useful in initially revealing the cellular processes underlying memory formation, which can then be targeted for further research.[128]

PP2A

Initial studies employed the drugs Okadaic Acid (OA) and Calyculin A (CalA), agents widely acknowledged to be specific and selective phosphatase inhibitors.[54] Both drugs disrupted retention from two different time points depending on the concentration of drug administered and the time relative to training at which administration took place.[17,224] At concentrations of OA that might be expected to selectively inhibit PP2A (0.5 nM per hemisphere), bilateral intracranial administration of either drug close to the time of training induced retention deficits from approximately 40-50 minutes post-training. This most probably indicates a role for PP2A in a process that normally occurs at or before this time relative to training and that is critical to a subsequent stage of memory formation.

The apparent requirement for PP2A was later shown to be confined to the left hemisphere of the chick. Administration of putative PP2A inhibitors to the right hemisphere was without effect on retention levels at any time tested, while administration to the left hemisphere mimicked the effects observed following bilateral administration (Bennett, Moutsoulas, Lawen & Ng, unpublished observations). These data are summarised schematically in Figure 3. The effects of PP2A inhibition were observed also to vary with the strength of reinforcement associated with the training experience. While left hemisphere inhibition of PP2A following strongly reinforced training resulted in significant memory loss (Fig. 3), left hemisphere inhibition following weakly reinforced training facilitated memory formation (Bennett, Moutsoulas, Lawen & Ng, unpublished observations).

Previous data from our laboratory have demonstrated that PKC inhibition has an amnestic effect following PAL which is evident by 25-30 minutes post training and which is also lateralized to the left hemisphere.[228] Transient changes in phosphorylation of the PKC substrate, growth associated protein of ~50 kD weight (GAP-43), are also observed to occur at around this time in critical areas of the chick brain,[1,226] and within a similar time frame following single-trial inhibitory-avoidance learning in the rat.[28] The amnestic effect of PP2A inhibition occurred soon after the effect observed following PKC inhibition. As PP2A is known to dephosphorylate GAP-43,[168] it is possible that PP2A may be a critical element in a proposed reinforcement-associated memory switch believed to require PKC activity.[143] Dephosphorylation of GAP-43 may be just as essential as its phosphorylation. The memory 'switch' is believed to determine whether information is retained beyond approximately 30 minutes post-training, but evidence supporting this speculation is circumstantial so far. Other candidate substrates for PP2A include a number of kinases that are dephosphorylated by this enzyme,[7] the protein called myristoylated alanine-rich C kinase substrate (MARCKS), which shows increased phosphorylation following imprinting in the chick;[173] the protein RC3/neurogranin, which shows enhanced phosphorylation during LTP;[30] and the protein tau, which shows altered dephos-

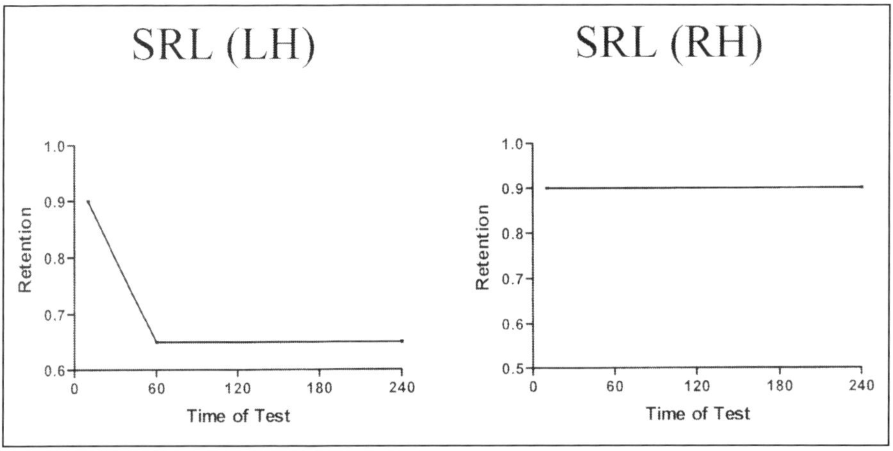

Figure 3. Schematic diagrams depicting the effects of inhibiting PP2A in one hemisphere of the chick following strongly-reinforced learning (SRL). Groups of chicks were trained to discriminate between a previously aversive red coloured bead and a previously pleasant blue coloured bead. They were then administered either OA (0.5 nM) or CalA (0.1 nM) into the left (LH) or right (RH) hemisphere and retention was tested at the specified time post-training. Groups of chicks that received either drug in the left hemisphere showed good retention, indicated by selective avoidance of a red coloured bead, only up until approximately 40-50 minutes post-training. All groups tested after this time showed significant retention deficits. Groups of chicks administered either drug in the right hemisphere showed good retention at all times of test.

phorylation in the brains of patients with Alzheimer's disease,[186] a disorder characterised by memory loss.

PP1

It remains difficult to elucidate roles for PP1 in complex systems, as all available cell-permeable inhibitors also curtail PP2A activity and most do so with greater potency. While some inhibitors, such as OA, differ in their potency towards the two enzyme classes in vitro,[63] it is not generally considered appropriate to generalize from in vitro to in vivo contexts as many factors can influence drug potencies in complex in vivo systems.[37] Nevertheless, previous studies have used differences in drug potencies to support arguments that either PP1 or PP2A is involved in a particular regulatory system in intact cells.[147,175,202] We have tentative evidence in support of a role for PP1 in a relatively early stage of memory formation based on such an approach.

In our early studies with OA and CalA, both drugs were found to produce two temporally distinct effects on memory formation.[17,224] The retention functions were statistically identical in each case and the two times of onset were relatively independent of time of administration or drug concentration within set ranges. The most parsimonious explanation for these results is that different cellular constituents, underlying different memory stages, might be inhibited by high and low drug concentrations. The later of the two effects, discussed in the preceding section, was most probably due to inhibition of PP2A as it was induced by very low drug concentrations (up to 0.5 nM OA) as well as the higher drug concentrations (100 nM OA) used in our later studies. It seems reasonable to attribute the alternative effect obtained in these experiments, with onset of retention deficits prior to 20 minutes post-training, to inhibition of PP1, since this effect was induced only by concentrations of OA sufficient to inhibit both PP1 and PP2A in vitro.

Of some interest is the later finding that the effect of inhibiting PP1, assuming this to be the case here, was confined to the right hemisphere of the chick, with no evidence of a role for PP1

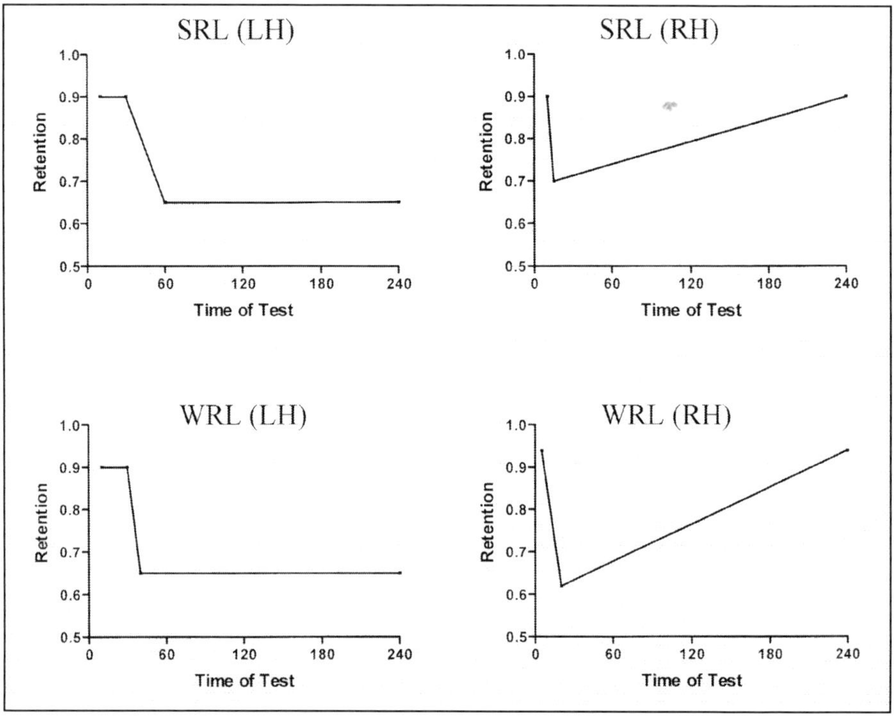

Figure 4. Schematic diagrams depicting the effects of inhibiting PP1 in one hemisphere of the chick following strongly-reinforced learning (SRL) or weekly reinforced learning (WRL). Trained groups of chicks were administered OA (100 nM) or CalA (100nM) into the left (LH) or right (RH) hemisphere and retention was tested at the specified time post-training. Groups of chicks that received either drug in the left hemisphere showed good retention, indicated by selective avoidance of a red coloured bead, only up until approximately 40-50 minutes post-training. All groups tested after this time showed significant retention deficits, most probably attributable to the effects of the drugs on PP2A (see Fig. 3). Groups of chicks administered either drug in the right hemisphere showed good retention only up until approximately 10-20 minutes post-training. Groups tested between 20 and 120 minutes post-training showed variable retention levels, with a gradual recovery being apparent. By 240 minutes post-training, retention levels in right hemisphere, drug-treated, chick groups were identical to those obtained from untreated control groups.

in left hemisphere memory processing being evident (Bennett, Moutsoulas, Lawen & Ng, unpublished observations). The left hemisphere retention deficits depicted in Figure 4 can be fully accounted for by the effects of the drugs on PP2A, as they were also obtained following administration of much lower drug concentrations (Fig. 3). More surprisingly, the effect on retention levels of PP1 inhibition in the right hemisphere was found to be both transient (Fig. 4) and independent of the level of reinforcement associated with the task (data not shown). Retention levels were reduced from before 20 minutes post-training through to after 60 minutes post weakly- or strongly-reinforced training, with drug-treated groups of chicks being statistically indistinguishable from control groups by 2-4 hours post-training. This pattern of results is unprecedented and, hence, difficult to explain. It was not observed when bilateral drug administration took place, but this is probably due to the effects of OA and CalA on PP2A in the left hemisphere, as described above.

It is possible that inhibition of PP1 in the right hemisphere may selectively affect retrieval of information from specific memory stages, rather than affecting memory formation per se. It is possible also that inhibition of PP1 may selectively affect only one or more of several informa-

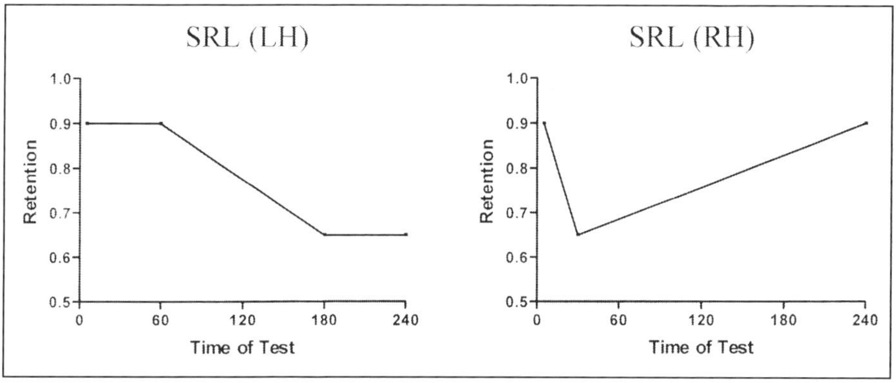

Figure 5. Schematic diagrams showing the effects of inhibiting PP2B in one hemisphere of the chick following strongly-reinforced learning (SRL). Chicks trained using the PAL task were administered either CyA (20 nM) or FK506 (20 nM) and tested for retention at a specified time post-training. Groups of chicks that received either drug in the left hemisphere (LH) showed good retention, indicated by selective avoidance of a red coloured bead, up until approximately 70-80 minutes post-training, an effect also obtained following administration of PPIase inhibitors (data not shown). Groups of chicks administered the drugs in the right hemisphere (RH) showed a pattern of retention deficits similar to those obtained following inhibition of PP1, with good retention for approximately 5-10 minutes, followed by a period of retention loss which had resolved completely by 180-240 minutes post-training.

tion traces that normally form in parallel, only one of which is expressed at any given time. Several additional experiments (Bennett, Moutsoulas, Lawen & Ng, unpublished observations) have provided preliminary support for the second of these conclusions but this has yet to be fully investigated. Were this to be the case, it would support compelling data from other learning paradigms which have recently challenged the traditional view that existing memory stages are sequentially dependent.[46,50,61,91,201]

PP2B

In our first series of studies using the PP2B inhibitor Cyclosporin A (CyA), we demonstrated that memory loss was observed following bilateral drug administration from approximately 70-80 minutes post-training.[16] We later showed (depicted in Fig. 5) that this effect was lateralized to the left hemisphere (Bennett, Moutsoulas, Lawen & Ng, unpublished observations), and that a temporally identical amnestic effect was obtained following administration of an alternative PP2B inhibitor, FK506.[14] Unexpectedly, similar data were obtained following administration of a number of pharmacological analogues which share with CyA and FK506 the ability to inhibit distinct classes of immunophilins, but which do not inhibit PP2B activity.[15] Immunophilins (also called peptidyl prolyl *cis/trans* isomerases or PPIases) catalyse protein folding reactions,[59] are enriched in brain tissue, and have important neuroprotective functions.[176] It was concluded, therefore, that our evidence was consistent with a role for these enzymes, rather than for PP2B, in a later stage of memory formation. Investigation of whether PP2B is also required for memory formation at or after this time awaits the development of more specific inhibitors.

In a subsequent series of studies we proposed that, if PP2B plays a role in memory via activation of PP1, as is suggested by leading theoretical models, then the temporal and spatial parameters of this involvement might be expected to be similar. We investigated this proposal by conducting further experiments with CyA and FK506, in which drug concentrations and times and location of drug administration were systematically varied. Consistent with our proposal, various treatments thought to act by inhibition of PP2B produced results almost identical to those produced by treatments thought to act by inhibition of PP1.[14] The requirement for

PP2B was shown to be localized to the right hemisphere; inhibition of either PP1 or PP2B in this hemisphere produced memory loss with an early time of onset; 10-20 minutes post-training for PP1 inhibition and 5-10 minutes post-training for PP2B inhibition. The retention loss with both classes of inhibitors was found to be transient, regardless of the level of reinforcement associated with the training experience, and both retention functions were very similar, with recovery of memory taking place by approximately 2-3 hours (Fig. 5).

This consistency is made more compelling by the fact that the pattern of retention deficits seen in both cases is so unusual. Few early drug effects on memory have been found to occur following right hemisphere drug administration, although this may be partially because many retention tests are conducted after the time at which memory recovered in our studies, so that transient effects might be overlooked. Recovery of memory following a prolonged period of retention loss is also rare following strongly-reinforced training and, as yet, has not been reported elsewhere following weakly-reinforced training. It seems reasonable to conclude, therefore, that at least one role for PP2B in memory formation in the chick may involve activation of PP1.

The exact nature of this role remains undetermined although, given the model of synaptic plasticity described previously, it is of interest that inhibition of CaMK-II in the right hemisphere of the chick also results in memory loss soon after training.[223] Unlike PP1 and PP2B inhibition, CaMK-II inhibition also produces memory loss in the left hemisphere and all retention deficits induced by CaMK-II appear to be permanent. It may be instructive that CaMK-II appears to have more than one, differentially localised, role in memory for a single-trial, step-down inhibitory avoidance task in rodents,[215] and that only some effects of CaMK-II inhibition in this species are permanent, while others can be attenuated by appropriate experimental protocols.[9] Additional work within the chick model is required to establish whether similar effects may be observed within this species.

Protein Phosphatases in Rodent Learning and Memory

Pharmacological studies examining phosphatase involvement in memory formation in rodents have emerged only recently. One of these found that rats infused with OA for several weeks developed learning difficulties.[4] The rats also demonstrated gross morphological changes, however, making it difficult to determine whether phosphatase inhibition was responsible for the learning deficits or whether these deficits may have reflected a secondary problem.[136] In another study, microinjection of OA into the dorsal hippocampus was found to impair spatial reference and working memory one day later in rats. The impairment was reported to be transient, having disappeared by the following day.[71] Chronic systemic administration of CyA at a concentration used therapeutically has been reported not to impair retention of information in rats.[21] Indeed, FK506 was able to ameliorate a discrimination learning impairment induced by chronic cerebral hypoperfusion in rats,[188] probably due to its well described neuroprotective effects.[65] Using a different approach, Ikegami and Inokuchi[80] found that infusion of antisense DNA against PP2B resulted in enhanced hippocampal-dependent contextual fear learning, although spatial learning performance on a water maze was unaffected.

Gene targeting approaches have also been applied to the issue of phosphatase involvement in memory formation in rodents although, thus far, these techniques have only directly addressed roles for the Ca^{2+}- dependent phosphatase, PP2B. As was described previously, a transgenic mouse strain (CN98) that overexpressed a truncated form of PP2B, under the control of a promoter molecule, in forebrain regions including the hippocampus, was developed by Mansuy et al.[122] and shown by Winder et al.[214] to exhibit only a truncated form of LTP. When hippocampal-dependent forms of learning were investigated in these mice, they were found to be significantly and selectively impaired relative to controls.[121] Just as the capacity for LTP in CN98 mice was reinstated by activation of PKA, however, so spatial learning performance could be improved by providing additional learning trials. This may indicate that PP2B is specifically involved in the transition from short-term memory stages to longer-term stages in the mouse. While this initially appears contradictory with the chick data, which suggested that the effect of PP2B inhibition was independent of the level of reinforcement associated with the training experience, it is of interest that memory loss induced by PP2B disruption in both species appears reversible. Another study found that expression of a PP2B inhibitory domain in the mouse brain reversibly facilitated LTP in vitro and in vivo, enhanced learning, and selectively strengthened short- and long-term memory for several hippocampal-dependent spatial and non-spatial tasks.[120] This again suggests that lowering PP2B activity has a beneficial effect on induction of LTP and some kinds of learning, although it is interesting that LTD induction was not affected, perhaps because PP2B activity was only partially inhibited. Another interesting aspect of this study was that, while PP2B inhibition was able to strengthen or prolong specific memory phases, it was not able to convert short-term memory into a more permanent representation. This again implies that PP2B may modulate rather than mediate memory formation in this species, a conclusion consistent with the data described previously.

A third study attempted to address this issue by reversibly but completely inhibiting a regulatory subunit of PP2B in the forebrain of adult mice.[221] This manipulation resulted in a significant shift in the LTP/LTD threshold, such that LTD was diminished and LTP enhanced. In addition, the mice exhibited a specific deficit in hippocampus-dependent working and episodic-like memory tasks that require single-trial learning, such as the delayed matching-to-place version of the Morris water maze and the working memory version of the 8-arm radial maze. Performance of the mice was normal in hippocampus-dependent reference memory tasks with multiple learning trials, such as contextual fear conditioning and a more standard version of the Morris water maze. A dissociation between the involvement of PP2B in different types of hippocampal-dependent memory tasks may partially explain the anomalous results obtained previously but, given the tentative status of all data available thus far, future research will be required to investigate this possibility. As was mentioned in relation to the role of PP2B in models of synaptic plasticity, it is also possible that different PP2B isoforms have distinct roles in memory formation that are obscured by present techniques. This possibility also awaits future investigation.

Conclusion

All existing evidence points to the view that information storage in biological organisms may rest upon a long and complex sequence of more-or-less transient events involving post-translational modifications to existing proteins. These events, which persist for at least two or more hours post-training in the chick, may act to store information while a decision is made as to whether permanent consolidation is appropriate. Our own research group has produced convincing evidence that each of five major kinase groups are required for permanent consolidation, each one playing a temporally specific role. Each of three classes of Ser/Thr phosphatases have also now been shown to be essential, suggesting that dephosphorylation may be as important as phosphorylation in ensuring that permanent storage takes place. Thus far, the pharmaco-behavioural results obtained by our group are broadly commensurate with more recent studies using gene technologies in rodent learning models. They are also consistent with the many studies implicating phosphatases in processes underlying various forms of synaptic plasticity. One might argue that an apparent inability to identify a kinase or phosphatase not implicated in learning and memory formation is problematic, since it implies a lack of specificity. The temporal specificity of the results obtained thus far, however, and the impressive similarities across studies from vastly different species and models of synaptic plasticity, is perhaps more consistent with the view that memory formation is indeed as complex as it is fascinating, with a carefully orchestrated cascade of events underlying our capacity to learn and retain information about events in our internal and external environments. While our present understanding of this cascade is limited, the advances made in the last decade are impressive, providing a solid scaffolding upon which future research can build.

References

1. Ali SM, Bullock S, Rose SPR. Phosphorylation of synaptic proteins in chick forebrain: changes with development and passive avoidance training. J Neurochem 1988; 50:1579-1587.
2. Allen PB, Hvalby O, Jensen V et al. Protein phosphatase-1 regulation in the induction of long-term potentiation: heterogeneous molecular mechanisms. J Neurosci 2000; 20:3537-3543.
3. Antoni FA, Smith SM, Simpson J et al. Calcium control of adenylyl cyclase: the calcineurin connection. Adv Second Messenger Phosphoprotein Res 1998; 32:153-172.
4. Arendt T, Holzer M, Fruth R et al. Paired helical filament-like phosphorylation of tau, deposition of beta/A4-amyloid and memory impairment in rat induced by chronic inhibition of phosphatase 1 and 2A. Neurosci 1995; 69:691-698.
5. Aszodi A, Muller U, Friedrich P et al. Signal convergence on protein kinase A as a molecular correlate of learning. Proc Natl Acad Sci USA 1991; 88:5832-5836.
6. Asztalos Z, von Wegerer J, Wustmann G et al. Protein phosphatase 1-deficient mutant Drosophila is affected in habituation and associative learning. J Neurosci 1993; 13:924-930.
7. Barnes GN, Slevin JT, Vanaman TC. Rat brain protein phosphatase 2A: an enzyme that may regulate autophosphorylated protein kinases. J Neurochem 1995; 64:340-353.
8. Barria A, Muller D, Derkach V et al. Regulatory phosphorylation of AMPA-type glutamate receptors by CaMKII during long-term potentiation. Science 1997; 276:2042-2045.
9. Barros DM. Stimulators of the cAMP cascade reverse amnesia induced by intra-amygdala but not intrahippocampal KN-62 administration. Neurobiol Learn Mem 1999; 71:94-103.
10. Barzilai A, Kennedy TE, Sweatt JD et al. 5-HT modulates protein synthesis and the expression of specific proteins during long-term facilitation in Aplysia sensory neurons. Neuron 1989; 2:1577-1586.
11. Baudry M. Synaptic plasticity and learning and memory: 15 years of progress. Neurobiol Learn Mem 1998; 70:113-118.
12. Bear MF, Abraham WC. Long-term depression in hippocampus. Ann Rev Neurosci 1996; 19:437-462.
13. Beattie EC, Carroll RC, Yu X et al. Regulation of AMPA receptor endocytosis by a signalling mechanism shared with LTD. Nat Neurosci 2000; 3:1291-1300.
14. Bennett PC, Schmidt L, Lawen A et al. Cyclosporin A, FK506 and rapamycin produce multiple, temporally distinct, effects on memory following single-trial, passive avoidance training in the chick. Brain Res 2002; 927:180-194.
15. Bennett PC, Singaretnam LG, Zhao WQ et al. Peptidyl-prolyl-cis/trans-isomerase activity may be necessary for memory formation. FEBS Letts 1998; 431:386-390.
16. Bennett PC, Zhao W, Lawen A et al. Cyclosporin A, an inhibitor of calcineurin, impairs memory formation in day-old chicks. Brain Res 1996; 730:107-117.
17. Bennett PC, Zhao WQ, Ng KT. Concentration-dependent effects of protein phosphatase (PP) inhibitors implicate PP1 and PP2A in different stages of memory formation. Neurobiol Learn Mem 2001; 75:91-110.
18. Bliss TVP, Gardner-Medwin AR. Long-lasting potentiation of synaptic transmission in the dentate area of the unanaesthetized rabbit following stimulation of the perforant path. J Physiol 1973; 232:357-374.
19. Bliss TVP, Lomo T. Long-lasting potentiation of synaptic transmission in the dentate area of the anaesthetized rabbit following stimulation of the perforant path. J Physiol 1973; 232:331-356.
20. Blitzer RD, Connor JH, Brown GP et al. Gating of CaMKII by cAMP-regulated protein phosphatase activity during LTP. Science 1998; 280:1940-1942.
21. Borlongan CV, Fujisaki T, Watanabe S. Chronic administration of cyclosporine A does not impair memory retention in rats. NeuroReport 1997; 8:673-676.
22. Brautigan DL. Flicking the switches: phosphorylation of serine/threonine protein phosphatases. Semin Cancer Biol 1995; 6:211-217.
23. Brennan PA. The effects of local inhibition of N-methyl-D-aspartate and AMPA/kainate receptors in the accessory olfactory bulb on the formation of an olfactory memory in mice. Neurosci 1994; 60:701-708.
24. Brown GP, Blitzer RD, Connor JH et al. Long-term potentation induced by theta frequency stimulation is regulated by a protein phosphatase-1-operated gate. J Neurosci 2000; 20:7880-7887.
25. Buxbaum JD, Dudai Y. A quantitative model for the kinetics of cAMP-dependent protein kinase (type II) activity. J Biol Chem 1989; 264:9344-9351.
26. Byrne JH, Kandel ER. Presynaptic facilitation revisited: state and time dependence. J Neurosci 1996; 16:425-435.
27. Byrne JH, Zwartjes R, Homayouni R et al. Roles of second messenger pathways in neuronal plasticity and in learning and memory. Insights gained from Aplysia. Adv Second Messenger Phosphoprotein Res 1993; 27:47-108.

28. Cammarota M, Paratcha G, de Stein L et al. B-50/GAP-43 phosphorylation and PKC activity are increased in rat hippocampal synaptosomal membranes after an inhibitory avoidance training. Neurochem Res 1997; 22:499-505.
29. Chan SF, Sucher NJ. An NMDA receptor signaling complex with protein phosphatase 2A. J Neurosci 2001; 21:7985-7992.
30. Chen SJ, Sweatt JD, Klann E. Enhanced phosphorylation of the postsynaptic protein kinase C substrate RC3/neurogranin during long-term potentiation. Brain Res 1997; 749:181-187.
31. Choi KW, Smith RF, Buratowski RM et al. Deficient protein kinase C activity in turnip, a Drosophila learning mutant. J Biol Chem 1991; 266:15999-606.
32. Clapham DE. Calcium signalling. Cell 1995; 80:259-268.
33. Clayton NS, Krebs JR. Lateralization in Paridae: comparison of a storing and a non-storing species on a one-trial associative memory task. J Comp Physiol [B] 1993; 171:807-815.
34. Cohen P. The discovery of protein phosphatases: from chaos and confusion to an understanding of their role in cell regulation and human disease. Bioessays 1994; 16:583-588.
35. Cohen P. Novel protein serine/threonine phosphatases: variety is the spice of life. Trends Biochem Sci 1997; 22:245-251.
36. Cohen P, Brewis ND, Hughes V et al. Protein serine/threonine phosphatases; an expanding family. FEBS Letts 1990; 268:355-359.
37. Cohen P, Holmes CF, Tsukitani Y. Okadaic acid: a new probe for the study of cellular regulation. Trends Biochem Sci 1990; 15:98-102.
38. Cooper DM, Mons N, Karpen JW. Adenylyl cyclases and the interaction between calcium and cAMP signalling. Nature 1995; 374:421-424.
39. Cory-Slechta DA. The impact of NMDA receptor antagonists on learning and memory functions. Psychopharm Bull 1994; 30:601-612.
40. Critz SD, Endo S, Shenolikar S et al. Protein phosphatase-1 inhibitors attenuate 5-HT modulated K+ currents in Aplysia sensory neurons. Soc Neursci Abstr 1991; 17:69.
41. Cummings JA, Mulkey RM, Nicoll RA et al. Ca^{2+} signaling requirements for long-term depression in the hippocampus. Neuron 1996; 16:825-833.
42. Dash PK, Moore AN. Characterization and phosphorylation of CREB-like proteins in Aplysia central nervous system. Mol Brain Res 1996; 39:43-51.
43. Dash PK, Hochner B, Kandel ER. Injection of the cAMP-responsive element into the nucleus of Aplysia sensory neurons blocks long-term facilitation. Nature 1990; 345:718-721.
44. Dosemeci A, Reese TS. Inhibition of endogenous phosphatase in a postsynaptic density fraction allows extensive phosphorylation of the major postsynaptic density protein. J Neurochem 1993; 61:550-555.
45. Dragunow M. A role for immediate-early transcription factors in learning and memory. Behav Genet 1996; 26:293-299.
46. Dubnau J, Tully T. Gene discovery in Drosophila: new insights for learning and memory. Ann Rev Neurosci 1998; 21:407-444.
47. Dunkley PR, Robinson PJ. Depolarization-dependent protein phosphorylation in synaptosomes: mechanisms and significance. Prog Brain Res 1986; 69:273-293.
48. Edwards TM, Rickard NS, Ng KT. Inhibition of guanylate cyclase and protein kinase g impairs retention for the passive avoidance task in the day-old chick. Neurobiol Learn Mem 2002; 77:313-326.
49. Eisenstein EM. Selecting a model system for neurobiological studies of learning and memory. Behav Brain Res 1997; 82:121-132.
50. Emptage NJ, Carew TJ. Long-term synaptic facilitation in the absence of short-term facilitation in Aplysia neurons. Science 1993; 262:253-256.
51. Endo S, Shenolikar S, Byrne JH. Characterization of a particulate protein phosphatase in neurons from Aplysia. Soc Neursci Abstr 1991; 17:69.
52. Endo S, Shenolikar S, Eskin A et al. Characterization of neuronal protein phosphatases in Aplysia californica. J Neurochem 1992; 58:975-982.
53. Fagnou DD, Tuchek JM. The biochemistry of learning and memory. Mol Cell Biochem 19951; 49-150:279-286.
54. Fernandez JJ, Candenas ML, Souto ML et al. Okadaic acid: useful tool for studying cellular processes. Curr Med Chem 2002; 9:229-262.
55. Figurov A, Boddeke H, Muller D. Enhancement of AMPA-mediated synaptic transmission by the protein phosphatase inhibitor calyculin A in rat hippocampal slices. Eur J Neurosci 1993; 5:1035-1041.
56. Fukunaga K, Kobayashi T, Tamura S et al. Dephosphorylation of autophosphorylated Ca2+/calmodulin-dependent protein kinase II by protein phosphatase 2C. J Biol Chem 1993; 268:133-137.

57. Fukunaga K, Muller D, Ohmitsu M et al. Decreased protein phosphatase 2A activity in hippocampal long-term potentiation. J Neurochem 2000; 74:807-817.
58. Funauchi M, Haruta H, Tsumoto T. Effects of an inhibitor for calcium/calmodulin-dependent protein phosphatase, calcineurin, on induction of long-term potentiation in rat visual cortex. Neurosci Res 1994; 19:269-278.
59. Galat A, Metcalfe SM. Peptidylproline cis/trans isomerases. Prog Biophys Mol Biol 1995; 63:67-118.
60. Gelperin A. Flies, genes, and memory engineering. Biol Bull 1996; 191:139-143.
61. Ghirardi M, Montarolo PG, Kandel ER. A novel intermediate stage in the transition between short- and long-term facilitation in the sensory to motor neuron synapse of aplysia. Neuron 1995; 14:413-420.
62. Ghosh A, Greenberg ME. Calcium signaling in neurons: molecular mechanisms and cellular consequences. Science 1995; 268:239-247.
63. Girault JA. Protein phosphorylation and dephosphorylation in mammalian central nervous system. Neurochem Int 1993; 23:1-25.
64. Gispen WH, Perumal R, Wilson JE et al. Phosphorylation of proteins of synaptosome-enriched fractions of brain during short-term training experience: the effects of various behavioural treatments. Behav Biol 1977; 21:358-363.
65. Gold BG. FK506 and the role of immunophilins in nerve regeneration. Mol Neurobiol 1997; 15:285-306.
66. Gold PE, McGaugh JL. A single-trace, two-process view of memory storage processes. In: Deutsch D, Deutsch JA, eds. Short-term memory. New York: Academic Press. 1975; 355-378.
67. Gomez-Puertas P, Martinez-Serrano A, Blanco P et al. Conditions restricting depolarization-dependent calcium influx in synaptosomes reveal a graded response of P96 dephosphorylation and a transient dephosphorylation of P65. J Neurochem 1991; 56:2039-2047.
68. Greengard P, Valtorta F, Czernik AJ et al. Synaptic vesicle phosphoproteins and regulation of synaptic function. Science 1993; 259:780-785.
69. Halpain S, Greengard P. Activation of NMDA receptors induces rapid dephosphorylation of the cytoskeletal protein MAP2. Neuron 1990; 5:237-246.
70. Halpain S, Girault JA, Greengard P. Activation of NMDA receptors induces dephosphorylation of DARPP-32 in rat striatal slices. Nature 1990; 343:369-372.
71. He J, Yamada K, Zou LB et al. Spatial memory deficit and neurodegeneration induced by the direct injection of okadaic acid into the hippocampus in rats. J Neural Trans 2001; 108:1435-1443.
72. Hebb DO. The organization of behaviour: a neuropsychological theory. New York: Wiley. 1949.
73. Herron CE, Malenka RC. Activity-dependent enhancement of synaptic transmission in hippocampal slices treated with the phosphatase inhibitor calyculin A. J Neurosci 1994; 14:6013-6020.
74. Heynen AJ, Abraham WC, Bear MF. Bidirectional modification of CA1 synapses in the adult hippocampus in vivo. Nature 1996; 381:163-166.
75. Hodgkiss JP, Kelly JS. Only 'de novo' long-term depression (LTD) in the rat hippocampus in vitro is blocked by the same low concentration of FK506 that blocks LTD in the visual cortex. Brain Res 1995; 705:241-246.
76. Huang EP. Synaptic plasticity: going through phases with LTP. Curr Biol 1998; 8:R350-R352.
77. Huang FL, Glinsmann WH. Separation and characterization of two phosphorylase phosphatase inhibitors from rabbit skeletal muscle. Eur J Biochem 1976; 70:419-426.
78. Hudmon A, Schulman H. Neuronal Ca2+/calmodulin-dependent protein kinase II: the role of structure and autoregulation in cellular function. Ann Rev Biochem 2002; 71:473-510.
79. Ichinose M, Byrne JH. Role of protein phosphatases in the modulation of neuronal membrane currents. Brain Res 1991; 549:146-150.
80. Ikegami S, Inokuchi K. Antisense DNA against calcineurin facilitates memory in contextual fear conditioning by lowering the threshold for hippocampal long-term potentiation induction. Neurosci 2000; 98:637-646.
81. Ikegami S, Kato A, Kudo Y et al. A facilitatory effect on the induction of long-term potentiation in vivo by chronic administration of antisense oligodeoxynucleotides against catalytic subunits of calcineurin. Brain Res Mol Brain Res 1996; 41:183-191.
82. Inagaki N, Ito M, Nakano T et al. Spatiotemporal distribution of protein kinase and phosphatase activities. Trends Biochem Sci 1994; 19:448-452.
83. Ingebritsen TS, Cohen P. The protein phosphatases involved in cellular regulation. 1. Classification and substrate specificities. Eur J Biochem 1983; 132:255-261.
84. Ito A, Hashimoto T, Hirai M et al. The complete primary structure of calcineurin A, a calmodulin binding protein homologous with protein phosphatases 1 and 2A. Biochem Biophys Res Commun 1989; 163:1492-1497.
85. Ito M. Cerebellar long-term depression: characterization, signal transduction, and functional roles. Physiol Rev 2001; 81:1143-1195.

86. Izquierdo I, McGaugh JL. Behavioural pharmacology and its contribution to the molecular basis of memory consolidation. Behav Pharmacol 2000; 11:517-534.
87. Izquierdo I, Medina JH. Memory formation: the sequence of biochemical events in the hippocampus and its connection to activity in other brain structures. Neurobiol Learn Mem 1997; 68:285-316.
88. Izquierdo I, Medina JH, Vianna MRM et al. Seperate mechanisms for short- and long-term memory. Behav Brain Res 1999; 103:1-11.
89. Kameyama K, Lee HK, Bear MF et al. Involvement of a postsynaptic protein kinase A sustrate in the expression of homosynaptic long term depression. Neuron 1998; 21:1163-1175.
90. Kandel ER, Schwartz JH. Molecular biology of learning: modulation of transmitter release. Science 1982; 218:433-443.
91. Kane NS, Robichon A, Dickinson JA et al. Learning without performance in PKC-deficient Drosophila. Neuron 1997; 18:307-314.
92. Kasahara J, Fukunaga K, Miyamoto E. Differential effects of a calcineurin inhibitor on glutamate-induced phosphorylation of Ca^{2+}/calmodulin-dependent protein kinases in cultured rat hippocampal neurons. J Biol Chem 1999; 274:9061-9067.
93. Kemp N, Bashir ZI. Long-term depression: a cascade of induction and expression mechanisms. Prog Neurobiol 2001; 65:339-365.
94. Kety SS. Brain catecholamines, affective states and memory. In: McGaugh JL, ed. The chemistry of mood, motivation and memory. New York, Plenum Press, 1972; 65-80.
95. Kirkwood A, Bear MF. Homosynaptic long-term depression in the visual cortex. J Neurosci 1994; 14:3404-3412.
96. Klee CB, Ren H, Wang X. Regulation of the calmodulin-stimulated protein phosphatase, calcineurin. J Biol Chem 1998; 273:13367-13370.
97. Kornhauser JM, Greenberg ME. A kinase to remember: dual roles for MAP kinase in long-term memory. Neuron 1997; 18:839-842.
98. Krasne FB, Glanzman D. What we can learn from invertebrate learning? Ann Rev Psychol 1995; 46:585-624.
99. Lee HK, Barbarosie M, Kameyama K et al. Regulation of distinct AMPA receptor phosphorylation sites during bidirectional synaptic plasticity. Nature 2000; 405:955-959.
100. Levenes C, Daniel H, Crepel F. Long-term depression of synaptic transmission in the cerebellum: cellular and molecular mechanisms revisited. Prog Neurobiol 1998; 55:79-91.
101. Levitan IB. Modulation of ion channels by protein phosphorylation and dephosphorylation. Ann Rev Physiol 1994; 56:193-212.
102. Liang KC, Hon W, Tyan YM et al. Involvement of hippocampal NMDA and AMPA receptors in acquisition, formation and retrieval of spatial memory in the Morris water maze. Chinese J Physiol 1994; 37:201-212.
103. Lindberg RA, Quinn AM, Hunter T. Dual-specificity protein kinases: will any hydroxyl do? Trends Biochem Sci 1992; 17:114-119.
104. Lisman JE. A mechanism for memory storage insensitive to molecular turnover: a bistable autophosphorylating kinase. Proc Natl Acad Sci USA 1985; 82:3055-3057.
105. Lisman JE. A mechanism for the Hebb and the anti-Hebb processes underlying learning and memory. Proc Natl Acad Sci USA 1989; 86:9574-9578.
106. Lisman JE, Zhabotinsky AM. A model of synaptic memory: a CaMKII/PP1 switch that potentiates transmission by organising an AMPA receptor anchoring assembly. Neuron 2001; 31:191-201.
107. Lisman J, Schulman H, Cline H. The molecular basis of CaMKII function in synaptic and behavioural memory. Nat Rev Neurosci 2002; 3:175-190.
108. Lu YF, Hayashi Y, Moriwaki A et al. FK506, a Ca^{2+}/calmodulin-dependent phosphatase inhibitor, inhibits the induction of long-term potentiation in the rat hippocampus. Neurosci Letts 1996; 205:103-106.
109. Lu YM, Mansuy IM, Kandel ER et al. Calcineurin-mediated LTD of GABAergic inhibition underlies the increased exitability of CA1 neurons associated with LTP. Neuron 2000; 26:197-205.
110. Lu YF, Tomizawa K, Moriwaki A et al. Calcineurin inhibitors, FK506 and cyclosporin A, suppress the NMDA receptor-mediated potentials and LTP, but not depotentiation in the rat hippocampus. Brain Res 1996; 729:142-146.
111. Luthi A, Chittajallu R, Duprat F et al. Hippocampal LTD expression involves a pool of AMPARs regulated by the NSF-GluR2 interaction. Neuron 1999; 24:389-399.
112. Lyn-Cook BD, Wilson JE. Effects of experiences on synaptic protein phosphorylation in vitro. Pharmacol Biochem Behav 1983; 18:949-952.
113. Lyn-Cook BD, Ruder FJ, Wilson JE. Regulation of phosphate incorporation into four brain phosphoproteins that are affected by experience. J Neurochem 1985.44:552-559.
114. Lynch G, Dunwiddie TV, Gribkoff VK. Heterosynaptic depression: a post-synaptic correlate of long-term potentiation. Nature 1977; 266:737-739.

115. Machlus B, Entingh D, Wilson JE et al. Brain phosphoproteins: the effect of various behaviors and reminding experiences on the incorporation of radioactive phosphate into nuclear proteins. Behav Biol 1974; 10:63-73.
116. Machlus BJ, Wilson JE, Glassman E. Brain phosphoproteins: the effect of short experiences on the phosphorylation of nuclear proteins of rat brain. Behav Biol 1974; 10:43-62.
117. MacMillan LB, Bass MA, Chang N et al. Brain actin-associated protein phosphatase 1 holenzymes containing spinophilin, neurabin, and selected catalytic subunit isoforms. J Biol Chem 1999; 274:35845-35854.
118. Malenka RC. Long-term depression: not so depressing after all. Proc Natl Acad Sci USA 1993; 90:3121-3123.
119. Malenka RC, Nicoll RA. Long-term potentiation—a decade of progress. Science 1999; 285:1870.
120. Malleret G, Haditsch U, Genoux D et al. Inducible and reversible enhancement of learning, memory and long-term potentiation by genetic inhibition of calcineurin. Cell 2001; 104:675-686.
121. Mansuy IM, Mayford M, Jacob B et al. Restricted and regulated overexpression reveals calcineurin as a key component in the transition from short-term to long-term memory. Cell 1998; 92:39-49.
122. Mansuy IM, Winder DG, Moallem TM et al. Inducible and reversible gene expression with the rtTA system for the study of memory. Neuron 1998; 21:257-265.
123. Maren S, Baudry M. Properties and mechanisms of long-term synaptic plasticity in the mammalian brain: relationships to learning and memory. Neurobiol Learn Mem 1995; 63:1-18.
124. Martin SJ, Grimwood PD, Morris RGM. Synaptic plasticity and memory: an evaluation of the hypothesis. Ann Rev Neurosci 2000; 23:649-711.
125. Mauelshagen J, Parker GR, Carew TJ. Dynamics of induction and expression of long-term synaptic facilitation in Aplysia. J Neurosci 1996; 16:7099-7108.
126. McGaugh JL. Time-dependent processes in memory storage. Science 1966; 153:1351-1358.
127. McGaugh JL. Involvement of hormonal and neuromodulatory systems in the regulation of memory storage. Ann Rev Neurosci 1989; 12:255-287.
128. McGaugh JL, Izquierdo I. The contribution of pharmacology to research on the mechanisms of memory formation. Trends Pharmacol Sci 2000; 21:208-210.
129. Merlevede W. Protein phosphatases and the protein phosphatases. Landmarks in an eventful century. Adv Prot Phosphatases 1985.1:1-18.
130. Micheau J, Riedel G. Protein kinases: which one is the memory molecule? Cell Mol Life Sci 1999; 55:534-548.
131. Millward TA, Zolnierowicz S, Hemmings BA. Regulation of protein kinase cascades by protein phosphatase 2A. Trends Biochem Sci 1999; 24:186-191.
132. Mons N, Guillou J-L, Jaffard R. The role of Ca2+/calmodulin-stimulable adenylyl cyclases as molecular coincidence detectors in memory formation. Cell Mol Life Sci 1999; 55:525-533.
133. Montoro RJ, Diaz-Nido J, Avila J et al. N-methyl-D-aspartate stimulates the dephosphorylation of the microtubule-associated protein 2 and potentiates excitatory synaptic pathways in the rat hippocampus. Neurosci 1993; 54:859-871.
134. Morishita W, Connor JH, Xia H et al. Regulation of synaptic strength by protein phosphatase 1. Neuron 2001; 32:1133-1148.
135. Moss SJ, Smart TG. Modulation of amino acid-gated ion channels by protein phosphorylation. Int Rev Neurobiol 1996; 39:1-52.
136. Muhder AK, Perry VH. Using okadaic acid as a tool for the in vivo induction of hyperphosphorylated tau. Neurosci 1998; 85:1329-1332.
137. Mulkey RM, Endo S, Shenolikar S et al. Involvement of a calcineurin/inhibitor-1 phosphatase cascade in hippocampal long-term depression. Nature 1994; 369:486-488.
138. Mulkey RM, Herron CE, Malenka RC. An essential role for protein phosphatases in hippocampal long-term depression. Science 1993; 261:1051-1055.
139. Muller D, Hefft S, Figurov A. Heterosynaptic interactions between LTP and LTD in CA1 hippocampal slices. Neuron 1995; 14:599-605.
140. Muzzio IA, Ramirez RR, Talk AC et al. Interactive contributions of intracellular calcium and protein phosphatases to massed-trials learning deficits in Hermissenda. Behav Neurosci 1999; 113:103-117.
141. Nestler EJ, Greengard P. Neuron-specific phosphoproteins in mammalian brain. Adv Cyclic Nucleotide Prot Phosphorylation Res 1984; 17:483-488.
142. Newton AC. Regulation of protein kinase C. Curr Opin Cell Biol 1997; 9:161-167.
143. Ng KT. Reinforcement, glucose metabolism and memory formation: a possible role for astrocytes. Aust J Psychol 1997; 49:156-163.
144. Ng KT, Gibbs ME, Crowe SF et al. Molecular mechanisms of memory formation. Mol Neurobiol 1991; 5:333-350.

145. Norman ED, Thiels E, Barrionuevo G et al. Long term depression in the hippocampus in vivo is assocaited with protein phosphatase-dependent alterations in extracellular-regulated kinase. J Neurochem 2000; 74:192-198.
146. O'Dell TJ, Kandel ER. Low-frequency stimulation erases LTP through an NMDA receptor-mediated activation of proteins phosphatases. Learn Mem 1994; 1:129-139.
147. Obara K, Yabu H. Dual effect of phosphatase inhibitors on calcium channels in intestinal smooth muscle cells. Amer J Physiol 1993; 264:C296-C301.
148. Oliet SH, Malenka RC, Nicoll RA. Two distinct forms of long-term depression coexist in CA1 hippocampal pyramidal cells. Neuron 1997; 18:969-982.
149. Orgad S, Dudai Y, Cohen P. The protein phosphatases of Drosophila melanogaster and their inhibitors. Eur J Biochem 1987; 164:31-38.
150. Pant HC, Veeranna. Neurofilament phosphorylation. Biochem Cell Biol 1995; 73:575-592.
151. Pasinelli P, Ramakers GM, Urban IJ et al. Long-term potentiation and synaptic protein phosphorylation. Behav Brain Res 1995; 66:53-59.
152. Pereira P, Ardenghi P, de Souza MM et al. Effects of infusions of the tyrosine kinase inhibitor radicicol into the hippocampus on short- and long-term memory of the inhibitory avoidance task. Behav Pharmacol 2001; 12:299-302.
153. Perumal R, Gispen WH, Wilson JE et al. Phosphorylation of proteins from the brains of mice subjected to short-term behavioral experiences. Prog Brain Res 1975; 42:201-207.
154. Price NE, Mumby MC. Brain protein serine/threonine phosphatases. Curr Opin Neurobiol 1999; 9:336-342.
155. Regehr WG, Tank DW. Dendritic calcium dynamics. Curr Opin Neurobiol 1994; 4:373-382.
156. Rickard NS, Poot AC, Gibbs ME et al. Both non-NMDA and NMDA glutamate receptors are necessary for memory consolidation in the day-old chick. Behav Neural Biol 1994; 62:33-40.
157. Riedel G. If phosphatases go up, memory goes down. Cell Mol Life Sci 1999; 55:549-553.
158. Riedel G, Wetzel W, Reymann KG. Comparing the role of metabotropic glutamate receptors in long-term potentiation and in learning and memory. Prog Neuro-Psychopharmacol Biol Psychiatry 1996; 20:761-789.
159. Rodnight R. Aspects of protein phosphorylation in the nervous system with particular reference to synaptic transmission. Prog Brain Res 1982; 56:1-25.
160. Rodnight R, Wofchuk ST. Roles for protein phosphorylation in synaptic transmission. Essays Biochem 1992; 27:91-102.
161. Rose SPR. God's organism? the chick as a model system for memory studies. Learn Mem 2000; 7:1-17.
162. Routtenberg A. Phosphoprotein regulation of memory formation: enhancement and control of synaptic plasticity by protein kinase C and protein F1. Ann N Y Acad Sci 1985; 444:203-211.
163. Rumelhart DE, McClelland JL, PDP Research Group. Parallel distributed processing: explorations in the microstructure of cognition. Cambridge, MIT Press, 1986.
164. Sacks DB, Mazus B, Joyal JL. The activity of calmodulin is altered by phosphorylation: modulation of calmodulin function by the site of phosphate incorporation. Biochem J 1995; 312:197-204.
165. Scanziani M, Malenka RC, Nicoll RA. Role of intercellular interactions in heterosynaptic long-term depression. Nature 1996; 380:446-450.
166. Schmidt HH, Pollock JS, Nakane M et al. Ca^{2+}/calmodulin-regulated nitric oxide synthases. Cell Calcium 1992; 13:427-434.
167. Searleman A, Herrmann D. Memory from a broader perspective. New York: McGraw-Hill Inc., 1994.
168. Seki K, Chen HC, Huang KP. Dephosphorylation of protein kinase C substrates, neurogranin, neuromodulin, and MARCKS, by calcineurin and protein phosphatases 1 and 2A. Arch Biochem Biophys 1995; 316:673-679.
169. Selcher JC, Weeber EJ, Varga AW et al. Protein kinase signal transduction cascades in mammalian associative conditioning. Neuroscientist 2002; 8:122-131.
170. Shaw CA, Lanius RA, van den Doel A. The origin of synaptic neuroplasticity: crucial molecules or a dynamical cascade? Brain Res Rev 1994; 19:241-263.
171. Shenolikar S. Protein phosphatase regulation by endogenous inhibitors. Semin Cancer Biol 1995; 6:219-227.
172. Sherrington C. The integrative action of the nervous system. Cambridge: Cambridge University Press, 1948.
173. Sheu FS, McCabe BJ, Horn G et al. Learning selectively increases protein kinase C substrate phosphorylation in specific regions of the chick brain. Proc Natl Acad Sci USA 1993; 90:2705-2709.
174. Skoulakis EM, Davis RL. Olfactory learning deficits in mutants for leonardo, a Drosophila gene encoding a 14-3-3 protein. Neuron 1996; 17:931-944.

175. Slack JR, Pockett S. Cyclic AMP induces long-term increase in synaptic efficacy in CA1 region of rat hippocampus. Neurosci Letts 1991; 130:69-72.
176. Snyder SH, Sabatini DM, Lai MM et al. Neural actions of immunophilin ligands. Trends Pharmacol Sci 1998; 19:21-26.
177. Soderling TR, Derkach VA. Postsynaptic protein phosphorylation and LTP. Trends Neurosci 2000; 23:75-80.
178. Souza DO, Elisabetsky E, Izquierdo I. Effect of various forms of training and stimulation on the incorporation of 32P into the nuclear phosphoproteins of the rat brain. Pharmacol Biochem Behav 1980; 12:481-486.
179. Squire LR. Memory and the hippocampus: a synthesis from findings with rats, monkeys, and humans. Psychol Rev 1992; 99:195-231.
180. Staubli U, Scafidi J. Studies on long-term depression in area CA1 of the anesthetized and freely moving rat. J Neurosci 1997; 17:4820-4828.
181. Strack S, Barban MA, Wadzinski BE et al. Differential inactivation of postsynaptic density-associated and soluble Ca^{2+}/calmodulin-dependent protein kinase II by protein phosphatases 1 and 2A. J Neurochem 1997; 68:2119-2128.
182. Suzuki T. Protein kinases involved in the expression of long-term potentiation. Int J Biochem 1994; 26:735-744.
183. Suzuki T. Messengers from the synapses to the nucleus (MSNs) that establish late phase of long-term potentiation (LTP) and long-term memory. Neurosci Res 1996; 25:1-6.
184. Sweatt JD, Kandel ER. Persistent and transcriptionally-dependent increase in protein phosphorylation in long-term facilitation of Aplysia sensory neurons. Nature 1989; 339:51-54.
185. Sweatt JD, Volterra A, Edmonds B et al. FMRFamide reverses protein phosphorylation produced by 5-HT and cAMP in Aplysia sensory neurons. Nature 1989; 342:275-278.
186. Szucs K, Ledesma MD, Dombradi V et al. Dephosphorylation of tau protein from Alzheimer's disease patients. Neurosci Letts 1994; 165:175-178.
187. Takaishi T, Saito N, Kuno T et al. Differential distribution of the mRNA encoding two isoforms of the catalytic subunit of calcineurin in the rat brain. Biochem Biophys Res Commun 1991; 174:393-398.
188. Tanaka K, Hori K, Wada-Tanaka N et al. FK506 ameliorates the discrimination learning impairment due to preventing the rarefaction of white matter induced by chronic cerebral hypoperfusion in rats. Brain Res 2001; 906:184-189.
189. Tavalin SJ, Colledge M, Hell JW et al. Regulation of GluR1 by the A-kinase anchoring protein 79 (AKAP79) signaling complex shares properties with long-term depression. J Neurosci 2002; 22:3044-3051.
190. Thiels E, Kanterewicz BI, Knapp LT et al. Protein phosphatase-mediated regulation of protein kinase C during long-term depression in the adult hippocampus in vivo. J Neurosci 2000; 20:7199-7207.
191. Thiels E, Norman ED, Barrionuevo G et al. Transient and persistant increases in protein phosphatase activity during long-term depression in the adult hippocampus in vivo. Neurosci 1998; 86:1023-1029.
192. Thiels E, Xie X, Yeckel MF et al. NMDA receptor-dependent LTD in different subfields of hippocampus in vivo and in vitro. Hippocampus 1996; 6:43-51.
193. Thompson RF, Kim JJ. Memory systems in the brain and localization of a memory. Proc Natl Acad Sci USA 1996; 93:13438-13444.
194. Tokuda M, Hatase O. Regulation of neuronal plasticity in the central nervous system by phosphorylation and dephosphorylation. Mol Neurobiol 1998; 17:137-156.
195. Tong G, Shepherd D, Jahr CE. Synaptic Desensitization of NMDA Receptors by Calcineurin. Science 1995; 267:1510-1512.
196. Torii N, Kamishita T, Otsu Y et al. An inhibitor for calcineurin, FK506, blocks induction of long-term depression in rat visual cortex. Neurosci Letts 1995; 185:1-4.
197. Traynelis SF, Wahl P. Control of rat GluR6 glutamate receptor open probability by protein kinase A and calcineurin. J Physiol 1997; 503:513-531.
198. Tsumoto T, Yasuda H. A switching role of postsynaptic calcium in the induction of long-term potentiation of long-term depression in visual cortex. Semin Neurosci 1996; 8:311-319.
199. Tully T. Discovery of genes involved with learning and memory: an experimental synthesis of Hirschian and Benzerian perspectives. Proc Natl Acad Sci USA 1996; 93:13460-13467.
200. Tully T, Boynton S, Brandes C et al. Genetic dissection of memory formation in Drosophila melanogaster. Cold Spring Harb Symp Quant Biol 1990; 55:203-211.
201. Vianna MRM, Izquierdo LA, Barros DM et al. Short- and long-term memory: differential involvement of neurotransmitter systems and signal transduction cascades. An Acad Bras Cienc 2000; 72:353-364.

202. Vinade L, Rodnight R. The dephosphorylation of glial fibrillary acidic protein (GFAP) in the immature rat hippocampus is catalyzed mainly by a type 1 protein phosphatase. Brain Res 1996; 732:195-200.
203. Vogel HJ. The Merck Frosst Award Lecture 1994. Calmodulin: a versatile calcium mediator protein. Biochem Cell Biol 1994; 72:357-376.
204. Walaas SI, Greengard P. Protein phosphorylation and neuronal function. Pharmacol Rev 1991; 43:299-349.
205. Wang JH, Stelzer A. Inhibition of phosphatase 2B prevents expression of hippocampal long-term potentiation. NeuroReport 1994; 5:2377-2380.
206. Wang LY, Orser BA, Brautigan DL et al. Regulation of NMDA receptors in cultured hippocampal neurons by protein phosphatases 1 and 2A. Nature 1994; 369:230-232.
207. Watanabe A. Protein phosphatase 1 regulation by inhibitors and targeting subunits. Proc Natl Acad Sci USA 2001; 98:3080-3085.
208. Weiler IJ, Hawrylak N, Greenough WT. Morphogenesis in memory formation: synaptic and cellular mechanisms. Behav Brain Res 1995; 66:1-6.
209. Wera S, Hemmings BA. Serine/threonine protein phosphatases. Biochem J 1995; 311:17-29.
210. Westphal RS, Tavalin SJ, Lin JW et al. Regulation of NMDA receptors by an associated phosphatase-kinase signaling complex. Science 1999; 285:93-96.
211. Whitechurch RA, Ng KT, Sedman GL. Tyrosine kinase inhibitors impair long-term memory formation in day-old chicks. Cog Brain Res 1997; 6:115-120.
212. Whitmarsh AJ, Davis RJ. Regulation of transcription factor function by phosphorylation. Cell Mol Life Sci 2000; 57:1172-1183.
213. Winder DG, Sweatt JD. Roles of serine/threonine phosphatases in hippocampal synaptic plasticity. Nat Rev Neurosci 2001; 2:461-474.
214. Winder DG, Mansuy IM, Osman M et al. Genetic and pharmacological evidence for a novel, intermediate phase of long-term potentiation suppressed by calcineurin. Cell 1998; 92:25-37.
215. Wolfman C, Fin C, Dias M et al. Intrahippocampal or intraamygdala infusion of KN62, a specific inhibitor of calcium.calmodulin-dependent protein kinase II, causes retrograde amnesia in the rat. Behav Neural Biol 1994; 61:203-205.
216. Woolf NJ. A structural basis for memory storage in mammals. Prog Neurobiol 1998; 55:59-77.
217. Xiao MY, Karpefors M, Gustafsson B et al. On the linkage between AMPA and NMDA receptor-mediated EPSPs in homosynaptic long-term depression in the hippocampal CA1 region of young rats. J Neurosci 1995; 15:4496-4506.
218. Yasuda H, Tsumoto T. Long-term depression in rat visual cortex is associated with a lower rise of postsynaptic calcium than long-term potentiation. Neurosci Res 1996; 24:265-274.
219. Yin JC, Tully T. CREB and the formation of long-term memory. Curr Opin Neurobiol 1996; 6:264-268.
220. Yoshimura Y, Sogawa Y, Yamauchi T. Protein phosphatase 1 is involved in the dissociation of Ca^{2+}/calmodulin-dependent protein kinase II from postsynaptic densities. FEBS Letts 1999; 446:239-242.
221. Zeng H, Chattarji S, Barbarosie M et al. Forebrain-specific calcineurin knockout selectively impairs bidirectional synaptic plasticity and working/episodic-like memory. Cell 2001; 107:617-629.
222. Zhabotinsky AM. Bistability in the Ca^{2+}/calmodulin-dependent protein kinase-phosphatase system. Biophys J 2000; 79:2211-2221.
223. Zhao WQ, Bennett P, Rickard NS et al. The involvement of Ca^{2+}/calmodulin-dependent protein kinase in memory formation in day-old chicks. Neurobiol Learn Mem 1996; 66:24-35.
224. Zhao W, Bennett P, Sedman GL et al. The impairment of long-term memory formation by the phosphatase inhibitor okadaic acid. Brain Res Bull 1995; 36:557-561.
225. Zhao WQ, Lawen A, Ng KT. Changes in phosphorylation of Ca^{2+}/calmodulin-dependent protein kinase II (CaMK-II) in processing of short-term and long-term memories after passive avoidance learning. J Neurosci Res 1999; 55:557-568.
226. Zhao WQ, Ng KT, Sedman G. Passive avoidance learning induced change in GAP43 phosphorylation in day-old chicks. Brain Res Bull 1995; 36:11-17.
227. Zhao WQ, Polya GM, Wang BH et al. Inhibitors of cAMP-dependent protein kinase impair long-term memory formation in day-old chicks. Neurobiol Learn Mem 1995; 64:106-118.
228. Zhao WQ, Sedman G, Gibbs ME et al. Effect of PKC inhibitors and activators on memory. Behav Brain Res 1994; 60:151-160.
229. Zhong Y, Budnik V, Wu CF. Synaptic plasticity in Drosophila memory and hyperexcitable mutants: role of cAMP cascade. J Neurosci 1992; 12:644-651.
230. Zhuo M, Zhang W, Son H et al. A selective role of calcineurin A_ in synaptic depotentiation in hippocampus. Proc Natl Acad Sci USA 1999; 96:4650-4655.

Chapter 4.8

Nitric Oxide

Kiyofumi Yamada and Toshitaka Nabeshima

Abstract

Among three isoforms of nitric oxide (NO) synthase, both neuronal and endothelial NO synthases may play an important role in learning and memory. NO production in the brain increases, in an activity-dependent manner, following an increase in intracellular Ca^{2+} levels. Electrophysiological studies revealed that the NO/cGMP pathway plays an important role as an intercellular messenger in the long-term potentiation and long-term depression, which is considered the cellular basis of learning and memory. Behavioral studies further support the role of NO in learning and memory. Collectively, the evidence suggests that NO plays a crucial role in certain forms of learning and memory formation. Furthermore, we believe that modulation of the NO/cGMP signaling pathway is a novel therapeutic strategy for at least some patients with cognitive impairments such as senile dementia.

Introduction

The cellular basis of learning and memory is an alteration in the effectiveness of synapses, which is called synaptic plasticity. Several studies have suggested a relationship between synaptic plasticity and memory.[16] Extensive study of the molecular mechanisms of long-term changes in electrophysiological responses at synapses, such as long-term potentiation (LTP)[9] and long-term depression (LTD),[40] has led to a better understanding of the cellular basis of learning and memory,[90,98] although the existence of a direct relation between LTP and memory has been questioned.[53,58]

Nitric oxide (NO) production in the brain was first reported by Garthwaite et al in 1988.[28] They demonstrated that the excitatory amino acid glutamate evokes the release of an endothelium-derived relaxing factor (EDRF)-like molecule, NO, in cerebellar slices, by activating N-methyl-D-aspartate (NMDA) receptors.[28] Subsequent studies have demonstrated that NO acts as an intercellular messenger in the brain and plays a crucial role in synaptic plasticity such as LTP and LTD, as well as learning and memory formation.[32]

In this chapter, we first give an outline of the regulation of NO synthesis in the brain, and then review electrophysiological and behavioral findings that imply a role for NO in learning and memory. Lastly, we describe the alterations of NO synthesis in the brain induced by learning and memory, and aging.

Regulation of NO Synthesis in the Brain

NO is synthesized from L-arginine by NO synthase (NOS) in a NADPH-dependent reaction. Several isoforms of NOS have been purified and molecularly cloned. Both neuronal NOS (nNOS) and endothelial NOS (eNOS) are calcium/calmodulin-dependent enzymes.[13,21] nNOS is constitutively expressed in certain populations of neurons in the CNS. eNOS is also constitutively expressed mainly in endothelial cells[59] but also in the CA1 pyramidal cells of the hippocampus.[23] In contrast to these two isoforms, inducible NOS (iNOS) is calcium-independent and expressed only after exposure to certain cytokines and/or bacterial endotoxins such as

From Messengers to Molecules: Memories Are Made of These, edited by Gernot Riedel and Bettina Platt.

lipopolysaccharide.[59,97] It is suggested that NO synthesized by iNOS plays a pathophysiological role in ischemic brain damage and other neuroinflammatory diseases.[38,81]

Studies both in vitro and in vivo have demonstrated that NO synthesis in neurons is stimulated by Ca^{2+} influx, which is induced by activation of ionotropic and metabotropic glutamate receptors. NO activates soluble guanylyl cyclase (sGC) which leads to the formation of cGMP.[13,21,88,89] Although relatively few neurons express NOS in the brain, the NO released could influence surrounding neurons over a wide area because it diffuses rapidly within spatial limits of approximately 0.3-0.4 mm.[87]

Another potential factor, which may modulate NO synthesis in the brain, is the availability of the precursor L-arginine in the NOS-containing neurons. L-Arginine is present in glial cells in the brain,[3] and its release was shown in cerebellar slices in vitro[30] and in the thalamus in vivo.[24] Systemic treatment with L-arginine, as well as direct infusion into the brain, produces a significant increase in NO synthesis in the brain.[74,88] In addition, an increase in extracellular L-arginine and NO metabolite (nitrite and nitrate) levels were observed when the glial function was impaired by treatment with the glial selective metabolic inhibitor fluorocitrate.[46,96] It is proposed that the glial toxin impairs the function of glial cells as a reservoir of L-arginine, leading to leakage of L-arginine from glial cells, which results in an enhancement of NO production by NOS-containing neurons.[96] Proposed mechanisms for the modulation of NO synthesis in the brain are illustrated in Figure 1.

Role of NO in LTP and LTD

LTP in the hippocampus refers to the phenomenon whereby a brief, high-frequency electrical stimulation of an excitatory pathway to the hippocampus produces a long-lasting enhancement in the strength of the stimulated synapse.[9] Pharmacologic as well as genetic blockade of NMDA receptors prevents the induction of hippocampal LTP without affecting normal synaptic transmission.[57,73,82] The expression of LTP is thought to involve, in part, a presynaptic increase in transmitter release, implying that postsynaptic neurons release retrograde messengers acting on presynaptic nerve terminals. NO is a potential candidate for such a retrograde messenger.[32] NO binds iron in the heme of sGC, altering the conformation to activate the molecule, which leads to formation of cGMP. Presynaptic injection of cGMP produces activity-dependent LTP in cultured hippocampal neurons. Based on experiments using a hippocampal cell culture system, the cellular mechanism of LTP has been proposed as follows: Tetanic stimulation causes Ca^{2+} influx through postsynaptic NMDA receptors and/or voltage-dependent calcium channels, thereby activating NOS in the postsynaptic neurons. NO then diffuses across the extracellular space to the presynaptic terminal, where it activates sGC and cGMP-dependent protein kinase (PKG) to produce an activity-dependent long-lasting enhancement of transmitter release[4,101] (Fig. 1). A recent study, however, revealed that cGMP-dependent protein kinases are not involved in LTP and that NO induces LTP through an alternative cGMP-independent pathway, possibly ADP-ribosylation in mice lacking cGMP-dependent protein kinases.[42]

Initial pharmacological studies suggested that NO is involved in LTP as a retrograde messenger[29,61,75] but there is no consensus on the role of NO in LTP[6,51,85] (Table 1). These discrepancies suggest that there are NO-dependent and independent forms of LTP. Alternatively, the results may be partly due to the stimulation pattern used to elicit LTP. High-frequency stimulation to induce LTP is different from neuronal activity under physiological conditions, and thus may lead to NO-independent forms of LTP. A recent study showed that the hippocampal LTP in vivo that was induced by a stimulation phase-locked with a theta rhythm, a more physiological stimulation pattern, is completely blocked by NOS inhibitors.[34] LTP in the hippocampus has been found to exhibit an early phase that is independent of protein and RNA synthesis (E-LTP) and a late phase that is reduced by inhibitors of those processes (L-LTP). Most studies into the role of NO in LTP have focused on E-LTP, not L-LTP. It has recently been demonstrated that NO contributes to L-LTP via the activation of sGC, PKG and CREB phosphorylation.[50]

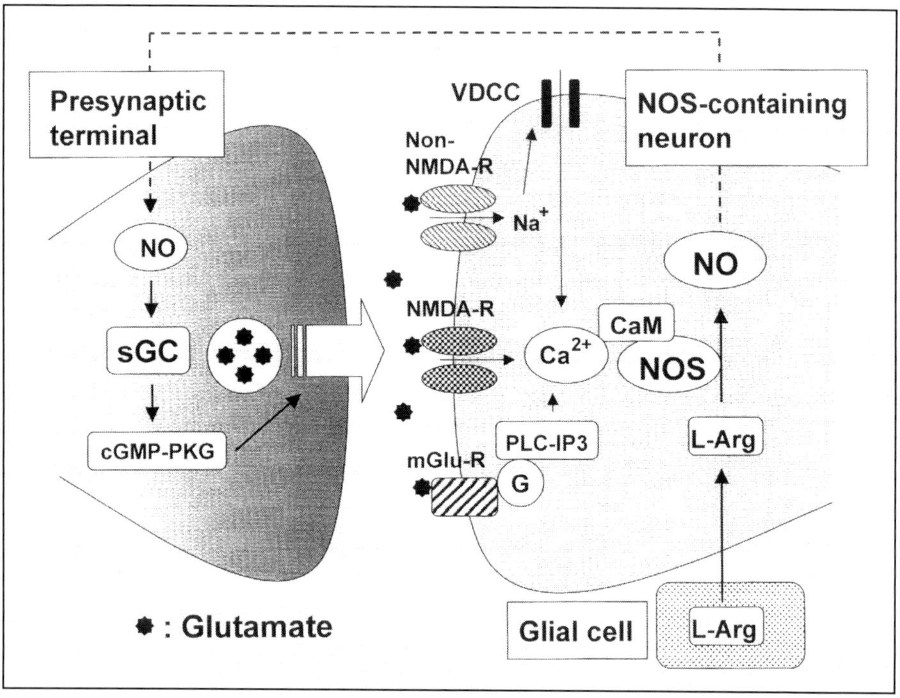

Figure 1. Regulation of NO synthesis in the brain and the role of NO in LTP. L-Arginine (L-Arg) is stored in glial cells. NO is synthesized by nNOS and/or eNOS from L-Arg in neurons. In the LTP, tetanic stimulation causes Ca^{2+} influx through postsynaptic NMDA and/or voltage-dependent calcium channels (VDCCs), thereby activating NOS in postsynaptic neurons. NO then diffuses across the extracellular space to the presynaptic terminal, where it activates sGC and cGMP-dependent protein kinase PKG to produce an activity-dependent long-lasting enhancement of neurotransmitter release.

LTP in the hippocampus is normal in mice with a targeted mutation of nNOS or eNOS,[62,78] but is eliminated in double mutants for nNOS and eNOS.[78] Inhibition of the membrane localization of eNOS results in an inhibition of LTP in the hippocampus.[41] It is shown that LTP in the CA1 subfield of the hippocampus of eNOS knockout mice was entirely absent under weak stimulation conditions although strong tetanic stimulation induced a robust LTP which was not blocked by NOS inhibitors.[86]

LTD in the cerebellum involves a persistent reduction of transmission efficacy at synapses from parallel fibers to Purkinje cells, which occurs when the parallel fibers are activated in conjunction with climbing fibers converging at the same Purkinje cells.[40] LTD in the cerebellum is thought to be a cellular basis of motor learning.[76] Stimulation of white matter of the cerebellum increases the NO concentration in the molecular layer. A NOS inhibitor N^G-monomethyl-L-arginine acetate (L-NMMA) and an NO scavenger hemoglobin inhibit LTD[77] (Table 1). LTD cannot be induced in Purkinje neurons of cerebellar slices from young adult nNOS knockout mice.[49]

The induction of LTP and LTD is not restricted to the hippocampus and cerebellum. Activity-dependent synaptic plasticity is induced in many other areas of the brain (Table 1). For instance, NO-dependent LTP was demonstrated in layer V of the auditory cortex[84] and the medial amygdaloid nucleus.[1] NMDA-independent, but NO-dependent LTP can be induced in the somatosensory cortex of mice, although it is not induced in eNOS knockout mice.[31] High-frequency stimulation of corticostriatal glutamatergic fibers induces LTD in striatal spiny neurons, and the NO/cGMP pathway plays a critical role in the corticostriatal LTD.[15]

Table 1. Role of NO in LTP and LTD

LTP/LTD	Effect	Drugs/Mutant Mice	References
LTP in the hippocampus	Inhibition	L-NA, L-NMMA,	61
		L-NAME,	75
		Hemoglobin,	29
		HMA	41
		7-NI, TRIM	34,50
	Temperature & age-dependent	L-NA, L-NAME, Hemoglobin	85
	Conditioning-dependent	L-NA	51
	No effect	L-NAME	6
	No change	nNOS mutant mouse	62
		eNOS mutant mouse	78
	Impairment	nNOS/eNOS double mutant mouse	78
		eNOS mutant mouse	86
LTP in the neocortex	Inhibition	L-NAME, L-NMMA	84
		Hemoglobin	31
	Impairment	eNOS mutant mouse	31
LTD in the cerebellum	Inhibition	L-NMMA, Hemoglobin	77
	Impairment	nNOS mutant mouse	49
LTD in the corticostriatal pathway	Inhibition	L-NAME, 7-NI	15
	Induction	SNAP	15

HMA: hydroxymyristic acid (a myristoylation inhibitor)
SNAP: S-nitroso-N-acetylpenicillamine
TRIM: 1-(2-trifluoromethylphenyl)imidazole

Role of NO in Memory Processes

As described above, NO plays an important role in LTP and LTD, suggesting that this intercellular messenger participates in some forms of learning and memory. This hypothesis has been tested in a number of studies with various behavioral tasks and NO-related agents (NOS inhibitors, NO donors, L-arginine, GC inhibitors, cGMP analogs, cGMP-dependent phosphodiesterases inhibitors). Some of these studies showed positive results supporting the involvement of NO in memory formation while others reported negative results[70,90] (Table 2). Accordingly, it is suggested that NO plays a role in certain forms of learning and memory but not others. Alternatively, nNOS and eNOS may have different roles in learning and memory processes. To test this hypothesis, the effects of specific inhibitors for each NOS isoform have been investigated in various behavioral tasks. Although mice with a targeted mutation of each NOS isoform (nNOS and eNOS knockout mice) are available, reports of behavioral experiments with these mutant mice on learning and memory are limited.[27]

Spatial Learning and Memory

Many studies have examined the role of NO in spatial learning and memory by using the water maze test,[5,11,18,25,36] but the results are not unequivocal. Several investigators have shown that systemic administration of the non-selective NOS inhibitor N^G-nitro-L-arginine methyl ester (L-NAME) as well as the nNOS-selective inhibitor 7-nitroindazole (7-NI) impairs performance in acquisition trials, in which the escape platform was submerged, although

Table 2. Role of NO in memory processes

Task	Effect	Drugs/Mutant Mice	References
Water maze	Impairment	L-NAME, 7-NI	18, 25, 36
	No effect	L-NAME, L-NA	5, 11
	Improvement	eNOS mutant mouse	27
Radial arm maze	Impairment	L-NA, L-NAME, 7-NI	12
		L-NMMA	92, 36, 79, 102, 55
3-Pannel runway	Impairment	L-NAME	63
Object recognition	Impairment	L-NAME	20
	State-dependent	L-NA	10
Operant conditioning	No effect	L-NAME	43
Y-maze	Impairment	L-NAME, 7-NI	93, 94, 52
14-Unit T-maze	Impairment	7-NI	39, 54
Habituation	Impairment	L-NAME	92
Passive avoidance	Impairment	L-NA, 7-NI, L-NMMA	35, 33, 7, 37, 26, 44
		Diphenyleneiodonium chloride	72
	No effect	L-NA, L-NAME	12, 5, 93
Olfactory memory	Impairment	L-NAME	12
	No effect	L-NA	14, 66
	Enhancement	Sodium nitroprusside,	67
		L-Arginine	68
Locomotor adaptation	Impairment	L-NMME, Hemoglobin	99
Eyeblink conditioning	Impairment	L-NAME	18

performance in cued trials, in which the platform was placed above the water, was not affected. These earlier studies suggest that NO plays a role in acquisition of spatial memory.[18,25,36]

In contrast, Bannerman et al[5] have shown that although L-NAME impairs spatial learning in the water maze task with multiple trials per day, it has no effect when only one trial per day is used. Based on these and other results, they have suggested that the impairment of performance caused by L-NAME is not due to any direct effect on the mechanism of spatial learning. Blokland et al[11] examined the role of hippocampal NO in spatial learning and reversal learning in the water maze test, because spatial learning is highly dependent on hippocampal function.[56] Local injection of N^G-nitro-L-arginine (L-NA) into the dorsal hippocampus before a daily training session transiently affected search behavior during the training, but failed to affect performance in the probe trials or reversal learning. Accordingly, it was concluded that hippocampal NO is not critically involved in place learning.[11]

The role of eNOS in learning and memory was investigated with eNOS knockout mice in the water maze test.[27] Unexpectedly, a clear improvement in performance was observed in the acquisition trials, although there was no difference between eNOS knockout mice and wild type mice in the cued trials. Thus, the performance of the mutant mice cannot be attributed solely to differences in sensorimotor capacities. Furthermore, eNOS knockout mice performed better in the probe trials and the reversal learning test. It remains to be determined whether the improvement depends on an increased spatial learning capacity per se or whether eNOS gene disruption induces changes in brain processes related to anxiety or reward which might play an important role in memory.[27] Pharmacologic studies with selective eNOS inhibitors will provide additional insight into the role of eNOS in learning and memory.

The role of NO in spatial learning and memory has also been investigated in a radial arm maze.[12,36,55,79,92,102] All these studies consistently demonstrated an inhibitory effect of NOS inhibitors on performance in the acquisition trials, suggesting a crucial role for NO in spatial memory formation. We demonstrated that L-NAME, but not D-NAME, impaired performance in a dose-dependent manner during the acquisition of the radial arm maze task, while it failed to affect performance in those rats that previously achieved the task.[92] A non-specific NOS inhibitor, L-NAME inhibited both spatial reference and working memory formation in rats whereas a specific nNOS inhibitor 7-NI impaired specifically spatial reference learning in a reference/working memory task without affecting working memory.[102] A working memory deficit in rats following the treatment with L-NAME was also demonstrated in the 3-pannel runway test[63] and the object recognition test.[20] However, a study using the operant conditioning test showed that inhibition of NOS activity induced by L-NAME does not interfere with the learning or retention of basic operant tasks that involve simple spatial or visual analysis.[43]

Daily administration of L-arginine increased the choice accuracy, by reducing the number of reference memory errors, in the late phase of training on the radial arm maze, although, during this period, NOS inhibitors had little effect.[102] Accordingly, it is possible that NO production in the brain changes depending upon the degree of memory formation, such that the level of production is high in the early phase of training, and then decreases with repeated training (Fig. 2). NOS inhibitors may cause an impairment of spatial learning in the early phase of training during which time NO production levels are high. The inhibitory effects of NOS inhibitors would disappear in the late phase of the training when the level of NO production decreases. Although systemic administration in rats of L-arginine increases NO production in the brain,[74,88] the effect may appear only when the level of production in the brain is low. It is likely, therefore, that L-arginine increases choice accuracy by increasing NO production in the late phase, but not the early phase of radial arm maze training.

In addition to the water maze and radial arm maze, several other tasks, including the Y-maze,[93,94] a 14-unit T-maze[39,52,54] and habituation tests,[92] have been utilized to assess the role of NO in spatial memory formation. It was demonstrated in a 14-unit T-maze test that learning impairment induced by the nNOS inhibitor 7-NI can be overcome by systemic administration of the NO donor, molsidomine.[54] Furthermore, an impairment of spontaneous alternation behavior in a Y-maze induced by systemic administration of NOS inhibitors such as L-NAME and 7-NI, which was associated with a significant decrease in cGMP contents in the hippocampus, was ameliorated by the intracerebroventricular injection of cGMP analogs.[52,93,94] These findings suggest that a NO/cGMP pathway in the hippocampus is responsible for spontaneous alternation behavior in the Y-maze.

Collectively, it appears that NO in the hippocampus may play a crucial role in certain forms of spatial learning but not others. It remains to be determined which NOS isoform is responsible for hippocampus-dependent spatial learning and memory.

Fear Memories

The passive avoidance task is widely used for the study of the mechanism of learning and memory because memory processes involved in the task can be manipulated at different phases, such as acquisition, consolidation/retention and retrieval. Since passive avoidance information is stored for a relatively long period, the memory involved in this task is considered long-term memory. In the chick, memory consolidation and/or retention in the passive avoidance task is impaired by treatment with NOS inhibitors[33,35] and enhanced by sodium nitroprusside,[71] which spontaneously releases NO. These results imply that NO is involved in long-term memory in the chick. Other groups have also shown in mice and rats that NOS inhibitors, administered immediately before or after the training, produce memory impairment.[7,26,37] A recent study showed that NO is involved in retention of the passive avoidance response through the modulation of the forebrain cholinergic system.[44] In contrast, other studies have failed to find any effect.[5,12,93] Therefore, further study is needed to confirm a possible role for NO in long-term memory involved in the passive avoidance task. One study suggested that the effects of NOS inhibitors on learning and memory are state-dependent, indicating that hippocampal NOS

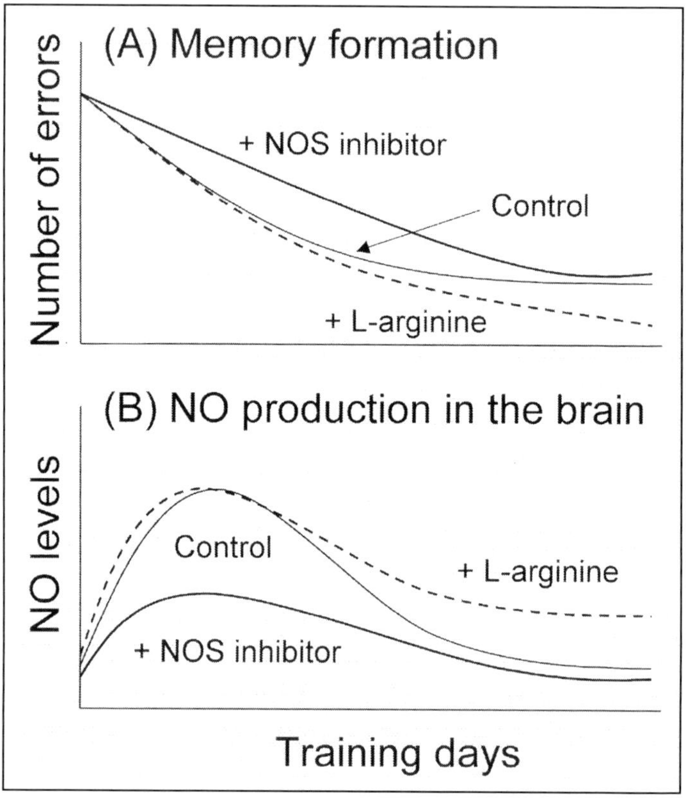

Figure 2. Schematic illustration of the hypothetical changes in NO production during memory formation. NO production in the brain changes depending upon the degree of memory formation, such that the level of production is high in the early phase of training, and then decreases with repeated training. NOS inhibitors cause an impairment of spatial learning in the early phase of training during which time NO production levels are high. The inhibitory effects of NOS inhibitors would disappear in the late phase of the training when NO production has decreased. The effect of L-arginine may appear only when the level of NO production in the brain is low. It is likely, therefore, that L-arginine increases choice accuracy by increasing NO production in the late phase, but not the early phase, of radial arm maze training.

inhibition leads to a change in the internal state of animals and this may affect the manner in which it performs in a cognitive task.[10] Another study[72] showed that day-old chicks trained on a single-trial passive avoidance task suffered significant memory loss from post-training intracranial administration of a selective inhibitor of eNOS, diphenyleneiodonium chloride, although administration of a selective nNOS or iNOS inhibitor at the same time had no effect on retention. Taken together, it is likely that eNOS plays a role in memory formation in the passive avoidance task, which is at least distinct from any role that may be played by nNOS.[72]

Olfactory Memory

The olfactory bulb is one of the brain structures involved in olfactory memory in animals. Since this structure possesses much activity of NOS, the role of NO in olfactory memory has been examined. Although one study showed impairment of olfactory memory following a systemic injection of L-NAME,[12] others did not observe any significant effect induced by a direct injection of L-NA into the accessory olfactory bulb.[14,66] Subsequently, it was shown that intrabulbar infusions of NO donors (sodium nitroprusside) and L-arginine induced formation of a pheromone-specific olfactory memory in the absence of mating.[67,68] Collectively, it is likely that NO in the olfactory bulb plays a role in olfactory memory.

Motor Learning

LTD in the cerebellum is considered a cellular basis of motor learning. Since NOS inhibitors impair LTD in the cerebellum, NO is considered to be involved in certain forms of motor learning.[40] For instance, NOS inhibitors and hemoglobin, which scavenges NO, abolish adaptive control of locomotion to perturbation, a form of motor learning, in the cat.[99] Cerebellar and brainstem structures play a critical role in the acquisition and retention of eyeblink conditioning.[80] Since L-NAME impairs acquisition of eyeblink conditioning in the rabbit as does selective lesion of the cerebellar cortex,[47] it is suggested that NO is involved in normal function or synaptic plasticity in the cerebellar cortex.[18] Furthermore, it is demonstrated that impairment of conditioned eyeblink response by L-NAME is accompanied by a retardation in the formation of conditioning-related activity in the interpositus nucleus.[2]

Learning and Memory-Associated Changes in NO Production in the Brain

As described above, accumulating evidence supports the hypothesis that NO plays a role in certain forms of learning and memory. If so, memory formation may be accompanied by an increase in NO production in the brain (Table 3). Papa et al[69] have shown that intense NADPH-diaphorase staining, a specific histochemical marker for neurons containing NOS[22] in the hippocampus is observed 2 hr, but not immediately, after exposure of rats to spatial novelty. The NOS inhibitor L-NA inhibits NADPH-diaphorase staining and behavioral habituation to spatial novelty.[69] In passive avoidance learning, intrahippocampal injection of L-NA immediately after training impairs memory retention, and NOS activity as well as cGMP content in the hippocampus increases immediately after training.[7,8] Chen et al[19] showed that NOS activity and nitrite levels in the hippocampus and cortex, and also the nitrite level in the cerebellum were significantly elevated one day after rats had learned a water-rewarded spatial alternation task in a Y-maze. Furthermore, immunohistochemistry with nNOS antibodies revealed that spatial learning and memory in the same water-rewarded spatial alternation task increased the number of NO-producing neurons in the dentate gyrus and frontal cortex.[100] Regarding olfactory memory formation, a stimulus-specific expression of nNOS mRNA was demonstrated, by using an in situ hybridization technique, in the female mouse accessory olfactory bulb.[65]

These results demonstrate that the levels of NO, as well as cGMP, increase after memory formation in the brain regions, which are considered to be important for learning and memory. Furthermore, the results provide additional evidence for a role of the NO/cGMP signaling pathway in learning and memory.

Conclusions

In this chapter, we first outlined the regulation of NO synthesis in the brain, and then reviewed electrophysiological and behavioral findings that imply a role for NO in learning and memory formation. Among three NOS isoforms, both nNOS and eNOS may play an important role in learning and memory. In NOS-containing neurons, NO production increases, in an activity-dependent manner, following an increase in intracellular Ca^{2+} levels. Electrophysiological studies revealed that the NO/cGMP pathway plays an important role as an intercellular messenger in the LTP and LTD, which is considered the cellular basis of learning and memory. Behavioral studies further support the role of NO in learning and memory. Collectively, the evidence suggests that NO plays a crucial role in certain forms of learning and memory formation.

It is important to examine whether compounds that can modulate the NO/cGMP signaling pathway have therapeutic potential for the treatment of patients with cognitive

Table 3. Learning and memory-associated changes in NO production in the brain

Task	Marker	Changes	Brain Region	References
Habituation to spatial novelty	NADPH-diaphorase staining	Increase	Hippocampus	69
Passive avoidance	NOS activity	Increase	Hippocampus	7
	cGMP	Increase	Hippocampus	8
Y-maze	NOS activity nitrite level	Increase Increase	Hippocampus/ cortex	19
	nNOS immunochemistry	Increase	Gentate gyrus/ Frontal cortex	100
Olfactory memory	nNOS mRNA	Increase	Accessory olfactory bulb	65

impairments such as aging-associated memory impairment. Some investigators suggested a decrease in NO synthesis in aged rat brain[17,60,83,91,95] whereas others suggested increased synthesis.[45,48] We believe that modulation of the NO/cGMP signaling pathway is a novel therapeutic strategy for at least some patients with cognitive impairments. Of note, Ohtsuka and Nakaya[64] reported an improving effect of oral administration of L-arginine on senile dementia. Other potential candidates include NO donors and inhibitors of cGMP-dependent phosphodiesterase.

References

1. Abe K, Watanabe Y, Saito H. Differential role of nitric oxide in long-term potentiation in the medial and lateral amygdala. Eur J Pharmacol 1996; 297:43-46.
2. Allen MT, Steinmetz JE. A nitric oxide synthase inhibitor delays the formation of learning-related neuronal activity in the cerebellar interpositus nucleus during rabbit eyelid conditoning. Pharmacol Biochem Behav 1996; 53:147-153.
3. Aoki E, Semba R, Mikoshiba K et al. Predominant localization in glial cells of free L-arginine. Immunocytochemical evidence. Brain Res 1991; 547:190-192.
4. Arancio O, Kiebler M, Lee CJ et al. Nitric oxide acts directly in presynaptic neuron to produce long-term potentiation in cultured hippocampal neurons. Cell 1996; 87:1025-1035.
5. Bannerman DM, Chapman PF, Kelly AT et al. Inhibition of nitric oxide synthase does not impair spatial learning. J Neurosci 1994; 14:7404-7414.
6. Bannerman DM, Chapman PF, Kelly AT et al. Inhibition of nitric oxide synthase does not prevent the induction of long-term potentiation in vivo. J Neurosci 1994; 14:7415-7425.
7. Bernabeu R, de Stein ML, Fin C et al. Role of hippocampal NO in the acquisition and consolidation of inhibitory learning. NeuroReport 1995; 6:1498-1500.
8. Bernabeu R, Schmitz P, Faillace MP et al. Hippocampal cGMP and cAMP are differentially involved in memory processing of inhibitory learning. NeuroReport 1996; 7:585-588.
9. Bliss TVP, Collingridge GL. A synaptic model of memory: long-term potentiation in the hippocampus. Nature 1993; 361:31-39.
10. Blokland A, Prickaerts J, Honig W et al. State-dependent impairment in object recognition after hippocampal NOS inhibition. NeuroReport 1998; 9:4205-4208.
11. Blokland A, de Vente J, Prickaerts J et al. Local inhibition of hippocampal nitric oxide synthase does not impair place learning in the Morris water escape task in rats. Eur J Neurosci 1999; 11:223-232.
12. Böhme GA, Bon C, Lemaire M et al. Altered synaptic plasticity and memory formation in nitric oxide synthase inhibitor-treated rats. Proc Natl Acad Sci USA 1993; 90:9191-9194.
13. Bredt DS, Snyder SH. Nitric oxide, a novel neuronal messenger. Neuron 1992; 8:3-11.
14. Brennan PA, Kishimoto J. Local inhibition of nitric oxide synthase activity in the accessory olfactory bulb does not prevent the formation of an olfactory memory in mice. Brain Res 1993; 619:306-312.
15. Calabresi P, Gubellini P, Centonze D et al. A critical role of nitric oxide/cGMP pathway in corticostriatal long-term depression. J Neurosci 1999; 19:2489-2499.

16. Castro CA, Silbert LH, McNaughton BL et al. Recovery of spatial learning deficits after decay of electrically induced synaptic enhancement in the hipppocampus. Nature 1989; 342:545-548.
17. Cha CI, Uhm MR, Shin DH et al. Immunocytochemical study of the distribution of NOS-immunoreactive neurons in the cerebral cortex of aged rats. NeuroReport 1998; 9:2171-2174.
18. Chapman PF, Atkins CM, Allen MT et al. Inhibition of nitric oxide synthesis impairs two different forms of learning. NeuroReport 1992; 3:567-570.
19. Chen J, Zhang S, Zuo P et al. Memory-related changes of nitric oxide synthase activity and nitrite level in rat brain. NeuroReport 1997; 8:1771-1774.
20. Cobb BL, Ryan KL, Frei MR et al. Chronic administration of L-NAME in drinking water alters working memory in rats. Brain Res Bull 1995; 38:203-207.
21. Dawson TM, Snyder SH. Gases as biological messengers: nitric oxide and carbon monoxide in the brain. J Neurosci 1994; 14:5147-5159.
22. Dawson TM, Bredt DS, Fotuhi M et al. Nitric oxide synthase and neuronal NADPH diaphorase are identical in brain and peripheral tissues. Proc Natl Acad Sci USA 1991; 88:7797-7801.
23. Dinerman JL, Dawson TM, Schell MJ et al. Endothelial nitric oxide synthase localized to hippocampal pyramidal cells: Implication for synaptic plasticity. Proc Natl Acad Sci USA 1994; 91:4214-4218.
24. Do K-Q, Binns KE, Salt TE. Release of the nitric oxide precursor, arginine, from the thalamus upon sensory afferent stimulation, and its effect on thalamic neurons in vivo. Neuroscience 1994; 60:581-586.
25. Estall LB, Grant SJ, Cicala GA. Inhibition of nitric oxide (NO) production selectively impairs learning and memory in the rat. Pharmacol Biochem Behav 1993; 46:959-962.
26. Fin C, Cunha CD, Bromberg E et al. Experiments suggesting a role for nitric oxide in the hippocampus in memory processes. Neurobiol Learn Mem 1995; 63:113-115.
27. Frisch C, Dere E, De Souza Silva MA et al. Superior water maze performance and increase in fear-related behavior in the endothelial nitric oxide synthase-deficient mouse together with monoamine changes in cerebellum and ventral striatum. J Neurosci 2000; 20:6694-6700.
28. Garthwaite J, Charles SL, Chess-Williams R. Endothelium-derived relaxing factor release on activation of NMDA receptors suggests role as intercellular messenger in the brain. Nature 1988; 336:385-388.
29. Haley JE, Wilcox GL, Chapman PF. The role of nitric oxide in hippocampal long-term potentiation. Neuron 1992; 8:211-216.
30. Hansel C, Batchelor A, Cuenod M et al. Delayed increase of extracellular arginine, the nitric oxide precursor, following electrical white matter stimulation in rat cerebellar slices. Neurosci Lett 1992; 142:211-214.
31. Haul S, Gödecke A, Schrader J et al. Impairment of neocortical long-term potentiation in mice deficient of endothelial nitric oxide synthase. J Neurophysiol 1999; 81:494-497.
32. Hawkins RD, Son H, Arancio O. Nitric oxide as a retrograde messenger during long-term potentiation in hippocampus. Prog Brain Res 1998; 118:155-172.
33. Hölscher C. 7-Nitro indazole, a neuron-specific nitric oxide synthase inhibitor, produces amnesia in the chick. Learn Mem 1994; 1:213-216.
34. Hölscher C. Nitric oxide is required for expression of LTP that is induced by stimulation phase-locked with theta rhythm. Eur J Neurosci 1999; 11:335-343.
35. Hölscher C, Rose SPR. An inhibitor of nitric oxide synthesis prevents memory formation in the chick. Neurosi Lett 1992; 145:165-167.
36. Hölscher C, McGlinchey L, Anwyl R et al. 7-Nitro indazole, a selective neuronal nitric oxide synthase inhibitor in vivo, impairs spatial learning in the rat. Learn Mem 1995; 2:267-278.
37. Huang A, Lee EHY. Role of hippocampal nitric oxide in memory retention in rats. Pharmacol Biochem Behav 1995; 50:327-332.
38. Iadecola C, Zhang F, Casey R et al. Inducible nitric oxide synthase gene expression in vascular cells after transient focal cerebral ischemia. Stroke 1996; 27:1373-1380.
39. Ingram DK, Spangler EL, Kametani H et al. Intracerebroventricular injection of N-nitro-L-arginine in rats impairs learning in a 14-unit T-maze. Eur J Pharmacol 1998; 341:11-16.
40. Ito M. Long-term depression. Annu Rev Neurosci 1989; 12:85-102.
41. Kantor DB, Lanzrein M, Stary SJ et al. A role for endothelial NO synthase in LTP revealed by adenovirus-mediated inhibition and rescue. Science 1996; 274:1744-1748.
42. Kleppisch T, Pfeifer A, Klatt P et al. Long-term potentiation in the hippocampal CA1 region of mice lacking cGMP-dependent kinases is normal and susceptible to inhibition of nitric oxide synthase. J Neurosci 1999; 19:48-55.
43. Knepper BR, Kurylo DD. Effects of nitric oxide synthase inhibitor N^G-nitro-L-arginine methyl ester on spatial and cued learning. Neurosci 1998; 83:837-841.
44. Kopf SR, Benton RS, Kalfin R et al. NO synthesis inhibition decreases cortical ACh release and impairs retention of a conditioned response. Brain Res 2001; 894:141-144.

45. La Porta CAM, Comolli R. Age-dependent modulation of PKC isoforms and NOS activity and expression in rat cortex, striatum and hippocampus. Exp Gerontol 1999; 34:863-874.
46. Largo C, Cuevas P, Somjen GG et al. The effect of depressing glial function in rat brain in situ on ion homeostasis, synaptic transmission, and neuron survival. J Neurosci 1996; 16:1219-1229.
47. Lavond DG, Steinmetz JE. Acquisition of classical conditioning without cerebellar cortex. Behav Brain Res 1989; 33:133-164.
48. Law A, Dore S, Blackshaw S et al. Alteration of expression levels of neuronal nitric oxide synthase and heme oxygenase-2 messenger RNA in the hippocampi and cortices of young adult and aged cognitively unimpaired and impaired Long-Evans rats. Neurosci 2000; 100:769-775.
49. Lev-Ram V, Nebyelul Z, Ellisman MH et al. Absence of cerebellar long-term depression in mice lacking neuronal nitric oxide synthase. Learn Mem 1997; 4:169-177.
50. Lu YF, Kandel ER, Hawkins RD. Nitric oxide signaling contributes to late-phase LTP and CREB phosphorylation in the hippocampus. J Neurosci 1999; 19:10250-10261.
51. Lum-Ragan JT, Gribkoff VK. The sensitivity of hippocampal long-term potentiation to nitric oxide synthase inhibitors is dependent upon the pattern of conditioning stimulation. Neurosci 1993; 57:973-983.
52. Mamiya T, Noda Y, Noda A et al. Effects of sigma receptor agonists on the impairment of spontaneous alternation behavior and decrease of cyclic GMP level induced by nitric oxide synthase inhibitors in mice. Neuropharmacol 2000; 39:2391-2398.
53. McEachern JC, Shaw CS. An alternative to the LTP orthodoxy: a plasticity pathology continuum model. Brain Res Rev 1996; 22:51-92.
54. Meyer RC, Spangler EL, Patel N et al. Impaired learning in rats in a 14-unit T-maze by 7-nitroindazole, a neuronal nitric oxide synthase inhibitor, is attenuated by the nitric oxide donor, molsidomine. Eur J Pharmacol 1998; 341:17-22.
55. Mizuno M, Yamada K, Olariu A et al. Involvement of brain-derived neurotrophic factor in spatial memory formation and maintenance in a radial arm maze test in rats. J Neurosci 2000; 20:2122-2130.
56. Morris RGM, Garrud P, Rawlins JNP et al. Place navigation impaired in rats with hippocampal lesions. Nature 1982; 297:681-683.
57. Morris RGM, Anderson E, Lynch GS et al. Selective impairment of learning and blockade of long-term potentiation by an N-methyl-D-aspartate receptor antagonist, AP-5. Nature 1986; 319:774-776.
58. Moser E, Mathiesen I, Andersen P. Association between brain temperature and dentate field potentials in exploring and swimming rats. Science 1993; 259:1324-1326.
59. Murphy S, Simmons ML, Agullo L et al. Synthesis of nitric oxide in CNS glial cells. Trends Neurosci 1993; 16:323-328.
60. Noda Y, Yamada K, Nabeshima T. Role of nitric oxide in the effect of aging on spatial memory in rats. Behav Brain Res 1997; 83:153-158.
61. O'Dell TJ, Hawkins RD, Kandel ER et al. Tests of the roles of two diffusible substances in long-term potentiation: evidence for nitric oxide as a possible early retrograde messenger. Proc Natl Acad Sci USA 1991; 88:11285-11289.
62. O'Dell T, Huang PL, Dawson TM et al. Endothelial NOS and the blockade of LTP by NOS inhibitors in mice lacking neuronal NOS. Science 1994; 265:542-546.
63. Ohno M, Yamamoto T, Watanabe S. Deficits in working memory following inhibition of hippocampal nitric oxide synthesis in the rat. Brain Res 1993; 632:36-40.
64. Ohtsuka Y, Nakaya J. Effect of oral administration of L-arginine on senile dementia. Am J Med 2000; 108:439.
65. Okera CO, Kaba H. Increased expression of neuronal nitric oxide synthase mRNA in the accessory olfactory bulb during the formation of olfactory recognition memory in mice. Eur J Neurosci 2000; 12:4552-4556.
66. Okera CO, Kaba H, Higuchi T. Failure of intrabulbar and peripheral administration of N–-nitro-L-arginine to prevent the formation of an olfactory memory in mice. Physiol Behav 1995; 58:387-391.
67. Okera CO, Kaba H, Higuchi T. Formation of an olfactory recognition memory in mice: reassessment of the role of nitric oxide. Neurosci 1996; 71:349-354.
68. Okera CO, Kaba H, Takahashi S et al. Intrabulbar infusions of (±)S-nitroso-N-acetylpenicillamine and L-arginine induce functional plasticity in the accessory olfactory bulb of female mice. Neurosci Lett 1996; 217:197-199.
69. Papa M, Pellicano MP, Sadile AG. Nitric oxide and long-term habituation to novelty in the rat. Annals NY Acad Sci 1994; 738:316-324.
70. Prast H, Philippu A. Nitric oxide as modulator of neuronal function. Prog Neurobiol 2001; 64:51-68.
71. Rickard NS, Ng KT, Gibbs ME. A nitric oxide agonist stimulates consolidation of long-term memory in the 1-day-old chick. Behav Neurosci 1994; 108:640-644.
72. Rickard NS, Gibbs ME, Ng KT. Inhibition of the endothelial isoform of nitric oxide synthase impairs long-term memory formation in the chick. Learn Mem 1999; 6:458-466.

73. Sakimura K, Kutsuwada T, Ito I et al. Reduced hippocampal LTP and spatial learning in mice lacking NMDA receptor _1 subunit. Nature 1995; 373:151-155.
74. Salter M, Duffy C, Garthwaite J et al. Ex vivo measurement of brain tissue nitrite and nitrate accurately reflects nitric oxide synthase activity in vivo. J Neurochem 1996; 66:1683-1690.
75. Schuman EM, Madison DV. A requirement for the intercellular messenger nitric oxide in long-term potentiation. Science 1991; 254:1503-1506.
76. Schweighofer N, Ferriol G. Diffusion of nitric oxide can facilitate cerebellar learning: A simulation study. Proc Natl Acad Sci USA 2000; 97:10661-10665.
77. Shibuki K, Okada D. Endogenous nitric oxide release required for long-term synaptic depression in the cerebellum. Nature 1991; 349:326-328.
78. Son H, Hawkins RD, Martin K et al. Long-term potentiation is reduced in mice that are doubly mutant in endothelial and neuronal nitric oxide synthase. Cell 1996; 87:1015-1023.
79. Suzuki Y, Ikari H, Hayashi T et al. Central administration of a nitric oxide synthase inhibitor impairs spatial memory in spontaneous hypertensive rats. Neurosci Lett 1996; 207:105-108.
80. Thompson RF, Krupa DJ. Organization of memory traces in the mammalian brain. Annu Rev Neurosci 1994; 17:519-549.
81. Tran MH, Yamada K, Olariu A et al. Amyloid β-peptide induces nitric oxide production in rat hippocampus: association with cholinergic dysfunction and amelioration by inducible nitric oxide synthase inhibitors. FASEB J 2001; 15:1407-1409.
82. Tsien JZ, Huerta PT, Tonegawa S. The essential role of hippocampal CA1 NMDA receptor-dependent synaptic plasticity in spatial learning. Cell 1996; 87:1327-1338.
83. Vallebuona F, Raiteri M. Age-related changes in the NMDA receptor/nitric oxide/cGMP pathway in the hippocampus and cerebellum of freely moving rats subjected to transcerebral microdialysis. Eur J Neurosci 1995; 7:694-702.
84. Wakatsuki H, Gomi H, Kudoh M et al. Layer-specific NO dependence of long-term potentiation and biased NO release in layer V in the rat auditory cortex. J Physiol London 1998; 513:71-81.
85. Williams JH, Li Y.-G, Nayak A et al. The suppression of long-term potentiation in rat hippocampus by inhibitors of nitric oxide synthase is temperature and age dependent. Neuron 1993; 11:877-884.
86. Wilson RI, Gödecke A, Brown RE et al. Mice deficient in endothelial nitric oxide synthase exhibit a selective deficit in hippocampal long-term potentiation. Neurosci 1999; 90:1157-1165.
87. Wood PL, Garthwaite J. Models for the diffusional spread of nitric oxide: implications for neuronal nitric oxide signaling and its pharmacological properties. Neuropharmacol 1994; 33:235-234.
88. Yamada K, Nabeshima T. Simultaneous measurement of nitrite and nitrate levels as indices of nitric oxide release in the cerebellum of conscious rats. J Neurochem 1997; 68:1234-1243.
89. Yamada K, Nabeshima T. Two pathways of nitric oxide production through glutamate receptors in the rat cerebellum in vivo. Neurosci Res 1997; 28:93-102.
90. Yamada K, Nabeshima T. Nitric oxide and cyclic GMP signaling pathway in learning and memory processes. Curr Top Pharmacol 1998; 4:77-86.
91. Yamada K, Nabeshima T. Changes in NMDA receptor/nitric oxide signaling pathway in the brain with aging. Microscop Res Tech 1998; 43:68-74.
92. Yamada K, Noda Y, Nakayama S et al. Role of nitric oxide in learning and memory and in monoamine metabolism in the rat brain. Br J Pharmacol 1995; 115:852-858.
93. Yamada K, Noda Y, Hasegawa T et al. The role of nitric oxide in dizocilpine-induced impairment of spontaneous alternation behavior in mice. J Pharmacol Exp Ther 1996; 276:460-466.
94. Yamada K, Hiramatsu M, Noda Y et al. Role of nitric oxide and cyclic GMP in the dizocilpine-induced impairment of spontaneous alternation behavior in mice. Neurosci 1996; 74:365-374.
95. Yamada K, Noda Y, Komori Y et al. Reduction in the number of NADPH-diaphorase-positive cells in the cerebral cortex and striatum in aged rats. Neurosci Res 1996; 24:393-402.
96. Yamada K, Senzaki K, Komori Y et al. Changes in extracellular nitrite and nitrate levels after inhibition of glial metabolism with fluorocitrate. Brain Res 1997; 762:72-78.
97. Yamada K, Komori Y, Tanaka T et al. Brain dysfunction associated with an induction of nitric oxide synthase following an intracerebral injection of lipopolysaccharide in rats. Neurosci 1999; 88:281-294.
98. Yamada K, Mizuno M, Nabeshima T. Role for brain-derived neurotrophic factor in learning and memory. Life Sci 2002; 70:735-744.
99. Yanagihara D, Kondo I. Nitric oxide plays a key role in adaptive control of locomotion in cat. Proc Natl Acad Sci USA 1996; 93:13292-13297.
100. Zhang S, Chen J, Wang S. Spatial learning and memory induce up-regulation of nitric oxide-producing neurons in rat brain. Brain Res 1998; 801:101-106.
101. Zhuo M, Hu Y, Schultz C et al. Role of guanylyl cyclase and cGMP-dependent protein kinase in long-term potentiation. Nature 1994; 368:635-639.
102. Zou L-B, Yamada K, Tanaka K et al. Nitric oxide synthase inhibitors impair reference memory formation in a radial arm maze task in rats. Neuropharmacol 1998; 37:323-330.

Chapter 5.1

CREB

Paul W. Frankland and Sheena A. Josselyn

Abstract

The cAMP Responsive Element Binding Protein (CREB) is an activity regulated transcription factor that modulates the transcription of genes with cAMP responsive elements (CRE) located in their promoter regions. A variety of signaling pathways converge to phosphorylate CREB at Ser133 and induce transcription. Here, we review the key features of CREB-dependent transcription and evaluate evidence suggesting that CREB plays a key role in different forms of plasticity in a wide range of species. The unifying feature of these studies is that manipulations of CREB function affect long-term, but not short-term, memory. This suggests that CREB-dependent transcription is required for the cellular events underlying long-term memory.

Introduction

Current psychological theories of memory propose that memory is not a unitary phenomenon. Rather, memory can be subdivided into qualitatively different memory systems, each of which is served by anatomically distinct brain regions.[104] A second distinction can be made within these systems. That is, memory can be subdivided into different phases, each of which can be distinguished from one another in terms of their temporal and biochemical properties.[32,78]

Using animal studies, at least two phases of memory have been identified: Short-Term Memory (STM) and Long-term Memory (LTM). Short-Term memory is induced rapidly and does not persist beyond a few hours. It involves transient changes in synaptic strength. These are mediated by covalent modifications of preexisting synaptic molecules, such as the phosphorylation or dephosphorylation of enzymes, receptors and/or ion channels. These post-translational modifications produce short-lasting changes in the efficacy of synaptic transmission.[106]

In contrast, Long-Term Memory lasts days or longer, and is thought to involve the growth and restructuring of synapses. There is extensive evidence from a wide variety of species that enduring changes underlying long-term memory require the synthesis of new proteins.[30,76] An essential feature of all of these studies is that administration of protein synthesis inhibitors at the time of learning specifically disrupts the formation of LTM, without affecting learning or STM.

The synthesis of most proteins is mediated by activity-regulated transcription factors. Studies in a wide variety of species have shown that the synthesis of proteins necessary for LTM formation is regulated, at least in part, by cAMP (cyclic adenosine 3',5'-monophosphate) responsive element binding protein (CREB). CREB is a transcription factor that modulates the transcription of genes with cAMP responsive elements (CRE) located in their promoter regions. Just as in studies examining the effects of protein synthesis inhibition on memory formation, a unifying feature of these studies is that manipulating CREB function affects only LTM, and not STM. The first studies showing that CREB plays a critical role in LTM

From Messengers to Molecules: Memories Are Made of These, edited by Gernot Riedel and Bettina Platt.

formation were conducted in *Aplysia* and *Drosophila.* Subsequent studies in vertebrate species (including, in particular, mice and rats) using a variety of molecular-genetic tools suggest that CREB has a highly conserved role in LTM formation.

Structure

CREB is a member of a family (CREB/ATF) of structurally similar, activity-regulated transcription factors. In mammals, at least three genes encode the CREB-like proteins, *CREB*, *CREM* (cAMP Response Element Modulator) and *ATF-1* (Activating Transcription Factor).[38,55,94] The mammalian CREB gene comprises 11 exons,[26,54,115] and alternative splicing generates the three major activator isoforms of CREB: α, Δ, and β.[11,47,118] Each of these is highly expressed in all tissues. In addition to these transcriptional activators, the CREB family also includes repressors of transcription. For example, the CREM gene codes at least four isoforms that repress CRE-dependent transcription: the CREM α, β and γ proteins as well as the inducible cyclic AMP early repressor (ICER).[37,82]

CREB regulates gene expression in response to a wide array of extracellular signals. In its inactive state, CREB is prebound as a dimer to CRE sites in the promoter regions of target genes. Neuronal stimulation may lead to the activation of CREB (via activation of various CREB kinases). In its activated form CREB binds CREB-binding protein (CBP); the recruitment of CBP links CREB directly and indirectly to other components of the basal transcription machinery, thus promoting transcription.[24]

Activation

A large number of signaling pathways converge on CREB, indicating that the transcriptional activity of CREB is regulated by a wide variety of extracellular signals.[31,77,100] Each of these pathways activate CREB via CREB kinases that phosphorylate CREB at serine 133 (Ser133). This is the critical residue for the transcriptional activity[46] since mutation of this residue to a nonphosphorylatable alanine (Ala) residue abolishes the transcriptional response to elevated cAMP levels.[46,83] Although CREB was initially identified as a transcription factor that responds to elevated levels of cAMP, it is now clear that CREB may be activated by three major signaling pathways (Fig. 1): 1) cAMP, 2) Ca^{2+}, and 3) growth factors.

1) cAMP: The activation of G-protein linked receptors (e.g., D1 receptors) leads to the increases in the second messenger cAMP via activation of adenylate cyclase.[44] Rises in levels of cAMP lead to the activation of protein kinase A (PKA) by dissociating the regulatory (R) from the catalytic (C) subunits. The C subunits of PKA passively translocate to the nucleus where they may phosphorylate CREB at Ser133.[5,29,50]

2) Ca^{2+}: Calcium is a pleiotropic second messenger that is activated via several different mechanisms following changes in membrane potential. Extracellular Ca^{2+} may enter the cytoplasm via ligand-gated ion channels of NMDA and AMPA receptors, or via voltage-gated calcium channels. In addition, Ca^{2+} may be released from intracellular stores.[100] Calcium signals are then transduced via a large number of different CREB kinases which include: CamKII, CaMKIV, RSK1-3 (via Ras-ERK), PKC and PKA.[10,29,33,75,101,107] The different kinetics of each of these pathways provides a mechanism for sustained CREB activation and CRE-mediated transcription. For example, activation of CaMKIV produces a wave of CREB phosphorylation with rapid on- and offset (lasting only minutes), whereas activation of the Ras-ERK-RSK2 pathway promotes a slower phase of CREB phosphorylation.[116] Furthermore, the distinct kinetic properties of these upstream regulatory pathways may allow CREB to compute information regarding the exact nature of the stimuli, perhaps allowing for specific stimuli (or patterns of stimulation) to be translated into specific patterns of gene expression. For example, recent data indicate that Ca^{2+} influx into neurons causes the phosphorylation of CREB at Ser142 and Ser143 (in addition to Ser133), and that CREB-induced transcription induced by this triple phosphorylation may not require the participation of CBP.[67] Therefore, Ca^{2+} influx promotes CREB-mediated transcription via a set of mechanisms that are distinct from those produced by other extracellular activation.

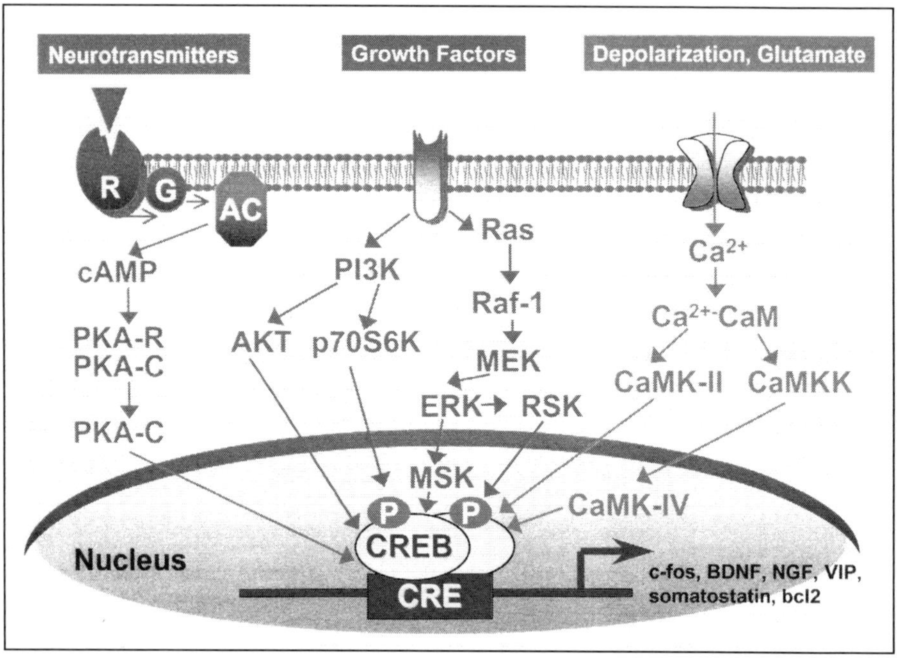

Figure 1. Activation of CREB by a multiple signaling pathways. In the first pathway, a neurotransmitter may bind to a receptor (R) that is linked to a G-protein (G), which leads to the increases in the second messenger cAMP via activation of adenylate cyclase (AC). Rises in levels of cAMP leads to the activation of protein kinase A (PKA) by dissociating the regulatory (R) from the catalytic (C) subunits. The C subunits of PKA passively translocate to the nucleus where they may phosphorylate CREB at Ser133. In the second pathway growth factors (such as NGF or BDNF) bind to and activate a Trk receptor. This, in turn, activates Ras and the downstream kinases Raf, MEK and ERK. Activated ERKs stimulate the activity of MSKs and RSKs which may then phosphorylate CREB at Ser133. In the third pathway, intracellular increases in Ca^{2+} which binds to calmodulin (CaM) which activates CaM kinases (CaMKII, CaMKIV, CaMKK) which may also phosphorylate CREB at Ser133.

3) Growth Factors: CREB mediates gene expression in response to a wide variety of growth factors, including nerve growth factor (NGF), fibroblast growth factor (FGF), epidermal growth factor (EGF) and brain-derived neurotrophic factor (BDNF) (see Brandner, this book). Signaling is then mediated by a large number of growth-factor-induced kinases. For example, NGF stimulation activates NGF receptors (tyrosine kinase receptor, Trk receptors) that stimulates guanine-nucleotide release factors (GRFs) that activate Ras, a small G protein. Activated Ras, in turn, stimulates the serine/threonine kinase, Raf, that triggers activation of MEK, and its targets, the ERK 1 /2 members of the MAPK family.[12] One downstream substrate of the Ras/ERK pathway is a 90 kDa ribosomal S-6 kinase-2 (RSK-2). Upon activation, both ERKs and RSKs translocate to the nucleus where they may phosphorylate CREB at Ser133.[23,36,117]

Just as phosphorylation of Ser133 seems to be critical for activation of CREB, dephosphorylation of this residue is important for inactivation of CREB. As with all other phosphoproteins, therefore, the level of CREB phosphorylation at Ser133 reflects a balance between the oppositional actions of kinases and phosphatases, such as protein phosphatase 1 and 2 (PP-1 and PP-2).[49] For example, dephosphorylation of CREB at Ser 133 may be initiated by the activation of calcineurin (PP-2B) by the Ca^{2+}-CaM pathway.[10] The transcriptional activity of phosphorylated CREB may also be actively suppressed by transcriptional repressors, such as CREM α, β and γ isoforms or ICER.[37,70,82]

The complexity of the pathways upstream from CREB may permit tight, fine-tuned regulation of CRE-mediated transcription, allowing it to produce distinct patterns of gene expression in response to different patterns of stimulation. For example, CREB activation may be moderated by phosphorylation events at sites other than Ser133 (e.g., Ser142 and/or Ser143), and also indirectly by phosphorylation or dephosphorylation of other components of the transcription machinery that CREB interacts with (e.g., CBP, POL II etc).

CREB and Electrophysiological Studies of Long-Term Plasticity in *Aplysia*

The withdrawal of the gill—an external respiratory organ— in the marine mollusk *Aplysia* can be produced by mechanical stimulation of either the siphon or mantle shelf. The reflex serves a defensive purpose: the retraction of the gill protects it from potential damage. This reflex exhibits a number of forms of plasticity. In particular, the sensitization of the withdrawal reflex—that is its enhancement following noncontingent shock applied to the tail of the animal—has been instrumental in the identification of many of the cellular and molecular mechanisms mediating synaptic and behavioral plasticity. The persistence of the reflex sensitization is related to the number of shocks applied to the tail: one shock produces a transient sensitization, lasting minutes, whereas 5 or more shocks produce a LTM lasting days or longer.[6,18,25,39,45,62,91] Long-term sensitization at the synaptic level can be studied in reduced preparations containing the sensory-motor synapse: short- and long-term facilitation of this synapse mediates the behavioral sensitization of the reflex.

The role of CREB in memory and plasticity has been studied in cocultured *Aplysia* sensory and motor neurons.[28] Injection of oligonucleotides with CRE sequences into cultured sensory neurons blocks long-term facilitation (LTF).[28] Presumably, these CRE-oligonucleotides act as competitive antagonists, trapping the CREB proteins needed for the transcriptional activation of genes that ultimately mediate LTF.[4,61] Moreover, a similar injection of a reporter gene driven by a CRE-containing promoter shows that stimuli that produce LTF also trigger CREB activation, while stimuli that do not produce LTF similarly do not trigger CREB activation.[61]

There are several CREB-like proteins in *Aplysia*. The CREB1 gene encodes three proteins (ApCREB1a, ApCREB1b and ApCREB1c) by alternative splicing.[7] The ApCREB1a shares structural and functional homology with CREB transactivators in mammals, while ApCREB1b resembles mammalian ICER, a repressor of CREB transcription. Injection of antibodies or antisense against CREB1a blocks LTF (but not short-term facilitation) while injection of phosphorylated ApCREB1a protein alone induces LTF.[7] Application of ApCREB1b blocks LTF while decreasing ApCREB1b function lowers the threshold for producing LTF.[7] ApCREB1c is a cytoplasmic protein that lacks a nuclear localization signal. Injection of unphosphorylated CREB1c followed by a single pulse of serotonin enhances STF and induces LTF. Therefore, this cytoplasmic form of CREB may play an important role not only in the modulation of CREB-mediated transcription necessary for LTF but also in STF.[7] *Aplysia* CREB2 is structurally unrelated to *Aplysia* CREB1 but shares some homology with mouse ATF4.[51] Decreasing ApCREB2 function decreases the threshold for producing LTF.[7] However, the precise mechanism underlying the effects that ApCREB2 exerts on LTF is unclear.

One neuron may participate in the storage of multiple memories. Therefore, activity-dependent changes must be synapse-specific so that the same neuron can encode multiple patterns of stimulation. Experiments using a single sensory neuron composed of two branches that contact two spatially separated motor neurons show that local application of serotonin onto a single synapse induces LTF that is specific to that branch.[22,73] This branch-specific LTF requires local protein synthesis (presumably at the synapse to be modified) as well as CREB activation in the nucleus of the presynaptic neuron. Repeated application of serotonin onto the cell body of the sensory neuron (rather than the branch) induces a transient, cell-wide LTF that does not persist beyond 48 hours. This transient LTF is CREB-dependent, but is not accompanied by synaptic growth. A similar pattern of transient LTF and no synaptic growth is produced by injection of phospho-CREB1 into the sensory neuron. In order for this transient LTF to become stable and for synaptic growth to appear, a single pulse of serotonin at either synapse is required. Thus, CREB-mediated transcription cooperatively induces synaptic changes in concert with local stimulation by serotonin, representing a mechanism by which individual synapses may be strengthened.

CREB and Memory in *Drosophila*

The molecular mechanisms underlying LTM have been successfully studied in *Drosophila* (or fruit flies). Learning in flies has been studied using an associative olfactory conditioning paradigm. Flies will learn to avoid a previously neutral odor that was paired with shock in favor of another odor that was not paired with shock in a T-maze.[112] Both forward and reverse genetic approaches have been used to study the involvement of CREB in memory in *Drosophila*.[112] Using a forward genetic approach, the progeny of flies that were treated with a mutagen were screened for learning and memory impairments. Two mutants identified by this screen were subsequently determined to have disruptions in Ca^{2+}/CaM-stimulated adenylate cyclase (*rutabaga*) and in cAMP-specific phosphodiesterase (*dunce*), both key enzymes in the regulation of intracellular levels of cAMP.[16,71,112]

Just as in other species, LTM (produced by multiple training trials) is dependent on protein synthesis.[113] Using a reverse genetics approach, Yin and colleagues[120] showed that disrupting CREB function in *Drosophila* blocks LTM produced by multiple training trials, suggesting that protein synthesis required for LTM is mediated, at least in part, by CREB. They found that transgenic over-expression of a CREB transcriptional repressor (dCREB2b) impairs LTM, but not STM, in this task. The finding that STM is intact indicates that the over-expression of this CREB repressor does not disrupt acquisition, and furthermore suggests that basic perceptual, motor, and motivational processes required for the task are intact in these flies.[120]

In species ranging from *Aplysia* to human, spaced training (training trials presented with intervening rest intervals) is more effective than massed training (the same number of training trials presented shorter intervening rest intervals) in producing LTM. The same is true in flies: multiple spaced training produces maximal LTM, whereas the same number of trials presented in a massed fashion produces strong STM but weak or no LTM. However, massed training alone is sufficient to produce maximal LTM if a CREB activator (dCREB2a) is over-expressed in transgenic flies prior to training. The over-expression of this CREB activator produces robust LTM following even just one training trial,[121] perhaps the fly equivalent of 'photographic' or 'flashbulb' memory.[119] Transgenic flies over-expressing a mutant activator, where Ser231 (similar to Ser133 of the mammalian CREB gene) was replaced by an Ala, do not show LTM after one training trial, indicating that phosphorylation of CREB at this residue is required for the enhancement of LTM.[121] Together, these results show the importance of CREB in LTM formation in *Drosophila* and, furthermore, suggest that CREB may be a rate-limiting component of this process.

CREB and LTM in Mammals

Targeted Disruption of CREB Function in a Mouse

The study of the role of CREB in mammalian memory was first made possible by the generation of a mouse in which the CREB gene was disrupted. A neomycin resistance (neo) gene was inserted into exon 2 of the CREB gene, which was believed to contain the translation initiation site for all CREB isoforms.[56] This neo insertion resulted in the loss of two main isoforms of CREB (α and Δ). However, the translation of a previously unknown CREB isoform (CREBβ) starts from exon 4. Therefore insertion of the neo gene into exon 2 did not disrupt the translation of this isoform; rather, in these CREB$^{\alpha\Delta}$ mice, expression of the CREBβ isoform is elevated.[11] The expression levels of CREM activator (τ) and repressor (α and β) isoforms were also increased in these mice. However, importantly, CREB-dependent transcription is decreased in these CREB$^{\alpha\Delta}$ mutant mice.[11] The homozygous deletion of all major CREB isoforms (α, β and Δ; CREB^{null}) is lethal.[96]

Since the CREB$^{\alpha\Delta}$ mice were generated, they have been exhaustively characterized at the behavioral level. Consistent with the effects of protein synthesis inhibition,[2,13,99] CREB$^{\alpha\Delta}$ mice exhibit normal STM but impaired LTM in several fear conditioning paradigms. For example, CREB$^{\alpha\Delta}$ mice show normal conditioned freezing to both tone and context when tested shortly (<1 hour) after training. However, both contextual and tone fear conditioning are impaired if these mice are tested 24 hours after training.[14,66] A similar pattern of results has been observed using a different assay of conditioned fear—fear-potentiated startle.[34]

A parallel set of findings has been observed in studies examining two forms of social learning in CREB$^{\alpha\Delta}$ mice. Rodents develop a preference for foods recently smelled on the breath of other rodents.[15,41,42] Memory for this socially transmitted food preference is normal in CREB$^{\alpha\Delta}$ mice when tested immediately following training. However, just as in fear conditioning, CREB$^{\alpha\Delta}$ mice are impaired when tested 24 hours following training.[42,66] The ability of rodents to remember conspecifics can be assessed in a social recognition task. Recognition is inferred from a decrease in the amount of time spent investigating a familiar (vs. unfamiliar) conspecific. Again, LTM, but not STM, for social recognition is disrupted in CREB$^{\alpha\Delta}$ mice.[65] Together with the fear conditioning data, these findings show that the CREB$^{\alpha\Delta}$ mutation specifically affects LTM, and not STM, in a variety of behavioral paradigms with widely varying performance demands. The extent of these impairments is influenced by gene dosage.[42] Further disruption of CREB function can be achieved by combining the CREB$^{\alpha\Delta}$ and CREB^{null} mutations to produce mice carrying only a single allele for the CREBβ isoform (CREB^{comb}). Memory impairments are more severe in these CREB^{comb} mice compared to the CREB$^{\alpha\Delta}$ mice carrying two alleles for the CREBβ isoform.

Drawing an intriguing parallel with the fly experiments, Kogan et al[66] showed that the LTM deficits in the CREB$^{\alpha\Delta}$ mice were rescued by increasing the spacing between training trials. This was true in three different forms of LTM: spatial (Morris water maze), contextual (fear conditioning) and social (socially transmitted food preference). These parallels suggest that the levels of activated CREB are rate-limiting for memory formation: The over-expression of the CREB activator (dCREB2a) in the transgenic flies removes the requirement for spaced training trials for LTM formation; Conversely, the reduced levels of CREB in the CREB$^{\alpha\Delta}$ mice necessitates multiple, spaced training, rather than fewer massed trials, to produced stable LTM.

One difficulty in the analysis of knockout mice in learning and memory studies is distinguishing between the effects of a given mutation on mnemonic vs. nonmnemonic processes. This problem is largely circumvented in the CREB$^{\alpha\Delta}$ and related mice since these mice show normal learning and STM. Therefore, compromising CREB function alone does not seem to have nonspecific effects on sensory, motor and motivational processes required for the acquisition and expression of learning. Rather, compromising CREB function appears to specifically affect the formation of LTM.

Gaining Temporal and Spatial Control of CREB Function in Mammals

One of the problems with traditional knockout approaches is that the target protein is deleted throughout development and in all tissues. For example, compensatory upregulation of the CREBβ and CREM isoforms complicates the analysis and interpretation of the CREB$^{\alpha\Delta}$ mice. Therefore, achieving both spatial and temporal control over gene expression has been one of the major goals, and three approaches have attempted to meet this challenge.

First, two studies have examined the effects of CREB antisense oligonucleotides on learning and memory in rats. Guzowski and McGaugh[48] examined acquisition in the hidden version of the water maze following injections of antisense against CREB mRNA directly into the dorsal hippocampus of rats. These injections disrupted acquisition in the hidden version of the water maze, a form of learning known to be dependent upon the hippocampus. Similar injections 2 days post-training had no effect on subsequent performance in the water maze, indicating that decreasing CREB function does not affect the expression of a previously consolidated memory. Acutely disrupting CREB function in the amygdala has also been shown to disrupt the development of a conditioned taste aversion.[69] Injections of antisense directed against CREB blocked long-term (3-5 days), but not short-term (2 hour), CTA memory. Sense control infusions, as well as infusions of antisense into brain regions (basal ganglia) not critical for plasticity underlying CTA, had no effect.

Second, a transgenic line of mice has been developed that inducibly expresses a CREB repressor ($\alpha CREB^{S133A}$).[63] The inducibliy of the system is produced by fusing the CREB repressor to a ligand-binding domain (LBD) of a human estrogen receptor with a G521R mutation (LBD^{G521R}). The activity of this mutated LBD is regulated not by estrogen but by the synthetic ligand, tamoxifen.[27,35,72] In the absence of tamoxifen, the LBD^{G521R}-$CREB^{S133A}$ fusion protein is bound to heat shock proteins and is therefore inactive.[35] However, administration of tamoxifen activates this inducible CREB repressor ($CREB^{IR}$) fusion protein, allowing it to compete with endogenous CREB and disrupt CRE-mediated transcription. This mouse has been used to dissect the role of CREB in potentially dissociable memory processes. By administering tamoxifen to activate the repressor in $CREB^{IR}$ transgenic mice at key time points in a fear conditioning protocol, the effects of acutely disrupting CREB function on 1) encoding or STM, 2) consolidation into LTM, 3) storage or maintenance, 4) retrieval were assessed. CREB is crucial for the consolidation of long-term conditioned fear memories, but not for encoding, storage or retrieval of these memories. While acute over-expression of a CREB repressor disrupts LTM, chronic over-expression of the same transgene throughout development has much milder effects.[93] The weaker effects associated with chronic over-expression of a CREB repressor (compared to conditional over-expression of this transgene in the $CREB^{IR}$ mice) may be due to upregulation or compensation through development. Alternatively, the weaker phenotype might be due to a milder disruption of CREB function in these mice: for example, transgene expression levels may not be sufficiently high to compete effectively with endogenous CREB.

A third approach has used viral vector-mediated gene transfer technology to manipulate CREB levels.[19-21] Josselyn et al[60] used herpes simplex viral vector-mediated gene transfer technology to specifically increase CREB expression in the amygdala of rats. This method exploits the natural ability of the herpes simplex virus to insert DNA into specific neuronal populations.[103] These rats were fear-conditioned using massed training that normally only produces STM but no or weak LTM for a light-shock pairing (Fig. 2). However, the over-expression of CREB in the amygdala neurons now results in normal LTM. These data are consistent with results in *Drosophila* showing that increasing CREB levels reduces the number of training trials required to produce LTM, or overcomes the requirement for trial spacing to produce LTM.[121]

Detecting CREB Activation During Learning

Complementary to approaches demonstrating that disruption of CREB function blocks the formation of LTM are those showing that CREB is activated following learning. These studies are invaluable since they provide a powerful synergy between systems and molecular approaches. They not only show that CREB-mediated transcription is critical for the formation of long-term memories, but they identify where and when these processes occur.

Activation of CREB leads to the transcription of genes with CRE sites in their promoter regions. Transgenic mice with a β-galactosidase reporter construct under the regulation of a CRE-containing promoter (CRE-LacZ) have been used to identify where in the brain learning-related CREB-mediated transcription occurs.[58,59] Following fear conditioning significant increases in CRE-dependent gene expression are observed in both the hippocampus and the amygdala, consistent with the idea that plasticity in these structures is critical for learning context-US and tone-US associations. In a clever control study Impey and colleagues showed that CRE-dependent gene expression related to tone-US associations was limited to the amygdala

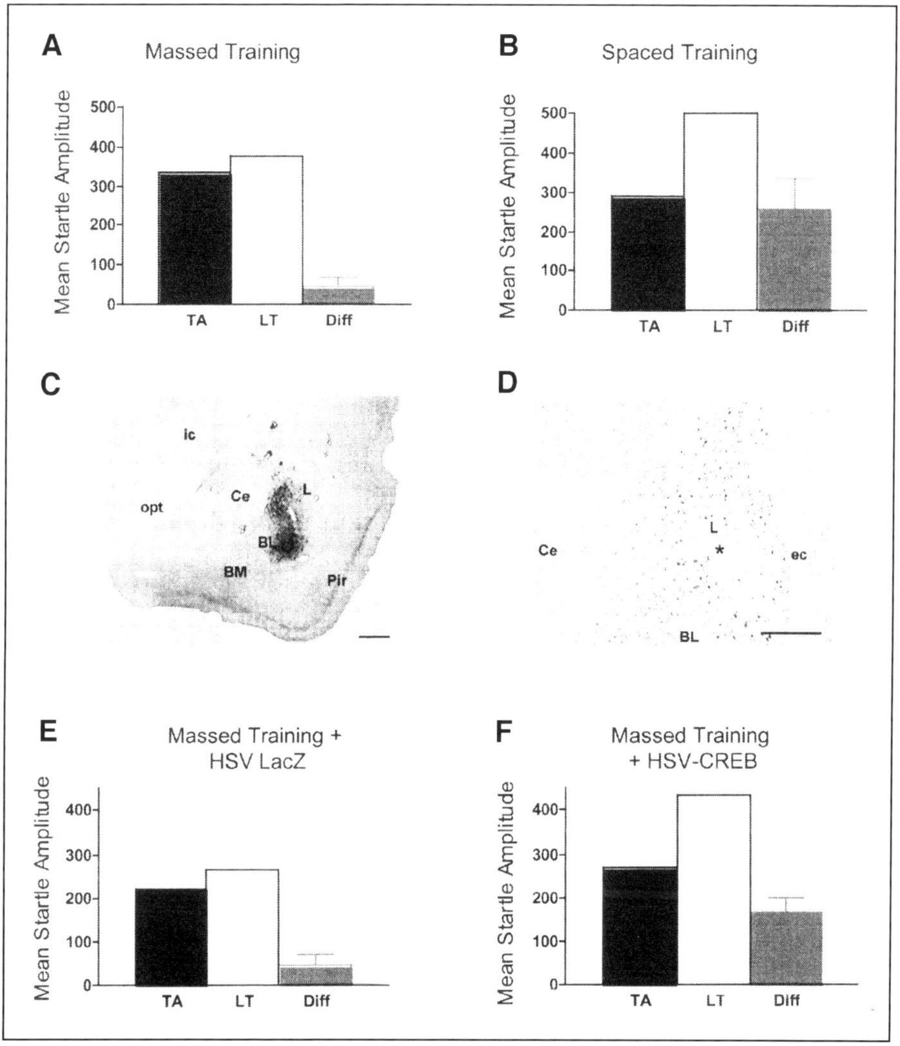

Figure 2. Effects of CREB over-expression in the amygdala on fear-potentiated startle. A) Massed training produces weak LTM, as assessed by mean fear-potentiated startle difference scores (difference between mean startle amplitude on light-tone (LT) trials from tone-alone (TA) trials). B) The same number of trials presented in a spaced fashion produces robust LTM. C) Infusion of HSV-LacZ herpes simplex viral vectors encoding LacZ (HSV-LacZ) into the basolateral amygdala produces high expression of β-galactosidase that is restricted to the basolateral amygdala. D) A high-power image of the amygdala following infusion of HSV-CREB showing over-expression of CREB that is restricted to the lateral nucleus of the amygdala. E) Infusion of HSV-LacZ into the amygdala does not change the weak LTM normally induced by massed training. F) Infusion of HSV-CREB into the amygdala prior to massed training enhances LTM.

using a latent inhibition protocol. To minimize the likelihood of the context becoming associated with shock, mice were pre-exposed to the training context for 12 hours prior to training. These procedures produced significant increases in CRE-dependent gene expression in the amygdala, but not the hippocampus. Consistent with this, when these mice were tested they only showed conditioned freezing when re-exposed to the tone, but not the context.

A second approach has been to use immunocytochemical procedures to detect learning-induced changes in levels of phosphorylated CREB (pCREB). For example, levels of pCREB are elevated in the olfactory bulbs following olfactory conditioning in neonate rat pups.[79] Consistent with the effects intra-amygdala infusions of CREB antisense oligonucleotides on the development of a conditioned taste aversion,[69] increases in pCREB levels are observed in the lateral nucleus of the amygdala following pairing of saccharin (CS) and LiCl-induced illness (US). Similar increases are not observed if the rats are exposed to the CS (saccharin) or US (LiCl) alone, indicating that activation of CREB is related to associative learning.

Several studies have examined pCREB levels in fear-motivated learning paradigms. Inhibitory avoidance training, for example, induces phosphorylation of CREB in the CA1 and Dentate Gyrus regions of the hippocampus.[9,17,88,108-110,114] These immunocytochemical data confirm similar findings using the CRE-reporter mouse.[59] Contextual fear conditioning increases pCREB levels in both the hippocampus and amygdala,[105] again consistent with the observations of Impey.[59]

The contribution of these studies is that they show that CREB activation is restricted to the brain regions that have been shown to critically mediate learning in each of these tasks. Furthermore, they allow us to characterize the time course of CREB activation following a learning event. Indeed, both contextual fear conditioning and inhibitory avoidance training appear to produce two waves of CREB activation:[8,105] pCREB levels are initially increased 0-30 minutes following training, and later 3-6 hours following training. These observations are consistent with the idea that LTM formation may involve multiple rounds of protein synthesis. For example, protein synthesis inhibition immediately following, or 4 hours following training, disrupts long-term contextual fear memories.[13] It is speculated that these later waves may be mediated by sustained PKA activity: In *Aplysia* CREB activation leads to the induction of a number of immediate response genes, including carboxy-terminal ubiquitin hydrolase. This hydrolase removes the regulatory subunit of PKA, allowing the kinase to become persistently active.[13]

CREB and Reconsolidation

Two studies showed that either CRE-dependent gene expression[59] or CREB activation[105] are detected in the amygdala following fear conditioning training. A third study has shown that pCREB levels are elevated in the amygdala following testing.[52] Therefore retrieval, as well as encoding, of fear memories initiates signaling cascades that culminate in CREB activation and presumably gene expression. These findings support the idea that memories are dynamic and modifiable entities.[81,85-87,98] That is, memory retrieval may induce a state of plasticity in which memories become labile before becoming stable again. The process of re-stabilization of the trace, or reconsolidation, following retrieval has been shown to be protein-synthesis dependent.[86] Consistent with the role for CREB in regulating gene expression required for initial consolidation of memories, recent data supports the role for CREB in regulating gene expression required for reconsolidation, implied by the Hall[52] study. Using the inducible CREB repressor transgenic mice, Kida et al[63] showed that acutely repressing CREB function following memory reactivation impairs the stability of memory. Although the exact molecular mechanisms mediating consolidation and reconsolidation may differ,[108] CREB appears to be necessary for both.

Conclusions

Much effort, using a wide variety of tools, has been focused on identifying the molecular mechanisms underlying learning and memory. Establishing that a particular molecule participates critically in these processes relies, it might be argued, on presentation of at least two types of evidence.[74,95] First, disruption of normal molecular function should interfere with memory formation. Second, activation of the molecule, in a predictable, region-specific manner, should be observed following learning. Reliance on evidence from one line of inquiry increases the potential for false-negative and false-positive results.[43,64] For example, targeted deletion of a particular molecule may cause learning impairments not because that molecule directly participates in processes critical for plasticity; rather, the loss of that molecule may produce more global disruption of cellular processes that indirectly impair the neuron's ability to respond appropriately to extracellular signals.

The case for the critical involvement of CREB in LTM is compelling since both types of evidence have been brought to bear on the problem. That is, disrupting CREB function, be it via the generation and testing of genetically-engineered mice or via the infusion of oligonucleotides, specifically disrupts LTM, but not learning. Secondly, studies using reporter mice or immunocytochemical approaches, have shown that CREB is activated following learning in a temporally- and region-restricted manner. In rodents, this conclusion is strengthened since these observations are drawn from a wide variety of tasks, each with widely varying stimulus properties and performance demands.

Similar manipulations of CREB function produce qualitatively similar effects in a wide variety of species including *Aplysia*, *Drosophila*, Chasmagnathus crab, honey bees, and song birds.[1,3,57,68,84,92,97,102,119] In humans, it is particularly noteworthy that the cognitive disabilities in several disorders appear to be directly related to disruption of CREB-mediated transcription. Mutations in RSK2, a protein kinase that activates CREB by phosphorylation at Ser133, are associated with Coffin-Lowry syndrome,[111] as well as nonspecific mental retardation.[80] For example, in tissue from Coffin-Lowry patients, reductions in RSK2-mediated CREB phosphorylation (following EGF stimulation) are linearly related to severity of cognitive deficits.[53] In addition, Rubinstein-Taybi syndrome, which is caused by a mutation of CBP—the cofactor that is essential for transcriptional activation of CREB—is associated with mental retardation.[90] Consistent with this, mice that are heterozygous for CBP mutation exhibit learning impairments.[89] These studies, along with those from sea slugs and flies, mice and rats, suggest an evolutionarily conserved role for CREB-transcription in role in long-term memory formation.

Acknowledgements

We thank Rui M. Costa for comments and discussion.

References

1. Abel T, Martin KC, Bartsch D et al. Memory suppressor genes: Inhibitory constraints on the storage of long- term memory. Science 1998; 279:338-341.
2. Abel T, Nguyen PV, Barad M et al. Genetic demonstration of a role for PKA in the late phase of LTP and in hippocampus-based long-term memory. Cell 1997; 885:615-26.
3. Alberini CM. Genes to remember. J Exp Biol 1999; 21:2887-2891.
4. Alberini CM, Ghirardi M, Metz R et al. C/EBP is an immediate-early gene required for the consolidation of long-term facilitation in Aplysia. Cell 1994; 76:1099-1114.
5. Bacskai BJ, Hochner B, Mahaut-Smith M et al. Spatially resolved dynamics of cAMP and protein kinase A subunits in Aplysia sensory neurons. Science 1993; 105:222-226.
6. Bailey CH, Montarolo P, Chen M et al. Inhibitors of protein and RNA synthesis block structural changes that accompany long-term heterosynaptic plasticity in Aplysia. Neuron 1992; 94:749-58.
7. Bartsch D, Ghirardi M, Skehel PA et al. Aplysia CREB2 represses long-term facilitation: Relief of repression converts transient facilitation into long-term functional and structural change. Cell 1995; 83:979-992.
8. Bernabeu R, Cammarota M, Izquierdo I et al. Involvement of hippocampal AMPA glutamate receptor changes and the cAMP/protein kinase A/CREB-P signalling pathway in memory consolidation of an avoidance task in rats. Braz J Med Biol Res 1997; 308:961-965.
9. Bevilaqua LR, Cammarota M, Paratcha G et al. Experience-dependent increase in cAMP-responsive element binding protein in synaptic and nonsynaptic mitochondria of the rat hippocampus. Eur J Neurosci 1999; 11:3753-3756.
10. Bito H, Deisseroth K, Tsien RW. CREB phosphorylation and dephosphorylation: A Ca(2+)- and stimulus duration-dependent switch for hippocampal gene expression. Cell 1996; 877:1203-1214.

11. Blendy JA, Kaestner KH, Schmid W et al. Targeting of the CREB gene leads to up-regulation of a novel CREB mRNA isoform. Embo J 1996; 155:1098-1106.
12. Blenis J, Chung J, Erikson E et al. Distinct mechanisms for the activation of the RSK kinases/MAP2 kinase/pp90rsk and pp70-S6 kinase signaling systems are indicated by inhibition of protein synthesis. Cell Growth Differ 1991; 26:279-285.
13. Bourtchouladze R, Abel T, Berman N et al. Different training procedures recruit either one or two critical periods for contextual memory consolidation, each of which requires protein synthesis and PKA. Learning Memory 1998; 5:365-374.
14. Bourtchouladze R, Frenguelli B, Blendy J et al. Deficient long-term memory in mice with a targeted mutation of the cAMP-responsive element-binding protein. Cell 1994; 791:59-68.
15. Bunsey M, Eichenbaum H. Selective damage to the hippocampal region blocks long-term retention of a natural and nonspatial stimulus-stimulus association. Hippocampus 1995; 56:546-556.
16. Byers D, Davis RL, Kiger Jr JA. Defect in cyclic AMP phosphodiesterase due to the dunce mutation of learning in Drosophila melanogaster. Nature 1981; 289:79-81.
17. Cammarota M, Bevilaqua LR, Ardenghi P et al. Learning-associated activation of nuclear MAPK, CREB and Elk-1, along with Fos production, in the rat hippocampus after a one-trial avoidance learning: Abolition by NMDA receptor blockade. Mol Brain Res 2000; 761:36-46.
18. Carew TJ, Castellucci VF, Kandel ER. An analysis of dishabituation and sensitization of the gill-withdrawal reflex in Aplysia. Int J Neurosci 1971; 2:79-98.
19. Carlezon Jr WA, Boundy VA, Haile CN et al. Sensitization to morphine induced by viral-mediated gene transfer. Science 1997; 277:812-814.
20. Carlezon Jr WA, Nestler EJ, Neve RL. Herpes simplex virus-mediated gene transfer as a tool for neuropsychiatric research. Crit Rev Neurobiol 2000; 141:47-67.
21. Carlezon Jr WA, Thome J, Olson VG et al. Regulation of cocaine reward by CREB. Science 1998; 282:2272-2275.
22. Casadio A, Martin KC, Giustetto M et al. A transient, neuron-wide form of CREB-mediated long-term facilitation can be stabilized at specific synapses by local protein synthesis. Cell 1999; 992:221-237.
23. Chen RH, Sarnecki C, Blenis J. Nuclear localization and regulation of erk- and rsk-encoded protein kinases. Mol Cell Biol 1992; 123:915-927.
24. Chrivia JC, Kwok RP, Lamb N et al. Phosphorylated CREB binds specifically to the nuclear protein CBP. Nature 1993; 365:855-859.
25. Cleary LJ, Lee WL, Byrne JH. Cellular correlates of long-term sensitization in Aplysia. J Neurosci 1998; 18:5988-5998.
26. Cole TJ, Copeland NG, Gilbert DJ et al. The mouse CREB (cAMP responsive element binding protein) gene: Structure, promoter analysis, and chromosomal localization. Genomics 1992; 134:974-982.
27. Danielian PS, White R, Hoare SA et al. Identification of residues in the estrogen receptor that confer differential sensitivity to estrogen and hydroxytamoxifen. Mol Endocrinol 1993; 72:232-240.
28. Dash PK, Hochner B, Kandel ER. Injection of the cAMP-responsive element into the nucleus of Aplysia sensory neurons blocks long-term facilitation. Nature 1990; 345:718-721.
29. Dash PK, Karl KA, Colicos MA et al. cAMP response element-binding protein is activated by Ca2+/calmodulin- as well as cAMP-dependent protein kinase. Proc Natl Acad Sci USA 1991; 811:5061-5065.
30. Davis HP, Squire LR. Protein synthesis and memory. Psychol Bull 1984; 6:518-559.
31. Deisseroth K, Tsien RW. Dynamic multiphosphorylation passwords for activity-dependent gene expression. Neuron 2002; 34:179-182.
32. Dudai Y. Consolidation: Fragility on the road to the engram. Neuron 1996; 173:367-370.
33. Enslen H, Sun P, Brickey D et al. Characterization of Ca2+/calmodulin-dependent protein kinase IV. Role in transcriptional regulation. J Biol Chem 1994; 269:15520-15537.
34. Falls WA, Kogan JH, Silva AJ et al. Fear-potentiated startle, but not prepulse inhibition of startle, is impaired in CREBalphadelta-/- mutant mice. Behav Neurosci 2000; 1145:998-1004.
35. Feil R, Brocard J, Mascrez B et al. Ligand-activated site specific recombination in mice. Proc Natl Acad Sci USA 1996; 93:10887-10890.
36. Finkbeiner S, Tavazoie SF, Maloratsky A et al. CREB: A major mediator of neuronal neurotrophin responses. Neuron 1997; 195:1031-1047.
37. Foulkes NS, Borrelli E, Sassone-Corsi P. CREM gene: Use of alternative DNA-binding domains generates multiple antagonists of cAMP-induced transcription. Cell 1991; 644:739-749.
38. Foulkes NS, Mellstrom B, Benusiglio E et al. Developmental switch of CREM function during spermatogenesis: From antagonist to activator. Nature 1992; 355:80-84.

39. Frost WN, Castelluci VG, Hawkins RD et al. The monosynaptic connections made by the sensory neurons of the gill– and siphon–withdrawal reflex participate in the storage of long–term memory for sensitization. Proc Nat Acad Sci USA 1985; 82:8266.
40. Galef Jr BG, Wigmore SW. Transfer of information concerning distant foods: A laboratory investigation of the 'information-centre' hypothesis. Anim Behav 1983; 31:748-758.
41. Galef Jr BG, Mason JR, Preti G et al. Carbon disulfide: A semiochemical mediating socially-induced diet choice in rats. Physiol Behav 1988; 42:119-124.
42. Gass P, Wolfer DP, Balschun D et al. Deficits in memory tasks of mice with CREB mutations depend on gene dosage. Learning Memory 1998; 5:274-288.
43. Gerlai R. Gene targeting: Technical confounds and potential solutions in behavioral brain research. Behav Brain Res 2001; 125:13-21.
44. Gilman AG. G proteins: Transducers of receptor-generated signals. Annu Rev Biochem 1987; 56:615-49.
45. Goelet P, Castellucci VF, Schacher S et al. The long and short of long-term memory - a molecular framework. Nature 1986; 322:419-422.
46. Gonzalez GA, Montminy MR. Cyclic AMP stimulates somatostatin gene transcription by phosphorylation of CREB at serine 133. Cell 1989; 594:675-680.
47. Gonzalez GA, Yamamoto KK, Fischer WH et al. A cluster of phosphorylation sites on the cyclic AMP-regulated nuclear factor CREB predicted by its sequence. Nature 1989; 337:749-752.
48. Guzowski JF, McGaugh JL. Antisense oligodeoxynucleotide-mediated disruption of hippocampal cAMP response element binding protein levels impairs consolidation of memory for water maze training. Proc Natl Acad Sci USA 1997; 946:2693-2698.
49. Hagiwara M, Alberts A, Brindle P et al. Transcriptional attenuation following cAMP induction requires PP-1-mediated dephosphorylation of CREB. Cell 1992; 701:105-113.
50. Hagiwara M, Brindle P, Harootunian A et al. Coupling of hormonal stimulation and transcription via the cyclic AMP-responsive factor CREB is rate limited by nuclear entry of protein kinase A. Mol Cell Biol 1993; 138:4852-4859.
51. Hai T, Liu F, Coukos W et al. Transcription factor ATF cDNA clones: An extensive family of leucine zipper proteins able to selectively form DNA-binding heterodimers. Genes and Dev 1989; 3:2083-2090.
52. Hall J, Thomas KL, Everitt BJ. Fear memory retrieval induces CREB phosphorylation and Fos expression within the amygdala. Eur J Neurosci 2001; 137:1453-1458.
53. Harum KH, Alemi L, Johnston MV. Cognitive impairment in Coffin-Lowry syndrome correlates with reduced RSK2 activation. Neurology 2001; 562:207-214.
54. Hoeffler JP, Meyer TE, Waeber G et al. Multiple adenosine 3',5'-cyclic monophosphate response element DNA-binding proteins generated by gene diversification and alternative exon splicing. Mol Endocrinol 1990; 46:920-930.
55. Hoeffler JP, Meyer TE, Yun Y et al. Cyclic AMP-responsive DNA-binding protein: Structure based on a cloned placental cDNA. Science 1988; 242:1430-1433.
56. Hummler E, Cole TJ, Blendy JA et al. Targeted mutation of the cAMP response element binding protein (CREB) gene: Compensation within the CREB/ATF family of transcription factors. Proc Natl Acad Sci USA 1994; 91:5647-5651.
57. Impey S, Goodman RH. CREB signaling—timing is everything. Sci STKE 2001; 82:E1.
58. Impey S, Mark M, Villacres EC et al. Induction of CRE-mediated gene expression by stimuli that generate long-lasting LTP in area CA1 of the hippocampus. Neuron 1996; 165:973-982.
59. Impey S, Smith DM, Obrietan K et al. Stimulation of cAMP response element (CRE)-mediated transcription during contextual learning. Nature Neurosci 1998; 1:595–601.
60. Josselyn SA, Shi C, Carlezon Jr WA et al. Long-term memory is facilitated by cAMP response element-binding protein overexpression in the amygdala. J Neurosci 2001; 21:2404-2412.
61. Kaang BK, Kandel ER, Grant SG. Activation of cAMP-responsive genes by stimuli that produce long-term facilitation in Aplysia sensory neurons. Neuron 1993; 103:427-435.
62. Kandel ER, Klein M, Bailey CH et al. Serotonin, cyclic AMP, and the modulation of the calcium current during behavioral arousal. In: Jacobs BL, Gelperin A, eds. Serotonin Neurotransmission and Behavior. Cambridge, MA: MIT Press, 1981:211-254.
63. Kida S, Josselyn SA, de Ortiz SP et al. CREB required for the stability of new and reactivated fear memories. Nature Neurosci 2002; 5:348-355.
64. Kim JJ, Baxter MG. Multiple brain-memory systems: The whole does not equal the sum of its parts. Trends Neurosci 2001; 246:324-330.
65. Kogan JH, Frankland PW, Silva AJ. Long-term memory underlying hippocampus-dependent social recognition in mice. Hippocampus 2000; 10:47-56.

66. Kogan JH, Frankland PW, Blendy JA et al. Spaced training induces normal long-term memory in CREB mutant mice. Curr Biol 1997; 71:1-11.
67. Kornhauser JM, Cowan CW, Shaywitz AJ et al. CREB transcriptional activity in neurons is regulated by multiple, calcium-specific phosphorylation events. Neuron 2002; 34:221-233.
68. Lamprecht R. CREB: A message to remember. Cell Mol Life Sci 1999; 554:554-563.
69. Lamprecht R, Hazvi S, Dudai Y. cAMP response element-binding protein in the amygdala is required for long- but not short-term conditioned taste aversion memory. J Neurosci 1997; 1721:8443-8450.
70. Laoide BM, Foulkes NS, Schlotter F et al. The functional versatility of CREM is determined by its modular structure. Embo J 1993; 123:1179-1191.
71. Levin LR, Han PL, Hwang PM et al. The Drosophila learning and memory gene rutabaga encodes a Ca2+/Calmodulin-responsive adenylyl cyclase. Cell 1992; 683:479-489.
72. Logie C, Stewart AF. Ligand-regulated site-specific recombination. Proc Natl Acad Sci USA 1995; 9213:5940-5944.
73. Martin KC, Casadio A, Zhu H et al. Synapse-specific, long-term facilitation of aplysia sensory to motor synapses: A function for local protein synthesis in memory storage. Cell 1997; 917:927-938.
74. Martin SJ, Grimwood PD, Morris RGM. Synaptic plasticity and memory: An evaluation of the hypothesis. Annu Rev Neurosci 2000; 23:649-711.
75. Matthews RP, Guthrie CR, Wailes LM et al. Calcium/calmodulin-dependent protein kinase types II and IV differentially regulate CREB-dependent gene expression. Mol Cell Biol 1994; 149:6107-6116.
76. Matthies H. In search of cellular mechanisms of memory. Prog Neurobiol 1989; 32:277-349.
77. Mayr B, Montminy M. Transcriptional regulation by the phosphorylation-dependent factor CREB. Nature Rev Mol Cell Biol 2001; 28:599-609.
78. McGaugh JL. Memory—a century of consolidation. Science 2000; 287:248-251.
79. McLean JH, Harley CW, Darby-King A et al. pCREB in the neonate rat olfactory bulb is selectively and transiently increased by odor preference-conditioned training. Learning Memory 1999; 6:608-618.
80. Merienne K, Jacquot S, Pannetier S et al. A missense mutation in RPS6KA3 (RSK2) responsible for nonspecific mental retardation. Nature Gen 1999; 221:13-14.
81. Miller RR, Matzel LD. Memory involves far more than 'consolidation'. Nature Rev Neurosci 2000; 1:214-216.
82. Molina CA, Foulkes NS, Lalli E et al. Inducibility and negative autoregulation of CREM: An alternative promoter directs the expression of ICER, an early response repressor. Cell 1993; 755:875-886.
83. Montminy M. Transcriptional regulation by cyclic AMP. Annu Rev Biochem 1997; 66:807-822.
84. Muller U. Prolonged activation of cAMP-dependent protein kinase during conditioning induces long-term memory in honeybees. Neuron 2000; 271:159-168.
85. Nadel L, Land C. Memory traces revisited. Nature Rev Neurosci 2000; 1:209-212.
86. Nader K, Schafe GE, LeDoux JE. Fear memories require protein synthesis in the amygdala for reconsolidation after retrieval. Nature 2000; 406:722-726.
87. Nader K, Schafe GE, LeDoux, JE. The labile nature of consolidation theory. Nature Rev Neurosci 2000; 1:216-219.
88. O'Connell C, Gallagher HC, O'Malley A et al. CREB phosphorylation coincides with transient synapse formation in the rat hippocampal dentate gyrus following avoidance learning. Neural Plast 2000; 74:279-289.
89. Oike Y, Hata A, Mamiya T et al. Truncated CBP protein leads to classical Rubinstein-Taybi syndrome phenotypes in mice: Implications for a dominant-negative mechanism. Hum Mol Genet 1999; 83:387-396.
90. Petrij F, Giles RH, Dauwerse HG et al. Rubinstein-Taybi syndrome caused by mutations in the transcriptional coactivator CBP [see comment[s]]. Nature 1995; 376:348-351.
91. Pinkser H, Castellucci VF, Kupfermann I et al. Habituation and dishabituation of the gill-withdrawal reflex in Aplysia. Science 1970; 167:1740-1742.
92. Pittenger C, Kandel E. A genetic switch for long-term memory. C R Acad Sci III 1998; 321:91-96.
93. Rammes G, Steckler T, Kresse A et al. Synaptic plasticity in the basolateral amygdala in transgenic mice expressing dominant-negative cAMP response element-binding protein (CREB) in forebrain. Eur J Neurosci 2000; 127:2534-2546.
94. Rehfuss RP, Walton KM, Loriaux MM et al. The cAMP-regulated enhancer-binding protein ATF-1 activates transcription in response to cAMP-dependent protein kinase A. J Biol Chem 1991; 26628:18431-18434.
95. Rose SP. What should a biochemistry of learning and memory be about? Neurosci 1981; 65:811-821.

96. Rudolph D, Tafuri A, Gass P et al. Impaired fetal T cell development and perinatal lethality in mice lacking the cAMP response element binding protein. Proc Natl Acad Sci USA 1998; 958:4481-4486.
97. Sakaguchi H, Wada K, Maekawa M et al. Song-induced phosphorylation of cAMP response element-binding protein in the songbird brain. J Neurosci 1999; 19:3973-3981.
98. Sara SJ. Strengthening the shaky trace through retrieval. Nature Rev Neurosci 2000; 1:212-213.
99. Schafe GE, Nadel NV, Sullivan GM et al. Memory consolidation for contextual and auditory fear conditioning is dependent on protein synthesis, PKA, and MAP kinase. Learning Memory 1999; 6:97-110.
100. Shaywitz AJ, Greenberg ME. CREB: A stimulus-induced transcription factor activated by a diverse array of extracellular signals. Annu Rev Biochem 1999; 68:821-861.
101. Sheng M, Thompson MA, Greenberg ME. CREB: A Ca(2+)-regulated transcription factor phosphorylated by calmodulin-dependent kinases. Science 1991; 252:1427-1430.
102. Silva AJ, Kogan JH, Frankland PW et al. CREB and memory. Annu Rev Neurosci 1998; 21:127-148.
103. Simonato M, Manservigi R, Marconi P et al. Gene transfer into neurones for the molecular analysis of behaviour: Focus on herpes simplex vectors. Trends Neurosci 2000; 23:183-90.
104. Squire LR. Memory and the hippocampus: A synthesis from findings with rats, monkeys, and humans. Psychol Rev 1992; 992:195-231.
105. Stanciu M, Radulovic J, Spiess J. Phosphorylated cAMP response element binding protein in the mouse brain after fear conditioning: Relationship to Fos production. Mol Brain Res 2001; 94:15-24.
106. Stork O, Welzl H. Memory formation and the regulation of gene expression. Cell Mol Life Sci 1999; 55:575-592.
107. Sun P, Enslen H, Myung PS et al. Differential activation of CREB by Ca2+/calmodulin-dependent protein kinases type II and type IV involves phosphorylation of a site that negatively regulates activity. Genes Dev 1994; 821:2527-2539.
108. Taubenfeld SM, Milekic MH, Monti B et al. The consolidation of new but not reactivated memory requires hippocampal C/EBPbeta. Nature Neurosci 2001; 4:813-818.
109. Taubenfeld SM, Wiig KA, Bear MF et al. A molecular correlate of memory and amnesia in the hippocampus. Nature Neurosci 1999; 2:309-310.
110. Taubenfeld SM, Wiig KA, Monti B et al. Fornix-dependent induction of hippocampal CCAAT enhancer-binding protein [beta] and [delta] Colocalizes with phosphorylated cAMP response element-binding protein and accompanies long-term memory consolidation. J Neurosci 2001; 21:84-91.
111. Trivier E, De Cesare D, Jacquot S et al. Mutations in the kinase Rsk-2 associated with Coffin-Lowry syndrome. Nature 1996; 384:567-570.
112. Tully T. Genetic dissection of learning and memory in Drosophila melanogaster. In: Madden J-IV, ed. Neurobiology of Learning, Emotion and Affect. New York: Raven Press Ltd, 1991:30-66.
113. Tully T, Preat T, Boynton SC et al. Genetic dissection of consolidated memory in Drosophila. Cell 1994; 79:35-47.
114. Viola H, Furman M, Izquierdo LA et al. Phosphorylated cAMP response element-binding protein as a molecular marker of memory processing in rat hippocampus: Effect of novelty. J Neurosci 2000; 20:RC112.
115. Waeber G, Meyer TE, LeSieur M et al. Developmental stage-specific expression of cyclic adenosine 3',5'-monophosphate response element-binding protein CREB during spermatogenesis involves alternative exon splicing. Mol Endocrinol 1991; 510:1418-1430.
116. Wu GY, Deisseroth K, Tsien RW. Activity-dependent CREB phosphorylation: Convergence of a fast, sensitive calmodulin kinase pathway and a slow, less sensitive mitogen-activated protein kinase pathway. Proc Natl Acad Sci USA 2001; 985:2808-28013.
117. Xing J, Ginty DD, Greenberg ME. Coupling of the RAS-MAPK pathway to gene activation by RSK2, a growth factor-regulated CREB kinase. Science 1996; 273:959-963.
118. Yamamoto KK, Gonzalez GA, Menzel P et al. Characterization of a bipartite activator domain in transcription factor CREB. Cell 1990; 604:611-617.
119. Yin JC, Tully T. CREB and the formation of long-term memory. Curr Opin Neurobiol 1996; 62:264-268.
120. Yin J, Wallach JS, Vecchio MD et al. Induction of a dominant-negative CREB transgene specifically blocks long-term memory in Drosophila melanogaster. Cell 1994; 79:49-58.
121. Yin JC, Del Vecchio M, Zhou H et al. CREB as a memory modulator: Induced expression of a dCREB2 activator isoform enhances long-term memory in Drosophila. Cell 1995; 81:107-115.

CHAPTER 5.2

Immediate-Early Genes

Jeffrey Greenwood, Pauline Curtis, Barbara Logan, Wickliffe Abraham and Mike Dragunow

Abstract

How long-term memories are formed in the brain is one of the principal targets of contemporary neuroscience research. This work is important from a fundamental perspective, because memory is a vital component of virtually all cognitive activity. It is also important from a clinical perspective since the early and most dramatic symptoms of Alzheimer's disease (AD) are an impairment of memory formation. Diseases such as AD can provide clues about which neurotransmitters and intracellular signalling pathways are involved in human memory formation. In this chapter, the importance of muscarinic cholinergic signalling in memory formation and AD will be discussed. The classical immediate early genes (IEGs) implicated in memory formation, and the signalling cascades which may regulate their activity will then be described. Finally, our recent work on the potential involvement of stress response immediate early genes in memory formation will be presented.

Introduction

The brain neurotransmitter acetylcholine has been shown to be critical for long-term memory storage[38,54] and drugs which activate muscarinic acetylcholine receptors improve memory by activating a system that is critical in memory formation (see refs. 36 and 46, and also Pepeu and Giovannini in this book for review). The best predictor of the cognitive decline in AD is the loss of acetylcholine markers in the brain reflecting death or atrophy of acetylcholine-producing neurons in the basal forebrain,[8,10] indicating that the memory loss in AD may be due to a reduction in acetylcholine signalling through muscarinic receptors on post-synaptic neurons. Cholinesterase inhibitors (which elevate acetylcholine in brain, such as tacrine) delay the progression of AD,[31] and their effects may be due to a combination of direct cognitive actions and enhanced acetylcholine signalling through postsynaptic M_1 muscarinic receptors, the latter having been shown to activate secretion of soluble amyloid precursor proteins[23] and reduce tau phosphorylation.[25,52] Although postsynaptic M_1 muscarinic receptors are not lost in AD brain,[37] there is a large impairment in signalling via the phosphoinositide (PI) system,[40] and this impairment in AD brain membranes is mimicked by application of amyloid-β-peptide, a peptide found in AD plaques and implicated strongly in AD causation,[44] to cultured neurons.[41] There is also a relationship between apolipoprotein E4 (APOE4), which is a risk factor for developing AD, and cholinergic systems such that the APOE4 allele is associated with greater loss of cholinergic markers and a reduced responsiveness to cholinergic agonists in AD.[50] APOE-deficient mice show a cholinergic impairment and memory formation problems that can be ameliorated by M_1 agonists,[24] and an impairment in muscarinic agonist-induced PI hydrolysis.[14] Thus, M_1 muscarinic receptor signalling is severely impaired in AD brain and this is likely to contribute to the memory formation problems seen in AD patients, as well as for the

From Messengers to Molecules: Memories Are Made of These, edited by Gernot Riedel and Bettina Platt.

poor efficacy of M_1 agonists in AD. Glutamate is also fundamental to memory formation and memory consolidation, and readers are referred to many excellent reviews on this topic in the literature (e.g., see ref. 48 and also Riedel et al in this book).

Learning Activates IEGs

Long-term changes in neuronal phenotype are required for the formation of long-term memories in the brain.[43] The discovery many years ago that the c-fos IEG, which codes for a transcription factor, was induced in neurons after various types of physiological and pathological stimulation generated great excitement in the neuroscience community because it provided the first indication that the neuronal genome was responsive to activation of membrane-bound receptors.[34] In particular, researchers interested in memory consolidation wondered whether c-fos might provide the link to the neuronal genome underlying the formation of long-term memory. Thus, neurotransmitters might cause permanent changes in neuronal phenotype by accessing the neuronal genome through induction of transcription factors (coded for by IEGs) such as c-Fos.[20,26,47] While this turned out not to be the case for c-fos,[16] other IEGs have emerged as better candidates. For example, it was shown that both glutamate, acting via N-methyl-D-aspartate (NMDA) receptors, and acetylcholine, acting via muscarinic receptors, induce the Krox 24 IEG product and transcription factor in neurons.[13,33,34,51,57,59] The induction of Krox 24 (but not c-Fos) via NMDA receptors predicts the permanence of synaptic changes in the long-term potentiation (LTP) model of memory storage,[2,5,51] implying that Krox 24 may be involved in inducing genes responsible for long-term memory storage.[9,15,17,18,49,61] These data have been well reviewed over the past few years.[3,18,61] Until recently, the link between krox 24 expression and LTP maintenance have been purely correlational. However, Jones et al[39] recently provided direct evidence for this relationship. They showed that the late, protein synthesis-dependent phase of LTP is absent in mice with a targeted disruption of krox 24. Furthermore, these mice exhibited long-term, but not short-term, memory deficits. Thus, Krox 24 appears to be a transcription factor that is vital for memory consolidation.

A Link between Cholinergic System and IEGs

Muscarinic receptor agonists enhance memory storage through a delayed action,[46] and muscarinic antagonists, which cause amnesia in humans,[1] block neuronal krox 24 expression,[33] suggesting that Krox 24 is an intermediate signalling molecule linking muscarinic receptor activation to memory-related gene expression. However, other transcription factors are also likely to act in concert with Krox 24. For example, phosphorylation of the transcription factor cyclic AMP response element binding protein (CREB) is also strongly implicated in long-term memory storage.[53] We recently discovered a novel signalling pathway linking muscarinic receptor agonists to CREB phosphorylation (maximal at 5-10 min) followed by Krox 24 induction (maximal 1 h) in human neuroblastoma cells.[19,27] Exactly how muscarinic receptor agonists activate CREB phosphorylation is unclear but the mechanism could involve one of several signalling cascades. Interestingly, we have discovered that Krox 24 induction downstream of muscarinic receptor activation is dependent on p42/44 mitogen activated protein (MAP) kinase activity, as it is blocked by the MEK inhibitor U0126, whereas CREB phosphorylation is unaffected by MAP kinase blockade.[27] Thus, the activation of muscarinic receptors leads to activation of both CREB and Krox 24, but these pathways appear to be parallel rather than sequential, with one (Krox 24) being dependent upon activation of the MAP kinase cascade while the other (CREB) is not. At present the functional implications of this are unclear. However, the ability of muscarinic agonists to activate two transcription factors, both of which play important roles in memory consolidation, might account for the importance of cholinergic signal transduction in learning and memory.

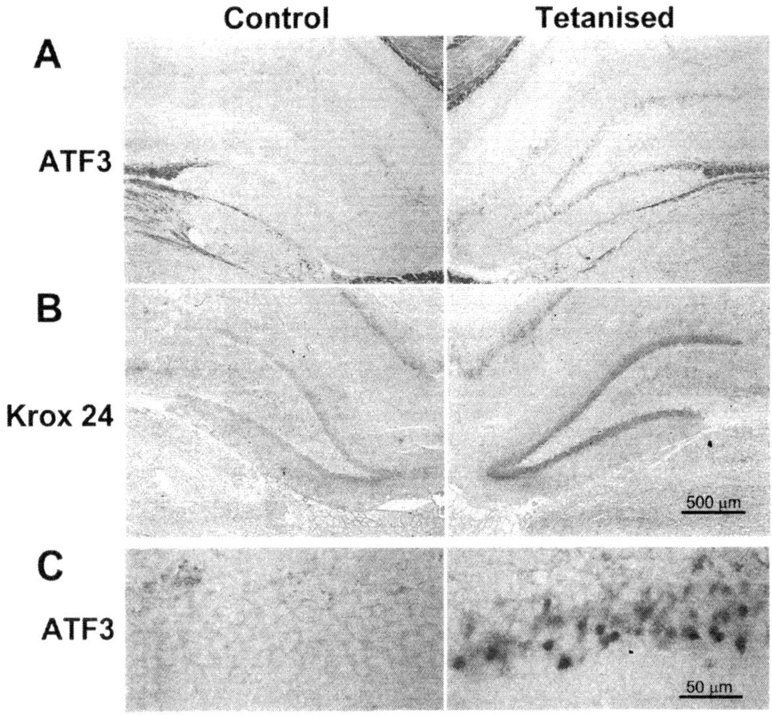

Figure 1. ATF3 is induced in dentate granule cells 2 hours after HFS. Immunohistochemical staining in dentate granule cells for ATF3 (A and C) and Krox 24 (B) in coronal brain sections 2 h following 50 T HFS. Control and tetanised hemispheres are indicated.

IEGs and Their Relation to Stress

In addition to classical IEGs such as krox 24 being involved in memory and LTP, we have recently discovered that transcription factors normally associated more with stress responses are also activated by LTP-inducing stimulation. In particular, we have been interested in how transcription factors such as Activating Transcription Factor 3 (ATF3) might be involved in LTP and hence memory processes. This interest stems in part from studies showing that c-Jun N-terminal kinase (JNK) activity, which is associated with some forms of cellular stress and apoptosis (see refs. 44 and 63, as well as 62), is also activated in the CNS by environmental stimulation.[65] Furthermore, one JNK target, c-Jun, is switched on in neurons during LTP.[5] ATF3 (also called LRF-1 in rat, LRG-21, CRG-5 and TI-241 in mouse) is a member of the ATF/CREB family of bZIP transcription factors but cannot be detected basally in neurons.[21,22,30,32,35] ATF3 has the characteristics of an IEG, with its induction being independent of protein synthesis.[22] The rapid induction of ATF3 by a number of physiological stress signals has led to its general designation as a stress response gene.[28,29] In the nervous system, for example, induction of ATF3 has been observed following peripheral nerve axotomy[55,58] and pentylenetetrazole-induced seizure activity.[12] This induction may be linked to Ca^{2+} influx, as ATF3 is also induced in SH-SY5Y neuroblastoma cells by calcium ionophore treatment.[7] In addition, there is evidence that induction of ATF3 is downstream of JNK activation.[11,29] The induction of ATF3 following seizures and the possible link to Ca^{2+} flux and JNK activation prompted us to ask if ATF3 is induced as a consequence of nonpathological synaptic activity.

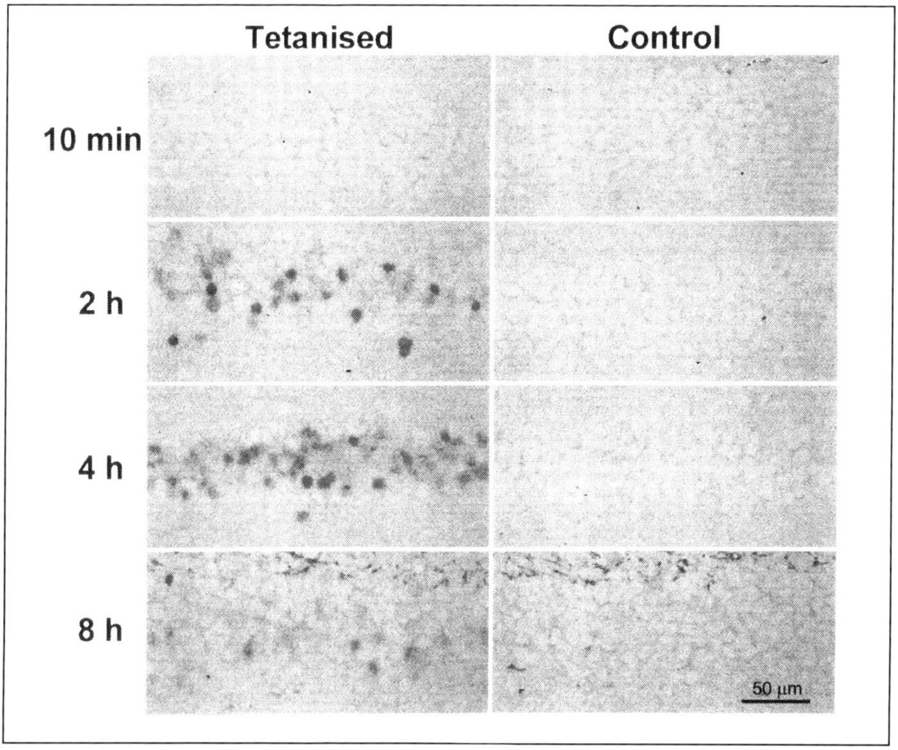

Figure 2. Time course for ATF3 induction in dentate granule cells following 50 T HFS. Brains were removed at the indicated times after the 50 T HFS protocol and coronal brain sections were immunostained for ATF3. Control and tetanised hemispheres are indicated.

We conducted experiments in freely moving adult rats which had bilateral chronically-indwelling stimulating and recording electrodes implanted suitable for recording perforant path-evoked field potentials in the dentate gyrus.[5] High-frequency stimulation (HFS, 400 Hz) was delivered to the perforant path of one hemisphere, with the contralateral hemisphere used as a nontetanised control. Brains were removed and frozen at defined times post-tetanisation (as described in detail previously in ref. 5), and 16 µm frozen coronal brain sections cut through the dorsal hippocampus were fixed and immunostained with antibodies to ATF3 (Santa Cruz sc-188) or Krox 24 (Santa Cruz sc-189). Antibody binding was visualised by peroxidase/ 3,3'-diaminobenzidine (DAB) staining.[63]

In these studies we observed ATF3 induction following HFS of the perforant path input to the dentate gyrus. A robust tetanisation protocol of 50 trains (50 T), delivered as described by Abraham et al,[5] which reliably induces LTP lasting weeks in the dentate gyrus,[6] resulted in ATF3 induction 2 h post-tetanisation in the dentate granule cell layer on the tetanised side, but not on the control side (Fig. 1A). This response coincided with the induction of Krox 24, which was strongly induced throughout the dentate granule cell layer of the tetanised hemisphere (Fig. 1B). However, the ATF3 expression appeared confined to the nuclei of a proportion of the granule cells (Fig. 1C) unlike the more general Krox 24 immunostaining. In sections that contained the path of electrode insertion, some ATF3 staining was evident in cells immediately surrounding the needle track for both hemispheres of the brain (results not shown), a result consistent with ATF3 being induced by tissue damage.[29]

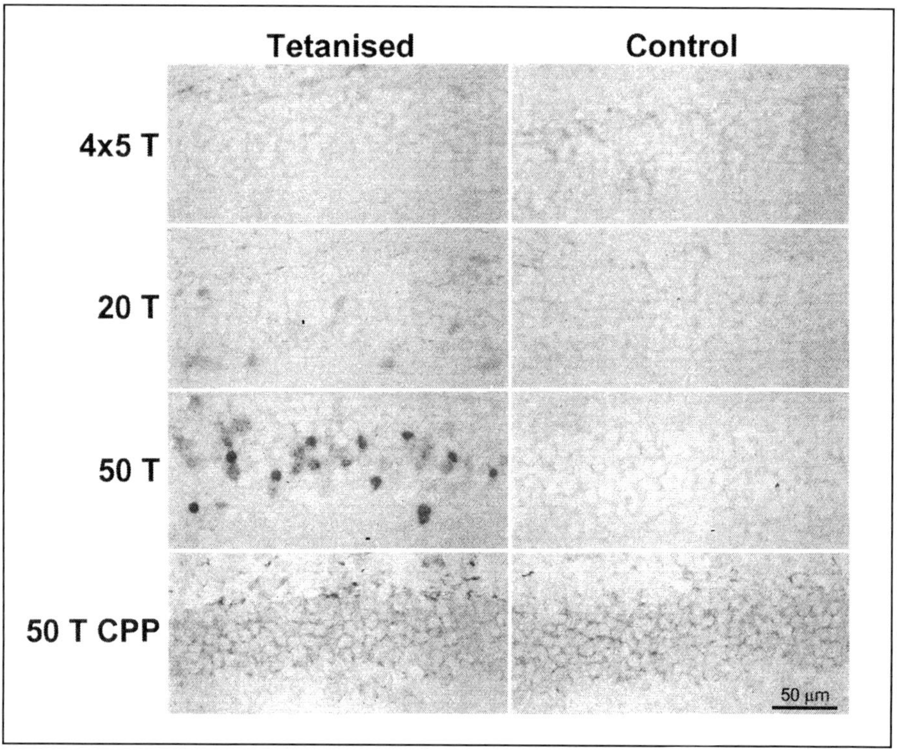

Figure 3. ATF3 induction in dentate granule cells varies with tetanisation protocol and is NMDA receptor dependent. Brains were removed 2 h after the indicated HFS protocols, and coronal brain sections were immunostained for ATF3. The NMDA receptor antagonist CPP (10 mg/kg) was injected i.p. 2 h prior to HFS (bottom panel). Control and tetanised hemispheres are indicated.

The time course of ATF3 expression in dentate granule cells was determined following completion of the 50 T HFS protocol (Fig. 2). No ATF3 immunostaining was present in dentate granule cells 10 min after completion of the 50 T protocol (4/4 animals). At 2 h post-tetanisation, 6/6 animals showed ATF3 immunostaining in a proportion of the dentate granule cells on the tetanised side. ATF3 expression was still present at 4 h post-tetanisation (6/8 animals), and was declining but still visible at 8 h post-tetanisation (2/3 animals). This time course of ATF3 expression is quite similar to that described for Krox 24, which shows a peak at 1-2 h and has returned to baseline by 8 h post-tetanisation.[51]

To determine whether the induction of ATF3 in dentate granule cells by HFS bore any relationship to LTP induction and stability, we compared ATF3 immunostaining at 2 h post-tetanisation following a range of tetanisation protocols (Fig. 3). As described above, the robust 50 T protocol reliably induced ATF3 expression (6/6 animals). Pretreatment with the NMDA receptor antagonist 3-(2-carboxypiperazin-4-yl)-propyl-1-phosphonic acid (CPP) completely blocked ATF3 induction 2 h following the 50 T protocol (Fig. 3, 4/4 animals), showing that the induction of ATF3, like LTP, is dependent on NMDA receptor activation. A 20 train protocol (20 T), which induces robust LTP that is however less stable than that induced by 50 T,[5,64] resulted in low but detectable ATF3 expression (2/3 animals). A spaced protocol of 20 trains administered in 4 sets of 5 trains with 10 min between each (4x5 T) did not result in ATF3 induction (4/4 animals). All of these protocols typically show substantial initial LTP,[4,5] indicating a poor correlation between LTP induction and ATF3 expression. This is also the case for Krox 24 induction, which shows a better correlation with LTP persistence than with the initial level of LTP.[51] However, the 4x5 T protocol typically produces more stable LTP than the condensed burst-type 20 T protocol,[4] which suggests that the strength of ATF3 induction is more closely related to the intensity of the HFS protocol than the stability of the resulting LTP.

Overall, we found from these experiments that protocols with fewer or less sustained bouts of tetanisation were less effective at inducing ATF3 expression. These results suggest that components of a stress response may be initiated in dentate granule cells following robust HFS protocols, possibly as a result of a large amplitude Ca^{2+} influx via NMDA receptors. Details of the signalling molecules involved in ATF3 induction in this system are not clear at present. However, we have preliminary evidence that the related transcription factor ATF2 is phosphorylated at Thr69/71 10 min after 50 T HFS (data not shown). Because the ATF3 promoter can be activated by ATF2/c-Jun heterodimers,[42] this result suggests that the induction of ATF3 after HFS may be mediated by activation of ATF2 and c-Jun. We have previously shown that ATF2 is phosphorylated in neurons undergoing apoptosis.[63] Furthermore, we hypothesised that Jun/ATF2 heterodimers cause apoptosis.[60] Clearly, dentate granule cells do not undergo apoptosis after LTP stimulation, but activation of stress-associated signalling cassettes in these neurons suggests some stress-related LTP processes. Whether these responses influence the characteristics of induced LTP is not clear at present.

Summary

In conclusion, recent data[39] provide support for the hypothesis, proposed many years previously[3] that the IEG krox 24 codes for a transcription factor (Krox 24) that links short-term neuronal events to long-term events required for consolidation of long-term memory storage. Although many other transcription factors are also induced by LTP stimulation, their role in memory processes is not as well defined. However, studies showing that LTP stimulation elicits not only changes in transcription factors classically associated with plasticity (e.g., Krox 24) but also in transcription factors linked historically to stress responses (e.g., ATF3, ATF2, c-Jun) increase the range of molecules associated with this model of memory formation. Whether this is because LTP represents a stress-response paradigm rather than a pure model of memory formation, or whether ATF3, ATF2 and c-Jun have multiple functions in addition to their presumed role in cell stress, is presently unclear.

Studies of IEG involvement in LTP and memory processes have revealed important signal transduction pathways that control the formation of long-term memories in the brain. Many challenges and opportunities lie ahead, including identifying the late-response genes activated by transcription factors such as Krox 24 (e.g., synapsin 1[56]) so that the molecular events involved in encoding the engram can be fully defined. The clinical implications of this work have yet to be fully realised but the hope is that drugs and/or gene therapy approaches, based upon these memory signalling cassettes, will one day provide a rational and effective approach to treating memory disorders such as AD.

Acknowledgements

Supported by grants from the Marsden Fund and the Health Research Council of New Zealand.

References

1. Aarsland D, Larsen JP, Reinvang I et al. Effects of cholinergic blockade on language in healthy young women. Implications for the cholinergic hypothesis in dementia of the Alzheimer type. Brain 1994; 117:1377-1384.
2. Abraham WC, Christie BR, Logan B et al. Immediate early gene expression associated with the persistence of heterosynaptic long-term depression in the hippocampus. Proc Natl Acad Sci USA 1994; 91:10049-10053.
3. Abraham WC, Dragunow M, Tate WP. The role of immediate early genes in the stabilization of long-term potentiation. Mol Neurobiol 1991; 5:297-314.
4. Abraham WC, Logan B, Greenwood JM et al. Induction and experience-dependent consolidation of stable long-term potentiation lasting months in the hippocampus. J Neurosci 2002; 22:9626-9634.

5. Abraham WC, Mason SE, Demmer J et al. Correlations between immediate early gene induction and the persistence of long-term potentiation. Neurosci 1993; 56:717-727.
6. Abraham WC, Mason-Parker SE, Williams J et al. Analysis of the decremental nature of LTP in the dentate gyrus. Mol Brain Res 1995; 30:367-372.
7. Adler EM, Fink JS. Calcium regulation of vasoactive intestinal polypeptide mRNA abundance in SH-SY5Y human neuroblastoma cells. J Neurochem 1993; 61:727-737.
8. Bartus RT, Dean III RL, Beer B et al. The cholinergic hypothesis of geriatric memory dysfunction. Science 1982; 217:408-414.
9. Beckmann AM, Wilce PA. Egr transcription factors in the nervous system. Neurochem Int 1997; 31:477-516.
10. Bierer LM, Haroutunian V, Gabriel S et al. Neurochemical correlates of dementia severity in Alzheimer's disease: Relative importance of the cholinergic deficits. J Neurochem 1995; 64:749-760.
11. Cai Y, Zhang C, Nawa T et al. Homocysteine-responsive ATF3 gene expression in human vascular endothelial cells: Activation of c-Jun NH_2-terminal kinase and promoter response element. Blood 2000; 96:2140-2148.
12. Chen BP, Wolfgang CD, Hai T. Analysis of ATF3, a transcription factor induced by physiological stresses and modulated by gadd153/Chop10. Mol Cell Biol 1996; 16:1157-1168.
13. Cole AJ, Saffen DW, Baraban JM et al. Rapid increase of an immediate early gene messenger RNA in hippocampal neurons by synaptic NMDA receptor activation. Nature 1989; 340:474-476.
14. De Sarno P, Jope RS. Phosphoinositide hydrolysis activated by muscarinic or glutamatergic, but not adrenergic, receptors is impaired in ApoE-deficient mice and by hydrogen peroxide and peroxynitrite. Exp Neurol 1998; 152:123-128.
15. Desjardins S, Mayo W, Vallee M et al. Effect of aging on the basal expression of c-Fos, c-Jun, and Egr-1 proteins in the hippocampus. Neurobiol Aging 1997; 18:37-44.
16. Douglas RM, Dragunow M, Robertson HA. High-frequency discharge of dentate granule cells, but not long-term potentiation, induces c-fos protein. Brain Res 1988; 464:259-262.
17. Dragunow M. Differential expression of immediate-early genes during synaptic plasticity, seizures and brain injury suggests specific functions for these molecules in brain neurons. In: Tolle TR, Zieglgansberger W, eds. Immediate-early genes in the CNS: More than just activity markers. Springer, New York: 1994:33-50.
18. Dragunow M. A role for immediate-early transcription factors in learning and memory. Behav Genet 1996; 26:293-299.
19. Dragunow M, Henderson C. An in vitro model system to investigate muscarinic receptor-mediated induction of the CREB and Krox 24 memory-related transcription factors. In: Keystone Symposium on Molecular Mechanisms of Alzheimer's Disease. Taos, New Mexico:1999:50.
20. Dragunow M, Robertson HA. Kindling stimulation induces c-fos protein(s) in granule cells of the rat dentate gyrus. Nature 1987; 329:441-442.
21. Drysdale BE, Howard DL, Johnson RJ. Identification of a lipopolysaccharide inducible transcription factor in murine macrophages. Mol Immunol 1996; 33:989-998.
22. Farber JM. A collection of mRNA species that are inducible in the RAW 264.7 mouse macrophage cell line by gamma interferon and other agents. Mol Cell Biol 1992; 12:1535-1545.
23. Farber SA, Nitsch RM, Schulz JG et al. Regulated secretion of beta-amyloid precursor protein in rat brain. J Neurosci 1995; 15:7442-7451.
24. Fisher A, Brandeis R, Chapman S et al. M1 muscarinic agonist treatment reverses cognitive and cholinergic impairments of apolipoprotein E-deficient mice. J Neurochem 1998; 70:1991-1997.
25. Genis I, Fisher A, Michaelson DM. Site-specific dephosphorylation of tau of apolipoprotein E-deficient and control mice by M1 muscarinic agonist treatment. J Neurochem 1999; 72:206-213.
26. Greenberg ME, Ziff EB, Greene LA. Stimulation of neuronal acetylcholine receptors induces rapid gene transcription. Science 1986; 234:80-83.
27. Greenwood JM, Dragunow M. Muscarinic receptor-mediated phosphorylation of cyclic AMP response element binding protein in human neuroblastoma cells. J Neurochem 2002; 82:389-397.
28. Hai T, Hartman MG. The molecular biology and nomenclature of the activating transcription factor/cAMP responsive element binding family of transcription factors: Activating transcription factor proteins and homeostasis. Gene 2001; 273:1-11.
29. Hai T, Wolfgang CD, Marsee DK et al. ATF3 and stress responses. Gene Expr 1999; 7:321-335.
30. Hai TW, Liu F, Coukos WJ et al. Transcription factor ATF cDNA clones: An extensive family of leucine zipper proteins able to selectively form DNA-binding heterodimers. Genes Dev 1989; 3:2083-2090.
31. Holford NH, Peace KE. Methodologic aspects of a population pharmacodynamic model for cognitive effects in Alzheimer patients treated with tacrine. Proc Natl Acad Sci USA 1992; 89:11466-11470.
32. Hsu JC, Laz T, Mohn KL et al. Identification of LRF-1, a leucine-zipper protein that is rapidly and highly induced in regenerating liver. Proc Natl Acad Sci USA 1991; 88:3511-3515.
33. Hughes P, Dragunow M. Activation of pirenzepine-sensitive muscarinic receptors induces a specific pattern of immediate-early gene expression in rat brain neurons. Mol Brain Res 1994; 24:166-178.

34. Hughes P, Dragunow M. Induction of immediate-early genes and the control of neurotransmitter-regulated gene expression within the nervous system. Pharmacol Rev 1995; 47:133-178.
35. Ishiguro T, Nakajima M, Naito M et al. Identification of genes differentially expressed in B16 murine melanoma sublines with different metastatic potentials. Cancer Res 1996; 56:875-879.
36. Izquierdo I, da Cunha C, Rosat R et al. Neurotransmitter receptors involved in post-training memory processing by the amygdala, medial septum, and hippocampus of the rat. Behav Neural Biol 1992; 58:16-26.
37. Jansen KL, Faull RL, Dragunow M et al. Alzheimer's disease: Changes in hippocampal N-methyl-D-aspartate, quisqualate, neurotensin, adenosine, benzodiazepine, serotonin and opioid receptors—an autoradiographic study. Neurosci 1990; 39:613-627.
38. Jerusalinsky D, Kornisiuk E, Izquierdo I. Cholinergic neurotransmission and synaptic plasticity concerning memory processing. Neurochem Res 1997; 22:507-515.
39. Jones MW, Errington ML, French PJ et al. A requirement for the immediate early gene Zif268 in the expression of late LTP and long-term memories. Nature Neurosci 2001; 4:289-296.
40. Jope RS, Song L, Powers RE. Cholinergic activation of phosphoinositide signaling is impaired in Alzheimer's disease brain. Neurobiol Aging 1997; 18:111-120.
41. Kelly JF, Furukawa K, Barger SW et al. Amyloid beta-peptide disrupts carbachol-induced muscarinic cholinergic signal transduction in cortical neurons. Proc Natl Acad Sci USA 1996; 93:6753-6758.
42. Liang G, Wolfgang CD, Chen BP et al. ATF3 gene. Genomic organization, promoter, and regulation. J Biol Chem 1996; 271:1695-1701.
43. Matthies H. In search of cellular mechanisms of memory. Prog Neurobiol 1989; 32:277-349.
44. Mattson MP. Cellular actions of beta-amyloid precursor protein and its soluble and fibrillogenic derivatives. Physiol Rev 1997; 77:1081-1132.
45. Mielke K, Herdegen T. JNK and p38 stress kinases—degenerative effectors of signal-transduction-cascades in the nervous system. Prog Neurobiol 2000; 61:45-60.
46. Mondadori C, Hengerer B, Ducret T et al. Delayed emergence of effects of memory-enhancing drugs: Implications for the dynamics of long-term memory. Proc Natl Acad Sci USA 1994; 91:2041-2045.
47. Morgan JI, Cohen DR, Hempstead JL et al. Mapping patterns of c-fos expression in the central nervous system after seizure. Science 1987; 237:192-197.
48. Newcomer JW, Krystal JH. NMDA receptor regulation of memory and behavior in humans. Hippocampus 2001; 11:529-542.
49. O'Donovan KJ, Tourtellotte WG, Millbrandt J et al. The EGR family of transcription-regulatory factors: Progress at the interface of molecular and systems neuroscience. Trends Neurosci 1999; 22:167-173.
50. Poirier J, Delisle MC, Quirion R et al. Apolipoprotein E4 allele as a predictor of cholinergic deficits and treatment outcome in Alzheimer disease. Proc Natl Acad Sci USA 1995; 92:12260-12264.
51. Richardson CL, Tate WP, Mason SE et al. Correlation between the induction of an immediate early gene, zif/268, and long-term potentiation in the dentate gyrus. Brain Res 1992; 580:147-154.
52. Sadot E, Gurwitz D, Barg J et al. Activation of m1 muscarinic acetylcholine receptor regulates tau phosphorylation in transfected PC12 cells. J Neurochem 1996; 66:877-880.
53. Silva AJ, Kogan JH, Frankland PW et al. CREB and memory. Annu Rev Neurosci 1998; 21:127-148.
54. Sirvio J. Strategies that support declining cholinergic neurotransmission in Alzheimer's disease patients. Gerontol 1999; 45:3-14.
55. Takeda M, Kato H, Takamiya A et al. Injury-specific expression of activating transcription factor-3 in retinal ganglion cells and its colocalized expression with phosphorylated c-Jun. Invest Ophthalmol Vis Sci 2000; 41:2412-2421.
56. Thiel G, Schoch S, Petersohn D. Regulation of synapsin I gene expression by the zinc finger transcription factor zif268/egr-1. J Biol Chem 1994; 269:15294-15301.
57. Tsiokas L, Watson M. Differential in vivo induction of immediate early genes by oxotremorine in the central nervous system of long- and short-sleep mice. Mol Pharmacol 1995; 47:272-282.
58. Tsujino H, Kondo E, Fukuoka T et al. Activating transcription factor 3 (ATF3) induction by axotomy in sensory and motoneurons: A novel neuronal marker of nerve injury. Mol Cell Neurosci 2000; 15:170-182.
59. von der Kammer H, Mayhaus M, Albrecht C et al. Muscarinic acetylcholine receptors activate expression of the EGR gene family of transcription factors. J Biol Chem 1998; 273:14538-14544.
60. Walton MR, Dragunow M et al. Is CREB a key to neuronal survival? Trends Neurosci 2000; 23:48-53.
61. Walton M, Henderson C, Mason-Parker S et al. Immediate early gene transcription and synaptic modulation. J Neurosci Res 1999; 58:96-106.
62. Walton M, MacGibbon G, Young D et al. Do c-Jun, c-Fos, and amyloid precursor protein play a role in neuronal death or survival? J Neurosci Res 1998; 53:330-342.
63. Walton M, Woodgate AM, Sirimanne E et al. ATF-2 phosphorylation in apoptotic neuronal death. Mol Brain Res 1998; 63:198-204.
64. Williams J, Dragunow M, Lawlor P et al. Krox20 may play a key role in the stabilization of long-term potentiation. Mol Brain Res 1995; 28:87-93.
65. Xu X, Raber J, Yang D et al. Dynamic regulation of c-Jun N-terminal kinase activity in mouse brain by environmental stimuli. Proc Natl Acad Sci USA 1997; 94:12655-12660.

Chapter 5.3

Protein Synthesis: I. Pharmacology

Oliver Stork and Hans Welzl

Abstract

The formation of long-lasting memory traces depends on the de novo synthesis of proteins. For more than 30 years substantial experimental evidence has been collected in species ranging from insects to mammals, in support of this hypothesis. A pharmacological approach to investigate the dependence of long-term memory formation on de novo protein synthesis is to administer drugs that prevent protein synthesis on the transcriptional or translational level. When injected during or after learning, these drugs block the development of long-term memory while leaving short-term memory unaffected. Recent research investigating the time course of protein synthesis following learning revealed the existence of two or probably more distinct time windows during which new proteins are synthesized in order to form an enduring memory trace. Another current topic addresses the question where in the neuron de novo protein synthesis takes place. Whereas a large part of the new proteins are synthesized in the soma, some of them are assembled specifically at those synapses whose modification of synaptic efficiency underlies memory formation.

Asking about the 'Where' and 'When' of Learning-Related Protein Synthesis

Over the past 30 years and more, a large body of evidence has been raised for an involvement of protein synthesis in the formation of long-term memory (LTM). The use of protein synthesis inhibitors (PSIs) has been fundamental for the development of this view. Short-term memory (STM), which is insensitive to protein synthesis inhibition, is now believed to be based on transient modifications of preexisting molecules, most importantly phosphorylation and dephosphorylation of enzymes, receptors and/or ion channels. For an alteration of synaptic efficiency beyond the scope of STM, however, the de novo synthesis of proteins appears to be indispensable. These proteins include transcriptional activators and regulators, neurotrophic factors, cytoskeletal and cell recognition molecules, to name but a few. In essence, it is believed that the protein synthesis during formation of LTM is required for structural changes of existing synapses or generation of new neuronal circuits through reactive sprouting and synaptogenesis.[4,22,67,83] A strong similarity is apparent to the regulation of growth during development and regeneration in the nervous system. All these processes are activity-dependent, require retrograde messengers and recruit cellular programs of growth including de novo protein synthesis to evoke changes in synaptic structure and function.[18,62,101]

The fact that LTM in vertebrates and invertebrates can be blocked by PSIs has long been the sole evidence for an involvement of protein synthesis.[29] In recent years, however, this hypothesis has obtained support from studies that employed molecular and genetic approaches, as the targeted disruption of transcription factors (e.g., c-fos, zif268, CREB) or specific effector genes (e.g., BDNF, NCAM, CamKII) disturbed the formation of LTM. It must be considered though that mutation of a gene may affect a variety of biological functions other than memory

From Messengers to Molecules: Memories Are Made of These, edited by Gernot Riedel and Bettina Platt.

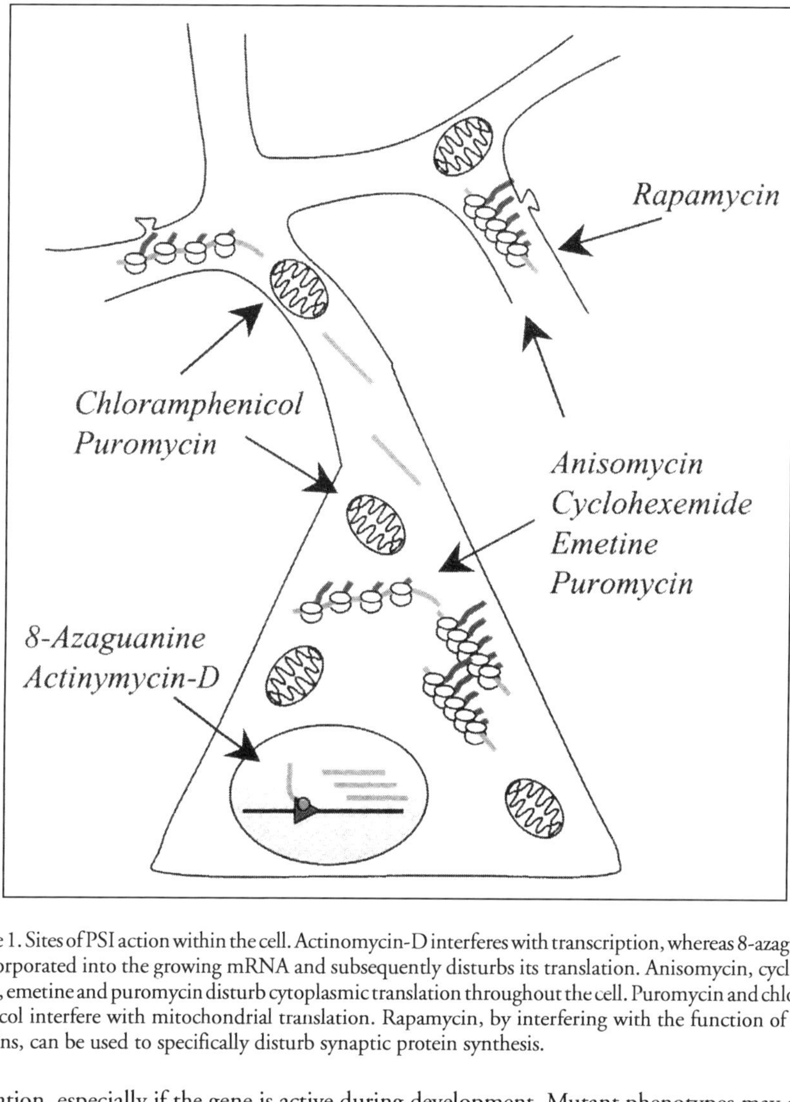

Figure 1. Sites of PSI action within the cell. Actinomycin-D interferes with transcription, whereas 8-azaguanin is incorporated into the growing mRNA and subsequently disturbs its translation. Anisomycin, cyclohex-imide, emetine and puromycin disturb cytoplasmic translation throughout the cell. Puromycin and chloram-phenicol interfere with mitochondrial translation. Rapamycin, by interfering with the function of TOR proteins, can be used to specifically disturb synaptic protein synthesis.

formation, especially if the gene is active during development. Mutant phenotypes may often be related to compensatory up-regulation or cis-activation of other genes[91] rather than the mutation itself, and are sensitive to variations in genetic background.[28,46] Some of these problems are overcome with inducible transgenic or knock out mutants,[110,118] with injection of anti-sense oligonucleotides or with virus-mediated gene-transfer.[61,131] However, it may still be difficult to clearly distinguish roles of the mutated molecules in STM and LTM as experimentally induced changes in gene expression require time and can hardly be controlled to occur in an exact time window before, during or after training.

Thus, while the tools are now available to investigate protein synthesis with unprecedented molecular specificity, PSIs become increasingly important to investigate the temporal and spatial aspects, in other words the "When? and Where?" of learning-related protein synthesis. They appear to be particularly powerful when studied in combination with or comparison to specific molecular interventions. In the following we describe the function of the most commonly used PSIs and their effect on memory formation in various paradigms.

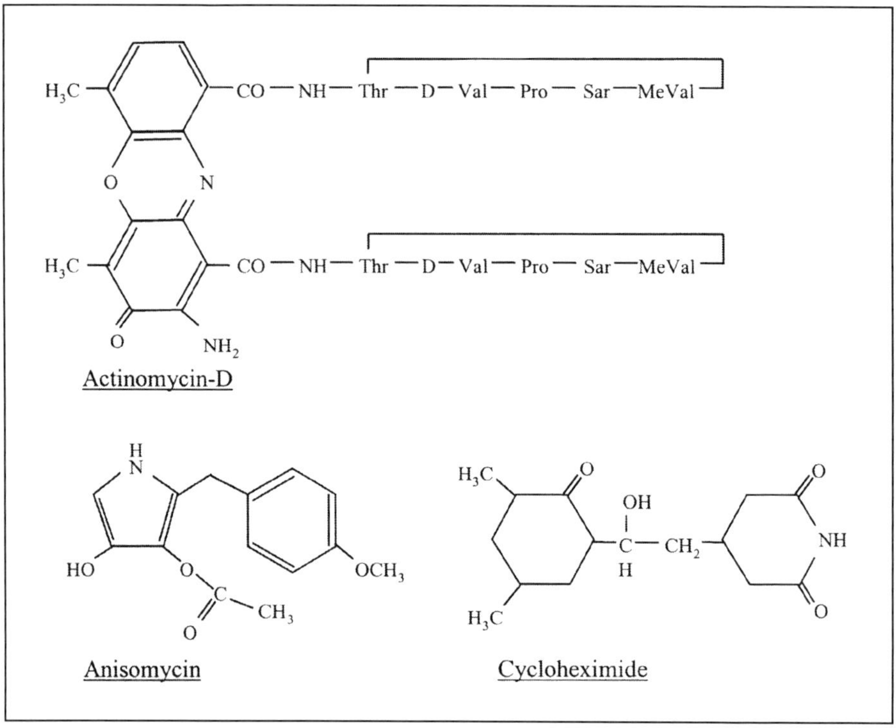

Figure 2. Chemical structure of the most commonly used PSIs in memory research, actinomycin-D, anisomycin and cycloheximide.

Inhibitors of Protein Synthesis

Transcription Inhibitors

Actinomycin-D (dactinomycin) and 8-azaguanine can be used to interfere with protein synthesis at the transcriptional level. Actinomycin-D is one of the best known antibiotics and used as a cytostatic agent in, e.g., Hodgkin's disease. It is produced by *Streptomyces antibioticus* and intercalates between GC base pairs of the DNA double strand without binding to single stranded RNA, and thus at low concentrations can inhibit transcription without interfering with translation or DNA replication.[99] Actinomycin-D does not cross the blood-brain barrier and, therefore, has to be injected intracranially when a blockade of transcription in the central nervous system is desired. 8-Azaguanine, which is identical to Pathocidin from *Streptomyces spectabilis*, is incorporated during transcription[43] and subsequently disturbs the translation of the affected mRNA. It also exerts influence on a variety of enzymes in cell metabolism and inhibits 5-phosphoribosylpyrophospate-amidotransferase, the key enzyme in purine synthesis. Unfortunately, doses of 8-azaguanine and actinomycin-D that are sufficient to substantially suppress cerebral protein synthesis cause rapid and irreversible systemic toxicity as well as necrosis.[21,121] Intracranial injections of very low doses (1µg) of actinomycin-D, on the other hand, have little effect on RNA synthesis but still attenuate retention even when injected as late as 24h after training.[121] This suggests that retention deficits may be the result of drug-induced brain damage and abnormalities in electrical activity. Thus, experiments with these transcription inhibitors should include anatomical and electrophysiological controls and have to be interpreted with caution.

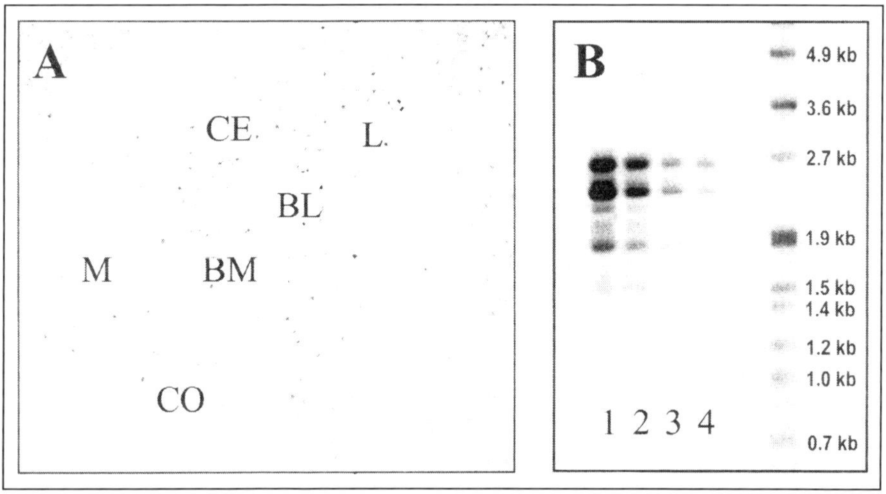

Figure 3. Gene expression during memory formation. (A) Induction of the immediate early gene c-fos in the amygdala, 30min after fear conditioning training. A number of strongly labeled cells can be found throughout the amygdala, whereas baseline c-fos expression levels (not shown) are hardly detectable. (B) Expression of a novel ubiquitination factor Praja1 in the amygdala, 6h after fear conditioning training. (1) Fear conditioned, (2) pseudo-trained and (3) fear memory retrieving mice, (4) naive controls. The graded increase across groups indicates fear-, stress- and learning-related regulation of gene-expression.[128] CE, central amygdala; L, lateral amygdala; BL, basolateral amygdala; MB, basomedial amygdala; M, medial amygdala and CO, cortical amydala.

Translation Inhibitors

Three classes of translation inhibitors have most commonly been used in memory experiments: puromycin (PURO), anisomycin (ANI), and various glutarimides (cycloheximide (CXM), emetine and acetoxycycloheximide). PURO is a nucleoside-antibiotic from *streptomyces alboniger* and inhibits protein synthesis at both the 70S and 80S, i.e., prokaryotic/mitochondrial and eukaryotic, ribosomes. Through its structural analogy to aminoacyl-tRNA, PURO can be incorporated into the growing peptide chain at its carboxyl end, which results in premature release of peptidyl-PURO fragments from the ribosomal complex. Unfortunately, such peptidyl-PURO fragments may by themselves have a long-lasting effect on cell function that causes amnesia.[29] In addition to inhibiting protein synthesis, PURO induces hippocampal seizures, swelling of mitochondria and disaggregation of ribosomes.[36] Due to its numerous side effects, PURO cannot be recommended to investigate the effects of protein synthesis inhibition on memory formation (for a discussion see ref. 29).

ANI is a pyrolidine antibiotic from *Streptomyces griseolus*. It inhibits the peptidyl transferase activity in eukaryotic ribosomes and thus interferes with peptide bond formation.[94] ANI is, at doses that successfully block retention, a fairly nontoxic protein synthesis inhibitor. Successive injections of ANI permit an inhibition of variable duration in the range of 2-8h, with the inactive form deacetyl-ANI serving as control compound.[36] However, it should be considered that ANI is a potent agonist of Jun-NH2 terminal kinase (JNK) as well as mitogen-activated kinase (MAPK), and can induce apoptosis.[15,136] CXM, which is isolated from *Streptomyces griseus,* interferes with the function of peptidyl transferase at the large subunit of the eukaryotic ribosome. However, CXM not only disturbs translation (initiation, translocation, and steps of elongation processes), but also DNA, rRNA and tRNA synthesis,[43,49] and in effective doses is far more toxic than ANI.[35] Doses inhibiting protein synthesis by 80% or more can cause sickness and, under stressful training conditions, even death. On the behavioral level, changes

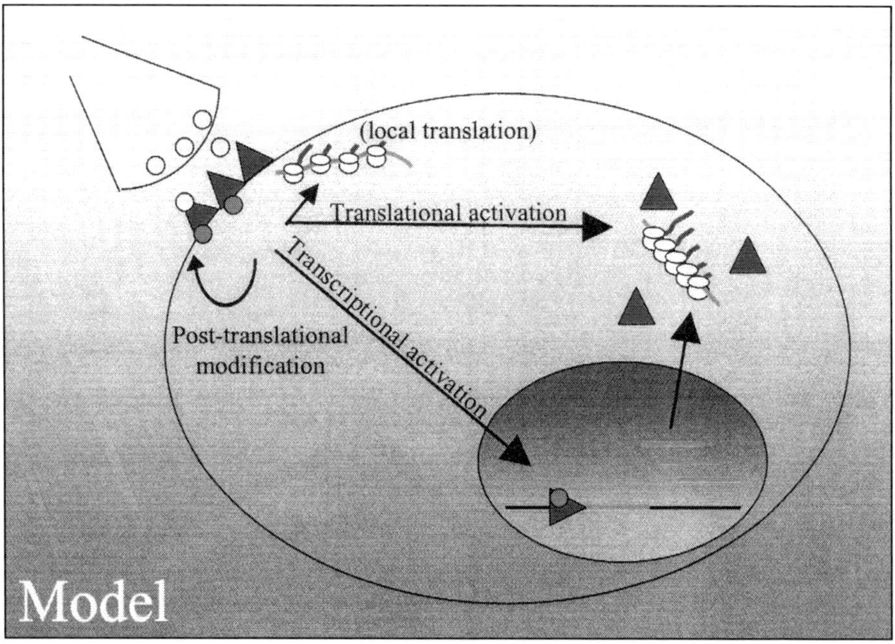

Figure 4. A model for protein synthesis during LTM formation. An incoming signal can produce several cellular responses: (1) post-translational modifications (e.g., phosphorylation) of proteins at activated synapses, responsible for STM, (2) a translational induction of protein synthesis at preexisting mRNA, to support ITM and possibly the generation of synaptic "tags", and (3) an induction of gene-expression through the activation of transcription factors, which are indispensable for LTM. All these processes appear to be activated in parallel during learning.

in locomotor activity occur that are not seen after ANI injections. Both protein synthesis inhibitors can cause diarrhea, but the signs are hardly noticeable after ANI injections.[36] Emetine is the main alkaloid component in *Radix ipecacuanhae* and inhibits protein synthesis through an interaction with the ribosomal protein S14 of the small eukaryotic ribosomal subunit. Emetine is very toxic and irritant and although considered one of the classical PSI antibiotics has only been used in few studies to interfere with memory formation.

Antibiotics that specifically interfere with the function of 70S ribosomes, such as the *Streptomyces venezuela* antibiotic chloramphenicol, may be used to inhibit prokaryotic and mitochondrial protein synthesis. Chloramphenicol binds to the 50S mitochondrial ribosome subunit and disturbs the mitochondrial peptidyl transferase reaction without affecting protein synthesis of eukaryotic ribosomes. Indirectly, however, chloramphenicol also affects nuclear and cytoplasmic gene expression and leads to the generation of new mitochondrial RNA species.[117] Moreover, the inhibition of mitochondrial protein synthesis may change energy provision or Ca^{2+} sequestration in the mitochondria. The effects of chloramphenicol treatment thus have to be interpreted with care, in particular as its blockade of LTM can be accompanied by profound changes in motor behavior.[42]

Rapamycin finally is a lipophilic macrolide, which has been employed to show synaptic protein synthesis in *Aplysia californica*.[16] Rapamycin is isolated from a strain of *Streptomyces hygroscopicus* indigenous to Easter Island ("Rapa Nui"). It interacts with the ubiquitous protein FKBP12 to specifically interfere with the TOR protein (target of rapamycin[98]). TOR modulates the phosphorylation of proteins that are involved in translation initiation and elongation of the peptide chain, and controls the abundance of the translation machinery in response to mitogen and growth factor stimulation. TOR localizes to synaptic sites and interacts with gephrin,[109] and is therefore thought to play a pivotal role in the control of synaptic protein synthesis.

In summary, a number of PSIs are available to interfere with different steps of protein synthesis during memory formation. However, possible unspecific effects have to be carefully controlled in experiments with PSIs. In fact, several hypotheses have been put forth to explain the amnesic effects of protein synthesis inhibitors by mechanisms other than their blockade of protein synthesis (e.g., by producing sickness and conditioned aversion, changes in locomotor activity, inhibition of steroidogenesis, and disturbance of catecholamine neurotransmission), but a number of experiments in different laboratories have provided results that make all these alternative explanations unlikely (for a review see ref. 29). Still, it should be considered that inhibitors of protein synthesis may affect behavioral performance not only through their interference with memory formation, but also by acting on learning-related processes involved in, e.g., attention or motivation. In addition to PSIs, the inhibitors of specific signal transduction pathways can be used to interfere with gene expression during memory formation. In fact, inhibition of the protein kinase A (PKA) and MAPK pathways, or blockade of glucocorticoid receptors have revealed striking similarity to the effects of PSIs.

Effects of Protein Synthesis Inhibitors on Memory Formation

Invertebrates

Learning in invertebrates has been extensively used to investigate cellular and molecular mechanisms of memory formation, including the role of protein synthesis in LTM. Invertebrates are capable of simple forms of learning that are controlled by less complicated and more accessible nervous systems than in vertebrates.

Aplysia californica

In the sea snail *Aplysia californica*, a simple gill-withdrawal reflex habituates with repeated stimulation (direct touch), but is sensitized by a noxious stimulation of the tail (for a review see ref. 32). The basic elements of this reflex can be simulated in vitro in a preparation consisting of sensory and motor neurons, and application of exogenous serotonin. Long-term facilitation has been shown to be synapse-specific and to require local protein synthesis, as well as RNA and protein synthesis in the cell body.[16,74] Long-term - but not short-term – in vivo sensitization or in vitro facilitation are prevented when mRNA or protein synthesis is inhibited during or up to 30min following treatments that normally induce the formation of a memory trace.[5,17,81] ANI, when applied around the time of such treatments (-1h to +2h) or 4-7h later, can also block structural changes that are associated with long-term sensitization and facilitation.[90] In addition to STM and LTM, an intermediate-term memory (ITM), which requires translational but not transcriptional activation, has been described.[48,75,129,130] The translational activation during this ITM seems to be independent of the soma and based on mRNA already existing at the synapse.

Hermissenda crassicornis

In this sea snail classical conditioning of the foot contraction reflex is achieved by pairing a light stimulus (conditioned stimulus; CS) with a rotation (unconditioned stimulus; US; for a review see ref. 32). Several different paradigms have been used on different nervous system preparations from *Hermissenda*. Although this makes a direct comparison of experimental data difficult, a number of key features, resembling those obtained in *Aplysia*, emerged. First, extensive conditioning causes structural changes of the neurons involved.[1] Second, three different stages of memory formation could be distinguished: STM,[25,26,97] which is independent of protein and RNA synthesis, ITM, which is dependent on translational but not transcriptional processes,[27] and LTM, which requires mRNA as well as protein synthesis for its formation.[25,26] And third, the requirement for protein synthesis is probably not restricted to the immediate post-training period, but continues for several hours, since prolonged and selective changes in RNA expression have been observed after extensive training (maximal at 24h and still detectable after four days[88]).

Insects

The fruitfly *Drosophila melanogaster* has been a long-time favourite with behavioral geneticists, due to the ease with which mutations can be created and their consequences are investigated (for a review see ref. 32). In a simple odor discrimination task, an anesthesia-resistant and CXM-insensitive memory can be induced that decays within 4 days after training. With spaced training, however, a LTM for odors is established that lasts for at least 7 days and can be blocked by application of CXM during training.[139] Few studies exist that investigated protein synthesis and memory formation in other insects. In honeybees (*Apis mellifera*), LTM for an odour-sucrose association is blocked by inhibition of either transcription or translation. The blocking effect can be detected 4 days after conditioning, if the drugs are injected 1h or 6h – but not 24h - after conditioning. Memory lasting for up to 2 days, however, is unaffected by the protein synthesis inhibition.[80,142,144] Inhibition of protein synthesis also prevents passive avoidance learning in the praying mantis (*Stagmatoptera biocellata*[59]). When tested after 24h, animals that were injected with CXM immediately after training act like naive animals. CXM injection 2h after training has no such effect.

Birds

The analysis of memory formation in birds has been particularly helpful to identify critical temporal phases of protein synthesis and the biochemical processes that occur after training.

Imprinting

When newly hatched chicks are briefly exposed to a distinct moving object, they become imprinted to it and the time spent near this object will be increased during future exposures. This imprinting process can be attenuated by immediate post trial injection of CXM.[47] Imprinting is accompanied by structural changes in the intermediate medial hyperstriatum ventrale, a brain structure known to be essential for different types of visual learning in chicks. However, the question whether protein synthesis inhibition also blocks these changes has not yet been investigated.[55]

Passive Avoidance Learning

Newly hatched chicks will also peck at water droplets that form at beads of different color and shape. When one of the beads is associated with a bitter tasting fluid, the chicks quickly learn to avoid pecking at beads of that color and shape.[103,105] Injection of ANI immediately before or up to 30min after initial training blocks the formation of LTM in this task.[37,133] Pretraining ANI injection can also prevent structural changes in the lobus parolfactorius that accompany the retention of passive avoidance.[120] A second time window of critical protein synthesis has been identified with ANI injections 4-5h post-training.[132] Inhibitors of protein fucosylation similarly prevent LTM formation when applied during this second time window, indicating the generation of glycoproteins that are essential for the learning process.[104,106]

Song Learning

When zebra finches listen to the song of a conspecific, neuronal activity and expression of an immediate early gene (zenk, synonym to zif268) is increased in the caudomedial striatum.[24,78,79,86] With repeated listening to the same song, neuronal activity declines and zenk expression disappears. However, novel calls or complex sounds reinduce neuronal activity and zenk expression. This 'stimulus-specific habituation' can last for up to two days.[23,24] Injection of CXM immediately after song presentation (0.5-3h) or at multiple subsequent periods (5.5-7h, 14-15h, and 17-18.5h) blocks the reduction in neuronal activity and prevents the decline of zenk expression.[23,24,140]

Mammals

Since the early 1960s a wealth of data has accumulated that supports a critical role of protein synthesis for LTM formation in mice and rats (for a review see ref. 29). The data have been collected in paradigms that range from classical conditioning to complex spatial learning tasks.

Fear Conditioning

In recent years, the involvement of protein synthesis in auditory cued and contextual fear conditioning has been extensively studied; this paradigm appears to be particularly well suited for molecular, cellular and biochemical analyses since a robust association of the auditory CS or the training context with the aversive US can be achieved within one brief training session. In different experiments with ANI, CXM and actinomycin-D it has been shown that the formation of such fear memory can be prevented with systemic, intra-ventricular or intra-amygdalar blockade of protein synthesis.[6,114,115,116,127] Contextual fear conditioning was further found to involve either one or two PSI-sensitive consolidation periods, depending on the intensity of training.[13] In the latter and in other studies, a striking similarity was observed to the effects of PKA and MAPK inhibition.[114,115] It has further been shown that fear memories require amygdalar protein synthesis also for reconsolidation after their retrieval.[84] Extinction of contextual fear memory, on the other hand, can take place in the presence of systemically injected PSIs.[65]

Conditioned Taste Aversion

Similar to fear conditioning, conditioned taste aversion learning is characterized by a brief training episode, in this case with a novel taste that is followed by the induction of sickness (e.g., through administration of lithium chloride). Consolidation of taste aversion memory can be prevented by intra-ventricular injections of CXM,[56,137] and by bilateral ANI injections into the gustatory cortex before or during trainingal.[108] Recent studies furthermore have shown that ANI injection into the gustatory cortex can also block the extinction of conditioned taste aversion.[10]

Avoidance Learning

It has long been established that administration of PSIs before or immediately after training prevents passive avoidance learning in rodents.[35,36,66,95,96,124] The same treatment does not affect STM, nor the performance during acquisition of the task. However, the amnesic action of protein synthesis inhibition was shown to decrease with the strength of training and to increase with longer duration of protein synthesis inhibition. The involvement of hippocampus and amygdala in this task is illustrated by the amnesic effects of local actinomycin-D and CXM injections, respecitively.[11] On the other hand, protein synthesis inhibition over a period of 6-8h is necessary to block active avoidance learning or the extinction of this task in mice.[33,34]

Discrimination Learning

Injections of CXM or ANI during and following training interfere with the establishment of LTM for an object discrimination task.[122] Moreover, rat olfactory discrimination learning in two different social situations was found to be blocked by protein synthesis inhibition.[70] ANI, however, does not prevent memory formation after olfactory discrimination learning,[124] possibly due to residual protein synthesis or the occurrence of protein-synthesis independent synaptic changes outlasting the training-test interval. Two critical waves of protein synthesis for brightness discrimination learning , one around training (between 10min pre and 80min post-) and one 4-6h post training,[50,51] could be identified with hippocampal ANI injections. Injections between these two time windows are without effect.

Spatial Learning

Few studies have investigated the effects of protein synthesis inhibition in spatial learning tasks. However, the formation of spatial LTM in the water maze can be disturbed by subcutaneous or intra-ventricular injection of ANI before, but not after each training session.[65,77] Extinction of spatial memory, in contrast, has proven insensitive to ANI.[65]

Electrophysiological Models

Long-term potentiation (LTP) and long-term depression (LTD) are electrophysiological models of synaptic plasticity that can be tested in vivo as well as in vitro. Physiological and molecular events that characterize LTP and LTD resemble those observed in learning paradigms, i.e., strong parallels exist between hippocampal LTP and spatial learning,[12,100] and amygdala LTP and fear conditioning.[73,102] Accordingly, a late phase of LTP (L-LTP), which is sensitive to PSIs in vitro and in vivo,[38,39,57,63,85,89,92,123] can be distinguished from a PSI-insensitive early phase (E-LTP). A similar protein synthesis-dependent late phase of cerebellar LTD has also been described.[69,71] The maintenance of LTP for up to 3h has been shown to require protein synthesis, but not mRNA synthesis,[92] suggesting that - as in ITM - proteins which can support a limited maintenance of LTP are synthesized from preexisting mRNA.

Principle Findings and Future Perspectives in Protein Synthesis Inhibitor Research

Investigations in both vertebrates and invertebrates have provided ample evidence for the dependence of LTM formation on de novo protein synthesis. Findings from protein synthesis inhibition are supported by recent molecular and genetic studies which interfere with the function of specific transcription factors during memory formation.[8,14,60,61,93,131,145,146] PSIs have been particularly valuable tools in determining spatial and temporal characteristics, i.e., the "Where? And When?" of learning-induced gene-expression.

The "When"

In essence, it seems that at least two critical waves of protein synthesis exist during memory consolidation; only when applied within these time windows are protein synthesis inhibitors able to block the formation of LTM.[13,50,51,104,132,134] Moreover, the finding that a limited LTM can still be induced in the presence of transcription inhibitors during an initial phase following training[92,130] has led to the concept of an intermediate-term memory (ITM), which is supported primarily through translation from preexisting mRNAs. Other studies have identified unique biochemical processes during this transitional phase of information storage, such as activation of PKA and calcineurin.[54,72] The biochemical processes that underly ITM and LTM appear to be activated in parallel; the induction of early gene transcription factors for example is essential for LTM, but not for ITM, and occurs within minutes after training.

Protein synthesis is further involved in the reconsolidation and continuous refinement of existing memories. Interestingly, opposite effects of PSIs were observed in different aversive learning tasks: ANI blocked the extinction of an active avoidance memory after systemic injection[34] and the extinction of conditioned taste aversion when infused to the insular cortex,[10] but strongly interfered with the reconsolidation of auditory cued fear memory after injection into the amygdala.[84] These observations may illustrate the complexity of memory reconsolidation and the contribution of different brain areas therein.

The "Where"

By disrupting memory consolidation with local injections of PSIs, it was in fact possible to identify brain areas that undergo neural plasticity following training, such as the amygdala during Pavlovian fear conditioning[114] or the gustatory cortex during aversive taste conditioning.[108] Moreover, PSIs may be used to address the intracellular localization of protein synthesis. It has also become evident in recent years that some critical protein constituents are synthesised synaptically during memory formation.[125,126] In fact, transport of mRNA into dendrites appears to be common for plasticity-related molecules including microtubule-associated protein2 (MAP2),[44,138] CamKII,[76] arg3.1,[68] brain spectrin,[45] trkB and BDNF.[135] Such synaptic protein synthesis may speed up local reorganization processes, allow localized and transcription-independent regulation of protein synthesis through retrograde messengers[82] and possibly provide new "synaptic tags".[16,40,41] As one of the best examples, CamKII mRNA is actively transported into dendrites and associates with synaptic polysomes upon depolarization.[3,76] Specific cis-acting elements in the 3' untranslated region of CaMKII mRNA are responsible for its localization and for its synaptic translation, which involves the cytoplasmic polyadenylation element binding protein, CPEB.[76,143] Initiation of translation, elongation of the peptide chain and the abundance of translation machinery components including CPEB are regulated by the TOR protein, which has become an exciting new target in memory research.[16]

PSIs have also been helpful for the identification of signal transduction pathways that control gene-expression during formation of LTM and to identify the factors that are induced. For instance, a striking parallel has been observed between protein synthesis inhibition and the effect of glucocorticoid inhibition during chick passive avoidance learning.[111,112] In Pavlovian fear memory, a temporal coincidence of PSI- with PKA- and MAPK inhibitor-sensitive phases of memory consolidation has been observed. e.g.,[113] As the PKA and MAPK pathways converge in CREB activation, these findings support the idea that CREB-mediated gene expression may be critical for fear memory formation.[14,61] Indeed, CREB activation in both vertebrates and invertebrates has been implicated in numerous other learning paradigms.[7,31,52,58,64,141]

Downstream of CREB, several genes are induced during a first PSI-sensitive phase immediately after training. In this phase, so-called immediate early gene transcription factors (e.g., c-fos, c-jun, zif268) and neurotrophic factors, the latter of which can initiate activity-dependent functional and structural reorganization in presynaptic and neighboring cells, are expressed.[2,18,30,53,78,79,107] Target genes of these transcriptional activators are expressed in later waves of protein synthesis and likely participate in the modification of synaptic structure and signal transduction.[9,19,20,87,119,128]

In summary, PSIs have been valuable tools for investigations on the brain structures and time windows that are critical for memory formation. They are used to dissect signalling pathways that control gene expression during learning and to identify the factors that are critical in these processes. The application of PSIs to questions of intracellular control mechanisms of synaptic plasticity, such as dendritic protein synthesis and synaptic tagging, and to the question of memory reconsolidation are promising new fields of research.

References

1. Alkon DL, Ikeno H, Dworkin J et al. Contraction of neuronal branching volume: An anatomical correlate of Pavlovian conditioning. Proc Natl Acad Sci USA 1990; 87:1611-1614.
2. Anokhin KV, Rose SPR. Learning-induced increase of immediate early gene messenger RNA in the chick forebrain. Eur J Neurosci 1991; 3:162-167.
3. Bagni C, Mannucci L, Dotti CG et al. Chemical stimulation of synaptosomes modulates alpha-Ca2+/calmodulin-dependent protein kinase II mRNA association to polysomes. J Neurosci 2000; 20:RC76.
4. Bailey CH, Kandel ER. Structural changes accompanying memory storage. Ann Rev Physiol 1993; 55:397-426.
5. Bailey CH, Montarolo P, Chen M et al. Inhibitors of protein and RNA synthesis block structural changes that accompany long-term heterosynaptic plasticity in Aplysia. Neuron 1992; 9:749-758.
6. Bailey DJ, Kim JJ, Sun W et al. Acquisition of fear conditioning in rats requires the synthesis of mRNA in the amygdala. Behav Neurosci 1999; 113:276-282.
7. Bartsch D, Casadio A, Karl KA et al. CREB1 encodes a nuclear activator, a repressor, and a cytoplasmic modulator that form a regulatory unit critical for long-term facilitation. Cell 1998; 95:211-223.

8. Bartsch DGM, Skehel PA, Karl KA et al. Aplysia CREB2 represses long-term facilitation: Relief of repression converts transient facilitation into long-term functional and structural change. Cell 1995; 83:979-992.
9. Barzilai A, Kennedy TE, Sweatt JD et al. 5-HT modulates protein synthesis and the expression of specific proteins during long-term facilitation in Aplysia sensory neurons. Neuron 1989; 2:1577-1586.
10. Berman DE, Dudai Y. Memory extinction, learning anew, and learning the new: Dissociations in the molecular machinery of learning in cortex. Science 2001; 291:2417-2419.
11. Berman RF, Kesner RP, Partlow LM. Passive avoidance impairment in rats following cycloheximide injection into the amygdala. Brain Res 1978; 158:171-188.
12. Bliss TVP, Collingridge GL. A synaptic model of memory: Long-term potentiation in the hippocampus. Nature 1993; 361:31-39.
13. Bourtchuladze R, Abel T, Berman N et al. Different training procedures recruit either one or two critical periods for contextual memory consolidation, each of which requires protein synthesis and PKA. Learn Mem 1998; 5:365-374.
14. Bourtchuladze R, Frenguelli B, Blendy J et al. Deficient long-term memory in mice with a targeted mutation of the cAMP-responsive element-binding protein. Cell 1994; 79:59-68.
15. Cano E, Hazzalin CA, Mahadevan LC. Anisomycin-activated protein kinases p45 and p55 but not mitogen-activated protein kinases ERK-1 and -2 are implicated in the induction of c-fos and c-jun. Mol Cell Biol 1994; 14:7352-7362.
16. Casadio A, Martin KC, Giustetto M et al. A transient, neuron-wide form of CREB-mediated long-term facilitation can be stabilized at specific synapses by local protein synthesis. Cell 1999; 99:221-237.
17. Castellucci VF, Blumenfeld H, Goelet P et al. Inhibitor of protein synthesis blocks long-term behavioral sensitization in the isolated gill-withdrawal reflex in Aplysia. J Neurobiol 1989; 20:1-9.
18. Castrén E, Pitkänen M, Sirviö Jetal. The induction of LTP increases BDNF and NGF mRNA but decreases NT-3 mRNA in the dentate gyrus. Neuro Report 1993; 4:895-898.
19. Cavallaro S, Schreurs BG, Zhao W et al. Gene expression profiles during long-term memory consolidation. Eur J Neurosci 2001; 13:1809-1815.
20. Cavallaro S, Meiri N, Yi C-L et al. Late memory-related genes in the hippocampus revealed by RNA fingerprinting. Proc Natl Acad Sci USA 1997; 94:9669-9673.
21. Chabner BA, Allegra CJ, Curt GA et al. Antineoplastic agents. In: Hardman JG, Limbird LE, Molinoff et al, eds. Goodman & Gilman's The Pharmacological Basis of Therapeutics. 9th ed. New York: McGraw-Hill, 1995:1233-1287.
22. Chang FF, Greenough WT. Transient and enduring morphological correlates of synaptic activity and efficacy change in the rat hippocampal slice. Brain Res 1984; 309:35-46.
23. Chew SJ, Mello C, Nottebohm F et al. Decrements in auditory responses to a repeated conspecific song are long-lasting and require two periods of protein synthesis in the songbird forebrain. Proc Natl Acad Sci USA 1995; 92:3406-3410.
24. Chew SJ, Vicario DS, Nottebohm F. Quantal duration of auditory memories. Science 1996; 274:1909-1914.
25. Crow T, Forrester J. Inhibition of protein synthesis blocks long-term enhancement of generator potentials produced by one-trial in vivo conditioning in Hermissenda. Proc Natl Acad Sci USA 1990; 87:4490-4494.
26. Crow T, Siddiqi V, Dash PK. Long-term enhancement but not short-term in Hermissenda is dependent upon mRNA synthesis. Neurobiol Learn Mem 1997; 68:343-350.
27. Crow T, Xue-Bian J-J, Siddiqi V. Protein synthesis-dependent and mRNA synthesis-independent intermediate phase of memory in Hermissenda. J Neurophysiol 1999; 82:495-500.
28. Crusio WE. Gene-targeting studies: New methods old problems. Trends Neurosci 1996; 19:186-187.
29. Davis HP, Squire LR. Protein synthesis and memory: A review. Psychol Bull 1984; 96:518-559.
30. Dragunow M, Hughes P, Mason-Parker SE et al. TrkB expression in dentate granule cells is associated with a late phase of long-term potentiation. Mol Brain Res 1997; 46:274-280.
31. Dubnau J, Tully T. Gene discovery in Drosophila: New insights for learning and memory. Annu Rev Neurosci 1998; 21:407-444.
32. Dudai Y. The Neurobiology of Memory. Oxford: Oxford University Press, 1989.
33. Flood JF, Bennett EL, Orme AE et al. Effects of protein synthesis inhibition on memory for active avoidance. Physiol Behav 1975; 14:177-184.
34. Flood JF, Jarvik ME, Bennett EL et al. Protein synthesis inhibition and memory for pole jump active avoidance and extinction. Pharmacol Biochem Behav 1977; 7:71-77.
35. Flood JF, Rosenzweig MR, Bennett EL et al. Influence of training strength on amnesia induced by pretraining injections of cycloheximide. Physiol Behav 1972; 9:589-600.

36. Flood JF, Rosenzweig MR, Bennett EL et al. The influence of duration of protein synthesis inhibition on memory. Physiol Behav 1973; 10:555-562.
37. Freeman FM, Rose SPR, Scholey AB. Two time windows of anisomycin-induced amnesia for passive avoidance training in the day-old chick. Neurobiol Learn Mem 1995; 63:291-295.
38. Frey U, Frey S, Schollmeier F et al. Influence of actinomycin D, a RNA inhibitor, on long-term potentiation in rat hippocampal neurons in vivo and in vitro. J Physiol (Lond) 1996; 490:703-711.
39. Frey U, Krug M, Reymann KG et al. Anisomycin, an inhibitor of protein synthesis, blocks late phases of LTP phenomena in the hippocampal CA1 region in vitro. Brain Res 1988; 452:57-65.
40. Frey U, Morris RGM. Synaptic tagging and long-term potentiation. Nature 1997; 385:533-536.
41. Frey U, Morris RGM. Synaptic tagging: Implications for late maintainance of hippocampal long-term potentiation. Trends Neurosci 1998; 21:181-188.
42. Fride E, Ben-Or S, Allweis C. Mitochondrial protein synthesis may be involved in long-term memory formation. Pharmacol Biochem Behav 1989; 32:873-878.
43. Gale EF, Cundliffe E, Reynolds PE et al. The Molecular Basis of Antibiotic Action. 2nd ed. London: John Wiley & Sons, 1981.
44. Garner CC, Tucker R, Matus A. Selective localization of messenger RNA for cytoskeletal protein MAP2 in dendrites. Nature 1988; 336:674-677.
45. Gelot A, Moreau J, Ben Ari Y et al. Alpha-brain spectrin mRNA belongs to the population of intradendritically transported mRNAs. Neuro Report 1996; 8:113-116.
46. Gerlai R. Gene-targeting studies of mammalian behavior: Is it the mutation or the background genotype? Trends Neurosci 1996; 19:177-181.
47. Gibbs ME, Lecanuet J-P. Disruption of imprinting by memory inhibitors. Anim Behav 1981; 29:572-580.
48. Ghirardi M, Montarolo PG, Kandel ER. A novel intermediate stage in the transition between short- and long-term facilitation in the sensory to motor synapse of Aplysia. Neuron 1995; 14:413-420.
49. Gokal PK, Cavanaugh AH, Thompson Jr EA. The effects of cycloheximide upon transcription of rRNA, 5S RNA, and tRNA genes. J Biol Chem 1986; 261:2536-2541.
50. Grecksch G, Matthies H. Two sensitive periods for the amnesic effect of anisomycin. Pharmacol Biochem Behav 1980; 12:663-665.
51. Grecksch G, Ott T, Matthies H. The effect of intrahippocampally applied anisomycin on the retention of brightness discrimination in rats. Behav Neural Biol 1980; 29:281-288.
52. Guzowski JF, McGaugh JL. Antisense ologideoxynucleotide-mediated disruption of hippocampal cAMP response element binding protein levels impairs consolidation of memory for water maze training. Proc Natl Acad Sci USA 1997; 94:2693-2698.
53. Hall J, Thomas KL, Everitt BJ. Fear memory retrieval induces CREB phosphorylation and Fos expression within the amygdala. Eur J Neurosci 2001; 13:1453-1458.
54. Hedge AN, Goldberg AL, Schwartz JH. Regulatory subunits of cAMP-dependent protein kinases are degraded after conjugation to ubiquitin: A molecular mechanism underlying long-term synaptic plasticity. Proc Natl Acad Sci USA 1993; 90:7436-7440.
55. Horn G, Bradley P, McCabe BJ. Changes in the structure of synapses associated with learning. J Neurosci 1985; 5:3161-3168.
56. Houpt TA, Berlin R. Rapid, labile, and protein synthesis-independent short-term memory in conditioned taste aversion. Learn Mem 1999; 6:37-46.
57. Huang Y-Y, Li X-C, Kandel ER. cAMP contributes to mossy fiber LTP by initiating both a covalently mediated early phase and macromolecular synthesis-dependent late phase. Cell 1994; 79:69-79.
58. Impey S, Mark M, Villacres EC et al. Induction of CREmediated gene expression by stimuli that generate long-lasting LTP in area CA1 of the hippocampus. Neuron 1996; 16:973-982.
59. Jaffé K. Effect of cycloheximide on protein synthesis and memory in praying mantis. Physiol Behav 1980; 25:367-371.
60. Jones MW, Errington ML, French PJ et al. A requirement for the immediate early gene Zif268 in the expression of late LTP and long-term memories. Nature Neurosci 2001; 4:289-296.
61. Josselyn SA, Shi C, Carlezon Jr WA et al. Long-term memory is facilitated by cAMP response element-binding protein overexpression in the amygdala. J Neurosci 2001; 21:2404-2412.
62. Kandel ER, O'Dell TJ. Are adult learning mechanisms also used for development? Science 1992; 258:243-245.
63. Krug M, Lossner B, Ott T. Anisomycin blocks the late phase of long-term potentiation in the dendate gyrus of freely moving rats. Brain Res Bull 1984; 13:39-42.

64. Lamprecht R, Hazvi S, Dudai Y. cAMP response element-binding protein in the amygdala is required for long- but not short-term conditioned taste aversion memory. J Neurosci 1997; 17:8443-8450.
65. Lattal KM, Abel T. Different requirements for protein synthesis in acquisition and extinction of spatial preferences and context-evoked fear. J Neurosci 2001; 21:5773-5780.
66. Lee EHY, Hung HC, Lu KT et al. Protein synthesis in the hippocampus associated with memory facilitation by corticotropin-releasing factor in rats. Peptides 1992; 13:927-937.
67. Lee KS, Schottler F, Oliver M et al. Brief bursts of high-frequency stimulation produce two types of structural change in rat hippocampus. J Neuro Physiol 1980; 44:247-258.
68. Link W, Konietzko U, Kauselmann G et al. Somatodendritic expression of immediate early gene is regulated by synaptic activity. Proc Natl Acad Sci USA 1995; 92:5734-5738.
69. Linden DJ. A protein synthesis-dependent late phase of cerebellar long-term depression. Neuron 1996; 17:483-490.
70. Malenfant SA, Barry M, Fleming AS. Effects of cycloheximide on the retention of olfactory learning and maternal experience effects in postpartum rats. Physiol Behav 1991; 49:289-294.
71. Manahan-Vaughan D, Kulla A, Frey U. Requirement of translation but not transcription for the maintenance of long-term depression in the CA1 region of freely moving rats. J Neurosci 2000; 20:8572-8576.
72. Mansuy IM, Mayford M, Jacob B et al. Restricted and regulated overexpression reveals calcineurin as a key component in the transition from short-term to long-term memory. Cell 1998; 92:39-49.
73. Maren S, Fanselow M. Synaptic plasticity in the basolateral amygdala induced by hippocampal formation stimulation in vivo. J Neurosci 1995; 15:7548-7564.
74. Martin KC, Casadio A, Zhu H et al. Synapse-specific, long-term facilitation of Aplysia sensory to motor synapses: A function for local protein synthesis in memory storage. Cell 1997; 91:927-938.
75. Mauelshagen J, Parker GR, Carew TJ. Dynamics of induction and expression of long-term synaptic facilitation in Aplysia. J Neurosci 1996; 16:7099-7108.
76. Mayford M, Baranes D, Podsypanina K et al. The 3'-untranslated region of CaMKII alpha is a cis-acting signal for the localization and translation of mRNA in dendrites. Proc Natl Acad Sci USA 1996; 93:13250-13255.
77. Meiri N, Rosenblum K. Lateral ventricle injection of the protein synthesis inhibitor anisomycin impairs long-term memory in a spatial memory task. Brain Res 1998; 789:48-55.
78. Mello C, Nottebohm F, Clayton D. Repeated exposure to one song leads to a rapid and persistent decline in an immediate early gene's response to that song in zebra finch telencephalon. J Neurosci 1995; 15:6919-6925.
79. Mello CV, Vicario DS, Clayton DF. Song presentation induces gene expression in the songbird forebrain. Proc Natl Acad Sci USA 1992; 89:6818-6822.
80. Menzel R, Manz G, Menzel R et al. Massed and spaced learning in honeybees: The role of CS, US, the intertrial interval, and the test interval. Learn Mem 2001; 8:198-208.
81. Montarolo PG, Goelet P, Castellucci VF et al. A critical period of macromolecular synthesis in long-term heterosynaptic facilitation in Aplysia. Science 1986; 234:1249-1254.
82. Morris BJ. Stabilization of dendritic mRNAs by nitric oxide allows localized, activity-dependent enhancement of hippocampal protein synthesis. Europ J Neurosci 1997; 9:2334-2339.
83. Moser M-B, Trommald M, Andersen P. An increase in dendritic spine densitiy on hippocampal CA1 pyramidal cells following spatial learning in adult rats suggests the formation of new synapses. Proc Natl Acad Sci USA 1994; 91:12673-12675.
84. Nader K, Schafe GE, Le Doux JE. Fear memories require protein synthesis in the amygdala for reconsolidation after retrieval. Nature 2000; 406:722-726.
85. Namgung U, Valcourt E, Routtenberg A. Long-term potentiation in vivo in the intact mouse hippocampus. Brain Res 1995; 689:85-92.
86. Nastiuk KL, Mello CV, George JM et al. Immediate-early gene responses in the avian song control system: Cloning and expression analysis of the canary c-jun cDNA. Mol Brain Res 1994; 27:299-309.
87. Nayak A, Zastrov DJ, Lickteig R et al. Maintenance of late-phase LTP is accompanied by PKA-dependent increase in AMPA receptor synthesis. Nature 1998; 394:680-683.
88. Nelson TJ, Alkon DL. Prolonged RNA changes in the Hermissenda eye induced by classical conditioning. Proc Natl Acad Sci USA 1988; 85:7800-7804.
89. Nguyen PV, Abel T, Kandel ER. Requirement of a critical period of transcription for induction of a late phase of LTP. Science 1994; 265:1104-1107.
90. O'Leary FA, Byrne JH, Cleary LJ. Long-term structural remodeling in Aplysia sensory neurons requires de novo protein synthesis during a critical time period. J Neurosci 1995; 15:3519-3525.

91. Olson EN, Arnold H-H, Rigby PWJ et al. Know your neighbors: Three phenotypes in null mutants of the myogenic bHLH gene MRF4. Cell 1996; 85:1-4.
92. Otani S, Marshall CJ, Tate WP et al. Maintenance of long-term potentiation in rat dentate gyrus requires protein synthesis but not messenegr RNA synthesis immediately post-tetanization. Neurosci 1989; 28:519-526.
93. Paylor R, Johnson R, Papaioannou V et al. Behavioral assessment of c-fos mutant mice. Brain Res 1994; 651:275-282.
94. Pestka S. Inhibitors of ribosome function. Ann Rev Microbiol 1971; 25:487-562.
95. Quatermain D, McEwen BS. Temporal characteristics of amnesia induced by protein synthesis inhibition: Determination by shock level. Nature 1970; 228:677-678.
96. Quinton EE, Kramarcy NR. Memory impairment correlates closely with cycloheximide dose and degree of inhibition of protein synthesis. Brain Res 1977; 131:184-190.
97. Ramirez RR, Gandhi CC, Muzzio IA et al. Protein synthesis-dependent memory and neuronal enhancement in Hermissenda are contingent on parameters of training and retention. Learn Mem 1998; 4:462-477.
98. Raught B, Gingras AC, Sonenberg N. The target of rapamycin (TOR) proteins. Proc Natl Acad Sci USA 2001; 98:7037-7044.
99. Reich E. Biochemistry of actinomycins. Cancer Res 1963; 23:1428-1441.
100. Richter-Levin G, Collingridge L, Bliss TV. Long-term potentiation and glutamate release in the dentate gyrus: Links to spatial learning. Behav Brain Res 1995; 66:37-40.
101. Rocamora N, Welker E, Pascual M et al. Upregulation of BDNF mRNA expression in the barrel cortex of adult mice after sensory stimulation. J Neurosci 1996; 16:4411-4419.
102. Rogan MT, Staubli U, LeDoux JE. Fear conditioning induces associative long-term potentiation in the amygdala. Nature 1997; 390:604-607.
103. Rose SPR. How chicks make memories: The cellular cascade from c-fos to dendritic remodelling. Trends Neurosci 1991; 14:390-397.
104. Rose SPR. Cell-adhesion molecules, glucocorticoids and long-term-memory formation. Trends Neurosci 1995; 18:502-506.
105. Rose SPR. God's Organism? The chick as a model system for memory studies. Learn Mem 2000; 7:1-17.
106. Rose SPR, Jork R. Long-term memory in the chick is blocked by 2-deoxygalactose, a fucose analogue. Behav Neural Biol 1987; 48:246-258.
107. Rosen JB, Fanselow M, Young SL et al. Immediate-early gene expression in the amygdala following footshock stress and contextual fear conditioning. Brain Res 1998; 796:132-142.
108. Rosenblum K, Meiri N, Dudai Y. Taste memory: The role of protein synthesis in gustatory cortex. Behav Neural Biol 1993; 59:49-56.
109. Sabatini DM, Barrow RK, Blackshaw S et al. Interaction of RAFT1 with gephyrin required for rapamycin-sensitive signaling. Science 1999; 284:1161-1164.
110. Saez E, No D, West A et al. Inducible gene expression in mammalian cells and transgenic mice. Curr Opin Biotechnol 1997; 8:608-616.
111. Sandi C, Rose SPR. Protein synthesis- and fucosylation-dependent mechanisms in corticosterone facilitation of long-term memory in the chick. Behav Neurosci 1992; 111:1098-1104.
112. Sandi C, Rose SP, Mileusnic R et al. Corticosterone facilitates long-term memory formation via enhanced glycoprotein synthesis. Neurosci 1995; 69:1087-1093.
113. Schafe GE, Atkins CM, Swank MW et al. Activation of ERK/MAP kinase in the amygdala is required for memory consolidation of pavlovian fear conditioning. J Neurosci 2000; 20:8177-8187.
114. Schafe GE, LeDoux JE. Memory consolidation of auditory Pavlovian fear conditioning requires protein synthesis and protein kinase A in the amygdala. J Neurosci 2000; 20:RC96.
115. Schafe GE, Nadel NV, Sullivan GM et al. Memory consolidation for contextual and auditory fear conditioning is dependent on protein synthesis, PKA, and MAP kinase. Learn Mem 1999; 6:97-110.
116. Schmaltz G, Marcant P. Transient aversion and long-lasting amnesia following cycloheximide injection in the rat. Physiol Behav 1983; 30:845-852.
117. Selwood SP, McGregor A, Lightowlers RN et al. Inhibition of mitochondrial protein synthesis promotes autonomous regulation of mtDNA expression and generation of a new mitochondrial RNA species. FEBS Lett 2001; 494:186-191.
118. Shockett P, Difillippantonio M, Hellman N et al. A modified tetracycline-regulated system provides autoregulatory, inducible gene expression in cultured cells and transgenic mice. Proc Natl Acad Sci USA 1995; 92:6522-6526.
119. Smirnova T, Laroche S, Errington ML et al. Transsynaptic expression of a presynaptic glutamate receptor during hippocampal long-term potentiation. Science 1993; 262:433-436.

120. Sojka M, Davies HA, Harrison E et al. Long-term increases in synaptic density in chick CNS after passive avoidance training are blocked by an inhibitor of protein synthesis. Brain Res 1995; 684:209-214.
121. Squire LR, Barondes SH. Actinomycin-D: Effects on memory at different times after training. Nature 1970; 225:649-650.
122. Squire LR, Barondes SH. Anisomycin, like other inhibitors of cerebral protein synthesis, impairs 'long-term' memory of a discrimination task. Brain Res 1974; 66:301-308.
123. Stanton PK, Sarvey JM. Blockade of long-term potentiation in rat hippocampal CA1 region by inhibitors of protein synthesis. J Neurosci 1984; 4:3080-3088.
124. Stäubli U, Faraday R, Lynch G. Pharmacological dissociation of memory: Anisomycin, a protein synthesis inhibitor, and leupeptin, a protease inhibitor, block different learning tasks. Behav Neural Biol 1985; 43:287-297.
125. Steward O. Molecular sorting in neurons: Cell biological processes that play a role in synapse growth and plasticity. In: Baudry M, Thompson RF, Davis JL, eds. Synaptic Plasticity. Cambridge MA: Bradford Book MIT Press, 1993:13-43.
126. Steward O, Schuman EM. Protein synthesis at synaptic sites on dendrites. Annu Rev Neurosci 2001; 24:299-325.
127. Stiedl O, Palve M, Radulovic J et al. Differential impairment of auditory and contextual fear conditioning by protein synthesis inhibition in C57BL/6N mice. Behav Neurosci 1999; 113:496-506.
128. Stork O, Stork S, Pape H-C et al. Identification of genes expressed in the amygdala during the formation of fear memory. Learn Mem 2001; 8:209-219.
129. Sutton MA, Carew TJ. Parallel molecular pathways mediate expression of distinct forms of intermediate-term facilitation at tail sensory-motor synapses in Aplysia. Neuron 2000; 26:219-231.
130. Sutton MA, Masters SE, Bagnall MW et al. Molecular mechanims underlying a unique intermediate phase of memory in Aplysia. Neuron 2001; 31:143-154.
131. Tischmeyer W, Grimm R, Schicknick H et al. Sequence-specific impairment of learning by c-jun antisense oligonucleotides. Neuro Report 1994; 5:1501-1504.
132. Tiunova A, Anokhin KV, Rose SPR. Two critical periods of protein and glycoprotein synthesis in memory consolidation for visual categorization learning in chicks. Learn Mem 1998; 4:401-410.
133. Tiunova A, Anokhin KV, Rose SPR et al. Involvement of glutamate receptors, protein kinases, and protein synthesis in memory for visual discrimination in the young chick. Neurobiol Learn Mem 1996; 65:233-243.
134. Tiunova A, Anokhin KV, Schachner M et al. Three time windows for amnestic effect of antibodies to cell adhesion molecule L1 in chicks. Neuro Report 1998; 9:1645-1648.
135. Tongiorgi E, Righi M, Cattaneo A. Activity-dependent dendritic targeting of BDNF and TrkB mRNAs in hippocampal neurons. J Neurosci 1997; 17:9492-9505.
136. Torocsik B, Szeberenyi J. Anisomycin affects both pro- and antiapoptotic mechanisms in PC12 cells. Biochem Biophys Res Commun 2000; 278:550-556.
137. Tucker A, Gibbs M. Cycloheximide-induced amnesia for taste aversion memory in rats. Pharmacol Biochem Behav 1976; 4:181-184.
138. Tucker RP, Garner C, Matus A. In situ localization of microtubule-associated protein mRNA in the developing and adult rat brain. Neuron 1989; 2:1245-1256.
139. Tully T, Preat T, Boynton SC et al. Genetic dissection of consolidated memory in Drosophila melanogaster. Cell 1994; 79:35-47.
140. Vicario DS, Nottebohm F. Quantal memory durations: Observations not reproduced. Science 1998; 279:1437.
141. Williams J, Dragunow M, Lawlor P et al. Krox20 may play a key role in the stabilization of long-term potentiation. Mol Brain Res 1995; 28:87-93.
142. Wittstock S, Kaatz H-H, Menzel R. Inhibition of brain protein synthesis by cycloheximide does not affect formation of long-term memory in honeybees after olfactory conditioning. J Neurosci 1993; 13:1379-1386.
143. Wu L, Wells D, Tay J et al. CPEB-mediated cytoplasmic polyadenylation and the regulation of experience-dependent translation of alpha-CaMKII mRNA at synapses. Neuron 1998; 21:1129-1139.
144. Wüstenberg D, Gerber B, Menzel R. Long- but not medium-term retention of olfactory memories in honeybees is impaired by actinomycin D and anisomycin. Eur J Neurosci 1998; 10:2742-2745.
145. Yin JC, Del Vecchio M, Zhou H et al. CREB as a memory modulator: Induced expression of a dCREB2 activator isoform enhances long-term memory in Drosophila. Cell 1995; 81:107-115.
146. Yin JC, Wallach J, Del Vecchio M et al. Induction of a dominant negative CREB transgene specifically blocks long-term memory in Drosophila. Cell 1994; 79:49-58.

CHAPTER 5.4

Protein Synthesis: II. New Proteins

Radmila Mileusnic

Abstract

The role of protein synthesis in long-term memory formation is still an area of intense scientific interest, which encompasses the study of mechanisms involved in gene expression and molecular mechanisms underlying synaptic plasticity. A number of low molecular weight compounds have been used to inhibit or enhance this fundamental cellular process. The pivotal role of protein synthesis in long-term memory formation suggest yet again that an understanding of how protein synthesis can be activated and regulated by events that ultimately lead to memory consolidation can lead to better understanding of the processes that keep our memories alive.

Introduction

Protein synthesis allows the remarkable capacity of nervous system to modify its neuronal architecture as a consequence of learning. The variety of protein molecules involved in this process encompasses very different type of proteins, from transcription factors and enzymes involved in neurotransmitter metabolism, to cell adhesion molecules.

Today, one takes a great risk when questioning the role of protein synthesis in memory formation. However, some readers may find it unusual that there is often a sign of scepticism regarding this dogma.[19,21] The main reason for the doubt is a paradoxical feature of proteins. Namely, proteins simultaneously encompass two rather different features: they are not stable molecules, yet they ensure continuity of our memories throughout our lifetime. Protein turnover replaces different types of proteins, including synaptic proteins, probably hundred times during our lifetime with no evident consequences for long-lasting synaptic modifications. Therefore, the question of how organisms store information for very long periods of time in spite of constant molecular turnover is one of the most captivating questions for many neuroscientists.

Bearing in mind the physical limits of a chapter, I have been 'forced' to omit many important papers that paved the way to a better understanding of the role of protein synthesis in memory formation. Thus, I would like to apologise to all of the zealous researchers whose name and work is not mentioned in this chapter.

A Brief History

The search for chemical substrates of learning and memory has a long history. Halstead[46] was perhaps the first to suggest that 'engrams' might be stored in "template" protein molecules in nerve cells. Correlational studies of Hydén focused on RNA and protein changes induced by training procedures, and implicated RNA changes in memory formation.[47-49] Although no replication of this type of experiment was ever reported from outside Hydén's laboratory, an unfortunate cul-de-sac stemming from this work was the hypothesis on memory transfer.[52,71] As a consequence, the conceptual model of "specific memories stored in specific molecules" was abandoned rather quickly and research shifted from "memory molecules" to biochemical and molecular processes that might result in modified synaptic interactions, either through (a)

From Messengers to Molecules: Memories Are Made of These, edited by Gernot Riedel and Bettina Platt.

structural changes, which inevitably involve proteins as building blocks of synaptic membranes or (b) process changes, which would, again, involve synthesis of protein molecules such as enzymes, neurotransmitters, receptors, etc.

Protein Synthesis Inhibitors

A reasonable test for the hypothesis that new proteins are necessary for long term memory was to substantially prevent the expression of new proteins using protein synthesis inhibitors. A vast literature has developed using different drugs and antibiotics that inhibit protein synthesis. Early studies[10,11,17] showed that actinomycin D, an antibiotic that in low concentrations inhibits transcription without appreciably affecting DNA replication could produce impairments in learning and memory. Results were criticised because of the severe toxicity of the drug. However, this objection has been overcome with the introduction of less toxic drugs such as anisomycin, which inhibit peptide bond synthesis. The most convincing data on the importance of protein synthesis for long-term memory formation came from work of Flood et al.[30,31,98] who demonstrated that the stronger the training and consequently the greater the amount of learning, the greater the duration of inhibition of protein synthesis had to be in order to produce amnesia.

However, there was some scepticism regarding the conclusion that treatment with protein synthesis inhibitors demonstrates that synthesis of new proteins is required for long-term memory. Importantly, it was not the concept that was in doubt, only whether antibiotics used as protein synthesis inhibitors demonstrate this notion (Table 1). Another reason for scepticism resided in the fact that, for example, puromycin produces effects on learning and memory that are qualitatively different from drugs such as acetocyloheximide, cycloheximide and anisomycin.[24] Namely, Flexner and Flexner[29] published data on the effect of NaCl which, when injected into the brains of mice some days after training, reversed the amnesia produced by puromycin given at the time of training, suggesting that puromycin did not prevented the long-term memory formation, but only its expression. On the other hand, Rosenbaum and colleagues,[97] while replicating this unusual finding of the Flexner's found that NaCl did not have an effect on amnesia produced by acetocycloheximide. Moreover, many other drugs 'attenuated' the amnestic

Table 1. Inhibitors of protein synthesis in eucaryotes most often used in research into the role of proteins in long-term memory formation

Inhibitors of Protein Synthesis	Processes Affected	Site of Action
Actinomycin D	Transcription and DNA replication	Prokaryotes and Eucaryotes: binds tightly to duplex DNA by intercalating phenoxazine ring between neighbouring base pairs in DNA
Puromycin	Elongation	Structural analogue of the 3'-terminal end of aminoacyl-tRNA. Enters aminoacyl site on the ribosome and is incorporated into the growing polypeptide chain that causes release of shortened polypeptide chains from the ribosome with puromycin attached.
Anisomycin	Translation (peptide bond formation)	Blocks peptide bond formation by binding to 60S ribosomal subunit. Concentration used for protein inhibition causes internucleosomal fragmentation and activates the stress-responsive pathways.
Cycloheximide and Acetocycloheximide	Elongation	Interferes with protein synthesis in eucaryotes by inhibiting peptidyl-transferase activity

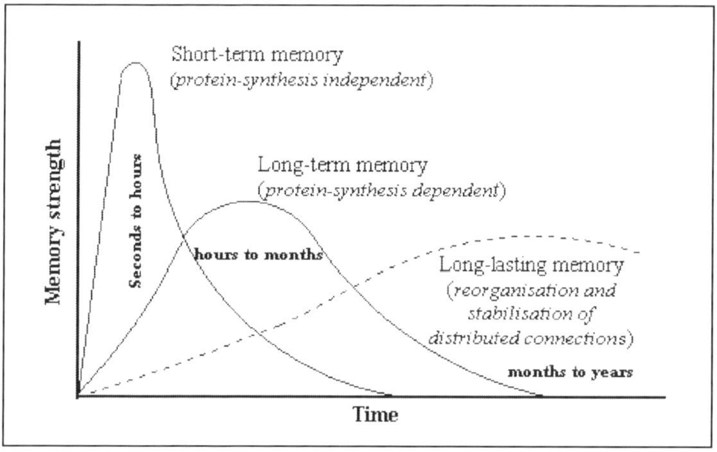

Figure 1. Different phases of memory consolidation (Adapted from ref. 75).

effect of protein synthesis inhibitors while not affecting the rate of protein synthesis.[24] For example, Flood et al.[32] found that amnesia induced by anisomycin could be attenuated by either amphetamine, strychnine or picrotoxin if these drugs were given at the time of training with no significant effect on the inhibition of protein synthesis induced by anisomycin. It seems that all these drugs act in this manner when their effect is opposed against the effect of amnestic agents such as protein synthesis inhibitors. The scepticism was nourished even further with the set of data showing that specific conditions such as stress, could bring back memories that were supposedly never formed.[67] Thus, one inescapable question arose: what are the contributors, of central or peripheral nature, to the process of memory formation?

De Wied[116] was among the first to suggest that hormones might influence the processes that underlie memory formation. It become obvious that the learning experience and the hormonal response have to occur within certain temporal limits for the hormonal response to influence the learning process and consequent memory formation. The list of hormones influencing learning processes is impressively long (see review by de Wied and Kovacs in this book). Many of them, such as vassopressine, affect learning through peripheral autonomic sequalae, while others, such as steroid hormones, modulate the process of gene expression.

The concept that memory storage is time-dependent process[74] brought new light on the idea that some form of neural activity must underlie the setting down of memory traces and that while the experience of events and the memories formed may be continuous, the underlying molecular processes are clearly discontinuous with defined time courses. This conceptual framework of McGaugh introduced post-training treatments as a new experimental approach in studies of different phases of memory consolidation. Among the earliest published data supporting the idea that the underlying molecular processes in memory consolidation are discontinuous with defined time courses are studies on goldfish by Agranoff et al.[3,5] They clearly demonstrate that animals injected intracranially with puromycin just before or within 30 min of a learning experience showed no problems with initial learning but had markedly impaired long-term memory. However, injections given more than 30 min after learning, did not prevent the formation of long-term memory. Thus, it was concluded that the most crucial difference between short- and long-term memory is that short-term memory is resistant to protein synthesis inhibitors while long-term memory is not (Fig. 1). From that time until today countless experiments confirmed that short-term memory is protein synthesis-independent, the formation of long-term memory is a time- and protein synthesis-dependent process, while the formation of long-lasting memory (consolidation of memory) is protein synthesis independent (for review see ref. 75).

Interim Summary

The large body of evidence on the effect of antibiotics on brain protein synthesis can be summarised, according to Squire and Davis,[106] into few basic findings: first, the effect of antibiotics on memory formation is a phenomenon found across different animal species and in wide variety of training paradigms (see also Strock and Welzl in this book). Second, in order to induce amnesia protein synthesis has to be inhibited up to a level of 90 - 95%, and finally, if inhibition is established 30 or more minutes after training, no amnesia develops. Despite the concise summary offered by Squire,[105] these studies left memory researchers with the already mentioned paradox: if protein synthesis is necessary for long-term memory then how could any other molecule that does not affect protein synthesis reverse the amnesia that was induced by drugs such as anisomycin? This paradox lead to the development of research into the role of drugs, such as amphetamine, that could substantially alter neural processing[67,98] and hormones, that could modify gene expression (for review see refs. 63, 72 and 73) thus altering the formation of long term-memories.

The Rate of Protein Synthesis

Changes in the rate of protein synthesis associated with learning have been extensively studied as potentially crucial factor in consolidating memories. Despite the fact that many experiments suggested that there may be a general increase in the rate of cerebral protein synthesis, incorporation of radiolabelled amino acids into brain proteins in trained animals relative to control ones in the period immediately following training[4,23,24,95] appeared to be affected by stress responses and as such, may be due to the release of the hormones known to respond during stress[25] rather than to consolidation. Moreover, the increased amino acid incorporation was often neither region-specific nor tissue specific - further evidence of a nonspecific response to activation.

The Posttranslational Modifications

The posttranslational modification of proteins attracted the attention of many laboratories because any process by which chemical groups can be added to or subtracted from the proteins very rapidly, resulting in major changes in the properties of the molecule, is likely to be utilised by neurons during the process of activity-dependent modifications.[4,23] Two types of covalent modification, very different in nature, attracted the most attention: phosphorylation and glycosylation. The process of cAMP-induced protein phosphorylation[59] offered fast, reliable, target-specific modification, an ideal candidate for signal transduction that could ultimately lead to modification of gene expression. Glycosylation, in contrast to phosphorylation, is a much slower process, but is capable of producing enormously heterogeneous and complex glycoprotein structures, making glycoproteins ideal candidates for intercellular interactions.

The first indication that protein phosphorylation was sensitive to behavioural manipulation came from the work of Machlus, Wilson and Glassman[65] on nuclear proteins, but subsequently emphasis has been shifted toward synaptic and membrane proteins.[14,38,55,62,100] It took neuroscience about 20 years after Wilson and Glassman's publication to 'rediscover' the role of activity-driven phosphorylation of transcription factors (TFs) as the crucial event in regulation of gene expression in the process of memory formation (for review see refs. 1, 18, 22, 69, 113 and 115).

The first indication that glycosylation of proteins is sensitive to behavioural manipulations came from work of Entingh et al.[27] who discovered changes in uridine metabolism following learning and suggested that uridine metabolites might be used as building blocks for glycoprotein synthesis. Two research groups, one in Magdeburg[68,88] and the other at The Open University[91,92,94,110] took another approach to investigate the role of glycoproteins in memory formation. Namely, they introduced a new 'tool' into the field: 2-D-galactose as competitor of fucose and consequent protein glycolsylation.

The 'Local' Protein Synthesis

When Fisher and Litvak[28] and Guiditta et al.[42] published their findings showing that axoplasmic proteins were labelled when isolated squid giant axon was incubated in vitro with radioactive amino acids, they set the stage for a number of studies indicating that translation might occur in cell compartments other than cell body. But, studies of glial-axonal protein transfer[60,61,114] provided strong suspicions of local protein synthesis and led to the belief that proteins were synthesised in periaxonal glial cells and secondarily transferred to the sub-adjacent axon by intercellular transfer. This view was generalised to all axons and became a commonly accepted view, despite evidence that the giant axon contained all of the prerequisites for such protein synthesis.[41,43,44,50] The idea of local protein synthesis was brought to light for the second time with the isolation of mitochondrial fractions from brain which contained a substantial portion of sheared off nerve terminals resealed into osmotically sensitive particles named synaptosomes.[20,45,87] This method allowed determination of whether presynaptic terminals contribute to protein synthesising activity[8] of neuronal cell bodies. Using two different inhibitors of protein synthesis, cycloheximide (inhibitor of ribosomal protein synthesis) and chloramphenicol (inhibitor of mitochondrial protein synthesis), Yellin et al.[117] confirmed that both cytoplasmic and mitochondrial systems of protein synthesis are present in synaptosomes.

The history of presynaptic protein synthesis is as turbulent as the history of protein synthesis. However, the convincing experimental evidence of Steward and Levy[108] which showed the existence of synapse-associated polyribosomal complexes (SPRCs) selectively localised in distal processes beneath postsynaptic sites on the dendrites and recent data indicating that presynaptic translational activity exists across different species cleared the clouds of suspicion and more importantly, lead to the belief that presynaptic protein synthesis is required for long-term plasticity changes.[66]

Present Time

Although the present times are characterised by breath-taking technological developments, time-dependent involvement of cellular processes enabling formation of lasting memories (consolidation theory proposed by Müller and Pilzecker in 1900 and revisited by McGaugh in ref. 74) is still shaping research into protein synthesis. Unfortunately, the contribution of Glassman and his two stage hypothesis of molecular cascade in memory consolidation, which postulated two distinct waves of protein synthesis during which an activator protein synthesised during phase 1 will act as an activator of the genes coding for phase 2 proteins,[39] is often forgotten.

One question that suddenly reemerged during the last four years and has lead to rather intensive arguments, that is the discussion of how stable long-lasting memories are after reactivation or retrieval. The experiments LeDoux and colleagues[83,84,103] aimed to test the stability of memory trace in terms of activation of molecular events. They showed that the same manipulations that could cause amnesia after initial learning, such as inhibition of protein synthesis around the time of training, can also lead to memory loss right after reactivation or retrieval. In other words, LeDoux and colleagues argue that any attempt to access memory when it is consolidated will bring the memory trace into a labile state making it vulnerable to the effects of inhibition of the same cellular and molecular processes that were critical for the original consolidation.[83,84] Hence the term reconsolidation entered the field yet again. The term reconsolidation as such is *contradiction in adjecto.* The problem is not of semantic nature but in the fact that anisomycin interfered with both new memories and reactivated memories, in other words, anisomycin rendered animals amnestic for the original learning task. On the other hand, the experiments of Morris, Anokhin, Sara and Taubenfeld contradicted the work of LeDoux's goup and showed that at the cellular level, NMDA receptors[104] or transcription factor C/EBPß[111] are not involved in retrieval of previously established memories. The stability of the memory trace during retrieval studied by the group of Anokhin,[64] using protein synthesis inhibitors in association with a reminder procedure, showed that administration of anisomycin and cycloheximide induced the development of temporary amnesia whose duration gradually

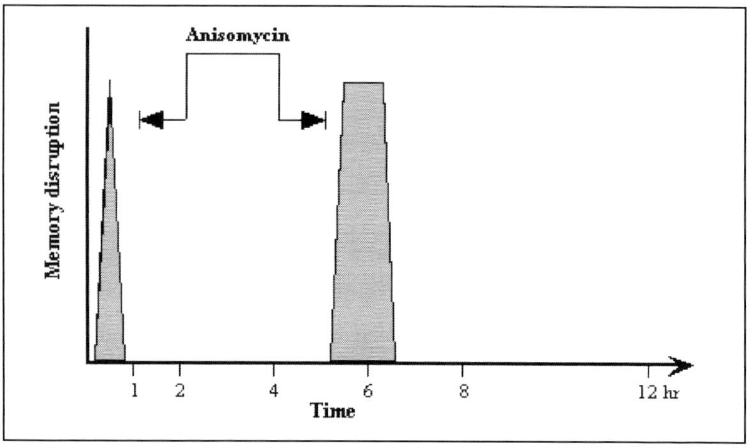

Figure 2. The time-windows of sensitivity of protein synthesis to antibiotics. The progress in imaging techniques, which can identify brain regions active during acquisition and retrieval, combined with visualisation of neural activity and gene expression could help us to elucidate the problem of 'reconsolidation'.

declined as the interval between training and reminder increased. The same conclusion was drawn from the experiments of Sara.[101,102] At a system level, hippocampus-dependent learning tasks have always been the 'darling' of research into memory consolidation. To identify whether hippocampal activity contributes to these processes independently, Riedel and colleagues used a novel method of inactivating synaptic transmission using a water-soluble antagonist of AMPA/kainate glutamate receptors and addressed the reconsolidation question by temporarily inactivating hippocampus prior to spatial memory test.[90] Their findings indicated that hippocampal neural activity is necessary for both encoding and retrieval of spatial memory and for either trace consolidation or long-term storage. On the other hand, experiments of McGaugh showed that other brain areas, such as amygdala, may have an important function in memory consolidation after a learning task has taken place.[76]

As for the *novelty* of the problem, it was in 1989 when Matthies[68] actually 'broke' the concept of protein synthesis independent phase of memory consolidation and showed the existence of the second wave of protein synthesis, which occurs 4-6 hr after learning experience, using a brightness discrimination task in rats (Fig. 2). The bimodal feature of protein synthesis was found in another extensively studied task, the one-trial passive avoidance task in chicks.[33] In spite of anatomical differences between rats and birds, the similarities in cellular processes between these two species is remarkable, from the very early events, such as involvement of glutamate receptors,[12,51] to the bimodal protein synthesis. Thus, the concept of 'stability' of long term memory, historically speaking, has already been challenged. Nevertheless, the reappearance of the consolidation problem emphasises, yet again, the complex nature of learning, consolidation, retrieval and extinction as well as our need to carefully study all these processes on both molecular and behavioural levels.

De Novo Protein Synthesis (with Paraphernaliae)

During the last two decades most of the research efforts have been directed toward discovering the sequence of events that will ultimately lead to activity-driven transcription and consequently protein synthesis. The major break-through was made by Kandel and colleagues[40,82] who unravelled the mechanism by which extracellular signals capable of inducing covalent modification of constitutive transcription regulators, mediated through second messengers, regulate the induction of gene expression (for review see refs. 9 and 54). Although inducible transcription factors, often called immediate early genes (IEGs), are not the focus of this chapter, one could not discuss protein synthesis without referring to IEGs (see also Greenwood et al. in this book). Many of the IEGs products are regulated by kinases, such as fos and CREB[6,7,13] or NF-kB[78] and are set in motion by a different state of organism almost on a minute-to-minute basis. Although it is obvious that the regulation of gene expression does not necessarily depend on *de novo* synthesis, but could be achieved by posttranslational modification of existing IEGs, it is important to emphasise that: (a) IEGs work only because of their combinatorial properties,[107] (b) CREB alone or any other transcription factor for that matter cannot be sufficient for initiation of DNA transcription or for any physiological process such as memory storage.

When a role for a particular protein is proposed on a basis of evidence from single-gene knockout approach, one should be aware of many interpretative difficulties precluding firm conclusions.[15,37,70] The reason for doubt is not solely due to the problems associated with knockout animals. When the idea that kinases can modify proteins already present within the synapse was proposed it was believed that these were synaptic transmission processes set in motion by the event itself.[99] Experiments in which kinase inhibitors were applied to the region of synapses showed that inhibition of phosphorylation ceases to be effective at about 1 hr after long-term potentiation (LTP) is initiated. However, one hour after the LTP tetanus or low frequency stimulation, the synthesis of Nf-kB and TFIIIa as well as p50 and p65 mRNA are increased.[78,79] The most probable reason for this rapid synthesis of transcription factors following phosphorylation events is that these proteins must be replenished.[77] Increased activation and/or synthesis of transcription factors could lead to more synthesis of target proteins as well as repressors and terminators of transcription[1] that would ultimately cause cessation of gene transcription. A good example of this complexity comes from work of Kinney et al.[58] describing the effect of decreased binding of hippocampal TFs to the E-box that is correlated with increased GAP-43 mRNA. All these events, rather different in nature, such as: transmembrane signalling, phosphorylation cascade, protein kinase activation, recruitment of transcription factors somehow merge their effect at the same time-point, 1-hr after the initial signal transduction event. Thus, one can assume that the transition from short-term memory to events that will trigger the consolidation is likely to occur at this point in time.

There is a time point at about 6 hr after the initial signal transduction event that may herald another transition point. Here, we find studies with widely disparate methods and animal models pointing to this time point.[53,58,7896] Across different species, model-systems and training tasks, memory consolidation occurs only after the initiation of the late-transcription phase that leads to the synthesis of a variety of proteins among which the cell adhesion proteins implicated in morphological changes at specific synapses (Fig. 3).

One could ask the question 'why search for glycoproteins'? Long-term remodelling must involve changes to the structure or geometry of synaptic or dendritic membranes, hence attention was drawn to glycoproteins, especially cell adhesion molecules (idea formulated as a general theory in ref. 26). And indeed, over the past 15 years there has been a remarkable convergence of evidence pointing to (a) activity-driven changes in the synthesis of cell adhesion molecules and (b) a key role for the cell adhesion molecules in the process of memory formation.

Early learning in the chick has proved a particularly fertile system in which to study the cellular and molecular processes of memory formation and consolidation. The protocol which has been most widely used for those studies is a one-trial passive avoidance learning task, based on the disposition of young chicks to peck spontaneously at small objects and to remember their characteristics for long periods. This model has all the experimental advantages of one-trial learning paradigms and since its first description by Cherkin in 1969[16] it has been widely used for the biochemical and pharmacological study of the molecular events involved in memory formation.

Following the pioneering work of Gibbs and Ng,[35,36] which utilised a pharmacological dissection procedure to identify biochemically sensitive periods in the minutes following training on this task, a combination of interventive and correlative studies has revealed a cascade of molecular processes occurring in defined brain regions, notably the left *intermediate medial hyperstriatum ventrale* and *lobus parolfactorius*. Briefly, within minutes of pecking at the bitter

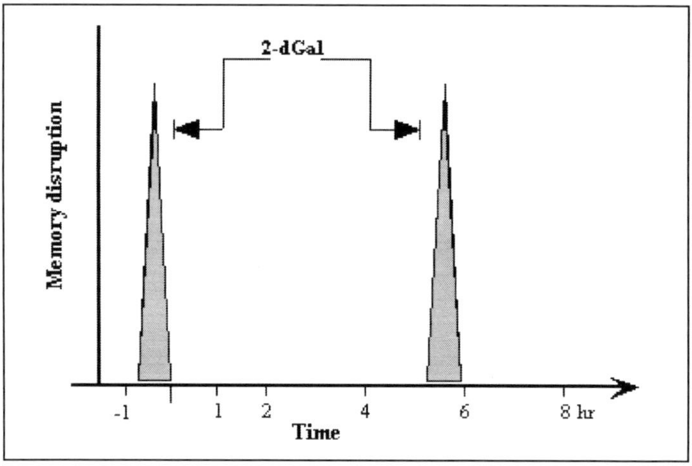

Figure 3. Pre and post-training intervals sensitive to 2-deoxy-Galactose.

bead, there is: (i) enhanced glutamate release, (ii) up-regulation of NMDA-sensitive glutamate receptors, and (iii) the opening of N-type conotoxin-sensitive calcium channels. These synaptic transients result in the activation of protein kinases and expression of IEGs such as c-fos and c-jun and consequently, the family of late genes coding for glycoproteins which, inserted into the pre and post-synaptic membranes, alter synaptic structure and connectivity (for review see ref. 93). Several aspects of this cascade and its time-dependencies are reminiscent of those occurring during and following LTP. However a one-trial task may not be typical of learning in general, because many instances of animal and human learning are based on the acquision of experience in a number of repeated trials, involving processes such as generalisation, categorisation and discrimination.

We have been able to identify a number of pre and post-synaptic membrane glycoproteins, which show enhanced fucose incorporation between 1 and 24 hr after training (presynaptic 50 kD; post-synaptic 33, 100-120 150-180 kD). This enhanced fucosylation was accompanied by increased activity of fucokinase. Moreover, the anti-metabolite 2-deoxygalactose, if injected around the time of training, and 4-6 hr after training , produced amnesia in animals tested 24hr later (for review see ref. 93).

What might be the significance of those 2 time-windows of sensitivity to the fucosylation inhibitor - presumably representing 2 waves of glycoprotein synthesis? A clue comes from manipulating the nature of the training experience: in a normal training protocol chicks peck at the bead which was made aversive by immersion in 100% methylanthranilate (MeA), and will avoid that bead for at least 48 hr subsequently. However, if the aversant is made weaker (10%), the avoidance response is initially as strong as for 100% MeA, but retention does not persist. Much more importantly, in the case of weak training fucosylation does not occur. These data led us to believe that, for the memory of passive avoidance training to endure, a functionally discrete second wave of neural activity, including glycoprotein, synthesis is required.

Two questions inevitably arise: (i) as even a single cell type uses multiple molecular mechanisms in adhering to the other cells (and to the extracellular matrix), the specificity of cell-cell adhesion must result from the integration of a number of different adhesion systems, some of which are associated with specialised cell junctions while others are not, and (ii) whether this cascade is unique to the specific case of one-trial learning in the chick or whether it is generalisable to other forms of avian and mammalian learning?

Our work on different synaptic transmembrane glycoproteins showed that the glycoproteins are recruited at different points in time and so are susceptible to blockade only at the time

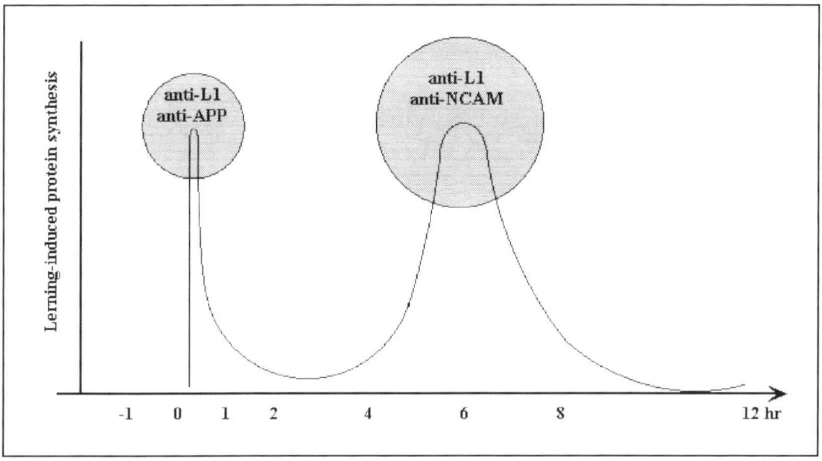

Figure 4. The antibody-induced onset of amnesia.

at which they are recruited. This conclusion was based on two different but complementary approaches: (i) an antibody approach, by which we were able to interfere with the expression of protein function by blocking the extracellular domain of protein in question (Fig. 4), and (ii) antisense approach, by which we were able to downregulate gene expression, hence to test the significance of de novo protein synthesis (Fig. 5).

L1, NCAM and APP antibodies were injected into the IMHV at various time-points before and after training. These groups received antibodies 30-min pretraining and 5.5 and 8-hr post-training. These times were similar to the time-windows of amnestic action observed with 2-d-Gal. Animals were tested 24 hr after training. Memory retention was impaired, compared with saline controls, (a) in the case of anti-L1 during both time-windows of protein synthesis; (b) in the case of anti-NCAM only during the second time-window, and (c) in the case of anti-APP only during the first time-window of protein synthesis. Since blocking the protein function by use of specific antibodies outside of specific time window were without effect, we addressed the question of the importance of de novo protein synthesis by use of synthetic oligodeoxynucleotides.

The two distinct time-windows of behavioural response to APP and NCAM downregulation confirmed that: (a) induction of de novo synthesis of different synaptic transmembrane glycoproteins occurs at different points in time after training, and (b) the specificity of cell-cell adhesion indeed results from the integration of a number of different adhesion systems.[80,81]

Conclusion

"... be always prepared to rewrite the encyclopaedia"

Umberto Eco, Serendipities, Chapter: The Force of Falsity, 1998

Although it is safe to say that protein synthesis leads to structural changes that are the physical substrate for long-term memory, there are time points in which the protein synthesis seems to be constitutive and independent of input. The experimental data that supports this notion are obtained from studies of transcriptional activation of GAP-43, which occurs 12, 24, 48 and 72 hr after LTP.[85,86] There is obviously a 2-3 day delay in promoter activation and mRNA synthesis. At this point we should start to doubt the prevailing view. What maintains memories for lifetime if proteins are synthesised, utilised, and finally replenished reflecting the recruitment of gene regulatory events that are input independent?

And, there are more questions to be asked. Namely, does dependence on the protein synthesis indicate that the proteins required for enduring synaptic modification are made on mRNAs

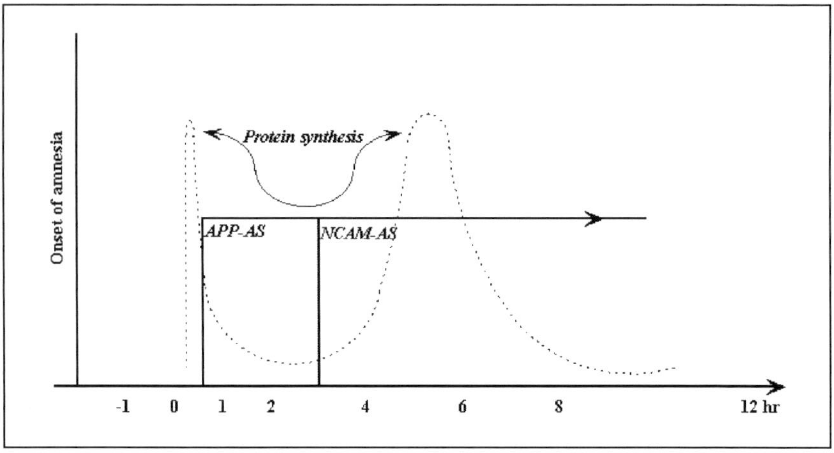

Figure 5. The APP- and NCAM-antisense-induced onset of amnesia.

that are present constitutively, or on mRNAs that are synthesised as a consequence of learning-induced transcriptional activation? Modification of existing proteins already positioned at synapses to be modified could potentially solve the problem of synaptic specificity,[34] because rapid changes in protein function can occur at the site of change. But, if the required proteins are made on constitutively expressed mRNAs, the precise mechanism of their increased translational activation should be synaptic activity. And if that is the case, what could be the 'pick-and chose' mechanism by which cells precisely regulate the translation of different mRNAs from the mix of mRNAS that are in place. If the required proteins are synthesised as a consequence of learning-induced transcriptional activation and the new gene expression is required, what is the nature of the process that allows signalling from the synapse to control the transcriptional activation of the neuron? What are the mechanisms by which the newly synthesised proteins essential for activity-dependent synaptic modification are selectively delivered to the synapses that are to be modified? How is this coordinated and how all of these molecules actually 'fit' into the process of memory consolidation?

The cellular mechanism for targeting newly synthesised mRNAs to synaptic sites on dendrites have been partially revealed by Steward and Worley[109] through studies on the intracellular transport and synaptic targeting of Arc (activity-regulated cytoskeleton associated protein). The synthesis of Arc mRNA is induced by patterns of synaptic activity that also includes LTP. Arc mRNA is rapidly transported into dendrites and the de novo synthesised arc protein is assembled into the synaptic junctional complex of the recently activated synapses. Moreover, the experiments with Arc antisense showed that preventing of Arc induction impairs consolidation of long-term memory thus suggesting that Arc protein induction is fundamental not only in marking neurons which have been sufficiently depolarised to activate NMDA receptors, but also in the stabilisation of activity-dependent processes.

The recent success of 'proteomics', the large-scale analysis of proteins, provided the basis for a better understanding of the interactive network of proteins within multiprotein complexes associated with neurotransmitter receptors and cell adhesion proteins involved in signalling. However, understanding the regulation of the flow of genetic information between the genome and 'proteome' is still uncharted territory. The first step in this flow is dependent on successful gene expression. The expression of gene product seems to depend on a complex network of ribonucleoproteins (mRNAPs) regulated by ELAV/Hu proteins.[112] This is particularly important for the expression of IEGs, which products could mediate the second wave of gene expression, in other words, the second wave of protein synthesis. Thus, the understanding of the properties of the mRNP expressed in neurons, their infrastucture and organised network, the approach called "ribonomics", could potentially provide research into memory formation with invaluable knowledge of the temporal and spatial regulation of activity-driven regulation of gene expression. Many of the neuronal mRNAs, to which ELAV/Hu proteins bind, encode transcription factors including CREB, ERG-1 and fos and appear to be packaged at distances far from the nucleus. Their localised expression in response to external stimuli may, when the transcription factors are transported back to nucleus, influence cellular events following stimulation of neurons.[5] Moreover, it seems to be that ELAV/Hu proteins are capable of regulating expression of clustered mRNP subset encoding proteins with related functional properties. The advantage of such a guarded regulatory pathway is that it could represent a controlled synapse-specific mode of direct activation of specific genes, and subsequent *de novo* synthesis of specific proteins, without involvement of multiple signal transduction cascade.[56,57] This mode of cellular response has the potential to be a selective mechanism by which dendritic branches, which received the appropriate input, activate *de novo* protein synthesis and modulate with great precision long-term modifications, the strength of the connections, underlying long-term memory formation.

References

1. Abel T, Lattal M. Molecular mechanisms of memory acquisition, consolidation and retrieval. Curr Opin Neurobiol 2001; 11:180-187.
2. Agranoff BW. Agents that block memory. In: Quarton GC et al. eds. The Neuroscience: A study programme. NewYork: Rockefeller Univ Press, 1967:756-764.
3. Agranoff BW. Learning and memory: Bochemical Approaches. In: Siegel GJ et al. eds. Basic Neurochemistry. Boston: Little Brown, 1981:801-820.
4. Agranoff BW, Davis RE, Brink JJ. Chemical study of memory fixation in goldfish. Brain Res 1966; 1, 303-309.
5. Albright TD et al. Neural Science: A century of progress and the mysteries that remain. Cell 2000; 100:S1-S55.
6. Anokhin KV, Rose SPR. Learning-induced increase of immediate early gene messenger RNA in the chick forebrain. Eur J Neurosci 1991; 3:162-167.
7. Anokhin KV et al. Effect of early experience on c-fos gene expression in the chick forebrain. Brain Res 1991; 544:101-107.
8. Austin L, Morgan IG. Incorporation of ^{14}C-labelled leucine into synaptosomes from rat cerebral cortex in vitro. J Neurochem 1967; 14:377-387.
9. Bailey CH, Bartsch D, Kandel ER. Toward a molecular definition of long-term memory storage. Proc Natl Acad Sci USA 1996; 93:13455-13502.
10. Barondes SH, Jarvik ME. The influence of actinomycin D on brain RNA synthesis and memory. J Neurochem 1964; 11:187-189.
11. Barraco RA, Stettner LJ. Antibiotics and memory. Psychol Bull 1976; 83:242-302.
12. Bourtchouladze R, Rose SPR. Memory formation in day old chicks requires NMDA but not nonNMDA glutamate receptors. Eur J Neurosci 1992; 4:533-538.
13. Bourtchouladze R et al. Deficient long-term memory in mice with a targeted mutation of the cAMP-responsive element-binding protein. Cell 1994; 79:59-68.
14. Browning M et al. Synaptic phosphoproteins: Specific changes after repetitive stimulation of the hippocampal slice. Science 1979; 203:60-62.
15. Brusa R. Genetically modified mice in neuropharmacology. Pharmacol Res 1999; 39:405-419.
16. Cherkin A. Kinetics of memory consolidation: Role of amnestic treatment parameters. Proc Natl Acad Sci USA 1969; 63:1094-1101.
17. Cohen HD, Barondes SH. Further studies on learning and memory after intracerebral injection of actinomycin D. J Neurochem. 1966; 13:207-211.
18. Curran T. Morgan JI. Memories of fos. BioEssays 1987; 7:255-258.
19. DeOrtiz SP, Arshavsky YI. DNA Recombination as a possible mechanism in declarative memory: A hypothesis. J Neurosci Res 2001; 63:72-81.
20. De Robertis E et al. On the isolation of nerve ending and synaptic vesicles. J Biophys Biochem Cytol 1961; 9:229-235.
21. Dietrich A, Been W. Memory and DNA. J Theor Biol 2001; 208:145-169.
22. Dragunow M. A role for immediate-early transcription factors in learning and memory. Behav Genet 1996; 26:293-299.
23. Dunn AJ. The Chemistry of learning and the formation of memory. In: Gispen WH, ed. Molecular and functional neurobiology. Amsterdam, Elsevier, 1976; 347-387.

24. Dunn AJ. Neurochemistry of learning and memory: An evaluation of recent data. Ann Rev Psychol 1980; 31:343-390.
25. Dunn AJ. Kramarcy NR. Effects of ACTH and related peptides on cerebral RNA and protein synthesis. Pharmacol Therap 1984; 12:353-372.
26. Edelman GM. Cell adhesion and the molecular processes of morphogenesis. Ann Rev Biochem 1985; 54:135-169.
27. Entingh D et al. Brain uridine monophosphate: Reduced incorporation of uridine during avoidance learning. Brain Res 1974; 70:131-138.
28. Fisher S, Litvak S. The incorporation of microinjected ^{12}C-aminoacid into TCA insoluble fraction of the giant axon in the squid. J Cell Physiol 1967; 70:69-74.
29. Flexner LB, Flexner JB. Intracerebral saline: Effect on memory of trained mice treated with puromycin. Science 1978; 159:330 -331.
30. Flood JF et al. The influence of duration of protein synthesis inhibition on memory. Physiol Behav 1973; 10:555 - 562.
31. Flood JF et al. Relation of memory formation to controlled amounts of brain protein synthesis. Physiol Behav 1975; 15:97-102.
32. Flood JF et al. Modification of anisomycin-induced amnesia by stimulants and depressants. Science 1978; 199:324 -326.
33. Freeman FM, Rose SPR. Two time windows of anisomycin-induced amnesia for passive avoidance training in the day-old chick. Neurobiol Learn Mem 1995; 63:291-295.
34. Frey U, Morris RGM. Synaptic tagging and long-term potentiation. Nature 1997; 385:533-536.
35. Gibbs ME, Ng K. Counteractive effects of norepinephrine and amphetamine on quabain-induced amnesia. Pharmacol Biochem Behav 1977; 6:533-537.
36. Gibbs ME, Ng K. Behavioural stages in memory formation. Neurosci Lett 1979; 13:279-283.
37. Gingrich JA, Hen R. The broken mouse: The role of development, plasticity and environment in the interpretation of phenotypic changes in knockout mice. Curr Opinion Neurobiol 2000; 10:146-152.
38. Gispen WH et al. Phosphorylation of proteins in synaptosome-enriched fraction of brain after a short-term training experience: The effects of various treatments. Behav Biol 1977; 21:358-363.
39. Glassman E. The biochemistry of learning. An evaluation of the role of RNA and proteins. Ann Rev Biochem 1969; 38:605-646.
40. Goelet P et al. The long and the short of long-term memory - a molecular framework. Nature 1986; 322:419-422.
41. Giuditta A, Cupello A, Lazzarini G. Ribosomal RNA in the axoplasm of the squid giant axon. J Neurochem 1980; 34:1757-1760.
42. Giuditta A, Dettbarn WD, Brzin M. Protein synthesis in the isolated giant axon of the squid. Proc Natl Acad Sci USA 1968; 59:1284-1287.
43. Giuditta A, Hunt T, Santella L. Messenger RNA in squid axoplasm. Biol Bull 1983; 165:526-528.
44. Giuditta A et al. Factors for protein synthesis in the axoplasm of the squid giant axon. J Neurochem 1977; 28:1393-1395.
45. Gray EG, Whittaker VP. The isolation of nerve endings from brain: An electron-microscopic study of cell fragments derived by homogenisation and centrifugation. J Anat 1962; 96:79-88.
46. Halstead WC. Cerebral mechanisms for behaviour. New York: Wiley, 1951.
47. Hydén H, Egyhazi E. Nuclear RNA changes of nerve cells during a learning experiment in rats. Proc Natl Acad Sci USA 1962; 48:1366-1373.
48. Hydén H, Egyhazi E. Changes in RNA content and base composition in cortical neurons in rats in a learning experiment involving a transfer of handedness. Proc Natl Acad Sci USA 1964; 52:1030-1035.
49. Hydén H, Lange PW. S-100 protein: Correlation with behaviour. Proc Natl Acad Sci USA 1970; 67:1959-1966.
50. Ingoglia NA et al. Incorporation of [^{3}H]amino acids into proteins in a partially purified fraction of axoplasm: Evidence for transfer RNA mediated, post-translational protein modification in squid giant axon. J Neurosci 1983; 3:2463-2473.
51. Izquierdo I, Medina JH. Memory formation: The sequence of biochemical events in the hippocampus and its connection to activity in other brain structures. Neurobiol Learn Mem 1997; 68:285-316.
52. Jacobson AL et al. Differential approach tendencies produced by injection of ribonucleic acid from trained rats. Science 1965; 150:636-637.
53. Jork R, Grecksch G, Matthies H. Impairment of glycoprotein fucosylation in rat hippocampus and the consequences on memory formation. Pharmacol Biochem Res 1986; 25:1137-1144.
54. Kandel ER, Pittenger C. The past, the future and the biology of memory storage. Philos Trans R Soc Lond B Biol Sci 1999; 354:2027-2052.
55. Kandel ER, Schwartz JH. Molecular biology of learning: Modulation of transmitter release. Science 1982; 218:433-443.
56. Keene J. Why is Hu where? Shuttling of early-response-gene messenger RNA subsets. Proc Natl Acad Sci USA 1999; 96:5-7.

57. Keene J. Ribonucleoprotein infrastructure regulating the flow of genetic information between genome and the proteome. Proc Nat Acad Sci USA 2001; 98:7108-7024.
58. Kinney WR et al. Prolong activation in E-box binding after a single systemic kainate injection: Potential relation to F1/GAP-43 gene expression. Mol Brain Res 1996; 38:25-36.
59. Kao JF, Greengard P. An adenosine3',5'-monophosphate-dependent protein kinase from Escherichia coli. J Biol Chem 1969; 244:3417- 3419.
60. Lasek RJ et al. Analysis of axoplasmic RNA from invertebrate giant axon. Nature New Biol 1973; 244:162-165.
61. Lasek JR et al. Transfer of newly synthesised proteins from Schwann cells to the squid giant axon. Proc Natl Acad Sci USA 1974; 71:1188-1192.
62. Levitan IB, Lemos JR, Novak-Hopper I. Protein phosphorylation and regulation of ion channels. Trends Neurosci 1983; 6:496-499.
63. Liang KC, Bennett C, McGaugh JL. Peripheral epinephrine modulates the effect of posttraining amygdala stimulation on memory. Behav Brain Res 1985; 15:93-100.
64. Litvin OO, Anokhin KV. Mechanisms of memory reorganisation during retrieval of acquired behavioural experience in chicks: The effects of protein synthesis inhibition in the brain. Neurosci Behav Physiol 2000; 30:671-678.
65. Machlus B, Wilson JE, Glassman E. Brain phosphoproteins: The effect of short experiences on phosphorylation of nuclear proteins of rat brain. Behav Biol 1974; 10:43-62.
66. Martin R. Ribosomes in peripheral and presynaptic domains of axon. In: Telkeen AW, Korf J, eds. Neurochemistry. New York: Plenum Press, 1997:661-665.
67. Martinez Jr JL. Endogenous modulators of learning and memory. In: Martinez JL, Kesner RP, eds. Learning and Memory. New York: Academic Press, 1981:127-155.
68. Matthies H. Plasticity in the nervous system - an approach to memory research. In: Amone Marsane C, Matties H, eds. Neural Plasticity and Memory formation. New York: Raven Press, 1982:1-15.
69. Mayford M, Kandel ER. Genetic approaches to memory storage. Trends Gen 1999; 15:463-470.
70. Mayford M, Abel T, Kandel ER. Transgenic approaches to cognition. Curr Opin Neurobiol 1995; 5:141-148.
71. McConnell JV. Memory transfer through cannibalism in planarians. J Neuropsychiatry Suppl 1962; 1:42-48.
72. McEwen B. Neuronal and cognitive effects of oestrogens. Introduction. Novartis Found Symp 2000; 230:1-6.
73. McEwen B. Effects of adverse experience for brain structure and function. Biol Psychiatry, 2000; 48:721-731.
74. McGaugh, JL. Time-dependent processes in memory storage. Science 1966; 153:1351-1358.
75. McGaugh JL. Memory - a century of consolidation. Science 2000; 287:248-251.
76. McGaugh JL, Cahill L, Roozendaal B. Involvement of the amygdala in memory storage: Interaction with other brain systems. Proc Natl Acad Sci USA 1996; 93:13508-13514.
77. Meberg PJ et al. Protein kinase C and F1/GAP-43 gene expression in hippocampus inversely related to synaptic enhancement lasting 3 days. Proc Natl Acad Sci USA 1993; 90:12050-12054.
78. Meberg PJ et al. Gene expression of the transcription factor NF-kB in hippocampus: Regulation by synaptic activity. Mol Brai Res 1996; 38:179-190.
79. Meberg PJ et al. MARCKS and protein H1/GAP-43 RNA in chick brain: Effects of imprinting. Mol Brain Res 1996: 35:149-156.
80. Mileusnic R, Lancashire C, Rose SPR. Sequence-specific impairment of memory formation by NCAM antisense oligodeoxynucleotides. Learning Memory 1999; 6:120-127.
81. Mileusnic R et al. APP is required during an early phase of memory formation. Eur J Neurosci 2000; 12:4487-4495.
82. Montarolo PG et al. A critical period for macromolecular synthesis in long-term heterosynaptic facilitation in Aplysia. Science 1986; 234:1249-1254.
83. Nader K, Schafe GE, LeDoux JE. Fear memories require protein synthesis in the amygdala for reconsolidation after retrieval. Nature 2000; 406:722-726.
84. Nader K, Schafe GE, LeDoux JE. The labile nature of consolidation theory. Nature Rev Neurosci 2000; 1:216-219.
85. Namgung U, Routtenberg A. Transcriptional and post-transcriptional replication of brain growth protein: Regional differentiation and induction of GAP-43. Europ J Neurosci 2000; 12:3214-3136.
86. Namgung U, Matsuyama S, Routtenberg A. Long-term potentiation activates GAP-43 promoter: Selective participation of hippocampal mossy cells. Proc Natl Acad Sci USA 1997; 94:11675-11680.
87. Petrushka E, Guiditta A. Electron microscopy of two subcellular fractions isolated from cerebral cortex homogenate. J Biophys Biochem Cytol 1959; 6:129-132.
88. Popov N et al. PAGE-autoradiography of fucose incorporation into rat hippocampal glycoproteins after acquisition of a brightness discrimination. BBA 1983; 42:763-776.

89. Rees HD et al. Effect of sensory stimulation on the uptake and incorporation of radioactive lysine into protein of mouse brain and liver. Brain Res 1974; 68:143-156.
90. Riedel G et al. Reversible neural inactivation reveals hippocampal participation in several memory processes. Nature Neurosci 1999; 2:898-905.
91. Rose SPR. Strategies in studying the cell biology of learning and memory. In: Squire LR, Butters N, eds. The Neuropsychology of memory. New York: Guilford Press, 1984:547-554.
92. Rose SPR. Cell adhesion molecules, glucocorticoids and memory. Trends Neurosci 1995; 18:502-506.
93. Rose SPR. God's organism: The chick as a model system for memory studies. Learning Memory 2000; 7:1- 17.
94. Rose SPR, Harding S. Training increase [^{3}H]-fucose incorporation in chick brain only if followed by memory storage. Neurosci 1984; 12:663-667.
95. Rose SPR, Haywood J. Experience, learning and brain metabolism. In: Davidson AN, ed. Biochemical correlates of brain structure and function. New York: Academic Press, 1977:249 – 292.
96. Rose SPR, Jork R. Long-term memory formation in chicks is blocked by 2-deoxygalacose, a fucose analog. Behav Neural Biol 1987; 48:246-258.
97. Rosenbaum M, Cohen HD, Barondes SH. Effect of intracerebral saline on amnesia produced by inhibitors of cerebral protein synthesis. Comm Behav Biol Part A 1968; 2:47 -50.
98. Rosenzweig MR, Bennett EL, Flood JL. Pharmacological modulation of formation of long-term memory. In: Adam G et al. eds. Brain and Behaviour. Adv Physiol Sci Vol 17 1981:101 - 111, London: Pergamon Press,
99. Routtenberg A. Anatomical localisation of phosphoprotein and glycoprotein substrate of memory. Prog Neurobiol 1979; 12:85-113.
100. Routtenberg A, Benson GE. In vitro phosphorylation of a 41.000-MW protein band selectively increased 24 hr after footshock or learning. Behav Neurol Biol 1980; 29:168-175.
101. Sara SJ. Retrieval and reconsolidation: Toward a neurobiology of remembering. Learning Memory 2000; 7:73-84.
102. Sara SJ. Strengthening the shaky trace through retrieval. Nature Rev Neurosci 2000; 1:212-213.
103. Schafe EG et al. Memory consolidation of Pavlovian fear conditioning: A cellular and molecular perspective. Trends Neurosci 2001; 24:540-546.
104. Steele RJ, Morris RGM. Delayed-dependent impairment of matching-to-place task with chronic and intrahippocampal infusion of the NMDA-antagonist D-AP5. Hippocampus 1999; 9:118-136.
105. Squire LR. Short-term memory as a biological entity. In: Deutsch D, Deutsch JA, eds. Short-term memory. New York: Academic Press, 1975:1-40.
106. Squire LR, Davis HP. The pharmacology of memory: A neurobiological perspective. Ann Rev Pharmacol. Toxicol 1981; 21:323 -356.
107. Struhl K. Mechanism of diversity in gene expression patterns. Neuron 1991; 7:177-181.
108. Steward O, Levy WB. Preferential localisation of polyribosomes under the base of dendritic spines of granule cells of the dentate gyrus. J Neurosci 1982; 2:84-91.
109. Steward O, Worley P. A cellular mechanism for targeting newly synthesised mRNAs to synaptic sites on dendrites. Proc Natl Acad Sci USA 2001; 98:70622-7068.
110. Sukumar R, Rose SPR, Bourgoyne RD. Incresed incorpotaion of [^{3}H]fucose into chick brain glycoproteins following training on a passive avoidance task. J Neurochem 1980; 34:1000-1006.
111. Taubenfeld SM et al. The consolidation of new but not reactivated memory requires hippocampal C/EBP. Nature Neurosci 2001; 4:813-818.
112. Tenenbaum SM et al. Identifying mRNA subsets in messenger ribonucleoprotein complexes by using cDNA. Proc Natl Acad Sci USA 2000; 97:14085-14090.
113. Tully T. Discovery of genes involved in learning and memory: An experimental synthesis of Hirschian and Benzerian perspective. Proc Natl Acad Sci USA 1996; 93:13460-13467.
114. Tytell M, Lasek JR. Glial polypeptides transferred into the squid giant axon. Brain Res 1984; 363:161-164.
115. Walton M et al. Immediate early gene transcription and synaptic modulation. J Neurosci Res 1999; 58:96-106.
116. Wied D de. Pituitary-adrenal system hormones and behaviour. In: Schmitt FO, Worden FG, eds. The neuroscience third study programme. Cambridge: MIT Press, 1974:653-666.
117. Yellin TO, Butler BJ, Stein HH. Inhibition of protein synthesis in mitochondria by cyclohexemide. Fed Proc 1967; 26:833-835.

CHAPTER 6.1

Learning-Induced Synaptogenesis and Structural Synaptic Remodeling

Yuri Geinisman, Robert W. Berry and Olga T. Ganeshina

Abstract

This chapter analyzes the results of quantitative electron microscopic studies of the vertebrate brain aimed at the elucidation of changes in synaptic unltrastructure that may underlie learning and memory. It has been reported that behavioral learning promotes new synapse formation, including both a net synaptogenesis, which causes a net gain in synapse number, and a specific synaptogenesis. The latter either accompanies a learning-related adult neurogenesis or leads to the formation of multiple-synapse boutons. Other data strongly suggest that behavioral learning also elicits a structural remodeling of existing synapses. This process is revealed morphologically as an increase in the number of perforated axospinous synapses and an enlargement of postsynaptic densities. Further research will show if the observed diversity of structural synaptic alterations reflects differences among various forms of learning and memory or among various consecutive processes underlying the formation, consolidation and long-term storage of each memory trace.

Introduction

Over a century ago, Ramón y Cajal[99] and Tanzi[114] postulated that changes in the number and structure of synaptic connections might underlie the establishment of long-term memories following behavioral learning. Since that time, cellular mechanisms of learning and memory have been thought to include the formation of new synapses and/or a structural remodeling of existing synapses. Numerous attempts have been made to verify this assumption by defining those modifications of synaptic ultrastructure in the vertebrate brain that result from the learning of new behaviors (for reviews see refs. 4, 6, 34 and 48). The review of the literature presented here examines a growing body of evidence for learning-induced synaptogenesis and synapse remodeling. The data reported so far are also related to the phenomena of adult neurogenesis, spine motility and transformation of postsynaptically silent synapses into functional ones. Finally, the role of these phenomena in shaping the patterns of learning-induced alterations in synaptic ultrastructure is discussed.

Patterns of Synaptogenesis Elicited by Behavioral Learning

Net Synaptogenesis Resulting from Learning

The search for the net synaptogenesis that may underlie the learning of new behaviors involves determining whether the process of learning is accompanied by a net gain in synapse number. Initial attempts to resolve this issue were based on the use of light microscopic preparations stained according to the Golgi method for counting dendritic spines, which usually synapse with axon terminals. However, since the Golgi method stochastically stains

From Messengers to Molecules: Memories Are Made of These, edited by Gernot Riedel and Bettina Platt.

only a small fraction of neurons, subsequent work over the past two decades has focused on the examination of electron microscopic preparations that reveal all the synapses present in a given tissue specimen. Samples of entire synaptic populations, including all morphological synaptic types, were taken from pertinent regions of the vertebrate brain and used for obtaining estimates of synapse number. Animals that had learned a given behavioral task were compared on this measure with controls that had no such learning experience. These electron microscopic studies, however, produced inconsistent results (Table 1).

Several studies found that behavioral learning promotes a net gain in synapse number (Table 1). Additionally, quantitative electron microscopic analyses of dendritic spines showed that their total number was increased in the molecular layer of the rat dentate gyrus as a consequence of either passive avoidance conditioning[91] or water maze training.[92] Similar results were obtained with the aid of confocal microscopy: an increase in the number of spines per unit length of basal, but not apical, dendrites of CA1 pyramidal cells was detected following spatial learning of rats in a complex environment.[82,83] Taken together, these data are consistent with the notion that the process of learning is accompanied by a net synaptogenesis.[4,6,48] At variance with such a notion is the observation of other studies that the learning of new behaviors does not cause a net change in synapse number (Table 1).

A number of factors might have contributed to the discrepancy in the results. One of these is that appropriate methods for synapse quantification were not available until recently, and most of the ultrastructural studies referred to in Table 1 used methodologically inadequate procedures. These included the identification of synapses in single (rather than in serial) ultrathin sections, sampling in selective (rather than in systematic, uniformly random) fashion, counting with two-dimensional (rather than with three-dimensional) probes, and estimating the numerical density of synapses (rather than their total number). Each of these procedures involves biases.[40,41] The direction and magnitude of the biases were not evaluated, which makes it difficult to interpret the data of the earlier electron microscopic studies. To circumvent these problems, a stereological method was designed for providing unbiased estimates of total synapse number in defined brain regions.[41]

This method was then used to explore the problem of whether the total number of synapses in the stratum radiatum of hippocampal subfield CA1 is altered by trace eyeblink conditioning.[40] The latter is a hippocampus-dependent form of associative learning[67,84,107] that is accompanied by increases in the synaptic responsiveness[40,96] and postsynaptic excitability[21,85] of CA1 pyramidal neurons. For conditioning, rabbits were given daily 80-trial sessions to a criterion of 80% conditioned responses in a session. During each trial, the conditioned stimulus (tone) and the unconditioned stimulus (corneal airpuff) were presented with an intervening trace interval of 500 msec. Brain tissue was taken for morphological analyses 24 hours after the last session. The results showed that the total number of synapses in the CA1 stratum radiatum was not changed in conditioned rabbits as compared to pseudoconditioned controls (Fig. 1A). No trend towards a conditioning-induced increase in total synapse number was observed (Fig. 1A), and the group means (± SEM) for pseudoconditioned and conditioned animals (23,267 ± 915 and 23,250 ± 869 synapses x 10^6, respectively) differed by only 0.07%. These data provide convincing evidence for the stability of total synapse number in the CA1 stratum radiatum 24 hours after acquisition of the trace eyeblink conditioned response.

If the discrepancy in the results of previous studies is due to their methodological limitations, it is reasonable to expect that the use of modern stereological methodology for synapse quantification following learning would yield consistent data. This, however, was not the case. Conflicting results were also obtained with the aid of the unbiased disector technique. Rusakov et al[101] found that spatial learning in the Morris water maze had no effect on the numerical density of synapses per unit volume of the rat CA1 stratum radiatum and dentate gyrus molecular layer. This finding is in agreement with our observation (Fig. 1A). In contrast, the work from the Greenough laboratory demonstrated that the learning of complex motor

Table 1. Results of synapse quantification based on analyses of samples taken from the entire synaptic population of a given brain area following behavioral learning

Behavioral Paradigm Used	Species and Brain Area Examined	Parameter Analyzed	Result	Reference
Brightness discrimination conditioning in a Y-maze	Rat hippocampus (subfield CA1)	N_A	Increase	Wenzel et al, 1980
Acquisition of male-like singing behavior by females treated with testosterone	Canary robustus archistralis nucleus	N	Increase	DeVoogt et al,1985
Passive avoidance conditioning	Chick hyperstriatum ventrale and lobus parolfactorius	N_V	Increase	Stewart et al,1987; Hunter & Stewart, 1989; Doubell & Stewart, 1993
Complex motor skill acquisition	Rat cerebellar and motor cortices	N_N	Increase	Black et al, 1990; Kleim et al, 1996; 1997
Visual imprinting	Chick hyperstriatum ventrale	N_A, N_V	No change	Bradley et al, 1981; Horn et al, 1985
Visual discrimination conditioning	Rabbit visual cortex	N_V	No change	Vrensen & Nunes Cardozo, 1981
One-way active avoidance conditioning	Rat dentate gyrus	N_A	No change	Van Reempts et al, 1992
Spatial learning in the Morris water maze	Rat hippocampus (subfield CA1) and dentate gyrus	N_V	No change	Rusakov et al, 1997
Trace eyeblink conditioning	Rabbit hippocampus (subfield CA1)	N	No change	Geinisman et al, 2000

Designations: N_A – synaptic numerical density per unit tissue area; N_V - synaptic numerical density per unit tissue volume; N_N – number of synapses per postsynaptic neuron; N – total synapse number.

skills increased the number of synapses per neuron in the rat motor and cerebellar cortex.[9,69,70] It appears, therefore, that the methodological limitations alone cannot account for the diversity of the data reported so far.

Another factor that may be of special importance in this respect is the transient nature of a learning-induced net gain in synapse number. In certain brain regions, such a change persists only for a limited period of time following behavioral training. This was clearly demonstrated in experiments involving passive avoidance conditioning of day-old chicks (for a review see ref. 112). Acquisition of the conditioned avoidance response was followed by an increase in the overall numerical density of axospinous synapses that was observed in the intermediate and medial hyperstriatum ventrale (IMHV) only at 1 hour, but not at 24 hours, after training.[22] On the other hand, the same change in the lobus parolfactorius (LPO) was detected 24 and 48 hours, but not 1 or 6 hours, post-training.[56] The different temporal sequence of alterations in the synaptic numerical density observed in the IMHV and LPO may be explained by differences in the involvement of these brain regions in task acquisition and consolidation. Lesion studies have indicated that the IMHV may be the site of initial registration of the memory trace, which is subsequently transferred to the LPO for long-term storage.[43] Similar observations were made more recently in experiments examining the effect of passive avoidance conditioning[91] or spatial learning[92] on the numerical density of dendritic spines in

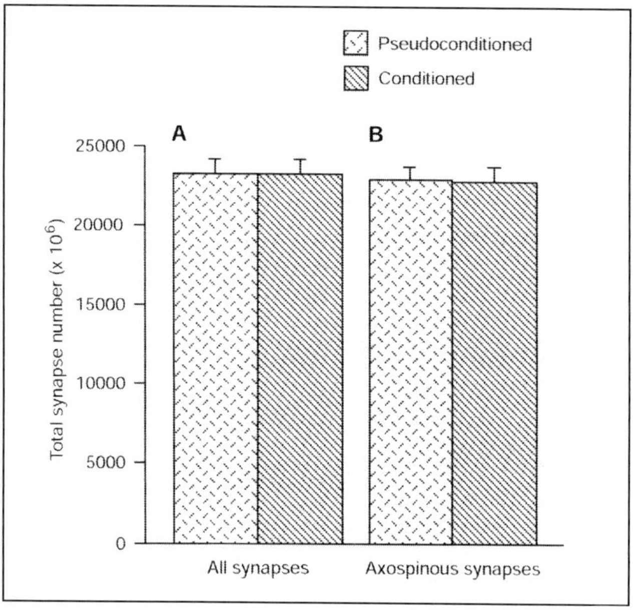

Figure 1. Total number of all synaptic contacts (A) and of axospinous synapses (B) in the CA1 stratum radiatum of pseudoconditioned and conditioned rabbits examined 24 hours after cessation of training (data are from ref. 34). Bars show group means ± SEM.

the molecular layer of the hippocampal dentate gyrus of adult rats. Following the cessation of training on either task, spine number was shown to increase at 6 hours and then return to control levels by 72 hours. The latter temporal pattern of ultrastructural changes is consistent with the concept that the hippocampal formation plays only a transitory role in the consolidation of memory (e.g., ref. 109). In the cerebellar cortex of adult rats, on the other hand, an increase in the number of synapses per Purkinje cell persists for at least 4 weeks after motor skill learning.[70] This suggests that the cerebellar cortex may be involved in a long-term retention of memories for motor skills.

The data described above underscore the necessity of examining various phases of the acquisition/consolidation process in order to reach a definitive conclusion of whether a net increase in synapse number is characteristic of a given form of behavioral learning. In most studies available in the literature, including the one from this laboratory,[40] synapses were quantified only at a single time point relative to behavioral acquisition. If quantitative ultrastructural analyses were performed at various time points along the acquisition/consolidation curve, the presence of additional synaptic contacts might have been detected as well. It is also conceivable, however, that some forms of learning may not involve a net gain in synapse number and, hence, a net synaptogenesis. Work from the laboratories of Rakic and Goldman-Rakic has provided evidence in favor of this supposition (for a review see ref. 11). Their estimates of synaptic numerical density in five major areas of the primate cerebral cortex reveal no ultrastructural sign of net synaptogenesis over the entire period of adulthood in spite of presumably continuous accumulation of long-term memories.

Even if the acquisition and retention of a given behavior does not result in a detectable net synaptogenesis, it is possible that new synapses may be nevertheless formed as a consequence of behavioral learning. In such cases, learning-induced synaptogenesis might be confined to re-arranging only a specific subset of synaptic connections in order to establish a memory trace. Examples of such specific synaptogenesis will be considered next.

Specific Synaptogenesis Related to Learning-Induced Adult Neurogenesis

It has been established that the production of new neurons in the vertebrate brain continues throughout adulthood, and there are observations suggesting that it may occur not only spontaneously but also as a consequence of behavioral learning (for reviews see refs. 47, 50, 75 and 102). These observations are especially relevant to the present discussion because to become functional, adult-born neurons have to form appropriate synaptic connections. This implies that a specific synaptogenesis involving newly generated neurons should accompany the process of adult neurogenesis.

Evidence for the existence of neurogenesis in the adult brain comes from studies employing the incorporation of tritiated thymidine, as well as from immunocytochemical studies using markers for the detection of proliferating cell progenies and their neuronal phenotype. Although these methodological approaches have serious limitations that make the interpretation of the results difficult (for reviews see refs. 44 and 98), there is a general consensus that new neurons are spontaneously added to several regions of the adult mammalian brain, most notably to the dentate gyrus and olfactory bulb. Adult neurogenesis has been most extensively studied in the dentate gyrus of the hippocampal formation, and the latter is known to mediate certain forms of learning and memory. Therefore, the following discussion is focused on data from the dentate gyrus.

Principal neurons of the dentate gyrus (granule cells) are generated in all adult mammalian species studied thus far, including rodents (e.g., see ref. 1, 5, 8 and 62), nonhuman primates[46,73] and humans.[18,28] New granule cells arise from their precursors residing in the subgranular zone of the dentate gyrus and migrate into the granule cell layer where they assume the morphological features characteristic of neighboring neurons.[14,61,62] The majority of the new granule cells die within two weeks of their birth.[14,45] However, adult-generated granule cells do display synapses,[61,62] receive functional synaptic inputs similar to those found in mature granule cells[118] and rapidly extend their axons through the mossy fiber tract to their natural target area, hippocampal subfield CA3.[52,81,110] This suggests that neurons newly generated in the adult dentate gyrus may become, at least temporarily, integral components of neural circuits in the hippocampal formation and participate in its functions related to learning and memory.[45,50,63]

Several experimental approaches have been used to test the validity of this suggestion. A number of studies have addressed the question of whether environmental complexity or aging alter adult hippocampal neurogenesis. Exposure of adult birds and rodents to enriched environments, which presumably offer more opportunities for learning than standard laboratory environments, increases the number of new hippocampal neurons. This does not affect the proliferation of progenitor cells, but rather promotes the survival of their progeny.[7,64,65,89] The enhanced survival of new granule cells in mice kept in an enriched environment is associated with improved spatial learning in the Morris water maze.[64,65,89] Conversely, the process of normal aging, which produces deficits in hippocampus-dependent forms of learning and memory (for a review see ref. 38), is accompanied by a dramatic reduction in the number of granule cells that are born in the dentate gyrus of aged mice and rats.[13,74,104]

Another approach was employed in experiments designed to determine if adult neurogenesis is modulated by specific learning experiences. An indication that behavioral learning may upregulate adult neurogenesis was initially provided by studies showing that the seasonal modification of song in birds coincides with the increased addition of new neurons to a forebrain vocal center involved in song learning.[2,68] A subsequent study of birds demonstrated that spatial learning associated with the first few exposures to the experience of storing and retrieving food augments the rate of neuronal recruitment into the hippocampus and hyperstriatum ventrale.[93] More recently, it has been found that learning the trace eyeblink conditioned response or the location of an invisible escape platform in the Morris water maze promotes the survival of granule cells born in the dentate gyrus of adult rats 1-2 weeks prior to training.[45] Additionally, the spatial learning ability of rats in the Morris water maze containing an invisible platform correlates with the extent of survival of adult-generated granule cells.[3] Unlike these hippocampus-dependent learning experiences, hippocampus-independent experiences, such as learning the delayed eyeblink conditioned response or the location of a visible and cued platform in the Morris water maze, have no effect on the number of new granule cells.[45] It should be noted here that functional integration of adult-born neurons into existing or newly developing circuitry occurs on a time-scale of several days. Thus an increased production of nerve cells resulting from an accelerated rate of adult neurogenesis would not have immediate functional consequences while a prolonged survival of functionally competent new neurons would.[45]

The results summarized above have been extended by the finding that the targeted destruction of the majority of newborn granule cells in adult rats by a toxin for proliferating cells (methylazoxymethanol acetate) impairs hippocampus-dependent trace eyeblink conditioning but does not affect delayed eyeblink conditioning, the latter of which is hippocampus-independent.[106] These observations suggest that adult-generated granule cells may play a role in the formation of hippocampus-dependent memories (see ref. 117). The hippocampal formation is considered to be especially important for the acquisition of associations between temporally or spatially discontiguous events[121] and for a transient storage of recently acquired memories.[109] It is possible that new granule cells, in spite of their limited lifespan, may support such major functions of the hippocampal formation. Since pyramidal cells can be newly generated in the adult hippocampus proper,[100] they may also mediate these hippocampal functions.

Newborn neurons in the adult brain may be much more modifiable and readily involved in synaptogenesis than older neurons. In order to survive and appropriately function, the adult-generated neurons that subserve learning and memory must make synaptic connections. Although such specific synaptogenesis associated with learning-induced adult neurogenesis may take place, it is difficult to document it. Markers for newly established synaptic contacts are not yet available for ultrastructural studies, which makes it necessary to demonstrate that learning-induced neurogenesis in a given brain region is accompanied by an increase in total synapse number. This alteration is not likely to be detected because synapses involving adult-born nerve cells constitute only a minute fraction of all synaptic contacts in a brain region. Quantification of the entire synaptic population is not necessary, however, for the detection of the other known form of specific synaptogenesis. This involves the formation of multiple synapses by single axonal boutons, which can be readily identified and selectively quantified as is described below.

Specific Synaptogenesis Related to Learning-Induced Formation of Multiple-Synapse Boutons

The term "multiple-synapse boutons" (MSBs) refers to those presynaptic axon terminals (boutons) that establish separate synaptic appositions with two or more postsynaptic elements instead of only one synaptic apposition with a single postsynaptic element. Earlier studies have reported that acquisition of complex motor skills increases the incidence of MSBs[29,59] as well as the overall number of synapses[9,69,70] in rat cerebellar and motor cortices. These data indicate that motor skill learning elicits both a specific synaptogenesis selectively producing additional MSBs and a net synaptogenesis. In light of the above observations, we explored the possibility that hippocampus-dependent associative learning, which does not alter total synapse number in the CA1 stratum radiatum[40] and hence does not involve a net synaptogenesis, nevertheless induces a specific synaptogenesis leading to the formation of MSBs, at least at the studied time point after training.

Electron micrographs obtained in our earlier study[40] were reanalyzed as described in detail elsewhere.[35] Trace eyeblink conditioned and pseudoconditioned rabbits were compared, the hippocampi being taken for morphological analyses 24 hours after the last training session. Unbiased stereological methods were used for obtaining estimates of the total number of MSBs in the CA1 stratum radiatum. Inspection of electron micrographs revealed that a typical MSB in this layer is a single presynaptic bouton that forms separate synapses with two spine heads (Fig. 2). Such MSBs can be unequivocally identified only in serial sections because in single

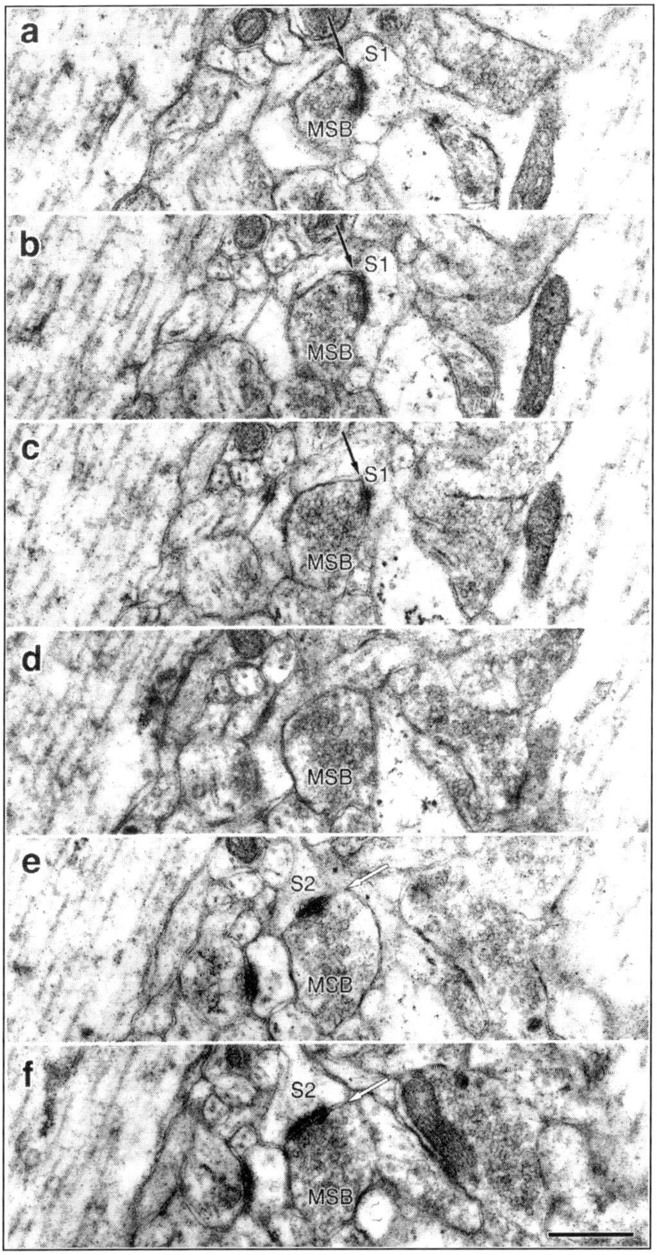

Figure 2. Electron micrographs of consecutive ultrathin sections (a-f) through the rabbit CA1 stratum radiatum demonstrating a typical multiple-synapse bouton (MSB). The bouton makes two synapses (black and white arrows), each one involving a separate dendritic spine (S1 or S2). Scale bar – 0.5 μm.

sections they usually exhibit a single synaptic profile involving only one spine (Fig. 2a-c,e,f) or no synaptic profile (Fig. 2d). MSBs synapsing with a dendritic shaft and a spine were encountered extremely rarely, in accordance with the observation that axodendritic synapses constitute only about 2% of the entire synaptic population of the rabbit CA1 stratum radiatum.[40] Therefore, only those MSBs that formed synapses exclusively with spines were quantified. The results showed that the mean total number of MSBs in the CA1 stratum radiatum was significantly increased in the group of conditioned rabbits ($1{,}236 \pm 43 \times 10^6$) as compared with the pseudoconditioned group ($1{,}047 \pm 28 \times 10^6$). Conditioned rabbits also had significantly more MSBs relative to untrained controls ($1{,}077 \pm 52 \times 10^6$), while the two control groups did not differ significantly from each other with respect to total MSB number.

Although MSBs in the rabbit CA1 stratum radiatum usually form synapses with two spines, some MSBs synapse with three or four spines. Therefore, we addressed the question of whether trace eyeblink conditioning alters the number of axospinous synapses per MSB and found that the mean numbers of axospinous synapses per MSB were the same (2.05) for the pseudoconditioned and conditioned groups. Another characteristic of MSBs is that they form both perforated synapses,[16,94] which exhibit a discontinuous profile of the postsynaptic density in at least one serial section, and nonperforated synapses that show a continuous postsynaptic density profile in all consecutive sections (Fig. 3). The perforated subtype has been implicated in synaptic plasticity associated with behavioral learning and hippocampal LTP (for reviews see refs. 34, 58 and 86). Our estimates of perforated synapse number per MSB showed, however, that the groups of pseudoconditioned and conditioned animals did not differ significantly on this measure.

The major finding of the cited study regarding an increase in total MSB number following trace eyeblink conditioning is in accord with those of earlier reports that motor skill learning induces the addition of MSBs.[29,59] Taken together, these results suggest that various forms of learning may promote MSB formation. Interestingly, this kind of morphological alteration is not unique to behavioral learning and to related phenomena such as hippocampal LTP[115] or an exposure to enriched environments.[60] Rather, the incidence of MSBs has been reported to increase as a consequence of various experimental manipulations that induce plasticity (for a review see ref. 35), indicating that the formation of MSBs may represent a general form of structural synaptic plasticity.

Individual MSBs in the CA1 stratum radiatum can synapse with spines arising from the same or different dendrites.[108,123] The LTP-induced increase in the proportion of activated boutons synapsing with two or more spines is essentially due to the formation of those MSBs that synapse with spines originating from the same dendrite.[115] However, we have been unable to reliably trace many multiple spines to their dendritic origins and to obtain representative samples for quantitative analyses. It is not known, therefore, whether the MSBs that are newly formed as a result of trace eyeblink conditioning make synapses with spines arising from the same dendrite. If this is the case, the strength of the conditioned synaptic input to target CA1 neurons may be amplified. If, however, the multiple postsynaptic spines synapsing with additional MSBs emanate from dendrites of different neighboring neurons, this may contribute to a synchronous activation of the latter and hence to the assembly of functional multineuronal units tuned to the synaptic input activated by conditioning stimulation. In either case, the effect of conditioning stimulation would be facilitated.

The results of our recent study and those reported in the literature suggest three models of structural plasticity that may underlie MSB formation after trace eyeblink conditioning (Fig. 4). For the purpose of simplicity, the following assumptions were incorporated into each model: 1) existing single-synapse boutons are transformed into MSBs by the conditioning and 2) the spines that are postsynaptic to newly formed MSBs originate from the same dendrite. The models also take into account our observation that trace eyeblink conditioning does not alter the total number of axospinous synapses in the CA1 stratum radiatum at the studied time point (24 hours) after acquisition of the conditioned response (Fig. 1B). This parameter would have been increased by about 2% if new axospinous synaptic contacts were added in the process of MSB formation. However, no trend towards such an increase was detected (Fig. 1B), suggesting that MSB formation in the conditioned animals may not be due to the recruitment of new axospinous synapses.

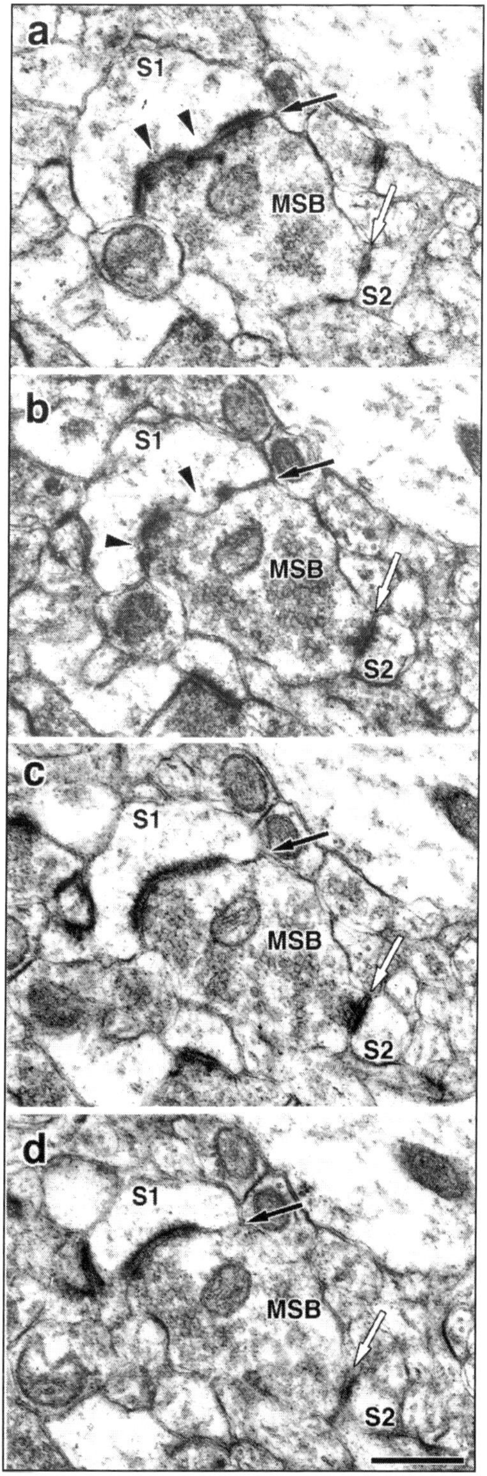

Figure 3. Electron micrographs of consecutive ultrathin sections (a-d) through the rabbit CA1 stratum radiatum demonstrating a multiple-synapse bouton (MSB) forming a perforated synapse (black arrows) with a large spine (S1) and a nonperforated synapse (white arrows) with a small spine (S2). The PSD of the large spine exhibits perforations (arrowheads) in some serial sections (a, b) whereas the PSD of the small spine shows no perforation in all sections. Scale bar – 0.5 μm.

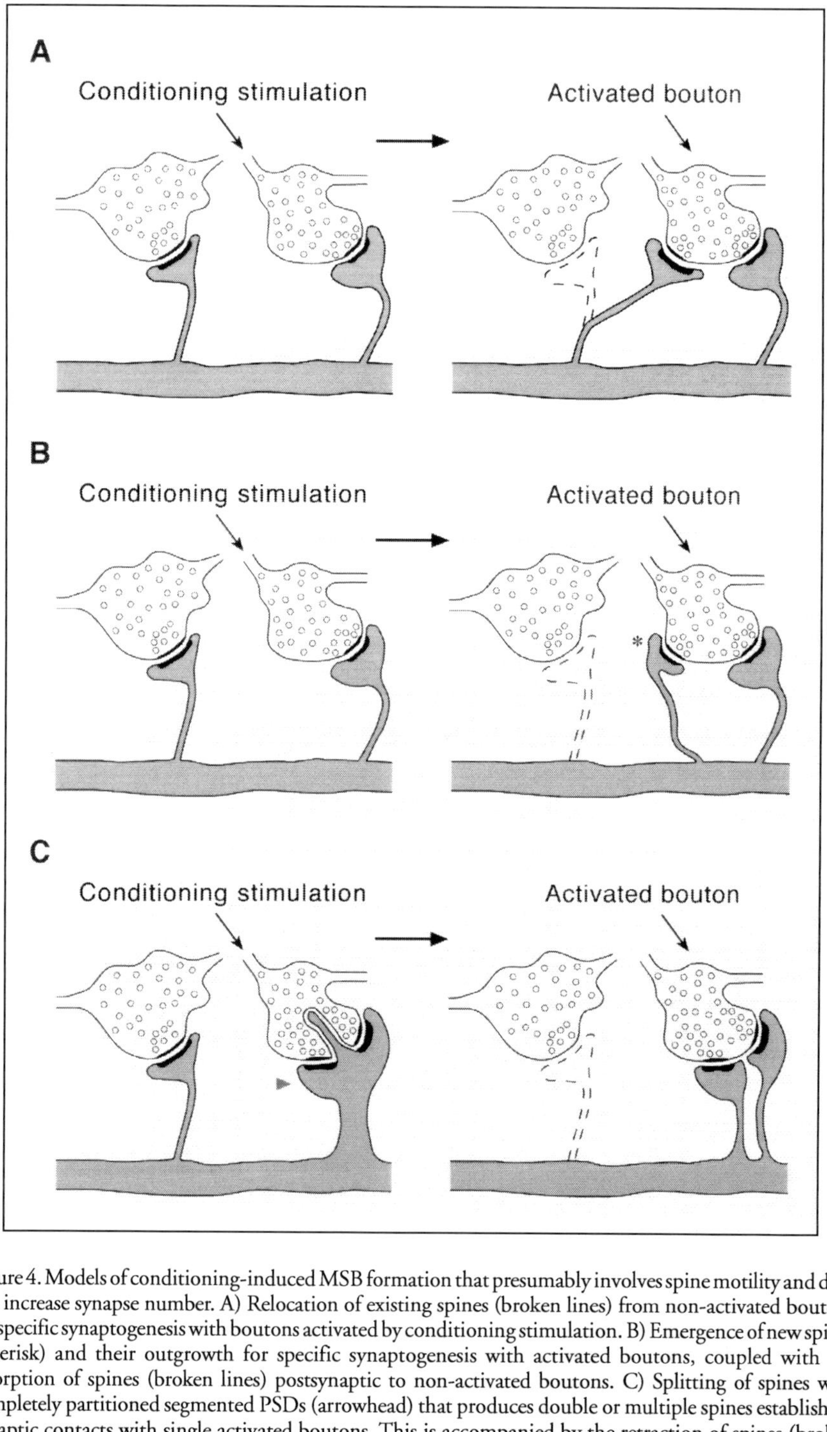

Figure 4. Models of conditioning-induced MSB formation that presumably involves spine motility and does not increase synapse number. A) Relocation of existing spines (broken lines) from non-activated boutons for specific synaptogenesis with boutons activated by conditioning stimulation. B) Emergence of new spines (asterisk) and their outgrowth for specific synaptogenesis with activated boutons, coupled with the resorption of spines (broken lines) postsynaptic to non-activated boutons. C) Splitting of spines with completely partitioned segmented PSDs (arrowhead) that produces double or multiple spines establishing synaptic contacts with single activated boutons. This is accompanied by the retraction of spines (broken lines) from non-activated boutons into parent dendrites.

Additionally, these models incorporate the recently discovered phenomenon of spine motility (for reviews see refs. 51, 80, 103 and 122). This phenomenon was established by labeling CA1 pyramidal neurons in cultured hippocampal slices with vital fluorescent markers and time-lapse two-photon imaging of their spines. Such experiments demonstrated that spines are highly dynamic structures constantly undergoing formation and resorption under normal conditions and that the process of new spine formation is markedly augmented by local high-frequency stimulation of dendrites, a manipulation which elicits LTP.[27,78] The ability of spines to rapidly elongate or retract is especially prominent during early postnatal development, but is retained to a certain degree after the maturation of CA1 pyramidal neurons in slices obtained from developing animals and maintained in culture.[17,24,31,72] Moreover, spines exhibiting larger and probably more mature synaptic contacts are no less motile than those forming smaller synapses.[111] A concordant movement of activated boutons and their postsynaptic spines[23] may also contribute to the spatial alignment of pre-and postsynaptic elements during MSB formation.

Accordingly, the first model (Fig. 4A) posits that, following trace eyeblink conditioning, some postsynaptic spines contacting non-activated boutons leave their presynaptic partners, relocate to boutons activated by conditioning stimulation and synapse with them. The second model (Fig. 4B) postulates that conditioning stimulation may induce the emergence of new spines and their outgrowth for a specific synaptogenesis with activated single-synapse boutons, probably in response to a signal emitted by the boutons. This model encompasses the retraction of some postsynaptic spines from non-activated boutons into parent dendrites, a process that keeps the total number of axospinous synapses constant. Finally, the third model (Fig. 4C) proposes that a specific synaptogenesis producing MSBs involves the splitting of large spines, which exhibit a PSD consisting of multiple segments separated from each other by complete spine partitions. The process of spine retraction maintaining the constancy of axospinous synapse number is also incorporated into the third model. This model was suggested by the findings of Toni et al[116] who examined the effect of hippocampal LTP on synaptic ultrastructure and observed a temporal coincidence between the disassembly of synapses involving such spines and the addition of MSBs that synapse with double spines arising from the same dendrite. However, the spatial arrangement of spines originating from the same dendrite and receiving synapses from the same MSB indicates that MSBs of this kind are unlikely to be formed by spine splitting.[30]

Further studies are needed to establish which of the three models is valid. In any event, our data described above demonstrate that trace eyeblink conditioning elicits specific synaptogenesis resulting in the formation of MSBs. Although the latter change does not require a net synaptogenesis to take place, it may facilitate the effect of conditioning stimulation.

Pattern of Structural Synaptic Remodeling Elicited by Behavioral Learning

Increase in the Number of Perforated Axospinous Synapses following Learning: A Possible Morphological Correlate of the Conversion of Synapses into More Efficacious Subtypes

An increase in the proportion or number of perforated axospinous synapses is perhaps the most notable and consistent change among activity-dependent alterations in synaptic ultrastructure (for a review see refs. 34, 58 and 86). There are reports in the literature that behavioral learning is also associated with such structural synaptic modification. The first indication that this may be the case came from the study of Greenough et al[49] showing that rats reared in a complex environment have more perforated synapses in the visual cortex than their counterparts kept in isolated conditions. The same effect was later found to be characteristic of visual discrimination conditioning.[120] Additionally, a significant correlation between the spatial learning ability of rats tested in an 8-arm maze and the number of perforated synapses per postsynaptic neuron was observed in the molecular layer of the dentate gyrus.[36] The numerical density of perforated synapses involving concave spines in the same synaptic layer was also estimated to increase following active avoidance conditioning.[119] It has been established that large perforated PSDs contain more AMPA receptors than do small nonperforated PSDs.[19] This finding implies that perforated synapses are more efficacious than nonperforated ones, and an augmentation of synaptic efficacy is believed to be essential for learning.[54,71,114]

A recent study from this laboratory demonstrated, however, that the total number of perforated axospinous synapses and their various morphological subtypes remained stable in the rabbit CA1 stratum radiatum 24 hours after trace eyeblink conditioning.[40] A plausible explanation for this negative finding is provided by data strongly suggesting that the addition of perforated synapses is an early and transitory event associated with the induction of the NMDA receptor-dependent form of hippocampal LTP,[39,115] which is widely regarded as a synaptic model of memory.[10] For example, the proportion of perforated synapses in the rat CA1 stratum radiatum increases at 30 min but returns to the control level at 60 min after potentiating stimulation of Schaffer collaterals in cultured hippocampal slices.[12,115] It is possible, therefore, that an increase in the number of perforated synapses was not observed by us 24 hours after trace eyeblink conditioning because it occurred at an earlier time point.

In any event, the formation of perforated axospinous synapses is a rapid process that is completed within 15 min or less.[87,115] In contrast, the assembly of new excitatory hippocampal synapses takes 1-2 hours following an initial contact of pre- and postsynaptic elements.[32] Comparison of the time frames of the two processes indicates that a rapid formation of perforated synaptic contacts results from a structural remodeling of existing synapses and not from synaptogenesis. In discussing such a remodeling, which may also occur as a result of behavioral learning, it is necessary to note that perforated axospinous synapses are morphologically heterogeneous and may be subdivided into several distinct subtypes (reviewed in ref. 33). This suggests that synaptic plasticity might be mediated by the conversion of some synaptic subtypes into others.[15,25,88]

Of special importance in this regard are the data demonstrating that an LTP-related increase in the number of perforated synapses is essentially due to the addition of their particular subtype distinguished by the presence of multiple, completely partitioned transmission zones.[37,116] In synapses belonging to this subtype (Fig. 5E), complete spine partitions provide barriers between two to four discrete transmission zones, each one being formed presynaptically by a separate axon terminal protrusion and delineated postsynaptically by a separate PSD segment.[33] It has been postulated that these synaptic contacts evolve from existing synapses.[33,34] The process is proposed to commence with an enlargement of small nonperforated synapses (Fig. 5A) and their conversion into atypically large ones (Fig. 5B). This is followed by the consecutive formation of perforated synapses that have initially a focal spine partition with a fenestrated PSD (Fig. 5C), then a sectional partition with a horseshoe-shaped PSD (Fig. 5D) and finally a complete partition(s) with a segmented PSD (Fig. 5E).

The structural features of the latter synaptic subtype suggest that it may be especially efficacious. Multiple transmission zones may function as independent units, provided that there is a barrier in the synaptic cleft preventing the diffusion of neurotransmitter and that each PSD segment is associated with an activated or newly inserted receptor cluster.[26,33] Under these conditions, an amplification of impulse transmission would be expected to take place. Mathematical modeling has also shown that the formation of multiple, completely partitioned transmission zones may facilitate synaptic transmission by altering calcium diffusion within the presynaptic bouton and enhancing thereby the probability of release.[42]

As was mentioned before, the number of axospinous synapses with multiple, completely partitioned transmission zones increases soon after LTP induction and then returns to the control level.[39,116] The reversal of the initial morphological change may reflect the transformation of such synapses into other synaptic subtypes. Three possibilities, which are not mutually exclusive,

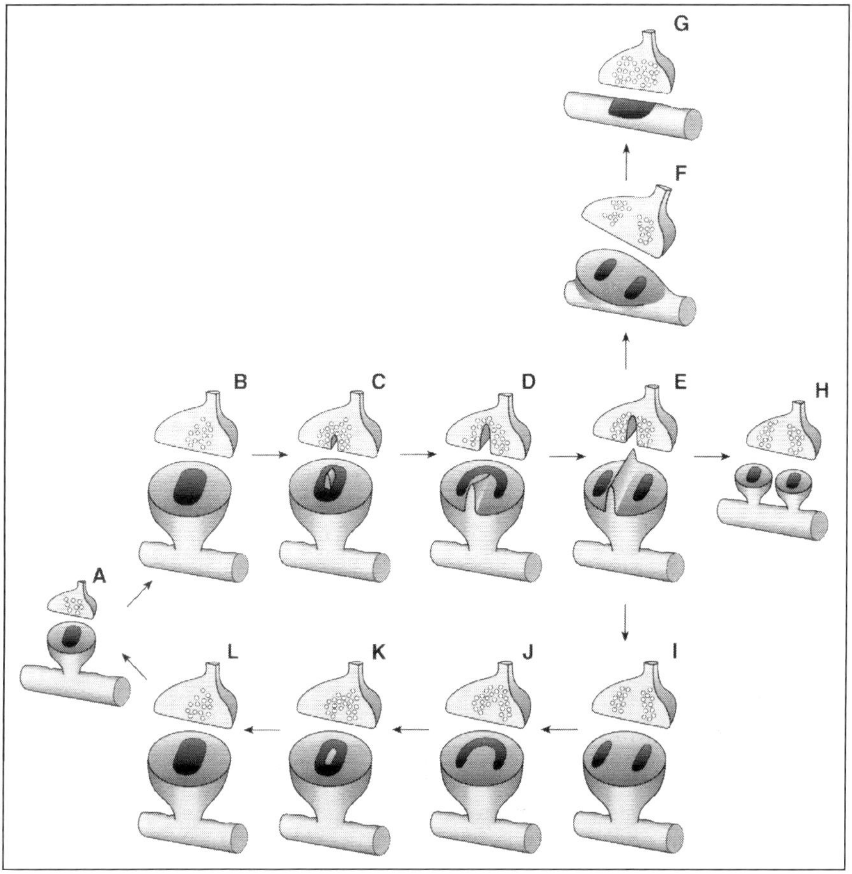

Figure 5. Diagram illustrating a hypothetical synapse restructuring that may underlie activation-dependent alterations in synaptic efficacy as explained in the text. The schematic shows the following synaptic subtypes: (A) typical (small) and (B, L) atypical (large) nonperforated axospinous synapses; perforated axospinous synapses that have (C) a focal spine partition and fenestrated PSD, (D) a sectional partition and horseshoe-shaped PSD, or (E) a complete partition(s) and segmented PSD; (F) an axospinous perforated synapse involving the postsynaptic spine that is partially retracted into a parent dendrite; (G) an asymmetrical axodendritic synapse; (H) two nonperforated axospinous synapses formed by a multiple-synapse bouton; perforated axospinous synapses that lack spine partitions and exhibit (I) a segmented, (J) horseshoe-shaped, or (K) fenestrated PSD. The sequence of synapse restructuring from A through B, C, D to E is proposed to be a rapid process that supports an initial maximal level of synaptic enhancement. The conversion of synapses from E through F to G and/or from E to H may underlie an enduring retention of synaptic enhancement at a relatively low level. Additionally, the consecutive transformation of synapses from E through I, J, K, L to A may account for the return of synaptic responses to the control level.

have to be considered in this respect. One of these is that some additional axospinous synapses with a segmented PSD and complete spine partitions (Fig. 5E) may be converted into asymmetrical axodendritic synapses (Fig. 5G). Such conversion is suggested by the observations that synapses of the latter kind are selectively increased in number during the maintenance phase of LTP and that there is a synaptic subtype (Fig. 5F), which appears to be transitional between axospinous and axodendritic junctions.[39] Another possibility alluded to in the preceding section was originally envisioned by Carlin and Siekevitz[15] who postulated that large segmented synapses (Fig. 5E) might split into small nonperforated ones (Fig. 5H). A temporal coincidence between the disassembly of synapses with multiple transmission zones and an increase in the proportion of MSBs after LTP induction[116] gives credence to this notion (but see ref. 30). Finally, it is also feasible that the transformation of additional synapses with multiple transmission zones into typical nonperforated synaptic contacts may be accomplished through the consecutive formation of perforated axospinous synapses that lack spine partitions and exhibit a segmented (Fig. 5I), horseshoe-shaped (Fig. 5J) and fenestrated (Fig. 5K) PSD.

These different patterns of the proposed remodeling of perforated synapses with multiple transmission zones (Fig. 5) may account for both the decay of synaptic responses (the sequence from E through I, J, K, L to A) and the sustained retention of a low level of synaptic enhancement (the sequences from E through F to G or from E to H) during LTP maintenance. Behavioral learning is accompanied, as is hippocampal LTP, by increases in the number of perforated axospinous synapses and of MSBs. It is tempting to speculate, therefore, that some forms of perforated synapse restructuring as outlined above may be characteristic of both phenomena.

Enlargement of Postsynaptic Densities following Learning: A Possible Morphological Correlate of the Conversion of Postsynaptically Silent Synapses into Functional Synapses

Recent work using time-lapse confocal imaging of hippocampal spines expressing a prominent PSD protein (PSD95) tagged with green fluorescent protein revealed that PSDs may rapidly (<15 min) expand or shrink.[79] It is not surprising, therefore, that previous electron microscopic studies indicated that the length of PSD profiles increases following acquisition of new behaviors.[20,55,112,119,120] In our experiments with trace eyeblink conditioning, we measured the length of PSD profiles on electron micrographs of consecutive sections through each synapse sampled from the CA1 stratum radiatum of the rabbit hippocampus to obtain estimates of the PSD area in conditioned and control animals.[40] These measurements showed that nonperforated axospinous synapses had a significantly larger PSD area in conditioned animals (30.3 ± 0.8 nm^2 x 10^3) than in pseudoconditioned (27.5 ± 0.9 nm^2 x 10^3) or unstimulated (26.1 ± 1.0 nm^2 x 10^3) controls.

The PSD contains signal transduction proteins, such as postsynaptic receptors and ion channels.[66,125] The recent discovery of "silent" hippocampal synapses leads to the suggestion that the conditioning-induced enlargement of nonperforated PSDs might reflect an addition of AMPA receptors. Electrophysiological experiments have revealed that a high proportion of synaptic contacts in the rat CA1 stratum radiatum exhibit functional NMDA receptors, but not functional AMPA receptors.[57,76] This makes such synapses postsynaptically silent, in that they do not generate a synaptic response to a release of a neurotransmitter, because NMDA receptor channels are blocked by extracellular magnesium at normal resting membrane potentials. Correspondingly, immunocytochemical studies have provided evidence for the existence of hippocampal synapses that exhibit only NMDA, but not AMPA, receptor immunoreactivity.[19,53,90,95,97,113] A lack of AMPA receptors, and not their inactive state, accounts for this phenomenon.[90,113] Notably, silent synapses acquire AMPA-type responses after LTP induction in the rat CA1 stratum radiatum,[57,76] indicating that they may be transformed into functional synaptic contacts due to an insertion of AMPA receptors into their PSDs. Based on these findings, it seems reasonable to hypothesize that the same synaptic modification may be induced by trace eyeblink conditioning (Fig. 6). This hypothesis appears to be inconsistent with the data showing that gene-targeted mice, which lack the AMPA receptor subunit GluR-A and have a reduced number of functional AMPA receptors, do not exhibit deficits in their spatial learning ability when tested in a water maze.[124] However, the trace eyeblink conditioning task taps the temporal, but not the spatial, domain of hippocampus-dependent memory function. Learning of the trace response may involve synaptic modifications that are different from those accompanying spatial learning.

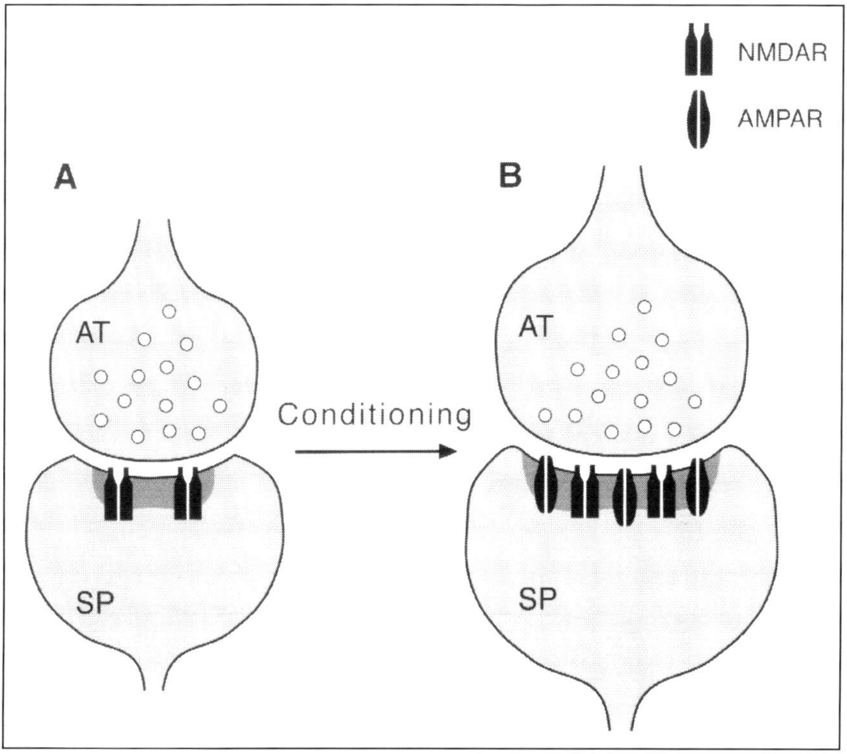

Figure 6. Diagram illustrating the hypothetical conversion of postsynaptically silent synapses into functional synapses associated with trace eyeblink conditioning. The schematic shows axospinous synaptic contacts between a presynaptic axon terminal (AT) and postsynaptic dendritic spine (SP). A. Nonperforated axospinous synapse with a small PSD that contains only NMDA receptors (NMDAR) but lacks AMPA receptors (AMPAR). Due to this, the synapse cannot generate a postsynaptic response and is postsynaptically silent. B. The PSD of this synapse is increased in size as a consequence of trace eyeblink conditioning. This structural modification is hypothesized to reflect the insertion of AMPAR, which makes the synapse functional.

Our hypothesis helps to explain why the enlargement of the PSD is a selective process that is characteristic only of nonperforated axospinous synapses and that does not involve any other synaptic subtype.[40] Especially relevant to this question are the data demonstrating that the ratio of AMPA to NMDA receptors is directly proportional to the PSD size in axospinous synapses from the rat CA1 stratum radiatum and that the AMPA receptor number regresses to zero when a PSD diameter is smaller than 180 nm.[90,97,113] These observations strongly suggest that the pool of silent axospinous synapses lacking AMPA receptors consists primarily of those synaptic junctions that have a relatively small, nonperforated PSD. Accordingly, our data show that only the smallest nonperforated PSDs, which probably lack AMPA receptors, are increased in their area by trace eyeblink conditioning (Fig. 7). Such a change may result from the insertion of AMPA receptors that are rapidly delivered to spines in response to synaptic NMDA receptor activation.[77,105] Provided that the spine delivery and insertion of AMPA receptors are triggered by associative learning, the observed enlargement of the smallest nonperforated PSDs may represent a structural correlate of the conversion of silent synapses into functional ones. Further ultrastructural studies using double labeling of AMPA and NMDA receptors are required to verify the validity of this supposition.

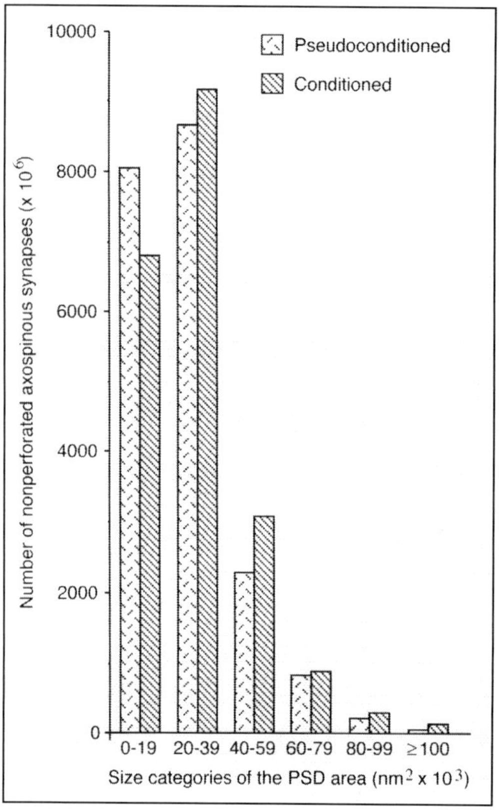

Figure 7. Comparison of pseudoconditioned and conditioned rabbits in terms of the distribution of nonperforated axospinous synapses according to the size of their PSD area (data from ref. 34). The total number of nonperforated synapses that had PSDs belonging to the smallest size category (PSD area < 20 $nm^2 \times 10^3$) was decreased in the conditioned group. Nonperforated synapses with PSDs that fell into all larger size categories were increased in number after conditioning. These data indicate that only the smallest nonperforated PSDs are enlarged in their area in conditioned rabbits.

Conclusions

Quantitative electron microscopic studies of the vertebrate brain have provided evidence that alterations in synaptic ultrastructure associated with learning and memory include both synaptogenesis and structural remodeling of synapses. A learning-induced net synaptogenesis manifested by an increase in the number of all synaptic contacts in a given brain region does not appear to be a ubiquitous phenomenon because it was detected in some studies, but not in others. Do these data tell us that only some varieties of learning and memory promote a net gain in synapse number or that such a gain takes place only during a certain phase of the acquisition/consolidation process?

Additionally, there are two known kinds of specific synaptogenesis, each one generating a special subset of synaptic connections that may be necessary for establishing a memory trace. One of these is confined to multiple-synapse bouton formation, which may complement a net synaptogenesis or occur independently of it. The addition of multiple-synapse boutons following learning may involve spine motility. Recent findings indicate that dendritic spines are highly dynamic structures capable of rapid motility. The neurobiological significance of this phenomenon remains unknown. Our models of multiple-synapse bouton formation raise the possibility that some existing or newly formed spines may relocate to single-synapse boutons activated by conditioning stimulation in order to synapse with them. The other kind of specific synaptogenesis, which may accompany learning-related adult neurogenesis, has not been directly demonstrated so far. Do both kinds of specific synaptogenesis represent a generalized phenomenon or are they unique for some particular forms of learning and memory?

Synapses have also been reported to undergo a learning-induced structural remodeling. The most demonstrative example of this is an enlargement of the PSD. Such a change was shown to selectively involve nonperforated axospinous synapses that had the smallest PSDs, usually lacking AMPA receptors, which probably made them postsynaptically silent. The increase in nonperforated PSD area was postulated to reflect the insertion of AMPA receptors and to represent a structural correlate of learning-associated conversion of postsynaptically silent synapses into functional ones. Is this hypothesis valid?

A learning-related increase in the proportion of perforated axospinous synapses is another alteration in synaptic ultrastructure resulting from structural remodeling of synapses: this alteration is completed within minutes, while the assembly of excitatory synapses takes at least 1 hour. It has been postulated that perforated axospinous synapses may evolve from nonperforated ones and then undergo a further restructuring. This would culminate in the formation of the perforated synaptic subtype distinguished by multiple transmission zones completely separated by spine partitions. The addition of synapses belonging to the latter subtype is known to occur early after the induction of hippocampal LTP, and it is possible that the same structural synaptic modification may be also associated with behavioral learning. Does the number of perforated synapses with multiple transmission zones increase soon after acquisition of new behaviors and is this change accompanied by a corresponding loss of nonperforated synapses?

The various learning-induced alterations in synaptic ultrastructure reported in the literature and reviewed here may be events that are interrelated or independent from each other. There are arguments in favor of the former possibility. For example, the enlargement of nonperforated PSDs promoted by learning may represent a step in the transformation of nonperforated axospinous synapses into perforated ones. Additionally, it is conceivable that specific synaptogenesis leading to MSB formation may be due to the splitting of axospinous synapses with multiple transmission zones. Such synapses transiently increase in number after LTP induction, and the reversal of this change coincides with MSB formation. Does behavioral learning elicit a similar sequence of structural synaptic alterations? The challenge of future research will be to address the questions posed above and to provide a better understanding of whether learning-associated net synaptogenesis, specific synaptogenesis and structural synapse remodeling are the links in a common chain of consecutive processes that underlie memory formation, consolidation and long-term storage.

Acknowledgements

Supported by grant 1 RO1 AG17139 from the National Institute on Aging.

References

1. Altman J, Das GD. Autoradiographic and histologic evidence of postnatal neurogenesis in rats. J Comp Neurol 1965; 124:319-335.
2. Alvarez-Buylla A, Kirn JR, Nottebohm F. Birth of projection neurons in adult avian brain may be related to perceptual or motor learning. Science 1990; 249:1444-1446.
3. Ambrogini P, Cuppini R, Cuppini C et al. Spatial learning affects immature granule cell survival in adult rat dentate gyrus. Neurosci Lett 2000; 286:21-24.
4. Andersen P, Soleng AF. Long-term potentiation and spatial training are both associated with the generation of new excitatory synapses. Brain Res Rev 1998; 26:353-359.
5. Angevine JB. Time of neuron origin in the hippocampal region. An autoradiographic study in the mouse. Exp Neurol Suppl 1965;2:1-70.
6. Bailey CH, Kandel ER. Structural changes accompanying memory storage. Annu Rev Physiol 1993; 55:397-426.
7. Barnea A, Nottebohm F. Seasonable recruitment of hippocampal neurons in adult free-ranging black-capped chickadees. Proc Natl Acad Sci USA 1994; 91:11217-11221.
8. Bayer S. Changes in the total number of dentate granule cells in juvenile and adult rats: a correlated volumetric and thymidine study. Exp Brain Res 1982; 46:315-323.
9. Black JE, Isaaks KR, Anderson BJ et al. Learning causes synaptogenesis, whereas motor activity causes angiogenesis, in cerebellar cortex of adult rats. Proc Natl Acad Sci USA 1990; 87:5568-5572.

10. Bliss TVP, Collingridge GL. A synaptic model of memory: long-term potentiation in the hippocampus. Nature 1993; 361:31-39.
11. Bourgeois J-P, Goldman-Rakic PS, Rakic P. Formation, elimination, and stabilization of synapses in the primate cerebral cortex. In: Gazzaniga MS, ed. Cognitive Neuroscience. A Handbook for the Field. 2nd Edition. Cambridge: MIT Press, 1999; 23-32.
12. Buchs P-A, Muller D. Induction of long-term potentiation is associated with major ultrastructural changes of activated synapses. Proc Natl Acad Sci USA 1996; 93:8040-8045.
13. Cameron HA, McKay RD. Restoring production of hippocampal neurons in old age. Nature Neurosci 1999; 2:894-897.
14. Cameron HA, Woolley CS, McEwen BS et al. Differentiation of newly born neurons and glia in the dentate gyrus of the adult rat. Neurosci 1993; 56:337-344.
15. Carlin PK, Siekevitz P. Plasticity in the central nervous system: do synapses divide? Proc Natl Acad Sci USA 1983; 80:3517-3521.
16. Cohen RS, Siekevitz. Form of the postsynaptic density. A serial section study. J Cell Biol 1978; 78:36-46.
17. Dailey ME, Smith SJ. The dynamics of dendritic structure in developing hippocampal slices. J Neurosci 1996; 16:2983-2994.
18. Del Bigio MR. Proliferative status of cells in adult human dentate gyrus. Microsc Res Tech 1999; 45:353-358.
19. Desmond NL, Weinberg RJ. Enhanced expression of AMPA receptor protein at perforated axospinous synapses. NeuroReport 1998; 9:857-860.
20. DeVoogt TJ, Nixdorf B, Nottenbohm F. Synaptogenesis and changes in synaptic morphology related to acquisition of new behavior. Brain Res 1985; 329:304-308.
21. Disterhoft JF, Coulter DA, Alkon DL. Conditioning-specific membrane changes of rabbit hippocampal neurons measured in vitro. Proc Natl Acad Sci USA 1986; 83:2733-2737.
22. Doubell TP, Stewart MG. Short term changes in the numerical density of synapses in the intermediate and medial hyperstriatum ventrale following one-trial passive avoidance learning. J Neurosci 1993; 13:2230-2236.
23. Dunaevsky A, Blazeski R, Yuste R et al. Spine motility with synaptic contact. Nature Neurosci 2001; 4:685-686.
24. Dunaevsky A, Tashiro A, Majewska A et al. Developmental regulation of spine motility in the mammalian central nervous system. Proc Natl Acad Sci USA 1999; 96:13438-13443.
25. Dyson SE, Jones DG. Synaptic remodelling during development and maturation: junction differentiation and slitting as a mechanisms of modifying connectivity. Dev Brain Res 1984;13:125-137.
26. Edwards FA. LTP—a structural model to explain the inconsistencies. Trends Neurosci 1995; 18:250-255.
27. Engert F, Bonhoeffer T. Dendritic spine changes associated with long-term synaptic plasticity. Nature 1999; 399:66-70.
28. Eriksson PS, Perfilieva E, Björk-Eriksson T et al. Neurogenesis in the adult human hippocampus. Nature Med 1998; 4:1313-1317.
29. Federmeier K, Kleim IA, Anderson BJ et al. Formation of double synapses in the cerebellar cortex of the rat following motor learning. Soc Neurosci Abstr 1994; 20:1435.
30. Fiala JC, Allwardt B, Harris KM. Dendritic spines do not split during hippocampal LTP or maturation. Nature Neurosci 2002; 5:297-298.
31. Fischer M, Kaech S, Knutti D et al. Rapid actin-based plasticity in dendritic spines. Neuron 1998; 20:847-854.
32. Friedman HV, Bresler T, Garner CC et al. Assembly of new individual excitatory synapses: time course and temporal order of synaptic molecule recruitment. Neuron 2000; 27:57-69.
33. Geinisman Y. Perforated axospinous synapses with multiple, completely partitioned transmission zones: probable structural intermediates in synaptic plasticity. Hippocampus 1993; 3:417-434.
34. Geinisman Y. Structural synaptic modifications associated with hippocampal LTP and behavioral learning. Cereb Cortex 2000; 10:952-962.
35. Geinisman Y, Berry RW, Disterhoft JF et al. Associative learning elicits the formation of multiple-synapse boutons. J Neurosci 2001; 21:5568-5573.
36. Geinisman Y, deToledo-Morrell L, Morrell F. Aged rats need a preserved complement of perforated axospinous synapses per hippocampal neuron to maintain good spatial memory. Brain Res 1986; 398:266-275.
37. Geinisman Y, deToledo-Morrell L, Morrell F et al. Structural synaptic correlate of long-term potentiation: formation of axospinous synapses with multiple, completely partitioned transmission zones. Hippocampus 1993; 3:435-446.

38. Geinisman Y, deToledo-Morrell L, Morrell F et al. Hippocampal markers of age-related memory dysfunction: behavioral, electrophysiological and morphological perspectives. Prog Neurobiol 1995; 45:223-252.
39. Geinisman Y, deToledo-Morrell L, Morrell F et al. Synapse restructuring associated with the maintenance phase of hippocampal long-term potentiation. J Comp Neurol 1996; 368:413-423.
40. Geinisman Y, Disterhoft JF, Gundersen HJG et al. Remodeling of hippocampal synapses following hippocampus-dependent associative learning. J Comp Neurol 2000; 417:49-59.
41. Geinisman Y, Gundersen HJG, Van der Zee E et al. Unbiased stereological estimation of the total number of synapses in a brain region. J Neurocytol 1996; 25:805-819.
42. Ghaffari-Farazi T, Liaw J-S, Berger TW. Consequence of morphological alterations on synaptic function. Neurocomputing 1999; 26/27:17-27.
43. Gilbert DB, Patterson TA, Rose SPR. Dissociation of brain sites necessary for registration and storage of memory for a one-trial passive avoidance task in the chick. Behav Neurosci 1991; 105:553-561.
44. Gould E, Gross CG. Neurogenesis in adult mammals: some progress and problems. J. Neurosci 2002; 22:619-623.
45. Gould E, Beylin A, Tanapat P et al. Learning enhances adult neurogenesis in the hippocampal formation. Nature Neurosci 1999; 2:260-265.
46. Gould E, Reeves AJ, Graziano MSA et al. Neurogenesis in the neocortex of adult primates. Science 1999; 286:548-552.
47. Gould E, Tanapat P, Hastings NB et al. Neurogenesis in adulthood: a possible role in learning. Tends Cogn Sci 1999; 3:186-192.
48. Greenough WT, Bailey CH. The anatomy of a memory: convergence of results across a diversity of tests. Trends Neurosci 1988; 11:142-147.
49. Greenough WT, West RW, DeVoogt TJ. Subsynaptic plate perforations: changes with age and experience in the rat. Science 1978; 202:1096-1098.
50. Gross CG. Neurogenesis in the adult brain: death of dogma. Nature Rev Neurosci 2000; 1:67-73.
51. Halpain S. Actin and the agile spine: how and why dendritic spines dance? Trends Neurosci 2000; 23:141-146.
52. Hastings NB, Gould E. Rapid extension of axons into CA3 region by adult-generated granule cells. J Comp Neurol 1999; 413:146-154.
53. He Y, Janssen WGM, Morrison JH. Synaptic coexistence of AMPA and NMDA receptors in the rat hippocampus: a postembedding immunogold study. J Neurosci Res 1998; 54:444-449.
54. Hebb DO. () The organization of behavior. New York: Wiley, 1975.
55. Horn G, Bradley P, McCabe BJ. Changes in the structure of synapses associated with learning. J Neurosci 1985; 5:3161-3168.
56. Hunter A, Stewart MG. A quantitative analysis of the synaptic development of the lobus parolfactorius of the chick (Gallus domesticus). Exp Brain Res 1989; 78:425-434.
57. Isaac JTR, Nicoll RA, Malenka RC. Evidence for silent synapses: implications for the expression of LTP. Neuron 1995; 15:427-434.
58. Jones DG, Harris RJ. An analysis of contemporary morphological concepts of synaptic remodeling in the CNS: perforated synapses revisited. Rev Neurosci 1995; 6:177-219.
59. Jones TA, Chu CJ, Grande LA et al. Motor skills training enhances lesion-induced structural plasticity in the motor cortex of adult rats. J Neurosci 1999; 19:10153-10163.
60. Jones TA, Klintsova AY, Kilman VL et al. Induction of multiple synapses by experience in the visual cortex of adult rats. Neurobiol Learn Mem 1997; 68:13-20.
61. Kaplan MS, Bell DH. Mitotic neuroblasts in the 9-day-old and 11-month-old rodent hippocampus. J Neurosci 1984; 4:1429-1441.
62. Kaplan MS, Hinds JW. Neurogenesis in the adult brain: electron microscopic analysis of light radioautographs. Science 1977; 197:1092-1094.
63. Kempermann G. Why new neurons? Possible functions for adult hippocampal neurogenesis. J. Neurosci 2002; 202:635-638.
64. Kempermann G, Kuhn HG, Gage FH. More hippocampal neurons in adult mice living in an enriched environment. Nature 1997; 386:493-495.
65. Kempermann G, Kuhn HG, Gage FH. Experience-induced neurogenesis in the senescent dentate gyrus. J. Neurosci 1998; 18:3206-3212.
66. Kennedy MB. Signal-processing machines at the postsynaptic density. Science 2000; 290:750-754.
67. Kim JJ, Clark RE, Thompson RF. Hippocampectomy impairs the memory of recently, but not remotely, acquired trace eyeblink conditioned responses. Behav Neurosci 1995; 109:195-203.
68. Kirn JR, Alvarez-Buylla A, Nottebohm F. Production and survival of projection neurons in a forebrain vocal center of adult male canaries. J Neurosci 1991; 11:1756-1762.

69. Kleim JA, Lussing E, Schwarz ER et al. Synaptogenesis and fos expression in the motor cortex of the adult rat after motor skill learning. J Neurosci 1996; 16:4529-4535.
70. Kleim JA, Vij K, Ballard DH et al. Learning-dependent synaptic modifications in the cerebellar cortex of the adult rat persist for at least four weeks. J Neurosci 1997; 17:717-721.
71. Konorski J. Conditioned Reflexes and Neuron Organization. Cambridge: Cambridge Univ Press, 1948.
72. Korkotian E, Segal M. Regulation of dendritic spine motility in cultured hippocampal neurons. J Neurosci 2001; 21:6115-6124.
73. Kornack DR, Rakic P. Continuation of neurogenesis in the hippocampus of the adult macaque monkey. Proc Natl Acad Sci USA 1999; 96:5768-5773.
74. Kuhn HG, Dickinson-Anson H, Gage FH. Neurogenesis in the dentate gyrus of the adult rat: age-related decrease of neuronal progenitor proliferation. J Neurosci 1996; 16:2027-2033.
75. Lee DW, Miyasato LE, Clayton NS. Neurobiological bases of spatial learning in the natural environment: neurogenesis and growth in the avian and mammalian hippocampus. NeuroReport 1998; 9:R15-27.
76. Liao D, Hessler NA, Malinow R. Activation of postsynaptically silent synapses during pairing-induced LTP in CA1 region of hippocampal slice. Nature 1995; 375:400-404.
77. Liao D, Scannevin RH, Huganir R. Activation of silent synapses by rapid activity-dependent synaptic recruitment of AMPA receptors. J Neurosci 2001; 21:6008-6017.
78. Maletic-Savatic M, Malinov R, Svoboda K. Rapid dendritic morphogenesis in CA1 hippocampal dendrites induced by synaptic activity. Science 1999; 283:1923-1927.
79. Marrs GS, Green SH, Dailey ME. Rapid formation and remodeling of postsynaptic densities in developing dendrites. Nature Neurosci 2001; 4:1006-1013.
80. Matus A. Actin-based plasticity in dendritic spines. Science 2000; 290:754-758.
81. Markakis EA, Gage F. Adult-generated neurons in the dentate gyrus send axonal projections to field CA3 and are surrounded by synaptic vesicles. J Comp Neurol 1999; 406:449-460.
82. Moser M-B, Trommald M, Andersen P. An increase in dendritic spine density on hippocampal CA1 pyramidal cells following spatial learning in adult rats suggests the formation of new synapses. Proc Natl Acad Sci USA 1994; 91:12673-12675.
83. Moser M-B, Trommald M, Egeland T et al. Spatial training in a complex environment and isolation alter the spine distribution differentially in rat CA1 pyramidal cells. J Comp Neurol 1997; 380:373-381.
84. Moyer JR Jr, Deyo RA, Disterhoft JF. Hippocampectomy disrupts trace eye-blink conditioning in rabbits. Behav Neurosci 1990; 104:243-252.
85. Moyer JR Jr, Thompson LT, Disterhoft JF. Trace eyeblink conditioning increases CA1 excitability in a transient and learning-specific manner. J Neurosci 1996; 16:5536-5546.
86. Muller D, Toni N, Buchs P-A. Spine changes associated with long-term potentiation. Hippocampus 2000; 10:596-604.
87. Neuhoff H, Roeper J, Schweizer M. Activity-dependent formation of perforated synapses in cultured hippocampal neurons. Eur J Neurosci 1999; 11:4241-4250.
88. Nieto-Sampedro MS, Hoff W, Cotman CW. Perforated postsynaptic densities: probable intermediates in synapse turnover. Proc Natl Acad Sci USA 1982; 79:5718-5722.
89. Nilsson M, Perfilieva E, Johansson U et al. Enriched environment increases neurogenesis in the adult dentate gyrus and improves spatial memory. J Neurobiol 1999; 39:569-578.
90. Nusser Z, Lujan R, Laube G et al. Cell type and pathway dependence of synaptic AMPA receptor number and variability in the hippocampus. Neuron 1998; 21:545-559.
91. O'Malley A, O'Connell C, Regan CM. Ultrastructural analysis reveals avoidance conditioning to induce a transient increase in hippocampal dentate spine density in the 6 h post-training period of consolidation. Neurosci 1998; 87:607-613.
92. O'Malley A, O'Connell C, Murphy KJ et al. Transient spine density increases in the mid-molecular layer of hippocampal dentate gyrus accompany consolidation of a spatial learning task in the rodent. Neurosci 2000; 99:229-232.
93. Patel SN, Clayton NS, Krebs JR. Spatial learning induces neurogenesis in the avian brain. Behav Brain Res 1997; 89:115-128.
94. Peters A, Kaiserman-Abramof IR. The small pyramidal neuron of the rat cerebral cortex. The synapses upon dendritic spines. Z Zellforsch 1969; 100:487-506.
95. Petralia RS, Esteban JA, Wang Y-X et al. Selective acquisition of AMPA receptors over postnatal development suggests a molecular basis for silent synapses. Nature Neurosci 1999; 2:31-36.
96. Power JM, Thompson LT, Moyer, JR Jr et al. Enhanced synaptic transmission in CA1 hippocampus after eyeblink conditioning. J Neurophysiol 1997; 78:1184-1187.

97. Racca C, Stephenson FA, Streit P et al. NMDA receptor content of synapses in stratum radiatum of the hippocampal CA1 area. J Neurosci 2000; 20:2512-2522.
98. Rakic P. Adult neurogenesis in mammals: an identity crisis. J Neurosci 2002; 22:614-618.
99. Ramón y Cajal S. Neue Darstellung vom histologischen Bau des Centralnervensystem. Arch Anat Physiol 1893; 319-428.
100. Rietze R, Poulin P, Weiss S. Mitotically active cell that generate neurons and astrocytes are present in multiple regions of adult mouse hippocampus. J Comp Neurol 2000; 424:397-408.
101. Rusakov DA, Davies HA, Harrison E et al. Ultrastructural synaptic correlates of spatial learning in rat hippocampus. Neuroscience 1997; 80:69-77.
102. Scharff C. Chasing fate and function of new neurons in adult brains. Curr Opin Neurobiol 2000; 10:774-783.
103. Segal M, Andersen P. Dendritic spines shaped by synaptic activity. Curr Opin Neurobiol 2000; 10:582-586.
104. Seki T, Arai Y. Age-related production of new granule cells in the adult dentate gyrus. NeuroReport 1995; 6:2479-2482.
105. Shi S-H, Hayashi Y, Petralia RS et al. Rapid spine delivery and redistribution of AMPA receptors after synaptic NMDA receptor activation. Science 1999; 284:1811-1816.
106. Shors TJ, Miesegaes G, Beylin A et al. Neurogenesis in the adult is involved in the formation of trace memories. Nature 2001; 410:372-376.
107. Solomon PR, Van der Schaaf EV, Thompson RF et al. Hippocampus and trace conditioning of the rabbit's classically conditioned nictitating membrane response. Behav Neurosci 1986; 100:729-744.
108. Sorra KE, Harris KM. Occurrence and three-dimensional structure of multiple synapses between individual radiatum axons and their target pyramidal cells in hippocampal area CA1. J Neurosci 1993; 13:3736-3748.
109. Squire LR, Zola SM. Amnesia, memory and brain systems. Philos Trans R Soc Lond Ser B 1997; 352:1663-1997.
110. Stanfield BB, Trice JE. Evidence that granule cells generated in the dentate gyrus of adult rats extend axonal projections. Exp Brain Res 1988; 72:399-406.
111. Star EN, Kwiatkowsky DJ, Murthy VN. Rapid turnover of actin in dendritic spines and its regulation by activity. Nature Neurosci 2002; 5:239-246.
112. Stewart MG, Rusakov DA. Morphological changes associated with stages of memory formation in the chick following passive avoidance training. Behav Brain Res 1995; 66:21-28.
113. Takumi Y, Ramírez-León V, Laake P et al. Different modes of expression of AMPA and NMDA receptors in hippocampal synapses. Nature Neurosci 1999; 2:618-624.
114. Tanzi E. I fatti i le induzione nell'odierna istologia del sistema nervoso. Riv Sperim Freniatr Med Leg 1893; 19:419-472.
115. Toni N, Buchs P-A, Nikonenko I et al. LTP promotes formation of multiple spine synapses between a single axon terminal and a dendrite. Nature 1999;402:421-425.
116. Toni N, Buchs P-A, Nikonenko I et al. Remodeling of synaptic membranes after induction of long-term potentiation. J Neurosci 2002; 1:6245-6251.
117. Van Praag H, Kempermann G, Gage FH. Running increases cell proliferation and neurogenesis in the adult mouse dentate gyrus. Nature Neurosci 1999; 2:266-270.
118. Van Praag, H, Schinder AF, Christie BR et al. Functional neurogenesis in the adult hippocampus. Nature 2002; 415:1030-1034.
119. Van Reempts J, Dikova M, Webrouck L et al. Synaptic plasticity in rat hippocampus associated with learning. Behav Brain Res 1992; 51:179-183.
120. Vrensen G, Nunes Cardozo J. Changes in size and shape of synaptic connections after visual training: an ultrastructural approach to synaptic plasticity. Brain Res 1981; 218:79-97.
121. Wallenstein GV, Eichenbaum H, Hasselmo ME. The hippocampus as an associator of discontiguous events. Trends Neurosci 1998; 21:317-323.
122. Wong WT, Wong ROL. Rapid dendritic movements during synapse formation and rearrangement. Curr Opin Neurobiol 2000; 10:188-124.
123. Yankova M, Hart SH, Woolley CS. Estrogen increases synaptic connectivity between single presynaptic inputs and multiple postsynaptic CA1 pyramidal cells: a serial electron-microscopic study. Proc Natl Acad Sci USA 2001; 98:3525-3530.
124. Zamanillo D, Sprengel R, Hvalby Ø et al. Importance of AMPA receptors for hippocampal synaptic plasticity but not for spatial learning. Science 1999; 284:1805-1811.
125. Ziff EB. Enlightening the postsynaptic density. Neuron 1997; 19:1163-1174.

Chapter 6.2

Cell Adhesion Molecules

Ciaran M. Regan

Abstract

The molecular cascade of events associated with hippocampal processing of information for long-term storage is a time-limited event. Learning sets in motion neural processes that continue to evolve after training, a phenomenon known as consolidation. The consolidation process has been proposed to involve the translation of neural activity into enduring synaptic change by a cascade of sequential molecular steps involving gene induction, increased protein synthesis and synaptic growth mediated, in part, by cell recognition systems. However, evidence for dendritic remodelling being a mechanism of memory consolidation is still in its infancy. There is no clear consensus as to whether spine or synapse frequency change accompanies memory formation. This may arise from a failure to consider temporal change in synaptic dynamics during information processing. By contrast, there is significant evidence that the temporal orchestration of cell adhesion molecule function is an integral process for memory consolidation. An attractive possibility is that transmembrane cell adhesion molecule expression is modulated in an activity-dependent manner that permits rapid alteration in synapse structure and/or efficacy. Evidence exists to correlate cell adhesion molecule function with spine and synapse formation following learning but also for their increased synaptic expression in the absence of dendritic remodelling. There appears to be a need to correlate the temporal dynamics of cell adhesion molecule function to synapse structure if a morphological correlate of learning is to be established.

Introduction

An important problem in the neurobiology of memory is whether cellular mechanisms of learning and memory include the formation of new synapses or the remodelling of existing ones. To elucidate this problem, numerous studies have examined alterations in the number and structure of hippocampal dendritic spines and synapses (see Geinisman et al in this book). This brain region, along with associated cortical structures in the temporal lobe, has been implicated in holding and processing information destined for consolidation as long term memory within the neocortex.[2] Moreover, this consolidation process has been demonstrated to involve a molecular cascade of events within the medial temporal lobe in which enhanced neural activity activates gene transcription, protein synthesis and synaptic reorganisation.[5,37,70,72] Activity-dependent synapse selection is attractive in its simplicity and most likely will provide a basis to understand neuroplastic events subservient to behavioural adaptation in the adult animal.

Most excitatory synapses in the adult brain are located on the bulbous heads of dendritic spines,[45] where they occur in vast numbers, estimated to be in the order of 10^{14} for the human cerebral cortex.[121] Dendritic filopodia are initially involved in the generation of shaft synapses, which later develop into spines.[28] This distinctive structure of dendritic spines is specialized by

From Messengers to Molecules: Memories Are Made of These, edited by Gernot Riedel and Bettina Platt. ©2004 Eurekah.com and Kluwer Academic / Plenum Publishers.

underlying microfilaments composed of actin that contrasts with the cytoplasmic microtubules that are dominant in the dendritic shaft.[68] The actin microfilaments are in close association with the postsynaptic density, a membrane specialization thought to be important for the clustering of postsynaptic receptors and ion channels and for the assembly of the postsynaptic signalling machinery.[42]

Cell adhesion molecule-regulated cell-cell recognition and/or signal transduction events play an integral role in dendritic spine development and synapse elaboration.[9] Cell-cell adhesion across the synaptic cleft is thought to hold the presynaptic active zone and the postsynaptic density in close register.[36] As yet, there is no clear consensus on the role of cell recognition systems in dendritic growth and synapse restructuring. Such molecules may function to promote dendritic growth and synapse formation and provide mechanisms that restrict elaboration of the dendritic arbor.[15] Moreover, in the adult hippocampus, region-specific cell genesis has been suggested to contribute to the fine-tuning of the neural structure throughout life by invoking a developmental replay of cell-cell recognition events.[55]

The challenge ahead lies in identifying the network of molecular interactions, including cell adhesion molecules (CAMs), that together produce the activity-induced morphological change, such as dendrite and synapse formation, as the greatest pitfall in research on activity-related plasticity is to overstate the relevance of single identified regulators.

Is Net Synapse Formation a Correlate of Learning?

Results from Studies Employing Long-Term Potentiation

Neural activity plays an important role in the emergence of new spines, the stabilization of existing spines, and changes in spine morphology. However, the extent to which dendritic motility depends on afferent innervation or synaptic activity is under debate. Much of this debate centres on the use of a cellular model of learning—long-term potentiation (LTP). The essence of LTP is a remarkably persistent enhancement of synaptic responses resulting from brief, repetitive activation of an excitatory afferent monosynaptic pathway by high frequency trains of electrical impulses.[11] LTP has been studied most intensively at excitatory hippocampal synapses formed by Schaffer collaterals on CA1 pyramidal cells or by perforant path fibres on dentate gyrus granule cells.

Recently developed imaging techniques that permit time-lapse observations have allowed an unprecedented understanding of dendritic spine dynamics. Studies using coordinated patterns of activity, such as the tetanic stimulation required for the induction of LTP, results in the growth of CA1 dendritic protrusions that are dependent on NMDA receptor-mediated neurotransmission.[27,65] Further, these dendritic spines have been shown to persist for >45 min in the absence of additional evoked activity.

Synaptic modifications are also reported to follow the induction of LTP in the perforant path. For example, following 4 daily tetanizations of the perforant path the ratio of perforated to non-perforated synapses in the mid-molecular layer of the dentate gyrus becomes increased by approximately 23% one hour after the last tetanization (Genisman; this book).[38,39] These studies suggest that spine splitting may be one of the mechanisms by which LTP is maintained at the synaptic level. Studies on activated synapses, identified by the detection of calcium precipitates in the spine apparatus, support this suggestion. Activated synapses appear transiently in the 30-60min post-tetanic period, exhibit perforations and, in the ensuing 2-3h period, an enduring three-fold increase in the frequency of activated CA1 presynaptic elements with at least two dendritic spines contacting the same axon terminal.[124]

By contrast, however, Sorra and Harris[110] have failed to observe new synapse formation at 2h following tetanization of the hippocampal CA1 region. Moreover, in separate studies utilizing a single tetanization of the perforant path, no change in spine frequency was observed in the outer-molecular layer of the dentate gyrus when examined 24h after potentiation.[105] These results suggest that LTP does not cause an overall formation of new synapses.

Alternatively, presynaptic terminals may move in concert with dendritic spines or synapses might be broken and reformed rapidly, in a matter of seconds to minutes, so that structural changes are relatively balanced across the dendritic arbor. This phenomenon has been observed for Purkinje cells in cerebellar slices and retinal ganglion cells in culture.[24,132]

Results from Studies Employing Behavioural Paradigms

A number of electron microscopic studies available in the literature report that the numerical density of synapses is increased in relevant areas of the vertebrate brain as a consequence of learning.[5,43] Change in hippocampal CA1 synapse numerical density has been observed at 1h following acquisition of a brightness discrimination task and in the intermediate hyperstriatum ventrale following avoidance conditioning in the chick.[20,130] In the hippocampal dentate gyrus, spine frequency changes occur at 6h following passive avoidance or water maze training that are transient and return to basal levels at 72h.[85,86] Motor skill learning also results in an increased synapse density in the rodent cerebellar cortex[10] that persists for at least 4 weeks.[59]

However, trace eyeblink conditioning in rabbits, a hippocampus-dependent task that is acquired over 5-10 days, failed to reveal a change in total CA1 synapse number when examined at 24h following the final conditioning session.[39] Moreover, extended water maze training, over a five day period, failed to elicit changes in either axospinous or axodendritic synapses in either the CA1 or dentate gyrus molecular layer at six days after the last training session.[97] Similar results have been obtained with rats subjected to a one-way active avoidance procedure over 3 days as no change in total synapse number in the molecular layer of the hippocampal denate gyrus could be observed.[127] Taken together, these results would suggest that the formation of long-term memory does not necessarily involve enduring synapse formation.

A possible reason for the disparity in these data sets may relate to the dynamic nature of the learning phenomenon. Stereological analyses performed at different time points during memory consolidation have suggested spine frequency changes to be transient in nature.[85,86] Similar studies have also provided evidence for temporally-regulated and region-specific changes in synapse number. Following avoidance conditioning in the chick, for example, transient synapse number increases are first observed in the intermediate hyperstriatum ventrale and, subsequently, in the lobus parolfactorius at 24h following training.[20,50]

Do Cell Adhesion Molecules Have a Role in Learning?

CAM Structure and Function

The spine surface exhibits an array of proteins many of which span the membrane thereby permitting bidirectional communication.[134] These proteins include ligand- and second messenger-gated ion channels, G-protein-coupled receptors and cell adhesion molecules that coexist in an organized but dynamic array in the postsynaptic density. Spectrin and spectrin-like proteins link these proteins to actin filaments that mingle with larger intermediate filaments in the spine head.

Analogous to the postsynaptic density, the presynaptic cytomatrix is organized into active zones that determine sites of synaptic vesicle fusion and recycling. These active zones are likely to spatially restrict proteins involved in vesicle docking and fusion, such as the SNAP (soluble N-ethylmaleimide-sensitive factor attachment protein) receptor complex and synaptotagamin. In addition, the active zone includes cytoskeletal proteins, such as members of the membrane-associated guanylate kinase (MAGUK) superfamily, as well as other multidomain proteins that are tightly associated with synaptic junctions. These are thought to function as adaptor proteins involved in localizing and assembling pre- and postsynaptic signalling complexes that also include cell adhesion molecules.[36]

The cell adhesion molecules (CAMs) localised to the synaptic complex include representatives from the cadherin, immunoglobulin and integrin superfamilies and, in addition, the neurexins and neuroligands.[9] The cadherins mediate Ca^{2+}-dependent cell-cell adhesion that is mainly *trans* homophilic. They have a single pass membrane domain that interacts with the actin cytoskeleton via a group of proteins termed the catenins. The integrins are formed from two non-covalently linked heterodimeric proteins that have single pass membrane domains that can link to actin by a variety of adaptor proteins, such as talin or vinculin. The extracellular domains of the integrins form a ligand binding site that requires divalent cations and interacts with defined matrix protein sequences (such as –Arg-Gly-Asp-).[4] The immunoglobulin superfamily is characterized by the presence of Ig-like domains and fibronectin repeats in their extracellular domains.

These adhesion molecules can interact in a hetero- or homophilic manner between cells (*trans* interactions) or in the same plane of the membrane (*cis* interactions) (Fig. 1). Many CAMs have single pass membrane domains, some of which interact with actin, or attach to the cell surface via a glycosylphosphatidylinositol linkage. The neurexins and neuroligands are located to the pre- and postsynaptic membranes, respectively, and mediate Ca^{2+}-dependent cell-cell adhesion through their extracellular N-terminal domains.

Unfortunately, the precise mechanism(s) of CAM signalling still remain to be resolved.[18,41,56,67,131]

CAMs and Learning-Induced Synaptic Plasticity

Across species and paradigms, numerous examples exist to support a role for CAMs in the synaptic plasticity that accompanies behavioural adaptation, however, these examples relate mainly to members of the immunoglobulin and integrin superfamilies. Targeted mutation of the *volado* gene product, an α-integrin subunit that is enriched in mushroom body neurons, impairs olfactory-avoidance learning in *Drosophila*.[44] Moreover, integrin receptor antagonists block the induction of LTP in the rat hippocampal formation[112] and intra-dentate injections of the integrin-associated protein impair avoidance conditioning.[49] With respect to the immunoglobulin superfamily, antibody interventive studies have implicated both NCAM and L1 to be necessary for the induction of LTP and avoidance conditioning and spatial learning paradigms.[1,21,64,87,92,102,103,123]

Disruption of CAM function does not always impair synaptic plasticity. For example, pretreatment of hippocampal slices with antibodies raised against the extracellular domain of either N-cadherin or E-cadherin have no effect on basal synaptic properties but significantly reduce LTP.[120] By contrast, deletion of cadherin-11 enhances LTP.[66] The basis for these opposing effects on LTP that are obtained by blocking N-cadherin or cadherin-11 is unknown but it suggests that different cadherins may have unique roles in synaptic signalling. An alternative explanation, as offered by Sanes and Lichtman,[100] is that many molecules are likely to be required to mediate the multi-step process of LTP that has been described to have at least four phases: initial, early, intermediate and late.[11] This aspect is further discussed below.

The case for members of the immunoglobulin superfamily being involved in the synaptic plasticity associated with memory formation is more substantial. Mice with a targeted null mutation in the L1 gene exhibit impaired spatial learning.[33] Moreover, adult mice with a targeted deletion of the NCAM gene exhibit behavioural abnormalities that include altered exploratory activity accompanied by increased anxiety and intermale aggression.[113,114] Not surprisingly, mice deficient for NCAM function fail to sustain LTP and exhibit an impaired spatial learning ability.[16,17] However, by crossing NCAM-deficient mice with those generated to over-express the NCAM180 isoform, the isoform enriched in synapses of postmitotic neurons,[89] Schachner and colleagues could rehabilitate the behavioural abnormalities induced by lack of NCAM expression.[115] This singular experiment clearly provides evidence of CAM function regulating behaviour at the level of the synapse.

Do Cell Adhesion Molecules Have a Temporal Role in Learning?

The molecular cascade of events associated with hippocampal processing of information for long-term storage is a time-limited event (Fig. 2).[58,72,135] Learning sets in motion neural processes that continue to evolve after training, a phenomenon known as consolidation. The consolidation process has been proposed to involve the translation of neural activity into

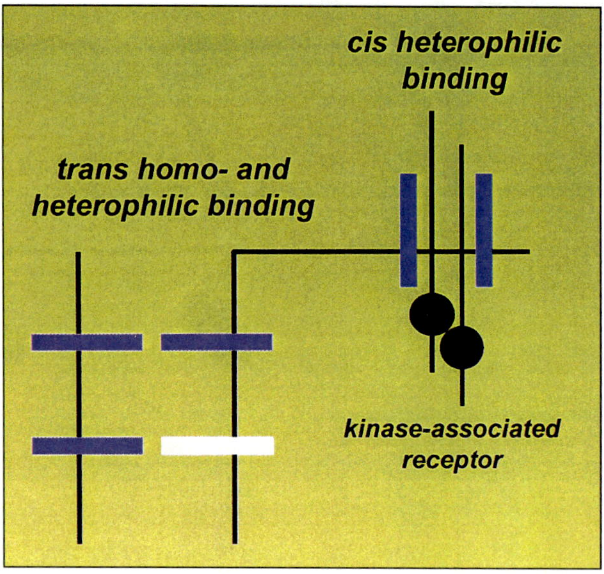

Figure 1. Proposed modes of CAM-CAM interactions and cell signalling. The blue and white bars represent individual CAMs that traverse the plasma membrane (black line). Trans homo- and heterophilic binding occurs between two apposing plasma membranes. By contrast, cis heterophilic binding involves CAM-induced clustering of transmembrane receptors with intrinsic kinase domains (black line with filled circle).

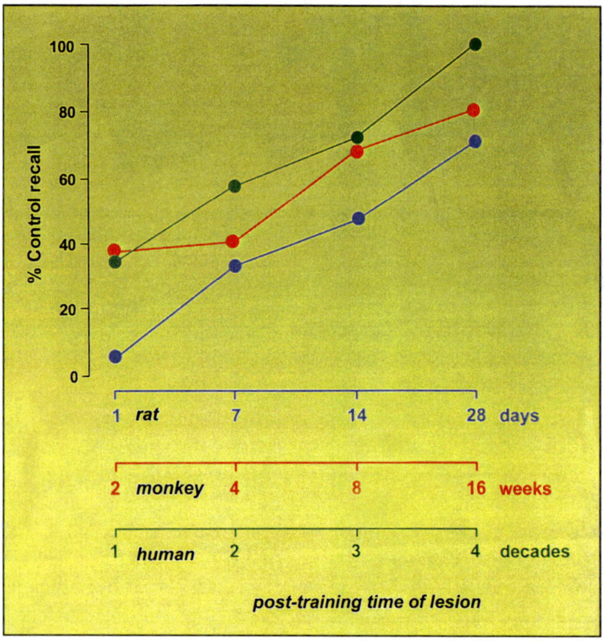

Figure 2. Temporal involvement of hippocampal formation in memory consolidation. This figure illustrates memory recall in rodents, primates and humans with hippocampal lesions at increasing times following paradigm acquisition. The data is expressed as percent recall in those with hippocampal lesions as compared to the control group. The data was adapted from information contained in refs. 58,117,135.

enduring synaptic change by a cascade of sequential molecular steps involving gene induction, increased protein synthesis and synaptic growth mediated, in part, by cell recognition systems.[6] As time passes, a more permanent memory develops which is independent of the hippocampal formation and most likely located in the neocortex. Such studies have contributed to a model of memory consolidation in which the medial temporal lobe memory system serves as a temporary reservoir prior to the eventual storage of long term memory within the cortex.[2]

At least two temporally and mechanistically distinct processes contribute to activity-dependent synaptic plasticity, which lasts from tens of minutes to hours or more. After its induction, LTP passes through a 30-60min period during which unknown chemistries make it progressively less vulnerable to disruption or reversal. For example, hypoxia of duration just sufficient to transiently block synaptic responses completely eliminates LTP when applied within the first minutes after induction but is without effect 30min later.[3] Also, LTP reversal is readily obtained with theta pattern stimulation when applied within the first minutes and becomes progressively less effective over 30min post-induction.[111]

Little is known about the neurochemical events responsible for consolidation. Protein and mRNA synthesis inhibitors are reported to dissipate LTP beginning several hours after induction.[80] Whether this reflects a need for newly synthesised molecules in order for potentiation to enter a late stage, as opposed to sustaining already established LTP, is an unresolved issue. In any event, the effects obtained with protein synthesis inhibitors develop too slowly to be part of the stabilisation process that begins within the first minute after induction.

Certain classes of adhesion receptors may be better candidates. Several of the more than 20 known integrins are expressed by hippocampal neurons and at least some of these are concentrated in the synapses[91] and many of these have been implicated in the stabilization of LTP. For example, function blocking antibodies to the $\alpha5$ or $\beta1$ integrin subunits have no effect on initial potentiation but significantly reduce it 45min later whereas antibodies to the α(v) or $\alpha2$ integrin subunits are without effect.[14] Moreover, function blocking with antibodies to the $\alpha3$ integrin subunit stabilize slow decay of LTP.[62] Members of the immunoglobulin superfamily also appear to be involved in the maintenance of LTP as judged by the increased expression of neuroplastin-65 and NCAM180 in the late phase following the induction of LTP.[104,109]

There is also evidence for the temporal involvement of CAMs in the consolidation processes that follow behavioural adaptation. For example, intraventricular administration of anti-L1 effectively blocks the acquisition of avoidance conditioning when administered just prior to training and, specifically, in the 5-8h and 15-18h post-training periods.[103,123] Similarly, anti-NCAM has been found to induce amnesia of avoidance conditioning and odour discrimination paradigms when administered in the 6-8h post-training period.[21,96,102] More recently, NCAM has been demonstrated to play a role in the acquisition of passive avoidance learning. Intraventricular infusions of a synthetic peptide ligand of NCAM (C3) strongly inhibited recall of a passive avoidance response in adult rats when given during training or in the 6-8 hours post-training period.[30] This peptide has an affinity for the IgI domain and the capability of disrupting NCAM-mediated neurite outgrowth in vitro.[93]

This unique amnesic action of the C3 peptide has also been related to disrupted NCAM internalization immediately following training. In the 3-4h period following training NCAM180, the synapse-associated isoform, becomes down-regulated in the dentate gyrus of the hippocampal formation. This effect is mediated by ubiquitination and prevented by C3 peptide administration during training. Thus, these findings indicate NCAM to be involved in the acquisition and in the later 6-8h post-training consolidation of a passive avoidance response in the rat. Moreover, this study provided the first in vivo evidence for CAM internalization in learning, an observation presaged by studies on ApCAM, the *Aplysia* NCAM homologue, that has been demonstrated to rapidly down-regulate and become internalized in an in vitro model of long-term sensitization of the gill and siphon-withdrawal reflex.[7,69] Internalization of CAMs may be a general mechanism for the dynamic modulation of synaptic plasticity during memory consolidation. Both NCAM and L1 are also endocytosed by clathrin-dependent pathways[54,73] and growth cone elaboration has been associated with endocytosis-dependent recycling of L1.[53]

Can Cell Adhesion Molecules Reveal Memory Pathway?

Studies of memory-related modifications in the mammalian brain are severely restricted by the difficulty of identifying the network in which specific memory-induced changes occur, and by the prospect that these cellular changes are too subtle to be distinguished from the background of previously acquired memories. This conundrum has, in part, been overcome by the development of a unique probe to a CAM glycosylation variant.

NCAM Polysialylation

A significant post-translational modification of NCAM involves the attachment of large homopolymers of α2,8 polysialic acid (PSA), a modification that is specific to NCAM in the mammalian brain (Fig. 3).[95,98] This post-translational modification of NCAM with PSA is believed to modulate NCAM-mediated homophilic adhesion and/or signal transduction events by virtue of its extensive negative charge .[47,131,133] By attenuating adhesion forces and modulating overall cell surface interactions, NCAM PSA has the potential to orchestrate dynamic changes in the shape and movement of cells, as well as their processes.[57,94]

The development of a monoclonal antibody to extended chains of PSA found on meningococcus group B polysaccharides[95] provided an unparalleled immunohistochemical tool for mapping PSA expression in the mammalian central nervous system. Moreover, a bacteriophage-derived endoneuraminidase (endo-N), that cleaves polysialosyl units associated with the K1 capsular antigen of certain *Escherichia coli* strains, provided the ideal control for these immunohistochemical procedures.[29,128] Both anti-PSA and endo-N require a minimum recognition size of at least 10-12 residues[46,71] and, as a consequence, only NCAM with extended chains of polysialic acid is recognised.

Using these tools the distribution of NCAM PSA in the adult brain has been mapped and found to be associated primarily with those brain regions that undergo structural reorganization and synaptic plasticity, such as the olfactory bulb, hypothalamus and hippocampal formation and its associated structures in the medial temporal lobe.[12,31,32,74,81,83,106] This convergent set of data has been used to suggest that NCAM PSA supports structural plasticity in adult nervous system.[57] NCAM PSA has also been implicated in activity-dependent synaptic remodelling. The hypothalmo-neurohypophysial system expresses high levels of NCAM PSA throughout life and undergoes extensive morphological synaptic plasticity in response to physiological stimuli that is dependent on the surface expression of polysialylated NCAM.[48,82] For example, microinjection of endo-N blocks cell surface expression of NCAM PSA and the synapse increase that occurs in response to lactation or osmotic stimulation.[122]

Consistent with this view is that removal of PSA with endoneuraminidase-N has been found to interfere with the induction and maintenance of hippocampal LTP and to produce spatial learning deficits in the Morris water maze paradigm.[8,76] Moreover, the frequency of NCAM polysialylated hippocampal neurons at the dentate infragranular border transiently increase at 10-12h following acquisition of either spatial or conditioned avoidance tasks.[22,31,77] These frequency changes are learning specific as they are not observed in animals rendered amnesic with either scopolamine or propofol.[23,84] Similar coincident frequency increase of polysialylated neurons occurs in layer II of the entorhinal and perirhinal cortex following spatial learning.[32,83]

Regulation of NCAM Polysialylation State

Addition of PSA to NCAM takes place in the trans Golgi compartment and is entirely catalyzed by the polysialyltransferases STX (ST8SiaII) and PST (ST8SiaIV).[26,60,61,63,79] PSA expression appears to be inducible as recent studies have demonstrated inhibition of calcium-independent PKCδ to be inversely related to NCAM polysialylation state both in vitro and in the hippocampal dentate gyrus during memory consolidation, an effect associated with the Golgi membrane fraction.[34,35] However, the role of polysialyltransferase in other forms of synaptic plasticity is not so clear-cut. Mice with a targeted deletion of the gene encoding PST-1 exhibit an impairment of Schaffer collateral-CA1 LTP but not mossy fibre-CA3 LTP.[26] These regional differences suggest that NCAM, but not PSA, is likely to be important for LTP in the hippocampal CA3 region. While PKC is known to be involved in learning and LTP, it is mainly the calcium-dependent isoforms that have been investigated and not in relation to the regulation of NCAM polysialylation.[125]

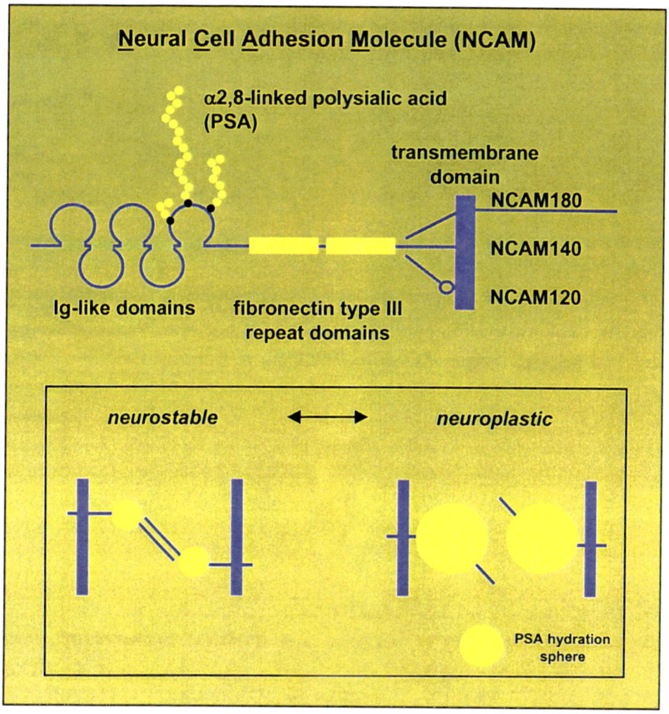

Figure 3. Structure and mode of NCAM-NCAM interactions.. The upper panel illustrates the basic structural features of NCAM. The three major splice variants – NCAM 180, NCAM 140 and NCAM 120 – associate with the membrane by a single transmembrane domain with extensive cytoplasmic domain, a single transmembrane with short cytoplasmic domain and by a glycosylphosphosphatidylinositol linkage, respectively. The lower panel illustrates the influence of increasing the PSA hydration sphere (yellow) on NCAM-NCAM interactions (dark blue line) and the degree of plasma membrane apposition (light blue line).

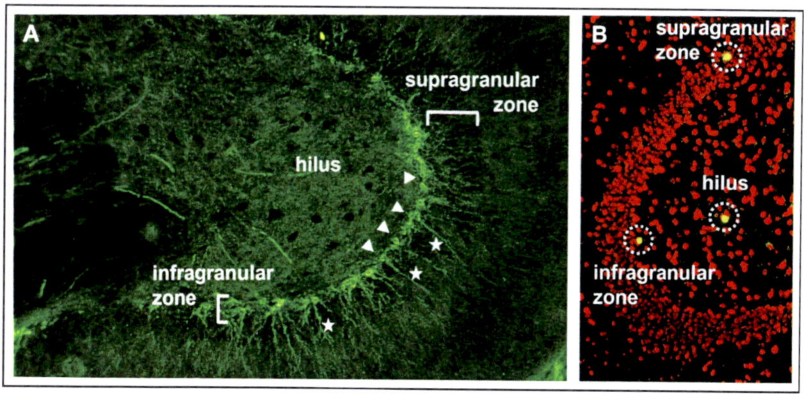

Figure 4. Distribution of polysialylated and BrdU-labelled neurons in the hippocampal dentate gyrus in the 12h post-training period following acquisition of a passive avoidance paradigm. Panel A illustrates (ls) PSA-labelled neurons at the infragranular layer (arrows) and their dendrites extending through the supragranular layer and into the molecular layer (stars). In Panel B the distribution of BrdU-labelled cells (circled) in the infra- and supra-granular layers and hilus of the hippocampal dentate gyrus is shown. In both panels the thickness of sections illustrated is 12μm. The images are adapted from ref.31.

It would appear, however, that polysialyltransferase activity is not regulated by enzyme levels alone but is directly controlled by a cellular signalling cascade that encompasses a PKCδ-dependent phosphorylation event involved in inhibition of polysialyltransferase activity.[35] The cellular signalling cascade appears to be NMDA-dependent as inhibition of this receptor system prevents the rapid decrease in NCAM PSA expression observed following electrical signalling.[13,78] Although cell activation seems to be an important regulatory factor in PSA synthesis other biosynthetic-independent mechanisms may also be involved. These most likely include the intracellular trafficking NCAM PSA by endocytotic cycling, however the molecular details of these mechanisms remain to be elaborated.[57]

The HNK1 Carbohydrate Epitope

NCAM PSA is not the only glycosylation mechanism associated with CAM function in the synaptic plasticity of memory consolidation. For example, many CAMs carry the HNK-1 carbohydrate structure, which was recognised first by a monoclonal antibody raised against human natural killer cells, hence, the acronym. The HNK-1 carbohydrate has been shown to consist of a minimal epitope of 3'-sulfated glucuronic acid attached to a lactosaminyl residue that is involved in the homophilic binding of NCAM.[101] The HNK-1 epitope also appears to be involved in synaptic plasticity associated with LTP and inhibitory avoidance learning in both fish and rodents.[87,99,116] However, the HNK-1 epitope lacks the advantage of NCAM PSA for defining the pathways associated with memory consolidation.

What about Neurogenesis in Learning?

Bromodeoxyuridine (BrdU) incorporation studies suggest that a proportion of the polysialylated neurons observed in the adult hippocampal dentate gyrus are recently generated[106] and necessary for odour discrimination.[40] Moreover, their depletion, following neurotoxic lesions, results in impaired trace conditioning of the eyeblink response, a hippocampal-dependent paradigm.[107] However, the learning-induced transient increase in polysialylated denate neuron frequency during the 12h post-training period does not appear to be associated with increased neurogenesis.[31] Indeed, the specificity of BrdU as a neurogenic marker has been recently questioned.[90] In contrast to ^{3}H-thymidine, BrdU is not a marker for cell division but rather a marker for DNA synthesis. This would explain the presence of BrdU-positive cells in layers other than those of the infragranular zone (Fig. 4).[31] Moreover, the cells of the infragranular zone are remarkably heterogenous[75] and, to date, there has been no attempt to relate BrdU or PSA labelling with their neurotransmitter phenotype. These polysialylated neurons receive synaptic input and are distinguished from the mature neurons, located in the superficial granule cell layer, by showing paired pulse facilitation and having a lower threshold for induction of LTP.[126,129] In my view, the most likely functional significance of learning-induced transient increases in the polysialylated neurons of the dentate infragranular zone is to facilitate the dendritic elaboration of recently acquired neurons in response to the acquisition of novel behavioural repertoires.

Perspective

The function of dendritic remodelling as a mechanism for the consolidation of information is still in its infancy. There is still no clear consensus as to whether spine or synapse frequency change accompanies memory formation. Aside from the consideration of differing tissue fixation and stereological procedures, the majority of studies have failed to take account of the fleeting nature of hippocampal information processing prior to its eventual consolidation within the neocortex.[117] Moreover, the diverse nature of the learning paradigms employed is also likely to result in temporal phase shifts of synaptic dynamics. For example, synapse frequency change occurs within minutes following the induction of LTP[124] as compared to the changes wrought at 6-8h following the acquisition of a spatial learning paradigm.[86] Moreover, 24h after the induction of LTP there is little evidence of spine frequency change or at 6 days following extensive training in the water maze.[97,104] Studies on spine and synapse frequency change following behavioural adaptation must also be cognisant of the separate pathways and sub-regions of the medial temporal lobe that are involved in the processing of information. Spine frequency change has been observed in the dentate gyrus but not CA1 region of the hippocampal formation following successful acquisition of tasks dependent on this brain region.[37,86]

The question of cell signalling from the extracellular environment to the nucleus of the neuron is central to how the brain modifies its structure and function to learn and remember. In this regard, ideal transmembrane signalling molecules should control the connections formed between neurons and govern cytoskeletal dynamics in a manner that regulates their morphology. CAMs are cell surface macromolecules that control cell-cell interactions during development of the nervous system by regulating such processes as neuronal adhesion and migration, neurite outgrowth, fasciculation, synaptogenesis and intracellular signalling.

There is now substantial evidence that CAMs are intimately associated with the molecular cascade that underpins the synaptic plasticity of memory consolidation and, moreover, their functions follow a defined sequence of events. The temporal involvement of CAMs in memory consolidation is dramatically illustrated by interventive studies using anti-L1 in chick avoidance conditioning.[103,123] In this paradigm L1 function is required at three discrete time periods over an 18h period. Again, with NCAM, and its PSA glycosylated variant, discrete periods of function are described for its role in acquisition, in a 6h post-training period that coincides with increased synapse production and, later at the 12h period.[21,22,30,31,32,77,83,96,102] The fact that antibodies are ineffective in the intervening periods further supports the notion that CAMs have specific temporal actions in the molecular cascade of memory consolidation. Moreover, these experiments are consistent with the need for their extensive involvement in the processing of information within the medial temporal lobe prior to its re-distribution to multiple neocortical sites.

What is unclear about the role of CAMs in memory consolidation, however, is their suggested contribution to synapse connectivity pattern that is proposed to be generated by activity-dependent spine growth and de novo synapse formation. Indeed, the significance of de novo synapse formation as a functional correlate of memory consolidation is far from clear, as some studies have failed to equate this morphological associate with learning.[37] Aside from the obvious consideration of differing tissue fixation procedures, failure to observe morphological change may relate to the temporal aspects of memory consolidation, as is exemplified by the CAM interventive studies. Although spine and synapse formation have been correlated with periods of CAM function in some studies,[22,85,86,102] others have failed to find an increased spine number but, rather, an increased frequency of synapses expressing greater levels of CAM.[104] In this regard, it is interesting to suggest that the increased NCAM labelling observed in the chick paradigm of avoidance conditioning, at a time sensitive to memory disruption with anti-L1 and anti-NCAM, may simply reflect selective retention of a synaptic population normally eliminated during the precocial development of this animal.[102,103,108]

What the future seems to hold is the need to associate increased CAM functions, such as signalling, and their role in de novo synapse formation. Too often CAM function has been extrapolated from in vitro studies and their role in the adult nervous system may be significantly different. This view is supported by the numerous models of targeted deletion of CAM function that have no significant developmental outcome but frequently exhibit learning deficits.[16,33] Another example that supports a potential role for CAMs in synapse remodelling, as opposed to de novo synapse formation, comes from studies of peptide ligands with an affinity for the homophilic binding domain of NCAM.[30] In vitro, such peptides induce neuritogenesis,[93] a function that might predict an inherent cognition-enhancing property if the model of de novo synapse formation has relevance to memory consolidation. By contrast, these peptides produce a profound amnesia when administered during training or in the 6-8h post-training period in which anti-NCAM and spine formation is observed to occur following training.[30] A more parsimonious interpretation could be that these peptide ligands disrupt pre- to postsynaptic signalling necessary for synapse strengthening.

An attractive possibility is that transmembrane CAMs transmit activity-dependent change of intracellular-signalling across the synapse hence modulating the strength of cell-cell binding and thereby rapidly altering synaptic structure and efficacy. Although the expression patterns of N-cadherin, NCAM and L1 can be regulated by distinct patterns of action potentials,[51,52] their adhesive interactions across the synaptic cleft and in synaptic function has remained elusive. If expression of cell adhesion molecules can be regulated by activity, then newly synthesised molecules may participate in the formation and modification of synapses. For example, at certain synapses in the central nervous system, pre- and postsynaptic adhesion is mediated by N-cadherin that, upon depolarisation, dimerizes and becomes markedly protease resistant.[119] Other in vitro studies have demonstrated activity-dependent strengthening of synapses to be dependent on integrins and/or the relative levels of post-synaptic, as opposed to pre-synaptic, NCAM.[14,19]

In conclusion, CAMs can no longer be viewed as the structural scaffold of the synapse but as active participants that are involved in all aspects of its plasticity, including trans-synapse signalling and structural modification. Understanding these trans-synapse signalling mechanisms will be crucial in relating extracellular adhesive interactions to gene transcription and protein synthesis that is required for long-term changes in synaptic function.

Acknowledgements

The author's work reviewed herein is currently supported by the EU 5th Framework Programme (QLRT-1999-02187) and basic research grants from Enterprise Ireland and the Health Research Board of Ireland.

References

1. Alexinsky T, Przybyslawski J, Mileusnic R et al. Antibody to day-old chick glycoprotein produces amnesia in adult rats. Neurobiol Learn Mem 1997; 67:14-20.
2. Alvarez P, Squire LR. Memory consolidation and the medial temporal lobe: A simple network model. Proc Natl Acad Sci USA 1994; 91:7041-7045.
3. Arai A, Larson J, Lynch G. Anoxia reveals a vulnerable period in the development of long-term potentiation. Brain Res 1990; 511:353-357.
4. Bahr BA, Lynch G. Purification of an Arg-Gly-Asp selective matrix receptor from brain synaptic plasma membranes. Biochem J 1992; 281:137-142.
5. Bailey CH, Kandel ER. Structural changes accompany memory storage. Annu Rev Physiol 1993; 55: 397-426.
6. Bailey CH, Bartsch D, Kandel ER. Toward a molecular definition of long-term memory storage. Proc Natl Acad Sci USA 1996; 93:13445-13452.
7. Bailey CH, Chen M, Keller F et al. Serotonin-mediated endocytosis of apCAM: An early step of learning-related synaptic growth in Aplysia. Science 1992; 256:645-649.
8. Becker CG, Artola A, Gerardy-Schahn R et al. The polysialic acid modification of the neural cell adhesion molecule is involved in spatial learning and hippocampal long-term potentiation. J Neurosci Res 1996; 45:143-152.
9. Benson DL, Schnapp LM, Shapiro L et al. Making memories: cell adhesion molecules in synaptic plasticity. Trends Cell Biol 2000; 10:473-482.
10. Black JE, Isaacs KR, Anderson BJ et al. Learning causes synaptogenesis, whereas motor activity causes angiogenesis, in cerebellar motor cortex of adult rats. Proc Natl Acad Sci USA 1990; 87:5568-5572.
11. Bliss TVP, Collingridge GL. A synaptic model of memory: long-term potentiation in the hippocampus. Nature 1993; 361:31-39.
12. Bonfanti L, Olive S, Poulain DA et al. Mapping of the distribution of polysialylated neural cell adhesion molecule throughout the central nervous system of the adult rat: An immunohistochemical study. Neurosci 1992 49:419-436.
13. Bouzioukh F, Tell F, Jean A et al. NMDA receptor and nitric oxide synthase activation regulate polysialylated neural cell adhesion molecule expression in adult brainstem synapses. J Neurosci 2001; 21:4721-4730.
14. Chun D, Gall CM, Bi X et al. Evidence that integrins contribute to multiple stages in the consolidation of long-term potentiation in rat hippocampus. Neurosci 2001; 105:815-829.
15. Cline HT. Dendritic arbor development and synaptogenesis. Curr Opin Neurobiol 2001; 11:118-126.

16. Cremer H, Lange R, Christoph A et al. Inactivation of the N-CAM gene in mice results size reduction of the olfactory bulb and deficits in spatial learning. Nature 1994; 367:455-459.
17. Cremer H, Chazal G, Carleton A et al. Long-term but not short-term plasticity at mossy fiber synapses is impaired in neural cell adhesion molecule-deficient mice. Proc Natl Acad Sci USA 1998; 95:13242-13247.
18. Crossin KL, Krushel LA. Cellular signaling by neural cell adhesion molecules of the immunoglobulin superfamily. Dev Dyn 2000; 218:260-279.
19. Dityatev A, Dityatev G, Schachner M. Synaptic strength as a function of post- versus presynaptic expression of the neural cell adhesion molecule NCAM. Neuron 2000; 26:207-217.
20. Doubell TP, Stewart MG. Short-term changes in the numerical density of synapses in the intermediate and medial hyperstriatum ventrale following one-trial passive avoidance training in the chick. J Neurosci 1993; 13:2230-2235.
21. Doyle E, Nolan P, Bell R et al. Intraventricular infusions of anti-neural cell adhesion molecules in a discrete post-training period impair consolidation of a passive avoidance response in the rat. J Neurochem 1992; 59:1570-1573.
22. Doyle E, Bell R, Regan CM. Hippocampal NCAM180 transiently increases sialylation during the acquisition and consolidation of a passive avoidance response in the adult rat. J Neurosci Res 1992; 31:513-523.
23. Doyle E, Regan CM. Cholinergic and dopaminergic agents which inhibit a passive avoidance response attenuate paradigm-specific increases in NCAM sialylation state. J Neural Transm 1993; 92:33-49.
24. Dunaevsky A, Blazeski R, Yuste R et al. Spine motility with synaptic contact. Nature Neurosci 2001 4:685-686.
25. Eckhardt M, Muhlenhoff M, Bethe A et al. Molecular characterisation of eukaryotic polysialyltransferase-1. Nature 1995; 373:715-718.
26. Eckhardt M, Bukalo O, Chazal G et al. Mice deficient in the polysialyltransferase ST8SiaIV/PST-1 allow discrimination of the roles of neural cell adhesion molecule protein and polysialic acid in neural development and synaptic plasticity. J Neurosci 2000; 20:5234-5244.
27. Engert F, Bonhoeffer T. Dendritic spine changes associated with hippocampal long-term synaptic plasticity. Nature 1999; 399:66-70.
28. Fiala JC, Feinberg M, Popov V et al. Synaptogenesis via dendritic filopodia in developing hippocampal area CA1. J Neurosci 1998; 18:8900-8911.
29. Finne J, Makela PH Cleavage of polysialosyl units of brain glycoproteins by a bacteriophage endosialidase. J Biol Chem 1985; 260:1265-1270.
30. Foley AG, Hartz BP, Gallagher HC et al. A synthetic peptide ligand of neural cell adhesion molecule (NCAM) IgI domain prevents NCAM internalisation and disrupts passive avoidance learning. J Neurochem 2000; 74:2607-2613.
31. Fox GB, O'Connell AW, Murphy KJ et al. Memory consolidation induces a transient and time-dependent increase in the frequency of neural cell adhesion molecule polysialylated cells in the adult rat hippocampus. J Neurochem 1995; 65:2796-2799.
32. Fox GB, Fichera G, Barry T et al. Consolidation of passive avoidance learning is associated with transient increases of polysialylated neurons in layer II of the rat medial temporal cortex. J Neurobiol 2000; 45:135-141.
33. Fransen E, D'Hooge R, van Camp G et al. L1 knockout mice show dilated ventricles, vermis hypoplasia and impaired exploratory patterns. Hum Mol Genet 1998; 7:999-1009.
34. Gallagher HC, Odumeru OA, Regan CM. Regulation of neural cell adhesion molecule polysialylation state by cell-cell contact and protein kinase C delta. J Neurosci Res 2000; 61:636-645.
35. Gallagher HC, Murphy KJ, Foley AG et al. Protein kinase C delta regulates neural cell adhesion molecule polysialylation state in the rat brain. J Neurochem 2001; 77:425-434.
36. Garner CC, Kindler S, Gundelfinger ED. Molecular determinants of presynaptic active zones. Curr Opin Neurobiol 2000; 10:321-327.
37. Geinisman Y. Structural synaptic modifications associated with hippocampal LTP and behavioral learning. Cerebral Cortex 2000; 10:952-962.
38. Geinisman Y, deToledo-Morrell L, Morrell F. Induction of long-term potentiation is associated with an increase in the number of axospinous synapses with postsynaptic densities. Brain Res 1991; 566:77-88.
39. Geinisman Y, Disterhoft JF, Gundersen HJG et al. Remodelling of hippocampal synapses following hippocampus-dependent associative learning. J Comp Neurol 2000; 417:49-59.
40. Gheusi G, Cremer H, McLean H et al. Importance of newly generated neurons in the adult olfactory bulb for odor discrimination. Proc Natl Acad Sci USA 2000; 97:1823-1828.
41. Giancotti FG, Ruoslahti E. Integrin signaling. Science 1999; 285:1028-1032.

42. Grant SGN, O'Dell TJ. Multiprotein complex signaling and the plasticity problem. Curr Opin Neurobiol 2001; 11:363-368.
43. Greenough WT, Bailey CH. The anatomy of memory: convergence of results across a diversity of tests. Trends Neurosci 1988; 11:142-147.
44. Groteweil MS, Beck CDO, Wu KH et al. Integrin-mediated short-term memory in Drosophila. Nature 1998; 391:455-460.
45. Harris KM, Kater SB. Dendritic spines: cellular specializations imparting both stability and flexibility to synaptic function. Annu Rev Neurosci 1994; 17:341-371
46. Hayrinen J, Bitter-Suermann D, Finne J. Interaction of meningococcal group B monoclonal antibody and its Fab fragment with alpha 2-8-linked sialic acid polymers: requirement of a long oligosaccharide segment for binding. Mol Immunol 1989; 26: 523-529.
47. Hoffman S, Edelman GM. Kinetics of homophilic binding by embryonic and adult forms of the neural cell adhesion molecule. Proc Natl Acad Sci USA 1983; 89:5762-5766.
48. Hoyk Zs, Parducz A, Theodosis DT. The highly sialylated isoform of the neural cell adhesion molecule is required for estradiol-induced morphological plasticity in the adult arcuate nucleus. Eur J Neurosci 2001; 13:649-656.
49. Huang A-M, Wang HL, Tang YP et al. Expression of integrin-associated protein gene associated with memory formation in rats. J Neurosci 1998; 18:4305-4313.
50. Hunter A, Stewart MG. Long-term increases in the numerical density of synapses in the chick lobus parolfactorius after passive avoidance training. Brain Res 1993; 605:251-255.
51. Itoh K, Stevens B, Schachner M et al. Regulated expression of the neural cell adhesion molecule L1 by specific patterns of neural impulses. Science 1995; 270:1369-1372.
52. Itoh K, Ozaki M, Stevens B et al. Activity-dependent regulation of N-cadherin in DRG neurons – differential regulation of N-cadherin, N-CAM and L1 by distinct patterns of action potentials. J Neurobiol 1997; 33:735-748.
53. Kamiguchi H, Lemmon V. Recycling of the cell adhesion molecule L1 in axonal growth cones. J Neurosci 2000; 20:3676-3686.
54. Kamiguchi H, Long KE, Pendergast M et al. The neural cell adhesion molecule L1 intteracts with the AP-2 adaptor and is endocytosed via the clathrin-mediated pathway. J Neurosci 1998; 18:5311-5321.
55. Kempermann G, van Pragg H, Gage FH. Activity-dependent regulation of neuronal plasticity and self repair. Prog Brain Res. 2000; 127:35-48.
56. Kennedy MB. Signal-processing machines at the post-synaptic density. Science 2000; 290:750-754.
57. Kiss JZ, Rougon G. Cell biology of sialic acid. Curr Opin Neurobiol 1997; 7:640-646.
58. Kim JJ, Fanselow M. Modality-specific retrograde amnesia of fear. Science 1992; 256:675-676.
59. Kleim JA, Vij K, Ballard DH et al. Learning-dependent synaptic modifications in the cerebellar cortex of the adult rat persist for at least four weeks. J Neurosci 1997; 17:717-721.
60. Kojima N, Yoshida Y, Tsuji S. A developmentally regulated member of the sialyltransferase family (ST8Sia II, STX) is a polysialic acid synthase. FEBS Letts 1995; 373:119-122.
61. Kojima N, Tachida Y, Yoshida Y et al. Characterisation of mouse ST8Sia II (STX) as a neural cell adhesion molecule-specific polysialic acid synthase. J Biol Chem 1996; 271:19457-19463.
62. Kramar EA, Bernard JA, Gall CM et al. Alpha3 integrin receptors contribute to the consolidation of long-term potentiation. Neurosci 2002; 110:29-39.
63. Livingston BD, Paulson JC. Polymerase chain reaction cloning of a developmentally regulated member of the sialyltransferase gene family. J Biol Chem 1993; 268:11504-11507.
64. Lüthi, A, Laurent, J-P, Figurov A et al. Hippocampal long-term potentiation and neural cell adhesion molecules L1 and NCAM. Nature 1994; 372:777-779.
65. Maletic-Savatic M, Malinow R, Svoboda K. Rapid dendritic morphogenesis in CA1 hippocampal dendrites induced by synaptic activity. Science 1999; 283:1923-1927.
66. Manabe T, Togashi H, Uchida N et al. Loss of cadherin-11 adhesion receptor enhances plastic changes in hippocampal synapses and modifies behavioral responses. Mol Cell Neurosci 2000; 15:534-546.
67. Maness PF, Beggs HE, Klinz SG et al. Selective neural cell adhesion molecule signaling by Src family tyrosine kinases and tyrosine phosphatases. Perspect Dev Neurobiol 1996; 4:169-181.
68. Matus A. Actin-based plasticity in dendritic spines. Science 2000; 290:754-758.
69. Mayford M, Barzilai A, Keller F et al. Modulation of an NCAM-related adhesion molecule with long-term synaptic plasticity in Aplysia. Science 1992; 256:638-644.
70. McGaugh JL. Memory – a century of consolidation. Science 2000; 287:248-251.
71. Michon F, Brisson J-R, Jennings JH. Conformational differences between linear α(2-8)-linked homo-oligosaccharides and the epitope of the group B Meningococcal polysaccharide. Biochemistry 1987; 26:8399-8405.

72. Milner B, Squire LR, Kandel ER. Cognitive neuroscience and the study of memory. Neuron 1998; 20:445-468.
73. Miñana R, Duran JM, Tomas M et al. Neural cell adhesion molecule is endocytosed via a clathrin-dependent pathway. Eur J Neurosci 2001; 13:749-756.
74. Miragall F, Kadman G, Husmann M et al. Expression of cell adhesion molecules in the olfactory system of the adult mouse: presence of the embyronic form of NCAM. Dev Biol 1988; 129:516-531.
75. Mott DD, Turner DA, Okazaki MM et al. Interneurons of the dentate-hilus border of the rat dentate gyrus: Morphological and electrophysiological heterogeneity. J Neurosci 1997; 17:3990-4005.
76. Muller D, Wang C, Skibo G et al. PSA-NCAM is required for activity-induced synaptic plasticity. Neuron 1996; 17:413-422.
77. Murphy KJ, O'Connell AW, Regan CM. Repetitive and transient increases in hippocampal neural cell adhesion molecule polysialylation state following multi-trial spatial training. J Neurochem 1996; 67:1268-1274.
78. Nacher J, Rosell DR, Alonso-Llosa G et al. NMDA receptor antagonist induces a long-lasting increase in the number of proliferating cells, PSA-NCAM-immunoreactive granule neurons and radial glia in the adult rat dentate gyrus. Eur J Neurosci 2001; 13:512-520.
79. Nakayama J, Fukuda M. A human polysialyltransferase directs in vitro synthesis of polysialic acid. J Biol Chem 1996; 271:1829-1832.
80. Nguyen PV, Abel T, Kandel ER. Requirement of a critical period of transcription for induction of a late phase of LTP. Science 1994; 265:1104-1107.
81. Ní Dhúill CM, Fox GB, Pittock SJ et al. Polysialylated neural cell adhesion molecule expression in the dentate gyrus of the human hippocampal formation from infancy to old age. J Neurosci Res 1999; 55:99-106.
82. Nothias F, Vernier P, von Boxberg Y et al. Modulation of NCAM polysialylation is associated with morphofunctional modifications in the hypothalamo-neurohypophysial system during lactation. Eur J Neurosci 1997; 9:1553-1565.
83. O'Connell AW, Fox GB, Murphy KJ et al. Spatial learning activates neural cell adhesion molecule polysialylation in a cortico-hippocampal pathway within the medial temporal lobe. J Neurochem 1997; 68:2538-2546.
84. O'Gorman DA, O'Connell AW, Murphy KJ et al. Nefiracetam prevents propofol-induced anterograde and retrograde amnesia in the rodent without compromising quality of anesthesia. Anesthesiology 1998; 89:699-706.
85. O'Malley A, O'Connell C, Regan CM. Ultrastructural analysis reveals avoidance conditioning to induce a transient increase in hippocampal dentate spine density in the 6h post-training period of consolidation. Neuroscience 1998; 87:607-613.
86. O'Malley A, O'Connell C, Murphy KJ et al. Transient spine density increases in the mid-molecular layer of hippocampal dentate gyrus accompany consolidation of a spatial learning task in the rodent. Neuroscience 2000; 99:229-232.
87. Pradel G, Schachner M, Schmidt R. Inhibition of memory consolidation by antibodies against cell adhesion molecules after active avoidance conditioning in zebrafish. J Neurobiol 1999; 39:197-206.
88. Pradel G, Schmidt R, Schachner M. Involvement of L1.1 in memory consolidation after active avoidance conditioning in zebrafish. J Neurobiol 2000; 43:389-403.
89. Pollerberg GE, Burridge K, Krebs KE et al. The 180-kD component of the neural cell adhesion molecule N-CAM is involved in cell-cell contacts and cytoskeleton-membrane interactions. Cell Tiss Res 1997; 250:227-236
90. Rakic P. Adult neurogenesis in mammals: an identity crisis. J Neurosci 2002; 22:614-618.
91. Rohrbough J, Groteweil MS, Davis RL et al. Integrin-mediated regulation of synaptic morphology, transmission and plasticity. J Neurosci 2000; 20:6868-6878.
92. Rønn LCB, Bock E, Linnemann D et al. NCAM-antibodies modulate induction of long-term potentiation in rat hippocampal CA1. Brain Res 1995; 677:145-151.
93. Rønn LCB, Olsen M, Østergaard S et al. Stimulation of neurite outgrowth by a synthetic NCAM ligand identified by means of combinatorial libraries. Nature Biotechnol 1999; 17:1000-1005.
94. Rougon G. Structure, metabolism and cell biology of polysialic acids. Eur. J Cell Biol 1993; 61:197-207.
95. Rougon G, Dubois C, Buckley N et al. A monoclonal antibody against meningococcus group B polysaccharides distinguishes embryonic from adult N-CAM. J Cell Biol 1986; 103:2429-2437.
96. Roullet P, Mileusnic R, Rose SPR et al. Neural cell adhesion molecules play a role in rat memory formation in appetitive as well as aversive tasks. Neuroreport 1997; 8:1907-1911.
97. Rusakov DA, Davies HA, Harrison E et al. Ultrastructural synaptic correlates of spatial learning in rat hippocampus. Neuroscience 1997; 80:69-77.

98. Sadoul R, Hirn M, Deagostini H et al. Adult and embryonic mouse neural cell adhesion molecules have different binding properties. Nature 1983; 304:347-349.
99. Saghatelyan AK, Gorissen S, Albert M et al. The extracellular matrix molecule tenascin-R and its HNK-1 carbohydrate modulate periosomatic inhibition and long-term potentiation in the CA1 region of the hippocampus. Eur J Neurosci 2000; 12:3331-3342.
100. Sanes JR, Lichtman JW. Can molecules explain long-term potentiation? Nature Neurosci 1999; 2:597-604.
101. Schachner M, Martini R. Glycans and the modulation of neural-recognition molecule function. Trends Neurosci 1995; 18:183-91.
102. Scholey AB, Rose SPR, Zamani MR et al. A role for the neural cell adhesion molecule in a late, consolidating phase of glycoprotein synthesis six hours following passive avoidance training in the young chick. Neuroscience 1993; 55:499-509.
103. Scholey AB, Mileusnic R, Schachner M et al. A role for a chicken homologue of the neural cell adhesion molecule L1 in consolidation of a passive avoidance task in the chick. Learn Memory 1995; 2:17-25.
104. Schuster T, Krug M, Hassan H et al. Increase in proportion of hippocampal spine synapses expressing neural cell adhesion molecule NCAM180 following long-term potentiation. J Neurobiol 1998; 37:359-372.
105. Schuster T, Krug M, Stalder M et al. Immunoelectron microscopic localization of the neural recognition molecules L1, NCAM, and its isoform NCAM180, the NCAM-associated polysialic acid, beta1 integrin and the extracellular molecule tenascin-R in synapses of the adult rat hippocampus. J Neurobiol 2001; 49:142-158.
106. Seki T, Arai Y. Highly polysialylated neural cell adhesion molecule (NCAM-H) is expressed by newly generated granule cells in the dentate gyrus of the adult rat. J Neurosci 1993; 13:2351-2358.
107. Shors TJ, Miesegaes G, Beylin A et al. Neurogenesis in the adult is involved in the formation of trace memories. Nature 2001; 410:372-376.
108. Skibo GC, Davies HA, Rusakov DA et al. Increased immunogold labelling of neural cell adhesion molecule isoforms in synaptic active zones of the chick striatum 5-6 hours after one-trial passive avoidance training. Neuroscience 1998; 82:1-5.
109. Smalla KH, Matthies H, Langnase K et al. The synaptic glycoprotein neuroplastin is involved in long-term potentiation at hippocampal CA1 synapses. Proc Natl Acad Sci USA 2000; 97:4327-32.
110. Sorra KE, Harris KM. Stability in synapse number and size at 2hr after long-term potentiation in hippocampal area CA1. J Neurosci 1998; 18:658-671.
111. Staubli U, Chun D. Factors regulating the reversibility of long term potentiation. J Neurosci 1996; 16:853-860.
112. Staubli U, Vanderklish P, Lynch G. An inhibitor of integrin receptors blocks long-term potentiation. Behav Neur Biol 1990; 53:1-5.
113. Stork O, Welzl H, Cremer H et al. Increased intermale aggression and neuroendocrine response in mice deficient for the neural cell adhesion molecule (NCAM). Eur J Neurosci 1997; 9:1117-1125.
114. Stork O, Welzl H, Wotjak CT. Anxiety and increased $5HT_{1A}$ receptor response in NCAM null mice. J Neurobiol 1999; 40:343-355.
115. Stork O, Welzl H, Wolfer D et al. Recovery of emotional behaviour in neural cell adhesion molecule (NCAM) mutant mice through transgenic expression of NCAM180. Eur J Neurosci 2000; 12:3291-3306.
116. Strekalova T, Wotjak CT, Schachner M. Intrahippocampal administration of an antibody against the HNK-1 carbohydrate impairs memory consolidation in an inhibitory learning task in mice. Mol Cell Neurosci 2001; 17:1102-1113.
117. Squire LR. Memory and the hippocampus: a synthesis from findings with rats, monkeys, and humans. Psychol Rev 2001; 99:195-231.
118. Squire LR, Barondes SH. Variable decay of memory and its recovery in cycloheximide-treated mice. Proc Natl Acad Sci USA 1972; 69:1416-1420.
119. Tanaka H, Shan W, Phillips GR et al. Molecular modification of N-cadherin in response to synaptic activity. Neuron 2000; 25:93-107.
120. Tang L, Hung CP, Schuman EM A role for the cadherin family of cell adhesion molecules in hippocampal long-term potentiation. Neuron 1998; 20:1165-1175.
121. Tang Y, Nyengaard JR, deGroot DMG et al. Total regional and global number of synapses in the human brain neocortex. Synapse 2001; 41:258-273.
122. Theodosis DT, Bonhomme R, Vitiello S et al. Cell surface expression of polysialic acid on NCAM is a prerequisite for activity-dependent morphological neuronal and glial plasticity. J Neurosci 1999; 19:10228-10236.

123. Tiunova A, Anokhin KV, Rose SPR. Three time windows for amnestic effect of antibodies to cell adhesion molecule L1 in chicks. Neuroreport 1998; 9:1645-1648.
124. Toni N, Buchs P-A, Nikonenko I et al. LTP promotes formation of multiple spine synapses between a single axon terminal and a dendrite. Nature 1999; 402:421-425.
125. van der Zee EA, Douma BRK. Historical review of research on protein kinase C in learning and memory. Prog Neuropsychopharm Biol Psychiat 1997; 21:379-406.
126. van Pragg H, Schinder AF, Christie BR et al. Functional neurogenesis in the adult hippocampus. Nature 2002; 415:1030-1034.
127. van Reempts J, Dikova M, Werbrouck L et al. Synaptic plasticity in rat hippocampus associated with learning. Behav Brain Res 1992; 51:179-83.
128. Vimr ER, McCoy RD, Vollger HF et al. Use of prokaryotic-derived probes to identify poly(sialic acid) in neonatal neuronal membranes. Proc Natl Acad Sci USA 1984; 81:1971-1975.
129. Wang S, Scott BW, Wojtowicz JM. Heterogenous properties of dentate granule neurons in the adult rat. J Neurobiol 2000; 42:248-257.
130. Wenzel J, Kammerer E, Kirsche W et al. Electron microscopic and morphometric studies on synaptic plasticity in the hippocampus of the rat following conditioning. J Hirnforsch 1980; 21:647-54.
131. Williams EJ, Furness J, Walsh FS et al. Activation of the FGF receptor underlies neurite outgrowth stimulated by L1, N-CAM, and N-cadherin. Neuron 1994; 13:583-594.
132. Wong WT, Faulkner-Jones BE, Sanes JR et al. Rapid dendritic remodeling in the developing retina: dependence on neurotransmission and reciprocal regulation by Rac and Rho. J Neurosci 2000; 20:5024-5036.
133. Yang P, Major D, Rutishauser U. Role of charge and hydration in effects of polysialic acid on molecular interactions on and between cell membranes. J Biol Chem 1994; 269:23039-23044.
134. Zhang W, Benson DL. Development and molecular organization of dendritic spines and their synapses. Hippocampus 2000; 10:512-526.
135. Zola-Morgan S, Squire LR. The primate hippocampal formation. Evidence for a time-limited role in memory storage. Science 1990; 250:288-290.

Chapter 7.1

Animal and Human Amnesia:

The Cholinergic Hypothesis Revisited

Robert Jaffard and Aline Marighetto

Identifying Memory Dysfunction

The net effect of an experimentally-induced or "naturally" occurring alteration in learning and memory is generally determined by the type of neurological dysfunction (from focal lesions to gene expression) and/or the nature of the learning task. Accordingly, memory systems are defined as distinct, but interactive, psychological and biological entities that still need specification. Within this framework, a full understanding of learning and memory covering the cellular/molecular, systemic and behavioural levels can be achieved only by experimental studies in animals. This, in turn, requires "animal models" asserting the structural congruence between sets of causally related neurobiological and behavioural variables. In the first part of this chapter, we examine the "cholinergic hypothesis of memory" formulated in the 80s, and then focus on cognitive ageing, giving an example of a possible alternative to the classical "pro-cholinergic" approach to age-related memory disorders. In doing so, the specific constraints on current animal research in learning and memory will be illustrated.

Acetylcholine and Memory: From a Key Neurotransmitter to the Functional Dynamics of Interactive Processes

The Cholinergic Hypothesis of Memory: Lesion Studies

By the 1960s, psychopharmacological experiments conducted in both animals and humans provided evidence that anticholinergic treatments produced a deficit in learning and memory performance. The hypothesis that these effects arose from the common dependency of a set of memory-related brain regions on acetylcholine was supported by the observations that these regions were rich in cholinergic elements. Investigations into the neuropathology of Alzheimer's disease finally showed that choline acetyltransferase activity (ChAT) was markedly reduced in brain tissue from these patients, and that this decrease was correlated with a loss of cognitive function. Together, these findings provided the basis for the "cholinergic hypothesis of memory".[4] Consequently, considerable effort has gone into investigating the behavioural effects of lesions of basal forebrain cholinergic neurons, targeting either the Nucleus Basalis Magnocellularis (NBM) that provides cholinergic innervation to the entire neocortex and amygdala, or the Medial Septal Nucleus (MS) and Vertical Diagonal Band (VDB) which, via the Fimbria/Fornix (FF), supplies almost the entire cholinergic innervation to the hippocampus.

By the 1990s most authors began to share the view that owing to a lack of specificity, the earlier interpretation of electrolytic or excitotoxic lesion studies in terms of support for the hypothesis that cholinergic neurons subserve learning and memory was generally unfounded.[21,33] More recently, immunotoxin IgG saporin, a powerful and selective tool with which to destroy cholinergic neurons, has been used in a number of experiments. Taken together, the results

From Messengers to Molecules: Memories Are Made of These, edited by Gernot Riedel and Bettina Platt.

have provided evidence that selective destruction of the NBM cholinergic neurons results in no—or only very mild—deficits in learning and memory as assessed by performance in spatial learning tasks such as the Morris water maze (MWM) in rats,[13] or by the delayed non-matching to sample (DNMTS) task in monkeys.[63] However, there is now strong evidence that the NBM cortical cholinergic projections support certain aspects of attentional function in both species.[54,63]

Selective destruction of MS cholinergic neurons has been reported to produce significant, although modest, learning and memory impairments in certain behavioural tasks that are sensitive to hippocampal lesions. Selective depletion of pre-synaptic cholinergic markers in the hippocampal formation induced by injection of 192 IgG saporin into the MS/VDB in rats has been shown to impair working memory both in the radial arm maze (RAM) and in the delayed non-matching to place (DNMTP) tasks using an operant chamber.[57,60,65] However, both the selectivity of the hippocampal cholinergic denervation and the delay-dependent effect of these impairments remain controversial. Recently, it has been suggested that extensive lesions of both the hippocampal (MS) and cortical cholinergic pathways (NBM) associated with great memory demands (i.e., remembering more than one simple location) are necessary to produce a delay-dependent deficit in the RAM task.[67] Although similar selective cholinergic deafferentation seems to result more consistently in delay-dependent impairing effects in the operant DNMTP than in the RAM task,[45,60] it is not clear whether these tasks assess the same mnemonic functions, as the delays used are generally much shorter in the former (tens of seconds) than in the latter task (minutes or hours). In contrast, infusions of 192 IgG saporin into the MS/VDB or even into both the MS/VDB and NBM failed to impair reference memory in either the MWM task or RAM tasks.[6,57,60] These reports indicate that removing cholinergic neurons that project to the hippocampus only produces, if any, mild learning and memory deficits in tasks that otherwise are clearly impaired by less selective MS neurotoxic lesions[22] or FF transections.[33] It is thus possible that damage to both the cholinergic and GABAergic neurons that project to the hippocampus is necessary to impair hippocampus-dependent learning and memory functions. Recent experiments showing that the previously observed cognitive-enhancing effect of muscarinic drugs targeted to the MS[44] may be better accounted for by increased impulse flow in the GABAergic than cholinergic septo-hippocampal pathways[68] emphasize the potential functional importance of this GABAergic component. Finally, since selective removal of cholinergic neurons in young rats did not reproduce the learning and memory impairments found in aged rats, and since these aged rats displayed a loss of GABAergic neurons in the septal region,[48] it may be that the deterioration of the cholinergic system is not sufficient to produce impairments. However, the fact that the severity of the behavioural impairment seen in aged animals is generally linked to the degree of septo-hippocampal cholinergic deficiency[23] suggests that this system is necessary for normal learning and memory function. In this respect, it remains possible that the sensitivity of the tasks that, up to now, have been used to assess the behavioural effects of a selective hippocampal cholinergic hypofunction is not sufficient to detect impairments. For example, it has recently been reported that rats injected with 192 IgG saporin into the MS displayed a significant bias toward the preferential use of an egocentric (versus allocentric) response strategy in the MWM task.[35] This would suggest that testing designs based on competing strategies[37] might provide a more sensitive tool to assess spatial memory dysfunction than the standard testing designs previously used. This also points to interactions existing between memory systems and to the possible involvement of cholinergic neurons in regulating such interactions.

Cholinergic Alterations Induced by Learning and Memory Testing

Based on the assumption that brain cholinergic neurons play a role in learning and memory, several studies have provided evidence that memory testing is associated with significant changes in pre-synaptic markers of cholinergic activity[10,51,66] as well as in alterations in the release of acetylcholine in hippocampal and/or cortical regions.[1,29,59] In a series of experiments using mice as subjects,[41] we showed that both reference-memory (RM) and working-memory (WM) testing in the RAM induced significant and long-lasting changes in hippocampal cholinergic activity using ex vivo measures of sodium-dependent high affinity choline uptake (SDHACU). Namely,

i. Both types of training induced an immediate increase in hippocampal SDHACU as compared to the "quiet" control condition.
ii. In the RM task, this immediate increase in SDHACU was followed by a decrease leading to a long-lasting (24 hours and 9 days) inhibition of this cholinergic marker. This secondary decrease in SDHACU occurred earlier with repetition of training, thereby leading to a shortening of the testing-induced cholinergic activation as RM training progressed.
iii. By contrast, in the WM task, SDHACU was still increased 24 hours after the last session of training.
iv. Finally, the amplitudes of both the immediate increase and subsequent secondary decrease in SDHACU were significantly related to the rate of acquisition and behavioural profile of learning in the RM task. Our set of results was interpreted with the aim of reconciling previous seemingly discordant data on training-induced changes in cholinergic activity. Specifically, our proposal was the following: The enhancement of hippocampal cholinergic transmission during training might facilitate the acquisition of a "relational kind" of information (sustaining WM and spatial mapping in RM), but to the detriment of simple associations (e.g., stimulus-response, or stimulus-reward that may also sustain RM performance). The subsequent post-training decrease and inhibition in cholinergic activity would facilitate the subsequent consolidation of the permanent (invariant) aspects of acquired information (e.g., information to be held in RM).

Subsequent pharmacological experiments[42] provided evidence that the long-lasting inhibition of hippocampal cholinergic activity subsequent to RM testing could be mediated, at least in part, by glutamatergic receptors located in the lateral septum (LS). Indeed, such receptors would, presumably through GABAergic interneurons, provide an inhibitory input to cholinergic cells in the medial area.[27] Therefore, we hypothesized that hippocampo-septal glutamatergic synapses in the LS could be the locus of an "LTP-like" mechanism sustaining the post-RM training inhibition of cholinergic cells, a phenomenon that should not be observed following WM testing. This hypothesis was confirmed. As compared to their controls (i.e., treadmill group), mice trained in the RM task exhibited a progressive and persistent enhancement in hippocampal-LS synaptic neurotransmission as training progressed, whereas an opposite change (i.e., a depression) occurred in mice trained in the WM task.[25,34] As previously observed for SDHACU, the magnitude of the RM training-induced enhancement of hippocampal-LS neurotransmission was correlated with discriminative performance. It is highly unlikely that the training-induced LTP- and LTD-like changes of the LS synapses are involved in the storage of specific information;[61] rather, they might "shape" the hippocampal-septal-hippocampal circuitry in the more appropriate configuration for coping with the requirement of the task. Specifically, the RM training-induced synaptic enhancement of hippocampal-septal synapses could be involved, through the rapid post-testing decrease in hippocampal acetylcholine release, in the consolidation of the to-be-learned trial-independent information (RM), a process which is not required for the trial-dependent WM performance. As mentioned above, the increase in hippocampal cholinergic activity might be necessary for the short-term maintenance of (a relational kind of) information during both RM acquisition and WM testing, whereas its immediate post-acquisition inhibition might make it possible to consolidate information hold in RM alone. The functionality of these training-induced "biphasic" changes in hippocampal cholinergic activity as well as its above-mentioned putative mechanisms are in fact congruent with data obtained with pharmacological approaches. Namely, it has been shown that increasing or decreasing cholinergic activity by pre-training infusion of drugs in the septal area (i.e., the NMDA receptor blocker AP5 or the GABAergic agonist muscimol, respectively) facilitated the maintenance of information in WM[50] and impaired acquisition of spatial RM in the MWM task.[49] Furthermore, using the same experimental design, Nagahara et al[49] showed that inducing inhibition of hippocampal cholinergic activity immediately after (rather than prior to) training actually facilitated (rather than impaired) subsequent retention.

Should the Cholinergic Hypothesis Be Re-Examined?

If the participation of the proposed "biphasic" mechanisms of the septo-hippocampal cholinergic neurons is to be confirmed in some forms of memory, many apparent discrepancies within cholinergic-related findings would be resolved, and the importance of the dynamical aspects would have to be taken into account. An example of such discrepancies is that, even though selective cholinergic lesions have, on the whole, failed to reproduce the impairing effect of ageing in the MWM task, it remains that the magnitude of this impairment is significantly correlated with the magnitude of cholinergic loss. It therefore seems that the cholinergic hypothesis of age-related cognitive decline deserves to be re-examined in the context of interactive and dynamical processes at both the neural (structures and circuits) and psychological (forms of memory) levels. It might indeed be that part of such a decline in cognitive abilities is related to alterations in the connective processes that sustain proper recruitment of a set of structures in the brain, as suggested by neuroimaging studies in humans.[7,12,47] The fact that in comparison with younger adults, aged animals displayed a different pattern of cholinergic responses to memory training and, in particular, that the long-lasting inhibition of hippocampal cholinergic activity subsequent to RM training was not observed in aged subjects.[10,32] is coherent with this view. Thus, if the plastic properties of glutamatergic synapses in the LS are really necessary for adjusting optimally the septo-hippocampal activity both as a function of the stage of memory formation and of the relative contribution of different information processing functions (e.g., "relational" vs. "simple associations"),[11] then the observation that aged mice display strong alterations in LS synaptic plasticity[24] should not be without consequences on their learning and memory capabilities. In this respect, alterations in LS synaptic plasticity and thus alterations in the capacity of the aged brain to configure the brain circuitry needed for optimal encoding, storage and retrieval of specific information, might be one of the numerous possible causes of the deficits observed at the behavioural level.[62]

Together, these data provide a potential framework in which to examine the neurobiological basis of cognitive ageing and, in particular, the possible involvement of the septo-hippocampal cholinergic pathway. However, due to the complexity of the system (both in terms of interactions between its components and of its temporal dynamics), it is not clear how it might be possible to correct the dysfunction by pharmacological means that could be used to alleviate age-related cognitive impairments in humans. Up to now, another problem that has not been considered is the validity of animal behavioural models used to assess the cognitive impairments occurring in senescence.[46] As an illustration of both issues, it has recently been shown that the enhancing effect of tacrine (an inhibitor of acetylcholinesterase) on spatial navigation performance (MWM task) in aged rats was blocked by non-spatial pre-training.[2] Indeed, this suggests either that targeting the cholinergic system of aged subjects is totally inefficient or that the cholinesterase inhibitor improves some major non-cognitive or procedural aspects of the task performance.[30,38] The latter possibility might explain why cholinergic agonists have been reported to be ineffective in reversing age-related deficits in tasks that need intensive pre-training (e.g., the DNMTP task).[14]

From Assessment to Alleviation of Age-Related Memory Impairments in Mice

Although the need for experimental studies of ageing in animals is obvious, the relevance of these studies depends on whether both the specific functions and biological systems targeted for study are appropriate models of human ageing. In this section, we describe how we tackled both issues using C57BL/6 mice as subjects.

Modelling Human Age-Related Memory Deficits

In humans, there is a consensus that declarative/explicit memory appears to be more vulnerable to deterioration in senescence than procedural/implicit memory.[56] Cohen[8] has identified two cardinal "non-verbal" characteristics of human declarative memory, i.e., its capacity to compare and contrast items in memory and to support the inferential use of memories in novel situations (flexibility). In contrast, procedural memory involves the facilitation of particular routines for which no such explicit comparisons are executed. Following on from the original reports by Eichenbaum et al,[16,17] we developed tasks to determine whether these different processes could be engaged or disengaged in mice simply depending on how the same items were presented (i.e., using a within-subject design).[40,43] The tasks consisted of unambiguous discriminations between arms of opposite valence in a radial-arm maze. As depicted in Figure 1, each experiment (either A or B) was designed in two-stages, with an initial learning phase followed by a test-phase. The only parameter which was varied among the successive stages was the way of presenting arms to mice, i.e., either in pairs (simultaneous discriminations) or one at a time (successive go-no-go discrimination). During the initial learning phase, most of the aged (21-23 months) mice were impaired in learning the simultaneous discriminations (B, stage 1) whereas all the aged subjects acquired the discrimination task as well as the adult (4-5 months) controls in a successive go-no-go design involving the same set of items (A, stage 1). When challenged with modified presentations of familiar items in the test phase, aged mice were impaired if two arms were presented to them simultaneously in a novel pairing, but not if they were presented one at a time, in a successive go-no-go procedure. Thus our results showed that the extent to which ageing could alter the ability to acquire (stage 1) or to use previously acquired spatial discriminations in novel situations (stage 2), strictly depends on the manner the discriminanda were presented to the subject. This set of data supports the conclusion that two forms of memory expression of the same acquired experience can be preferentially triggered through a change in the way discriminanda are presented. Specifically, even though two-choice tasks can be theoretically solved on the basis of elemental associations, it appears that, at least in the present (and other) specific conditions,[43] presentation of discriminanda may encourage the use of explicit comparisons thereby requiring relational representations of past experience. In this sense, the dissociation observed in aged mice is reminiscent of the dissociation between implicit and explicit expression of the same piece of previously acquired material in human amnesic subjects.[55]

In contrast with most data from experiments carried out using the MWM,[5] results from the experiments described above did not provide evidence that impaired and unimpaired individuals could be distinguished within the population of aged mice. In fact, in a study conducted in a large population of F-344 rats tested in the MWM from 1.5 to 26 months, Lindner[38] did not find any evidence among aged rats of a bimodal distribution of performance in the spatial learning versions of the task.

Assessing Similarities of Memory Impairments in Senescent and Hippocampal Lesioned Subjects

Within a neuropsychological framework, the cognitive decline in non-demented aged individuals grossly parallels that observed in medial temporal lobe amnesia[9] and is thought to originate mainly in deficient hippocampal processing. In this view (i) aged mice should display alterations in hippocampal functioning when confronted with problems in the RAM tasks described above, and (ii) hippocampal damage in adult mice should result in the same selective deficit as that seen in aged mice. When confronted with selected RAM problems in which they were able to perform at the same (above-chance) level as adults, aged animals indeed displayed an overall reduction of the testing-induced increase in the expression of the immediate early gene cFos (for review, see Greenwood et al, this book) in the whole septo-hippocampal system (SHS). Importantly, this diminished activation was accompanied by a loss of correlation between Fos activity levels in the connected sub-regions of the SHS.[69] Ibotenate hippocampal

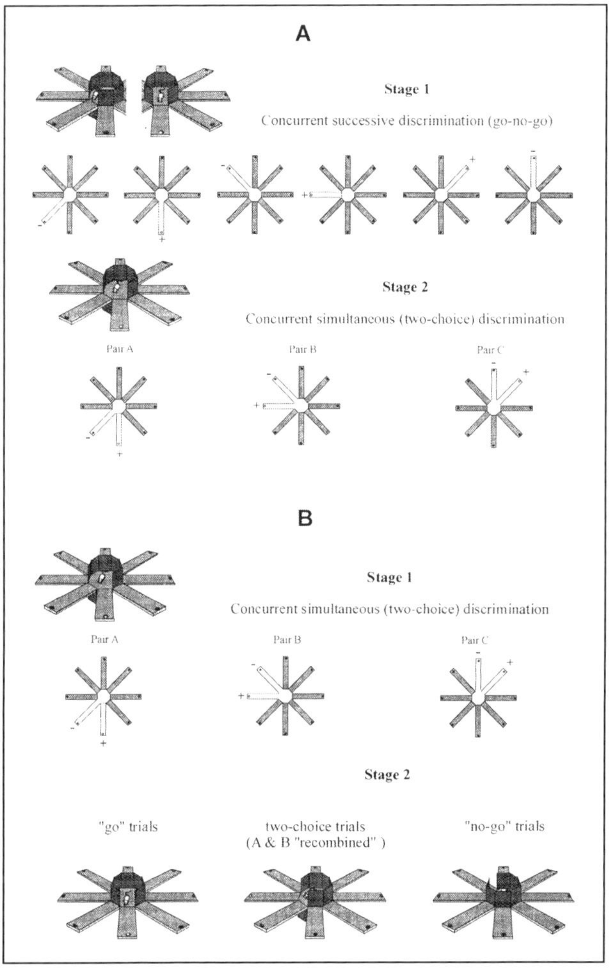

Figure 1. Two-stage paradigms of RAM discrimination tasks (adapted from ref. 40). Each mouse was separately assigned six adjacent arms. Out of these, three served as positive (baited) arms and the remaining three served as negative (not baited) arms. Design A: In stage 1, the six arms were always presented one at a time using a go-no-go discrimination procedure (in each trial, the door to only one arm was open). Go-no-go discrimination was indexed by a ratio between the median latency to enter negative arms and positive arms. Each mouse was trained until reaching a predetermined criterion of go-no-go discrimination and then transferred to stage 2. Mice failing to reach the criterion within 360 trials were dropped from further testing. The discrimination problems presented in stage 2 were between the same arms as in stage 1: the reward contingency of the discriminanda remained unchanged, but their presentation was modified. The six arms were now grouped into, and presented as, three adjacent pairs A, B and C. In each trial, the mouse was confronted with access to two adjacent arms with opposing valence and allowed to visit only one of them. Accuracy was measured by percentage correct, i.e., choice of the positive arm within a pair. Design B: In stage 1, the six arms were grouped into the three pairs A, B and C (exactly as design A, stage 2) . A mouse was considered to reach criterion performance when its choice accuracy was above 75% correct for two consecutive sessions and then transferred to stage 2. Again mice failing to reach the acquisition criterion within 360 trials were rejected. In stage 2, the presentation of the arms only was modified. Two choice arms of opposing reward valence, each taken from a different discrimination problem featured in stage 1, were presented either one at a time or simultaneously as members of a novel pairing. These corresponded to "go" or "no-go" trials and "recombined" two-choice discrimination, respectively.

lesions in adult mice nicely reproduced the selective behavioural impairment previously observed in aged mice.[19] Namely, hippocampus-lesioned mice performed as intact mice in learning the location of reward versus non-reward when arms were presented one at a time (see Figure 1, A stage 1). However, as aged mice, they were unable to translate this knowledge into an efficient approach to the rewarded arm in the two-choice (simultaneous) discrimination (i.e., A stage 2).

The selective deficit seen in both aged mice and hippocampal-lesioned mice in the RAM tasks described above is unlikely to be due to confounding (non-specific) changes (affect, motivation, perception or motor control), as all the basic requirements of the task were largely identical in the successive stages. In contrast, in-depth analysis of the performance of rats in the MWM[38] has revealed significant relationships between non-cognitive and or non-mnemonic factors (e.g., swim speed, thigmotaxia, performance in the cued platform version of the task) and measures of cognitive functions (i.e., swim distances in the spatial learning versions). This suggests that swim distances in the MWM may be affected by non-specific factors.[31] Moreover, it is unlikely that presenting arms one at a time (see A, stage 1 in Figure 1) in the RAM task is easier than presenting them by pairs (i.e., B, stage 1), as the speed of acquisition of adult mice was very similar in both situations. This is also an important point since it has been suggested that since the visible platform task in the MWM is easier to solve than the hidden platform task, this experimental design may be biased towards false specificity.[26] Whatever the case, two (multiple)-stage testing designs[17] such as the RAM tasks seem to provide a valuable tool for dissociating impaired from unimpaired forms of memory expression and, as we have shown, to support the hypothesis that the selective cognitive deficit of senescent mice stems from hippocampal dysfunction.

The anomalous performance seen in both aged mice and hippocampal-lesioned mice resembles the deficit observed in rats with fornix lesions in the studies of Eichenbaum et al[16,17] on odour-guided discrimination tasks. It has been proposed that in both animals and humans, this deficit may stem from alterations in forming relations between separately experienced cues (i.e., relational representation),[15,36] a process that would critically depend on the functional integrity of the hippocampus here conceived of as an associator of discontiguous (spatial and/or temporal) items.[64] It is therefore conceivable that one basic dysfunction underlying learning and memory impairments in senescence is the difficulty in prolonging neural activity subserving the representation of a cue.[53] Such a dysfunction may account for age-related impairments observed in a variety of tasks such as Pavlovian conditioning in the trace paradigm,[58] delayed response or delayed recognition tasks,[14] and in tasks that require the encoding and storage of relationships between discontiguously perceived cues or events to guide performance (i.e., contextual information, cognitive maps).

Alleviating the Selective Age-Related Memory Deficit

Most of the pharmacological strategies traditionally used to improve age-related cognitive decline are designed to restore the deficiency of neurotransmission of one (e.g., cholinergic) or several (e.g., cholinergic and serotoninergic) specific types. Alternatively, other compounds (e.g., Nerve Growth Factor) are used to improve the "general health" of neurons and thus to maintain their normal cellular functions as they age. Recently, experiments were carried out to determine whether normalising a broad profile of brain gene expression in aged mice to pre-senescent (adult) levels would improve their associated cognitive deficits. The rationale was the following. First, it has been reported in mice that ageing is associated with a reduction (20-30 %) in the levels of mRNA for brain retinoid acid nuclear receptors (i.e., RAR and RXR) and in particular that this age-related reduction is susceptible to reversal by acute systemic administration of retinoic acid (RA).[18] Second, it is now well established that these nuclear receptors regulate the expression of a number of genes coding for neural proteins involved in synaptic plasticity (e.g., neurogranin, NMDA receptors, synaptophysin etc.), for cholinergic-specific proteins, and for neurotrophic factors (for references, see ref. 20). All of

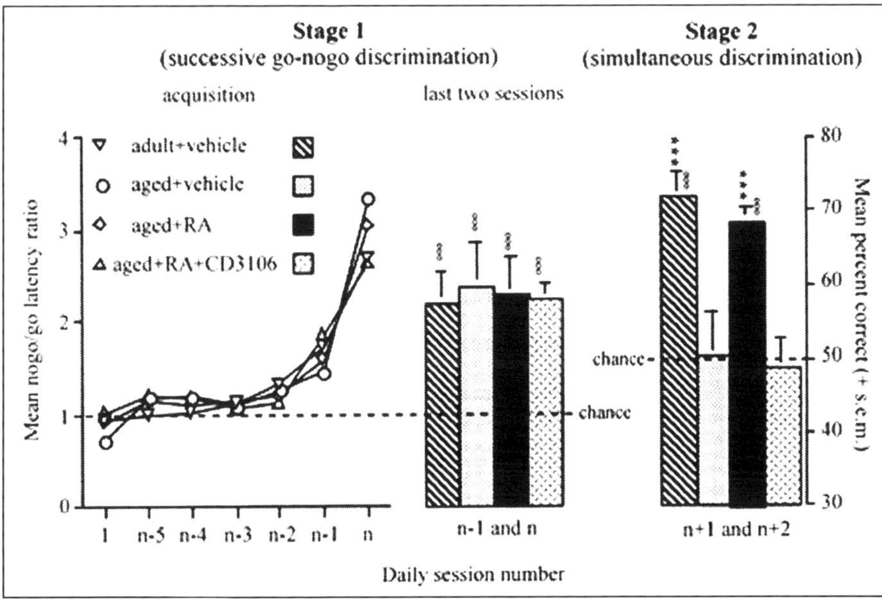

Figure 2. Behavioural data from ref. 20. Four groups of mice (i.e., adult mice treated with vehicle, aged mice treated with vehicle, aged mice treated with RA and aged mice treated with RA and the antagonist of retinoid nuclear receptors, CD3106) were trained in our RAM discrimination tasks (design A as described in figure 1). Learning curves in stage 1 (successive go-no-go discriminations) were identical among the four groups. Each group of mice reached about the same mean level of no-go/go latency ratio in the last two sessions of stage 1. Conversely, between-groups differences emerged in stage 2 when the same arms were combined into three pairs. While aged mice treated either with vehicle or a mixture of RA and CD3106 behaved as if they were naive, failing to transfer their acquired preference into the choice of the positive arm within a pair, aged mice treated with RA displayed nearly the same above chance level of accuracy as seen in adult controls.

these constitute a potential therapeutic target for attenuating cognitive deterioration. Using the RAM tasks we[20] showed that restoring the brain (and hippocampal) levels of retinoid receptors and the expression of certain specific associated target genes [in particular, that of the Ca2+-sensitive calmodulin-binding protein neurogranin (RC3)] by administration of RA specifically alleviates the selective deficit of aged mice (see Fig. 2). Conversely, decreasing brain levels of brain retinoid receptors (and expression of RC3) by 20-30 % in adult mice using a vitamin A deprived diet was demonstrated to produce the same selective cognitive deficit as seen in aged mice.[70] Given that retinoid receptors play a major role in the foetal development of the nervous system, these findings are, in a sense, in line with the general principle that "the signals transduced by cells during growth and physiologic activity are the same as those that become overloaded during pathological events and ageing".[39]

Conclusion

We have shown in this chapter that processes more complex than that suggested by the hypothesis linking age-related memory loss to cholinergic deficiency have to be considered. First of all, we have underlined the need to optimise the development of animal behavioural models providing closer analogy with -and greater sensitivity and selectivity to- the cognitive deficits known to date to accompany human senescence. Second, we have emphasised the need to recognise that ageing cannot be reduced to a neurobiological dysfunction at a single locus, but rather that it involves a set of dysfunctions, including compensatory changes whose functional relevance lies at the system-property level. In this respect, it must be noted that both observations are not inherent to ageing studies, as they concern all research investigating the problem of relating brain function to memory. Returning to the ageing issue, we tentatively propose that rather than compensating a single identified dysfunctional target, a better strategy would be to attempt to globally re-establish cellular homeostasis. Although the strategy of using RA administration seems efficient (see Figure 2), we are at present unable to pinpoint the protein product(s) of the retained-activated target genes that are responsible for the observed improvement in cognitive function in senescent animals. Yet it has long been a dream of neurobiologists to reduce complex cognitive function to a simple molecular device (but see ref. 3 and 52), as well as to equate cognition (a psychological construct) with measures of behavioural performance assessed by training and testing techniques currently in vogue.[28] As already widely remarked, no one has ever measured learning and memory. They can only be inferred from careful behavioural analysis specifying what, how, and why, and not simply by stating that "something" has been learned. Such are the prerequisites for thoughtful integration of the neurobiological level with the behavioural level.

References

1. Acquas E, Wilson C, Fibiger HC. Conditioned and unconditioned stimuli increase frontal cortical and hippocampal acetylcholine release:effects of novelty, habituation and fear. J Neurosci 1996; 16:3089-3096.
2. Aura J, Riekkinen PJ. Pre-training blocks the improving effect of tetrahydroaminoacridine and D-cycloserine on spatial navigation performance in aged rats. Eur J Pharmacol 2000; 390:313-318.
3. Bannerman DM, Good MA, Butcher SP et al. Distinct components of spatial learning revealed by prior training and NMDA receptor blockade. Nature 1995; 378:182-186.
4. Bartus RT, Dean RL, Pontecorco, MJ et al. The cholinergic hypothesis: a historical overview, current perspective, and future directions. Ann NY Acad Sci 1985; 444:332-358.
5. Baxter MG, Gallagher M. Neurobiological substrates of behavioral decline: models and data analytic strategies for individual differences in aging. Neurobiol Aging 1996; 17:491-495.
6. Baxter MG, Bucci DJ, Gorman LK et al. Selective immunotoxic lesions of basal forebrain cholinergic cells: effects on learning and memory in rats. Behav Neurosci 1995; 109:714-722.
7. Cabeza R, McIntosh AR, Tulving E et al. Age-related differences in effective neural connectivity during encoding and recall. NeuroReport 1997; 8:3479-3483.
8. Cohen NJ. Preserved learning capacity in amnesia: evidence for multiple memory systems. In: Squire LR, Butters N, eds. Neurospsychology of memory. New York, Guilford Press, 1984; 83-103.
9. Craik FIM, Jennings JM. Human memory. In: Craik FIM, Salthouse TA, eds. Handbook of aging and cognition. Erlbaum NJ, Hillsdale, 1992; 51-83.
10. Decker MW, Pelleymounter MA, Gallagher M. Effects of training on a spatial memory task on high affinity choline uptake in hippocampus and cortex in young adult and aged rats. J Neurosci 1988; 8:90-99.
11. Desmedt A, Garcia R, Jaffard R. Differential modulation of changes in hippocampal-septal synaptic excitability by the amygdala as a function of either elemental or contextual fear conditioning in mice. J Neurosci 1998; 18:480-487.
12. D'Esposito M. New answers to old questions. Curr Biol 1999; 9:939-941.
13. Dornan WA, McCampbell AR, Tinkler GP et al. Comparison of site-specific injections into the basal forebrain on water maze and radial arm maze performance in the male rat after immunolesioning with 192 IgG saporin. Behav Brain Res 1996; 82:93-101.
14. Dunnett SB, Evenden JL, Iversen SD. Delay-dependent short-term memory deficits in aged rats. Psychopharmacol 1988; 96:174-180.
15. Eichenbaum H. How does the brain organize memories? Science 1997; 277:330-332.
16. Eichenbaum H, Fagan A, Mathews P et al. Hippocampal system dysfunction and odor discrimination learning in rats: Impairment or facilitation depending on representational demands. Behav Neurosci 1988; 102:3531-3542
17. Eichenbaum H, Mathews P, Cohen NJ. Further studies of hippocampal representation during odor discrimination learning in rats. Behav Neurosci 1989; 103:1207-1216.
18. Enderlin V, Pallet V, Alfos S et al. Age-related decreases in mRNA for brain nuclear receptors and target genes are reversed by retinoic acid treatment. Neurosci Lett 1997; 229:125-129.
19. Etchamendy N, Desmedt A, Cortes-Torrea C et al. Contrasting effects of selective hippocampal lesion on different memory expressions of spatial discriminations in mice. Hippocampus 2003; 13:197-211.

20. Etchamendy N, Enderlin V, Marighetto A et al. Alleviation of a selective age-related relational memory deficit in mice by pharmacologically induced normalization of brain retinoid signaling. J Neurosci 2001; 21:6423-6429.
21. Everitt BJ, Robbins TW Central cholinergic systems and cognition. Ann Rev Psychol 1997; 48:649-684.
22. Gallagher M, Colombo PJ. Ageing: the cholinergic hypothesis of cognitive decline. Curr Opinion Neurobiol 1995; 5:161-168.
23. Gallagher M, Nagahara AH, Burwell RD. Cognition and hippocampal systems in aging: animal models. In: McGaugh JL, Weinberger N, Lynch G eds. Brain and Memory: modulation and mediation of neuroplasticity. New York: Oxford University Press, 1995; 103-126.
24. Garcia R, Jaffard R. Age-related changes in inhibition and long-term potentiation in the lateral septum in mice. Brain Res 1993; 620:229-236.
25. Garcia R, Vouimba RM, Jaffard R. Spatial discrimination learning induces LTP-like changes in the lateral septum of mice. NeuroReport 1993; 5:329-332.
26. Gerlai R. Behavioral tests of hippocampal function: simple paradigms complex problems. Behav Brain Res 2001; 125:269-277.
27. Giovannini MG, Mutolo D, Bianchi L et al. NMDA receptor antagonists decrease GABA outflow from the septum and increase acetylcholine outflow from the hippocampus: a microdialysis study. J Neurosci 1994; 14:1358-65.
28. Grant SG, O'Dell TJ, Karl KA et al. Impaired long-term potentiation, spatial learning, and hippocampal development in fyn mutant mice. Science 1992; 258:1903-1910.
29. Hironaka N, Tanaka K, Izaki Yet al. Memory-related acetylcholine efflux from rat prefrontal cortex and hippocampus: a microdialysis study. Brain Res 2001; 901:143-150.
30. Hoh TE, Cain DP. Fractionating the nonspatial pretraining effect in the water maze task. Behav Neurosci 1997; 111:1285-1291.
31. Huerta PT, Scearce KA, Farris SM et al. Preservation of spatial learning in fyn tyrosine kinase knockout mice. NeuroReport 1996; 7:1685-1689.
32. Jaffard R, Marighetto A, Lebrun C et al. Cholinergic alterations induced by training in hippocampus in adult and aged mice: involvement of septal noradrenergic and glutamatergic synapses. Neurosci Res Comm 1993; S13, 27-30.
33. Jaffard R, Micheau J. Central cholinergic systems, learning and memory. In: Delacour J, ed. The memory system of the brain. Adv Ser Neurosci, World Scientific 1994;4:389-430
34. Jaffard R, Vouimba RM, Marighetto A et al. Long-term potentiation and long-term depression in the lateral septum in spatial working and reference memory. J Physiol 1996; 90:939-941.
35. Janis LS, Glasier MM, Fulop Z et al. Intraseptal injections of 192 IgG saporin produce deficits for strategy selection in spatial-memory tasks. Behav Brain Res 1998; 90:23-34.
36. Johnson MK. MEM: mechanisms of recollection. J Cog Neurosci 1992; 4:268-280.
37. Kim JJ, Baxter MG. Multiple brain-memory systems: the whole does not equal the sum of its parts. Trends Neurosci 2001; 24:324-330.
38. Lindner MD. Reliability, distribution, and validity of age-related cognitive deficits in the Morris water maze. Neurobiol Learn Mem 1997; 68:203-220.
39. Malik MA, Blusztajn JK, Greenwood CE. Nutrients as trophic factors in neurons and the central nervous system: role of retinoic acid. J Nutr Biochem 2000; 11:2-13.
40. Marighetto A, Etchamendy N, Touzani K et al. Knowing which and knowing what: a potential mouse model for age-related human declarative memory decline? Eur J Neurosci 1999; 11:3312-3322.
41. Marighetto A, Micheau J, Jaffard R. Relationships between testing-induced alterations of hippocampal cholinergic activity and memory performance on two spatial tasks in mice. Behav. Brain Res 1993; 56:133-144.
42. Marighetto A, Micheau J, Jaffard R. Effects of intraseptally injected glutamatergic drugs on hippocampal sodium-dependent-high-affinity choline uptake in "naive" and "trained" mice. Pharmacol Biochem Behav 1994; 49:689-699.
43. Marighetto A, Touzani K, Etchamendy N et al. Further evidence for a dissociation between different forms of mnemonic expressions in a mouse model of age-related cognitive decline: effects of tacrine and S 17092: a novel prolyl endopeptidase inhibitor. Learning Memory 2000; 7:159-169.
44. Markowska AJ, Olton DS, Givens B. Cholinergic manipulations in the medial septal area: age-related effects on working memory and hippocampal electrophysiology. J Neurosci 1995; 15:2063-2073.
45. McAlonan GM, Dawson GR, Wilkinson LO et al. The effects of AMPA-induced lesions of the medial septum and vertical limb nucleus of the diagonal band of Broca on spatial delayed non-matching to sample and spatial learning in the water maze. Eur J Neurosci 1995; 7:1034-1049.

46. McDonald MP, Overmier JB. Present imperfect: a critical review of animal models of the mnemonic impairments in Alzheimer's disease. Neurosci Biobehav Rev 1998; 22:99-120.
47. McIntosh AR, Sekuler AB, Penpeci C et al. Recruitment of unique neural systems to support visual memory in normal aging. Curr Biol 1999; 9:1275-1278.
48. Miettinen R, Sirvio J, Riekkinen P Sr et al. Neocortical, hippocampal and septal parvalbumin- and somatostatin-containing neurons in young and aged rats: correlation with passive avoidance and water maze performance. Neurosci 1993; 53:367-378.
49. Nagahara AH, Brioni JD, McGaugh JL. Effects of intrseptal infusion of muscimol on inhibitory avoidance and spatial learning: differential effects of pretraining and posttraining administration. Psychobiol 1992; 20:198-204.
50. Puma C, Bizot JC. Intraseptal infusions of a low dose of AP5, a NMDA receptor antagonist, improves memory in an object recognition task in rats. Neurosci Lett 1998; 248:83-86.
51. Raaijmakers WGM. High affinity choline uptake in hippocampal synaptosomes and learning in the rat. In: Ajmone-Marsan C, Matthies H, eds. Neuronal plasticity and memory formation. New York, Raven Press, 1982; 373-385.
52. Rampon C, Tang YP, Goodhouse J et al. Enrichment induces structural changes and recovery from nonspatial memory deficits in CA1 NMDAR1-knockout mice. Nature Neurosci 2000; 3:238-244.
53. Rawlins JNP. Associations across time: The hippocampus as a temporary memory store. Behav Brain Sci 1985; 8:479-496.
54. Sarter M, Bruno JP. Cognitive functions of cortical acetylcholine: towards a unifying hypothesis. Brain Res Rev 1997; 23:28-46.
55. Schacter DL. Implicit expressions of memory in organic amnesia: learning of new facts and associations. Hum Neurobiol 1987; 6:107-118.
56. Schugens MM, Daum I, Spindler M et al. Differential effects of aging on explicit and implicit memory. Aging, Neuropsychol Cog 1997; 4:33-44.
57. Shen J, Barnes CA, Wenk GL et al. Differential effects of selective immunotoxic lesions of medial septal cholinergic cells on spatial working and reference memory. Behav Neurosci 1996; 110:1181-1186.
58. Solomon PR, Beal MF, Pendlebury WW. Age-related disruption of classical conditioning: a model systems approach to memory disorders. Neurobiol Aging 1988; 9:535-546.
59. Stancampiano R, Cocco S, Cugusi C et al. Serotonin and acetylcholine release response in the rat hippocampus during a spatial memory task. Neurosci 1999; 89:1135-1143.
60. Torres EM, Perry TA, Blokland A et al. Behavioral, histochemical and biochemical consequences of selective immunolesions in discrete regions of the basal forebrain cholinergic system. Neurosci 1994; 63:95-122.
61. Urban IJ, Ontskul A, Croiset G et al. A long-lasting increase and decrease in synaptic excitability in the rat lateral septum are associated with high and low shuttle box performance, respectively. Behav Brain Res 1995; 68:173-183.
62. Vouimba RM, Garcia R, Jaffard R. Opposite effects of lateral septal LTP and lateral septal lesions on contextual fear conditioning in mice. Behav Neurosci 1998; 112:875-884.
63. Voytko ML, Olton DS, Richardson RT et al. Basal forebrain lesions in monkeys disrupt attention but not learning and memory. J Neurosci 1994; 14:167-186.
64. Wallenstein GV, Eichenbaum H, Hasselmo ME. The hippocampus as an associator of discontiguous events. Trends Neurosci 1998; 21:317-323.
65. Walsh TJ, Herzog CD, Gandhi C et al. Injection of 192-saporin into the medial septum produces cholinergic hypofunction and dose-dependent working memory deficits. Brain. Res 1996; 726:69-79.
66. Wenk G, Hepler D, Olton D. Behavior alters the uptake of (3H) choline into acetyl-cholinergic neurons of the nucleus basalis magnocellularis and medial septal area. Behav Brain Res 1984; 13:129-138.
67. Wrenn CC, Lappi DA, Wiley RG. Threshold relationship between the lesion extent of cholinergic basal forebrain in the rat and working memory impairment in the radial maze. Brain Res 1999; 847:284-298.
68. Wu M, Shanabrough M, Leranth C et al. Cholinergic excitation of septohippocampal GABA but not cholinergic neurons: implications for learning and memory. J Neurosci 2000; 20:3900-3908.
69. Touzani K, Marighetto A, Jaffard R. Fos imaging reveals ageing-related changes in hippocampal response to radial maze discrimination testing in mice. Eur J Neurosci 2003; 17:628-640.
70. Etchamendy N, Enderlin V, Marighetto A et al. Vitamin A deficiency and relational memory deficit in adult mice: Relationships with changes in brain retinoid signalling. Behav Brain Res 2003; 145:37-49.

CHAPTER 7.2

Aging and the Calcium Homeostasis

Wendy W. Wu and John F. Disterhoft

Abstract

Normal brain aging is associated with physiological alterations in Ca^{2+} homeostasis and deficits in learning and memory. The hippocampus, a structure critical for proper learning and memory functions, is frequently implicated in aging-related learning deficits. Consistent with the "Ca^{2+} hypothesis of aging", there are many reports of aging-related changes in Ca^{2+} signaling cascades in hippocampal pyramidal neurons, including enhancement in the Ca^{2+}-dependent AHP caused by an enhanced sI_{AHP}. Previous experiments from our laboratory have shown that the size of the AHP and the sI_{AHP} is inversely correlated to the acquisition of hippocampus-dependent tasks and positively correlated to the aging process (i.e., the AHP and the sI_{AHP} are reduced during learning and enhanced during aging). We thus hypothesize that the AHP and the sI_{AHP} are critically involved in learning and age-related learning deficits. The fact that the sI_{AHP} receives neuromodulation from many transmitter systems important for learning and sensitive to aging lends further support for its role in age-related learning deficits. Our data suggest that the sI_{AHP} is a good candidate to be an important link between age-related changes in Ca^{2+} homeostasis and learning deficits.

Introduction

Neuronal activity is accompanied with a rapid rise in the level of cytosolic Ca^{2+}, which functions as a second messenger to mediate a wide range of cellular responses such as neurotransmitter release, changes and maintenance in cytoarchitecture, gene transcription, and activation of enzyme cascades. The pivotal role of Ca^{2+} in these and other cellular processes dictates a need for precise homeostatic control. Because the transmembrane gradient for Ca^{2+} is far greater than those of other physiologically relevant ions, a small change in membrane Ca^{2+} permeability can result in a large difference in the level of cytoplasmic Ca^{2+}. Under normal conditions, the level of intracellular free Ca^{2+} is carefully regulated to maintain a sharp concentration gradient to ensure rapid signal transduction for subsequent events (see also Sun and Alkon, this book).

The Ca^{2+} signal in response to neuronal activation is derived from many sources and modified by different mechanisms: Ca^{2+} influx into the cytoplasm from the extracellular space, Ca^{2+} release from intracellular stores, cytoplasmic Ca^{2+} buffering, and Ca^{2+} clearance systems (that either extrude Ca^{2+} from the cell or accumulate Ca^{2+} into internal stores). Changes in all of these processes have been implicated in aging. Despite negative feedback and compensatory mechanisms, the summed effect of the age-related dysregulation in Ca^{2+} homeostasis is an enhanced free cytosolic Ca^{2+} level in aging neurons.[60,98,118]

Prolonged exposure to elevated levels of Ca^{2+} can lead to irreversible cell damage and even cell death.[14] Based on the toxic effects of excess Ca^{2+}, the "Ca^{2+} hypothesis of brain aging" was proposed to account for age-related changes in neuronal function and viability in aging. The "Ca^{2+} hypothesis" has received considerable support since its inception, including studies demonstrating age-related changes in Ca^{2+}-dependent synaptic plasticity,[53,71,72,75,103] and others

From Messengers to Molecules: Memories Are Made of These, edited by Gernot Riedel and Bettina Platt.

showing an age-related increase in vulnerability to Ca^{2+}-dependent neurotoxicity and neurodegeneration.[12,63,89] It is believed that dysregulation in Ca^{2+} homeostasis in the hippocampus, a brain region critical for learning and memory, contributes to compromised behavioral/learning deficits associated with aging.

Much work has focused on determining the sources for Ca^{2+} dysfunction and alterations in Ca^{2+}-mediated events in aging tissues. However, less is known about how these physiological changes translate into compromised learning and memory. To address the relationship between age-related Ca^{2+} dysregulation and learning deficits, our laboratory has performed a series of experiments using hippocampal preparations from young and aging animals trained in hippocampus-dependent learning tasks. We have identified several age- and/or learning-related changes in Ca^{2+}-mediated processes in the hippocampus, including changes in Ca^{2+} influx, the Ca^{2+}-dependent postburst afterhyperpolarization (AHP), as well as two of the AHP currents (the I_{AHP} and sI_{AHP}). Furthermore, we have shown that changes in the AHP and the sI_{AHP} are inversely correlated with ability to acquire hippocampus-dependent tasks. The AHP currents are critical determinants of neuronal excitability. Thus, our data indicate that regulation of neuronal excitability by the AHP currents are important events that occur during learning, and further suggest that changes in the AHP currents are likely to play a role in aging-related learning deficits. This chapter focuses on studies performed in our laboratory illustrating the involvement of the AHP and its underlying currents in the cellular mechanisms of learning and memory. We review briefly 1) age-related changes in Ca^{2+} homeostasis; 2) two hippocampus-dependent learning tasks, trace eyeblink conditioning and Morris water maze training, that are useful to examine learning and aging-related learning deficits; and 3) changes in the AHP and the sI_{AHP} of hippocampal pyramidal neurons in relation to compromised Ca^{2+} regulation, as well as to learning, in aging animals. We have compelling evidence suggesting that changes in the AHP, and in particular, the sI_{AHP}, could link altered Ca^{2+} homeostasis in aging to changes in neural function and learning deficits. Some of the topics addressed in this chapter are covered in somewhat greater depth in two recent reviews.[127,128]

Altered Ca^{2+} Homeostasis in Aging

A major source of Ca^{2+} for signal transduction comes from the extracellular Ca^{2+} pool. Extracellular Ca^{2+} can enter the cell via voltage-gated Ca^{2+} channels (VGCCs) and ligand-gated receptors. In aging CA1 hippocampal pyramidal neurons, Ca^{2+} action potentials are larger[88] and have a longer plateau phase that is mainly attributable to an enhanced L-type Ca^{2+} influx (Fig. 1).[71] This enhanced L-type Ca^{2+} influx is partially the result of an increase in the functional L-type Ca^{2+} channel density.[12,37,113] Whether influxes via other VGCCs exhibit similar age-related alterations is unclear. Ca^{2+} influx via ligand-gated receptors —alpha-amino -3-hydroxy- 5-methyl-4-isoxazolepropionate (AMPA)/kainate receptors, N-methyl-D-aspartate (NMDA) receptors, and nicotinic acetylcholine receptors— has also been implicated in aging.[7,8,32,33,57,81,99,100] Altogether, these studies point to an enhanced Ca^{2+} influx in aging hippocampal pyramidal neurons.[71,88]

Another major source of Ca^{2+} comes from the release from the intracellular Ca^{2+} store—the endoplasmic reticulum (ER).[10] ER Ca^{2+} release can occur via two pathways: 1) the ryanodine receptor (RyR) pathway, in which Ca^{2+} binds to RyRs on ER and initiates store release in a process termed Ca^{2+}-induced-Ca^{2+}-release (CICR); and 2) the inositol-triphosphate (IP_3) pathway, in which IP_3 and Ca^{2+} activate IP_3 receptors on ER, causing Ca^{2+} to be released from IP_3 stores, which then activates the RyR-sensitive store to cause CICR. Store release in the hippocampus has not been quantified in the context of aging. Nonetheless, ER Ca^{2+} content has also been shown to increase with age,[35] and the caffeine-induced rise of intracellular Ca^{2+} appears to be higher in aging neurons.[40] There are also reports of age-related and region specific changes in IP_3Rs in aging rats.[5,62] Together, these data suggest that Ca^{2+} release from internal stores is impacted by the aging process.

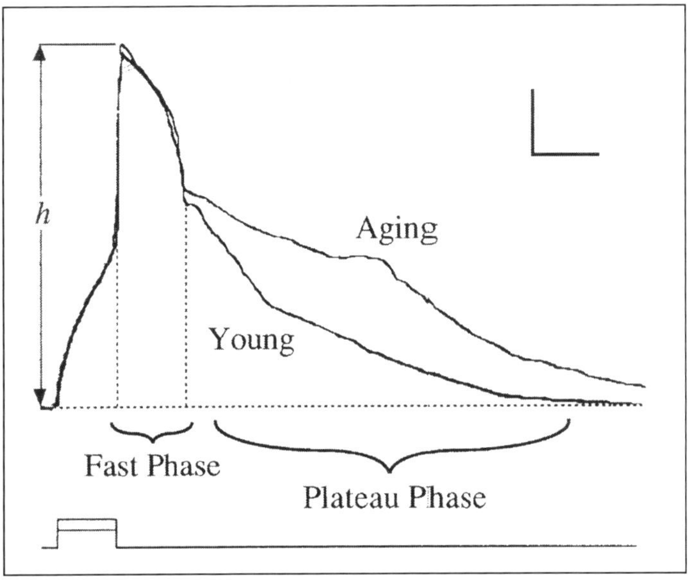

Figure 1. Age-related enhancement in Ca^{2+} influx in CA1 pyramidal neurons. The calcium action potential consisted of an initial fast phase followed by a slower plateau phase. There was no difference in the peak amplitude of the initial fast phase between young and aging neurons. The amplitude and duration of the slower plateau phase was significantly larger in aging neurons. [Reprinted from ref. 69].

Aging is also associated with significant, albeit region-specific changes in neuronal Ca^{2+} buffering capacity. The expression of many Ca^{2+} binding proteins that may act as buffers is down-regulated in certain brain regions of aged animals. For example, in the hippocampus, expressions of both mRNA and protein[47] for calbindin-D28K, a major cytosolic Ca^{2+} binding protein, are reduced in aging. The number of interneurons positive for parvalbumin, calbindin, or calretinin is also decreased in aging (see ref.104; ref.90 showed no decrease for parvalbumin-positive interneurons in the hippocampus but a decrease for the calbindin-D48K-positive interneurons). Conceivably, age-related alteration in neuronal Ca^{2+} buffering capacity, compounded with a decrease in the inhibitory drive in the hippocampus, can lead to different hippocampal network activities and contribute to age-related cognitive deficits.

The decay of depolarization-induced Ca^{2+} transients is prolonged in aging neurons,[40] indicating an impaired buffering and/or Ca^{2+} extrusion mechanism. The clearance of Ca^{2+} load after neuronal activity and maintenance for a low resting Ca^{2+} level is performed by several membrane-located and intracellular systems.[118] The extrusion of Ca^{2+} to the extracellular space is mediated by membrane Ca^{2+}-ATPase and the Na^+/ Ca^{2+} exchanger. In addition, Ca^{2+} is removed from the cytoplasm by sequestration into the ER via the sarco/endoplasmic reticulum Ca^{2+}-ATPase (SERCA) type of receptors, and into the mitochondria via uniporters. Both Ca^{2+} extrusion and sequestration mechanisms are impaired in aging. For example, Ca^{2+} effluxes through the Na^+/ Ca^{2+} exchanger and Ca^{2+}-ATPase are markedly reduced in old rats.[61,65] ER store loading is also substantially reduced in aging neurons, suggesting deficiencies in the activity of the SERCA type receptors.[122] Similarly, the amount of Ca^{2+} bound to mitochondria,[61] as well as mitochondrial Ca^{2+} uptake, is decreased in aged brains—a result associated with a decrease in the activity of the mitochondrial Ca^{2+} uniporter.[98] Together, these results suggest that, during neuronal activation, aging neurons are exposed to high elevations of Ca^{2+} for a longer period of time.

Altered Ca^{2+} Homeostasis and Age-Related Learning Deficits

The various changes in Ca^{2+} homeostatic processes point toward an increased intracellular Ca^{2+} level in aging, especially under conditions of high Ca^{2+} loads—a situation that may exacerbate neuronal vulnerability to excitotoxicity. The same mechanisms involved in Ca^{2+} dysregulation in aging have also been implicated in learning. For example, the density of the L-type Ca^{2+} channels is negatively correlated with performance in Morris water maze,[113] supporting a role for Ca^{2+} influx in learning. Our laboratory has also demonstrated that nimodipine, an L-type Ca^{2+} channel blocker, facilitated learning in the trace eyeblink conditioning task in aging rabbits,[21] further linking changes in the L-type Ca^{2+} influx with aging-related learning deficits. D-cycloserine, a partial coagonist to the NMDA receptor glycine site, improved the acquisition of trace eyeblink conditioning task in both young and aging rabbits, suggesting that an age-related decrease in NMDAR-mediated Ca^{2+} influx affects learning ability.[114,115]

Alterations in Ca^{2+}-Mediated Plasticity in Aging: Implications for Learning

There are several forms of activity-dependent plasticity in the hippocampus thought to underlie certain forms of memory. In nearly all cases, the occurrence of these phenomena involves Ca^{2+}-mediated cascades. Disruptions of these Ca^{2+}signals after neural activation prevent plastic changes from occurring, and in some cases have been shown to hamper learning in animals as well.[6,48,82,121] These results are consistent with the view that Ca^{2+} dysregulation contributes to age-related learning deficits.

Modulation of the postburst AHP, a hyperpolarizing voltage shift from the resting membrane potential that occurs after a burst of action potentials, is an example of an activity-dependent plasticity that is readily observed after learning and occurs normally during aging. The AHP is a compensatory mechanism for the cell to prevent Ca^{2+} overload. When Ca^{2+} concentrations reach high levels, Ca^{2+}-dependent K^+ currents underlying the AHP are activated. Together, these outward currents hyperpolarize the membrane, thus limiting further firing in response to sustained excitation in a process called spike frequency adaptation, or accommodation. Once activated, the AHP currents exert powerful negative influence on neuronal excitability and therefore regulate subsequent Ca^{2+} signals.

Our laboratory and others have identified age- and learning-related changes in the AHP and its underlying currents.[70,72,77,91,116] In the following sections, we review the paradigms that were used to establish an involvement of the AHP and its underlying currents in learning and in aging. Our data have led us to propose that changes in the AHP, partially attributable to changes in the sI_{AHP}, link Ca^{2+} dysregulation in aging neurons with learning deficits.

Paradigms Used to Study Age-Related Learning Deficits

The hippocampus is critically involved in learning and memory.[17] Hippocampal lesions in humans and animals cause severe deficits in the ability to transfer information from short-term to long-term stores, thus preventing the formation of new memories.[108] The hippocampus, as well as learning and memory processes that depend on proper hippocampal function, is particularly vulnerable to the aging process.[38,87] Aging animals and humans have shown an impairment in acquiring hippocampus-dependent learning tasks while they are not impaired in versions of the task that do not require the hippocampus.[41,42,116]

Two of the hippocampus-dependent tasks that our laboratory has used to address the relationship between age-related changes in Ca^{2+} dynamics and learning are trace eyeblink conditioning[68,107,124] and Morris water maze learning.[66,67] Successful acquisition of these tasks requires a properly functioning and intact hippocampus.[15,27,64,67,68,107,124]

Trace Eyeblink Conditioning

Trace eyeblink conditioning is a hippocampus-dependent temporal paradigm that we have used to characterize learning- and age-related associative learning deficits in rabbits, rats, and humans. In trace eyeblink conditioning, the subject must learn to associate a conditioning stimulus (CS), with a behaviorally significant unconditioned stimulus (US), and give a conditioned response (CR). An empty trace period intervenes between CS offset and US onset, requiring the subject to form a short-term memory of the CS in order to predict US onset successfully and give a CR timed appropriately to avoid the US. Trace eyeblink conditioning taps into the role of the hippocampal system in forming temporal associations, making the hippocampus necessary for acquiring this task.[17,105,123] In addition, trace eyeblink conditioning is impaired in aging rabbits,[21,31,106,111,116] aging rats,[41] and aging humans.[42] Thus, this task is useful for analyzing the cellular mechanisms underlying aging-related learning deficits.

Morris Water Maze

Morris water maze learning is a paradigm designed to target hippocampus-dependent spatial learning, and is sensitive to deficits accompanying lesions of the hippocampus.[27,67] During this task, the animals are placed in a circular tank filled with opaque water, and are required to swim to a hidden platform in order to escape. Successful acquisition of this task requires the animals to remember the placement of the platform location by using extramaze cues to escape the water quickly and effectively. Acquisition of the Morris water maze requires that spatial associations between the platform location and the surrounding environment be formed, making the hippocampus necessary for this task.[123] In addition, learning in the Morris water maze is impaired in aging rats.[27,93,94]

Learning-Related Changes in Hippocampal CA1 Pyramidal Neurons—Postsynaptic Excitability Increases in Learning

The AHP and accommodation are reduced in rabbit CA1 and CA3 hippocampal pyramidal neurons after the acquisition of the trace eyeblink conditioned response (Fig. 2).[69,70,117] These biophysical changes are learning-induced, as they are not observed in neurons of pseudoconditioned controls (which receive unpaired presentations of the same tone conditioned stimulus and air puff unconditioned stimulus), naïve controls, and animals that were trained but failed to acquire the task. They are postsynaptic, as they are evoked by intracellular current injection, and persist after blocking Na^+ spike-dependent synaptic transmission.[19] More importantly, changes in the AHP and accommodation return to baseline after acquisition and initial consolidation,[70,117] consistent with the hypothesis that the hippocampus functions as an intermediate storage buffer during learning.[17,22,39]

Eduction in the AHP is a general phenomenon that occurs in learning, and is not species-related or task-specific. In rat CA1 pyramidal neurons, the AHP was also reduced following the acquisition of the trace eyeblink conditioning task[78] or the spatial water maze learning.[77] Other laboratories have also reported reductions in the AHP in hippocampal pyramidal neurons after radial arm maze training[28] and in rat piriform cortex pyramidal neurons after odor discrimination operant-conditioning.[95] Together, these results strongly suggest that an increase in postsynaptic neuronal excitability, caused by a reduction in the AHP, is a conserved cellular mechanism underlying learning across species.

Components of the AHP Altered in Learning

The AHP can be separated into fast, medium, and slow components (fAHP, mAHP, sAHP, respectively) based on kinetic and pharmacological criteria.[96,110] The currents underlying these components are four classes of outward K^+ currents (I_C, I_M, I_{AHP}, and sI_{AHP}), and the time course of the AHP is in part regulated by a mixed cationic current, I_Q/I_h.[3,34,50,55,79,109] Three of the four K^+ currents, I_C, I_{AHP}, and the sI_{AHP}, are Ca^{2+}-dependent and have been implicated in learning. In the invertebrate *Hermissenda*, I_C, part of the fAHP, is reduced after classical conditioning.[4] Current-clamp data from our laboratory suggest that for rats and rabbits, changes in the AHP after learning associative tasks are mainly driven by changes in the currents underlying the mAHP (I_M and I_{AHP}) and the sAHP (I_{AHP} and sI_{AHP}).[69,70,116,117] Using whole-cell

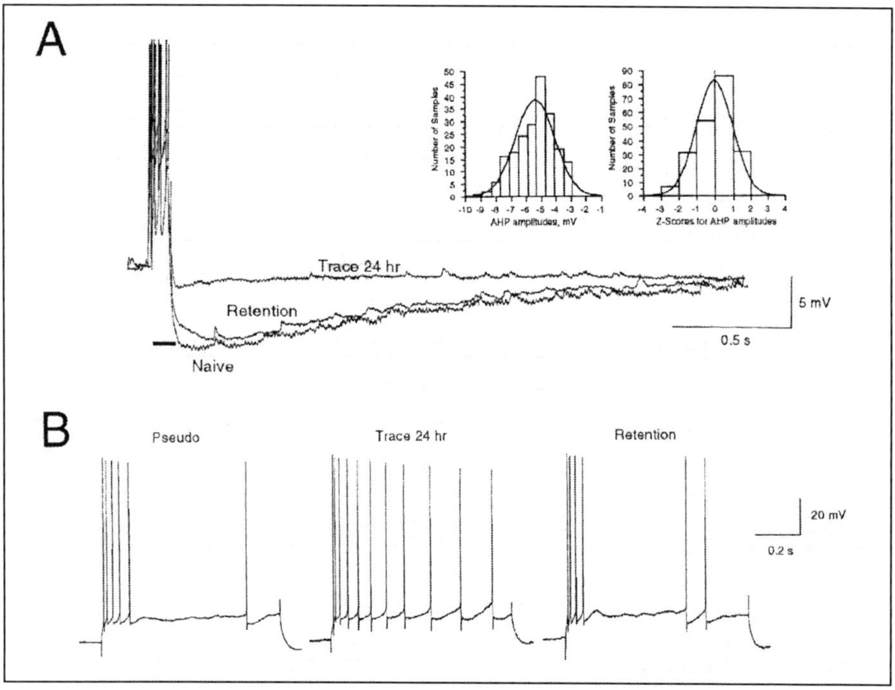

Figure 2. Acquisition of trace eyeblink conditioning task increased excitability of CA1 pyramidal neurons. A) Voltage traces showing an overlay of representative AHP recordings in neurons from naïve (Naive) and trace eyeblink conditioned rabbits (Trace). The learning-related AHP reduction is transient, as AHP recordings taken 14 days after the rabbits reached behavioral criterion (Retention) showed no change as compared to those from naïve animals. The AHPs from naive rabbits followed a normal distribution, as demonstrated in the frequency distribution and z-score graphs (insets). B) Typical accommodation responses in CA1 pyramidal cells from rabbits. During the 800 ms of depolarizing stimulus, trace eyeblink conditioned animals fired more action potentials in comparison to the pseudoconditioned animals. Since the AHP regulates the degree of accommodation, the reduction in accommodation is also transient. Accommodation measurements taken 14 days after the rabbits reached behavioral criterion (Retention) showed no change from those from pseudoconditioned animals. [Reprinted from ref. 70].

voltage-clamp techniques, we have subsequently demonstrated that in CA1 pyramidal neurons from rats trained in the water maze task, both the I_{AHP} and the sI_{AHP} are reduced.[77]

Postsynaptic Excitability in CA1 Hippocampal Pyramidal Neurons Decreases in Aging: Implications for Age-Related Learning Deficits

Acquisition of the trace eyeblink conditioning response is impaired in aging rats and rabbits.[41,45,116] Interestingly, the AHP and accommodation are enhanced in CA1 neurons of rabbits and rats at ages that show learning deficits (Fig. 3).[49,54,69,70] Although many aging animals failed to acquire the trace eyeblink conditioned response,[41,116] those that did learn also showed a reduction in the AHP (Fig. 4).[69] Together, these data revealed an inverse relationship between the AHP and learning ability, and suggest that AHP enhancement in aging is involved in aging-related learning deficits.

Nimodipine reduces the L-type Ca^{2+} influx in aging hippocampal pyramidal neurons,[71] thereby reducing the AHP[72] and restoring excitability to levels closely resembling those of young adult CA1 neurons. Administering nimodipine to aging rabbits facilitated the acquisition of trace eyeblink conditioning,[21] suggesting that the effect of this drug might be mediated

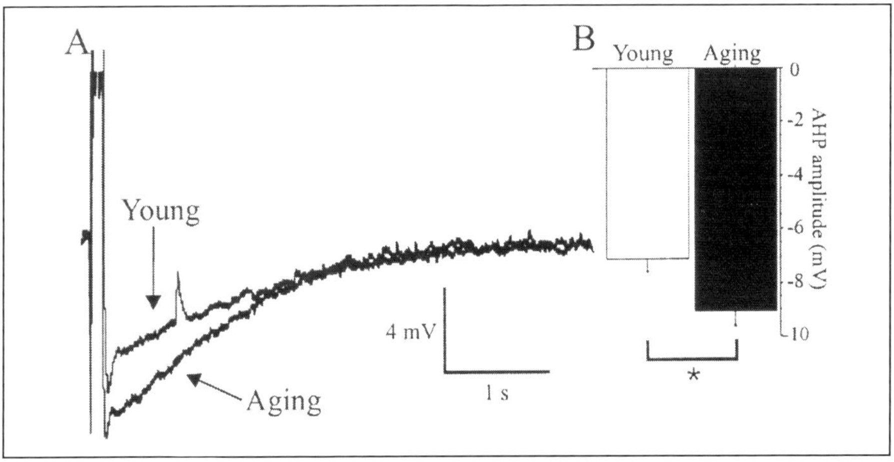

Figure 3. Age-related enhancement in the AHP. A) An overlay of representative AHP traces recorded from neurons of young and aging rabbits. B) Mean AHP amplitude was greater in aging neurons than in young neurons (mean ± SEM; unpaired t test; *p < 0.05). [Reprinted from ref. 91].

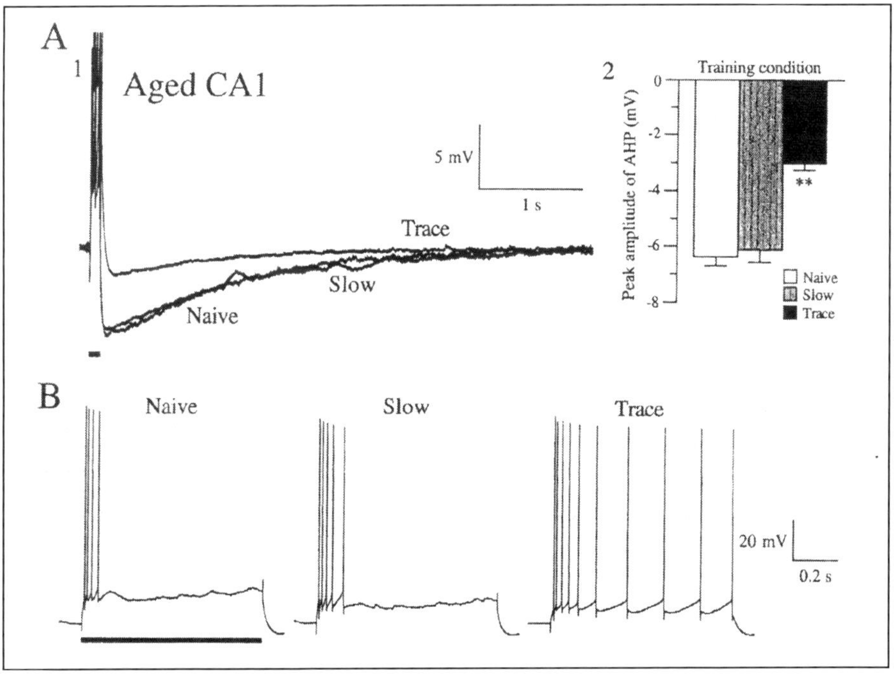

Figure 4. Acquisition of trace eyeblink conditioning task increased excitability of CA1 pyramidal neurons from aging animals. A) 1. Overlay of voltage traces showing representative AHP recordings in neurons from aging naive (Naive), aging slow-learning (Slow), and aging trace-conditioned (Trace) rabbits. 2. Mean effects of trace eyeblink conditioning on AHP amplitude in aging rabbit CA1 neurons. After learning, the AHP was significantly reduced compared with naive and slow-learning aging controls. B) Typical accommodation responses in CA1 pyramidal cells from aging naive (Naive), aging slow-learning (Slow), and aging trace-conditioned (Trace) rabbits. [Reprinted from ref. 69].

by its secondary effect on the AHP. Cholinergic treatments that facilitated learning in aging animals were also found to reduce the AHP[45,46,80,125] and the sI_{AHP} (ref. 92) without directly affecting Ca^{2+} dynamics, further supporting the idea that the effects of these treatments were mediated by reducing the AHP and the sI_{AHP}.

Our data show that two of the AHP currents, the I_{AHP} and the sI_{AHP}, are enhanced in aging neurons (Fig. 5).[91] Previous studies have shown that the apamin-sensitive I_{AHP} accounts for only a small percentage (~20%) of the total AHP.[79,109] Thus by inference, most of the enhancement in the AHP in aging is attributable to the sI_{AHP}. This deduction is fairly consistent with the reduced neuronal excitability seen in aging hippocampal pyramidal neurons,[72] because the sI_{AHP} is the main determinant of neuronal excitability.[96,110] Bath applications of nimodipine also caused a quantitatively greater reduction in the sI_{AHP} in hippocampal pyramidal neurons of aging animals in comparison to that of the young animals,[91] consistent with enhanced L-type Ca^{2+} influx in aging. However, the residual sI_{AHP} in bath-applied nimodipine was still larger in aging neurons than in young neurons, suggesting that other mechanisms also contribute to an enhancement of this current in aging.

Mechanisms Underlying Aging-Related Enhancement in the sI_{AHP}

The sI_{AHP} is a Ca^{2+}-dependent K^+ current. So far, Ca^{2+} influx from the extracellular pool and Ca^{2+} release from intracellular stores have both been shown to activate the sI_{AHP}. Furthermore, Ca^{2+} buffering/Ca^{2+} clearance systems have also been shown to affect the time course of the sI_{AHP}. In hippocampal pyramidal cells, Ca^{2+} influxes from both the VGCCs and ligand-gated receptors (such as NMDA and AMPA)[51,91,101] have both been shown to generate the sI_{AHP}. In particular, Ca^{2+} influxes from both the L- and the N-type channels have greater influences on the sI_{AHP}, as bath applications of nifedipine, nimodipine, and ω-conotoxin GVIA all cause partial reductions of this current (see ref. 112; ref. 102 showed no contribution for the N-type channels). The P/Q-type Ca^{2+} channels probably contribute only a small amount to the total Ca^{2+} influx, if at all, as ω-agatoxin IVA had no effect on the sI_{AHP}.[112]

Ca^{2+} released from intracellular stores is also important for the sI_{AHP}. In cultured hippocampal neurons, blocking this store release with ryanodine causes a reduction in the sI_{AHP}.[102] Similarly in CA1 and CA3 neurons, bath applications of ryanodine and thapsigargin, which deplete the store, also reduced the AHP and the sI_{AHP}.[102,112,119,120] The decay for the sI_{AHP} is regulated by the extent of cytoplasmic Ca^{2+} buffering,[52,131] and Ca^{2+} clearance mechanisms.[52] The deficits in these two mechanisms are consistent with a prolonged AHP and sI_{AHP} tail in aging neurons.

The sI_{AHP} also receives modulation from a variety of neurotransmitters, neuropeptides, and neuromodulators.[110,128] Many of these neurotransmitter systems that modulate the sI_{AHP} are sensitive to aging, raising a possibility that the enhanced sI_{AHP} we observed in aging hippocampal pyramidal neurons results from differences in neuromodulation.

The various neurotransmitters and neuromodulators suppress the sI_{AHP} through protein kinase activities.[1,29,36,73,83-85,120] For example, in CA1 pyramidal neurons, the effects of cholinergic transmission on the sI_{AHP} in CA1 pyramidal neurons are mainly mediated through the muscarinic receptors[92] by the activation of PKC,[2,18,58] and/or Ca^{2+}/calmodulin-dependent kinase II,[2,73,85] along with a phosphatase.[44] Glutamatergic transmission also reduces the AHP currents—I_M and I_{AHP}/sI_{AHP}—by activating metabotropic glutamate receptors (mGluRs) and the subsequent PKC cascade.[1,13,16,20,59] Monoamines such as dopamine, noradrenaline, histamine, and serotonin, as well as neuropeptides such as VIP, CGRP, or CRF all affect the sI_{AHP} by activating the cAMP/PKA pathway.[36,56,83,84]

Under normal conditions, the sI_{AHP} is maintained by a balance of the activities between kinases and phosphatases.[86] Many of these kinases such as PKC and CaMKII, as well as phosphatases such as PP1 and PP2A, that are known to modulate the sI_{AHP} depend on Ca^{2+} influx for activation. Thus, phosphorylation of the sI_{AHP} channels is likely to be altered with age as well. It has previously been suggested that the enhanced L-type Ca^{2+} influx in aging might

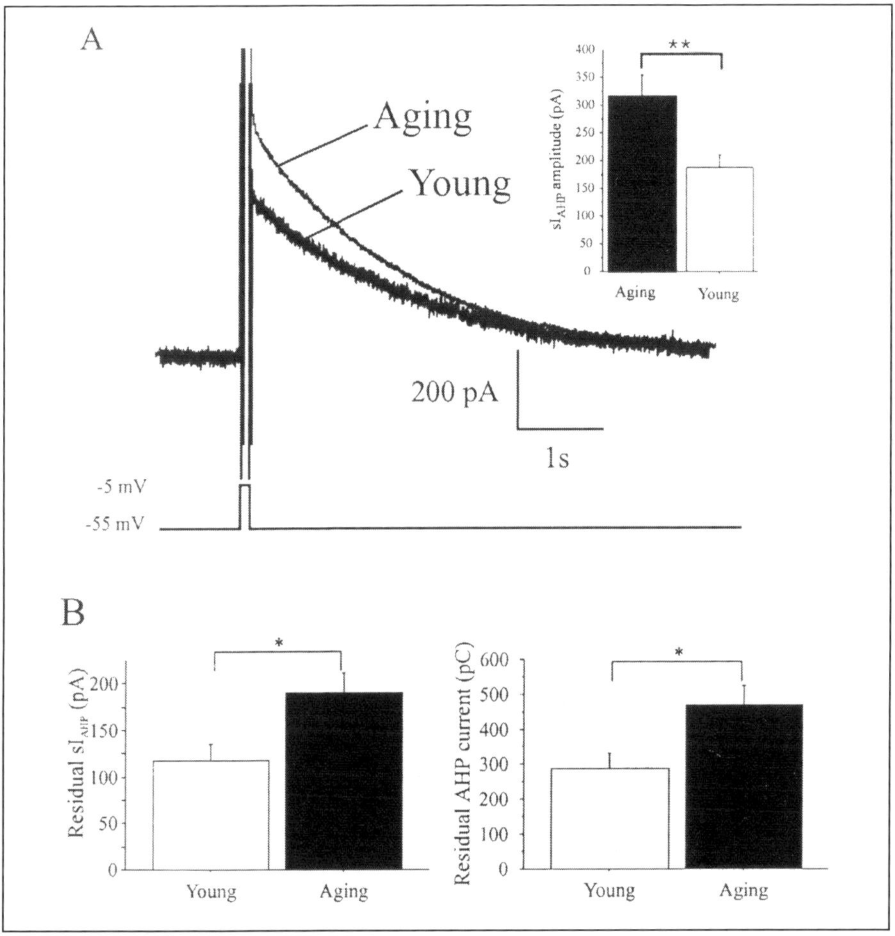

Figure 5. Age-related enhancement in the sI_{AHP}. A) Representative sI_{AHP} recordings from CA1 pyramidal neurons of young and aging rabbits. The sI_{AHP} amplitude is greater in aging neurons than in young neurons (mean ± SEM; unpaired t test; $^{**}p < 0.01$). B) In saturating concentration of nimodipine (10 μM), the sI_{AHP} recorded from neurons of aging rabbits is still significantly larger than that of young rabbits, suggesting that the enhancement of sI_{AHP} in aging is not solely due to an enhanced L-type Ca^{2+} influx. [Modified from ref. 91].

preferentially activate protein phosphatases relative to protein kinases.[25] A study examining the effect of phosphorylation on regulating synaptic strength has indeed demonstrated a shift in the balance of kinase and phosphatase activities in aging hippocampal neurons toward the direction of higher phosphatase activities.[75] This shift may partially explain why the sI_{AHP} is enhanced in aging neurons, since the effect of kinases has been shown to suppress this current.

Another possibility that might contribute to an enhanced sI_{AHP} is an increase in the functional sI_{AHP} channel density. In perfusate containing nimodipine, the sI_{AHP} was still significantly larger in aging neurons than in young neurons,[91] indicating that an enhanced Ca^{2+} influx alone or the resultant shift in kinase/phosphatase activities is insufficient to account for the enhanced sI_{AHP} in aging. It is known that increases in intracellular Ca^{2+} concentration can activate various signaling pathways that drive new gene expression essential for neuronal development, survival, and plasticity.[23,126] The route of Ca^{2+} entry determines which signaling pathways are activated, and thus plays a critical role in specifying the cellular response to Ca^{2+}. For example, L-type Ca^{2+} influx is particularly effective in activating transcription factors such as CREB and MEF-2.[23] Given the numerous changes in the Ca^{2+} signaling mechanisms in aging, it is conceivable that the profile of gene expression alters with age as well. As the sI_{AHP} is a compensatory mechanism to counter neuronal Ca^{2+} overload, it is possible that the functional channel density for the sI_{AHP} channels is upregulated in aging.

sI_{AHP} As a Link Between Age-Related Changes in Ca^{2+} Homeostasis and Learning

Depending on the route of entry, Ca^{2+} can differentially modulate the sI_{AHP} by activating different kinase cascades and affecting subsequent plastic changes. The interaction between the Ca^{2+} signaling cascade and the glutamatergic system exemplifies the complex nature of the regulation of the sI_{AHP} and neuronal excitability. In hippocampal pyramidal neurons, NMDAR-mediated Ca^{2+} influx can activate the sI_{AHP} in the absence of action potentials.[51] However at synapses, this Ca^{2+} influx shows associative features in that it becomes supralinear when it occurs with the pairing of an action potential and EPSPs.[43,129,130] Likewise, repetitive activation of mGluRs can induce large increases in intracellular Ca^{2+} level at the proximal dendrite when paired with backpropagating action potentials.[74] Given the putative locations of the sI_{AHP} channels,[9,11,97] the NMDA receptor- and mGluR-mediated Ca^{2+} transients might be especially important in regulating the sI_{AHP} by strategically increasing local dendritic Ca^{2+} levels and activating local kinase cascades. Furthermore, in hippocampal pyramidal neurons, the Ca^{2+} transient evoked by stronger stimulation is dependent on VGCCs but independent of NMDAR, whereas the Ca^{2+} transient evoked by subthreshold stimulation is independent of voltage-gated Ca^{2+} channels and dependent on NMDA receptors. Thus, differential modulation of the sI_{AHP} by Ca^{2+} channels and the glutamatergic system can be adjusted in an activity-dependent manner.

The exact biochemical steps that lead to changes in the sI_{AHP} in learning hippocampus-dependent tasks and in aging are not totally understood. What is clear, however, is that numerous neuromodulators can alter both sI_{AHP} and higher brain functions. For example, activation of mGluRs has been shown to reduce the AHP and the sI_{AHP}.[1,13,16,20,59] However, by activating PKC, mGluRs can also prevent activation of β-adrenergic receptors, which couple to adenylyl cyclase, from blocking the sI_{AHP}.[76] Since the AHP modulates neuronal responsiveness, cross talk between PKC and the adenylyl cyclase pathway is likely to have physiological consequences. The interference with the β-adrenergic response by mGluRs suggests that under physiological conditions, mGluRs can exert dominance over β-adrenergic receptors in a task-specific manner.

Furthermore, it is likely that these mechanisms also involve Ca^{2+} signaling and second messenger systems that were previously implicated in other forms of synaptic plasticity. For example, kinases known to modulate the sI_{AHP}—PKC, PKA, and CaMKII—are also important for the induction of LTP. Pharmacological manipulations that facilitated LTP have also been shown to reduce the AHP,[16] suggesting that the AHP and its underlying currents can serve as an adjustable gain control, variably hyperpolarizing and shunting synaptic potentials arising in the apical dendrites and controlling the induction of LTP.[97] A recent study confirmed this hypothesis, demonstrating that steady state activation of the sI_{AHP} dampens temporal summation of the EPSPs as well as speeds up their decay rate.[51] Accordingly, a reduction in the AHP of CA1 pyramidal neurons during learning can allow further synaptic plasticity to occur at critical synapses, and an enhanced AHP in aging can hamper the formation of further plasticity important for learning and memory.[24,30,97]

References

1. Abdul-Ghani MA, Valiante TA, Carlen PL et al. Tyrosine kinase inhibitors enhance a Ca(2+)-activated K+ current (IAHP) and reduce IAHP suppression by a metabotropic glutamate receptor agonist in rat dentate granule neurones. J Physiol 1996; 496:139-144.
2. Agopyan N, Agopyan I. Effects of protein kinase C activators and inhibitors on membrane properties, synaptic responses, and cholinergic actions in CA1 subfield of rat hippocampus in situ and in vitro. Synapse 1991; 7:193-206.
3. Alger BE, Sim JA, Brown DA. Single-channel activity correlated with medium-duration, Ca-dependent K current in cultured rat hippocampal neurones. Neurosci Lett 1994; 168:23-28.
4. Alkon DL. Calcium-mediated reduction of ionic currents: A biophysical memory trace. Science 1984; 226:1037-1045.
5. Araki T, Kato H, Kanai Y et al. Age-dependent changes in second messenger and rolipram receptor systems in the gerbil brain. J Neural Transm Gen Sect 1994; 97:135-147.
6. Balschun D, Wolfer DP, Bertocchini F et al. Deletion of the ryanodine receptor type 3 (RyR3) impairs forms of synaptic plasticity and spatial learning. Embo J 1999; 18:5263-5273.
7. Barnes CA. Normal aging: Regionally specific changes in hippocampal synaptic transmission. Trends Neurosci 1994; 17:13-18.
8. Barnes CA, Rao G, Shen J. Age-related decrease in the N-methyl-D-aspartateR-mediated excitatory postsynaptic potential in hippocampal region CA1. Neurobiol Aging 1997; 18:445-452.
9. Bekkers JM. Distribution of slow AHP channels on hippocampal CA1 pyramidal neurons. J Neurophysiol 2000; 83:1756-1759.
10. Berridge MJ. Neuronal calcium signaling. Neuron 1998; 21:13-26.
11. Bowden SE, Fletcher S, Loane DJ et al. Somatic colocalization of rat SK1 and D class (Cav 1.2) L-type calcium channels in rat CA1 hippocampal pyramidal neurons. J Neurosci 2001; 21:RC175,1-6.
12. Campbell LW, Hao SY, Thibault O et al. Aging changes in voltage-gated calcium currents in hippocampal CA1 neurons. J Neurosci 1996; 16:6286-6295.
13. Charpak S, Gahwiler BH, Do KQ et al. Potassium conductances in hippocampal neurons blocked by excitatory amino-acid transmitters. Nature 1990; 347:765-767.
14. Choi DW. Calcium and excitotoxic neuronal injury. Ann N Y Acad Sci 1994; 747:162-171.
15. Clark RE, Squire LR. Classical conditioning and brain systems: The role of awareness. Science 1998; 280:77-81.
16. Cohen AS, Coussens CM, Raymond CR et al. Long-lasting increase in cellular excitability associated with the priming of LTP induction in rat hippocampus. J Neurophysiol 1999; 82:3139-3148.
17. Cohen NJ, Eichenbaum H. Memory, amnesia, and the hippocampal system. Cambridge: MIT Press, 1993.
18. Cole AE, Nicoll RA. The pharmacology of cholinergic excitatory responses in hippocampal pyramidal cells. Brain Res 1984; 305:283-290.
19. Coulter DA, Lo Turco JJ, Kubota M et al. Classical conditioning reduces amplitude and duration of calcium-dependent afterhyperpolarization in rabbit hippocampal pyramidal cells. J Neurophysiol 1989; 61:971-81.
20. Desai MA, Conn PJ. Excitatory effects of ACPD receptor activation in the hippocampus are mediated by direct effects on pyramidal cells and blockade of synaptic inhibition. J Neurophysiol 1991; 66:40-52.
21. Deyo RA, Straube K, Disterhoft JF. Nimodipine facilitates associative learning in aging rabbits. Science 1989; 243:809-811.
22. Disterhoft JF, Thompson LT, Moyer Jr JR et al. Calcium-dependent afterhyperpolarization and learning in young and aging hippocampus. Life Sci 1996; 59:413-420.
23. Dolmetsch RE, Pajvani U, Fife K et al. Signaling to the nucleus by an L-type calcium channel-calmodulin complex through the MAP kinase pathway. Science 2001; 294:333-339.
24. Dunwiddie TV, Taylor M, Heginbotham LR et al. Long-term increases in excitability in the CA1 region of rat hippocampus induced by beta-adrenergic stimulation: Possible mediation by cAMP. J Neurosci 1992; 12:506-517.
25. Foster TC, Norris CM. Age-associated changes in Ca(2+)-dependent processes: Relation to hippocampal synaptic plasticity. Hippocampus 1997; 7:602-612.
26. Gallagher M, Burwell R, Burchinal M. Severity of spatial learning impairment in aging: Development of a learning index for performance in the Morris water maze. Behav Neurosci 1993; 107:618-626.
27. Gallagher M, Holland PC. Preserved configural learning and spatial learning impairment in rats with hippocampal damage. Hippocampus 1992; 2:81-88.

28. Gant JC, Glasier C, Lea P et al. Age-dependent changes in Ca^{2+}-mediated excitability in rat CA1 neurons correlate with changes in radial arm maze learning. Soc Neurosci Abstr 1999; 25:888,105.
29. Gerber U, Sim JA, Gahwiler BH. Reduction of potassium conductances mediated by metabotropic glutamate receptors in rat CA3 pyramidal cells does not require protein kinase C or protein kinase A. Eur J Neurosci 1992; 4:792-797.
30. Giese KP, Peters M, Vernon J. Modulation of excitability as a learning and memory mechanism: A molecular genetic perspective. Physiol Behav 2001; 73:803-810.
31. Graves CA, Solomon PR. Age-related disruption of trace but not delay classical conditioning of the rabbit's nictitating membrane response. Behav Neurosci 1985; 99:88-96.
32. Gray R, Rajan AS, Radcliffe KA et al. Hippocampal synaptic transmission enhanced by low concentrations of nicotine. Nature 1996; 383:713-716.
33. Guan ZZ, Zhang X, Ravid R et al. Decreased protein levels of nicotinic receptor subunits in the hippocampus and temporal cortex of patients with Alzheimer's disease. J Neurochem 2000; 74:237-243.
34. Halliwell JV, Adams PR. Voltage-clamp analysis of muscarinic excitation in hippocampal neurons. Brain Res 1982; 250:71-92.
35. Hanahisa Y, Yamaguchi M. Brain microsomal calcium accumulation in rats with increasing age: Involvement of thapsigargin-sensitive Ca2+-ATPase. Int J Mol Med 1999; 4:627-631.
36. Haug T, Storm JF. Protein kinase A mediates the modulation of the slow Ca(2+)-dependent K(+) current, I(sAHP), by the neuropeptides CRF, VIP, and CGRP in hippocampal pyramidal neurons. J Neurophysiol 2000; 83:2071-2079.
37. Herman JP, Chen KC, Booze R et al. Up-regulation of alpha1D Ca2+ channel subunit mRNA expression in the hippocampus of aged F344 rats. Neurobiol Aging 1998; 19:581-587.
38. Jack Jr CR, Petersen RC, Xu Y et al. Rates of hippocampal atrophy correlate with change in clinical status in aging and AD. Neurology 2000; 55:484-489.
39. Kim JJ, Clark RE, Thompson RF. Hippocampectomy impairs the memory of recently, but not remotely, acquired trace eyeblink conditioned responses. Behav Neurosci 1995; 2:195-203.
40. Kirischuk S, Pronchuk N, Verkhratsky A. Measurements of intracellular calcium in sensory neurones of adult and old rats. Neuroscience 1992; 50:947-951.
41. Knuttinen MG, Gamelli AE, Weiss C et al. Age-related effects on eyeblink conditioning in the F344 x BN F1 hybrid rat. Neurobiol Aging 2001; 22:1-8.
42. Knuttinen MG, Power JM, Preston PR et al. Awareness in classical differential eyeblink conditioning in young and aging humans. Behav Neurosci 2001; 115:747-757.
43. Kovalchuk Y, Eilers J, Lisman J et al. NMDA receptor-mediated subthreshold Ca(2+) signals in spines of hippocampal neurons. J Neurosci 2000; 20:1791-1799.
44. Krause M, Pedarzani P. A protein phosphatase is involved in the cholinergic suppression of the Ca(2+)-activated K(+) current sI(AHP) in hippocampal pyramidal neurons. Neuropharmacol 2000; 39:1274-1283.
45. Kronforst-Collins MA, Moriearty PL, Ralph M et al. Metrifonate treatment enhances acquisition of eyeblink conditioning in aging rabbits. Pharmacol Biochem Behav 1997; 56:103-110.
46. Kronforst-Collins MA, Moriearty PL, Schmidt B et al. Metrifonate improves associative learning and retention in aging rabbits. Behav Neurosci 1997; 111:1031-1040.
47. Krzywkowski P, Potier B, Billard JM et al. Synaptic mechanisms and calcium binding proteins in the aged rat brain. Life Sci 1996; 59:421-428.
48. Kubota M, Murakoshi T, Saegusa H et al. Intact LTP and fear memory but impaired spatial memory in mice lacking Ca(v)2.3 (alpha(IE)) channel. Biochem Biophys Res Commun 2000; 282:242-248.
49. Kumar A, Foster TC. 17beta-Estradiol Benzoate Decreases the AHP Amplitude in CA1 Pyramidal Neurons. J Neurophysiol 2002; 88:621-626.
50. Lancaster B, Adams PR. Calcium-dependent current generating the afterhyperpolarization of hippocampal neurons. J Neurophysiol 1986; 55:1268-1282.
51. Lancaster B, Hu H, Ramakers GM et al. Interaction between synaptic excitation and slow afterhyperpolarization current in rat hippocampal pyramidal cells. J Physiol 2001; 536:809-823.
52. Lancaster B, Zucker RS. Photolytic manipulation of Ca2+ and the time course of slow, Ca(2+)-activated K+ current in rat hippocampal neurons. J Physiol 1994; 475:229-239.
53. Landfield PW, Pitler TA, Applegate MD. The effects of high Mg2+-to-Ca2+ ratios on frequency potentiation in hippocampal slices of young and aged rats. J Neurophysiol 1986; 56:797-811.
54. Landfield PW, Pitler TA. Prolonged Ca2+-dependent afterhyperpolarizations in hippocampal neurons of aged rats. Science 1984; 226:1089-1092.
55. Maccaferri G, Mangoni M, Lazzari A et al. Properties of the hyperpolarization-activated current in rat hippocampal CA1 pyramidal cells. J Neurophysiol 1993; 69:2129-2136.

56. Madison D, Nicoll RA. Actions of noradrenaline recorded intracellularly in rat hippocampal CA1 pyramidal neurons, in vitro. J Physiol 1986; 372:221-244.
57. Magnusson KR. The aging of the NMDA receptor complex. Front Biosci 1998; 3:70-80.
58. Malenka RC, Nicoll RA. Dopamine decreases the calcium-activated afterhyperpolarization in hippocampal CA1 pyramidal cells. Brain Res 1986; 379:210-215.
59. Mannaioni G, Marino MJ, Valenti O et al. Metabotropic glutamate receptors 1 and 5 differentially regulate CA1 pyramidal cell function. J Neurosci 2001; 21:5925-5934.
60. Martinez A, Vitorica J, Satrustegui J. Cytosolic free calcium levels increase with age in rat brain synaptosomes. Neurosci Lett 1988; 88:336-342.
61. Martinez-Serrano A, Blanco P, Satrustegui J. Calcium binding to the cytosol and calcium extrusion mechanisms in intact synaptosomes and their alterations with aging. J Biol Chem 1992; 267:4672-4679.
62. Martini A, Battaini F, Govoni S et al. Inositol 1,4,5-trisphosphate receptor and ryanodine receptor in the aging brain of Wistar rats. Neurobiol Aging 1994; 15:203-206.
63. Mattson MP. Calcium as sculptor and destroyer of neural circuitry. Exp Gerontol 1992; 27:29-49.
64. McGlinchey-Berroth R, Carrillo MC, Gabrieli JD et al. Impaired trace eyeblink conditioning in bilateral, medial-temporal lobe amnesias. Behav Neurosci 1997; 111:873-882.
65. Michaelis ML, Johe K, Kitos TE. Age-dependent alterations in synaptic membrane systems for Ca2+ regulation. Mech Ageing Dev 1984; 25:215-225.
66. Morris RGM. Spatial localization does not require the presence of local cues. Learning Motivation 1981; 12:239-260.
67. Morris RGM, Garrud P, Rawlins JNP et al. Place navigation impaired in rats with hippocampal lesions. Nature 1982; 297:681-683.
68. Moyer Jr JR, Deyo RA, Disterhoft JF. Hippocampectomy disrupts trace eye-blink conditioning in rabbits. Behav Neurosci 1990; 104:243-252.
69. Moyer Jr JR, Power JM, Thompson LT et al. Increased excitability of aged rabbit CA1 neurons after trace eyeblink conditioning. J Neurosci 2000; 15:5476-5482.
70. Moyer Jr JR, Thompson LT, Disterhoft JF. Trace eyeblink conditioning increases CA1 excitability in a transient and learning-specific manner. J Neurosci 1996; 16:5536-5546.
71. Moyer Jr JR, Disterhoft JF. Nimodipine decreases calcium action potentials in rabbit hippocampal CA1 neurons in an age-dependent and concentration-dependent manner. Hippocampus 1994; 4:11-18.
72. Moyer Jr JR, Thompson LT, Black JP et al. Nimodipine increases excitability of rabbit CA1 pyramidal neurons in an age- and concentration-dependent manner. J Neurophysiol 1992; 68:2100-2109.
73. Müller W, Petrozzino JJ, Griffith LC et al. Specific involvement of Ca(2+)-calmodulin kinase II in cholinergic modulation of neuronal responsiveness. J Neurophysiol 1992; 68:2264-2269.
74. Nakamura T, Barbara JG, Nakamura K et al. Synergistic release of Ca2+ from IP3-sensitive stores evoked by synaptic activation of mGluRs paired with backpropagating action potentials. Neuron 1999; 24:727-737.
75. Norris CM, Halpain S, Foster TC. Alterations in the balance of protein kinase/phosphatase activities parallel reduced synaptic strength during aging. J Neurophysiol 1998; 80:1567-1570.
76. Nouranifar R, Blitzer RD, Wong T et al. Metabotropic glutamate receptors limit adenylyl cyclase-mediated effects in rat hippocampus via protein kinase C. Neurosci Lett 1998; 244:101-105.
77. Oh MM, Gamelli AE, Wu WW et al. Morris Watermaze learning enhances neuronal excitability of CA1 hippocampal pyramidal neurons in rats. Soc Neurosci Abstr 2001; 27.
78. Oh MM, Knuttinen MG, Gamelli AE et al. Enhanced Neuronal excitability of CA1 hippocampal pyramidal neurons is observed after trace eyeblink conditioning in rats. Soc Neurosci Abstr 1999; 25:84.
79. Oh MM, Power JM, Disterhoft JF. Apamin increases neuronal excitability of CA1 hippocampal pyramidal neurons from rabbits in vitro. Neurosci Res Comm 2000; 27:135-142.
80. Oh MM, Power JM, Thompson LT et al. Metrifonate increases neuronal excitability in CA1 pyramidal neurons from both young and aging rabbit hippocampus. J Neurosci 1999; 19:1814-1823.
81. Pagliusi SR, Gerrard P, Abdallah M et al. Age-related changes in expression of AMPA-selective glutamate receptor subunits: Is calcium-permeability altered in hippocampal neurons? Neurosci 1994; 61:429-433.
82. Pak JH, Huang FL, Li J et al. Involvement of neurogranin in the modulation of calcium/calmodulin-dependent protein kinase II, synaptic plasticity, and spatial learning: A study with knockout mice. Proc Natl Acad Sci 2000; 97:11232-11237.
83. Pedarzani P, Storm JF. PKA mediates the effects of monoamine transmitters on the K+ current underlying the slow spike frequency adaptation in hippocampal neurons. Neuron 1993; 11:1023-1035.

84. Pedarzani P, Storm JF. Dopamine modulates the slow Ca(2+)-activated K+ current IAHP via cyclic AMP-dependent protein kinase in hippocampal neurons. J Neurophysiol 1995; 74:2749-2753.
85. Pedarzani P, Storm JF. Evidence that Ca/calmodulin-dependent protein kinase mediates the modulation of the Ca2+-dependent K+ current, IAHP, by acetylcholine, but not by glutamate, in hippocampal neurons. Pflugers Arch 1996; 431:723-728.
86. Pedarzani P, Krause M, Haug T et al. Modulation of the Ca2+-activated K+ current sIAHP by a phosphatase-kinase balance under basal conditions in rat CA1 pyramidal neurons. J Neurophysiol 1998; 79:3252-3256.
87. Petersen RC, Jack Jr CR, Xu YC et al. Memory and MRI-based hippocampal volumes in aging and AD. Neurology 2000; 54:581-587.
88. Pitler TA, Landfield PW. Aging-related prolongation of calcium spike duration in rat hippocampal slice neurons. Brain Res 1990; 508:1-6.
89. Porter NM, Thibault O, Thibault V et al. Calcium channel density and hippocampal cell death with age in long-term culture. J Neurosci 1997; 17:5629-5639.
90. Potier B, Krzywkowski P, Lamour Y et al. Loss of calbindin-immunoreactivity in CA1 hippocampal stratum radiatum and stratum lacunosum-moleculare interneurons in the aged rat. Brain Res 1994; 661:181-188.
91. Power JM, Wu WW, Sametsky E et al. Age-related enhancement of the slow outward calcium-activated potassium current in hippocampal CA1 pyramidal neurons in vitro. J Neurosci 2002; 22:7234-7243.
92. Power JM, Oh MM, Disterhoft JF. Metrifonate decreases sI(AHP) in CA1 pyramidal neurons in vitro. J Neurophysiol 2001; 85:319-322.
93. Rapp PR, Gallagher M. Preserved neuron number in the hippocampus of aged rats with spatial learening deficits. Proc Natl Acad Sci 1996; 93:9926-9930.
94. Rasmussen T, Schliemann T, Sorensen JC et al. Memory impaired aged rats: No loss of principal hippocampal and subicular neurons. Neurobiol Aging 1996; 19:143-147.
95. Saar D, Grossman Y, Barkai E. Reduced after-hyperpolarization in rat piriform cortex pyramidal neurons is associated with increased learning capability during operant. Eur J Neurosci 1998; 10:1518-1523.
96. Sah P. Ca(2+)-activated K+ currents in neurones: Types, physiological roles and modulation. Trends Neurosci 1996; 19:150-154.
97. Sah P, Bekkers JM. Apical dendritic location of slow afterhyperpolarization current in hippocampal pyramidal neurons: Implications for the integration of long-term potentiation. J Neurosci 1996; 16:4537-4542.
98. Satrustegui J, Villalba M, Pereira R et al. Cytosolic and mitochondrial calcium in synaptosomes during aging. Life Sci 1996; 59:29-34.
99. Segovia G, Porras A, Del Arco A et al. Glutamatergic neurotransmission in aging: A critical perspective. Mech Ageing Dev 2001; 122:1-29.
100. Seguela P, Wadiche J, Dineley-Miller K et al. Molecular cloning, functional properties, and distribution of rat brain alpha 7: A nicotinic cation channel highly permeable to calcium. J Neurosci 1993; 13:596-604.
101. Shah MM, Haylett DG. K(+) Currents generated by NMDA receptor activation in rat hippocampal pyramidal neurons. J Neurophysiol 2002; 87:2983-2989.
102. Shah M, Haylett DG. Ca(2+) channels involved in the generation of the slow afterhyperpolarization in cultured rat hippocampal pyramidal neurons. J Neurophysiol 2000; 83:2554-2561.
103. Shankar S, Teyler TJ, Robbins N. Aging differentially alters forms of long-term potentiation in rat hippocampal area CA1. J Neurophysiol 1998; 79:334-341.
104. Shetty AK, Turner DA. Hippocampal interneurons expressing glutamic acid decarboxylase and calcium-binding proteins decrease with aging in Fischer 344 rats. J Comp Neurol 1998; 394:252-269.
105. Solomon PR. Temporal versus spatial information processing theories of hippocampal function. Psych Bull 1979; 86:1272-279.
106. Solomon PR, Pendelbury WW. A model systems approach to the study of disorders of learning and memory. Neurobiol Aging 1994; 15:283-286.
107. Solomon PR, Van der Schaaf ER, Thompson RF et al. Hippocampus and trace conditioning of the rabbit's classically conditioned nictitating membrane response. Behav Neurosci 1986; 5:729-744.
108. Squire LR. Memory and Brain. New York: Oxford Univ Press, 1987.
109. Stocker M, Krause M, Pedarzani P. An apamin-sensitive Ca2+-activated K+ current in hippocampal pyramidal neurons. Proc Natl Acad Sci 1999; 96:4662-4667.
110. Storm JF. Potassium currents in hippocampal pyramidal cells. Prog Brain Res 1990; 83:161-187.
111. Straube KT, Deyo RA, Moyer Jr JR et al. Dietary nimodipine improves associative learning in aging rabbits. Neurobiol Aging 1990; 11:659-661.

112. Tanabe M, Gahwiler BH, Gerber U. L-Type Ca2+ channels mediate the slow Ca2+-dependent afterhyperpolarization current in rat CA3 pyramidal cells in vitro. J Neurophysiol 1998; 80:2268-2273.
113. Thibault O, Landfield PW. Increase in single L-type calcium channels in hippocampal neurons during aging. Science 1996; 272:1017-1020.
114. Thompson LT, Disterhoft JF. Age- and dose-dependent facilitation of associative eyeblink conditioning by D-cycloserine in rabbits. Behav Neurosci 1997; 111:1303-1312.
115. Thompson LT, Moskal JR, Disterhoft JF. Hippocampus-dependent learning facilitated by a monoclonal antibody or D-cycloserine. Nature 1992; 359:638-641.
116. Thompson LT, Moyer Jr JR, Disterhoft JF. Trace eyeblink conditioning in rabbits demonstrates heterogeneity of learning ability both between and within age groups. Neurobiol Aging 1996; 17:619-629.
117. Thompson LT, Moyer Jr JR, Disterhoft JF. Transient changes in excitability of rabbit CA3 neurons with a time course appropriate to support memory consolidation. J Neurophysiol 1996; 76:1836-1849.
118. Toescu EC, Verkhratsky A. Parameters of calcium homeostasis in normal neuronal ageing. J Anat 2000; 197:563-569.
119. Torres GE, Arfken CL, Andrade R. 5-Hydroxytryptamine4 receptors reduce afterhyperpolarization in hippocampus by inhibiting calcium-induced calcium release. Mol Pharmacol 1996; 50:1316-1322.
120. Torres GE, Chaput Y, Andrade R. Cyclic AMP and protein kinase A mediate 5-hydroxytryptamine type 4 receptor regulation of calcium-activated potassium current in adult hippocampal neurons. Mol Pharmacol 1995; 47:191-197.
121. Tsien JZ, Huerta PT, Tonegawa S. The essential role of hippocampal CA1 NMDA receptor-dependent synaptic plasticity in spatial memory. Cell 1996; 87:1327-1338.
122. Verkhratsky A, Shmigol A, Kirischuk S et al. Age-dependent changes in calcium currents and calcium homeostasis in mammalian neurons. Ann N Y Acad Sci 1994; 747:365-381.
123. Wallenstein GV, Eichenbaum H, Hasselmo ME. The hippocampus as an associator of discontiguous events. Trends Neurosci 1998; 21:317-323.
124. Weiss C, Bouwmeester H, Power JM et al. Hippocampal lesions prevent trace eyeblink conditioning in the freely moving rat. Behav Brain Res 1999; 99:123-132.
125. Weiss C, Preston AR, Oh MM et al. The M1 muscarinic agonist CI-1017 facilitates trace eyeblink conditioning in aging rabbits and increases the excitability of CA1 pyramidal neurons. J Neurosci 2000; 20:783-790.
126. West AE, Chen WG, Dalva MB et al. Calcium regulation of neuronal gene expression. Proc Natl Acad Sci 2001; 98:11024-11031.
127. Wu WW, Disterhoft JF. Calcium homeostasis and learning deficits in ageing. In: Mattson M, ed. Advances in Cell Aging and Gerontology. New York: Elsevier, 2002.
128. Wu WW, Oh MM, Disterhoft JF. Age-related biophysical alterations of hippocampal pyramidal neurons: Implications for learning and memory. Ageing Res Rev 2002; 1:181-207.
129. Yuste R, Majewska A, Cash SS et al. Mechanisms of calcium influx into hippocampal spines: Heterogeneity among spines, coincidence detection by NMDA receptors, and optical quantal analysis. J Neurosci 1999; 19:1976-1987.
130. Yuste R, Majewska A, Holthoff K. From form to function: Calcium compartmentalization in dendritic spines. Nature Neurosci 2000; 3:653-659.
131. Zhang L, Pennefather P, Velumian A et al. Potentiation of a slow Ca(2+)-dependent K+ current by intracellular Ca2+ chelators in hippocampal CA1 neurons of rat brain slices. J Neurophysiol 1995; 74:2225-2241.

Index

A

B

C

H

I

K

L

M

N

O

P

R

S

T

V

W